Carbohydrate Chemistry and Biochemistry

Structure and Mechanism

WITHDRAWN
FROM
UNIVERSITIES
AT

Carbohydrate Chemistry and Biochemistry

Structure and Mechanism

Michael L. Sinnott
School of Applied Sciences, University of Huddersfield, Huddersfield, UK
(Formerly Professor of Chemistry, University of Illinois at Chicago, USA)

RSC Publishing

Cover design and computer graphics assistance by
Gilbert E. M. Sinnott

ISBN: 978-0-85404-256-2

A catalogue record for this book is available from the British Library

Reprinted 2008

Published by The Royal Society of Chemistry,
Thomas Graham House, Science Park, Milton Road,
Cambridge CB4 0WF, UK

Registered Charity Number 207890

For further information see our web site at www.rsc.org

Preface

Carbohydrates are important because of biology. Of the six Kingdoms of life (Eubacteria, Archaebacteria, Protista, Fungi, Plantae and Animalia), carbohydrate polymers are a main component of the cell wall of organisms in the first five. Other carbohydrate polymers, such as glycogen and starch, are energy storage materials. Simple sugars, and their phosphate esters, are the working materials of primary metabolism. Finally, it is now clear that interactions of receptors with the carbohydrate portions of glycoproteins and glycolipids are they key to many recognition phenomena.

This book aims to provide a graduate level text on structure and mechanism in carbohydrate chemistry and biochemistry for beginning researchers. In addition to research workers new to carbohydrates in academic research departments, the work is intended to be useful for researchers and practitioners in the carbohydrate-processing industries, such as pulp and paper, textiles and food.

It was originally commissioned as a constituent volume in a series on Physical Organic Chemistry by the late Professor Andrew Williams, in his capacity as Series Editor. With Andy's sad and untimely death, the book is now published as a stand-alone volume. I hope the work is worthy of his memory.

By "structure" is meant not only the disposition of covalent bonds, but also conformation, at the small molecule and at the macromolecular level. By "mechanism" is meant a description of reactive intermediates and transition states – evidence-based curly arrows. It is with some reservation that the word "carbohydrate" is in the title, as *strictu sensu* carbohydrate biochemistry includes much of primary metabolism, such as glycolysis or the Calvin cycle. These reactions are in general not covered; there are several excellent curly-arrow-based biochemistry texts available.

With the exception of carbohydrate-binding modules attached to glycosidases, the book does not cover those interactions of carbohydrates with proteins that do not lead to covalency changes. The thinking behind the omission is that the interactions are weak and often polyvalent, and that the free energy changes involved are below the resolution of physical organic chemistry, even if detailed qualitative descriptions of some interactions are possible. Indeed, given the powerful forces available to bind a sugar to a protein – monosaccharide dissociation constants in the nM region are attainable – one could argue that the main mechanistic question glycobiology has to answer is not "Why does such and such a protein bind such and such a sugar", but "Why is the interaction so weak?"

The non-covalent interactions of polysaccharides with themselves are covered, as the regular, repeating structures that result (helices, ribbons and so forth) can be rationalised by the normal methods of conformational analysis.

I have tried to cover the principles of carbohydrate reactivity and to give an outline of their application in various technologies. In this respect, organic synthesis is regarded as a technology, alongside starch processing or papermaking: principles behind key reactions are covered, but not their detailed application.

New researchers in the carbohydrate area have first degrees from a range of subjects in the physical, biological and material sciences. My experiences of graduate teaching on two continents is that for students of varied backgrounds a condensed, telegraphic recapitulation of the relevant basics, particularly structures and equations, is essential. The inclusion of basic background in this way has meant, of course, that the book was not dependent on companion volumes in the proposed series. Thus, in Chapter 5, I include a fair amount on the enzyme kinetics of the glycosyl-transferring enzymes: even though enzyme kinetics are covered in good undergraduate biochemistry courses, they will be unfamiliar to chemists. Likewise, in Chapter 1, I include for reference much of the structural chemistry of carbohydrates with which graduate chemists should be familiar, but which will be foreign to people with degrees in the polysaccharide processing technologies. I have also tried to include enough background about the complex techniques applied to carbohydrates (such as X-ray crystallography and NMR) that readers of this book will be able to assess the usefulness and relevance of papers in the primary literature to their own research work.

Each chapter thus has "how to" sections interspersed with literature reviews which cover what is known up to late 2006. The treatment aims to be conceptually, but not experimentally, comprehensive: where essentially the same experiment has been done in a number of systems, only one or two are covered in detail.

Carbohydrate nomenclature has been treated in some detail, as it is so complex and counter-intuitive that researchers often get it wrong themselves in print (a particular egregious example is that methyl β-D-glucopyranoside is the mirror image of methyl β-L-glucopyranoside, but the preferred conformation of the former is $^{4}C_{1}$ and of the latter $^{1}C_{4}$). At the same time, I have tried to use structural nomenclature as it actually appears in the contemporary research literature, although for key compounds (such as *N*-acetylneuraminic acid) I also give IUPAC names, despite a suspicion that their rigorous use is confined to members of IUPAC nomenclature committees.

After some discussion, the Ingold system of mechanistic nomenclature (A1, S_N2, *etc.*) has been adopted, in preference to the Guthrie–Jencks system

($D_N + A_N$, *etc.*) recommended by IUPAC,[i] which, though, for important reactions is given in a foortnote. The Jencks–Guthrie system is in my view superior, as it can indicate preassociation reactions and proton transfers, which are particularly important in carbohydrate chemistry, in a way that the older Ingold system cannot, but unfortunately in the 15 years since its formulation, the Jencks–Guthrie system has not found widespread acceptance.

Illustrations with organic structural formulae have been used widely. In depicting sugar rings, Mills formulae have been used if the conformation is not known with confidence, otherwise conformational structures are drawn. Ribbon diagrams of various carbohydrate-active proteins with known structures are available from protein databases. Their reproduction in this book would have increased its cost and, at a time when all beginning researchers have Web access, only marginally increased its usefulness. If the reader wants to know what, say, a GH6 cellulase looks like, he or she should go to CAZy, the wonderful resource for the whole scientific community started by Bernard Henrissat and maintained by him and Pedro Coutinho, and click on the links in the 3D column. The protein can then be viewed in various downloadable viewers.

Finally, I should like to thank the many colleagues who have courteously answered my questions and/or allowed me to reproduce diagrams from their papers: Andrew Bennet, Harry Brumer, David Crich, Gideon Davies, Harry Gilbert, the late Kirill Neustroev, Andrew Laws, Anthony Serianni, Junji Sugiyama and Tuula Teeri. Particular thanks are due to David and Gideon for their patience in answering my many, sometimes elementary, questions so promptly.

Michael L. Sinnott
Wigan

[i] R. D. Guthrie and W. P. Jencks, *Acc. Chem. Res.* 1989, **10**, 343.

Contents

Chapter 2 Conformations of Monosaccharides

Chapter 3 Nucleophilic Substitution at the Anomeric Centre

CHAPTER 1

Structures of the Open-chain Forms of Reducing Sugars, and their Carbonyl Group Reactions

1.1 DEFINITIONS AND STRUCTURES

Historically, carbohydrates were defined as substances with the empirical formula $C_n(H_2O)_m$. The common sugars such as glucose and fructose ($n = m = 6$), or sucrose ($n = 12$, $m = 11$) fit this formula, but nowadays the convention is to regard as a carbohydrate a polyhydroxyaldehyde or polyhydroxy ketone with the classical formula, a molecule closely related to it, or oligomers or polymers of such molecules. Their study evolved as a separate sub-discipline within organic chemistry for practical reasons – they are water-soluble and difficult to crystallise – so that their manipulation demanded different sets of skills from classical "natural products" such as terpenes, steroids, alkaloids, *etc.*

The term "monosaccharide" refers to a carbohydrate derivative possessing a single carbon chain; "disaccharide" and "trisaccharide" refer to molecules containing two or three such monosaccharide units joined together by acetal or ketal linkages. "Oligosaccharide" and "polysaccharide" refer to larger such aggregates, with "a few" and "many" monosaccharide units, respectively. Current usage seems to draw the distinction between "few" and "many" at around 10 sugar units.

The everyday term "sugar" is used to describe mono-, di-, and the lower oligosaccharides, particularly when unprotected.

All common sugars have an unbranched carbon chain, and are referred to as trioses (3 carbons) (Figure 1.1), tetroses (4), pentoses (5), hexoses (6), heptoses (7), octoses (8) and nonoses (9). Depending on whether the carbonyl group is an aldehyde or a ketone, the monosaccharides are referred to as aldoses or ketoses, although IUPAC nomenclature prefers "-uloses", as in "hexuloses", for ketoses. It is amazing to realise that the stereochemistry of glucose and related aldohexoses was deduced by Emil Fischer in the 1891,[1] less than 20 years after van't Hoff and Le Bel's postulation that the valencies of carbon were directed towards the apices of a tetrahedron. Fischer had only two physical parameters with which to characterise compounds (optical

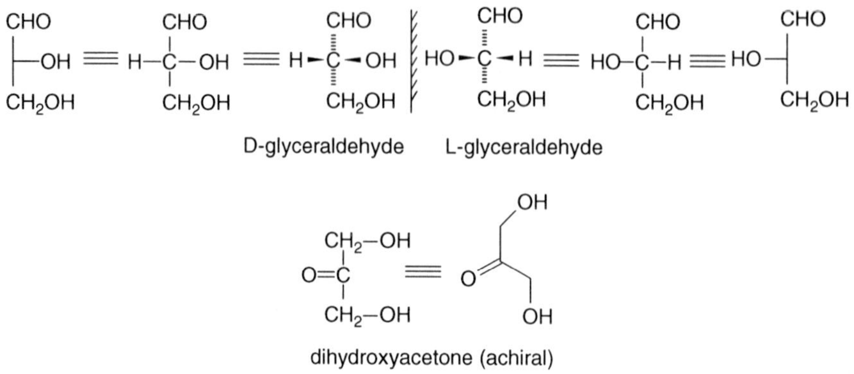

D-glyceraldehyde L-glyceraldehyde

dihydroxyacetone (achiral)

Figure 1.1 Triose structures. The two central structures for glyceraldehyde enantiomers are written according to the Mills convention (wedges out of the paper, dashed lines going in). The next structures are the full Fischer convention and the outer structures an abbreviated Fischer convention in which the hydrogen and central carbon atoms are implicit.

rotation and melting point), but nonetheless succeeded by rigorous stereochemical logic.

There are three trioses – a ketose (dihydroxyacetone), and the two enantiomeric forms of glyceraldehyde. In the carbohydrate field the *R,S* system of designating stereochemistry of a chiral centre, used elsewhere in organic chemistry, becomes very cumbersome, and Fischer's original conventions, as modified by Rosanoff[2] are still used.

The first Fischer convention is that the four bonds to a chiral centre are written horizontally and vertically, with the vertical bonds going into the plane of the paper and the horizontal bonds coming out. Such Fischer structures can be rotated through 180° in the plane of the paper, but not through 90°, since then the convention would be violated.

The second Fischer (strictly, Fischer-Rosanoff) convention is that if the main carbon chain is written vertically, with the carbon of highest oxidation state (in the case of glyceraldehyde, the CHO group) at the top, then if the bottom asymmetric centre has the OH on the right hand side, the configuration is D; if it is on the left, it is L. This rule for designating the absolute configuration as D or L also applies to sugars with more than one asymmetric centre.[i]

---

[i] Although the etymology is the same (*dextro* (right in Greek) and *laevo* (left)), this upper case D and L, denoting absolute configuration, should not be confused with the lower case d and l, used in the older literature in place of the current (+) and (−) to denote the sign of the optical rotation. The nomenclature arose because of Fischer's guess that d (dextrorotatory) glyceraldehyde had the D configuration; the assumption was shown to be correct by stereochemical correlations and X-ray diffraction studies, which exploited the phase changes introduced by atoms scattering X-rays of wavelength close to their X-ray absorption bands; ordinary X-ray crystallography cannot distinguish between enantiomers (J. M. Bijvoet, A. F. Peerdeman and A. J. van Bommel, *Nature*, 1951, **168**, 271).

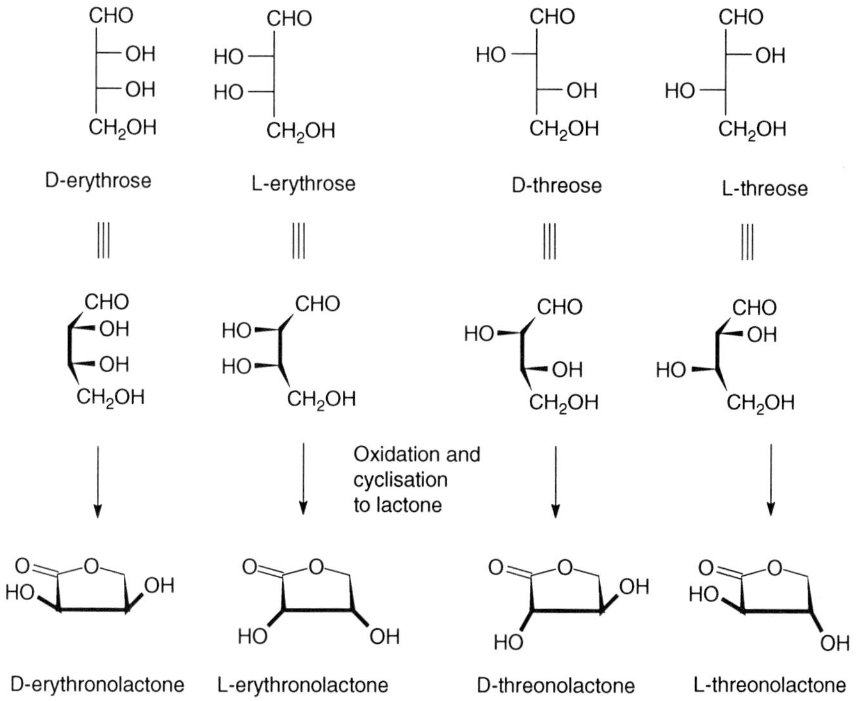

Figure 1.2 Aldotetroses and derivatives, illustrating diastereoisomerism.

The aldotetroses have two non-equivalent chiral centres, so there are four (2^2) stereoisomers, two pairs of enantiomers. Enantiomers have identical chemical and physical properties, except where they interact with agents which are themselves chiral. However, because the distances between atoms in different diastereomers differ, they have different chemical and physical properties with achiral reagents. Thus, erythrose and threose have different melting points, and reactivities to achiral reagents. This can be grasped intuitively by considering the oxidation of these sugars to the corresponding lactones: in threonolactone the hydroxyl groups are *trans*, in erythronolactone *cis* (Figure 1.2).

The carbonyl groups of aldoses and ketoses, like other aldehydes and ketones with electron-withdrawing substituents, are hydrated in aqueous solution; participation of one of the hydroxyl groups, to give an internal cyclic hemiacetal also occurs, and indeed largely predominates where relatively strain-free five and six-membered rings result. Aqueous solutions of aldoses and ketoses thus consist of a slowly equilibrating mixture of various species, of which the unhydrated, straight chain structures shown in this section are a very minor proportion. Compositions of equilibrated solutions of reducing monosaccharides are given later in Table 1.1.

If the two ends of the carbon chain of erythrose are made identical (for example, by reducing the aldehyde to a CH_2OH group with $NaBH_4$), then the resulting alditol (erythritol) is achiral, since it contains a mirror plane. Erythritol

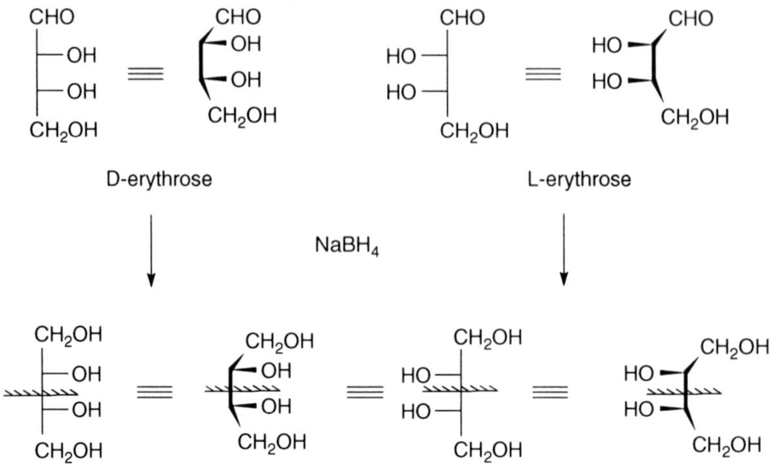

Figure 1.3 Erythritol is a *meso* form. The mirror plane in the various representations of the compound are shown.

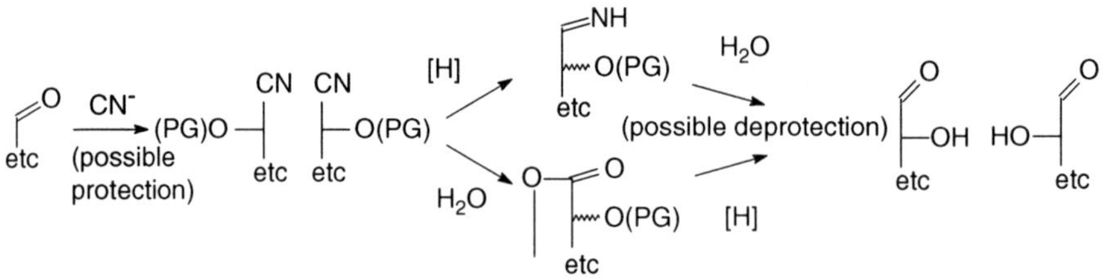

Figure 1.4 General outline of the various versions of the Kiliani reaction. *PG* = *protecting group*.

is termed a *meso* form, and is internally compensated (*i.e.* one asymmetric centre is *R*, another *S*). (Figure 1.3).

Extension of the carbon chain of an aldose from the carbonyl by one unit at a time can be carried out fairly readily by the Kiliani reaction.[3] A cyanhydrin is formed by addition of cyanide ion, followed by reduction and hydrolysis (in either order); historically, the sugar was unprotected, and the cyanohydrin was hydrolysed to the sugar lactone, and then reduced with sodium amalgam (Figure 1.4). Because a new asymmetric centre is formed, two epimeric sugars result (epimers are diastereomers that differ in the configuration of only one carbon).

The derivation of the aldopentoses from the aldotetroses is shown in Figure 1.5. Aldoses are numbered from the carbonyl group – thus ribose differs from arabinose in the configuration at C2 and xylose in the configuration at C3. Carbons and attached hydrogens and oxygens are referred to as C1, C2 C6, H1, H3, H4, O2, O4 *etc.*

Ketoses have one asymmetric centre fewer than the corresponding aldoses, and therefore according to IUPAC nomenclature should be named from the

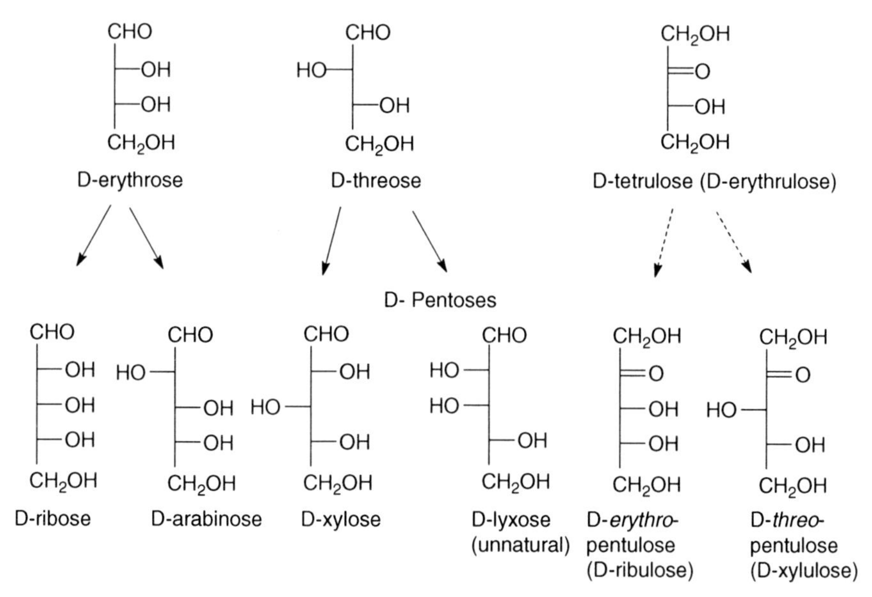

Figure 1.5 Structures of the pentoses. Whilst the aldopentoses can in practice be derived from the aldotetroses by the Kiliani reaction, no such simple route is available for the ketoses.

aldose of one fewer carbon. In practice, however, in the areas in which they are important ketopentoses are named from the common aldose of the same number of carbon atoms. Thus, the key reaction in carbon fixation in green plants is the reaction of ribulose 1,5-bisphosphate (not *erythro*-pentulose 1,5-bisphosphate) with carbon dioxide.

Figure 1.6 sets out the structures of the aldohexoses and the aldopentoses from which they can be derived by the Kiliani reaction. Aldohexoses have four non-equivalent asymmetric centres, and hence there are 16 (2^4) stereoisomers, 8 pairs of enantiomers.

The stereochemistry of the alditols formed by reduction of the aldehyde group is instructive. In the pentose series, xylitol and ribitol are *meso* forms, whereas D-arabinitol and D-lyxitol are identical. In the hexose series, allitol and galactitol are *meso* forms, and D-altritol and D-talitol are identical, as are D-glucitol and L-gulitol.

Natural 2-ketohexoses, with the relative configurations of the three asymmetric centres corresponding to all four aldopentoses are known, and have trivial names D-psicose (= D-*ribo*-hexulose), D-fructose (= D-*arabino*-hexulose), L-sorbose (= L-*xylo*-hexulose) and D-tagatose (= D-*lyxo*-hexulose) (Figure 1.7).

Monosaccharide units derived from oxidised or reduced sugars are important components of polysaccharides. The most important deoxy sugars are L-rhamnose (6-deoxy-L-mannose) and L-fucose (6-deoxy-L-galactose). Uronic acids are sugars in which the 6 position has been formally oxidised to a carboxylic acid, and form the major basis for anionic polysaccharides. Important

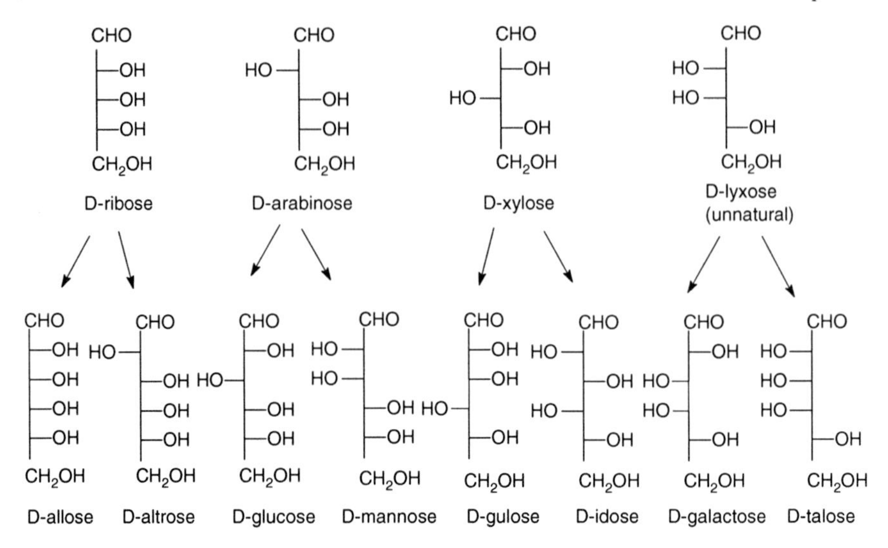

Figure 1.6 Structures of aldohexoses and their parent pentoses.

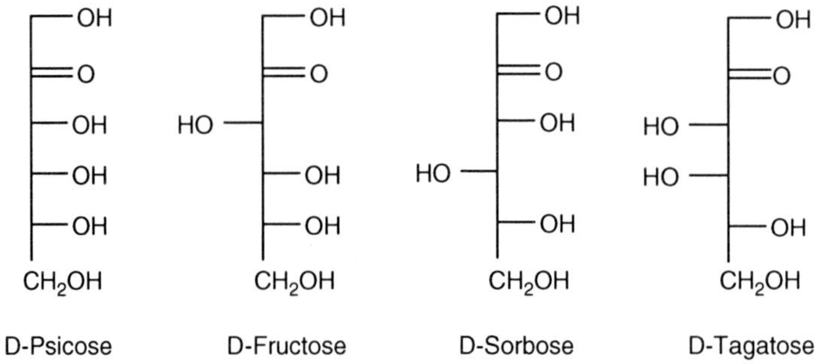

D-Psicose D-Fructose D-Sorbose D-Tagatose

Figure 1.7 2-Ketohexoses.

uronic acids are D-glucuronic, D-galacturonic and D-mannuronic. In some anionic polysaccharides, of which alginate is the classic example, the polysaccharide is biosynthesised with one uronic acid and then position 5 is epimerised enzymically in the polymer. Thus, alginate contains D-mannuronic acid residues, and L-guluronic acid residues from their epimerisation. Likewise, L-iduronic acid residues arise from epimerisation of D-glucuronic acid residues in mammalian polysaccharides such as heparin (Figure 1.8).

Other important components of naturally-occurring oligo- and polysaccharides are D-mannosamine, D-glucosamine, and D-galactosamine, molecules in which the the 2-OH of the parent aldose has been replaced by $-NH_2$. These sugars often occur as their N-acetyl derivatives, N-acetyl mannosamine, *etc.* IUPAC names for these sugars[4] (*e.g.* 2-amino-2-deoxy D-glucose and 2-acetamido-2-deoxy D-glucose for glucosamine and N-acetyl glucosamine,

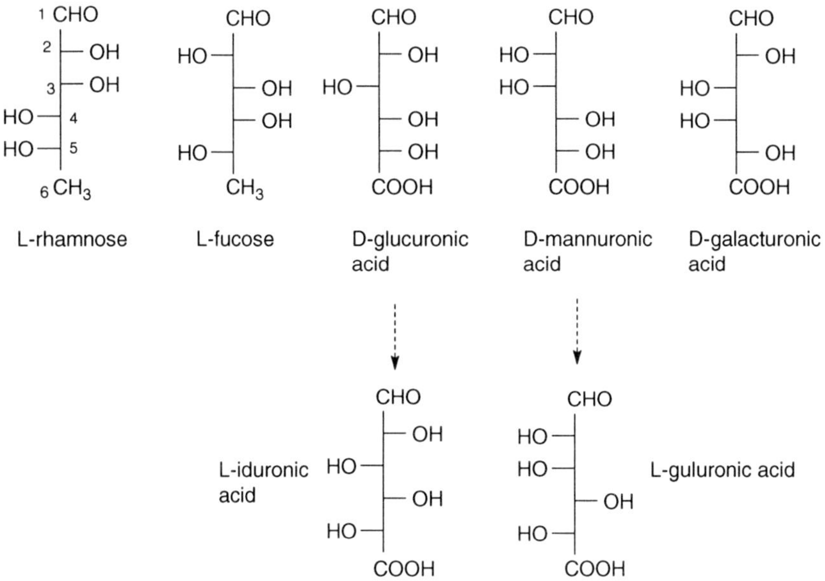

L-rhamnose L-fucose D-glucuronic acid D-mannuronic acid D-galacturonic acid

L-iduronic acid L-guluronic acid

Figure 1.8 Important reduced and oxidised monosaccharide units.

respectively) are encountered in the literature of preparative organic chemistry, but are infrequent elsewhere.

Hugely important in the interactions of cells of organisms in the animal kingdom with each other, and with the cells of their microbial parasites (such as the influenza virus or cholera bacterium) are the sialic acids, nine-carbon acidic sugars based on various O- and N-acylation patterns of neuraminic acid. N-acetyl neuraminic acid is biosynthesised from N-acetyl mannosamine 6-phosphate and phosphoenol pyruvate, acting as masked enolate. Other 3-deoxy-2-ulosonic acids, such as Kdo, important in the bacterial glycocalyx, are biosynthesised analogously (Figure 1.9).

IUPAC nomenclature unfortunately obscures this biosynthetic origin. Sugars with more than four asymmetric centres are numbered with the first four asymmetric centres (which need not necessarily be contiguous) *from the top* (in the Fischer-Rosanoff convention) named after the appropriate hexose, and the remaining asymmetric centres *at the bottom* named *glycero-*, *erythro-*, *etc*. Thus, N-acetyl neuraminic acid becomes 5-acetamido-3,5-dideoxy-D-*glycero*-D-*galacto*-non-2-ulosonic acid, and Kdo, 3-deoxy-D-*manno*-oct-2-ulosonic acid.

1.1.1 *R,S* Stereochemical Designations in Carbohydrate Chemistry – Prochirality

With more complex molecules than straight-chain aldoses, particularly those containing tertiary or quaternary chiral centres, the Fischer-Rosanoff convention founders on the impossibility of deciding reliably what is the main chain.

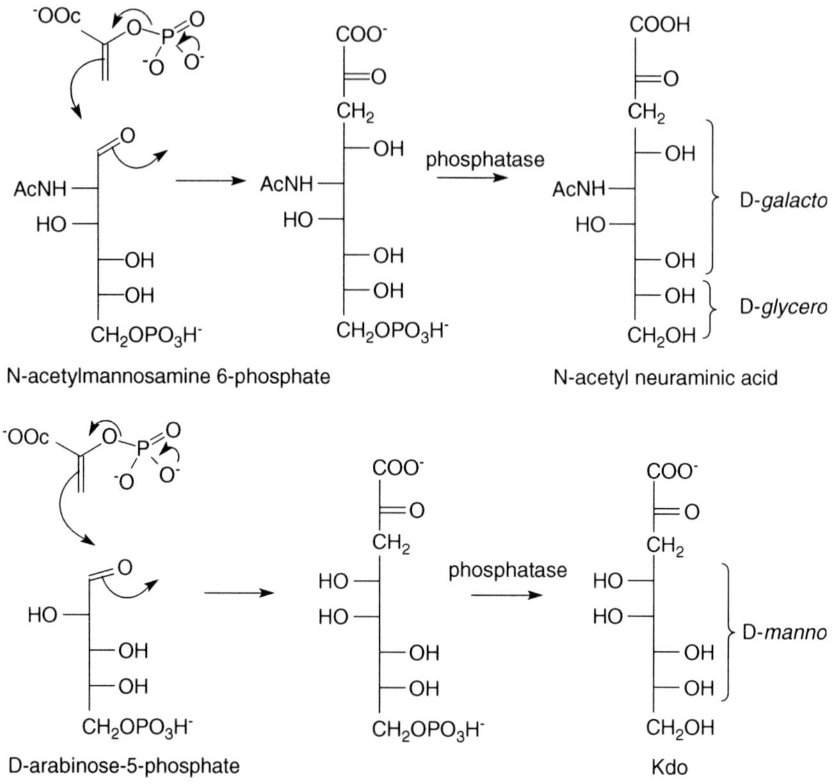

Figure 1.9 Schematic of the biosynthesis of N-acetyl neuraminic acid and Kdo. The synthetases will have catalytic machinery for opening the ring forms of the sugar, and an enzyme nucleophile preassociated to attach the phosphoryl group of phosphoenol pyruvate (free metaphosphate is not formed).

R. S. Cahn, C. K. Ingold and V. Prelog[5] proposed the system now universally in use outside carbohydrate chemistry.

The stereochemical designation of a tetrahedral chiral centre is determined by the priorities of the four substituents. The convention applies to all tetrahedral centres, not just carbon. Priorities in the first instance are determined by atomic weight (not atomic number – the convention works for molecules chiral by isotopic substitution). The chiral centre is viewed along the line of the bond from the central atom to the substituent of lowest priority, with the central atom in front. If the priorities of the three remaining substituents **decrease clockwise** the stereochemistry is *R* (for *rectus,* right in Latin), otherwise it is *S* (for *sinister*, left).

If two or more atoms in the first coordination sphere have the same atomic mass, then priorities are decided by the priorities of atoms attached to them. Thus, -CH_2OH (carbon bearing one oxygen and two hydrogens) has a higher

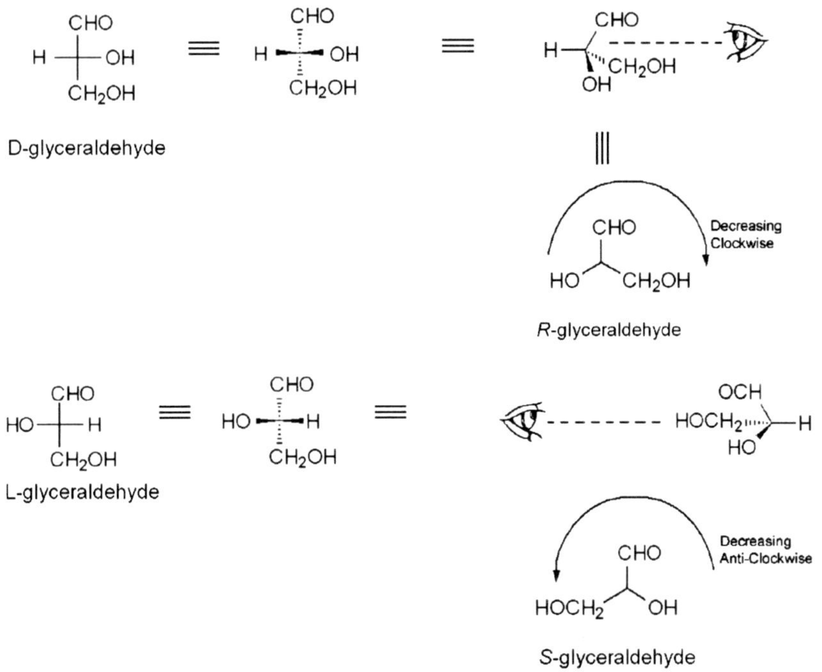

Figure 1.10 Application of Cahn-Ingold-Prelog (*R, S*) Convention to D-Glyceraldehyde.

priority than -CH_3 (carbon bearing three hydrogens). The relative priorities of two RCHOH groups attached to a chiral centre are determined by the priorities of the two R groups, in the third coordination sphere, and so on.

There is one non-intuitive rule – doubly- and triply-bonded substituents count two and three times, respectively. Thus -CHO (carbon attached to hydrogen and two oxygens) has a higher priority than -CH_2OH (carbon attached to two hydrogens and one oxygen).

Figure 1.10 illustrates how D-glyceraldehyde has the *R* configuration.

The *R, S* system becomes very complex when there are chains of >CHOH groups, which is why the Fischer-Rosanoff system is still used in carbohydrate chemistry. However, the Cahn-Ingold-Prelog system has to be used to distinguish between prochiral groups.

Prochirality was discovered in 1948 when A. G. Ogston,[6] a metabolic biochemist, found that the two carboxymethyl groups of citric acid were not metabolically equivalent. They are in enantiomeric environments, the *proR* – CH_2COOH group "seeing" the other three substituents in one sense, and the *proS* in the other. They are therefore transformed differently by all chiral reagents, in Ogston's case the enzymes of the Krebs cycle. The definition of *proR* and *proS* is very straightforward – a *proR* group is one, which on substitution with a heavier isotope would give the R configuration. This is

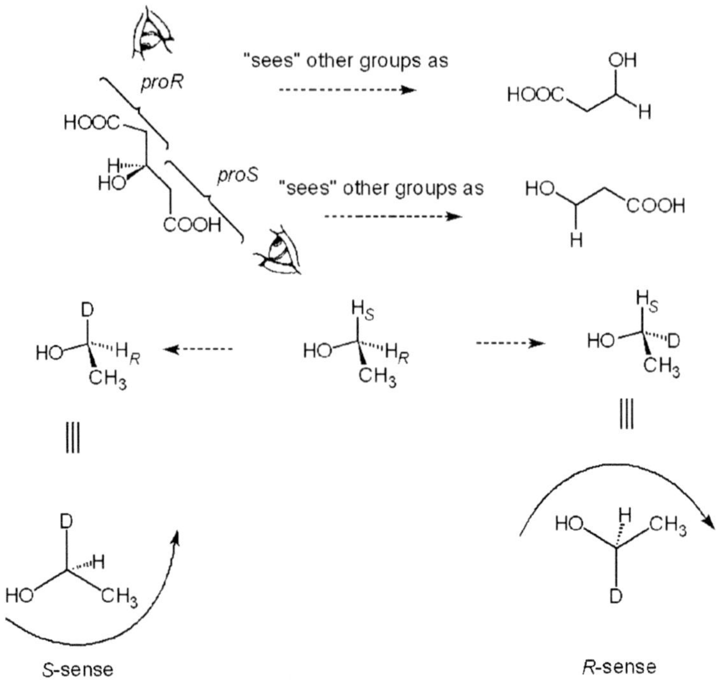

Figure 1.11 Enantiotopic groups. Top, enantiopic carboxymethyl groups of citric acid, showing their enantiomeric environments, and enantiopic hydrogens of ethanol, showing their *proR* and *proS* environments.

illustrated for the enantiotopic hydrogens of ethanol as well as for citric acid in Figure 1.11.[ii]

If prochiral centres of the type Ca_2bc are in a molecule which is already chiral, then the two groups have different properties, and are said to be diastereotopic (since isotopic substitution produces diastereomers). Notably, they have in general different NMR chemical shifts. The two diastereotopic hydrogens of terminal methylene groups, and the 3-hydrogens of ketodeoxyacids such as Kdo are the most important examples. The *pro-R* and *pro-S* designations

---

[ii] Glycerol has two enantiopic CH_2OH groups, which make the central carbon atom chiral if they are differently substituted, as they usually are in glycerides. Unfortunately a completely different, Fischer-based convention applies in lipid chemistry. Glycerol is written in the Fischer convention with 2-OH on the left (*i.e.* L); by definition OH1 is then at the top and OH3 at the bottom, so that the numbering, not a stereochemical designator, defines stereochemistry. The two compounds below (*sn*-1-acyl glycerol and *sn*-3-acyl glycerol) are enantiomers (*sn* = **s**ystematic **n**omenclature).

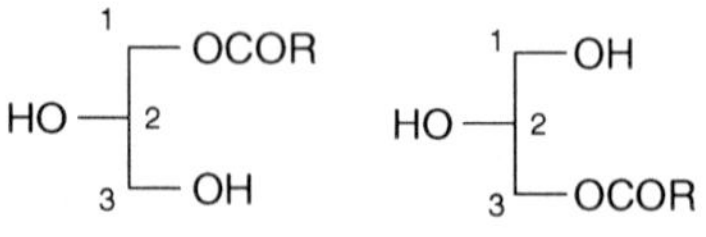

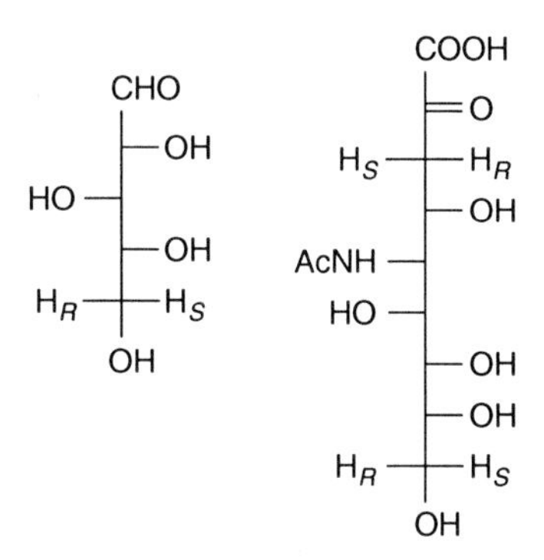

Figure 1.12 Fischer representation of the diastereotopic hydrogens in D-xylose (right) and N-acetylneuraminic acid (left).

still apply. These are shown (for D-xylose and N-acetyl neuraminic acid) in Figure 1.12.

1.2 NUCLEOPHILIC ADDITIONS TO CARBONYL GROUPS OF SUGARS

1.2.1 Additions of Water, and OH Groups of the Same Sugar

1.2.1.1 Equilibria. Equilibrium constants for the addition of alcohols to aldehydes are in the molar range,[7] but the up to 10^8 M entropic advantage possessed by intramolecular reactions compared to their intermolecular counterparts[8] means that reducing sugars exist predominantly as cyclic hemiacetals or hemiketals. The cyclic forms are termed *furanose* if the ring is five-membered and *pyranose* if the ring is six-membered, after the unsaturated heterocyles furan and pyran. The strain in four-membered rings means that they are present in negligible quantities, but seven-membered rings contribute 1.6% to the equilibrium composition of aqueous idose,[9] where pyranose forms are conformationally destabilised (see Chapter 2), and are important with 5-deoxy D-*ribo*-hexose,[10] where pyranose forms cannot form (perversely, these seven-membered ring forms are called septanoses, rather than oxepinoses, by analogy with pyranose and furanose).

Addition of an internal hydroxyl group generates a new asymmetric centre, and results in the cyclic hemiacetals existing as two diastereomers. Diastereomers differing only in their configuration at the masked carbonyl group of a sugar are termed anomers (from the Greek ανομερια, a place which is up – actually, the name of a real mountain village in Kefallinia). Anomers are distinguished by the designators α and β; originally the α anomer was defined as the more dextrorotatory one in the D series, and the more laevorotatory one in the L series; later the α-pyanose or furanose form was defined as the one with the OH on the right-hand-side in the D series and the left-hand-side in the L series, when written in the Fischer convention (Figure 1.13).

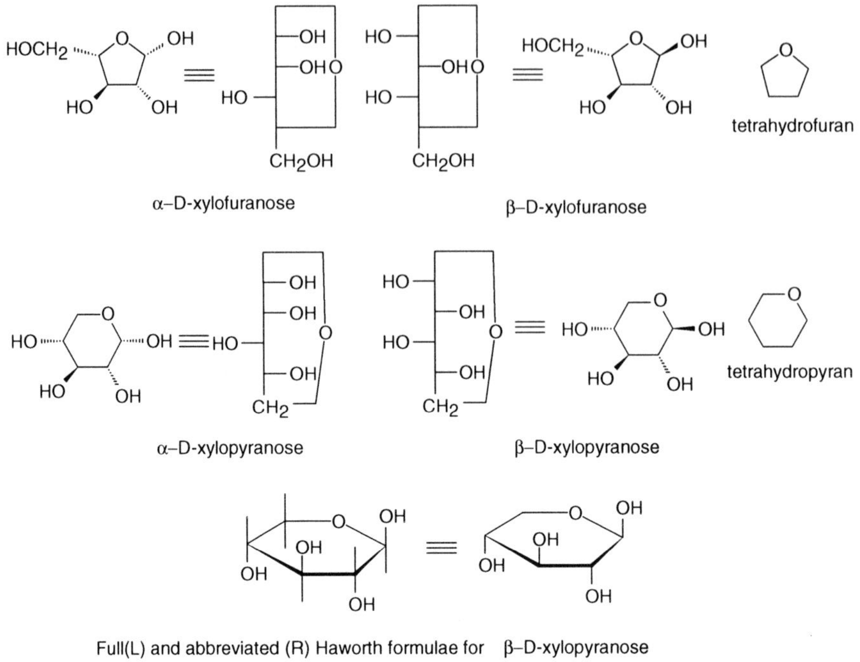

Figure 1.13 Fischer and Mills representations of α- and β-xylofuranose and α- and β-xylopyranose. Also shown are full and abbreviated Haworth representations of β-D-xylopyranose.

Fischer representations of α and β xylofuranose and xylopyranose are given in Figure 1.13, together with the corresponding Mills representations (in which the ring is considered to be in the plane of the paper, so that wedges represent bonds coming out of it, and hatchings, bonds which go into it). Fischer representations of sugar rings put sharp angles in the C–O bonds in the ring, thus defying modern convention, which would understand a non-existent CH_2 group at the angle. In general Mills representations for representing sugar rings are to be preferred, unless the conformation is known with confidence, when conformational representations can be used.

Also used are Haworth formulae, in which an imaginary flat ring is viewed in perspective. These have nothing to recommend them, but are quite widespread in the chemistry literature before the 1990s, and persist in the technology literature.

Competing with intramolecular hemiacetal formation in aqueous solutions of reducing sugars is hydration, intermolecular addition of water to form a gem-diol. Addition of water or alcohols to carbonyl groups is favoured by electron withdrawing substituents attached to the carbonyl group. The lower the degree of s character in the carbon orbital forming a bond with an electron-withdrawing substituent, the easier it is to satisfy that substituent's electron

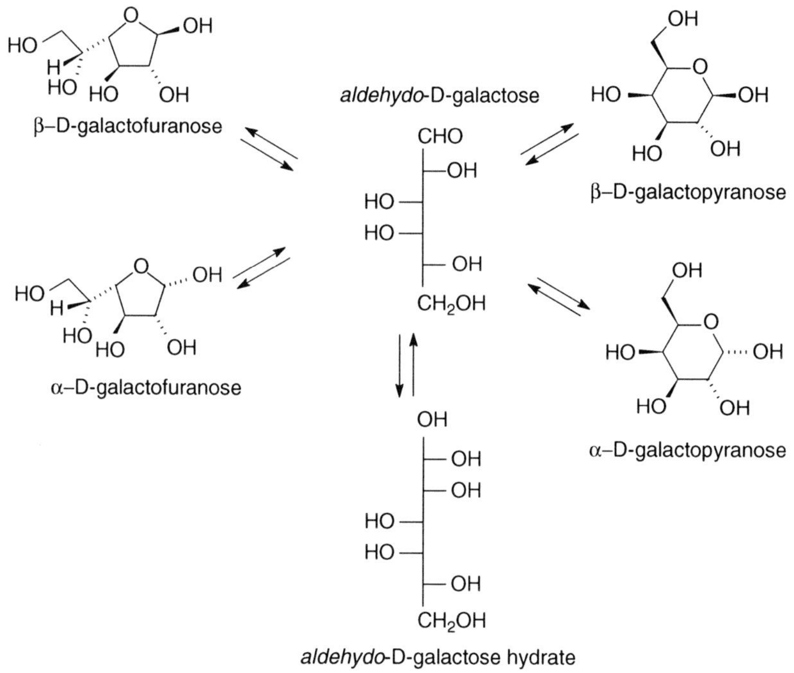

Figure 1.14 All detectable forms of D-galactose in aqueous solution.

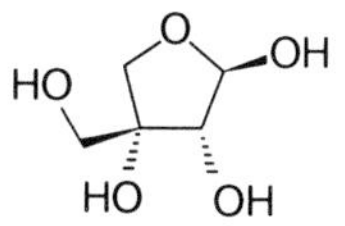

Figure 1.15 β-D-Apifuranose.

demand. Therefore, rehybridisation of the carbonyl carbon from sp^2 to sp^3, as on hydration, is favoured by electron-withdrawing substituents.

The multiple structures available to a simple sugar (galactose) are illustrated in Figure 1.14.

Ring formation from either of the two diastereotopic hydroxymethyl groups of the one important branched-chain sugar, apiose (straight chain formula $(CH_2OH)_2COH$ CHOH-CHO) is in principle possible, and creates two asymmetric centres, at the anomeric carbon and another at the 3 position. Current, *ad hoc*, usage is to describe the structure in Figure 1.15 as β-D-apifuranose.

Table 1.1 gives the equilibrium compositions of the common sugars, in some cases at several temperatures. The following trends are apparent:

(i) Straight-chain aldoses are much more hydrated than straight-chain ketoses (see especially glyceraldehyde and dihydroxyacetone), but, as

Table 1.1 Equilibrium composition of solutions of reducing sugars.

Sugar	*Temperature (°C)*	*α-Furanose*	*β-Furanose*	*α-Pyranose*	*β-Pyranose*	*Aldehydo*	*Hydrate*	*Comments*	*Ref.*
Glycolaldehyde	–	–	–	–	–	4	70	At 0.1 M – rest dimer	67
DL-Glyceraldehyde	24	–	–	–	–	2	37	At 1.0 M– rest dimer	68
DL-Glyceraldehyde	45	–	–	–	–	6	44		68
DL-Glyceraldehyde	75	–	–	–	–	16	53		68
Dihydroxyacetone	22	–	–	–	–	80	20		69
D-Erythrose	36	25	63	–	–	–	12		70
D-Erythrose	24	ND	ND	–	–	1	11		68
D-Erythrose	75	ND	ND	–	–	5	6		68
D-Threose	28	48	35	–	–	1	16		68
D-Threose	45	51	36	–	–	3	10		68
D-Arabinose	31	3	1	60	35	–	–		71
D-Lyxose	31	2	1	70	28	–	–		71
D-Lyxose	44	(3.5)	(3.5)	69	27	–	–	α*f*, β*f* not distinguished	71
D-Ribose	31	7	13	21	59	–	–		71
D-Ribose	44	9	15	17	59	–	–		71
D-Xylose	31	(<1)	(<1)	36	63	–	–	Furanoses detected only	71
	44	(<1)	(<1)	37	62	–	–		
D-Allose	24	3	5	15	76	–	–		71
D-Allose	30	3.0	5.3	14.7	77.1	0.003	0.006	1-^{13}C-enriched	72
D-Allose	44	4	6	17	73	–	–		71
D-Altrose	30	18.6	13.4	26.9	41.0	0.014	0.079	1-^{13}C-enriched	72
D-Altrose	44	18	15	29	37	–	–		71

D-Galactose	15	1.2	2.5	33	64	–	–		73
	25	1.8	3.1	32	62	–	–		73
	30	2.3	3.7	31.2	62.8	0.006	0.046	1-^{13}C-enriched	72
	35	4	3	29	64	–	–		71
2-Fluoro-2-deoxy-D-galactose	35	1.0	2.2	41.0	55.7	–	–		74
3-Fluoro-3-deoxy-D-galactose	35	0.7	1.6	40	58	–	–		74
D-Glucose	30	0.11	0.28	37.6	62.0	0.004	0.006	1-^{13}C-enriched	72
D-Glucose	31	–	–	38	62	–	–		70
D-Glucose	44	0.14	–	37	63	–	–		71, 75 (*f*, 43 °C)
D-Gulose	30	0.94	3.04	12.2	83.7	0.006	0.08	1-^{13}C-enriched	72
D-Idose	30	12.14	16.12	33.7	37.4	0.09	0.07	1-^{13}C-enriched	72
D-Idose	37	12.3	15.2	37.9	32.9	α,β sept-	anose1.6		9
D-Idose	44	13	19	37	31	–	–		71
D-Mannose	30	0.64	0.24	66.2	32.8	0.004	0.022		72
D-Mannose	44	0.6	0.3	66	34	–	–		71, 76 (*f*, 36 °C)
D-Talose	30	17.9	11.1	42.2	28.7	0.03	0.05		72
	44	17	14	37	32	–	–		71
D-Fructose	27	4	21	Trace	75	–	–		77
D-Fructose	85	11	33	Trace	66	–	–		77
1-Deoxy-D-fructose	85	10	13	7	44	26			78
L-Sorbose	27	2	–	98	–	–	–		77
L-Sorbose	85	9	–	91	–	–	–		77
D-Tagatose	27	1	4	79	16	–	–		77
D-Psicose	27	39	15	22	24	–	–		77
N-Acetylneuraminic acid	25	–	–	5	95	–	–		79

would be expected for an electrically neutral dissociative equilibrium,[iii] hydration decreases with temperature.

(ii) Furanose forms of a given sugar are favoured over pyranose forms by increase in temperature, probably because of the extra internal rotation possible in furanose forms.

(iii) α/β pyranose ratios are temperature-insensitive.

(iv) Open-chain forms of the sugar are favoured by removal of hydroxy groups (see the contrast between fructose and 1-deoxy fructose).

The differing proportions of furanose and pyranose forms for each sugar arise from conformational effects, discussed in Section 2.8.

1.2.1.2 Kinetics of Mutarotation. When a crystalline aldose or ketose, consisting of only one form of the sugar, is dissolved in water, it is converted to the equilibrium mixture given in Table 1.1. This conversion takes place, for the most part, on the time-scale of minutes to hours, and is associated with a change in the optical rotation. For the first century in the history of carbohydrate chemistry (until the advent of ^{13}C NMR in the 1970s) optical rotation was the only instrumental technique that could be used to monitor the reactions of sugars in water.[iv] Therefore much of the basic physical organic chemistry of mutarotation was established using polarimetry.

In general, if a crystalline sugar on dissolving in water produces a solution which changes exponentially to an equilibrium value, then the equilibrium composition of that sugar (Table 1.1) will be virtually exclusively the two pyranoses. Thus, glucose, xylose and mannose show such "simple" mutarotation curves. If the mutarotation curve is not a simple exponential, it is called "complex". Such curves are seen for sugars which have appreciable quantities of furanose forms at equilibrium – the curve for ribose in fact goes through a maximum. Using 1-^{13}C labelled sugars it is possible using NMR saturation transfer experiments to measure the rate of conversion of the acyclic form to the cyclic forms (and the hydrate, if present).[11] A saturating rf pulse is applied to the C1 resonance of the acyclic form. During this pulse the open form is converted to the cyclic forms by ring closure chemistry, and depending on the length of the pulse, more and more of the cyclic forms find themselves with a C1 atom which is magnetically saturated. It turns out that the kinetics of loss of intensity in the ring form resonances are simply first order: if I_t is the intensity of the resonance of a particular cyclic form after a saturating pulse of duration t applied to the resonance of the acyclic form,

[iii] Entropy changes for such equilibria are dominated by the translational partition function, since translational energy is quantised in much smaller units than rotational or vibrational energy. Reactions in which there is an increase in the number of translational modes are thus favoured by increasing temperature.

[iv] ^{1}H NMR was available from the 1960s, but required spectra to be run in D_2O, the UV transitions of sugars were obscured by atmospheric oxygen, and their IR bands by water solvent.

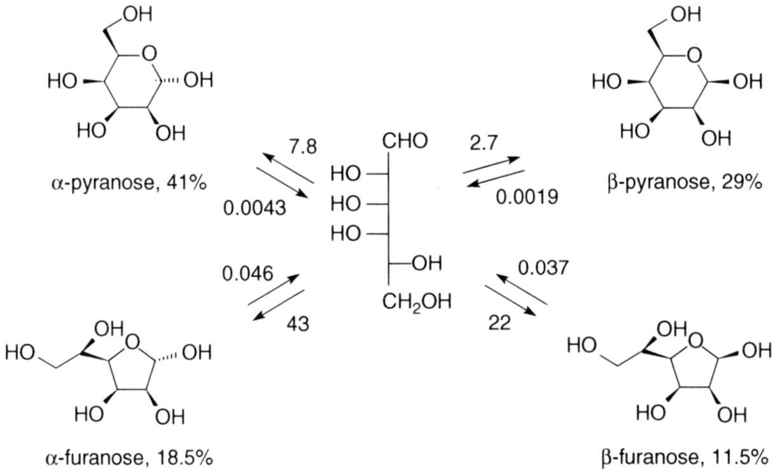

α-pyranose, 41%

β-pyranose, 29%

α-furanose, 18.5%

β-furanose, 11.5%

Figure 1.16 Rate constants (s^{-1}) for the mutarotation of talose, in 50 mM sodium acetate buffer, pH 4.0, at 28°.

and I_∞ intensity after a saturating pulse of infinite duration, the rate constant for ring closure is given by equation 1.1.

$$-kt = \ln(I_t - I_\infty) \quad (1.1)$$

The rate constants for ring opening are obtained from the position of equilibrium. The procedure can be expanded by two-dimensional techniques, which have been used to determine all the microscopic rate constants for the mutarotation of talose (Figure 1.16).[12] The greater rate constants for closure to furanose forms, despite these being less thermodynamically stable than pyranose forms, is a general phenomenon which governs not only mutarotation but also the Fischer glycoside synthesis.

For many years there was discussion over whether the mutarotation of galactose was "simple" or "complex" – the proportions of furanose and straight chain forms are higher than for glucose, but not as high as ribose.[13] The answer, of course, depends on the precision of the polarimeter used, and its ability to detect slight deviations from exponential time-courses of rotation, with galactose mutarotation measured on modern polarimeters capable of recording to the nearest millidegree being clearly complex.

The differential equations describing complex mutarotations resist explicit solution, and are usually solved numerically. Much mechanistic work has been done, however, on simple mutarotations monitored polarimetrically. Such reactions are can be treated as reversible first-order reactions, with the mutarotation rate constant corresponding to the sum of the ring-opening reactions of the two pyranose forms, which are the rate determining steps in each direction (Equation 1.2).

$$k_{\text{mut}} = k^{\alpha}_{\text{open}} + k^{\beta}_{\text{open}} \quad (1.2)$$

1.2.1.3 General Acid and Base Catalysis of Mutarotation. Mutarotation is catalysed by acids and bases, with the base-catalysed reaction being generally more efficient. However, both $[H_3O^+]$ and $[OH^-]$-catalysed processes for glucose mutarotation are negligible between pH 4 and 6, where there is a dominant uncatalysed, "water" reaction. Glucose mutarotation was one of the first reactions for which the phenomenon of *general* acid and base catalysis was described.[14] The experimental manifestation phenomenon is that the rate is dependent on the concentration of *undissociated* acids or *unprotonated* bases at constant pH. It is thus distinct from *specific* acid or base catalysis, in which the rate is proportional to $[H_3O^+]$ or $[OH^-]$ only. In molecular terms, general acid or base catalysis appears when the fully protonated (or deprotonated) substrate is "too unstable to exist", *i.e.* would have an estimated lifetime shorter than a molecular vibration, so that the proton is transferred as bonds between other atoms are broken. A particular example is discussed in detail in Section 3.7.3.

Experimentally, general acid and/or base catalysis is manifested by a dependence of first-order rate constant on the concentration of aqueous buffer, at constant buffer ratio and hence constant pH. Altering buffer concentration of course alters the ionic environment, and this by itself can have kinetic consequences. The problem is circumvented in two ways. The first is to use low concentrations of buffer and swamping concentrations (1–2 M) of an inert electrolyte such as $NaClO_4$ or KCl, the argument being that the *proportional* change in the ionic environment caused by changing the buffer concentration is small enough to be neglected. The second is to maintain ionic strength constant. The concept of ionic strength, defined as $\frac{1}{2}\Sigma c_i z_i^2$, where c_i is the concentration of the *i*th ion which has a unit charge of z_i, comes from the Debye-Hückel theory of electrolytes, which applies only to very dilute solutions. However, it is often applied to ionic solutions with electrolyte concentrations in the molar range, even in the presence of organic cosolvents. Not surprisingly, spurious observations of buffer catalysis when "ionic strength" is maintained constant under these conditions occur regularly.[15] The ever-present possibility of specific salt effects is minimised if inert electrolyte and catalysing buffer systems involve only 1:1 electrolytes.

With these provisos, buffer catalysis is generally measured from gradients of plots of observed first-order rate constants against buffer concentration; the separation of the second-order "buffer catalysis" constant into catalytic constants for acid and basic components of the buffer is achieved by secondary plots of these second-order constants against buffer ratio. This is illustrated in Figure 1.17.

In general, it is found that mutarotation is catalysed by acids and bases, and also exhibits a "water" reaction, so that we can write equation 1.3

$$k_{\text{mut}} = k_0 + \Sigma k_{\text{HA}}[\text{HA}] + \Sigma k_{\text{B}}[\text{B}] \quad (1.3)$$

where the solvated proton and hydroxide ion are included in HA and B respectively.

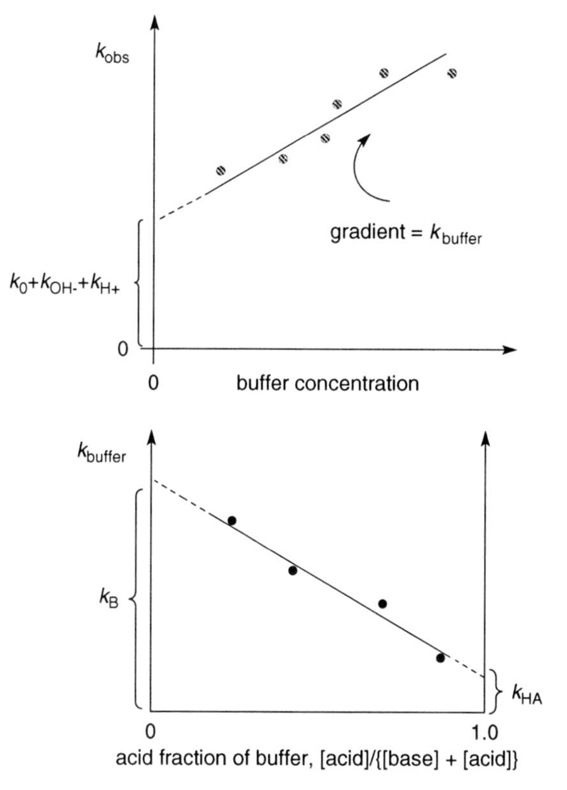

Figure 1.17 Cartoon of the observation of buffer catalysis and separation of the catalytic constants into acid and base components. The intercepts on a series of primary (LHS) plots permit a pH-rate profile at zero buffer concentration to be constructed, and hence the values of k_o, the water-catalysed reaction, k_{OH^-}, the hydroxide-ion catalysed reaction, and k_{H^+}, the proton-catalysed reaction, to be extracted.

When the appropriate statistical corrections are made, the logarithms of the catalytic constants k_{HA} and k_B are linear functions of the pK_a values of HA and BH^+, respectively. This is the Brønsted catalysis law. We can write equation 1.4 for general acid catalysis and equation 1.5 for general base catalysis, where p is the number of acid sites and q is the number of basic sites in the catalyst. These statistical factors arise in the LHS of equations 1.4 and 1.5 because each single encounter between the catalyst and substrate has a p- or q-fold higher chance of being reactive than one with a unidentate molecule. The statistical factors on the RHS arise because the operative K_a is the microscopic one, of the individual catalysing group, rather than the macroscopic one.

Thus, for a carboxylic acid as a general acid or carboxylate as a general base $p = 1$ and $q = 2$: in the acid dissociation equilibrium there is only one OH group to lose a proton from, but two basic sites to recombine, so the OH groups of carboxylic acids are more acidic by a factor of 2 compared with phenols of the same macroscopic pK_a. Likewise, for $H_2PO_4^-$ acting as a general acid or HPO_4^{2-} acting as a general base, p = 2, q = 3.

$$\log_{10}(k_{HA}/p) = \text{constant} + \alpha \log_{10}(qK_a/p) \quad (1.4)$$

$$\log_{10}(k_B/q) = \text{constant} - \beta \log_{10}(qK_a/p) \quad (1.5)$$

The Brønsted parameters α and β measure the degree of proton donation or removal. They have the values α = 0.26, β = 0.36 for the mutarotation of glucose.[14] These numbers will be a weighted average for ring opening of the two anomers of glucose. As the parameters α and β approach 1.0, it becomes increasingly difficult to detect the general acid- or base-catalysed pathway in the presence of a proton- or hydroxide-catalysed pathway. Consider an attempt to detect general acid catalysis by trichloroacetic acid (pK_a 1.4 {K_a = 0.04}, p = 1, q = 2), at pH 1.4. We take the pK_a of the proton as $-\log_{10}55$ (water is 55 M in itself, and $K_a = [H^+][A^-]/[HA]$, only in this case $[H^+] \equiv [HA]$), and p = 3, q = 0.[v] Putting these values separately into equation 1.4, and subtracting, results in equation 1.6.

$$\log_{10}\left(k_{H_3O^+}/3\right) - \log_{10}(k_{TCA}) = \alpha \log_{10}(55/3) - \alpha \log_{10}(2 \times 0.04) = 2.35\alpha \quad (1.6)$$

If the proton-catalysed reaction proceeds with $k_{obs} = 1 \times 10^{-4}\,s^{-1}$ (a typical easily-followed rate) at pH 1.4, then $k_{H_3O^+} = 2.5 \times 10^{-3} M^{-1} s^{-1}$. If α = 0.9, $k_{TCA} = 6.4 \times 10^{-6}\,M^{-1}\,s^{-1}$. Therefore, to observe a 10% increase in k_{obs}, about the smallest that can be reliably detected, 1.6 M trichloroacetic acid, or 3.2 M 1:1 buffer, would be required. If, on the other hand, α = 0.1, equation 1.6 gives $k_{TCA} = 4.9 \times 10^{-4}\,M^{-1}\,s^{-1}$, and only 20 mM trichloroacetic acid would be required to produce a 10% increase in k_{obs}.

The Brønsted catalysis law was the first linear free energy relationship to be discovered. Such relationships occur in many areas of chemistry and biochemistry, and correlate the effects of a series of small structural changes on a rate constant and an equilibrium constant, or two rate constants. They are established by the existence of linear plots of the logarithms of rate constants against equilibrium constants or rate constants of a different reaction. The homologous changes might be in the structure of a small molecule – in the classic Brønsted treatment, the structure and hence pK_a of an acid or base – or single amino acid changes in the structure of an enzyme.

Homologous changes in small molecules are generally used to probe electron demand at a reaction centre. Classic Brønsted relationships are particularly simple, since they correlate rates and equilibria for a single reaction (aqueous K_a values being necessarily proportional to equilibrium constants for proton

[v] This is somewhat controversial – some authors consider the solvated proton just that, so that p = 1.

transfer to bases other than water). α is usually considered to represent the degree of proton transfer, the reasoning being as follows. If a series of changes in acid structure have only half the effect on the rate of the catalysed reaction that they do on acid strength, *i.e.* $\alpha = 0.5$, then the change in electron demand at the acid group is half that which occurs during complete ionisation of the acid. It is then generally assumed that this is equivalent to the proton being half transferred at the transition state. Values of α and β in classic Brønsted plots thus usually lie between 0 and 1, although examples are known, usually involving acids whose deprotonation results in delocalisation of charge well away from the original proton-bearing atom, of values outside this range. Thus, deprotonation of primary nitroalkanes by pyridine exhibits an α value of 1.89, and changes in the structure of secondary nitroalkanes can even result in effects on kinetic and thermodynamic acidity in opposite senses (negative α values).[16]

Linear free energy relationships are extrathermodynamic – they are not predicted by thermodynamics, and their foundation is experimental. One theoretical treatment of proton transfer, indeed, predicts that Brønsted plots should be shallow parabolas, not straight lines. The Marcus theory of electron transfer, originally formulated for simple electron transfers between transition metal complexes,[17] considers the distortions that have to be made in the donor or acceptor during the transfer. It makes the simplest possible assumptions, that the energy of each complex varies as the square of its displacement from the energy minimum (*i.e.* restoring force proportional to displacement, Hooke's law), and that the energy barrier is given by the height of the point of intersection of the two parabolas. The success of Marcus theory for electron transfer reactions, and has led to its extension to proton and even methyl transfer.

As applied to proton transfers,[18] the theory takes the form of equations 1.7 and 1.8, where k_{diff} is the diffusion-controlled encounter rate. $\Delta G^{\ddagger}$ is the free energy of activation defined by equation 1.8, ΔG^{o} being the standard equilibrium free energy change and $\Delta G^{\ddagger}{}_{o}$ the "intrinsic barrier", the free energy of activation when $\Delta G^{o} = 0$, a constant of the particular type of reaction.

$$k_{\mathrm{obs}} = k_{\mathrm{diff}} \exp(-\Delta G^{\ddagger}/RT) \tag{1.7}$$

$$-\Delta G^{\ddagger} = -\Delta G^{\ddagger}_{o}(1 + (\Delta G^{\circ}/4\Delta G^{\ddagger}_{o}))^{2} \tag{1.8}$$

Equation 1.8 describes a shallow parabola which is observed, more or less, for deprotonations of carbonyl compounds,[18] but not for amine-catalysed deprotonation and ring opening of benzisoxazoles (to o-cyanophenolates),[19] where Brønsted plots against pK_{a}s of both amines and benzisoxazoles are strictly linear. It seems to the author that the chemical physics of the transfer of electrons, light particles with de Broglie wavelengths similar to molecular dimensions, should be very different from that of the transfer of heavier, better spatially defined species, and that the linearity or otherwise of Brøsted plots depends on the precise characteristics of the intramolecular potential energy curve. Bell, for example, found that intersecting Morse potential curves gave less curved Brønsted plots than intersecting parabolas.[20]

For proton transfers unaccompanied by any rearrangements of heavy atoms, $\Delta G^{\ddagger}_{o} < \Delta G^{o}$, so that the dominating term in equation 1.7 is k_{diff}. This has the value of $10^{10}\,M^{-1}s^{-1}$ in water at ambient temperatures, so that equation 1.7 predicted what had been previously found experimentally, that all thermodynamically downhill proton transfers occur at the diffusion rate. Necessarily, therefore, all the corresponding uphill proton transfers occur at a rate or $10^{10-\Delta pKa}$, where ΔpK_a is the difference between the acidities of donor and acceptor. Brønsted plots consisting of a horizontal portion of $\log_{10}k_{HA} = 10$ intercepting with a portion of gradient –1 at $\Delta pK_a = 0$ are diagnostic of rapid proton transfer, and are termed Eigen plots.[vi] (The frequently-encountered statement that proton transfer between electronegative atoms is diffusion-controlled is economical with the truth, since proton transfers from phenyl acetylenes[21] and HCN[22] are also diffusion-controlled. In the deprotonated forms of both these carbon acids the charge lies in the sp orbital previously forming the C–H bond, whereas with the so-called "anomalous" acids (carbonyl compounds and nitroalkanes) charge of the anion is delocalised away from the carbon from which the proton is removed).

For reactions in aqueous solution, it is best to use experimental aqueous pK_a values in linear free energy correlations for determining, *e.g.* the sensitivity of reactions to nucleophile or base strength or leaving group nucleofugacity. For reactions in non-aqueous solution there are many variations on the classic Hammett approach.[23] This defined a σ value for a meta or para substituent in a benzene ring (σ_x) as the acid-strengthening effect of the substitution on the ionisation of benzoic acid (*i.e.* ΔpK_a, acid strengthening/electron withdrawing being positive). The ρ value of a reaction was then the gradient of a plot of the $\log_{10}k_{obs}$ against σ, and was a measure of the change in electron demand at the transition state (equation 1.9). Such a plot is of course equivalent to a plot of $\Delta G^{\ddagger}$ for the reaction against ΔG^{o} for the ionisation of the appropriate benzoic acids.

$$\sigma_x = \log_{10}(K_{a(x)}/K_{a(0)}); \quad \log_{10}(k_x/k_0) = \rho\sigma \qquad (1.9)$$

Some of the modifications in the basic Hammett equation proposed in the subsequent 70 years, such as the use of different σ_p constants where there is direct conjugation between the reaction centre and the substituent (σ^+ or σ^-, depending on whether positive or negative charge is being generated), have stood the test of time. Less happy have been attempts to separate conjugative and inductive effects with two-parameter equations, the popular Yukawa-Tsuno equation (1.10) in particular suffering from the scaling of the "resonance" r parameter by the ordinary ρ value, although the (neglected) Young-Jencks

---

[vi] From Manfred Eigen, Nobel Laureate in 1967 for his work on rapid reactions, including proton transfers. Unrelated, except by a quirk of German surname coinage, to eigenvalue, eigenvector *etc.* in quantum mechanics (eigen = "own, characteristic").

modification[24] (equation 1.11) does not suffer from this flaw.

$$\log_{10}(k/k_0) = \rho(\sigma + r(\sigma^+ - \sigma)) \quad (1.10)$$

$$\log_{10}(k/k_0) = \rho\sigma^n + \rho^r(\sigma^+ - \sigma^n) \quad (1.11)$$

The Taft adaptation of equation 1.9 to aliphatic systems (with substituent parameters σ_I) does not distinguish between through-space and through-bond effects. Very recently it has been convincingly shown that this can mislead: effects of substituents in carbohydrate frameworks depend on their orientation.[25] Many supposed inductive effects in aliphatic systems are in fact the simple effects of electrostatic dipoles, as discussed in detail in the carbohydrate context in Chapters 3 and 6.

1.2.1.4 Kinetic Isotope Effects on Mutarotation

1.2.1.4.1 Origin of kinetic isotope effects.[26] By virtue of the Born-Oppenheimer approximation, which states that the motions of electrons and of nuclei can be considered separately, the potential energy surface of a molecule does not vary with the isotopic nature of nuclei moving on it. This makes isotopic substitution a uniquely valuable probe of mechanism, since, unlike electronic substituent effects, the probe does not alter what is being investigated.

In general isotopic substitution alters reaction rate according to the Bigeleisen equation,[27] which is given in simplified form as equation 1.12

$$k_L/k_H = \text{MMI} \times \text{EXC} \times \text{ZPE} \quad (1.12)$$

EXC is the ratio of ratios of partition functions associated with different occupancy of the various vibrational levels consequent upon isotopic substitution, in ground and transition states, for both isotopic species. Since at ordinary temperatures organic molecules are practically exclusively in their zeroth vibrational states anyway, in practice at ordinary temperatures EXC = 1. MMI is a similar ratio of ratios of partition functions for translation and rotation: it can be neglected for hydrogen isotopes, but for other elements can be significant. The most important term is ZPE, which arises from the different zero point energies of isotopic species. For each vibrational mode, there is a zero point energy of $\frac{1}{2}h\nu$, ν being the classical vibration frequency, $(\kappa/m)^{\frac{1}{2}}/2\pi$, where m is the reduced mass and κ is the bond force constant. (For hydrogen isotope effects most of the motion of a bond vibration is in the hydrogen, so that m can be equated with the mass of the hydron,[vii] but for other vibrations this is not so – see Section 3.10.4). The Born-Oppenheimer approximation requires κ to not change on isotopic substitution. Heavier isotopomers have less zero point energy, and therefore more energy is required to break or weaken the bond.

Zero point energy kinetic isotope effects lie exclusively in the energy of activation, so that rate-ratios have the form of equation 1.13, where $\Delta\Delta$(ZPE) is

---

[vii] A general term for any isotope of hydrogen. In formulae given the symbol L, meaning "H, D or T, as appropriate".

the difference between isotopic species in the difference in zero point energy between ground and transition state. This leads directly to equation 1.14, describing the temperature-dependence of these semi-classical effects

$$k_L/k_H = \exp(-\Delta\Delta(\text{ZPE})/\text{RT}) \quad (1.13)$$

$$\text{T} \ln (k_L/k_H) = \text{constant} \quad (1.14)$$

With the further approximation that the rest of the molecule is heavy compared to the hydron, it is seen that the zero point energy of a deuterated species is $1/\sqrt{2}$, and that of a tritiated species $1/\sqrt{3}$, that of a protiated species. Consequently, at ordinary temperatures, the Swain-Schaad relation (equation 1.15) holds.

$$k_H/k_T = (k_H/k_D)^{1.44} \quad (1.15)$$

The foregoing applies to both primary kinetic isotope effects, where the bond to the site of isotopic substitution is made or broken, and secondary isotope effects, where it is weakened or strengthened.

1.2.1.4.2 Hydrogen tunnelling. Aliphatic C–H stretching vibrations are associated with an infrared absorbance at $2800\,\text{cm}^{-1}$. If we approximate aliphatic C–L bonds as diatomics X–L, with X infinitely heavy, then the zero point energy[viii] of X–H is $16.8\,\text{kJ mol}^{-1}$, of the X–D bond is $16.8/\sqrt{2}$ (11.9) kJ mol^{-1}, and of the X–T bond $16.8/\sqrt{3}$ (9.7) kJ mol^{-1}. Therefore the maximum possible kinetic isotope effects, corresponding to complete rupture of X–L at the transition state, with no compensating bonding elsewhere, arises from the difference in activation energy equal to differences in zero point energy. Thus $k_H/k_D = e^{4.9/RT}$, or 7.3, and $k_H/k_T = e^{7.1/RT}$, or 17.8 at 25 °C. However, experimental values well in excess of these are sometimes observed, particularly in enzyme systems – abstraction of the key substrate hydron in lipoxygenase action is associated with a k_H/k_D value of 86 at room temperature.[28] The reason appears to be quantum mechanical tunnelling, which permits a particle to cross an energy barrier even when it does not have enough energy to surmount it. Tunnelling probability has a very high dependence on mass, which accounts for the greater than classical values of isotope effects. Tunnelling is also favoured by high, narrow barriers.

Whilst kinetic isotope effects in excess of the above semiclassical theoretical maxima are an indication of tunnelling, they can in principle have other causes (for example, if the approximation of lumping the rest of the molecule into a notional atom X is abandoned, then bending frequencies as well as stretching frequencies have to be considered and maximal semi-classical isotope effects double). Other evidence is therefore required before tunnelling can be

---

[viii] $\frac{1}{2}$Nhυ per mole, where N is Avogadro's number (6.022×10^{23}), h is Planck's constant (6.626×10^{34} J s) and υ is the frequency, in this case the speed of light ($3 \times 10^{10}\,\text{cm s}^{-1}$) $\times\, 2800\,\text{cm}^{-1}$, or $8.4 \times 10^{13}\,\text{s}^{-1}$.

established. Like all quantum-mechanical processes, tunnelling probability is temperature-independent. Such a temperature-independent contribution in principle results in non-linear Arrhenius plots, with ln$k$ approaching the temperature-independent tunnelling contribution at high values of 1/T. Over small temperature ranges, this is experienced as an apparently anomalously low value of the pre-exponential factor A, with A_H/A_D values <0.5 being considered diagnostic for tunnelling.

It has been suggested that protein motions in enzyme-substrate complexes play a catalytic role by transiently altering the shapes of energy barriers to permit tunnelling ("vibrationally assisted tunnelling")[29] but the experimental basis for this appears insecure.[30]

1.2.1.4.3 Solvent isotope effects. Much mechanistic information can be obtained about reactions involving proton transfer from solvent kinetic isotope effects, particularly in solvents of mixed isotopic composition. For practical reasons work is essentially confined to H/D effects, especially those in water. Unlike ordinary primary hydrogen isotope effects, solvent isotope effects have to take into account a host of exchangeable sites, subject to equilibrium as well as kinetic isotope effects. A key concept is that of the fractionation factor, φ, [31]which is the deuterium occupancy of a site in a 1:1 H_2O/D_2O mixture: more formally it is defined by equation 1.16[ix]

$$\varphi = \frac{F(1-n)}{(1-F)n} \tag{1.16}$$

where n is the atom fraction of deuterium and F the atom fraction in the site in question, which can be measured by, *e.g.* measurement of NMR chemical shifts as a function of n in H_2O/ D_2O mixtures. Fractionation factors essentially measure the steepness of the potential well in which the hydron finds itself, with deuterium accumulating in steep potential wells, and thus minimising the zero point energy of the system.[x] The fractionation factors of alcohols, amines, and carboxylic acids are close to unity, but the fractionation factors of RSL are ~0.5, and of L_3O^+, 0.69. The fractionation factor of L in OL^- is 1.25, but the three lone pairs on oxygen are all hydrogen bonded to a water molecule, and have fractionation factors of 0.7.

Specific acid catalysed processes go faster in D_2O than in H_2O by a factor of 2-3; in the pre-equilibrium process, the fractionation factor of the (fully) transferred hydron changes from $\varphi=0.69$ to somewhere between 0.69 and 1.0, whilst the two remaining hydrons change from $\varphi=0.69$ to $\varphi=1.00$. The value of k_{D_2O}/k_{H_2O} thus lies between $(0.69)^{-3}$ (3.0) and $(0.69)^{-2}$ (2.1).

[ix] This treatment assumes the rule of the geometric mean, which requires the equilibrium constant $[HOD]^2/[H_2O][D_2O]$ to be exactly 4.0. In fact it is not, (W. A. Van Hook, *J. Chem. Soc., Chem. Commun.* 1972, 479) and small corrections have to be applied.

[x] Since k, which measures the steepness of the potential well, remains unchanged on isotopic substitution, the zero point energy of a site occupied by a deuteron is smaller than one occupied by a proton by a factor of $1/\sqrt{2}$.

Measurements in mixtures of H_2O and D_2O give information about the number of protons changing their fractionation factors between ground state and transition state. The analysis is conducted in terms of the Gross-Butler equation, whose modern form[32] is given as equation 1.17, where k_n is the rate constant at atom fraction deuterium n, k_0 that in pure H_2O, and φ_I^{TS} and φ_I^{GS} are the fractionation factors of the ith exchangeable proton in the transition state and the ground state, respectively.

$$\frac{k_n}{k_0} = \frac{\prod(1 - \mathrm{n} + \mathrm{n}\varphi_I^{TS})}{\prod(1 - \mathrm{n} + \mathrm{n}\varphi_I^{GS})} \tag{1.17}$$

Clearly, if the fractionation factor does not change between ground state and transition state, terms in the numerator and denominator cancel. If all the ground state hydrons have fractionation factors near unity, the denominator itself becomes unity. In this case the shape of the plot of k_n/k_0 against n (often called a "proton inventory") immediately gives an indication of the number of protons transferred.[33] In the case of the acetate-catalysed mutarotation of tetramethyl[xi] glucose,[34] the proton inventory is linear, suggesting that one proton only alters its fractionation factor. The proton inventory for the water reaction was bowed down (concave downwards), indicating alteration of fractionation factors of at least two protons (Figure 1.18).

Improvements in instrumentation have permitted a comprehensive study of all kinetic isotope effects in the uncatalysed (water) mutarotation of glucose itself.[36] The solvent isotope effect was slightly smaller than for tetramethyl glucose ($k_{H_2O}/k_{D_2O} = 3.0$), but the proton inventory again indicated a many-proton mechanism. The other kinetic isotope effects in the system, 1-^{13}C, 5-^{18}O, and 1,2, 3, 4,5 and 6-^{2}H, were measured by the isotopic quasi-racemate method (see Section 3.10, which also gives the equilibrium hydrogen isotope effects for glucose mutarotation). Finally, the absolute rate constants for closure of *aldehydo* glucose to the two anomers ($k_{ald\rightarrow\alpha}$ and $k_{ald\rightarrow\beta}$) were determined by ^{13}C NMR polarisation transfer. It was therefore possible to determine isotope effects for the $\alpha\rightarrow\beta$ reaction and the $\beta\rightarrow\alpha$ reaction from equations 1.18 and 1.19 (EIE = equilibrium isotope effect; KIE = kinetic isotope effect).

$$\mathrm{EIE} = \mathrm{KIE}_{\alpha\rightarrow\beta}/\mathrm{KIE}_{\beta\rightarrow\alpha} \tag{1.18}$$

$$\mathrm{KIE}_{obs} = \frac{\mathrm{KIE}_{\alpha\rightarrow\beta} + \mathrm{KIE}_{\beta\rightarrow\alpha}(k_{ald\rightarrow\alpha}/k_{ald\rightarrow\beta})}{1 + (k_{ald\rightarrow\alpha}/k_{ald\rightarrow\beta})} \tag{1.19}$$

Observed anomeric ^{13}C and 5-^{18}O kinetic effects (1.004 and 1.015 respectively) were both less than the quilibrium effects (defined as $[\beta]_L[\alpha]_H/[\beta]_H[\alpha]_L$) of 1.007 and 0.975, respectively, indicating little changing of bonding at these

[xi] Early workers were quite concerned that the fractionation factors of sugar hydroxyl groups would change during reaction, so much work was done on tetramethyl glucose.

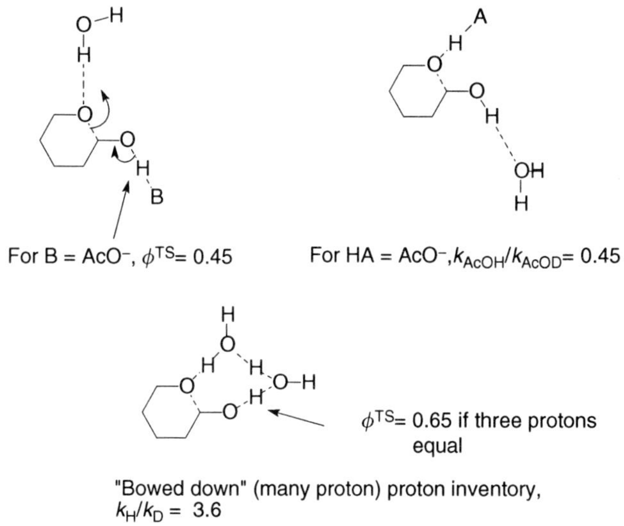

For B = AcO$^-$, ϕ^{TS}= 0.45

For HA = AcO$^-$, k_{AcOH}/k_{AcOD}= 0.45

ϕ^{TS}= 0.65 if three protons equal

"Bowed down" (many proton) proton inventory, k_H/k_D = 3.6

Figure 1.18 Transition states for the acetate-catalysed, acetic acid-catalysed and water reactions in the mutarotation of tetramethyl glucose. The additional waters for the acetate and acetic acid reactions are drawn to indicate solvation, rather than a change in bonding that would alter fractionation factors. Isotope effects are taken from ref 34 and fractionation factors calculated from their data using 1.0, rather than 1.23, for the fractionation factor of the anomeric hydroxyl. The latter was based on an implausible, equilibrium isotope effect of 4.1 on the acid dissociation constant of tetramethyl glucose.[35]

positions. An observed α-deuterium effect of 1.10, compared with an equilibrium effect of 1.04, though, indicated relatively significant changes. Quantum mechanical calculations supported a structure for both α- and β- transition states in which the O1H had been substantially deprotonated by a single water molecule, but the C1–O5 bond remained substantially intact, the significant α-deuterium effect arising from an inductive effect on the deprotonation, rather than a rehybridisation.

1.2.1.5 Mutarotation of 5-thioglucose. The mutarotation of 5-thioglucose presents some interesting features.[37] As expected from the lower basicity of sulfur compared with oxygen, mutarotation of thioglucose is much less sensitive to acid than mutarotation of glucose itself, with the acid-catalysed part of the pH-rate profile only becoming apparent below pH 1.5. Although a "water" reaction can just be observed around pH 2, it is nothing like the broad minimum between pH 4 and 6 seen with glucose itself (and its rate at 25.0 °C, $1.8 \times 10^{-7}\,s^{-1}$, is much lower than that for glucose, $3.7 \times 10^{-4}\,s^{-1}$). Base catalysis, though, is comparably efficient. The thiol group of the straight-chain form of the sugar can undergo disulfide exchange with a suitable thiol reagent. It was expected that, as the concentration of the thiol reagent was increased, a limiting rate of reaction, corresponding to rate of ring opening,

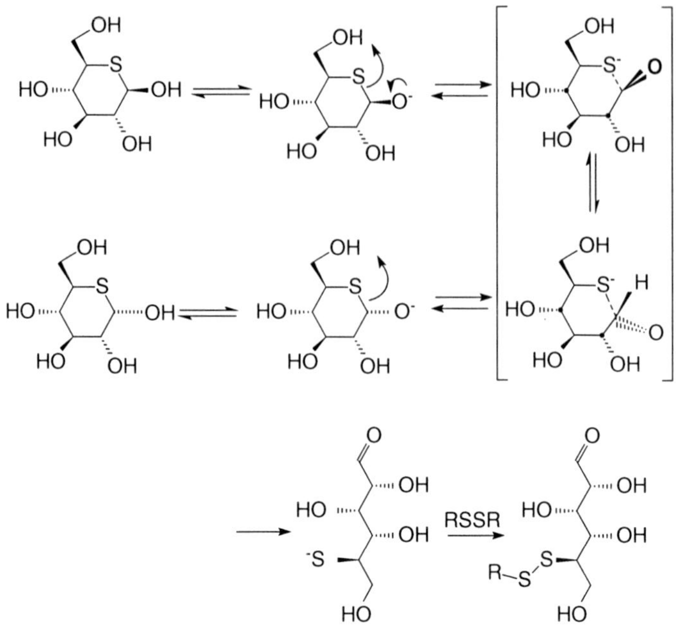

Figure 1.19 Mechanism of mutarotation and reaction with thiol reagents of 5-thioglucose.

would be observed. In fact thioglucose mutarotation is several hundredfold too fast under basic conditions for it to proceed through an open chain *aldehydo* sugar, and a thiolate–aldehyde complex, which can mutarotate but not react with thiol reagents, is thought to exist. Such a species would be similar to the oxocarbenium ion–thiophenolate complex whose dissociation is thought to be rate determining in the hydrolysis of pMeOC$_6$H$_4$CH(OEt)SPh.[38](Figure 1.19)

1-Thio aldoses are configurationally stable at the anomeric centre, since mutarotation would require the transient generation of a very unstable thioaldehyde: C=S bonds are particularly unstable because of poor 2p-3p orbital overlap (see Section 3.9, Figure 3.20).

1.2.1.6 Synchronous Catalysis of Mutarotation? In Figure 1.18 the proton transfers in the water-catalysed mutarotation of glucose are drawn as synchronous; this is for simplicity, there being no unambiguous evidence about whether the several protons changing their fractionation factors are in a circle or not. Measurements in water or other hydroxylic solvents cannot throw light on this point, or indeed whether the donation of a proton to O5 is concurrent with the removal of a proton from O1. In a classic, landmark experiment Swain and Brown[39] showed that in benzene solution the mutarotation of tetramethyl glucose was first order in both pyridine as a base catalyst and phenol as a an acid catalyst, *i.e.* equation 1.20

$$k_{obs} = k_3[\text{phenol}][\text{pyridine}] \qquad (1.20)$$

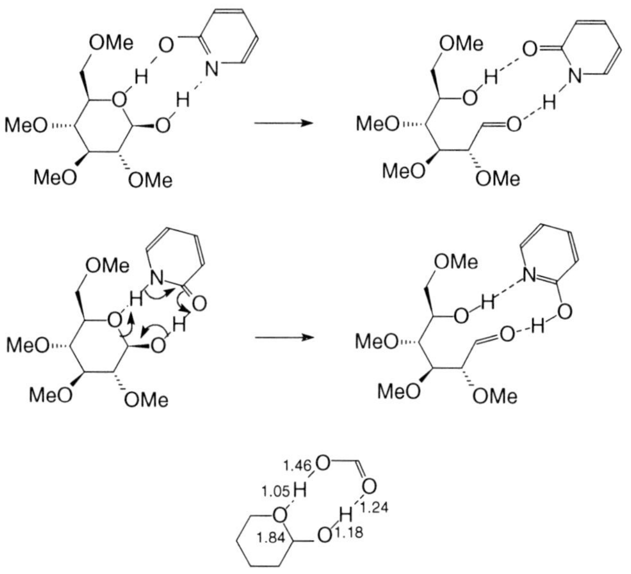

Figure 1.20 Possible transition states in the catalysis of tetramethyl glucose mutarotation by 2-pyridone, and calculated transition state for formic acid catalysis of 2-hydroxytetrahydropyran ring opening. In the calculated transition state, the proton is largely transferred to the endocyclic oxygen, the endocyclic C–O bond has started to break, but the hydroxyl proton is not transferred to the catalyst, *i.e.* the reaction is concerted but not synchronous.

The reaction proceeded very slowly in the presence of either pyridine or phenol by itself, thereby establishing that proton removal and proton donation occur concurrently. Dramatic rate enhancements are achieved by incorporation of acidic and basic sites into the same molecule, so the reaction becomes bimolecular rather than termolecular: 1 mM 2-pyridone was as effective as 100-fold greater concentrations of phenol or pyridine[40] (Figure 1.20).

Two mechanisms can be written for the pyridone-catalysed reactions, depending on whether the ring opening catalyst is the pyridone, or its minor 2-hydroxypyridine tautomer (Figure 1.20). Microscopic reversibility requires the ring closure reaction to use the opposite catalyst tautomer to the ring opening reaction.

In benzene solution mutarotation is catalysed by a range of bifunctional proton-transfer catalysts, including carboxylic acids[41] and the nucleoside cytosine.[42] *Ab initio* calculations on formic acid catalysis of the mutarotation of 2-hydroxytetrahydropyran indicate strong coupling of the double proton transfer to endocyclic C–O bond cleavage.[43]

1.2.1.7 Mutarotases (For a treatment of enzymological terms and concepts see Chapter 5). Mutarotases ensure that the mutarotation of sugars *in vivo* is much faster than spontaneous, and thus couples various transport systems and metabolic pathways. The relatively non-specific aldose mutarotases isolated from various sources by early workers would seem to be the proteins now termed galactose 1-epimerases, which have a very relaxed specificity.

The structures of three galactose-1-epimerases are known: the discrete enzymes from humans,[44] and the bacterium *Lactococcus lactis,*[45] and the mutarotase domain of the UDP-galactose 4-epimerase/galactose 1-epimerase of the yeast *Saccharomyces cerevisiae,* in which two enzyme active sites are formed by a single polypeptide.[46] All three enzymes possess the same protein fold, a β sandwich, with the active site in a shallow cleft, although the bacterial enzyme is a dimer and the human enzyme as isolated a monomer. There are three potentially catalytic groups in the vicinity of the C1: two histidines and a glutamate (His96, His170, and Glu304 in the *L. lactis* numbering). Mutation of His 175 (equivalent to His 170) in the homologous *Escherichia coli* enzyme eliminated activity, but the effect of mutation of His 104 (equivalent to His 96) was somewhat less pronounced;[47] in the *L. lactis* enzyme Glu 304 and His 170 mutants were devoid of activity, but His 96 mutants had some activity.[48] In the complex of the human enzyme with β-D-galactopyranose, Glu 307 ($\approx$300 in *L. lactis*) is ideally suited to act as the catalytic base and His 176 ($\approx$170 in *L. lactis*) as the catalytic acid, although electron density for the complex of galactose and the dimeric bacterial enzyme showed both galactopyranose anomers were bound. The three homeomorphs galactose, arabinose and D-fucose all bound in the same fashion to *L. lactis* enzyme, and had k_{cat}/K_m values around $2 \times 10^5\,M^{-1}\,s^{-1}$ (well below diffusion), whilst the three homeomorphs glucose, xylose and quinovose were bound in a completely different fashion, and had k_{cat}/K_m values 1.3 – $2 \times 10^4\,M^{-1}\,s^{-1}$; it was reasonably surmised that the complexes observed with the *gluco* configured sugars were non-productive.[49]

Detailed mechanistic studies have been carried out with the *E. coli* enzyme, which is more active than the *L. lactis* enzyme, with k_{cat}/K_m for galactose $4.6 \times 10^6\,M^{-1}\,s^{-1}$ and glucose $5 \times 10^5\,M^{-1}\,s^{-1}$.[50] The value for galactose is sufficiently close to the diffusion limit that glucose, whose ten-fold lower rate establishes that pH studies will not be complicated by diffusional effects, was used as substrate. k_{cat} was invariant with pH, but $\log_{10} k_{cat}/K_m$ showed a distorted bell-shaped profile, with the acid limb having a gradient of +1 and the alkaline limb a gradient of −2, governed by pK values of 6.0 and 7.6. The profile for the His 104Gln mutant was sigmoid, with loss of activity to acid governed by a pK of 5.8. With the *E. coli* enzyme it was possible to measure the low activity of the glutamate mutant (Glu 309Gln) as a function of pH, which was a symmetrical bell-shape, with the acid limb governed by a pK of 6.2 and 8.7 on the alkaline side. These data establish quite nicely that three acid/base groups, not two, are involved in catalysis, with different groups involved in proton transfer to and from the α and β forms of the sugar. Compatible with the X-ray structures, and data for the non-enzymic reaction indicating protonation and deprotonation of OH1 are more important than analogous processes with OH5, is the idea that Glu 307 is the acid-base to the β-OH, His 175 to the α-OH and His104 to O5. The acid limb of the wild-type asymmetrical profile corresponds to protonation of Glu 307, with the overall rate being

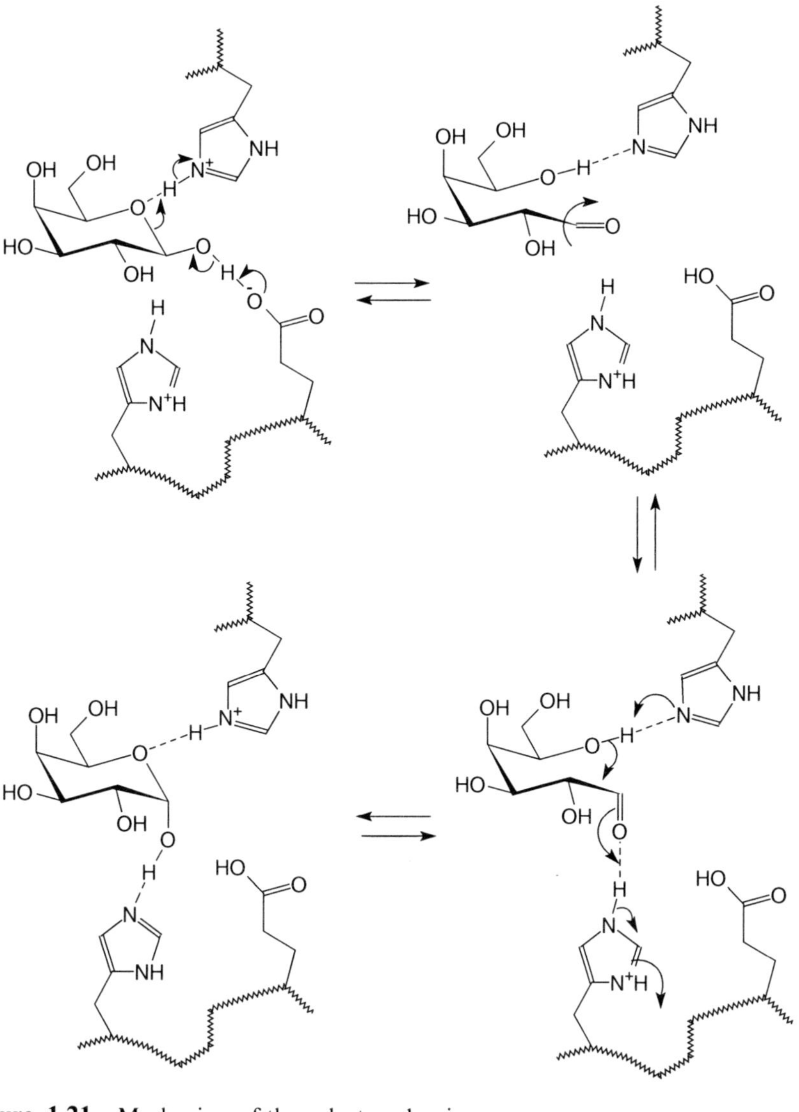

Figure 1.21 Mechanism of the galactose 1-epimerases.

limited by ring opening. The two-proton basic limb corresponded to the deprotonation of both histidines, and ring closure becoming rate determining (Figure 1.21).

The use of NMR techniques, rather than polarimetry, to monitor mutarotation, as with the spontaneous reaction, allowed accelerations of particular conversions of sugars to be directly monitored. Saturation difference measurements indicated that three proteins from *Escherichia coli,* RbsD, FucU (*sic*) and YiiL, of previously unknown function, had mutarotase activity.[51] RbsD interconverted the β-pyranose and β-furanose forms of ribose, without any catalysed conversions to or from α- furanose or α-pyranose forms, and was termed a "pyranase". FucU was an L-fucose mutarotase which had some "pyranase" ability. YiiL proved to be a rhamnose mutarotase, with a completely different structure to the galactose mutarotatases. The complex of L-rhamnose with the enzyme revealed a single histidine capable of interacting with the ring oxygen, and the acid/base at O1 to be Tyr 18: in support of such a role for Tyr 18, the Tyr18Phe mutant was devoid of activity.[52]

1.2.2 Reaction with Low Molecular Weight Alcohols – the Fischer Glycoside Synthesis

The reaction of solutions of reducing sugars in low molecular weight alcohols under acid catalysis to give a range of glycosides is known as the Fischer glycoside synthesis.[53] The mechanism[54] is generally accepted to be analogous to the mutarotation, in that the *aldehydo* sugar is intercepted by solvent to give an acyclic hemiacetal, which is then cyclised to (predominantly) furanosides. The reaction can be stopped after a short time, but if allowed to continue the furanosides open to give acyclic acetals which open and close to eventually give high yields of the thermodynamically most stable product, the glycopyranosides. The anomerisation of the pyranosides, like acid-catalysed glycoside hydrolysis, can proceed largely by the same mechanism as acid-catalysed hydrolysis of glycosides, via a cyclic oxocarbenium ion (in a complex rather than free – see Chapter 3). Anomerisation of the pyranosides eventually yields the most stable, axial anomer (Figure 1.22).

Key pieces of evidence for these mechanisms are:

(i) The failure to isolate the dimethyl acetal in the early stages of the reaction, before furanosides are formed.

(ii) The presence of nucleophilic assistance in the ring closure of the dimethyl acetal to the furanoside, and hence, by microscopic reversibility, in the ring opening reaction of furanosides. This very prescient discovery (in 1965) involved two sets of experiments. Treatment of the dimethyl acetals of glucose and galactose with acid in dilute aqueous solution led to the production of glucofuranosides in >98% yield and galactofuranosides in 71% yield, but the rate of loss of glucose dimethyl acetal was > 10 times faster than the rate of loss of galactose dimethyl acetal. Any sort of unimolecular generation of acyclic oxocarbenium ions would have proceeded at essentially the same rate for glucose and galactose.[55] Logarithms of rate constants for acid-catalysed hydrolysis of a series of dimethyl acetals $RCH(OMe)_2$ gave a good correlation with σ_I for substituents in R, but points for aldose dimethyl acetals lie above the correlation line.[56]

(iii) The acid-catalysed methanolysis of ethyl 5-O-methyl glucofuranosides, and the acid-catalysed ethanolysis of methyl 5-O-methyl glucofuranosides, which cannot form pyranosides, give exclusively the transfer product of opposite configuration in the initial stages of the reaction

not stereospecific; exchange with retention

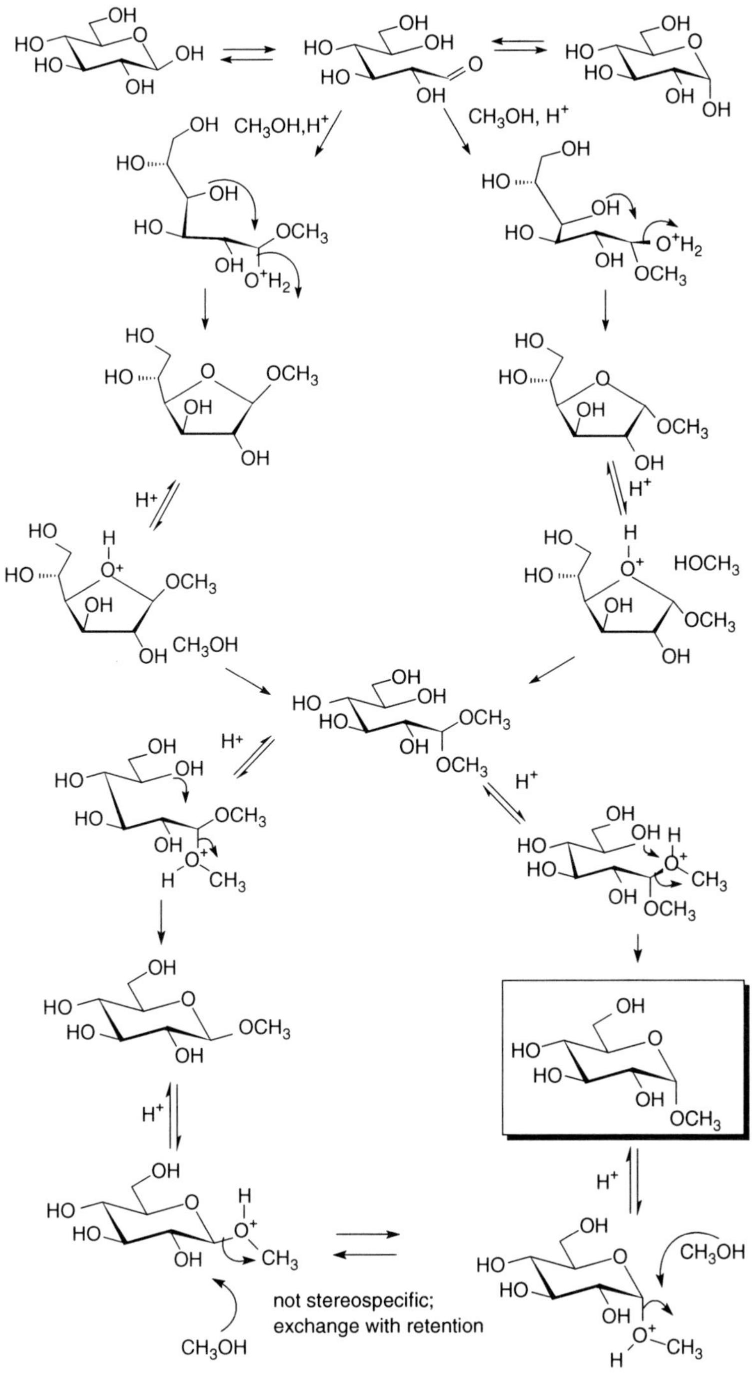

Figure 1.22 The Fischer glycoside synthesis, illustrated by reaction of methanol and glucose. The ring opening and closing reactions are shown as stereospecific, by analogy with transglycosylations, whereas the displacements at C1 of the pyranosides is nonstereopsecific.

(e.g, methyl β-glucofuranoside gives only ethyl α-glucofuranoside), whilst the starting material does not epimerise[57] The 5-O-methyl group prevents the formation of pyranosides, and the absence of any epimerisation of starting material rules out the intermedicacy of any acyclic oxocarbenium ion with a real lifetime. Whilst a series of stereospecific S_N2 reactions with preservation of the ring are compatible with these data, the non-stereospecific nature of the transglycosidation in six-membered rings, where reaction occurs with preservation of the ring, makes this unlikely.

(iv) The acid-catalysed methanolysis of ethyl 4-O-methyl glucopyranosides, and the acid-catalysed ethanolysis of methyl 4-O-methyl glucopyranosides, which cannot give furanosides, gives an initial slight preference for inversion, and also a distinct preference for attack at the α-face of the pyranose ring.[58] Epimerisation of the starting material is again not observed. This is behaviour very reminiscent of, for example, solvolysis of glucopyranosyl derivatives which cannot ring open,[59] and suggests epimerisation of the pyranosides via oxocarbenium ion-counterion-leaving group complexes.

The acid used in the Fischer glycoside synthesis was originally HCl (which made the kinetic data reproducible only with difficulty, as in methanol HCl is slowly consumed to form methyl chloride). Nowadays, ion-exchange resins are used preparatively, whilst the data on 4- and 5-O-methylated glycosides were obtained with camphor sulfonic acid.

In contrast to the thermodynamic products of the reaction of low-molecular-weight alcohols with reducing sugars being axial pyranosides, the thermodynamic products of the reaction of thiols RSH with reducing sugars are the open-chain dithioacetals (Figure 1.23).[60]

1.2.3 Formation, Anomerisation, and Hydrolysis of Glycosylamines

If reducing sugars are reacted with ammonia or primary amines in neutral solution in water or alcohols, glycosylamines are produced. The reaction appears to involve initial formation of a Schiff base and then ring closure. The reactions are complex, and in the synthesis direction give rise to a range of products, so that reliable mechanistic information is sparse. In the reverse sense, the hydrolysis of glycosylamines, the conclusions of the 40-year-old work of Capon and Connett still stand.[61] The key intermediates are an acyclic Schiff base and its cation (Figure 1.24). The hydrolysis of N-aryl glucosylamines is preceded by their rapid anomerisation to an equilibrium mixture of 10% α and 90% β. The hydrolysis is formally general acid catalysed, but, by analogy with the mechanism of hydrolysis of simple Schiff bases, this formal general acid catalysis is likely to be the kinetically equivalent general base catalysis of the attack of water on the protonated Schiff base.[62]

The pH (or H_o – see Chapter 3, footnote xi) dependences of the rates of hydrolysis of glycosylamines are bell-shaped, with the maximimum at a pH

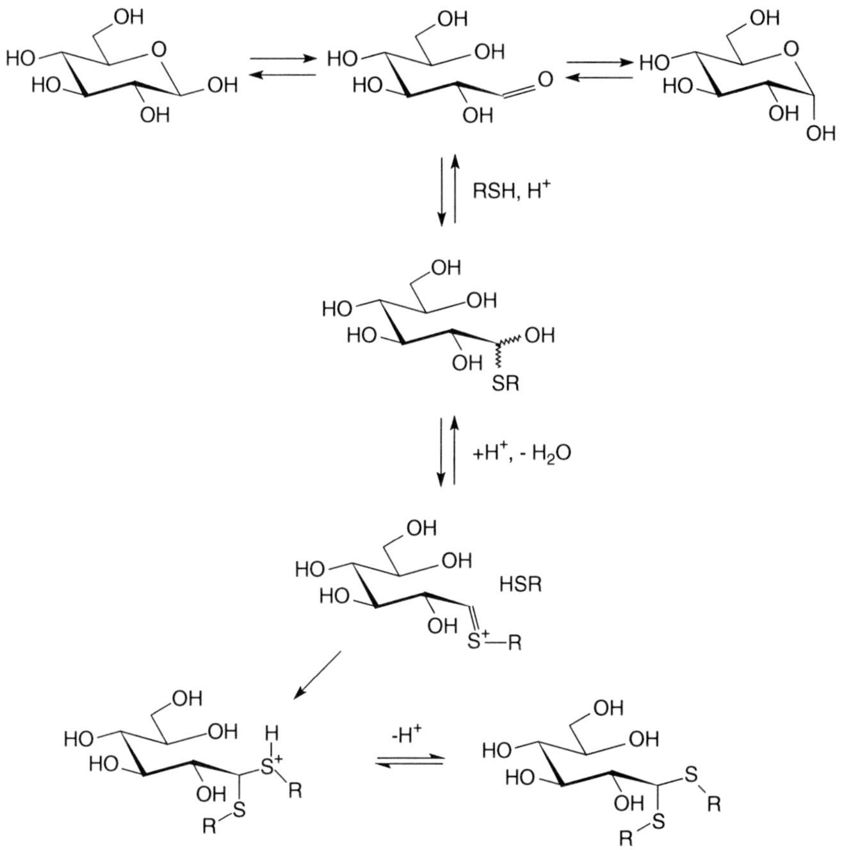

Figure 1.23 Formation of straight-chain dithioacetals with thiols. Reactions are probably preassociated S_N2 reactions similar to the reactions in the Fischer glycoside synthesis, but the thiacarbenium ion is drawn to emphasise that the stability of the dithioacetal reflects, not only the weaker basicity of sulfur as compared to oxygen, but also its lesser ability to stabilise a carbenium ion centre by conjugation (see Section 1.3).

or H_o around 4–5 units more negative than the pK_a of the parent amine.[54] The alkaline limb of the bell arises from from increased conversion of the Schiff base to its more reactive conjugate acid as the pH is lowered. The acid limb of the bell probably arises from the protonation of potential general base catalysts, for removal of the proton from the carbinolamine, which makes the hydrolysis of the carbinolamine (rather than its formation) rate-determining.

The mutarotation of glycosylamines has been studied more in methanolic solutions than in water, because any intermediate from addition of this solvent to the acyclic Schiff base cannot yield *aldehydo* sugar. As would be expected from the mechanisms of Figure 3.23, increased acid strength results in faster mutarotation[63] (Figure 1.24).

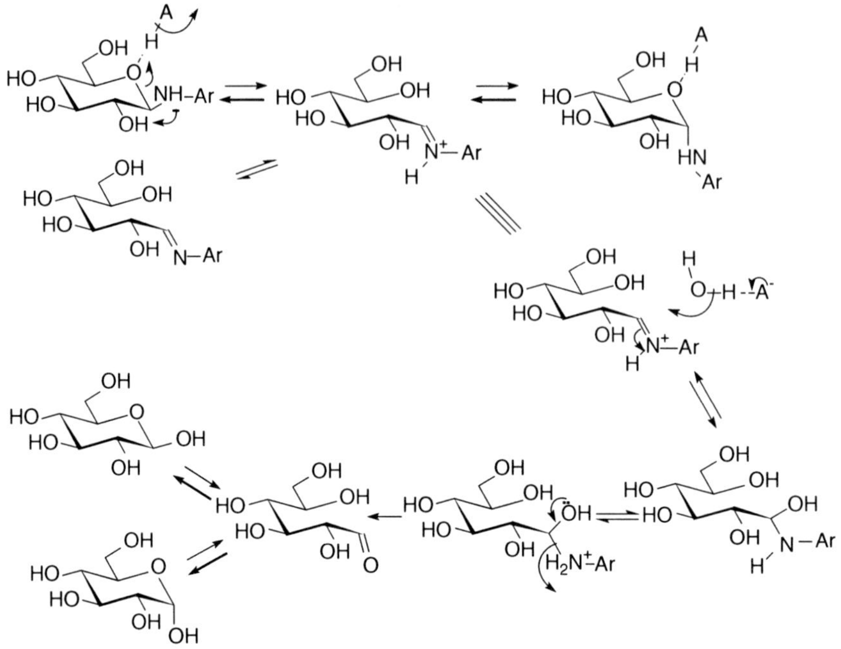

Figure 1.24 Mutarotation and hydrolysis of glycosylamines, illustrated with N-aryl glucosylamines.

Glycosylamines from simple secondary amines, such as N-glycosyl piperidine, appear to behave like glycosylamines derived from primary amines and ammonia.[54] Where, however, formation of a cationic Schiff base would involve disruption of an aromatic sextet (as with glycosyl imidazoles, pyrazoles and purines) the glycosylamines are configurationally stable.

Studies of the reaction of aldoses with the classic carbonyl group reagents hydroxylamine, semicarbazide and hydrazine indicated that the reactions were as expected for such condensations preceded by a ring-opening step. The reactions showed the usual bell-shaped pH dependence as the rate determining step changed from formation to decomposition of the tetrahedral intermediate. Below pH 4.6, the reaction is buffer-catalysed (the author surmises, general acid catalysis of the departure of OH from XNH-CRHOH appearing as general base catalysis of the decomposition of XNH_2^+-CRHOH). The overall rate varied with the nature of the sugar broadly according to the proportion of aldehyde at equilibrium (Table 1.1); however, opening of the aldose ring was partly rate-determining in some cases, as anomers of the same sugar showed small kinetic differences.[64]

1.3 CYCLITOLS

Although not strictly sugars, polyhydroxylated carbocylic rings have many of the non-anomeric reactions of carbohydrates proper, and because of the

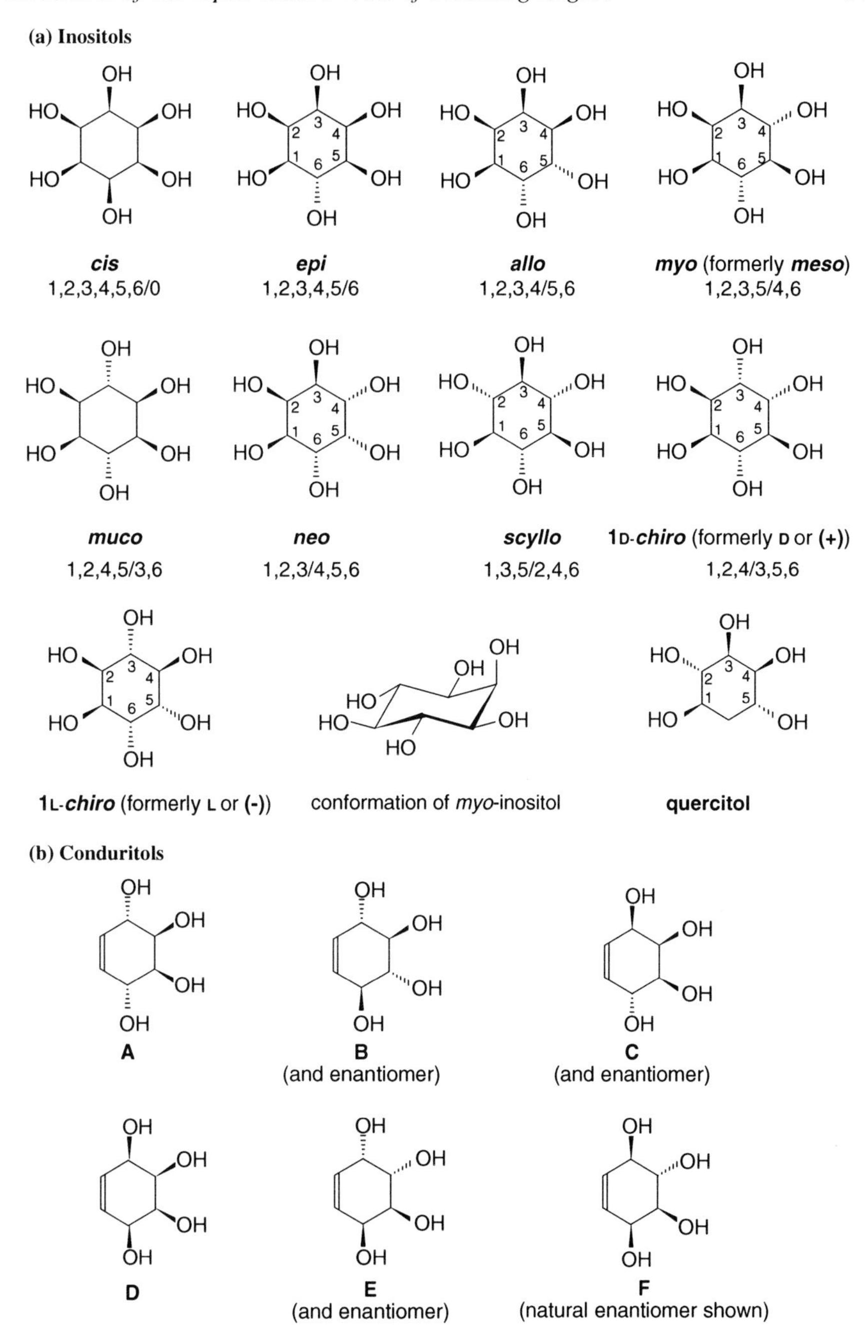

Figure 1.25 (a) Structures of the inositols and related compounds. The numbering system, as well as the sequence of numbers for description as a cyclohexane hexol, is indicated. (b) Structures of the conduritols.

similarities of handling techniques (they are water-soluble and frequently only sparingly soluble in organic solvents) are often regarded as "honorary" carbohydrates. Where naturally occurring, they are biosynthesised by internal aldol reactions from carbohydrate precursors.

The basic IUPAC system of nomenclature[65] is straightforward: the compounds are named as cycloalkane polyols, with a slash separating substituents on each side of the ring; thus, 1,2,3/4,5 cyclopentane pentol has three adjacent hydroxyls on one side of the ring, and two on the other (there are four cyclopentane pentols, all achiral).

IUPAC cyclitol nomenclature is rarely used for cyclohexane hexols, with the older system based on "inositol" for the parent cyclohexane hexol, and stereochemistry denoted by a Graeco-Latin prefix, being used in the biochemical literature. There are eight inositols, only one of which (*chiro*) is chiral. The rules for carbon numbering and for designation of stereochemistry are complex, and examples of frequently-encountered systems are given in Figure 1.25 (a).

Trivial names are also given to tetrahydroxylated cyclohexenes, conduritol. These are given letters in the sequence of their discovery.[66] Only conduritols A and F occur naturally.

The commonest naturally-occurring inositol is *myo*-inositol, and its structure is readily memorised because in the preferred conformation only one hydroxyl group is axial. Its 1-L phosphate ester is biosynthesised from its isomer glucose-1-phosphate by an oxidation-intramolecular aldol condensation-reduction sequence, see Section 6.8.4.

REFERENCES

1. E. Fischer, *Ber.*, 1891, **24**, 1836.
2. M. A. Rosanoff, *J. Am. Chem. Soc.*, 1906, **28**, 114.
3. For a recent protocol, and a discussion of the reaction see N. Adje, F. Vogeleisen, and D. Uguen, *Tetrahedron Lett.*, 1996, **37**, 5893.
4. www.chem.qmul.ac.uk/iupac/2carb/index.html.
5. R. S. Cahn, C. K. Ingold and V. Prelog, *Experimentia*, 1956, **12**, 81.
6. A. G. Ogston, *Nature*, 1948, **162**, 963.
7. K. B. Wiberg, K. M. Morgan and H. Maltz, *J. Am. Chem. Soc.*, 1994, **116**, 11067.
8. M. I. Page and W. P. Jencks, *Proc. Natl. Acad. Sci. USA*, 1971, **68**, 1678.
9. T. B. Grindley and V. Gulasekharam, *J. Chem. Soc., Chem. Commun.*, 1978, 1073.
10. Z. Pakulski and A. Zamojski, *Pol. J. Chem.*, 1995, **69**, 912.
11. A. S. Serianni, J. Pierce, S.-G. Huang and R. Barker, *J. Am. Chem. Soc.*, 1982, **104**, 4037.
12. J. R. Snyder, E. R. Johnston and A. S. Serianni, *J. Am. Chem. Soc.*, 1989, **111**, 2681.
13. P. W. Wertz, J. C. Garver and L. Anderson, *J. Am. Chem. Soc.*, 1981, **103**, 3916.
14. (a) T. M. Lowry and G. F.Smith, *J. Chem. Soc.*, 1927, 2539; (b) J. N. Brønsted and E. A. Guggenheim, *J. Am. Chem. Soc.*, 1927, **49**, 2554.

15. (a) P. Salomaa, A. Kankaanperä and M. Lahti, *J. Am. Chem. Soc.*, 1971, **93**, 2084; (b) J. L. Hogg and W. P. Jencks, *J. Am. Chem. Soc.*, 1976, **98**, 5643; (c) M. Lahti and A. Kankaanperä, *Acta Chem. Scand.*, 1972, **26**, 2130.
16. F. G. Bordwell, J. E. Bartmess and J. A. Hautala, *J. Org. Chem.*, 1978, **43**, 3107.
17. http://nobelprize.org/nobel_prizes/chemistry/laureates/1992/marcuslecture.pdf.
18. J. Wirz, *Pure Appl. Chem.*, 1998, **70**, 2221.
19. D. S. Kemp and M. L. Casey, *J. Am. Chem. Soc.*, 1973, **95**, 6670.
20. R. P. Bell, *J. Chem. Soc., Faraday Trans. 2*, 1976, **72**, 2088.
21. A. C. Lin, Y. Chiang, D. B. Dahlberg and A. J. Kresge, *J. Am. Chem. Soc.*, 1983, **105**, 5380.
22. R. A. Bednar and W. P. Jencks, *J. Am. Chem. Soc.*, 1985, **107**, 7117.
23. L. P. Hammett, *Chem. Rev.*, 1935, **17**, 125.
24. P. R. Young and W. P. Jencks, *J. Am. Chem. Soc.*, 1979, **101**, 3288.
25. H. Heligsø and M. Bols, *Acc. Chem. Res.*, 2006, **39**, 259.
26. L. Melander and W. H. Saunders, *Reaction Rates of Isotopic Molecules*, John Wiley and Sons, New York, 1980.
27. (a) J. Bigeleisen and M. G. Meyer, *J. Chem. Phys.*, 1947, **15**, 261; (b) J. Bigeleisen and M. Wolfsberg, *Adv. Chem. Phys.*, 1958, **1**, 15.
28. (a) M. H. Glickman, J. S. Wiseman and J. P. Klinman, *J. Am. Chem. Soc.*, 1994, **116**, 793; (b) C. C. Hwang and C. B. Grissom, *J. Am. Chem. Soc.*, 1994, **116**, 795.
29. J. Basran, M. J. Sutcliffe and N. S. Scrutton, *Biochemistry*, 1999, **38**, 3218.
30. K. M. Doll, B. R. Bender and R. G. Finke, *J. Am. Chem. Soc.*, 2003, **125**, 10877.
31. R. P. Bell, *The Proton in Chemistry*, 2nd edn, Chapman and Hall., London, 1973.
32. V. Gold, *Adv. Phys. Org. Chem.*, 1972, **7**, 259.
33. K. S. Venkatasubban and R. L. Schowen, *CRC Crit. Rev. Biochem.*, 1984, **17**, 1.
34. H. H. Huang, R. R. Robinson and F. A. Long, *J. Am. Chem. Soc.*, 1966, **88**, 1866.
35. Y. Pocker, *Chem. Ind.*, 1960, 560.
36. B. E. Lewis, N. Choytun, V. L. Schramm and A. J. Bennet, *J. Am. Chem. Soc.*, 2006, **128**, 5049.
37. C. E. Grimshaw, R. L. Whistler and W. W. Cleland, *J. Am. Chem. Soc.*, 1979, **101**, 1521.
38. J. L. Jensen and W. P. Jencks, *J. Am. Chem. Soc.*, 1979, **101**, 1476.
39. C. G. Swain and J. F. Brown, *J. Am. Chem. Soc.*, 1952, **74**, 2534.
40. C. G. Swain and J. F. Brown, *J. Am. Chem. Soc.*, 1952, **74**, 2538.
41. P. R. Rony, *J. Am. Chem. Soc.*, 1968, **90**, 2824.
42. C. Melander and D. A. Horne, *Tetrahedron Lett.*, 1997, **38**, 713.
43. S. Murpurgo, M. Brahimi, M. Bossa and G. O. Morpurgo, *Phys. Chem. Chem. Phys.*, 2000, **2**, 2707.
44. J. B. Thoden, D. J. Timson, R. J. Reece and H. M. Holden, *J. Biol. Chem.*, 2004, **279**, 23431.
45. J. B. Thoden and H. M. Holden, *J. Biol. Chem.*, 2002, **277**, 20854.

46. J. B. Thoden and H. M. Holden, *J. Biol. Chem.*, 2005, **280**, 21900.
47. J. A. Beebe and P. A. Frey, *Biochemistry*, 1998, **37**, 14989.
48. J. B. Thoden, J. Kim, F. M. Raushel and H. M. Holden, *Protein Sci.*, 2003, **12**, 1051.
49. J. B. Thoden, J. Kim, F. M. Raushel and H. M. Holden, *J. Biol. Chem.*, 2002, **277**, 45458.
50. J. A. Beebe, A. Arabshahi, J. G. Clifton, D. Ringe, G. A. Petsko and P. A. Frey, *Biochemistry*, 2003, **42**, 4414.
51. K.-S. Ryu, C. Kim, I. Kim, S. Yoo, B.-S. Choi and C. Park, *J. Biol. Chem.*, 2004, **279**, 25544.
52. K.-S. Ryu, J.-I. Kim, S.-J. Cho, D. Park, C. Park, H.-K. Cheong, J.-O. Lee and B.-S. Choi, *J. Mol. Biol.*, 2005, **349**, 153.
53. E. Fischer, *Chem. Ber.*, 1893, **26**, 2400.
54. B. Capon, *Chem. Rev.*, 1969, **69**, 407.
55. B. Capon and D. Thacker, *J. Am. Chem. Soc.*, 1965, **87**, 4199.
56. B. Capon and D. Thacker, *J. Chem. Soc(B)*, 1967, 1322.
57. K.-J. Johansson, P. Konradsson and Z. Trumpakaj, *Carbohydr. Res.*, 2001, **332**, 33.
58. P. J. Garegg, K.-J. Johansson, P. Konradsson, B. Lindberg and Z. Trumpakaj, *Carbohydr. Res.*, 2002, **337**, 517.
59. M. L. Sinnott and W. P. Jencks, *J. Am. Chem. Soc.*, 1980, **87**, 2026.
60. E.g. (a) O. Kölln and H. Redlich, *Synthesis*, 1995, 1383; (b) M. Funabashi, S. Arai and M. Shinohara, *J. Carbohydr. Chem.*, 1999, **18**, 333.
61. B. Capon and B. E. Connett, *J. Chem. Soc.*, 1965, 4492.
62. E. H. Cordes and W. P. Jencks, *J. Am. Chem. Soc.*, 1963, **85**, 2843.
63. K. Smiataczowa, J. Kosmalski, A. Nowacki, M. Czaja and Z. Warnke, *Carbohydrate Res.*, 2004, **339**, 1439.
64. J. W. Haas and R. E. Kadunce, *J. Am. Chem. Soc.*, 1962, **84**, 4910.
65. http://www.chem.qmul.ac.uk/iupac/cyclitol/.
66. M. Desjardins, M.-C. Lallemand, S. Freeman, T. Hudlicky and K. A. Abboud, *J. Chem. Soc., Perkin Trans. 1*, 1999, 621.
67. G. C. S. Collines and W. O. George, *J. Chem. Soc.(B)*, 1971, 1352.
68. S. J. Angyal and R. G. Wheen, *Austr. J. Chem.*, 1980, **33**, 1001.
69. L. Davis, *Bioorg. Chem.*, 1973, **2**, 197.
70. A. S. Serriani, E. L. Clark and R. Barker, *Carbohydr. Res.*, 1979, **72**, 79.
71. S. J. Angyal and V. A. Pickles, *Austr. J. Chem.*, 1972, **25**, 1695.
72. Y. Zhu, J. Zajicek and A. S. Serriani, *J. Org. Chem.*, 2001, **66**, 6244.
73. P. W. Wertz, J. C. Garver and L. Anderson, *J. Am. Chem. Soc.*, 1981, **103**, 3916.
74. J. N. Barlow and J. S. Blanchard, *Carbohydr. Res.*, 2000, **328**, 473.
75. C. Williams and A. Allerhand, *Carbohydr. Res.*, 1977, **56**, 173.
76. D. J. Wilbur, C. Williams and A. Allerhand, *J. Am. Chem. Soc.*, 1977, **99**, 5450.
77. S. J. Angyal and D. S. Bethell, *Austr. J. Chem.*, 1976, **29**, 1249.
78. S. J. Angyal, G. S. Bethell, D. E. Cowley and V. A. Pickles, *Austr. J. Chem.*, 1976, **29**, 1239.
79. J. C. Wilson, D. I. Angus and M. von Itzstein, *J. Am. Chem. Soc.*, 1995, **117**, 4214.

CHAPTER 2

Conformations of Monosaccharides

2.1 DIFFERENCES BETWEEN CONFORMATIONAL ANALYSIS OF CARBOHYDRATES AND OTHER ORGANIC MOLECULES

The gas-phase conformations of small molecules can now be computed with some accuracy by *ab initio* methods. In the case of non-polar molecules, whose conformations are unlikely to alter much in solution in non-polar solvents, computational calculations are often the method of first choice in determining conformation.

Carbohydrates, however, are very polar molecules. This means that their gas-phase conformations, which can be computed, are likely to be largely determined by electrostatic and intramolecular hydrogen bonding interactions. They are therefore likely to be of questionable relevance to the conformation of the same molecules in polar solvents, particularly water. At the time of writing (early 2007) there does not appear to be a force field which can accurately predict the conformations of carbohydrates in water; by contrast, Angyal's instability factors, a purely empirical way of estimating carbohydrate conformations, are still used.[1] The major problem with computational approaches appears to be nearly free rotation about the C–O bonds of hydroxyl groups, which gives rise to a very asymmetric potential which is difficult to handle with computational economy.

In addition to steric effects and electrostatic effects, the conformational analysis of carbohydrates requires two stereoelectronic effects to be taken into consideration: the *gauche* effect and the anomeric effect, in its various manifestations. Both of these effects can be considered as aspects of no-bond resonance.

2.2 THE *GAUCHE* EFFECT

As is well known, *trans*-butane has about 4 kJ mol^{-1} less internal energy than *gauche*-butane and is therefore favoured in the gas phase at ordinary temperatures. However, 1,2-difluoroethane prefers the *gauche* conformation, with an internal energy difference of 7.3 kJ mol^{-1} between it and the *trans* conformation.[2] The effect is seen in other, less clear-cut cases with oxygen substituents. The origin of the effect is considered to lie in the same phenomena that are

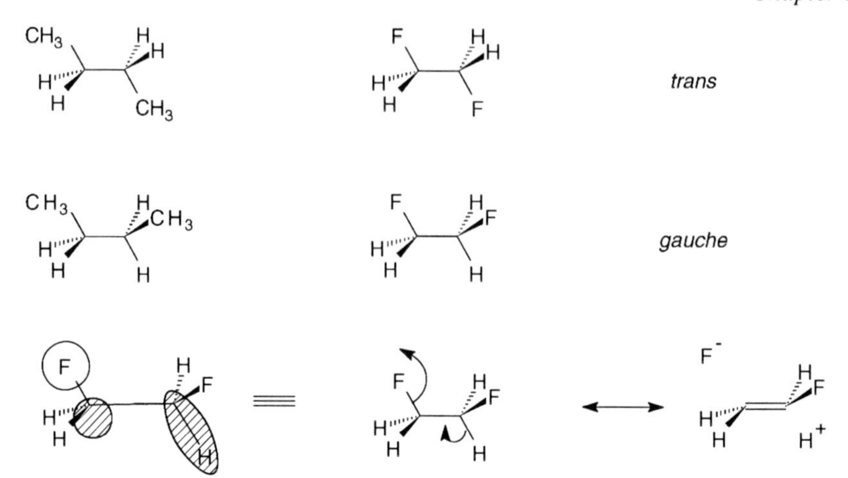

Figure 2.1 Preferred conformations of butane and 1,2-difluoroethane, illustrating the *gauche* effect. In the *gauche* conformation each C–F σ* orbital can overlap with a C–H σ orbital on the vicinal carbon, corresponding to no-bond resonance as shown. Such no-bond resonance would be disfavoured if the C–F bonds were *trans*, since it would remove electron density from an electronegative element. The effect is strong enough in this case to overcome the electrostatic repulsion between the two C–F dipoles, which favours the *trans* form.

responsible for the great stabilisation of carbenium ions β-substituted with silyl groups,[3] namely overlap of σ orbitals with vicinal σ* orbitals, as shown in Figure 2.1.[4,5]

2.3 CONFORMATIONS OF ACYCLIC SUGARS

In general, the conformations of alditol chains are determined by the tendency of the carbon backbone to adopt the extended zig-zag conformation of polyethylene, with all the four-carbon units being in the conformation of *trans*-butane. Because of the *gauche* effect, there is no large preference for OH groups to be *trans* or *gauche* and this zig-zag conformation is adopted by mannitol and galactitol. However, such a conformation involves a 1,3-parallel interaction between the OH groups of glucitol (Figure 2.2). This is strongly disfavoured by electrostatics, so that glucitol adopts a so-called *sickle* conformation.[6,7]

2.4 DESCRIPTION OF THE CONFORMATIONS OF SUGAR RINGS

The currently used terms for describing sugar conformations in general derive from a mathematical study of Cremer and Pople,[8] who built on the treatment of furanose rings of Altona and Sundaralingam.[9] Cremer and Pople showed rigorously that the conformation of a ring with x atoms could be described rigorously by $x - 3$ spatial coordinates (this, of course, neglects any substituents

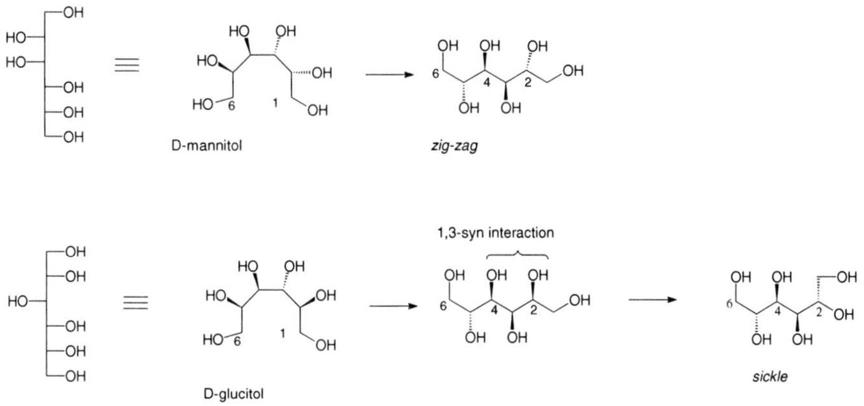

Figure 2.2 Conformations of D-glucitol and D-mannitol. Note the apparent toleration of vicinal hydroxyl groups being either *gauche* or *trans*, but the prohibition on 1,3-*syn* interactions of hydroxyl groups.

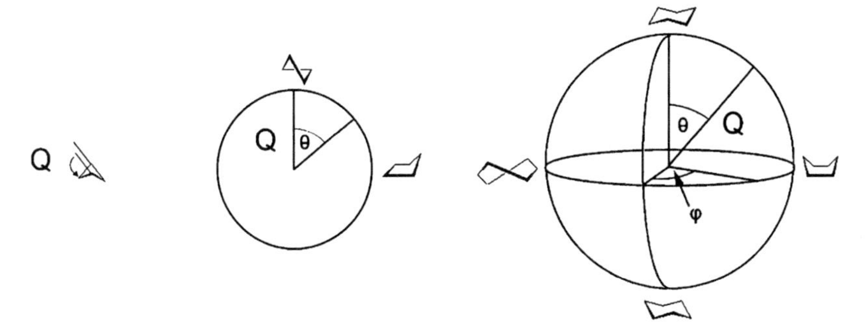

Figure 2.3 Cremer–Pople treatment of ring conformations. Left, cyclobutane conformations are completely defined by degree of pucker; centre, cyclopentane conformations require specification of the degree of pucker and a pseudorotational angle; right, cyclohexane conformations require specification of the degree of pucker and two angles; for conformations on the equator ($\theta = 90°$) φ becomes a pseudorotational angle similar to that for cyclopentane.

on the x atoms). They realised that if these coordinates were polar rather than Cartesian, then ring conformations could be described in ways which corresponded to chemical intuition. They accordingly defined a puckering parameter Q, essentially the degree to which the atoms departed from the mean plane of the ring, as the radius, and $x - 4$ angles which described the position on a multi-dimensional surface of this radius.

Although four-membered rings are rare in carbohydrate chemistry, they represent the simplest case for the Cremer–Pople treatment, $x = 4$, so that their conformation is described uniquely by the puckering amplitude, *i.e.* the amount by which the fourth atom is displaced from the plane defined by the remaining three (Figure 2.3).

For five-membered rings ($x = 5$), only two parameters, the puckering parameter and an angle, are required. The internal angle of a pentagon (108°) is very close to the tetrahedral angle (109.5°), but in the planar forms of five-membered rings all the substituents are eclipsed. Puckering of the ring results in two low-energy forms, the envelope, in which one ring atom lies out of a plane containing the remaining four, and the twist, in which two adjacent atoms lie above and below the plane defined by the remaining three. There are 10 envelope conformations and 10 twist conformations, which can be interconverted by rotations about single bonds. This gives the appearance of the out-of-plane atoms rotating round the ring, and, in the case of cyclopentane itself, looks equivalent to the ring rotating about its vertical axis. The origin of the term "pseudorotation" for the interconversions of twist and envelope conformations lies in this feature of the geometry of cyclopentane itself, but is extended to systems where the ring atoms are not equivalent and the ring does not appear to rotate.

In the case of six-membered rings, a puckering parameter and two angles are required; therefore, all conformations at a given degree of pucker can be described by their position on a sphere. The chair conformations are located at the poles, and the twist and classical boat conformations round the equator. The conformations round the equator form a pseudorotational itinerary similar to that seen with five-membered rings: rotations about single bonds can give the appearance of a cyclohexane molecule on the skew/boat itinerary rotating, and the term is extended to systems where ring atoms are non-equivalent. In middle latitudes on the sphere of six-membered ring conformations are found the half-chair and envelope conformations that are transition states for cyclohexane, but the half-chair (like the boat conformer on the equator) is a stable conformer for cyclohexene.

In addition to an envelope conformation similar to that seen with five-membered rings, but with five, rather than four, contiguous atoms coplanar, six-membered ring conformations are discussed in terms of chair, boat, skew and half-chair conformations. The half-chair conformation has four contiguous ring atoms coplanar, with the remaining two above and below the plane so defined. The well-known chair conformation of cyclohexane has a six-fold axis of rotation–inversion and perfectly staggered substituents about each C–C bond, although since the C–O bond is somewhat shorter than the C–C bond, this perfect geometry is not fully transferred to pyranosides. Also well known is the boat conformation, with two orthogonal mirror planes, despite its being at a local energy maximum for saturated systems. The skew conformation, with a two-fold axis of symmetry, is at a local energy minimum.

The conformations of furanose and pyranose rings are described qualitatively by italicised letters *T*, *E*, *C*, *H*, *S* and *B* for twist, envelope, chair, half-chair, skew and boat, respectively; *T* refers only to furanose rings and *C*, *H*, *S* and *B* only to pyranose rings, but *E*, describing conformations with just one ring atom out of the plane of the remainder, can refer to both pyranoses and furanoses. The particular conformation is specified by superscripts and subscripts referring to atoms above and below a defined plane; thus $^4C_1$ (the commonest preferred conformation of hexopyranose derivatives) refers to a chair conformation with carbon 4 above the plane and carbon 1 below.

The reference planes in the case of E and H define themselves. The reference plane in B is defined not by either of the planes containing four contiguous atoms, but by the plane defining the bottom of the boat. "Above" the plane is defined as the direction from which the numbered carbon atoms *increase* in a *clockwise* direction and thus corresponds to the commonest representation of D-sugars in Haworth or conformational representations. Several planes containing three ring atoms can be drawn through T conformers and containing four ring atoms through C and S conformers. Where there is a choice of planes, that plane which gives the lowest superscripts and subscripts is chosen as the reference. Thus, the $^4C_1$ conformation (reference plane C2, C3, C5, O) is called such, rather than $^2C_5$ (reference plane C3, C4, O5, C1).

Unfortunately, the rule that "above" the definition plane of a particular conformer is the direction from which the carbon atom numbers increase round the ring has the effect of reversing conformational designations between enantiomers – the mirror image of α-D-glucopyranose in the $^4C_1$ conformation is α-L-glucopyranose in the $^1C_4$ conformation.

In the case of furanosides, especially nucleosides, Altona and Sundaralingam's original symbols for the pseudorotational angle P and pucker τ_m (or sometimes φ_m) are used; $P = 0°$ is defined as the $^3T_2$ conformation. Figure 2.4 shows the pseudorotational itinerary, with a nucleoside (sugar D-ribose) as an example. Nucleoside chemists refer to conformations with $0 \leq P \leq 36°$ as "N"

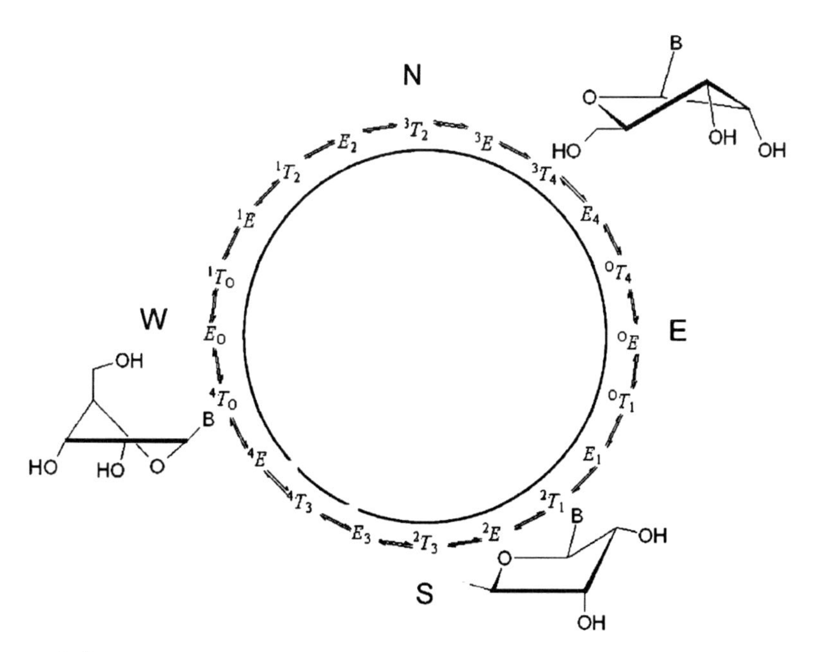

Figure 2.4 Conformational itinerary of a furanose ring.

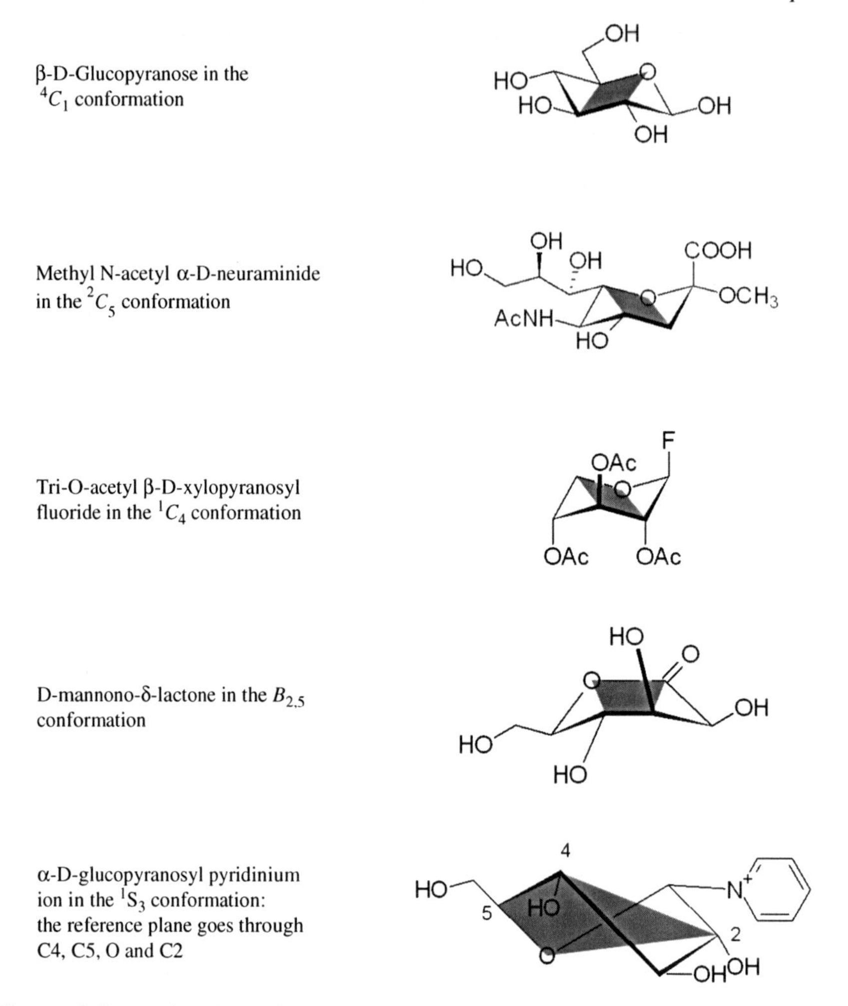

Figure 2.5 Preferred conformations of some pyranose derivatives, showing the reference plane.

(for "Northern", although more properly NNE) and with $144 \leq P \leq 180°$ as "S" (for "Southern", really SSE), as these are the two commonest conformations adopted by the ribose ring. Replacement of a nucleoside base with an OMe group did not alter conformational preferences.[10]

Figure 2.5 shows the preferred conformations of some pyranosyl derivatives. The examples are chosen both to illustrate the principles by which the conformations are named and also to illustrate the operation of various effects which determine the conformations. The conformations are all necessarily found on the surface of the Cremer–Pople sphere; the inter-relations of the various conformations in the northern and southern hemispheres are set out in Figure 2.6.

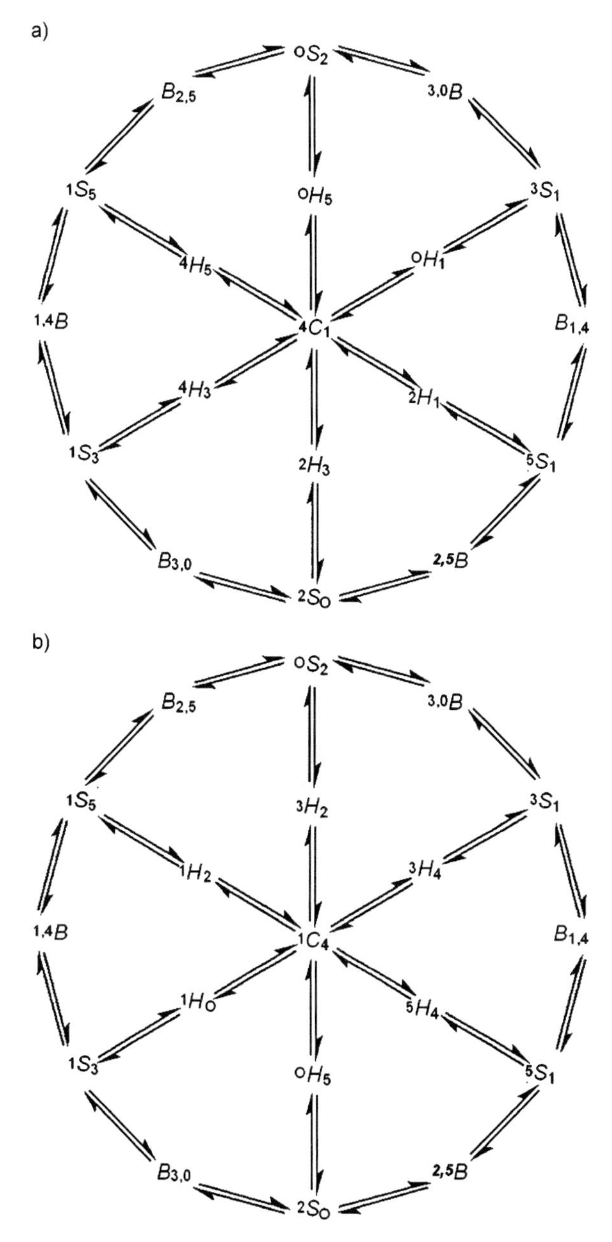

Figure 2.6 Conformations on the surface of the Cremer–Pople sphere for a pyranose ring: (a) Northern hemisphere; (b) Southern hemisphere.

In addition to the familiar axial and equatorial description of substituents in chair conformations, substituents in non-chair conformers of six-membered rings also have their descriptors: there are four isoclinal, four pseudoaxial and four pseudoequatorial bonds in skew cyclohexane and four pseudoaxial, four

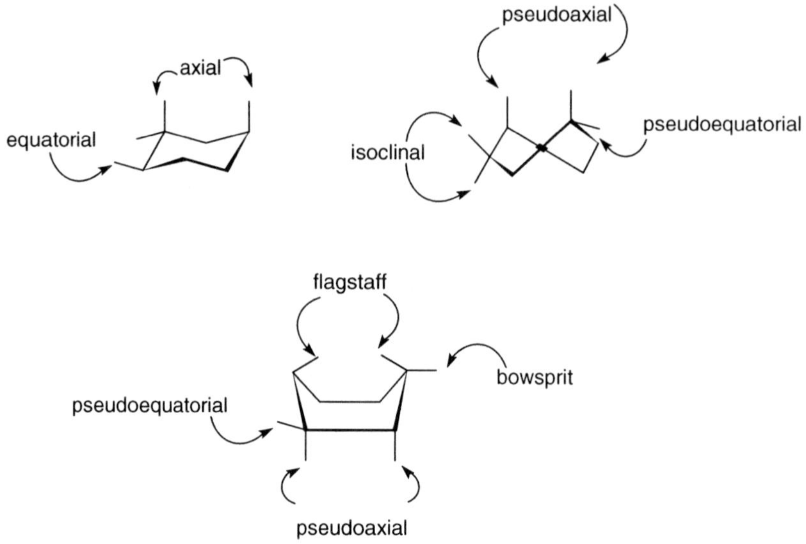

Figure 2.7 Designations of bonds in chair and non-chair six-membered rings.

pseudoequatorial, two bowsprit and two flagstaff bonds in boat cyclohexane (Figure 2.7).

2.5 ANALYSIS OF CARBOHYDRATE CONFORMATION AND CONFIGURATION BY NMR – THE KEY ROLE OF THE KARPLUS EQUATION

Most of our knowledge of the conformation of sugar rings in solution has come from NMR, in particular from spin–spin coupling constants. The coupling is a through-bond effect that can be envisaged as a minute unpairing of the electrons in the bond to pair up with a nucleus with a spin. The unpairing is transmitted through valence electrons, depending on the symmetries of their overlaps, to another nucleus. Coupling falls off sharply with the number of bonds through which it has to be transmitted, ^{1}H–^{1}H coupling through four bonds only being detectable with particular favourable geometries ("W coupling").

Three-bond coupling constants ^{3}J are particularly informative, as their magnitude depends sinusoidally on dihedral angles, with, in general, maxima at dihedral angles of 0° and 180° (*i.e.* synperiplanar and antiperiplanar disposition). Karplus[11] derived eqn (2.1) for three-bond coupling of protons. This is illustrated in Figure 2.8.

$$^{3}J = A + B\cos\varphi + C\cos2\varphi \qquad (2.1)$$

It is often forgotten that the magnitudes of *A*, *B* and *C* depend not only on the nuclei coupled, but also on the particular system. In particular, coupling

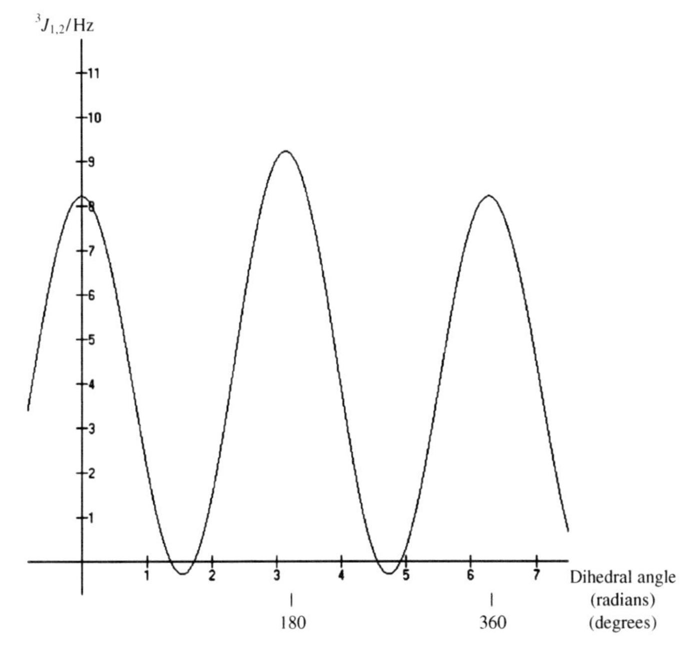

Figure 2.8 Three-bond proton–proton coupling constants as a function of dihedral angle, according to eqn (2.1), illustrated with Karplus's original values of $A=4.22$, $B=-0.5$ and $C=4.5$, applicable to hydrocarbons.

constants are reduced by electron-withdrawing substituents: there have been many attempts to correct for this explicitly. The most successful is a five-parameter approach for three-bond proton–proton coupling constants by Altona and colleagues.[12] In practice, coupling constants usually are interpreted empirically, by reference to other closely related molecules.

Some caution must be exercised before a peak splitting is identified with a coupling constant. This is in general only the case where the chemical shift difference ($\Delta\sigma$) between the coupled nuclei, expressed in Hz rather than ppm, is significantly (say, 10-fold) larger than the coupling constant, J, so that the spectra are first order. As $\Delta\sigma$ approaches the value of the coupling constant, the inner lines of multiplets increase at the expense of the outer lines. Thus, a so-called AX two-proton system, where the chemical shift difference is large compared with the coupling constant, appears as two clean doublets (*i.e.* four equal-intensity lines). However, as $\Delta\sigma$ approaches J, the two inner lines increase in intensity relative to the two outer ones, so that in a one-dimensional spectrum, a so-called AB system with $\Delta\sigma \sim J$ can be mistaken for the 1:4:4:1 methylene quartet of an ethyl group. Since the important parameter is $\Delta\sigma$ in Hz, increase of field strength, which increases while J remains constant, can simplify spectra. At 60 MHz, although anomeric protons can often be distinguished, other ring protons form an unanalysable complex multiplet, whereas at 600 MHz most splitting systems of the ring protons of sugars are first order, although even at this field (14.1 T) there are exceptions.

Although many pyranoid systems adopt one conformation predominately, and coupling constants can be assigned to this conformation, furanoid systems are generally mobile, and more than one conformation is adopted. In systems in the "fast exchange" region, chemical shifts and coupling constants are weighted averages. The requirement for fast exchange is that the rate constant for interconversion of observed molecules is much greater than chemical shift differences (in Hz) between them. This is generally true of furanosyl systems at accessible temperatures, because the energy barriers to interconversion of conformers are so low. The significant barriers to interconversion of chair six-membered rings, however, mean that pyranosyl systems which significantly occupy both chair conformers may exhibit broadened spectra as the temperature is lowered and the "intermediate exchange" region approached. In CS_2 solution, the single CH_3 proton resonance of the axial and equatorial methoxyl groups of 2,2-dimethoxytetrahydropyran begins to broaden at −80 °C; a barrier to chair–chair interconversion of 36 kJ mol^{-1} can be calculated [interestingly, this is less than the barriers for chair–chair interconversion of cyclohexane (44 kJ mol^{-1}) and 1,1-dimethoxycyclohexane (45 kJ mol^{-1}), indicating that the anomeric effect (Section 2.6), while it affects preferences, if anything reduces conformational barriers[13]]. In the "slow exchange" regime, two distinct spectra of the different conformers are observed, in the same way as different anomers, *etc.*

Pyranosyl derivatives in chair conformations are the most amenable to analysis of three-bond proton–proton coupling by the Karplus equation, eqn (2.1) and Figure 2.8. It is seen that a dihedral angle of ~60°, (1.047 radians) corresponding to axial–equatorial or equatorial–equatorial coupling, gives rises to a fairly low coupling constant, whereas axial–axial couplings, corresponding to a dihedral angle of 180°, give rise to a maximum coupling constant.

In general, the anomeric protons of sugar derivatives resonate at the lowest field of all the ring protons, since they are attached to carbons with two electron-withdrawing oxygen substituents. Axial hydrogens resonate at higher field than equatorial hydrogens, but even axial anomeric hydrogens can usually be readily identified. More troublesome in the case of water-soluble materials is that the proton resonance from residual HOD in D_2O commonly comes between the expected resonances of axial and equatorial anomeric protons and if care is not taken to exclude moisture during sample preparation, may obscure them.

The $^3J_{H1,H2}$ coupling alone can unambiguously identify anomeric configuration, if the conformation can be predicted with confidence. Thus, in the D-*gluco* series, all the substituents except the anomeric are equatorial in the $^4C_1$ conformation, so that the preference for this conformation can only be overcome by bulky, charged aglycones (Figure 2.5); for ordinary oxygen aglycones, therefore, a 1–2 Hz splitting of the H1 resonance indicates an α configuration and a 7–8 Hz splitting a β configuration; as discussed above, the axial H1 of β-configured glycoside resonates at higher field (lower δ). The technique can be applied to galactopyranosides and xylopyranosides, but not to mannopyranosides, since H2 is now equatorial in the $^4C_1$ conformation, so that 60° dihedral angles, and

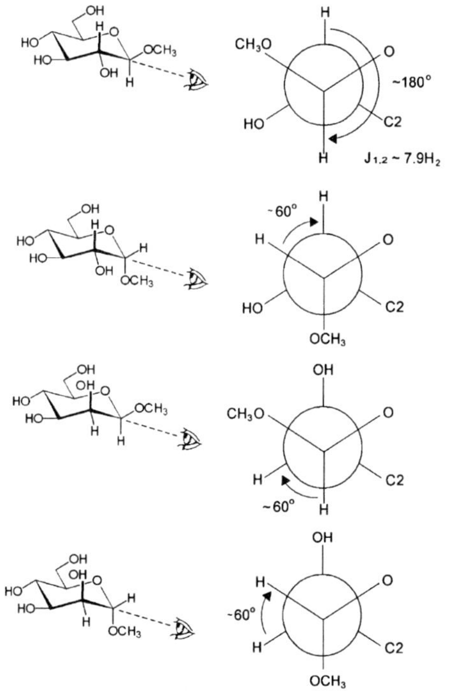

Figure 2.9 Predicted H1–H2 dihedral angles for β- and α-glucopyranosides (top) and β- and α-mannopyranosides (bottom).

attendant low splittings, are observed with both mannopyranoside anomers (Figure 2.9). Anomeric assignments of mannopyranosides may be based on the chemical shift of the anomeric proton if both anomers are available, but otherwise are uncertain; one-bond ^{13}C–^{1}H couplings at the anomeric centre are more reliable ($^{1}J_{^{13}C,H}\,{}^{1}J_{13C-H}$ for α glycosides is around 170 Hz, compared to 160 Hz for their anomers[14]).

If the anomeric configuration of a glycosyl derivative is known with confidence, then three-bond proton coupling constants can enable the conformation to be determined. Thus, the bulk of the pyridinium ring in α-D-xylopyranosylpyridinium ions, and the necessity for extensive solvation of the positive charge, constrain the ring in the $^{1}C_{4}$ conformation, as shown by the proton–proton coupling constants displayed in Figure 2.10. Note the four-bond (or W) coupling observed between H3 and H1. Also displayed are the couplings from methyl β-D-galactopyranoside; note how they are all lower than would be predicted for a perfect chair and from Figure 2.8 and also that very similar dihedral angles can give appreciably different couplings.

2.6 THE ANOMERIC EFFECT

Many carbohydrate conformations follow simply from the same considerations that govern alicyclic conformations, that bulky substituents prefer equatorial

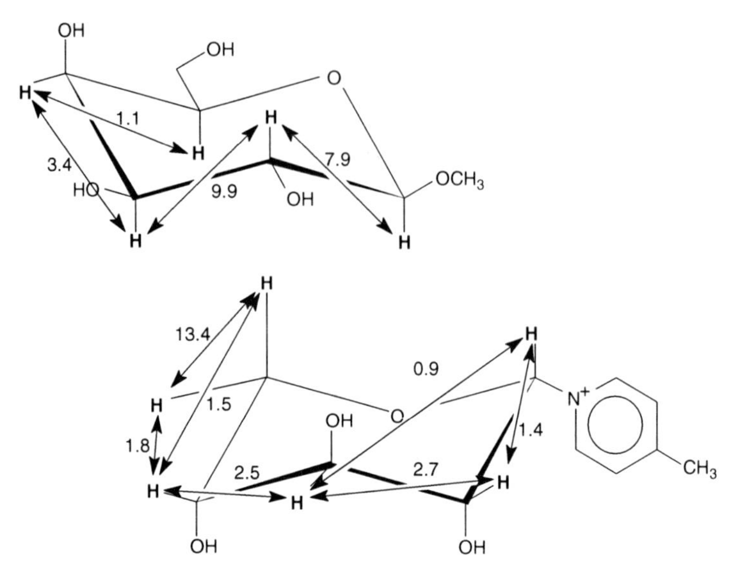

Figure 2.10 Proton–proton coupling constants in α-D-xylopyranosyl-4-methylpyridinium ion[15] and methyl-β-D-galactopyranoside. Note the differences between coupling constants associated with very similar dihedral angles and the lower couplings experienced when the proton-attached carbon atoms have multiple electron-withdrawing groups.

and pseudo-equatorial orientations and that aligned electrostatic dipoles involve large energetic cost. Many, but not all: β-D-xylopyranosyl fluoride adopts the $^1C_4$ conformation in which all four ring substituents are axial, as does its tri-*O*-acetyl derivative. In preparative carbohydrate chemistry, it is common to make protected glycopyranosyl halides by reaction of the protected sugar with a highly acidic solution of the appropriate hydrogen halide – the axial halide is the major thermodynamic product.[i] In the case of cellobiosyl fluoride, 1–2% only of the β-compound appears to be present at equilibrium.[19]

The tendency of electronegative substituents at C1 of a pyranose ring to adopt an axial orientation is termed the anomeric effect.[20] The name has been extended to the tendency of the C–O dihedral angle of X–C–O–R fragments to adopt a conformation in which X is antiperiplanar to a lone pair on the oxygen, when X is an electron-withdrawing group. There have been two explanations of the origin of the anomeric effect, one based on classical (pre-quantum) electrostatics and the other on frontier orbital theory. The consensus that appears

---

[i] The classic way of making per-*O*-acetylglycopyranosyl bromides ("acetobromo sugars") is acetylation of glucose, galactose, mannose, xylose, *etc*., with acetic anhydride and perchloric acid, followed by reaction of the products with HBr generated *in situ* from red phosphorus, bromine and water[16]; chlorides are made similarly, with the acetylation solution saturated with HCl; fluorides can be made from the anomeric mixture of fully acetylated sugars in liquid HF or, less hazardously, pyridinium poly(hydrogenfluoride).[17] Per-*O*-acetylglycosyl iodides can be made with anhydrous HI in dichloromethane at low temperature.[18]

to have emerged is that frontier orbital interactions are the main origin of the effect, but that classical electrostatics plays some role.

The frontier orbital explanation is that there is overlap between a lone pair on oxygen and the σ* orbital of the C–X bond, which is efficient (in the case of a pyranose ring) when the C–X substituent is axial, but not when it is equatorial.

To understand this, it is necessary to be clear about what the conventional representations of the lone pair electrons on oxygen do and do not represent. In a fragment R–O–R, if the ROR angle is 109.5°, then the lone pairs can be represented as being in two orthogonal sp^3 orbitals, drawn in the conventional skittle shape; however, the outline of the skittle represents a contour of ψ, not electron density; the small and large lobes of the skittle in fact have opposite signs. If these two sp^3 orbitals are both doubly occupied, then the resulting electron distribution has the shape of a large, blunt sp^3 orbital, with a maximum in charge density between the two individual sp^3 orbitals.[21] The same electron density, however, results from considering the lone pairs as occupying an sp and a p orbital; in fact, the lone pairs on oxygen can be represented as two sp^3 orbitals, one p and one sp orbital or anything in between: the same electron density, which is the experimentally measurable quantity, results (Figure 2.11).

The σ* orbital of a C–X bond, is, like the p-type lone pair on oxygen, composed of two approximately equal lobes. The overlap between the p-type lone pair and the σ* orbital of an axial C–X bond in a pyranose ring is relatively efficient [the dihedral angle about C1–O5 bond described by the axis of the p-type lone pair and the σ* orbital will be about 30°; p–p overlap of this type varies approximately as the square of the cosine of the dihedral angle and $\cos^2(30°) = 0.75$]. While the overlap of the sp lone pair with an equatorial C–X bond is geometrically optimal, the electrons in an sp orbital are held very much closer to the nucleus than those of a p orbital and are much less readily available for donation. The anomeric effect, on the frontier orbital model, thus corresponds to "no bond resonance" in the sense shown, but only in the axial position (Figure 2.12).

Most of the features of the effect can also be rationalised by classical electrostatic considerations. The C–X bond will be associated with a dipole, usually with its negative end towards X. The ring oxygen atom will be

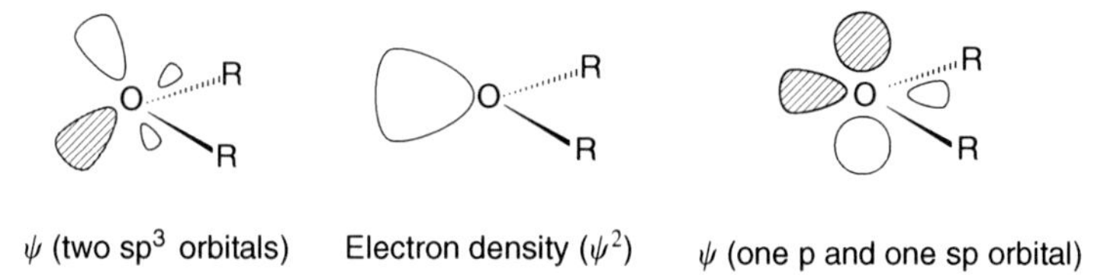

Figure 2.11 Cartoons of contours of ψ for the equivalent representations of oxygen lone pairs in an ether as in two sp^3 orbitals or one p and one sp orbital and an electron density contour. Note that hybrid sp^n orbitals do not have a node at the nucleus, as a pure p orbital does, but rather one "behind" the nucleus, between it and the small lobe.

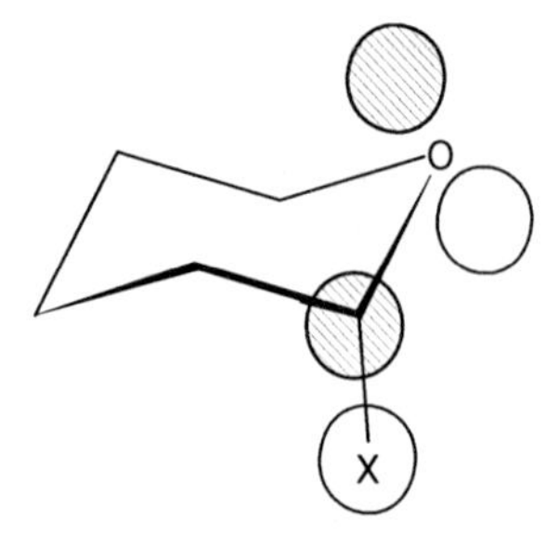

Overlap of p-type lone pair with σ* orbital of an axial substituent.
This corresponds to "no-bond resonance" in the sense shown

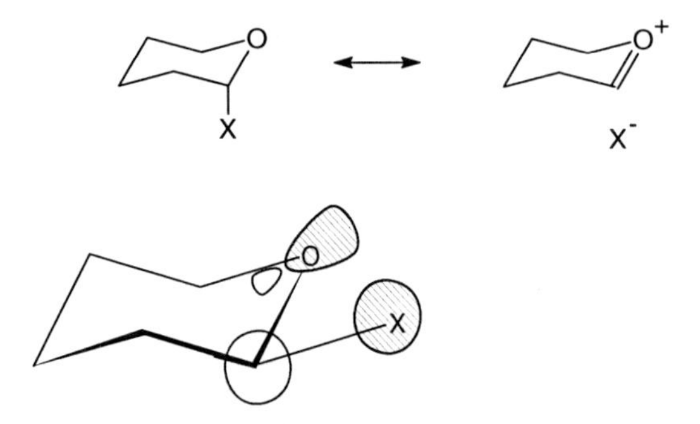

Overlap of sp-type lone pair with the σ* orbital of an equatorial substituent. The sp lone pair is much closer to the nucleus, so the overlap is much less important, even though the sp lone pair and the C-X σ* orbital are exactly eclipsed.

Figure 2.12 Frontier orbital rationalisation of the anomeric effect.

associated with a region of negative charge, both by virtue of the lone pairs on oxygen[ii] (Figure 2.11) and also by virtue of the dipoles associated with the C1–O5 and C5–O5 bonds, which have a resultant in the plane defined by C5, O5 and C1 and approximately bisecting the C5–O5–C1 angle. This resultant dipole will exactly eclipse the C–X dipole if the C–X bond is equatorial, but not if it is axial (Figure 2.13).

The no-bond resonance picture of the anomeric effect predicts that the proportion of axial conformer in mobile 2-aryloxytetrahydropyrans should increase as electron-withdrawing substituents are introduced into the aryl group and the electron demand of the C–X bond increases This will have the effect, however, of decreasing the negative charge on the exocyclic oxygen and according to the electrostatic model the anomeric effect should decrease. In fact,

[ii] If the two lone pair sp^3 orbitals on oxygen are drawn in the customary skittle shapes, the major lobes in a Newman projection look like two rabbit's ears. The direct electrostatic effect of lone pair electrons was therefore dubbed the "rabbits' ears" effect, but failed to find acceptance, as much from the overtones of Beatrix Potter as from the fact that the shape of electron density resembled not two, but one ear (Figure 2.11).

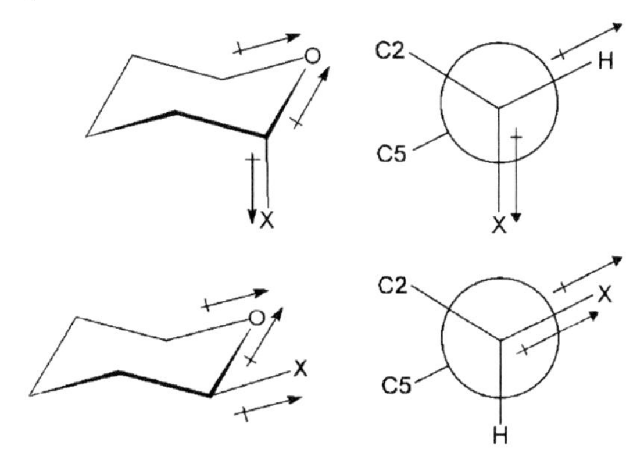

Figure 2.13 Electrostatic explanation of the anomeric effect.

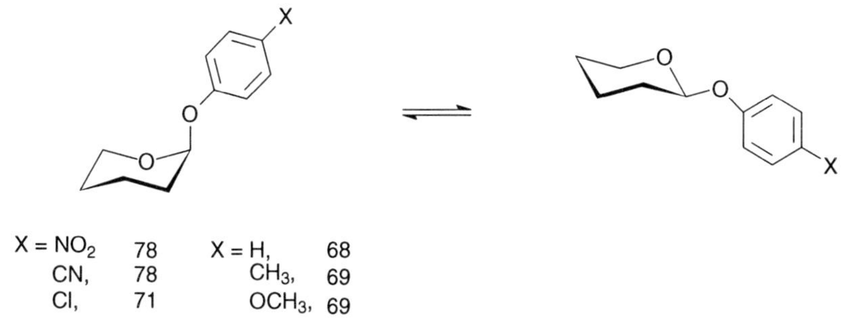

Figure 2.14 Proportions of axial conformer of 2-aryloxytetrahydropyrans in deuterochloroform solution.[22] The conformational equilibrium is invariant with X in cyclohexane (84–87% axial).

as is shown in Figure 2.14, there is a small effect in $CDCl_3$ solution, which disappears in cyclohexane. Since $CDCl_3$ is a weakly hydrogen-bonding solvent, it could be that hydrogen-bonding reduces the charge on the exocyclic oxygen atom, allowing the frontier orbital effect to be experienced unmasked by the electrostatic effect, as it is in cyclohexane solution.

The no-bond resonance model of the anomeric effect predicts that, in a series of axial aryloxytetrahydropyran derivatives, the intracyclic bond should shorten and the extracyclic bond should lengthen as the parent phenol becomes more acidic and the ion-paired canonical form; any effect should be much smaller in the equatorial case. Careful X-ray crystallographic studies[23] of a series of tetrahydropyranyl ethers (structures in Figure 2.15) indeed showed that for seven such axial compounds, with the $\mathrm{p}K_\mathrm{a}$ of ROH spanning 8 pK units, the lengths (Å) of the extracyclic bond (x) and the intracyclic bond (n) were given by eqns (2.2) and (2.3), respectively:

$$x = 1.493 - 0.006495\mathrm{p}K_\mathrm{a} (r = 0.985) \quad (2.2)$$

$$n = 1.364 + 0.003639\mathrm{p}K_\mathrm{a} (r = 0.939) \quad (2.3)$$

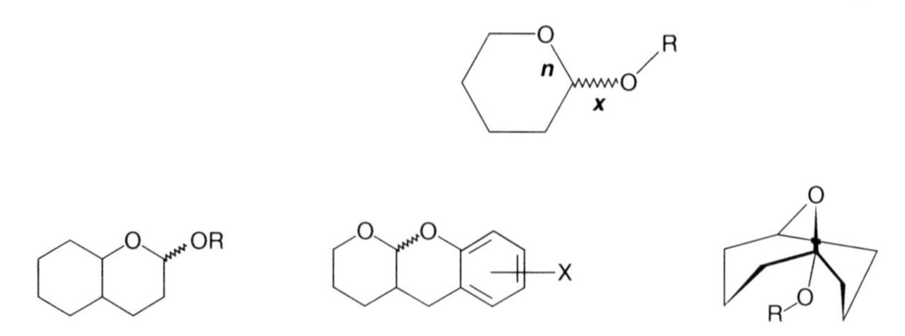

Figure 2.15 Tetrahydropyranyl structures used in the investigation of the anomeric effect by X-ray crystallography.

However, similar, but not as pronounced, trends were also seen with seven equatorial tetrahydropyranyl ethers, with a pK_a range of 13 units:

$$x = 1.456 - 0.00476pK_a (r = 0.945) \tag{2.4}$$

$$n = 1.394 + 0.00214pK_a (r = 0.911) \tag{2.5}$$

The picture was made yet more ambiguous by the fact that the equivalent bond lengths in seven α-D-glucopyranosides (over an aglycone acidity range of 12 pK units) showed no such correlation for n and only a weak one for x:

$$x = 1.427 - 0.00102pK_a \ (r = 0.636) \tag{2.6}$$

Such a correlation is to be expected in the absence of the frontier orbital interactions of Figure 2.12, since the C–OR bond in a variety of structures lengthens with increasing acidity of ROH in a wide variety of structures, with a sensitivity comparable to that seen in the tetrahydropyranyl systems.[24]

The key piece of evidence for the rationalisation of the anomeric effect in terms of classical electrostatic dipole interactions has been the apparent existence of a reverse anomeric effect. Reversal of the sense of the dipole of the C–X bond, so that X now carries a net positive rather than negative charge, should, on the electrostatic picture, result in X having a more than ordinary tendency to be equatorial. The effect was first observed in glycosyl pyridinium salts[25] and, in what at first sight seemed to be an elegant experiment, the steric demand of the bulky pyridinium salt was apparently eliminated as the source of the effect by the observation that tri-*O*-acetyl-α-D-xylopyranosylimidazole changed from 65% to >95% equatorial on addition of excess acid.[26] Protonation of imidazole will have only a minor effect on its steric requirements, since protonation takes place at a remote site[iii].[27]

[iii] The A values for imidazole and imidazolium in cyclohexyl systems, measured directly, are both 9.2 ± 0.4 kJ mol^{-1}, but measurement of the difference in K_a directly indicates that the imidazolium A value is about 0.4 kJ mol^{-1} higher.

However, NMR titration of a mixture of the anomeric glucopyranosylimidazoles in a variety of solvents established conclusively that the α-anomer was more basic than the β-anomer by around 0.3 pK units in D_2O, but less in less polar solvents.[28] This is exactly the opposite of what is predicted by any reverse anomeric effect, but is what would be predicted on the frontier orbital picture of the normal anomeric effect of Figure 2.12, since protonation would increase the electron demand of the anomeric substituent. The additional hydroxymethyl group of glucopyranosylimidazoles, ensures that they, unlike xylopyranosylimidazoles, remain in the $^4C_1$ conformation on protonation. On the electrostatic model, therefore, the protonated α-anomer is expected to be destabilised and the protonated β-anomer stabilised, which should make the β-anomer more basic.

The elegant NMR approach by Perrin *et al.*[28] enables the ratio to be measured without any absolute measurement of pH. Equation (2.7) follows from the thermodynamic box of Figure 2.16, where K_a refers to an acid dissociation constant and K_e to an anomeric equilibrium constant (=[β]/[α]):

$$K_a^{\beta} K_e^{+} = K_a^{\alpha} K_e^{0} \tag{2.7}$$

The difference in anomeric equilibrium between protonated and neutral glucosylimidazoles can thus be measured as an acid dissociation constant ratio (*i.e.* pK_a difference).

Proton transfers are normally rapid on the NMR time-scale at ordinary temperatures. Therefore, the chemical shift of all the nuclei in part-protonated glucosylimidazoles will be a weighted average of the chemical shifts for protonated and unprotonated species. If both anomers are present in the same solution, then, irrespective of the absolute value of [H^+], K_a^{α}/K_a^{β} will be

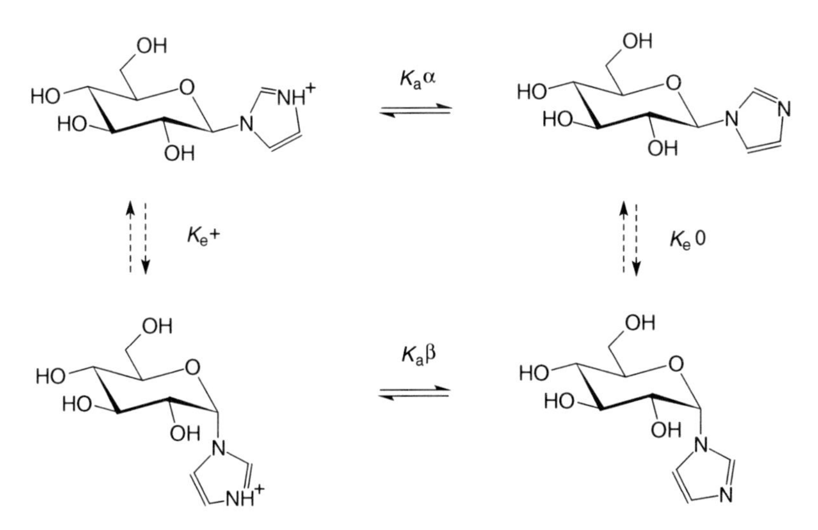

Figure 2.16 Thermodynamic box for protonation and equilibration of glucopyranosylimidazoles.

$[\alpha H^+][\beta]/[\alpha][\beta H^+]$. The method thus permits measurement of ratios of acid dissociation constants (differences in pK_a values) in any medium, without the necessity for measuring any absolute pH and thus can be used in all solvents, provided that proton transfer is fast in them. Equation (2.8) holds, where an unsubscripted δ refers to the observed chemical shift of a reporter nucleus, δ_0 and δ_+ to its value when the substrate is fully deprotonated and fully protonated, respectively, and the superscripts to the anomer. The acid dissociation constant can thus be obtain from the gradient of a linear plot obtained from chemical shifts obtained during an NMR titration.

$$(\delta^{\beta} - \delta_0^{\beta})(\delta_+^{\alpha} - \delta^{\alpha}) = (K_a^{\alpha}/K_a^{\beta})(\delta^{\alpha} - \delta_0^{\alpha})(\delta_+^{\beta} - \delta^{\beta}) \quad (2.8)$$

The ratio K_a^{α}/K_a^{β} is, however, solvent dependent, increasing from 0.52 in D_2O to 0.80 in CD_3OD and to 0.97 in $(CD_3)_2SO$, consistent with a small solvent-dependent, electrostatic interaction favouring protonation of the β-anomer, which becomes more important as the solvent becomes less polar.

Similar experiments with tetra-*O*-methylglucosylanilines in acetonitrile indicated that here the protonation of the β-anomer was slightly favoured, with K_a^{α}/K_a^{β} varying[29] from 1.6 to 2.1, depending on the aniline substituents. However, measurements of the analogous parameter for the corresponding axial and equatorial 4-*tert*-butylcyclohexylanilines revealed that protonation of the equatorial isomer was favoured by a slightly larger factor (3.5–3.1), indicating the existence of steric hindrance to solvation of the axial epimers and a small normal anomeric effect in the protonated glucosylanilines.[30] The greater basicity of β- compared with α-glucosylamines is likewise attributed to such steric hindrance of solvation. Glucosylanilines, like most glycosylamines, are not configurationally stable, but the NMR titration method requires only that the interconversion be slow on the NMR time-scale.

The anomeric effect appears to be particularly strong when there is an axial substituent on position C2 (this was at one stage called the Δ2 effect). This would appear to be an electrostatic effect, the C2–O dipole opposing the C–X dipole of an equatorial substituent and thus reinforcing the anomeric effect (Figure 2.17).

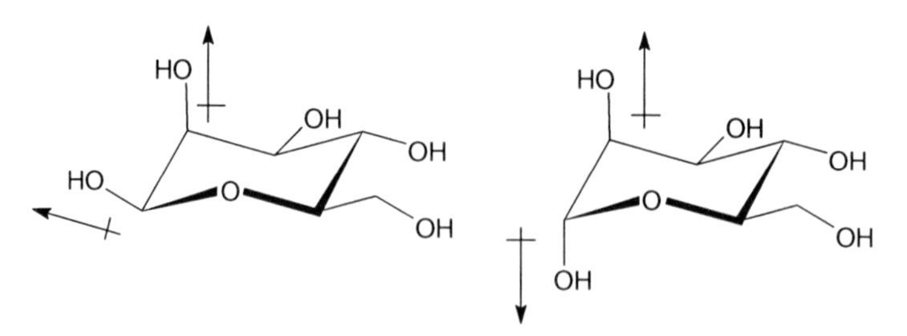

Figure 2.17 Enhancement of the anomeric effect in D-mannopyranose. The D-mannopyranose ring is rotated by 180° about a vertical axis from the normal viewpoint.

2.7 CONFORMATIONAL FREE ENERGIES IN PYRANOSES

The length of the C–O bond in ethers is generally taken as 1.43 Å,[24] whereas the length of an unstrained C–C bond is around 1.53 Å.[31] This makes tetrahydropyran slightly less puckered than cyclohexane. In cyclohexane chemistry, conformational preferences are discussed in terms of *A* values for substituents (the free energy difference between the axial and equatorial chair conformers of a monosubstituted cyclohexane). These *A* values for cyclohexane[32,33] cannot be applied directly to tetrahydropyran,[34,35] even in the absence of an anomeric effect, not least because they vary with the position of substitution, even with non-polar substituents which do not exert an anomeric effect (Table 2.1).

For 2-aryloxysubstituents, which do exert an anomeric effect, Kirby and Williams[36] combined their own data for cyclohexane[37] with that of Ouedrago and Lessard for aryloxytetrahydropyrans[38] to derive eqn (2.9) for the strength of the anomeric effect of a 2-phenoxysubstituent:

$$\text{Anomeric effect/kJ mol}^{-1} = (7.1 \pm 0.8) - (0.22 \pm 0.08)\text{p}K_a \qquad (2.9)$$

A values, however, which refer to non-polar solvents such as CS_2 or $CDCl_3$, are not particularly useful for the medium to which most carbohydrate research applies, water. In the 1960s, by measurement of conformational energy differences of a wide range of substances, Angyal and co-workers produced a series of empirical instability factors which could predict the conformations of pyranose rings.[1,39] These interaction energies are set out in Table 2.2; they are destabilisation energies, so the anomeric effect factor is added to the equatorial conformer or epimer.

The existence of a destabilising interaction between *gauche* vicinal oxygen substituents seems at variance with the results from studies of acyclic sugars, but may be a solvation effect.

An example of the power of these instability factors is their prediction that epimerisation of β-D-mannopyranose at C5 to give α-L-gulose will change the conformational preference of the ring from $^4C_1$ to $^1C_4$. A closely analogous epimerisation occurs during biosynthesis of alginate (Chapter 4), whose

Table 2.1 *A* values (kJ mol^{-1}) for cyclohexane and tetrahydropyran.

Substituent	*Cyclohexane*[32,33]	*Tetrahydropyran*[34,35]
2-CH_3	7	12
3-CH_3		6
4-CH_3		8
2-$COOCH_3$	5.5	5.8
3-$COOCH_3$		2.5
2-CH_2OH		12
3-CH_2OH		3.3
2-Cl	2.2	<-7.5
3-Cl		2.8
4-Cl		1.3

Table 2.2 Interaction energies ($kJ\,mol^{-1}$) in carbohydrate systems.

Axial H–axial O (OH, OAc, *etc.*)	1.9
Axial H–axial C (CH3 or CH2OH)	3.8
Axial O–axial O	6.3
Axial O–axial C	10.5
Vicinal O–vicinal O (e–e or a–e only, not a–a)	1.5
Vicinal O–vicinal C (e–a or e–e only, not a–a)	1.9
Anomeric effect, O1 and O2 equatorial	2.1
Anomeric effect, O1 equatorial, O2 axial	4.2

β-D-mannopyranose, $^4C_1$		β-D-mannopyranose, $^1C_4$		α-L-gulopyranose, $^4C_1$		α-L-gulopyranose, $^1C_4$	
3 × $O–O_{vic}$	4.5	2 × $O–O_{vic}$	3.0	3 × $O–O_{vic}$	4.5	2 × $O_{ax}–H_{ax}$	3.8
1 × $O–C_{vic}$	1.9	1 × $O_{ax}–H_{ax}$	1.9	1 × $O_{ax}–H_{ax}$	1.9	2 × $O–O_{vic}$	3.0
1 ×$O_{ax}–H_{ax}$	1.9	2 ×$C_{ax}–O_{ax}$	21.0	1 × $O–C_{vic}$	1.9	1 × $O–C_{vic}$	1.9
1 × anomeric, $O2_{ax}$	4.2	1 × $O_{ax}–O_{ax}$	6.3	2 × $C_{ax}–H_{ax}$	7.6		
				1 × anomeric, $O2_{ax}$	4.2		
Total	**12.5**	**Total**	**32.2**	**Total**	**20.1**	**Total**	**8.7**

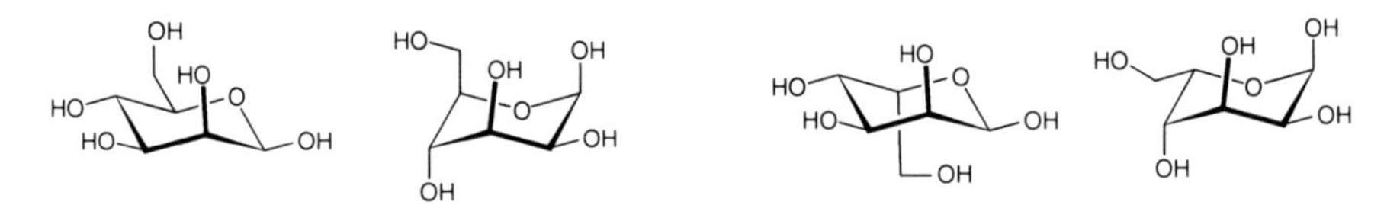

Figure 2.18 Use of instability factors to predict the conformation of β-D-mannopyranose and α-L-gulopyranose.

properties are critically dependent on alternating sequences of poly β(1→4)-mannuronic acid and poly-α(1→4)-guluronic acid, in which the pyranose rings of the latter adopt the alternative chair conformation (Figure 2.18).

2.8 RATIONALISATION OF THE COMPOSITION OF AQUEOUS SOLUTIONS OF REDUCING SUGARS (SEE TABLE 1.1)

Easiest to understand is that those sugars which have all substituents except the anomeric OH equatorial in the pyranose form (glucose, xylose) have few if any furanoses at equilibrium: five-membered rings are generally less stable than six-membered rings (though they are formed faster). Glucofuranose and xylofuranose additionally have a *cis* interaction between the 3OH and the 4-substituent, which further destabilises furanose forms. Pyranose forms of galactose and arabinose are destabilised by a single axial OH at position 4, whereas the furanose forms have all the substituents *trans* to each other; furanose forms are therefore more abundant than with glucose and xylose. There is a single axial OH at position 3 in ribopyranose and allopyranose and accordingly furanose forms of these sugars are also more abundant, although it

β-D-xylopyranose and -xylofuranose (R = H) and β-D-glucopyranose and -glucofuranose (R = CH_2OH)

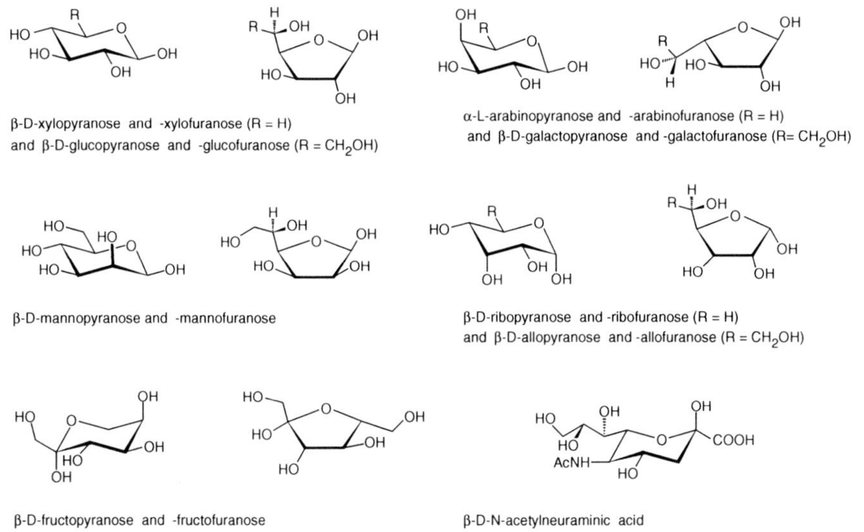

α-L-arabinopyranose and -arabinofuranose (R = H) and β-D-galactopyranose and -galactofuranose (R= CH_2OH)

β-D-mannopyranose and -mannofuranose

β-D-ribopyranose and -ribofuranose (R = H) and β-D-allopyranose and -allofuranose (R = CH_2OH)

β-D-fructopyranose and -fructofuranose

β-D-N-acetylneuraminic acid

Figure 2.19 Conformations of representative pyranose and furanose forms of common reducing sugars.

is not clear why they should be more abundant than in arabinose and galactose. Mannofuranose has the substituents at C2, C3 and C4 all *cis*, so that despite the axial 2OH in mannopyranose, furanose forms are present in only trace amounts.

D-Fructose is D-arabinose with the anomeric hydrogen replaced by a hydroxymethyl group; furanose forms are therefore present for the same reason as with arabinose.

The approximately 60:40 ratio of equatorial to axial pyranose anomers of glucose, xylose, galactose and arabinose[iv] illustrates the operation of the anomeric effect: the *A* value for hydroxyl (Table 2.1) predicts an equatorial/axial preference in cyclohexane of about 5.4:1. The enhancement of the anomeric effect by an axial 2OH in mannopyranoses results in α-mannopyranose becoming the predominant tautomer. By contrast, an axial 3-OH, as in ribose and allose, results in 1,3-diaxial clashes with an axial anomeric OH group and an equatorial anomeric OH group being favoured more than in, say, glucose.

In ketopyranoses, the reluctance of carbon substituents to become axial acts in the same direction as the anomeric effect, so that the OH-axial pyranose anomer is overwhelmingly predominant, as in fructopyranose and *N*-acetylneuraminic acid (Figure 2.19).

[iv] Because of the change of "bottom" reference asymmetric carbon, α- not β-arabinopyranose has the anomeric OH equatorial (formal removal of the C5 CH_2OH from β-D-galactopyranose gives α-L-arabinopyranose).

2.9 CONFORMATIONS OF HYDROXYMETHYL GROUPS

The conformations about the C5–C6 bond in hexopyranoses can be dealt with in several ways. In crystal structures of small molecules, the C4–C5–C6–O6 dihedral angle, χ, can be determined reasonably exactly and is often quoted (see Section 4.1). In some polymers χ can be determined, but in others (most notably some forms of cellulose) there is disorder about the C5–C6 bond. In solution, it is fairly easy to obtain vicinal coupling constants between H5 and both $H6_R$ and $H6_S$. However, rotation about C5–C6 is sufficiently fast that the system is in the fast-exchange region and an average value of χ comes out of the Karplus equation, with the nature of the averaging that produced it not clear. Some authors consider the system in terms of a rapid equilibrium between three perfectly staggered conformers, termed *gg* (O6 *gauche* to both O5 and C4), *gt* (O6 *gauche* to O5 and *trans* to C4) and *tg* (O6 *trans* to O5 and *gauche* to C4).[v] These are illustrated in Figure 2.20. Other authors consider the equilibrium in terms of a continuum of rotamers of smoothly varying energy.[40]

In the *gg* and *gt* conformations, the two oxygens are *gauche* to each other and are thus stabilised relative to the *tg* conformation by the *gauche* effect; additionally, there is the possibility of weak hydrogen bonding by OH6 to the lone pairs on the ring oxygen. It is therefore unsurprising that (on the three perfect conformer model), the *tg* conformation is relatively uncommon.

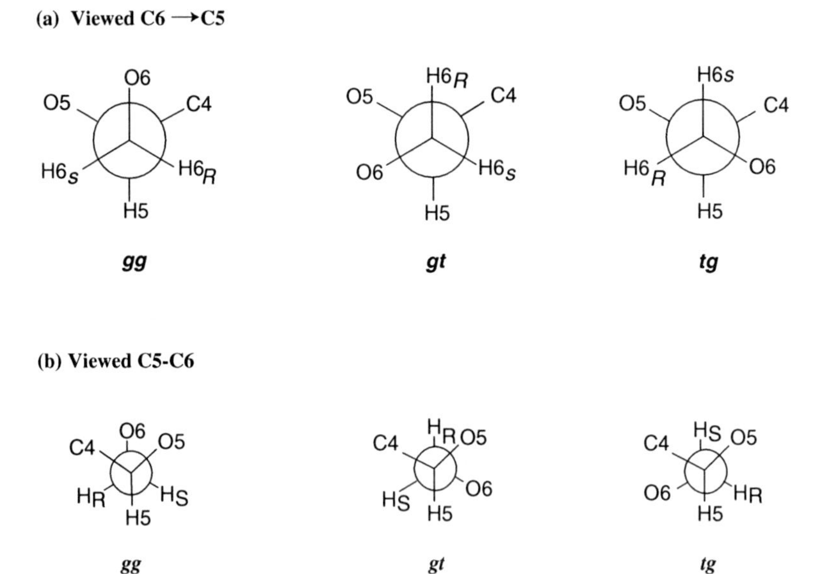

Figure 2.20 Three perfectly staggered rotamers about the exocyclic C5–C6 of aldohexopyranoses.

[v] Note that the relationship of O6 to O5 is described before its relationship to C4.

2.10 CONFORMATIONS OF SEPTANOSIDES

Septanosides are rarely, if ever, encountered in Nature, but the ability of hydroxylated azepanes (azacycloheptanes) to inhibit glycosyl-transferring enzymes prompted an investigation of the conformational preferences of some septanosides.[41] The Cremer–Pople treatment of a seven-membered ring requires three angles in addition to the puckering parameter and loses its visual usefulness. It appears, nonetheless, that low-energy conformers are confined to positions on a pseudorotational itinerary involving twist-chair and chair conformers. Chair conformers are described by the one atom above and the two below the "seat" of the chair, as in the $^{5}C_{12}$ conformation sketched in Figure 2.21. As might be expected from the fully eclipsed C1–C2 bond in such a conformation, it is at a local energy maximum and distorts to the twist-chair (*TC*) conformation. The reference plane in *TC* conformations is defined by three contiguous ring atoms, with the ring atoms above and below the reference plane, superscripts or subscripts. Substitution patterns can enforce the occupancy of a single conformer, as in the $^{5,6}TC_{3,4}$ conformation adopted exclusively by methyl α-D-*glycero*-D-idoseptanoside, in which the anomeric effect is fully expressed. It appears that methyl β-D-*glycero*-D-guloseptanoside, whilst occupying the $^{5,6}TC_{3,4}$ conformation predominantly, can also occupy the $^{6,O}TC_{4,5}$ conformation to a significant extent (Figure 2.21).

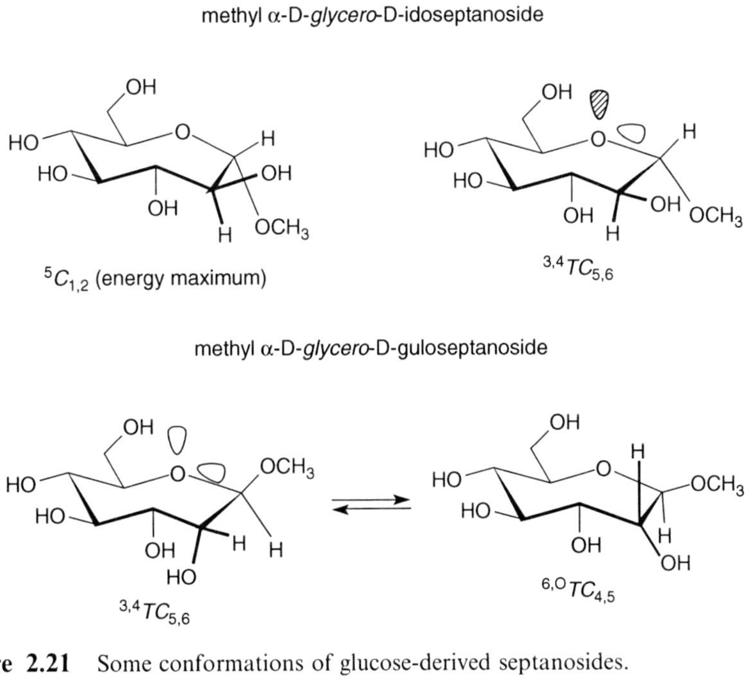

methyl α-D-*glycero*-D-idoseptanoside

$^{5}C_{1,2}$ (energy maximum) $^{3,4}TC_{5,6}$

methyl α-D-*glycero*-D-guloseptanoside

$^{3,4}TC_{5,6}$ $^{6,O}TC_{4,5}$

Figure 2.21 Some conformations of glucose-derived septanosides.

REFERENCES

1. S. J. Angyal, *Angew. Chem. Int. Ed. Engl.*, 1969, **8**, 157.
2. D. Friesen and K. Hedberg, *J. Am. Chem. Soc.*, 1980, **102**, 3987.
3. J. B. Lambert, Y. Zhao, R. W. Emblidge, L. A. Salvador, X. Y. Liu and E. C. Chelius, *Acc. Chem. Res.*, 1999, **32**, 183.
4. A. J. Kirby and A. H. Williams, *ACS Symp. Ser.*, 1993, **539**, 53.
5. B. M. Pinto and R. M. Y. Leung, *ACS Symp. Ser.*, 1993, **539**, 126.
6. Y. Israeli and C. Detellier, *Carbohydr. Res.*, 1997, **297**, 201.
7. A. Schouten, J. A. Kanters, J. Kroon, S. Comini, P. Looten and M. Mathouthli, *Carbohydr. Res.*, 1998, **312**, 131.
8. D. Cremer and J. A. Pople, *J. Am. Chem. Soc.*, 1975, **97**, 1354.
9. C. Altona and M. Sundaralingam, *J. Am. Chem. Soc.*, 1972, **94**, 8205.
10. J. B. Houseknecht, C. Altona, C. M. Hadad and T. L. Lowary, *J. Org. Chem.*, 2002, **67**, 4647.
11. M. Karplus, *J. Chem. Phys.*, 1959, **30**, 11.
12. C. A. G. Haasnoot, F. A. A. M. de Leeuw and C. Altona, *Tetrahedron*, 1980, **36**, 2783.
13. C. L. Perrin and O. Nuñez, *J. Chem. Soc., Chem. Commun.*, 1984, 333.
14. K. Bock and C. Pedersen, *J. Chem. Soc., Perkin Trans. 2.*, 1974, 293.
15. L. Hosie, P. J. Marshall and M. L. Sinnott, *J. Chem. Soc., Perkin Trans. 2.*, 1984, 1121.
16. M. Bárczai-Marcos and F. Kőrösy, *Nature*, 1950, **165**, 359.
17. J. Jünnemann, J. Thiem and C. Pedersen, *Carbohydr. Res.*, 1993, **249**, 91.
18. S. M. Chervin, P. Abado and M. Koreeda, *Organic Letters*, 2000, **2**, 369.
19. D. Becker, K. S. H. Johnson, A. Koivula, M. Schülein and M. L. Sinnott, *Biochem. J.*, 2000, **345**, 315.
20. The effect was first proposed by J. T. Edward, whose account of the historical background is published in *ACS Symp. Ser.*, 1993, **539**, 1.
21. A. Streitweiser and P. H. Owens, *Orbital and Electron Density Diagrams*, Macmillan, New York, 1973, p. 120.
22. M. J. Cook, T. J. Howe and A. Woodhouse, *Tetrahedron Lett.*, 1988, **29**, 471.
23. A. J. Briggs, R. Glenn, P. G. Jones, A. J. Kirby and P. Ramaswamy, *J. Am. Chem. Soc.*, 1984, **106**, 6200.
24. F. H. Allen and A. J. Kirby, *J. Am. Chem. Soc.*, 1984, **106**, 6197.
25. R. U. Lemieux and A. R. Morgan, *Can. J. Chem.*, 1965, **43**, 2205.
26. H. Paulsen, Z. Györgydeák and M. Friedemann, *Chem. Ber.*, 1974, **107**, 1590.
27. C. L. Perrin, M. A. Fabian and K. B. Armstrong, *J. Org. Chem.*, 1994, **59**, 5246.
28. C. L. Perrin, M. A. Fabian, J. Brunckova and B. K. Ohta, *J. Am. Chem. Soc.*, 1999, **121**, 6911.
29. C. L. Perrin and K. B. Armstrong, *J. Am. Chem. Soc.*, 1993, **115**, 6825.
30. C. L. Perrin and J. Kuperman, *J. Am. Chem. Soc.*, 2003, **125**, 8846.
31. L. S. Bartell and J. K. Higginbotham, *J. Chem. Phys.*, 1965, **42**, 851.

32. A. Streitweiser, C. H. Heathcock and E. M. Kosower, *Introduction to Organic Chemistry*, Macmillan, New York, 4th edn, 1992.
33. E. L. Eliel, K. D. Hargrave, K. M. Pietrusiewicz and M. Manoharan, *J. Am. Chem. Soc.*, 1982, **104**, 3635.
34. H. Booth, K. A. Khedhair and S. A. Readshaw, *Tetrahedron*, 1987, **43**, 4699.
35. T. H. Lowry and K. S. Richardson, *Mechanism and Theory in Organic Chemistry*, HarperCollins, New York, 3rd edn, 1987, p. 139.
36. A. J. Kirby and N. H. Williams, *J. Chem. Soc., Chem. Commun.*, 1992, 1286.
37. A. J. Kirby and N. H. Williams, *J. Chem. Soc., Chem. Commun.*, 1992, 1285.
38. A. Ouedrago and J. Lessard, *Can. J. Chem.*, 1991, **69**, 474.
39. S. J. Angyal, *Aust. J. Chem.*, 1968, **21**, 2737.
40. K. Bock and J. Ø. Duus, *J. Carbohydr. Chem.*, 1994, **13**, 513.
41. M. P. DeMatteo, S. Mei, R. Fenton, M. Morton, D. M. Baldisseri, C. M. Hadad and M. W. Peczuh, *Carbohydr. Res.*, 2006, **341**, 2927.

CHAPTER 3

Nucleophilic Substitution at the Anomeric Centre

Two-electron chemistry at the anomeric centre is governed by electron release from the adjacent ring oxygen atom. At one mechanistic extreme, glycosyl cations are stable enough to be solvent-equilibrated intermediates, whose fates are independent of their method of generation; at the other, nucleophiles such as azide or appropriately-positioned intramolecular nucleophiles attack the anomeric centre in unambiguous S_N2[i] reactions. Most nucleophilic substitutions at anomeric centres fall between these two extremes and involve reactions in which bond making and bond breaking take place within the same solvent shell, but whose transition states have a high buildup of positive charge on the anomeric centre and the ring oxygen. Such reactions appear to be characterised by plateaux on free-energy surfaces, rather than the deep valleys and narrow passes of traditional transition state theory. Comparatively small variations in solvent, nucleophile (and any attendant general base) and leaving group (and any attendant general acid or electrophile) can therefore have large effects on the path taken by the system. However, because Natural Selection will progressively stabilise only one reaction pathway of an enzyme-catalysed reaction, the free energy surfaces for enzymic glycosyl transfers are probably more "classical"; at least, it is often easier to predict the consequences of a change in substrate or enzyme structure on enzyme reactions than to make analogous predictions about their non-enzymic counterparts.

Ions of the general structure $R_1R_2C{=}O^+{-}R_3$, such as the glycosyl cations formed on S_N1 reactions of glycosyl derivatives, have been given various names. When tricovalent, six-electron carbon cations $>C^+-$ were first proposed as reaction intermediates in the 1930s, they were given the name "carbonium ions". However, in the 1960s, a series of hypervalent ions such as CH_5^+ were discovered and it was pointed out that, by analogy with other "onium" ions, such as ammonium, oxonium, *etc*., these should really have the name "carbonium" and the tricovalent, six-electron species should be called "carbenium" ions, by analogy with nitrenium ions, $>N^+$. Ions of the type $R_1R_2C{=}O^+{-}R_3$ were therefore variously called oxocarbonium ions, oxocarbenium ions and oxacarbenium ions

[i] D_N A_N reaction in the IUPAC (Guthrie–Jencks) system.

(the difference between "oxo-" and "oxa-" being that "oxo" considers oxygen to have replaced a hydrogen and "oxa-" to have replaced a carbon). "Oxocarbenium ion" seems to be the commonest usage and will be adopted here.

3.1 STEREOCHEMISTRY OF OXOCARBENIUM IONS

Most efficient overlap of a filled lone-pair orbital on oxygen with an empty p orbital on the adjacent carbon atom occurs when the lone pair is in a pure p orbital. Therefore, the oxygen atom is sp^2 hybridised and the electron-deficient carbon, the oxygen and the three substituents R_1, R_2 and R_3 are all coplanar. Direct NMR studies of the methoxymethyl cation $H_2C{=}O^+{-}CH_3$[1] in "Magic Acid" (a mixture of the strong Lewis acid SbF_5, low-basicity diluents such as SO_2 SO_2ClF or SO_2F_2 and sometimes very strong protic acids such as fluorosulfonic or trifluoromethanesulfonic acid, used in direct studies of carbocations[2]) indicated that the two methylene hydrogens became equivalent by two routes, depending on the solvent composition (Figure 3.1). Rotation of the methyl group out of the plane was favoured by solvation, in part because of partial bond formation to the sp^2 carbon from solvent components. In methoxybenzyl cations, π-donation from the aromatic system likewise favoured rotation over inversion, with rotation rates, rates, as expected, correlating with σ^+ for the aromatic substituent.[3] Rotation is the predominant route in the gas phase, but is disfavoured by solvation: early work with "Magic Acid" gave activation parameters $\Delta H^{\ddagger} = 50\,\mathrm{kJ\,mol^{-1}}$, $\Delta S^{\ddagger} = -17\,\mathrm{J\,K^{-1}\,mol^{-1}}$.[4]

Incorporation of the structural grouping $R_1R_2C{=}O^+{-}R_3$ into a ring restricts the range of permitted conformations of that ring, confining furanose rings to either the ${}^3E$ or E_3 conformations and pyranose rings to the ${}^4H_3$ and ${}^3H_4$ half-chair conformations and the ${}^{2,5}B$ and $B_{2,5}$ classical boat conformations (see Figure 3.1). Mechanistic studies on non-enzymic reactions have so far ignored all but the ${}^4H_3$ half-chair conformation, but recent studies on Family 11 xylosidases have indicated that they act through oxocarbenium ion-like transition states in which the pyranose ring is in the ${}^{2,5}B$ conformation; likewise, Family 26 mannosidases constrain the sugar ring to the $B_{2,5}$ conformation for reaction.[5]

The p_π–p_π interaction in an oxocarbenium ion is mimicked by sugar lactones and their conformations can be taken as a guide; Horton and co-workers[6] predicted reaction of mannosidases via the $B_{2,5}$ conformation, on the basis of this being the exclusive conformation of mannono-δ-lactone.

3.2 LIFETIMES OF INTERMEDIATES

Any real intermediate must, by definition, live long enough to experience several molecular vibrations–otherwise it is not an intermediate, but a microstate on a reaction pathway. The period of a C–H or O–H bond stretching vibration is around 1 ps[ii]–other vibrations are slower. Therefore, no species

---

[ii] ps = picosecond, 10^{-12} s.

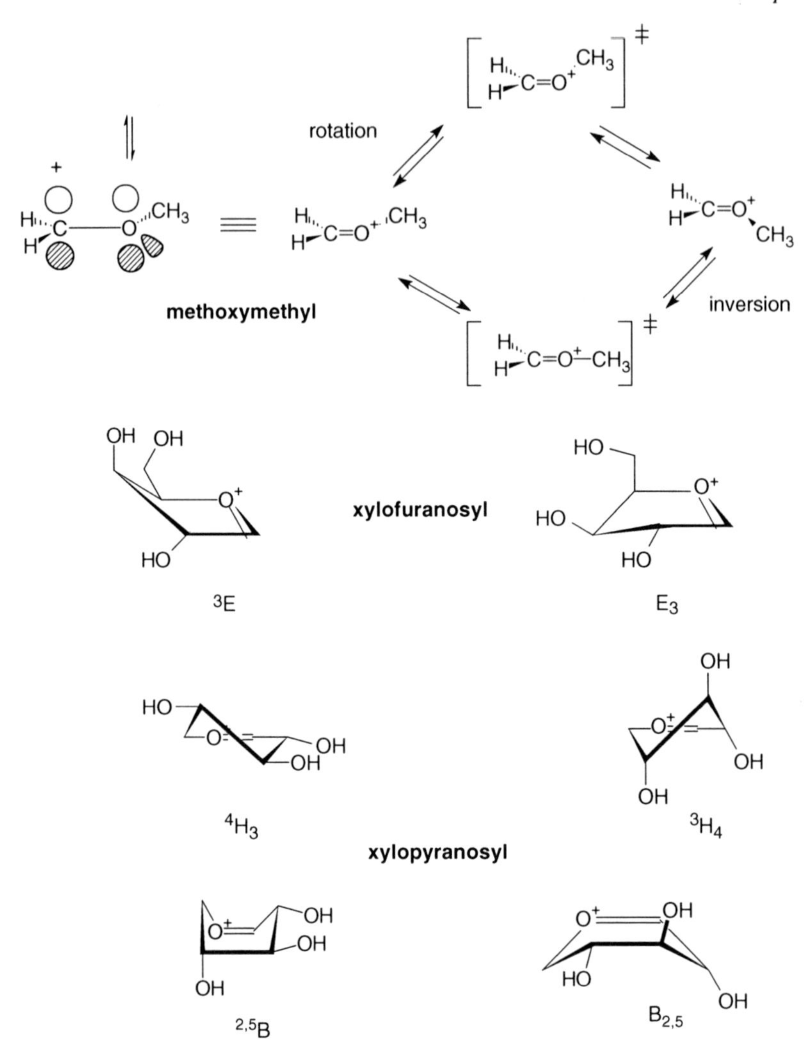

Figure 3.1 Stereochemistry of oxocarbenium ions. Top: the methoxymethyl cation, showing overlap between an empty p orbital on carbon and a p-type lone pair of electrons on oxygen and also the two mechanisms for isomerisation. Rotation involves breaking the π bond, via a perpendicular transition state, whereas during inversion the ion remains planar and the oxygen atom undergoes a process similar to the inversion of ammonia. Centre: the two permitted conformations of a xylofuranosyl cation. Bottom: the four permitted conformations of a xylopyranosyl cation.

with an estimated lifetime of less than ~1 ps can be meaningfully invoked: the intermediate is "too unstable to exist". In the case of acid or base catalysis, a protonated or deprotonated species that becomes too unstable to exist will necessarily involve the protonating acid or deprotonating base in the rate-determining transition state, so that general acid or base catalysis will occur. In the case of nucleophilic substitution reactions, supposedly unimolecular (S_N1 or $D_N + A_N$) reactions involving a species with a lifetime less than 1 ps will in fact be bimolecular, even if the degree of bond formation from the incoming nucleophile is low and the degree of bond breaking to the departing leaving group is high (a situation written D_N $^*A_N$ in the IUPAC system). This now appears to be the situation for nucleophilic reactions at anomeric centres of sugars, although until the 1980s, they were almost universally written as involving discrete oxocarbenium ions [a notable exception, though, was Capon and Thacker's discovery[7] that acid-catalysed ring closures of acylic sugar dimethyl acetals was 30–40-fold faster than would be predicted from rates of hydrolysis of dimethyl acetals $RCH(OCH_3)_2$, with R having the same electron demand as the polyhydroxymethylene chain of a sugar].

3.2.1 The Jencks Clock

Oxocarbenium ions of general formula $Ar–C(CH_3)=O^+–CH_3$ exhibit unusual behaviour in their partitioning between the strong nucleophile SO_3^{2-} and water;[8] absolute preferences for sulfite dianion (10^1–$10^3\,M^{-1}$) are 5–7 orders of magnitude lower than expected on the basis of the relative nucleophilicities of bisulfite and water, but the ratio of products displays a high dependence on the electronic properties of Ar [the oxocarbenium ions are solvent equilibrated, since the same distribution of products is obtained by acid-catalysed hydrolysis of $ArC(OCH_3)_2CH_3$ or acid-catalysed hydration of $ArC(OCH_3)=CH_2$]. The reason is that reactions of the cation with solvent water are activation limited, but reactions with bisulfite are limited by diffusion of oxocarbenium ion and bisulfite. The rate of the latter, estimated at $5 \times 10^9\,M^{-1}\,s^{-1}$ in water at 25 °C, enables the lifetimes of the ions $Ar–C(CH_3)=O^+–CH_3$ to be estimated (they vary from 140 ns[iii] for Ar=*p*-methoxyphenyl to 3 ns for Ar=*m*-bromophenyl). Similar ions could be observed directly in sulfuric acid and their reaction on dilution into aqueous sulfuric acid monitored by UV spectroscopy: extrapolation to pure water gave lifetimes in agreement with the trapping experiments.[9]

A linear relationship was observed between the logarithms of the rate constants for addition of water to $Ar–C(CH_3)=O^+–CH_3$ and addition of SO_3^{2-} to $ArCOCH_3$; extrapolation of this correlation to formaldehyde gave an estimated lifetime of the methoxymethyl cation of ~1 fs.[iv] Thus, the reactions of methoxymethyl derivatives in water were required to be bimolecular, even though the methoxymethyl cation was a stable, observable species in "Magic Acid". The requirement was confirmed (see Section 3.3).

[iii] ns = nanosecond, 10^{-9} s.
[iv] fs = femtosecond, 10^{-15} s.

The Jencks clock was developed into a technique whereby the lifetimes of a whole range of oxocarbenium and carbenium ions of moderate stability could measured. Azide ion replaced thiosulfate, as the products were non-ionic and more stable, although the diffusional rate constant for anion–cation recombination of $5 \times 10^9\,M^{-1}\,s^{-1}$ in water at 25 °C still applied,[10] and was later confirmed directly for simple carbocations by laser flash photolysis.[11] Simple analysis of ROH and RN_3 obtained from R^+ in the presence of x M azide enabled k_0, the rate of reaction of R^+ with water, to be calculated from the equation

$$[\mathrm{ROH}]/[\mathrm{RN_3}] = 10^9 k_0/5x \tag{3.1}$$

Generation of R^+ from the covalent azide itself and common ion inhibition of this process by added azide enabled the product analysis to be dispensed with completely.[12] Applying the steady-state approximation, $d[R^+]/dt=0$, to the reaction scheme of eqn (3.2) results in the rate law given by eqn (3.3). Inversion of this expression gives eqn (3.4), which describes a linear plot, from which k_0 can be calculated from slope and intercept.

$$\mathrm{RN_3} \underset{k_-[\mathrm{N_3^-}]}{\overset{k_+}{\rightleftharpoons}} \mathrm{R^+} \xrightarrow{k_0} \mathrm{ROH} + \mathrm{N_3^-} \tag{3.2}$$

$$k_{obs} = \frac{k_+ k_0}{k_0 + k_-[\mathrm{N_3^-}]} \tag{3.3}$$

$$\frac{1}{k_{obs}} = \frac{1}{k_+} + \frac{k_-[\mathrm{N_3^-}]}{k_0 k_+} \tag{3.4}$$

In the 1-phenylethyl system, S_N1 and S_N2 pathways could occur in the same solution: concerted reactions occurred when the carbenium ion did not have a meaningful lifetime in the presence of a nucleophile.[13] This finding had important applications to the reactions of glycosyl derivatives in water, where the reality or otherwise of glycosyl cation intermediates is now known to depend on the leaving group.

3.3 THE METHOXYMETHYL SYSTEM

The predictions that the methoxymethyl cation was too unstable to exist was confirmed by the discovery of unmistakably bimolecular substitution reactions at the methylene group of two methoxymethyl systems, methoxymethyl 2,4-dinitrophenolate[14] and *N*-(methoxymethyl)-*N*,*N*-dimethylanilinium ions.[15] Although the reactions were bimolecular, quantitative analysis of their rates by means of various linear free energy relationships indicated there was little bonding of either leaving group or nucleophile to the reaction centre at the transition state, resulting in the methoxymethyl moiety carrying a pronounced positive charge. Thus, rates of attack by various nucleophiles on the

2,4-dinitrophenolate showed a very weak quantitative dependence on Swain–Scott[v] nucleophilicity,[16] with points badly scattered about a trend line of gradient 0.2–0.3,[14] indicating that bond formation was only about one-quarter as developed as in nucleophilic displacements of methyl bromide. Likewise, the value of β_{nuc}[vi] for attack of a series of primary amines on the *N*-(methoxymethyl)-*N*,*N*-dimethyl-*m*-nitroanilinium ion was only 0.14.[15] The neutral leaving group system permitted values of β_{lg}[vii] of −0.89 for the attack of water and −0.7 for the attack of propylamine to be obtained. Values of β_{eq}[viii] for this displacement are not available, but it is clear from the value of β_{eq} of ~1.5 for the quaternisation of substituted pyridines with methyl iodide[17] that at the transition state there is little bonding to either leaving group or nucleophile.

Such "exploded" transition states, in which the reaction centre carries a pronounced positive charge, are the commonest type of nucleophilic displacements at acetal centres. The reactions are unambiguously bimolecular, but in quantitative measures of transition state structure they differ only slightly from those of true S_N1 reactions. Thus, the value of β_{lg} for the hydrolysis of aryloxytetrahydropyrans, in which the solvent-equilibrated tetrahydropyranyl cation is a true intermediate, is −1.18, modestly but significantly more negative than the value of −0.82 obtained for hydrolysis of methoxymethyl esters of the type CH_3OCH_2OCOR.[18]

Nucleophilic displacements at the methoxymethyl methylene and indeed nucleophilic substitutions in general can be considered in terms of a More O'Ferrall–Jencks diagram in which the progress of bond making to the nucleophile and bond breaking to the leaving group were represented on two orthogonal axes, although the inability of carbon to accommodate more than eight electrons in its valence shell means that the northwest half of the diagram is prohibited (Figure 3.2) (a similar treatment of acyl transfer does not operate under this constraint, of course, so in addition to an acylium ion, $RC{\equiv}O^+$, analogous to the carbenium ion in the southeast corner, a tetrahedral intermediate in the northwest corner occurs).

For the simple cleavage of a single bond, with no other bonds formed or broken, it is possible to plot graphically the energy of the system as a function of the bond being cleaved. In the same way, energy can be plotted on an axis perpendicular to the plane of Figure 3.2, in the same way as contours on a map. The transition state is then the structure at the highest energy point of the

---

[v] The Swain–Scott relationship, $\log_{10}(k/k_0) = ns$, defines a linear free energy relationship. The nucleophilicity parameter for a given nucleophile is defined by $n = \log_{10}(k/k_0)$, where k is the rate of reaction of the nucleophile with methyl bromide and k_0 the rate of reaction of methyl bromide with methanol. Originally the relationship was formulated for reactions in methanol, but later corrections enabled n parameters to be used in water. The parameter s measures the sensitivity of a reaction to nucleophilic push, and is the gradient of a plot of $\log_{10}k$ for a series of nucleophiles against n. By definition, s for S_N2 reactions of methyl bromide is 1.00, and n for methanol is 0.00.

[vi] The gradient of the plot of $\log_{10}k$ against the pK_a of the nucleophile.

[vii] The gradient of the plot of $\log_{10}k$ against the pK_a of the leaving group.

[viii] The gradient of the plot of $\log_{10}K_{eq}$ against the pK_a of the leaving group or nucleophile, according to context.

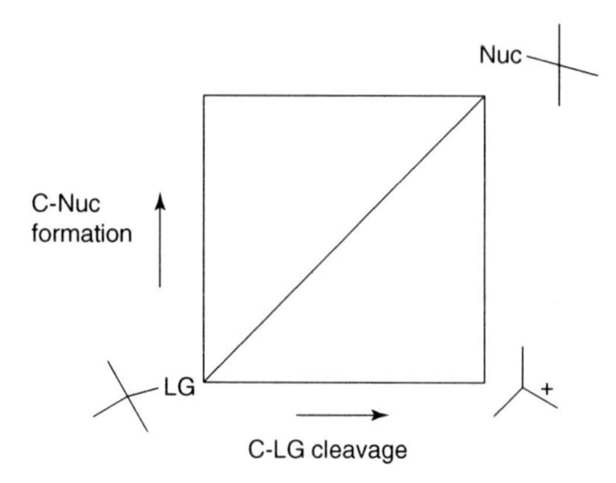

Figure 3.2 More O'Ferrall–Jencks Diagram for nucleophilic displacements at saturated carbon.

"pass" leading from reactants to products. A true S_N1 reaction corresponds to reaction along the southern and eastern edges of the diagram; an S_N2 reaction on a substrate such as a methyl derivative, where there is little charge buildup, to a reaction along the SW–NE diagonal. The reactions of alkoxyalkyl derivatives involve a transition state at the southeast corner of the diagram, but not a fully fledged oxocarbenium ion.

3.3.1 α-Hydrogen Isotope Effects in the Methoxymethyl System

The methoxymethyl system also permitted the limitations of the then widespread use of α-hydrogen isotope effects as a measure of "S_N1 character" to be understood. When a carbon bearing a hydrogen isotope changes hybridisation from sp^3 to sp^2, an out-of-plane bending vibration at $1340\,cm^{-1}$ weakens to $800\,cm^{-1}$. Since at ordinary temperatures only the ground vibrational energy state is occupied, this entails a loss of zero point energy, proportionally greater for the lighter isotope (Figure 3.3).[19] Weakening of a bond can be regarded as attenuated version of its cleavage (see Section 1.2.1.4.1).

α-Deuterium equilibrium and kinetic isotope effects of 1.00–1.30 (and α-tritum effects of 1.00–1.46, since the Swain–Schaad relationship,[20] $k_H/k_T = (k_H/k_D)^{1.44}$, holds for these zero-point energy effects) are observed. An incoming nucleophile restricts the motion of the vibrating hydrogen, effectively increasing the bonding force constant and reducing the effects. However, k_H/k_D per deuteron for the attack of I^-, Br^-, Cl^- and F^- on the *N*-(methoxymethyl)-*N*, *N*-dimethyl-*m*-nitroanilinium ion are 1.18, 1.16, 1.13 and 0.99, respectively, the opposite order from nucleophilicities and presumed "S_N1 character".[15] The effects are determined by the essentially mechanical nature of the restriction of the motion of the α-hydrogen: the "softer", more polarisable the nucleophile, the less effective it is at increasing effective force constants. α-Hydrogen isotope effects can therefore only be used as a measure of "S_N1 character" if the same nucleophilic atom is used (effects per deuterium for hydroxide, phenoxide and acetate are 1.07, 1.08 and 1.07, respectively).

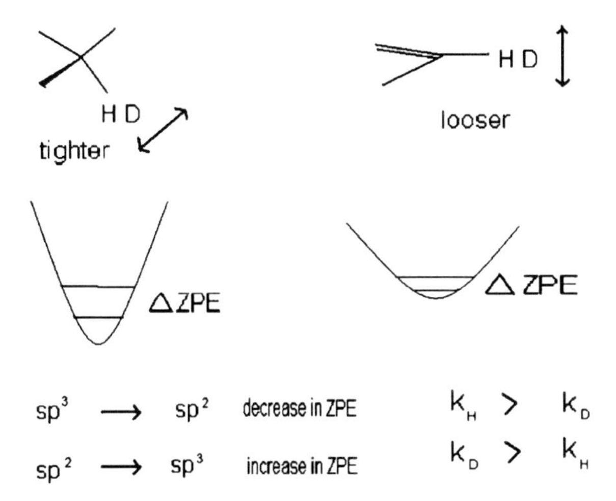

Figure 3.3 Origin of α-hydrogen isotope effects.

3.4 GEMINAL EFFECTS AND DEVELOPMENT OF CONJUGATION

The use of the Jencks clock has enabled a number of oxocarbenium ion and related systems to be analysed in detail. It is clear that the intuitive expectation that the more stable the oxocarbenium ion, the slower it reacts with water and the faster it is formed, are complicated by two further phenomena, geminal interactions in the ground state, and imbalance between σ bond cleavage and development of conjugation.

3.4.1 Geminal Effects

On the frontier orbital picture, the anomeric effect (see Section 2.6) arises because of a stabilising interaction between a ring oxygen lone pair and the σ* orbital of the C1–O1 bond; likewise, the *exo* anomeric effect (see Section 4.5) arises because of a stabilising interaction between a O1 lone pair and the σ* orbital of the C1–O5 bond. Such stabilising n–σ* interactions are cumulative, and in the limit are held to account for the inertness of carbon tetrachloride and the relative inertness of orthocarbonates, $(RO)_4C$,[21] although the interpretation has been questioned on theoretical grounds.[22] Whatever the validity of the explanation, the effect is experimentally undeniable. Methoxymethyl fluoride has a similar solvolytic reactivity to *p*-methoxybenzyl fluoride, but that of methoxymethyl chloride is 12 000 times greater than that of *p*-methoxybenzyl chloride.[23] Efficient stabilising n–σ* interactions between C–O and C–F can only occur in the methoxymethyl system, but not between C–O and C–Cl bonds in either system, because of the different sizes of the orbitals involved.

3.4.2 Development and Loss of Conjugation

p-Methoxybenzyl cations can be accorded honorary oxocarbenium ion status by virtue of a canonical structure which puts the positive charge on the oxygen

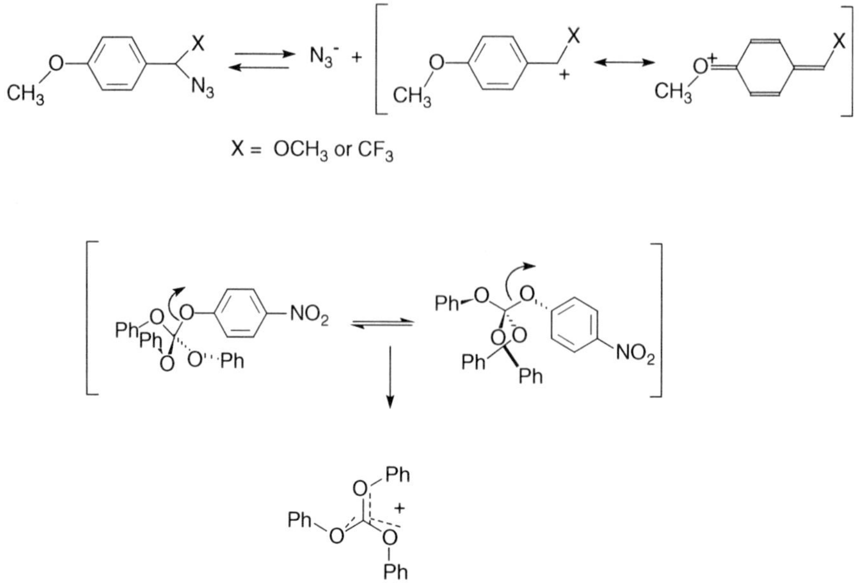

$X = OCH_3$ or CF_3

Figure 3.4 Oxocarbenium ion systems demonstrating non-synchronous σ and π bond cleavage and formation.

(Figure 3.4). They can be generated in aqueous solution from the azides, and their lifetimes estimated by common ion suppression of hydrolysis, as described in Section 3.2.1. Remarkably, replacement of an α-methoxy group by an α-trifluoromethyl group decreases the lifetime of the oxocarbenium ion by merely a factor of 2,[24] despite decreasing the rate of its generation from the azide by 10^{14}. The reason is that in the direction of carbenium ion generation, development of conjugation of the positive charge into the aromatic system lags behind C–N bond cleavage; at the transition state for solvolysis, therefore, the trifluoromethyl group exerts much of the effect it had in saturated systems. Conversely, the major contributor to the activation energy required for reaction of the cation comes from that necessary to break the partial π bonds between the aromatic ring and the substituents, which will be little influenced by the trifluoromethyl group. Development of conjugation in a nascent carbenium ion lagging behind σ bond cleavage is very similar to the development of conjugation in nascent nitroalkane anions lagging behind C–H cleavage during deprotonation, so that deprotonation is slow (the "nitromethane anomaly"). Aryl orthocarbonates $(ArO)_3COAr'$ provide a similar system, where the crowded propeller-shaped ground states mean that cleavage of C–OAr′ is ahead of the development of the planarity permitting full delocalisation of the positive charge in the trioxocarbenium fragment.[25]

There is some evidence (see Section 3.7) that the development of double bond character between C1 and O5 in actual sugars may occur with varying degrees of synchronicity. In a description of nucleophilic substitution at an oxocarbenium centre, therefore, description of the transition state really

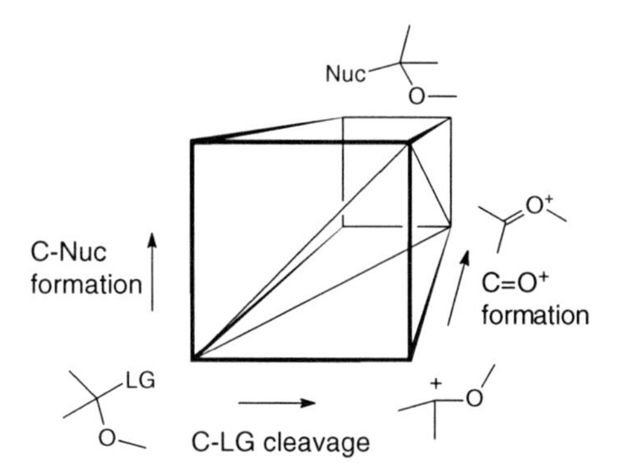

Figure 3.5 Generalised three-dimensional More–O'Ferrall–Jencks diagram for nucleophilic substitution at oxocarbenium ion centres. Because carbon cannot have more than eight electrons, only the front south-east tetrahedron is accessible, as shown.

requires the addition of an extra axis, representing development of C–O double bond character, to the More–O'Ferrall–Jencks diagram of Figure 3.2 (Figure 3.5).

3.5 SPONTANEOUS HYDROLYSIS OF GLYCOSYL DERIVATIVES

By spontaneous is meant the loss of a leaving group from a glycosyl or similar derivative in a process independent of acid or electrophiles, and dependent only on nucleophilic assistance from solvent or internal nucleophiles. The spontaneous hydrolysis of 2-(*p*-nitrophenoxy)tetrahydropyran was discovered in 1970[26] and a β_{lg} value of -1.18 for a range of such compounds was measured.[18] The measurement of the rate of similar processes for 2-aryloxytetrahydrofurans,[27] faster by only a factor of ~2, showed that ring size *per se* had a limited effect on reactivity.

3.5.1 Departure of Anionic Oxygen Leaving Groups from Sugars

Similar losses of alkoxide or phenoxide anions were observed with sugars themselves. The rates of spontaneous hydrolysis of various 2,4-dinitrophenyl β-aldopyranosides[28,29] could be used to examine the effect of glycone substitution pattern. These data complemented those from the hydrolyses of α- and β-aldopyranosyl phosphates [30,31] under acidic conditions, which were known to proceed via C–O rather than P–O cleavage, with departure of $H_2PO_4^-$. At elevated temperature (220 °C), departure of the unactivated leaving group methoxide from methyl glycosides of glucopyranose and ribofuranose could be observed.[32] The following trends emerged:

(i) In pyranosides, loss of an equatorial oxygen leaving group is invariably faster than loss of an axial leaving group, by a factor which varies over the range 1.2–3, depending on the system. This is true even for

Table 3.1 ρ_I values for the spontaneous hydrolyses of equatorial 2,4-dinitrophenyl aldopyranosides.

Position	*Glucose series*	*Galactose series*	*Other series*
6	–2.2	–2.5	
5	–6.3	–5.4	
4	–5.1	–3.1	
3	–2.9	–3.5	–2.7 (allose)
2	–8.3	–8.7	–10.7 (mannose)

loss of methoxide from the methyl glucopyranosides, where the α/β rate ratio is 2.5.[32] Ground-state stabilisation or destabilisation by the anomeric effect obviously plays a part here, because for the spontaneous hydrolysis of α- and β-glucopyranosyl fluorides the factor increases to about 18 at 50 °C.[33]

(ii) Table 3.1 gives the ρ_I values calculated from the three substituents F, OH and I for positions 2, 3, 4 and 6[29], or the four substituents H, CH_2OH, CH_3 and CH_2Cl at position 5[28], for the spontaneous hydrolysis of equatorial 2,4-dinitrophenylglycopyranosides. The ρ_I values for positions 2 and 5 suggest that charge is distributed evenly between C1 and O5 at the transition state. However, Namchuk *et al.*[29] performed a Kirkwood–Westheimer analysis (a calculation based on simple electrostatics) of the effects at positions 6, 4, 2 and 3 and obtained good agreement with experiment if they assumed most of the positive charge resided on O5. Such analysis assumes the effects are electrostatic (*i.e.* through-space). The treatment in terms of σ_I, moreover, neglects the proven effect of the orientation of the OH and other substituents on their effective electron withdrawing power (see point iv). When this is taken into account, by use of "stereoelectronic" substituent constants, σ_S, defined as $-\Delta pK_a$[ix] of the piperidine brought about by a particular substituent at a particular site in a particular axial or equatorial orientation, then an excellent linear relationship is observed between $\log_{10}k$ for spontaneous hydrolyses of dinitrophenylglycosides and $\Sigma\sigma_S$ ($r = 0.99$; gradient 1.07).[34] The value of the gradient (-1.07) indicates, however, that the effective charge at C1 of the sugar is slightly in excess of that introduced by protonation of an amine.

(iii) The deactivating nature of most sugar substituents means that for practical reasons few β_{lg} values have been measured, but loss of phenoxides from the anions of aryl *N*-acetyl-α-neuraminides is governed by $\beta_{lg} = -1.3_0$,[35] and a value of -1.2_4 for their anomers,[36] within experimental error the same as that for hydrolysis of aryloxytetrahydropyrans; in all three cases solvent-equilibrated oxocarbenium ions are intermediates (see Section 3.11.1).

[ix] So that σ_S is in the same sense–positive, electron-withdrawing–as conventional σ constants.

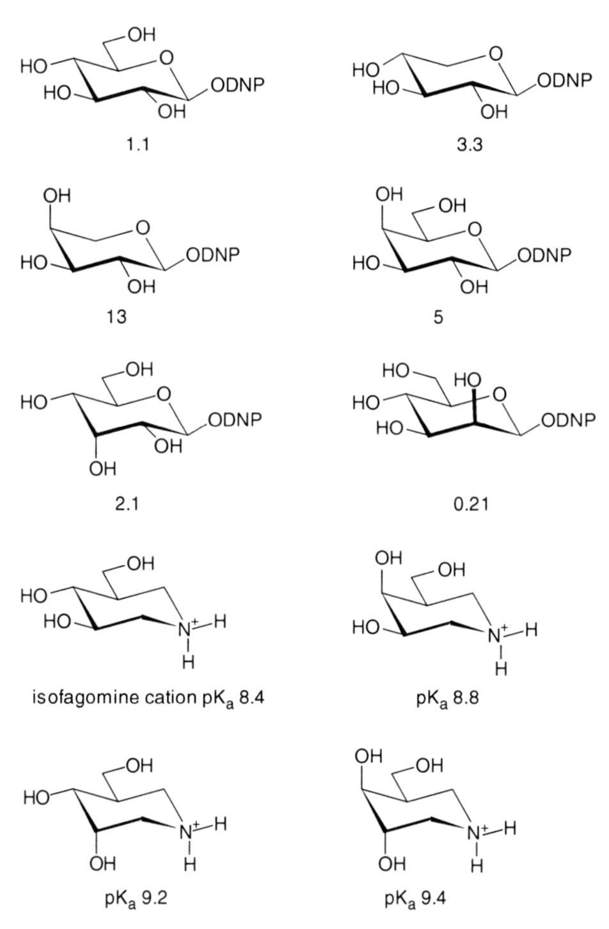

Figure 3.6 Effect of hydroxyl epimerisation. Top, relative rates of hydrolysis of 2,4-dinitrophenylglycosides, taken from Ref. 28 or extrapolated from Ref. 29 Bottom, effects of hydroxyl epimerisation in the isofagomine skeleton on amine basicity, taken from Ref. 37.

(iv) The dependence of rate on sugar configuration is not simple (Figure 3.6). At one time it was thought that galactopyranosides hydrolysed faster than glucopyranosides because there was a release of torsional (conformational) strain from axial OH groups as the system moved towards the half-chair or boat conformation of the oxocarbenium ion. While such effects may play a minor role, the very different effects of axial OH groups at positions 2, 3 and 4 mean they cannot by themselves be an explanation. If, on the other hand, the effects of OH groups are essentially electrostatic, then the *galacto* compound hydrolyses faster because the C4–O4 dipole is pointing with its negative end away from the positive charge on O5 in the *gluco* series and approximately at right-angles in the *galacto* series. A compelling argument in favour of this explanation is that very similar trends are observed in the pK_as of hydroxylated piperidinium ions related to the alkaloid isofagomine. Change in orientation of the dipole of a hydroxyl group by epimerisation of a hydroxyl group has exactly parallel effects on the pK_as of the isofagomine and the rate of acid-catalysed hydrolysis of a glycoside, to the extent of giving rise to a linear free energy relationship. In both cases a positive centre is stabilised or destabilised by electrostatic interaction with a dipole. If the logarithm of the ratio of rates of acid catalysed hydrolysis of a pair of glycosides, epimeric at one of the ring hydroxyls, is plotted against the pK_a difference caused by the analogous epimerisation of isofagomine, a straight line of gradient 0.86 is obtained.[37]

(v) Conclusive evidence, at least in terms of the *galacto*/*gluco* difference, that the effect arises from electrostatics comes from the fact that it is preserved in acid-catalysed hydrolyses even of glycosides with an additional methyl group at C4 (Figure 3.7), so that steric effects, if anything, would be reversed (the instability factor for CH_3 is greater than for OH).[38]

(vi) The spontaneous hydrolysis of ribofuranosides is about an order of magnitude faster than that of glucopyranosides of the same anomeric configuration;[32] the extra torsional strain in furanosides may be a reason.

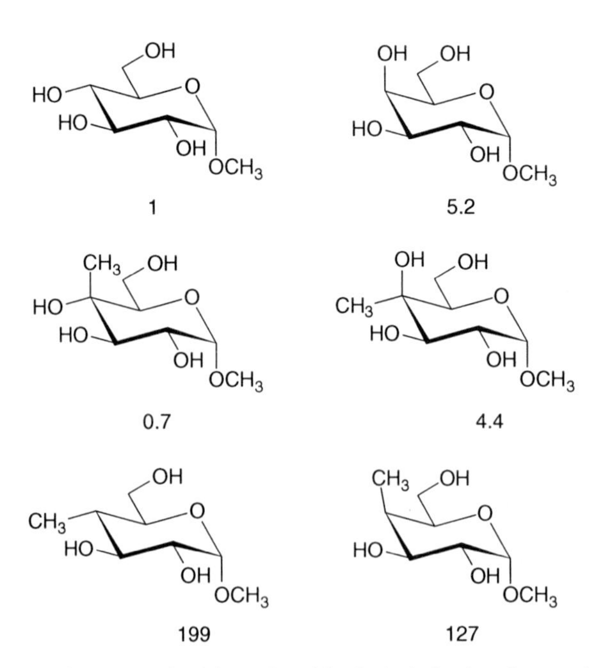

Figure 3.7 Relative rates of acid-catalysed hydrolysis in the *gluco* and *galacto* series. The fact that the additional methyl group does not alter the *gluco*/*galacto* ratio indicates that the faster rates of hydrolysis in the galacto series are not steric in origin.

3.5.2 Departure of Pyridines

Detailed studies of the spontaneous hydrolyses of β-D-galactopyranosyl,[39] α-D-glucopyranosyl and α-D-xylopyranosyl,[40] 2-deoxy-β-D-glucopyranosyl,[41] 2-deoxy-α-D-glucopyranosyl,[42] pyridine ring-substituted analogues of NAD^+[43] and *N*-acetyl-α-D-neuraminylpyridinum salts[44] have been performed. All processes are characterised by large negative β_{lg} values (-1.2_7, -1.0_6, -1.2_8, -1.2_7, -1.0, -0.8_4, -1.0 and -1.2_2, respectively), indicating nearly complete cleaving of the bond to the leaving group at the transition state (the value for the *N*-acetyl-neuraminylpyridinium salts is independent of whether or not the carboxylic acid is protonated).

The effects of substitution at position on hydrolysis of NAD^+ analogues, in both the *ara* series[45] (six points) and the *ribo* series (five points),[46] gave values of ρ_I of -6.7 and -7.0, respectively (Figure 3.8). These values are lower than those observed for spontaneous hydrolysis of 2,4-dinitrophenyl pyranosides (Table 3.1), suggesting slightly different mechanisms (see Section 3.6).

The rates of hydrolysis and the preferred conformation of the sugar ring, for a series of glycosyl 4-bromoisoquinolinium ions, are given in Figure 3.9. The hydrolysis of the arabinofuranosyl derivative,[47] with an all-*trans* substitution pattern, is around an order of magnitude faster than the hydrolysis of the β-D-glucopyranosyl derivative, much the same as the *gluco*/*ribo* ratio observed in spontaneous hydrolysis of the methyl glycosides. Important information, however, comes from the anomeric rate ratios in the pyranosyl series. The bulk of the pyridine moiety, and any contribution from an electrostatic reverse anomeric effect, mean that in these studies the leaving group is a conformational determinant. Its preference for an equatorial orientation forces the xylose ring of α-D-xylopyranosylpyridinium ions into the $^1C_4$ conformation and α-D-glucopyranosylpyridinium ions into the $^1S_3$ conformation, as shown in Figure 3.9; at the same time it reinforces the normal preference for the $^4C_1$ conformation of β-glucopyranosyl derivatives.

The anomeric rate ratios in hydrolysis of glucopyranosyl- and xylopyranosylpyridinium ions were one of the first pieces of evidence against the "theory of stereoelectronic control", more informatively also termed the "antiperiplanar lone pair hypothesis (ALPH)". As originally advanced, the theory stated that C–X bond cleavage from a carbon substituted with one or more heteroatoms bearing lone pairs of electrons was favoured if the (sp^3) lone pairs were antiperiplanar (dihedral angle of 180°) to the breaking bond.[48,49] The ideas

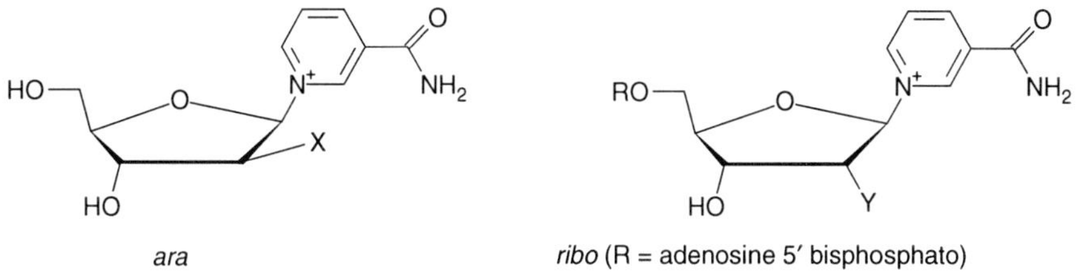

ara *ribo* (R = adenosine 5′ bisphosphato)

Figure 3.8 Effects of sugar ring substitution on departure of pyridine.

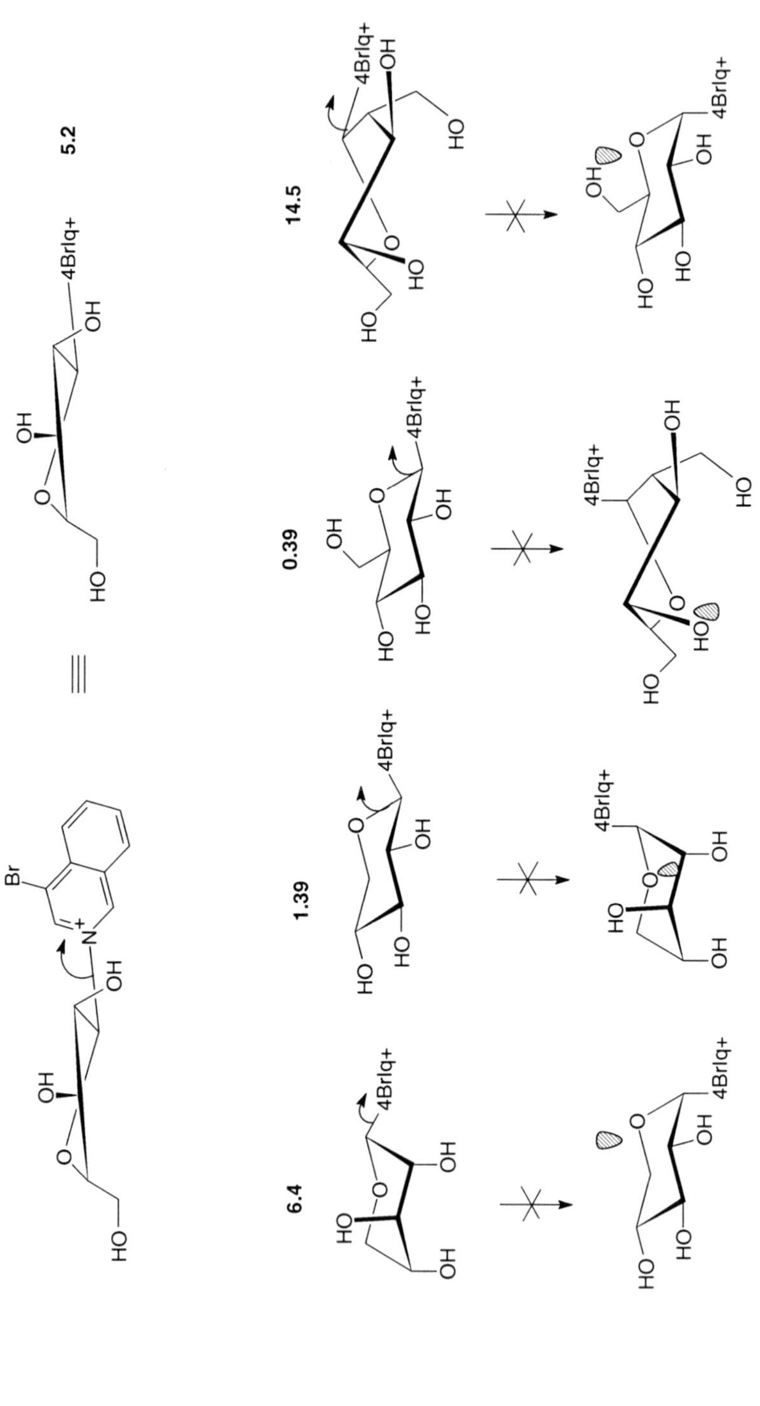

Figure 3.9 Relative rates of hydrolysis of glycosyl pyridinium ions.

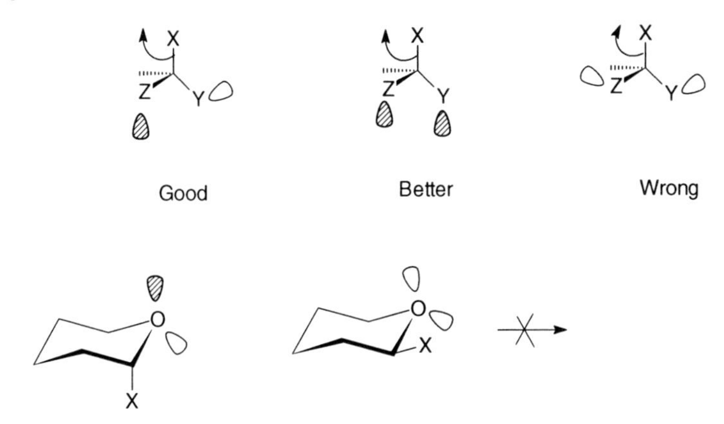

Figure 3.10 Ideas behind the "theory of stereoelectronic control" or "antiperiplanar lone pair hypothesis (ALPH)" Antiperiplanar (sp^3) lone pairs are shown shaded.

are illustrated in Figure 3.10. As a statement about reactive ground-state conformations, ALPH violated the Curtin–Hammett principle[x] in systems in rapid conformational equilibrium and is often recast as a statement about transition states.

According to ALPH, therefore, loss of an axial leaving group from a pyranose ring should be more favoured than loss of an equatorial group, whereas it was known, even 30 years ago, that the reverse was the case. It was accordingly asserted that "α-glycosides must hydrolyse via their ground-state conformation whereas β-glycosides must first assume a boat conformation in order to fulfil the stereoelectronic requirement" (Ref. 48, p. 39). The implications of this assertion in respect of glycopyranosylpyridinium ions, which all have equatorial leaving groups, is set out in Figure 3.9. Using Angyal's instability factors and other experimental conformational data, it was calculated that the "stereoelectronic requirement" would predict an α/β ratio for the xylopyranosylpyridinium ions of 150 and one of 10^8 for glucopyranosylpyridinium ions, much greater than the ratios in fact observed.

The data which had been held to support ALPH are now thought to arise from either least motion effects, which are minimised in reactions with early or late transition states,[50] or systematic bias in the choice of system. In the decomposition of tetrahedral intermediates, an upper limit of 8 kJ mol^{-1} has been placed on any antiperiplanar lone pair effect in any system.[51] Although discredited, the ideas are superficially appealing and even at the time of writing (2007) are occasionally invoked by the non-specialist.

[x] The Curtin Hammett principle states that, for systems in rapid conformational equilibrium, rates are determined only by the free energy difference between the conformational ground state and the transition state; nothing can be said about high-energy conformations occupied en route to the transition state.

3.6 LIFETIMES OF GLYCOSYL CATIONS IN WATER AND BIMOLECULAR DISPLACEMENTS AT THE ANOMERIC CENTRE

Extrapolation of direct measurements, by azide trapping, of the lifetimes of methoxy-substituted alkyl cations[12] suggested that the glucopyranosyl cation in water would have a lifetime of 1 ps: long enough to have a real existence but too short for it to be solvent equilibrated. Unambiguous bimolecular reactions on α-glucopyranosyl fluoride were observed, but only with anionic nucleophiles: no products from attack by pyridine were detected. The bimolecular reactions were otherwise S_N2 reactions of the sort exhibited in the methoxymethyl system, with a Swain–Scott *s* of 0.18.[52] When substrates were changed to 2-deoxy α- and β-D-glucopyranosylpyridinium ions, so that the leaving group was neutral, however, rather different behaviour was observed.[41,42] The reactions had the following characteristics:

(i) Although kinetically bimolecular, the second-order constants for the attack of anionic nucleophiles did not correlate with nucleophilicity–the Swain–Scott *s* for both α and β ions was zero.
(ii) The presence of anions promoted both hydrolysis and nucleophilic substitution reactions with the anion, in a way which varied with the nature of the anion.
(iii) In the presence of 2 M azide ion, the predominant product in both cases was inverted azide. Different product distributions were observed with the α- and β-pyridinium salts, most tellingly with methanol as a third nucleophilic component.

The reactions were not therefore straightforward bimolecular reactions. It was proposed that the rate-determining step was the formation of a solvent-separated ion-pair molecule complex of the type N_3^-//Glcp^+.Py, where the slashes refer to solvent molecules. Inverted azide was formed in a post-rate-determining solvent reorganization step. All reaction, whether hydrolysis or substitution, went through this or an analogous complex with another anion. The intimate ion pair N_3^-Glcp^+, with an estimated "lifetime" of 10^{-19} s, was too unstable to exist by 5–6 orders of magnitude (the estimate was based on equilibrium constants for encounter complex formation, extrapolated lifetimes of the glucosyl cation in water and the N^+ value for azide[53]).[xi] Where azide was the only anion present, approximately equal quantities of hydrolysis and substitution products were formed: N_3^-//Glcp^+.Py partitioned between solvent reorganisation to give azide and solvent capture to give hydrolysis product. On the basis of the directly measured rate constant (10^{11} s^{-1}) for loss of intervening solvent between solvent-separated pairs of stable ions, the estimated lifetime of

---

[xi] This makes the argument less circular, since the uncertainty in the lifetimes of the glucosyl cation produced by various estimates ranged over, at most, an order of magnitude. The Ritchie N_+ parameter for a given nucleophile governs its relative reactivity to cationic electrophiles – surprisingly, all nucleophiles have the same relative reactivates to a whole range of cations, so that N_+ is simply defined as $\log(k/k_0)$.

the 2-deoxyglucosyl cation in the N_3^-//Glc*p*$^+$.Py complex was estimated as 10 ps; the 4-fold effect of a 2-hydroxy group on oxocarbenium ion lifetimes measured by azide trapping in simple systems then indicates the glucopyranosyl cation itself has a lifetime slightly in excess of 2.5 ps.

These reactions with a neutral leaving group are somewhat different from the "exploded" but otherwise conventional S_N2 reactions observed with α-glucopyranosyl fluoride. Accordingly, the effects of 2-substituents appear to be somewhat smaller with a nicotinamide leaving group (Figure 3.8) than with dinitrophenolate (Table 3.1).

If the oxocarbenium ion is stable enough and a nitrogen leaving group acidic enough, then spontaneous cleavage of a neutral *N*-glycoside, similar to the hydrolysis of *O*-glycosides, can be observed, as with the pH-independent hydrolysis of 2′-deoxy-β-D-ribofuranosides of uracil, thymine and 5-bromodeoxyuracil.[54]

3.7 ACID-CATALYSED HYDROLYSIS OF GLYCOSIDES

3.7.1 Specific Acid Catalysis

Studies of the mechanism of acid-catalysed hydrolysis of glycosides date back over 50 years,[55] well before doubts were raised about whether glycosyl cations had a real existence. Even at the time, though, it was noticed that the absolute value of kinetic parameters varied slightly, depending on whether hydrochloric, sulfuric or perchloric acid was used. It seems likely that reactions proceed through some sort of solvent-separated ion pair encounter complex, similar to that shown to be an intermediate in the hydrolysis of deoxyglucosylpyridinium salts. The electronic properties of the sugar residues in such complexes are, as we have seen, very close to those postulated for glycosyl cations, so that we can broadly concur with the original authors in considering them as such.

Convenient rates for most glycosides are only obtained with approximately molar concentrations of acid even at moderately elevated temperatures (50–100 °C). Much early work was focused on trying to determine whether solvent was involved as a nucleophile, using as a criterion the dependence of rate on h_0, a parameter designed to measure the ability of the solution to protonate a neutral base.[xii] In dilute aqueous solution, it was identical with

---

[xii] H_0 for aqueous (and indeed non-aqueous) solutions was measured from the ionisation ratios of a series of structurally similar weakly basic, electrically neutral indicators (nitroanilines were widely used). To give a simplified example, the pK_a of *m*-nitroaniline (2.47) could be measured in aqueous solution from ionisation ratios, measured by UV spectroscopy, and pH meter readings, via the Henderson–Hasselbach equation, $\mathrm{pH} = \mathrm{p}K_a + \log_{10}([\mathrm{B}]/[\mathrm{BH}^+])$. The glass electrode of pH meters is unreliable below about pH 1.5, but an H_0 values of the medium of ~1.0 could be measured from the ionisation ratio of *m*-nitroaniline in strongly acid solutions. From this new value of H_0, a pK_a of *p*-nitroaniline (1.0) could be estimated from its ionisation -ratio, $[\mathrm{B}]/[\mathrm{BH}^+]$. The H_0 scale could be extended further into the acidic region from this pK_a and spectroscopically measured ionisation ratios of *p*-nitroaniline. It could then be used to determine the pK_a of, say, 3,5-dinitroaniline. The whole process could then be repeated with nitroanilines of decreasing basicity. The H_0 scale was adjusted computationally so that overlapping plots of $\log_{10}([\mathrm{B}]/[\mathrm{BH}^+])$ against H_0 for a whole series of indicators had gradients as close to 1 as possible. The scale, of course, assumes that thermodynamic non-ideality affects all neutral indicators similarly.

Table 3.2 Bunnett w parameters for the acid catalysed hydrolyses of some glycosides.

Glycoside	w	*Glycoside*	w
tert-Butyl β-D-glucopyranoside	−5.0	Methyl α-D-glucopyranoside	+1.7
Methyl 2-deoxy-β-D-glucopyranoside	−1.6	Methyl β-D-glucopyranoside	+1.6
Methyl 2-deoxy-α-D-glucopyranoside	−1.8	Phenyl α-D-glucopyranoside	+3.0
Sucrose	−0.6	Phenyl β-D-glucopyranoside	−1.4
Methyl 2-deoxy-β-D-mannopyranoside	+0.4	Lactose	+3.0
		Maltose	+3.0

$[H_3O^+]$, but much above 0.5 M increased far more rapidly ($-\log_{10}h_0$ was given the symbol H_0 and becomes pH in low concentrations of acid). According to the Zucker–Hammett hypothesis, the rates of $A1$ reactions (those involving only protonated substrate in the rate-determining step) are proportional to h_0 and those of $A2$ reactions (those involving both protonated substrate and a water molecule) to $[H_3O^+]$.[xiii] Acid catalysis of glycosides generally follows h_0.[55] A more sophisticated is treatment is that of Bunnett [eqn (3.5)], where c is a constant:

$$\log k_{obs} + H_0 = w \log_{10} a_{H_2O} + c \tag{3.5}$$

Generally, values of w are negative for $A1$ reactions, and positive for $A2$ reactions; the actual values for various glycosides[56] (Table 3.2) are very much in line with contemporary ideas of glucosyl cations being on the border of a real existence. The large negative value for the hydrolysis of *tert*-butyl β-D-glucoside reflects the fact that this compound hydrolyses with alkyl-oxygen fission to give a probably solvent-equilibrated *tert*-butyl cation.[57] Tertiary alkyl glycosides are crowded, though, and even bridgehead tertiary glycosides (such as 1-adamantyl β-D-glucopyranoside), which give unstable carbenium ions, still hydrolyse faster than their uncrowded analogues.

3.7.2 Site of Productive Protonation

Glycosides are asymmetrical acetals, and in principle protonation could occur on either oxygen, to give a cyclic or acyclic oxocarbenium ion (Figure 3.11). The pathway taken will depend on the particular system, although in practice all common pyranosides and most furanosides hydrolyse by the exocyclic C–O fission.

The balance between exocyclic and endocyclic fission is governed by four factors:

(i) *Acidity of the aglycone*. Electron-withdrawing groups in the aglycone will disfavour protonation of the exocyclic oxygen atom, but make the

[xiii] In the IUPAC system of reaction mechanism representation, $A1$ is termed $A_e + D_N + A_N$, and $A2$, $A_e + A_N D_N$.

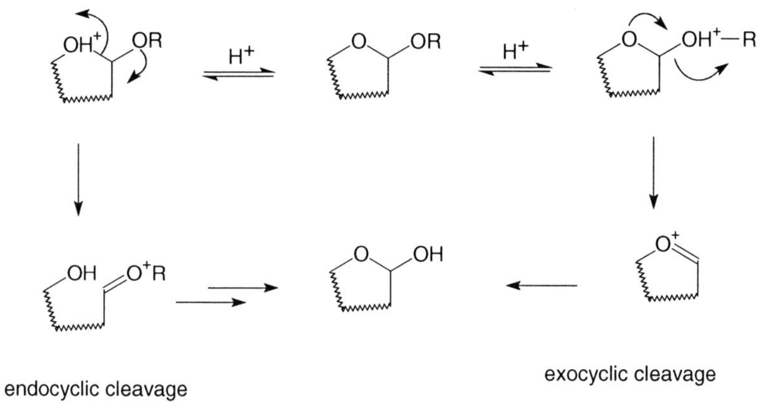

Figure 3.11 Exocyclic and endocyclic cleavage during the hydrolysis of glycosides.

ROH a better leaving group once protonated. It is reasonable to assume that the transition state for C–O cleavage in the RHS pathway of Figure 3.11 is late, like the transition states for hydrolysis of pyridinium salts, so that overall β_{lg} values for the RHS pathway will be close to zero. However, electron-withdrawing substituents in R will disfavour the protonation of the endocyclic oxygen atom and also make the acyclic oxocarbenium ion less stable. The LHS pathway will therefore show strongly positive β_{lg} values.

(ii) *Ring size*. It has been established from studies of mutarotation (Section 1.2.1.2) that closure to furanose rings is often faster than closure to pyranose rings, even though the pyranose rings are thermodynamically more stable. It follows that opening of furanose rings is faster than that of pyranose rings and that therefore the LHS pathway is more likely for furanosides.

(iii) *Temperature*. Exocyclic cleavage (RHS, Figure 3.11) generates two fragments, whereas after endocyclic cleavage the two fragments remain associated. The dominance of the translational partition function means that exocyclic cleavage exhibits positive entropies of activation and endocyclic cleavage negative entropies of activation.[xiv] In the case of competing endo- and exocyclic cleavage, raising the temperature will therefore favour exocyclic cleavage.

(iv) *Ring substitution pattern*. Endocyclic cleavage is favoured by eclipsed substituents on the ring.

[xiv] Many authors tend to take as a standard state the particular acid concentration they use, and treat the reaction as unimolecular. This is, of course, strictly incorrect; data should be referred to a common acid concentration, adopted as standard state, for comparisons of $\Delta S^{\ddagger}$. In practice, use of acid concentrations close to 1.0 M means that trends can be seen even without this correction.

The exocyclic cleavage pathway is taken by all common pyranosides; evidence includes:

(i) ^{18}O kinetic isotope effects > 1.00 (Table 3.3).
(ii) β_{lg} values for aryl glycopyranosides of 0.27 ± 0.04 for β-glucosides, 0.09 ± 0.02 for β-xylosides, 0.01 ± 0.02 for α-glucosides and 0.03 ± 0.01 for α-mannosides.[58]
(iii) A linear relationship (gradient 1.0) between logarithms of rates of acid-catalysed hydrolysis of methyl glycosides and rates of spontaneous hydrolysis of 2,4-dinitrophenyl glycosides of the same glycone, for a series of eight β-glycosides, (four *gluco* and *galacto* homeomorphs).[28]

Entropies of activation in the hydrolysis of various series of alkyl aldofuranosides show a sharp increase at a particular level of aglycone acidity, corresponding to a change from endocyclic to exocyclic cleavage. A similar discontinuity within an aglycone series is seen in the effect of added solvents. In the α-D-arabinofuranosyl case, with all the substituents staggered, this occurs between ethyl (exocyclic) and isopropyl (endocyclic). In both the α- and β-D-xylofuranosyl series, with a *cis* relationship between the C4 hydroxymethyl group and OH-3, the changeover occurs between 2-chloroethyl (exocyclic) and 2-methoxyethanol (endocyclic).[59] The changeover points in other aldofuranosyl series are difficult to interpret.

There is, of course, no *a priori* reason why aldopyranosides with a sufficiently strained ring should not react by endocyclic cleavage and such a system has been constructed (Figure 3.12).[60]

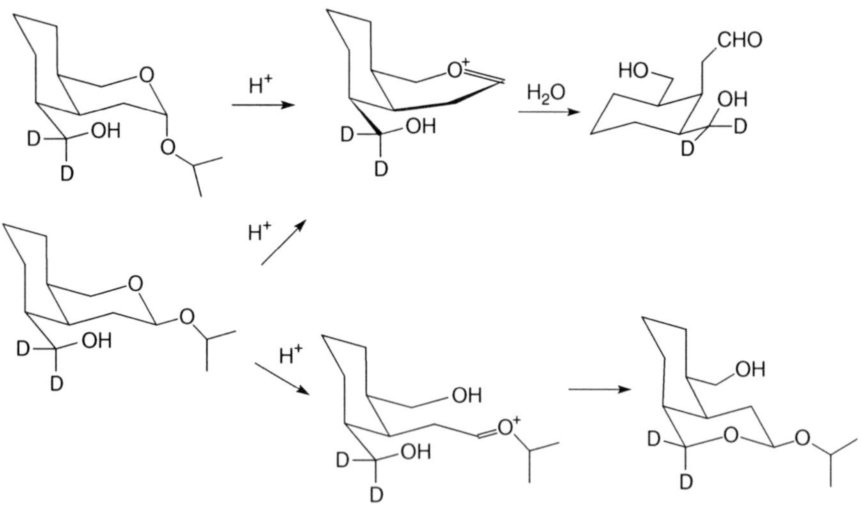

Figure 3.12 "Honorary" pyranoside showing both endocyclic and exocyclic fission. The axial anomer only shows scrambling of the deuterium label, indicating an endocyclic pathway.

The fact that more ring opening was seen in the equatorial than the axial anomer has led to the suggestion that ALPH was operating. In the preferred exocyclic conformations, "O1" of the equatorial anomer has an sp^3 lone pair antiperiplanar to the "C1–O5" bond, but the axial anomer does not. However, the equatorial anomer requires a conformation change for ALPH to apply to exocyclic fission, whereas the axial anomer does not. This is illustrated in Figure 3.13. There is also a least motion explanation for these small and elusive apparent effects consistent with ALPH (Figure 3.13b).

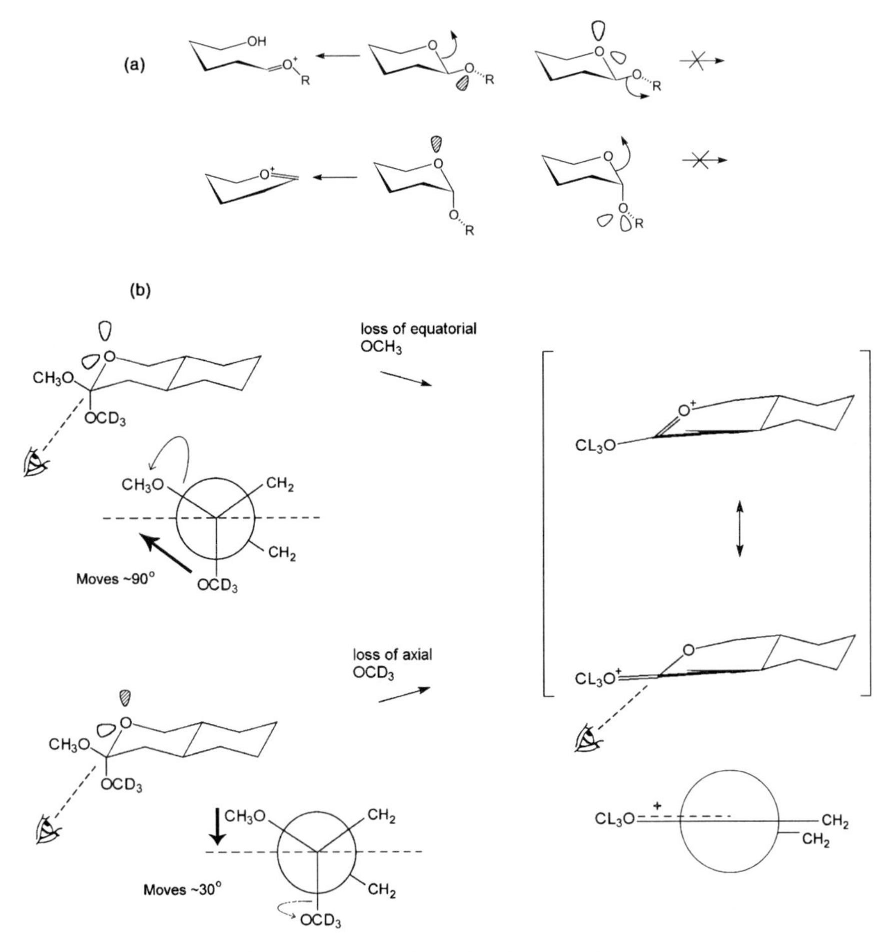

Figure 3.13 (a) ALPH favours endocyclic cleavage in acid-catalysed hydrolysis of equatorial, but not axial, pyranosides. (b) Alternative principle of least motion explanation of ALPH effects, shown here in an ortho-ester, where a small effect may exist.[61] On generation of a pyranosyl cation, the "anomeric" hydrogen has to change its position relative to the rest of the molecule much more in the axial than the equatorial anomer.

3.7.3 Differences in Structure–Reactivity Patterns in Acid-catalysed and Spontaneous Hydrolyses–Effect of the Pre-equilibrium Protonation

Generally, electronic effects on spontaneous hydrolyses and acid-catalysed hydrolyses with the same glycone are essentially identical, as shown by the gradient of 1.0 in linear free energy relationship between acid-catalysed hydrolyses of methyl glycosides and spontaneous hydrolyses of their 2,4-dinitrophenyl analogues.[28] Likewise, a ρ_I value of between −8 and −9 can be calculated from rates of acid-catalysed glycosides and their 2-deoxy and 2-aminodeoxy analogues,[62] the same as the values in Table 3.1. The electronic effect on the pre-equilibrium protonation is therefore minor compared with the effect on oxocarbenium ion stability; one can compare the roughly 1000-fold greater rates of hydrolysis of 2-deoxyglycosides with the difference in pK_a values of ethylamine (10.8) and ethanolamine (9.5).

Whereas the acid-catalysed hydrolysis of equatorial alkyl glycosides is faster than that of their axial analogues, this reactivity preference is reversed in the case of aryl glycosides, the factor favouring the axial anomer being 4 for phenyl glucopyranosides and 10 for *p*-nitrophenyl glucopyranosides. It would appear that formation of the solvent-separated ion pair–molecule complex, which we assume is the rate-limiting step in the hydrolysis, by analogy with the results for pyridinium salts, is likewise favoured when a neutral, bulky group leaves from the α face of a pyranose ring. There is circumstantial evidence that the 2-OH may bring about steric acceleration of departure of α-glycones in these systems. Removal of the 2-OH increases the rate of hydrolysis of β-D-glucopyranosyl-4-bromoisoquinolinium ion by the expected factor of $\sim 10^3$ (680 at 65 °C), but of its anomer by a factor of only 97.[42]

3.7.4 Acid-catalysed Hydrolysis of Nucleosides

Structures of common nucleosides whose acid-catalysed hydrolysis has been studied are shown in Figure 3.14.

The lone pair at N9 of a purine is involved in aromatic bonding, and so is not available to provide electronic "push" for ring-opening reactions. Consequently, adenosine and guanosine and their 2′-deoxy derivatives hydrolyse by a similar mechanism to *O*-pyranosides: pre-equilibrium protonation at N7, the non-glycosylated imidazole nitrogen, followed by essentially unimolecular cleavage of the glycone/aglycone bond, probably, by analogy with 2-deoxyglucosyl pyridinium ions, leading to some sort of solvent-separated anion–oxocarbenium ion-molecule complex. As expected for such a mechanism, removal of the 2′-OH results in the reaction going around 10^3-fold faster (1000-fold in the case of adenosine[63] and 500-fold in the case of guanosine[64]). These features result in the well-known acid-catalysed depurination of DNA.

The plots of $\log_{10}k_{obs}$ versus pH (or H_0) are linear,[63] of gradient −1.0 for guanosine, 2′-deoxyguanosine and 2′-deoxyadenosine, over most of the accessible range, passing through the pK_a of N7 (~ 3) without detectable inflection

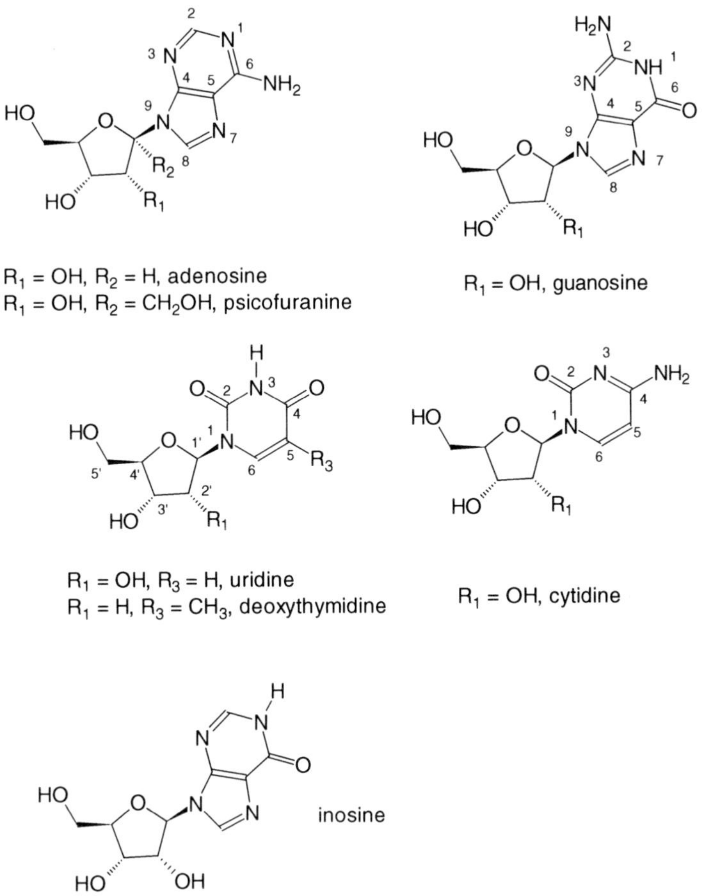

Figure 3.14 Structures of common nucleosides whose acid-catalysed hydrolysis has been studied. Adenosine, guanosine and cytidine are three of the four common nucleosides in RNA and their 2′-deoxy derivatives in DNA, whereas uridine is found only in RNA and 2′-deoxythymidine in DNA. Psicofuranine is an antibiotic and is not a common constituent of nucleic acids. Inosine is a commonly used substrate in investigations of enzymic ribosyl transfer.

or curvature.[64] This is expected for leaving groups with many protonation states in a reaction with $\beta_{lg} = 0$; in the reaction scheme of eqn (3.6), $k_1/K_{a_1} = k_2/K_{a_2}$. The analysis was confirmed by studies of 7-methylguanosine, 1,7-dimethylguanosine and 2′-deoxy-7-methylguanosine, which all hydrolysed in a pH-independent fashion until high acidity, when the rates were similar to those of the unmethylated compounds. In very concentrated perchloric acid, a levelling off of the rate of methylated and unmethylated substrates alike could be detected at $H_0 \approx -2.5$, corresponding to complete conversion of methylated and unmethylated substrates similar to the dication.[65]

$$S + H^+ \underset{K_{a_1}}{\rightleftharpoons} SH^+ \underset{\downarrow k_1}{} + H^+ \underset{K_{a_2}}{\rightleftharpoons} SH_2^{2+} \underset{\downarrow k_2}{} \quad (3.6)$$

Psicofuranine likewise hydrolyses via both mono- and dication, although in this case a point of inflection around pH = pK_a of N7 is discernible in the $\log_{10}k_{obs}$ versus pH plots.[66]

The lone pairs at N1 of a pyrimidine are less involved in aromatic bonding than the lone pairs at N9 of a purine, and the pyrimidine ring is anyway much more electrophilic, so that, as discussed in Section 3.6, pyrimidine anions can depart in pH-independent processes. The acid-catalysed hydrolysis of 2′-deoxypyrimidine nucleosides with C1′–N1 cleavage can occur (at a lower rate than hydrolysis of 2′-deoxypurine nucleosides), but the presence of ring-expansion and anomerisation products in small amounts in the products of acidic hydrolysis of 2′-deoxythymidine and 2′-deoxyuridine (but not 2′-deoxycytidine or 2′-deoxy-5-bromouridine) indicates that the N1 lone pair of pyrimidines may be sufficiently available to provide a minor, glycosylamine-like pathway (see Section 1.2.3).[67] The acidity dependence of the hydrolysis of 2′-deoxycytidine is different from that of purine nucleosides, two sections of the $\log_{10}k_{obs}$ versus pH (or H_0) plots of gradient −1.0 being joined by a horizontal region around pH 2.0, corresponding to reaction of the monocation.[68] Pyrimidine nucleosides themselves are so deactivated to C1′–N1 cleavage that in aqueous acid they hydrolyse by water attack on the pyrimidine ring, shown by an *inverse* α-tritium kinetic isotope effect at C6 ($k_T/k_H = 1.2$), as expected from rate-determining nucleophilic attack of water at this position.[69]

The idea that in purine nucleoside hydrolysis C1′–N9 fission is overwhelmingly favoured, but that in pyrimidine nucleoside hydrolysis endocyclic C′1–O4′ fission is finely balanced with C1′–N1 fission, is illustrated by the contrasting effects of a 3′,5′-phosphate ester bridge (Figure 3.15).[xv] This slows the depurination of 2′-deoxyadenosine by a factor of 2500, but increases the rate of *N*-glycosyl bond fission by factors of 760 and 260 in the case of thymidine and uridine, respectively, which hydrolyse by ring opening[70] (the system is complicated in the case of the unreactive cyclic nucleosides by concurrent phosphate ester hydrolysis). The 3′,5′-cyclic phosphate moiety forms part of a chair-conformation ring, to which the furanose ring is fused *trans*. Formation of the ^3E or E_3 conformation of a furanosyl cation with this fusion involves considerable strain. However, if the glycosylamine (ring opening) pathway is available, as for pyrimidines, this is markedly sterically accelerated.

[xv] 3,5-Cyclic purine nucleoside monophosphates (cAMP and cGMP) are important second messengers in metabolic regulation (*i.e.* they carry messages within the cell, triggered by extracellular hormones).

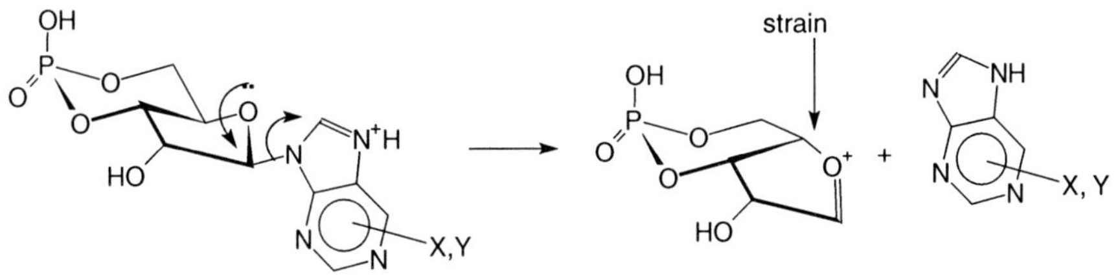

Purine nucleoside hydrolysis decelerated by bridging with 3′, 5′-cyclic phosphate

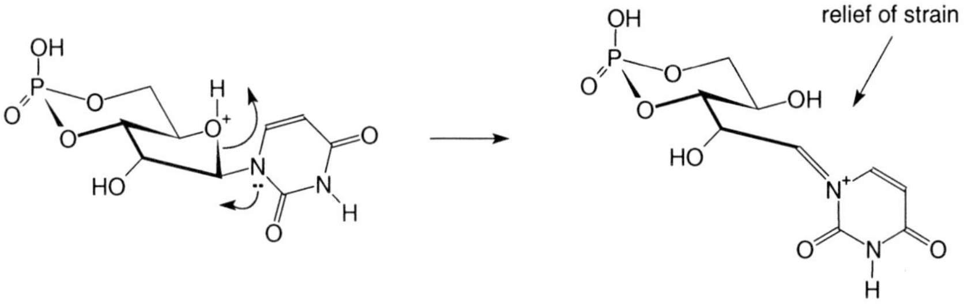

Pyrimidine nucleoside hydrolysis accelerated by bridging with 3′, 5′-cyclic phosphate

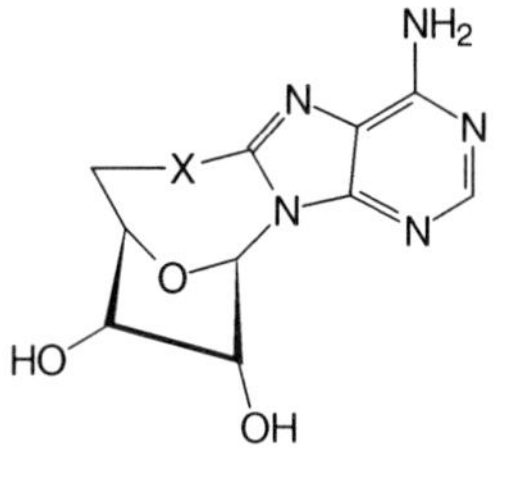

X = S, O, NH, nothing;
Small (<10-fold) deceleration on bridging

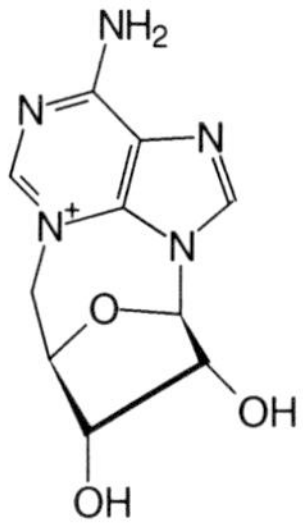

Large deceleration
(inductive effect)

Figure 3.15 Effect of bridging on acid lability of nucleosides. 3′,5′-Bridging with a cyclic phosphate disfavours formation of ribofuranosyl cations but favours ring opening by steric acceleration: 5′,8-bridging has a small effect but 5′,3-bridging in the case of purines has a large rate-retarding effect because of the electronic effect of the quaternary nitrogen.

Direct bridging from position 5′ to position 8 of purine nucleosides has a small rate-retarding effect (a factor of 3–9); similar factors are seen if O, NH or S bridges are inserted.[71] The 29 000-fold lower reactivity of 5′-N3-cycloadenosine is therefore overwhelmingly the electronic effect of putting a quaternary nitrogen at position 5 (*cf.* ρ_I for position 6 of pyranosides, Table 3.1).

3.7.5 Intermolecular General Acid Catalysis of Glycoside Hydrolysis

A glycosidase, lysozyme, was the first enzyme whose X-ray structure was solved, and the mechanism of action proposed at the time (1966) postulated general acid catalysis by Glu35: the proton was transferred from Glu35 as the C–O bond of the substrate was broken. An example of intramolecular general acid catalysis (the accelerated hydrolysis of salicyl β-D-glucopyranoside) was already known,[72] but the lysozyme structure prompted a search for further model systems.

Analysis of the schematic free energy profile for an A1 reaction suggested that general acid catalysis would be observed in acetal systems whose structures raised the energy of the conjugate acid of the substrate or the ground state, but lowered the energy of the oxocarbenium ion.[73] Although this analysis was correct in its outcome, it relied heavily on the Hammond postulate, which is now considered to be not universally valid. Exactly the same predictions are made if general acid catalysis is considered to occur when the protonated substrate in the A1 reaction becomes too unstable to exist, *i.e.* has an estimated lifetime less than 1 ps. Instability of the fully protonated substrate is favoured by low basicity (acidic leaving groups), oxocarbenium ion stability, or steric crowding, which can be relieved on C–O cleavage. The acetal C–O bond in such systems cleaves as the proton is being transferred, rather than subsequently.

2-(*p*-Nitrophenoxy)tetrahydropyran provides such an example.[26] It exhibits a spontaneous reaction with $k(30\,°C)$ of $8.8 \times 10^{-5}\,s^{-1}$. Such spontaneous hydrolyses have a β_{lg} value of -1.18.[18] The pK_a of *p*-nitroanilinium ion (acid $ArNH_3^+$) is 1.0[74] and that of *p*-nitroaniline (acid $ArNH_2$) is 18.9.[75] If we assume the difference in pK_a values of protonated *p*-nitrophenol (acid $ArOH_2^+$) and *p*-nitrophenol (acid ArOH) is the same as that between the analogous aniline and anilinium ion (17.9), then protonated 2-(*p*-nitrophenoxy)tetrahydropyran ether is estimated to decompose with a rate constant of $8.8 \times 10^{-5} \times 10^{(1.18 \times 17.9)}\,s^{-1}$ or $1.1 \times 10^{17}\,s^{-1}$, *i.e.* it is far too unstable to exist. The hydrolysis of *p*-nitrophenyl tetrahydropyranyl ether does indeed exhibit general acid catalysis with a Brønsted α of 0.5.[26]

The general acid-catalysed hydrolysis of acetals is a so-called class n reaction[xvi] and can be represented on a More O'Ferrall–Jencks diagram, in which Brønsted α (horizontal axis) represents proton transfer and a (normalised) Hammett ρ (vertical axis) measures C–O bond cleavage (Figure 3.16). Since the effect of leaving group ability is easy to measure, a diagonal axis running NW–SE is also defined. This two-dimensional diagram assumes, of course, that development of conjugation is synchronous with C–O cleavage. Reactions with $\beta_{lg}=0$ lie along a line joining the SW and NE corners.

The effect of altering oxocarbenium ion stability, acid strength and leaving group ability has been measured for dialkylacetals of the type ArCH(OR)(OR′),[76] and also for arylbenzaldehyde acetals of the type PhCH

[xvi] n for "nucleophile", since acid/base catalysis is applied to the nucleophilic/leaving group centre, as distinct from class e reactions, where acid/base catalysis is applied to the electrophile.

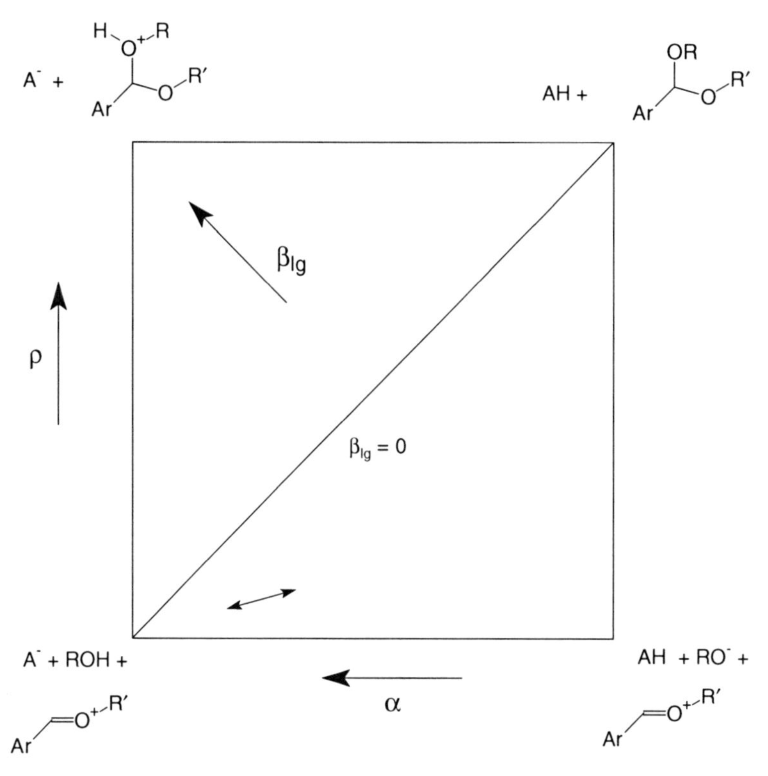

Figure 3.16 More O'Ferrall–Jencks diagram for the general acid catalysis of an acetal. The transition state position and reaction coordinate direction are those deduced for benzaldehyde alkyl acetals of acidic alcohols or phenols.

(OMe)OAr and ArCH(OMe)OPh.[77] Increasing the stability of the oxocarbenium ion corresponds to a lowering of the southern edge of the diagram and results in a decrease in α, whereas decreasing the strength of the catalysing acid results in an increase in ρ. Increasing leaving group ability corresponds to tipping the surface about the SW–NE diagonal, lowering the SE corner. The β_{lg} plots are non-linear, going through a minimum; one explanation is that diffusional separation of the ternary complex of oxocarbenium ion, leaving group and catalysing acid has becomes rate determining, another that the transition state moves to cross the $\beta_{lg} = 0$ line. It is possible to deduce a reaction coordinate direction from the cross-correlation coefficients;[78] this is shown qualitatively in Figure 3.14.

With some highly reactive systems,[79] decomposition of the intermediate hemiacetal,[80] rather than of the acetal itself, may become rate determining. The problem illustrates the general uncertainty created by following kinetic processes using only one instrumental signal, in these cases UV changes.

If the generation of an oxocarbenium ion is general acid catalysed, then, by microscopic reversibility, its reaction with alcohols should be general base catalysed. This has been observed in reactions of alcohols with *p*-$MeOC_6H_4CH^+Me$,[81] which by virtue of its quinonoid resonance forms can be accounted as an honorary oxocarbenium ion.

The intermolecular general acid catalysis of the hydrolysis of glycosides themselves, as would be expected from the lower stabilities of the oxocarbenium ions involved, has not been fully characterised. An anomalous temperature dependence of the solvent D_2O and leaving-group ^{18}O kinetic isotope effects in the acid-catalysed hydrolysis of *p*-nitrophenyl β-D-glucopyranoside, together with observation of general acid catalysis by trifluoroacetic acid at 45.0 but not 75.0 °C, indicates that the mode of acid catalysis changes with temperature, the general acid catalysed route, involving immobilisation of a molecule of catalysing acid, becoming less favoured the higher the temperature.[82]

The hydrolysis of α-D-glucopyranosyl fluoride is catalysed by both acids and bases, general base catalysis (of the attack of water) occurring in the case of phosphate and phosphonate dianions and general acid catalysis by monoanions of those dianions which showed general base catalysis.[83] The Brønsted parameters are small ($\alpha = 0.15$, $\beta = 0.06$). The molecular mechanisms involve acid catalysis by hydrogen bonding (rather than complete proton transfer) to departing fluoride and slight proton removal from nucleophilic water, made observable by the enhancement of these processes when the catalyst is an anion and can interact electrostatically with the partial positive charge at C1.

3.7.6 Intramolecular General Acid Catalysis of Glycoside Hydrolysis

Intramolecular general acid catalysis is normally detected in the first instance by a horizontal portion on the plot of $\log_{10}k_{obs}$ versus pH, governed by the acid dissociation constant of the substrate. Thus, the hydrolyses of various salicyl acetals (including the β-D-glucopyranoside below pH ≈ 10,[84] where a base-catalysed process occurs) obey the rate law of eqn (3.7), where k_0 is the first-order rate constant for hydrolysis of the neutral molecule and k_a is the second-order rate constant for the acid-catalysed hydrolysis of the neutral molecule:

$$\log_{10}k_{obs} = \log_{10}\{[k_0/(1 + K_a/10^{-pH})] + [10^{-pH}k_a/(1 + K_a/10^{-pH})]\} \quad (3.7)$$

The appearance of the plot of $\log_{10}k_{obs}$ versus pH is of two parallel lines of gradient −1 joined by an inflection, which, if the acid catalysis is efficient enough, can become a large horizontal region.

Although sometimes other practices are adopted, it is best to plot the decadic logarithm of a parameter against pH rather than the parameter itself. In the present case, the parameter is an observed first-order rate constant, but the comment is general and applies, for example, to dissociation or various kinetic parameters in enzyme kinetics. The reasons are twofold. First, errors in kinetic parameters tend to be (roughly) constant as percentages, so that least-squares fitting of the parameter, rather than its logarithm, to the rate law would grossly overweight points at high absolute values of the parameter. Second, specific acid- and base-catalysed processes are readily recognised by slopes which are at the limit integers (usually +1 or −1).

Rate laws, however, can only establish the composition of the rate-determining transition state, not its structure, and in particular they cannot distinguish between tautomers. 2′-Naphthyl β-D-glucuronide has a similar rate law to salicyl β-glucoside, but the molecular events giving rise to it are different (Figure 3.17).[85] If k_0 is the rate constant for the pH-independent part of the

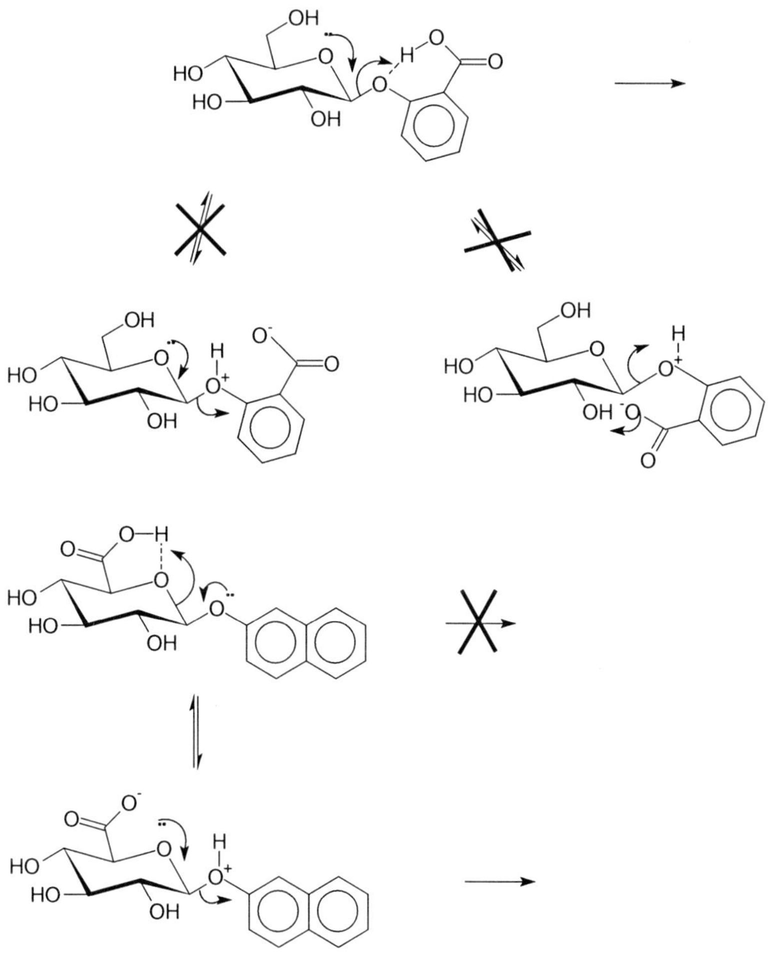

Figure 3.17 Molecular mechanisms giving rise to enhanced hydrolysis rates of glycosides with carboxylic acid groups. In the case of salicyl β-glucoside, frontside nucleophilic attack is stereoelectronically prohibited and distinction between specific acid catalysis of the hydrolysis of the anion and intramolecular general acid catalysis was made on the basis of solvent isotope effects in related systems.[86]

profile and K_a the acid dissociation constant of the glucuronyl carboxylate, then $k_{a-} = k_0/K_a$, where k_{a-} is the second-order rate constant for specific acid catalysis of the anion. If logarithms of second-order rate constants for acid-catalysed hydrolyses of a series of C5-substituted 2′-naphthyl β-D-xylopyranosides are plotted against σ_I of the C5 substituent, $\log k_{a-}$ falls exactly where predicted by σ_I of COO^-. The lability of 2′-naphthyl β-glucuronide in weak acid is therefore the result of the inductive effect of the carboxylate group: the same mechanism dictates the lability of all glycosides of uronic acids in weak acid. A similar phenomenon gives rise to the lability of sialic acid glycosides: in this case distinction was made between two kinetically equivalent pathways on the basis of β_{lg} values.[35]

The effectiveness of intramolecular catalysis is quantitated by the effective molarity of the reaction, the ratio of the first-order rate constant for the intramolecular reaction to the second-order rate constant for a sterically and electronically analogous reaction.[87] Effective molarities have a theoretical maximum of 10^8 M,[88] but in small molecule systems rarely come anywhere near this.

Estimates of effective molarities in general acid catalysis of acetals require the analogous intermolecular reaction to be observable and therefore, as we have seen in the previous section, for the oxocarbenium ion to be particularly stable. The termolecular processes corresponding to the hydrolysis of salicyl β-glucoside,[72] 2-methoxymethoxybenzoic acid[89] and 8-methoxymethoxy-1-(*N*,*N*-dimethylamino)naphthalene,[90] involving a nucleophile in addition to the catalysing acid, cannot be detected. Effective molarities can be calculated in the systems shown in Figure 3.18, which generate solvent-equilibrated oxocarbenium ions.

Most intramolecular proton transfers have effective molarities of only in the tens; the efficiency of the salicyl systems (I–IV, Figure 3.18) is based partly in a tight hydrogen bond in salicylate anion, which raises the pK_a to 12.95, three units higher than that of phenol.[95] The leaving oxygen and catalysing carboxylic acid are conjugated in the salicyl system, but similar efficiencies were obtainable in non-conjugated systems V–VII.

The electronic effects of substituents in the leaving group and the general acid in electronically independent systems such as V–VII have not yet been examined, but there have been concordant attempts to analyse two salicyl systems according to the equation

$$\log(k/k_0) = \rho_1\sigma_1 + \rho_2\sigma_2 \qquad (3.8)$$

Analysis of pH-independent rates of hydrolysis of substituted 2-methoxymethoxy benzoic acids[96] and salicyl-substituted versions of VII[92] gave for the phenol $\rho_1 = 0.89$ and 0.93, respectively. These values correspond to a β_{lg} value of ~ -0.4, which is commonly observed with many glycosidase-catalysed hydrolyses of aryl glycosides. The low values for the carboxylic acid ρ_2 (0.02 and ~ 0.2) are more problematic, as they should be the same as the Brønsted α; perhaps a strong hydrogen bond in the ground state is part of the cause.

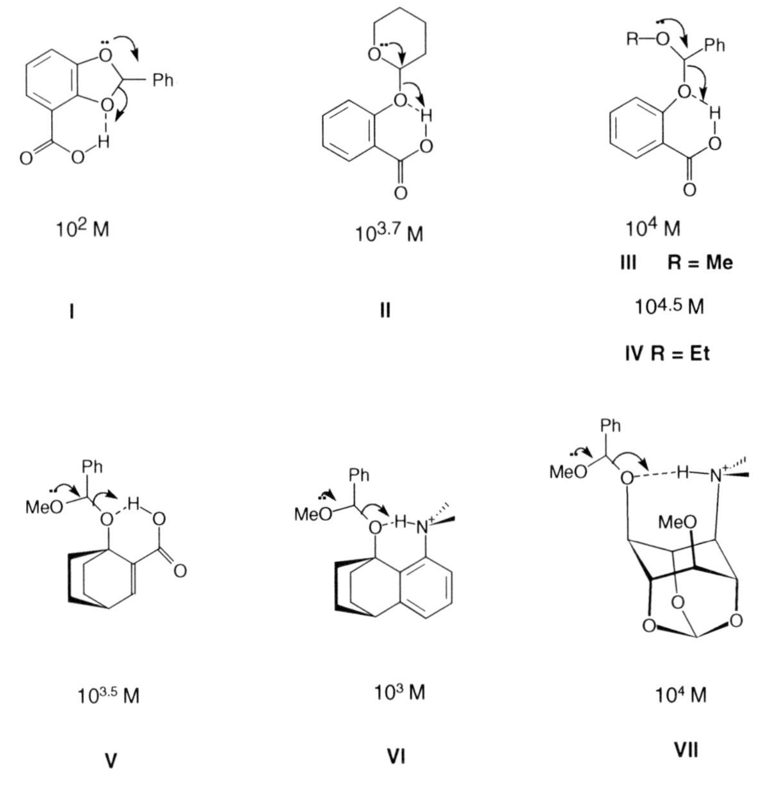

Figure 3.18 Effective molarities in the intramolecular hydrolysis of various acetals. Original references: I,[86] II and III,[91] IV[92] and V, VI and VII.[93] The kinetically equivalent nucleophilic pathway (*cf.* Figure 3.17) can be rejected on stereoelectronic grounds (the 6-*endo*-tet cyclisations violate Baldwin's rules), but nonetheless the methoxymethyl ester of salicylic acid was synthesised and shown not to be formed in the hydrolysis of 2-methoxymethoxybenzoic acid, but to be stable under the reaction conditions.[94]

3.8 ELECTROPHILIC CATALYSIS OF GLYCOSIDE HYDROLYSIS

Electrophilic catalysis of the departure of halogens in the century-old Koenigs–Knorr reaction[97] is implicit in the use of heavy metal bases such as silver oxide and mercuric cyanide, but the first demonstration of electrophilic catalysis in water (in the hydrolysis of the β-glucoside of 8-hydroxyquinoline by first-row transition metals ($Cu^{II} \gg Ni^{II} > C^{II}$)) was by Clark and Hay in 1973.[98] The observations were expanded to the more conveniently followed (because more labile) benzaldehyde methylacetals[99] or tetrahydropyranyl derivatives[100] of 8-hydroxyquinoline, whose hydrolysis is now known to give solvent-equilibrated oxocarbenium ions (Figure 3.19). Surprisingly, however, the observation of electrophilic catalysis of glycoside hydrolysis itself was not picked up by paper

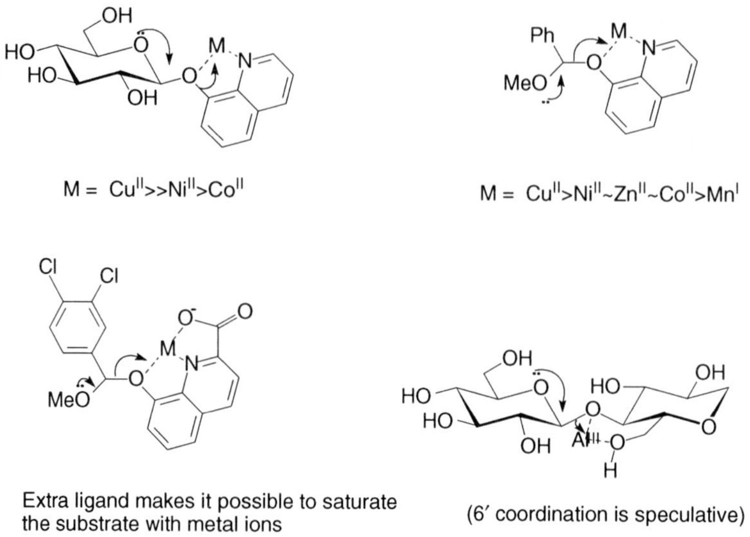

M = Cu^{II}>>Ni^{II}>Co^{II}

M = Cu^{II}>Ni^{II}~Zn^{II}~Co^{II}>Mn^{I}

Extra ligand makes it possible to saturate the substrate with metal ions

(6′ coordination is speculative)

Figure 3.19 Glycoside and acetal hydrolyses catalysed by electrophiles.

technologists until 30 years later, despite its obvious relevance to the problem of paper permanence.

The mechanical strength of paper sheets depends on the degree of polymerisation (*DP*) of the cellulose chains in the microfibrils; chemical pulping usually brings this down to about 1700 from the biosynthetic 10^4 and paper becomes unusably friable when the *DP* falls further to below 400. Many old paper documents, particularly those from the nineteenth and early twentieth centuries, have become unusably fragile. Although oxidative processes can contribute, it has long been thought that the main process in loss of cellulose *DP* under archival conditions is the aqueous hydrolysis of the cellulose (paper is 5–7% water). A particular problem is "rosin–alum" sizing, invented around 1800 and widespread by the 1840s, which is responsible for the paradox that nineteenth and early twentieth century documents are often in worse condition than those from the seventeenth and eighteenth centuries. During manufacture, paper for printing and writing is given a hydrophobic surface, so that ink does not "wick" as it does on blotting paper. "Rosin–alum" sizing uses a mixture of diterpene acids ("rosin") and "papermakers' alum" [hydrated aluminium sulfate, not $KAl(SO_4)_2.12H_2O$]. The paper from such treatment has a pH of 3.5–5.5 and the idea was so firmly entrenched that protons from the aluminium hydration sphere were the depolymerisation catalyst that massive programmes of document deacidification, such as much of the stock of the US Library of Congress, were set in train. In fact, a steep dependence of k_{obs} for hydrolysis of 1,5-anhydrocellobiitol on the concentration of added Al^{III} at constant pH demonstrates the culprit is in all likelihood aluminium acting as an electrophile, not protons.[101]

3.9 HYDROLYSIS OF THIOGLYCOSIDES AND THIOACETALS

The mechanistic consequences of replacing an oxygen atom in an acetal by a sulfur atom are dictated by three phenomena:

(i) Sulfur is "softer", more polarisable than oxygen, so that assistance of leaving group departure by Brønsted acids (whether via general or specific acid catalysis) is less ready than with oxygen and assistance by electrophiles, particularly heavy metal ions, is more ready.

(ii) 2p–3p orbital overlap is less efficient than 2p–2p orbital overlap, so conjugative stabilisation of thiocarbenium ions is less efficient than that of oxocarbenium ions (Figure 3.20). Measurements of this effect uncomplicated by others are rare, one of the few being a 40-fold lower rate of methanolysis of 2,3,4-tri-*O*-acetyl-5-thio-5-deoxy-α-D-xylopyranosyl bromide than of its oxygen analogue.[102]

(iii) The C–S bond seems to be more stable to spontaneous, uncatalysed heterolytic cleavage, without an electrophile, than the C–O bond, even if the pK_a values of the leaving alcohol and thiol are the same. Again, quantitative data in simple systems are lacking, but a lower limit of 10^4 could be placed on the relative rates of acetolysis of *tert*-butyl 2,4,6-trinitrophenolate (picrate) and *tert*-butyl 1-thio-2,4,6-trinitrophenolate.[103]

In acid-catalysed hydrolysis of a hemithioacetal, $R_1R_2CH(SR_3)(OR_4)$, in general, other things being equal, C–O cleavage is slightly favoured over C–S cleavage, the greater ease of protonation of oxygen just offsetting the poorer

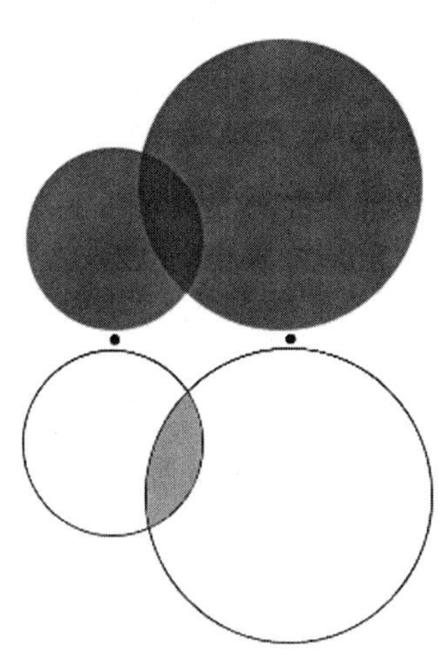

Figure 3.20 2p–3p orbital overlap is less efficient than 2p–2p orbital overlap. The overlap integral is governed by the proportion of electron density in each orbital that overlaps, not the size of the area shown overlapping in this cartoon.

conjugation by sulfur.[104] However, conformational effects can over-ride this preference, with phenyl 1-thio-β-D-glucopyranoside being hydrolysed by C–S fission some 20-fold slower than its oxygen analogue.[105] Where the conformational preference is for ring opening, as in the hydrolysis of methyl 1-thio-α-D-ribopyranoside, where a 1,3-diaxial interaction between the 3-OH and the MeS groups destabilises the $^4C_1$ conformation, anomerisation and ring contraction accompany hydrolysis, indicating initial endocyclic C–O fission.[106] Methyl 5-thio-α- and β-xylopyranosides hydrolyse, as expected, with exocyclic fission, some 10 times faster than their oxygen analogues: apparently the lesser inductive effect of sulfur, which increases the basicity of O1, more than offsets the poor 2p–3p orbital overlap.[107]

In accord with the low fractionation factors of S–H bonds (see Section 1.2.1.4.3), k_{D_2O}/k_{H_2O} for the acid-catalysed hydrolyses of 2-(substituted-thiophenoxy)tetrahydropyrans is lower (1.25) than would be expected for a specific acid-catalysed reaction with an oxygen leaving group (the reaction proceeds by sulfur–oxygen fission because of the electron-withdrawing aryl group on sulfur). The Hammett ρ value for this process (0.96) is more positive than for the acid-catalysed hydrolysis of glycosides, suggesting that after complete protonation of the sulfur, the C–S bond is less cleaved than the C–O bond in analogous reactions. A similar ρ value (0.88) is found for the Hg^{II}-catalysed hydrolysis.[108]

Spontaneous departure of thiophenoxide from hemithioacetals of the type ArCH(OEt)SPh gives rise to a "water" reaction whose rate-limiting step appears to be the departure of thiophenoxide from an ion pair;[104a,109] the ability to form such a species is reminiscent of the thiolate–carbonyl complexes detected in the mutarotation of 5-thioglucose (see Section 1.2.1.5).

3.10 HEAVY ATOM AND REMOTE HYDROGEN KINETIC ISOTOPE EFFECTS IN GLYCOSYL TRANSFER

3.10.1 Measurement of Small Isotope Effects

Whereas primary hydrogen kinetic isotope effects, solvent deuterium kinetic isotope effects (see Section 1.2.1.4) and (at a pinch) α-deuterium kinetic isotope effects (see Section 3.3.1) are large enough to be measured by direct comparison, non-hydrogen ("heavy") kinetic isotope effects, and the kinetic effects arising from hydrogen isotopic substitution at remote sites require specialised techniques. Isotope ratios in starting material or product can be measured by NMR, mass spectrometry or radio counting. All three methods depend on the enrichment of product and starting material in the isotope of interest. If R_0 is the isotopic ratio in the starting material and R_P and R_R are the isotopic ratios of the product and starting material, respectively, at a fractional conversion F, then the kinetic isotope effect is given by the equations[110]

$$\mathrm{KIE} = \frac{\ln(1-F)}{\ln[(1-F)R_R/R_0]} \tag{3.9}$$

$$\mathrm{KIE} = \frac{\ln(1 - F)}{\ln[1 - (FR_{\mathrm{P}}/R_0)]} \tag{3.10}$$

A feature of eqn (3.9) which has only been recognised in the last decade is that at very high conversions ($F \rightarrow 1$), $R_{\mathrm{R}}/R_0 \rightarrow \infty$. This makes small isotope effects measurable by conventional NMR integration if small amounts of starting material are isolated at very high conversion.[111] Thus, a 3% effect will raise the isotope ratio in the remaining reactant by 20%, if the unreacted starting material is isolated at 99.8% conversion. Obviously, therefore, large amounts of substrate are required. Moreover, internal standards have to be used for the NMR integration, otherwise reliance on methods such as UV absorbance, which are linear functions of the degree of reaction, would merely displace the precision problem to the measurement of F.

Mass spectrometric analysis of isotope ratios is in principle very accurate (isotope ratio mass spectrometers were developed for geochemical investigations), but the analytical problem comes from possible fractionation of isotopes in the workup procedure. Traditionally, isotope ratio mass spectrometry measurements of kinetic isotope effects have been confined to reactions which give a small molecule product (such as CO_2), which can be directly introduced to the mass spectrometer.

It is possible to count the ratio of the disintegrations of two β-emitters of different energies, such as ^{14}C and ^{3}H, with very high precision. This fact can be exploited to measure both ^{14}C and ^{3}H effects themselves and effects of other, non-radioactive isotopes. For measurement of a ^{3}H effect, the substrate is synthesised with ^{3}H at the site of interest and ^{14}C at a site where there is no expected isotope effect. Fractionation of ^{3}H at the site of interest then results in alteration of the $^{14}C/^{3}H$ ratio according to eqns (3.9) and (3.10). Obviously, the isotopes can be reversed, with ^{14}C at the site of interest and ^{3}H at a remote site, although because ^{3}H for ^{1}H is a large change, great care has to be taken that secondary ^{3}H effects do not mask the ^{14}C one. Isotope effects due to stable isotopes can be measured analogously, with one substrate with ^{15}N at natural abundance labelled with ^{3}H and its isotopomer with enriched ^{15}N at the site of interest labelled with remote ^{14}C.

Small isotope effects on reactions of sugars have also been measured by the isotopic quasi-racemate method.[188] The optical rotation of a solution containing as nearly equal as possible concentrations of the unlabelled sugar in the unnatural (usually L) series and its enantiomer labelled in high enrichment at the site of interest in the natural series is followed as a function of time (Figure 3.21). If concentrations are exactly the same (and the isotope effect on optical rotation can be neglected), optical activity builds up by virtue of the kinetic isotope effect and then disappears. If the effect is small, δk is the rate constant difference between light and heavy isotopomers, k the rate constant and $\Delta\alpha$ the optical rotation change on complete reaction of one isomer, a maximum of optical activity, at time $1/k$ and of magnitude $\delta k\Delta\alpha/2.72k$, is observed. More generally, the data can be fitted to the equation

$$\alpha_t = A\mathrm{e}^{-kt} + B\mathrm{e}^{-kt/C} + \alpha_\infty \tag{3.11}$$

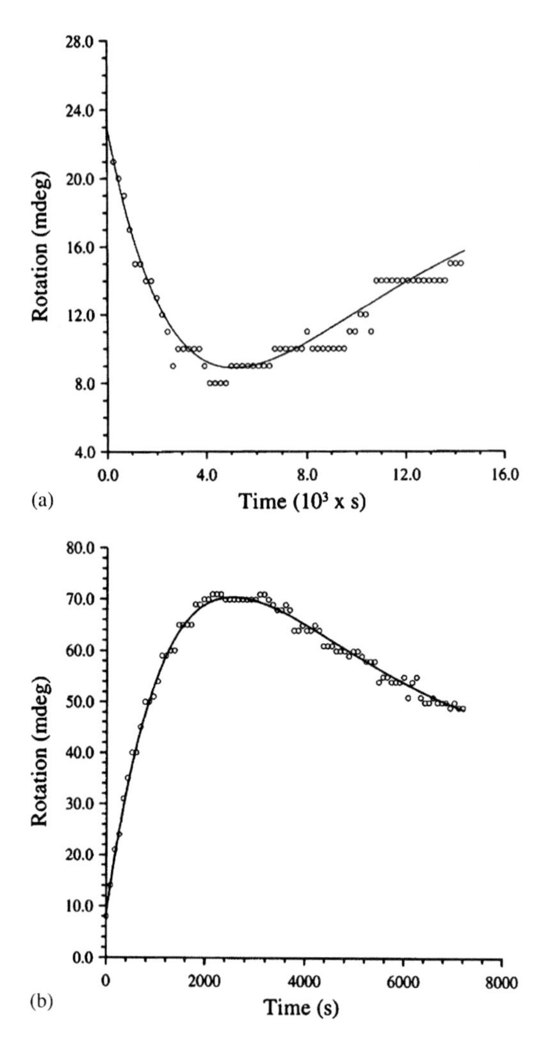

Figure 3.21 Time courses of the hydrolysis of an isotopic quasi-racemate in two acid-catalysed hydrolyses of glycosides. Both figures, taken from Ref. 107, by kind permission of Prof. A. J. Bennet and the American Chemical Society, refer to a mixture of labelled D-sugar and unlabelled L-sugar ($10\,\mathrm{mg\,mL^{-1}}$ each), show the least-squares best fit and display only every fourth data point for clarity. (a) Ring ^{18}O effect for methyl β-xylopyranoside; (b) anomeric ^{13}C effect for methyl 5-thio-α-xylopyranoside. See Table 3.3 for numerical values of the effects, which are in opposite senses.

Fits to such an expression with all parameters as variables are very ill-conditioned and it is advisable to measure A in a separate experiment, constrain k to within 10% of its separately measured value, ensure that $\mathrm{B} \approx -\mathrm{A}$ and only analyse time courses which are obviously, by visual inspection, not simple exponentials (preferably they should have definite maxima and minima).

3.10.2 Inductive and Steric Effects of Isotopes of Hydrogen

The relative change in mass between the isotopes of hydrogen is a large one, and secondary effects of 1–3% are often observed which have uncertain origins. These effects are largest with tritium, which makes tritium a poor choice for a "remote label" in ratio counting determinations of isotopic enrichments. Any sort of partitioning or non-covalent interaction, such as formation of an ES complex in an enzyme-catalysed reaction, can experience significant hydrogen isotope effects. Indeed, baseline separation of $(C_6H_5)CO$ and $(C_6D_5)CO$ can be obtained by gas chromatography on capillary columns.[112]

Deuterium (and *a fortiori* tritium) behaves as if it is slightly smaller and somewhat more electron-donating than hydrogen. An explanation, which rationalises both the small size and electron-donating inductive effect, is based on the anharmonicity of the C–L bond vibration. The heavier L, the lower is its zero point energy and hence, because of the anharmonicity, the nearer the centre of electron density of the C–L bond gets to the potential energy minimum (Figure 3.22).

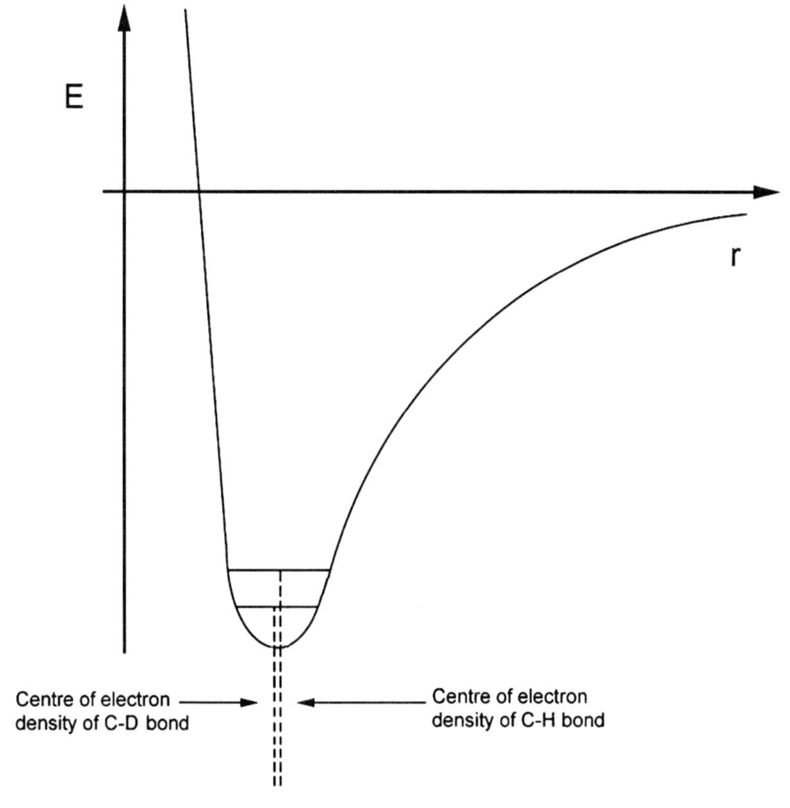

Figure 3.22 Cartoon of a C–L potential energy curve, illustrating the bond stretch–anharmonicity explanation of the inductive and steric effects of deuterium and tritium. The anharmonicity and zero-point energies are greatly exaggerated.

The inductive effective of deuterium is unarguable: direct measurement by NMR[xvii] of the pK_a differences between deuterated and protonated carboxylic acids gave[113] 0.034 for LCOOH, 0.0134 for L_3CCOOH, 0.0077 for L_3CCH_2COOH and 0.0046 for $L_3CCH_2CH_2COOH$, *i.e.* the inductive effect falls off with distance in much the same way as a conventional inductive effect. It is likewise cumulative; ΔpK_a for C_6L_5COOH is 0.0099, for *o*-C_6H_4LCOOH 0.0020 and for *m*- and *p*-C_6H_4LCOOH 0.0018.

Steric isotope effects are less clear cut, possibly because many small effects considered to be steric are in fact a mix of steric and stereoelectronic. Early work on the racemisation of optically active biphenyls gave a value of k_{H6}/k_{D6} of 0.85 for a 2,2′-dimethylbiphenyl also containing a 6,6′ ethylene bridge (4,5-dimethyl 9,10 dihydrophenanthrene),[114] and 1-deuteriocyclohexane prefers the deuterium-axial conformation by 25 J mol^{-1},[115] but the preference decreases next to a heteroatom.[116] Effects of deuterium substitution of carbon-bound protons in glucose on the anomeric equilibrium in water cannot be simply rationalised by a single effect; the equilibrium isotope effect (defined as $[\beta]_H[\alpha]_D/[\beta]_D[\alpha]_H$) being 1.043 for H1, 1.027 for H2, 1.027 for H3, 1.001 for H4, 1.036 for H5 and 0.998 for H6,6′.[117]

3.10.3 β-Hydrogen Kinetic Isotope Effects

β-Hydrogen kinetic isotope effects (effects due to deuterium or tritium substitution vicinal to the reaction centre) give information on both charge buildup and conformation and are thus particularly valuable for studies of mechanisms of glycosyl transfer. They have their origin in hyperconjugative weakening of the vicinal C–L bond as σ electrons flow into an electron-deficient vicinal p-orbital (Figure 3.23). This overlap is optimal when the C–L bond is parallel to the axis of the p-orbital and zero when they are orthogonal. We can thus write eqn (3.12), in which the observed effect $(k_H/k_D)_{obs}$ is related to the dihedral angle θ between the C–L bond and the electron-deficient p-orbital, the maximum possible hyperconjugative contribution to the effect, $(k_H/k_D)_{max}$, and an inductive effect, $(k_H/k_D)_i$ [*mutatis mutandis*, eqn (3.12) also applies to tritium effects].

$$\ln(k_H/k_D)_{obs} = \cos^2\theta \ln(k_H/k_D)_{max} + \ln(k_H/k_D)_i \tag{3.12}$$

β-Deuterium effects on glycosyl transfer are in the range $0.98 < (k_H/k_D)_{obs} < 1.15$ and are particularly informative in two cases. First, where they are inverse or zero and from other evidence it can be deduced the reaction centre is electron-deficient, then in the reactive conformation the C–L bond and the electron-deficient p-orbital must be approximately orthogonal. Second, where there are two diastereotopic β-hydrons which can be independently labelled (as with neuraminides), their relative magnitudes can immediately give conformational

[xvii] Measurement of the chemical shift difference between protonated and deuterated acids as a function of pH, essentially as described by eqn (2.8), but for isotopomers, not stereoisomers.

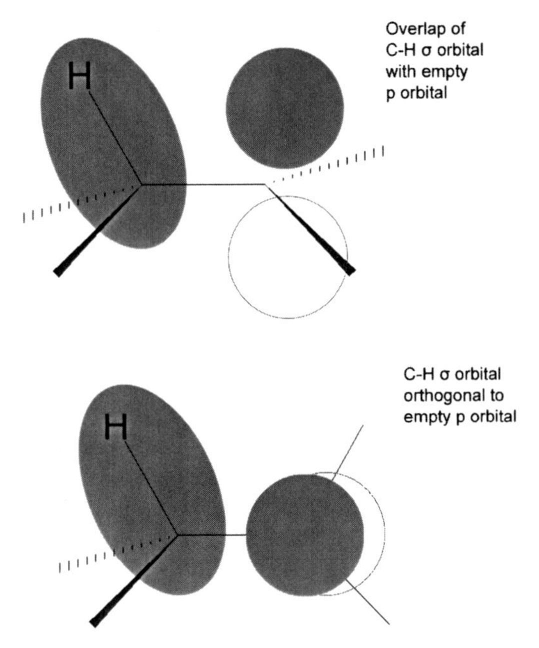

Figure 3.23 Hyperconjugative origin of β-hydrogen effects.

information, and, if the small correction $\ln(k_H/k_D)_i$ can be estimated from effects at other remote sites, eqn (3.12) can be solved explicitly without any assumptions.

3.10.4 Heavy Atom Kinetic Isotope Effects

Whereas substitution of protium by deuterium or tritium results in the near doubling or tripling of the reduced mass of a vibration,[xviii] substitution of ^{12}C by ^{13}C or the radioisotope ^{14}C, ^{16}O by ^{17}O or ^{18}O, or ^{14}N by ^{15}N produces much smaller changes in reduced mass. Primary kinetic isotope effects arising from such isotopic substitutions nonetheless do appear to be substantially zero point energy effects. Maximum zero point energy effects (k_L/k_H[xix] at 25 °C) for breaking bonds to carbon are 1.25 for $^{12}C/^{13}C$, 1.14 for $^{15}N/^{14}N$ and 1.19 for $^{16}O/^{18}O$. In practice, effects are much smaller than this, largely because bonds are not completely broken at the transition state. With these small effects, the effect of change of mass on molecular moments of inertia cannot be neglected. Such "ponderal" kinetic isotope effects have a lower dependence on temperature than zero point effects, for which $T\ln(k_L/k_H)$ = constant (equation 1.14) holds generally, T being the absolute temperature.

[xviii] For a diatomic molecule, the reduced mass (m^*) is $m_1m_2/(m_1+m_2)$. This means that most of the motion of an X–H bond is in the hydrogen. For bonds linking atoms whose masses are more nearly matched, the proportional change in m^* is much lower. Thus, if the C–O bond is treated as a diatomic molecule, the reduced mass of ^{12}C–^{16}O is 6.86, but that of ^{12}C–^{18}O is 7.2.

[xix] k_{Light}/k_{Heavy}.

The relative importance of non-zero-point contributions to heavy atom effects also makes exact comparisons of results with different isotopes of the same element difficult. There is no counterpart of the Swain–Schaad equation (equation 1.15) for isotopes of hydrogen, although for isotopes of carbon, the intuitive expectation that ^{14}C effects would be around double ^{13}C effects was confirmed; eqn (3.15) held for a series of effects, with $1.6 < r < 2.1$:[118]

$$\frac{\ln(k_{12}/k_{14})}{\ln(k_{12}/k_{13})} = r \tag{3.13}$$

^{13}C effects are particularly amenable to determination by the Singleton method, since the $I = \frac{1}{2}$ ^{13}C isotope is present at a natural abundance of 1.11%.[119] At the reaction centre ^{13}C effects appear to indicate the symmetry of an S_N transition state, being low for S_N1 reactions but high for S_N2 reactions, particularly those with little bonding to leaving group or nucleophile but charge delocalisation away from the reaction centre.

Because ^{17}O is present in such low abundance (0.037%), there has up to now been no motive for measuring ^{17}O rather than the larger ^{18}O effects (^{18}O is present in 0.20% abundance), but this could change because ^{17}O is the only oxygen isotope with a nuclear spin ($I = 5/2$), so that it is the only one to which the Singleton method can be applied directly. ^{18}O effects at the anomeric centre can distinguish unambiguously between endocyclic and exocyclic cleavage during acid-catalysed hydrolysis of glycosides: *p*-nitrophenyl α-arabinofuranoside-1-^{18}O (exocyclic cleavage) gives a value of k_{16}/k_{18} of 1.023 at 80 °C, whereas isopropyl α-arabinofuranoside-1-^{18}O (endocyclic cleavage) gives a value of k_{16}/k_{18} of 0.988, in accord with a decrease in bonding to O1 in exocyclic cleavage and an increase in endocyclic cleavage.[120] The direct effect is, however, far from the theoretical maximum, in part because in an acid-catalysed process bonding (to hydrogen) is gained. In the absence of acid catalysis (as in the spontaneous hydrolysis of *p*-nitrophenyl *N*-acetyl-α-neuraminide,[121] for which $k_{16}/k_{18} = 1.053$), larger effects are seen.

^{15}N isotope effects in carbohydrate chemistry have only been measured in the leaving group, by remote label methods. However, the properties of this stable isotope (0.365% abundance, $I = \frac{1}{2}$) would seem to make the Singleton method attractive in the future. The acid-catalysed hydrolysis of AMP gives $k_{14}/k_{15} = 1.03$;[122] N7, rather than the leaving N9, is protonated, so the effect of the loss of C–N bonding is not offset by the gain of N–H bonding, and, for a one-unit mass difference, the effect is comparatively large.

3.10.5 Transition State Structure Determination from Multiple Kinetic Isotope Effects

Each individual kinetic isotope effect places certain constraints on the transition state structure. Many such effects thus enable the structure to be located with some precision, and for the past 20 years computer software packages, starting with BEBOVIB,[123] a bond-order, bond-energy vibrational analysis package, and progressing to VIBE, CTBI and ISOEFF,[124] which incorporate quantum chemical calculations, have become available which will calculate transition state geometry and bond orders. What still presents problems are the input parameters: subtle effects of anharmonicity or solvation can be computed if they are known, but they are often unknown. When using such packages it is always good to check that the results correspond to chemical intuition: as late as 1994, using BEBOVIB-IV, it was possible to fit an extended set of isotope effects on the enzymic hydrolysis of α-glucosyl fluoride to a model in which total bond order to carbon increased at the transition state.[125]

Table 3.3 sets out kinetic isotope effects for a series of hydrolyses of glycosyl derivatives. A number of features emerge:

(i) Leaving group ^{18}O effects for acid-catalysed hydrolyses of methyl glycoside are direct, constant and low, in accord with the specific-acid-catalysed nature of the processes and their late transition states: the (major) zero point energy portion of the effect arises from the nearly complete exchange of C1–O1 vibrations for O1–H vibrations.

(ii) Anomeric ^{13}C effects for the pyranosides fit nicely into the picture of hydrolyses with neutral leaving groups proceeding through essentially S_N1 transition states giving rise to small effects and reactions with anionic leaving groups being S_N2 and giving rise to large effects. The clean S_N2 reaction of α-glucopyranosyl fluoride with azide ion gives a particularly large effect. The moderate anomeric ^{14}C effect for AMP, equivalent to the ^{13}C effect for glucosyl fluoride hydrolysis, suggests that it is the charge of the atom immediately attached to the anomeric carbon, rather than the charge of the whole leaving group, which determines the S_N1/S_N2 mechanistic change–AMP is protonated on N7, not N9. This is confirmed by the much lower ^{14}C effect for hydrolysis of NAD^{+}, where the leaving group is a pyridine, than for AMP. The effects are higher for the methyl 5-thioxylosides than for methyl xylosides or glucopyranosides because of the well-known poor 3p–2p overlap, which makes $>C{=}S^{+}-$ weaker (and hence lower frequency) than $>C{=}O^{+}-$.

(iii) α-Deuterium effects are small because of the inductive effect of deuterium and differ between anomers in a way quantitatively predictable from the isotope effect on anomeric equilibrium of 1.043 (see Section 3.10.2).

(iv) γ-Deuterium effects in the *xylo* series are closer to unity in the 5-thioxylose system than xylose itself; since they arise from an axial and equatorial hydron there will be a minimal conformational component, the effects will be largely inductive. They therefore support the idea of poor 3p–2p orbital overlap in the thiocarbenium ion.

(v) Ring ^{18}O effects for acid-catalysed hydrolysis of methyl glucosides are very low compared with both the bimolecular reactions of glucosyl

Table 3.3 Kinetic isotope effects for hydrolysis of various glycosyl derivatives. The isotopic quasi-racemate method was used for all measurements except those for AMP, where the remote label method was used.

Substrate	α-*D*	β-*D*	*Leaving group* ^{18}O	*Anomeric* ^{13}C	*Ring* ^{18}O	γ-*D*
α-Glc*p*OMe, H^+, 80 °C[188]	1.13(7)	1.07(3)	1.02(6)	1.007[189]	0.996(5)	0.98(7)
β-Glc*p*OMe, H^+, 80 °C[188]	1.08(9)	1.04(5)	1.02(4)	1.01(1)[189]	0.99(1)	0.97(1)
α-Xyl*p*OMe, H^+, 80 °C[107]	1.12(8)	1.09(8)	1.02(3)	1.006	0.98(3)	0.98(6) (d_2)
β-Xyl*p*OMe, H^+, 80 °C[107]	1.09(8)	1.04(2)	1.02(3)	1.006	0.97(8)	0.96(7) (d_2)
α-*S*-Xyl*p*OMe, H^+, 80 °C[107]	1.14(2)	1.06(1)	1.02(7)	1.03(1)	N/A	1.00 (d_2)
β-*S*-Xyl*p*OMe, H^+, 80 °C[107]	1.09(4)	1.018(5)	1.03(5)	1.02(8)	N/A	0.98(6) (d_2)
α-Glc*p*F, 80 °C[33]	1.14(2)	1.06(7)	N/A	1.03(2)	0.98(4)	
β-Glc*p*F, 50 °C[33]	1.08(6)	1.03(0)	N/A	1.017	0.98(5)	
α-Glc*p*F+2 M NaN_3, 50 °C[33]	1.16(9)			1.08(5)		
NAD^+[190]	1.19(4) (T)	1.11(4) (T)	1.02(0) (^{15}N)	1.016(^{14}C)	0.98(8)	1.00(0) (T)
AMP[122]	1.12(3)	1.07(7)	1.03(0) (^{15}N)	1.04(4) (^{14}C)		

fluorides and effects with xylose. It would appear that the hydroxymethyl group is precluding development of conjugation, possibly in the axial case via a conformation change of the protonated substrate to the $^{1}S_3$ conformation of the α-glucosylpyridinium ion, but the requirement for a nucleophile stops such a change in the reactions of the fluoride.

3.11 HYDROLYSES OF KETOSIDES

Oxocarbenium ion formation from ketosides is expected to occur more readily than from aldosides, because of possible hyperconjugative and inductive stabilisation of charge. The acid-catalysed hydrolysis of sucrose has been studied for a century and half, and played a key role in establishing the concept of a Brønsted acid (the stronger the acid, the better it catalysed the "inversion" of sucrose – the hydrolysis to a fully mutarotated equimolar mixture of glucose and fructose is associated with a change of rotation at the sodium D line from positive to negative), but the most recent mechanistic work seems to be determination of a Bunnett w value indicative of an A1 process in 1961 (see Table 3.2). This author would guess that the fructofuranosyl cation has a lifetime similar to that of the 2-deoxyglucopyranosyl cation.

3.11.1 Hydrolysis of Sialic Acid (Neuraminic Acid) Derivatives

The sialic acid structure incorporates a carboxylate at the anomeric centre, which can be inductively electron donating in its deprotonated form and inductively electron withdrawing when protonated, as well as deoxygenation of the position next to the anomeric centre. These features make sialosyl cations more stable than simple aldopyranosyl cations, and hydrolysis and solvolysis reactions have been investigated using the substrates shown in Figure 3.24.

Solvolysis of both α- and β-NeuNAcOPNP in aqueous ethanol yielded NeuNAcOEt in the same α:β ratio of 1 : 9[36] as was obtained for the NeuNAc-OMe formed by solvolysis of α-NeuNAcPy$^+$ in aqueous methanol.[126] These data indicate that all three derivatives form a solvent-equilibrated cation which collapses to an anomeric mixture of substitution products whose composition is dictated by the relative stabilities of the anomers (*cf.* the anomeric equilibrium of NeuNAcOH, Table 1.1); the α-configured NeuNAc derivatives are destabilised by the axial carboxylic acid group. This group brings about very remarkable anomeric reactivity differences: the spontaneous hydrolysis of α-NeuNAcOPNP is about 110 times faster than that of β-NeuNAcOPNP, despite the β_{lg} values for α- and β-NeuNAcOAr being the same (−1.32[35] and −1.24,[36] respectively. Likewise, the pK_a of the carboxylic acid group in α-NeuNAcOPNP is 2.86, but that in β-NeuNAcOPNP, 1.58, attributed to steric hindrance to solvation of the axial carboxylate.[36] A complication of the reactions in aqueous alcohols is that small amounts (3–14%) of the elimination

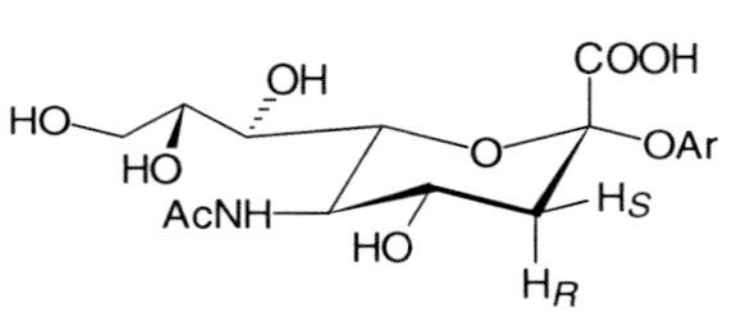

Aryl N-acetyl α–neuraminides (α–**NeuNAcOAr**)(natural stereochemistry)

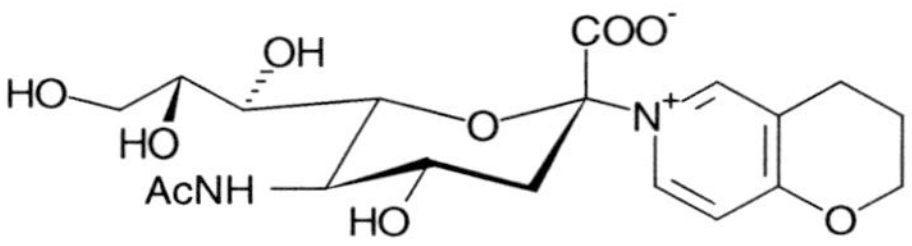

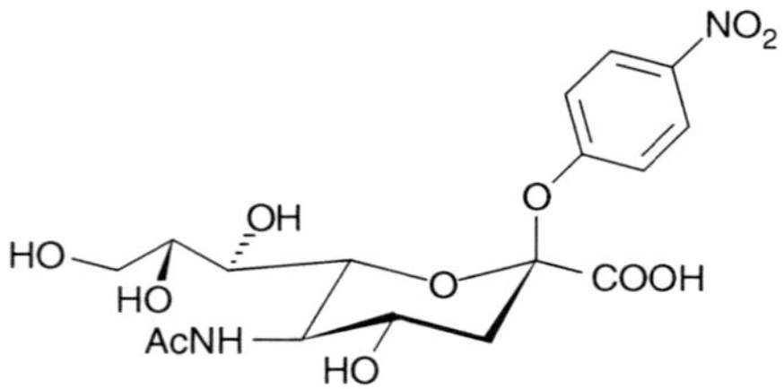

p-Nitrophenyl N-acetyl β-neuraminide (β-**NeuNAcOPNP**)

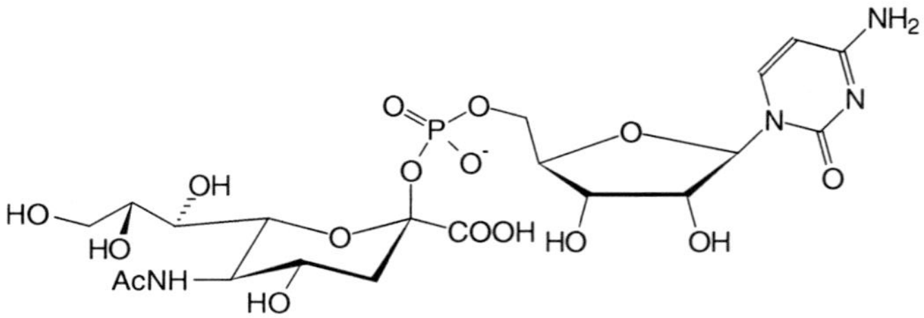

An N-acetyl α-neuraminyl pyridinium salt
(**α-NeuNAcPy+**)

CMP - NeuNAc

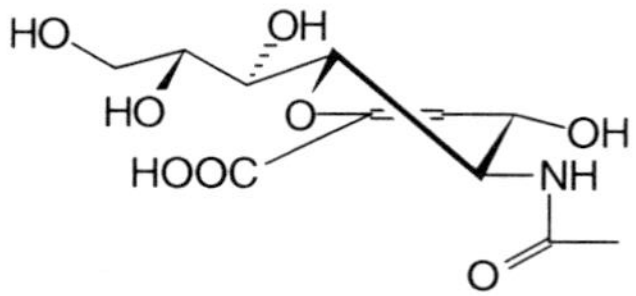

2,3-Dehydro-N-acetyl neuraminic acid (**DANA**)

Figure 3.24 Substrates for studies of sialoside hydrolysis.

product DANA are concurrently produced, although not from α-NeuNAcOPNP in pure water.[xx]

These results seem to be at variance with those with another leaving group, CMP, where significant quantities of DANA are produced and approximately equal amounts of α- and β-NeuNAcOMe are produced in the presence of methanol.[127] It is possible that with a negative leaving group with a poorly delocalised charge (unlike *p*-nitrophenolate), concurrent reactions through ion pairs and/or S_N2 transition states are occurring, giving largely inversion, whereas solvent-equilibrated sialosyl cations give a preponderance of β-NeuNAcOMe.

Both α- and β-NeuNAcOPNP hydrolyse through four processes: the spontaneous departure of the phenolate from the substrate anion, an acid-catalysed process of the neutral molecule, a base-catalysed process of the anion and an apparently spontaneous process of the neutral molecule. This last process was shown to be the kinetically equivalent acid-catalysed hydrolysis of the anion of β-NeuNAcOAr by a β_{lg} value of 0.14,[36] and of the anion of α-NeuNAcOAr by a β_{lg} value of 0.00 and a solvent deuterium kinetic isotope effect of 0.96,[35] when any intramolecular general acid catalysis would give rise to negative β_{lg} values and direct solvent isotope effects.

The measurement of individual *proS* and *proR* β-deuterium effects for α-NeuNAcOPNP, and also the inductive effect of deuterium substitution at position 4, enabled eqn (3.12) to be solved for the acid-catalysed hydrolysis of the neutral molecule, the acid-catalysed hydrolysis of the anion and the spontaneous reaction. An isotope effect k_H/k_D of 1.000 for the *proS* and position 4 hydrons for the spontaneous and acid-catalysed reactions of the anion indicated the inductive correction in eqn (3.12) was zero. The *proR* effects of 1.07_5 and 1.08_6 then lead immediately to a value of the *proR* hydron–breaking C–O bond dihedral angle (θ) of 30° and values of $(k_H/k_D)_{max}$ of 1.09_8 and 1.11_5 for the spontaneous and acid-catalysed processes of the anion, respectively. The acid-catalysed hydrolysis of neutral α-NeuNAcOPNP, by contrast, has a *proS* effect of 1.02_5 and a *proR* effect of 1.06_8, indicating a less flattened transition state in which $\theta > 30°$, The value of 0.97_6 for the effect at C4 is too high to be wholly inductive [implausible values of $(k_H/k_D)_{max}$ result if $\ln(k_H/k_D)_i$ is set equal to 3ln0.976], but even in the absence of any inductive correction a $(k_H/k_D)_{max}$ of 1.19 results, indicating a much more electron-deficient reaction centre.

That the ionisation of a geminal carboxylate should have such a large effect was puzzling: initially, neighbouring group participation to give some sort of α-lactone was tentatively invoked, but this is incompatible with the subsequent stereochemical results. A possible resolution comes from the ^{14}C kinetic isotope effects measured for CMP-NeuNAc hydrolysis, an unremarkable primary effect of 1.030 at position 2 being surpassed by a very large secondary one (1.037) at

---

[xx] DANA arises from an *E*1 reaction or an *E*1-like *E*2 reaction. Elimination is probably much more ready in sialosyl derivatives than simple sugars because of the stability of the conjugated product.

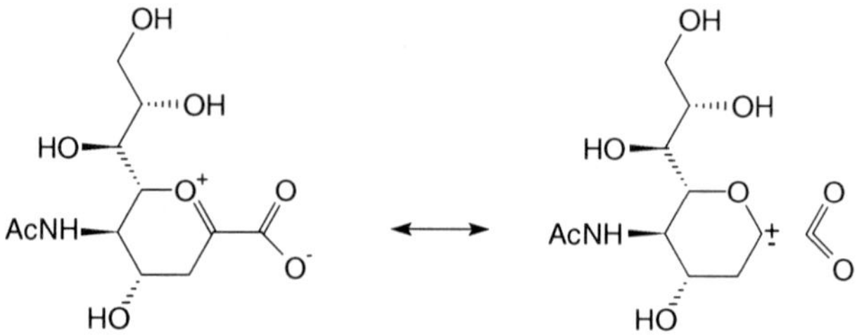

Figure 3.25 "Carbene resonance" possibly associated with the sialosyl cation.

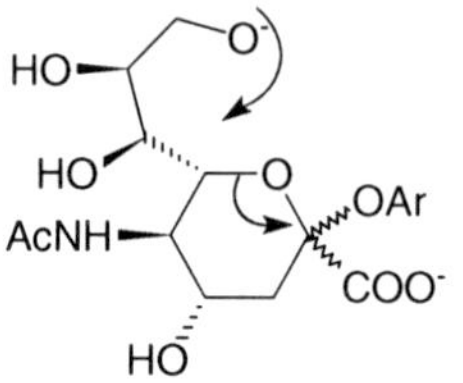

Figure 3.26 Possible base-catalysed hydrolysis mechanism of NeuNAcOAr.

position 1[127] "carbene resonance" in the sense shown in Figure 3.25 explains both sets of results.

At high pH, both α-and β-NeuNAcOAr liberate ArO^- by a base-catalysed process which, however, shows only a weak dependence on the pK_a of ArOH ($\beta \approx -0.2$), is not subject to a primary deuterium kinetic isotope effect at C3, and does not result from nucleophilic attack on the aromatic ring, as shown by ^{18}O labelling. A possible, though far from proven, mechanism, is that of Figure 3.26.

3.12 NEIGHBOURING GROUP PARTICIPATION IN GLYCOSIDE HYDROLYSES

Much information of a qualitative nature is available about participation by various protecting groups in nucleophilic displacements at the anomeric centre during glycoside synthesis, but quantitative information is confined to participation by acetamido, carboxylate and the ionised hydroxyl groups of the sugar itself.

3.12.1 Participation by Acetamido

The early work on participation by the acetamido group was motivated by its apparent relevance to lysozyme action. Lysozyme itself is now known to act through the canonical retaining mechanism (see Figure 5.32), but GH18 and GH20 chitinases and hexosaminidases do indeed act through an amide participation mechanism.

p-Nitrophenyl 2-acetamido-2-deoxy-β-D-glucopyranoside (β-D-GlcNAc*p*-OPNP) was shown to undergo a pH-independent hydrolysis in 1967,[128] but the intermediacy of the presumed oxazoline was not demonstrated directly until work with β-D-GlcNAc*p*F a decade later, when methanolysis of the oxazoline was shown to give β-D-GlcNAc*p*OMe exclusively.[129] The rate enhancement, compared with the simple glucoside, depended on the leaving group and medium: spontaneous hydrolysis of β-D-GlcNAc*p*F is ~400 times faster than that of β-D-Glc*p*F, but with 2,4-dinitrophenolate as leaving group the ratio falls to 40 (although it rises again to 500 in acetolysis).[130] Distortion of the pyranose ring is necessary for the carbonyl oxygen of the acetamido group to displace the β leaving group in an in-line fashion (such distortion has been directly observed in a chitinase; see section 5.9.6.7) and the greater nucleophilic assistance to the fluoride may reflect the stronger anomeric effect with fluoride making this readier (Figure 3.27). The greater assistance in acetolysis than hydrolysis appears to be a simple polarity effect: the positive charge is spread over more atoms with amide participation, which is thus favoured relative to a transition state with a more localised charge in less polar solvents.

Amide participation appears to be about as effective in specific acid-catalysed hydrolysis (of the methyl glycosides) as in spontaneous hydrolysis (of the 2,4-dinitrophenyl glycosides).[28] There is some evidence for simultaneous general acid catalysis and amide participation in the hydrolysis of β-D-GlcNAc*p*-O*o*C_6H_4COOH.[84]

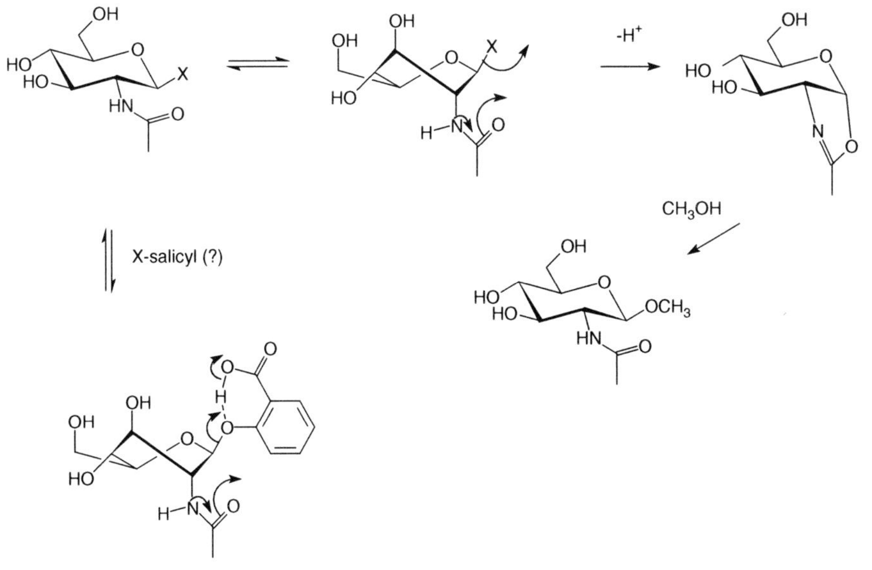

Figure 3.27 Participation by 2-acetamido groups in glucoside hydrolysis. The conformational pre-equilibrium is drawn by analogy to the ES complexes of GH 18 and GH20 chitinases and hexosaminidases.

3.12.2 Participation by Carboxylate and Phosphate. Electrostatic Catalysis?

Given the canonical mechanism of retaining *O*-glycopyranosidases, it is surprising that there is no system involving an actual sugar in which nucleophilic participation by a carboxylate has been demonstrated. The closest is such process is participation by phosphate. The tetraphosphate of β-D-Glc*p*OPNP hydrolyses about 100-fold faster than its anomer above pH 9; part of this acceleration may well be due to nucleophilic attack by the 2-phosphate in a cyclic five-membered transition state, but disentangling the kinetic behaviour of a system which, even if tautomers having any electrically neutral phosphate groups are neglected, has 5 macroscopic and 12 microscopic protonation states, was not attempted.[131] Another complication is that, as observed with polysaccharides (Chapter 4), sulfation of vicinal equatorial OH groups can change the conformation[132] and the effect of isoelectronic phosphates is likely to be similar.

Participation in the departure of anionic leaving groups from acetal centres that otherwise would give rise to solvent-equilibrated oxocarbenium ions is illustrated in Figure 3.28.

Systems based on bis-salicylacetals, $RCH(O\textit{o}C_6H_4COOH)_2$, show bell-shaped pH–rate profiles compatible with combined intramolecular general acid–nucleophilic catalysis, and rate enhancements of 3×10^9 compared with their dimethyl esters.[136] However, when the carboxyl group becomes electronically insulated from the oxocarbenium ion fragment and the leaving group, this factor falls[137] to 4×10^4. At the time the systems in Figure 3.29 were constructed, however, the now-discredited ion-pair model for the

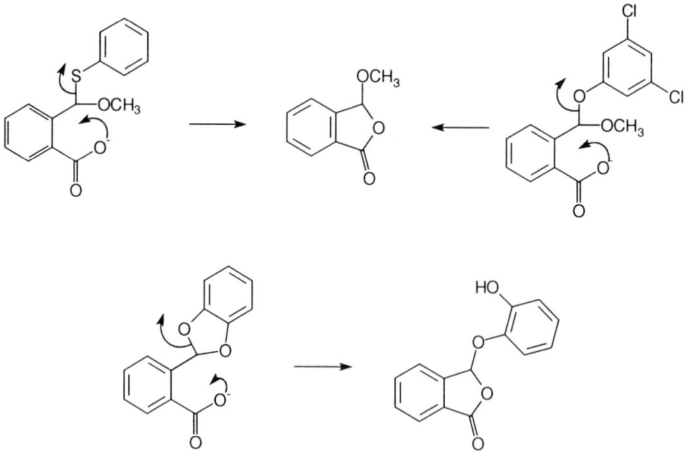

Figure 3.28 Nucleophilic participation by carboxylate in the departure of various leaving groups from mixed acetals of phthalic hemialdehyde Rate enhancements of these spontaneous processes of 100 for 3,5-dichlorophenolate departure,[133] 22 for thiophenolate departure[134] and 20 for catechol monoanion departure[135] were estimated from the rates of the *p*-phthalic derivatives; intermolecular general acid catalysis of the catechol was also accelerated.

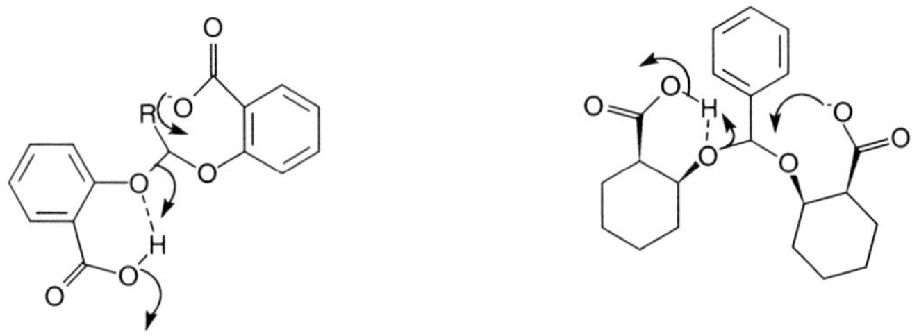

Figure 3.29 Systems showing synchronous intramolecular general acid and nucleophilic catalysis.

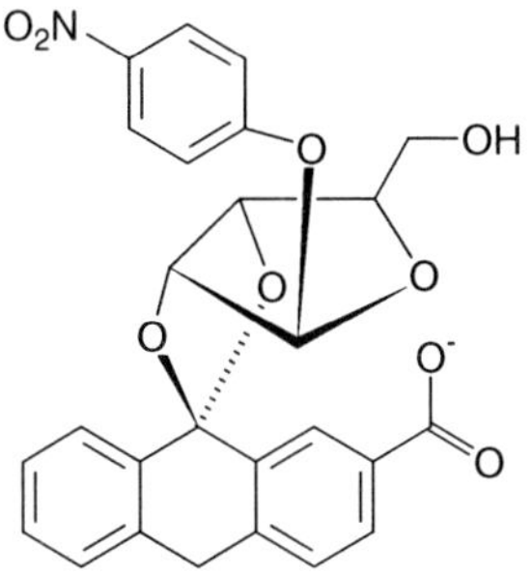

Figure 3.30 Model system showing electrostatic, but not nucleophilic, catalysis.

glycosyl–enzyme intermediate in lysozyme had such a hold on the thinking of workers in the field that the original papers describing these systems refer to electrostatic rather than nucleophilic catalysis, even though there was no attempt to distinguish between the two.

Electrostatic effects, however, can be important and rationalise the effects of hydroxyl epimerisation on rates of glycoside hydrolysis (Section 3.5.1). Many glycosidase hydrolase families have not two but three catalytic groups, the function of the third one being electrostatic in some sense. Czarnik's system[138] (Figure 3.30), based on calculations which suggested that apposition of the negative charge of a carboxylate to the lone pairs of the ring oxygen would promote oxocarbenium ion formation, is particularly relevant, since nucleophilic participation is sterically impossible, yet loss of *p*-nitrophenolate from the anion is 860 times faster (at 100 °C) than from the analogous compound with the carboxylate removed.

3.12.3 Participation by Ionised Sugar Hydroxyls – Base-catalysed Hydrolysis of Glycosides

Aryl glycosides, particularly those of acidic glycones, react readily in alkali, to the extent that deprotection of 2,4-dinitrophenyl glycosides is experimentally demanding. Most pathways involve participation by ionised hydroxyl groups. The effect of the aryl substituent on the rates of hydrolysis of aryl α- and β-glucopyranosides gave Hammett ρ values of 2.8_1[139] and 2.4_8,[140] respectively, corresponding to β_{lg} values of -1.2_8 and -1.1_3, with the α-compounds hydrolysing about two orders of magnitude more slowly than their anomers. Isolated products are predominantly 1,6-anhydroglucose,[62] although small amounts of glucose, and molecular rubble derived from it, such as isosaccharinic acids, are formed, by direct attack of hydroxyl. Removal (or methylation) of the 2-OH makes the reactivity of deoxy-β-glucopyranosides similar to those of the α-anomers. That 2-deoxy-β-glycopyranosides are not completely inert suggests that the less reactive systems may hydrolyse by direct attack of OH^- on the anomeric centre.

These data indicate that the reactions of the β-anomers proceed through the mechanism in Figure 3.31, in which the react initially to give a 1,2-anhydro sugar (whose protected derivatives are comparatively stable compounds). This has only access to the $^5H_4$, $^4H_5$, $^{3,O}B$ and $B_{3,O}$ conformations, since the epoxide ring ensures the coplanarity of C3, C2, C1 and O5. In the $^5H_4$ and $B_{3,O}$ conformations, the ionised 6-OH is ideally placed to open the 1,2-epoxide. The reactive conformations of the glycosides themselves are probably somewhere on the skew-boat pseudorotational itinerary around $^OS_2$.

Aryl α-glucopyranosides react by direct displacement of the substituent at C1 by the ionised 6-OH, although this cannot happen without a change in ring conformation. At very high base concentrations (~1.5 M NaOH), the 6-OH should be completely ionised and the reaction rate should not increase further with increasing base concentration; however, above 1.5 M NaOH, the rate of hydrolysis of phenyl α-glucopyranoside increases in a more than first-order fashion,[142] suggesting a direct displacement by hydroxide ion at the anomeric centre, which, however, is of minor importance under less extreme conditions. Galactosides of both anomeric configurations behave as their glucosyl compounds.

Aryl pentopyranosides are hydrolysed in base, but give a range of products from reaction of the reducing sugar with base.[62]

The dihedral angle of close to 180° between C1–OR and C2–OH in α-mannopyranosides in their preferred $^4C_1$ conformation makes participation of the ionised 2-OH in displacements of the aglycone particularly ready: α-D-Man*p*OPh is 20 times as reactive to base as β-D-Glc*p*OPh. Generally, aryl α-mannosides are between two and three orders of magnitude more reactive to base than their β-anomers (Figure 3.32). Very large negative β_{lg} values (-1.2_1 and -1.4_3) for aryl α- and β-mannopyranosides, respectively, indicate the C1–OAr bond has largely cleaved at the transition state.[143] The production of significant quantities of 1,6-anhydromannose from phenyl β-mannopyranoside suggests participation by the 4-OH.[62]

The near −180° dihedral angle between the participating ionised 2-OH and the leaving group is found in *trans*-furanosides in addition to α-mannopyranosides, which are accordingly much more base-labile. Extensive data for phenoxide leaving groups are not available,[144] but the base-catalysed departure of pyridine leaving groups from the α-L-arabinofuranosyl system is around 1600 times faster than from the β-D-galactopyranosyl system at 25 °C,[47,145] in line with the base-lability of NAD^+.[146] Studies of methyl glycosides under

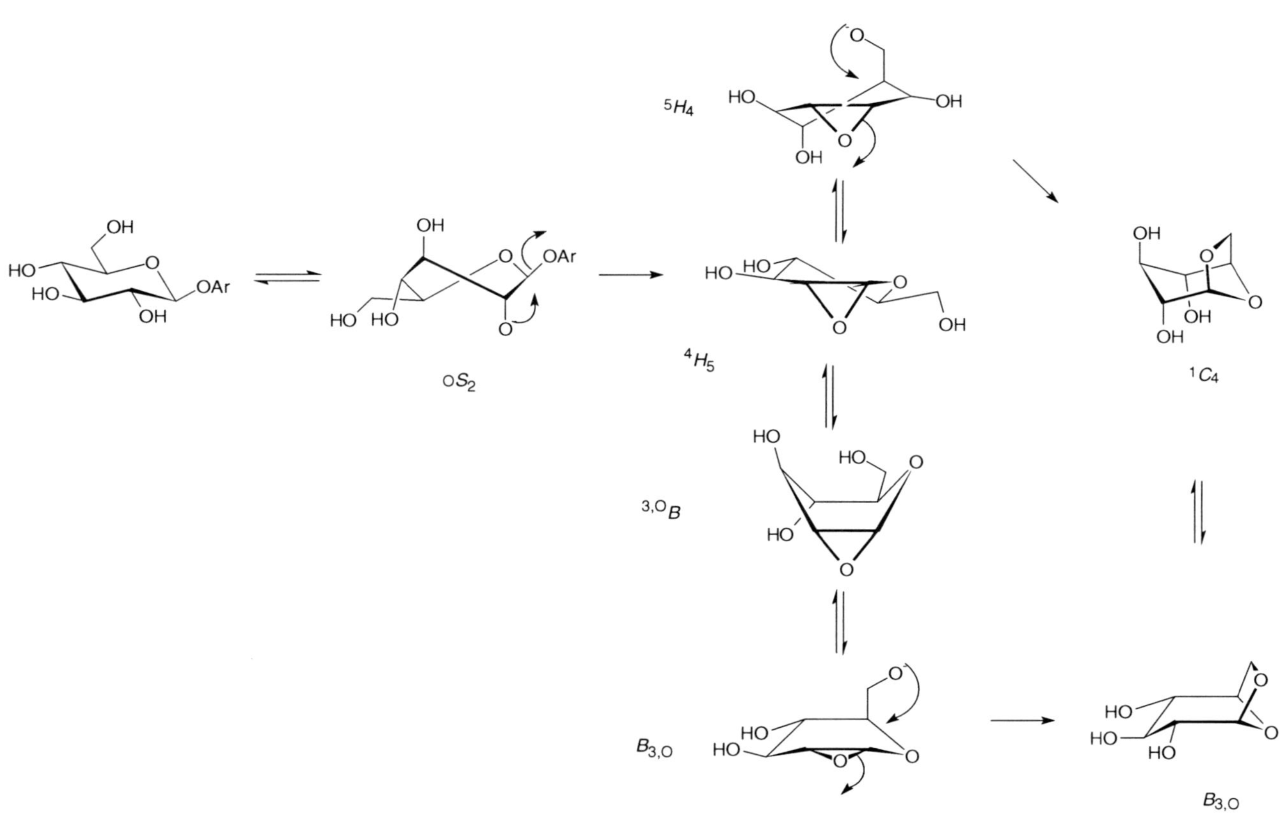

Figure 3.31 Hydrolysis of aryl β-glucopyranosides in alkali. They cannot react from the ground-state $^4C_1$ conformation, but the reactive conformations shown, although reasonable, are speculative. The two conformations of the 1,6-anhydroglucose product shown are both occupied and in the solid state give rise to different crystal forms.[141]

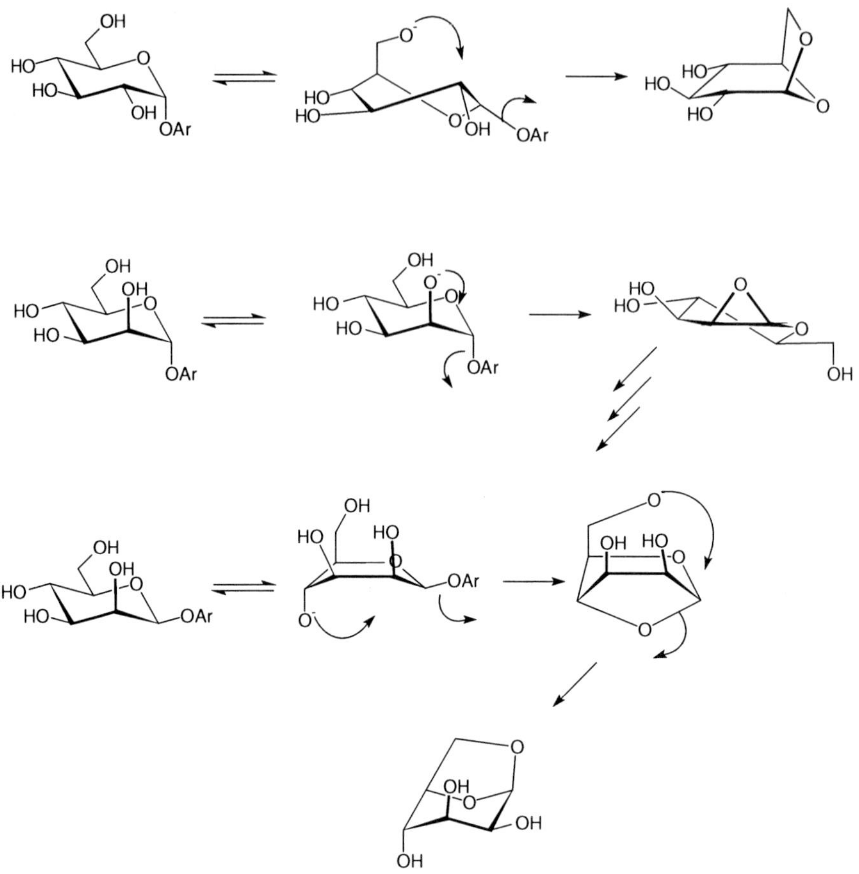

Figure 3.32 Base-catalysed hydrolyses of aryl α-glucopyranosides and aryl α- and β-mannopyranosides.

conditions mimicking the Kraft process for pulping of wood chips (170 °C, 10% aqueous NaOH) showed that furanosides with a 2-OH *trans* to the aglycone were hydrolysed 5–20 times faster than the pyranosides of the same sugar and anomeric configuration.[147] Small amounts of methyl arabinopyranosides found in the products of base-catalysed hydrolysis of methyl α-L-arabinofuranoside suggest the incursion of the ring-opening mechanism in Figure 3.33.[148]

Not all apparently hydrolytic cleavages of glycosyl derivatives in base in fact involve reaction at the anomeric centre. The anomeric anion of a reducing sugar ($pK_a \approx 12$) can be displaced from disaccharide alditols such as maltitol and cellobiitol by an ionised hydroxyl group under the violent conditions of Kraft pulping,[62] and the base-lability of *p*-hydroxybenzyl glycosides arises from the generation of a quinone methide.[62] Moreover, electron-deficient aromatic systems used as leaving groups can be attacked at the aromatic moiety by OH^-. This appears to be largely the case for nucleosides, although complex ring opening and reclosing reactions of the aromatic rings can regenerate certain

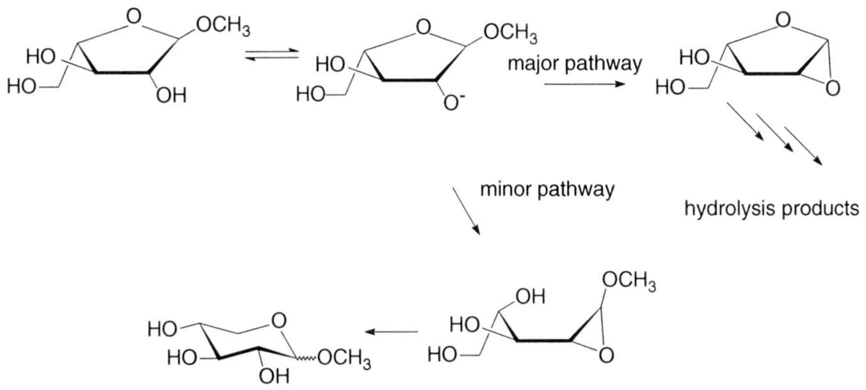

Figure 3.33 Partial ring opening of a furanoside ring in base.

purines and give the appearance of an attack at the anomeric centre.[149] Likewise, attack on the pyridinium rings of α-glucosyl- and α-xylopyranosyl-pyridinium salts is a minor pathway of their alkaline hydrolysis.[40] Glycosides of *o*- and *p*-nitrophenol can react via formation of cyclic Meisenheimer complexes which permit the aryl group to migrate round the sugar ring and nitrophenoxide to be produced in $E1_{CB}$-like reactions.[150] Therefore, monitoring the appearance of nitrophenolate will overestimate the rate of the anomeric centre reaction for these acidic phenols. The very negative measured β_{lg} values for alkaline hydrolysis of aryl glycosides (more negative than for spontaneous hydrolysis) may therefore be an artefact of other pathways for the generation of anions of acidic phenols such as nitrophenol (Figure 3.34).

3.13 REACTIONS IN ORGANIC MEDIA

Most organic solvents are less polar than water, so that if intimate ion pairs of glycosyl cations and anions are "too unstable to exist" in water, *a fortiori* they have no real existence in organic solvents and mechanistic proposals which invoke them are simply in error. Even the reality of the ion–molecule complexes shown to exist in water must be questioned. Nonetheless, many literature sources draw intimate ion pairs and even solvent-equilibrated oxocarbenium ions, in the absence of any direct evidence.

3.13.1 Solvolyses

In early work on the methanolysis of tetra-*O*-methyl α-D-glucopyranosyl and -mannopyranosyl chlorides, it was shown that the reactions were cleanly first order whether followed by titration of the acid produced or by polarimetry.[151] The *gluco* chloride gave exclusively the β-methyl glucoside, the *manno* an anomeric mixture. Reactions of the *gluco* but not *manno* chloride with thiophenoxide ion were cleanly bimolecular and gave exclusive inversion, as expected.

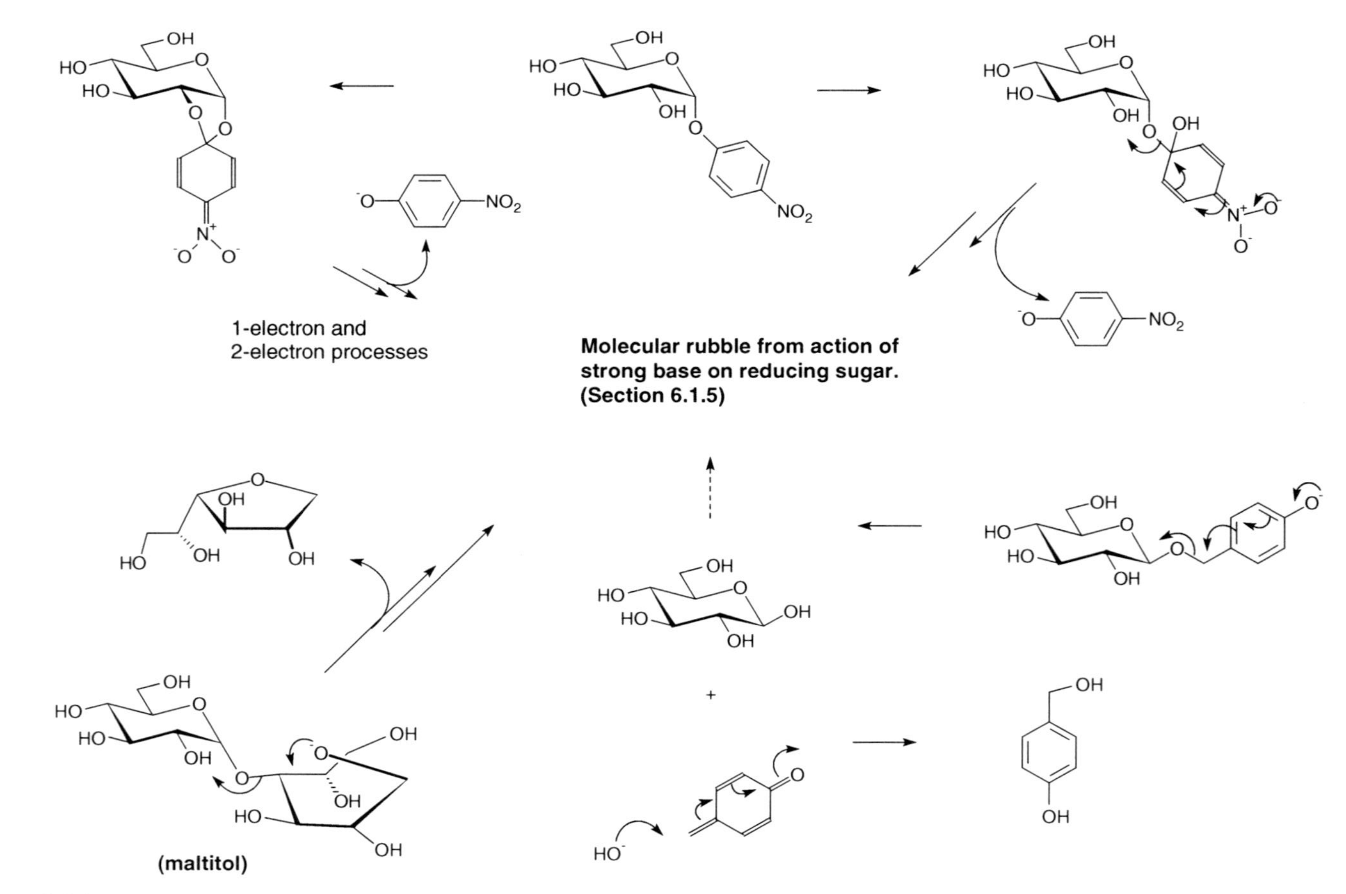

Figure 3.34 Some pathways for base-catalysed glycoside cleavage not involving reaction at the anomeric centre.

Since the solvolyses were accelerated by small quantities of added sodium methoxide only to the same degree as by other sodium salts, it was thought that the solvolyses were S_N1. The effect of addition of salts of the anion of the solvent was widely used as a mechanistic criterion at the time (1960s), with the absence of an effect of small ($\sim$0.1 M) amounts of lyate ion being considered a good indicator of an S_N1 reaction. In fact, if the reactions are S_N2, but with "exploded" transition states which result in low β_{nuc} values, no such effect of added lyate ion is to be expected. If we assume that the pK_a values of MeOH and MeO^+H_2 differ by approximately the same 17.4 units as H_2O and H_3O^+,[xxi] then a β_{nuc} value of 0.1 would result in bimolecular rate constants for attack by MeOH and MeO^- differing by a factor of 50. Methanol is 25 M in itself, so 0.1 M methoxide would result in a rate increase of only 20%, difficult to see above an ordinary salt effect.

In solvolyses of acetyl-protected halides, neighbouring group participation where leaving group and 2-acetoxy group were *trans* could be demonstrated:[152] such participation remains the bane of β-mannopyranoside synthesis. Under certain conditions, the dioxolanyl cations can be intercepted by alcohols to give orthoesters. Solvolysis of tetra-*O*-acetyl-α-D-glucopyranosyl bromide in various alcohols uncovered a pathway for formation of apparently retained glycoside product–epimerisation of the starting material by bromide ion liberated in the early stages of the reaction.[153] This pathway probably accounts for the formation of orthoesters from *cis*-2-acyloxyglycosyl bromides under certain conditions, although *syn* attack may be involved to some extent (see below). Pathways for reaction of *cis* and *trans* acyl glycosyl halides are given in Figure 3.35.

The solvolyses of a number of α- and β-glucopyranosyl derivatives in mixtures of ethanol and trifluoroethanol brought to light features not readily understood in terms of classical S_N1 and S_N2 chemistry.[154] For all leaving groups except fluoride, the products were largely, but not exclusively, inverted, with the preference for inversion being greater for the β-anomer of a particular substrate pair. However, both retained and inverted products showed a several-fold preference for the ethyl glycosides, giving β_{nuc} in the approximate range 0.2–0.3, even for retention. This pattern of behaviour was observed with protonated phenol as a leaving group; phenol is less nucleophilic than even trifluoroethanol, so no reversibly formed glucosyl cation–leaving group complexes are involved. The data therefore compel the model for transition states of Figure 3.36, in which the α-face preference arises from the ability of the 2-OH to stabilise an attacking alcohol molecule either by hydrogen bonding and/or electrostatically. Preferences for attack by ethanol rather than trifluoroethanol are unlikely to be simply nucleophilic, since derived β_{nuc} values are far too large. One possibility is that substrate molecules whose hydroxyl groups are solvated by ethanol rather than trifluoroethanol are more reactive because the hydroxyl groups are less electron withdrawing.

[xxi] $K_a(H_2O) = [H^+][OH^-]/[H_2O] = 10^{-15.7}$, since water is 55 M in itself, and $[H^+][OH^-] = K_w = 10^{-14}$; $K_a(H_3O^+) = [H^+][H_2O]/[H_3O^+] = 55$ M.

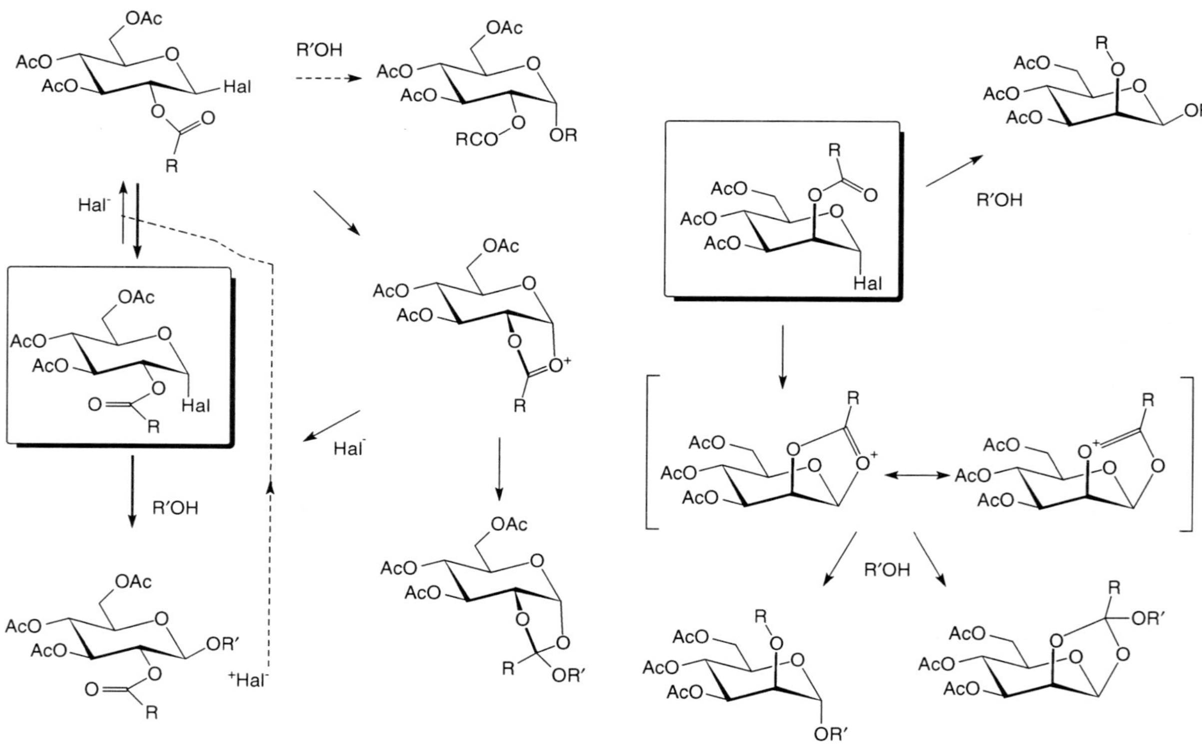

Figure 3.35 Reactions of acylated glycosyl halides under solvolysis conditions. The two canonical forms for the dioxanyl cation are drawn only with mannose and only the predominant *exo* forms of the orthoester products are shown.

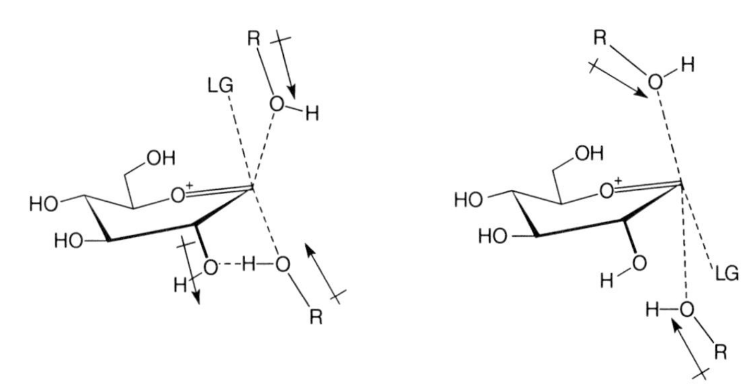

Figure 3.36 Qualitative transition states for solvolysis of glucopyranosyl derivatives. The dotted lines are meant to convey a minimal degree of covalent bonding: kinetic isotope effect data for hydrolysis of α-glucosyl fluoride were fitted to a transition state structure with bond orders to leaving group and nucleophile of only 0.001.

This model, and the well-known hydrogen bonding requirements of fluoride ion, permit the behaviour of the fluorides to be interpreted. Both anomeric fluorides showed a chemoselectivity for retention of close to unity and an inversion/retention ratio for trifluoroethyl glycoside formation of 0.13; neat trifluorethanolysis of α-glucosyl fluoride yields 88% retained product. The fluoride ion hydrogen bonds to the most acidic component of the solvent and then performs an internal return reaction in which the leaving group $F^- \cdots HOCH_2CF_3$ recombines with the cationic fragment on oxygen. This explanation was confirmed by the major retained product of solvolysis of α-glucosyl fluoride in an equimolar mixture of ethanol, trifluoroethanol and phenol being phenyl α-glucoside, derived from the most acidic component of the mixture. Such internal return reactions are well known in solvolysis chemistry – chiral *exo*-norbornyl tosylate racemises 2–3 times faster than it acetolyses[155]–and are perhaps best regarded as the system moving over a plateau on a free energy surface with the carbenium fragment carrying almost a full positive charge (Figure 3.37).

Most mechanistic thinking in the area of glycoside synthesis is shaped by a seminal paper by Lemieux *et al.*, in which an array of solvent-separated and intimate ion pairs is invoked, in heroic defiance of Occam's razor[xxii].[156,157] The key discovery was that, whereas tetraacetyl α-D-glucopyranosyl bromide in a mixture of phenol and pyridine gave the β-pyridinium salt exclusively, in anhydrous pyridine anomeric mixtures were produced, and addition of tetraethylammonium bromide gave exclusively the α-anomer (Figure 3.38). Similar

[xxii] A literal translation of William of Oakham's words is "It is vain to do with more what can be done with fewer", not the commonly quoted "Entities are not to be multiplied without necessity".

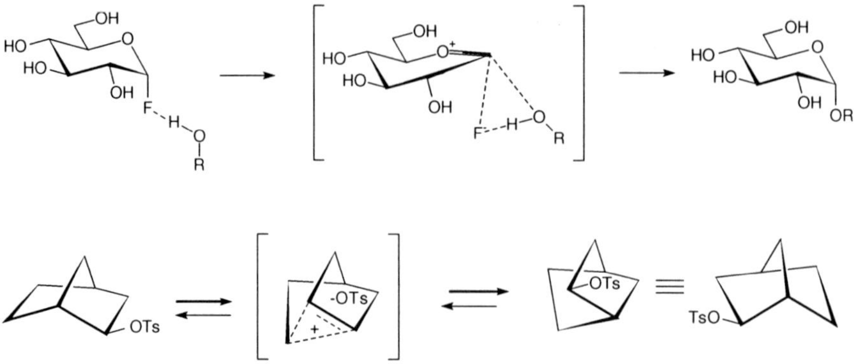

Figure 3.37 Internal return in the solvolysis of α-glucosyl fluoride and in a classical example from solvolysis chemistry, the acetolysis of *exo*-norbornyl *p*-toluenesulfonate.

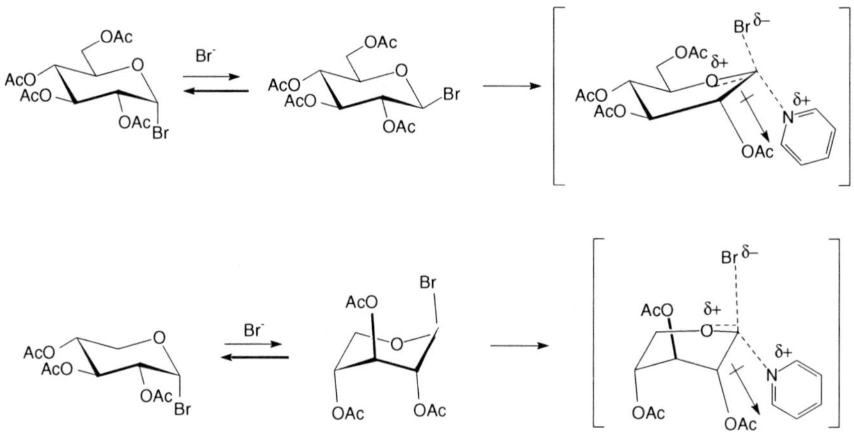

Figure 3.38 Reaction of acetylated glycosyl halides in pyridine.

results were obtained in the *galacto* and *xylo* series, the tendency to give acetylated α-xylopyranosylpyridinium ion being particularly pronounced. These results can be rationalised by simple S_N2 reactions in which the relative selectivities for bromide and pyridine are on the two faces of the sugar ring are different, pyridine being a more effective nucleophile, relative to bromide, on the α-face. An obvious reason for this would be that the developing positive charge on the pyridine is stabilised electrostatically by the negative end of the C2–O2 dipole. β-D-Glucopyranosyl and -galactopyranosyl bromides are in the $^4C_1$ conformation (although they probably react through skew conformations), whereas β-D-xylopyranosyl bromide is in the $^1C_4$ conformation, in which leaving group and C2 substituent are *trans* diaxial. In accord with this explanation, α-2-deoxyglucopyranosylpyridinium salts cannot be synthesised by the halide ion epimerisation route.[42] The role of the phenol in the production of the β-pyridinium salts appears to be to hydrogen bond to the bromide ion and make it less nucleophilic and therefore unable to carry out the epimerisation.

Whatever the objections to Lemieux *et al.*'s mechanistic rationale, though, the pyridinolysis results did lead to the first general synthesis of α-glycopyranosides. The starting point was the axial glycosyl halide (readily available as a thermodynamic product because of the anomeric effect), a source of halide ions, the glycosyl acceptor and a non-nucleophilic weak base. As alcohols are less nucleophilic than pyridine, they cannot compete with acyloxy neighbouring groups at C2 and benzyl or other weakly/non-participating groups are employed in the synthesis of 1,2-*cis*-glycosides.

3.13.2 Synthesis of Glycosides

The importance of oligosaccharide fragments in biological molecular recognition has, over the last 30 years, prompted the exposition of a bewildering array of methods for constructing glycosidic linkages. Most of these methods involve nucleophilic attack by a derivative of the aglycone ("glycosyl acceptor") on the anomeric centre of a sugar derivative ("glycosyl donor"). However, glycosidation protocols are frequently put forward in the manner of the recipes of celebrity chefs, with the evidence for any mechanism too often "anecdotal or circumstantial",[158] and propounded in ignorance of physical organic studies. It is possible, though, to discern some rational patterns in contemporary protocols.

3.13.2.1 Reaction of Phenoxides with Glycosyl Halides in Organic and Aqueous–Organic Solutions. The first glycoside to be chemically synthesised was phenyl β-D-glucopyranoside, by reaction of sodium phenoxide with 2,3,4,6-tetra-*O*-acetyl-α-D-glucopyranosyl chloride (acetochloroglucose) in ethanol (the basic conditions used removed the acetyl protecting groups from the sugar).[159] The use of an aqueous solvent followed in 1909,[160] with Mauthner[161] in 1915 adding the final refinement of change of leaving group to bromide and homogeneous solution (aqueous acetone). Mauthner's conditions are still in use and the likely reason they work is that the reaction goes through the familiar "exploded" S_N2 transition states. Like any sort of Schotten–Baumann reaction, the Michael glycosylation (as it is now called[162]) depends on the low β_{nuc} of the reaction of interest for its success.

Consider a glycosyl donor, GlyZ, reacting exclusively with the anions of two potential acceptors, HX and HY, whose acid dissociation constants are K_X and K_Y. For simplicity we consider the solution to be buffered at a pH well below the lower of pK_X or pK_Y. We can then write

$$[X^-]/[Y^-] = K_X[HX]/[HY]K_Y \tag{3.14}$$

For simplicity we also assume that reaction of GlyZ takes place only with the anions, so that eqn (3.15) holds, where k_X and k_Y are the second-order rate constants for reaction of X^- and Y^- with GlyX:

$$[GlyX]/[GlyY] = k_X[HX]K_X/k_Y[HY]K_Y \quad (3.15)$$

However, by the definition of β_{nuc}, eqn (3.16) also holds:

$$\log(k_X/k_Y) = \beta_{nuc}\log(K_Y/K_X) \quad (3.16)$$

so that we can write

$$\begin{aligned}\log([GlyX]/[GlyY]) &= (1-\beta_{nuc})\log(K_X/K_Y) + \log([HX]/[HY]) \\ &\equiv (1-\beta_{nuc})\Delta pK + \log([HX]/[HY])\end{aligned} \quad (3.17)$$

Consequently, with a β_{nuc} of 1.0, products are produced in the ratio of the acceptors, but as β_{nuc} approaches zero, more and more of the product is derived from the *stronger* acid. Nucleophilic displacements at anomeric centres are characterised by low β_{nuc} values. Hence it is possible to obtain good yields of aryl glycosides in aqueous media.

The Michael glycosidation gives decreasing yields as the pK_a of the phenol falls below 6, probably because of interception of the glycosyl halide by water (as distinct from hydroxide) in preference to phenolate. Good yields are obtained, however, in dipolar aprotic solvents, sometimes very simply and crudely (di- and trinitrophenyl glycosides can be made by boiling the phenol with the acetohalo sugar in acetone containing anhydrous potassium carbonate).[163]

The stereochemistry of the reaction of halide with phenolate is inversion. In the case of the formation of 1,2-*cis*-glycosides, competing neighbouring group participation by the substituent on C2 in the starting halide is more important in aqueous media, and to make aryl α-glucopyranosides and galactopyranosides it is necessary to start from the β-halide and use the pre-formed phenolate in dipolar aprotic solvents: likewise, α-mannopyranosyl halides in hexamethylphosphoramide give aryl β-mannopyranosides.[164]

3.13.2.2 Leaving Groups. Generally, there are two approaches to activation of the anomeric position to nucleophilic substitution by nucleophiles weaker than phenoxide. The first is to use an anomeric mixture or an equilibrating system and rely on other features such as neighbouring group participation to control stereochemistry. The approach was first exemplified by the Helferich fusion techniques for making aryl glycosides – fusion of fully acetylated sugars with phenols in the presence of *p*-toluenesulfonic acid gave protected β-glycosides[165] whereas use of anhydrous zinc chloride gave their anomers.[166] The α-glycosides are the thermodynamic product because of the anomeric effect, so that they are formed predominantly under more vigorous conditions: the β-anomers are formed first because of neighbouring group participation, although the conditions are vigorous enough to ensure epimerisation of the anomeric acetates but not the aryl glycoside products. Lemieux *et al.*'s synthesis of α-glycosides discussed in Section 3.13.1 is another example.

A second approach is to use other factors to control the stereochemistry of the leaving group and rely on Walden inversion to control the stereochemistry of the product. As we have seen, this may not give complete stereochemical control in unfavourable cases and it is often advisable to have a "backup" phenomenon that would give the same stereochemistry. The use of acetohalo sugars to give 1,2-*trans* diequatorial glycosides is the classic example, as in the Koenigs–Knorr reaction: because of the anomeric effect, the axial acetohalo sugars are produced as the thermodynamic products from reaction of the hydrogen halide in anhydrous acetic acid with the acetylated sugar. The 1,2-*cis*-glycosyl halides then react either react directly with Walden inversion or through the dioxanyl cation formed by participation of the 2-acyloxy group, in both cases giving the desired product.

Of course, in the furanoside series the anomeric effect cannot be used so predictably to control stereochemistry, so that stereocontrol of furanoside synthesis is less well understood than that of pyranoside synthesis.

Halides (except fluoride) have the disadvantage that they are themselves nucleophilic and can epimerise the starting halide. Promoters are often added that serve both to activate the halide leaving group and to complex it once departed and prevent further reaction. Examples are the silver salts (carbonate or oxide) used in the classic Koenigs–Knorr reaction with chlorides and bromides: Cl^-, Br^- and I^- are "soft" nucleophiles which complex strongly with the lower B group metals by virtue of the overlap of the full d shells on the metals with the empty ones on the halide. The practice has been to use solid silver oxide or silver carbonate to also neutralise the acid formed in the glycosylation. If homogeneous reaction conditions are desired, the Helferich catalyst of mercuric cyanide and mercuric bromide[167] can be used, but only on a laboratory scale, as the base is cyanide ion which gives toxic HCN. Fluorides have recently been used as glycosyl donors: Sn^{II} and the lanthanide ions[168] form tight complexes with F^-, and its electrophilic salts are used as activators. Promoters can be very exotic: cyclopentadienylhafnium dichloride is also used as a fluoride activator. Glycosyl iodides can be used as glycosyl donors, but only for α-glycoside synthesis.[162] This is not surprising, since iodide is one of the most powerful nucleophiles available, and the product of a Lemieux halide-catalysed glycosylation is produced whether intended or not.

The glycosyl trichloroacetimidates introduced by Schmidt and Kinzy are a great advance (Figure 3.39).[169] Because of the greater acidity of the anomeric hydroxyl than ring hydroxyls, fully acylated sugars can be selectively deprotected at the anomeric position by, for example, hydrazine.[170] Reaction of the anomeric anion under kinetically controlled conditions yields the equatorial trichloroacetimidate, as the relatively unhindered equatorial hydroxyl is a better nucleophile towards the bulky trichloroacetonitrile (the success of the method may depend on the large size of the chorine atoms rendering the electrophilic carbon effectively neopentyl like). However, if trichloroacetimidate formation is made reversible (for example, by treatment with sodium hydride, which generates the ion), the axial trichloroacetimidate, favoured by the anomeric effect, predominates. Allowing derivatives of general type

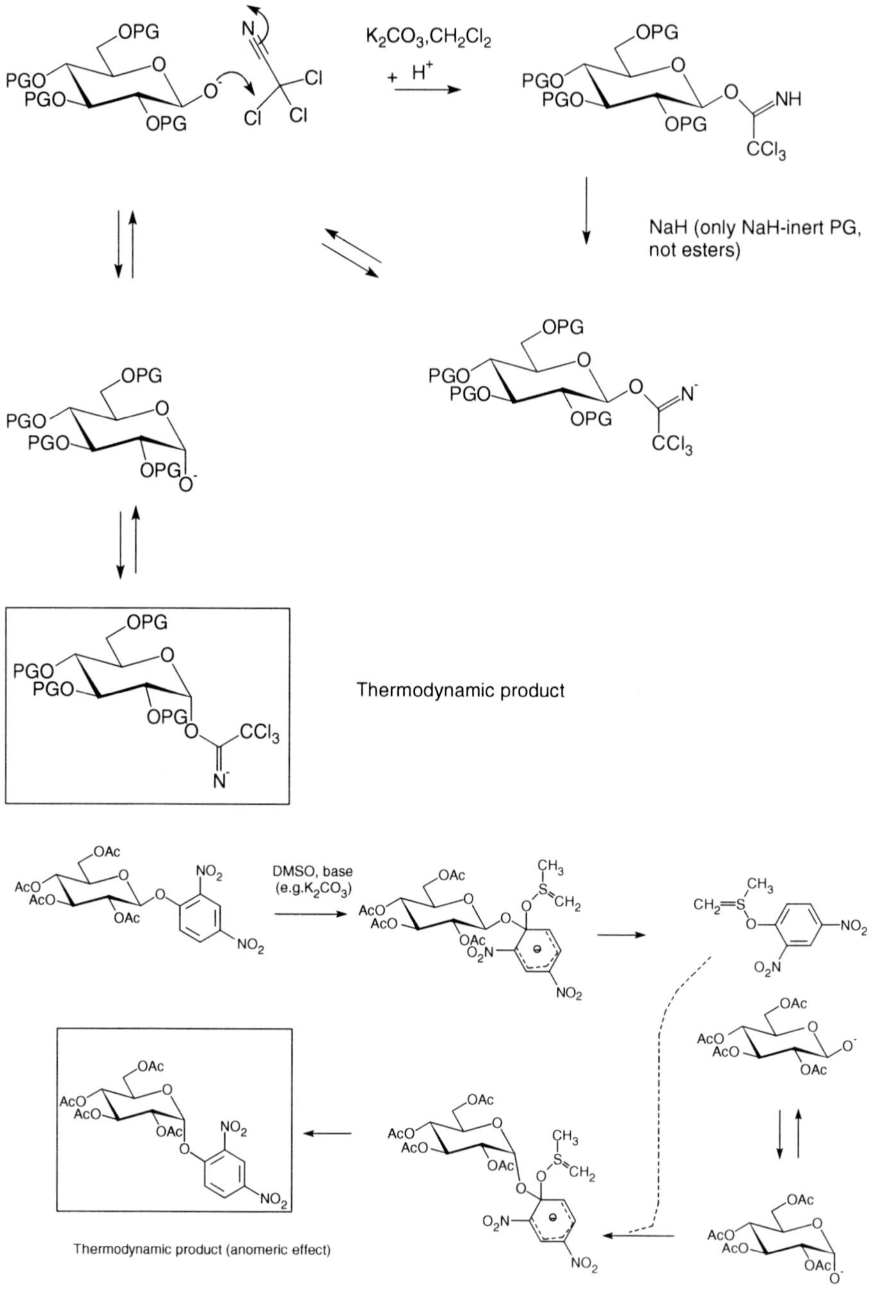
K2CO3,CH2Cl2
+ H+
NaH (only NaH-inert PG, not esters)
Thermodynamic product
DMSO, base
(e.g.K2CO3)
Thermodynamic product (anomeric effect)

Figure 3.39 Formation of anomeric pairs of glycosyl trichloroacetimidates and 2,4-dinitrophenolates. Anhydrous caesium carbonate is also used for trichloroacetimidate formation. In the 2,4-dinitrophenolate epimerisation, nucleophilic attack by oxygen rather than sulfur is proposed since sp^2-hybridised carbon is a "hard" electrophile.

Gly–O–X to reach anomeric equilibrium by reversible cleavage of the O–X bond is also the major route to 2,4-dinitrophenyl α-glycopyranosides.[171]

The trichloroacetimidate anion is not a good enough leaving group to depart by itself, although phenols acidic enough to protonate the leaving group, such as 3,4-dinitrophenol,[172] need no additional acid catalyst. Common activators are $F_3B.OEt_2$ and $(CH_3)_3SiOSO_2CF_3$ ("TMS triflate").[173] Stereochemistry is determined by Walden inversion and/or neighbouring group participation; with a weakly participating group such as benzyloxy at C2, β-mannopyranosides can be made.[173]

Trichloroacetimidates have to be made just before use, but other glycosyl donors can be made in latent form, and then activated during the glycosylation reaction. Two oxygen leaving group precursors based on this principle are vinyl glycosides[174] and 4-pentenyl glycosides (Figure 3.40).[175] Vinyl glycosides are made as allyl glycosides and then isomerised with an organometallic catalyst: treatment with the acceptor and a Lewis acid probably results in *C*-protonation and departure of the leaving group as a ketone. The pentenyl glycosides are activated by the addition of bromine, attack of O1 on the bromonium ion yields a trialkyloxonium ion and the effective leaving group is an ether.[xxiii]

Glycosyl phosphites, activated by catalytic amounts of trifluoromethyl triflate, are also glycosyl donors with oxygen leaving groups.[176]

Thioglycosides and their sulfoxide oxidation products and selenoglycosides have been used as latent glycosyl donors. The thioglycosides are activated by "soft" electrophiles such *N*-bromo- or *N*-iodosuccinimide (with trifluoromethanesulfonic acid) or salts of the interesting hypermethylated disulfide Me_2S^+–SMe.[173] The immediate glycosidation precursors appear to have sulfur leaving groups such as sulfenyl halides (from Gly–S^+Hal–R) or dialkyl disulfides (from Gly–S^+R–SMe).

This situation does not apply to sulfoxides, made by mild oxidation of the sulfides. They are usually activated with trifluoromethanesulfonic anhydride in the presence of an acid acceptor such as 2,6-di-*tert*-butylpyridine, which is sterically prohibited from being a nucleophile. The originators of the method considered it to produce a solution, stable at low temperatures, of protected glycosyl cations. The reality is that the glycosylating agent is the covalent glycosyl triflate,[177] overwhelmingly in the axial orientation because of the powerful anomeric effect of this very electronegative leaving group. The reaction of a 4,6-benzylidene-protected mannosyl donor, generated from the sulfoxide and Tf_2O, with a typical carbohydrate acceptor was shown to exhibit an α-deuterium kinetic isotope effect of 1.33 at −60 °C,[178] the usual temperature of such couplings. It corresponds to a value of 1.23 at 25 °C, at the upper end of

---

[xxiii] The rate of spontaneous hydrolysis of α-D-glucopyranosyl-3-bromopyridinium ion, with a leaving group pK_a of 2.8, is $2 \times 10^{-7}\,s^{-1}$ at 25 °C. The pK_a of a protonated tetrahydrofuran will be between −2 and −3. S_N1-like reactions involving departure of a neutral leaving group are relatively solvent insensitive, although they are somewhat accelerated in less polar solvents. With a β_{lg} of −1 the lifetime of the trialkyloxonium salt is thus estimated to be on the order of seconds.

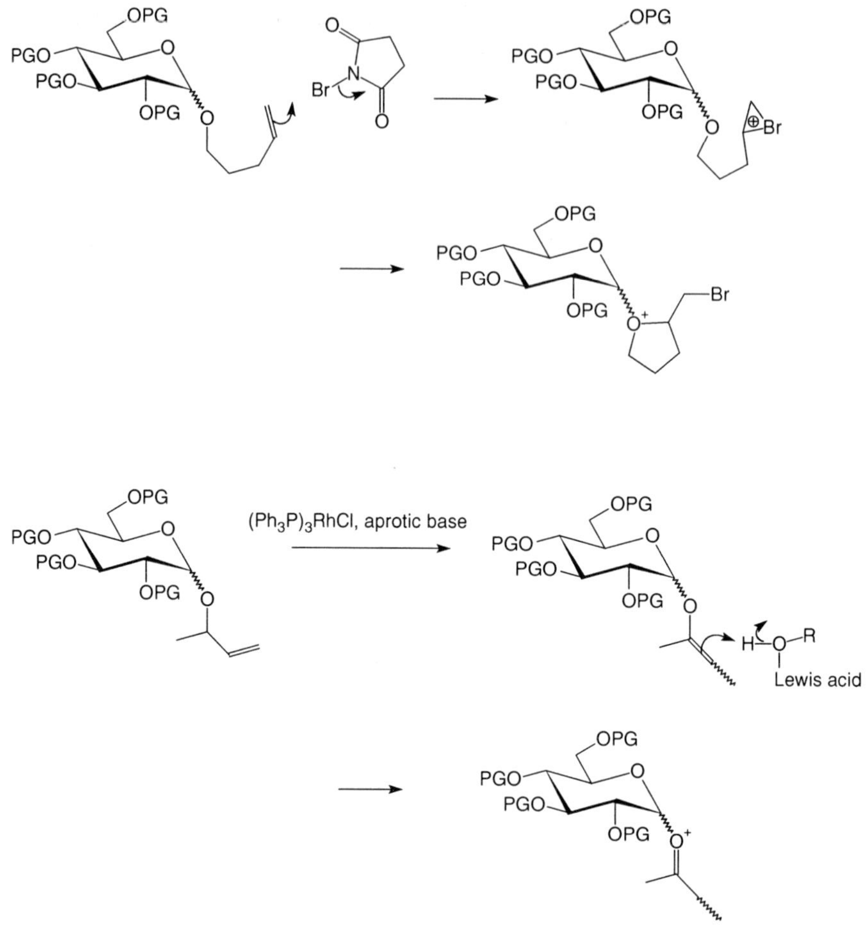

Figure 3.40 Use of pentenyl glycosides and vinyl (allyl) glycosides as glycosylation donor precursors.

the range of such effects observed with the familiar "exploded" transition states at oxocarbenium ion centres.

Additions of an electrophilic species E–X to glycals result in the electrophile becoming attached to position 2, and X to the anomeric position. Electrophiles where E can be converted to an amino or acetamido group (such as iodine azide equivalents) or reduced off (such as ArSeCl), and X is a halogen which can be displaced nucleophilically are particularly popular in glycosidation reactions starting from glycals (Figure 3.41). However, if there is a potential leaving group on position 3, the Ferrier rearrangement (see Section 6.7.3) may take place. Epoxidation of a glycal results in a protected 1,2-anhydro sugar (the same as are intermediates in base-catalysed glycoside hydrolysis), which ring strain makes a good glycosylating agent. The availability of neutral epoxidation reagents, such as dimethyldioxetane, made the route apparently attractive for

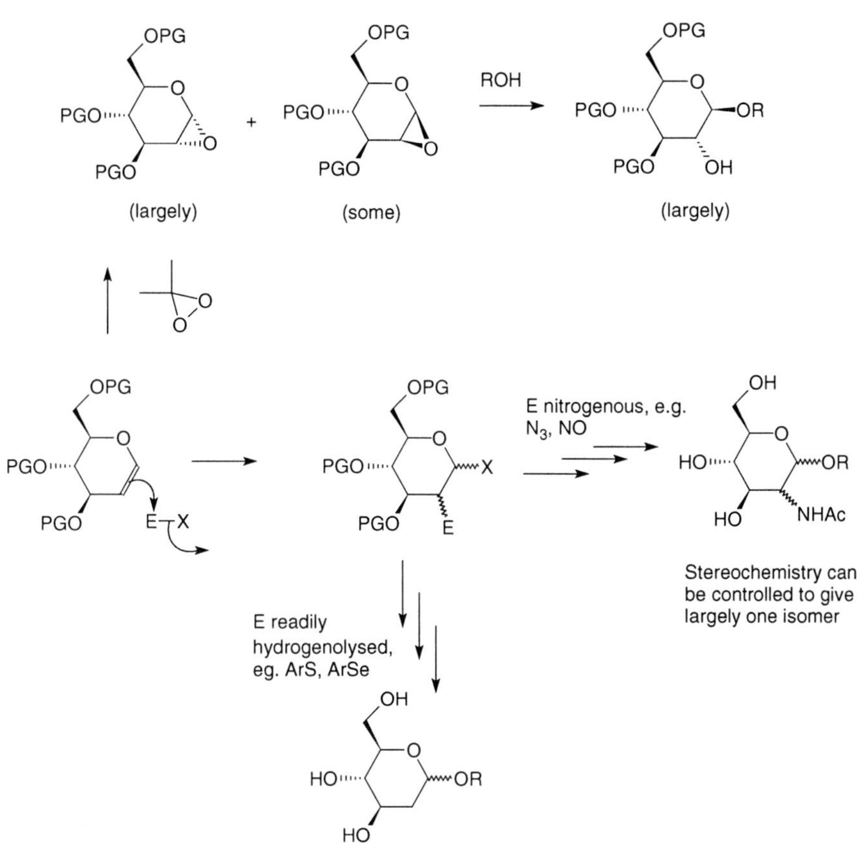

Figure 3.41 Glycosylations from protected glycals. PG=protecting group, E=electrophile.

solid-phase synthesis,[179] but the reality turned out to be that the initial epoxidation was not stereospecific, *manno* and *gluco* sugars being both produced from glucal.[180]

3.13.2.3 Effect of Protecting Groups. Not surprisingly, the electronic nature of protecting groups affects the rate of glycosidations, with the more electron-withdrawing groups reacting more slowly. Although competitive data pertaining to the bromine-mediated hydrolysis of pentenyl glycosides are available,[181] quantitative data on simple systems are sparse: acetolysis of tetra-*O*-acetyl-2,4-dinitrophenyl β-D-galactopyranoside is about 10-fold slower than that of the deprotected compound.[182] Benzyl and other alkyl protecting groups are accelerating, acyl groups decelerating. Obviously, the degree of activation or deactivation ("armed" or "disarmed") of the glycosyl donor can be manipulated by substituents in the protecting groups and complex syntheses in which selectivity is based of the effects of protection on reactivity are possible.[183]

In a 4,6-benzylidene group, the conformation about C5–C6 is constrained to *tg*, so that the positive end of the C6–O6 dipole points towards the developing positive charge on O5. Protection by cyclic acetals which enforce this conformation is thus strongly deactivating.

The group protecting the substituent at C2 may or may not participate as an intramolecular nucleophile, depending on its stereochemistry and its identity. Generally, acyloxy groups will participate if *trans* to the leaving group, although they can become more or less effective as nucleophiles, depending on the intermolecular competition. The nucleophilicity of acyloxy groups can be manipulated, *e.g.* by introducing nitro groups into benzoyl derivatives. Ethers and acetals are generally regarded as non-participating, but benzyl protection of mannose O2 does not guarantee formation of β-mannosides from α-mannosyl donors, and it seems likely that some degree of participation occurs.

The orthoesters derived from dioxanyl cations (Figure 3.35) were intensively investigated as glycosyl donors in synthesis in the Soviet Union,[184] but their use seems to have abated. Generally, the more basic the conditions, the more likely they are to form, suggesting attack on a cation stabilised with two oxygen atoms exhibits a larger β_{nuc} than the preassociation-type reactions at the anomeric centre. A combination of two masking strategies–orthoester and pentenyl glycoside–is possible.[185]

An acetamido group at C2 will, of course, efficiently participate in reactions at C1 if *trans* to the leaving group; if this is a halide, any epimerisation from free halide ion will also make *cis*-acylamido halides very unstable. Since Cl^- is a weaker nucleophile than Br^- for the epimerisation step, the most widely used fully acetylated GlcNAc donor is the chloride, not the bromide. Many workers in fact prefer to preclude any possibility of participation by working with the azide or with the nitrogen protected as an imide such as phthalimide, not an amide, during the coupling reaction.

The 4,6-benzylidene group restricts the conformational possibilities of the pyranose ring: skew conformations are restricted by the requirement for the O4–C4–C5–C6 dihedral angle to be 60° or less and $^4C_1$ is the only chair conformation possible. This is important in the synthesis of β-mannopyranosides. Skew conformations which promote the participation of O2 in displacements of α-leaving groups, even by normally "non-participating" protecting groups such as carbonate (see Figure 6.32), seem to be precluded. For this reason, 4,6-benzylidene protection (or similar) is a key feature of practical syntheses of β-mannopyranosides based on nucleophilic substitution at the anomeric centre (Figure 3.42).[186]

3.13.2.4 Effect of Solvent. The usual solvent for synthetic glycosidation reactions is dichloromethane. Acetonitrile is occasionally used, but it may be nucleophilic enough that *N*-glycosyl derivatives are formed transiently: the pK_a of protonated acetonitrile is –10.1.[187] A calculation similar to that in footnote xxiii suggests that an *N*- glycosyl nitrilium ion would have a lifetime of around microseconds in water at 25 °C, and obviously longer at the usual glycosidation temperature of –60 °C.

Increase of electron-withdrawing inductive (through-bond) and field (through-space) effects of OH by acylation

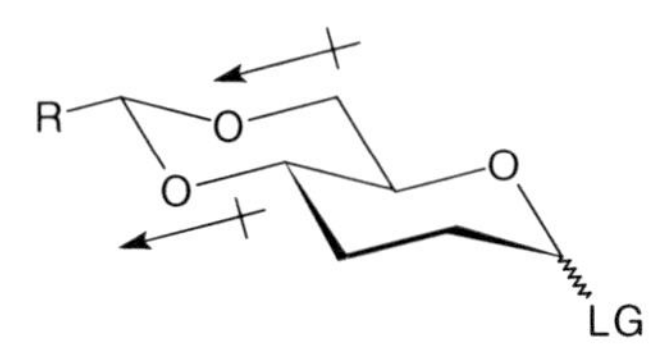

4,6-Acetalisation increases electron-withdrawing through space effect of 6-OH by constraining C5-C6 to *tg* conformation, with positive end of C6-O6 dipole pointing towards O5 and C1

Figure 3.42 Deactivation of glycosyl donors by protecting groups.

REFERENCES

1. D. Cremer, J. Gauss, R. F. Childs and C. Blackburn, *J. Am. Chem. Soc.*, 1985, **107**, 2435.
2. (a) G. A. Olah, S. Prakash and J. Sommer, *Superacids*, Wiley, New York, 1985; (b) R. J. Gillespie, *Acc. Chem. Res.*, 1968, **1**, 202.
3. C. Blackburn, R. F. Childs, D. Cremer and J. Gauss, *J. Am. Chem. Soc.*, 1985, **107**, 2242.
4. D. Fărcaşiu, J. J. O'Donnell, K. B. Wiberg and M. Matturo, *J. Chem. Soc., Chem. Commun.*, 1979, 1124.
5. G. J. Davies, V. M.-A. Ducros, A. Varrot and D. L. Zechel, *Biochem. Soc. Trans.*, 2003, **31**, 523.
6. Z. Wałaszek, D. Horton and I. Ekiel, *Carbohydr. Res.*, 1982, **106**, 193.
7. B. Capon and D. Thacker, *J. Chem. Soc. B*, 1967, 1322.
8. P. R. Young and W. P. Jencks, *J. Am. Chem. Soc.*, 1977, **99**, 8238.
9. R. A. McClelland and M. Ahmad, *J. Am. Chem. Soc.*, 1978, **100**, 7031.
10. J. P. Richard, M. E. Rothenburg and W. P. Jencks, *J. Am. Chem. Soc.*, 1984, **106**, 1361.
11. R. A. McClelland, V. M. Kanagasabapathy, N. S. Banait and S. Steenken, *J. Am. Chem. Soc.*, 1991, **113**, 1009.
12. T. L. Amyes and W. P. Jencks, *J. Am. Chem. Soc.*, 1989, **111**, 7888.
13. J. P. Richard and W. P. Jencks, *J. Am. Chem. Soc.*, 1984, **106**, 1383.
14. G.-A. Craze, A. J. Kirby and R. Osborne, *J. Chem. Soc., Perkin Trans. 2*, 1978, 357.

15. B. L. Knier and W. P. Jencks, *J. Am. Chem. Soc.*, 1980, **102**, 6789.
16. C. G. Swain and C. B. Scott, *J. Am. Chem. Soc.* 1953, **75**, 141; modification to water, J. Koivurinta, A. Kyllönen, L. Leonen, K. Valaste and J. Koskikallio, *Suom. Kemistil.* 1974, 239.
17. M. Sawada, Y. Takai, C. Chong, T. Hanafusa, S. Misumi and Y. Tsuno, *Tetrahedron Lett.*, 1985, **26**, 5065.
18. G.-A. Craze and A. J. Kirby, *J. Chem. Soc., Perkin Trans. 2*, 1978, 354.
19. A. Streitweiser, R. H. Jagow, R. C. Fahey and S. Suzuki, *J. Am. Chem. Soc.*, 1958, **80**, 2326.
20. C. G. Swain, E. C. Stivers, J. F. Reuwer and L. J. Schaad, *J. Am. Chem. Soc.*, 1958, **80**, 5885.
21. N. Narasimhamurthy, H. Manohar, A. G. Samuelson and J. Chandrasekhar, *J. Am. Chem. Soc.*, 1990, **112**, 2937.
22. K. B. Wiberg and P. R. Rablen, *J. Am. Chem. Soc.*, 1993, **115**, 614.
23. J. P. Richard, T. L. Amyes and D. J. Rice, *J. Am. Chem. Soc.*, 1993, **115**, 2523.
24. T. L. Amyes and J. P. Richard, *J. Chem. Soc., Chem. Commun.*, 1991, 200.
25. P. Kandanarachchi and M. L. Sinnott, *J. Am. Chem. Soc.*, 1994, **116**, 5601.
26. T. H. Fife and L. H. Brod, *J. Am. Chem. Soc.*, 1970, **92**, 1681.
27. H. Lönnberg and V. Pohjola, *Acta Chem. Scand., Ser. A*, 1976, **30**, 669.
28. D. Cocker and M. L. Sinnott, *J. Chem. Soc., Perkin Trans. 2*, 1975, 1391.
29. M. Namchuk, J. D. McCarter, A. Becalski, T. Andrews and S. G. Withers, *J. Am. Chem. Soc.*, 2000, **122**, 1270.
30. J. V. O'Connor and R. Barker, *Carbohydr. Res.*, 1979, **73**, 227.
31. S. G. Withers, M. D. Percival and I. P. Street, *Carbohydr. Res.*, 1989, **187**, 43.
32. R. Wolfenden, X. Lu and G. Young, *J. Am. Chem. Soc.*, 1998, **120**, 6814.
33. Y. Zhang, J. Bommuswamy and M. L. Sinnott, *J. Am. Chem. Soc.*, 1994, **116**, 7557.
34. M. Bols, X. Liang and H. H. Jensen, *J. Org. Chem.*, 2002, **67**, 8970.
35. M. Ashwell, X. Guo and M. L. Sinnott, *J. Am. Chem. Soc.*, 1992, **114**, 10158.
36. V. Dookhun and A. J. Bennet, *J. Am. Chem. Soc.*, 2005, **127**, 7456.
37. H. H. Jensen, L. Lyngbye and M. Bols, *Angew. Chem. Int. Ed.*, 2001, **40**, 3447.
38. H. H. Jensen and M. Bols, *Org. Lett.*, 2003, **5**, 3419.
39. C. C. Jones, M. L. Sinnott and I. J. L. Souchard, *J. Chem. Soc., Perkin Trans. 2*, 1977, 1191.
40. L. Hosie, P. J. Marshall and M. L. Sinnott, *J. Chem. Soc., Perkin Trans. 2*, 1984, 1121.
41. X. Huang, C. Surry, T. Hiebert and A. J. Bennet, *J. Am. Chem. Soc.*, 1995, **117**, 10614.
42. J. Zhu and A. J. Bennet, *J. Am. Chem. Soc.*, 1998, **120**, 3887.
43. F. Schuber, P. Travo and M. Pascal, *Bioorg. Chem.* 1979, **8**, 83; C. Tarnus and F. Schuber, *Bioorg. Chem.* 1987, **15**, 31.

44. D. T. H. Chou, J. N. Watson, A. A. Scholte, T. J. Borgford and A. J. Bennet, *J. Am. Chem. Soc.*, 2000, **122**, 8357.
45. A. L. Handlon and N. Oppenheimer, *J. Org. Chem.*, 1991, **56**, 5009.
46. A. L. Handlon, C. Xu, H. M. Muller-Steffner, F. Schuber and N. J. Oppenheimer, *J. Am. Chem. Soc.*, 1994, **116**, 12087.
47. M. L. Sinnott and W. S. S. Wijesundera, *Carbohydr. Res.*, 1985, **136**, 357.
48. P. Deslongchamps, *Stereoelectronic Effects in Organic Chemistry*, Pergamon Press, Oxford, 1983.
49. A. J. Kirby, *The Anomeric Effect and Related Stereoelectronic Effects at Oxygen*, Springer-Verlag, Berlin, 1983; A. J. Kirby, *Acc, Chem. Res.* 1984, **17**, 305.
50. M. L. Sinnott, *Adv. Phys. Org. Chem.*, 1987, **24**, 113.
51. C. L. Perrin, *Acc. Chem. Res.*, 2002, **35**, 28.
52. N. S. Banait and W. P. Jencks, *J. Am. Chem. Soc.*, 1991, **113**, 7951.
53. C. D. Ritchie, *Can. J. Chem.*, 1986, **64**, 2239.
54. R. Shapiro and S. Kang, *Biochemistry*, 1969, **8**, 1806.
55. For a review of the literature prior to 1969, see B. Capon, *Chem. Rev.* 1969, **69**, 407; a later review of acetal and ketal hydrolysis in general is given by E. H. Cordes and H. G. Bull, *Chem. Rev.*, 1974, **74**, 581.
56. J. F. Bunnett, *J. Am. Chem. Soc.*, 1961, **83**, 4978.
57. D. Cocker, L. E. Jukes and M. L. Sinnott, *J. Chem. Soc., Perkin Trans. 2*, 1973, 190.
58. Recalculated from (cited) literature data by M. L. Sinnott, in *The Chemistry of Enzyme Action*, ed. M. I. Page, Elsevier, Amsterdam, 1984, p. 411. Hammett ρ values are calculated in the original papers, but the involvement of substituted benzoic acids seems a needless complication, when experimental pK_a values of phenols are readily available.
59. (a) H. Lönnberg, A. Kankaanperä and K. Haapakka, *Carbohydr. Res.*, 1977, **56**, 277; (b) H. Lönnberg and A. Kulonpää, *Acta Chem. Scand., Ser. A*, 1977, **31**, 306; (c) H. Lönnberg and L. Valtonen, *Finn. Chem. Lett.*, 1978, 209.
60. J. L. Liras, V. M. Lynch and E. V. Anslyn, *J. Am. Chem. Soc.*, 1997, **119**, 8191.
61. O. E. Desvard and A. J. Kirby, *Tetrahedron Lett.*, 1982, **23**, 4163.
62. B. Capon, *Chem. Rev.*, 1969, **69**, 407.
63. E. R. Garrett and P. J. Mehta, *J. Am. Chem. Soc.*, 1972, **94**, 8532.
64. J. A. Zoltewicz, D. F. Clark, T. W. Sharpless and G. Grahe, *J. Am. Chem. Soc.*, 1970, **92**, 1741.
65. J. A. Zoltewicz and D. F. Clark, *J. Org. Chem.*, 1972, **37**, 1193.
66. E. R. Garrett, *J. Am. Chem. Soc.*, 1960, **82**, 827.
67. J. Cadet and R. Teoule, *J. Am. Chem. Soc.*, 1974, **96**, 6517.
68. R. Shapiro and M. Danzig, *Biochemistry*, 1972, **11**, 23.
69. J. Prior and V. Santi, *J. Biol. Chem.*, 1984, **259**, 2429.
70. M. Oivanen, M. Rajamäki, J. Varila, J. Hovinen, S. Mikhailov and H. Lönnberg, *J. Chem. Soc., Perkin Trans. 2*, 1994, 309.

71. A. Karpeisky, S. Zavgorodny, M. Hotokka, M. Oivanen and H. Lönnberg, *J. Chem. Soc., Perkin Trans. 2*, 1994, 741.
72. B. Capon, *Tetrahedron Lett.*, 1963, **4**, 911.
73. E. Anderson and B. Capon, *J. Chem. Soc. B*, 1969, 1033.
74. D. D. Perrin, *Dissociation Constants of Organic Bases in Aqueous Solution*, Butterworths, London, 1972.
75. R. A. Cox and R. Stewart, *J. Am. Chem. Soc.*, 1976, **98**, 488.
76. J. L. Jensen, L. R. Herold, P. A. Lenz, S. Trusty, V. Sergi, K. Bell and P. Rogers, *J. Am. Chem. Soc.*, 1979, **101**, 4672.
77. B. Capon and K. Nimmo, *J. Chem. Soc., Perkin Trans. 2*, 1975, 1113.
78. D. A. Jencks and W. P. Jencks, *J. Am. Chem. Soc.*, 1977, **99**, 7948.
79. T. H. Fife and E. Anderson, *J. Am. Chem. Soc.*, 1971, **93**, 1701.
80. J. L. Jensen, A. B. Martines and C. L. Shimazu, *J. Org. Chem.*, 1983, **48**, 4175.
81. R. Ta-Shma and W. P. Jencks, *J. Am. Chem. Soc.*, 1986, **108**, 8040.
82. A. J. Bennet, A. J. Davis, L. Hosie and M. L. Sinnott, *J. Chem. Soc., Perkin Trans. 2*, 1987, 581.
83. N. S. Banait and W. P. Jencks, *J. Am. Chem. Soc.*, 1991, **113**, 7958.
84. D. Piszkiewicz and T. C. Bruice, *J. Am. Chem. Soc.*, 1968, **90**, 2156.
85. B. Capon and B. C. Ghosh, *J. Chem. Soc. B*, 1971, 739.
86. B. Capon, M. I. Page and G. H. Sankey, *J. Chem. Soc., Perkin Trans. 2*, 1972, 529.
87. A. J. Kirby, *Adv. Phys. Org. Chem.*, 1980, **17**, 183.
88. M. I. Page and W. P. Jencks, *Proc. Natl. Acad. Sci. USA*, 1971, **68**, 1678.
89. B. M. Dunn and T. C. Bruice, *J. Am. Chem. Soc.*, 1970, **92**, 2410.
90. A. J. Kirby and J. M. Percy, *J. Chem. Soc., Perkin Trans. 2*, 1989, 907.
91. T. H. Fife and E. Anderson, *J. Am. Chem. Soc.*, 1971, **93**, 6610.
92. C. Buffet and G. Lamaty, *Recl. Trav. Chim.*, 1976, **95**, 1.
93. C. J. Brown and A. J. Kirby, *J. Chem. Soc., Perkin Trans. 2*, 1997, 1081.
94. B. Capon, M. C. Smith, E. Anderson, R. H. Dahm and G. H. Sankey, *J. Chem. Soc. B*, 1969, 1038.
95. A. J. Kirby, *Acc. Chem. Res.*, 1997, **30**, 290.
96. G.-A. Craze and A. J. Kirby, *J. Chem. Soc., Perkin Trans. 2*, 1974, 61.
97. W. Koenigs and E. Knorr, *Chem. Ber.*, 1901, **34**, 957.
98. C. R. Clark and R. W. Hay, *J. Chem. Soc., Perkin Trans. 2*, 1973, 1943.
99. T. J. Przystas and T. H. Fife, *J. Am. Chem. Soc.*, 1980, **102**, 4391.
100. T. J. Przystas and T. H. Fife, *J. Chem. Soc., Perkin Trans. 2*, 1987, 143.
101. J. Baty and M. L. Sinnott, *Chem. Commun.*, 2004, **1**, 866.
102. R. L. Whistler and T. van Es, *J. Org. Chem.*, 1963, **28**, 2303.
103. M. L. Sinnott and M. C. Whiting, *J. Chem. Soc. B*, 1971, 965.
104. (a) J. L. Jensen and W. P. Jencks, *J. Am. Chem. Soc.*, 1979, **101**, 1476; (b) F. Guinot and G. Lamaty, *Tetrahedron Lett.*, 1972, **13**, 2569.
105. C. Bamford, B. Capon and W. G. Overend, *J. Chem. Soc.*, 1962, **1**, 5138.
106. C. J. Clayton, N. A. Hughes and S. A. Saaed, *J. Chem. Soc. C*, 1967, 644.
107. D. Indurugalla and A. J. Bennet, *J. Am. Chem. Soc.*, 2001, **123**, 10889.
108. L. R. Fedor and R. S. Murty, *J. Am. Chem. Soc.*, 1973, **95**, 8407.

109. J. P. Ferraz and E. H. Cordes, *J. Am. Chem. Soc.*, 1979, **101**, 1488.
110. L. Melander and W. H. Saunders, *Reaction Rates of Isotopic Molecules*, Wiley, New York, 1980, p. 95.
111. D. A. Singleton and A. A. Thomas, *J. Am. Chem. Soc.*, 1995, **117**, 9357.
112. T. Holm, *J. Am. Chem. Soc.*, 1994, **116**, 8803.
113. T. Pehk, E. Kiirend, E. Lippmaa, U. Ragnarsson and L. Grehn, *J. Chem. Soc., Perkin Trans. 2*, 1997, 445.
114. R. E. Carter and L. Melander, *Adv. Phys. Org. Chem.*, 1973, **10**, 1.
115. F. A. L. Anet and M. Kopelevich, *J. Am. Chem. Soc.*, 1986, **108**, 1355.
116. F. A. L. Anet and M. Kopelevich, *J. Chem. Soc., Chem. Commun.*, 1987, 595.
117. B. E. Lewis and V. L. Schramm, *J. Am. Chem. Soc.*, 2001, **123**, 1327.
118. D. J. Miller, R. Subramanian and W. H. Saunders, *J. Am. Chem. Soc.*, 1981, **103**, 3519.
119. J. K. Lee, A. D. Bain and P. D. Berti, *J. Am. Chem. Soc.*, 2004, **126**, 3769.
120. A. J. Bennet, M. L. Sinnott and W. S. S. Wijesundera, *J. Chem. Soc., Perkin Trans. 2*, 1985, 1233.
121. M. Ashwell, M.L. Sinnott and Y. Zhang, *J. Org. Chem.*, 1994, **59**, 7539.
122. D. Parkin and V. L. Schramm, *Biochemistry*, 1987, **26**, 913.
123. L. B. Sims and D. E. Lewis in E. Buncel and C. C. Lee, Eds, *Isotopes Organic Chemistry* Elsevier, New York, 1984, Vol. 6, p. 161.
124. V. Anisimov and P. Paneth, *J. Math. Chem.*, 1999, **26**, 75.
125. Y. Tanaka, W. Tao, J. S. Blanchard and E. J. Hehre, *J. Biol. Chem.*, 1994, **269**, 32306.
126. T. L. Knoll and A. J. Bennet, *J. Phys. Org. Chem.*, 2004, **17**, 478.
127. B. A. Horenstein and M. Bruner, *J. Am. Chem. Soc.*, 1996, **118**, 10371.
128. D. Piszkiewicz and T. C. Bruice, *J. Am. Chem. Soc.*, 1967, **89**, 6237.
129. F. W. Ballardie, B. Capon, W. M. Dearie and R. L. Foster, *Carbohydr. Res.*, 1976, **49**, 79.
130. D. Cocker and M. L. Sinnott, *J. Chem. Soc., Perkin Trans. 2*, 1976, 618.
131. P. Camilleri, R. F. D. Jones, A. J. Kirby and R. Strömberg, *J. Chem. Soc., Perkin Trans. 2*, 1994, 2085.
132. H. P. Wessel and S. Bartsch, *Carbohydr. Res.*, 1995, **274**, 1.
133. T. H. Fife and T. J. Przystas, *J. Am. Chem. Soc.*, 1977, **99**, 6693.
134. T. H. Fife and T. J. Przystas, *J. Am. Chem. Soc.*, 1980, **102**, 292.
135. B. Capon and M. I. Page, *J. Chem. Soc., Perkin Trans. 2*, 1972, 2057.
136. E. Anderson and T. H. Fife, *J. Am. Chem. Soc.*, 1973, **95**, 6437.
137. T. H. Fife and T. J. Przystas, *J. Am. Chem. Soc.*, 1979, **101**, 1202.
138. X. M. Cherian, S. A. Van Arman and A. W. Czarnik, *J. Am. Chem. Soc.*, 1988, **110**, 6566.
139. R. L. Nath and H. N. Rydon, *Biochem. J.*, 1954, **57**, 1.
140. A. N. Hall, S. Hollingshead and H. N. Rydon, *J. Chem. Soc.*, 1961, 4290.
141. V. N. Nikitin, I. Yu. Levdik and M. A. Ivanov, *J. Struct. Chem.*, 1968, **9**, 901.
142. H.-Z. Lai and D. E. Ontto, *Carbohydr. Res.*, 1979, **75**, 51.

143. S. Kyosaka, S. Murata and M. Tanaka, *Chem. Pharm. Bull (Tokyo)*, 1983, **31**, 3902.
144. Kinetic studies on the basic hydrolysis of *p*-acetylphenyl 3- and 5-*O*-methyl->-D-xylofuranoside were reported by H. Lönnberg, *Finn. Chem. Lett.* 1978, 213 and support the idea of greater base-lability of *trans-O*-furanosides.
145. C. C. Jones, M. L. Sinnott and I. J. L. Souchard, *J. Chem. Soc., Perkin Trans. 2*, 1977, 1191.
146. B. M. Anderson and C. D. Anderson, *J. Biol. Chem.*, 1963, **238**, 1475.
147. J. Janson and B. Lindberg, *Acta Chem. Scand.*, 1960, **14**, 2051.
148. J. Janson and B. Lindberg, *Acta Chem. Scand., Ser. B*, 1982, **36**, 277.
149. (a) H. Lönnberg and P. Lehikoinen, *J. Org. Chem.*, 1984, **49**, 4964; (b) P. Lehikoinen and H. Lönnberg, *Chem. Scr.*, 1986, **26**, 103.
150. (a) D. Horton and A. E. Luetzow, *J. Chem. Soc. D, Chem. Commun.*, 1971, 79; (b) C. S. Tsai and C. Reyes-Zamora, *J. Org. Chem.*, 1972, **37**, 2725.
151. A.-J. Rhind-Tutt and C. A. Vernon, *J. Chem. Soc.*, 1960, 4637.
152. B. Capon, P. M. Collins, A. A. Levy and W. G. Overend, *J. Chem. Soc.*, 1964, 3242.
153. L. R. Schroeder, J. W. Green and D. G. Johnson, *J. Chem. Soc. B*, 1966, 447.
154. M. L. Sinnott and W. P. Jencks, *J. Am. Chem. Soc.*, 1980, **102**, 2026.
155. S. Winstein and D. Trifan, *J. Am. Chem. Soc.*, 1952, **74**, 1147.
156. R. U. Lemieux, K. B. Hendricks, R. V. Stick and K. James, *J. Am. Chem. Soc.*, 1975, **97**, 4056.
157. B. Russell, *A History of Western Philosophy*, George Allen and Unwin, London, 2nd edn, 1961, p. 462.
158. L. G. Green and S. V. Ley, in *Carbohydrates in Chemistry and Biology*, ed. B. Ernst, G. W. Hart and P. Sinay, Wiley-VCH, Weinhein, 2000, Vol. 1, p. 427.
159. A. Michael, *Am. Chem. J.*, 1879, **1**, 307.
160. E. Fischer and K. Raske, *Chem. Ber.*, 1909, **42**, 1465.
161. F. Mauthner, *J. Prakt. Chem.*, 1915, **91**, 174.
162. K. J. Jensen, *J. Chem. Soc., Perkin Trans. 1*, 2002, 2219.
163. H. G. Latham, E. L. May and E. Mosettig, *J. Org. Chem.*, 1950, **15**, 884.
164. W. Nerinckx, unpublished work, 2006.
165. (a) B. Helferich and K. Iloff, *Hoppe Seylers Z. Physiol. Chem.*, 1933, **221**, 252; (b) J. Conchie, G. A. Levvy and C. A. Marsh, *Adv. Carbohydr. Chem.*, 1957, **12**, 157.
166. (a) B. Helferich and E. Schmitz-Hillebrecht, *Chem. Ber*, 1933, **66**, 378; (b) B. Helferich and D. V. Kashelikar, *Chem. Ber.*, 1957, **90**, 2094.
167. (a) B. Helferich and K. L. Bettin, *Chem. Ber.*, 1961, **94**, 1159; (b) B. Helferich and W. Ost, *Chem. Ber.*, 1962, **95**, 2616.
168. S. Hosono, W.-S. Kim, H. Sasai and M. Shibasaki, *J. Org. Chem.*, 1995, **60**, 4.
169. R. R. Schmidt and W. Kinzy, *Adv. Carbohydr. Chem. Biochem.*, 1994, **50**, 21.

170. G. Excoffier, D. Gagnaire and J.-P. Utille, *Carbohydr. Res.*, 1975, **39**, 368.
171. L. A. Berven, D. Dolphin and S. G. Withers, *Can. J. Chem.*, 1990, **68**, 1859.
172. Y. Zhao, C. J. Chany, P. F. G. Sims and M. L. Sinnott, *J. Biotechnol.*, 1997, **57**, 181.
173. F. Barresi and O Hindsgaul, *J. Carbohydr. Chem.*, 1995, **14**, 1043.
174. G.-J. Boons and S. Isles, *J. Org. Chem.*, 1996, **61**, 4262.
175. (a) B. Fraser-Reid, U. E. Udodong, Z. Wu, H. Ottosson, J. R. Merritt, C. S. Rao, C. Roberts and R. Madsen, *Synlett*, 1992, 927; (b) B. Fraser-Reid, J. R. Merritt, A. L. Handlon and C. W. Andrews, *Pure Appl. Chem.*, 1993, **65**, 779.
176. H. Kondo, S. Aoki, Y. Ichikawa, R. L. Halcomb, H. Ritzen and C.-H. Wong, *J. Org. Chem.*, 1994, **59**, 864.
177. D. Crich and S. Sun, *J. Am. Chem. Soc.*, 1997, **119**, 11217.
178. D. Crich and N. S. Chandrasekera, *Angew. Chem. Int. Ed.*, 2004, **43**, 5386.
179. S. J. Danishefsky, K. F. McClure, J. T. Randolph and R. B. Ruggeri, *Science* 1993, **260**, 1307; P. H. Seeburger and S. J. Danishefsky, *Acc. Chem. Res.* 1998, **31**, 685.
180. C. M. Timmers, G. A. van der Marel and J. H. van Boom, *Recl. Trav. Chim.*, 1993, **112**, 609.
181. D. R. Mootoo, V. Date and B. Fraser-Reid, *J. Am. Chem. Soc.*, 1988, **110**, 2662.
182. L. E. Jukes, BSc thesis, part II, University of Bristol, 1972.
183. B. Fraser-Reid, Z. F. Wu, U. E. Udodong and H. Ottosson, *J. Org. Chem.*, 1990, **55**, 6068.
184. A. F. Bochkov and G. E. Zaikov, *The Chemistry of the O-Glycosidic Bond*, Pergamon Press, Oxford, 1979.
185. B. Fraser-Reid, S. Grimme, M. Piacenza, M. Mach and U. Schlueter, *Chem. Eur. J.*, 2003, **9**, 4687.
186. D. Crich, H. Li, Q. Yao, D. J. Wink, R. D. Sommer and A. L. Rheingold, *J. Am. Chem. Soc.*, 2001, **123**, 5826.
187. A. Streitweiser, C. H. Heathcock and A. M. Kosower *Introduction to Organic Chemistry*, Macmillan, New York, 4th edn, 1992, p. 314.
188. A. J. Bennet and M. L. Sinnott, *J. Am. Chem. Soc.*, 1986, **108**, 7287.
189. Same figure obtained by Berti *et al.*, 119, using the Singleton method.
190. P. J. Berti and V. L. Schramm, *J. Am. Chem. Soc.*, 1997, **119**, 12069.

CHAPTER 4

Primary Structure and Conformation of Oligosaccharides and Polysaccharides

4.1 INTRODUCTION – DEPICTION AND ISOLATION OF POLYSACCHARIDES

The central dogma of modern molecular biology is that structural information flows in the sense DNA → RNA → protein. Over the last 20 or so years, it has become clear that much fine-tuning of the cell involves carbohydrates, whether by themselves, or attached to proteins in glycoproteins, or to lipids in glycolipids, so we can now write of most information flowing DNA → RNA → protein → oligo- and polysaccharides. Of course, there are minor pathways in which information flows in the reverse sense (*e.g.* reverse transcriptases), but the main path of information transfer helps to explain why, at a time when automated DNA sequencers can produce sequences of hundreds of base-pairs in a day, determination of the structure of a polysaccharide with a five-sugar repeat remains a major undertaking. The root cause of the problem is that molecular complexity increases down the information stream. DNA and RNA are linear and based on a four-letter code (although bases, particularly in RNA, can be extensively modified). Proteins are almost always linear and based on a 20- or 21-letter code,[i] but protein sequencing is so much less efficient than DNA sequencing that the usual modern practice is to determine enough protein sequence to be able to isolate mRNA and determine the sequence of its complementary DNA.

It is difficult to see how molecular biology could come up with an analogous strategy, of going up the information stream and sequencing something easier, for polysaccharides. Although oligosaccharides, particularly those attached to proteins, have definite structures, many polysaccharides have an element of randomness. Determination of the primary sequence of a polysaccharide therefore remains a matter of frontal assault on the problem by wet chemistry, mass spectrometry and multi-dimensional NMR. (The determination of the

[i] 20 without selenocysteine, coded for by UGA (in RNA), normally a stop codon.

sequences of oligosaccharides attached to proteins or lipids in mammalian systems is greatly simplified because of the limited range of structures found – many sugars can be identified by their chromatographic mobilities, much as amino acids in protein hydrolysates are identified by their position on the ion-exchange chromatogram of a classical amino acid analyser.)

Carbohydrate primary structures are written in much the same way as protein structures used to be, using a fairly transparent three-letter code. Ara, Xyl and Rib stand for arabinose, xylose and ribose and Glc, Man, Gal, Ido, Gul and Fru for glucose, mannose, galactose, idose, gulose and fructose, respectively. The 6-deoxy sugars rhamnose and fucose have their own symbols (Rha and Fuc), but the uronic acids are indicated by the letter A after the three-letter code – GalA is galacturonic acid. Likewise, 2-amino-2-deoxy sugars are denoted by N – GalN is galactosamine. Amino sugars are often acetylated, so GalNAc stands for 2-acetamino-2-deoxygalactose. Occasionally UA is used for uronic acid, as in GalUA for galacturonic acid.

Because a number of sugars – galactose, rhamnose and fucose in particular – occur in polysaccharides in both D and L forms, it is usual to precede the sugar code by D or L. The ring form is designated by an italicised *f* or *p*. Finally, the stereo- and regiochemistry of the glycosidic linkage is denoted by α or β before the residue whose anomeric configuration is being designated and the regiochemistry by numbers of the form ($n \rightarrow m$), where n is the number of the anomeric carbon and m is the oxygen to which it makes a glycosidic linkage. Thus, lactose is β-D-Gal*p*-(1→4)-D-Glc and chitobiose β-D-GlcNAc*p*-(1→4)-D-GlcNAc. Linkages between anomeric centres should be denoted by double-headed arrows – sucrose is α-D-Glc*p*-(1↔2)-β-D-Fru*f*. However, the use of both single- and double-headed arrows is not universal, many authors preferring to use a comma, as non-standard symbols such as single- and double-headed arrows do not transfer between word-processing programs.[ii]

Sialic acids are denoted by Neu and Kdo by Kdo: NeuAc denotes *N*-acetylneuraminic acid and NeuGc. KDN (sometimes Kdn) is 3-deoxy-D-*glycero*-D-*galacto*-non-2-ulopyranosonic acid CNeuNAC with the 5 M+AC replaced by OH *N*-glycolylneuraminic acid. Apiose is denoted Api, with the hydroxymethyl branch not forming the furanose ring denoted the 3′-position.

In few areas of natural product chemistry does the injunction "first, catch your hare"[iii] so succinctly summarise the main difficulties. Whereas a battery of well-established techniques can be used for the purification of a protein from biological materials, each sub-field, and even each laboratory, in the carbohydrate purification and sequencing area tends to adopt its own extraction procedures. Polysaccharides *in vivo* are often heterogeneous with respect to molecular weight, which restricts the usefulness of size-exclusion chromatography. Some polysaccharides are covalently attached to other macromolecular

[ii] Microsoft Word being particularly bad.

[iii] Anon, but commonly misattributed to recipes for jugged hare in Isabella Beeton's *Household Management* (1861) or Hannah Glasse's *The Art of Cookery Made Plain and Easy* (1747).

components of biological tissue, while others, such as pectin, may have regions of radically different chemical structure in the same large polysaccharide molecule. Fragments are often solubilised by treatment with enzymes or by chemical treatments. Splitting polysaccharides into smaller fragments is anyway often desirable if extensive NMR studies are to be performed.

Some polysaccharides carry negative charges, by virtue of their uronic acid components and/or modification of OH groups as sulfate esters or cyclic acetals of pyruvic acid ($CH_3COCOOH$); these can often be precipitated by cationic surfactants such as cetyltrimethylammonium bromide. They can also be fractionated on weak anion exchangers. For neutral polysaccharides, elaborate precipitation protocols involving initial removal of proteins and nucleic acids have to be used – a useful precipitant of the polysaccharide is 50% ethanol.

The problem is greatly simplified in the case of glycoproteins, where all the standard techniques of protein isolation and purification can be used and the oligosaccharide fragments liberated from the purified protein. The sugar chains of O-linked glycoproteins, in which a glycosidic bond is formed between the reducing end of the oligosaccharide and the oxygen of a serine or threonine residue of the protein, can be cleaved off in mild base. The reaction appears to be an $E1_{CB}$-like $E2$ reaction, with negative charge delocalised into the peptide carbonyl group, which leaves a dehydroserine or dehydrothreonine residue in the protein and liberates the reducing end of the oligosaccharide. The oligosaccharides from N-linked glycoproteins are often cleaved off by treatment with endoglycosidase H, which cleaves the reducing end chitobiosyl unit between the GlcNAc residues, leaving one still attached to the protein (illustrated in Figure 4.1).

Since, unlike nucleic acids and peptides, sugars do not normally possess a UV chromophore, monitoring liquid chromatographic separations can be difficult. Traditionally, point-by-point assays of low-pressure chromatograms using chromogenic reagents such as anthrone or phenol were performed. HPLC of oligosaccharides became possible in conjunction with refractive index detectors; however, for obvious reasons only isocratic elution can be performed, and because only small changes in refractive index are brought about by fairly large concentrations of sugars, high concentrations have to be used. Reducing sugars can be subjected to HPLC with a normal UV detector if they are first converted to glycosylamines with a chromophoric aglycone, but of course the method will not work with non-reducing sugars.

The recent advent of high-performance anion-exchange chromatography coupled to pulsed amperometric detection (HPAEC–PAD) has revolutionised the separation of oligosaccharides. The anion exchanger is usually a quaternary ammonium residue and the chromatogram is run at alkaline pH, so that fractionation is essentially on the basis of the pK_a of the sugar hydroxyls and their number. For this reason, reducing sugars, possessing in the anomeric hydroxyl a more acidic OH than others, are better separated than non-reducing sugars, although the performance with even these can be impressive. Detection involves making the solution very alkaline (M-level NaOH) and making a gold electrode anodic, at a potential high enough to oxidise RO^- groups to $RO^\bullet$, for

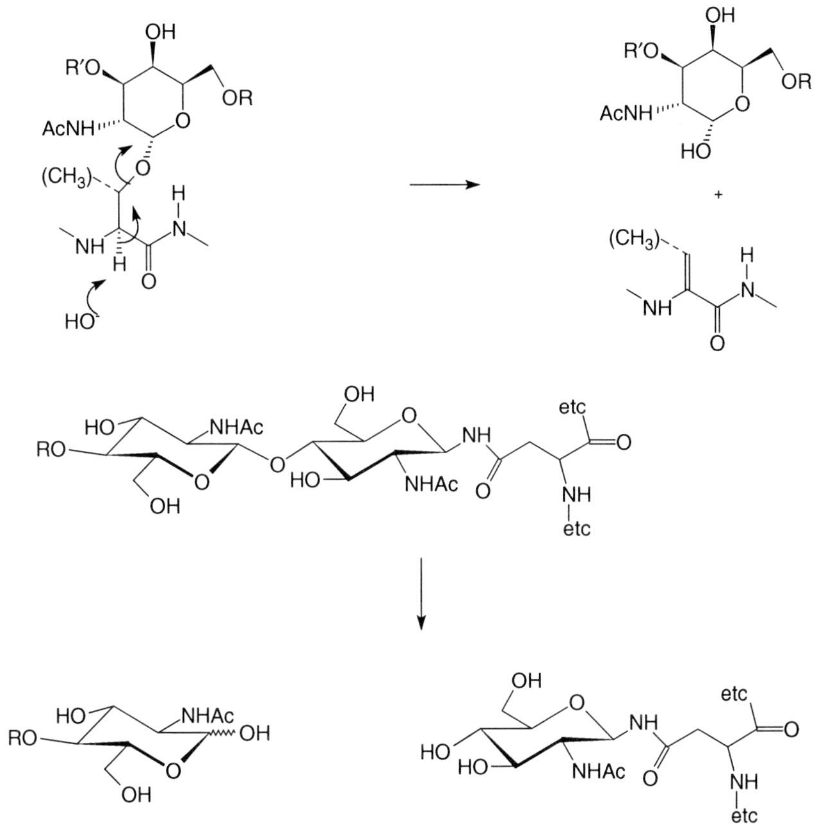

Figure 4.1 Isolation of oligosaccharides and polysaccharides. Top, base-catalysed elimination of O-linked glycoproteins; one of the common *O*-glycan linkages (α GalNAc*p*) is shown. Bottom, endoglycosidase H-catalysed cleavage of N-linked glycoproteins.

a fraction of a second. The resulting small current can be detected with such sensitivity that micrograms of carbohydrate can be quantitated in the overall chromatogram. The carbohydrate so detected is destroyed by the oxidation and alkaline pH, but the detection method is so sensitive that stream splitters can be employed and most of the eluent recovered.

4.2 DETERMINATION OF STRUCTURE AND CONFORMATION OF OLIGO- AND POLYSACCHARIDES[1]

4.2.1 Determination of Constituent Sugars and Substitution Pattern

The polymer is hydrolysed to its constituent sugars, the acid hydrolysate reduced to the alditols with $NaBH_4$, the alditols acetylated and then analysed by gas–liquid chromatography (GLC), using a stationary phase that separates components essentially on the basis of their boiling points. This technique continues to be used because it is so well established, even though HPAED–PAD is now as sensitive as GLC. If the polysaccharide contains carboxylic acid groups, these are reduced to the alcohols. They can be reduced before methylation, by initial treatment with a carbodiimide and reduction of the complex by $NaBD_4$ to $-CD_2OH$ groups. Mass spectrometry coupled to GLC (GC–MS) can then distinguish alditol acetates derived from uronic acids from those derived from aldoses. Alternatively, the methyl esters of uronic acids can be reduced after methylation. $NaBD_4$ is also used for the reduction of the aldoses after the hydrolysis if GC–MS is going to be performed, since it permits the ends of the alditols to be distinguished in mass spectral fragmentation patterns.

Distinction between the D and L forms of sugars is made by acid-catalysed formation of glycosides with chiral 2-butanol or 2-octanol; the acetylated or trimethylsilylated glycosides with D and L sugars will be diastereomers and thus have different GLC retention times.

With the constituent sugars determined, the pattern of substitution is determined by methylation analysis. The polysaccharide is fully methylated at the polymer level, and, if it still contains carboxylic acid groups, the methyl esters are reduced using $LiBH_4$. The methylated polymers are then hydrolysed (2 M aqueous CF_3COOH is widely used) and the methylated sugars analysed. This is usually done by first reducing them to methylated alditols, acetylating and then analysing by GLC or GC–MS.[2] The concept behind methylation analysis is simple – non-reducing termini give rise to fully methylated sugars, sugars in the main chain not bearing branches will have one position unmethylated and branch points will have an unmethylated hydroxyl for each sugar chain attached. Uronic acids will appear as 6-acetylated alditols.

Simple linear polymers of even moderate degree of polymerisation will in practice give only the main chain methylated sugars – amylose [(4)-α-D-Glc*p*-(1→)$_n$] and cellulose [(4)-β-D-Glc*p*-(1→)$_n$] both give substantial quantities of 1,4,5-triacetyl-2,3,6-trimethylglucitol and very little else (Figure 4.2). By themselves, however, the methylation patterns are compatible either with the correct structures of 1→4-linked glucopyranose units or with 1→5-linked glucofuranose units. However, the unlikelihood of four-membered ring sugars makes the almost exclusive production of 1,3,5-triacetyl-2,4,6-trimethylglucitol from methylation analysis of laminarin [(3)-β-D-Glc*p*-(1→)$_n$] a definite proof, both of the 1→3 linkage between residues and their pyranoside nature.

If there is significant branching – as with amylopectin, which has 1→6 branches as well as a preponderance of α-(1→4) linkages – effectively equal quantities of 2,3-dimethyl-1,4,5,6-tetraacetylglucitol, from the branch point and 1,5-diacetyl-2,3,4,6-tetramethylglucitol, from the non-reducing ends, will be produced. The production of 1,5-diacetyl-2,3,4,6-tetramethylglucitol from the non-reducing ends establishes that these are in the pyranose form. However, methylation analysis gives no information about the size of the chain attached to the branch point – the methylation data are compatible both with occasional single α-D-Glc*p*-(1→6) residues or the true Christmas-tree like structure with long α-D-Glc*p*-(1→4) chains attached to each branch point.

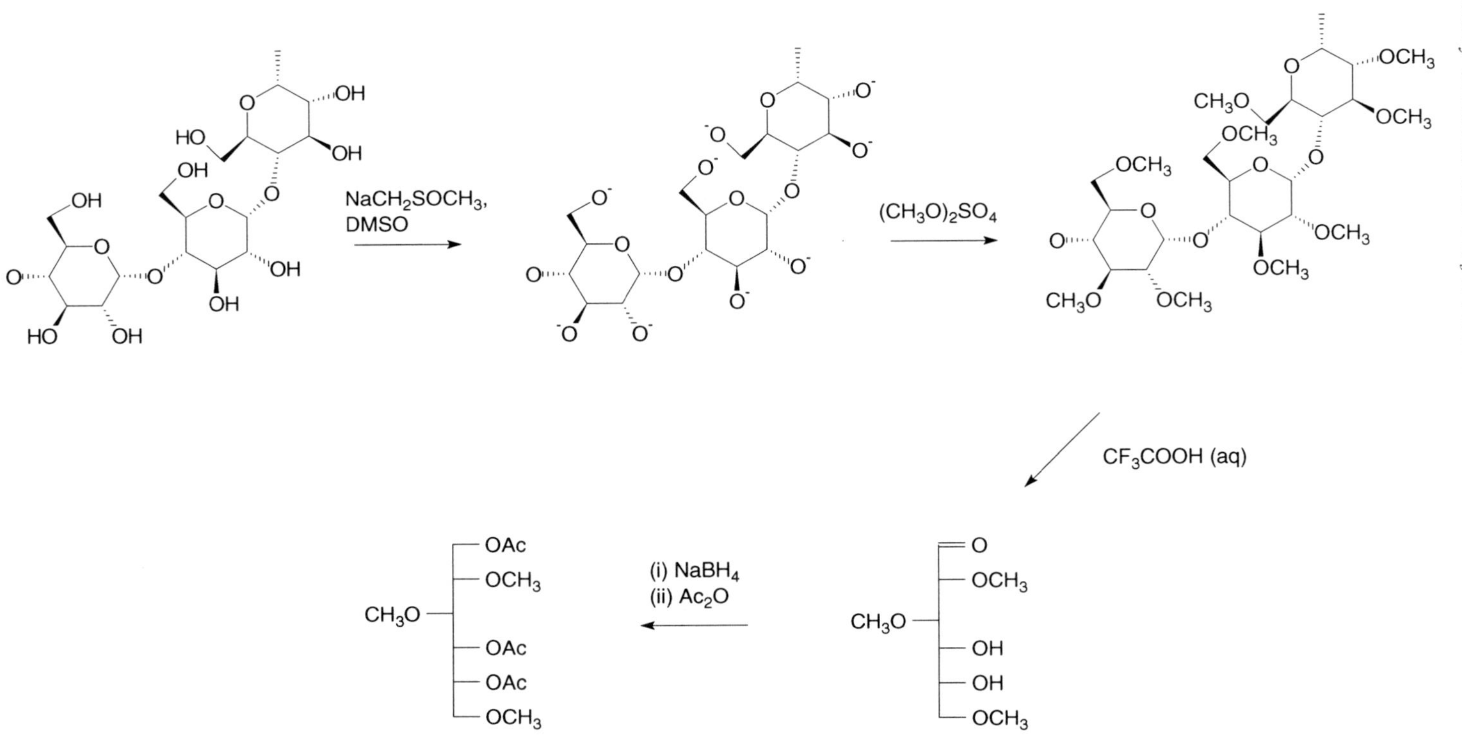

Figure 4.2 Methylation analysis of amylose.

The methylation technique, although conceptually simple, was for many years compromised by the difficulty of fully methylating a polysaccharide – incomplete methylation would give rise to structures with bogus branch points. Hakomori[3] solved the problem by dissolving the polymer in anhydrous dimethyl sulfoxide and then treating with an excess of a solution of its anion,[iv] which deprotonated all the OH and AcNH groups (the pK_a of dimethyl sulfoxide is 28.5); subsequent reaction with dimethyl sulfate gives the fully methylated polymer.

Even today, the resolution of the method is not good and occasional branch points can be missed. This is particularly the case where occasional branch points giving rise to long branches are involved, as with pectin, where they showed up only in the atomic force microscopy of single molecules.[4] Similarly, examination of single molecules of amylose by atomic force microscopy reveals occasional branch points (1 residue in 10^2), well below the resolution of methylation analysis.[5]

4.2.2 Mass Spectrometry

Classical electron ionisation mass spectrometry, in conjunction with GLC, is the routine method of confirming the gross structure of the sugar monomers. There are tables of standard spectra, but the fragmentation patterns can be understood from those of the product of the standard methylation product of a non-reducing terminal glucose residue, 1,5-di-*O*-acetyl 2,3,4,6-tri-*O*-methyl-1-deuteroglucitol, molecular ion (Figure 4.3). The initial product of electron ionization is the molecular ion, the radical cation. This can cleave in a number of ways, with loss of a neutral fragment; the equipment only detects ions. What is observed depends on the stability of the oxocarbenium ion formed. Thus, an ion m/z 45 is seen from C5–C6 cleavage, with the methoxylated fragment splitting from the rest of the molecule, which becomes a neutral radical; the reverse cleavage in which the positive charge would be carried by the acetoxylated carbon is not observed. However, both ions are seen from C2–C3 cleavage and also C3–C4 cleavage, giving rise to the fragment ions shown. These can fragment further by loss of neutrals such as methanol or acetic acid, so that, for example the m/z 205 ion loses acetic acid to give an m/z 145 ion. The effectiveness of these processes for the m/z 269 ion expected from C1–C2 cleavage presumably accounts for its lack of prominence.

The development of methods for ionising molecules that are far less drastic than the traditional route of vaporising them, and then bombarding them with electrons from an electron gun, has enabled the mass spectra of underivatised carbohydrates to be studied. Moreover, tandem techniques have enabled ions to be selected, made to undergo collision-induced dissociation by the presence of an inert gas such as argon or helium and the fragments examined. The

---

[iv] Generated by reaction of NaH with anhydrous DMSO, under carefully controlled conditions to avoid a runaway reaction.

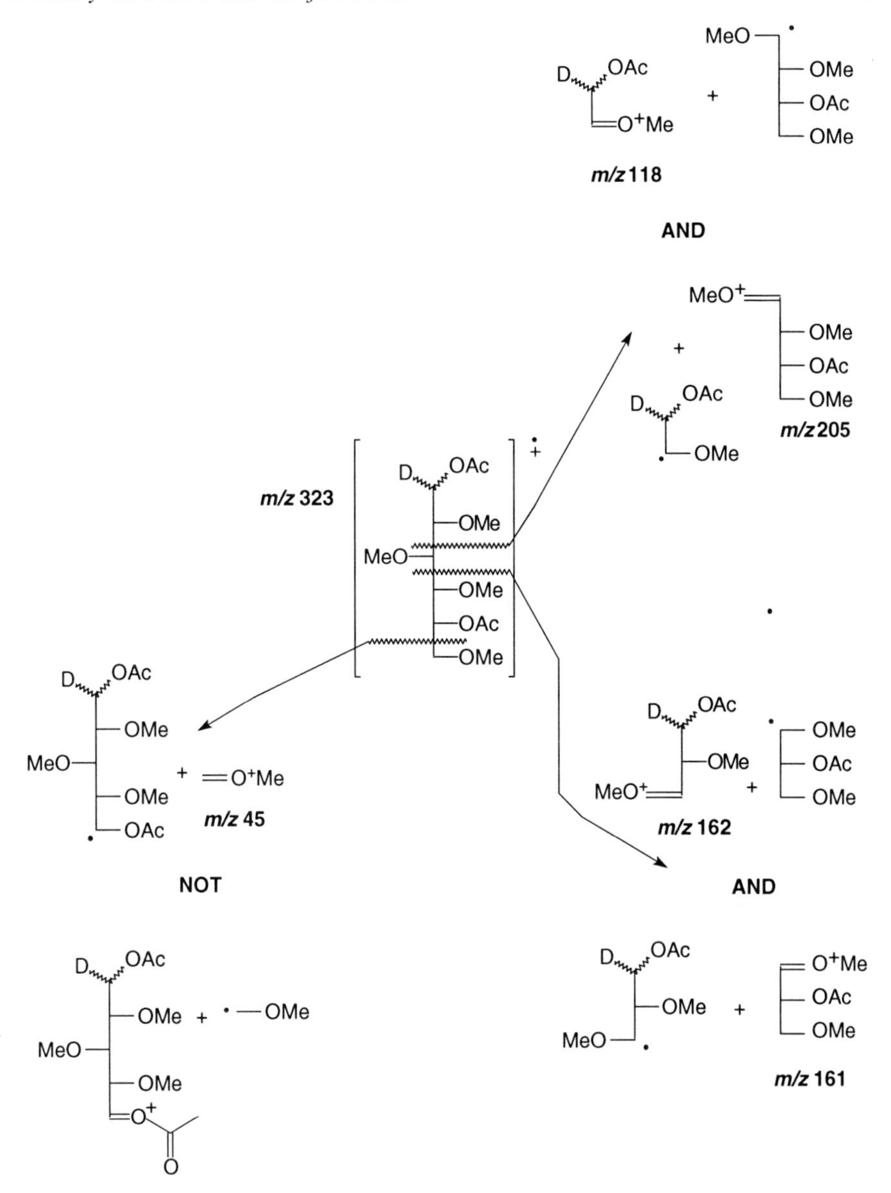

Figure 4.3 Electron ionization cleavage patterns of a partially acetylated methylated alditol.

process can be repeated several times, modern equipment being capable of $(MS)^{12}$. Much of the initial work on carbohydrates was done with glycoproteins and glycolipids,[6] where the structural possibilities were more limited than with polysaccharides *per se* and a list of *m*/*z* values and intensities correspondingly more informative.

Three "soft" ionisation methods are in use for carbohydrates, fast atom bombardment (FAB), electrospray ionisation (ESI) and matrix-assisted laser desorption/ionisation (MALDI). FAB is the oldest and involves directing a high-energy beam of Cs^+ ions or Xe atoms at the sample dissolved in a non-volatile solvent such as *m*-nitrobenzyl alcohol. The atoms sputter the sample and matrix; $[M+H]^+$ or $[M+Na]^+$ ions are commonly observed. With an upper limit of M^+ of about 2000, FAB is not that soft, and is usually used for small oligosaccharides; it has the further disadvantage that the sample is prepared and then directly introduced into the mass spectrometer, so that it cannot be combined with liquid chromatography.

In ESI, a stream of small, highly charged droplets is produced by passing the solution to be analysed (usually, but not necessarily, aqueous) through a nozzle at a potential of ~3.5 kV into the mass spectrometer. As the droplets shrink by evaporation, the electrostatic potential at the surface becomes sufficiently high to eject positive ions. The technique is most widely used with proteins, which have many basic sites. A whole series of multiply charged ions are produced, and as the mass analyser measures mass-to-charge ratios rather than simply mass, the crude spectra have to be deconvoluted with some sophistication. With carbohydrates, though, singly charged ions tend to be produced, with $[M+Na]^+$ as common as $[M+H]^+$.[7]

MALDI occurs from a solution of the sample in a UV-absorbing matrix: 2,5-dihydroxybenzoic acid is the current favourite for carbohydrate analysis.[8] A pulse of laser light directed on the matrix gives rise to both positive and negative ions, which can be further analysed. In the positive mode, $[M+Na]^+$ is commoner than $[M+H]^+$.

The cleavage patterns of oligosaccharides are discussed in terms of a specialised nomenclature,[9] in which cleavages leaving the charge on the non-reducing end are described by letters at the beginning of the alphabet and those leaving it on the reducing end by letters at the end of the alphabet (Figure 4.4). Cleavages across the ring are denoted by A or X, with the bonds cleaved indicated by superscripts, using the lowest number (oxygen counting as zero). Cleavages of the glycosidic linkage are B and Y, with cleavage to glycosyl cations (charge at non-reducing end) being B-type. C and Z cleavages are those of the bond joining the glycosidic oxygen to a non-anomeric carbon. Finally, subscripts describe the residue cleaved, with numbering starting from the non-reducing end for A-, B- and C-type cleavages and at the reducing end for X-, Y- and Z-type cleavages.

4.2.3 Diffraction by Single Crystals, Crystal Powders and Fibres

The various types of diffraction of importance in determining polysaccharide (and protein) conformation (neutron and electron diffraction in addition to the more familiar X-ray diffraction) can be understood from the simple physics of a diffraction grating. Imagine light of wavelength λ incident on a diffraction grating of spacing *d*. [We will consider Fraunhofer diffraction (Figure 4.5), where the bulk of the beam passes through the diffraction grating unscattered,

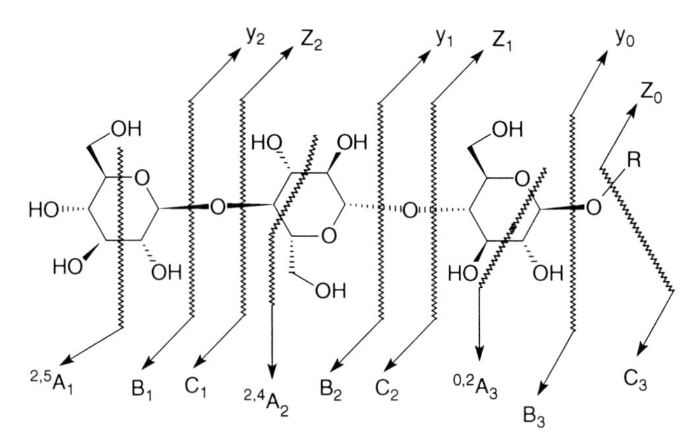

Figure 4.4 Nomenclature for mass spectral fragmentation of oligosaccharides.

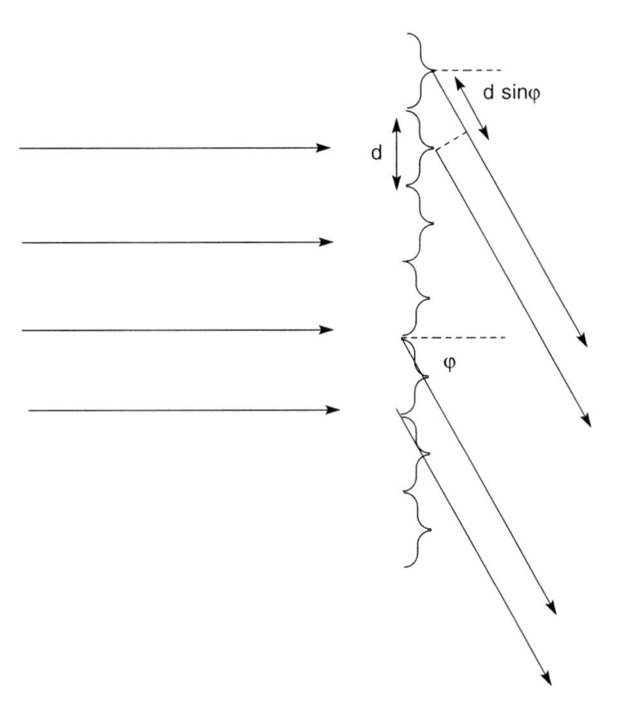

Figure 4.5 Classical Fraunhofer diffraction. (Standard trigonometry equation: $d\cos(90° - \varphi) = d\sin\varphi$)

as this more closely corresponds to various diffraction experiments.] Light is scattered each part of the diffraction grating. The pathlength difference between wavelets scattered from the same part of adjacent repeat units is $d\sin\varphi$, where φ is the angle of scatter. For the wavelets to reinforce each other, this distance must be an integral number of wavelengths, *i.e.* eqn. (4.1) must hold, where n is an integer:

$$d \sin \varphi = n\lambda \tag{4.1}$$

It can be shown that the longer the array of lines on the diffraction grating, the more the waves from light which does not obey this condition cancel each other out. If light with a range of values of λ is incident on the grating, then the light is split up into a spectrum, the familiar rainbow colours seen, for example, on a CD. If the light is monochromatic, then just a series of lines corresponding to various values of n is observed. These are the various orders of diffraction.

Now imagine a two-dimensional grating, as in Figure 4.6. There are lines of letter Rs which do not correspond to those defining the unit cell (the smallest asymmetric unit). The situation is analogous to that in a carefully planted forest, where if the trees are all planted in rows and the distance between trees in

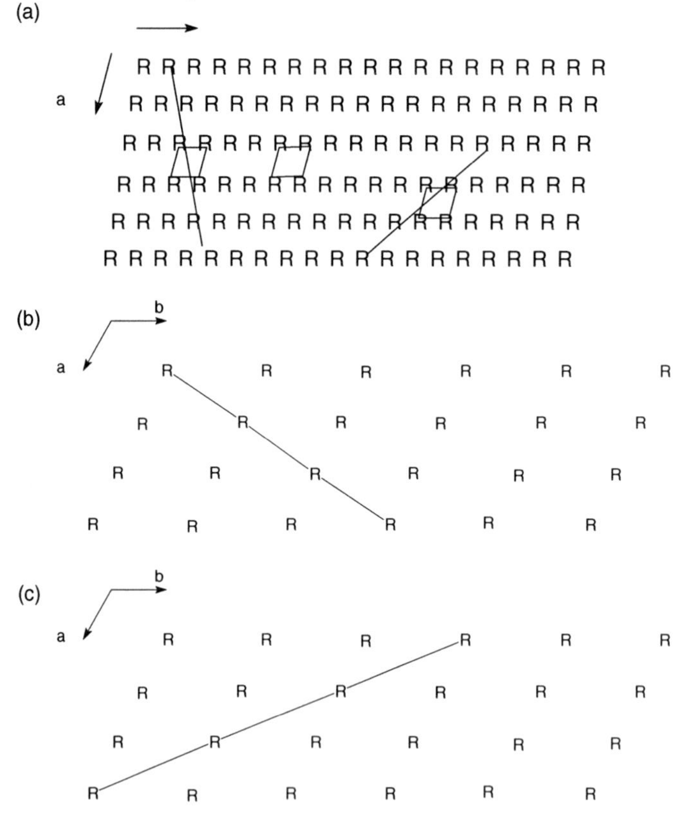

Figure 4.6 (a) Two-dimensional array, illustrating a two-dimensional unit cell and how lines with a regular repeat need not be parallel to an axis, (b) The $\bar{1}$,1 (or 1,$\bar{1}$) line of a two dimensional array. (c) The 1,1 (or $\bar{1}$,$\bar{1}$) line of a two dimensional array.

a row and between rows is a round number of metres, then one can often mistake a certain line of sight for a row of trees. These new lines will act just like a diffraction grating, as well as the horizontal and near-vertical rows with which the lattice is made up. The simplest repeating unit that defines the layout is called the unit cell – illustrated by the rhombuses of Figure 4.6(a), in which the letter R represents a lattice point (or tree, to continue the forest analogy). The Miller index of a lattice line (or plane, in three dimensions) is the reciprocal of the intercept on the unit cell axis: the Miller indices corresponding to the intercepts on the a, b, and c axes are given the symbols h, k, l respectively. Thus, the LH line drawn bisecting the unit cell in Figure 4.6(a) has $|k| = 2$, and the RH one $|h| = 2$. Lines (or planes) parallel to an axis have Miller indices of zero. In order to ensure Miller indices are integers, negative Miller indices are employed, so that a line cutting an axis 2/3 of the way along the unit cell has a Miller index of –3. The operation of the sign convention means that the h, k line is identical to the $-h$, $-k$ line, or in three dimensions, the h, k, l plane is identical to the $-h$, $-k$, $-l$ plane. Conventionally, a negative Miller index is written with the negative sign over the digit, rather than in front, *e.g.* $\bar{1}$ rather than –1.

The diffraction pattern recorded on a film from a planar array irradiated with monochromatic light would be a series of spots, corresponding to the various orders of diffraction from the regular lines comprising the equivalent of various diffraction gratings.

Crystals correspond to a diffraction grating in three dimensions. The unit cells now have axes a, b and c and the Miller indices of a particular plane are h, k, l. The condition for diffraction is understood from Bragg's[v] demonstration that scattering from a series of planes with Miller indices h, k, l was equivalent to reflection from those same planes. If the angle between the plane and the incident beam is θ, then for reflected beams, initially in-phase, to also be in-phase after reflection, the extra distance travelled by the reflected beam must be a whole number of wavelengths, *i.e.* eqn. (4.2) should hold (Figure 4.7):

$$n\lambda = 2d_{hkl} \sin \theta \tag{4.2}$$

(In fact, for any value of n, we can choose the Miller indices to take account of n, so that the Bragg equation is normally written $\lambda = 2d_{hkl}\sin\theta$.)

Just as a two-dimensional array can only be constructed by packing together polygons of certain shapes and symmetries (*e.g.* regular triangles, parallelograms and regular hexagons, but not regular pentagons or other polygons), so in three dimensions the three-dimensional shapes that can be packed together are limited to the 7 crystal systems. The unit cells of the seven crystal systems contain just one lattice point (in Figure 4.6, each corner "R" is shared between four unit cells). However, certain types of symmetry be more conveniently treated with unit cells containing 2, 3, or four lattice points, and this gives rise to the 14 Bravais lattices. The three most important from the perspective of

---

[v] Lawrence Bragg, the father, not William, the son, both X-ray crystallographers at Cambridge.

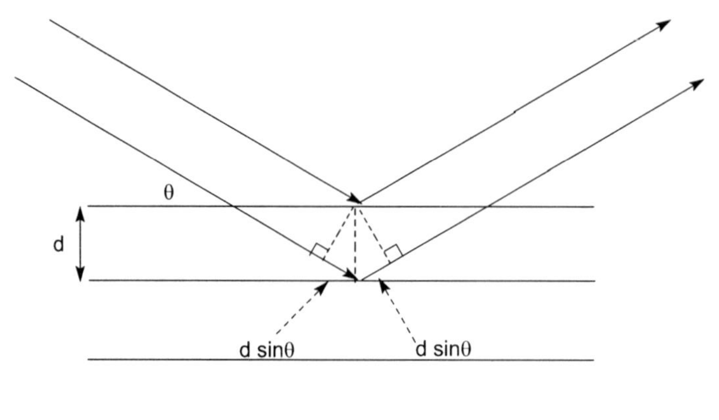

Figure 4.7 Cartoon illustrating how Bragg reflections from various lattice planes.

carbohydrate polymers are the triclinic, the monoclinic and the orthorhombic. The triclinic unit cell is a parallelepiped and is the simplest that can exist (an older term was primitive); no two sides have equal lengths or two angles that are equal. Two of the three angles defining the monoclinic unit cell are right-angles, but none of the three sides have the same length. All the angles defining the orthorhombic unit cell are right-angles, but all the sides are of different lengths.

Crystal unit cells are defined by the lengths of the three sides, a, b and c, and the magnitude of the three angles, α, β and γ: α is the angle between the b and c sides, β that between a and c and γ that between a and b. Where, and pointing in what direction, the axes should be placed is defined by convention: the current standard convention (at least for polysaccharides[10]) is illustrated in Figure 4.8, with the origin at the back LHS corner, the c axis pointing up, the a axis to the viewer and γ for the trigonal and monoclinic structures obtuse.

The crystal plane ($2h$, $2k$, $2l$) has half the spacing of the (h, k, l) plane and thus the conditions for fulfilment of the Bragg equation (equation 4.2) are the same for first order diffraction from the ($2h$, $2k$, $2l$) plane as for second order diffraction from the (h, k, l) plane, and usually all observed diffractions are treated as first order.

There are various types of and outcomes from, diffraction experiments on solids at the molecular level, summarised in Table 4.1. By far the most important for determining molecular structure in general are X-rays, although important information about polysaccharides has come from the diffraction of electrons and thermal neutrons. (From the De Broglie relationship, $\lambda = h/p$, the wavelengths of electrons, which are light, become commensurate with atomic dimensions only when p is made very large by accelerating electrons through, typically, around 100 kV; by contrast, the 1828-fold heavier neutrons achieve similar wavelengths with only thermal energy.)

Historically the oldest is the Laue method, in which "white" X-rays containing many frequencies impinge on a stationary single crystal. The result is a

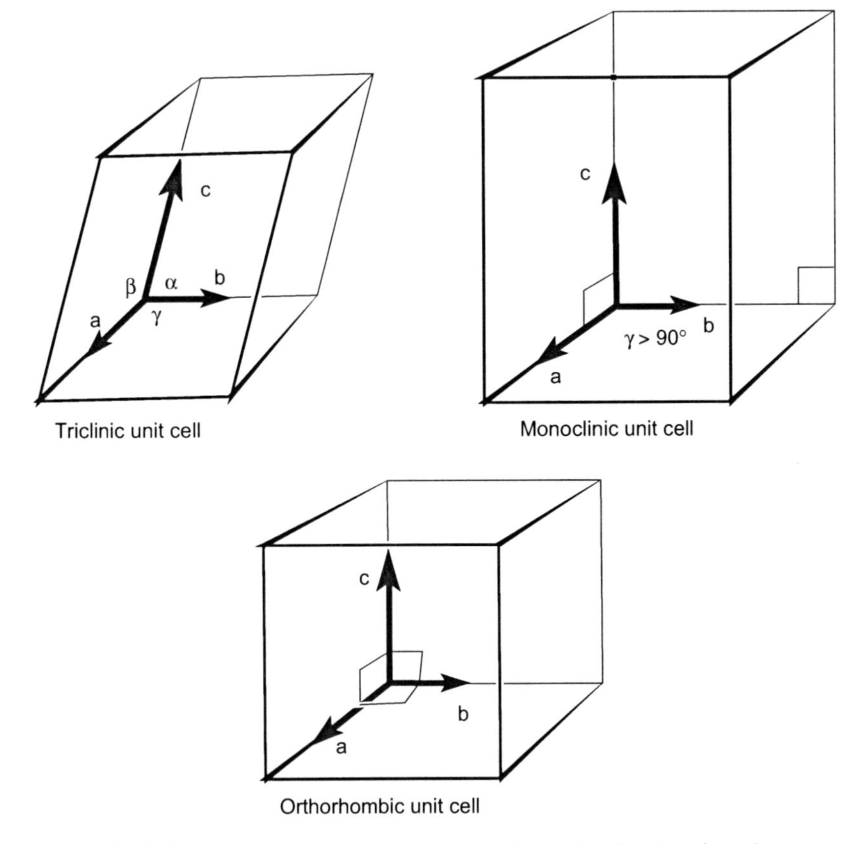

Figure 4.8 Triclinic, monoclinic and orthorhombic unit cells, showing placement of axes.

Table 4.1 Various configurations of diffraction experiments.

Source	*Sample*	*Rotation?*	*Pattern*	*Comments*
Continuous, "white"	Single crystal	No	Spots	Laue diffraction pattern
Monochromatic	Single crystal	No	Few spots if hit a diffraction plane	–
Monochromatic	Single crystal	Yes	Spots	"Standard" conditions
Monochromatic	Randomly oriented crystallites	No	Rings	Powder diffraction pattern
Monochromatic	Fibres	No	Spots on layer lines	If linear polymer molecules completely out-of-register, only one layer line

series of diffraction spots in arcs and circles. The X-rays produced by a synchrotron are "white" and one might expect that the Laue method would experience a revival in protein crystallography (see Section 5.6) but, except for a few studies in which the progress of a reaction was monitored in the protein crystal (see Section 5.12.2.6), the technical difficulties, such as correcting for the intensity spectrum of the incident light and the response spectrum of the detectors, and also differentiating between various orders of diffraction, make it unsuitable for structure determination.

If a beam of monochromatic[vi] X-rays is directed at a stationary single crystal, the chance of the Bragg condition [eqn. 4.2] being fulfilled with any given crystallographic plane is small. It is therefore routine to rotate (or sometimes oscillate) single crystals about one crystallographic axis during data accumulation. In this way, the familiar pattern of spots with (on average) decreasing intensity with distance from the unscattered beam is produced; the spots furthest away from the beam are scattered by the planes which are closest together. The spacing of the crystal planes producing the most distant spots which can still be measured defines the resolution of the experiment.

If an array of crystals are randomly oriented with respect to an incident monochromatic beam, then a series of concentric rings are produced. The spacings of the rings indicates the lattice dimensions and their sharpness the "degree of crystallinity" of the sample. Rather than record a series of concentric rings on a photographic plate, the traces are now instrumentalised (Figure 4.9).

Many water-soluble polysaccharides can be induced to orient themselves in a single direction by annealing and pulling spun fibres in a certain direction,[11] and many natural linear polymers (such as cellulose or chitin) are biosynthesised partly crystalline.

If we define the c (or z) direction as the direction of the fibre, then the polymer molecules are aligned parallel to some extent in this direction. The molecules will of course be a constant width, and even if they are completely out of register with each other in the c direction, the regularities of packing in the a and b directions will make a fibre much like a diffraction grating: Irradiation with a pencil of monochromatic X-rays perpendicular to the fibre will therefore result only in diffraction in the a, b plane. As the polymer molecules become in register with each other in the c direction, diffraction will take place in this plane too, and increasingly X-rays will be diffracted in a cone whose apex is the intersection of the X-ray beam and the fibre, rather than a fan in the ab plane. The fibre is not one big crystal, though, and regions which are increasingly ordered in the c direction are not in register with each other, so it is not necessary to rotate the fibre.

Fibre diffraction is one of the oldest of X-ray diffraction techniques: examination of fibrous proteins gave the α helix and of fibrous DNA the famous double helix.

[vi] X-rays are produced by the bombardment of a metal target with electrons *in vacuo*: electrons are knocked out of the K shell and the resultant refilling of the "hole" by outer-shell electrons results in sharp emission spectra, which by use of filters can be made effectively monochromatic. A commonly used radiation is Cu Kα.

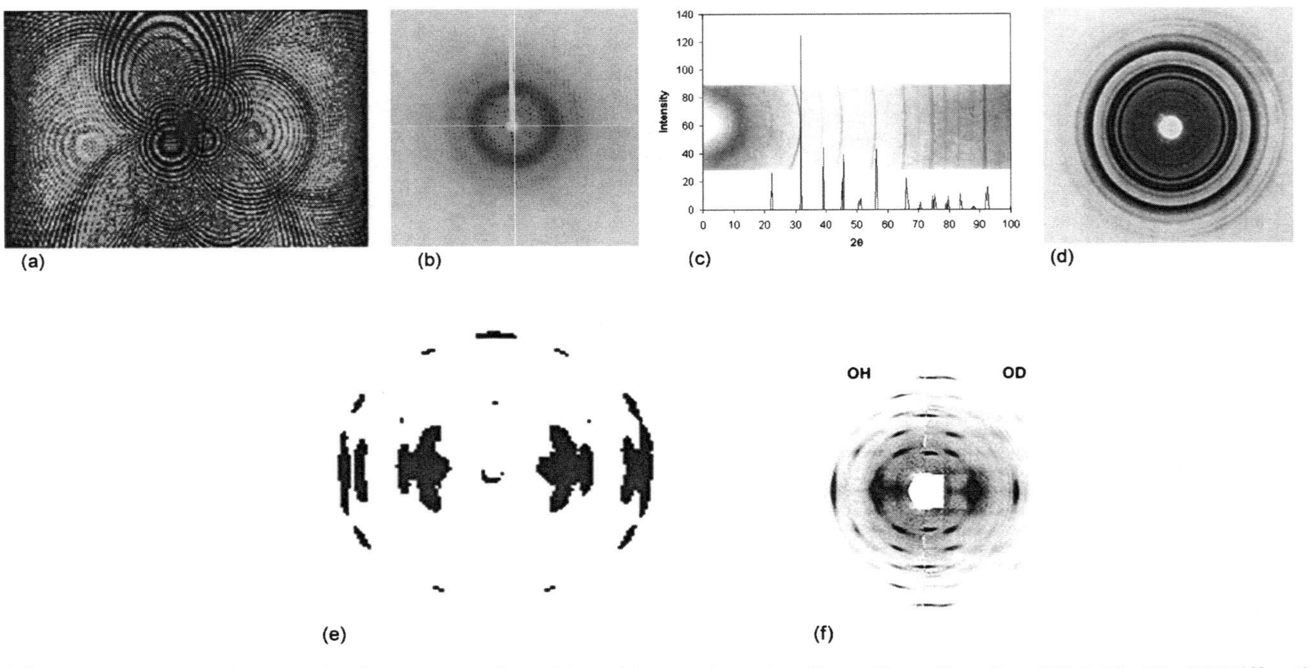

Figure 4.9 (a) X-ray Laue photograph of a concanavalin A/glucoside complex taken from *Chem. Soc. Rev.*, 2004, **33**, 548. (b) Diffraction of a lysozyme crystal under "standard" conditions. Note the large number of spots. Taken, with permission, from biop.ox.ac.uk/www/lj2001/ garman/garman_2a.html. (c) Powder diffraction pattern of a highly crystalline metallic catalyst. Note the sharp lines and the modern, digitised trace. Taken, with permission, from www.doitpoms.ac.uk/tlplib/xray-diffraction/printall.php (d) Power diffraction pattern of starch. Note the few, fuzzy rings indicating overall poor crystallinity (e) Fibre diffraction pattern of a hyaluronate salt. The very few diffraction spots not in the *a,b* plane (horizontal line through the undiffracted beam) indicates that the hyaluronate chains are only poorly in register with each other along the fibre direction. Taken, with permission, from http://www.glycoforum.gr.jp/science/hyaluronan/HA21/HA21E.html) (f) Fibre diffraction pattern formed by scattering of neutrons by (LHS) O-protiated cellulose II and (RHS) O-deuterated cellulose II. Reproduced from reference 35 by kind permission of Dr. Henri Chanzy.

X-rays are scattered by electron density – the more electrons an atom has, the more intensely it scatters. Neutrons, however, are scattered by a parameter of the atomic nucleus, which is different for different isotopes (but shows no general trend with nuclear mass). Hydrogen and deuterium have very different (but large) neutron scattering cross-sections and therefore neutron diffraction, which requires access to an atomic reactor, is used where location of hydrogen atoms is critical.

As discussed earlier in connection with the Fischer convention for depicting stereochemistry, conventional diffraction studies cannot distinguish chirality. A consequence of this for fibrous polymers is that, even if one knows the unit cell and contents, one cannot tell which way the polymer chain is running by standard X-ray (or other) diffraction. Both cellulose and chitin [β-D-GlcNAc*p*-(1→4)]$_n$ can adopt monoclinic crystal forms (and cellulose also a triclinic one), in which the polysaccharide chains are parallel and run parallel to the *c* axis. Fibre X-ray diffraction patterns, however, were compatible with either a "parallel up" structure, in which, in the monoclinic unit cell of Figure 4.8, the chains running from non-reducing to reducing end also ran from bottom to top of the figure; the parallel down structure was the converse. Another way of expressing this is that for each saccharide unit in the "parallel up" structure the *c* coordinate of O5 was larger than that of C5, whereas for the "parallel down" structure the *c* coordinate of O5 was smaller than that of C5. An ingenious solution has been devised by Sugiyama and co-workers, using a combination of electron diffraction and microscopy.[12] The following explanation is taken from their own description of their work on *β*-chitin (Figure 4.10).[13] Two methods are used. In the first, a chitin *β* microcrystal is selected and rotated on the microscope stage until the *c* axis is parallel to the axis of a part of the stage which it is possible to tilt. The crystal is adjusted until an incident beam of electrons produces a set of diffraction spots from the $(1\bar{1}0)$ plane (which a practised observer would recognise). If the *c* axis is in the direction shown, then a tilt of (as it happens, with this particular chitin polymorph) –27° will produce the pattern of spots from diffraction in he [010] plane, whereas a rotation of 28° will produce the diffraction pattern from the [100] plane. In the second method, the direction of tilt is perpendicular to the *c* axis. The microcrystal is manipulated until the reflections from the $[1\bar{2}0]$ plane appear. Tilting in one direction brings reflections from the $[0\bar{1}3]$ plane and in the other the [013] plane. It is generally advisable to label one end of the microcrystal (*e.g.* by treatment with an *exo*-acting enzyme or by direct staining of the reducing ends), as crystal habit may make identification of crystallographic direction difficult, and do both experiments on crystals from the same preparation.

4.2.4 The Role of Fourier Transforms

Two techniques which are used extensively in the determination of polysaccharide structure and conformation are unusually configured X-ray crystallography and NMR spectroscopy. Both techniques rely on the availability of

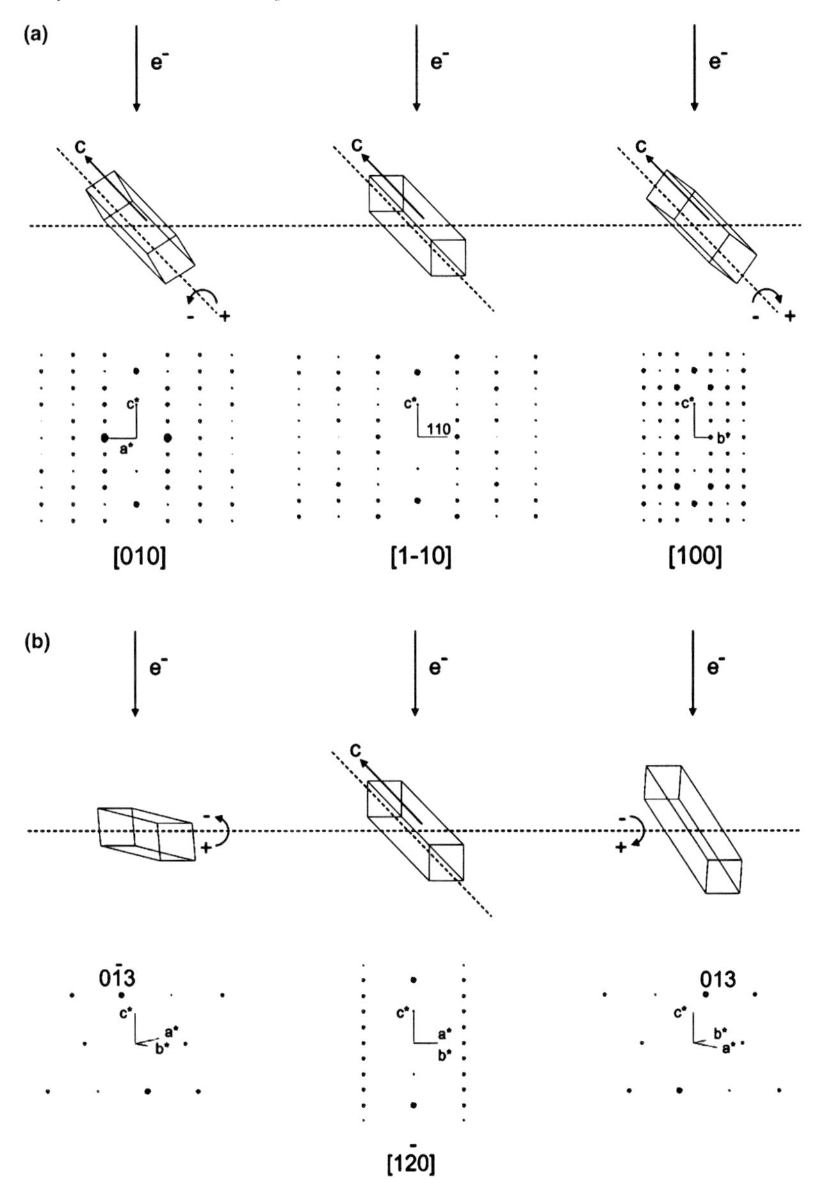

Figure 4.10 The Sugiyama tilt technique for determining the direction of a polymeric chain which aligns with the *c* axis. (Reproduced from Ref. 13 with permission from the Biochemical Society)

powerful computers and the key to their success is the Fourier transform (FT) theorem. This theorem states that any periodic function can be expressed as the sum of a series of sinusoidal functions with frequencies which were integral multiples of a fundamental frequency (v, $2v$, $3v$, $4v$, ...) with different amplitudes and phases. Mathematically, any sinusoidal wave can be written as in eqn. (4.3), where W is the perturbation in whatever medium the wave is in, A the amplitude, λ the wavelength and φ the phase:

$$W = A\cos[(2\pi x/\lambda + \varphi)] \tag{4.3}$$

This is illustrated in Figure 4.11, which also illustrates how different phases can alter the appearance of a periodic function.

Each component of the Fourier series, once its position is known in the sequence, needs be specified by only two parameters, amplitude and phase. A consequence is that complex wave forms can be represented by plots of intensity (*i.e.* A^2) against frequency and vice versa. The Fourier transform, which converts one to the other, is usually written as in eqn. (4.4), where ω is the frequency:

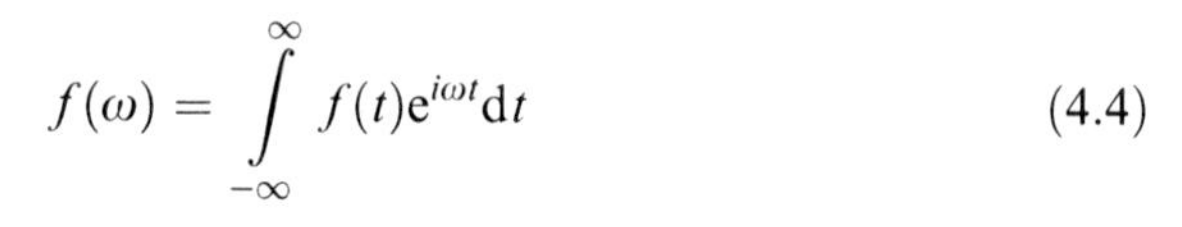

$$f(\omega) = \int_{-\infty}^{\infty} f(t)e^{i\omega t}dt \tag{4.4}$$

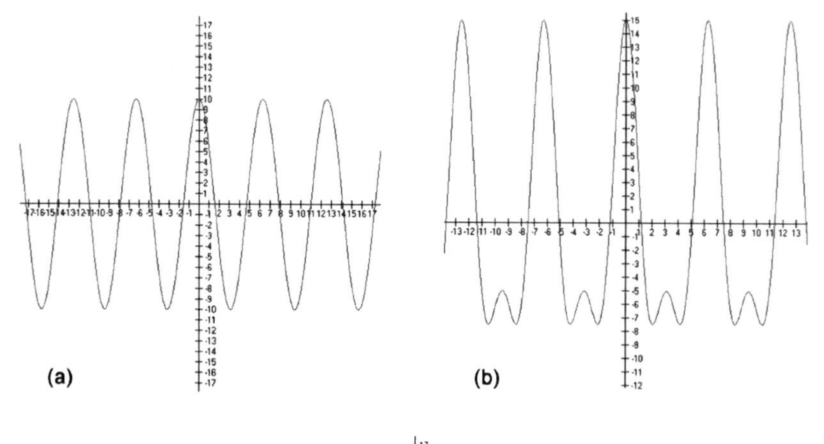

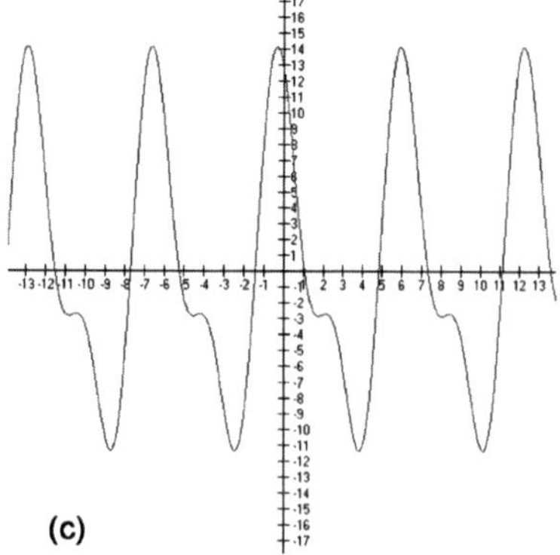

Figure 4.11 (a) A simple cosine wave (plot of $y = 10\cos x$); (b) summation of two cosine waves (plot of $y = 10\cos x + 5\cos 2x$; (c) effect of phase [plot of $y = 10\cos x + 5\cos(2x + 1)$] (the x-axis is in radians; $180° = \pi = 3.14159$ rad).

The principle can be understood if an NMR spectrum is considered: just one line corresponds to the absorption or emission of radiofrequency radiation of just one frequency, whereas say a 1.2.1 triplet will correspond to three frequencies with amplitudes 1:$\sqrt{2}$:1. Simple Fourier transform NMR spectroscopy involves giving a short burst of radiofrequency power [the maximum duration is given by Heisenberg uncertainty in the frequency which encompasses all resonances of interest, eqn. (4.5), but the pulse is usually much shorter than this]. The signal initiated by this burst of power then decays exponentially, but during the decay the wave-form is digitised and stored. Figure 4.12 illustrates the effect of adding an exponential decay turn to the three-component wave produced by a 1:2:1 triplet, when the graph looks much like an FID (free induction decay) displayed on the monitor of an FT NMR spectrometer. When the signal has decayed to an appropriate value (and there is a whole art in deciding this), the system is pulsed again. When sufficient signals have been accumulated (the signal-to-noise ratio increases as the square root of the number of accumulations), the Fourier transform is performed.

$$h \approx \Delta Et = h\Delta\nu t; t \sim 1/\Delta\nu \tag{4.5}$$

The power of FT NMR is that one is not confined to a single exciting pulse. One can have several pulses with various durations, delays and phases in order to edit a one-dimensional spectrum. Or one can have an array of pulses with a variable evolution time and then perform the Fourier transform with respect to both the evolution time and the decay of the FID, generating a two-dimensional spectrum whose output is a contour plot. With very powerful machines (>600 MHz, ^{1}H) it is even possible to perform the Fourier transform in three dimensions, with two evolution times. These pulse sequences are known by (usually arch) acronyms such as COSY, INADEQUATE, *etc.*, and modern NMR machines are supplied with the hardware and software to perform the commoner experiments already installed. It is not necessary to understand fully the spin physics behind such sequences in order to use them, but the basic viewpoint used in their description is worth grasping.

This involves observing the various magnetic fields in a rotating set of coordinates. When a magnetic field is applied to a spinning nucleus, it experiences a torque and precesses about the applied static magnetic field. One can imagine the magnetic moments of all the nuclei in the sample added together in a magnetisation vector precessing around the applied, static magnetic field (by convention, the z direction of a set of Cartesian coordinates). The viewpoint adopted is that of an observer on the z axis who is rotating at the precession frequency (the Larmor frequency, which it can be shown is the frequency of the radiowave absorbed or emitted by transitions between the energy levels of the nucleus in the magnetic field). The magnetisation in the rotating coordinates thus appears to be static. When an rf field is applied this can be decomposed into two oppositely rotating components; in the rotating coordinate framework the important component is the static one on the x axis. This applies a torque to the main magnetisation, which starts to move in the yz plane. When it has moved 90° ($\pi/2$ in the radian units used in NMR) it is static along the y axis in

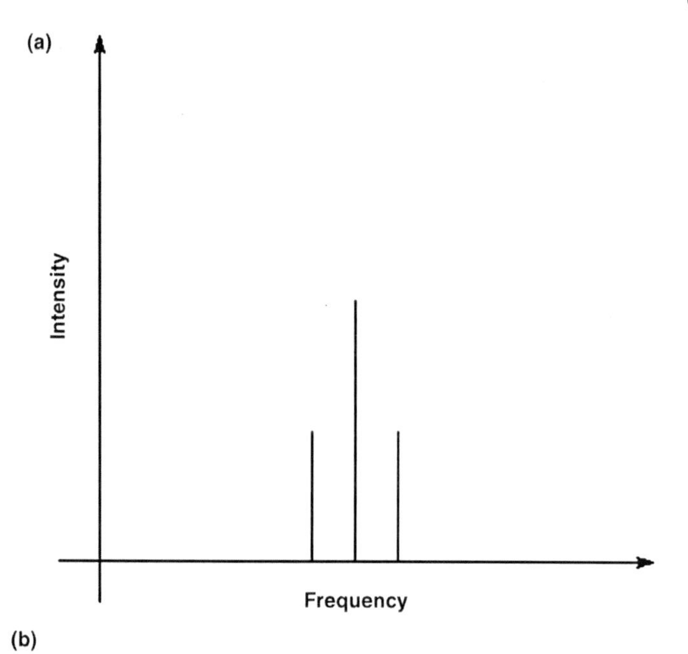

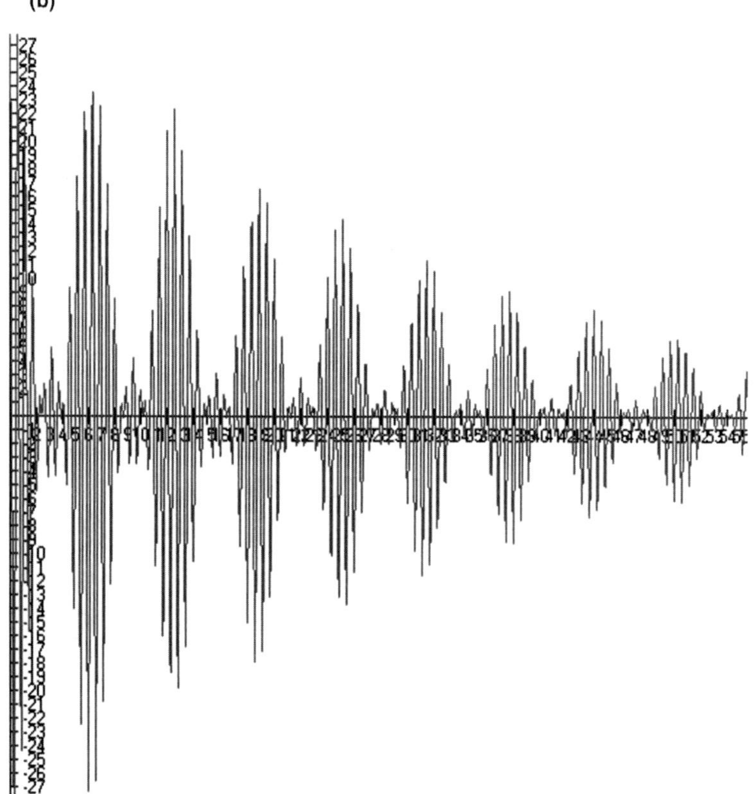

Figure 4.12 (a) Line spectrum of a 1:2:1 triplet and (b) its Fourier transform with decay of the signal superimposed [$y = (10\cos 10x + 14.14\cos 11x + 10\cos 12x)\text{-}\exp(-x/30)$].

the rotating frame but precessing about B_0 in the laboratory frame. If the radiofrequency power is now switched off, the precession generates an rf signal which is detected as an FID and slowly decays. On the other hand, if the rf pulse lasts until the magnetisation has travelled through 180°, no signal is observed when the power is switched off as there is no precession that would generate such a signal. For this reason the exciting pulse in simple FT NMR is $\pi/2$ (Figure 4.13).

In FT NMR, one starts off with a series of Fourier components and generates a spectrum, whereas in X-ray crystallography the transformation is in the reverse sense. The raw data of an X-ray diffraction experiment are a two-dimensional array of diffraction intensities, much like an oddly plotted spectrum in which x and y axes are replaced by the angles of a polar coordinate system. The desired information is the detailed shape of the repeating electron distribution in three dimensions in the crystal, to which the molecular structure can be fitted. The main problem in reconstructing the electron density distribution is that, as incoherent X-radiation has been used to obtain the diffraction pattern (that is to say, each X-ray quantum has its waves "out of step" with the others), phase information has been lost. A comparison of Figure 4.11b and c shows what a drastic effect different phases can have on the appearance of the repeat unit, even though the waves have the same intensity. This "phase problem" has been solved in various areas of crystallography: for small molecules, where X-ray crystallography is now a routine service, the solution

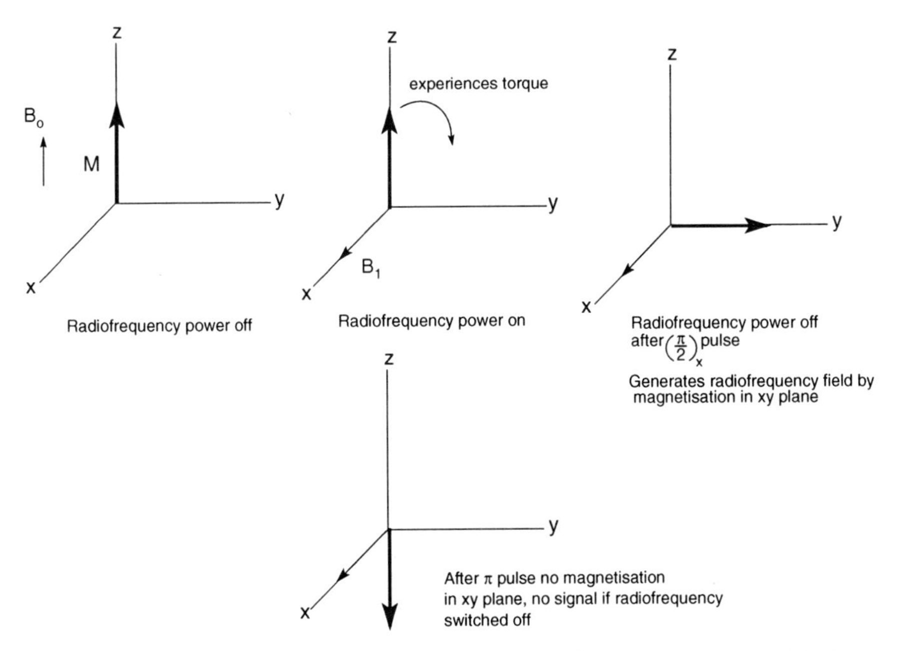

Figure 4.13 Meaning of a "$\pi/2$" pulse. B_0 is the applied static magnetic field (from the instrument magnet) and B_1 one of two components of the applied rf field.

has come from computers powerful enough to make enough guesses at the phases that the X-ray diffraction patterns can be fitted to experiment directly (direct methods). In protein crystallography the technique of isomorphous replacement is still used: small, intensely scattering heavy atoms (such as Hg) are introduced into one or two places in the protein structure by soaking the crystals in solutions of the appropriate derivatives. The changes in diffraction intensities then permit the phases to be estimated on the assumption the rest of the scattering density is not changed.

The reconstruction of the conformation of a polysaccharide from a fibre diffraction diagram is a much less sophisticated business and basically involves model-building according to known rules of valency and to maximise productive contacts such as hydrogen bonding.[14]

4.2.5 Use of NMR Pulse Sequences in the Determination of Sequence – an Example

Once the constituent sugars of a polysaccharide are known and one has some idea of their substitution pattern, conditions for NMR investigations are established. Sometimes the purified polysaccharide will, with some coaxing (*e.g.* higher than ambient temperature, pH optimisation for non-neutral polysaccharide), give sharp spectra. More often, though, particularly as many polysaccharides are gums and mucilages, the microscopic viscosity will be too high – the polysaccharide molecules will not undergo enough thermal motion to average out local magnetic fields and broad spectra will result. It is then necessary to degrade the polysaccharide partially to smaller fragments, *e.g.* by ultrasonic irradiation. The danger with this procedure is that labile "decoration" on the main polysaccharide (such as *O*-acetyl groups or cyclic acetals of pyruvic acid) may be removed.

One-dimensional proton and carbon spectra are often immediately informative. Because in general anomeric protons and carbons resonate at lower field than the rest of the molecule (from the electron-withdrawing effect of two oxygens), the number of saccharide units in the repeat can often be deduced by simply counting anomeric resonances. Here anomeric ^{13}C resonances, with their wider dispersion (δ 95–109 ppm), are more useful than anomeric hydrogen resonances, which can overlap with each other and, in the case of the upfield axial anomeric hydrogens, with other resonances, even at very high fields. In cases where individual anomeric proton resonances are well separated, the α- and β-pyranosidic linkages can be seen. Acetyl groups can also often be identified, and also the methyl groups of reduced sugars like rhamnose.

The sensitivity of ^{13}C is inconveniently low, and in the conventional ^{1}H-decoupled ^{13}C spectrum the excitation of the proton resonances increases the strength of the ^{13}C signal in addition to simplifying it. However, rather than simply applying rf power to the protons, a series of pulses to both protons and carbons results in Distortionless Enhancement by Polarisation Transfer (DEPT) and gives one-dimensional ^{13}C spectra with usefully higher sensitivity than from proton decoupling. Additionally, the one-dimensional spectrum displays ^{13}C resonances of carbons with an odd number of protons attached as sharp lines above the baseline, whereas carbons with two protons attached are sharp lines below the baseline.

The use of the various pulse sequences will be illustrated with respect to the neutral polysaccharide produced by a key bacterium in the manufacture of yoghurt, *Lactobacillus delbrueckii* var. *bulgaricus*[vii,viii].[15] The polysaccharide has the structure given in Figure 4.14.

Complete hydrolysis gave galactose and glucose in a 4:3 ratio and methylation analysis gave 1,5-diacetyl-2,3,4,6-tetramethylgalactitol, 1,3,5-triacetyl-2,4,6-trimethylglucitol, 1,4,5-triacetyl-2,3,6-trimethylglucitol, 2,4-dimethyl-1,3,5,6-tetracetylgalactitol and 6-methyl-1,2,3,4,5-pentacetylglucitol in the molar ratio 3:1:1:1:1. The tetramethylgalactitol indicated terminal galactopyranose, 1,3,5-triacetyl-2,4,6-trimethylglucitol indicated 3-substituted glucopyranose and 2,4-dimethyl-1,3,5,6-tetracetylgalactitol indicated 3,6-disubstituted galactopyranose, but methylation analysis could not distinguish between furanose and pyranose rings for the other two residues.

The DEPT spectrum (Figure 4.15) immediately confirms the seven-saccharide repeat, since there are seven anomeric resonances; such a confirmation of the results of methylation analysis by NMR is important, because incomplete methylation can give the appearance of additional branch points. Moreover, the hydroxymethylene carbons at high field show inverted peaks, which identifies them as the C6 carbons of the various residues.

The next task is to assign the signals to individual saccharide units and here multi-dimensional techniques are essential. The first historically was COSY, a homonuclear shift correlation technique with a simple pulse sequence [$(\pi/2)_x$–evolve–$(\pi/2)_x$–acquire]. Fourier transformation of the accumulated FIDs with respect to both the acquisition time and the evolution time results in a three-dimensional contour map in which a diagonal mountain range (usually presented SW–NE) is accompanied by smaller hills in the surrounding plains. The projection of the three-dimensional contours on either the x or y axis gives the ordinary one-dimensional spectrum. The hills ("cross peaks") arise when peaks in the one-dimensional spectrum are strongly coupled together: if a peak on the diagonal at $x=y=a$ is strongly coupled to a peak at $x=y=b$, then cross peaks arise at $x=a$, $y=b$ and $x=b$, $y=a$. It is therefore often possible to follow a set of three-bond hydrogen couplings round a sugar ring, starting from

---

[vii] I thank Drs. Lindsay Harding and Andrew Laws, University of Huddersfield, for provision of these spectra.

[viii] The binary (Linnean) names for organisms are written in italics, with the first name the genus and the second the particular species in that genus. The name is written in full the first time it appears, the genus name subsequently being abbreviated to its first letter (unless organisms from two genera with the same initial letter are being discussed). Although Prokaryota and Archaea may be simple organisms, an inoculum is not the same as a sample of a pure organic chemical and will contain many variants. When a variant becomes a different species is, with these simple organisms, a very fine point and with the advent of rapid DNA sequencing, classical topological, nutritional and staining classifications are being overturned regularly. A notable example is a workhouse cellulolytic microorganism, which started out as *Pseudomonas fluorescens* subspecies *cellulosa* and is now *Cellvibrio japonicus*.

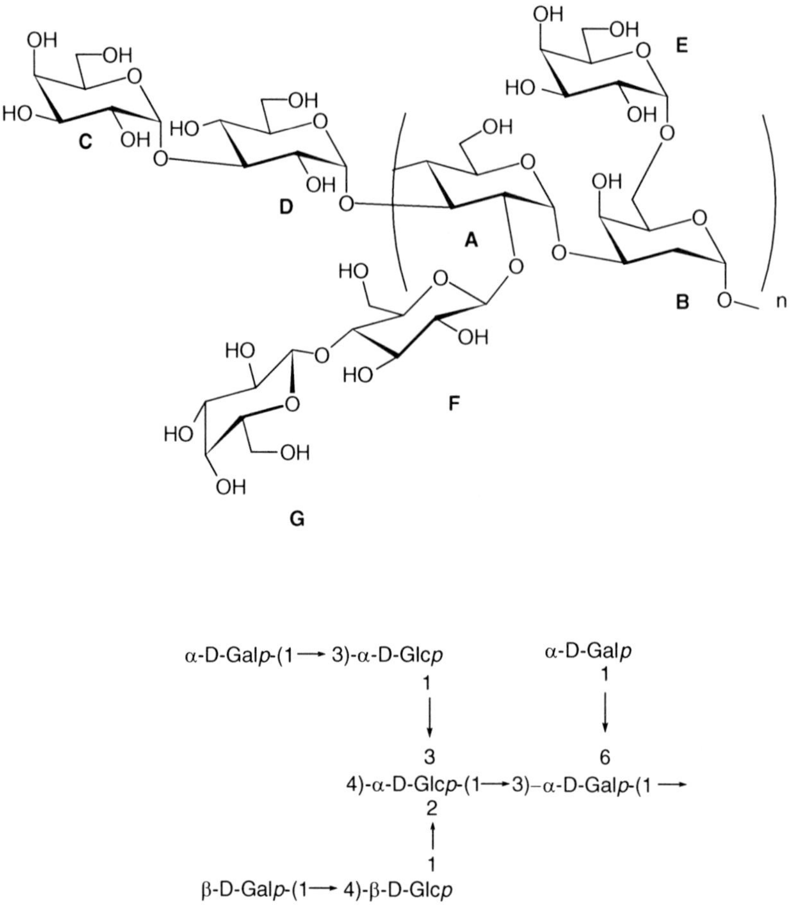

α-D-Gal*p*-(1→ 3)-α-D-Glc*p* α-D-Gal*p*
1 1
↓ ↓
3 6
4)-α-D-Glc*p*-(1→3)–α-D-Gal*p*-(1 →
2
↑
1
β-D-Gal*p*-(1→ 4)-β-D-Glc*p*

Figure 4.14 Structure of the extracellular polysaccharide of *L. delbrueckii* var. *bulgaricus*.

the anomeric resonance. COSY spectra have a mirror plane on the SW–NE diagonal and spectra are often symmetrised to increase the signal-to-noise ratio. The digital resolution of COSY spectra is often less than that of 1D spectra and the fine structure of the cross peaks can be lost, but if it is not, the fine structure is a combination of the fine structure of the two coupled resonances. Thus, in the case of a β-D-glucopyranoside, the 1D H2 resonance is approximately a 1:2:1 triplet, because of nearly equal *trans*-diaxial coupling to H1 and H3. In the usual form of COSY spectrum, the cross peaks between H1 and H3 have six components, equal doublets when viewed along the lines connecting the cross peak to H1 on the diagonal, but 1:2:1 triplets when viewed along the lines connecting H2 to the cross peaks. With a colour printer it is possible to represent COSY spectra with the sign of the cross peak still present.

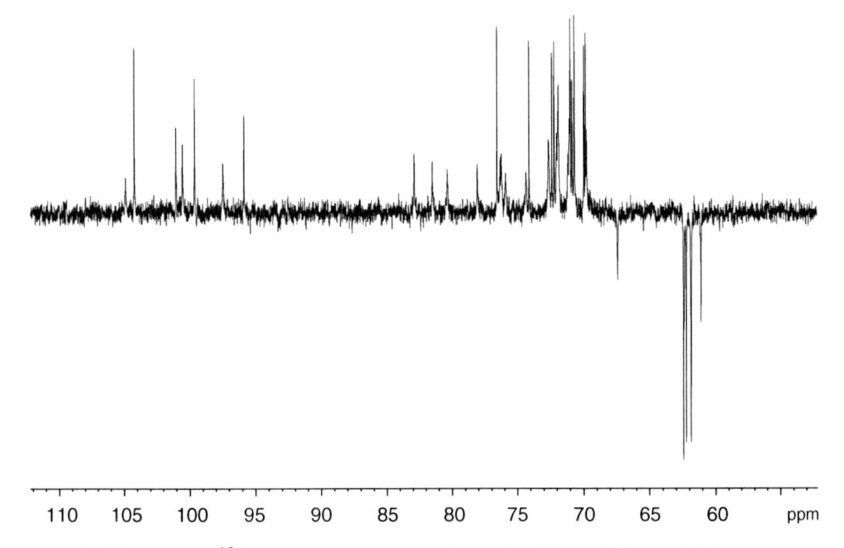

Figure 4.15 DEPT ^{13}C spectrum of the polysaccharide in Figure 4.14.

There have been many refinements to the original 1971 COSY sequence in addition to symmetrisation; the version commonly used today, DQF-COSY (for Double Quantum Filtered) introduces a third (orthogonal) $(\pi/2)$ pulse immediately after the second of the classic COSY sequence. This has the effect of suppressing all singlets (including, particularly usefully, the solvent). A DQF-COSY spectrum of the polysaccharide in Figure 4.14 is given as Figure 4.16.

A problem with the various versions of COSY is that the size of the cross peaks increase with coupling constant, so that detection of proton–proton couplings over more than three bonds is a rarity. A homonuclear technique which visualises all the couplings of a system is TOCSY, in which the second $(\pi/2)_x$ pulse in the COSY sequence is replaced by a spin lock [*i.e.* basic sequence $(\pi/2)_x$–evolve–spin lock–acquire]. This gives cross peaks for all the spins in the spin system, not just those which are directly coupled to each other, but is otherwise similar to a COSY spectrum (including the symmetry about the diagonal, which, however, can be exploited by the instrument software before presentation). Thus, an anomeric proton at $x = a$, $y = a$ can in principle display cross peaks along both the lines $x = a$ and $y = a$, for H2, H3, H4, H5, H6$_R$ and H6$_S$. In practice, however, polarisation transfer to remoter nuclei takes time and the spin-lock time determines the peaks actually observed (Figure 4.17).

The final step in the assignment of all resonances is correlating the ^{13}C peaks with the proton peaks. This can often be done from chemical shift alone for the anomeric resonances, but not for other ring atoms. The most important pulse sequence here is HSQC (Heteronuclear Single Quantum Correlation), which involves six pulses to protons and four to ^{13}C. The spectrum is now a non-symmetrical map with a peak at each carbon attached to a proton; the projection on the ^{13}C axis is the ^{13}C DEPT spectrum and on the proton axis the ordinary proton spectrum (Figure 4.18).

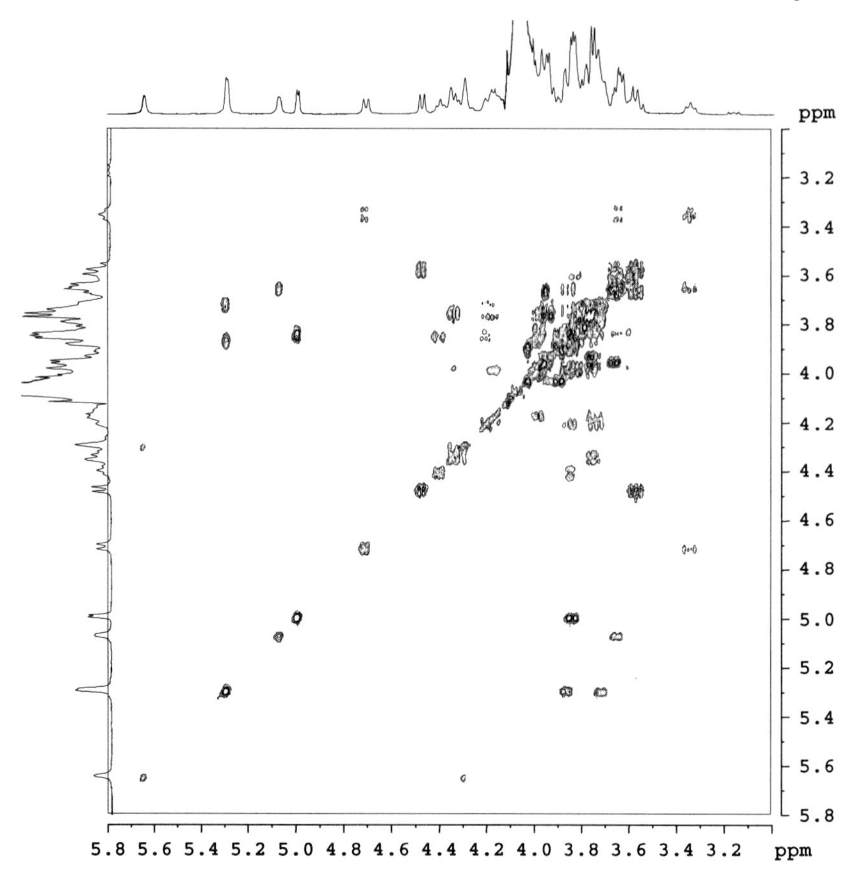

Figure 4.16 DQF-COSY spectrum of the polysaccharide in Figure 4.14.

With the foregoing techniques, it was possible to assign all the resonances to each monosaccharide unit; the chemical shifts of H5 and C5 for residues A and F established their ring form, left ambiguous as between pyranose and furanose by methylation analysis, as pyranose. It then remained to determine how the units are put together. DEPT, COSY, TOCSY and HSQC all exploit through-bond ("scalar") coupling between nuclei. It would in principle be possible to determine connectivity between monosaccharide units by detecting scalar coupling between anomeric protons and the ring carbon of the next saccharide unit. Techniques such as HMBC [Heteronuclear Multiple Bond Correlation (spectroscopy)] and INEPT (a similar sequence to DEPT) can do this in principle, but in practice their use is confined to polysaccharides with simple repeats (such as the two-unit repeats with which the techniques were first explored[16,17]). Rather, through-space interactions, in particular the nuclear Overhauser effect (NOE), are exploited. Irradiation (to saturation) of one nucleus will alter the Boltzmann distribution of a nearby nucleus in a magnetic field. The commonest mechanism is simply an interaction between the magnetic dipoles of the nuclei through space, when the strength of the interaction is

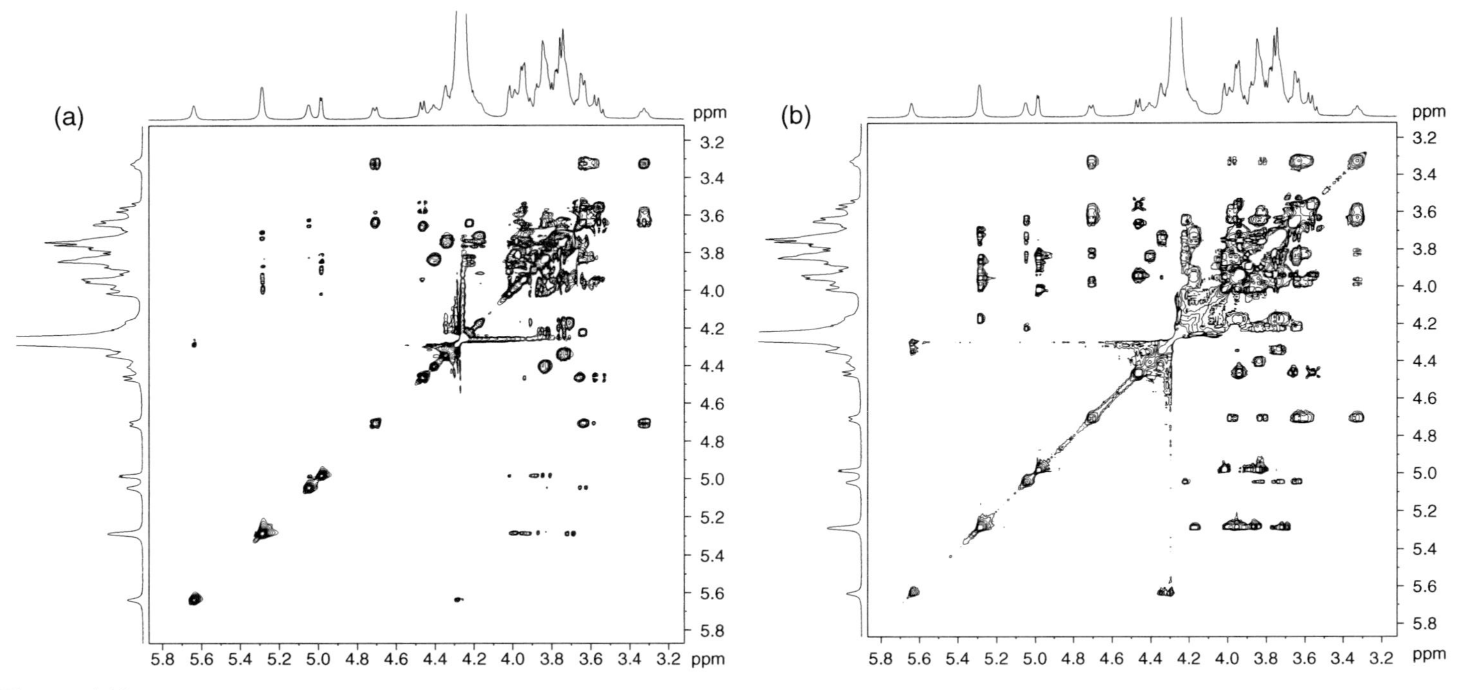

Figure 4.17 TOCSY spectra of the polysaccharide shown in Figure 4.14, (a) with a 60 ms lock (b) a 210 ms lock.

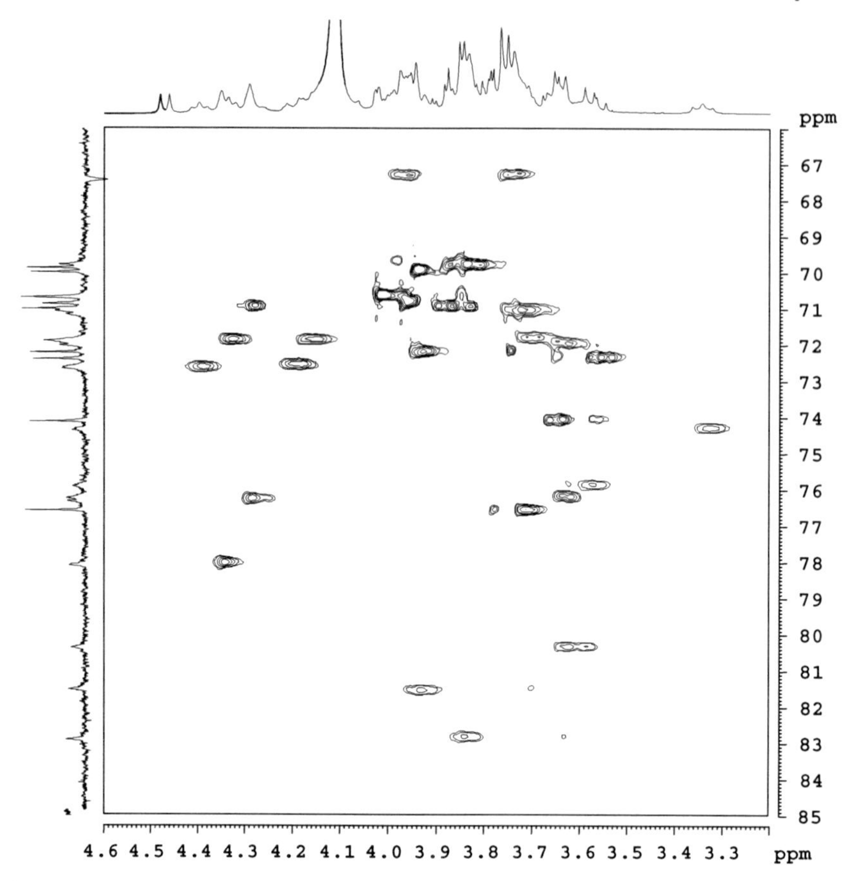

Figure 4.18 HSQC spectrum of the non-anomeric ring protons and carbons of the polysaccharide shown in Figure 4.14. Note the lowest-field methylene carbon at δ 67.3, corresponding to the glycosylated 6-position of residue B.

inversely proportional to the sixth power of the distance between them. The pulse sequence which gives rise to two-dimensional correlation spectra is ROESY (Rotating-frame Overhauser Effect SpectroscopY) and cross peaks are associated with nuclei less than about 5 Å between them (Figure 4.19). The pulse sequence is similar to that in TOCSY, and indeed with injudicious settings of machine parameters in either experiment, misleading cross peaks due to the operation of the other phenomenon can be seen (NOESY, the standard pulse sequence for determining nuclear Overhauser effects, gives poor results with carbohydrates because of their slow tumbling times).[18]

4.2.6 High-resolution Solid-state ^{13}C NMR – CP-MAS

NMR spectra of solids are generally broad, because each nucleus in the solid sample experiences slightly different magnetic fields, due to the different positions of other magnetic nuclei. In most mobile liquids, molecular tumbling

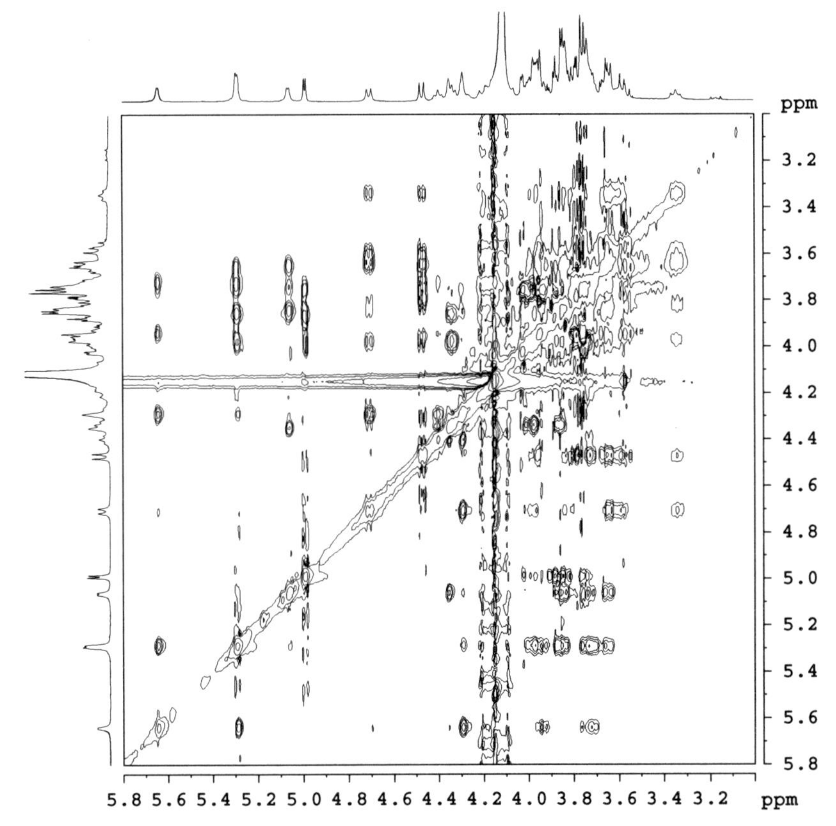

Figure 4.19 ^1H–^1H ROESY spectrum of the polysaccharide in Figure 4.14. The structurally informative Overhauser peaks are between anomeric protons and the hydrogens of the next residue.

is fast compared with the lifetime of spin states, so that the local fields are averaged out and sharp lines result. Rigorous analysis of a system of interacting nuclear spins indicates that the interaction energies due to chemical shielding, heteronuclear spin interaction and homonuclear spin interaction all have terms multiplied by $(3\cos^2\theta - 1)$, where θ is the angle between the magnetic dipole and the applied magnetic field. At the "magic angle", $\theta = 54° 44'$, $3\cos^2\theta = 1$, these terms disappear. If the sample is rotated very fast (120 000–1 200 000 rpm: sophisticated engineering is involved) about an axis making an angle of $54° 44'$ with the applied field, magnetic dipolar interactions from along the other two orthogonal axes can be averaged out, much as in a liquid. Magic angle spinning (MAS) can thus give high resolution spectra of solids. A refinement, with nuclei such as ^{13}C which are present only in low abundance, so that there are negligibly few ^{13}C–^{13}C interactions, is cross-polarisation (CP), similar to an INEPT experiment. Essentially, rf fields are applied so as to excite both proton and ^{13}C resonances for a locking time, then the ^{13}C rf field is switched off (the proton field continuing, for spin decoupling) and the ^{13}C signal followed. The outcome of a CP-MAS experiment is a high-resolution proton-decoupled ^{13}C spectrum, similar to one in solution.

4.2.7 Atomic Force Microscopy

In the investigation of surfaces and macromolecules on surfaces, enormous advances have recently come from atomic force microscopy (AFM), and it is now possible to visualise the conformation of individual polysaccharide molecules,[19,20] and also to measure directly the forces between them,[21] as well as the forces responsible for conformational transitions of a single sugar unit.[22] At the heart of the equipment is a probe consisting of a tip of hard material (silicon or silicon nitride) attached to a cantilever with a very low deflection constant, so that it can be "bent" by atomic-scale non-covalent forces (Figure 4.20). The tip can be pyramidal, tetrahedral or conical, with conical tips giving the highest resolution (tip diameter about 50 Å). Laser light is shone on the back of the cantilever and reflected back to a photodetector. This creates a record of the deflection and also permits, via the generation of a voltage applied across a piezoelectric ceramic supporting the sample, the movement of the sample.

The atomic force microscope can be configured in several ways, the most obvious (contact mode) merely involving scanning the tip over the sample at regular intervals, rasper fashion, and recording the deflection. Because of the proportionately very large capillary forces that arise from a contamination layer, imaging of polysaccharides in direct contact mode is carried out under a solvent

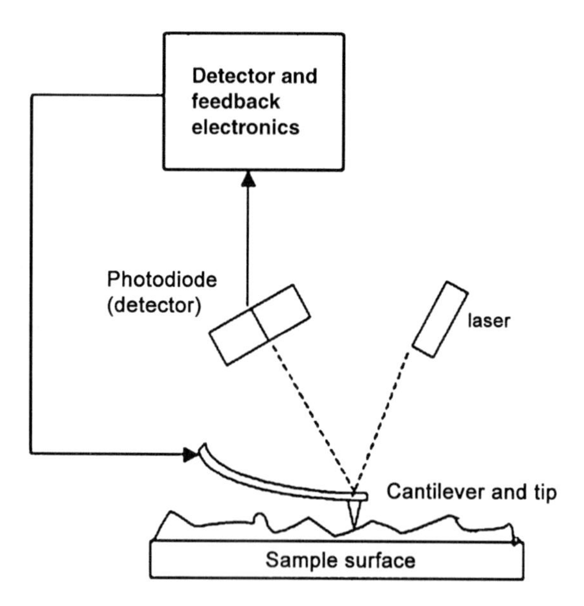

Figure 4.20 Cartoon of atomic force microscope. The sample mount is usually on a piezoelectric crystal, which can be moved by a voltage from the data-processing module, in response to the signal from the cantilever.

such as water or butanol. A refinement that has been used for most imaging of polysaccharides is the constant force mode: the piezoelectric feedback circuits are used to raise or lower the sample to maintain a constant deflection of the tip. This mode was successfully used to image the surface of the cellulose of *Valonia ventricosa* cellulose.[23] Here, however, the highest resolution was obtained in so-called "deflection" or "error" mode: the feedback to the piezoelectric ample support was switched on, but the deflection of the cantilever also recorded, so that two images were produced: one was the normal height image and the other was created from the mismatch between where the tip was and where the feedback electronics had calculated it to be, *i.e.* from the gradient of surface features rather than their height. This technique proved sufficiently powerful that on the surface of a *Valonia* crystal it was possible to see suggestions of individual hydroxymethyl groups.[24] Since the smallest tips available have at tip radius of 50 Å, it was suggested that the result arose from the contact of atomic-scale "whiskers" protruding from the tip. Other contact modes involve measuring both lateral and vertical displacements of the tip with a four-quadrant photoelectric detector. An AFM image of amylose is shown in Figure 4.21.

In the tapping mode, the cantilever assembly is vibrated at or near its resonance frequency (200–400 kHz), but with a high amplitude (20–100 nm). When this vibration encounters a surface, the amplitude decreases and the piezoelectric base can be set to move so that the amplitude is constant or to modulate the cantilever oscillation in various ways so that the change in the main oscillation is detected by the electronics of the system. Tapping mode has the advantage that damage to the sample is avoided (as is damage to the tip) and is the method of choice for imaging biological materials in air. Studies of hyaluronic acid[25] and scleroglucan (a fungal polysaccharide with a laminarin [(3)-β-D-Glc*p*-(1→)$_n$] main chain decorated on every third or so residue with a single β-D-glucopyranosyl residue on O6) used this mode.[26]

The ability of the tip to pick up one end of a biopolymer while the other still remains bound to the base has permitted direct measurements of forces

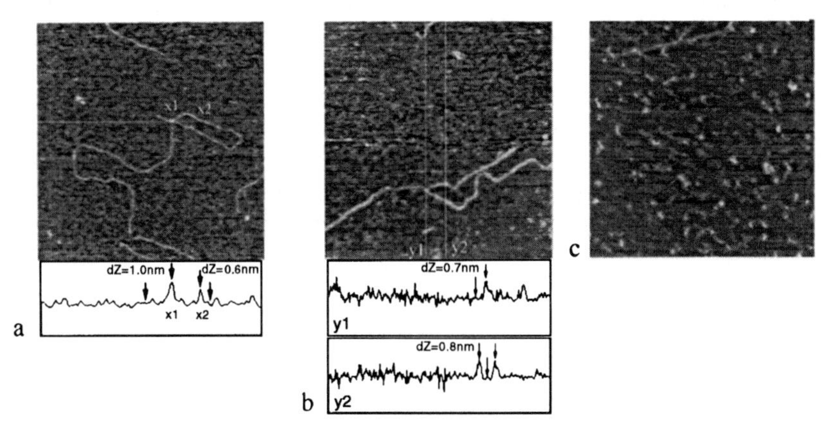

Figure 4.21 AFM image of amylose, taken from Ref. 5, reprinted with permission from Elsevier.

within molecules and between them. In the first experiments,[27] dextran ([Glc*p*-α-(1→6)]$_n$) was derivatised with epoxyalkanethiols and bound to a gold surface with Au–S–CH_2CHOH–CH_2–O linkages. The bound polymer was then carboxymethylated and reacted with the protein streptavidin, which binds the coenzyme biotin very tightly. Biotin was then attached to the AFM tip and a "fly-fishing" mode of operation of the cantilever enabled the tip to pick up a single dextran molecule, the experiment being conducted under water. When the AFM tip was raised at a constant rate, the dextran molecule originally obeyed Hooke's law, with a force constant of $670 \pm 100\,\mathrm{pN\,Å^{-1}}$, but at some point at appeared to stiffen, with a new constant of $1700\,\mathrm{pN\,Å^{-1}}$. In the high force constant region the resistance to the motion of the cantilever tip was coming from a conversion of $^4C_1$ chairs to a skew or boat conformation. It later transpired that simply depositing a polysaccharide on glass and pressing an AFM tip into it gave enough adhesion that force–extension curves could be measured,[28] although the original careful experiments with unambiguous modes of attachment[27] were necessary to establish what was being recorded. Experiments with carboxymethylated cellulose, carboxymethylated amylose and pectin confirmed the assignment of the high force constant region to forcible distortion of pyranose rings out of the $^4C_1$ conformation.[29] Carboxymethylcellulose, a 1,4-diequatorially linked polymer, showed no conformational transitions, as the 1,4-diequatorial arrangement is the one with the longest effective length of the anhydroglucose unit. Carboxymethylamylose, based on a linear chain of α-(1→4) glucopyranose units, showed one such transition, corresponding, as with the dextran, to a change to a region where resistance to the applied force came from conversions of $^4C_1$ conformations to skew conformations (probably $^0S_2$, since this gives the greatest separation of 1,4-*cis* substituents). The nanomechanical properties of pectin (see Section 4.7.3.2) are governed by the main chain, which is an α-(1→4)-galacturonan, *i.e.* 1,4-diaxially linked. Three regions of the force–extension curve are observed, the first being the usual straightening of the chain, the second the conversion of the $^4C_1$ chairs to a boat and the third the conversion of the boat to the $^1C_4$ chairs in which the pyranose rings are diequatorially linked. The anhydroglucose unit in the first transition (1,4-diaxial $^4C_1$ to $^0S_2$) lengthens by 0.37 Å and in the second ($^0S_2$ to $^1C_4$) by 0.06 Å.

Modification of the atomic force microscope with feedback loops enabled force–extension curves for individual polysaccharide molecules to be measured under constantly increasing force (rather than constant displacement), in a nanoscale version of the INSTRON tester for measuring paper strength.[30]

4.3 DESCRIPTION OF OLIGOSACCHARIDE AND POLYSACCHARIDE CONFORMATION

With rare exceptions (one discussed above), the monosaccharide units, particularly pyranose units, of oligosaccharides and polysaccharides can be regarded as rigid. The conformation of any inter-saccharide linkage can then be described by a small number of dihedral angles. Where two rings are joined directly by an oxygen atom, only two angles are required, φ to describe the conformation about the anomeric centre–oxygen bond and ψ to describe that between the intersaccharide oxygen atom and the next ring. Where the glycosidic linkage is to the hydroxymethyl group of a hexopyranose, a third angle, χ, is required.[ix]

It is not easy to determine the conformation of an inter-saccharide linkage from numerical values of φ and ψ in the research literature, since several different conventions are used,[x] and additionally there can be errors in their application. In the case of methyl β-lactoside (Figure 4.3), the IUPAC–IUBMB rules[31] define φ as GalO5–GalC1–GlcO4–GlcC4 and ψ as GalC1–GlcO4–GlcC4–GlcC3.[xi] The IUPAC system, which specifies the galactose ring oxygen as one fixed point for determination of φ, and the next *lowest* numbered glucose ring carbon to the attachment-point for determination of ψ, has the advantage that it can be applied to ketosides and also glycosidic linkages to tertiary centres. Unfortunately, another heavy atom-based system, which would specify ψ as GalC1–GlcO4–GlcC4–GlcC5, *i.e.* a reference point is to the next *highest* carbon to the attachment point, is in widespread use.[32,33,34] This text will use the conventional lower case φ and ψ for dihedral angles in the IUPAC convention,[35] and upper case (Φ, Ψ)for the system where the reference point for ψ is the next highest carbon to the attachment point, as in Refs. 33 and 34.

Whereas the heavy atom-based conventions are favoured by fibre crystallographers (since fibre X-ray diffraction cannot detect hydrogens), hydrogen-based systems, in which φ is defined as GalH1–GalC1–GlcO4–GlcC4 and ψ as GalC1–GlcO4–GlcC4–GlcH4, are favoured by NMR spectroscopists, who measure coupling constants, which are converted to dihedral angles via Karplus-type equations. Sometimes authors using this definition will add an "H", superscripted[2] (*e.g.* φ^{H}) or subscripted (*e.g.* ψ_{H}),[36] but sometimes not;[37] this text will use a subscripted H.

Dihedral angles are conventionally measured in the sense written, *i.e.* for φ_{H} (defined as GalH1–GalC1–GlcO4–GlcC4) one looks along GalC1→GlcO4 and the angle is that between the GalH1–GalC1 and GlcO4–GlcC4 bonds in the Newman projection. Likewise for ψ_{H}, defined as GalC1–GlcO4–GlcC4–GlcH4, one looks along GlcO4→GlcC4[xii] and ψ_{H} is the angle between the GalC1–GlcO4 and the GlcC4–GlcH4 in the Newman projection. Positive angles are

---

[ix] For such things as polysialic acid, with 2→8 linkages, additional dihedral angles have of course to be defined.

[x] The author has encountered five, but there may be more.

[xi] Primes (*e.g.* O5′–C1′–O4–C4 for φ) are more commonly used to specify dihedral angles, but there is no agreement on which sugar (galactose or glucose in this case) should carry the prime (Ref. 2 favours galactose and Ref. 31 glucose).

[xii] Because of the properties of helices, one could also define ψ_{H} as GlcH4–GlcC4–GlcO4–GalC1 and look along GlcC4→GlcO4 and get the same signs and magnitudes of dihedral angles. J. Lehmann (*Carbohydrates: Structure and Biology*, transl. A. H. Haines, Thieme, Stuttgart, 1998, p. 35) does just this.

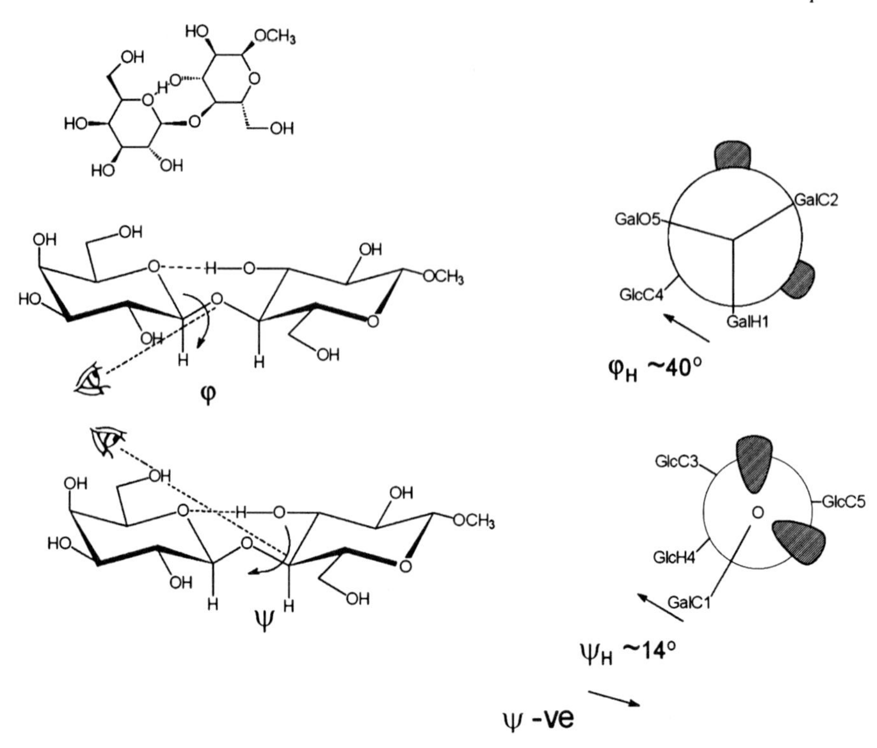

Figure 4.22 Definition of dihedral angles about the glycosidic bond in methyl β-lactoside. This is drawn in the preferred conformation in water, deduced from the following three-bond coupling constants via the appropriate Karplus equations[38] (the prime referring to the glucose moiety): ${}^3J_{C4',H1} = 3.8$ Hz, ${}^3J_{C2,C4'} = 3.1$ Hz, defining φ, and ${}^3J_{C1,H4'} = 4.9$ Hz, ${}^3J_{C1,C3'} = 0$ Hz, ${}^3J_{C1,C5'} = 1.6$ Hz, defining ψ. The disaccharide unit in cellulose[35,xiv] and other 1→4 diequatorially linked polysaccharides adopts a very similar conformation.

those in which, in the Newman projection, one moves from the "front" bond to the "rear" bond in a clockwise direction (Figure 4.22); to avoid angles greater than 180°, authors usually quote negative (anticlockwise) angles in their text.[xiii]

A fourth convention, mercifully confined to some of the older, but still current, fibre diffraction literature, defined dihedral angles as those between two planes going through O4, C4 and C1 of each residue;[39] the description in terms of bond vectors[40] is equivalent. These are broadly similar to φ_H and ψ_H (and would be identical if the plane were a mirror one on a substituted

[xiii] That does not stop them quoting angles from 0 to 360° in three-dimensional energy contour plots, though – possibly because mapping software cannot cope with negative coordinates.

[xiv] Reference 35 adopts the IUPAC system and gives $\varphi = -95°$ and $\psi = 92°$, corresponding, if C1, O4 and C4′ have perfect tetrahedral angles (which is unlikely), to $\varphi_H = 25°$ and $\psi_H = 28°$.

cyclohexane). This text will use the closest Cyrillic equivalents to φ and ψ (ф and ш) for angles described using this convention.

Although it is not possible to convert accurately dihedral angles described according to non-IUPAC convention into the IUPAC convention without knowing all the bond angles of all the atoms involved, a qualitative idea can be obtained if it is assumed that all angles are exactly tetrahedral. IUPAC-conforming dihedral angles calculated in this way are underlined and are always approximate (*e.g.* $\underline{\varphi} \approx 90°$).

φ and ψ can be determined by various X-ray or neutron diffraction studies on single crystals (in the case of oligosaccharides) or fibres (in the case of polysaccharides). In the case of solution conformations of oligosaccharides, both angles can in principle be determined from three-bond NMR coupling constants and Karplus-type relationships, although in practice with the methyl β-lactoside in Figure 4.1, this involved elaborate preparative chemistry and biochemistry to make specifically ^{13}C-labelled sugars and a 600 MHz NMR machine.[38] Also, coupling constants, which only give $\cos^2\varphi$ or $\cos^2\varphi$, cannot determine the sign of the dihedral angle and a value of ψ_H of –14° for methyl lactoside would fit the NMR data equally well (and is more likely because of the C5 $\cdots$ HO3′ bond). Where a molecule adopts essentially a single conformation, coupling constants do not vary with temperature. Conversely, if the barriers to rotation are low, so that coupling constants alter with temperature, further knowledge of the averaging process is required. The importance of solvation is illustrated by the fact that benzyl β-lactoside has a radically different conformation in the gas phase[41] ($\varphi_H = 180°$, $\psi_H = 0°$).

χ is defined in the IUPAC system as (glycosylated oxygen)–C_n–C_{n-1}–C_{n-2}, so that it has a value of 180° for a perfect *gt* conformation, 60° for a perfect *gg* conformation and –60° for a perfect *tg* conformation (Figure 2.20). In this text lower case is used (χ) for the IUPAC-conforming angle and upper case (X) for another convention defining the angle as O5–C5–C6–O6[xv].[33,34]

It is common to present the results of conformational calculations on a disaccharide unit as a map with energy contours and φ and ψ as orthogonal axes: an example is shown in Figure 4.23. Also, various experimental determinations of φ and ψ for a linkage can be plotted as individual points, much in the manner of Ramachandran plots[42] in protein chemistry.[xvi] However, whereas a single X-ray crystal structure determination of a protein can give hundreds, if not thousands, of points on a Ramachandran plot, meaningful Ramachandran-like scatter diagrams cannot be obtained from a single polysaccharide – cellulose II only gives two points,[6] for example.

[xv] In references 33,34 and 97, the symbol χ' is used for the IUPAC-defined angle and χ for O5–C4–C5–C6.

[xvi] The dihedral angles about the N–C_α and C_α–CO and bonds of each amino acid are given the symbols φ and ψ, so that each residue contributes one point to the plot. Since peptide bonds are transoid and planar and essentially rigid, protein secondary structure is defined by the values of φ and ψ for each amino acid.

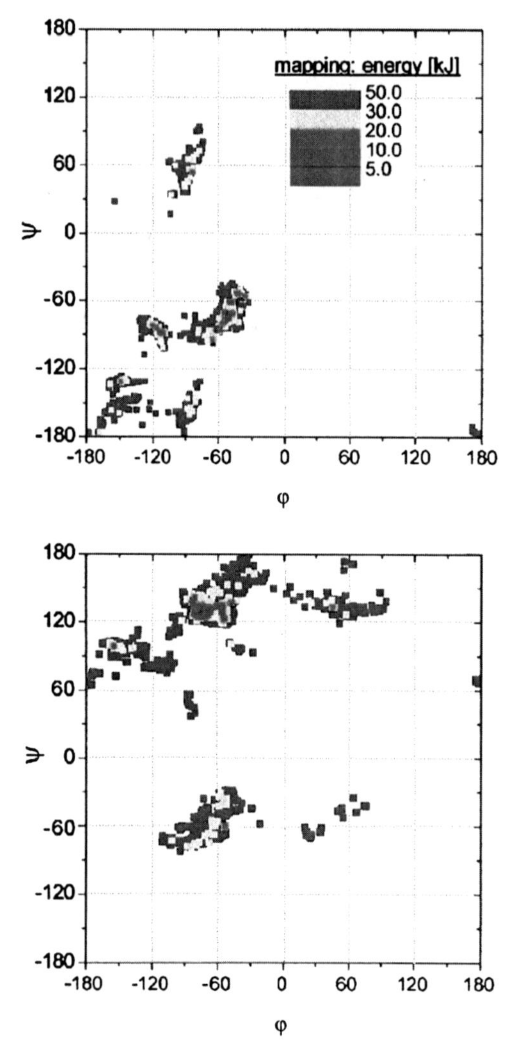

Figure 4.23 Conformational energy map of (top) the α-Fuc*p*-(1→3)-GlcNAc link and (bottom) the β-Gal-(1→4)-GlcNAc link according to the Monte Carlo calculations of Azurmendi et al;[43] reproduced with permission of John Wiley and Sons, Inc.

4.4 THE *EXO*-ANOMERIC EFFECT

It is widely observed that preferred conformations of glycosides with equatorial glycosidic C–O bonds have values of φ_H in the general range 45–25° ($\varphi = -75$ to –95°) and those of axial glycosides around –30° ($\varphi = 90°$). The electronic or electrostatic interactions that determine these preferences are the same as those governing the anomeric effect and were termed the *exo*-anomeric effect by Lemieux and co-workers.[44] Whereas the no-bond resonance explanation for the anomeric effect considers the key interaction to be overlap of a lone pair on O5[xvii] with the σ* orbital of the exocyclic C1–O1 bond, the analogous explanation of the *exo*-anomeric effect considers the overlap of a lone pair on O1 with the σ* orbital of the endocyclic C1–O5 bond. Likewise, whereas the electrostatic explanation for the anomeric effect considers the interaction between the C1→O1 dipole and the resultant of the C1→O5 and C5→O5 dipoles, the electrostatic explanation of the effect considers the interaction between the C1→O5 dipole and the resultant of the C1→O1 and R→O1 dipoles, where R is the aglycone. By analogy with the anomeric effect, it seems likely that no-bond resonance effects are dominant, but that electrostatic effects also play a part. Figure 4.24 shows the three perfectly staggered rotamers about the C1–O1 bond of an equatorial pyranoside, and, in Newman projection, the orbital and electrostatic interactions. It is noteworthy that in cellulose and methyl lactoside, φ_H is in fact smaller than 60°, corresponding to an imperfectly staggered conformation, but one in which the p-type lone pair on O1 has more efficient overlap with the C1–O5 σ* orbital.

The rotamer with $\varphi_H = 180°$ is favoured by the *exo*-anomeric effect but disfavoured by ordinary steric interactions: the clashes between R and the axial substituent at C2 (even if only hydrogen) are what they would be if R were an axial substituent at C4.

Similar considerations apply to the *exo*-anomeric effect in axial pyranosides (Figure 4.25). If the O1 lone pair electrons are regarded as being in a p and an sp orbital, then a if φ_H narrows to −30°, the p-type lone pair on O1 is exactly aligned with the C1–O5 σ* orbital in the right-hand-side conformation, in the same way as with equatorial pyranosides.

As with equatorial pyranosides, the $\varphi_H \approx 180°$ rotamer is disfavoured by ordinary steric interactions, as the group R is right under the α face of the pyranose ring. As with equatorial pyranosides also, the electrostatic explanation of the anomeric effect indicates the same two rotamers are favoured as the non-bond resonance explanation: in the rotamer with $\varphi_H \approx -60°$, in the Newman projection the resultant of the R→O1 and C1→O1 dipoles is aligned exactly with the C1→O5 dipole.

The very existence of an *exo*-anomeric effect, however, was brought into question by Kishi and co-workers' studies of *C*-glycosides, in which the glycosidic oxygen atoms of common glycosides and di- and oligosaccharides had been replaced by methylene (CH_2) groups. Whether carbon analogues of ethyl α- and β-glucopyranosides,[45] isomaltose and gentiobiose (*i.e.* α-D-Glc*p*-(1→6)-D-Glc*p* and β-D-Glc*p*-(1→6)-D-Glc*p*),[46] 1→4 disaccharides[47] or even "carbon trisaccharides",[48] they all had the same preferred conformations as their oxygen analogues, at least as estimated by nuclear Overhauser effect measurements. A methylene group, of course, should have no lone pair effects and very minimal electrostatic ones.

[xvii] Numbering is for an aldopyranoside.

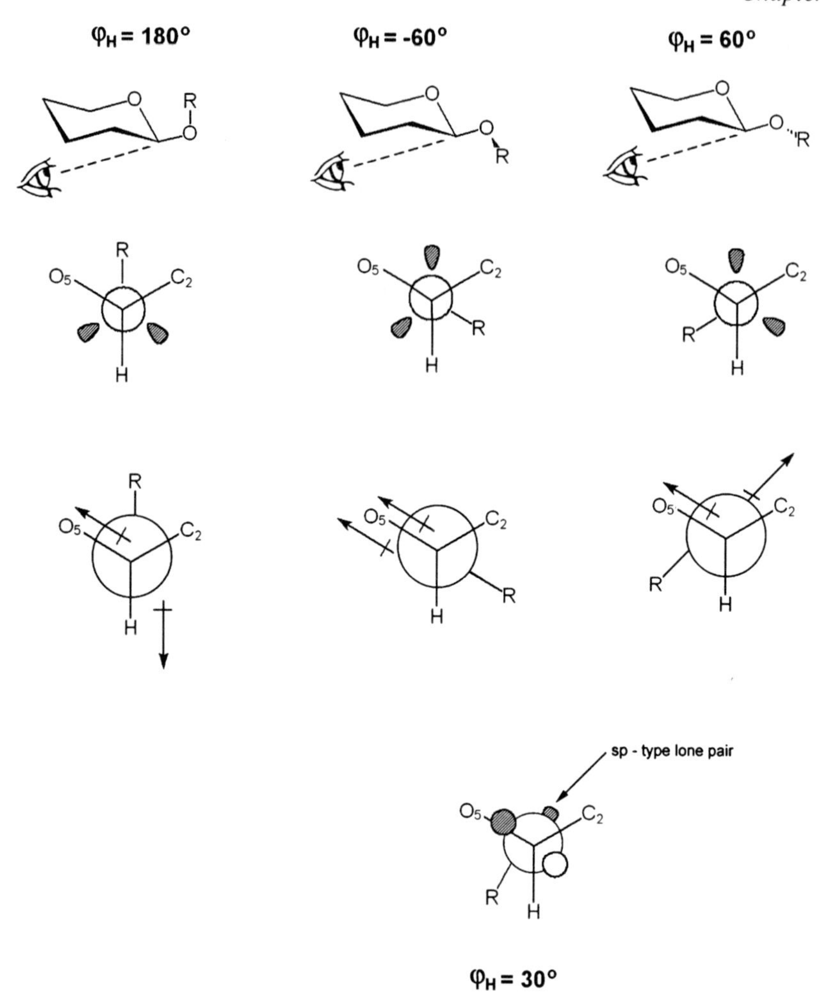

Figure 4.24 The *exo*-anomeric effect in equatorial pyranosides. The second row (first row of Newman projections) depicts the lone pair electrons on O1 as in sp^3 orbitals and the third row dipole interactions. The final Newman projection illustrates how, if φ_H narrows to 30°, the p-type lone pair on O1 is exactly aligned with the C1–O5 σ^* orbital.

4.5 POLYSACCHARIDES IN SOLUTION

The rheological[xviii] behaviour of aqueous solutions of some polysaccharides, such as bacterial *exo*-polysaccharides and vertebrate glycosaminoglycans, is the reason for their biosynthesis in the first place, with bacterial *exo*-polysaccharides protecting the bacterium, and glycosaminoglycans acting as lubricants and

[xviii] Mechanical, in connection with flow, from the Greek ρεος, stream.

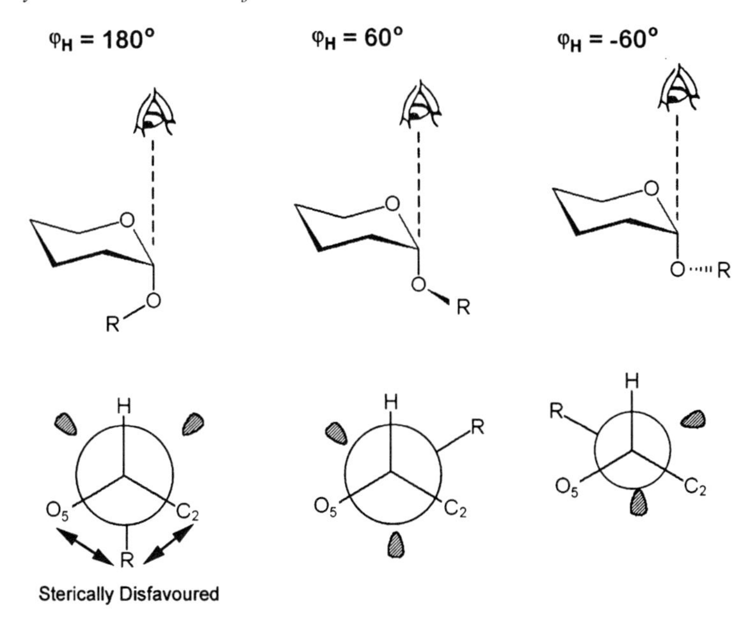

Figure 4.25 Operation of the *exo*-anomeric effect in axial pyranosides.

mechanical shock-absorbers. The rheological behaviour of polysaccharides is also important because it is exploited commercially in foodstuffs, health care products and some industrial processes such as papermaking – basically, in any application where a jelly- or gloop-like consistency is required and the thickening agent must be seen as harmless.

In addition to highly viscous solutions, polysaccharides can also form gels, defined as materials which, though largely liquid by weight, have some capacity to resist shear force and maintain their shape. The gelling agent forms networks whose cavities enclose the solvent. It remains widely thought that gel formation is associated with formation of double helices by the polysaccharide chains, although for particular polymers, the lateral association of single helices, as securely demonstrated for alginate, is increasingly favoured.

Two theoretical models for polysaccharides in aqueous solution come from polymer physics:

(i) The random coil, in which linear polymers are modelled by rigid rods, connected at each end by universal joints infinitely flexible with respect to two orthogonal rotations. This is sometimes called a "random flight" chain, as the individual rods undergo Brownian motion with respect to each other. The model rarely applies to polysaccharides with direct glycosidic linkages between rings, but is more successful with polysaccharides linked through side-chains, *e.g.* hexopyranoses linked (1→6).

(ii) The "wormlike chain" model, sometimes called the Kratky–Porod model, in which the polymer chain is continuously deformable, but resists deformation, particularly over short distances.[xix] A key concept is the "persistence length", L_p, defined by eqn. (4.6), where k_B is Boltzmann's constant, κ the bending elasticity of the polymer (angular deformation per length per unit force, *i.e.* dimensions of $\mathcal{M}\mathcal{T}^{-2}$[xx]) and T the absolute temperature:

$$L_p = \kappa/k_B T \tag{4.6}$$

The relatively small region of allowable values of φ and ψ in polysaccharides linked between rings makes the wormlike chain model realistic for them. The polymer behaves as a random flight one over contour lengths $S \gg L_p$, but as a stiff rod if $S \ll L_p$.[49] Persistence lengths can be fairly large: 350–450 Å for xanthan gum, 80–100 Å for alginate or hyaluronate.[50]

Many polysaccharides are anionic, either because their building blocks are anionic (uronic acids, sialic acids, KDO, *etc.*) or because of substitution with sulfate, phosphate or pyruvate groups (the last as cyclic acetals – see Section 4.6.10.3.1). Naturally positively charged polysaccharides seem confined to chitosan (Section 4.6.1.1.3), but neutral polysaccharides such a starch or cellulose are modified by the introduction of amine-based substituents for industrial use as cationic polymers. The titration curve of a weak polyacid (such as hyaluronan) is extended over many pH units, because the electrostatic effect of negative charge makes the introduction of further negative charge by deprotonation progressively more difficult. It is conventional to describe the ionisation in terms of an effective pK_a at each point of the titration curve [eqn. (4.7), related to the Henderson–Hasselbach equation], where α is the degree of ionisation, related to the titration ratio α_n ([titrant added]/c_p, where c_p is the molar concentration of ionisable protons) by eqn. (4.8):

$$pK_a = pH - \log[\alpha/(1-\alpha)] \tag{4.7}$$

$$\alpha = \alpha_n + 10^{-pH}/c_p \tag{4.8}$$

The electrostatic effect has been treated explicitly for polymers with one type of ionising group,[51] and for two.[52] Pronounced discontinuities or abnormal shapes of pK_a versus α plots are a reliable indication of a change in conformation as the charge increases.

The effect of charge on polysaccharide conformation, although very important, has not been successfully treated in the general case.[53] It is often minimised by performing studies in high concentrations of salt.

[xix] The phrase "wormlike chain" can imply, to readers familiar with earthworms, a *segmented* chain, which is exactly the wrong picture. Something like the stem of a vine or creeping plant would be more appropriate.

[xx] To avoid confusion with M for molar, L for length and T for absolute temperature, dimensions are written in Corsiva font.

4.5.1 Separation on the Basis of Molecular Size

Gel permeation chromatography (GPC), sometimes called size-exclusion chromatography (SEC), is based on a very simple principle: a gel is constructed with a narrow range of pore sizes and packed into a chromatography column. If a polymer of infinite molecular weight is applied at the top of the column, it cannot fit into any of the pores and is eluted in volume V_0, the excluded volume. A small molecule, on the other hand, can fit into all the pores and is this eluted in volume V_t, the total volume of the column. Fractionation according to molecular size thus occurs between elution volumes V_0 and V_t with molecules eluted in order of decreasing size.

The first SEC material to become available commercially was Sephadex, made by cross-linking a bacterial β-(1→6)glucan with epichlorohydrin (1-chloro-2,3-epoxypropane) which was confined to aqueous solution. A wide range of SEC materials are now available, based on various polymer chemistries and compatible with a range of organic solvents.

Although fractionation occurs on the basis of molecular size, not molecular weight, by careful choice of molecular weight standards it is often possible to obtain fairly accurate molecular weights. The simplest form of the technique is particularly powerful with globular proteins. Sephadex or similar columns are a standard way of determining the oligomeric state of a protein, after the monomer molecular weight has been determined by an electropherogram under denaturing conditions.

Over the fractionation range of a GPC column, the relationship between elution volume V_E and molecular weight is approximately logarithmic, so eqn. (4.9) applies:

$$V_E = A - B \log M_r \tag{4.9}$$

The problem, of course, comes from the implicit assumption that the gel matrix has no specific interactions with the soluble polymer, and that the relationship between effective volume and molecular weight is the same for the polysaccharide of interest and the standards. A recent development has been to place instruments which measure molecular weight at the exit of a GPC column, so that the column is used only for fractionation, and a full molecular weight distribution of a polydisperse polymer can be obtained. Viscometers and light-scattering monitors can be so employed,[54] as can on-line electrospray mass spectrometers.[55] The last technique is particularly powerful, since the masses determined by the mass spectrometer are absolute.

Undecorated linear (1→4) diequatorial polysaccharides such as cellulose tend to be insoluble in water, so any attempt at determining a molecular weight distribution first has to solubilise them. A common method is complete reaction with phenyl isocyanate (Ph–N=C=O) in warm (~80 °C) pyridine which converts all the hydroxyl groups of cellulose into "carbanilate" (cell–O–CO–NH–Ph) groups.[56] SEC in an organic solvent such as tetrahydrofuran[57] can then, with appropriate standards, give the molecular weight distribution.

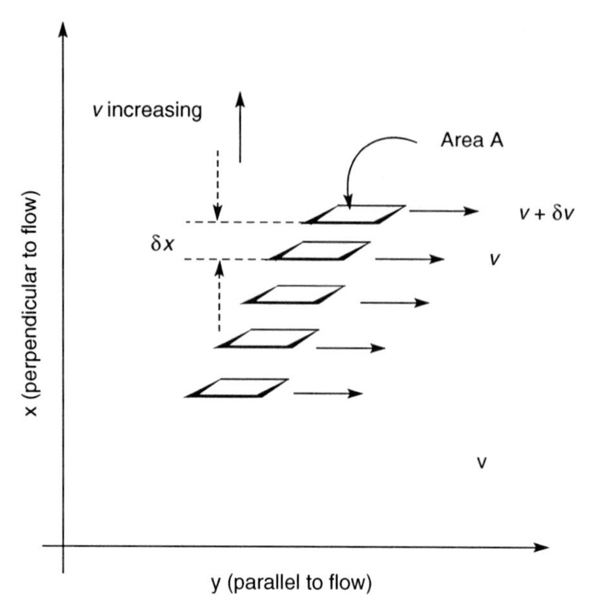

Figure 4.26 Laminar flow and viscosity.

4.5.2 Rheological Properties of Polysaccharides

Rheology details the behaviour of liquid-like materials under the influence of mechanical stresses and covers viscous and viscoelastic behaviour.

Viscosity is the resistance presented by a liquid to external forces subjecting it to flow. Laminar flow only is considered in polysaccharide rheology – as the name implies, the velocity in laminar flow increases monotonically with distance away from the edge of the vessel, pipe, *etc.* (in turbulent flow the flow rate will have local maxima and minima) (Figure 4.26). Imagine two hypothetical plates parallel to the direction of flow and the sides of the container, separated vertically by a distance δx. One plate will be moving faster than the plate below it and will experience a force F because of viscous forces. This force will be in proportional to the area of the plane, so we can define a sheer stress of F/A, where A is the area of the plate. The force will bring about a difference of velocity δv between two adjacent plates separated by δx and we can define a sheer strain rate, usually denoted γ, as $\delta v/\delta x$, in the limit $\mathrm{d}v/\mathrm{d}x$ (with dimensions of $\mathscr{T}^{-1}$). Kinetic or dynamic viscosity, η, is defined by eqn. (4.10) and therefore has dimensions of pressure × time or $\mathscr{M}\mathscr{L}^{-1}\mathscr{T}^{-1}$,[xxi,58]:

$$\eta = \frac{\text{shear stress}}{\gamma} = \frac{F}{A} \times \frac{\mathrm{d}x}{\mathrm{d}v} \tag{4.10}$$

[xxi] *Kinematic* viscosity is the viscosity divided by the density.

Most pure liquids and solutions of small molecules show a simple proportionality between sheer stress and sheer strain rate. Such liquids are known as Newtonian, after their discoverer. Solutions of polysaccharides usually show Newtonian behaviour only below a certain critical concentration, c^*, at which the chains start to interact. Above this concentration, except at very low sheer strain rates, viscosity decreases rapidly with sheer strain rate until at very high values of γ it becomes constant. This is not a small effect: with xanthan gum (see Section 4.6.10.3.1), the fall in viscosity is $\sim 10^6$-fold. The dependence of viscosity on sheer strain rate is often given by

$$\eta = \eta_\infty + \frac{\eta_0 - \eta_\infty}{(\tau\gamma)^m} \tag{4.11}$$

where η_0 is the viscosity at zero shear, η_∞ that at infinite shear, τ is the shear at half-maximal viscosity and m is a parameter characteristic of the system. It can be readily determined by a plot of $\log \eta$ against $\log \gamma$, which in practice has an extensive linear portion of gradient $-m$ between η_0 and η_∞ (Figure 4.27). Newtonian behaviour is exhibited when $m = 1$; if $m < 1$, shear thinning (as with most polysaccharides) is observed, whereas if $m > 1$, shear thickening takes place. Shear thickening is usually associated with slurries, in which under conditions of high shear empty cavities are created. Slurries of uncooked maize starch ("cornflour") in water are obvious domestic examples.

The reason for shear thinning is straightforward: above the critical concentration c^*, the polysaccharide chains start to interact with each other, but when placed under shear, the linear molecules align with the flow, reducing resistance.

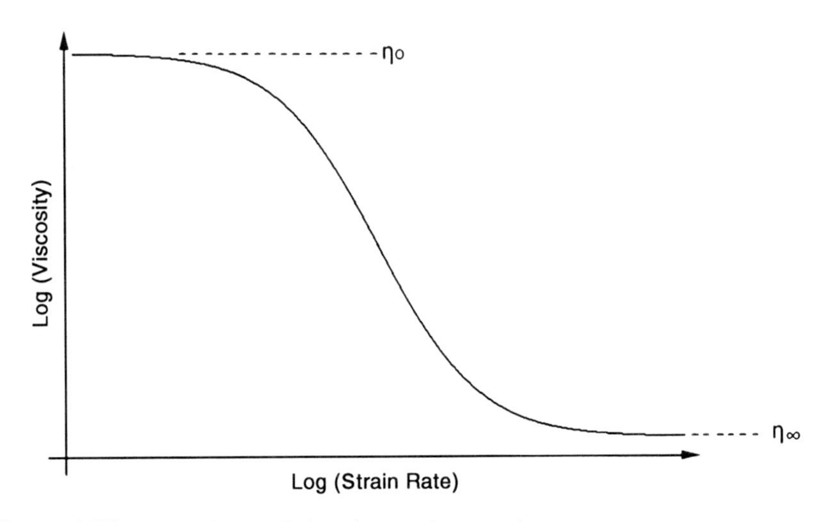

Figure 4.27 Dependence of viscosity on shear strain rate.

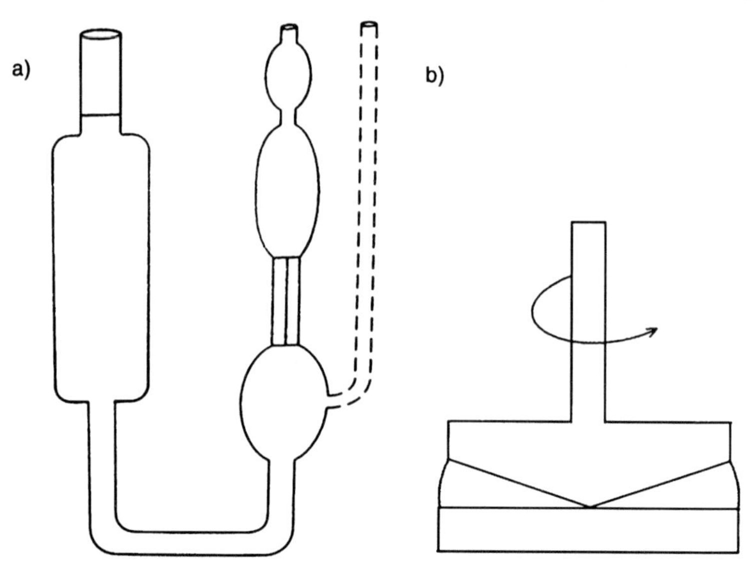

Figure 4.28 (a) Ubbelohde and (b) cone-and-plate viscometers.

Shear thinning (and shear thickening) are instantaneous, in contrast to thixotropic behaviour, which occurs when a response occurs well after a perturbation.[xxii]

Traditionally, viscosities (in the Newtonian range) were measured by timing the flow of a given volume of liquid (V) through a capillary of length l and radius r under a pressure p, when the viscosity is given by Poiseuille's law:

$$\eta = \frac{\pi t r^4 p}{8Vl} \tag{4.12}$$

The Ostwald and Ubbelohde viscometers, a series of glass bulbs which connected to a standard capillary, and the proportionality between t and η made relative measurements of viscosity under conditions of low shear rate fairly easy, particularly as thermostatting the whole device was simple (viscosity shows an Arrhenius-type temperature dependence, $\eta = A\mathrm{e}^{-E/RT}$, with typical E values around 5 kJ mol^{-1}). For measurements of viscosity as a function of shear stress rate, various electrical devices, which rotate a disc of various configurations in a liquid container and measure torque, are available. A useful configuration of such devices is the "cone and plate" type, in which a solid cone of angle θ rad makes contact with a flat, stationary plate (Figure 4.28). As the cone is rotated, the liquid between the cone and plate is subjected to a constant shear

[xxii] *E.g.* the yearly liquefaction of the blood of San Gennaro, after the phial containing it has been handled by the Bishop of Naples (L. Garlaschelli, F. Ramaccini and S. Della Sala, *Chem. Br.*, 1990, **30**, 123).

rate of Ω/θ s^{-1}, where Ω is the rate of rotation of the cone (in radians per second).[59]

In discussions of the behaviour of polymer solutions, the following viscosity parameters can be used, where η_0 is the viscosity of the solvent and η the observed viscosity at a polymer concentration c:

$$\text{Relative viscosity (dimensionless)} = \eta_r = \eta/\eta_0 \tag{4.13}$$

$$\text{Specific viscosity (dimensionless)} = \eta_{sp} = \eta_r - 1 = (\eta - \eta_0)/\eta_0 \tag{4.14}$$

$$\text{Reduced viscosity [dimensions } \mathscr{L}^3\mathscr{M}^{-1} \text{ (reciprocal concentration)]} \\ = \eta_{red} = \eta_{sp}/c \tag{4.15}$$

$$\text{Intrinsic viscosity} = [\eta] = \text{limit of } \eta_{red} \text{ as } c \text{ approaches zero} \tag{4.16}$$

The relationship between observed viscosity and intrinsic viscosity depends on the volume occupied by the polymer chains (dependent on the first power of their concentration) and the interactions between polymer chains (in the dilute and semi-dilute regions, dependent on the second power of their concentration). The resulting equation, eqn. (4.17), is known as the Huggins equation, with k_H being the (dimensionless) Huggins coefficient, which measures chain–chain interaction:

$$\eta_{sp} = [\eta]c + k_H[\eta]^2c^2 \tag{4.17}$$

The more polymer chains interact with the solvent, the less they will interact with each other and, for neutral polysaccharides in water or anionic polysaccharides in salt solutions, k_H lies between 0.3 and 05.[54]

Addition of further terms in higher powers of c and $[\eta]$ to eqn. (4.15), to represent higher order interactions, will obviously improve the fit of the data. Viscosities of hyaluronan (Section 4.6.10.1.1) and xanthan (Section 4.6.10.3.1) obey the equation[54]

$$\eta_{sp} = [\eta]c + k_H[\eta]^2c^2 + B[\eta]^nc^n \tag{4.18}$$

Considerable effort has been expended on elucidating the dependence of $[\eta]$ on polymer molecular structure. The relation between intrinsic viscosity and molecular weight, M_r, is given by the Mark–Houwink equation, eqn. (4.19); in the case of polydisperse samples, the average depends on the value of α: if it is 1.0, the number-average molecular weight is produced, whereas if $\alpha < 1.0$, the usual case with polysaccharides, the average will be below the number average.

$$[\eta] = KM_r^{\alpha} \tag{4.19}$$

For polysaccharides in water, α is usually around 0.7, above the values for a freely coiling chain, but well below that for a stiff rod (1.8).[59]

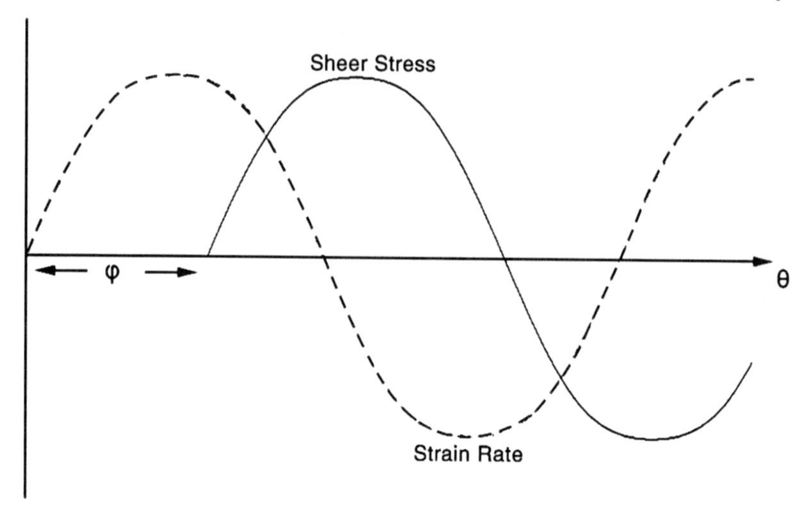

Figure 4.29 Variation of G' and G'' with shear rate.

Above c^*, the behaviour of polymer solutions can become viscoelastic: some of the displacement of the material when a small stress is applied is reversed when the stress is removed. Viscolelasticity can be readily characterised if it is linear, *i.e.* if the stress is sufficiently small that the elastic response is Hookean and the viscous response is Newtonian.[60] Viscoelasticity is normally characterised by applying a small, sinusoidally oscillating strain to the sample and monitoring peak stress, peak strain and the phase shift, δ, between the input and output signal. The stress per unit strain gives the complex modulus G^*. This is made up of the storage modulus G', arising form the elastic storage of energy, and the loss modulus, G'', arising from the dissipation of energy from the viscous response to stress (Figure 4.29). The relation between the two moduli is given by eqn. (4.20); a perfectly elastic response means that the strain occurs immediately on the application of the stress, so that $\delta = 0$.

$$G' = G^* \cos\delta; \quad G'' = G^* \sin\delta \tag{4.20}$$

If G' and G'' are studied as a function of frequency (a technique sometimes known as mechanical spectrometry), it is usually found that G' increases sharply with frequency, to a new plateau value, whereas the response of G'' is less pronounced and it may even go through a shallow maximum and then decrease. The frequency at which G' is half-maximal corresponds approximately to τ, the shear stress rate at which the velocity is half-maximal. This is easily understood on a molecular basis, since τ is the rate at which the polymer molecules can reorientate themselves. As a modulus, G has dimensions of pressure ($\mathcal{M}\mathcal{L}^{-1}\mathcal{T}^{-2}$).[xxiii]

[xxiii] The mathematics of treating sine waves is simplified by the introduction of imaginary numbers: strictly G^* is complex, $G^* = G' + iG''$.

4.5.3 Laser Light Scattering and Related Techniques

The electron clouds of a molecule irradiated with visible light will orient themselves to some extent in line with the electric vector of the light. If the light is plane polarised, scattered light is visible only in one plane, and if the dimensions of the molecule are small compared with the wavelength of the light, the Rayleigh equation, eqn. (4.21), holds:

$$\frac{I}{I_0} = \frac{16\pi^4\alpha^2 \sin^2\theta}{\lambda^4 r^2} \qquad (4.21)$$

where λ is the wavelength of the light, I_0 the intensity of the incident light, I the intensity of the light scattered through an angle θ, observed at a distance r from the observer, and α the polarisability of the molecule. The inverse fourth power dependence on the wavelength of sunlight scattered by the Earth's atmosphere explains why the sky is blue, as shorter wavelengths are scattered preferentially. Scattering is more intense from solutions than from gases, and can give an idea of the size and shape of the scattering solute molecules.

The parameter Ir^2/I_0 at a given value of θ is called the Rayleigh ratio, R_θ. Modern multi-angle laser light scattering equipment (MALLS) simultaneously measures scattered intensity at several values of θ. Therefore, a typical static light-scattering experiment involves the measurement at a matrix of values of θ and c the polymer concentration.

For an ideal solution of small particles, Kc/R_θ, where c is the polymer concentration and K is a system constant depending only on the solvent and θ, becomes the reciprocal of the weight-average molecular weight. For non-ideal solutions, molecular properties are expanded as a power series in concentration. Corrections can also be made for complications that arise when the size of the polymer becomes comparable to the wavelength of the light used (internal interference).

Data for the matrix of values of R_θ as a function of values of both θ and c is conventionally visualised in a Zimm plot (Figure 4.30). As with families of double-reciprocal plots in steady-state enzyme kinetics, parameters are now extracted computationally,[61] but presented graphically. The functions, however, are not simple and need not be linear. The ordinate is Kc/R_θ, but the abscissa is $\sin^2(\theta/2) + kc$, where c is the concentration of the polymer and k is an arbitrary constant chosen so that the y intercept of families of plots at constant c or constant θ is constant.

Before computerisation, the weight-average molecular weight was obtained from the y intercept:

$$\lim_{c\to 0,\theta\to 0} \frac{Kc}{R_\theta} = \frac{L}{M_w} \qquad (4.22)$$

and the radius of gyration, S, from the ratio of slope to intercept of the extrapolated curve at $c = 0$. From molecular weight and radius of gyration, some indication of shape is obtainable. Note that the molecular weight

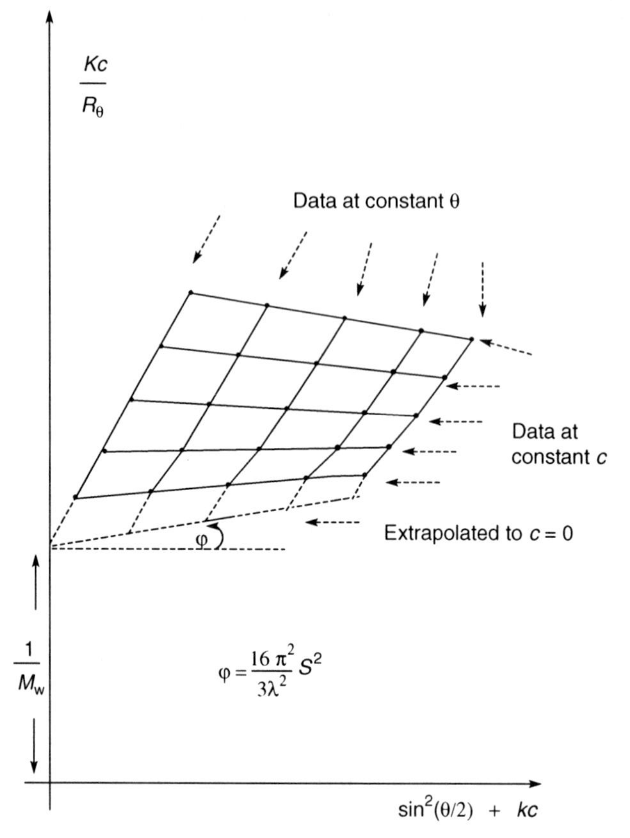

Figure 4.30 Schematic Zimm plot.

calculated on a polydisperse sample is the weight-average molecular weight, $\Sigma n_i M_i^2 / \Sigma n_i M_i$, where there are n_i molecules of molecular weight M_i. This averaging dramatically weights the average towards the higher molecular weights,[xxiv] and makes the technique particularly vulnerable to aggregation or high molecular weight contamination.

Various other scattering techniques have been used in the recent past to determine the size and shape of polysaccharides. Dynamic light scattering analysis is based on the changes in frequency and phase of laser light scattered by particles and macromolecules undergoing Brownian motion. Its main advantage in the study of hydrocolloids is its ability to detect small amounts of highly aggregated material – the output of modern instruments is a plot of scattering intensity (as a percentage) against the hydrodynamic radius, plotted logarithmically.[62] Dynamic light scattering is widely used to analyse

[xxiv] The number-average molecular weight of a polymer, 90% of whose molecules have $M = 10^5$ and 10% of whose molecules have $M = 10^6$, is 1.9×10^5, but the weight-average molecular weight is three times larger, 5.7×10^5.

suspensions of cellulose fibres and fillers in papermaking. Application of an electric field gradient across the same cell as is used to analyse Brownian motion can permit the zeta potential (an electrophoretic mobility parameter for macroscopic particles, equivalent to average surface charge) to also be measured.

Small-angle X-ray and neutron scattering have been used in the past to obtain an estimate of polymer shape. X-rays scattered from different parts of a macromolecule can interfere when the scattering angle is very small ($\sim 1°$), to yield, from solutions, a plot of intensity against scattering angle like a much compressed and smudged-out version of an X-ray powder diagram from a polycrystalline sample. Neutron scattering of various types can give shape, charge and dynamic information, but requires an atomic pile as a radiation source.

4.5.4 Chiroptical Methods

Since the monomer units of polysaccharides are themselves chiral, formation of chiral units on the macromolecular scale, such as double and single helices, will occur with only one chirality, left- or right-handed (strictly, this is true only for a given set of conditions, but in practice it is rare for the same polysaccharide forming both right- and left-handed helices even under different conditions). Formation of such macromolecular chiral species results in a large change in the interaction of the material with polarised light.

The most usual way in which such interaction is monitored is by the rotation of plane polarised light as it traverses a solution. In order to understand the language in which the other chiroptical techniques are described, it is useful to consider the general mechanism of the rotation of plane polarised light.

Figure 4.31 represents a beam of plane polarised light directed at the observer. By definition, the oscillation of the electric vector of plane-polarised light is confined to a line which can be regarded as the diameter of a circle; the oscillation is sinusoidal and so is zero at its extremities. But the electric vector oscillating along the diameter is equivalent to two electric vectors, of constant scalar magnitude, rotating in opposite senses. Thus, a beam of plane polarised light is equal to two beams, of equal intensity, of circularly polarised light. If these two beams travel through the sample at the same speed v, their (formal) recombination on exiting the sample will give the same plane of polarisation as on entering it. However, if one of the circularly polarised beams travels faster than the other, on recombination the plane of polarisation will have been rotated.

Such a difference in speeds is manifested in different refractive indices and arises from the different polarisabilities of matter with a right- or left-handed helical screw sense. If only dissolved molecules can contribute to the helical screw sense experienced by the circularly polarised light, the observed rotation at a given wavelength will be given by the following familiar expression:

$$[\alpha]_{\lambda}^{T} = \frac{100\alpha}{cl} \tag{4.23}$$

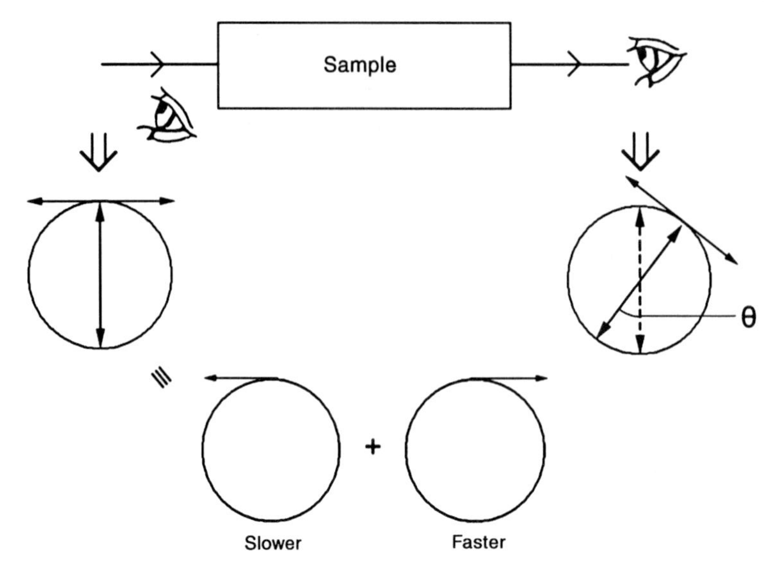

Figure 4.31 Rotation of plane polarised light. Resolution into two beams of oppositely circularly polarised light is shown.

where $[\alpha]_\lambda^T$ is the specific rotation at a temperature T °C and wavelength λ, calculated as defined by a perverse and memorably unmemorable convention whereby l is the pathlength in decimetres and c is the concentration in grams per 100 mL. The molecular rotation $[\varphi]$, useful in considerations of structure, is $M_r[\alpha]/100$.

$[\varphi]$ depends closely on wavelength, the variation of $[\varphi]$ with λ being known as the optical rotatory dispersion (ORD). At values of λ well above λ_0, the wavelength of maximum absorption of the nearest chromophore, the Drude equation, eqn. (4.24), holds:

$$[\varphi] = \frac{k}{\lambda^2 - \lambda_0^2} \tag{4.24}$$

However, as λ_0 is approached, the behaviour of $[\varphi]$ becomes complex, as shown in Figure 4.32. The presence of the a sharp maximum adjacent to a sharp minimum is known as the Cotton effect, with the effect said to be positive if the maximum in $[\varphi]$ or $[\alpha]$ is at higher wavelength than the minimum.

For a single chromophore, the ORD curve crosses the wavelength axis at the wavelength of maximum absorption. The extinction coefficient for right- and left-handed circularly polarised light is different for a chiral chromophore, and a plot of this difference is known as the circular dichroism (CD) spectrum. Sometimes circular dichroism is defined in a straightforward manner, as the difference between the extinction coefficients of right- and left-handed circularly polarised light [eqn. (4.25): I_0 is the intensity of light incident on, and I that

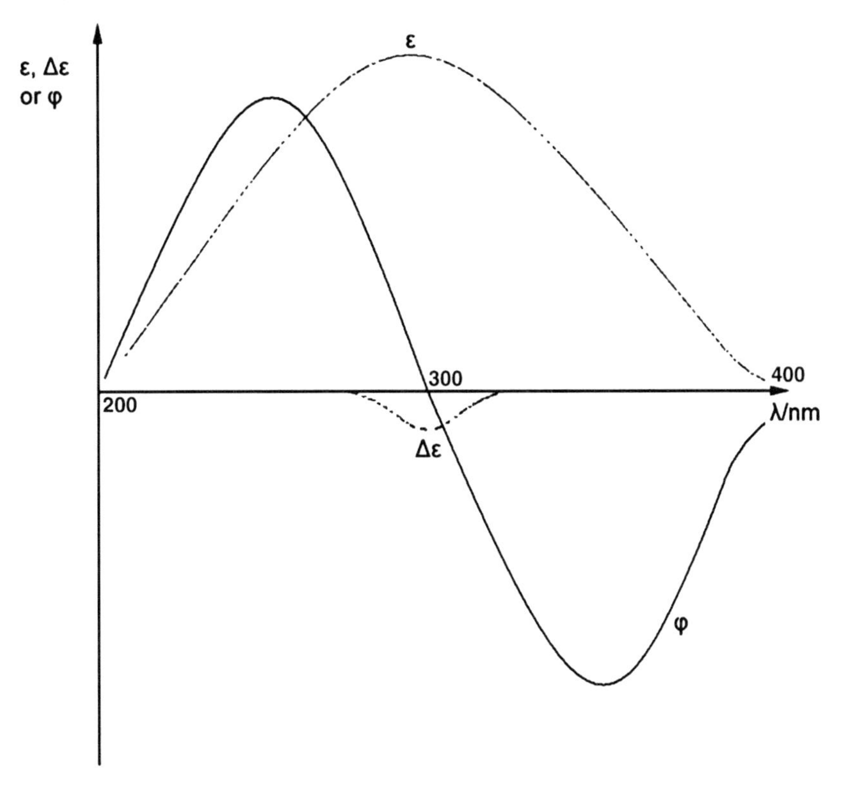

Figure 4.32 Cotton effect ORD and CD spectra. A negative Cotton effect and hence a negative value of Δε are shown.

transmitted through, a cell of pathlength L (cm) containing sample at a molar concentration C:

$$\Delta\varepsilon = \varepsilon_L - \varepsilon_R; \ \varepsilon = \frac{\log_{10}(I_0/I)}{CL} \tag{4.25}$$

With the sign conventions above, the sign of the Cotton effect is the same as that of the circular dichroism spectrum (Figure 4.32).

Modern CD instrumentation measures the difference in absorbance of right-add left-handed circularly polarised light directly, so CD spectra are often reported as Δε, with units the same as for ε itself. Sometimes, however, for historical reasons, the circular dichroism of a sample is recorded as ellipticity. The origin of the term lies in the fact that if one of the circularly polarised beams emerging from the sample is attenuated relative to the other, on recombination the two circularly polarised beams form elliptically polarised light, not plane polarised light. The ellipticity θ, is $\tan^{-1}(b/a)$ where b is the minor and a the major axis of the ellipse. The conventional molar ellipticity $[\theta]$

is expressed in $^{\circ}\mathrm{cm}^2\,\mathrm{mol}^{-1}\,\mathrm{dmol}^{-1}$ and is related to the conventionally expressed $\Delta\varepsilon$ by

$$[\theta] = 3298\Delta\varepsilon \tag{4.26}$$

Chromophores often exploited in CD studies of polysaccharides are the carboxylates, whose n–π* transitions occur at 208 nm. CD is more sensitive, but otherwise gives much the same information as ORD, and indeed to a first approximation the shape of the ORD curve is that of (minus) the first differential with respect to wavelength of the CD curve.

4.6 SOME IMPORTANT POLYSACCHARIDES

From the structures of polysaccharides containing only one type of sugar in one type of linkage, it is possible to infer general qualitative guidelines governing the structures of polysaccharides. Much of the behaviour of homopolysaccharide derivatives with "decoration" of various sorts attached to the basic structure can also be rationalised. Polysaccharides with different linkages to the same sugar are less common, but the behaviour of amylopectin and β-(1→3),β-(1→4)-glucan is largely understood.

Despite the experimental difficulties in deducing them, the sequences of a very large number of heteropolysaccharides are known from many sources. An area particularly fertile in novel structures is that of the bacterial glycocalyx, the extracellular polysaccharides and lipopolysaccharides on the outside of bacteria, which is under biological selection pressure to be as diverse as possible and thus resist attempts by the bacterium's competitors to dissolve away its protective coat with various enzymes. Those heteropolysaccharides for which there is some information on the relation between structure and properties are, unsurprisingly, those which are important to the human body or have industrial applications, and only these will be covered here.

The oligosaccharides derived from polysaccharides often have opaque trivial names: a list of common oligosaccharides is given in Table 4.2.

4.6.1 1→4-Linked Diequatorial Pyranosides

Most of the polysaccharides of the plant cell wall are based on 1,4-linked diequatorial pyranosides (Figure 4.33); structural polysaccharides of invertebrate animals have the same motif. The solid-state structures and solution conformations of these polysaccharides and their oligosaccharides are governed by a hydrogen bond in the sense O5 ··· HO3′ between the 3-OH and the ring oxygen of the preceding glucose: this results in a value of φ of around –95°, close to the value required for optimal overlap of the p-type lone pair on the glycosidic oxygen with the C1–O5 σ* orbital; ψ is more variable, but with fibrous polysaccharides is around $\psi \approx 96°(\Psi \approx -140°)$. In some fibrous polysaccharides, the O3′H ··· O5 hydrogen bond is bifurcated, with a second acceptor (O1 or O6), but the O3′H ··· O5 distance is always the shorter.

Table 4.2 Some oligosaccharides.

Name	*Structure*	*Comments*
Allolactose	β-D-Gal*p*-(1→6)-D-Glc	
Cellobiose, -triose, *etc.*	(4)-β-D-Glc*p*-(1→)$_n$	
Gentianose	β-D-Glc*p*-(1→6)-α-D-Glc(1↔2)-β-D-Fru*f*	Non-reducing
Gentiobiose	β-D-Glc*p*-(1→6)-D-Glc	
Isomaltose	α-D-Glc*p*-(1→6)-D-Glc	
Kestose	β-D-Fru*f*-(2→1 or 6)-β-D-Fru*f*-(2↔1)-α-D-Glc*p*	Formed by GH 32 transglycosylation
Kojibiose, -triose, *etc.*	(2)-α-D-Glc*p*-(1→)$_n$	
Lacto-*N*-triose I	β-D-Gal*p*-(1→3)-β-D-GlcNAc*p*-(1→3)-D-Gal	
Lacto-*N*-triose II	β-D-GlcNAc*p*-(1→3)-β-D-Gal*p*-(1→4)-D-Glc	Milk oligosaccharide
Lacto-*N*-tetraose	β-D-Gal*p*-(1→3)-β-D-GlcNAc*p*-(1→3)-β-D-Gal*p*-(1→4)-D-Glc	
Lacto-*N*-neotetraose	β-D-Gal*p*-(1→4)-β-D-GlcNAc*p*-(1→3)-β-D-Gal*p*-(1→4)-D-Glc	
Laminaribiose, -triose, *etc.*	(3)-β-D-Glc*p*-(1→)$_n$	
Lactose	β-D-Gal*p*-(1→4)-D-Glc	
Lactulose	β-D-Gal*p*-(1→4)-D-Fru	
Maltose, -triose, *etc.*	(4)-α-D-Glc*p*-(1→)$_n$	Note maltose is a disaccharide
Mannobiose, -triose, *etc.*	(4)-β-D-Man*p*-(1→)$_n$	In the medical literature other stereo- and regioisomers also called mannobiose, *etc.*
Melibiose	α-D-Gal*p*-(1→6)-D-Glc	
Melizitose	α-D-Glc*p*-(1→3)-α-D-Glc*p*-(1↔2)-β-D-Fru*f*	3-α-D-Glucopyranosylsucrose
Planteose	α-D-Gal*p*-(1→6)-β-D-Fru*f* (2↔1)-α-D-Glc*p*	Non-reducing
Primeverose	β-D-Xyl*p*-(1→6)-D-Glc	
Raffinose	α-D-Gal*p*-(1→6)-α-D-Glc*p*-(1↔2)-β-D-Fru*f*	Non-reducing
Stachyose	α-D-Gal*p*-(1→6)-α-D-Gal*p*-(1→6)-α-D-Glc*p*-(1↔2)-β-D-Fru*f*	Non-reducing
Sucrose	α-D-Glc*p*-(1↔2)-β-D-Fru*f*	Common sugar. Non-reducing
Trehalose	α-D-Glc*p*-(1↔1)-α-D-Glc*p*	Naturally occurring isomer, properly α,α-trehalose, α,β- and β,β-isomers known. Non-reducing
Turanose	α-D-Glc*p*-(1→3)-D-Fru*f*.	
Xylobiose, -triose, *etc.*	(4)-β-D-Xyl*p*-(1→)$_n$	

Figure 4.33 Key conformational motif in 1,4-diequatorial pyranosides. Shown in both conformational and Mills representations.

1,4-Diequatorially linked polypyranose structures form fibres when the polysaccharides are undecorated – indeed, oligosaccharides beyond a degree of polymerisation of ~6 become almost water insoluble. When, however, the backbone chain is substituted ("decorated"), as it is in the plant hemicelluloses, the decoration interferes with the parallel packing of the chains and much more soluble polymers result.

4.6.1.1 Undecorated, Fibrous, 1→4-Diequatorial Polysaccharides

4.6.1.1.1 Cellulose. Cellulose is the major structural component of plant cell walls, forming the fibres in the natural composite which is lignocellulose. It forms about 30% of the carbon in the biosphere, about 10^9 tonnes being formed and hydrolysed per year.[63] It is generally considered to have been discovered by Anselme Payen in 1838,[64] and the details of its structure were painfully deduced over the next century. Methylation analysis gave overwhelmingly 2,3,6-tri-*O*-methylglucose, a result compatible with (1→4)-pyranosidic and (1→5)-furanosidic linkages in rings or infinite chains. Acetolysis of cellulose gave ~50% octaacetylcellobiose, the structure of which was rigorously established by "wet" chemical methods, confirming the β-(1→4) nature of a majority of the linkages[65] (furanosidic linkages were anyway unlikely because of the relative acid stability of the polymer). Ring structures were eliminated by the isolation (in pre-chromatographic times!) of 0.6% 2,3,4,6-tetra-*O*-methylglucose from the non-reducing end of methylated cellulose.[66] Early fibre X-ray work suggested (correctly) that the chains of β-(1→4)-glucan were flat and linear, a shape they could take if rings were in the conformation now known as $^4C_1$ and each ring was rotated through 180° about the axis of the chain with respect to its predecessor:[67] conformational analysis was thereby introduced into carbohydrate chemistry before it was introduced into mainstream organic chemistry. The degree of polymerisation (*DP*) of native cellulose is quoted by secondary sources as around 10^4,[68] but the media used to solubilise cellulose can give spuriously low or spuriously high figures. Derivatisation as the carbanilate, already discussed, runs the risk of hydrolysing the occasional glucan link. The traditional use of an octahedrally coordinated transition metal to form a chelate with the 2- and 3-OH groups of the anhydroglucose unit has been gradually improved. The original "cuprammonium" solution (a solution of Cu^{II} and ammonia), which gave the first reasonably reliable ultracentrifuge data,[69] was a one-electron oxidant and therefore could lead to loss of *DP* (see Chapter 7). Progressive improvements on the theme (*e.g.* Cu^{II}–, Ni^{II}– and

Cd^{II}–en (ethylenediamine, $NH_2CH_2CH_2NH_2$)) complexes led to the present Cd–tren system [tren = $N(CH_2CH_2NH_2)_3$] (Figure 4.34a), which is not at all redox active and will solubilise celluloses, such as those from bacteria or cotton, which give the highest observed *DP* in well-behaved solution of 9700.[70]

Some clear "solutions" of cellulose, however, are in fact micellar. Clear, apparently purely physical solutions of cellulose are formed in *N*-methylmorpholine *N*-oxide (NMNO)[71] and anhydrous dimethylacetamide–LiCl,[72] but they can give rise to anomalously high *DP* values, because of association of the glucan chains in so-called "fringe micelles", in which a group of chains associate chain-to-chain but the ends are largely random chain (Figure 4.34b and c).

It has recently been found that salts which melt at or near room temperature, so-called ionic liquids, can form physical solutions of cellulose[73] and starch.[74] 1-*N*-Butyl-3-methylimidazolium chloride dissolved plant and bacterial cellulose with no apparent loss of *DP*, and cellulose in the resulting solutions was much more readily derivatised to various esters than in the solid (Figure 4.34d).[75] The same applied to 1-*N*-allyl-3-methylimidazolium chloride;[76] in both solvents, ^{13}C NMR indicated that the cellulose chains were disordered in solution.[77] Studies

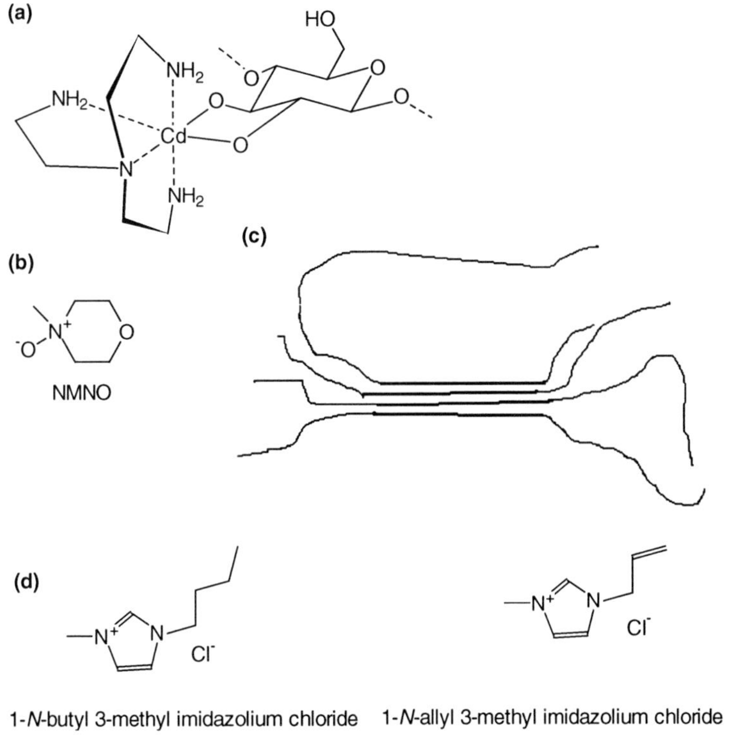

Figure 4.34 Solubilisation of cellulose. (a) By Cd–tren; (b) *N*-methylmorpholine *N*-oxide; (c) fringe micelles formed in apparently simple solutions; (d) ionic liquids capable of solubilising cellulose and other polysaccharides.

of NMR relaxation of ^{13}C, ^{35}Cl and ^{37}Cl nuclei in 1-*N*-butyl-3-methylimidazolium chloride solutions of model compounds indicated hydrogen bonding of each sugar hydroxyl to a chloride ion,[78] suggesting a possible mechanism for the dissolving power of these salts and also for the ready derivatisation of the solubilised carbohydrate. (Measurements of colligative properties, which would have revealed whether the polysaccharide chains were isolated or in a fringe micelle, are not possible in ionic liquids.)

Determination of cellulose conformation was less straightforward, partly because biosynthesised cellulose is only partly crystalline. A typical cellulose fibre[79] will consist of crystalline and amorphous regions, with the same chain wandering through several such regions. The crystallites are small enough that most glucan chains will be at a surface, and the validity of treating the material as two distinct phases has been questioned.[80] The chains are associated into microfibrils, about $(1–5) \times 10^4$ Å^2 in cross-section, containing several hundred glucan chains and several crystallites; the axis of the microfibril and the axis of the fibre are rarely parallel, the angle between the two (microfibril angle) having a great bearing on the mechanical properties of the fibre. The problems of the poor crystallinity of cellulose from abundant sources cannot be solved by dissolution of the cellulose and recrystallisation, because biosynthesised cellulose appears to be thermodynamically metastable, and deposition of cellulose, either from physical solution or by decomposition of its xanthate derivative, as in the viscose process, produces a different polymorph, cellulose II. Attempts to determine the crystal structure of cellulose I have therefore had to focus in the first place upon finding a suitable source.

A couple of fibre X-ray studies[81] suggested a monoclinic unit cell with parallel glucan chains for cellulose I, but the structural breakthrough came when it was found that the cellobiohydrolase II (Cel6A) from *Trichoderma reesei*, which had been shown to be an *exo*-cellulase, acting from the non-reducing end, was found to sharpen only one end of the microcrystals of the highly crystalline cellulose I from the marine tunicate *Valonia ventricosa*.[82] If an *exo*-acting enzyme acted only on one end of the needle-like crystals, then the chains in cellulose I had to run parallel to each other, rather than antiparallel. Soon thereafter, solid-state NMR indicated that solid cellulose had two anomeric ^{13}C resonances, rather than just the one, as expected from the chemical structure.[83] Moreover, the intensities of these resonances varied with the source, with one resonance being predominant in celluloses from primitive organisms (bacteria, algae), whereas in celluloses from higher plants the two resonances were of comparable intensity. Natural cellulose is a mixture of two forms, cellulose Iα, with a triclinic unit cell, and cellulose I_β, with a monoclinic cell, the unit cell dimensions being confirmed by electron diffraction.[84] Cellulose Iα is the thermodynamically more stable form, as shown by its conversion to cellulose I_β by heat annealing at 260 °C.[85]

Recent electron,[86] X-ray and neutron diffraction work has established the structures for both cellulose I_α[34] and I_β[33]. Microcrystalline samples were obtained by mild acid hydrolysis of the cellulose walls of the freshwater alga *Glaucocystis nostochinearum* and the tunicate *Halocynthia roretzi* and were oriented in the direction of the *c* axis. In both crystalline forms cellobiose is the repeat unit, with a strong intrachain hydrogen bond from 3-OH to the preceding ring O5 maintaining a value of $\Phi = \varphi$ between −89° and −99° and Ψ between −147° and −138° ($\underline{\psi} \approx$ 93–102°) and hydrogen bonding between the O6 and the O2 of the preceding residue giving a range of χ between −83° and −70° over the four types of glucose units involved (two in each allomorph) (Figure 4.35). All glucose units therefore adopt the normally disfavoured *tg* conformation about C5–C6, but the O5 ··· HO3′ hydrogen bond reinforces the conformation about C1–O1 that favoured is by the *exo*-anomeric effect.

The inter-chain hydrogen bonding and packing of the chains in the crystal are slightly different in the two forms. The unit cells are shown in Figure 4.36, with one cellobiose unit in the triclinic form and two in the monoclinic form. The glucan chains in both cellulose I_α and cellulose I_β form extended sheets by hydrogen bonding to their neighbours, but the sheets themselves are held together only by van der Waals interactions and weak C–H ··· O bonds. A cartoon of the stacking of the glucan chains is given in Figure 4.37.

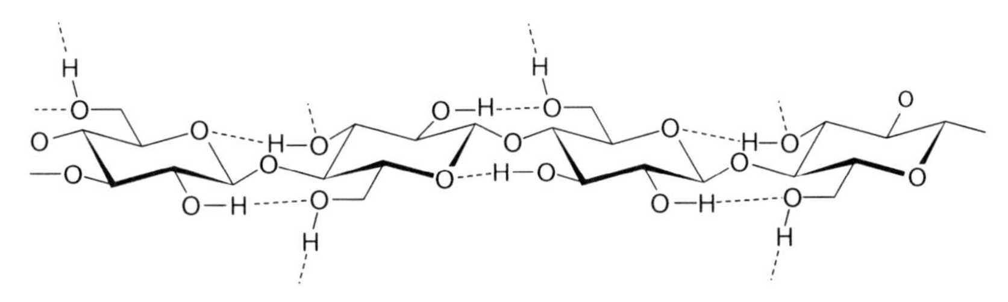

Figure 4.35 Hydrogen bonding and conformation of an individual glucan chain in cellulose I allomorphs.

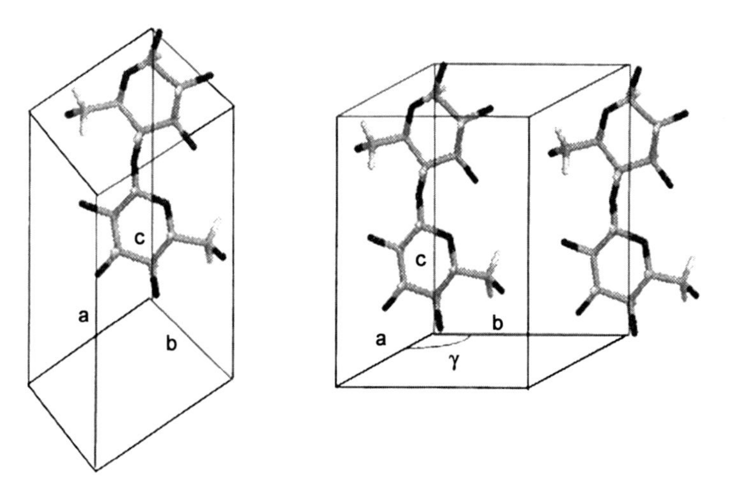

Figure 4.36 Unit cells of cellulose I_α (left) and cellulose I_β (right); reproduced from ref. 86 by kind permission of Prof. J. Sugiyama.

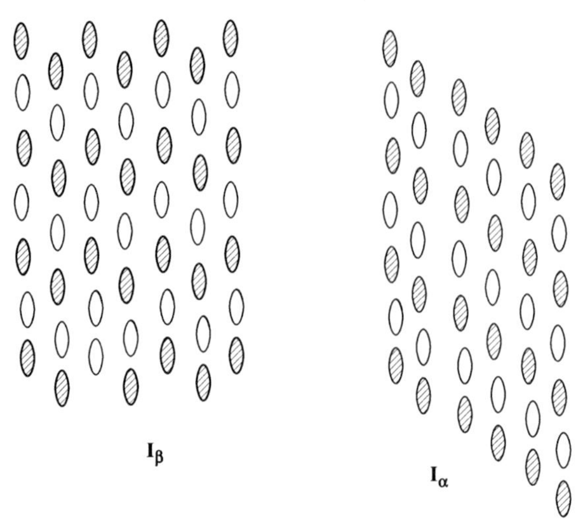

Figure 4.37 Cartoon of packing of cellulose chains in the two forms of cellulose I. The view is perpendicular to both the *c* (fibre) axis and the hydrogen-bonded planes. The shaded and unshaded ellipses refer to the two directions of the glucose units.

In the triclinic form, there is only one glucan chain, but the two glucose units in it are not equivalent and one is slightly more puckered than the other, in accord with detailed CP-MAS NMR studies.[87] In the monoclinic form there are two non-equivalent chains, one of which goes through the centre of the unit cell (centre) and another of which goes through the crystallographic origin (origin), although the individual glucose units within each type of chain are identical. There are two hydrogen-bonded sheets in the monoclinic form (Figure 4.38).

In both forms the neutron diffraction study revealed a well-defined intramolecular O5 ··· DO3′ bond, with the deuterium in a single position, across all linkages (although it is bifurcated to O1 as well in alternate linkages in cellulose I_α). Deuterium atoms forming other hydrogen bonds were disordered and could be considered as belonging to either of two networks. However, it is possible that this disorder is a property only of the deuterated celluloses, with the protiated celluloses having a single structure. In some tightly hydrogen-bonded systems, the deuterium moves in a double potential well, but protium in a single one, since the barrier between the two wells is lower than the zero point energy of the protiated system but higher than that of the deuterated one.[88] Nonetheless, the partially occupied networks appear to all involve intrachain hydrogen bond between O2 and O6 of the next residue, although their types differ; sometimes they are in the sense O2(*n*)–H ··· O6(*n* – 1), where the 6-OH forms a bifurcated hydrogen bond to O2 and O3 of an adjacent chain; at other times they are in the sense O2(*n*)–H (bifurcated) to O1(*n*) and O6(*n* – 1); in this case the 6-OH forms an unbifurcated hydrogen bond to O3 of an adjacent chain.

Microdiffraction tilt experiments with crystalline samples of cellulose I_α and I_β, in combination with staining of the reducing ends of highly crystalline specimens,

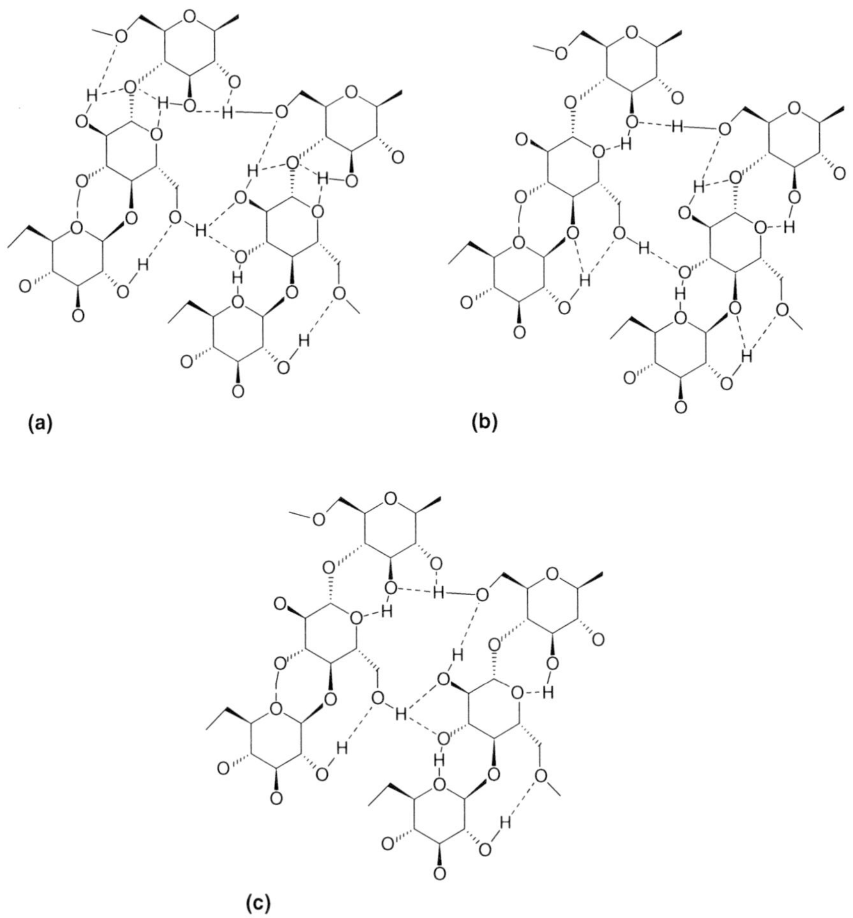

Figure 4.38 Most occupied hydrogen bonding networks in (a) cellulose I_{α}, (b) the origin chain of cellulose I_{β} and (c) the centre chain of hydrogen I_{β}. The less occupied networks involve variations on the same themes of bicoordinate/tricoordinate hydrogen.

established that both forms of cellulose had the "parallel up" structure.[86] The same Sugiyama tilt experiments (Figure 4.10) could be used to establish the direction of the crystalline *c* axis in cellulose microfibrils emerging from the cellulose synthase apparatus of *Acetobacter aceti*. These were pointing away from the bacterium. Therefore the nascent cellulose chains grow from their non-reducing end.

The topological problem of how polymer biosynthesis that proceeds by the addition of monosaccharide units[xxv] to the non-reducing end can be initiated

---

[xxv] For several years a mechanism involving simultaneous transfer of two glucose residues (I. M Saxena, R. M. Brown, M. Fevre, R. A. Geremia and B. Henrissat, *J. Bacteriol.* 1995, **177**, 1419) to form a cellobiose unit was favoured, but can now be discounted because of the absence of more than one nucleotide sugar binding domain in cellulose synthases.

was solved by the elucidation of the role of the most common plant sterol, sitosterol (Figure 4.39).[89] This first forms a β-glucoside, and then the glucan chain is built up until it reaches a critical size, at which point an endoglucanase cleaves the chain, leaving a sitosterol cellooligosaccharide glucoside free to start a new chain. The key GH 9 endoglucanase was actually found genetically, as a deficiency caused the KORRIGAN mutation in *Arabidopsis thaliana*, the only higher plant for which the genome has been sequenced.[90]

The biosynthesis of a metastable polymorph of a polymer would be expected to, and does, require a complex apparatus.[91] Attempts to make cellulose *in vitro* without this apparatus intact yield callose, a β-(1→3),β-(1→4)-glucan with a preponderance of β-(1→3) linkages, although if care is taken a "loose" form of cellulose I can be produced.[92] The apparatus in plants appears to be a rosette-like organisation of spinnerets (Figure 4.40), each one of which appears to produce a single glucan chain; the glucan chains must crystallise close to the mouths of the spinnerets. In the bacterium *Acetobacter xylinum,* though, the cellulose exits from a slit-like microstructure and there is a report of a mutant strain that gave cellulose I or cellulose II according to the medium viscosity.[93]

The structure of cellulose II has been a longstanding puzzle, with the X-ray and neutron diffraction evidence favouring an antiparallel structure[35] difficult to reconcile with biochemical and chemical evidence suggesting cellulose I and cellulose II are similar. Enzymic polymerisation of cellobiosyl fluoride, and also recrystallisation of cellulose I, yield cellulose II.[94] A key problem has been mercerisation, the conversion of cellulose I to cellulose II by sodium hydroxide, without apparent macroscopic disintegration,[xxvi] although more recent reports have described the complete macroscopic reordering of flax fibres on complete conversion to cellulose II.[95] Parallel structures for cellulose II have also been proposed on theoretical grounds.[96] The monoclinic antiparallel structure favoured by neutron and X-ray diffraction involves two chains running in opposite senses per unit cell; the adoption of the preferred *gt* conformation by all the hydroxymethyl groups provides a ready explanation for the stability of the this allomorph ($\chi = -165°$ and $-175°$ for origin and centre chains, respectively). The conformation of the glycosidic linkage is much like other cellulose allomorphs, with a key role again played by the O3-H $\cdots$ O5′ hydrogen bond, ($\varphi = -95°$ and $-91°$, $\psi = 92°$ and $89°$ for origin and centre chains, respectively), although because of the *gt* conformation of the hydroxymethyl group, the hydrogen bond is now bifurcated to O6′. Hydrogen bonds between origin chains are all of the type O2–H $\cdots$ O6′, whereas those between centre chains are of the type O2 $\cdots$ H–O6′. Additionally, there are hydrogen bonds between the origin and centre layers. Hydrogen bonding in cellulose II is illustrated in Figure 4.41.

On treatment with liquid ammonia or mono-, di- or triamines, the crystal structures of both cellulose I and cellulose II open out to form cellulose III_I and

[xxvi] So-called because it was discovered by the Lancashire textile chemist John Mercer in 1844, who found that passing cotton thread through a bath of caustic soda gave the resultant cloth an attractive sheen.

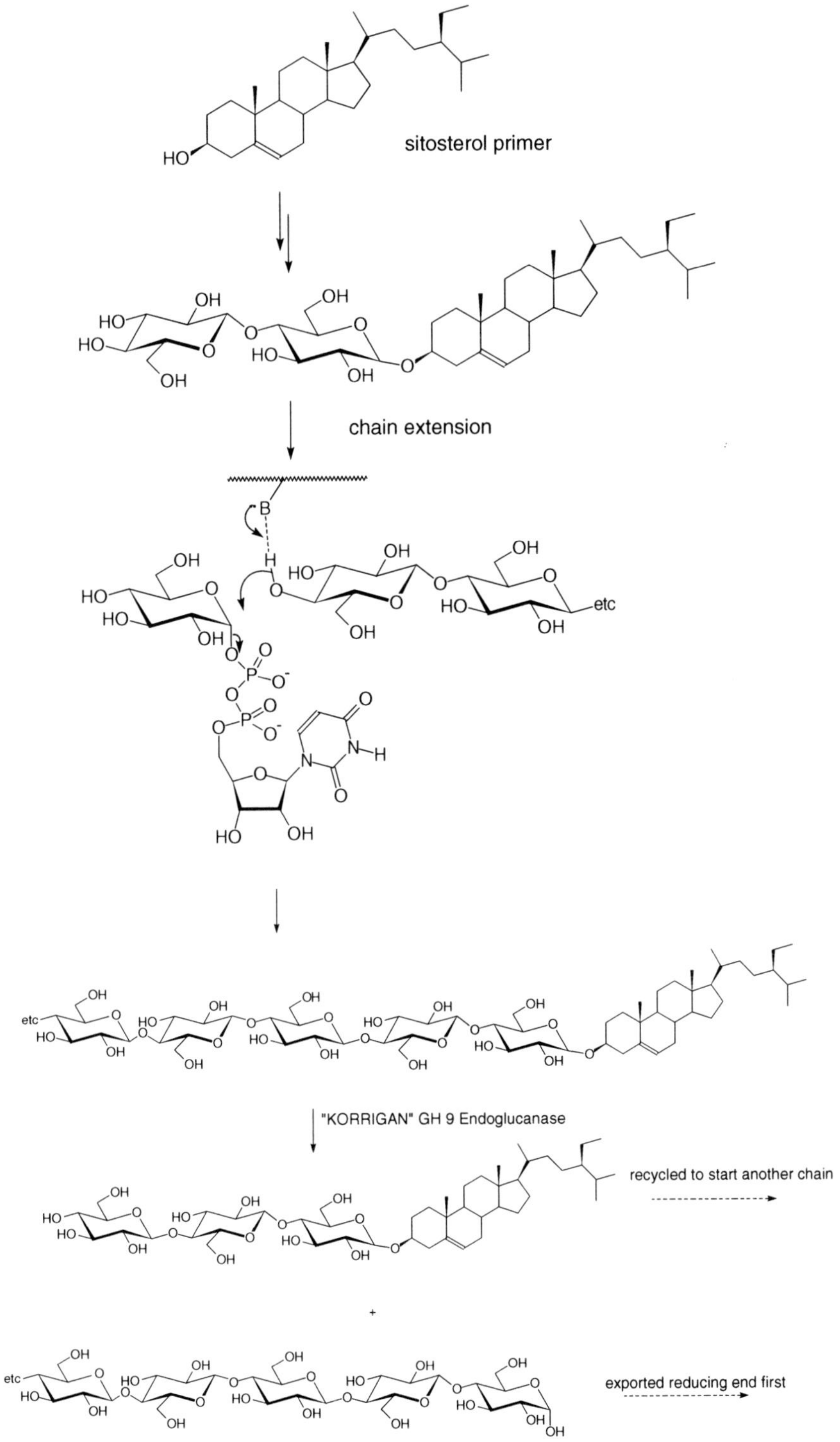

sitosterol primer

chain extension

"KORRIGAN" GH 9 Endoglucanase

recycled to start another chain

exported reducing end first

Figure 4.39 Biosynthesis of cellulose.

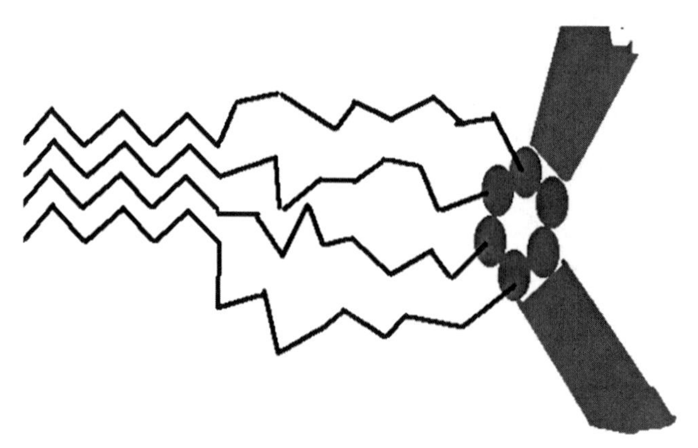

Figure 4.40 Cartoon of the organisation of the plant cellulose biosynthetic apparatus.

cellulose III_{II}, respectively (Figure 4.42). Cellulose III polymorphs survive removal of ammonia or amines, but revert back to their respective parent on hydrothermal treatment. Cellulose III_I has a one-chain monoclinic unit cell, with $\varphi = -92°$, $\psi = 93°$ and $\chi = 163°$, close to the chain conformations of cellulose II.[97]

A combination of methods has recently apparently solved the mercerisation paradox.[98] As discussed above, the cellobiohydrolase Cel6A of *T. reesei* will sharpen the non-reducing ends of the cellulose I microcrystals of *V. ventricosa*. At the same time, a method of labelling the reducing ends of the microcrystals was found. This exploited the very powerful affinity of a protein called streptavidin for the cofactor biotin. Biotin was linked to a hydrazide and this was reacted with the reducing end of the glucan chains and the imine reduced with sodium cyanoborohydride, covalently attaching biotin (Figure 4.43). The crystal suspensions were then treated with streptavidin to which colloidal gold particles has been attached. The gold showed up in electron micrographs because of its high number of electrons. In Cel6A-treated *V. ventricosa* microcrystals, the labelled end was always the blunt, unsharpened end. Microfibrils which had been vigorously treated with sodium borohydride were not labelled, as after this treatment they possessed no reducing end.

After mercerisation, the cellulose II microcrystals were labelled; however, the labels were seen to be attached to both ends of the microcrystals. When *V. ventricosa* cellulose I was converted to cellulose III_I by treatment with supercritical ammonia, only one end of the microcrystals was labelled, demonstrating its parallel structure. The key to the mechanism of mercerisation was provided by microdiffraction experiments, when it was found that diffraction by one layer of cellulose I microfibrils gave two sets of 110 diffraction spots, corresponding to the cellulose chains going in opposite directions. Mercerisation of *V. ventricosa* cellulose I resulted in complete reordering of the surface of the cellulose crystal on the micrometre scale. The process involved interdigitation of the cellulose chains of microcrystals in close proximity to each other in

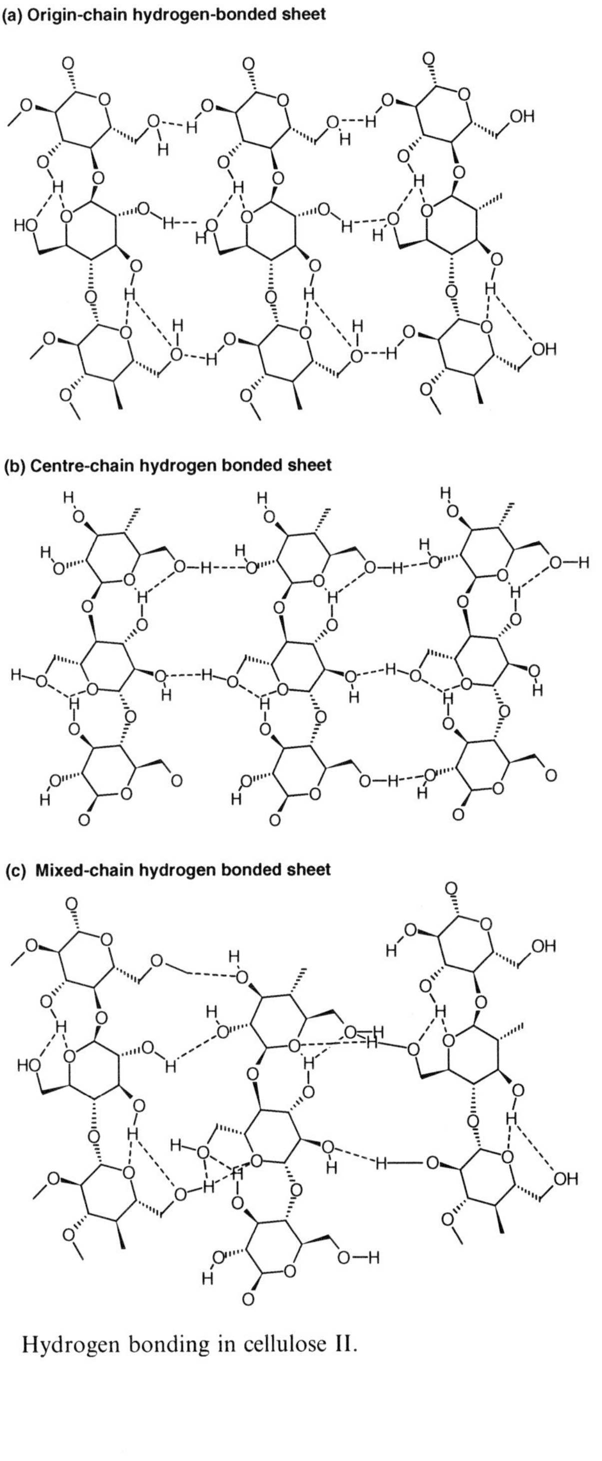

(a) Origin-chain hydrogen-bonded sheet

(b) Centre-chain hydrogen bonded sheet

(c) Mixed-chain hydrogen bonded sheet

Figure 4.41 Hydrogen bonding in cellulose II.

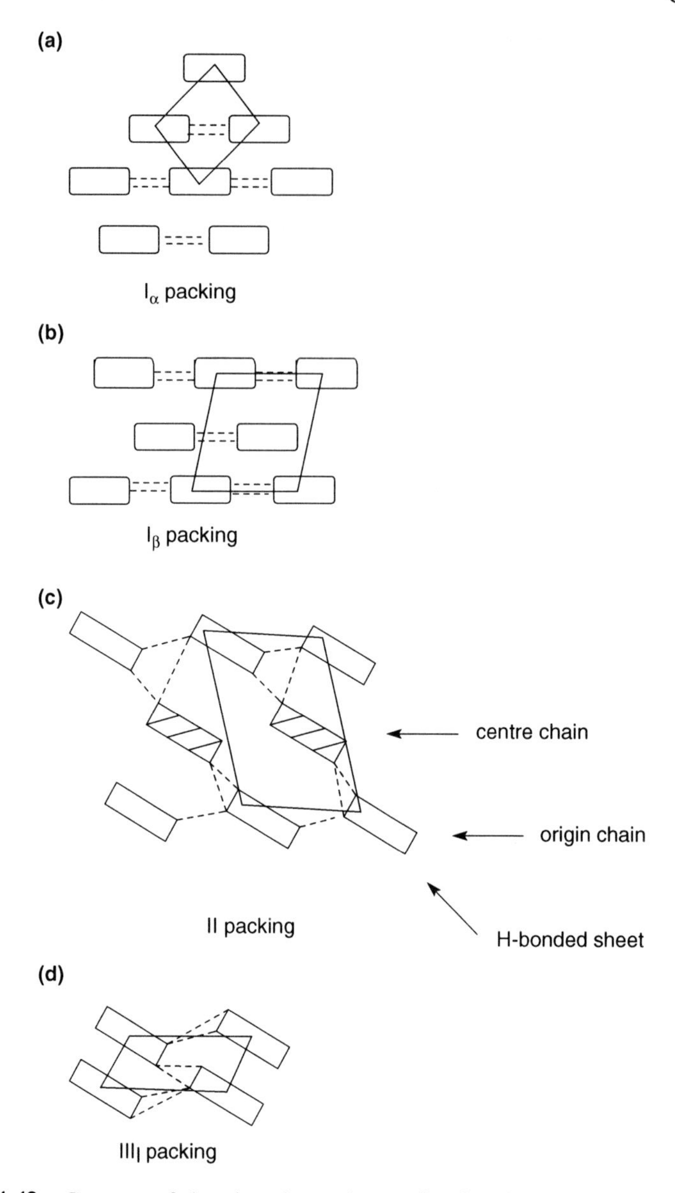

Figure 4.42 Cartoon of the view down the *c* axis of various cellulose polymorphs: shaded rectangles represent glucan chains in the opposite sense and dotted lines hydrogen bonds.

the cell wall, which have opposite polarity (Figure 4.44). With the smaller microcrystals of a higher plant (ramie) cellulose I, the results were not as clear cut, but suggested the presence of microcrystalline regions with opposite chain polarity in this cellulose also.

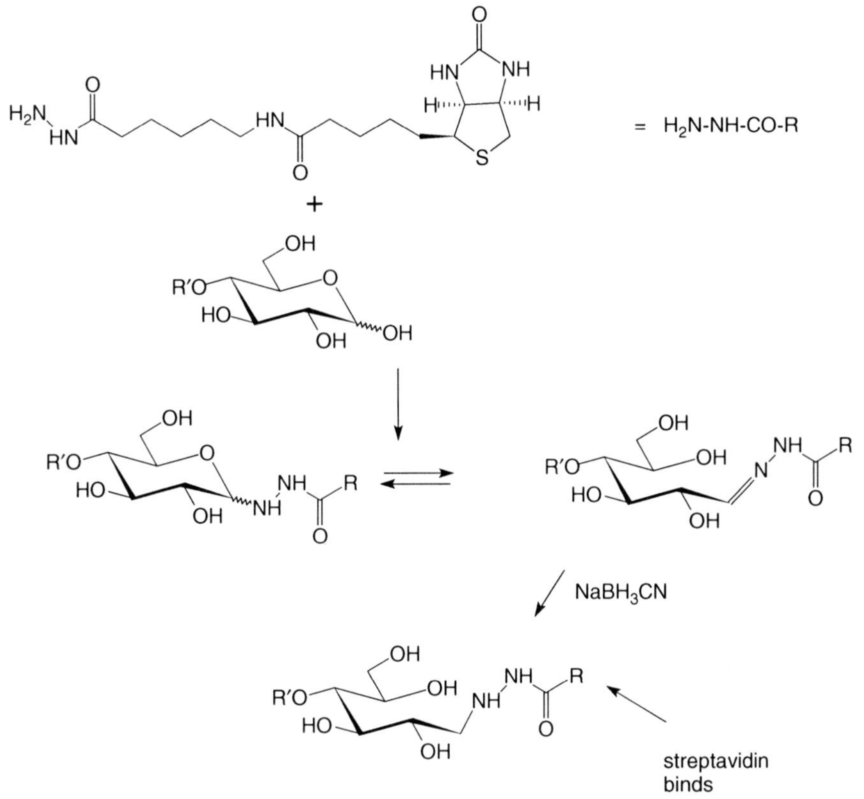

= H_2N-NH-CO-R

$NaBH_3CN$

streptavidin binds

Figure 4.43 Chemistry of polysaccharide reducing-end labelling.

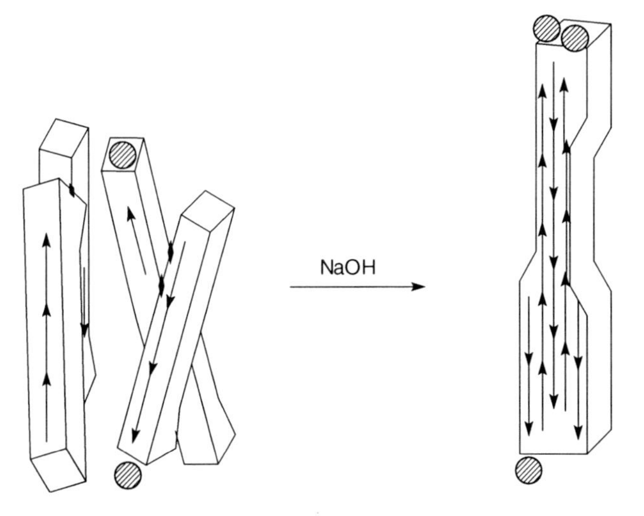

Figure 4.44 Interdigitation mechanism of mercerisation.

4.6.1.1.2 Chitin. Chitin, the structural polysaccharide of arthropods and one of the structural polysaccharides of fungi, plays a role in the marine biosphere similar to that played by cellulose in the terrestrial – in the older literature it is called "animal cellulose". The "rain" of particles from the benthic layer of the seas to the ocean floor is largely chitin and globally is said to consist of "billions" of tonnes of chitin from just one subclass of zooplankton, the copepods.[99] Its primary structure is [β-D-GlcNAc*p*-(1→4)]$_n$, *i.e.* cellulose with all the 2-OH groups replaced by acetamido groups. Chitin is found in close association with proteins and calcium carbonate, which have to be dissolved away before a native *DP* can be estimated, so the resultant figures are even more problematical than with cellulose, with the higher figures being associated with more careful extraction procedures.[100] Very large quantities of chitin are produced as waste by the shellfish industry and much of the research on the material has been motivated by a desire to find a remunerative use for it.

Three crystalline allomorphs have been described, α, β and γ, with structures available only for the first two. These are based on X-ray and electron diffraction, so that hydrogen atoms are not located.

β-Chitin has a structure reminiscent of cellulose I_α, but with a monoclinic rather than a triclinic unit cell (Figure 4.45).[101] Fibre X-ray studies on the β-chitin from *Oligobrachia ivanovi* revealed one polysaccharide chain and two sugar units, per unit cell, with the usual pronounced O3′–H ··· O5 hydrogen bond enforcing values of ϕ = ɯ = 23–24° ($\underline{\varphi} \approx -97°$, $\underline{\psi} \approx 90°$). Major differences from cellulose I allomorphs, however, are that the hydroxymethyl groups are in the *gg* conformation and that the interchain hydrogen bonds are between these hydroxyls and the amide carbonyls (the amide plane being approximately perpendicular to the sugar ring). Probably there are no intersheet hydrogen bonds, though in the absence of neutron diffraction data to locate hydrogen atoms, this conclusion is necessarily tentative.

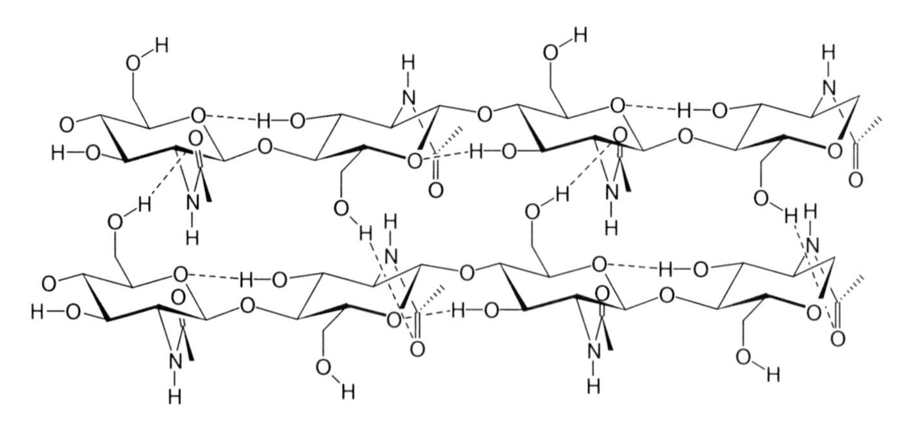

Figure 4.45 β-Chitin.

The biosynthesis of chitin oligosaccharides (in the zebrafish and the terrestrial bacterium *Rhizobium* spp.) occurs by processive transfer of GlcNAc residues from UDPGlcNAc to the non-reducing end; whatever the initial anchor (*cf.* sitosterol in cellulose biosynthesis), the *p*-nitrophenyl glycoside of GlcNAc can replace it.[102] The configuration of chitin in the unit cell was established as "parallel up" by Sugiyama "tilt" microdiffraction experiments on the β-chitin produced by the vestimentiferan *Lamellibrachia satsuma*, similar to those performed with cellulose.[13] The *c* axis in the crystalline chitin emerging from diatoms *Thalassiosira* spp. was established by tilt diffraction experiments as going away from the organism, thereby establishing that the chitin was being exported reducing end first.[103] Therefore, chitin is biosynthesised from the non-reducing end.

α-Chitin is the commonest chitin allomorph. X-ray studies of deproteinised lobster tendon indicated an antiparallel arrangement of crystal chains in an orthorhombic unit cell.[104] Biosynthesis of fibrous polymers with an antiparallel molecular arrangement presents obvious topological problems, but the absence of any cross-specificity between α- and β-chitin, among chitinases active on crystalline chitin[105] or chitin binding domains,[106] makes it most unlikely the antiparallel structure is incorrect. The observation, in the yeast *Mucor rouxii*, of a subcellular vesicle termed a chitosome, which could produce chitin microfibrils *in vivo*, suggests a mechanism whereby the chains are first produced in a soluble of liquid-crystalline form in the vesicle and crystallise into antiparallel α-chitin when the contents of the vesicle are "extruded" by cytosis out of the cell.[107]

The dihedral angles about the glycosidic bonds in α-chitin were the same as those in β-chitin,[xxvii] but in α-chitin there are two chitin chains in the unit cell. To account for an X-ray reflection which should not have been diffracted by crystals of the derived space group, it was proposed that the hydroxymethyl group was disordered. Subsequently, however, it was shown that this reflection was a double diffraction artefact.[108] The detailed structure of α-chitin is therefore uncertain.

4.6.1.1.3 Chitosan. Treatment of demineralised shellfish waste with hot 50% sodium hydroxide for several hours results in *N*-deacetylation of the chitin of which it is largely composed, without significant depolymerisation. The extent of *N*-deacetylation depends on the severity of the treatment: to fully deacetylate a pure form of chitin (crab tendon) three sequential 2 h treatments with 67% sodium hydroxide at 110°C were required.[109] Chitosan was at first thought to be a semisynthetic product, but more recently partly deacetylated chitins have been found in the cell walls of some fungi,[110] although the distribution of the remaining *N*-acetyl groups is non-random, unlike chemically produced chitosan. Unsurprisingly, given the natural occurrence of the polymer, "chitosanases" exist and are found in glycoside hydrolase families 8, 46, 75 and 80. However, even in the family that contains only chitosanases (46), the reactive amino

---

[xxvii] Although described differently, a negative sign being placed before ψ in the α-chitin paper. If the sign in the β-chitin paper is correct, $\psi \approx -143°$, very close, as expected, to the figure for cellulose.

group appears, from the crystal structure of an enzyme that cleaves both βGlcN-(1→4)GlcN and βGlcN-(1→4)GlcNAc linkages,[111] to play no covalent role in the action of a standard inverting glycohydrolase.

The availability of a positively-charged hydrophilic biopolymer has opened the way to many potential practical applications, such as removal of metal ions from waste-water and applications that involve interactions with negatively charged polysaccharides, whether soluble or insoluble. In papermaking, for example, the possibility exists that chitosan could replace the expensive synthetic and semisynthetic amine polymers, currently in use to interact with the negatively charged cellulose fibres as flocculation or wet-strength agents.[112]

In its hydrated, cationic form, chitosan adopts helical conformations with eight residues per five left-hand or three right-hand turns,[113] in four-fold helical conformations of dimeric units. The anhydrous, fully deacetylated chitosan adopts a crystalline form much like that of cellulose II,[109] with $\varphi = \Phi = -98°$, $\Psi = -148°$ ($\underline{\psi} \approx 92°$), and the hydroxymethyl group adopting the *gt* conformation ($X = 60°$), and two antiparallel chains per unit cell (although this is orthorhombic, not monoclinic as in cellulose II).

4.6.1.2 Decorated 1→4-Diequatorially Linked Polysaccharides – the Plant Hemicelluloses. The molecules are much smaller than cellulose ($DP \approx 10^2$), but, insofar as their conformations are known, they are dominated by a hydrogen bond from the 3-OH of one sugar to the ring oxygen of the preceding sugar. They owe their unfortunate name to a nineteenth-century belief that they were precursors of cellulose. The term "hemicellulose" is vague, the classical definition being that they are extractable from the lignocellulose composite of plant cell walls by cold dilute sodium hydroxide. Most of our knowledge of wood hemicelluloses derives from work undertaken for the pulp and paper industry, where a key distinction is between softwood (wood from gymnosperms, generally conifers) and hardwood (wood from angiosperms, generally deciduous).

The detailed molecular structure of the hemicelluloses varies from species to species and even from part to part of the same plant, but a common feature is that the main chain is "decorated" with side groups, some of which can contain a carboxylic acid group. The "decoration" stops the hemicellulose molecules aggregating: undecorated xylan or mannan are nearly as insoluble as cellulose.

The major hemicellulose of grasses and cereals (Graminaceae) and of hardwoods is based on xylan (Figure 4.46).[114] The basic xylan structure is like that of cellulose, although without the O2H ··· O6′ hydrogen bond of cellulose, the hydrogen-bonded O5–C1–O1–C4′–C3′–O3′ ring by itself dictates a left-handed helix with three monosaccharide residues per turn, with $\varphi = -65°$, $\psi = 135°$, rather than the two-fold helix of cellulose (Figure 4.47).[115] This helical structure persists even when the xylose chain binds to certain xylanases,[116] although the viscosity behaviour of arabinoxylan is that of a semi-flexible coil, with a persistence length of 60–80 Å, independent of degree of substitution.[117] In grasses the 2- and 3-positions of the xylose units are sparsely and randomly substituted with α-L-arabinofuranosyl groups, 4-*O*-methylglucuronic acid

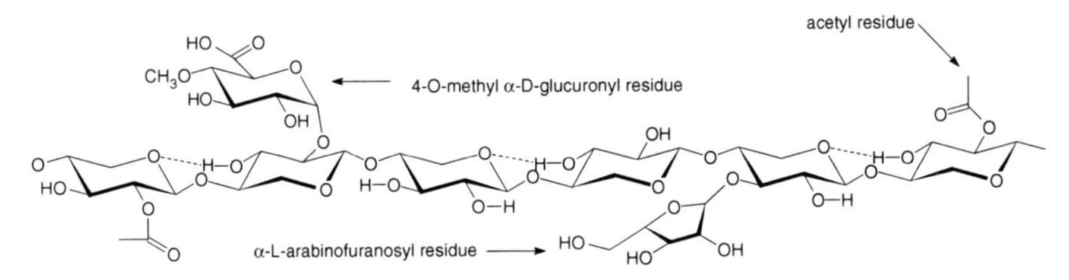

Figure 4.46 Xylan from grasses. The major hemicellulose of hardwoods lacks the arabinofuranosyl groups and the 4-*O*-methyl-α-glucuronosyl residues are attached only to O2 of xylose. The minor hemicellulose of softwoods lacks the acetyl groups, but again the 4-*O*-methyl-α-glucuronosyl residues are attached only to O2 of xylose.

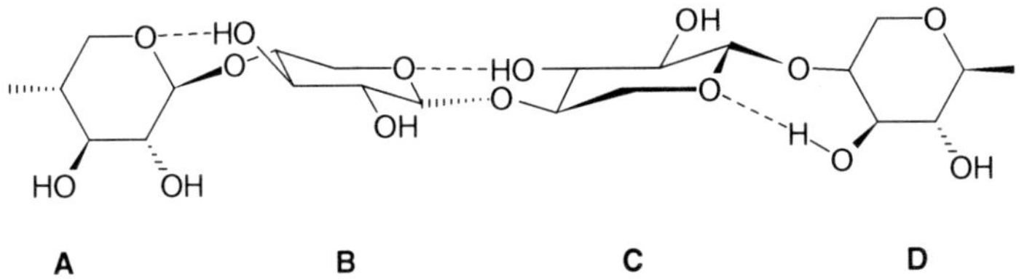

Figure 4.47 Cartoon of the three-fold helix of xylan. Residues A and D are in the plane of the paper, residue B below it and residue C above it.

groups and acetyl groups and the three-fold left-handed helical structure persists in these "decorated" xylans.[118] O5 of the arabinofuranosyl group is sometimes esterified with ferulic acid (*m*-methoxy-*p*-hydroxycinnamic acid), which on oxidation can form lignin-type links with other feruloyl residues or bulk lignin, and therefore provide a covalent link between hemicellulose and lignin.

Glucuronoxylan (15–30%) is the major hemicellulose of hardwoods such as birch, eucalyptus and poplar/aspen. It is similar to grass xylan, but lacks the arabinofuranosyl decoration (Figure 4.44).

Although some xylan-based hemicellulose (non-acetylated arabinoglucuronoxylan, 7–10%) is found in softwood, its major hemicelluloses have a main chain of glucomannan, which can be considered as cellulose with configuration at C2 of about three out of four of the glucose units inverted. The decoration consists of α-D-galactopyranosyl units at C6 and acetyl groups,[119] when the hemicellulose is known as galactoglucomannan (15–25%). Glucomannan itself forms a minor component (2–5%) of hardwood hemicellulose, where it is found acetylated (Figure 4.48).[120]

The basic conformational motif of these molecules appears to be the two-fold helix, maintained by intra-chain O5 ··· HO3′, familiar from cellulose. Electron diffraction studies on single crystals of mannan I revealed an antiparallel arrangement[121] in an orthorhombic unit cell containing four equivalent mannoses. The polymer chain adopted a two-fold helix with $\varphi = -81°$, $\Psi = -161°$ ($\underline{\psi} \approx 79°$) and $\chi = 161°$. In addition to the intra-chain O5 ··· HO3′ hydrogen

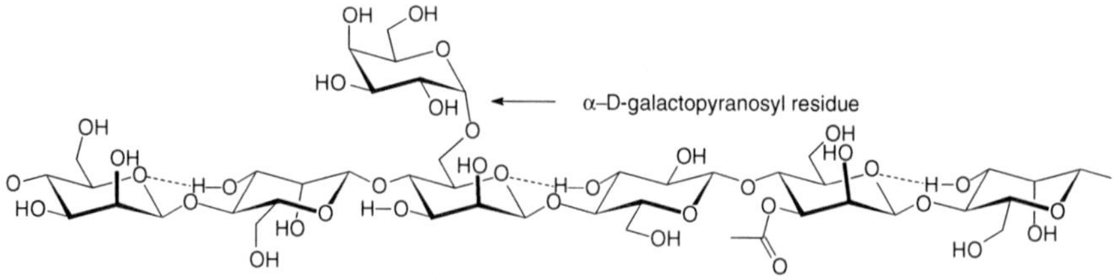

Figure 4.48 Softwood galactoglucomannan. Sometimes a distinction is made between galactoglucomannans, with a sugar ratio Man:Glc:Gal of ~3:1:1, and (galacto) glucomannans, where the ratio is more like 4:1:0.1.

bond, the chains have intrachain O2′-H ··· O5 hydrogen bonds. Konjac glucomannan, a material with a mannose:glucose ratio of 1:1.6, adopts the crystal conformation of mannan II, with an antiparallel arrangement of mannan chains, of which there are four (four mannobiose units) per orthorhombic unit cell. The dihedral angles are reported as $\varphi_H = 23.5$, $\psi_H = -25°$, $X = 57°$ ($\underline{\varphi} \approx -97°$, $\underline{\psi} \approx 145°$, $\underline{\chi} \approx 180°$), *i.e.* the hydroxymethyl groups adopt the *gt* conformation.[122] The glucose and mannose positions appeared isomorphous, *i.e.* positions in the crystal structure could be occupied by either Man or Glu. The α-galactopyranosyl branches of a galactomannan (guaran gum) do not appear alter the conformation of the main chain in a major fashion: an X-ray fibre study[123] gave $\varphi = -98°$, $\Psi = -143°$ ($\underline{\psi} \approx 97°$), $\chi = 158°$ for the main chain and $\varphi = 149°$, $\psi = -111°$, $\chi = 116°$ for the Gal-α-(1→6) side-chains: φ for these crowded side-chains is not dictated by the *exo*-anomeric effect.

A crucially important plant polysaccharide based on a β-(1→4)-glucan backbone is xyloglucan, in which the six positions are decorated with α-D-xylopyranosyl residues and which is often then further elaborated. This has the capacity to absorb very strongly to cellulose fibrils and cross-link them; the cellulose/xyloglucan network forms the primary cell wall of plant cells. When the plant cell expands the xyloglucan is cleaved by xyloglucan endo-transglycosylase, a GH 16 enzyme (see Section 5.9.6.6) to form a glycosyl enzyme which does not hydrolyse, but remains stable until it encounters another xyloglucan "loose end" with which it can recombine and form part of an expanded primary cell wall "net". The ability of the xyloglucan to bind irreversibly to cellulose fibres has opened up the possibilities of many applications from cellulose composites to security papers.[124]

Because of the importance of this polysaccharide, the various substitutions of the backbone glucan have their own nomenclature:[125]

G: Backbone glucose only

X: Backbone glucose with α-D-Xyl*p*-(1→6) branch

L: Backbone glucose with β-D-Gal*p*-(1→2)-α-D-Xyl*p*-(1→6) branch

F: Backbone glucose with α-L-Fuc*p*-(1→2)-β-D-Gal*p*-(1→2)-α-D-Xyl*p*-(1→6) branch

J: Backbone glucose with α-L-Gal*p*-(1→2)-β-D-Gal*p*-(1→2)-α-D-Xyl*p*-(1→6) branch

S: Backbone glucose with α-L-Ara*f*-(1→2)-α-D-Xyl*p*-(1→6) branch

T: Backbone glucose with β-L-Ara*f*-(1→3)-α-L-Ara*f*-(1→2)-α-D-Xyl*p*-(1→6) branch

A: Backbone glucose with α-D-Xyl*p*-(1→6) branch and α-L-Ara*f*-(1→2) branch

B: Backbone glucose with α-D-Xyl*p*-(1→6) branch and β-D-Xyl*p*-(1→2) branch

As might be expected from such a highly branched structure, the gummy polysaccharide has no definite solution conformation. Light scattering experiments on the polymer itself[126] suggested a random coil and NMR studies on the heptasaccharide XXXG[127] indicated several occupied minima on Ramachandran φ,ψ plots.

4.6.1.3 Conformationally Mobile, Originally 1→4 Diequatorially Linked Polysaccharides. The most important member of this class of polysaccharides is alginate, a polysaccharide produced by seaweed and bacteria and widely used in the food industry. It is biosynthesised initially as a β-(1→4)-linked polymannuronic acid, by sequential transfer from GDP mannuronate.[128] Some of the individual mannuronate residues are then epimerised at C5 to yield α-L-guluronate residues. The $^4C_1$ conformation of these guluronate residues is disfavoured by the axial OH on C2, the anomeric effect at C1 and above all by the now-axial carboxylate at C5; the $^1C_4$ conformation, with the intersaccharide linkages now 1,4-diaxial, is adopted by the guluronate residues.[129]

There are a range of mannuronan epimerases (at least seven in *Azotobacter vinelandii*, for example)[130] which produce different patterns of mannuronate and guluronate residues. They are part of a superfamily of lyases and epimerases,[131] of which the mannuronate lyases/epimerases were the first to be identified,[132] whose active sites stabilise carboxyl enolates which can either expel a β-substituent (lyases) or be reprotonated on the opposite face of the α-carbon (epimerases).[133] There are three basic structures: mannuronate blocks (MMMMM), guluronate blocks (GGGGG) and regions of strictly alternating sequence (MGMGM).[134] However, the alginate produced by *Pseudomonas aeruginosa*, a source much studied because of its ability to produce alginate in the lungs of cystic fibrosis patients and, in effect, drown them, has no G blocks and contains some acetyl groups at positions 2 and 3.[135]

Fibre X-ray analysis of polymannuronic acid indicates that it adopts a "cellulose-like" two-fold helix form, described as a ribbon, with two disaccharide units per orthorhombic unit cell and the carboxylic acid group approximately perpendicular to the plane of the sugar ring,[136] whereas its salts crystallise in a left-handed three-fold helix,[137] Molecular modelling suggests a two-fold and three-fold helices are similar in energy, so that slight differences in conditions will alter the pitch of the helix.[138] Fibre X-ray studies of polyguluronic acid[139] showed it to adopt a two-fold helix with the plane of the sugar rings perpendicular to the direction of the helix and a hydrogen bond between the axial 3-OH and the carboxylic acid, which again was approximately perpendicular to the plane of the sugar ring. NMR studies on salts of guluronic acid oligosaccharides indicated that salt formation did not in this case alter conformation and were able to locate the dihedral angles about the glycosidic link: $\varphi = \Phi = -95°$ ($\varphi_H = 25°$) and $\Psi = -147°$ ($\psi_H = -27°$).[140]

Alginate is important because of its ability to form rigid, stable gels on addition of divalent metal ions, particularly Ca^{2+} (binding efficiency $Mg^{2+} \ll Ca^{2+} < Sr^{2+} < Ba^{2+}$).[141] The polysaccharides in solution as behave as semi-flexible chains,[142] but Ca^{2+} enables them to associate by being coordinated to guluronate residues of different chains.[143] Whole stretches of guluronate are capable of doing this, with multiple binding of Ca^{2+} by adjacent guluronate residues in a cooperative manner – once one Ca^{2+} has joined two residues in different polyguluronate chains together, subsequent association becomes progressively easier.[144] The resemblance of the junction region to the boxes in which eggs are sold led to the term "egg-box model". An experimental determination of the coordination of Ca^{2+} in the array of egg-boxes derived from polyguluronate has never been performed and theoretical molecular modelling studies[145] do not even give a definite answer as to whether the ordering of the chains in the junction zones is parallel or antiparallel, although electron microscopy of alginate junction zones suggests that lateral association of polyguluronate chains beyond the dimer is rare.[146] Nonetheless, the existence of interactions with Ca^{2+} was plausibly inferred from experiments which monitored the changes in circular dichroism on gelation.[xxviii] The titration curves for Ca^{2+} and Mn^{2+} binding to alginate were not dependent on the M/G ratio and showed saturation at about 0.2 mol divalent metal per saccharide:[147] the model of guluronate metal binding in Figure 4.49, if lateral aggregation stops at the dimer, predicts a ratio of 0.25, with every other carboxylate uncoordinated.

The ^{13}C anomeric resonances in alginate are sufficiently dispersed that the frequency of the eight possible mannuronate and guluronate triads (MMM, MMG, MGM, GMM, GGM, GMG, MGG, GGG) can be analysed from the intensity of the signals from the central saccharide of the triad.[148] A study of the effect of paramagnetic Mn^{2+} ions on the ^{13}C NMR spectrum of alginate showed,[149] as expected for electrostatic binding to individual carboxylate groups, only large effects on C6 and C5 of mannuronate blocks. Guluronate blocks showed effects on C1, in line with binding of metal cations in the "holes" in the two-fold helix and coordination by O5 and O1 (Figure 4.49(b)). Remarkably, however, the guluronate residues in the sequence GM were more strongly affected than even those in polyguluronate blocks. The conformation of strictly alternating blocks was suggested to be that of Figure 4.50, with Mn^{2+} coordinated by O5 and O3 of the guluronate and O4 and O3 of mannuronate. Lateral aggregation of such structures would explain why bacterial alginate, which lacks G blocks, can still form gels.

If gels are formed from mixtures of G blocks, M blocks and strictly alternating MG blocks, which are not covalently connected, and the unimmobilised saccharides then leached out from the resultant gel and analysed by NMR, the

---

[xxviii] The chromophore being the n-π^* transition of the carboxylates.

composition of the gel can be obtained by difference. In this way, it has recently been shown that MG blocks and mixed MG and G block aggregates, in addition to G block aggregates, can be formed.[150]

4.6.2 1→4-Linked Equatorial–axial Pyranosides

The *exo*-anomeric effect enforces a value of φ around 90° for anomeric axial linkages. This ensures a proclivity towards helical structures (as distinct from the ribbons of 1,4-diequatorially linked polysaccharides), as in starch and glycogen. β-(1→4)-Linked galactans, 1,4 axial–equatorial in the other sense, are known as hemicellulose components of compression and tension wood,[151] the storage polysaccharides of lupins[152] and as arabinogalactans[153] attached to the rhamnogalacturonan I component of pectin, but have not been subject to conformational studies: they appear to be biosynthesised from UDP-Gal.[154]

4.6.2.1 Starch. Starch is the reserve polysaccharide of many species of plants, and occurs as granules of diameter 0.5–100 μm, depending on the source. Tubers such as potatoes tend to have larger granules than cereals such as maize, although the size-distribution within a given species is large.[155] Starch granules have a high degree of organisation; viewed through crossed polaroids under visible light, a pattern similar to a Maltese cross is seen.[156] The granule is partly crystalline, with semicrystalline bands alternating with amorphous regions to give a series of concentric spheroids with a periodicity of 120–400 nm, which in section resemble growth rings in trees.[157] At the centre of at least some starch granules is a central cavity (hilum) connected to the surface by pores.[158]

Starch is generally considered as an intimate mixture of two components, amylose, a linear chain of α-(1→4)-glucopyranose units and amylopectin, with a significant number of α-(1→6) branches.[xxix] Amylose gives the familiar blue–purple colour with iodine, whereas amylopectin gives a brown colour. Amylopectin is crystalline, whereas amylose is generally considered to be amorphous. Amylose forms both the amorphous spheroids with a large periodicity (120–400 nm) in the starch granule and may intercalate with short crystalline amylopectin regions to form the lamellae with a periodicity of ~9 nm within the semicrystalline portions of the granule.[159] Most starch granules are a mixture of 75–80% amylopectin and 20–25% amylose, but in the "waxy starches", which are highly crystalline, the amylopectin content can be essentially 100%,[160] whereas in some high-amylose strains the amylopectin content falls to 20–30%.[161]

4.6.2.1.1 Amylose. Although amylose has historically been considered to be unbranched, occasional branching is observed by AFM (see Figure 4.21(b)).[5] Its *DP* depends on source; cereal amyloses generally have $DP < 10^3$ (500–1000)

[xxix] The discussion in Section 4.2.1 of the methylation analysis is historically simplified: the structure was established before GLC (methylated sugars had to be separated by fractional distillation) and Hakamori methylation (so there was a long debate over whether the 2,3-dimethylglucose arose from branch points or incomplete methylation).

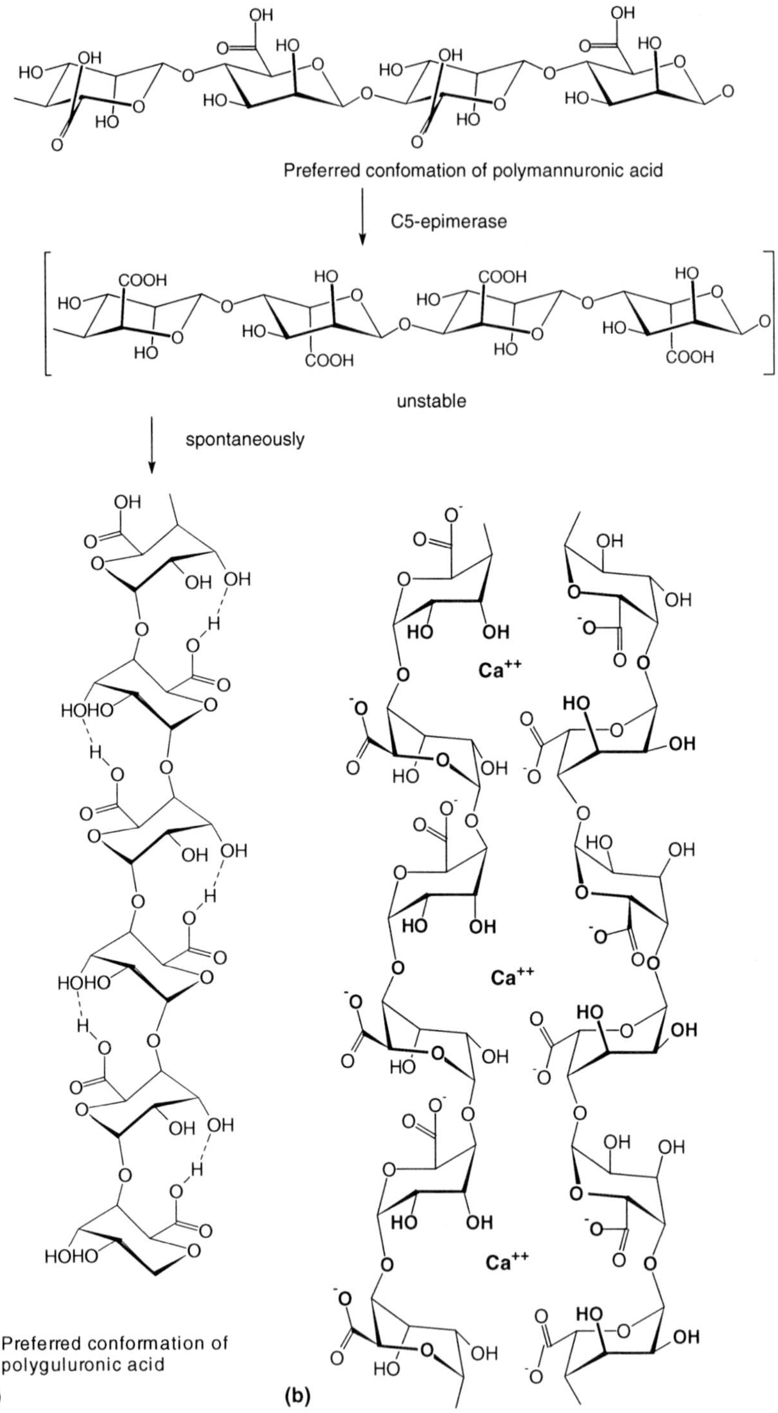

Preferred confomation of polymannuronic acid

C5-epimerase

unstable

spontaneously

Ca++

Ca++

Ca++

Preferred conformation of polyguluronic acid

(a)

(b)

Figure 4.49 (*Continued*)

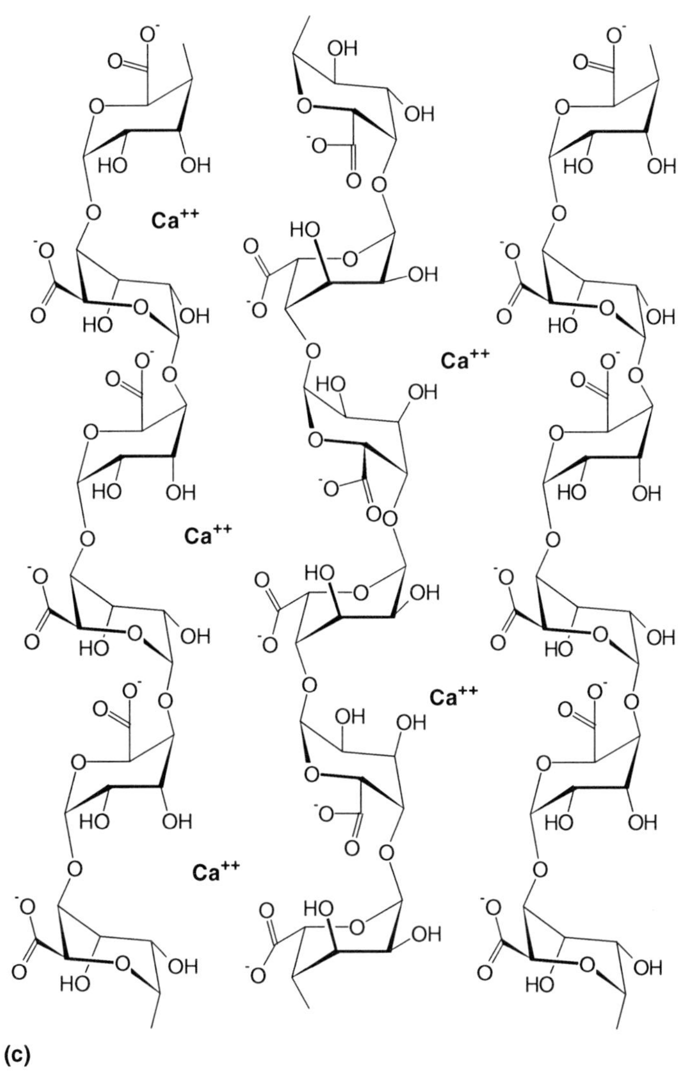

Figure 4.49 (a) Polymannuronic acid in its cellulose-like two-fold helical conformation and conversion to polyguluronic acid. (b) Polyguluronate in its two-fold helical conformation and a possible chain association via the "egg-box" model. The atoms likely to be coordinating the chelated Ca^{2+} are shown in bold. Antiparallel chains are drawn in line with modelling studies, but there is no definite experimental evidence. (c) Possible "layering" of egg-box structures.

and tuber starches $DP > 10^3$ (1–4000).[162] A key to an understanding of amylose conformation, both amylose itself and the unbranched chains of amylopectin is the cycloamyloses, sometimes called cyclodextrins or Schardinger dextrins. These are cyclic structures of α-(1→4)-linked D-glucopyranose units produced by cyclodextrin glucanotransferases of various *Bacillus* spp.:

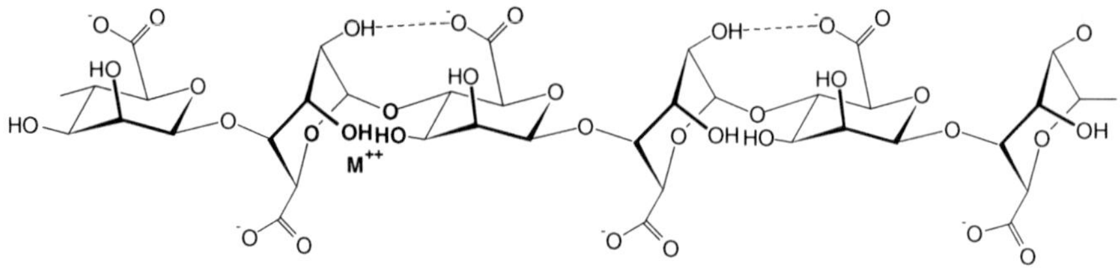

Figure 4.50 Possible conformation of strictly alternating mannuronate–guluronate blocks, showing possible metal ion coordination: coordinating oxygen atoms are in bold.

although very large rings are formed initially, the final products are those with 6, 7 and 8 glucose units, called α-, β- and γ-cyclodextrins or cyclohexaamylose, cycloheptaamylose and cyclooctaamylose, of which cycloheptaamylose is the most stable.[163] These molecules have hydrophobic interiors and form host–guest complexes with hydrophobic organic molecules. Although amylose is amorphous, in the presence of a wide range of organic molecules it forms structures closely related to the cyclodextrins, V-amyloses. In the V-amyloses, the amylose chains adopt a left-handed[xxx] helix with 6, 7 or 8 residues per turn, depending on the organic molecule – with glycerol, for example, it forms a six-membered helix (Figure 4.51).[164] V-amylose can be formed in the starch granule by accommodation of the long fatty acid chain of lipids in the central hydrophobic cavity.[165]

The parallel between the cyloamyloses and V-amyloses helped to explain the well-known blue colour of starch–iodine complexes: a series of intensely coloured complexes of cyclohexaamylose, iodine and metal iodides had chains of iodine atoms in the central cavity of stacks of cyclohexaamyloses.[166] The blue colour was thought to arise from charge transfer interactions amongst the I_2 molecules and the linear I_3^- and I_5^- ions trapped in the central cavity.

$^3J_{CH}$ values from the NMR spectra of a series of maltooligosaccharides [α-(1→4)-linked glucose oligomers], gave values for φ_H and ψ_H[167] after parameterisation of Karplus-type equations from observed values of $^3J_{CH}$ for α-, β- and γ-cyclodextrins in solution and their X-ray crystal structures. Values of φ_H were narrowly clustered (–28 to –22°) in DMSO, but in water varied from –54.5° for maltose to –33° ($\underline{\varphi} \approx 87°$) for maltoheptaose. A perfect overlap of the O1 p-type lone pair with the C1–O5 σ* orbital, with all atoms exactly tetrahedral, would require a φ_H value of –30°. The values of ψ_H clustered between 0° and –16° in DMSO, but completely reversed from +70.5° for maltose to –25° for maltohexaose in water, suggesting substantial disorder in water, in indirect conformation of the X-ray results on starch granules. In neither solvent is a 2-OH–3′-OH hydrogen bond, observed with methyl β-maltoside,[168] possible.

[xxx] A left-handed helix is one which, when looked at from the end, requires an *anticlockwise* motion around the turns of the helix to take a point *away* from the observer.

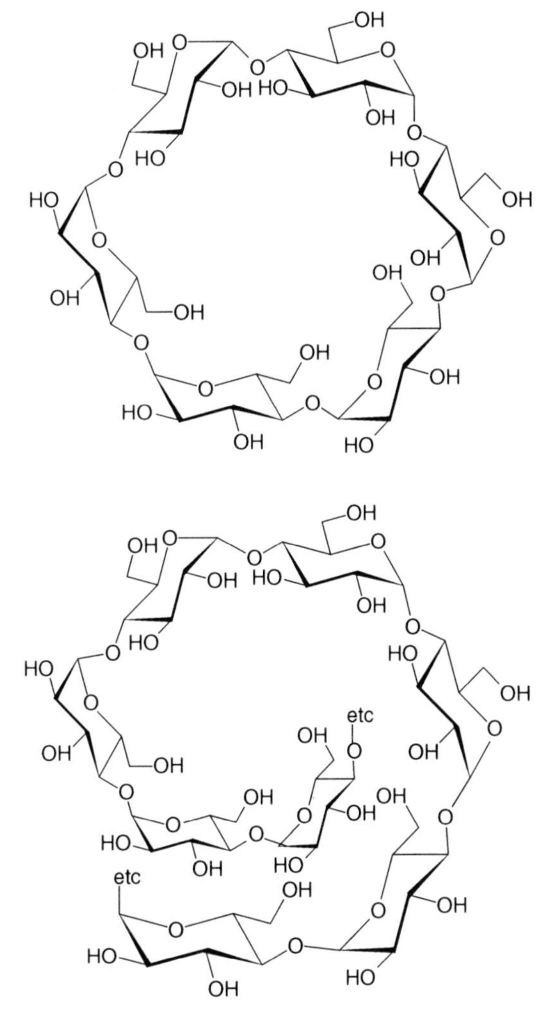

Figure 4.51 Relationship between cyclohexaamylose and a V-amylose with six residues per turn.

Where the degrees of freedom are reduced, however, as in a cycloamylose with 26 residues,[169] V-amylose structures can be observed in the absence of guests, at atomic resolution. The two helical regions in this molecule involve the left-handed helix with six residues per turn seen in the complex with glycerol; it is possible to detect the 2-OH–3′-OH hydrogen bond between adjacent residues,[xxxi] and cross-helix bifurcated hydrogen bonds between $O3_n$ and $O2_n$ and

[xxxi] Confusingly labelled $O(3)_n$ $O(2)_{n+1}$ in Ref. 169; if the chemical convention of numbering from the non-reducing terminus is adopted, they are $O2_n$–$O3_{n+1}$.

$O6_{n+6}$. Values of φ vary between 91° and 115° and of ψ between 97° and 131° (Figure 4.52).

Because of their hydrophobic interior and biologically innocuous constitution, cycloamyloses find many commercial uses for the delivery of hydrophobic molecules, be they flavourings in foods, pharmaceuticals, cosmetics or agrochemicals; they are also often used simply as phase-transfer catalysts.[163] Cyclohexaamylose is shown in Figure 4.51 viewed from one end of the hydrophobic cavity: another representation is that of Figure 4.53, which emphasises the truncated conical nature of cyclodextrins and the fact that the hydroxymethyl groups surround one opening of the hydrophobic interior and the 2-OH and 3-OH to the other (wider) opening.

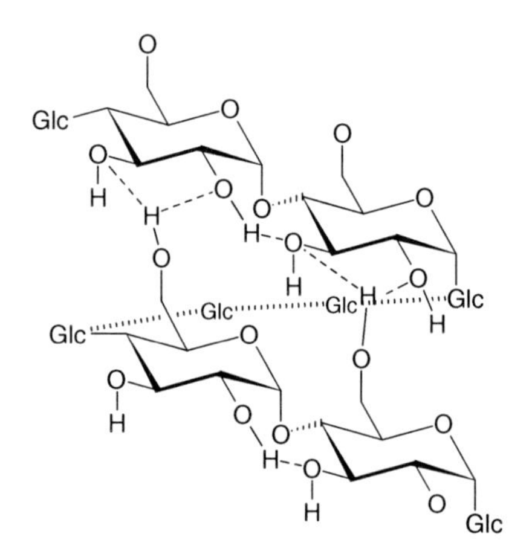

Figure 4.52 Cartoon of hydrogen bonding in V-amylose. The hydroxymethyl groups adopt the *gg* conformation. The locations of the hydrogens forming hydrogen bonds are speculative.

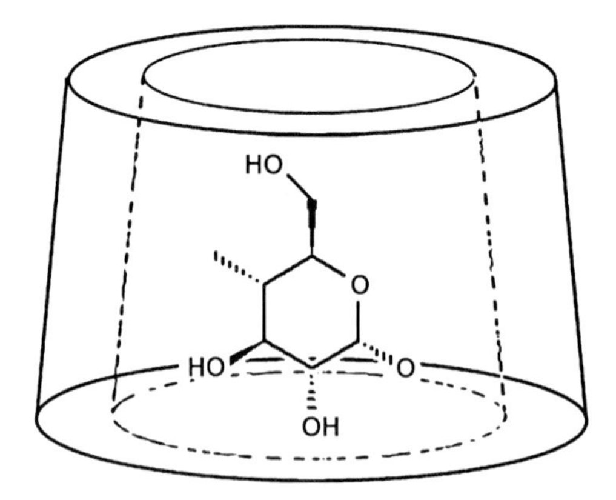

Figure 4.53 Small cycloamylose structure, showing the approximate orientation of the glucose residues.

4.6.2.1.2 Derivatisation of cycloamyloses and catalysis. The ability of cyclodextrins to act as hosts for small organic molecules has led to a continuing investigation of their potential as catalysts. In 1966, the discovery that they acted as catalysts in the hydrolysis of aryl esters, with the aryl group binding in the cavity and first transferring the acyl group to a sugar hydroxyl, was very remarkable.[170] Such a mechanism was reminiscent of that adopted by the serine proteinases (Section 6.5.3.2.1) and the analogy was reinforced by the Michaelis–Menten kinetics (see Section 5.2) followed. Unfortunately, the term "enzyme mimic" was coined and the idea that enzyme-like rate accelerations could be achieved, simply by hanging various types of functionality on the cyclodextrin framework, has survived to this day. It is synthetically easy to selectively modify the 6-OH groups of the cyclodextrins, and enzyme-like kinetics with modest rate enhancements continue to be reported, such as a factor of 7 in the deprotonation of 4-*tert*-butyl-α-nitrotoluenc by 6-amino-6-deoxycyclodextrins,[171] and an apparently more impressive 7000 in the hydrolysis of nitrophenyl glycosides by the 6-dicyanohydrins.[172]

However, for the last quarter of a century, the consensus amongst enzymologists, largely through an influential review by Jencks,[173] has been that most of the catalytic power of those enzymes not employing complex cofactor chemistry derives from transition state interactions between parts of the substrate and parts of the enzyme not involved in covalency changes, and that therefore small-molecule mimics of these enzymes will only ever produce small accelerations. A rate enhancement even of 7000 is unimpressive when that of real glycosidases is $\sim 10^{17}$.

Much more successful has been the tailor-making of catalysts combining cyclodextrins with transition metals or other heavy elements with multiple valencies.[174] The cyclodextrins envelope labile catalytic centres and/or provide a binding site. Enzyme mimicry is more plausible where the specialist chemistry of a cofactor is crucial. Thus, a 2 2′-ditelluro-bis-cycloheptaamylose in its oxidised form had a k_{cat}/K_m for oxidation of 3-carboxy-4-nitrobenzenethiol only 600-fold less than that of glutathione peroxidase[175] oxidising glutathione – but even here the "mimicry" is not close, as the enzyme uses a selenium, not a tellurium, cofactor.

4.6.2.1.3 Amylopectin. Amylopectin has significant numbers of α-(1→6) branches and these branches are not randomly distributed, but occur in clusters,[176] giving rise to the lamellar structure with the 9 nm periodicity observed from diffraction studies. The existence of (1→6) branches means that whereas a single molecule has many non-reducing ends, it has only one reducing end, probably linked to a protein similar to glycogenin (see Section 5.12.2.3). Amylopectin (Figure 4.54) is a very large molecule, with estimated M_r ranging from 10^6 to 10^8 ($C_6H_{10}O_5 = 182$).[177] The cluster structure arises from a combination of three types of chains:[178]

(1) *A-chains*: these are short (20–40 glucose units) and have no branches. They form double helices with adjacent terminal, unbranched α-(1→4)-glucan chains. The helices are usually taken as left-handed, but

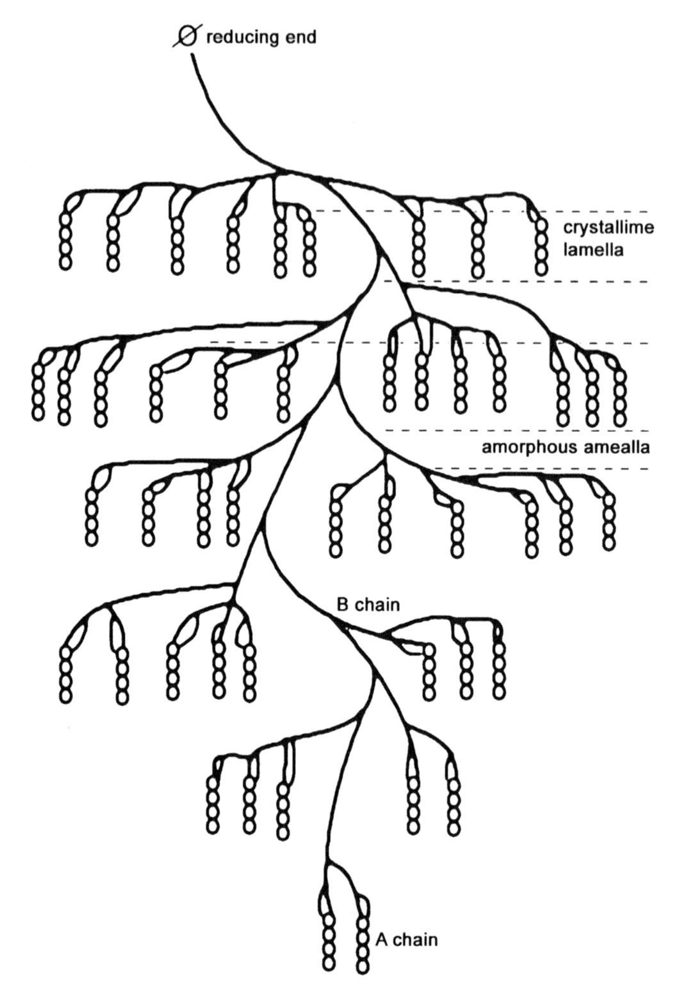

Figure 4.54 Amylopectin model.

no incontrovertible evidence exists on the point and right-handed helices have been proposed for B-type starch.[179]

(2) *B-chains*: these have several α-(1→6) branches and are longer.

(3) *C-chain*: this is the B-type chain which contains the single reducing end in the molecule. Although definite evidence for amylose is lacking, the closely related glycogen is known to be built up on the protein glycogenin (in GT 8 – see Section 5.12.2.3), which first glycosylates one of its own Tyr residues and then puts the first few glucose units.[180]

This model arose from the action of various enzymes on amylopectin. Much of the work was done several decades ago with wild-type enzymes purified by the standards of the time.[181] Some specificity differences between enzymes now appear to be quantitative than qualitative, with different names being given to various proteins on the insecure basis of fairly minimal values of $\Delta\Delta G^{\ddagger}$ for competing catalysed reactions. The following discussion uses the contemporary nomenclature for subsites of polysaccharide-degrading enzymes (see Section 5.5). Starch-degrading enzymes which give information on amylopectin are:

(i) β-Amylase (GH 14, inverting), which has −2, −1 and +1 subsites. The −2 subsite can only accommodate an unsubstituted glucose unit and the −1, +1 only an α-(1→4)-linked glucose unit. The enzyme therefore yields β-maltose and a polymer ("β-limit dextrin") with "stubs" of 1–3 glucose units left at the branch points; the 3-glucose stubs hint at a +2 subsite with definite specificities.[182]

(ii) Various GH 13 (retaining) enzymes with various designations. "α-Amylase" appears to be the name given to enzymes with a large number (up to 9) of positive and negative subsites which will only accommodate α-(1→4)-linked glucose units. At low loadings such enzymes preferentially hydrolyse the amorphous, unbranched regions of the B and C chains between the lamellae, but given time will also attack the double-helical A chains. "Isoamylase" seems to be merely an α-(1→6)-glucanase with no (or very promiscuous) positive subsites, rather than something with specificity: it will hydrolyse the phenolic glycosidic link in glycogenin.[183] Pullulanase hydrolyses pullulan (4)-α-D-Glc*p*-(1→4)-α-D-Glc*p*-(1→6)-α-D-Glc*p*-$(1\rightarrow)_n$ at the 1,6-linkages, but the +1 subsite will accommodate 4-substituted glucose units, so that branch points are completely removed.

(iii) Glycogen phosphorylase (GT 35, Section 5.12.2.7) removes non-reducing ends from glycogen, starch, *etc.*, one unit at a time, but its active site tunnel, enclosing subsites −1 through +4, dictates that it leaves maltotetraosyl residues at both the 4- and 6-positions at a branch point. The product of extensive phosphorolysis is a "limit dextrin", in which the shortest α-(1→6) branch is a maltotetraosyl unit.

(iv) Glycogen debranching enzyme takes the branch point with two maltotetraosyl residues attached, left by glycogen phosphorylase and converts it to a linear maltooctaosyl unit and a molecule of glucose. The monomeric rabbit muscle enzyme is very large (165 kDa) and consists of two domains, an N-terminal GH 13 α-glucan transglycosylase which transfers an α-maltotriosyl residue to from the 4-position of an α-(1→6)-linked glucose to the 4-position of an α-(1→4)-linked glucose and an inverting[184] glucosidase which removes the "stub" of the (1→6) branch as β-glucopyranose. The two activities can operate independently (Figure 4.55).[185] The family of the C-terminal glucosidase domain is uncertain: site-directed mutagenesis experiments identified two catalytic aspartates, which suggests GH 49, but closest (but still remote) sequence similarity was seen with a GH 13 (retaining) oligo-(1→6)-glucosidase;[186] the domain remains unclassified. The transglucanase activity can produce large cyclodextrins (11–50 glucose units per ring).[187]

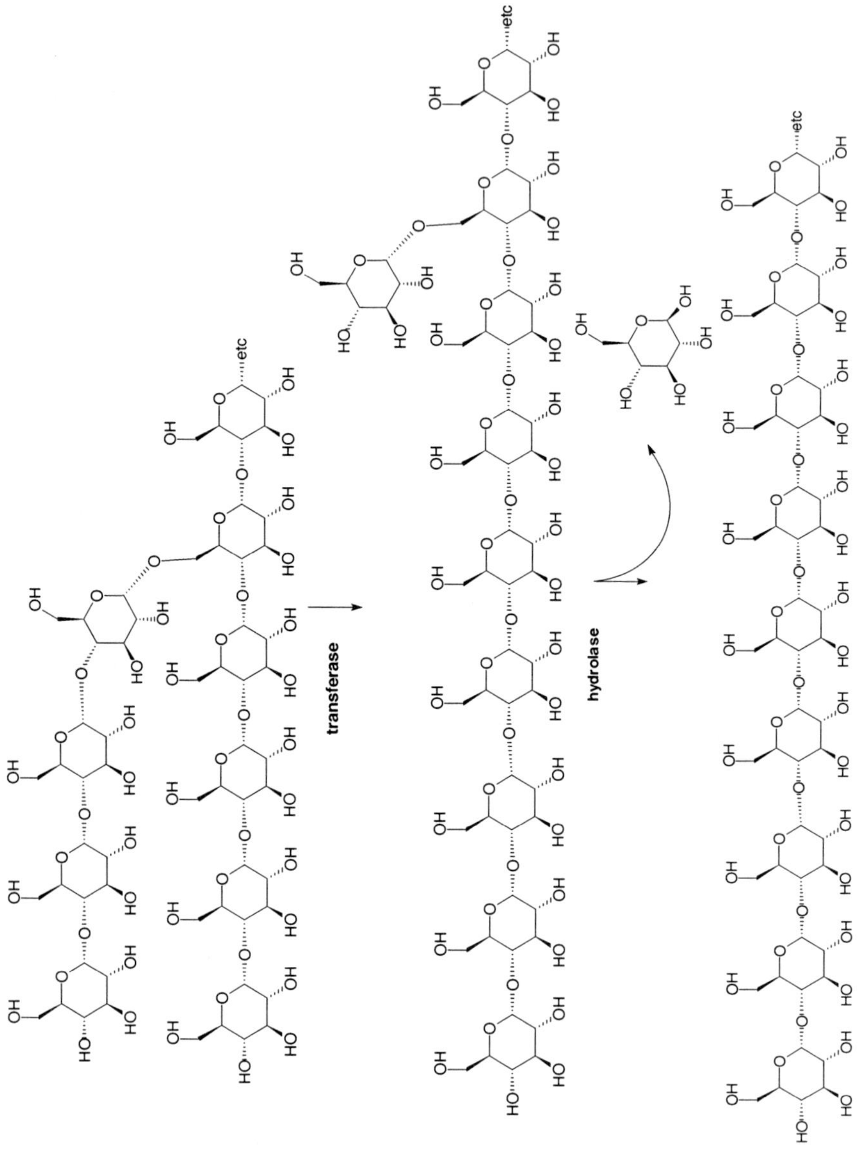

transferase

hydrolase

Figure 4.55 Action of debranching enzyme.

(v) Glucoamylase is a GH 15 (inverting) enzyme with a strict *exo* action, which liberates β-glucose from the non-reducing termini. It is comparatively non-specific with respect to its +1 subsite, and so will hydrolyse both α-(1→4)- and α-(1→6)-linked glucose. It will, of course, eventually turn starch into glucose, although it will slow down considerably when it encounters crystalline regions in which the non-reducing ends are masked.

The packing of the A-chain double helices of amylopectin in the crystalline lamellae can occur in two ways. In the low-water starches found in cereals ("A-type"[xxxii]) there are two parallel double helices with six residues per turn in the monoclinic unit cell, which contains 12 glucose residues and four water molecules:[188] because of symmetry considerations there are only three types of glucose molecule in the unit cell, two of which have $\varphi = 92°$, $\psi = -153°$ and the third $\varphi = 85°$, $\psi = -143°$; in all three glucose molecules the hydroxymethyl group adopts the *gg* conformation. The double helices are held together by O2–O′6 hydrogen bonds. The conformation of the polysaccharide chains is thus similar to that in V-amylose. In the starches found in tubers ("B-type"), the crystal structure is more open and hydrated, but again is based on a left-handed parallel double helix with six glucose residues per turn.[189] In this case the hexagonal unit cell again has 12 glucose units, but because of symmetry considerations there are only two sorts of glucose units, but within experimental error the two sets of dihedral angles are identical ($\varphi = 84°$, $\psi = -144°$). Both hydroxymethyl groups adopt the *gg* conformation and the helices are held together with O2–O6′ hydrogen bonds. The main differences between the two forms appear to be the 36 water molecules in the unit cell of "B-type" amylopectin helices and the chain length. "B-type" helices are associated with longer A chains (30–44 glucose units) and "A-type" helices with shorter ones (23–29 glucose units).[190]

The nature of the levels of organisation of the starch granule between the molecular and the microscopic "tree rings" remain uncertain. One model arranges the lamellae of double helices into super-helices (helices of helices),[191] whereas another regards the diffraction spots that gave rise to the idea of helices of helices to have arisen from liquid crystalline-type order.[192] The latter idea gave rise to the term "blocklets" containing several lamellae.

4.6.2.1.4 Biosynthesis of starch. A definitive picture of the biosynthesis of starch is not yet available.[193] It is widely thought that it proceeds by stepwise addition of α-(1→4)-linked glucose units from ADP glucose to the non-reducing end of a growing α-(1→4)-glucan chain by family GT 5 glycosyl transferases, and that elaboration of the polymer to yield amylopectin requires the coordinated action of starch branching enzymes and starch debranching enzymes of family GH 13 to create the structure in Figure 4.56. The widespread

[xxxii] Not to be confused with A-chains, which occur in all three types of starch, A, B and C.

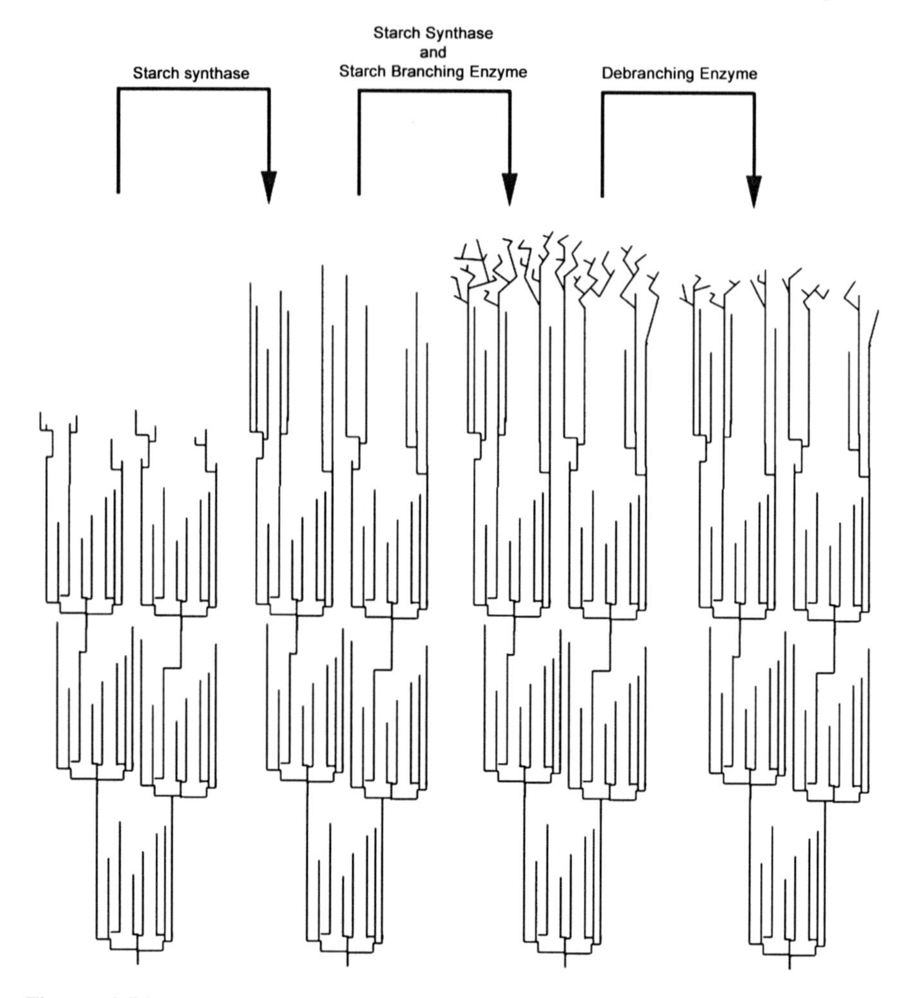

Figure 4.56 Model of starch biosynthesis.

assumption of an autocatalytic protein primer similar to glycogenin rests on the parallel between plant starch and animal glycogen biosynthesis, which in other respects is inexact – animal glycogen synthetases are in family GT 3 and plant (and bacterial) starch synthetases are in GT 5.

Moreover, the general picture continues to be disputed in its fundamentals. If the primer plus non-reducing end transfer mechanism is adopted, the results of the following pulse–chase experiment[194] require *ad hoc* assumptions about the accessibility of various portions of the whole starch granule to added precursors. Starch granules were incubated with ADP–[^{14}C]glucose for a given time (the "pulse" part of the pulse–chase experiment). Half of them were then incubated for the same time with unlabelled ADP glucose (the "chase" part of the pulse–chase experiment). Both types of starch granules were then treated identically with *exo*-acting enzymes, β-amylase (with only –1 and –2 negative subsites) or glucoamylase (with only –1 negative subsites). The radioactivity liberated either as glucose or as maltose was the same for starch granules subjected to the "hot" pulse and the "cold" chase or just the "hot" pulse. On the non-reducing end extension model, the non-reducing ends of the granules subjected to the cold chase should have had very little radioactivity. Such experiments with organelles (rather than pure enzymes) are, however, vulnerable to various endogenous metabolic processes in the organelle. Similar pulse–chase experiments in the same laboratory, on glucose incorporation in cellulose being synthesised by *Acetobacter xylinum*, also indicated, incorrectly,[86] glucan chain extension from the reducing end.[195]

Despite these experiments, the balance of evidence favours non-reducing end extension of at least amylose. Certain mutants (*waxy* mutants) of cereals produce waxy starches with little or no amylose, with the *waxy* mutant of maize being grown commercially. The defect has been located to a single gene, that for granule-bound starch synthetase (GBSS),[196] the only synthetase found exclusively within the granule. Because of the necessity of generating lipid intermediates, it is not possible for the biosynthetic machinery of a lipid-linked pathway (necessary for chain extension from the reducing end) to be located in a single enzyme. Amylose is biosynthesised with amylopectin, the increasing amylose decreasing amylopectin crystallinity in the lamellae.[197]

The various soluble starch synthetases extend α-(1→4)-glucans, starting from chain lengths of various sizes. SS1, for example, appears to have an extended starch-binding-domain at the C-terminus and transfer glucose residues up to a chain length of about 10.[198] The three other soluble starch synthetases found in cereals (SSIIa, SSIIb and SSIII) appear to make longer chains between clusters: there is a possibility that branches may have to be introduced before some of them operate.

The enzymes that catalyse starch branching are GH 13 transglycosylases, which act in much the same way as the GH 13 module of starch debranching enzyme, but in this case act, in the physiological direction, to transfer malto-oligosaccharyl residues of various sizes from O4 to O6. They operate via formation of a β-maltooligosaccharyl enzyme intermediate. Two main classes have been discovered, BEI and BEII, with BEII transferring the shorter chains.[199] BEII is, however, further subdivided into BEIIa and BEIIb, which have slightly different substrate specificities and locations.

Surprisingly, starch biosynthesis (Figure 4.56) also involves debranching enzymes, whose elimination results in reduced starch yield and the production of a highly branched, glycogen-like polymer, phytoglycogen.[200] They are of two types, the isoamylase-type and the pullulanase type,[201] both hydrolysing α-(1→6) branches, but with the pullulanase being specific for maltotriose in sites –3 to –1 and + 1 to + 3; isoamylase will not hydrolyse pullulan. There is as yet some discussion as to whether the enzymes remove small branches which would interfere with the packing of the starch granule or are merely scavenging malformed oligosaccharides.[193a]

The only naturally occurring covalent modification of starch (apart from the hypothetical link to primer) is phosphorylation. In potato, phosphorylation of starch at O3 and O6 is carried out by a dikinase,[202] which transfers the β (central) phosphate of ATP to O3 or O6 of the glucan, the γ (terminal) phosphate appearing as inorganic phosphate. Phosphorylation occurs on fairly long linear stretches of α-(1→4)-glucan, whether in amylose itself or the long A-chains of potato amylopectin. It interferes with chain packing and so increases the ability of starches to swell. The phosphate content of potato starch, already high (0.08% by weight), increases by a factor of 5 when high-amylose starches are engineered by interfering with the glucan branching enzyme.

4.6.2.1.5 Interaction of starch and water – cooking and retrogradation[203]. When starch is heated in water, the amylose largely dissolves and the double helices in the amylopectin A chains break apart and become disordered. A gel is formed, but on cooling slowly the glucan chains crystallise in new patterns, a process called retrogradation. Traditionally, amylose is separated from amylopectin by gently stirring the starch granules at 50–80 °C, centrifuging and then adding an organic compound such as 1-butanol whose hydrophobic portion can fit in the tunnel of V-amylose (Figure 4.51). The amylose–butanol complex precipitates and the amylose can be regenerated by washing out the organic compound with a more volatile solvent, such as acetone. The very different behaviour of amylose and amylopectin towards water (they are, in polymer terminology, incompatible) can lead under various conditions to phase separation into two solutions, one containing very largely amylose and the other somewhat enriched in amylopectin.[204]

A widely cited description of the events that occur when raw starch is cooked, termed gelatinisation uses liquid crystal theory to describe the changes in the various types of order in the starch granule.[205] These types of order are, in order of increasing size of the periodicity, molecular, the amylopectin lamellae with a 9 nm periodicity, the "growth rings" and the shape of the granules. The amylopectin lamellae with the 9 nm spacing (Figure 4.54) are considered to form smectic liquid crystals, with the lamellae ordered in two dimensions and disordered in a third and the dimensions of the liquid crystals governed by the growth rings.

The gelatinisation was monitored simultaneously by several techniques, corresponding to different hierarchies of order on different length scales, Solid-state ^{13}C NMR[206] (see Section 4.2.6) monitored order at the molecule level, viz., the double helices. Powder X-ray diffraction (see Section 4.2.5) monitored the crystallinity of the blocklets of double helices. Small-angle X-ray and neutron scattering monitored the periodicity of the growth rings [as θ in eqn. (4.2) decreases, at constant λ, d increases, so at small angles X-ray diffraction will monitor larger periodicities].[207] A comparison of small-angle scattering by neutrons when the starch was placed in H_2O–D_2O mixtures of varying composition enabled the water content of the various scattering structures to be determined.[207] Differential scanning calorimetry (DSC) enabled the energy changes to be monitored. DSC involves measuring the difference in electrical energy required make the sample and an inert reference conform to the same programme of increasing temperature. When the sample undergoes an order–disorder transition (such as melting), it requires more energy to match

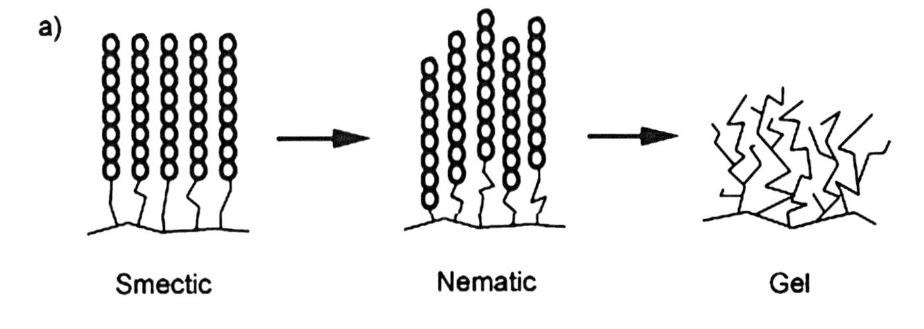

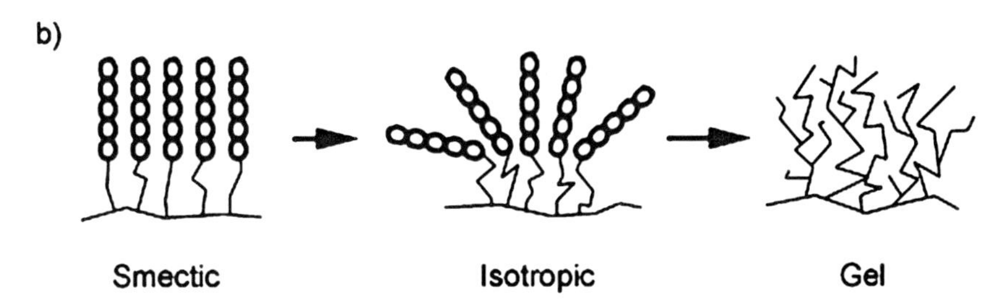

Figure 4.57 Donald model of starch gelatinisation on cooking. (a) in limiting water (b) in excess water.

the temperature of the reference, whereas a disorder–order event, such as crystallisation, it requires less. The following pictures emerged of the changes taking place during starch gelatinisation (Figure 4.57), when the process was and was not, limited by water availability:

(i) In the presence of excess water, the amorphous regions of the granule expand as they are hydrated.
(ii) As the heating is continued, a single DSC transition is observed, associated with the A-chain double helices breaking loose from their crystalline packing and, in a fast, non-rate-determining process, unwinding from the loose ends to form coils.
(iii) When heated with limiting water, the A-chain double helices first come loose from their crystallites. The shorter double helices in A-type starch can then persist as an effectively homogeneous solution/melt until the helix–coil transition occurs. The longer double helices of B-type starch, when they come loose from the crystallite, maintain one-dimensional order as a nematic liquid crystalline phase until the helix–coil transition occurs. Two DSC transitions are therefore seen during processing of starch under conditions of limiting water.

Retrogradation is the spontaneous reassociation of starch components after cooking, particularly when the suspension is cooled. It is of major practical importance, being responsible, for example, for the change in the texture of bread as it becomes stale.[208]

Retrogradation is not the reverse of the initial processing: whereas in native starch the amylose is amorphous, in retrogradation it is the first component to crystallise, followed by the slower crystallisation of amylopectin.[209] As would be expected for a crystallisation, retrogradation depends on the conditions of the original treatment and the presence of other polymers. Depending on the conditions, amylose can crystallise on retrogradation in a form similar to V-amylose,[210] or, if the amylose leachate of a starch granule is allowed to crystallise, as large double-helical aggregates,[211] which can only be redissolved with difficulty. Amylose appears to interfere with crystallisation of amylopectin: the amylopectin of waxy starches (containing little or no amylose) recrystallises faster than that of normal starches.[212] The amylopectin helices appear to recrystallise as the A form.[213] The potential for altering gelatinisation and retrogradation by alteration of starch structure by deletion, suppression or introduction of biosynthetic enzymes is enormous, but is only starting to be explored.[214]

4.6.3 1→4-Diaxially Linked Pyranosides

4.6.3.1 Marine Galactans. Galactans, which may be sulfated or even sulfated and ring closed, are made by a number of marine organisms. Most of them have mixed linkages and are discussed in Section 4.6.9.2, but a 3-*O*-sulfated 4-linked homo-L-galactan produced by tunicates has been characterised.[215]

4.6.3.2 Pectin. The "glue" which keeps plant cells together, and appears to play many other roles, is a complex polysaccharide called pectin. Its existence has been recognised ever since the discovery of jam making (fruit pectin causes jams and jellies to set). Early work on the carbohydrate chemistry of plant cells had a catch-all term "pectic substances", which included pectin but also many other gel- and slime-forming polysaccharides, such as xyloglucan. Modern usage confines the word "pectin" to a series of polysaccharides based on poly-α-(1→4)-galacturonan. The only carbohydrate component of the so-called "smooth" regions of pectin is homogalacturonan, irregularly decorated by methyl ester formation of the carboxylates and acetylation of O2 and O3.[216] On the other hand, "hairy" regions of pectin are extensively ramified with a range of radically different sugars and polysaccharides.[217] These "hairy" regions are considered in terms of two structures, rhamnogalacturonan I and rhamnogalacturonan II (Figure 4.58). The second name is a misnomer: rhamnogalacturonan II, although it has an occasional rhamnose residue, has the same poly-α-(1→4)-galacturonan main chain as the "smooth" regions of pectin, in contrast to the main chain of rhamnogalacturonan I, which has the disaccharide repeat α-D-GalA*p*-(1→2)-α-L-Rha*p*-(1→4) and is a true rhamnogalacturonan.

Determination of the molecular weight of native pectin presents the usual problems of large, fragile biomolecules, which need only a few chemical events to drastically lower the measured molecular weight, and these are compounded

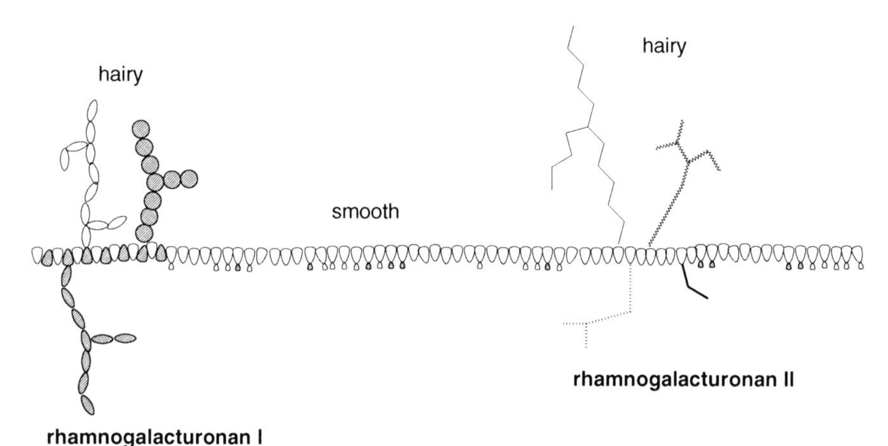

Figure 4.58 Cartoon of the "smooth" and "hairy" regions of pectin, on the assumption that rhamnogalacturonans and homogalacturonans are glycosidically linked. An impression of the acetylation and methylation patterns of the homogalacturonan is given.

by pectin's semi-solubility. Pectin for academic study is usually extracted with buffers and/or calcium chelators; more drastic conditions involving aqueous acid are used industrially.[218] Gently extracted pectin has an MW in excess of 0.2 MDa and homogalacturonan, rhamnogalacturonan I and rhamnogalacturonan II cannot be separated by gel permeation chromatography, leading to the widespread belief that pectin is one big polysaccharide.[219] This is established for at least homogalacturonan and rhamnogalacturonan II,[220] which have a common backbone, but data might be skewed by borate cross-links between rhamnogalactronan II moieties and dehydroferuloyl cross-links between rhamnogalacturonan I moieties (see below).

4.6.3.2.1 Homogalacturonan. Naturally occurring homogalacturonan is randomly acetylated on O2 and O3, and methylated on the carboxylate: in some pectins the degree of methylation can exceed 50%. Highly methylated pectins form weak gels, particularly in acid when the carboxylates are protonated, but in neutral solution strong gels are formed by low-methylated pectin in the presence of Ca^{2+}. Commercial pectin is often demethylated. A recent innovation is demethylation with ammonia, which leaves the strongly hydrogen bonding $-CONH_2$ groups in the polymer.[218]

Fibre X-ray studies indicated a 3_1 right-handed helical structure for polygalacturonic acid and sodium polygalacturonate ($\varphi = 73–80°$, $\Psi = 90–97°$; three residues per turn):[221] this is illustrated in Figure 4.59. The three-fold helix accommodated junction zones, both between neutral pectins, and when Ca^{2+}-mediated.[222] However, modelling and NMR of 1,4-α-galacturonobiose[223] indicated that a two-fold helical structure, the similar to the guluronate blocks in alginate, was very close to the 3_1 helix in energy (Figure 4.60). Such a structure is the near-mirror image (but for the stereochemistry of O3) of the

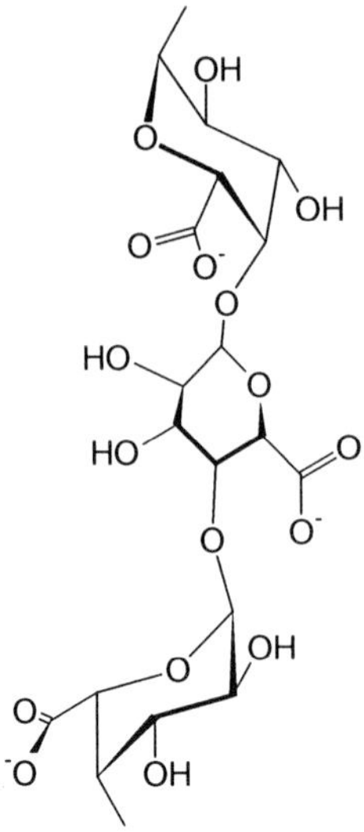

Figure 4.59 Three-fold helix of polygalacturonate.

two-fold helix of the polyguluronate regions of alginate, suggesting a parallel between the gelation of low-methyl pectin and of alginate.

Modification of homoglucuronan with single β-D-xylopyranosyl residues at O3, leading to a Xyl:Gal ratio between 0.5 and 1, is considered to lead to a different pectic polysaccharide, xylogalacturonan.[224,225] Heavily xylosylated rhamnogalacturonans are associated with plant cell detachment.[226]

4.6.3.2.2 Rhamnogalacturonan I. The backbone of rhamnogalacturonan is now considered to be a strictly alternating α-D-Gal*p*A-(1→2)-α-L-Rha*p*-(1→4) structure.[227] Despite the radical difference in primary structure, calculations suggest that the 1,2-diaxially linked rhamnose unit will not radically change the conformation of the polygalacturonan, so the possibility exists that the rhamnogalacturonan I backbone can form the linkage region of gels.[228] Its Gal*p*A units may be acetylated, but not esterified.[229]

To O4 of the rhamnose residues are connected a range of oligo- and polysaccharide chains, which depend on plant source.[230] Chief amongst them are (1→5)-α-L-arabinans and β-D-galactans, although xylogalacturonan and (1→3)-α-L-arabinan are known.[231] The (1→5)-α-L-arabinans may have (1→3) branch points. Single sugar residues such as α-L-Fuc*p*, β-D-Gal*p*A, β-D-Glc*p*A and 4-OMe-β-D-Glc*p*A may also be present.[219] It is becoming increasingly clear that this decoration is part of the mechanical and physiological machinery of the plant. CP-MAS and other solid-state NMR techniques indicated that the galactan and arabinan side-chains of rhamnogalacturonan 1, examined *in situ* in the plant cell wall, are the most mobile parts of the pectin molecule, with the arabinan, with three degrees of rotational freedom (corresponding to φ, ψ and χ), being more mobile than the galactans, although the longer galactans were nonetheless mobile.[232] The (1→5)-α-L-arabinan of stomata (the "ports" in plant

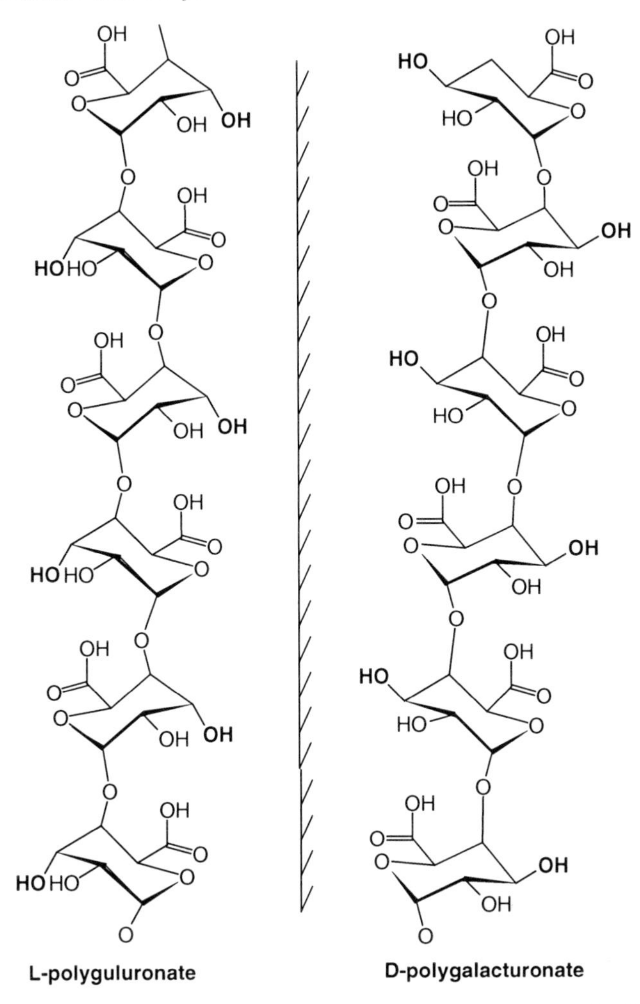

Figure 4.60 Near-enantiomorphism of two-fold helical structures of poly-L-guluronate and poly-D-galacturonate. The 3-OH (bold) destroys the symmetry.

tissue which permit entry and exit of gases) plays a role in their opening and closing, since in the absence of the arabinan the stomata are stuck open.[233] The β-D-galactan attached to rhamnogalacturonan I appears to be largely (1→4)-linked,[234] although it may have (1→6) branch points and the occasional (1→3) link in the main chain.[235] Arabinogalactans are of the Type I variety,[153] with a β-(1→4) main chain and short, sometimes single, arabinofuranose residues on O3.

The α-L-Ara*f*-(1→5)-arabinan, of *DP* 45–80, covalently attached to rhamnose residues, can be cleaved and isolated, and forms reasonably well-defined crystallites, which, however, cannot be aligned for fibre X-ray. Recent powder X-ray work[236] has indicated that the polymer crystallises in a monoclinic unit cell with one chain per unit cell. It adopts a single two-fold helix with the

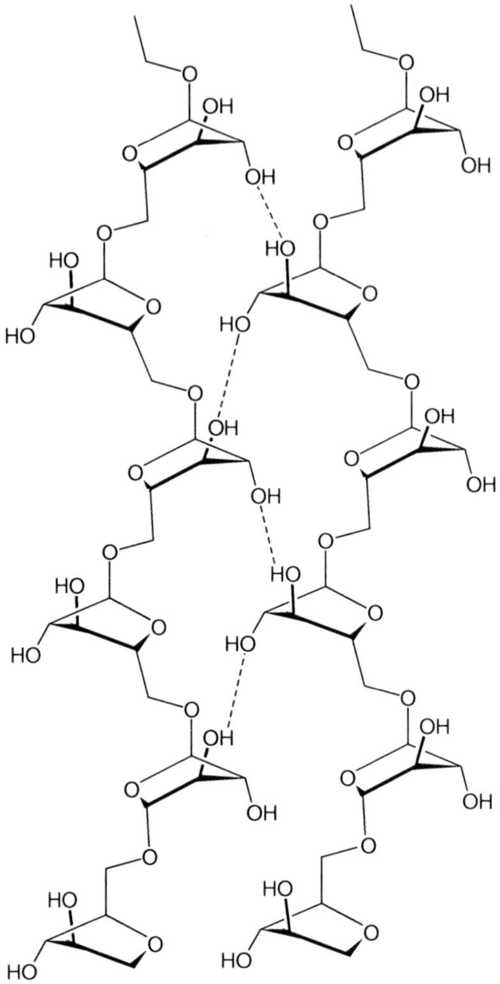

Figure 4.61 Conformation of crystalline arabinan. The possible intrachain hydrogen bonds are shown, but in the absence of neutron structures it is not possible to tell in which sense they are.

furanose rings in the *E*2 conformation (Figure 4.61). Parallel chains can form interchain hydrogen bonds O2 and O3.

The best fit to the data are obtained with $\psi = -174°$ and $\chi^{xxxiii} = 179°$, with $\underline{\varphi} \approx +81°$,[xxxiii] in accord with the operation of the *exo*-anomeric effect. The two-fold helices can aggregate by interchain OH2–OH3 bonds.

[xxxiii] In fact given the symbol ω. At the same time the dihedral angle about C1–O5′ was defined as C2–C1–O5′–C5′, which took a value of −159°.

In many plant species, the O5 of α-L-Ara*f* residues in plant cells walls is esterified with ferulic acid. One-electron oxidation results in cross-linking, both of feruloylated polysaccharides to each other, as with the terminal α-L-Ara*f* residues esterified with ferulic acid in the arabinan of rhamnogalacturonan I,[237] and to lignin (Figure 4.62).

4.6.3.2.3 Rhamnogalacturonan II. In rhamnogalacturonan II, the poly α-(1→4)-guluronan has four side-chains, labelled A–D, attached. Their relative location is not known with complete certainty, although a likely distribution is that in Figure 4.63.[238,239]

The structure contains a number of uncommon sugars, including Kdo (Figure 1.9), apiose (Figure 1.15), Dha (2-keto-3-deoxy-D-*lyxo*-heptulosaric acid) and aceric acid, AceA (3-carboxy-5-deoxy-L-xylose) (Figure 4.64). Side-chain structures are shown in Figures 4.63, 4.65 and 4.66.

It can be seen that both A and B chains are highly crowded, except for the apiosyl residues which connect them to the galacturonan main chain. These apiofuranosyl residues play a key role in changes of morphology of the plant cell, because they can be cross-linked with borate. The cross-linking was first observed with isolated rhamnogalacturonan II,[240] but its physiological relevance was established by a mutant of *A. thaliana,* in which the L-fucosyl residues of the branches were replaced by L-galactosyl residues. This lowered the formation constant for the borate cross-link, so that the plants were dwarfed. When excess borate was provided, the plants grew normally.[241] There appears to be a preference for cross-linking of two A side-chains,[240] although mixed AB cross-links are possible.[242] The tetrahedral borate of the cross-link is chiral and so two diastereomeric borate cross-links involving apiose can be envisaged (Figure 4.67).

4.6.3.2.4 Biosynthesis and biodegradation of pectin. Little is known about pectin biosynthesis, other than that it takes place in the cytoplasm and the product is then exported. Of the estimated minimal 53 genes,[243] only one [a rhamnogalacturonan II α-(1→3)-xylosyltransferase] had been unambiguously identified in early 2005,[244] although less definitive candidates for genes for a couple of side-chain transferases and a main-chain galacturonyl transferase (archly named *QUASIMODO1*) have been found in *A. thaliana.*[245]

Little is known about the L-arabinofuranosyl transfer machinery. A gene in the genome of *A. thaliana* has been identified as a membrane-bound arabinofuranosyl transferase by its membrane-insertion sequence and the fact that mutants in the gene led to expression of mRNA, but phenotypes that were nonetheless defective in arabinan side-chains of their pectin.[246] Attempts to use UDP-α-L-arabinopyranose as arabinofuranose donor in crude extracts, presumably in an attempt to uncover any mutase similar to UDP galactopyranose mutase (see Section 5.12.4), merely resulted in the incorporation of pyranose units into the arabinan.[247] However, an enzyme using this donor has been shown to add the terminal arabinopyranosyl residue to the 3-position of the terminal arabinose of the 1,5-arabininan of rhamnogalacturonan I.[248] [By contrast, the D-arabinofuranosyl transfer machinery that produces the

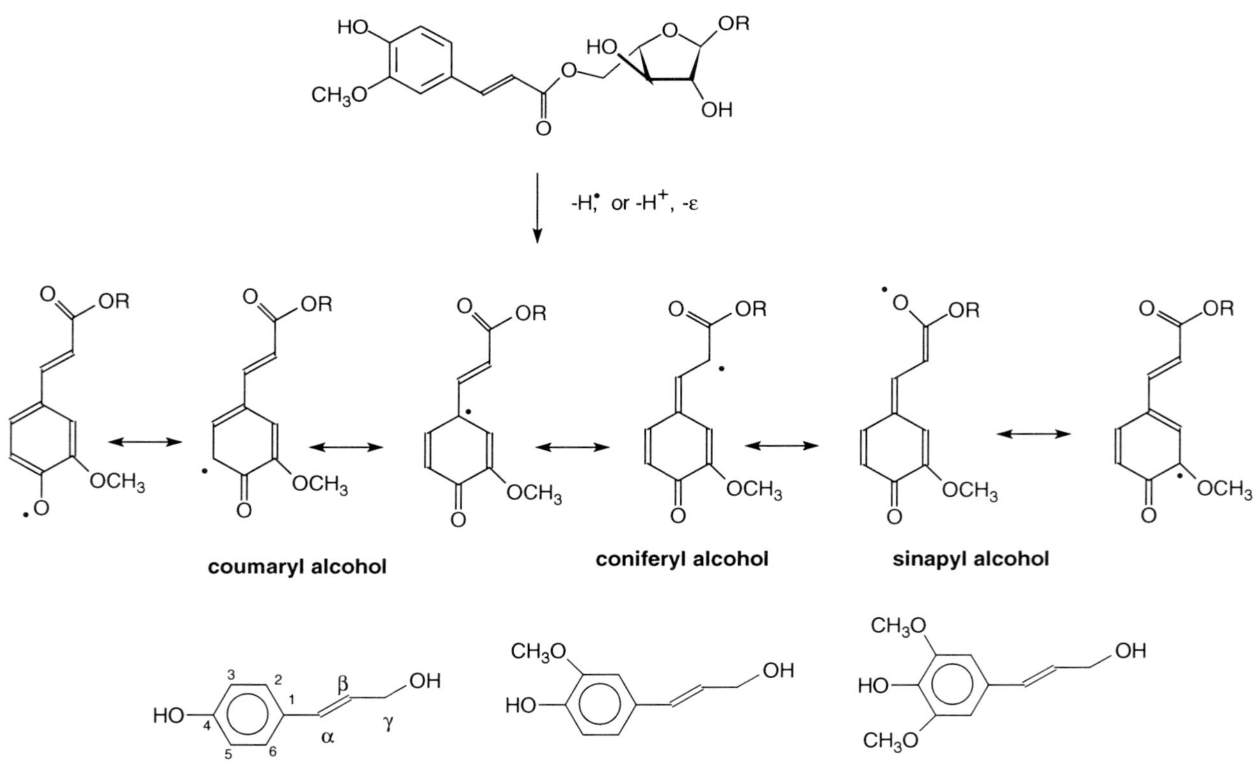

Figure 4.62 One-electron oxidation of feruloyl groups. The radicals can couple at all possible spin sites, although β-O4 is the most frequent. Lignin, the matrix material of plant cells, is an oxidative polymer of three related alcohols, *p*-coumaryl, coniferyl and sinapyl alcohol, formed by coupling of radicals at all possible spin sites. The feruloyl side-chains of pectin and hemicellulose can participate in this polymer formation, generating covalent links between polysaccharides and lignin.

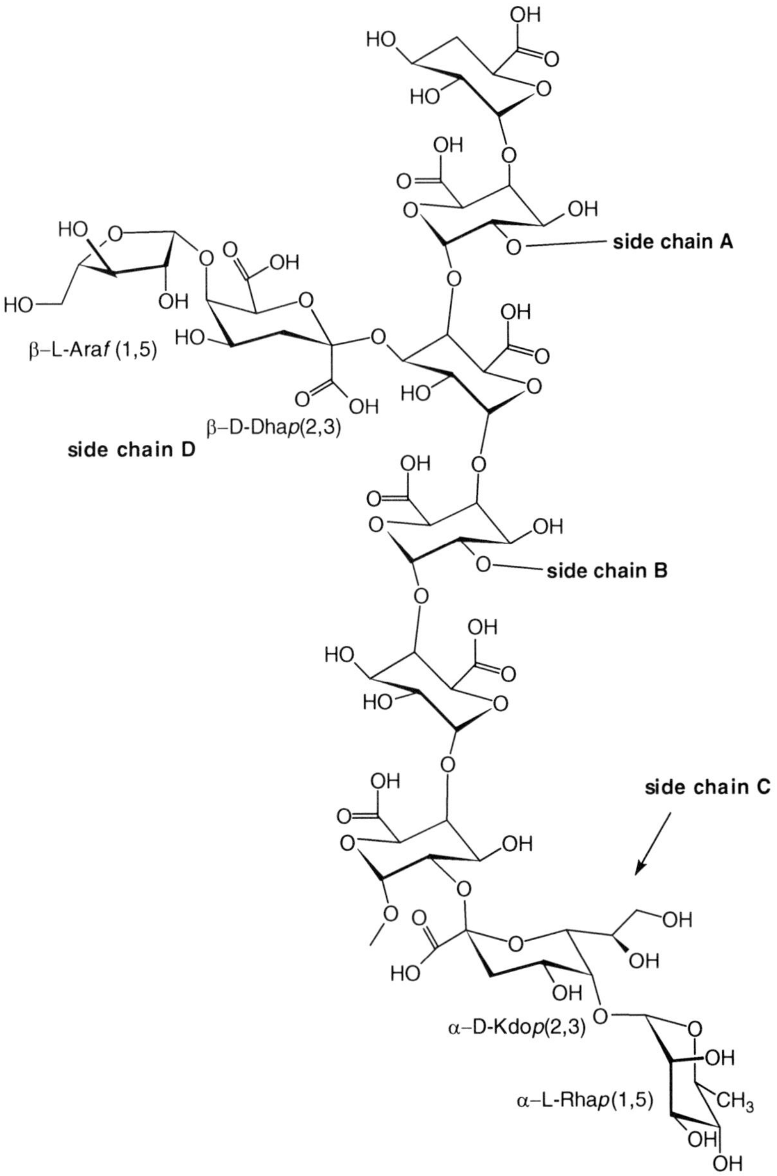

Figure 4.63 Structure of rhamnogalacturononan II.

D-arabinogalactan of *Mycobacterium tuberculosis* is known to involve conversion of 5-phosphoribosyl-α-pyrophosphate to 5-phospho-β-ribosyldecaprenyl phosphate by a membrane-bound enzyme, which has been cloned, sequenced and functionally expressed.[249] The sugar attached to the decaprenyl glycosyl donor (not dolichol, since the sugar-proximal prenyl unit is not reduced) is then

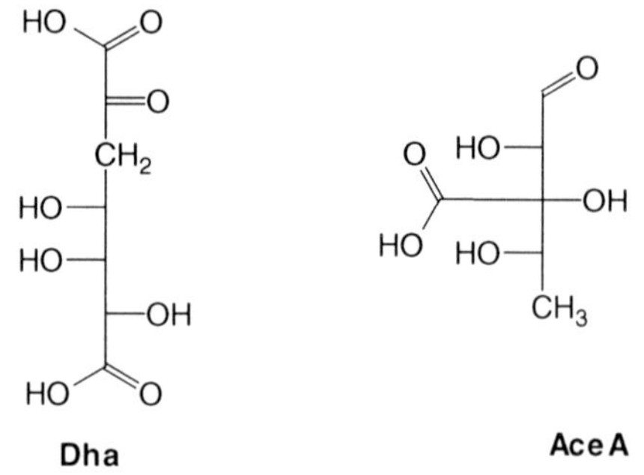

Dha **Ace A**

Figure 4.64 Straight-chain Fischer structures for the pectin sugars Dha (2-keto-3-deoxy-*lyxo*-heptulosaric acid) and aceric acid (3-carboxy-5-deoxy-L-xylose).

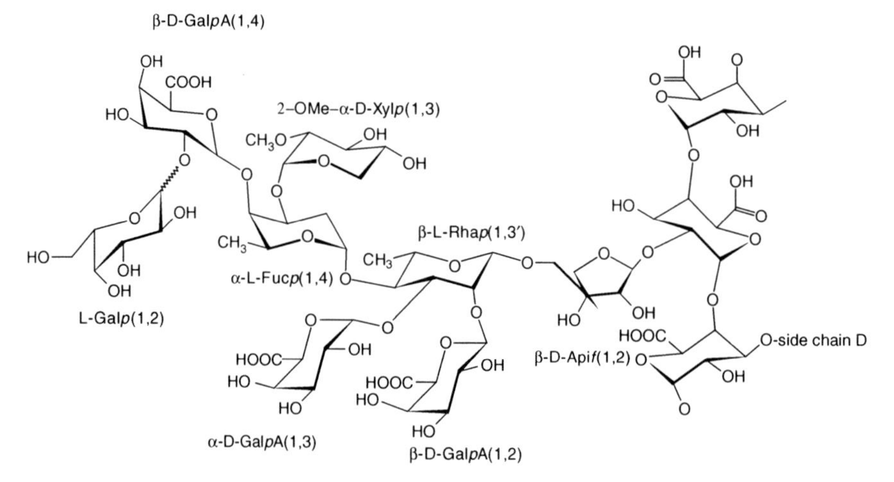

Figure 4.65 Structure of rhamnogalacturonan II A side-chain, shown with three sugars of the α-(1→4)-galacturonan backbone.

phosphorylated and epimerised at C4 and C3 to give the α-D-arabinofuranosyldecaprenylphosphoryl donor.]

The degradation of the pectin main chain is catalysed not only by pectinases, which hydrolyse the glycosidic link, but also by lyases, which carry out and $E1_{CB}$-like elimination across C4 and C5, leaving a 4-deoxy-4,5-dehydrogalacturonic acid at the new non-reducing end (see Section 6.7.4).

Since invasion of plant pathogens is inevitably associated with degradation of the cell wall, α-(1→4)-linked galacturonan oligosaccharides, unsurprisingly, play a key role in the mobilisation of the plant defences. They also act during normal growth an development, acting in the opposite sense to plant growth hormones during stem elongation and promoting the generation of ethylene during fruit ripening.

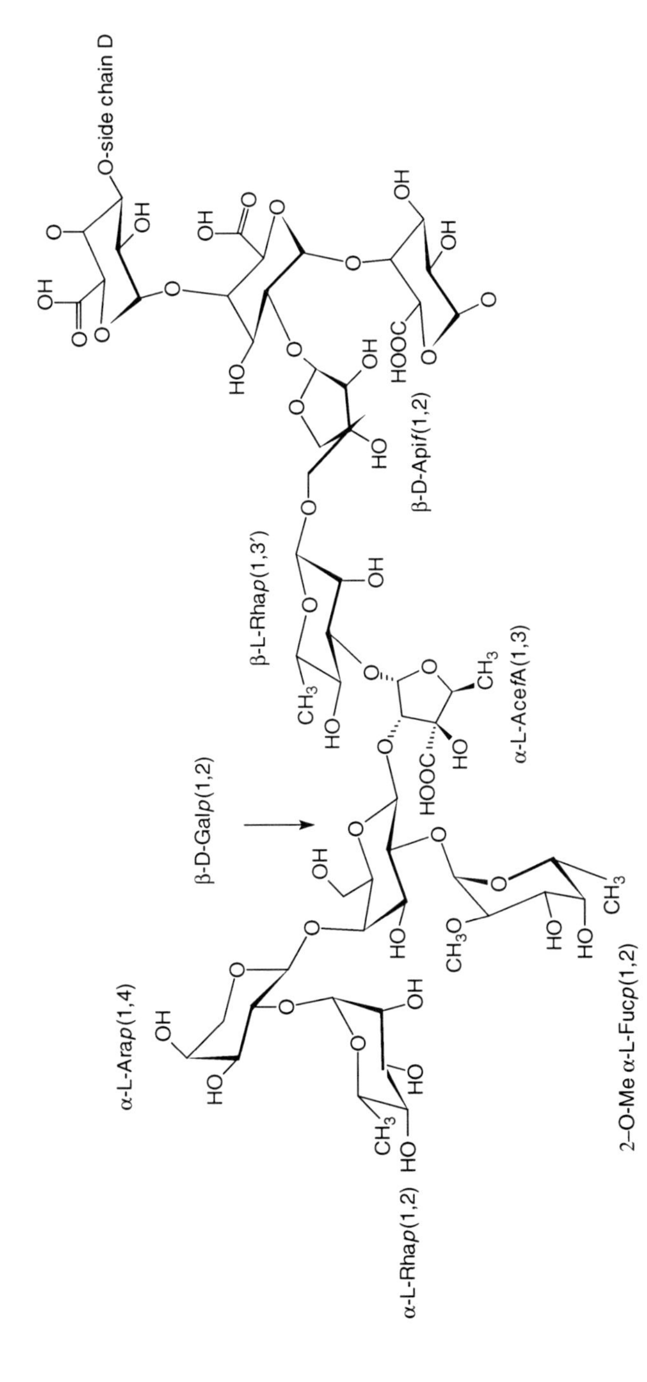

Figure 4.66 Structure of rhamnogalacturonan II B side-chain, shown with three sugars of the α-(1→4)-galacturonan backbone.

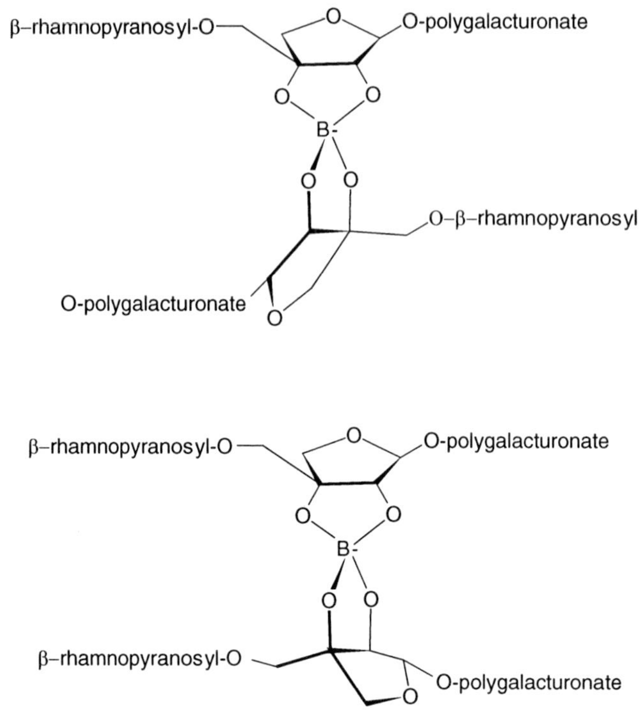

Figure 4.67 Two stereochemistries about the tetrahedral borate cross-links of rhamnogalacturonan II.

The softening of fruit during ripening results from cleavage of the pectin main chain, and is associated with production of galacturonan-cleaving enzymes. So that they remain firm during handling and transport, commercially grown tomatoes are picked green and ripened artificially in an atmosphere of ethylene. The Flavor-Savr tomato was developed in an attempt to make tomatoes remain firm while they ripened naturally. A gene coding for an "antisense" mRNA to the mRNA of an endogenous polygalacturonase was introduced, so that when the polygalacturonase gene was transcribed a duplex was formed between the sense and antisense mRNA and polygalacturonase mRNA was not translated into enzyme.

4.6.4 1,3-Diequatorially Linked Pyranosides

The operation of the *exo*-anomeric effect and the steric restrictions consequent on β-(1→3) linkages appear to strongly influence such polymers to adopt a wide-amplitude six-fold right-handed helix, which can be single or triple. A triple, right-handed helix for the β-(1→3)-xylan which formed the structural polysaccharide of some green algae was identified by fibre X-ray.[250] Its dihedral angles of $\varphi_H = 34°$, $\psi_H = 10°$ ($\underline{\varphi} \approx -86°$ $\underline{\psi} \approx -110°$) are very similar to those found by the same technique for a right-handed triple helical structure of the β-(1→3)-glucan in its hydrated native form $\varphi_H = 35°$, $\psi_H = 23°$ ($\underline{\varphi} \approx -86°$, $\underline{\psi} \approx -97°$), [251] and also for the dried polymer.[252]

4.6.4.1 β-(1→3)-Glucans. A range of names is given to naturally occurring β-(1→3)-glucans. Curdlan is the undecorated polymer with $DP \approx 450$ produced by the bacterium *Alcaligenes faecalis*; a similar polysaccharide produced by the eukaryote *Euglenia gracilis* is termed paramylon. The polysaccharide with a few β-(1→6) branches, 75% of them single glucose residues,[253] is produced by *Laminaria* spp. (seaweeds), commercially by *L. digitata*, and termed laminarin. Despite the linkage inhomogeneity of laminarin, the series of β-(1→3)-linked glucose residues is termed laminaribiose, -triose, *etc*. Fungal β-(1→3)-glucans with single β-(1→6) branches have been investigated because of their medical effects. Lentinan, from the shi-itake mushroom, *Lentinus edodes*, has a 7-glucose repeat with single β-(1→6) branches on residues 2 and 4 of the 5-glucose main chain.[254] Schizophyllan (produced by the fungus *Schizophyllum commune*) and scleroglucan (produced by members of the fungal genus *Sclerotium*) are two very similar polysaccharides, with a β-Glc*p*-(1→6) substitution on every third glucose residue of the β-(1→3) chain.

All the foregoing polysaccharides adopt six-fold right-handed helices, whether single or triple, which appear to be enforced by the *exo*-anomeric effect and the steric restrictions imposed on ψ by the β-(1→3) linkage. Interconversion between the various conformations of β-(1→3)-glucans is slow and, considering that covalency changes are not involved, can require surprisingly drastic conditions.

Thus curdlan, the undecorated β-(1→3)-glucan, has three forms. Six-fold right-handed single helices are formed by form I, obtained by brief heating of native curdlan (form II) at 70 °C in a large excess of water;[255] the conformation adopted by its triacetate is similar.[256] The single helix form of curdlan can also be produced from hydration of lyophilised DMSO solutions, but simple lyophilisation (from the gel or aqueous, DMSO or aqueous NaOH solution) gives a random chain;[257] in aqueous solution a random coil is formed by curdlan in high ($< \sim 0.2$ M) solutions of alkali, probably because of deprotonation of sugar OHs, but the single helical conformation is regained on neutralisation.[258]

Thus, regeneration of curdlan form II requires heating with water at 120 °C, [251] and form III requires intensive drying of form II.[252] The conformation changes can be readily monitored by CP-MAS ^{13}C NMR (chemical shift differences of up to 8 ppm being observed).[259] CP-MAS, of course, once signals have been assigned, can then analyse specimens which are mixtures of the two forms.[260]

The β-(1→3)-glucans and their conformations are of interest because of their antitumour, antiviral and immunomodulating activities. Their presence in various mushrooms used as traditional medicaments in East Asia makes them one of the very few such medicaments to have a proven, beneficial effect.[xxxiv]

[xxxiv] Sympathetic-magic-based traditional Chinese and Japanese medicine is usually worse than useless, as illustrated by the 100% infant mortality ($n = 9$) inflicted on the family of Emperor Meiji, until its practitioners were displaced by Western-trained doctors in the 1880s (D. Keane, *Emperor of Japan: Meiji and His World, 1852–1912*, Columbia University Press, 2002).

Binding to lipooligosaccharide receptors and thereby modulating the release of nitric oxide,[261,262] is the suggested mechanism. The single helical form of the molecule appears to be the active form.[263,257]

Lentinan,[264] scleroglucan[265] and schizophyllan[266] all adopt triple helical conformations, essentially similar to the triple helical conformation of unsubstituted curdlan, in aqueous solution. In these triple helices, the β-(1→6) single glucose units prevent aggregation and crystallisation of the triple helices, and networks of polysaccharide are formed by imperfect registration of the individual glucan chains in the helices.[267] The conversion of triple helices to single helices requires the complete unwinding of the triple helix: any portions of triple helix left after, say, DMSO or NaOH treatment will merely act as a template for triple helix reformation.[268] The triple helix does not have the central cavity of the single helix and is therefore much more stable: it is stabilised by interchain bifurcated hydrogen bonds, as illustrated in Figure 4.68. The instability of the single helical form permits it to encapsulate and solubilise carbon nanotubules,[269] in much the same way as amylose encapsulates much smaller single aliphatic chains.

4.6.4.2 β-(1→3)-Galactans. The unsubstituted β-(1→3)-galactan chain is not found in Nature, but it forms the backbone of a number of highly substituted polymers. Plant arabinogalactan proteins are proteins which are heavily *O*-glycosylated on Ser, Thr or hydroxyproline, with main-chain carbohydrate being very largely a β-(1→3)-galactan, but with very many and varied other glycosylation patterns. The arabinogalactan protein from spruce (*Picea abies*), for example, is acidic,[270] as it contains β-D-glucuronyl and 4,6-*O*-pyruvyl

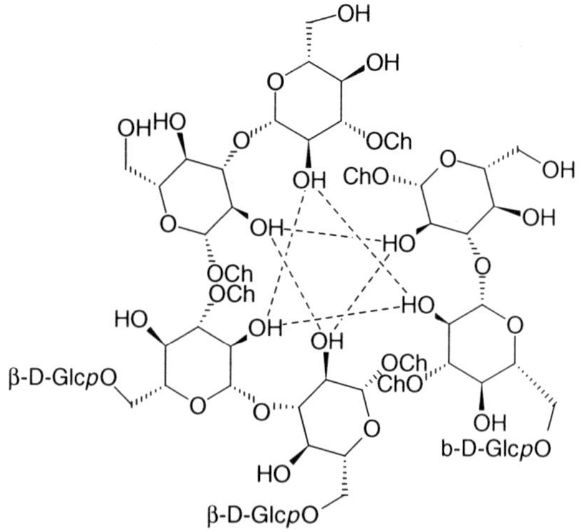

Figure 4.68 Cartoon of the end view of a β-(1→3)-glucan triple helix, showing the interchain hydrogen bonds between the 2-OH groups. Each 2-OH probably forms one donor and one acceptor hydrogen bond, but the locations of the protons are not known. Ch = rest of the β-(1→3)-glucan chain. Note how any substituent on O6 will stick out from the coiled "rope" of β-(1→3)-glucan chains.

residues. The details of the structure and function of these molecules, for which "protein" is a misnomer as they are 90–97% carbohydrate, are still unknown.[271] Plant type II arabinogalactans have a β-(1→3)-galactan backbone, [272] with substitution of β-D-galactopyranosyl and α-L-arabinopyranosyl residues at O6 [the type I arabinogalactans are β-(1→4)-linked and are encountered attached to rhamnogalacturonan 1 – see Section 4.6.3.2.2].

The larch family, which is unique in comprising deciduous conifers, is also unique in the hemicelluloses, which are not based on 1,4-diequatorial polysaccharides, but are largely arabinogalactans with a highly branched β-(1→3)-galactan backbone.[273] The branches consist of single (1→6)-linked β-D-galactopyranosyl or α-L-arabinofuranosyl residues two to three (1→6)-linked β-D-galactopyranosyl residues. (1→3)-Linked β-D-arabinopyranosyl residues have also been found in the terminal position of the side-chains.

4.6.5 (1→3)-Linked Axial–Equatorial Pyranosides

Extended polymers with this stereochemistry are confined to α-(1→3)-glucans. An integral part of the cell wall of fission yeasts has been shown[274] to consist of a linear, largely α-(1→3), glucan with the structure [α-Glc*p*-(1→3)]$_m$[α-Glc*p*-(1→4)]$_n$[α-Glc*p*-(1→3)]$_m$[α-Glc*p*-(1→4)]$_n$OH, where $m \approx 120$ and $n \approx 12$. Likewise, an unbranched α-(1→3)-glucan has been found in the Japanese mushroom rheishi *(Ganoderma lucidum)*.[275] However, the α-(1→3)-glucan attracting most attention has been that produced by oral bacteria such as *Streptococcus mutans* and *Streptococcus sobrinus* from sucrose, which contributes to the formation of dental plaque and therefore increases the risk of caries. "Dextran" in this context means a polymer produced from sucrose by transfer of the glucosyl residue by a GH 70 transglycosylase, not a Leloir enzyme; the thermodynamic driving force, as with the biosynthesis of "levan" by oral bacteria, is the steric crowding of the sucrose molecule. The polymers produced have a sucrose residue at the reducing end.[276] Because of the relaxed specificity of the +1 acceptor site of the biosynthetic enzyme, polymers with various linkages are made, depending on the bacterial strain – thus *Leuconostoc* spp. produce mainly α-(1→6)-linked glucans. The GH 70 transglycosylase of one such species[277] could be induced to produced a mixed polymer with α-(1→2) and α-(1→3) linkages simply by mutation of a single amino acid in the +1 site to lysine. The balance of glycosidic linkage formation can be altered even by the occupancy of higher numbered positive subsites of the transglycosylase, one of the enzymes from *S. sobrinus* making[278] α-(1→6) linkages in the presence of short acceptors, but α-(1→3) linkages in the presence of pre-existing α-(1→3)-glucan.

Despite the linkage inhomogeneity of the native polymer, however, by an acetylation–alignment–deacetylation sequence it was possible to obtain orientated fibres of α-(1→3)-glucan from *S. salivarius* suitable for X-ray analysis (Figure 4.69).[279] Although the crystallographic repeat is the disaccharide, there is very little difference between the two residues which are close to a two-fold helix. The near two-fold helices are packed in antiparallel sheets, which have

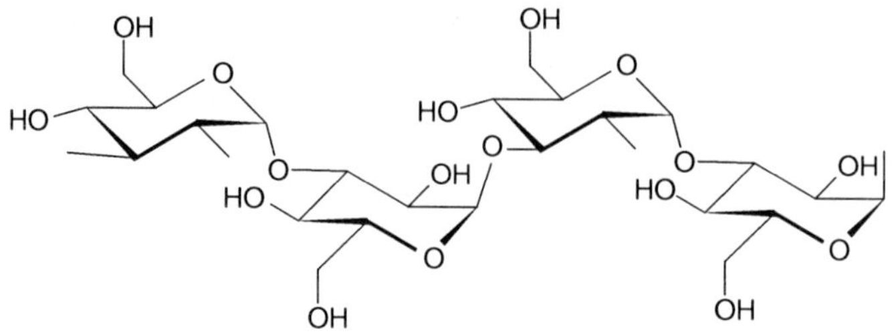

Figure 4.69 Ordered conformation of α-(1→3)-glucan.

close to the maximum number of hydrogen bonds. Dihedral angles (which differed between the two residues by only a couple of degrees: the average is quoted) are $\varphi_H = -17^\circ$ ($\underline{\varphi} \approx 103^\circ$), in accord with the *exo*-anomeric effect, $\psi_H = -16^\circ$ ($\underline{\psi} \approx -136^\circ$) and $X = -60^\circ$ ($\underline{\chi} \approx 60^\circ$), the hydroxymethyl groups being in the *gg* conformation. The solution macromolecular behaviour of rheishi α-(1→3)-glucan, however, while indicating a conformation more extended that amylose or curdlan, showed no longer scale order, suggesting that the chain was largely flexible.[275]

4.6.6 (1→2) Pyranosidic Homopolymers

β-(1→2)-Cyclic glucans are virulence factors for a number of bacteria. In the nitrogen-fixing plant symbionts *Rhizobium* spp. they are associated with attachment of the bacteria to root hairs and nodule formation: in the infectious organism *Agrobacterium tumefaciens*, the workhorse microorganism for introducing foreign DNA into dicotyledonous plant species, they are associated with attachment to plant cells. Their usual size is 17–22 glucose units.

The NMR spectrum of the eicosa (20-mer) β-(1→2)-cyclic glucan is identical with that of β-glucopyranose, so that on the NMR time-scale all residues are equivalent: the conformation in Figure 4.70 represents at least the NMR-average conformation.[280]

In *A. tumefaciens*, larger rings (*DP* 24–30) can be produced in the presence of β-(1→2)-cyclic glucans of the usual *DP* 17–22: the normal products appear to act as specific inhibitors of the cyclisation activity.[281]

The cyclic glucans of *DP* 17–22 are also produced as virulence factors by *Brucella* spp., mammalian bacterial pathogens causing undulant fever in humans and abortion in cattle. In *B. abortus*, a single glycosyl transferase, transferring β-glucose from UDPGlc,[282] initiates the reaction by forming a glucosylated enzyme, elongates the polysaccharide and then, when the glucan chain has reached the required length, cyclises it. The C-terminal domain is homologous to cellodextrin phosphorylase, which is in GH 84, not a GT family (see Chapter 5).[283]

Although individual (1→2) linkages occur in a wide range of heteropolymers, the only other biologically important homopolymer with this linkage appears

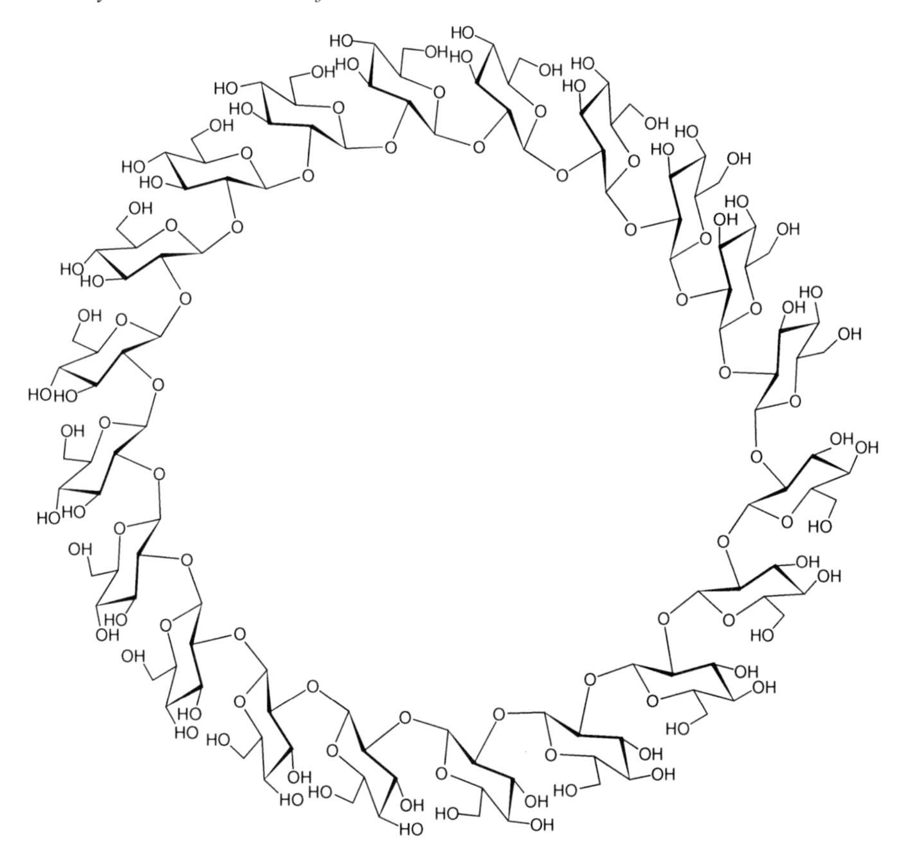

Figure 4.70 Cyclo[β-Glc*p*-(1→2)]$_{20}$.

to be a β-(1→2)-mannan of *DP* 4–40, produced in large amounts (80–90% of total carbohydrate) by the infectious stage of the single-celled human parasite *Leishmania mexicana*.[284] Chemically synthesised β-(1→2)-mannan has a disordered structure manifested in the NMR spectrum; the sharply decreasing yield in the coupling reactions for *DP* 4 and above support calculations of a "collapsed, disordered" structure of the polymer.[285]

4.6.7 Pyranosidic Homopolymers Without Direct Linkages to the Ring

The extra degree of freedom of (1→6)-linked hexopyranosides, and *a fortiori* the two extra degrees of freedom of α-(2→8)-linked polysialic acid, mean that these polymers tend to behave as freely coiling chains, rather than be semi-structured in solution.

The extracellular dextran, native *DP* 30 000–250 000, produced by *Leuconostoc mesenteroides*, which contains largely α-(1→6) linkages (the proportion of other linkages varying with the strain) is industrially important.[286] *L. mesenteroides* is an environmental bacterium, first isolated from sauerkraut, which in the presence of sucrose produces dextran by transglycosylation of glucopyranose residues from sucrose, catalysed by a GH 70 enzyme. The dextran cross-linked with epichlorohydrin (1-chloro-2,3-epoxypropane), provided the first gel-permeation material, Sephadex.[287] When swollen in water, the "holes" in the gel of cross-linked dextran were such that, when the gel was packed in a chromatographic column, fractionation on the basis of molecular size was possible (see Section 4.5.1). Native dextran behaves as a nearly ideal statistical coil.[288]

β-(1→6)-Glucans form part of the cell wall of yeasts,[289] where they interconnect mannoprotein, glucan and chitin.[290] Their biosynthesis occurs at the cell surface, not at the Golgi membrane, and two genes associated with their synthesis are in family GH 16, rather than a GT family.[291] This suggests formation by a transglycosylase, as with α-(1→6)-glucans. There are no reports of the conformational preferences of the polymer, probably because it is a free coiling chain.

α-(1→6)-Mannans are associated with yeasts. Chains of *DP* 30–100 are attached, in place of the shorter oligosaccharides of typical N-linked glycoproteins, to the cell wall proteins of *Saccharomyces* spp. The mannan is produced by a range of glycosyl transferases (not transglycosylases), which also produce α-(1→2) and α-(1→3) branches.[292] The yeasts *Trichophyton mentagrophytes* and *T. rubrum*, causative agents of human skin conditions such as athletes' foot and dhobi itch, produce a mannan which has an α-(1→6) backbone and single α-(1→2)-mannopyranose units every fourth mannose unit in the main chain,[293] and which partly suppresses the host immune response to the infection. A phytotoxic bacterium (*Pseudomonas syringae*) produces a mannan[294] with the same α-(1→6) backbone, but with much more frequent α-(1→2) branches, not only of single mannoses, but also of disaccharides and trisaccharides with α-(1→2) linkages. A broadly similar polysaccharide is also produced by the workhorse host for recombinant glycoproteins, *Pichia pastoris*, though some β-mannose residues were apparently detected in the branches.[295]

Polysialic acids (Figure 4.71) are found throughout Nature, outside the plant kingdom. The commonest is an α-(2→8)-linked polymer of *N*-acetylneuraminic acid, produced by bacteria as a food reserve and environmental protectant:[296] in mammals it is attached to neural cell adhesion molecules and modulates cell shape.[297] Other polymers are also known, including an α-(2→8)-linked polymer of *N*-glycolylneuraminic acid in mammals,[298] the α-(2→9)-linked isomer found in a nerve cell line and neuroinvasive strains of *Neisseria memingitidis*,[299] an alternating α-(2→8)-, α-(2→9)-polymer produced by a pathogenic strain of *E. coli*,[300] and an α-(2→5′)-poly-*N*-glycolylneuraminic acid from the coat of sea urchin eggs, in which the glycosyl bond is between C2 and the primary hydroxyl of the N-linked glycolyl group.[301] The jelly of the eggs of salmonid fish (salmon, trout, char, *etc.*) contains a range of α-(2→8)-linked polysialic acids, poly(NeuNAc), poly(NeuGc), mixed polymer chains of the two residues and poly(KDN) (Figure 4.71).[302]

Polysialic acids can be biosynthesised by either GH 33 transialidases, from existing glycosidic bonds,[296] or from CMPNeuNAc in a Leloir pathway.[300]

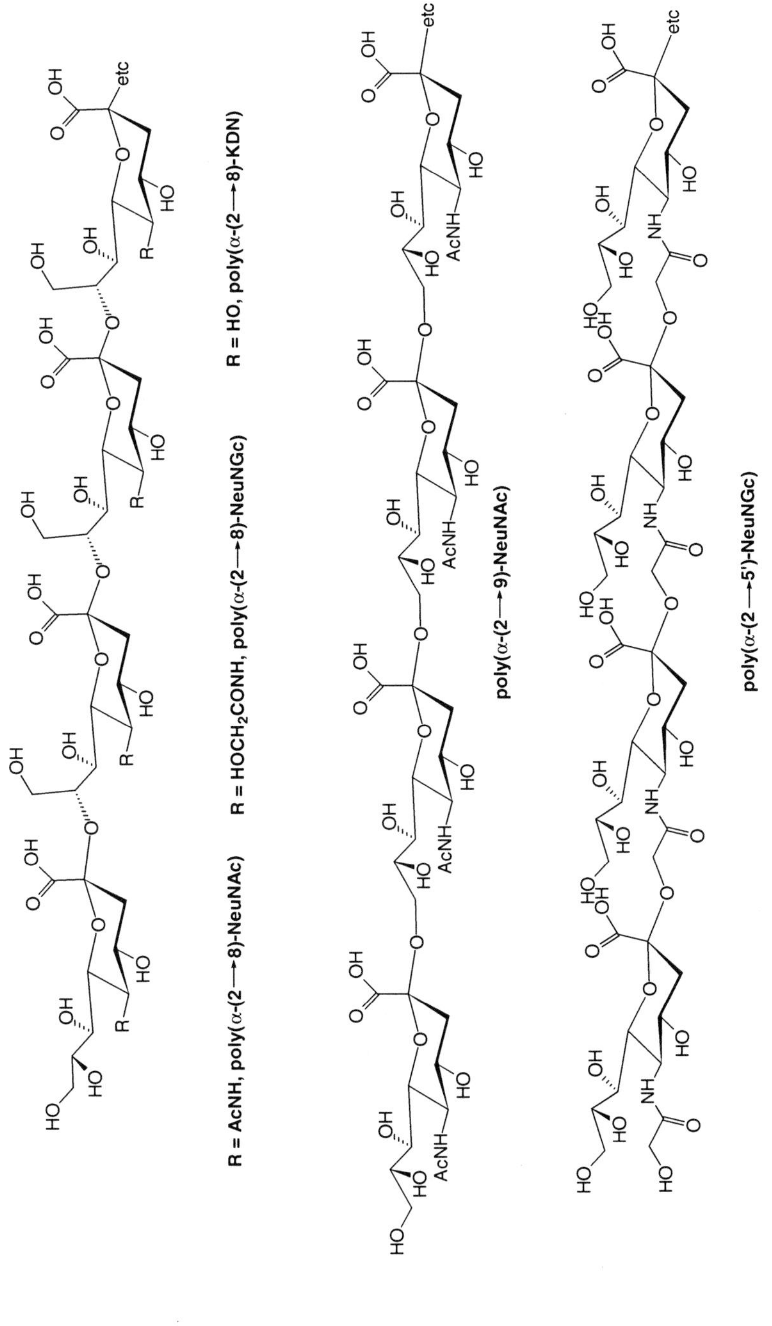

R = AcNH, poly(α-(2→8)-NeuNAc) R = $HOCH_2CONH$, poly(α-(2→8)-NeuNGc) R = HO, poly(α-(2→8)-KDN)

poly(α-(2→9)-NeuNAc)

poly(α-(2→5′)-NeuNGc)

Figure 4.71 Polysialic acids.

The proximity of the C1 carboxylate of residue *n* to the hydroxyl groups of the glycerol side-chain of $n+1$ makes the lactonisation of polysialic acid in acid solution very ready and it can occur on, for example, freeze-drying or even on standing in buffer at the acid end of the physiological pH range (*e.g.* pH 3.5). Lactonisation of O8-linked polymers occurs on O9, rather than O7,[303] as expected for a less hindered primary OH. Lactonisation of O9-linked polymers occurs on O8,[304] since lactonisation on O7 would generate a seven-membered ring.

The lactonisation of poly-α-(2→8)-NeuAc is neither temporally nor regiochemically random (Figure 4.72).[305] Temporally, three stages can be identified: stage 1, when a few internal lactonisations take place; stage 2, when all the internal linkages are lactonised, leaving the two external linkages unlactonised; and stage 3, when all linkages are lactonised. The lactonisation-patterns of partly lactonised oligomers support the model, with the tetramer being lactonised between residues 2 and 3 and the pentamer between residues 2 and 3 or 3 and 4. The analysis of lactonisation patterns is made possible by the inertness of lactonised links to sialidases.

Lactonised linkages are much less labile to hydrolysis of the glycosidic link in weak acid than normal intersialic acid linkages – as considered in Section 3.11.1, ionisation of the 1-carboxylate provides an electron donating (or at least less electron-withdrawing) group to stabilise the sialosyl cation-like transition states.

The non-enzymic hydrolysis of poly-α-(2→8)-NeuAc at pH 3–4 and 37°C is very much faster than that of the dimer, the apparently terminal products of hydrolysis being the dimer and trimer.[306] The rate increases with chain length, with the octamer being about twice as reactive as the tetramer. Only the internal bond of the tetramer is cleaved, with the dimer being formed exclusively. The alternating poly-α-(2→8), α-(2→9)-NeuNAc behaved similarly and α-(2→8)-NeuNGc and poly-α-(2→8)-KDN are comparably labile, whereas the polymer becomes very stable on reduction of the carboxylate groups to CH_2OH.

Intramolecular general acid catalysis, in polymer conformations in which the carboxylic acid residue of the $n+2$ or possibly $n-2$ residue is favourably placed, appear to be the reason for this behaviour. In order to provide an epitope recognised by antibodies, at least 10 α-(2→8)-NeuNAc units are required. NMR studies confirmed this, also indicating the conformation of external linkages was different from that of internal linkages and uncovering a very wide dispersion of chemical shifts for equivalent atoms in residues in different parts of the polymer.[307] Inter-residue NOEs in solutions of [α-(2→8)-NeuNAc]$_3$OH were between 3H$_R$ of residue 1 and H8 of residue 2 and between H7 of residue 2 and H9 of residue 3, clearly indicating different conformations about the two linkages of the oligosaccharide. A wide, nine-residue helix was proposed as one component of an array of conformations adopted by this essentially flexible polysaccharide,[308] which accounted for the immunogenicity of the longer oligosaccharides. However, if the observed increase in hydrolysis-rate of the polymer of 1–2 orders of magnitude arose from partial ($<10\%$)

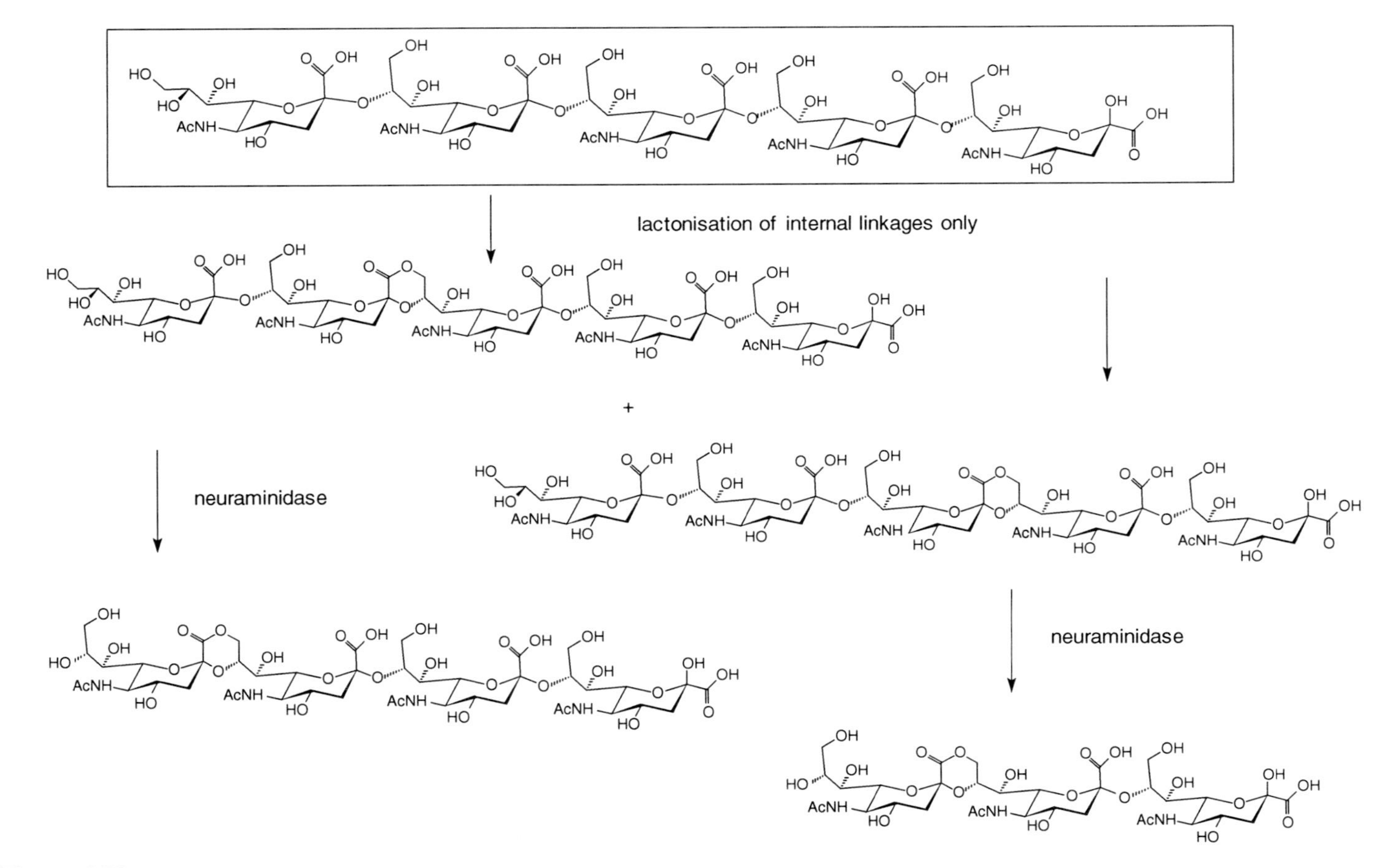

Figure 4.72 Lactonisation of α-(2→8)-NeuAc pentamer.

occupancy of conformations in which remote carboxylates were placed in suitable positions to act as intramolecular general acids, their effective molarity in such a conformation must be much higher than in the model systems used hitherto to investigate the phenomenon (see Figure 3.18).

The linkages in the polymer which are particularly susceptible to acid hydrolysis (such as the internal link in the tetramer) are those which are also particularly susceptible to lactonisation, but lactonisation renders the linkage relatively unreactive to acid hydrolysis. Because of this competition, in weakly acid solution the products observed with any given oligomer will vary with exact pH and temperature in a complex fashion.[309]

4.6.8 Furanosidic Homopolymers

4.6.8.1 Inulin. Inulin is a linear polymer of β-(1→2)-D-fructofuranose, terminating in a βDFru*f*-(2↔1)αDGlc*p* unit. It is an important reserve carbohydrate in the roots of certain species of plant, such as dahlia, chicory and Jerusalem artichokes, and is found in lower amounts in other plants.[310] Humans have no inulinase activity in the gut, so dietary inulin is metabolised by intestinal flora[xxxv].[311] Plant inulins are comparatively small, with *DP* < 100, but much larger fungal and bacterial inulins are known.

From Figure 4.73, it can be seen that the polymer chain of inulin does not involve the sugar ring, and can be regarded as a polyoxyethylene chain, with a sugar ring attached in a spiro fashion on one of the two carbons of the ethylene unit. The conformational determinants associated with relatively rigid sugar units therefore do not operate, and inulin behaves like a freely coiling chain in solution, although small (average *DP* 17) inulin fragments can be crystallised.[312] A ^{13}C NMR study of molecular motion in solutions of various polysaccharides showed that inulin was much the most mobile of those examined.[313] NMR also failed to detect helical conformations in inulin polymers up to *DP* 9.[314] Reports that the very high molecular weight inulin (*DP* 60 000) of *Aspergillus sydowi*[315] had, according to the results of low-angle laser light and X-ray scattering, gel permeation chromatography and viscometry,[316] an unusual compact structure, in which the polymer appeared to be curled up into a ball, were subsequently shown to arise from branch points in the primary structure.[317] Not only the *A. sydowi* inulin, but also that made from sucrose by recombinant levansucrase of *Streptococcus mutans*, was shown by methylation analysis to have 5–7% (2→6) branch points. The compact nature of the branched polymer is therefore unsurprising.

Inulin is not biosynthesised by a Leloir pathway involving a nucleotide mono- or diphospho sugar, but by transglycosylating enzymes of GH 32 and GH 68 acting on sucrose. Only two enzymes are involved, a sucrose–sucrose 1-fructosyl transferase (1-SST), which produces the primer trisaccharide 1-kestose, and fructan 1-fructosyltransferase (1-FFT), which exchanges

[xxxv] With production of gas and social embarrassment.

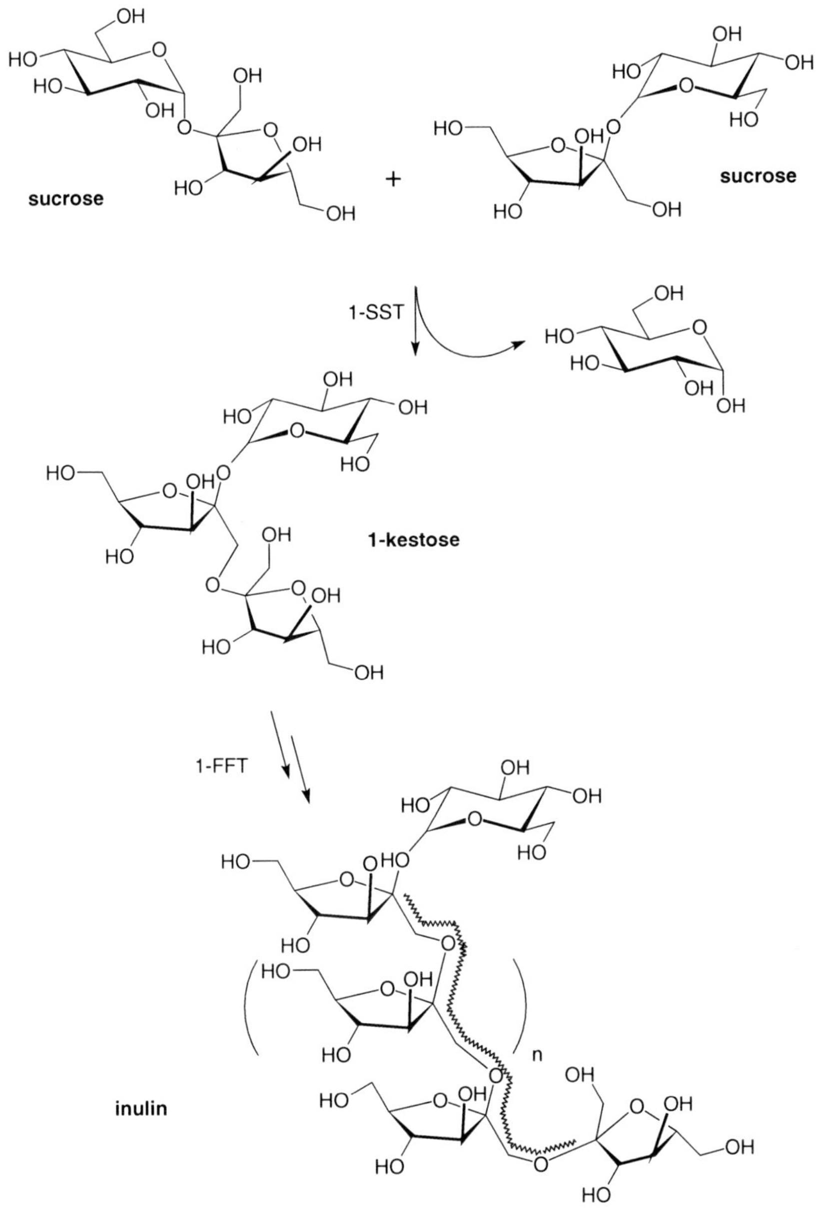

sucrose + sucrose

1-SST

1-kestose

1-FFT

inulin

Figure 4.73 Biosynthesis of inulin from sucrose. The wavy line illustrates the polymer main chain.

fructofuranosyl moieties between fructan chains. The two-enzyme hypothesis[318] was unequivocally proved when genes for 1-SST and 1-FFT from the globe artichoke (*Cynara scolymus*), which produced particularly long (*DP* up to 200) inulin molecules, were introduced into potato.[319] The potato tubers, which normally do not synthesise inulin, then synthesised the full spectrum of insulin molecules observed in the globe artichoke.

The thermodynamic driving force for fructan biosynthesis is the sterically strained nature of the crowded glycosidic linkage in sucrose.

4.6.8.2 Phleins, Levans and Fructans. The strictly β-(2→6)-linked polymer of fructofuranoside units attached to sucrose is termed phlein. There appear to be no experimental investigations of its structure, although some theoretical work with the primitive AM1 basis set suggested a helix for the β-(2→6)-linked fructofuranose heptamer.[320]

Phlein is biosynthesised in a manner analogous to inulin, but only the one transglycosylase, sucrose:fructan 6-fructosyltransferase (6-SFT), appears to be involved in both chain initiation and extension.[321] Cloning of the 6-SFT gene from barley into tobacco, whose wild-type produces no fructans of any description, resulted in the tobacco plants producing 6-kestose and phleins (Figure 4.74), up to *DP* 10.[322] Cloning of the 6-SFT into chicory, a plant which made only inulin-type fructans, resulted in the production of large, mixed (2→1)- and (2→6)-fructans, with many branch points.

In the microbiological literature, "levan" is used for β-(2→6)-fructans, in place of "phlein". Levans are made as extracellular polysaccharide by a wide range environmental bacteria, most notoriously the oral bacterium *Streptococcus mutans* (see Section 4.6.5). The levansucrases of oral bacteria, which make levan from sucrose, are GH 68 transglycosylases, rather than Leloir enzymes. The word "levan" can include polysaccharides with a significant number of (2→1) linkages.[323] Practically, bacterial levans tend to be of much higher *DP* than plant phleins (and the structure at the reducing end consequently less certain): the levan from *Lactobacillus reuteri*, for example, has a *DP* of 10^3.[324]

Fructan is a general term for fructose polymers.

4.6.9 Polysaccharides from One Sugar but with more than One Linkage in the Main Chain

Pullulan,[325] a bacterial *exo*-polysaccharide produced by *Aureobasidium pullulans,* a black yeast, is a polymer of maltotriosyl units linked α-(1→6), (4)-α-D-Glc*p*-(1→4)-α-D-Glc*p*-(1→6)-α-D-Glc*p*-$(1\rightarrow)_n$, with a wide range of industrial uses. Depending on the culture conditions, its *DP* (monomer glucose) is 10^3–10^4. It appears to act as a random coil in aqueous solution, although with preparations with $DP > 200$ some evidence of transient ordering is obtained. The trimer repeat is first assembled from UDPGlc as a phospholipid-linked intermediate, α-Glc*p*-(1→4)-α-Glc*p*-(1→6)-Glc*p*-phosphate-lipid, which is subsequently polymerised.

β-(1→3),β-(1→4)-Glucan is an important hemicellulose of grasses, with *DP* up to 1200: a similar polymer, called lichenan, is produced by some lichens.[326] Precipitation of β-(1→3),β-(1→4)-glucan which has dissolved from the barley grain during malting and mashing results in beer becoming cloudy on storage, and this commercial inconvenience underlies a large proportion of the work done on the polymer and enzymes to hydrolyse it.[327] Around 90% of the

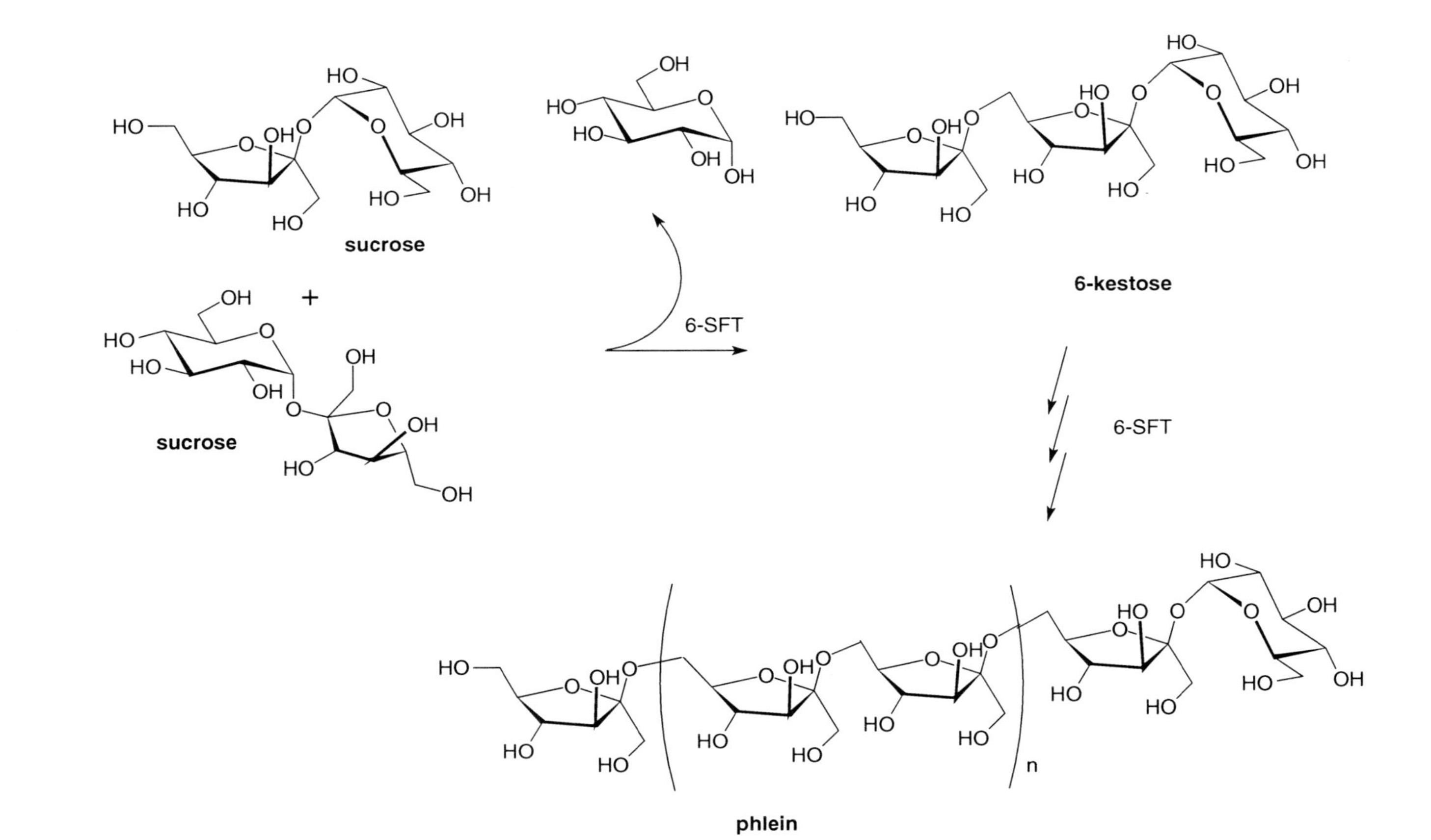

Figure 4.74 Phleins and their biosynthesis.

β-(1→3),β-(1→4)-glucan of barley grains consists of regions in which cellotriosyl or cellotetraosyl units are joined by single β-(1→3) linkages.[328] In aqueous solution, barley β-(1→3),β-(1→4)-glucan forms fringed micelles in which chains are associated sideways.[329] A β-(1→3),β-(1→4)-glucan with strictly alternating β-(1→3) and β-(1→4) linkages has been synthesised from α-laminobiosyl fluoride and a glycosynthase (see Section 5.9.3) in which the nucleophilic Glu of a GH 16 β-(1→3),β-(1→4)-glucanase has been mutated to Ala:[330] the glycosynthase constructs only β-(1→4) linkages. The resulting polymer, as expected from its regular structure and the crystallinity of both β-(1→3)- and β-(1→4)-glucans, is highly crystalline, being produced as spherules of platelet crystals. The unit cell dimensions, measured by X-ray diffraction, suggest an extended conformation like cellulose, but more open.

4.6.10 Heteropolysaccharides with Several Sugars in the Main Chain

4.6.10.1 Glycosaminoglycans. Glycosaminoglycans are a range of anionic polysaccharides in animal connective tissue. They are based on four different disaccharide repeats, with one of the saccharides being a 2-acetamido-2-deoxy sugar. In the case of heparan and chondroitin, the reducing end is attached to a protein through an oligosaccharide link of general formula (sugar)-β-GlcA-(1→3)-β-Gal*p*-(1→3)-β-Gal*p*-(1→4)-β-Xyl*p*-(1→Serine) sequence (see Section 5.12.1.9 for the enzyme that adds the GlcA).

4.6.10.1.1 Initial polymer chains and their biosynthesis. The four basic repeats are as follows (Figure 4.75):

Heparan: β-GlcA*p*-(1→4)-α-GlcNAc*p*-(1→4). The β-GlcA and α-GlcNAc are added sequentially, from UDPGlcA and UDPGlcNAc, by a bifunctional transferase which is a single polypeptide chain.[331,332] Necessarily, since a reducing-end primer is involved, chain extension occurs at the non-reducing end. The "sugar" in the linkage region above is the first α-GlcNAc-(1→4) (for the enzyme that transfers the residue, see Section 5.12.2.9).

Chrondroitin: β-GlcA*p*-(1→3)-β-GalNAc*p*-(1→4). Initiating enzyme(s) transfer a βGalNAc residue to O4 of GlcA in the key β-GlcA-(1→3)-β-Gal*p*-(1→3)-β-Gal*p*-(1→4)-β-Xyl*p*-1→Serine region of the common core protein, committing the core protein to chondroitin, rather than heparan, biosynthesis.[333] An enzyme which is a single polypeptide chain then adds GlcA and GalNAc units sequentially from UDPGlcA and UDPGalNAc, respectively.[334]

Hyaluronan: β-GlcA*p*-(1→3)-β-GlcNAc*p*-(1→4). The biosynthesis proceeds, apparently without any primer requirement (unlike heparan and chondroitin), by stepwise addition of UDPGlcA and UDPGlcNAc, from the non-reducing end.[335] As with heparan, the synthetases are single polypeptides which catalyse both transfer reactions.[336]

Keratan: β-Gal*p*-(1→4)-β-GalNAc*p*-(1→3). Oligosaccharide chains (20–50 residues) derived from this polymer form part of biantennary N-linked glycoproteins (keratin sulfate type I, such as corneal keratan sulfate) and shorter stretches, fucosylated, as part of O-linked glycoproteins (keratin sulfate type II, such as

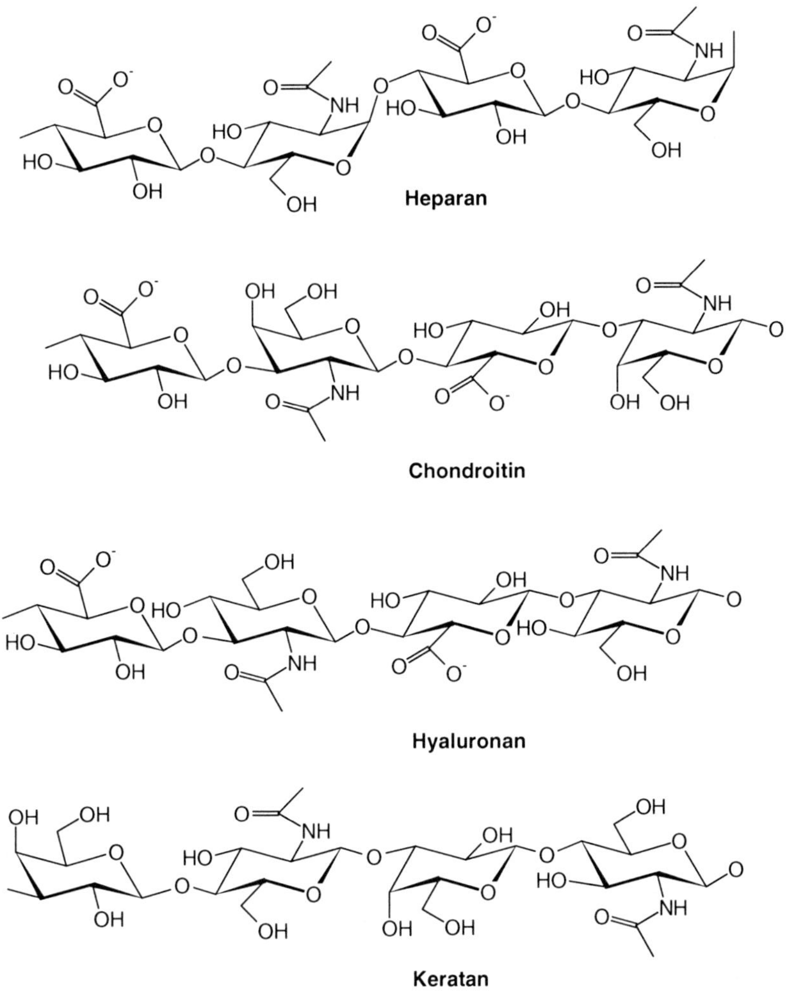

Heparan

Chondroitin

Hyaluronan

Keratan

Figure 4.75 Two repeat units of the first-biosynthesised polysaccharide chains of four main types of glycosaminoglycans.

articular cartilage). The chain extension appears to involve discrete transferases (rather than a single polypeptide chain).[337] The polymer is only known sulfated on the 6-OH of the galactosamine and less frequently of galactose.[338]

The heparan, chondroitin and hyaluronan synthetases, although one polypeptide chain, have two active sites corresponding to the two glycosyl transfers catalysed. Kinetic studies are technically very difficult, but the point was established in the case of the hyalurononan and chondroitin synthetases by the construction of chimeric proteins. *Pasteurella multocida* produces both chondroitin and hyaluronan by the action of two-domain single polypeptide synthetases. The chimeric protein from the N-terminal half of the hyaluronan synthetase and the C-terminal half of the chondroitin synthetase produced only hyaluronan, whereas the chimeric protein from the N-terminal half of chondroitin synthetase and the C-terminal half of hyaluronan synthetase produced only chondroitin. Clearly, the hexosaminyl transferase domain of both enzymes lies in the N-terminal domain.[339]

4.6.10.1.2 Hyaluronan. Hyaluronan occurs undecorated as a molecule of *DP* $\approx$ 6000,[340] which is one of the lubricants of load-bearing joints in mammals – it is a major component of synovial fluid. In water, the approximate two-fold helix formed by the ionised disaccharide unit, enforced by $O5_n \cdots HO4_{n+1}$ and $O2H_n \cdots O{=}C$ hydrogen bonds across the β-(1→3) link, and $O5_n \cdots HO3_{n+1}$ and $N_n\text{–}H \cdots {}^{-}OOC\text{–}_{n+1}$ hydrogen bonds across the β-(1→4) linkages, curves into a shallow helix; in DMSO solutions the hydrogens forming these hydrogen bonds give sharp ^1H resonances (Figure 4.76).[341] These helices persist in water and can associate in an antiparallel fashion via interchain hydrogen bonds and bonding between hydrophobic patches, as manifested by ^{13}C NMR.[342] The double helices can then aggregate laterally. It is suggested that these solution aggregates, which are formed only from longer hyaluronan, can disperse and reform readily, thus providing the "bounce" needed for hyaluronan function as lubricant and cushion in the joints. The sharp increase in intrinsic viscosity [eqn. (4.16)] as solutions of sodium hyaluronate are diluted is in accord with the dissociation of these aggregates.[343] Hyaluronan can be induced to crystallise in an extended conformation[344] as potassium and sodium salts in an orthorhombic unit cell with two antiparallel single helices per unit cell. One potassium form has a shallow left-handed single helix with four disaccharide residues per turn.[345] Across the β-(1→4)-glycosidic linkage, $\varphi = \Phi = -71°$, $\Psi = -110°$ and across the β-(1→3)-glycosidic linkage, $\varphi = \Phi = -58°$, $\Psi = -103°$; *X* for the GlcNAc hydroxymethyl was 152° (*gt* conformation). In a more compact "sinuous hyaluronate" potassium form, these angles changed to: across the β-(1→4)-glycosidic linkage, $\varphi = \Phi = -77°$, $\Psi = -100°$, and across the β-(1→3)-glycosidic linkage, $\varphi = \Phi = -64°$, $\Psi = -103°$; *X* for the GlcNAc hydroxymethyl was −178° (*gt* conformation).[346] After much coaxing, the polysaccharide was induced to crystallise also in a double helical form.[347]

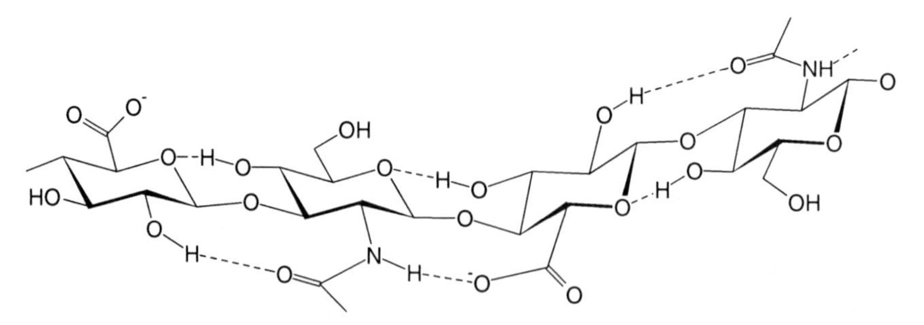

Figure 4.76 Solution conformation of hyaluronan. Note the combination of the two conformational features imposed by two types of glycosidic bonds: the two-fold helix imposed by the β-(1→4) link (cf. cellulose) and shallow helix of the disaccharide units typical of a β-(1→3) link (*cf.* curdlan).

4.6.10.1.3 Sulfation and epimerisation. Chondroitin can occur unmodified, but it can also occur as chondroitin 4-sulfate and chondroitin 6-sulfate, according to the predominant position of the *N*-acetylgalactosamine sulfated. Various regions can be further sulfated, with non-reducing termini being sulfated at both the 4- and 6-positions on the galactosamine and the 2-position of the uronic acid. The uronic acid can be epimerised, with regions of the polymer containing only residues of C5-epimerised uronic acid, α-L-iduronic acid, being termed dermatan sulfate. (The epimerisation of the uronic acid at the reference asymmetric carbon is the reason the designation of the bond changes from β to α, although of course the absolute stereochemistry at C1 does not change: the situation is exactly analogous to that with alginate, where C5 epimerisation changes β-D-ManU to α-L-GulU.) Unlike hyaluronan, chondroitin and dermatan are covalently linked to protein in proteoglycans. Heparan and keratan can also be *O*-sulfated and the heparan/heparin family is also *N*-deacetylated, *N*-sulfated.

The SO_3^- is transferred from the biological sulfation agent 3-phosphoadenosine 5′-phosphosulfate (PAPS) (Figure 4.77). The initial reaction in PAPS biosynthesis is an enzymic S_N2 reaction at the α-phosphorus of ATP, which occurs with inversion of configuration with α-thio-ATP, in which P_α is chiral.[348] The reaction is so unfavourable thermodynamically, however, that metabolically it has to be dragged to product both by the exergonic hydrolysis of pyrophosphate and the consumption of another molecule of ATP in the phosphorylation of the 3′-position position of adenosine phosphosulfate. The stereochemistry at sulfur of transfers from PAPS is considered to be inversion, by analogy with phosphoryl transfers from ATP,[349] and the structure of a PAP–vanadate complex, in which the trigonal bipyramidal V^V mimics the putative in-line sulfatyl transfer transition state.[350] Various PAPS-using glycosaminoglycan sulfotransferases have been isolated and characterised,[351] and heterologous expression systems for several members of the family of Golgi enzymes which 6-sulfate Gal, GalNAc or GlcNAc have been worked out.[352] There is one crystal structure of a carbohydrate sulfotransferase, that of the sulfotransferase domain of a heparan *N*-deacetylase *N*-sulfotransferase, and the structure of the putative active site supports, but does not prove, an in-line transfer mechanism with inversion.[353]

Sulfation beyond one sulfate group per two carbohydrate residues may change the conformation of the sugar rings (Figure 4.78). The effect is particularly marked where two vicinal equatorial OH groups can be sulfonated. Methyl 2,3,4-tri-*O*-sulfonyl-β-glucuronide adopts a range of non-chair conformations, as does methyl 2,3,4,6-tetra-*O*-sulfonyl-β-glucoside, but their α-anomers, stabilised by the anomeric effect, remain in the $^4C_1$ conformation.[354] The effect is an electrostatic one: sulfamation (converting OH to neutral OSO_2NH_2 groups) does not change the conformation. Chemical sulfonation of chondroitin to the level of four sulfates per sugar dimer result in the GlcA residue adopting the $^1C_4$ conformation at 30 °C and the $^2S_0$ conformation at 60 °C.[355]

The glycosaminoglycan epimerases are part of a superfamily of lyases and epimerisases whose active sites stabilise carboxyl enolates, similar to the mannuronate epimerases in alginate biosynthesis.[133]

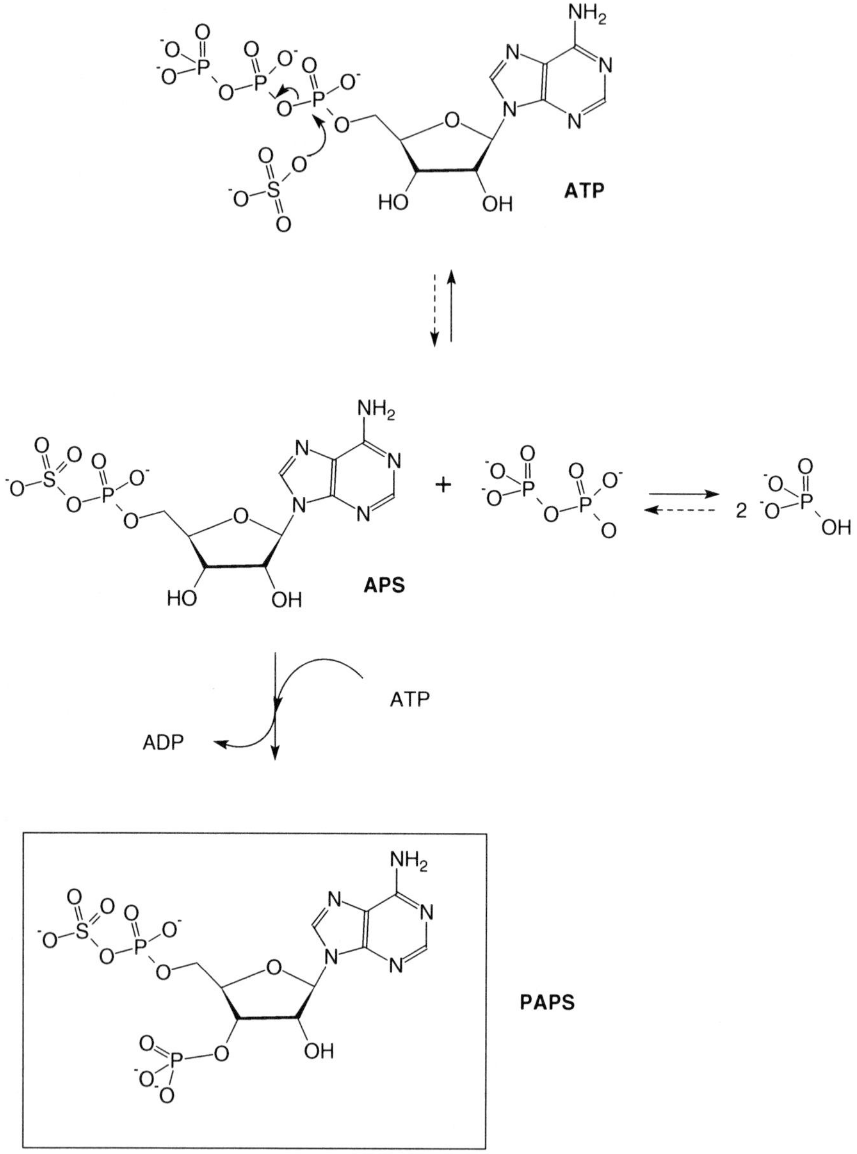

Figure 4.77 The biological sulfation agent PAPS and its biosynthesis.

4.6.10.1.4 Chondroitin, dermatan and their sulfates. Chondroitin sulfate proteoglycans have been localized to cartilage, bone, cornea and intervertebral disc, whereas dermatan sulfate is found in skin and aorta. In addition to its mechanical role, appears to have a host of regulatory functions.[356]

Chondroitin 4-sulfate and its uronic acid C5-epimerised isomer, dermatan 4-sulfate, occur not as free polysaccharides but as substituents on a core protein.

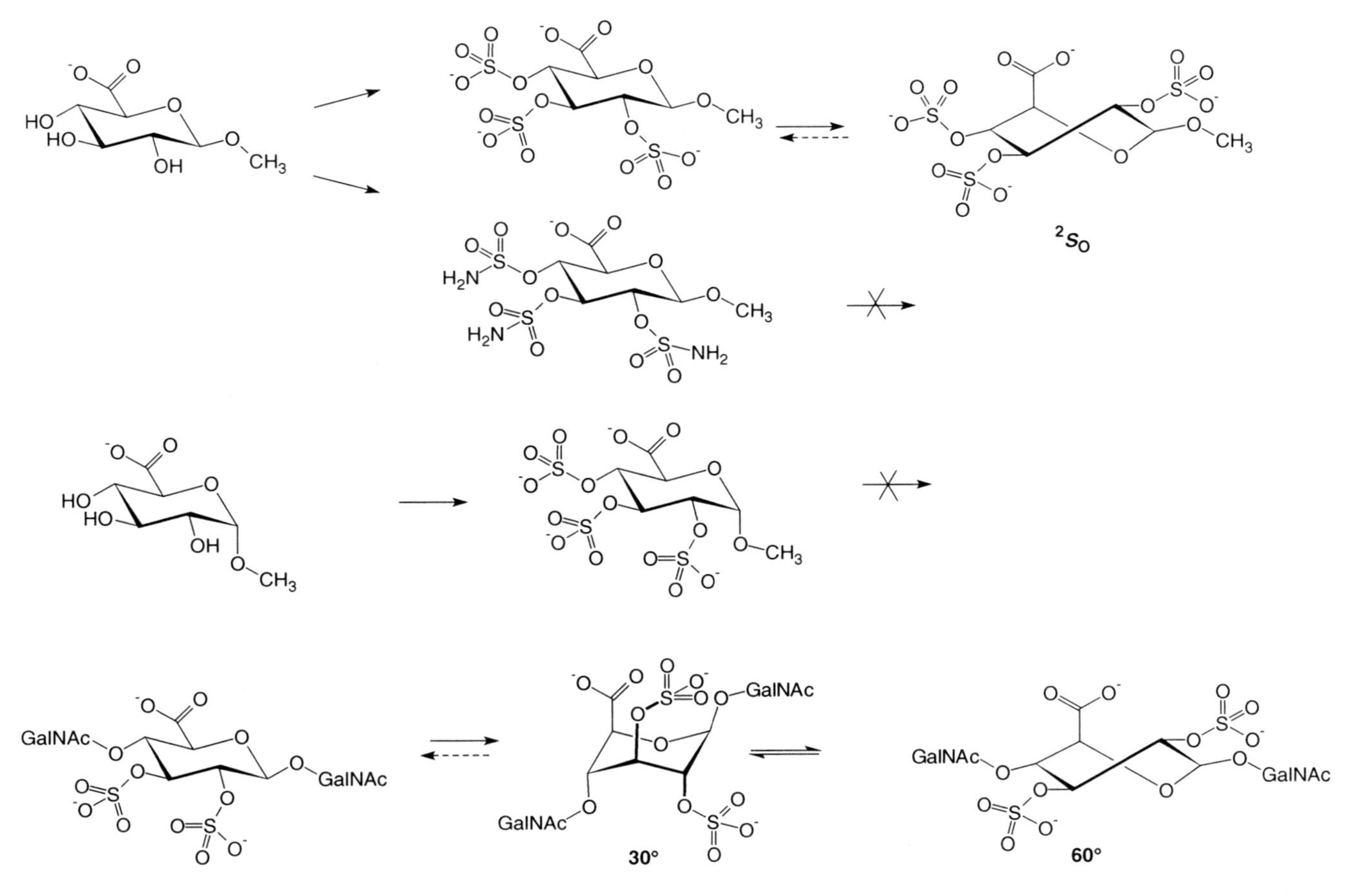

Figure 4.78 Conformational options for a sulfated glucuronic acid residue, as monomer and polymer. Sulfation of vicinal diequatorial residues is sufficiently disfavoured by electrostatics that it can alter conformation, but the effect is not large enough to overcome a methoxyl anomeric effect.

The general arrangement is seen best with aggrecan, the major load-bearing proteoglycan in the extracellular matrix of all cartilaginous tissues. The ~300 kDa core protein is substituted with ~100 chondroitin sulfate and, in some species, keratan sulfate glycosaminoglycan chains. It associates non-covalently with hyaluronan and the ~45 kDa link glycoprotein to form high molecular weight aggregates (>200 MDa). In cartilage, these aggregates form a densely packed, hydrated gel that is enmeshed within a network of reinforcing collagen fibrils. Atomic force microscopy[357] permits the layout of the aggrecan molecule to be seen directly – it is clear that the glycosamino glycan chains are extended (Figure 4.79).

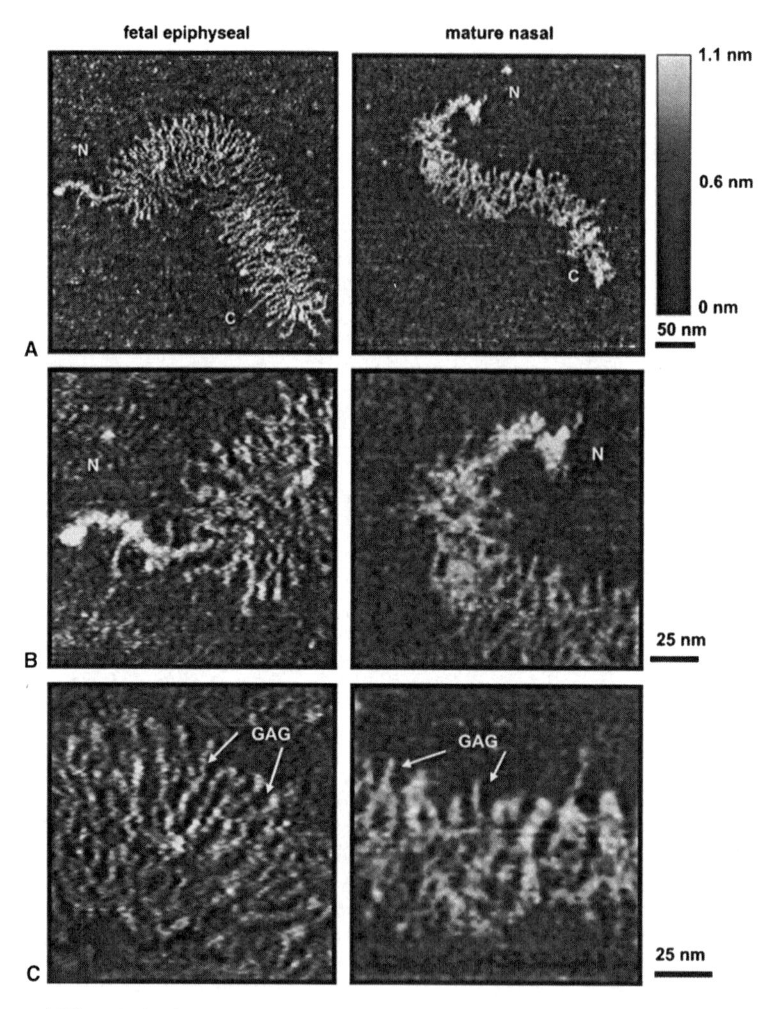

Figure 4.79 A single aggrecan molecule visualised by atomic force microscopy. (Dr Christine Ortiz and Elsevier are thanked for permission to reproduce this figure from Ref. 357.)

Chondroitin 4-sulfate can exist is shallow three-fold left-handed helices in the presence of Na^+ or K^+: dihedral angles are (average for the two salts): across the β-(1→4)-glycosidic linkage, $\varphi = \Phi = -83°$, $\Psi = -132°$, and across the β-(1→3)-glycosidic linkage, $\varphi = \Phi = -86°$, $\Psi = 132°$.[358] In the presence of Ca^{2+}, a two-fold helix is enforced by the chelation of Ca^{2+} by the 4-sulfate on GalNAc and the GlcA carboxylate, as well as the familiar $O5_n \cdots HO3_{n+1}$ hydrogen bond across the β-(1→4) linkage (Figure 4.80). Refinement of the fibre X-ray diffraction pattern of the Ca^{2+} salt[359] indicated that across the β-(1→4)-glycosidic linkage, $\varphi = \Phi = -98°$, $\Psi = -174°$, and across the β-(1→3)-glycosidic linkage, $\varphi = \Phi = -80°$, $\Psi = -107°$: χ for the GalNAc hydroxymethyl was $-146°$. The carboxylate and acetamido groups are roughly perpendicular to the planes of the sugar rings. This general shape agreed with solution NMR measurements of the sodium salt.[360]

Molecular modelling suggests that the conformations of chondroitin and its 6-sulfate should be very similar, with that of chondroitin 4-sulfate somewhat different because of the restrictions on ψ imposed by the sulfate group for the β-(1→3) linkage.[361]

The conformation of dermatan 4-sulfate, in which C5 of the glucuronic acid has been epimerised, is still unsettled. Fibre X-ray data fitted models in which the α-L-IdoA residue was in the $^4C_1$ conformation better than those in which it was in the $^1C_4$ conformation.[362] Models containing skew conformers were not considered and NMR studies have since established that in D_2O solutions of dermatan 4-sulfate, the α-L-IdoA residues are in the $^1C_4$ or $^2S_0$ conformations.[360] Molecular dynamics simulations have also supported a $^2S_0$ conformation for the iduronate in dermatan sulfate.[363]

4.6.10.1.5 Heparan sulfate and heparin. These two connective tissue polysaccharides are first biosynthesised as the linear polymer [β-GlcA*p*-(1→4)-α-GlcNAc*p*-(1→4)]$_n$, on the primer (sugar)-β-GlcA-(1→3)-β-Gal*p*-(1→3)-β-Gal*p*-(1→4)-β-Xyl*p*-1→Serine. They are then partly modified by the following reactions, which occur biosynthetically in the order given (Figure 4.81):[364]

(i) *N*-Deacetylation followed by PAPS-dependent *N*-sulfation of some of the GlcNAc residues, to yield GlcNS residues. So far, four *N*-deacetylases/*N*-sulfotransferases have been identified. They are

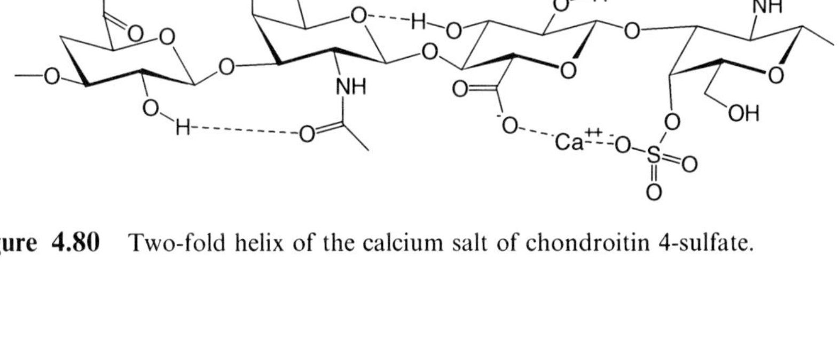

Figure 4.80 Two-fold helix of the calcium salt of chondroitin 4-sulfate.

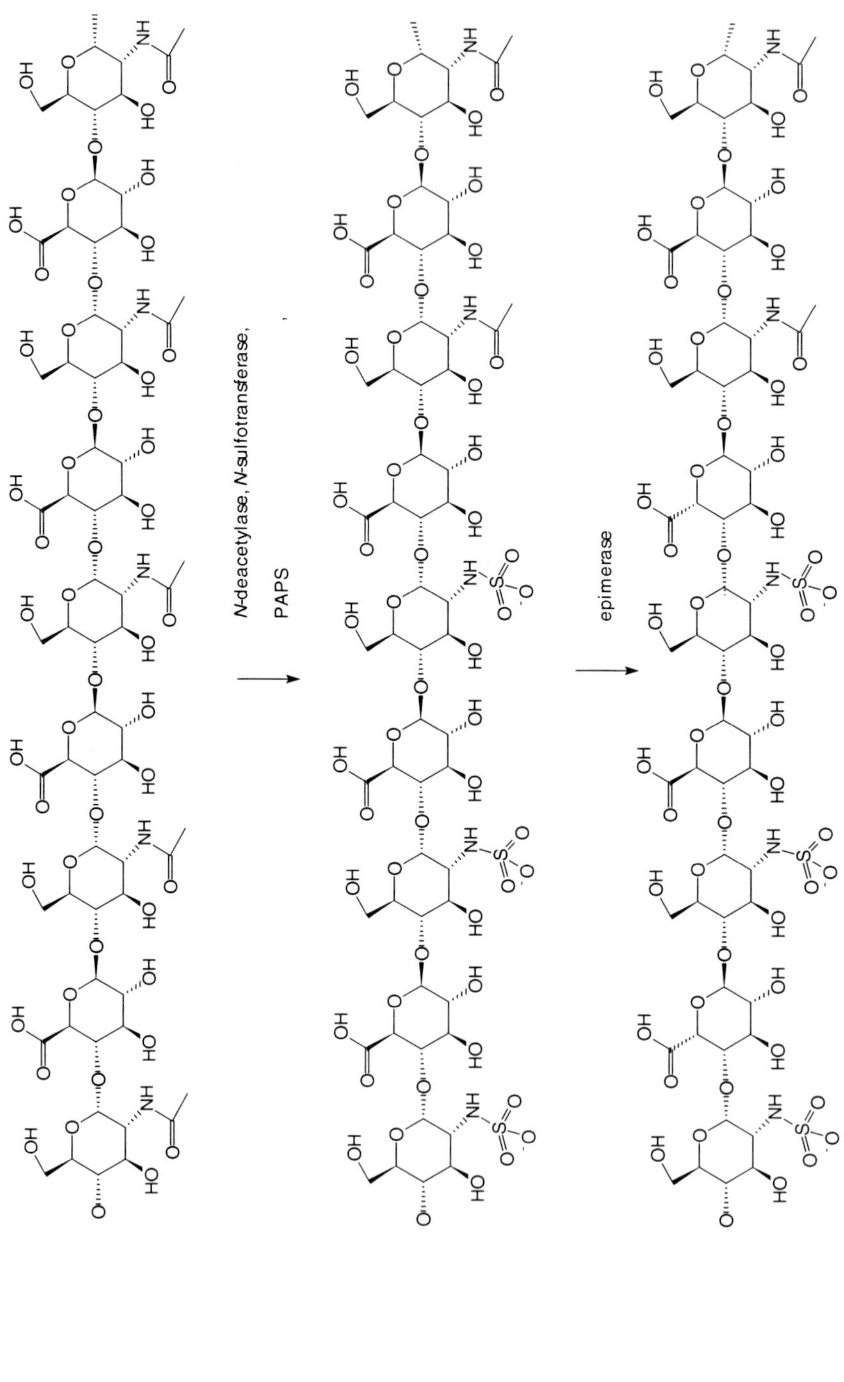

N-deacetylase, *N*-sulfotransferase,
PAPS

epimerase

Figure 4.81 *(Continued)*

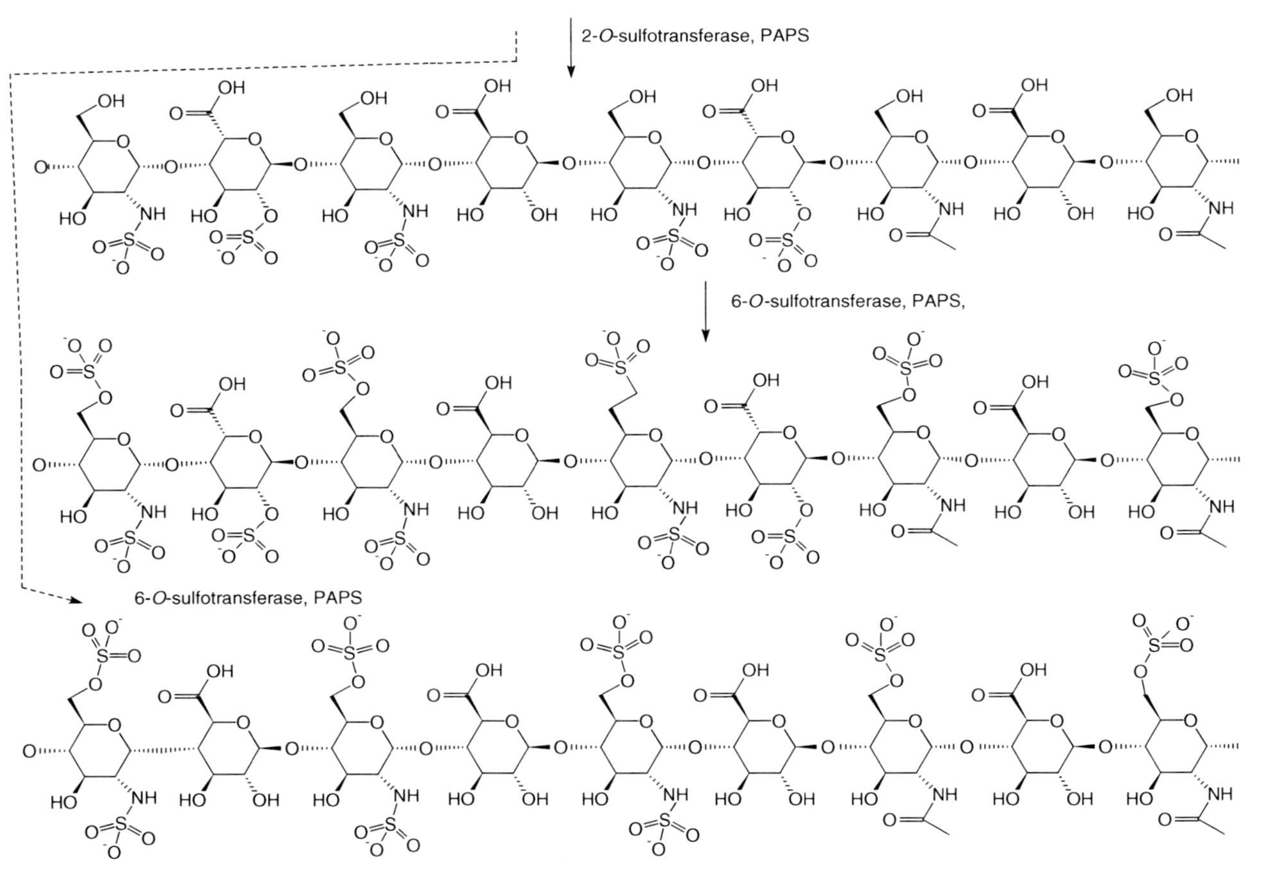

Figure 4.81 Modification of first-biosynthesised heparan. The transformations are schematic, and 3-*O*-sulfation of the GlcNAc residues is also observed.

single polypeptides with separate deacetylase and sulfotransferase domains,[365] which can be expressed and are active independently.[366] The presence of the two activities on a single polypeptide explains why "heparosan", the *N*-deacetylated polymer, does not accumulate, although there must be some escape of polymer chains from the enzyme, since the occasional GlcN residue is encountered, in certain sequences.[367]

(ii) Epimerisation of β-GlcA*p* residues with a GlcNS residue in the −2 subsite.[xxxvi] Attachment of GlcNS to O4 of the βGlcA*p* residue to be epimerised at C5 is a necessary, but not a sufficient, condition for the epimerisation, so that unepimerised GlcA residues occur in α-GlcNS*p*-(1→4)-β-GlcA*p*-(1→4)-α-GlcNAc*p*-(1→4) or α-GlcNS*p*-(1→4)-β-GlcA*p*-(1→4)-α-GlcNS*p*-(1→4) sequences.[368]

(iii) Partial *O*-sulfation on O6 of the GlcNS residues and O2 of the α-L-IdoA*p* residues. Any GlcNS residue can be 6-*O*-sulfated, but 2-*O* sulfation of α-L-IdoA*p* requires a GlcNS or GlcNAc residue, rather than a 6-sulfated-GlcNS residue in the + 1 subsite of the transferase. This suggests that 2-*O*-sulfation of iduronate precedes 6-O-sulfation of GlcNS. Moreover, if epimerisation at C5 of the uronic acid residues were to attain equilibrium, there would be a preponderance of GlcA residues in the polymer, whereas in heparin α-L-IdoA*p* predominate. It is suggested 2-*O*-sulfation after epimerisation is one way in which the equilibrium could be pulled over to iduronate.

(iv) Occasional 3-sulfation of GlcNS or GlcNS6S.

There are quantitative differences in the chemical constitution of heparan sulfate and heparin polysaccharides, but qualitative differences in their biological location and core protein.[332,364] Heparin is biosynthesised only in connective tissue mast cells and attached only to a unique protein core (serglycin). The initial polysaccharide chains have a *DP* (dimer) of 250–400, but can be cleaved subsequently to give an array of heparin polysaccharides. Heparan sulfate, by contrast, can be synthesised on an array of proteins in various cells, with two major subfamilies of heparan sulfate proteoglycans, the syndecans and glypicans, which carry fewer and shorter polysaccharide chains.

Quantitatively, heparin is much more modified than heparan sulfate, with up to 90% of some heparins (*e.g.* bovine lung) consisting of the disaccharide α-L-IdoA*p*-(2S)-(1→4)-α-D-GlcNS*p*-(6S), where "S" indicates sulfate. Heparan sulfate has a block structure, with regions of unmodified GlcNAc and GlcA residues (called "NA domains") coexisting in the same polymer with highly modified, heparin-like stretches ("NS domains") and mixed, lightly modified regions between the two ("NA/NS transition domains").

[xxxvi] Using the polysaccharide enzyme nomenclature (Figure 5.17) first developed for polysaccharide hydrolases, in which the sugar residue at which the reaction takes place is −1, with positive numbers increasing towards the reducing end and negative numbers towards the non-reducing end.

The series of local structures that can be formed by various combinations of the modification reactions is very large [there are 24 possible structures that can be formed by the enzymes described above acting on the single disaccharide sequence β-GlcA*p*-(1→4)-α-GlcNAc*p*-(1→4)] and it is now clear that local sequences are methods of transferring information.[369] Investigation of the ability of various heparin sequences to bind to proteins dates back to 1935, when heparin started to be used to clinically as a blood anticoagulant. The minimal sequence which will have an effect on blood clotting is α-D-GlcN(S or Ac)*p*-(6S)-β-D-GlcA*p*-(1→4)-α-D-GlcNS*p*-(6S,3S)-(1→4)-α-L-IdoA*p*-(2S)-(1→4)-α-D-GlcNS*p*-(6S).[370]

Because of the information content of individual heparin and heparan sulfate sequences, considerable effort has gone into specialised sequencing technologies. The readily available heparinases are lyases, which yield a $\Delta^{4,5}$-unsaturated sugar at the +1 residue, rather than hydrolases, but which have the disadvantage that the C5 stereochemistry of the uronic acid residue is lost. The deamination chemistry of the 2-acetamido and 2-sulfamido groups is also exploited to bring about chain cleavage (Figure 4.82).

On treatment with nitrosating agents such as nitrous acid, amines, RNH_2, yield alkane diazonium ions, $R{-}N^+{\equiv}N$, which are particularly susceptible to nucleophilic displacement. It is sometimes claimed that carbenium ions R^+ are intermediates. However, it is most unlikely that the electron-deficient simple secondary cations that would result from deamination of a hexosamine have a real existence in water. In water, the methanediazonium ion $CH_3{-}N^+{\equiv}N$ has a lifetime of 0.3 s,[371] and synchronous fragmentation of triazenes to R–N=N–NHAr to R^+, N_2 and X^- is not important even when R^+ is the stabilised *p*-methoxybenzyl cation,[372] although it takes place in glacial acetic acid when R is a simple secondary alkyl group.[373] In water, a highly polar solvent, diazonium ions are stabilised with respect to the transition states for their S_N1 decomposition, in which the positive charge is more delocalised.

Reaction of heparin with nitrous acid at pH ≈ 1.5 results in deamination of the *N*-sulfato but not *N*-acetyl groups.[374] The $-NHSO_3H$ groups will be largely in their monoanionic form (the pK_a of sulfamic acid is 0.99[375]) and will form *N*-nitrososulfamates. Simple *N*-alkyl-*N*-nitrososulfamic acids, as their potassium salts, are stable, crystalline materials, which give deamination-type products around pH 1.[376] The decomposition is buffer-catalysed and a reasonable mechanism is either loss of sulfur trioxide in a reaction S_N1 at sulfur or, by analogy with the mechanism for deamination reactions of *N*-nitrosocarboxamides, nucleophilic attack at sulfur (Figure 4.83).

Carbohydrate equatorial 2-diazonium ions undergo a ring contraction, in which the C1–O5 bond, antiperiplanar to the leaving nitrogen, undergoes a 1,2-shift to give a five-membered ring (Figure 4.83). Whether the lone pair electrons on oxygen, in the ground state pointing away from the reaction centre, are involved, is a difficult question. Their participation would result in a discrete oxonium ion as an intermediate and is disfavoured by both the strain and drastic conformation changes involved. The product of the 1,2-shift is commonly written as an acyclic oxocarbenium ion, which adds water to yield a

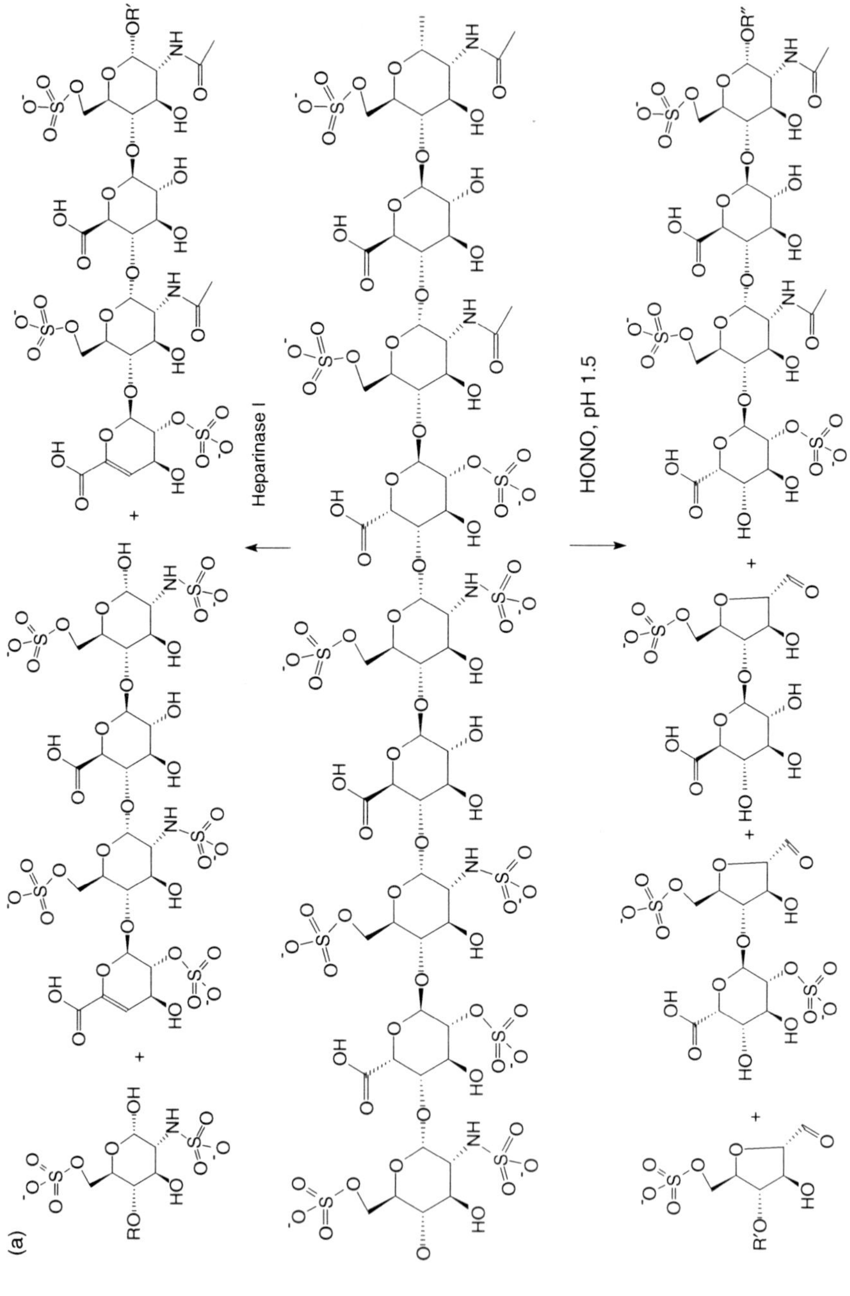

(a)

Heparinase I

HONO, pH 1.5

Figure 4.82 (*Continued*)

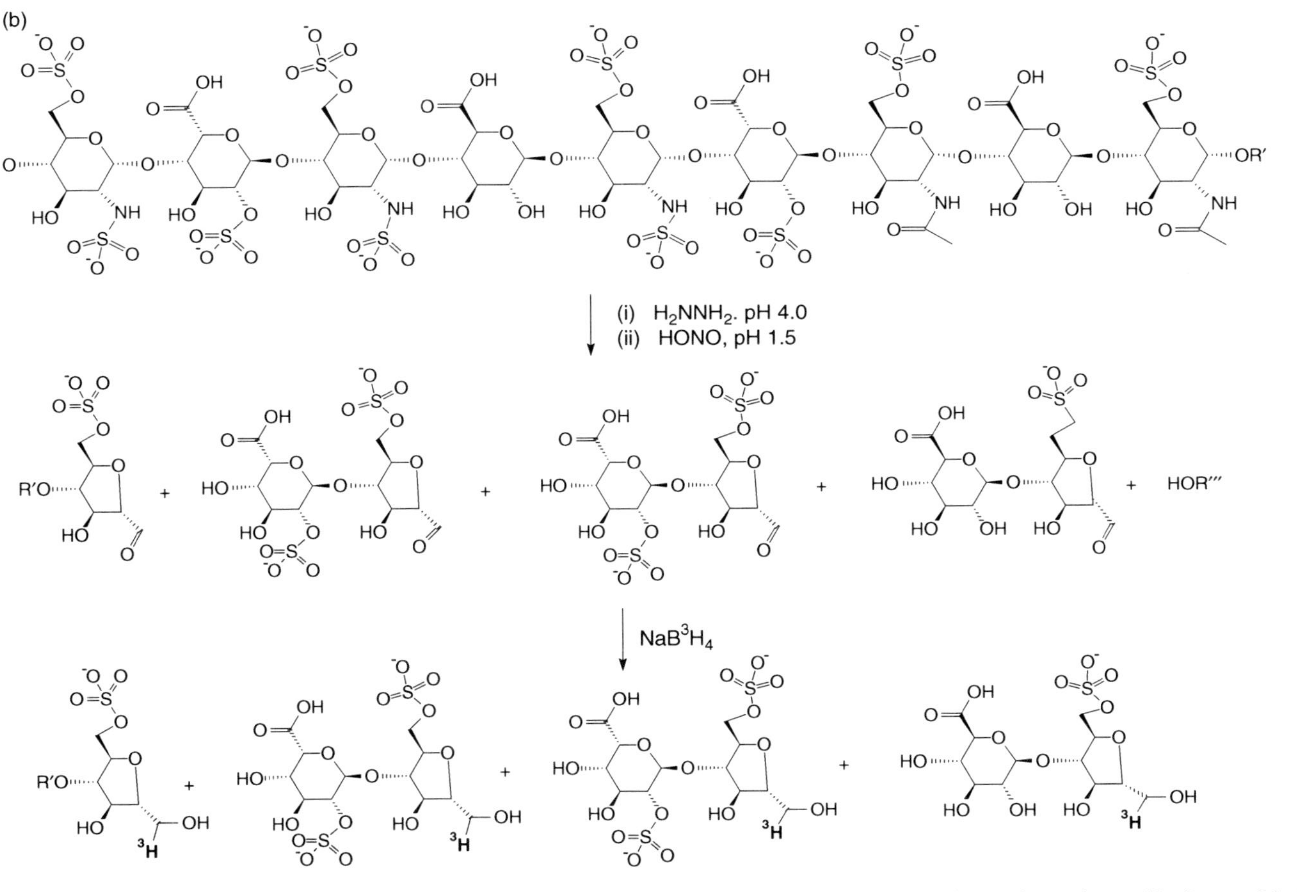

Figure 4.82 (a) Lyase and nitrous acid treatment of a representative fragment of heparin. The heparinase shown is specific for α-L-IdoAp(2S) residues, but other heparinases with broader specificities, which will act on GlcA residues, are known. (b) Complete deamination of a representative fragment of heparin, followed by labelling and reduction.

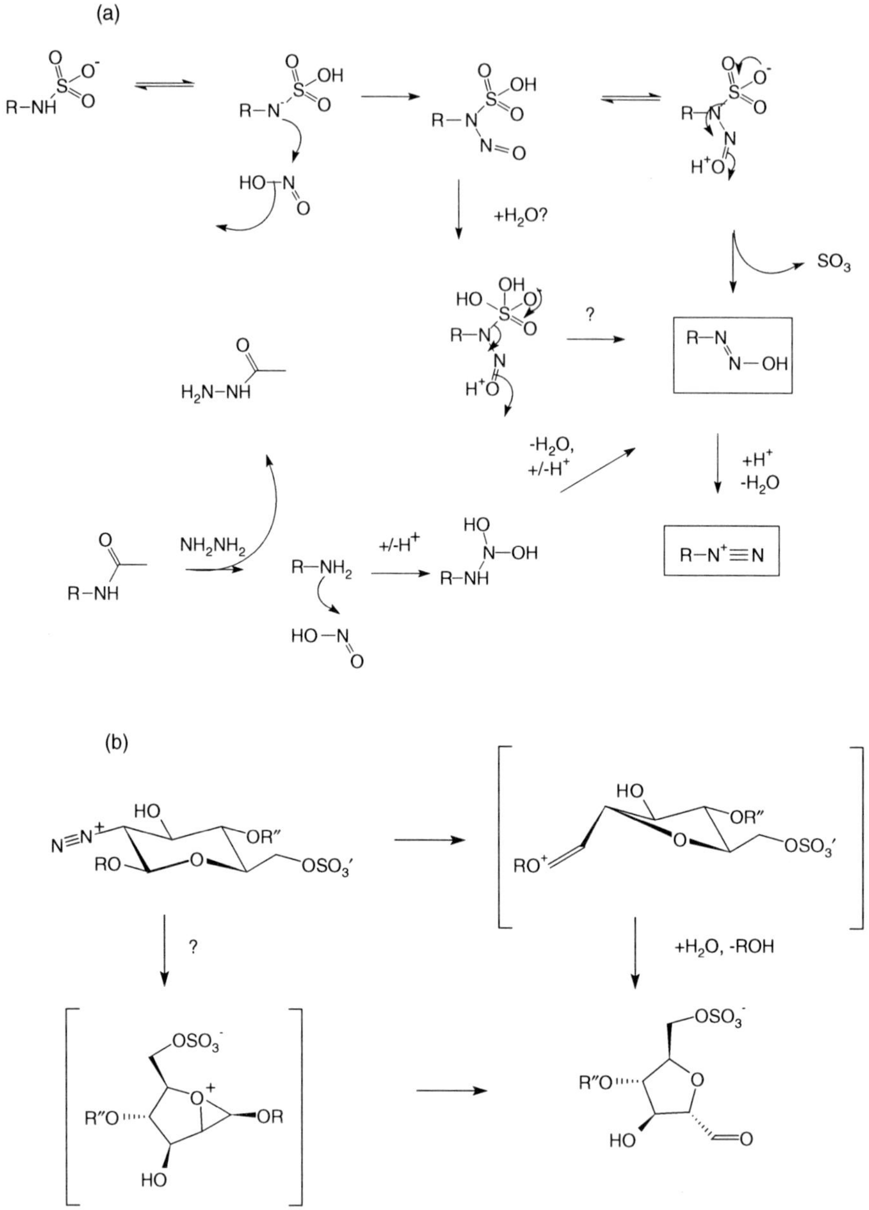

Figure 4.83 Mechanism of deamination of equatorial hexosamines and derivatives. (a) Generation of diazonium ions; (b) bond shift.

hemiacetal and then a new reducing centre, in a 2,5-anhydro sugar. However, the acyclic oxocarbenium ions as written are very probably "too unstable to exist" and there are unlikely to be any solvent-equilibrated intermediates between the diazonium ion and the hemiacetal.

In order to identify the new reducing end, the products from nitrous acid treatment are reduced by NaB^3H_4. Cleavage at both *N*-acetyl and *N*-sulfato sites can be achieved by initial de-*N*-acetylation with hydrazine (at pH 4.0) prior to nitrous acid treatment.[364] The combined treatment yields only disaccharides, but the *O*-sulfation pattern is preserved.

Just as the primary structure of heparin and heparan sulfate has a wealth of fine detail, depending on its exact provenance and function, so the exact conformation of a stretch of polysaccharide depends on its exact location in the chain. The key to this conformational flexibility lies in iduronic acid residues, which can adopt either the $^1C_4$ or the $^2S_0$ conformation (the glucosamine-derived residues are firmly in the $^4C_1$ conformation, as are glucuronic acid residues). Interpretation of vicinal proton–proton coupling constants of IdoA residues in terms of an equilibrium between just $^1C_4$ and $^2S_0$ conformations suggests the equilibrium changes from 60:40 for internal IdoA residues to 40:60 for terminal residues (Figure 4.84(a)).[364,377]

X-ray diffraction patterns from heparan sulfate indicated a helical repeat length of 1.86 nm, consistent with a two-fold helix of the disaccharide unit;[378] similar patterns from crystallised heparin, from a source (bovine lung) in which >90% of the disaccharide residues were the trisulfated α-D-GlcNS*p*(6S)-(1→4)-α-L-IdoA*p*(2S)-(1→4) residue. The 1.65–1.73 nm repeat observed for macromolecular heparin was consistent with a two-fold helical structure with the IdoA in the $^1C_4$ conformation (Figure 4.85).[379] This structure was virtually identical with one model derived from NOE and molecular modelling, but the NMR data was also compatible with a different structure in which the IdoA residue was in the $^2S_0$ conformation.[380] The major difference between these structures lay in the disposition of the sulfates, with the model with $^1C_4$ iduronates having a "three on one side, then three on the other" arrangement and the model with $^2S_0$ iduronates having the sulfates more evenly dispersed. An NMR study of the hexasaccharide $\Delta^{4,5}$β-D-GlcA*p*-(1→4)[xxxvii]-α-D-GlcNS*p*(6S)-(1→4)-α-L-IdoA*p*(2S)-(1→4)-α-D-GlcNS*p*(6S)-(1→4)-α-L-IdoA*p*(2S)-(1→4)-D-GlcNS*p*(6S)-OH [*i.e.* based on three of the standard α-D-GlcNS*p*(6S)-(1→4)-α-L-IdoA*p*(2S)-(1→4) disaccharide units, with the non-reducing end dehydrated by a heparinase] demonstrated that the IdoA units did indeed adopt both $^1C_4$ and $^2S_0$ conformations, although change of the IdoA conformation affected neither helix pitch nor dihedral angles (Figure 4.84(b)).[381] Surprisingly, the degree of *O*-sulfation or *N*-sulfation or acetylation did not alter the equilibrium between the $^1C_4$ and $^2S_0$ conformations of the IdoA units markedly, with the sole exception of the structure α-D-GlcNAc*p*[or α-D-GlcNAc*p*(6S)]-(1→4)-α-L-IdoA*p*(2S), where the IdoA residue was exclusively in the $^1C_4$ conformation.[382]

There are now many X-ray crystal structures of proteins complexed with heparin fragments. Despite the possibility of distortion from the free solution conformation by interaction of the protein, in fact the points for the α-D-GlcNS*p*(6S)-(1→4)-α-L-IdoA*p*(2S) glycosidic link on the pseudo-Ramachandran

---

[xxxvii] According to IUPAC nomenclature, the first residue is an α-L-erythrose derivative.

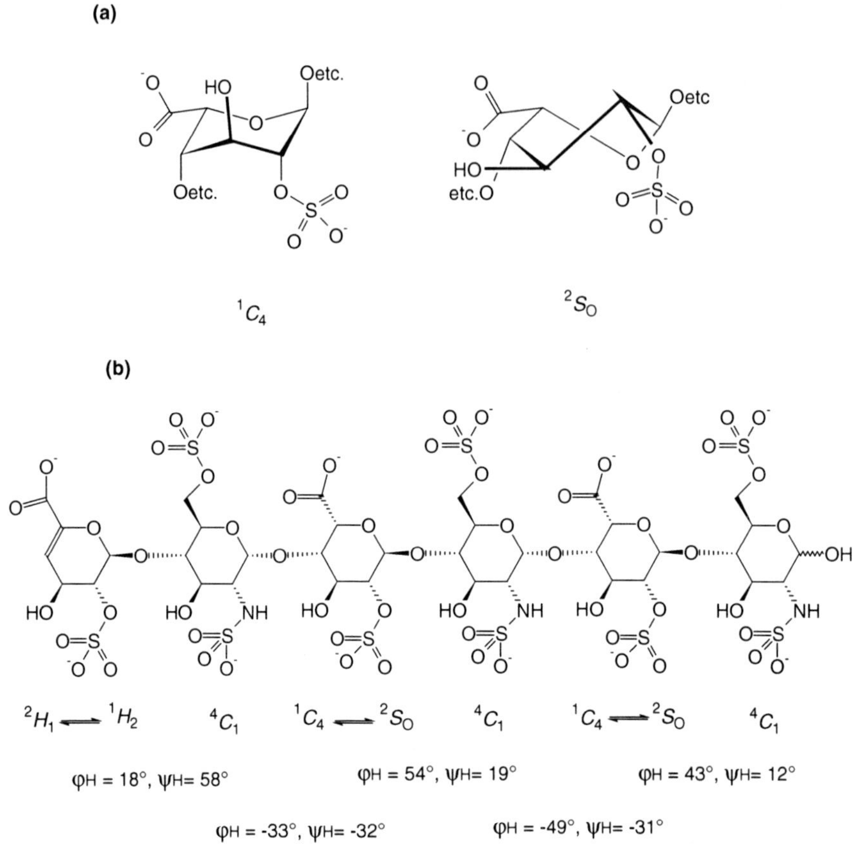

Figure 4.84 (a) ${}^{1}C_4$ and ${}^{2}S_0$ conformation of an α-L-iduronosyl residue. (b) Dihedral angles for $\Delta^{4,5}$β-D-GlcA*p*-(1→4)-α-D-GlcNS*p*(6S)-(1→4)-α-L-IdoA*p*(2S)-(1→4)-α-D-GlcNS*p*(6S)-(1→4)-α-L-IdoA*p*(2S)-(1→4)-D-GlcNS*p*(6S)-OH. Taken from Ref. 381.

(φ and ψ) plot are closely clustered round the free solution values, as are the points for the α-L-IdoA*p*(2S)(1→4)-α-D-GlcNS*p*(6S) glucosidic link (Figure 4.86).[383]

4.6.10.2 Marine Galactans. The galactans isolated from red seaweeds are important commercially but, as could be expected from the difficulties of doing molecular biology on organisms which grow slowly in habitats which are difficult to duplicate in the laboratory, little is known of their biosynthesis.[384] The basic structure is of alternating 3-substituted β-D-glucopyranose and 4-substituted α-galactopyranose units in a chain of about 10^3 monosaccharides. Two families of marine galactans are in use. In the carrageenan type, isolated from Atlantic seaweed genera such as *Chondrus*, *Eucheuma* and *Gigartina* and named after the Irish town of Carragheen, the α-galactose unit is D. In the agar type, isolated from Pacific and Indian ocean genera such as *Gelidium* and

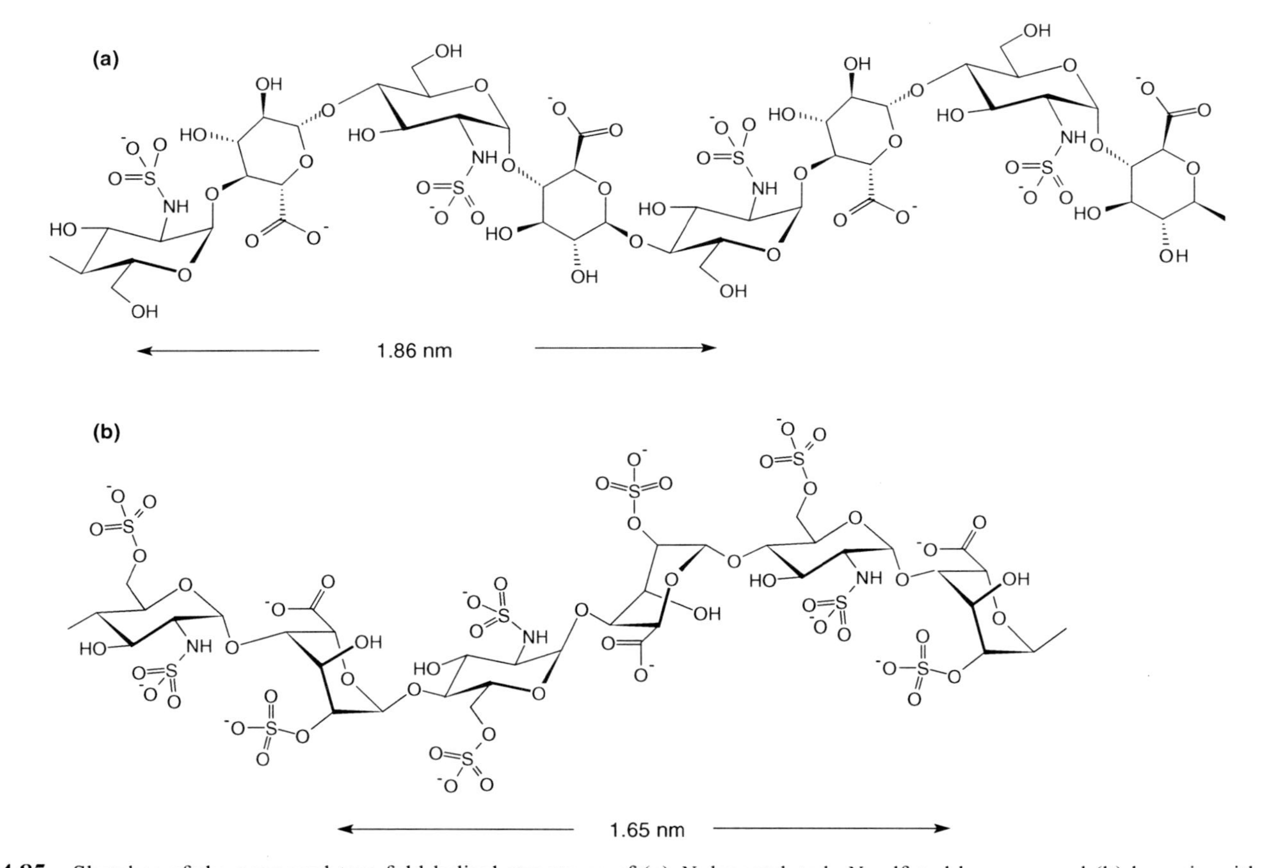

Figure 4.85 Sketches of the proposed two-fold helical structures of (a) *N*-deacetylated, *N*-sulfated heparan and (b) heparin with the IdoA residues in the $^{1}C_{4}$ conformation.

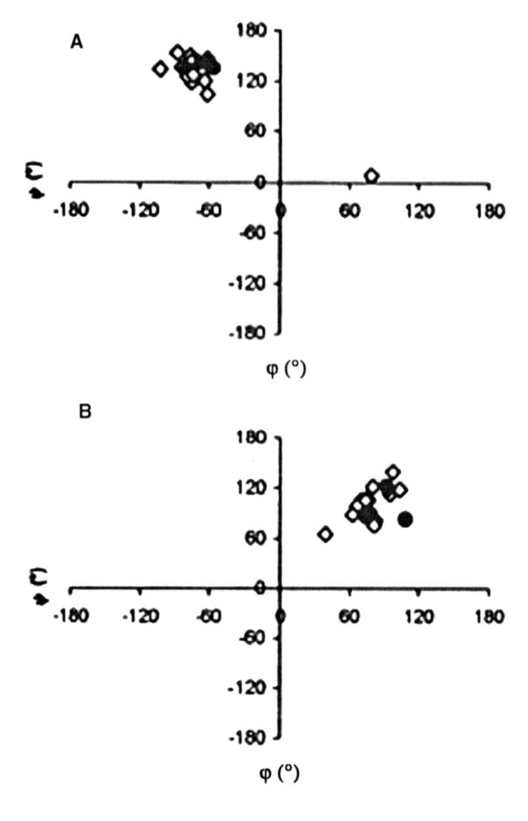

Figure 4.86 Distribution of glycosidic dihedral angles in protein-bound heparin fragments: (a) α-D-GlcNS*p*(6S)-(1→4)-α-L-IdoA*p*(2S); (b) α-L-IdoA*p*(2S)-(1→4)-α-D-GlcNS*p*(6S). We thank Dr. Barbara Mulloy for permission to reproduce this from Ref. 383.

Gracilaria, the α-galactose unit is L. UDP-D-Gal*p* and GDP-L-Gal*p* appear to be the biosynthetic glycosyl donors of the two sugar enantiomers.[384]

The first-formed galactan structure is then sulfated at various positions, presumably by PAPS-dependent enzymes. Sulfation at the 6-position of the α-galactose ring provides a good leaving group at a primary position and this can be displaced by the 3-OH of the same residue, giving a 3,6-anhydro bridge which constrains the α-D-galactose residue to the $^1C_4$ conformation. [The 3,6-anhydro-α-L-galactose is constrained to the mirror-image conformation, which strict adherence to the perversities of carbohydrate nomenclature (see Section 2.4) dictates is described as $^4C_1$ (the conformation without the bridge in the L series being $^1C_4$).] In the case of the agar-type polysaccharides, 3,6-anhydro bridge formation is enzyme catalysed.[386] Methylation of these polymers is also known.[387] The unusual L-galactoside linkage makes agar a valuable gelation agent in microbiology, since few bacteria produce enzymes to hydrolyse it.

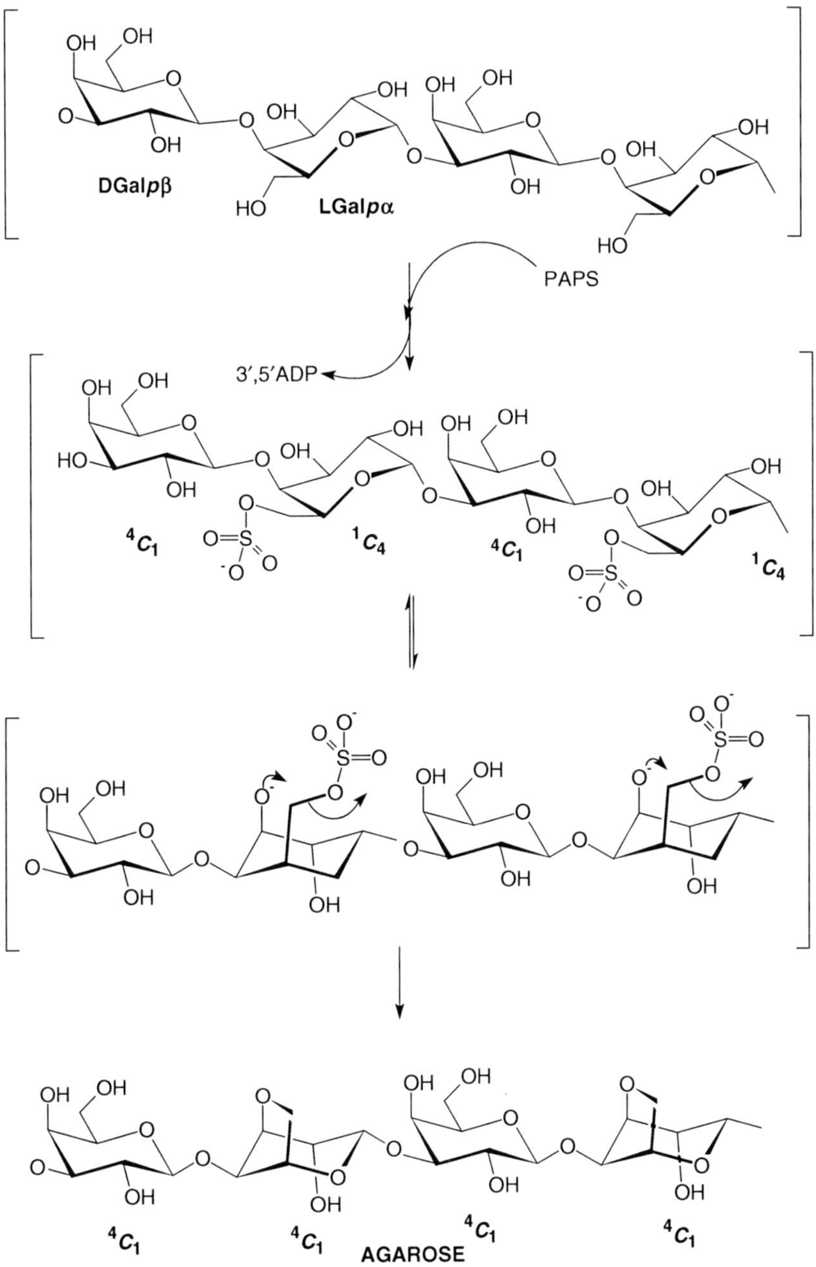

Figure 4.87 Agarose and its probable biosynthesis.

Moreover, the absence of charge makes it an excellent medium in which to run electrophoresis gels of DNA fragments (Figure 4.87).

Agarose powder dissolves in water around 70 °C to give a clear solution, which on cooling forms rigid gels around 45 °C. These gels do not melt again until ~70 °C. The thermal hysteresis involved can be monitored by several physical techniques, most notably chiroptical measurements (Section 4.5.4). Models involving formation of double helices at the junction zones were advanced to explain the hysteresis: gel formation had to involve a loose polymer winding itself round its partner a number of times before cooperative gelation could begin; likewise, melting had to involve an end becoming loose and spontaneously unwinding. The invocation of double helices, however, was permissive rather than compelling, and heavy reliance was placed 30–40 years ago on fibre X-ray studies. However, the diffraction patterns of agarose films produced at near-ambient temperature were poor, and the half-dozen or so splodges on typical X-ray diffraction diagrams of the period merely permitted, rather than compelled, double helical structures. It has since become possible to obtain good diffraction diagrams of agarose, which compel a single helix.[388] Several single helical models are compatible with the (many) diffraction spots, although the contracted conformation with a "fat" helix has values of $\underline{\varphi}$ at variance with the *exo*-anomeric effect [$\sim -125°$ for the (1→4) linkage and $\sim -30°$ for the (1→3) linkage]. The shallow single left-handed helices have values of $\phi = 30-40°$ ($\underline{\varphi} \approx -80$ to $-90°$) and ш $= 0-20°$ ($\underline{\psi} \approx 120-140°$) for the β-(1→4) linkage and $\phi = 45-57°$ ($\underline{\varphi} \approx -63$ to $-75°$) and ш $= -12-10°$ ($\underline{\psi} \approx -108-130°$) for the α-(1→3) linkage.

Detailed considerations of the structure of the carrageenans (Figure 4.88) is often compromised because the polysaccharides have been subjected to harsh extraction conditions, in particular treatment with strong alkali, which will promote formation of the 1,6-anhydro unit from 6-sulfated residues anyway. It does appear, though, that the 3,6-anhydro bridge also occurs naturally in the carrageenans, since enzymes have evolved to hydrolyse the consequent linkage.[389] Carrageenans can also be pyruvylated and methylated. Any particular sample will contain a range of structures and what crystallises during attempts to create fibres suitable for X-ray analysis may not be representative of the majority molecular species.

The structure of the carrageenans is considered in terms of ideal disaccharide repeats, given a Greek letter, which are displayed in Table 4.3.

κ-Carrageenan forms rigid gels, but the polymer chain conformations involved remain a matter of debate, in much the same way as agarose chain conformations. NMR studies in dilute solution[390] suggest that the chain conformation is disordered, with the polymer adopting several related conformations with low energy barriers between them. A persistent local structure – the exchangeable hydrogens of which can be identified in DMSO solution – is that given in Figure 4.89. ι-Carrageenan forms semi-liquid gels and again the polysaccharide can be crystallised in several forms. It appears, however, that gel formation does not arise from multiple helices, but from single chains cross-linked with cations. Many polymorphic forms of ι-carrageenan are formed, depending on the cation.[391] The crystal structure of the sodium salt has been determined twice,[392,393] with, unlike agarose, the modern structure agreeing with the 1970s structure: the chains were packed in three-fold right-handed parallel double helices, which were themselves packed three to a unit cell, in the sense ↑↑↓, with,

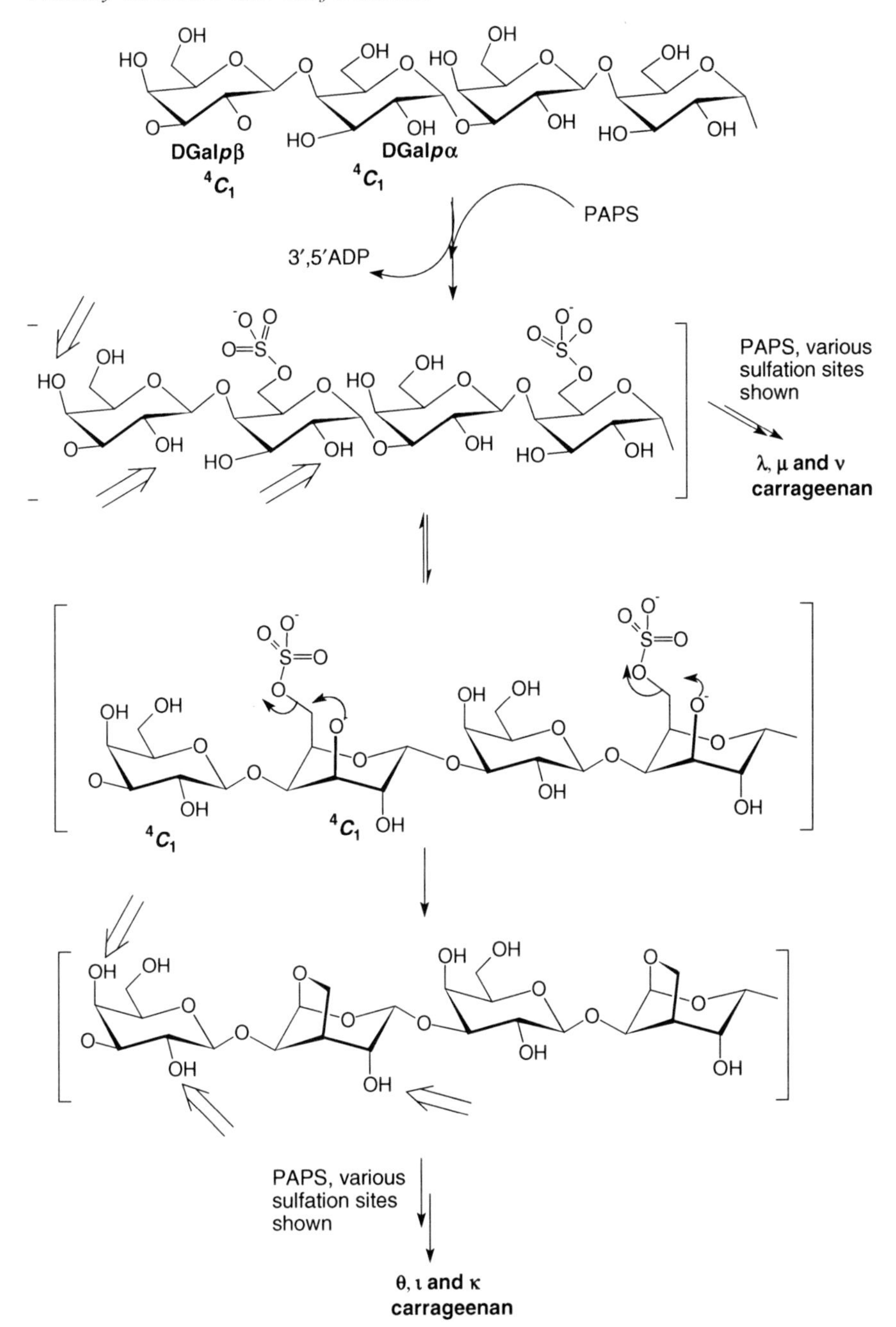

Figure 4.88 Structures and probable biosynthesis of the carrageenans.

Table 4.3 Structures of carrageenans.

Carrageenan type	*Sulfation positions of 3-substituted β-Gal*	*Sulfation positions of 4-substituted α-Gal*	*3,6-Anhydro ring in 4-substituted α-Gal?*
θ (theta)	2	2	Yes
ι (iota)	4	2	Yes
κ (kappa)	4	None	Yes
λ (lambda)	2	2,6	No
μ (mu)	4	6	No
ν (nu)	4	2,6	No

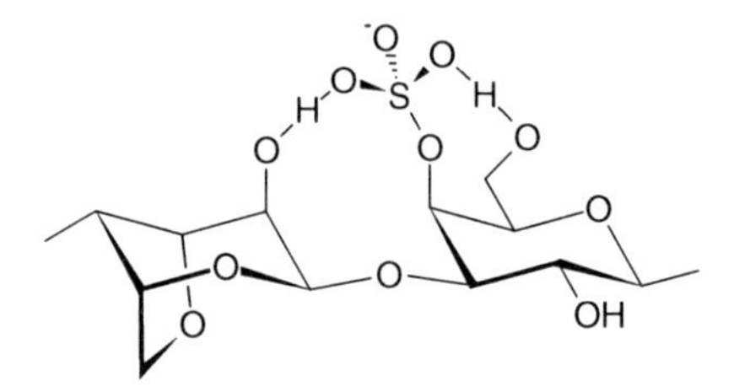

Figure 4.89 Persistent hydrogen bonds in κ-carrageenan.

as expected, sulfate groups on the outside of the helices. The dihedral angles differed little between salts across the β-(1→4) link: $\varphi = \Phi = -88°$, $\Psi = 107°$ ($\psi \approx -133°$); across the α-(1→4) link they were more variable: for the Na^+ salt, $\varphi = \Phi = 60°$, $\Psi = 77°$; for the Ca^{2+}salt, $\varphi = \Phi = 75°$, $\Psi = 99°$ ($\psi \approx -141°$). The hydroxymethyl group adopts the *gt* conformation about C5–C6. λ-Carrageenan does not form gels and is disordered.

4.6.10.3 Industrially and Commercially Important Bacterial exo-*Polysaccharides.* Two families of bacterial polysaccharides are of interest to various industries because of their rheological properties. The xanthan family possesses a β-(1→4)-glucan main chain, with every other glucose substituted with an oligosaccharide branch,[394] whereas the gellan family is based around a tetrasaccharide main chain repeat, β-D-Glc*p*-(1→4)-β-D-GlcA*p*-(1→4)-β-D-Glc*p*-(1→4)-α-L-Rha*p*-(1→3), with various branches and occasionally conservative substitutions in the main chain, such as α-L-Man*p* for α-L-Rha*p*.[394]

4.6.10.3.1 Xanthan family. Xanthan gum is produced by the bacterial plant pathogen *Xanthomonas campestris*, and appears to be associated with its pathogenicity. It has a very high native MW (4×10^6) and is of interest because of its enormous range of shear thinning (Figure 4.90) and its rapid and complete recovery after shearing.[396] Initial use was in secondary oil recovery, since mud slurries in its dilute solutions did not sediment, because of its high viscosity at low shear, but could, because of the low viscosity at high shear, be forced into rock formations. It is now used in a range of industrial and domestic products where stability of suspensions and absence of thixotropy are important.

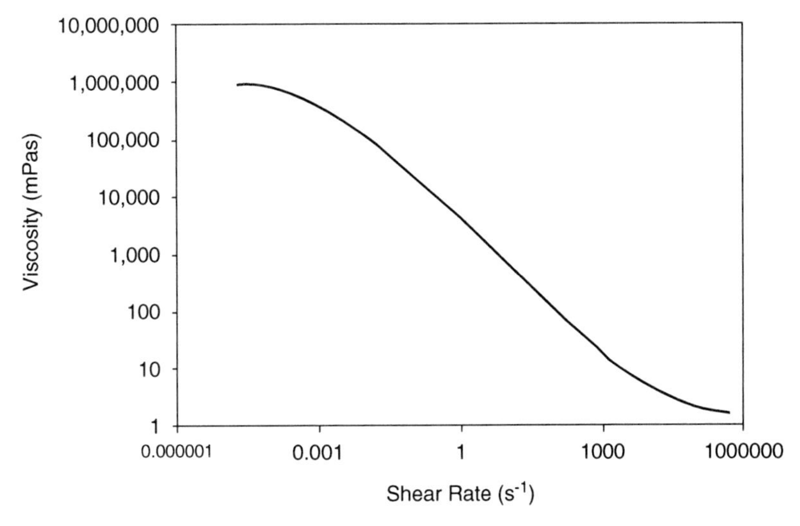

Figure 4.90 Flow curve of 0.5% xanthan solution. Reproduced, with permission, from Ref. 396.

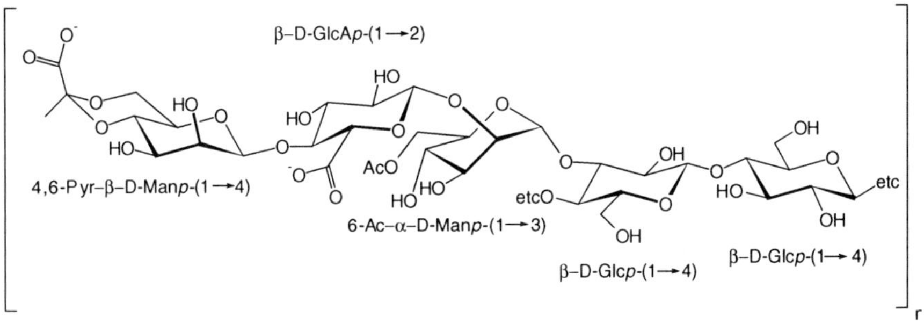

Figure 4.91 Structure of xanthan gum.

Xanthan has the pentasaccharide repeat shown in Figure 4.91,[397] which is first assembled as a polyprenyl pyrophosphate moiety; the lipid is established as polyprenol, rather than the reduced polyprenol dolichol (see Figure 5.2b) by the liberation of neutral sugar by catalytic reduction, followed by alkaline phosphatase treatment (allylic pyrophosphate esters being hydrogenolysed off).[398] The side-chain sugars are transferred from GDPMan and UDPGlcA and the internal mannose acetylated by acetyl-CoA.[399] The acetylation of the internal mannose occurs after the transfer of GlcA.[400] The pyruvyl moiety is then added from phosphoenol pyruvate[401] by a transferase which has been identified.[402] The assembly of the pentasaccharide can be disrupted by a mutation in the gene which transfers GlcA from UDP GlcA to the 2-OH of the inner mannose, but the polymerisation reaction to a main chain still occurs, producing a cellulose-type polymer with an α-(1→3)-mannose branch on every other glucose residue.[403] Normally, the polymerisation requires the whole repeat, including pyruvate decoration, to be assembled, after which the polysaccharide is assembled from the reducing end.[404] Sometimes the terminal pyruvate is replaced by a 6-*O*-acetyl group.[394,397]

The introduction of the pyruvate moiety creates a new asymmetric centre (at C2 of the pyruvic acid). From NMR chemical shift data in model compounds, it was possible to deduce that in xanthan gum, as in other polysaccharides where O4 and O6 of glucose or mannose formed part of a six-membered pyruvate ketal ring, the stereochemistry was *S*, whereas in polysaccharides in which the pyruvate acetal bridged O4 and O6 of galactose, the stereochemistry was *R*.[405] In both cases the methyl group is equatorial (Figure 4.92(a)). (Occasionally five-membered pyruvate acetals, as on O2 and O3 of galactose, are encountered in Nature: in such cases, without a clear distinction between axial and equatorial substituents, NOE enhancements have to be relied on).[406] The mechanism of the pyruvylating enzyme is not known, but that in Figure 4.92(b), differing from the established mechanism of carboxyvinyl transferases, which transfer the $CH_2{=}C{-}COOH$ group from phosphoenol pyruvate to alcohols,[407] only in the last step, which is an addition rather than an elimination, seems plausible.

Native xylan gum polymer exists in an ordered conformation which is broken on heating and a random coil formed: on coiling another ordered conformation, different to the native, is formed. The native form is a single helix with a persistence length estimated at 45 nm from solution measurements: heating in dilute solution causes the irreversible change to a double-helical form with a persistence length of 125 nm.[408] AFM measurements of arrays of single molecules indicated a number-average contour length of 1651 nm for the single helix, which decreased to 450 nm in the presence of salt, which induced the single to double helix transition.[409] Persistence lengths for the double helical form, estimated by AFM on the assumption that the molecules on a mica surface were in thermodynamic equilibrium with their surroundings, were between 141 and 417 nm: clearly, any additional stiffness in the polymer caused by its adsorption on mica will be reflected in a higher persistence length than in free solution.

A structurally similar polysaccharide is acetan, made by *Acetobacter xylinum,* in which the substituted glucose of the cellulose main chain is 6-*O*-acetylated and the terminal acetylated or pyruvylated mannose of the side-chain is replaced the trisaccharide α-L-Rha*p*-(1→6)-β-D-Glc*p*-(1→6)-α-D-Glc*p*-(1→4) (Figure 4.93).[410] The biosynthetic pathway for the two polysaccharides is very similar[411] and, as with xanthan, the elaboration of the oligosaccharide repeat can be arrested by deletion of the gene for a transferase which adds a side-chain sugar. In the case of acetan, deletion of *aceP*, coding for a β-(1→6)-glucosyl transferase, produced a polysaccharide with a pentasaccharide repeat.[394] Fibre X-ray structures suggest that single 5-fold helical structure of the two polysaccharides is similar.[412]

4.6.10.3.2 Gellan family. *Sphingomonas elodea*[xxxviii] produces an acylated version of a polysaccharide,[413] which forms weak gels: on deacylation it forms

[xxxviii] Known previously as *Auromonas elodea*.

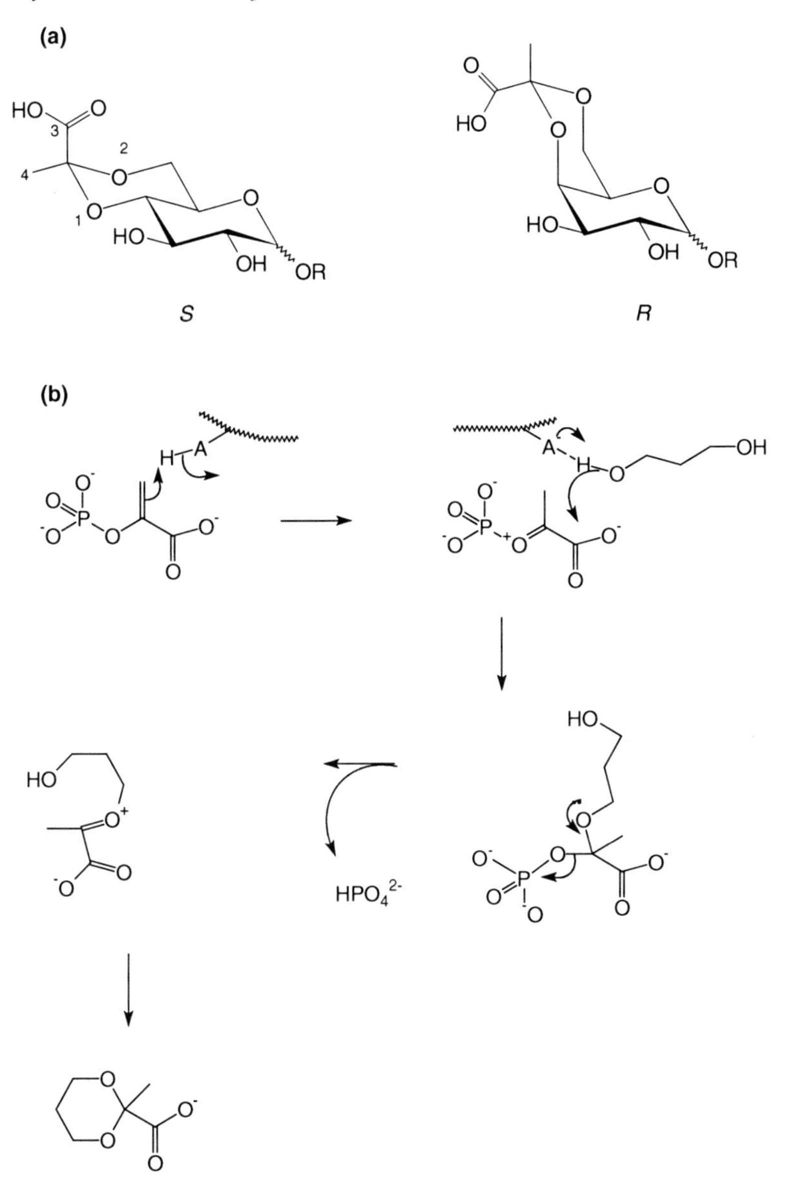

Figure 4.92 (a) Stereochemistry of six-membered pyruvate acetals; in both cases the methyl group is equatorial. The priorities of the groups attached to pyruvate C2 are indicated. (b) Possible mechanism of the pyruvylation reaction.

stiff, brittle gels. The deacylated material can be induced to crystallise in oriented fibres.[414] The primary structure is based on a tetrasaccharide repeat, [3-β-D-Glc*p*-(1→4)-β-D-GlcA*p*-(1→4)-β-D-Glc*p*-(1→4)-α-L-Rha*p*-(1→)]$_n$ with the first glucose residue (residue A) partly acetylated on position 6 and fully

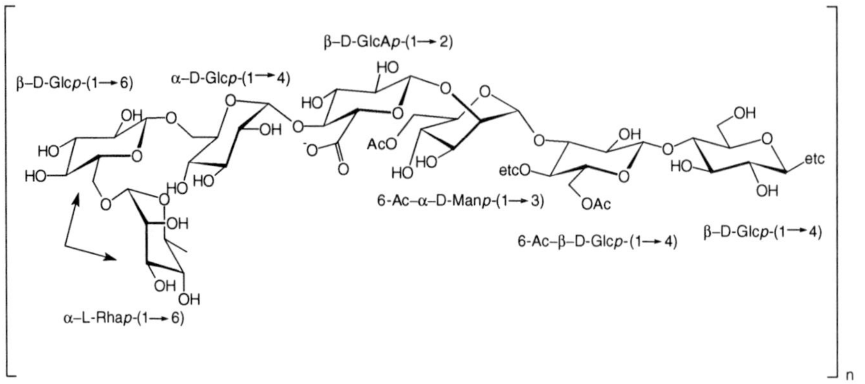

β-D-GlcAp-(1→2)
β-D-Glcp-(1→6) α-D-Glcp-(1→4)
6-Ac-α-D-Manp-(1→3)
6-Ac-β-D-Glcp-(1→4) β-D-Glcp-(1→4)
α-L-Rhap-(1→6)

Figure 4.93 Structure of acetan. Genetically modified polysaccharide, in which the two arrowed residues are missing, can be obtained by deleting a glycosyltransferase.

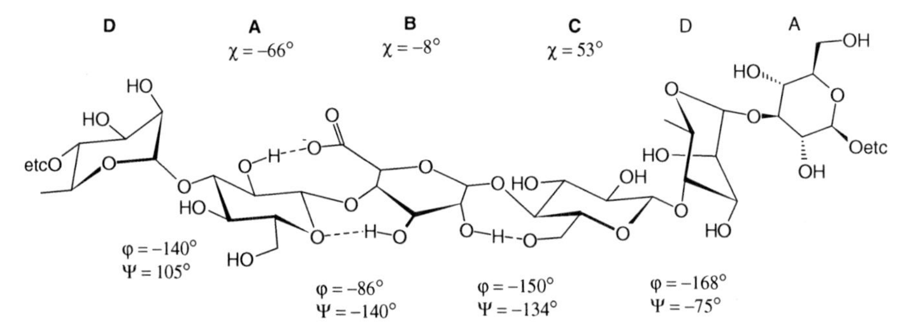

Figure 4.94 Half a pitch (six residues) of a single gellan chain helix in its parallel double-helical conformation.

esterified with L-glyceric acid on position 2.[415] The structure adopted by the chain in the deacylated material is a three-fold helix (*i.e.* with 12 monosaccharide residues per turn) with a pitch of 5.64 nm; all the monosaccharide residues are in the $^4C_1$ conformation. Two parallel helices intertwine, with the carboxylates acting as interhelix hydrogen-bond acceptors. The O5 (Glc) ··· HO3 (GlcA) hydrogen bond between residues A and B (Figure 4.94) is standard for 1→4-linked diequatorial pyranosides such as cellulose and xylan and the $O2_n$–H ··· O6 hydrogen bond of cellulose I is paralleled by a similar one between residues A and B (with the carboxylate of GlcA as acceptor) and an identical one between residues B and C, which puts the hydroxymethyl group of residue C close to the *tg* conformation.

Related polysaccharides include welan, with an α-L-Man*p* or α-L-Rha*f* side chain on the second glucose O3 residue C and rhamsan, with αβ-D-Glc*p*-(1→6)-α-D-Glc*p*-(1→6) side-chain on residue A, and other polysaccharides merely given numbers by their discoverers. The variant of rhamsan with a

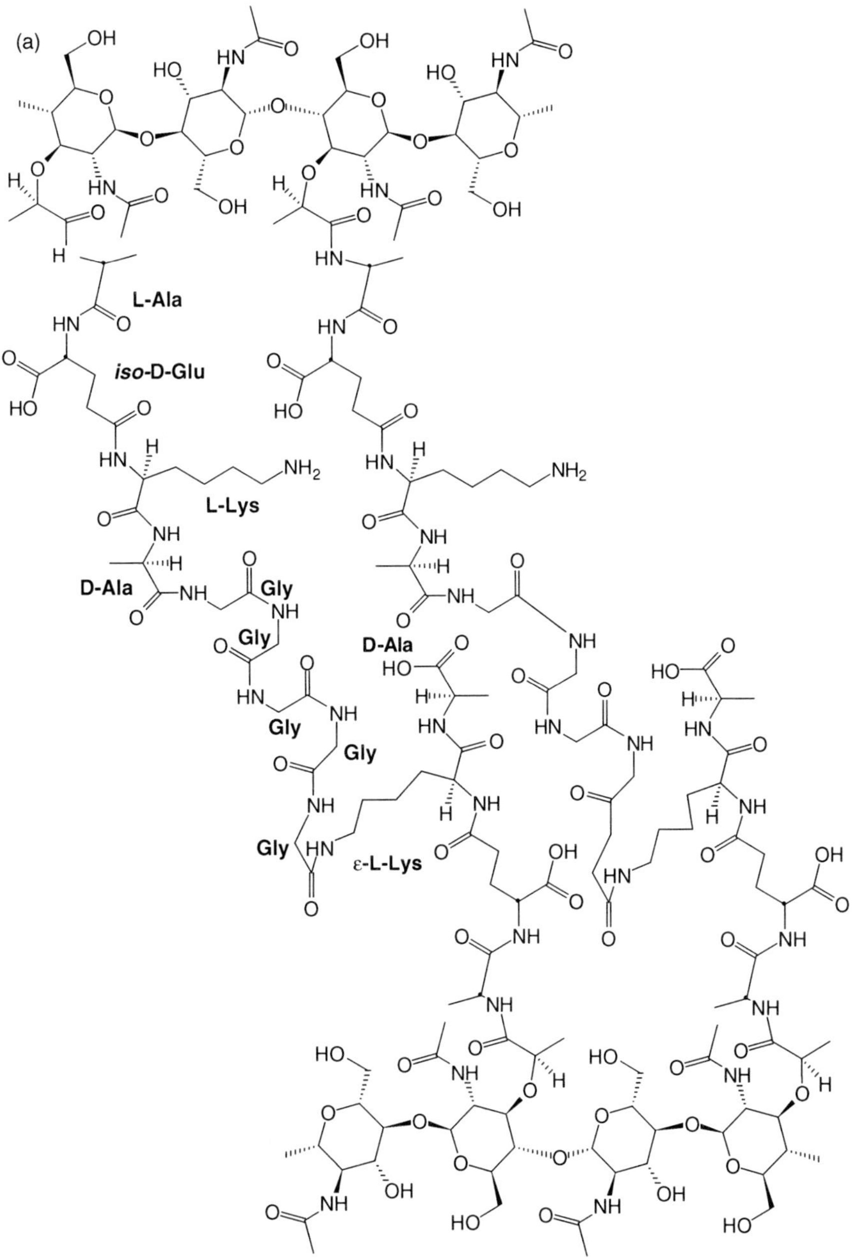

Figure 4.95 Peptidoglycan of (a) Gram-positive and (b) Gram-negative bacteria.

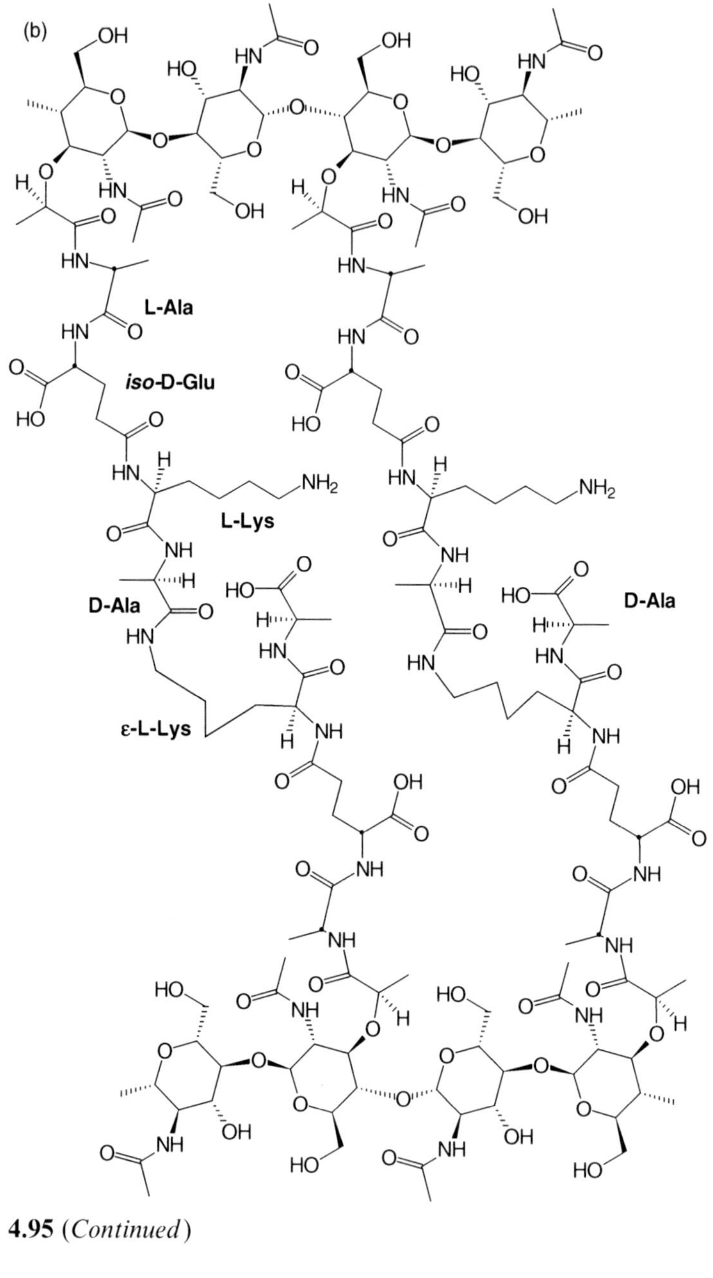

Figure 4.95 (*Continued*)

2-deoxy-GlcA*p* unit as residue B of the main chain, RMDP17, forms similar helices to gellan, with the substituents on the outside of the helices – this may well be general for this family of polysaccharides,[395] in which there is continuing interest, because of the possibility of "tweaking" the polymer properties for better food technological properties.[416]

4.6.10.4 Bacterial Cell Wall Peptidoglycans and Related Material. Polysaccharides form important components of the large macromolecules which form the cell walls of bacteria. As the chemistry of the polysaccharide components cannot be considered in isolation, and coverage of molecular interactions in the bacterial cell wall would require a large text in itself, the area will only be covered only sketchily: there are excellent texts available.[417]

Bacteria are traditionally divided into those which can be stained by the Gram stain ("Gram positive") and those which are not. The chromophoric ingredient in Gram stain is the very stable delocalised carbenium ion of Crystal Violet, $(Me_2NpC_6H_4)_3C^+$. Bacteria whose outer surface consists a thick layer of peptidoglycan bind this chromophoric cation because of the charge-transfer interactions that are possible, whereas Gram-negative bacteria, which present glycolipids on their outer surface, are not stained. Gram-negative bacteria do, however, also have a thin layer of peptidoglycan inside the glycolipid.

Both types of peptidoglycans (Figure 4.95) are forms of decorated chitin, in which O3 of every other GlcNAc residue carries a D-lactic acid ether side-chain, which in turn forms amide links to a pentapeptide. The D-3-lactyl GlcNAc residue is termed *N*-acetylmuramic acid and sometimes abbreviated NAM (GlcNAc is then abbreviated NAG). The carboxylate also forms a peptide bond to L-Ala, the initial amino acid of the pentapeptide peptide which forms part of the cross-link; other residues in sequence are D-Glu, connected in an *iso* fashion, so that the α-carboxylate remains free, L-Lys and D-Ala. In Gram-positive bacteria, the cross-link is a pentapeptide of five glycine units from the D-Ala to the ε-amino group of the Lys of another decorated glycan; in Gram-negative bacteria, the link is made directly.

Gram-positive bacteria also have additional anionic polymers (secondary cell-wall polymers, SCWPs), which are covalently attached to the muramic acid residues of the peptidoglycan and attach cell-surface proteins. They can be of the teichoic acid type, linear copolymers of a sugar alcohol and phosphoric acid, or the teichuronic acid type, in which the anionic component is uronic acids.[418] The commonest teichoic acids are based on 1,3-linked glycerol or 1,5-linked ribitol.

REFERENCES

1. An excellent set of practical protocols in use today is given by R. Stenitz, P.-E. Jansson and G. Widmalm, *A Practical Guide to the Structural Analysis of Carbohydrate*, http://www.casper.organ.su.se.
2. H. Björndal, C. G. Hellerquist, B. Lindberg and S. Svensson, *Angew. Chem. Int. Ed.*, 1970, **9**, 610.
3. S. Hakomori, *J. Biochem. (Tokyo)*, 1964, **55**, 205.
4. A. N. Round, A. J. MacDougall, S. G. Ring and V. J. Morris, *Carbohydr. Res.*, 1997, **303**, 251.
5. A. P. Gunning, T. P. Giardina, C. B. Faulds, N. Juge, S. G. Ring, G. Williamson and V. J. Morris, *Carbohydr. Polym.*, 2003, **51**, 177.
6. A. Dell, *Adv. Carbohydr. Chem. Biochem.*, 1987, **45**, 19.

7. A. Reis, M. A. Coimbra, P. Domingues, A. J. Ferrer-Correia and M. R. M. Domingues, *Carbohydr. Polym.*, 2004, **55**, 401.
8. D. J. Harvey, *Int. J. Mass. Spectrom.*, 2003, **226**, 1.
9. B. Domon and C. E. Costello, *Glycoconjugate J.*, 1988, **5**, 397.
10. T. Imai, T. Watanabe, T. Yui and J. Sugiyama, *Biochem. J.*, 2003, **374**, 755.
11. E. D. T. Atkins and W. Mackie, *Biopolymers*, 1972, **11**, 1685.
12. M. Koyama, W. Helbert, T. Imai, J. Sugiyama and B. Henrissat, *Proc. Natl. Acad. Sci. USA*, 1997, **94**, 9091.
13. E.-L. Hult, F. Katouno, T. Uchiyama, T. Watanabe and J. Sugiyama, *Biochem. J.*, 2005, **388**, 851.
14. (a) I. A. Nieduszynski, in *Polysaccharides – Topics in Structure and Morphology*, ed. E. D. T. Atkins, Macmillan, Basingstoke, 1985, p. 107; (b) D. H. Isaac, in *Polysaccharides – Topics in Structure and Morphology*, ed. E. D. T. Atkins, Macmillan, Basingstoke, 1985, p. 141.
15. L. P. Harding, V. M. Marshall, Y. Hernandez, Y. Gu, M. Maqsood, N. McLay and A. P. Laws, *Carbohydr. Res.*, 2005, **340**, 1107.
16. A. Bax, M. F. Summers, W. Egan, N. Guirgis, R. Schneerson, J. B. Robbins, F. Ørskov, I. Ørskov and W. F. Vann, *Carbohydr. Res.*, 1988, **173**, 53.
17. F.-P. Tsui, W. Egan, M. F. Summers, R. A. Byrd, R. Schneerson and J. B. Robbins, *Carbohydr. Res.*, 1988, **173**, 65.
18. J. Breg, D. Romijn, J. F. G. Vliengthart, G. Strecker and J. Montreuil, *Carbohydr. Res.*, 1988, **183**, 19.
19. A. R. Kirby, A. P. Gunning and V. J. Morris, *Biopolymers*, 1996, **38**, 355.
20. T. M. McIntire and D. A. Brant, *Biopolymers*, 1997, **42**, 133.
21. T. E. Fischer, P. E. Marszalek, A. F. Oberhauser, M. Carrion-Vazquez and J. M. Fernandez, *J. Physiol.*, 1999, **520**, 5.
22. Z. Lu, W. Nowak, G. Lee, P. E. Marszalek and W. Yang, *J. Am. Chem. Soc.*, 2004, **126**, 9033.
23. A. A. Baker, W. Helbert, J. Sugiyama and M. J. Miles, *J. Struct. Biol.*, 1997, **119**, 129.
24. A. A. Baker, W. Helbert, J. Sugiyama and M. J. Miles, *Biophys. J.*, 2000, **79**, 1139.
25. M. K. Cowman, M. Li and E. A. Balazs, *Biophys. J.*, 1998, **75**, 2030.
26. A. K. Vuppu, A. A. Garcia and C. Vernia, *Biopolymers*, 1997, **42**, 89.
27. M. Rief, F. Oesterhelt, B. Heymann and H. E. Gaub, *Science*, 1997, **275**, 1295.
28. P. E. Marszalek, A. F. Oberhauser, Y.-P. Pang and J. M. Fernandez, *Nature*, 1998, **396**, 661.
29. P. E. Marszalek, Y.-P. Pang, H. Li, J. El Yazal, A. F. Oberhauser and J. M. Fernandez, *Proc. Natl. Acad. Sci. USA*, 1999, **96**, 7894.
30. P. E. Marszalek, H. Li, A. F. Oberhauser and J. M. Fernandez, *Proc. Natl. Acad. Sci. USA*, 2002, **99**, 4278.
31. http://www.chem.qmul.ac.uk/iupac/misc/psac.html; *Biochemical Nomenclature and Related Documents*, Portland Press London, 2nd edn, 1992, pp. 177–179.

32. Discussed by C. A. Stortz and A. S. Cerezo, *Carbohydr. Res.*, 2003, **338**, 1679.
33. Y. Nishiyama, P. Langan and H. Chanzy, *J. Am. Chem. Soc.*, 2002, **124**, 9074.
34. Y. Nishiyama, J. Sugiyama, H. Chanzy and P. Langan, *J. Am. Chem. Soc.*, 2003, **125**, 14300.
35. P. Langan, Y. Nishiyama and H. Chanzy, *J. Am. Chem. Soc.*, 1999, **121**, 9940.
36. G. L. Strati, J. L. Willett and F. A. Momany, *Carbohydr. Res.*, 2002, **337**, 1833.
37. F. Cloran, I. Carmichael and A. S. Serianni, *J. Am. Chem. Soc.*, 1999, **121**, 9843.
38. M. L. Hayes, A. S. Serianni and R. Barker, *Carbohydr. Res.*, 1982, **100**, 87.
39. P. R. Sundarajan and V. S. R. Rao, *Biopolymers*, 1969, **8**, 305.
40. S. A. Foord and E. D. T. Atkins, *Int. J. Biol. Macromol.*, 1980, **2**, 193.
41. R. A. Jockusch, R. T. Kroemer, F. O. Talbot, L. C. Snoek, P. Çarçabal, J. P. Simons, M. Havenith, J. M. Bakker, I. Compagnon, G. Meijer and G. von Helden, *J. Am. Chem. Soc.*, 2004, **126**, 5709.
42. See any standard text of protein structure, *e.g.* C. I. Brandén and J. R. Tooze, *Introduction to Protein Structure,* Garland Publishing, New York, 1991, p. 8.
43. H. F. Azurmendi, M. Martin-Pastor and C. A. Bush, *Biopolymers*, 2002, **63**, 89.
44. R. U. Lemieux, A. A. Pavia, J. C. Martin and K. Watanabe, *Can. J. Chem.*, 1969, **47**, 4427.
45. P. G. Goekjian, T.-C. Wu and Y. Kishi, *J. Org. Chem.*, 1991, **56**, 6412.
46. P. G. Goekjian, T.-C. Wu, H.-Y. Kang and Y. Kishi, *J. Org. Chem.*, 1991, **56**, 6422.
47. Y. Wang, P. G. Goekjian, D. M. Ryckman, W. H. Miller, S. A. Babirad and Y. Kishi, *J. Org. Chem.*, 1992, **57**, 482.
48. T. Haneda, P. G. Goekjian, S. H. Kim and Y. Kishi, *J. Org. Chem.*, 1992, **57**, 490.
49. J.-B. Fournier, *Continuum Mech. Thermodyn.*, 2002, **14**, 241.
50. M. Rinaudo, *Polym. Bull.*, 1992, **27**, 585.
51. G. S. Manning, *J. Phys. Chem.*, 1981, **85**, 870.
52. S. Paoletti, R. Gilli, L. Navarini and V. Crescenzi, *Glycoconjugate J.*, 1997, **14**, 513.
53. M. Rinaudo, *Biomacromolecules*, 2004, **5**, 1155.
54. M. Rinaudo, *Food Hydrocolloids*, 2001, **15**, 433.
55. M. J. Deery, E. Stimson and C. G. Chappell, *Rapid Commun. Mass Spectrom.*, 2001, **15**, 2273.
56. B. F. Wood, A. H. Conner and C. G. Hill, *J. Appl. Polym. Sci.*, 1986, **32**, 3703.
57. K. M. Kleman-Leyer, N. R. Gilkes, R. C. Miller and T. K. Kirk, *Biochem. J.*, 1994, **302**, 463.

58. S. F. Sun, *Physical Chemistry of Macromolecules*, Wiley, New York, 1994, p. 129.
59. G. Robinson, S. B. Ross-Murphy and E. R. Morris, *Carbohydr. Res.*, 1982, **107**, 17.
60. J. W. Goodwin, in *Industrial Water-Soluble Polymers*, ed. C. A. Finch, Royal Society of Chemistry, Cambridge, 1996, p. 28.
61. See. *e.g.*, http://www.wyatt.com/solutions/software/analysis_a2.cfm.
62. See, *e.g.*, http://www.malvern.co.uk.
63. M. P. Coughlan, *Biochem. Soc. Trans.*, 1985, **13**, 405.
64. A. Payen, *C. R. Acad. Sci.*, 1838, **7**, 1052.
65. H. Friese and K. Hess, *Liebigs Ann.*, 1927, **456**, 38.
66. W. N. Haworth and H. Machemer, *J. Chem. Soc.*, 1932, 2270.
67. W. J. Astbury and M. M. Davies, *Nature*, 1944, **154**, 84.
68. E. Sjöström, *Wood Chemistry*, Academic Press, San Diego, 2nd edn, 1993.
69. N. Gralén and T. Svedberg, *Nature*, 1943, **152**, 625.
70. K. Saalwächter, W. Burchard, P. Klüfers, G. Kettenbach, P. Mayer, D. Klemm and S. Dugarmaa, *Macromolecules*, 2000, **33**, 4094.
71. L. Schulz, B. Seger and W. Burchard, *Macromolecular Chem. Phys.*, 2000, **201**, 2008.
72. T. Röder, A. Potthast, T. Rosenau, T. Kosmsa, T. Baldinger, B. Morgenstern and O. Glatter, *Macromol. Symp.*, 2002, **190**, 151.
73. R. P. Swatloski, S. K. Spear, J. D. Holbrey and R. D. Rogers, *J. Am. Chem. Soc.*, 2002, **124**, 4974.
74. D. G. Stevenson, A. Biswas, J. Jane and G. E. Inglett, *Carbohydr. Polym.*, 2007, **67**, 21.
75. K. Schlufter, H. P. Schmauder, S. Dorn and T. Heinze, *Macromol. Rapid Commun.*, 2006, **27**, 1670.
76. H. Zhang, J. Wu, J. Zhang and J. S. He, *Macromolecules*, 2005, **38**, 8272.
77. J. S. Moulthrop, R. P. Swatloski, G. Moyna and R. D. Rogers, *Chem. Commun.*, 2005, 1557.
78. R. C. Remsing, R. P. Swatloski, R. D. Rogers and G. Moyna, *Chem. Commun.*, 2006, 1271.
79. S. J. Eichhorn, C. A. Baillie, N. Zafeiropoulos, L. Y. Mwaikambo, M. P. Ansell, A. Dufresne, K. M. Entwistle, P. J. Herrera-Franco, G. C. Escamilla, L. Groom, M. Hughes, C. Hill, T. G. Rials and P. M. Wild, *J. Mater. Sci.*, 2001, **36**, 2107.
80. (a) R. H. Atallah, in *Trichoderma reesei Cellulases and Other Hydrolases*, ed. T. Reinikainen and P. Suominen, Akateeminen Kirjakauppa, Helsinki, 1993, p. 25; (b) R. H. Attalla and D. L. VanderHart, *Solid State Nucl. Magn. Reson.* 1999, **15**, 1.
81. (a) A. Sarko and R. Muggli, *Macromolecules*, 1974, **7**, 486; (b) K. H. Gardner and J. Blackwell, *Biopolymers*, 1974, **13**, 1975.
82. H. Chanzy, B. Henrissat, H. Vuong and M. Schülein, *FEBS Lett.*, 1983, **153**, 113.
83. R. H. Atalla and D. H. VanderHart, *Science*, 1984, **223**, 283.
84. J. Sugiyama, R. Vuong and H. Chanzy, *Macromolecules*, 1991, **24**, 4168.

85. J. Sugiyama, J. Persson and H. Chanzy, *Macromolecules*, 1991, **24**, 2461.
86. M. Koyama, W. Helbert, T. Imai, J. Sugiyama and B. Henrissat, *Proc. Natl. Acad. Sci. USA*, 1997, **94**, 9091.
87. H. Kono, S. Yunoki, T. Shikano, M. Fujiwara, T. Erata and M. Kawai, *J. Am. Chem. Soc.*, 2002, **124**, 7506.
88. M. M. Kreevoy and T. M. Liang, *J. Am. Chem. Soc.*, 1980, **102**, 3315.
89. L. Peng, Y. Kawagoe, P. Hogan and D. Delmer, *Science*, 2002, **295**, 147.
90. D. R. Lane, A. Wiedemeier, L. Peng, H. Höfte, S. Vernhettes, T. Desprez, C.H. Hocart, R. J. Birch, T. I. Baskin, J. E. Burn, T. Arioli, A. S. Betzner and R. E. Williamson, *Plant Physiol.*, 2001, **126**, 278.
91. M. S. Doblin, I. Kurek, D. J. Wilk and D. P. Delmer, *Plant Cell. Physiol.*, 2002, **43**, 1407.
92. J. Lai-Kee-Him, H. Chanzy, M. Müller, J.-L. Putaux, T. Imai and V. Bulone, *J. Biol. Chem.*, 2002, **277**, 36931.
93. H. Shibezaki, M. Saito, S. Kuga and T. Okano, *Cellulose*, 1998, **5**, 165.
94. S. Kobayashi, L. J. Hobson, J. Sakamoto, S. Kimura, J. Sugiyama, T. Imai and T. Itoh, *Biomacromolecules*, 2000, **1**, 168.
95. E. Dinand, M. Vignon, H. Chanzy and I. Heaux, *Cellulose*, 2002, **9**, 7.
96. (a) R. J. Marhofer, S. Reising and J. Brickmann, *Ber. Bunsen Ges. Phys. Chem.*, 1996, **100**, 1350; (b) L. M. Kroon-Batenburg, B. Bouma and B. J. Kroon, *Macromolecules*, 1996, **29**, 5695.
97. M. Wada, H. Chanzy, Y. Nishiyama and P. Langan, *Macromolecules* 2004, 37, 8548. This paper reverts to the IUPAC definition for φ and ψ, as in Ref. 35, but continues to use upper-case Φ and Ψ, as Refs. 33 and 34 do for their non-IUPAC definitions.
98. N.-H. Kim, T. Imai, M. Wada and J. Sugiyama, *Biomacromolecules*, 2006, **7**, 274.
99. C. Yu, A. M. Lee, B. L. Bassler and S. Roseman, *J. Biol. Chem.*, 1991, **266**, 24260.
100. G. A. F. Roberts, *Chitin Chemistry*, Macmillan, London, 1992.
101. K. H. Gardner and J. Blackwell, *Biopolymers*, 1975, **14**, 1581.
102. E. Kamst, J. Bakkers, N. E. M. Quaedvlieg, J. Pilling, J. W. Kijne, B. J. J. Lugtenberg and H. P. Spaink, *Biochemistry*, 1999, **38**, 4045.
103. J. Sugiyama, C. Boisset, M. Hashimoto and T. Watanabe, *J. Mol. Biol.*, 1999, **286**, 247.
104. R. Minke and J. Blackwell, *J. Mol. Biol.*, 1978, **120**, 167.
105. A. L. Svitil, S. M. Ní Chadhain, J. A. Moore and D. L. Kirchman, *Appl. Environ. Microbiol.*, 1997, **63**, 408.
106. L. Chamoy, M. Nicolaï, J. Ravaux, B. Quennedy, F. Gaill and J. Delachambre, *J. Biol. Chem.*, 2001, **276**, 8051.
107. (a) J. Ruiz-Herrera and S. Bartnicki-Garcia, *Science*, 1974, **186**, 357; (b) J. Ruiz-Herrera, V. O. Sing, W. J. Van der Woude and S. Bartnicki-Garcia, *Proc. Natl. Acad. Sci. USA*, 1975, **72**, 2706.
108. Y. Saito, T. Okano, H. Chanzy and J. Sugiyama, *J. Struct. Biol.*, 1995, **114**, 218.

109. T. Yui, K. Imada, K. Okuyama, Y. Obata, K. Suzuki and K. Ogawa, *Macromolecules*, 1994, **27**, 7601.
110. T. Fukamizo, T. Ohkawa, K. Sonoda, H. Toyoda, T. Nishiguchi, S. Ouchi and S. Goto, *Biosci. Biotech. Biochem.*, 1992, **56**, 1632.
111. J. Saito, A. Kita, Y. Higuchi, Y. Nagata, A. Ando and K. Miki, *J. Biol. Chem.*, 1999, **274**, 30818.
112. M. Laleg and I. Pikilik, *Nordic Pulp Paper Res. J.*, 1992, **7**, 174.
113. K. Ogawa and S. Inukai, *Carbohydr. Res.*, 1987, **160**, 425.
114. (a) K. C. B. Wilkie, *Adv. Carbohydr. Chem. Biochem.*, 1979, **36**, 215; (b) K. C. B. Wilkie, *ChemTech*, 1983, **13**, 306; (c) R. C. Sun, X. F. Sun and I. Tomkinson, *ACS Symp. Ser.*, 2004, **864**, 2.
115. E. D. T. Atkins, in *Xylans and Xylanases*, ed. J. Visser, G. Beldman, M. A. Kusters-van Someren and A.G.J.Voragen, *Progress in Biotechnology*, Vol. 7, Elsevier, Amsterdam, 1992, p. 39.
116. L. Lo Leggio, J. Jenkins, G. W. Harris and R. W. Pickersgill, *Proteins Struct. Funct. Genet.*, 2000, **41**, 362.
117. G. Dervilly-Pinel, J.-F. Thibault and L. Saunier, *Carbohydr. Res.*, 2001, **330**, 365.
118. T. Yiu, K. Imada, N. Shibuya and K. Ogawa, *Biosci. Biotech. Biochem.*, 1995, **59**, 965.
119. Recent example (spruce) with modern sequencing techniques: T. Hannuksela and C. H. du Penhoat, *Carbohydr. Res.*, 2004, **339**, 301.
120. A. Teleman, M. Nordström, M. Tenkanen, A. Jacob and O. Dahlman, *Carbohydr. Res.*, 2003, **338**, 525.
121. H. Chanzy, S. Pérez, D. P. Miller, G. Paradoni and W. T. Winter, *Mancromolecules*, 1987, **20**, 2407.
122. T. Yui, K. Ogawa and A. Sarko, *Carbohydr. Res.*, 1992, **229**, 41.
123. (a) R. Chandrasekaran, A. Radha and K. Okuyama, *Carbohydr. Res.*, 1998, **306**, 243; this fitted the X-ray intensities to a structure with β-galactosyl residues and was corrected in; (b) R. Chandrasekaran, W. Bian and W. Okuyama, *Carbohydr. Res.*, 1998, **312**, 219.
124. H. Brumer, Q. Zhou, M. J. Baumann, K. Carlsson and T. T. Teeri, *J. Am. Chem. Soc.*, 2004, **126**, 5715.
125. S. C. Fry, W. S. York, P. Albersheim, A. Darvill, T. Hayashi, J.-P. Joseleau, Y. Kato, E. P. Lorences, G. A. MacLachlan, M. McNeil, A. J. Mort, J. S. G. Reid, H. U. Seitz, R. R. Selvendran, A. G. J. Voragen and A. R. White, *Phys. Plant.*, 1993, **89**, 1.
126. D. Ieiri, T. Kawamura, M. Mimura, H. Urakawa, Y. Yuguchi and K. Kajiwara, *Sen-I Gakkaishi*, 2003, **59**, 93.
127. C. Picard, J. Gruza, C. Derouet, C. M. G. C. Renard, K. Mazeau, J. Koca, A. Imberty and C. Hervé du Penhoat, *Biopolymers*, 2000, **54**, 11.
128. U. Remminghorst and B. H. A. Rehm, *Appl. Environ. Microbiol.*, 2006, **72**, 298.
129. H. Grasdalen, B. Larsen and O. Smidsrød, *Carbohydr. Res.*, 1981, **89**, 179.
130. B. I. G. Svanem, G. Skjåk-Bræk, H. Ertesvåg and S. Valla, *J. Bacteriol*, 1999, **181**, 68.

131. D. R. J. Palmer, B. K. Hubbard and J. A. Gerlt, *Biochemistry*, 1998, **37**, 14350.
132. P. Gacesa, *FEBS Lett.*, 1987, **212**, 199.
133. M. E. Tanner and G. L. Kenyon, in *Comprehensive Biological Catalysis*, ed. M. L. Sinnott, Academic Press, London, 1998, Vol. II, p. 7.
134. A. Haug, B. Larsen and O. Smidsrød, *Carbohydr. Res.*, 1974, **32**, 217.
135. V. Sherbrock-Cox, N.J. Russell and P. Gacesa, *Carbohydr. Res.*, 1984, **135**, 147.
136. E.D.T. Atkins, I.A. Nieduszynski, W. Mackie, K.D. Parker and E.E. Smolko, *Biopolymers*, 1973, **12**, 1865.
137. W. Mackie, *Biochem. J.*, 1971, **125**, 89P.
138. I. Braccini, R. P. Grasso and S. Pérez, *Carbohydr. Res.*, **317**, 119.
139. E.D.T. Atkins, I.A. Nieduszynski, W. Mackie, K.D. Parker and E.E. Smolko, *Biopolymers*, 1973, **12**, 1879.
140. W. Mackie, S. Perez, P. Rizzo, F. Taravel and M. Vignon, *Int. J. Biol. Macromol.*, 1983, **5**, 329.
141. R. Kohn, *Pure Appl. Chem.*, 1975, **42**, 371.
142. H. Zhang, H. Zheng, Q. Zhang, J. Wang and M. Konno, *Biopolymers*, 1998, **46**, 395.
143. G. T. Grant, E. R. Morris, D. A. Rees and P. J. C. Smith, *FEBS Lett.*, 1973, **32**, 195.
144. E. R. Morris, D. A. Rees, D. Thom and J. Boyd, *Carbohydr. Res.*, 1978, **66**, 145.
145. I. Brancini and S. Pérez, *Biomacromolecules*, 2001, **2**, 1089.
146. O. Smidsrød, *Faraday Discuss. Chem. Soc.*, 1975, **57**, 263.
147. C. K. Siew, P. A. Williams and N. W. G. Young, *Biomacromolecules*, 2005, **6**, 963.
148. H. Grasdalen, B. Larsen and O. Smidsrød, *Carbohydr. Res.*, 1981, **89**, 179.
149. N. Emmerichs, J. Wingender, H. -C. Flemming and C. Mayer, *Int. J. Biol. Macromol.*, 2004, **34**, 73.
150. I. Donati, S. Holtan, Y. A. Mørch, M. Borgogna, M. Dentini and G. Skjåk-Bræk, *Biomacromolecules*, 2005, **6**, 1031.
151. A series of a papers largely in *Svensk Papperstidning – Nordisk Cellulosa* with T. E. Timell as corresponding author: H. R Schreude, W. A. Cote and T. E. Timell, 1966, **69**, 641; C. M. Kuo and T. E. Timell, 1969, **72**, 703; K. J. Jiang and T. E. Timell, 1972, **75**, 592; see also H. Meier, *Acta Chem. Scand.*, 1962, 16, 2275.
152. M. S. Buckeridge and J. S. G. Reid, *Planta*, 1994, **192**, 502.
153. S. W. A. Hinz, R. Verhoef, H. A. Schols, J.-P. Vincken and A. G. J. Voragen, *Carbohydr. Res.*, 2005, **340**, 2135.
154. N. Geshi, B. Jorgenson, H. V. Scheller and P. Ulvskov, *Planta*, 2000, **210**, 622.
155. C. Martin and A. M. Smith, *Plant Cell*, 1995, **7**, 971.
156. D. J. Gallant, B. Bouchet, A. Buléon and S. Pérez, *Eur. J. Clin. Nutr.*, 1992, **46**, S3.
157. M. Yamaguchi, K. Kainuma and D. French, *J. Ultrastruct. Res.*, 1979, **69**, 249.

158. K. C. Huber and J. N. BeMiller, *Cereal Chem.*, 1997, **74**, 537.
159. P. J. Jenkins and A. M. Donald, *Int. J. Biol. Macromol.*, 1995, **17**, 315.
160. J. G. Sargeant, *Starch/Stärke*, 1982, **34**, 89.
161. W. Banks, C. T. Greenwood and D. D. Muir, *Stärke*, 1974, **26**, 289.
162. I. Hanashiro and Y. Takeda, *Carbohydr. Res.*, 1998, **306**, 421.
163. E. M. M. Del Valle, *Process Biochem.*, 2004, **39**, 1033.
164. S. H. D. Hulleman, W. Helbert and H. Chanzy, *Int. J. Biol. Macromol.*, 1996, **18**, 115.
165. K.R. Morgan, R. H. Furneaux and N. G. Larsen, *Carbohydr. Res.*, 1995, **276**, 387.
166. M. Noltemeyer and W. Saenger, *J. Am. Chem. Soc.*, 1980, **102**, 2710.
167. H. Sugiyama, T. Nitta, M. Horii, K. Motohashi, J. Sakai, T. Usui, K. Hisamichi and J. Ishiyama, *Carbohydr. Res.* 2000, **325**, 177. These authors appear to confuse psi (ψ) with the alternative form of lower-case phi (φ) and tabulate values of ø and "φ".
168. F. Goldsmith, S. Sprang and R. Fletterick, *J. Mol. Biol.*, 1982, **156**, 411.
169. K. Gessler, I. Usón, T. Takaha, N. Krauss, S. M. Smith, S. Okada, G. M. Sheldrick and W. Saenger, *Proc. Natl. Acad. Sci. USA*, 1999, **96**, 4246.
170. M. L. Bender, R. L. Van Etten and G. A. Clowes, *J. Am. Chem. Soc.*, 1966, **88**, 2319.
171. L. Barr, C. J. Easton, K. Lee and S. F. Lincoln, *Org. Biomol. Chem.*, 2005, **3**, 2990.
172. F. Ortega-Caballero, J. Bjerre, S. S. Laustsen and M. Bols, *J. Org. Chem.*, 2005, **70**, 7217.
173. W. P. Jencks, *Adv. Enzymol.*, 1975, **43**, 219.
174. (a) V. T. D'Souza, *Supramol. Chem.*, 2003, **15**, 221; (b) E. E. Karakhanov, A. L. Maksimov, E. A. Runova, Y. S. Kardasheva, M. V. Terenina, T. S. Buchneva and A. Ya. Guchkova, *Macromol. Symp.*, 2003, **204**, 159.
175. Z. Dong, J. Liu, S. Mao, X. Huang, B. Yang, X. Ren and G. Luo, *J. Am. Chem. Soc.*, 2004, **126**, 16395.
176. (a) Z. Nikuni, *Starch/Stärke*, 1978, **30**, 105; his cluster model was originally reported in the Japanese-language *Science of Cookery*, 1969, **2**, 6; (b) D. French, *Denpun Kagaku J. Jpn. Soc. Starch Sci.*, 1972, **19**, 8.
177. (a) H. F. Zobel, *Starch/Stärke*, 1988, **40**, 44; (b) T. Aberle, W. Burchard, W. Vorwerg and S. Radosta, *Starch/Stärke*, 1994, **46**, 329; (c) Y.-H. Chang, J.-H. Lin and S.-Y. Chang, *Food Hydrocolloids*, 2006, **20**, 332; (d) S. Mali, L. B. Karam, L. P. Ramos and M. V. E. Grossmann, *J. Agric. Food. Chem.*, 2004, **52**, 7720.
178. P. J. Jenkins and A. M. Donald, *Int. J. Biol. Macromol.*, 1995, **17**, 315.
179. A. Sarko and H.-C. H. Wu, *Starch/Stärke*, 1978, **30**, 73.
180. M. D. Alonso, J. Lomako, W. M. Lomako and W. J. Whelan, *FEBS Lett.*, 1994, **342**, 38.
181. Literature to ~1988 reviewed by D. J. Manners, *Carbohydr. Polym.*, 1989, **11**, 87.
182. S.-H. Yun and N. K. Matheson, *Carbohydr. Res.*, 1993, **243**, 307.
183. J. Lomako, M. Lomako and W. J. Whelan, *Carbohydr. Res.*, 1992, **227**, 331.

184. C. Braun, T. Lindhorst, N. Madsen and S.G. Withers, *Biochemistry*, 1996, **35**, 5458.
185. T. E. Nelson, E. Kolb and J. Larner, *Biochemistry*, 1969, **8**, 1419.
186. A. Nakayama, K. Yamamoto and S. Tabata, *J. Biol. Chem.*, 2001, **276**, 28824.
187. M. Yanase, H. Takata, T. Takaha, T. Kuriki, S. M. Smith and S. Okada, *Appl. Environ. Microbiol.*, 2002, **68**, 4233.
188. A. Imberty, H. Chanzy, S. Pérez, A. Buléon and V. Tran, *J. Mol. Biol.*, 1988, **201**, 365.
189. A. Imberty and S. Pérez, *Biopolymers*, 1988, **27**, 1205.
190. S. Hizukuri, *Carbohydr. Res.*, 1985, **141**, 295.
191. G. T. Oostergetl and E. F. J. van Bruggen, *Carbohydr. Polym.*, 1993, **21**, 7.
192. T. A. Waigh, A. M. Donald, F. Heidelbach, C. Riekel and M. J. Gidley, *Biopolymers*, 1999, **49**, 91.
193. (a) A. M. Smith, *Curr. Opin. Plant. Biol.*, 1999, **2**, 223; (b) M. G James, K. Denyer and A. M Myers, *Curr. Opin. Plant. Biol.*, 2003, **6**, 215; (c) S. Jobling, *Curr. Opin. Plant. Biol.*, 2004, **7**, 210.
194. R. Mukerjea and J. F. Robyt, *Carbohydr. Res.*, 2005, **340**, 2206.
195. N. S. Han and J. F. Robyt, *Carbohydr. Res.*, 1998, **313**, 125.
196. M. Shure, S. Wessler and N. Fedoroff, *Cell*, 1983, **35**, 225.
197. P. J. Jenkins and A. M. Donald, *Int. J. Biol. Macromol.*, 1995, **17**, 315.
198. P.D. Commuri and P.L. Keeling, *Plant J.*, 2001, **25**, 475.
199. Y. Takeda, H.-P. Guan and J. Preiss, *Carbohydr Res.*, 1993, **240**, 253.
200. R. A. Burton, H. Jenner, L. Carrangis, B. Fahy, G. B. Fincher, C. Hylton, D. A. Laurie, M. Parker, D. Waite, S. van Wegen, T. Verhoeven and K. Denyer, *Plant J.*, 2002, **31**, 97.
201. A. Kubo, N. Fujita, K. Harada, T. Matsuda, H. Satoh and Y. Nakamura, *Plant Physiol.*, 1999, **121**, 399.
202. G. Ritte, J. R. Lloyd, N. Eckermann, A. Rottmann, J. Kossmann and M. Steup, *Proc. Natl. Acad. Sci. USA*, 2002, **99**, 7166.
203. Review with emphasis on rice: G. E. Vandeputte and J. A. Delcour, *Carbohydr. Polym.*, 2004, **58**, 245.
204. S. Kim and J. L. Willett, *Starch/Stärke*, 2004, **56**, 29.
205. T. A. Waigh, M. J. Gidley, B. U. Komanshek and A. M. Donald, *Carbohydr. Res.*, 2000, **328**, 165. The undefined acronym "WAXS" in this paper stands for wide-angle X-ray scattering or powder X-ray diffraction.
206. D. Cooke and M.J. Gidley, *Carbohydr. Res.*, 1992, **227**, 103.
207. P. J. Jenkins and A. M. Donald, *Carbohydr. Res.*, 1998, **308**, 133.
208. There is a large food technology literature on retrogradation: for recent reviews see (a) R. Hoover, *Carbohydr. Polym.*, 2001, **45**, 253; (b) R. Hoover, *Food Rev. Int.*, 1995, **11**, 331; (c) A. Abd. Karim, M. H. Norziah and C. C. Seow, *Food. Chem.*, 2000, **71**, 9.
209. K. S. Lewen, T. Paeschke, J. Reid, P. Molitor and S. J. Schmidt, *J. Agric. Food. Chem.*, 2003, **51**, 2348.
210. G. F. Fanta, F. C. Felker and R. L. Shogren, *Carbohydr. Polym.*, 2002, **48**, 161.

211. J.-L. Jane and J. F. Robyt, *Carbohydr. Res.*, 1984, **132**, 105.
212. A. M. Donald, K. L. Kato, P. A. Perry and T. A. Waigh, *Starch/Stärke*, 2001, **53**, 504.
213. F. Lionetto, A. Maffezzoli, M.-A. Ottenhof, I. A. Farhat and J. R. Mitchell, *Starch/Stärke*, 2005, **57**, 16.
214. A. Blennow, B. Wischmann, K. Houborg, T. Ahmt, K. Jørgensen, S. Balling Engelsen, O. Bandsholm and P. Poulsen, *Int. J. Biol. Macromol.*, 2005, **36**, 159.
215. R. S. Aquino, A. M. Landeira-Fernandez, A. P. Valente, L. R. Andrade and P. A. S. Mourão, *Glycobiology*, 2005, **15**, 11.
216. J. Visser and A. G. J. Voragen eds., *Pectins and Pectinases Progress in Biotechnology*, **14**, Elsevier, Amsterdam, 1996.
217. T. P. Kravtchenko, M. Penci, A. G. J. Voragen and W. Pilnik, *Carbohydr. Polym.*, 1993, **20**, 195.
218. http://www.herbstreith-fox.de/en/forschung_und_entwicklung/index.htm. Excellent summaries of technical aspects of pectin production and use by a commercial pectin supplier (in English).
219. B. L. Ridley, M. A. O'Neill and D. Mohnen, *Phytochemistry*, 2001, **57**, 929. This review gives the structures of the rhamnogalacturonan I arabinan and galactan as both (1→3), whereas the arabinan is largely (1→5) and the galactan (1→4).
220. T. Ishii and T. Matsunaga, *Phytochemistry*, 2001, **57**, 969.
221. M. D. Walkinshaw and S. Arnott, *J. Mol. Biol.*, 1981, **153**, 1055.
222. M. D. Walkinshaw and S. Arnott, *J. Mol. Biol.*, 1981, **153**, 1075.
223. S. Cros, C. Hervé du Penhoat, N. Bouchemal, H. Ohassan, A. Imberty and S. Pérez, *Int. J. Biol. Macromol.*, 1992, **14**, 313.
224. M. M. H. Huisman, C. T. M. Fransen, J. P. Kamerling, J. F. G. Vliegenthart, H. A. Schols and A. G. J. Voragen, *Biopolymers*, 2001, **58**, 279.
225. A. Le Goff, C. M. G. C. Renard, E. Bonnin and J.-F. Thibault, *Carbohydr. Polym.*, 2001, **45**, 325.
226. W. G. T. Willats, L. McCartney, C. G. Steele-King, S. E. Marcus, A. Mort, M. Huisman, G.-J. van Alebeek, H. A. Schols, A. G. J. Voragen, A. Le Goff, E. Bonnin, J.-F. Thibault and J. P. Knox, *Planta*, 2004, **218**, 673.
227. M. McNeil, A. G. Darvill and P. Albersheim, *Plant. Physiol.*, 1980, **66**, 1128.
228. M. Broadhurst, S. Cros, R. Hoffman, W. Mackie and S. Pérez, , in J. Visser and A. G. J. Voragen, eds., *Pectins and Pectinases*, Elsevier, Amsterdam, 1996, p. 517.
229. P. Komalavilas and A. J. Mort, *Carbohydr. Res.*, 1989, **189**, 261.
230. P. Lerouge, M. A. O'Neill, A. G. Darvill and P. Albersheim, *Carbohydr. Res.*, 1993, **243**, 359.
231. R. Oechslin, M. V. Lutz and R. Amadò, *Carbohydr. Polym.*, 2003, **51**, 301.
232. M.-A. Ha, R. J. Viëtor, G. D. Jardine, D. C. Apperley and M. C. Jarvis, *Phytochemistry*, 2005, **66**, 1817.

233. L. Jones, J. L. Milne, D. Ashford and S. J. McQueen-Mason, *Proc. Natl. Acad. Sci. USA*, 2003, **100**, 11783.
234. N. W. H. Cheetham, P. C.-K. Cheung and A. J. Evans, *Carbohydr. Polym.*, 1993, **22**, 37.
235. S. W. A. Hinz, R. Verhoef, H. A. Schols, J.-P. Vincken and A. G. J. Voragen, *Carbohydr. Res.*, 2005, **340**, 2135.
236. S. Janaswamy and R. Chandrasekaran, *Carbohydr. Res.*, 2005, **340**, 835.
237. S. Levigne, M.-C. Ralet, B. Quéméner and J.-F. Thibault, *Carbohydr. Res.*, 2004, **339**, 2315.
238. S. Vidal, T. Doco, P. Williams, P. Pellerin, W. S. York, M. A. O'Neill, J. Glushka, A. G. Darvill and P. Albersheim, *Carbohydr. Res.*, 2000, **326**, 277.
239. B. L. Reuhs, J. Glenn, S. B. Stephens, J. S. Kim, D. B. Christie, J. G. Glushka, E. Zablackis, P. Albersheim, A. G. Darvill and M. A. O'Neill, *Planta*, 2004, **219**, 147.
240. T. Ishii, T. Matsunaga, P. Pellerin, M. A. O'Neill, A. Darvill and P. Albersheim, *J. Biol. Chem.*, 1999, **274**, 13098.
241. M. A. O'Neill, S. Eberhard, P. Albersheim and A. G. Darvill, *Science*, 2001, **294**, 846.
242. T. Matoh, M. Takasaki, M. Kobayashi and K. Takabe, *Plant Cell. Physiol.*, 1998, **39**, 483.
243. D. Mohnen, in *Carbohydrates and Their Derivatives Including Tannins, Cellulose and Related Lignins*, ed. S. D. Barton, H. Nakanisho, O. Meth-Cohn and B. M. Pinto, Elsevier, Amsterdam, 1999, p. 498.
244. J. Egelund, M. Skjøt, N. Geshi, P. Ulvskov and B. L. Petersen, *Plant Physiol.*, 2004, **136**, 2609.
245. C. Orfila, S. Oxenbøll Sørensen, J. Harholt, N. Geshi, H. Crombie, H.-N. Truong, J. S. G. Reid, J. P. Knox and H. V. Scheller, *Planta*, 2005, **222**, 613.
246. J. Harhot, J. K. Jensen, S. O. Sørensen, C. Orfila, M. Pauly and H. V. Schneider, *Plant Physiol.*, 2006, **140**, 49.
247. K. J. Nunan and H. V. Scheller, *Plant Physiol.*, 2003, **132**, 331.
248. T. Ishii, T. Konishi, Y. Ito, H. Ono, M. Ohnishi-Kameyama and I. Maeda, *Phytochemistry*, 2005, **66**, 2418.
249. H. Huang, M. S. Scherman, W. D'Haeze, D. Vereecke, M. Holsters, D. C. Crick and M. R. McNeil, *J. Biol. Chem.*, 2005, **280**, 24539.
250. A. Veluraja and E. D. T. Atkins, *Carbohydr. Polym.*, 1987, **7**, 133.
251. C. T. Chuah, A. Sarko, Y. Deslandes and R. H. Marchessault, *Macromolecules*, 1983, **16**, 1375.
252. Y. Deslandes, R. H. Marchessault and A. Sarko, *Macromolecules*, 1980, **13**, 1466.
253. S. M. Read, G. Currie and A. Bacic, *Carbohydr. Res.*, 1996, **281**, 187.
254. T. Sasaki and N. Takasuka, *Carbohydr. Res.*, 1976, **47**, 99.
255. K. Okuyama, A. Otsubo, Y. Fukuzawa, M. Ozawa, T. Harada and N. Kansai, *J. Carbohydr. Chem.*, 1991, **10**, 645.

256. K. Okuyama, Y. Obata, K. Noguchi, T. Kusaba, Y. Ito and S. Ohno, *Biopolymers*, 1996, **38**, 557.
257. H. Saitô, Y. Yoshioka, N. Uehara, J. Aketagawa, S. Tanaka and Y. Shibata, *Carbohydr. Res.*, 1991, **217**, 181.
258. H. Saitô, T. Ohki and T. Sasaki, *Biochemistry*, 1977, **16**, 908.
259. H. Saitô, M. Yokoi and Y. Yoshioka, *Macromolecules*, 1989, **22**, 3892.
260. H. Saitô, Y. Yoshioka, M. Yokoi and J. Yamada, *Biopolymers*, 1990, **29**, 1689.
261. A. G. Ljungman, P. Leanderson and C. Tagesson, *Environ. Toxicol. Pharmacol.*, 1998, **5**, 273.
262. E. Wakshull, D. Brunke-Reese, J. Lindermuth, L. Fisette, R. S. Nathans, J. J. Crowley, J. C. Tufts, J. Zimmerman, W. Mackin and D. S. Adams, *Immunopharmacology*, 1999, **41**, 89.
263. S.-H. Young, W.-J. Dong and R. R. Jacobs, *J. Biol. Chem.*, 2000, **275**, 11874.
264. L. Zhang, X. Zhang, Q. Zhou, P. Zhang and X. Li, *Polym. J.*, 2001, **33**, 317.
265. T. Yanaki, T. Kojima and T. Norisuye, *Polym. J.*, 1981, **13**, 1135.
266. T. Sato, T. Norisuye and H. Fujita, *Carbohydr. Res.*, 1981, **95**, 195.
267. A. K. Vuppu, A. A. Garcia and C. Vernia, *Biopolymers*, 1997, **42**, 89.
268. B. H. Falch, A. Elgsaeter and B. J. Stokke, *Biopolymers*, 1999, **50**, 496.
269. M. Numata, M. Asai, K. Kaneko, A.-H. Bae, T. Hasegawa, K. Sakurai and S. Shinkai, *J. Am. Chem. Soc.*, 2005, **127**, 5875.
270. S. Karácsonyi, V. Pätoprstý and M. Kubačkovà, *Carbohydr. Res.*, 1998, **307**, 271.
271. Reviews: (a) J. Sommer-Knudsen, A. Bacic and A. Clarke, *Phytochemistry*, 1998, **47**, 483; (b) N. I. Rumyantseva, *Biochemistry (Moscow)*, 2005, **70**, 1073 (*Biokhimiya*, 2005, **70**, 1301).
272. E. Luonteri, C. Laine, S. Uusitalo, A. Teleman, M. Siika-aho and M. Tenkanen, *Carbohydr. Polym.*, 2003, **53**, 155.
273. (a) M. Manley-Harris, *Carbohydr. Polym.*, 1997, **34**, 243; (b) G. R. Ponder and G. N. Richards, *Carbohydr. Polym.*, 1997, **34**, 251.
274. C. H. Grün, F. Hochstenbach, B. M. Humbel, A. J. Verkleij, J. H. Sietsma, F. M. Klis, J. P. Kamerling and J. F.G. Vliegenthart, *Glycobiobiology*, 2005, **15**, 245.
275. J. Chen, L. Zhang, Y. Nakamura and T. Norisuye, *Polym. Bull.*, 1998, **41**, 471.
276. N. W. H. Cheetham, M. E. Slodki and G. J. Walker, *Carbohydr. Polym.*, 1991, **16**, 341.
277. K. Funane, T. Ishii, H. Ono and M. Kobayashi, *FEBS Lett.*, 2005, **579**, 4739.
278. H. Mukasa, H. Tsumori and A. Shimamura, *Carbohydr. Res.*, 2001, **333**, 19.
279. K. Ogawa, K. Okamura and A. Sarko, *Int. J. Biol. Macromol.*, 1981, **3**, 31.
280. L. Poppe, W. S. York and H. van Halbeek, *J. Biomol. NMR*, 1993, **3**, 81.
281. G. Williamson, K. Damani, P. Devenney, C. B. Faulds, V. J. Morris and B. J. H. Stevens, *J. Bacteriol.*, 1992, **174**, 7941.

282. A. E. Ciocchini, M. S. Roset, N. Iñón de Iannino and R. A. Ugalde, *J. Bacteriol.*, 2004, **186**, 7205.
283. N. Iñón de Iannino, G. Briones, M. Tolmasky and R. A. Ugalde, *J. Bacteriol.*, 1998, **180**, 4392.
284. J. E. Ralton, T. Naderer, H. L. Piraino, T. A. Bashtannyk, J. M. Callaghan and M. J. McConville, *J. Biol. Chem.*, 2003, **278**, 40757.
285. D. Crich, A. Banerjee and Q. J. Yao, *J. Am. Chem. Soc.*, 2004, **126**, 14930.
286. www.sigmaaldrich.com.
287. J. Porath and P. Flodin, *Nature*, 1959, **183**, 1657.
288. A. M. Basedow and K. H. Ebert, *J. Polym. Sci., Polym. Symp.*, 1979, **66**, 101.
289. B. M. Humbel, M. Konomi, T. Takagi, N. Kamasawa, S. A. Ishijima and M. Osumi, *Yeast*, 2001, **18**, 433.
290. R. Kollár, B. B. Reinhold, E. Petráková, H. J. C. Yeh, G. Ashwell, J. Drgonová, J. C. Kapteyn, F. M. Klis and E. Cabib, *J. Biol. Chem.*, 1997, **272**, 17762.
291. R. C. Montijn, E. Vink, W. H. Müller, A. J. Verkleij, H. Van Den Ende, B. Henrissat and F. M. Klis, *J. Bacteriol.*, 1999, **181**, 7414.
292. J. Stolz and S. Munro, *J. Biol. Chem.*, 2002, **277**, 44801.
293. K. Ikuta, N. Shibata, J. S. Blake, M. V. Dahl, R. D. Nelson, K. Hisamichi, H. Kobayashi, S. Suzuki and Y. Okawa, *Biochem. J.*, 1997, **323**, 297.
294. M. M. Corsaro, A. Evidente, R. Lanzetta, P. Lavermicocca and A. Molinaro, *Carbohydr. Res.*, 2001, **330**, 271.
295. E. Vinogradov, B. O. Petersen and J. Ø. Duus, *Carbohydr. Res.*, 2000, **325**, 216.
296. E. R. Vimr, K. A. Kalivoda, E. L. Deszo and S. M. Steenbergen, *Microbiol. Mol. Biol. Rev.*, 2004, **68**, 132.
297. S. Inoue and Y. Inoue, *J. Biol. Chem.*, 2001, **276**, 31863.
298. C. Sato, K. Kitajima, S. Inoue and Y. Inoue, *J. Biol. Chem.*, 1998, **273**, 2575.
299. S. Inoue, G. L. Poongodi, N. Suresh, H. J. Jennings and Y. Inoue, *J. Biol. Chem.*, 2003, **278**, 8541.
300. G.-J. Shen, A. K. Datta, M. Izumi, K. M. Koeller and C.-H. Wong, *J. Biol. Chem.*, 1999, **274**, 35139.
301. S. Kitazume, K. Kitajima, S. Inoue, F. A. Troy, J.-W. Cho, W. J. Lennarz and Y. Inoue, *J. Biol. Chem.*, 1994, **269**, 22712.
302. C. Sato, K. Kitajima, I. Tazawa, S. Inoue, Y. Inoue and F. A. Troy, *J. Biol. Chem.*, 1993, **268**, 23675.
303. M. R. Lifely, A. S. Gilbert and C. Moreno, *Carbohydr. Res.*, 1981, **94**, 193.
304. M. R. Lifely, A. S. Gilbert and C. Moreno, *Carbohydr. Res.*, 1984, **134**, 229.
305. Y. Zhang and Y. C. Lee, *J. Biol. Chem.*, 1999, **274**, 6183.
306. A. E. Manzi, H. H. Higa, S. Diaz and A. Varki, *J. Biol. Chem.*, 1994, **269**, 23617.

307. F. Michon, J.-R. Brisson and H. J. Jennings, *Biochemistry*, 1987, **26**, 8399.
308. J.-R. Brisson, H. Baumann, A. Imberty, S. Pérez and H. J. Jennings, *Biochemistry*, 1992, **31**, 4996.
309. T.-P. Yu, M.-C. Cheng, H.-R. Lin, C.-H. Lin and S.-H. Wu, *J. Org. Chem.*, 2001, **66**, 5248.
310. R. H. F. Beck and W. Praznick, *Starch/Stärke*, 1986, **38**, 391.
311. A. Fuchs, *Starch/Stärke*, 1987, **39**, 335.
312. I. André, J. L. Putaux, H. Chanzy, F. R. Taravel, J. W. Timmermans and D. de Wit, *Int. J. Biol. Macromol.*, 1996, **18**, 195.
313. M. Tylianakis, A. Spyros, P. Dais, F. R. Taravel and A. Perico, *Carbohydr. Res.*, 1999, **315**, 16.
314. J.H. Liu, A. L. Waterhouse and N. J. Chatterton, *J. Carbohydr. Chem.*, 1994, **13**, 859.
315. T. Harada, S. Suzuki, H. Taniguchi and T. Sasaki, *Food Hydrocolloids*, 1993, **7**, 23.
316. S. Kitamura, T. Hirano, K. Takeo, M. Mimura, K. Kajiwara, B. T. Stokke and T. Harada, *Int. J. Biol. Macromol.*, 1994, **16**, 313.
317. D. Wolff, S. Czapla, A. G. Heyer, S. Radosta, P. Mischnick and J. Springer, *Polymer*, 2000, **41**, 8009.
318. J. Edelman and T. G. Jefford, *New Phytol.*, 1968, **67**, 517.
319. E. M. Hellwege, S. Czapla, A. Jahnke, L. Willmitzer and A. G. Heyer, *Proc. Natl. Acad. Sci. USA*, 2000, **97**, 8699.
320. S. Gonta, M. Utinans, O. Neilands and I. Vīna, *J. Mol. Struct. Theochem.*, 2004, **710**, 61.
321. N. Sprenger, K. Bortlik, A. Brandt, T. Boller and A. Wiemken, *Proc. Natl. Acad. Sci. USA*, 1995, **92**, 11652.
322. N. Sprenger, L. Schellenbaum, K. van Dun, T. Boller and A. Wiemken, *FEBS Lett.*, 1997, **400**, 355.
323. E. J. Yoon, S.-H. Yoo, J. Cha and H. G. Lee, *Int J. Biol.Macromol.*, 2004, **34**, 191.
324. G. H. Van Geel-Schutten, E. J. Faber, E. Smit, K. Bonting, M. R. Smith, B. Ten Brink, J. P. Kamerling, J. F. G. Vliegenthart and L. Dijkhuizen, *Appl. Environ. Microbiol.*, 1999, **65**, 3008.
325. K. I. Shingel, *Carbohydr. Res.*, 2004, **339**, 447.
326. P. S. Stinard and D. J. Nevins, *Phytochemistry*, 1980, **19**, 1467.
327. A. Planas, *Biochim. Biophys. Acta*, 2000, **1543**, 361.
328. J. R. Woodward, G. B. Fincher and B. A. Stone, *Carbohydr. Polym.*, 1983, **3**, 207.
329. A. Grimm, E. Krüger and W. Burchard, *Carbohydr. Polym.*, 1995, **27**, 205.
330. M. Faijes, T. Imai, V. Bulone and A. Planas, *Biochem. J.*, 2004, **380**, 635.
331. T. Lind, F. Tufaro, C. McCormick, U. Lindahl and K. Lidholt, *J. Biol. Chem.*, 1998, **273**, 26265.
332. U. Lindahl, M. Kusche-Gullberg and L. Kjellén, *J. Biol. Chem.*, 1998, **273**, 24979.
333. T. Uyama, H. Kitagawa, J. Tanaka, J. Tamura, T. Ogawa and K. Sugahara, *J. Biol. Chem.*, 2003, **278**, 3072.

334. H. Kitagawa, T. Uyama and K. Sugahara, *J. Biol. Chem.*, 2001, **276**, 38721.
335. S. Bodevin-Authelet, M. Kusche-Gullberg, P. E. Pummill, P. L. DeAngelis and U. Lindahl, *J. Biol. Chem.*, 2005, **280**, 8813.
336. P. L. DeAngelis, *Cell. Mol. Life Sci.*, 1999, **56**, 670.
337. J. L. Funderburgh, *Glycobiology*, 2000, **10**, 951.
338. M. Peña, C. Williams and E. Pfeiler, *Carbohydr. Res.*, 1998, **309**, 117.
339. W. Jing and P. De Angelis, *Glycobiology*, 2003, **13**, 661.
340. N. Adnan and P. Ghosh, *Inflam. Res.*, 2001, **50**, 294.
341. J. E. Scott, F. Heatley and W. E. Hull, *Biochem. J.*, 1984, **220**, 197.
342. J. E. Scott and F. Heatley, *Proc. Natl. Acad. Sci. USA*, 1999, **96**, 4850.
343. M. Rinaudo, M. Milas, N. Jouon and R. Borsali, *Polymer*, 1993, **34**, 3710.
344. E. D. T. Atkins, C. F. Phelps and J. K. Sheehan, *Biochem. J.*, 1972, **128**, 1255.
345. A. K. Mitra, J. K. Sheehan and S. Arnott, *J. Mol. Biol.*, 1983, **169**, 813.
346. A. K. Mitra, S. Raghunathan, J. K. Sheehan and S. Arnott, *J. Mol. Biol.*, 1983, **169**, 829.
347. S. Arnott, A. K. Mitra and S. Raghunanthan, *J. Mol. Biol.*, 1983, **169**, 861.
348. H. Zheng and T. S. Leyh, *J. Am. Chem. Soc.*, 1999, **121**, 8692.
349. G. Lowe, in *Comprehensive Biological Catalysis*, ed. M. L. Sinnott, Academic Press, London, 1998, Vol I, p. 627.
350. Y. Kakuta, E. V. Petrotchenko, L. C. Pedersen and M. Negishi, *J. Biol. Chem.*, 1998, **273**, 27325.
351. E.g. S. Ohtake, H. Kimata and O. Habuchi, *J. Biol. Chem.*, 2005, **280**, 39115.
352. J. R. Grunwell, V. L. Rath, J. Rasmussen, Z. Cabrilo and C. R. Bertozzi, *Biochemistry*, 2002, **41**, 15590.
353. Y. Kakuta, T. Sueyoshi, M. Negishi and L. C. Pedersen, *J. Biol. Chem.*, 1999, **274**, 10673.
354. H. P. Wessel and S. Bartsch, *Carbohydr. Res.*, 1995, **274**, 1.
355. T. Maruyama, T. Toida, T. Imanari, G. Yu and R. J. Linhardt, *Carbohydr. Res.* 1998, **306**, 35. The sketch of the uronic acid in the $^{2}S_{O}$ conformation is of L-altruronic acid.
356. T. E. Hardingham and A. J. Fosang, *FASEB J.*, 1992, **6**, 861.
357. L. Ng, A. J. Grodzinsky, P. Patwari, J. Sandy, A. Plaas and C. Ortiz, *J. Struct. Biol.*, 2003, **143**, 242.
358. R. P. Millane, A. K. Mitra and S. Arnott, *J. Mol. Biol.*, 1983, **169**, 903.
359. J. J. Cael, W. T. Winter and S. Arnott, *J. Mol. Biol.*, 1978, **125**, 21.
360. J. E. Scott, F. Heatley and B. Wood, *Biochemistry*, 1995, **43**, 15467.
361. M.A. Rodríguez-Carvajal, A. Imberty and S. Pérez, *Biopolymers*, 2003, **69**, 15.
362. A. K. Mitra, S. Arnott, E. D. T. Atkins and D. H. Isaac, *J. Mol. Biol.*, 1983, **169**, 873.
363. A. Almond and J. K. Sheehan, *Glycobiology*, 2000, **10**, 329.
364. D. L. Rabenstein, *Nat. Prod. Rep.*, 2002, **19**, 312.

365. J. Aikawa, K. Grobe, M. Tsujimoto and J. D. Esko, *J. Biol. Chem.*, 2001, **276**, 5876.
366. M. B. Duncan, M. Liu, C. Fox and J. Liu, *Biochem. Biophys. Res. Commun.*, 2006, **339**, 1232.
367. J. Van Den Born, K. Gunnarsson, M. A. H. Bakker, L. Kjellén, M. Kusche-Gullberg, M. Maccarana, J. H. M. Berden and U. Lindahl, *J. Biol. Chem.*, 1995, **270**, 31303.
368. Å. Hagner-McWhirter, U. Lindahl and J.-P. Li, *Biochem. J.*, 2000, **347**, 69.
369. H.-J. Gabius, H.-C. Siebert, S. André, J. Jiménez-Barbero and H. Rüdiger, *ChemBioChem*, 2004, **5**, 741.
370. M.-C. Bourin and U. Lindahl, *Biochem. J.*, 1993, **289**, 313.
371. J. F. McGarrity and T. Smyth, *J. Am. Chem. Soc.*, 1980, **102**, 7303.
372. C. C. Jones, M. A. Kelly, M. L. Sinnott, P. J. Smith, G. T. Tzotzos, *J. Chem. Soc., Perkin Trans. 2*, 1982, 1655.
373. H. Maskill, R. M. Southam and M. C. Whiting, *Chem. Commun.*, 1965, **496**.
374. S. Radoff and I. Danishefsky, *J. Biol. Chem.*, 1984, **259**, 166.
375. J. A. Dean, *Handbook of Organic Chemistry*, McGraw-Hill, New York, 1987.
376. E. H. White, M. Li and S. Lu, *J. Org. Chem.*, 1992, **57**, 1252.
377. W.-L. Chuang, M. D. Christ, J. Peng and D. L. Rabenstein, *Biochemistry*, 2000, **39**, 3542.
378. E. D. T. Atkins and T. C. Laurent, *Biochem. J.*, 1973, **133**, 605.
379. E. D. T. Atkins and I. A. Nieduszynski, *Adv. Exp. Med. Biol.*, 1975, **52**, 19.
380. B. Mulloy, M. J. Foster, C. Jones and D. B. Davies, *Biochem. J.*, 1993, **293**, 849.
381. D. Mikhailov, R. J. Linhardt and K. H. Mayo, *Biochem. J.*, 1997, **328**, 51.
382. B. Mulloy, M. J. Forster, C. Jones, A. F. Drake, E. A. Johnson and D. B. Davies, *Carbohydr. Res.*, 1994, **255**, 1.
383. B. Mulloy and M. J. Forster, *Glycobiology*, 2000, **10**, 1147.
384. B. Kloareg and R.S. Quatrano, *Oceanogr. Mar. Biol. Annu. Rev.*, 1988, **26**, 259.
385. F. Goulard, M. Diouris, E. Deslandes and J.-Y. Floc'h, *Eur. J. Phycol.*, 1999, **34**, 21.
386. J. A. Hemmingson, R. H. Furneaux and H. Wong, *Carbohydr. Res.*, 1996, **296**, 285.
387. Q. Zhang, H. Qi, T. Zhao, E. Deslandes, N. M. Ismaeli, F. Molloy and A. T. Critchley, *Carbohydr. Res.*, 2005, **340**, 2447.
388. S. A. Foord and E. D. T. Atkins, *Biopolymers*, 1989, **28**, 1345.
389. G. Michel, W. Helbert, R. Kahn, O. Dideberg and B. Kloareg, *J. Mol. Biol.*, 2003, **334**, 421.
390. M. Bosco, A. Segre, S. Miertus, A. Cesàro and S. Paoletti, *Carbohydr. Res.*, 2005, **340**, 943.
391. S. Janaswamy and R. Chandrasekaran, *Carbohydr. Polym.*, 2005, **60**, 499.

392. S. Arnott, W. E. Scott, D. A. Rees and G. G. A. McNab, *J. Mol. Biol.*, 1974, **90**, 253.
393. S. Janaswamy and R. Chandrasekaran, *Carbohydr. Res.*, 2001, **335**, 181.
394. I. J. Colquhoun, A. J. Jay, J. Eagles, V. J. Morris, K. J. Edwards, A. M. Griffin and M. J. Gasson, *Carbohydr. Res.*, 2001, **330**, 325. This paper, concerned with a truncated version of acetan which does not contain rhamnose, gives the terminal rhamnose of the full polysaccharide as D, which is probably a typo: all other references to acetan give the configuration as L.
395. W. Bian, R. Chandrasekaran and M. Rinaudo, *Carbohydr. Res.*, 2002, **337**, 45.
396. R. C. Clark and B. Lockwood, in *Industrial Water-Soluble Polymers*, ed. C. A. Finch, Royal Society of Chemistry, Cambridge, 1996, p. 52.
397. P.-E. Jansson, L. Kenne and B. Lindberg, *Carbohydr. Res.*, 1977, **45**, 275.
398. L. Ielpi, R. O. Couso and M. A. Dankert, *FEBS Lett.*, 1981, **130**, 253.
399. L. Ielpi, R. O. Couso and M. A. Dankert, *Biochem. Int.*, 1983, **6**, 323.
400. M. Barreras, P. L. Abdian and L. Ielpi, *Glycobiology*, 2004, **14**, 233.
401. L. Ielpi, R. O. Couso and M. A. Dankert, *Biochem. Biophys. Res. Commun.*, 1981, **102**, 1400.
402. M. P. Marzocca, N. E. Harding, E. A. Petroni, J. M. Cleary and L. Ielpi, *J. Bacteriol.*, 1991, **173**, 7519.
403. A. A. Vojnov, D. E. Bassi, M. J. Daniels and M. A. Dankert, *Carbohydr. Res.*, 2002, **337**, 315.
404. L. Ielpi, R. O. Couso and M. A. Dankert, *J. Bacteriol.*, 1993, **175**, 2490.
405. P. J. Garegg, P.-E. Jansson, B. Lindbergh, F. Lindh, J. Lönngren, I. Kvarnström and W. Nimmich, *Carbohydr. Res.*, 1980, **78**, 127.
406. C. Jones, *Carbohydr. Res.*, 1990, **198**, 353.
407. K. A. Gruys and J. A. Sikorski, in *Comprehensive Biological Catalysis*, ed. M. L. Sinnott, Academic Press, London, 1998, Vol. I, p. 273.
408. L. Chazeau, M. Milas and M. Rinaudo, *Int. J. Polym. Anal. Charact.*, 1995, **2**, 21.
409. T. A. Camesano and K. J. Wilkinson, *Biomacromolecules*, 2001, **2**, 1184.
410. C. Ojinnaka, A. J. Jay, I. J. Colquhoun, G. J. Brownsey, E. R. Morris and V. J. Morris, *Int. J. Biol. Macromol.*, 1996, **19**, 149.
411. N. I. de Iannino, R. O. Couso and M. A. Dankert, *J. Gen. Microbiol.*, 1988, **134**, 1731.
412. V. J. Morris, G. J. Brownsey, P. Cairns, G. R. Chilvers and M. J. Miles, *Int. J. Biol. Macromol.*, 1989, **11**, 326.
413. (a) M. A. O'Neill, R. R. Selvandran and V. J. Morris, *Carbohydr. Res.*, 1983, **124**, 123; (b) P.-E. Jansson, B. Lindberg and P. A. Sandford, *Carbohydr. Res.*, 1983, **124**, 135.
414. R. Chandrasekaran, R. P. Millane, S. Arnott and E. D. T. Atkins, *Carbohydr. Res.*, 1988, **175**, 1.
415. M. S. Kuo, A. Dell and A. J. Mort, *Carbohydr. Res.*, 1986, **156**, 173.

416. M. T. Nickerson, A. T. Paulson and R. A. Speers, *Food Hydrocolloids*, 2004, **18**, 783.
417. E.g. L. M. Prescott, J. P. Harley and D. A. Klein, *Microbiology*, McGraw-Hill, New York, 6th edn, 2004; R. H. Garrett and C. M. Grisham, *Biochemistry,* Harcourt Brace, Fort Worth, 1996.
418. C. Schäffer and P. Messner, *Microbiology – SGM*, 2005, **151**, 643.

CHAPTER 5

Enzyme-catalysed Glycosyl Transfer

5.1 TYPES OF ENZYME-CATALYSED GLYCOSYL TRANSFER

Enzymic glycosyl transfer proceeds through transition states similar to those described for non-enzymic glycosyl transfer, in which nucleophile and leaving group interact weakly with a reaction centre which carries a high degree of positive charge. The reactions are catalysed by enzymes, which are globular proteins whose three-dimensional structure is determined by the sequence of L-amino acids that make up the condensation polymer which is the protein. (An *aide-memoire* to the structures of individual amino acids and the commonest elements of protein structure is provided in the Appendix). The substrate is bound to the protein at the active site, catalysis occurs and products are released. The fit between enzyme and substrate is generally not of the "lock and key" type first proposed by Emil Fischer,[1] which would imply an essentially rigid protein and substrate. Rather, as first proposed by Koshland,[2] there are changes in conformation of both enzyme and substrate(s) which have evolved to maximise catalysis – the "induced fit" model. The conformation changes may be subtle and minor or may involve substantial movement of whole lobes of protein.

Where there are two substrates, there are two fundamentally different kinetic mechanisms. In the first, the so-called "ping pong" mechanism, one substrate is bound, is partly transformed at the active site (often with a loss of molecular fragment) and then the second substrate binds and the product is released. In the ternary complex mechanism, by contrast, both substrates have to bind at the active site before any catalysis occurs, after which products are released. The ternary complex mechanism is called "sequential" in most texts on enzyme kinetics, because the substrates bind in sequence, but here the term will be avoided because the difference from the ping-pong mechanism is not self-evident.

Enzymes that transfer glycosyl groups to water are known as glycosidases or glycoside hydrolases. The glycosyl transfer can occur with either retention or inversion of configuration. Koshland[3] first to applied physical organic chemical reasoning to glycosidase catalysis. He realised that the inversion of configuration was universal in bimolecular displacements at saturated carbon, so that those glycosidases, which gave the product sugar initially in the same anomeric configuration as the substrate proceeded by a double displacement mechanism, whereas those that gave the initial sugar product in the inverted configuration proceeded by a single displacement mechanism[i] (Figure 5.1). The stereochemistry of action of a glycosyl-transferring enzyme is its most fundamental characteristic.

In the majority of retaining glycoside hydrolases, the nucleophile X is a carboxylate group from an Asp or Glu residue of the protein. The natural leaving group in *O*-glycoside hydrolases usually has a high pK_a (around 14 if it is another sugar) and so a group which can act as a proton donor (*i.e.* general acid catalyst of aglycone departure), again usually Asp or Glu of the protein, is usually involved. Likewise, water is a poor nucleophile, so a carboxylate group which can act as a general base catalyst (proton acceptor) is usually involved with inverting *O*-glycosidases. An analysis of the X-ray crystal structures of a large number of retaining and inverting *O*-glycosidases led to the generalisation that retaining glycosidases have the two closest oxygen atoms of their twin-carboxylate catalytic machinery 5.5 Å apart, whereas in inverting glycosidases, where a water molecule, in addition to the reactive sugar, had to be accommodated between the catalytic carboxylates, the distance increased to ~9.5 Å.[4] As more data became available, it became clear that the distance for retaining glycosidases was fairly constant, whereas that for inverting enzymes ranged between 6 and 12 Å.[5]

The glycosyl-enzyme intermediates in "retaining", double displacement glycosidases can often be intercepted by acceptors other than water; in some enzymes this transglycosylation capability has evolved into the primary physiological function of the enzyme. They are then known as "transglycosylases" and act, inevitably, by a ping-pong mechanism.

Glycosyl transferases, by contrast, are a different type of enzyme entirely. They use activated glycosyl donors and the nucleophilic displacement involves departure of a phosphate or pyrophosphate ester (Figure 5.2). Physiologically, glycosyl transferases act in the anabolic direction and are responsible for the biosynthesis of polysaccharides, glycoproteins and glycolipids. This is sometimes called the Leloir pathway, after L. F. Leloir, the Nobel laureate who discovered the role of the nucleotide diphospho sugars. A form of physiological control can occur when the same sugar can occur as two activated forms, as with UDPGlc and ADPGlc. Some polysaccharides, usually simple ones such as starch, chitin or cellulose, are built up one sugar at a time by direct transfer of the sugar to the non-reducing end of the growing polysaccharide. The repeat units of more complex polysaccharides can be built up as phosphate esters or diesters of long-chain polyisoprenes, which are anchored in the hydrophobic cell membranes. The polysaccharide-chain is then built up as a hydroxyl of the newly synthesised repeat unit displaces a dolichol phosphate from the growing

[i] In strict logic, inversion must arise from an odd number and retention from an even number of displacements. In biological practice, however, it is easier for natural selection to develop a protein which stabilises one or two transition states, than one which stabilises 3, 5, 7, ... or 4, 6, 8, ... transition states, as the case may be. The same argument makes it unlikely that simple glycoside hydrolases which work by a ring-opening mechanism will ever be found.

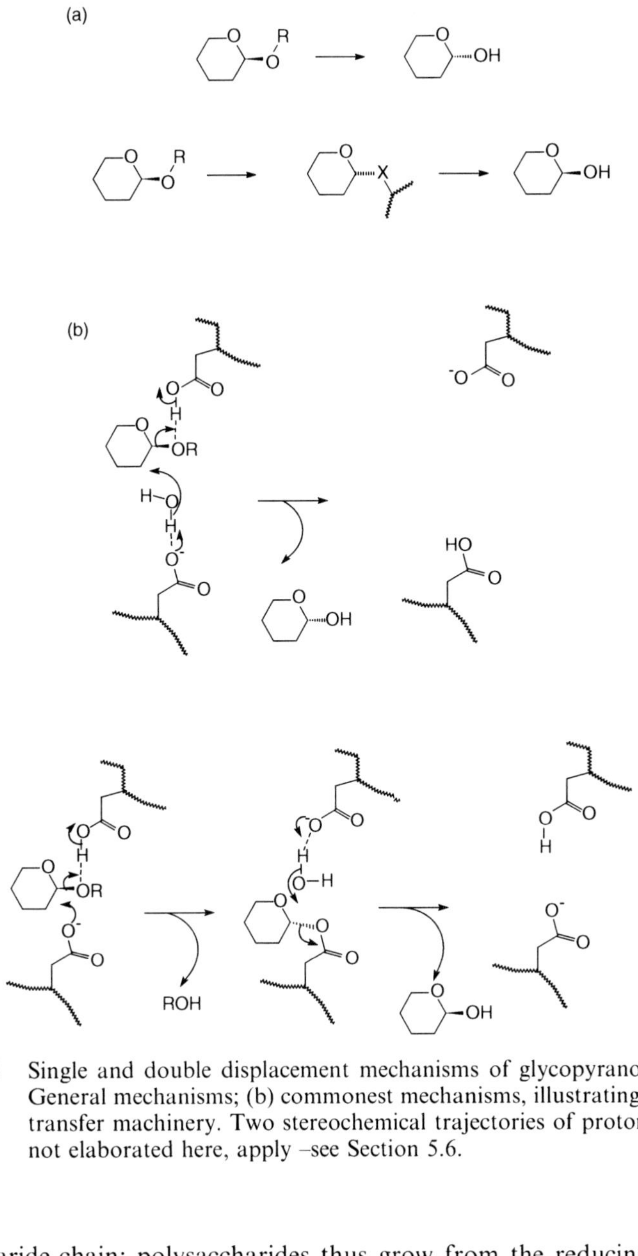

Figure 5.1 Single and double displacement mechanisms of glycopyranosidases. (a) General mechanisms; (b) commonest mechanisms, illustrating the proton transfer machinery. Two stereochemical trajectories of proton donation, not elaborated here, apply –see Section 5.6.

polysaccharide-chain; polysaccharides thus grow from the reducing end and require two dolichol chains anchored in the lipid membrane.

The advance of DNA sequencing technology has meant that the sequences of many glycosyl-transfer enzymes are known, either from cDNA sequences or the sequences of whole genomes. Starting around a dozen years ago, initially by the use of the manual technique of hydrophobic cluster analysis,[6] and later by the use of advanced sequence-similarity searching programs, first glycosidase and transglycosylase sequences and later glycosyl transferases, carbohydrate esterases, polysaccharide lyases and carbohydrate binding modules have been classified into families by Henrissat. They are displayed on a website continually updated by Henrissat and Coutinho:[7] currently (June 2007) there are ~110 GH and ~90 GT families. The classification into sequence families is

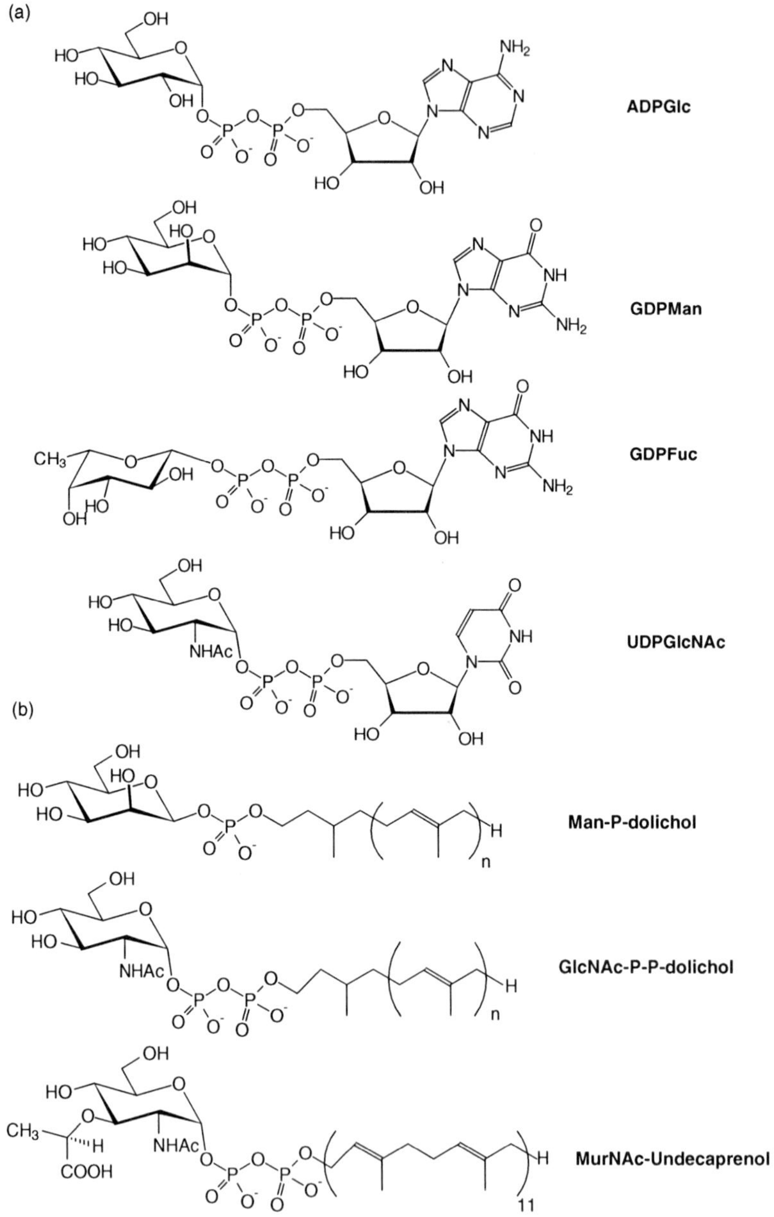

(a)

ADPGlc

GDPMan

GDPFuc

UDPGlcNAc

(b)

Man-P-dolichol

GlcNAc-P-P-dolichol

MurNAc-Undecaprenol

now the backbone of our understanding of structure and function of these enzymes, for the following reasons:

(i) Enzymes within the same family have the same basic protein fold.
(ii) Enzymes within the same family have the same basic chemical mechanism (although differences in detail, particularly in respect of acid–base catalysis, are known).
(iii) Enzymes within the same family have the same stereochemistry of hydrolysis.

Although there is an occasional recruitment of metal ions as structurally stabilising elements (*e.g.* GH[ii] 35 sialidases require Ca^{2+}) or metals and small organic molecules as part of the acid/base machinery, except for GH 4 enzymes there are no cofactor requirements: catalysis is wholly carried out by the protein itself.

In some cases, families are further grouped into clans. The largest such clan is the glycosyl hydrolase Clan A (clan GH-A), all of which have a protein fold of eight alternating α-helices and β-sheets, giving the $(\beta/\alpha)_8$ sructure, sometimes called a "TIM barrel" because it was first encountered with triose phosphate isomerase.

The glycosyl hydrolase and glycosyl transferase sequence classification is so far, unfortunately, restricted to enzymes in which both leaving group atom and nucleophile atom are oxygen[iii] This includes some, but not all, *O*-glycoside phosphorylases. The metabolically crucial glycogen and starch phosphorylases are afforded "honorary" glycosyl transferase status as GT[iv] Family 35, as they appear to have a mechanism and protein fold similar to the "retaining" glycosyl

Figure 5.2 Activated glycosyl donors in the biosynthesis of glycosidic bonds. (a) Some nucleotide diphospho sugars. The commonest form of glycosyl activation is by UDP. The unusual equatorial leaving group in GDPFuc (and other 6-deoxy sugar donors such as TDPRha) arises because the biosynthetic pathway involves extensive modification of common nucleotide diphospho sugars, GDPFuc being biosynthesised from GDPMan via inversion of stereochemistry at C3, C4 and C5 – see Section 6.8.2). For the structure of CMPNeuNAc, see Figure 3.24. (b) Activated sugar donors with hydrophobic, polyisoprene cell-membrane anchors. The undecaprenol-activated compounds are found in bacteria, whereas dolichols are found in eukaryotes. Dolichol is a range of compounds (n = 14–24); note the reduction of the last isoprene unit, which will arrest any unwanted S_N1 hydrolysis by alkyl–oxygen fission.

[ii] GH = glycoside hydrolase.
[iii] Although, of course, in enzyme reactions the difference between nucleophile and leaving group is somewhat artificial: enzymes catalyse reactions in both directions and fairly often, as with phosphorylases, the physiological direction of chemical flux is apparently in the thermodynamically disfavoured direction, small concentrations of the thermodynamically disfavoured component being "mopped up" by efficient enzymes further along the pathway.
[iv] GT = glycosyl transferase.

transferases: conversely, GT Family 36 (cellobiose phosphorylase) has been abolished, as it appears that it has much in common with the inverting GH Family 94.

Some phosphorylase reactions are illustrated in Figure 5.3.

Despite their absence from the current sequence classification, much is known about the mechanisms of *N*-glycoside formation, hydrolysis and phosphorolysis at the anomeric centre of D-ribofuranose. The enzymes physiologically are involved in nucleoside and nucleotide biosynthesis and degradation: some work at the nucleic acid polymer level to remove or add modified nucleic acid bases.

5.2 STEREOCHEMISTRY AND STEADY-STATE KINETICS OF ENZYMIC GLYCOSYL TRANSFER

Most measurements of glycosidase kinetics[v] are carried out under steady-state conditions. Substrate is in large excess over enzyme and the reaction is monitored on a time-scale that is long compared with the reciprocals of the rate constants for individual molecular events, so that changes in the concentrations of various liganded[vi] and unliganded forms of the enzyme can be set to zero. If only one substrate is involved and the active sites are independent, eqn. (5.1), the Michaelis–Menten equation, holds:

$$v = \frac{k_{cat}[E]_0[S]}{K_m + [S]} = \frac{V_{max}[S]}{K_m + [S]} \tag{5.1}$$

where $[E]_0$ is the total concentration of enzyme and v the rate of the reaction at a substrate concentration [S]. This is the same algebraic form as the expression for heterogeneous catalysis with a single reactant, which obeys the Langmuir isotherm. It describes a rectangular hyperbola with axes $x = V_{max}$, $y = -K_m$. In biochemical practice, a range of initial rates are usually measured, at various values of [S] flanking K_m, often $3 \times K_m$ to $K_m/3$, taken as the initial substrate concentration.[vii] The reasons for this are the lability of many enzymes, particularly in dilute solution, and the ever-present possibility of inhibition by products. If these complications are known definitely not to be present, however, time courses of enzyme reactions are useful. Inspection of eqn. (5.1)

---

[v] Detailed kinetic studies of glycosyl transferases are rare, because the enzymes are usually membrane-bound.

[vi] A liganded enzyme is one which has substrate, product or fragments thereof covalently or non-covalently attached. "Ligand" is a general term for a small molecule which binds to a particular protein.

[vii] Before the widespread availability of non-linear curve-fitting software, the individual parameters K_m and k_{cat} were separated by means of a Lineweaver–Burk plot, in which $1/v$ was plotted against 1/[S]. The plot should be a straight line with a y intercept of $1/V_{max}$, x intercept of $-1/K_m$ and gradient of K_m/V_{max}. Lineweaver–Burk plots are still occasionally encountered, but give very poor estimates of parameters, as relatively inaccurate points at low values of [S] are grossly overweighted.

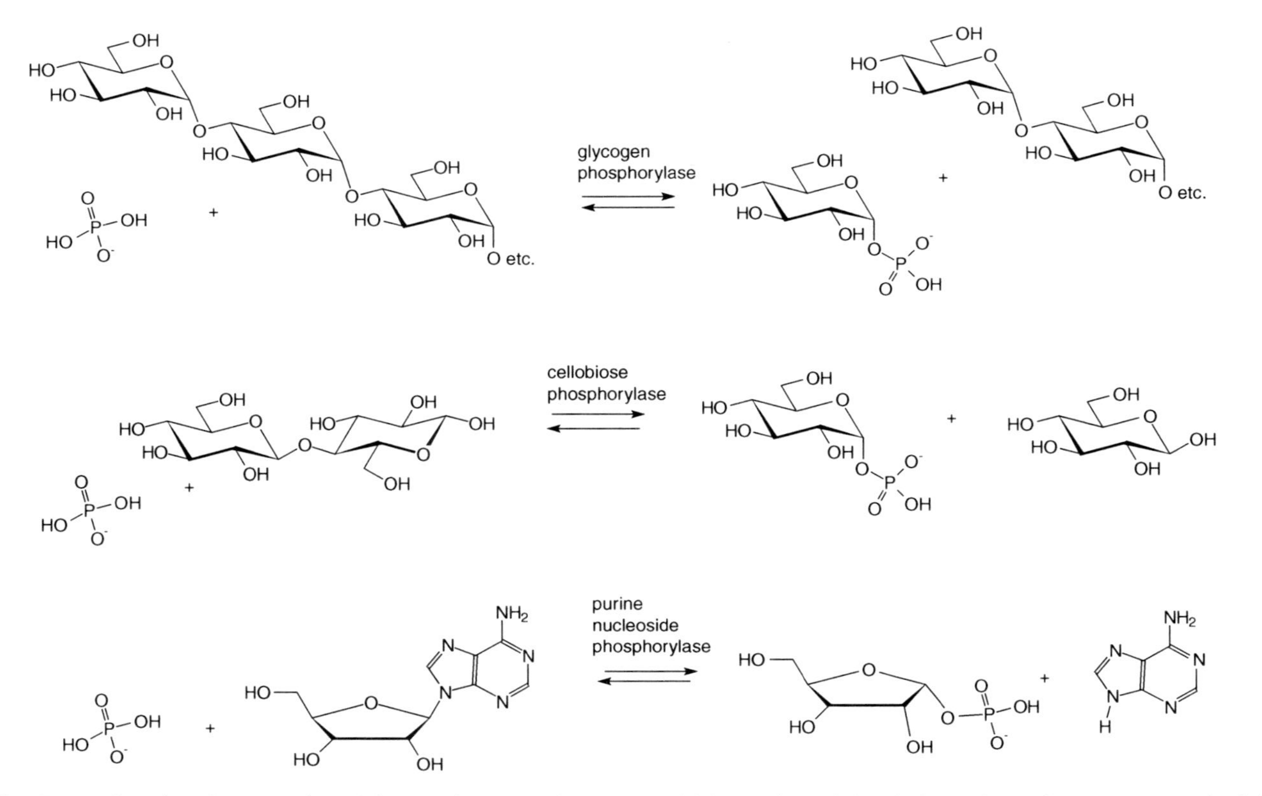

Figure 5.3 Some phosphorylase reactions. The reactions as written are stoichiometric and thus independent of pH. However, both inorganic phosphate and glycosyl phosphates ionise over physiological pH ranges, so if the equilibrium constant is quoted in terms of total glycosyl phosphate and total inorganic phosphate, it is pH dependent.

$$\mathrm{E} + \alpha\mathrm{GlyOR} \underset{k_{-1}}{\overset{k_{+1}}{\rightleftharpoons}} \mathrm{E.}\alpha\mathrm{GlyOR} \xrightarrow{k_{+2}} \mathrm{E} + \beta\mathrm{GlyOH}$$

Scheme 5.1 Minimal kinetic scheme for an inverting glycosidase: k_{+1} is the bimolecular rate constant for enzyme–substrate (ES) combination, k_{-1} is the unimolecular rate constant for loss of the substrate from the enzyme and k_{+2} is the first-order rate constant for the chemical step in the ES (Michaelis) complex. Product loss is assumed to be fast and isomerisations of the Michaelis complex to be either non-existent or rapid and reversible.

reveals it has two limiting regions: one where [S] $\gg K_m$ and the reaction is zero order and the other where [S] $\ll K_m$ and the reaction is first order, with a rate constant V_{max}/K_m.

Great care must be exercised before attributing molecular significance to values of k_{cat} and K_m. The minimal kinetic mechanism for an inverting glycosidase is that of Scheme 5.1, which leads to the expressions for k_{cat}, K_m and k_{cat}/K_m in eqns (5.2), (5.3) and (5.4).

$$k_{cat} = k_{+2} \tag{5.2}$$

$$K_m = \frac{k_{-1} + k_{+2}}{k_{+1}} \tag{5.3}$$

$$\frac{k_{cat}}{K_m} = \frac{k_{+2}k_{+1}}{k_{-1} + k_{+2}} \tag{5.4}$$

Even with this minimal scheme, it is clear that K_m can be equated with the dissociation constant of the ES complex, K_s, only if $k_{-1} \gg k_{+2}$, *i.e.* the substrate comes off the enzyme many times for every occasion it is transformed or the "commitment to catalysis" (defined in Section 5.4.4.) is zero. On the other hand, if $k_{-1} \ll k_{+2}$, every molecule that binds to the enzyme is transformed, the substrate is said to be "sticky", the "commitment to catalysis" is unity and k_{cat}/K_m becomes equal to k_{+1}, the rate of enzyme–substrate combination. This is usually the diffusion limit, so that absolute values of k_{cat}/K_m approaching $10^7\,\mathrm{M^{-1}\,s^{-1}}$, particularly if they do not vary with substrate, are a sign that the diffusion limit is being reached. The value of $10^7\,\mathrm{M^{-1}\,s^{-1}}$ is much reduced from the value of $5 \times 10^9\,\mathrm{M^{-1}\,s^{-1}}$ for the diffusional rate of azide–cation combination encountered in Chapter 3 because proteins are large molecules, and only diffusion to the active site is productive. Diffusion-limited substrate binding can be verified by a decrease of k_{cat}/K_m with increasing microscopic viscosity.[viii]

[viii] But of course the physical chemists' standby of sucrose should not be used, because at the concentrations required (~M) it will bind to any carbohydrate-active enzyme.

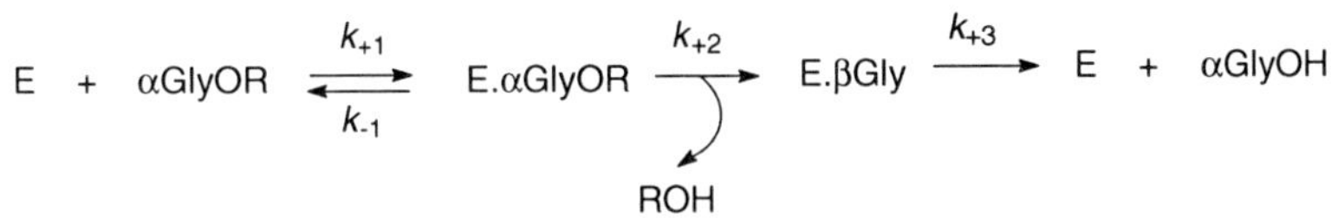

$$\mathrm{E} + \alpha\mathrm{GlyOR} \underset{k_{-1}}{\overset{k_{+1}}{\rightleftharpoons}} \mathrm{E.}\alpha\mathrm{GlyOR} \xrightarrow[\text{ROH}]{k_{+2}} \mathrm{E.}\beta\mathrm{Gly} \xrightarrow{k_{+3}} \mathrm{E} + \alpha\mathrm{GlyOH}$$

Scheme 5.2 Minimal kinetic scheme for a retaining glycosidase. The double displacement reaction involves two enzyme–substrate complexes, the second of which is commonly (but not always) a glycosylated enzyme.

The minimal kinetic scheme for a retaining glycosidase is given in Scheme 5.2. This scheme gives rise to expressions for $k_{cat,}$ K_m and k_{cat}/K_m as shown in eqns (5.5), (5.6) and (5.7).

$$k_{cat} = \frac{k_{+2}k_{+3}}{k_{+2} + k_{+3}} \tag{5.5}$$

$$K_m = \frac{k_{-1} + k_{+2}}{k_{+1}} \frac{k_{+3}}{k_{+2} + k_{+3}} \tag{5.6}$$

$$\frac{k_{cat}}{k_m} = \frac{k_{+1}k_{+2}}{k_{-1} + k_{+2}} \tag{5.7}$$

Importantly, the expression for k_{cat}/K_m does not contain k_{+3}, so that leaving group effects, for example, are usually measured with respect to this parameter; in general, whatever the kinetic mechanism, k_{cat}/K_m contains only terms up to and including the first irreversible step.

There are important and non-intuitive features of the kinetics of the system that appear when $k_{+2} \gg k_{+3}$:

(i) Even if the substrate is not in the least sticky ($k_{-1} \gg k_{+2}$), so that the first term in eqn. (5.6) becomes K_s, the dissociation constant for the E.αGlyOR complex, the K_m value is lower than K_s by a factor of $k_{+3}/(k_{+3}+k_{+2})$. Fast first steps followed by slow second steps therefore result in low K_m values.

(ii) Addition of a better nucleophile than water (methanol often serves) results in k_{+3} in eqns (5.5) and (5.6) being replaced by a term $k_{+3}+k_{+4}$[MeOH], so that a linear increase with [MeOH] is seen with both k_{cat} (for production of ROH) and K_m, if $k_{+2} \gg k_{+3}$.

Interception of the glycosyl-enzyme intermediate with another nucleophile such as methanol is an example of transglycosylation; if there is a binding site for the acceptor, the reaction scheme becomes that in Scheme 5.3.

The rate law for such a transglycosylation becomes that of eqn. (5.8); the constants are defined as in eqns (5.9)–(5.13). If there is no concurrent hydrolysis, as occurs with some transglycosylases such as the GH 16 xyloglucan endotransglycosylases, then putting $k_{+3}=0$ generates the rate law for a so-called "ping pong" reaction [eqn. (5.14)].

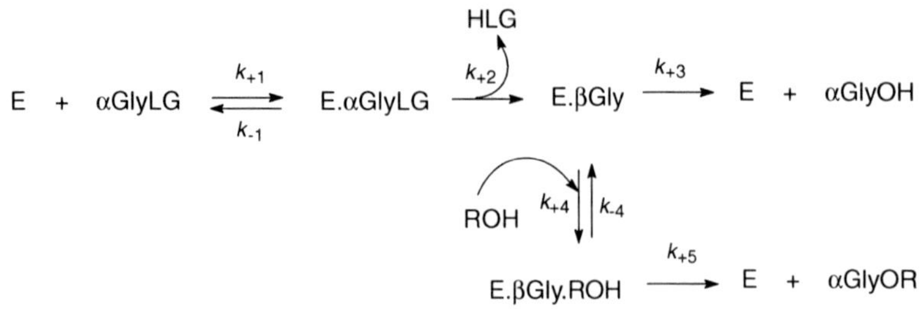

$$\mathrm{E} + \alpha\mathrm{GlyLG} \underset{k_{-1}}{\overset{k_{+1}}{\rightleftharpoons}} \mathrm{E.\alpha GlyLG} \xrightarrow[\text{HLG}]{k_{+2}} \mathrm{E.\beta Gly} \xrightarrow{k_{+3}} \mathrm{E} + \alpha\mathrm{GlyOH}$$

$$\mathrm{E.\beta Gly} + \mathrm{ROH} \underset{k_{-4}}{\overset{k_{+4}}{\rightleftharpoons}} \mathrm{E.\beta Gly.ROH} \xrightarrow{k_{+5}} \mathrm{E} + \alpha\mathrm{GlyOR}$$

Scheme 5.3 Hydrolysis and transglycosylation of a retaining glycosidase. LG = leaving group, to emphasise that activated donors such as nitrophenyl glycosides or fluorides are often used.

$$\frac{v}{[E]_0} = \frac{[\mathrm{GlyLG}]k_{\mathrm{cat1}} + [\mathrm{GlyLG}][\mathrm{ROH}]k_{\mathrm{cat2}}}{K_{\mathrm{m}} + [\mathrm{GlyLG}][\mathrm{ROH}] + [\mathrm{ROH}]K_A + [\mathrm{GlyLG}]K_{\mathrm{B}}} \tag{5.8}$$

$$k_{\mathrm{cat1}} = \frac{k_{+2}k_{+3}(k_{+5} + k_{-4})}{k_{+4}(k_{+2} + k_{+5})} \tag{5.9}$$

$$k_{\mathrm{cat2}} = \frac{k_{+2}k_{+5}}{k_{+2} + k_{+5}} \tag{5.10}$$

$$K_{\mathrm{m}} = \frac{k_{+3}(k_{+2} + k_{-1})(k_{+5} + k_{-4})}{k_{+1}k_{+4}(k_{+2} + k_{+5})} \tag{5.11}$$

$$K_{\mathrm{A}} = \frac{k_{+5}(k_{+2} + k_{-1})}{k_{+1}(k_{+2} + k_{+5})} \tag{5.12}$$

$$K_{\mathrm{B}} = \frac{(k_{+5} + k_{+4})(k_{+2} + k_{+3})}{k_{+4}(k_{+2} + k_{+5})} \tag{5.13}$$

$$\frac{v}{[\mathrm{E}]_0} = \frac{k_{\mathrm{cat}}[\mathrm{GlyLG}][\mathrm{ROH}]}{K_{\mathrm{A}}[\mathrm{ROH}] + K_{\mathrm{B}}[\mathrm{GlyLG}] + [\mathrm{GlyLG}][\mathrm{ROH}]} \tag{5.14}$$

Inversion of this expression gives

$$k_{\mathrm{cat}}[\mathrm{E}]_0/v = K_{\mathrm{A}}/[\mathrm{GlyLG}] + K_{\mathrm{B}}/[\mathrm{ROH}] + 1 \tag{5.15}$$

Therefore, a set of Lineweaver–Burk plots of $1/v$ against $1/[\mathrm{S}]$ for either substrate, at various fixed concentrations of the other, gives a set of parallel lines, of gradient $K_{\mathrm{A}}/k_{\mathrm{cat}}[\mathrm{E}]_0$.

It sometimes occurs that the acceptor molecule is a second molecule of glycosyl donor, GlyLG. The kinetic scheme is that of Scheme 5.3, except that

"GlyLG" replaces "ROH". The rate law then becomes that of eqn. (5.16), where v refers to rate of generation of LGH:

$$\frac{v}{[\mathrm{E}]_0} = \frac{k_{\mathrm{cat}}[\mathrm{S}] + k_{\mathrm{cat2}}[\mathrm{S}]^2}{K_{\mathrm{m}} + [\mathrm{S}] + \dfrac{[\mathrm{S}]^2}{K_{\mathrm{m2}}}} \tag{5.16}$$

If the rate of transglycosylation becomes zero, then one has a special case of substrate inhibition, in that a second molecule of substrate binds to the glycosyl-enzyme and does not react further, and eqn. (5.17) holds:

$$\frac{v}{[\mathrm{E}]_0} = \frac{k_{\mathrm{cat}}[\mathrm{S}]}{K_{\mathrm{m}} + [\mathrm{S}] + \dfrac{[\mathrm{S}]^2}{K_{\mathrm{is}}}} \tag{5.17}$$

K_{is} in this case is not, however, the dissociation constant of the second molecule of substrate from the ES_2 complex, but $K_d[(k_{+2}+k_{+3})/k_{+2}]$, where K_d is the dissociation constant of the glycosyl-enzyme intermediate.

In the case of enzymes working via a ternary complex mechanism, we have two extreme cases. The easiest to comprehend is the rapid equilibrium random mechanism (Scheme 5.4): this is the mechanism where the chemistry is most likely to be rate determining and kinetic isotope effects or structure-reactivity correlations are likely to be mechanistically informative. Enzymes acting on their physiological substrates at optimal pH are likely to show a degree of preference for one or the other substrate binding first, but they can often be induced to revert to a rapid equilibrium random mechanism by the use of non-optimal substrates or pH.

K_B and K_A are dissociation constants of the two complexes with a single substrate.[ix] The upper and lower pathways for the formation of the ternary complex are parts of a thermodynamic box, so the overall termolecular dissociation constant of the E.ROH.GlyLG complex is $\alpha K_A K_B$, where α is a measure of the cooperativity of binding between the two substrates. If $\alpha = 1$, obviously

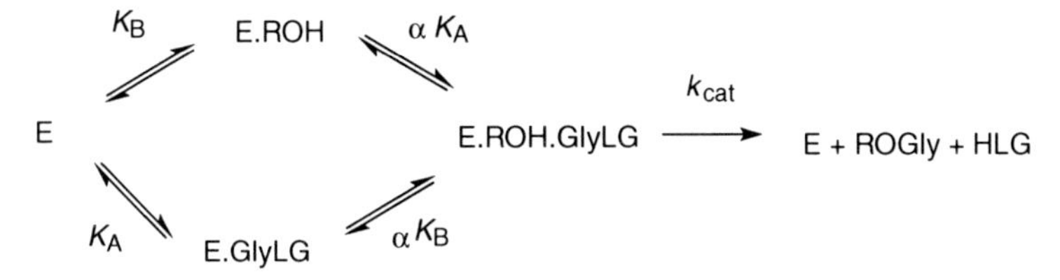

Scheme 5.4 Rapid equilibrium random mechanism for a two-substrate enzyme, illustrated for a glycosyl transfer.

[ix] In texts on enzyme kinetics, "A" is usually the first-binding substrate, a convention adhered to here; it does not mean "acceptor".

binding of one substrate in its subsite site is unaffected by binding of the other; conversely, if $\alpha = 0$, the system is infinitely cooperative and complexes with a single ligand are effectively not formed. The rate law is that of eqn. (5.18). Unless $\alpha = 0$, double reciprocal plots in which one substrate is varied at a constant concentration of the other are not parallel, but all meet at a point. In the case of a set of plots of $1/v$ against 1/[GlyLG] at various concentrations of [ROH], they meet at $1/v = (1 - \alpha)/V_{max}$, $1/[\text{GlyLG}] = -1/K_A$:

$$\frac{v}{[\text{E}]_0} = \frac{k_{cat}[\text{GlyLG}][\text{ROH}]}{\alpha K_A K_B + \alpha K_B[\text{GlyLG}] + \alpha K_A[\text{ROH}] + [\text{GlyLG}][\text{ROH}]} \qquad (5.18)$$

The other extreme form of ternary complex mechanism occurs when the second substrate will only bind after the first substrate has bound – for example, the binding of the first substrate may induce a conformation change which opens up the active site for the binding of the second substrate. Under such circumstances, rapid equilibrium will be observed only for very poor substrates. The kinetic scheme for a system in which glycosyl donor binds first, as is common for glycosyl transferases, is given in Scheme 5.5 and the rate law in eqn. (5.19).

$$\frac{v}{[\text{E}]_0} = \frac{k_{cat}[\text{GlyLG}][\text{ROH}]}{K_A K_B + K_B[\text{GlyLG}] + [\text{GlyLG}][\text{ROH}]} \qquad (5.19)$$

Double reciprocal plots with the first-binding substrate as the variable, at various concentrations of the second, therefore give a family of lines intercepting on the abscissa at $1/v = 1/V_{max}$ (Figure 5.4).

A number of key metabolic glycosyl-transferring enzymes, most notably glycogen phosphorylase, do not show a simple hyperbolic (Michaelis–Menten) variation of rate with substrate concentration. They are termed allosteric enzymes. The physical situation that can give rise to such plots is usually an oligomeric enzyme, where binding in one active site of the oligomer alters the affinity of the remaining sites, although even monomeric enzymes[8] undergoing slow conformational changes can show such behaviour. If binding at one site increases the affinity (or activity) of the others, the cooperativity is positive, whereas if binding at one site reduces binding or activity at the others, the cooperativity is negative. If we imagine a positively cooperative, a negatively cooperative and a Michaelis–Menten system with the same V_{max} and substrate concentration ($[\text{S}]_{1/2}$) at which $v = V_{max}/2$, then the v *versus* [S] curve for a positively cooperative system is below the Michaelis–Menten hyperbola at $[\text{S}] < [\text{S}]_{1/2}$ and above it at $[\text{S}] > [\text{S}]_{1/2}$, whilst the v *versus* [S] curve for a negatively cooperative system is above the Michaelis–Menten hyperbola at $[\text{S}] < [\text{S}]_{1/2}$ and below it at $[\text{S}] > [\text{S}]_{1/2}$. (Figure 5.5)

$$\text{E} + \text{GlyLG} \overset{K_A}{\rightleftharpoons} \text{E.GlyLG} + \text{ROH} \overset{K_B}{\rightleftharpoons} \text{E.ROH.GlyLG} \xrightarrow{k_{cat}} \text{E} + \text{GlyOR} + \text{HLG}$$

Scheme 5.5 Ordered equilibrium ternary complex mechanism.

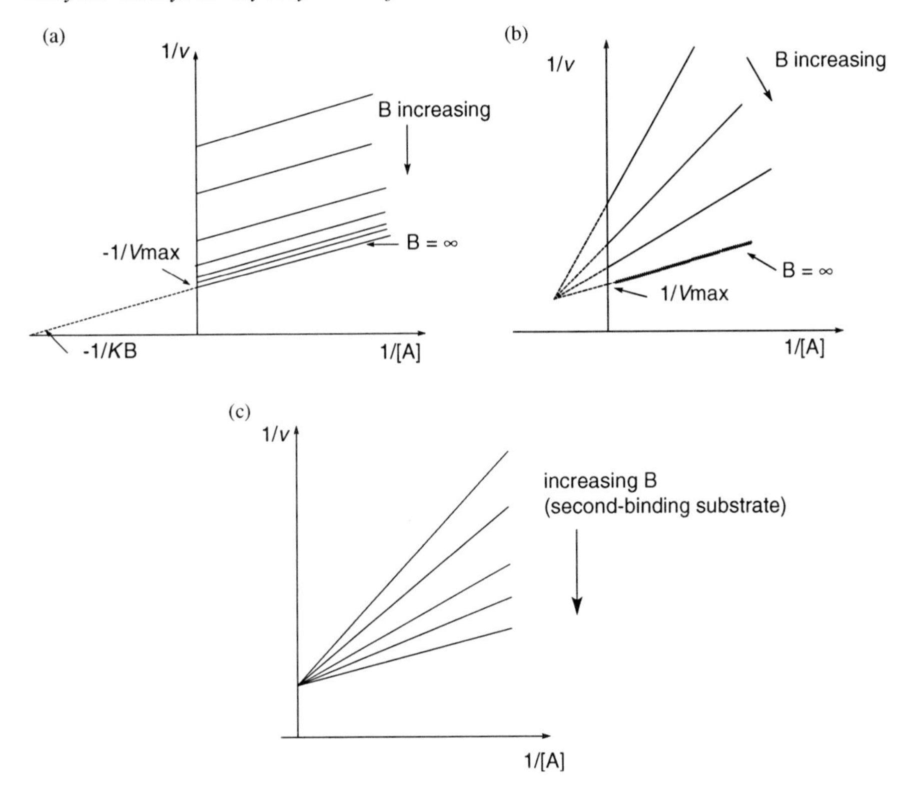

Figure 5.4 Patterns of double reciprocal plots for two substrate reactions. (a) Ping-pong. The double reciprocal plots for one substrate in the presence of fixed concentrations of the other are parallel, no matter which substrate reacts first. (b) Rapid equilibrium random. (c) Rapid equilibrium ordered, first binding substrate only.

It is conventional to describe the degree of cooperativity in terms of a Hill coefficient, n_H, defined by eqn. (5.20):

$$\log \frac{v}{V_{max} - v} = n_H \log[S] - \log K' \tag{5.20}$$

This equation has its origin in a rigorously derived equation for an infinitely cooperative equilibrium system with n subsites (*i.e.* only E and ES_n complexes exist), for which eqn. (5.21) holds:

$$\frac{v}{V_{max}} = \frac{[S]^n}{K + [S]^n} \tag{5.21}$$

If n_H has a value of 1, the system is non-cooperative; if it has a value equal to the number of binding sites, the system is infinitely cooperative. Intermediate values of n_H are best interpreted as a rough measure of the degree of

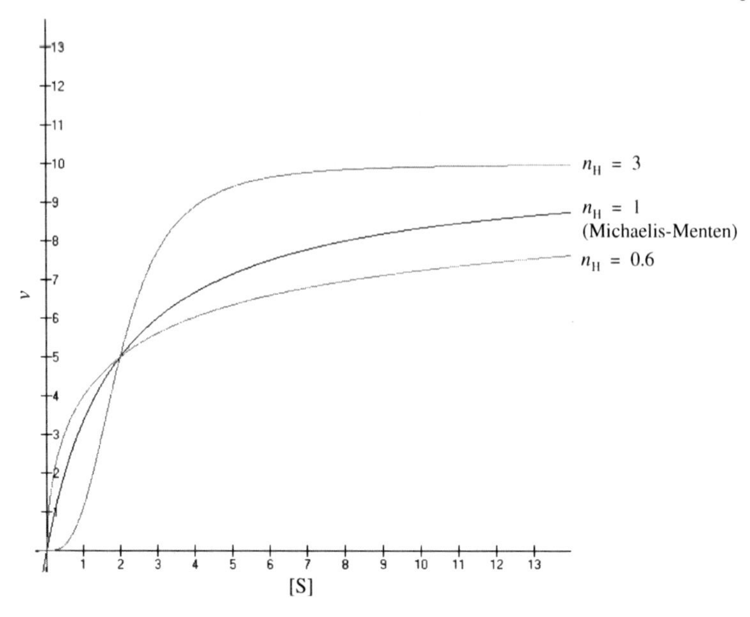

Figure 5.5 Plots of v versus [S] for a Michaelis–Menten enzyme ($n_H = 1$), a negatively cooperative one ($n_H = 0.6$) and a positively cooperative one ($n_H = 3$), all plotted with $V_{max} = 10$ and $[S]_{1/2}$ (K_m in the Michaelis–Menten case) = 2. Note how the higher the degree of cooperativity, the smaller is the rate below [S] = 2 and the higher the rate above it.

cooperativity {in fact, if partially liganded forms of the enzyme are significant, eqn. (5.20) is not linear at high and low [S]}.

5.3 REVERSIBLE INHIBITION

5.3.1 Competitive Inhibition

Many sugar-like molecules, or even molecules with a few correctly disposed hydroxyl groups [including, notoriously, the common buffer component Tris,[9] $(CH_2OH)_3CNH_2$)] will bind to the active site of glycosidases, and there block binding of substrate. If the inhibitor (I) competes for the active site, so that there are EI and ES complexes, but no ESI complexes, the inhibition is said to be competitive. As might be intuitively expected, the inhibitor has the apparent effect of increasing substrate K_m, whilst leaving V_{max} unchanged. Equation 5.22 holds:

$$v = \frac{V_{max}[S]}{[S] + K_m\left(1 + \frac{[I]}{K_i}\right)} \qquad (5.22)$$

K_i is, in the first instance, the dissociation constant of the EI complex. Binding and release of inhibitor is often considered to be fast compared with the other steps in the reaction, but frequently this is not so. One reason is a simple mass action effect: if the fastest possible rate of binding is the diffusional rate of ligand–enzyme combination, $10^7\,M^{-1}\,s^{-1}$, and K_i is in the nM range, then the "off" rate is $10^{-2}\,s^{-1}$, implying that establishment of binding equilibrium takes several minutes. These considerations mean that, without an X-ray structure of the EI complex, it is very difficult to distinguish between tight non-covalent binding and covalent binding; the binding of a non-covalent inhibitor with K_i in the pM range is effectively irreversible from mass action considerations alone ($k_{off} \approx 1\,\text{day}^{-1}$, if the enzyme survives that long).

If a poor substrate, rather than a completely inert ligand, is used to inhibit the hydrolysis of a good substrate, then the K_i measured is the K_m, not the K_s, of the poor substrate. This is because, from the point of view of the hydrolysis of the good substrate, the important parameter is the total amount of enzyme tied up in various ES complexes with the poor substrate, not the amount tied up in the first, non-covalent one. The apparent powerful inhibition of retaining glycosidases by glycals arises from this phenomenon: a 2-deoxyglycosyl-enzyme intermediate is formed much faster than it is hydrolysed, giving rise to a low K_m.[10]

An important consequence of the K_i for one substrate measured by its inhibition of the hydrolysis of a second being its K_m is that the ratio of rates of two competing substrates depends only on their k_{cat}/K_m ratio, at all substrate concentrations. This is shown in eqn. (5.23). It is particularly germane to the measurement of kinetic isotope effects from the changes in isotopic enrichment in substrate or product; such measurements *always* give the effect on k_{cat}/K_m.

$$\frac{v_1}{v_2} = \frac{k_{cat1}[E]_0[S_1]\left\{[S_2] + K_{m2}\left(1 + \dfrac{[S_1]}{K_{m1}}\right)\right\}}{k_{cat2}[E]_0[S_2]\left\{[S_1] + K_{m1}\left(1 + \dfrac{[S_2]}{K_{m2}}\right)\right\}} = \frac{k_{cat1}K_{m2}[S_1]}{K_{m1}k_{cat2}[S_2]} \tag{5.23}$$

Slow binding and even slower release are often found with inhibitors with K_i well above nM levels, and this is usually interpreted in terms of slow protein conformational changes or loss of water, particularly to accommodate a poorly analogous "transition state analogue"[11] (see Section 5.3.2). Such systems are described by Scheme 5.6.

On addition of enzyme to a mixture of substrate and slow-binding inhibitor, one sees an exponential approach to a steady-state rate (if an increase in absorbance is followed, the analytical form of the curve is the same as for a pre-steady-state "burst" – see Figure 5.38). If K_i (fast) is the K_i calculated from the very initial rates and K_i (slow) from the steady-state rate, then the relation between the two constants is given by eqn. (5.24). The exponential describing the approach to the steady state depends in a complex way on the concentration

$$E + I \rightleftharpoons EI \underset{k_{off}}{\overset{k_{on}}{\rightleftharpoons}} EI'$$

Scheme 5.6 Slow binding and release of competitive inhibitors: two EI complexes are formed.

of inhibitor and substrate,[10] so that it is often convenient to measure K_i (fast) and K_i (slow) in the usual way and measure k_{off} directly from the exponential approach to the steady-state rate when enzyme pre-incubated with an excess of inhibitor is diluted into a large volume of substrate.

$$K_i(\text{slow}) = K_i(\text{fast})[1 + (k_{on}/k_{off})] \tag{5.24}$$

5.3.2 Transition State Analogues and Adventitious Tight-binding Inhibitors

Consider the schematic free energy profile for a single substrate, single product enzyme reaction and its analogous non-enzymic counterpart (Figure 5.6). The free energy differences are those dictated by the van't Hoff isotherm and transition state theory (R = gas constant, h = Planck's constant, κ = Boltzmann's constant, T = absolute temperature). From the very fact that the enzyme is a catalyst, $\Delta G^{\ddagger}_{uncat} \gg \Delta G^{\ddagger}_{cat}$. It is possible to define a free energy difference ΔG°_{TS}, which is the binding energy of the transition state. However, since the enzyme is a catalyst, so that $\Delta G^{\ddagger}_{uncat} \gg \Delta G^{\ddagger}_{cat}$, it follows that $\Delta G^{\circ}_{TS} \gg \Delta G^{\circ}_{S}$, ΔG°_{P}. That is, the enzyme binds the transition state of the catalysed reaction much more tightly than it binds substrates or products. If transition states can possess a state function such as Gibbs free energy (*i.e.* a thermodynamic variable whose value is independent of pathway; this is a big "if", and in some very reactive systems the assumption breaks down[12]), it follows immediately that the enzyme binds the transition state more tightly than substrate *by the same factor by which it catalyses the reaction*. Therefore, we can write eqn. (5.25), the *K*s being dissociation constants:

$$k_{cat}/k_{uncat} = K_S/K_{TS} \tag{5.25}$$

The idea of transition state binding has proved very fruitful in understanding enzymic catalysis in general,[13] as researchers have looked for specific transition state interactions between enzyme and substrate. To the author, though, the frequently encountered statement that "enzymes work by binding transition states", without elaboration, is a tautology – as we have seen, the concept of transition state binding follows immediately from elementary transition state theory.

If enzymes bind transition states more tightly than substrates, then they should bind molecules which resemble transition states in shape and/or charge more tightly than they bind substrates or products. The concept of "transition state analogue" was first applied to glycosyl transfer by Leaback,[14] who

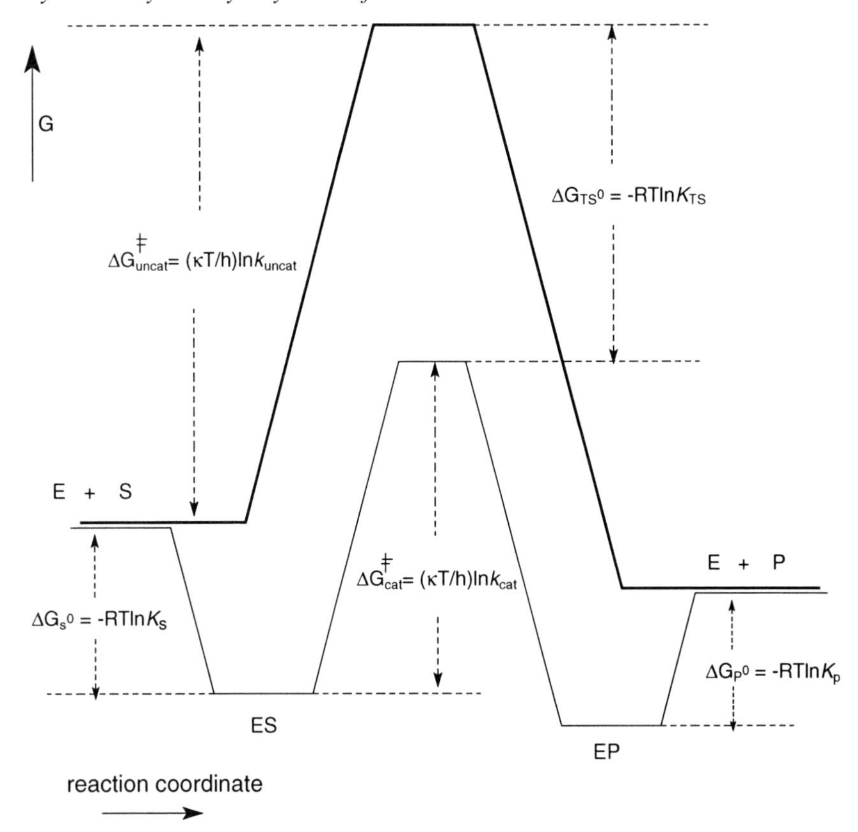

Figure 5.6 Schematic free-energy profile for a single substrate, single product enzyme reaction and its analogous uncatalysed reaction.

reasoned that the long-standing puzzle of the powerful inhibition of glycosidases by aldonolactones[15] could be explained by the resemblance of the lactones to the shape, and to some extent the charge, of a glycosyl cation (Figure 5.7). However, if the lactones were transition state analogues, the analogy was remote: spontaneous cleavage of a β-(1 → 4) pyranose link has a rate constant of $0.9 \times 10^{-15}\,s^{-1}$ at 25 °C,[16] whereas k_{cat} values for enzymes cleaving such linkages were $\sim 10^2\,s^{-1}$. It follows that an exact analogue of the transition state should be bound around 10^{17}-fold more tightly than the substrate, whereas lactones had K_i values in the μM region or a mere 10^3-fold lower than a typical substrate. Why α-glycosidases did not bind lactones tightly was also unexplained.

In the late 1970s and early 1980s, a series of basic inhibitors of glycosidases were investigated (see Figures 5.8 and 5.9). The paradigmal compound was the naturally occurring nojirimycin, 5-amino-5-deoxyglucose.[17] The *manno*[18] and *galacto*[19] analogues were also later isolated as natural products. However, these reducing sugar analogues are unstable, and undergo a series of slow

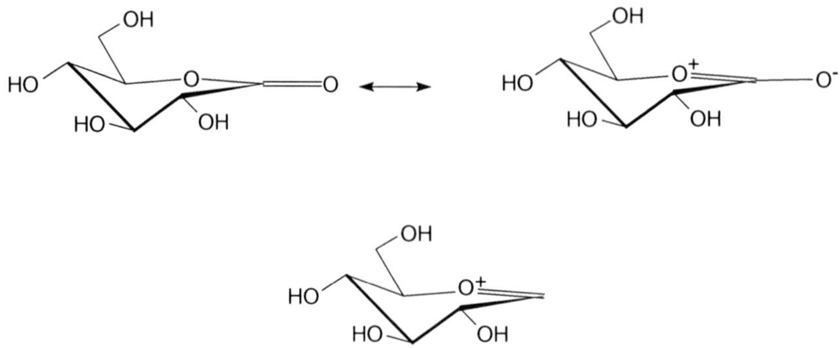

Figure 5.7 Analogy between aldonolactones and oxocarbenium ions.

anomerisation and dehydration reactions. Their reduced analogues, such as deoxynojirimycin, also inhibit the appropriate glycosidases and the iminoalditol structure has become a key structural motif in the design of glycosidase and glycosyl transferase inhibitors [indeed, the synthesis and investigation of 1,5-iminogalactitol ("deoxygalactostatin") predated the isolation of the natural product[20]]. The protonated forms of such structures, and also of natural products such as castanospermine and swainsonine, were considered to mimic oxocarbenium ions (or oxocarbenium ion-like transition states), with the protonated nitrogen atom mimicking the positive oxygen.

This cannot be the whole story, however, as strong inhibition of pyranosidases is observed with a whole range of *five-membered* iminoalditols,[21] including most notably the natural product swainsonine,[22] an α-mannosidase inhibitor. Likewise, isofagomine, in which the positive site mimics C1 rather than O5 of a glucopyranosyl cation, is a good inhibitor of β-glucosidases, but is less effective against α-glucosidases (although it does inhibit).[23] The seven-membered ring analogue of isofagomine, 1;6-anhydro-1,6-imino-D-glucitol, also inhibits a range of glycosidases.[24]

The active sites of most glycosidases contain two or usually three catalytic carboxylates, and it became clear that specific interactions with these carboxylates accounted for much of the enhanced binding of basic inhibitors. Such interactions, perhaps enhanced by fortuitous substrate-mimicking interactions of hydroxyl groups with the protein, had to be responsible for the fact that acyclic polyhydroxyamines were mM inhibitors of glucoamylase[25] and yeast α-glucosidase,[26] and the inhibition of several glycosidases by simple imidazoles, without a sugar recognition site,[27] as well as the perpetually rediscovered inhibition by Tris buffer. The presence of a basic site in place of O1 in a sugar derivative, as in glycosylamines and related compounds, had been shown by Legler[28] to enhance binding, presumably by formation of a salt bridge with a catalytic proton donor on the enzyme. Such interactions were clearly at the root of the inhibitory power of mannostatin, trehalozin and acarbose[29] (although in the last case, the constraining of the carbocyclic ring to a half-chair

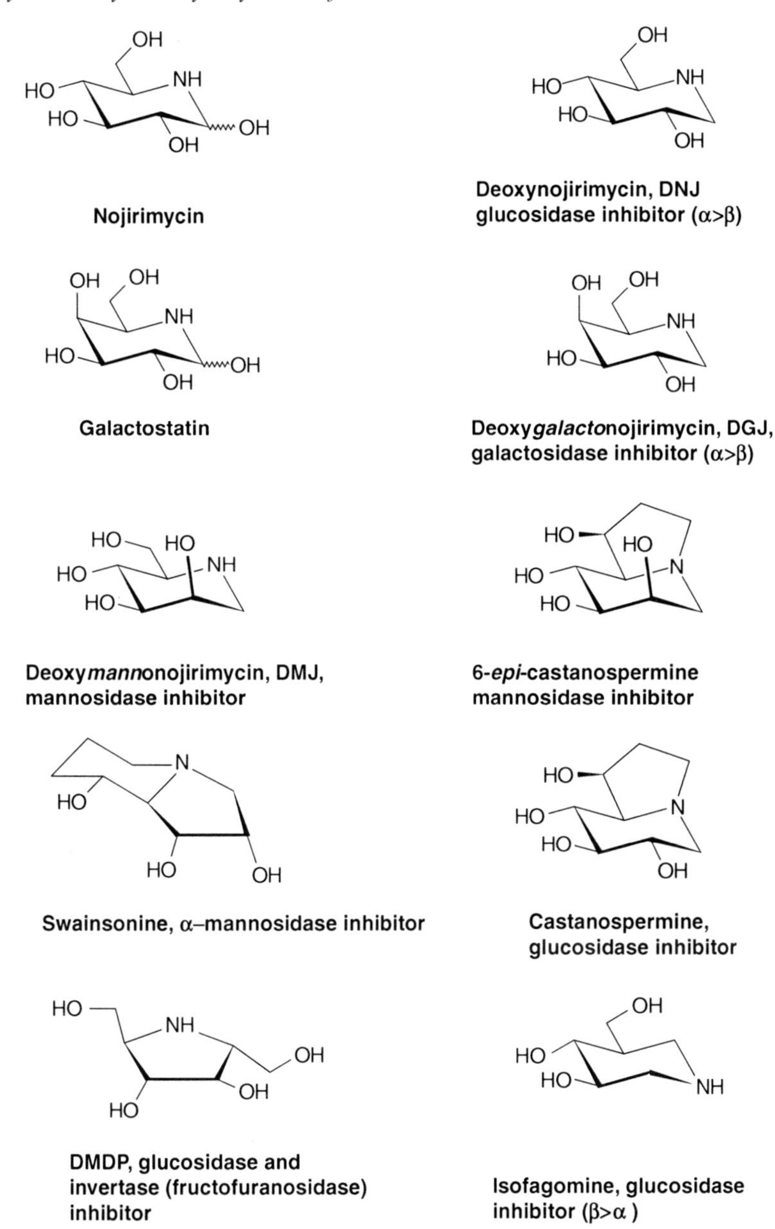

Figure 5.8 Glycosidase inhibitors with endocyclic basic nitrogen.

conformation by the double bond confused matters until it was realised that the half-chair in question was equivalent to a pyranose $^{2}H_{3}$, rather than the $^{4}H_{3}$ conformation of an oxocarbenium ion). A recent study of hydroxylated *N*-benzylcyclopentylamines has shown the expected correlation between configuration and inhibitory selectivity.[30]

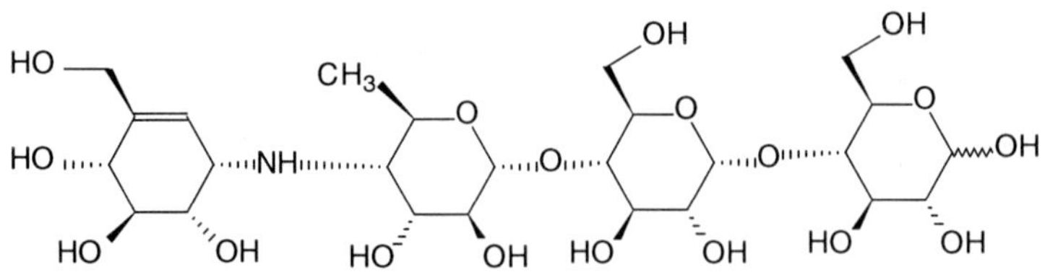

acarbose
(amylase and glucosidase inhibitor)

Related α-glucosidase inhibitors:

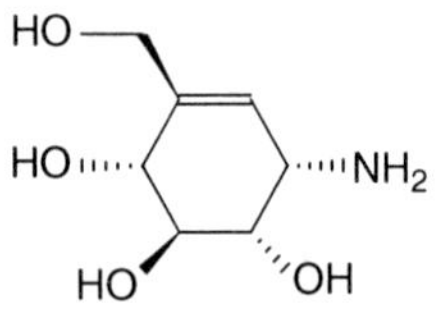

valienamine

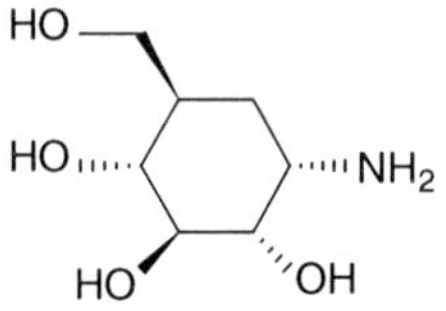

validamine

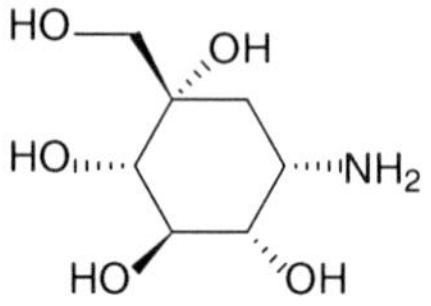

valiolamine

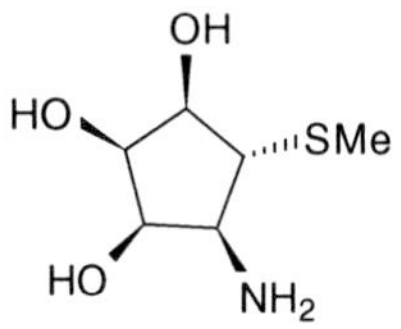

mannostatin (inhibits β-mannosidase)

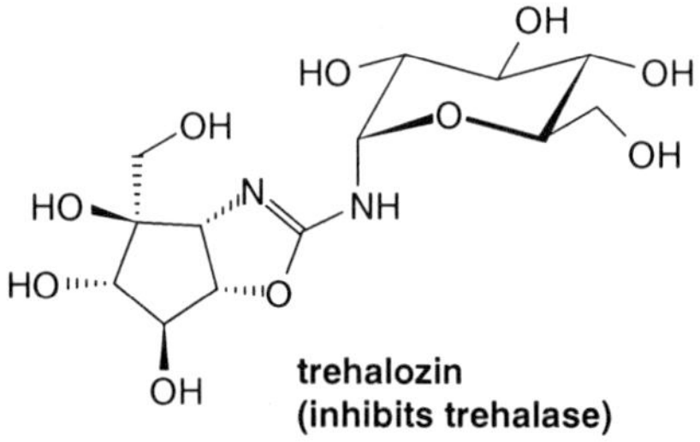

trehalozin (inhibits trehalase)

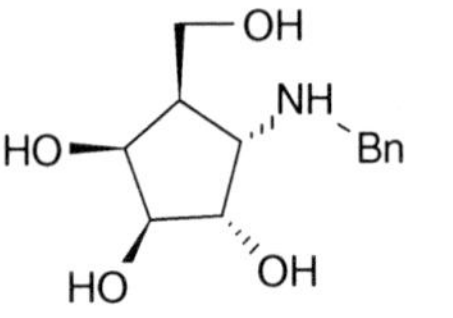

inhibits α galactosidase

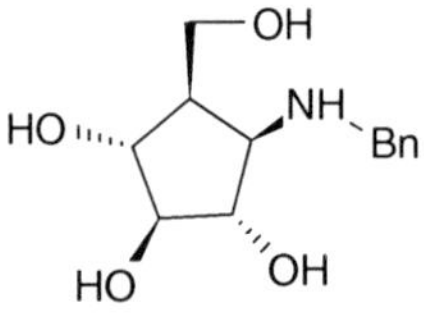

inhibits β-glucosidase

Figure 5.9 Glycosidase inhibitors exploiting interactions with enzyme acid catalytic groups.

Therefore, as was understood 20 years ago, there are three types of interaction which can be exploited in designing glycosidase inhibitors: similarity to the shape of the transition state, similarity to the charge distribution of the transition state and salt bridge/hydrogen bonding interactions with catalytic carboxylates.[28]

In mammals, the oligosaccharide of *N*-glycoproteins is transferred as a tetradecasaccharide from a tetradecasaccharyldolichol pyrophosphate to the nitrogen of protein Asn residues (see Section 5.12.1.9)[31] and the N-linked oligosaccharide is then processed by various "trimming" glycosidases, which remove some of the sugar residues and leave the protein being "correctly addressed" for its role in the cell. The trimming glycosidases therefore became important therapeutic targets, in the treatment of cancer and of viral infections. Unhappily for an understanding of glycosidase mechanism, the result has been an enormous, confused literature on glycosidase inhibition,[32] from which it is too often impossible to retrieve the experimental facts, let alone draw mechanistic conclusions.[x] Three common failings are:

(i) Impurity of inhibitor. Enzymes are labile materials and the by-products of organic syntheses can be well-known enzyme inactivators (Michael acceptors, heavy metal salts, alkylating agents and so forth). Particularly if enzymes are assayed as a single point in a stopped assay, inactivation by dirt in a synthetic product can be mistaken for the desired reversible inhibition. Continuous assays and analytical purity of the inhibitor are needed.

(ii) Impurity of enzyme. Commercial enzymes are often specified in terms of catalytic activity and absence of other activities, not electrophoretic purity. Plant seed enzymes, such as almond β-glucosidase[33] and coffee bean α-galactosidase, are notoriously mixtures of isoenzymes with very different catalytic properties – single K_i values quoted for such indeterminate mixtures of isoenzymes are meaningless. Enzymes should be pure as catalysts: the ideal is a recombinant enzyme expressed in a host with no equivalent endogenous enzyme. Even electrophoretic purity, particularly under standard denaturing conditions (polyacrylamide gel in the presence of the detergent sodium dodecyl sulfate, SDS-PAGE), which separates essentially on monomer molecular weight, can provide false reassurance – for example, the cellulolytic fungus *Trichoderma reesei* excretes at last half a dozen enzymes which hydrolyse β-(1 → 4) glucan links,[34] all monomers with an M_r about 60 kDa and all with radically different properties.

(iii) Wrong measurement. Much of the literature on glycosidase inhibition quotes IC_{50} values. IC_{50} – the inhibitor concentration at which the enzyme reaction is slowed to half its original rate – is a physiologists' parameter, useful in examining metabolism with physiological concentrations of the natural substrate of an enzyme (or indeed cell extracts). In the case of a competitive inhibitor, its value depends on the substrate concentration and K_m, and also the K_i [eqn. (5.22)]; if these are not both given, the measurement is meaningless.

---

[x] The outcome in terms of clinically useful drugs has also been meagre: Relenza, a designed sialidase inhibitor, is in use and *N*-butyldeoxynojirimycin, an inhibitor of one of the trimming α-glucosidases, is in clinical trial against HIV/AIDS.

K_i values are measured from initial rates at a range of substrate and inhibitor concentrations; it is usual to measure v for a matrix of values of [S] and [I], with [S] ranging from $\sim 0.7 \times K_m$ to $\sim 3 \times K_m$ and [I] ranging from $\sim 0.7 \times K_m$ to $\sim 3 \times K_m$. Around 25 initial rate measurements is a minimum: if it is not desired to publish a pleasing double reciprocal plot, these could arise from unreplicated measurements at each of five values of [S] and [I], although very pretty plots result from thrice-replicated measurements at each of three well-chosen values of [S] and [I]. The use of microtitre plates is a mixed blessing: automated plate-readers reduce the tedium of the measurements and hugely economise on inhibitor, substrate and enzyme, but as their use depends on a single point stopped assay, they should only be used when it is established that neither the inhibitor nor substrate causes non-linearity in the assay.

Even a properly determined K_i at a single pH can be uninformative, as the pH dependences of catalysis and of inhibition can differ, not least because the pK_a values of iminoalditol inhibitors lie between 6 and 8. Ideally, a plot of both pK_i and $\log(k_{cat}/K_m)$ against pH is needed. Most glycosidases show a bell-shaped dependence of k_{cat}/K_m on pH, corresponding to ionisations of free enzyme (see Section 5.4 2); more often than not the identification of the alkaline limb with deprotonation of the catalytic acid and the acid limb with protonation of the nucleophile in retaining glycosidases or the general base in inverting glycosidases is largely correct. In the case of an α-L-arabinofuranosidase of the fungus *Monlinea fructigena* (AFIII), the acid limb was not observed (it is probably at very low pH, since the enzyme had evolved to work in rotting fruit) and a sigmoid variation of k_{cat}/K_m on pH was observed [eqn. (5.26), with $K_E = 10^{-5.9}$].

$$\frac{k_{cat}}{K_m} = \frac{A}{1 + K_E/[H^+]} \tag{5.26}$$

The binding of two iminoalditol inhibitors showed bell-shaped profiles, with the acid limb corresponding to the acid dissociation of the enzyme (constant K_E) and the basic limb to the deprotonation the inhibitor, acid dissociation constant K_I [eqn. (5.27)].

$$\frac{1}{K_i} = \frac{A}{1 + [H^+]/K_E + K_I/[H^+]} \tag{5.27}$$

This is exactly what one would expect for binding of deprotonated enzyme to protonated inhibitor. The pH of maximal inhibition (~ 7) is well removed from the pH optimum of catalysis (<5). However, as with all pH variation studies (see the discussion of kinetic equivalence, Section 3.7.4), information is provided on the composition of the complex, not its structure. If we set $B = AK_E/K_I$, then eqn. (5.28) holds, giving the appearance of the binding of protonated enzyme to deprotonated inhibitor. pH variation studies cannot by themselves make the distinction between the two modes of binding.[35]

$$\frac{1}{K_i} = \frac{B}{1 + [H^+]/K_I + K_E/[H^+]} \tag{5.28}$$

As with in the arabinofuranosidase work, a pH study of isofagomine binding to a GH 1 β-glucosidase from *Thermotoga maritima* indicated that the complex contained one proton located on either the inhibitor, the acid catalyst or the nucleophile.[36] In this case the acid limb of catalysis of the k_{cat}/K_m–pH plot was accessible (pK_a 4.75), which resulted in a pH optimum for inhibition of 7.6, in contrast to maximal catalytic activity at pH 5.85. X- ray studies of the complex of the enzyme and inhibitor showed that a strong hydrogen bond was formed between the amine nitrogen and the nucleophilic carboxylate. Such interaction had earlier been proposed to explain the tighter binding of isofagomine-type inhibitors to β-glycosidases and of deoxynojirimycin-type inhibitors to α-glycosidases, at least if the enzymes are retaining. In the preferred case, a salt bridge can be made with the nucleophilic carboxylate,[37] *i.e.* X in Figure 5.1. In support of this explanation, a hydroxylated piperazine of the *gluco* configuration binds equally tightly to α- and β-glucosidases[38] (Figure 5.10).

Careful studies of analogous oxazines with *T. maritima* GH 1[39] and an endoglucanase (Cel5A from *Bacillus agaradhaerens*) revealed that here the pH optimum of catalysis corresponded with the pH optimum of inhibition. The oxazine pK_a lies well below the catalytic optimum pH. Again, however, the complex contained one proton on the array of inhibitor, nucleophile and acid catalyst, although in this case it resided on one of the two enzyme catalytic groups.

Heightman and Vasella attempted to improve the binding of a range of glycosidase inhibitors by incorporating a fused five-membered aromatic ring, containing 2–4 nitrogen atoms, which constrained the pyranose ring to the same conformations as an oxocarbenium ion (Figure 5.11).[37] Enzymes from GH 1, 2, 5 and 10 were inhibited if there was a nitrogen lone pair on the atom equivalent to O1 of the substrate, but not if there were a carbon atom; moreover, the inhibition increased with the basicity of the nitrogen heterocycle. With enzymes from the other GH families tested, inhibition was weak. Examination of X-ray crystal structures of complexes between these inhibitors and various glycosidases and of the enzymes themselves complexed with substrate–analogues led to remarkable conclusions about the trajectory of proton donation in glycosidases. Largely for typographical reasons, proposed mechanisms of β-glycosidases had been written with the proton donor above the glycosidic oxygen atom: in the conformation preferred by the *exo*-anomeric effect (Section 4.4), this involved proton donation into the "vertical" sp^3 lone pairs. However, it appears this does not happen: proton donation is to the lone pair *anti* to the C1–O5 bond. The other trajectory appears to be proton donation to a lone pair *syn* to the C1–O5 bond: to do this, the C1–O1 dihedral angle has to adopt a conformation disfavoured by the *exo*-anomeric effect or steric effects.

The question of whether a molecule is a true transition state analogue or merely an adventitious tight binder can be addressed by preparing a series of site-directed mutants of the enzyme in question, and measuring k_{cat}/K_m against (ideally the natural) substrate and K_i for inhibitor for each mutant. For a true transition state analogue, there should be a linear relationship between $\log(k_{cat}/K_m)$ and $\log K_i$, as a linear relationship should exist between free energy of binding of the inhibitor

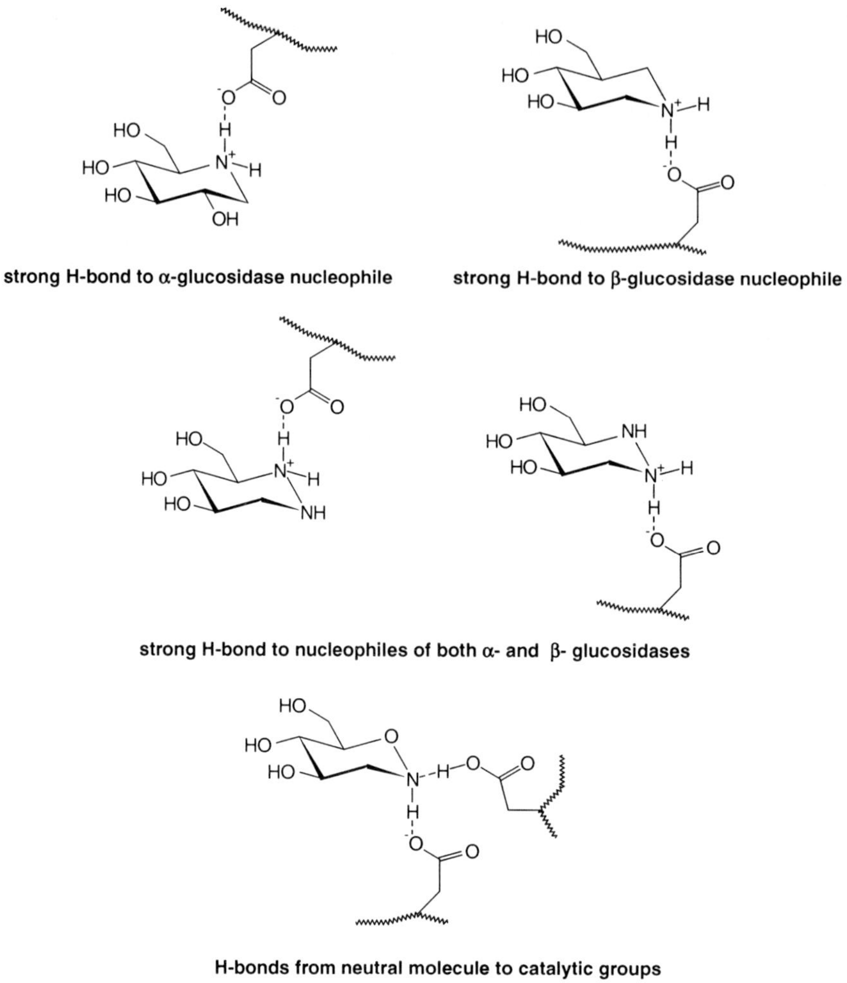

strong H-bond to α-glucosidase nucleophile **strong H-bond to β-glucosidase nucleophile**

strong H-bond to nucleophiles of both α- and β- glucosidases

H-bonds from neutral molecule to catalytic groups

Figure 5.10 Interactions of isofagomine-type and deoxynojirimycin-type inhibitors with retaining glucosidases. The piperazine can make either type of interaction and inhibits both α- and β-glucosidases comparably powerfully. The oxazine appears to be unprotonated and oxazine-based inhibitors have inhibition optima close to the optimum for catalysis.

and free energy of binding of the transition state. The labour involved in such experiments has restricted their use to a study of a series of mutants of the GH 13 transglycosylase, cyclodextrin transglycosylase, binding to acarbose and acarbose and deoxynojirimycin binding to GH 15 glucoamylase.[40]

The relationship held for acarbose binding to the GH 13 transglycosylase, but with a slope of only 0.4 and a better correlation for the non-natural substrate α-glucosyl fluoride than for the "more natural" substrate α-maltotriosyl fluoride.[41] A less onerous version of the same idea was exploited with measurements of the binding of *manno*- and *gluco*-tetrazoles to glycosidases of

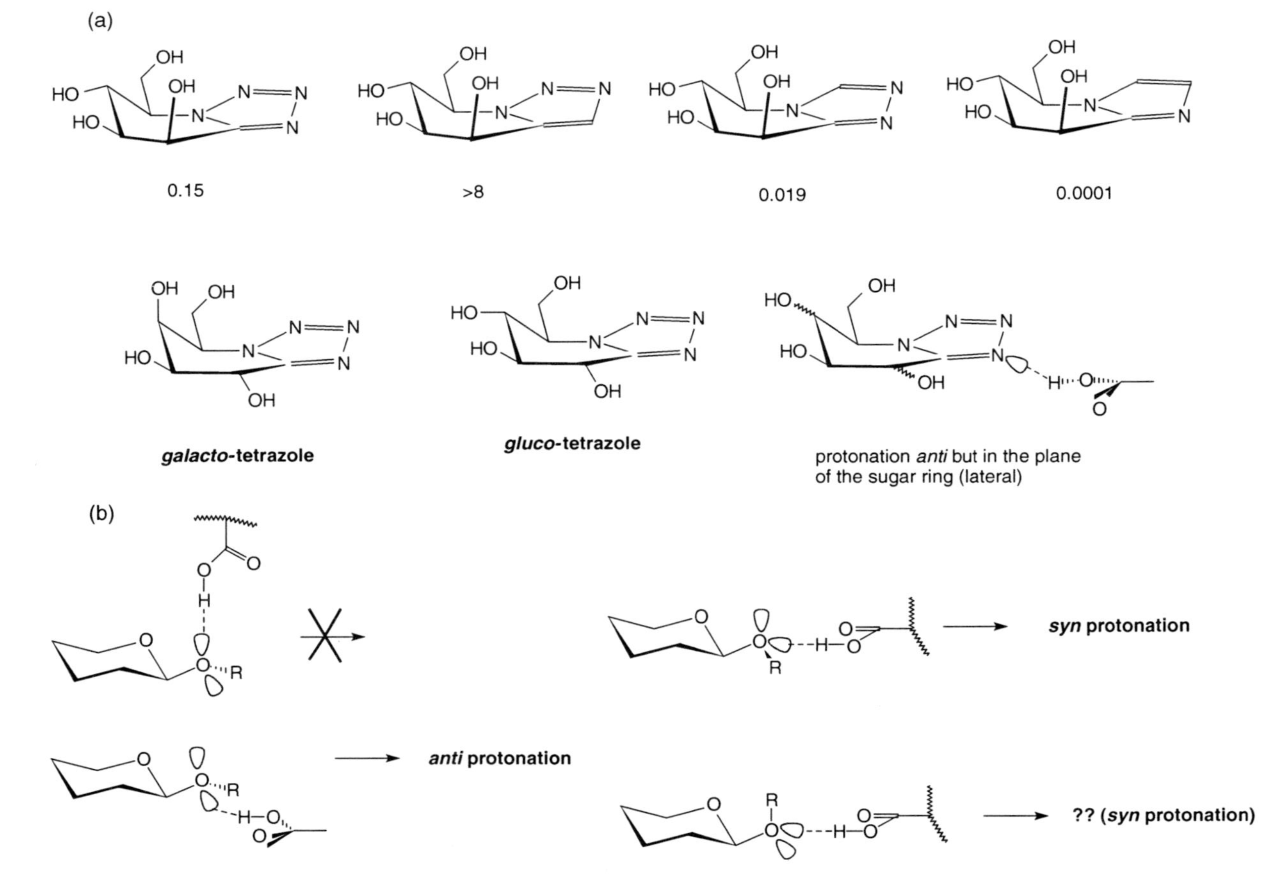

Figure 5.11 (a) Structures of substituted piperidines annulated with five-membered azoles and K_i values (mM) for the *manno* series binding to snail β-mannosidase. (b) *Syn* and *anti* protonation of an equatorial leaving group.

relaxed specificity, so that each bound both tetrazoles: the plot of pK_i for the tetrazole against $\log(k_{cat}/K_m)$ of the corresponding *p*-nitrophenyl glycoside ranged over five orders of magnitude and was linear with a gradient of 1.[42]

Very recently, in the first thorough study, the hydrolysis of o-nitrophenyl xylobioside by ten mutants of the GH10 xylanase/glycanase Cex from *Cellulomonas Document1fimi* was correlated with K_i for the binding of five xylobiose analogues. The three inhibitors in which the atom corresponding to C1 was sp^2 hybridised showed excellent correlations between -log K_i and log k_{cat}/K_m ($r = 0.97$–0.99), whereas both the isofagomine-type inhibitor ($r = 0.77$) and the deoxynojirimycin-type inhibitor ($r = 0.89$), in which the atom corresponding to C1 was sp^3 hybridised, showed worse correlations, even though, in the case of the isofagomine-type inhibitor, absolute values of K_i were the lowest observed.

It has been suggested that true transition state analogues should be distinguished from adventitious tight binders by their favourable enthalpy, rather than entropy, of interaction with the enzyme.[43] The difference between the thermodynamics of binding of isofagomine (enthalpy-driven) and deoxynojirimycin (entropy-driven) to *T. maritima* GH 1 β-glucosidase then seemed to indicate that isofagomine was the transition state analogue and deoxynojirimycin was the adventitious inhibitor; however, this may simply be a consequence of the strong hydrogen bond in the isofagomine case, where deoxynojirimycin binding is modulated by ordering of water molecules.

5.3.3 Anticompetitive Inhibition

If an inhibitor binds only to the glycosyl-enzyme intermediate of a retaining glycosidase, then uncompetitive (sometimes more informatively called anticompetitive) inhibition is observed. The rate law is that of eqn. (5.29). It should be noted that, like the analogous case of substrate inhibition described by eqn. (5.17), K_i is $K_d[(k_{+2}+k_{+3})/k_{+2}]$, where K_d is the dissociation constant of the glycosyl-enzyme intermediate.

$$v = \frac{V_{max}[S]}{K_m + [S]\left(1 + \frac{[I]}{K_i}\right)} \quad (5.29)$$

Inhibition can be analysed visually by double reciprocal plots (Figure 5.12). In the case of competitive inhibition, $1/v$ *versus* $1/[S]$ plots at various values of [I] give a series of lines all intersecting at $1/[S] = 0$, $1/v = 1/V_{max}$, whereas in the case of anticompetitive inhibition, the $1/v$ *versus* $1/[S]$ plots at various values of [I] are all parallel. The fascination of double reciprocal plots was such that a further form of inhibition, where the lines all intersected at $1/v = 0$, $1/[S] = -1/K_m$ was defined and confusingly termed non-competitive inhibition; the physical situation, rarely encountered in practice, was that inhibitor bound equally to E and ES. The rate law is that of eqn. (5.30). In practice, inhibition which is neither cleanly competitive nor cleanly uncompetitive is best described by the "mixed inhibition" eqn. (5.31).

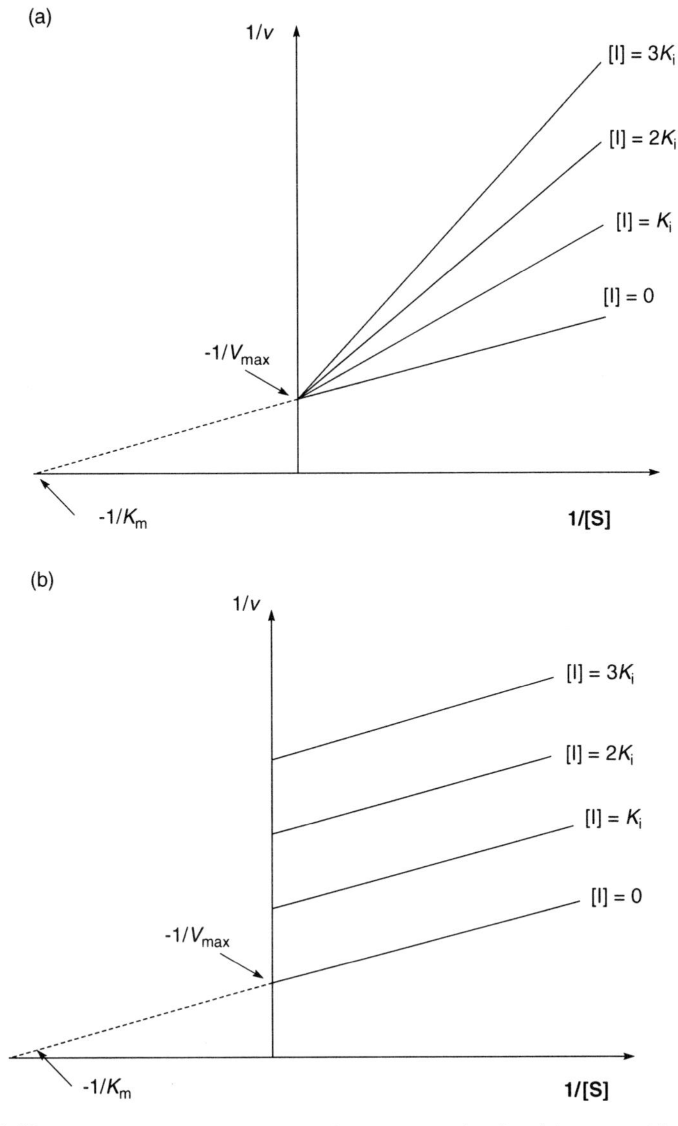

Figure 5.12 Double reciprocal plots, drawn to scale, for (a) competitive and (b) anticompetitive inhibition.

$$v = \frac{V_{\max}[\mathrm{S}]}{K_{\mathrm{m}}\left(1 + \frac{[\mathrm{I}]}{K_{\mathrm{i}}}\right) + [\mathrm{S}]\left(1 + \frac{[\mathrm{I}]}{K_{\mathrm{i}}}\right)} \tag{5.30}$$

$$v = \frac{V_{\max}[\mathrm{S}]}{K_{\mathrm{m}}\left(1 + \frac{[\mathrm{I}]}{K_{\mathrm{i}}}\right) + [\mathrm{S}]\left(1 + \frac{[\mathrm{I}]}{\alpha K_{\mathrm{i}}}\right)} \tag{5.31}$$

5.4 DETERMINATION OF THE MECHANISM OF ENZYMIC GLYCOSYL TRANSFER – MODIFICATION OF TOOLS FROM SMALL-MOLECULE PHYSICAL ORGANIC CHEMISTRY AND NEW TOOLS FROM PROTEIN CHEMISTRY

Many of the tools of physical organic chemistry can be applied to enzyme mechanisms, but their application requires an awareness of two properties of enzymes: that they are labile polyelectrolytes, and that their kinetics usually involve a number of sequential kinetic steps, which evolution tends to make of similar rates, at least for the natural substrate. To offset these restrictions on the use of standard physical-organic tools, protein X-ray crystallography permits the observation of bound substrate or substrate–analogue in its reactive ground-state conformation, thus circumventing the Curtin–Hammond prohibition on observing reactive ground-state conformations, that applies to small molecule reactions.

5.4.1 Temperature Dependence of Rates and Equilibria

Whereas energies of activation for a rate process and standard enthalpies for equilibria involving small molecules can be calculated with confidence from Arrhenius or van't Hoff plots of ln*k* or ln*K* against 1/T [eqns (5.32) and (5.33), respectively, where *R* is the gas constant], the situation with enzymes is not so clear cut.

$$\ln k = \ln A - E_a/RT \tag{5.32}$$

$$\ln K = \Delta S^{\ominus}/R - \Delta H^{\ominus}/RT \tag{5.33}$$

Natural selection will have made all the steps in an enzyme turnover of its natural substrate roughly similar in rate at its working temperature, but that does not mean that they have the same activation energies. Indeed, with the discovery of organic cosolvents (such as dimethyl sulfoxide), which permitted enzymes to be active[44] and the media still liquid well below 0 °C, the field of cryoenzymology was founded: it often became possible to follow different kinetic steps of an enzyme simply by slowly raising the temperature of an enzyme-substrate solution.[45] Over such extended temperature ranges, Arrhenius plots may not be linear, simply because each segment refers to a different chemical process – this appears to be the case for GH 2 (*lacZ*) β-galactosidase from *Escherichia coli*,[46] but not a β-glucosidase from the extreme thermophile *Caldocellum saccharolyticum*, which exhibits a linear Arrhenius plot between 90 °C and −75 °C.[47]

A second reason for caution is that in enzyme reactions the heat capacity change may be appreciable, which means that enthalpy changes themselves are temperature dependent: one can define a standard heat capacity change and a heat capacity of activation as in eqns (5.34) and (5.35):

$$\frac{\partial \Delta H^{\ominus}}{\partial T} = \Delta C_p^{\ominus} \tag{5.34}$$

$$\frac{\partial \Delta H^{\ddagger}}{\partial T} = \Delta C_{\mathrm{p}}^{\ddagger} \tag{5.35}$$

The existence of a very large $\Delta C_{\mathrm{p}}^{\ominus}$ is one possible reason for the value of $\Delta H^{\ominus}$ of $+58.6\,\mathrm{kJ\,mol^{-1}}$ obtained by application of the van't Hoff equation, eqn. (5.33), to K_{i} measured for the inhibition of almond β-glucosidase by isofagomine[48] which differed even in sign from the value measured directly by isothermal titration calorimetry.[36]

"Optimum temperatures" of enzymes are often quoted in the literature. For the most part these are an artefact of two opposing phenomena and single-point stopped assays. The rate of the enzyme reaction itself increases with increasing temperature according to eqn. (5.32), as does the rate of enzyme denaturation. Enzyme denaturation generally has a very high E_{a}, so at some temperature the rate of enzyme denaturation becomes significant. In a series of single-point stopped assays at different temperatures, therefore, at some point the denaturation of the enzyme during the course of the assay overtakes the increase in rate of the catalysed reaction, giving rise to an "optimum temperature".

5.4.2 Effect of Change of pH

Whereas the effect of changing pH on a reaction involving only small molecules can be immediately informative, the interpretation of the effects of pH on enzyme reactions is more complex, since enzymes have many ionising groups and often react by a series of chemical steps with similar rates.

The classical treatment of Alberty and Bloomfield[49] assumed that all proton transfers were rapid, the substrate was not "sticky" and the reaction proceeded through one protonic form of the enzyme. The ionisations then seen in the plot of $k_{\mathrm{cat}}/K_{\mathrm{m}}$ against pH represented only ionisations of free enzyme and free substrate. Some enzymes have sigmoidal ionisations, attributable to the loss of one proton, either with the protonated form active [eqn. (5:26)] or with the deprotonated form active [eqn. (5.36)], but the commonest form of variation is the bell-shaped curve [eqn. (5.37)]:

$$\frac{k_{\mathrm{cat}}}{K_{\mathrm{m}}} = \frac{A}{1 + [\mathrm{H}^+]/K_{\mathrm{A}}} \tag{5.36}$$

$$\frac{k_{\mathrm{cat}}}{K_{\mathrm{m}}} = \frac{(k_{\mathrm{cat}}/K_{\mathrm{m}})_{\mathrm{max}}}{1 + [\mathrm{H}^+]/K_{\mathrm{B}} + K_{\mathrm{A}}/[\mathrm{H}^+]} \tag{5.37}$$

Many glycosidases exhibit such a pH–rate profile and the temptation to equate K_{A} with the dissociation constant of the acid-catalytic group and K_{B} with the dissociation constant of the base or nucleophile frequently proves irresistible. Even in an Alberty–Bloomfield system, however, the equation should be not made unthinkingly. K_{A} and K_{B} are *macroscopic* dissociation constants, which are related to the *four* microscopic dissociation constants of the two-carboxylate system (Figure 5.13) by eqns (5.38) and (5.39) (since Figure 5.13

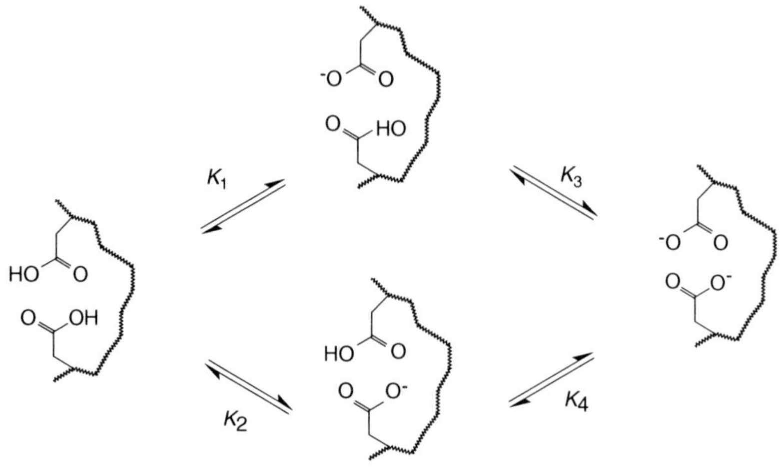

Figure 5.13 Protonation states of a glycosidase with two active site carboxylates.

describes a thermodynamic box, $K_1K_3 = K_2K_4$, only three dissociation constants have to be determined individually for a full description of the system).

$$K_B = K_1 + K_2 \tag{5.38}$$

$$K_A = K_3K_4/(K_3 + K_4) \tag{5.39}$$

If the bottom carboxylate (for example, a nucleophile), is very much more acidic than the top carboxylate (for example, an acid catalyst), then $K_2 \gg K_1$, $K_B = K_2$ and $K_3 \gg K_4$ so that $K_A = K_4$ and the commonly held perception that the acid limb of a bell-shaped pH–rate profile for a glycosidase represents protonation of the catalytic nucleophile/base and the basic limb the deprotonation of an acid catalyst is not incorrect. However, it is not necessarily correct either. Simple pH studies can never tell anything about the structure of the species involved, only their composition, so they cannot distinguish between catalysis by the two tautomeric forms of the monoprotonated species. If we substitute $K_B = K_2$ and $K_A = K_4$ in eqn. (5.38) and then multiply the top and bottom of the right-hand side by K_4/K_2 we obtain eqn. (5.40), which is the form that would be expected for the most acidic group acting as the catalysing acid and the least acidic as the catalysing base, *i.e.* catalysis by the top, not bottom, tautomer in Figure 5.13. K_2/K_4 approximates to the tautomerisation constant of the monoprotonated species.

$$\frac{k_{cat}}{K_m} = \frac{(k_{cat}/K_m)_{max}}{1 + [H^+]/K_2 + K_4/[H^+]} = \frac{K_2/K_4(k_{cat}/K_m)_{max}}{K_2/K_4 + [H^+]/K_4 + K_2/[H^+]} \tag{5.40}$$

Such "reverse protonation" mechanisms, in which the enzyme acts through a minor tautomer of the catalytically active species, are occasionally encountered. The GH 11 retaining β-xylanase from *Bacillus circulans* has an M_r of only 20 kDa, so it was possible to examine the ionisation state of the two catalytic glutamates directly by growing the organism on ^{13}C-labelled glutamic acid and following NMR chemical shifts. All the four microscopic ionisation states of Figure 5.10 could be observed, but the ionisations of the macroscopic pH–k_{cat}/K_m profile were indeed largely accounted for by K_2 and K_4.[50] However, in a mutant in which an "auxiliary" active site carboxylate had been mutated to its electrical neutral carboxamide, the correspondence between macroscopic and microscopic protonation disappeared and the mutant enzyme worked by a reverse protonation mechanism.[51]

The requirement of the Bloomfield–Alberty treatment, that all enzyme flux goes through a single protonation state is obviously, with a polyelectrolyte, going to be a gross approximation; indeed, most enzyme systems which give apparently clean bell-shaped pH profiles can, on closer examination (particularly with mutation of a major contributor to the ionisation curve to a non-ionising group) be shown to be more complex.[52] Ionisations which influence, but do not completely control, the pH behaviour of catalysis must be commonplace, but not often noticed because of the low precision of most experimental pH–rate profiles.

Most enzyme kinetic studies assume that proton transfer is fast compared with catalysis, but this is also not necessarily so. It has long been known that bimolecular rate constants for proton transfer between electronegative atoms follow an "Eigen curve", with the rate in the thermodynamically favourable direction being diffusion controlled (10^{10} M^{-1} s^{-1} at ambient temperature) and in the thermodynamically unfavourable direction being $10^{10-\Delta pK}$ M^{-1} s^{-1}, where ΔpK is the difference in acidity between acid and base (see Section 1.2.1.3).[53] If we consider a group with a pK_a of ~ 10 (such as a cysteine thiol, a lysine NH_3^+ or a tyrosine OH) in solution in pure water (55 M, pK_a -1.7), it is seen that the rate of deprotonation rate by water is only 1 s^{-1}, below the k_{cat} values for most enzymes. The rate of deprotonation by OH^- (thermodynamically downhill) is only 100 s^{-1} even at pH 6 ($[OH^-] = 10^{-8}$ M). Buffering improves matters considerably: in 0.1 M sodium acetate (p$K_{HOAc} = 4.7$), the deprotonation rate becomes 2×10^4 s^{-1}.

Slow proton transfer makes possible the occurrence of "iso" mechanisms – mechanisms in which the form of the enzyme released after catalysis is different to that at the start of the cycle. A candidate would be any inverting glycosidase, which is released with the acid group deprotonated and the basic group protonated [Figure 1(b)], although no example in the glycosyl transfer area has yet been demonstrated (the best example is proline racemase,[54] in which two cysteines act, one as an acid and the other as a base).

The interpretation of the pH dependence of the separated parameters k_{cat} and K_m presents even more problems than that of their ratio. An additional problem here is the "kinetic pK_a", in which a pH–rate profile of a parameter which is a blend of two or more steps shows a dependence on an ionisation of a certain pK_a, even though nothing has in fact ionised. As an illustration, consider the expression for k_{cat} for a two-step glycosidase [eqn. (5.5)] and make k_{+3} pH independent with a rate of 100 s^{-1} and k_{+2} proportional to $[OH^-]$, with a rate of 10^{pH-5}. Inserting these values into eqn. (5.5) (and dividing through by

10^{pH-5}) gives eqn. (5.41):

$$k_{cat} = \frac{100}{1 + [H^+]/10^{-7}} \quad (5.41)$$

which is exactly what would be expected for a single chemical step dependent on the deprotonation of a group of pK_a 7.

5.4.3 Determination of Stereochemistry

Reaction stereochemistry is the most fundamental piece of mechanistic information about a glycosyl transfer. In the case of glycosyl transferases and transglycosidases, the stereochemistry is simply determined by the structures of stable reactants and products. Reducing sugars, however, mutarotate. Many, but not all, retaining glycosidases have transferase activity and in this case circumventing the problem of mutarotation of the product sugar by carrying out the hydrolysis in solutions of potential acceptors can give transfer products which no longer mutarotate, and can be analysed at leisure. For example, the α-L-arabinofuranosidases of *Monilinia fructigena* were shown to act with retention by carrying out the hydrolysis of *p*-nitrophenyl α-L-arabinofuranosidase in the presence of methanol and identifying the product as methyl the α-L-arabinofuranoside by glc of its trimethylsilyl ether.[55] Although it is always advisable to characterise transfer products, their very existence can be taken as evidence for retention. Were an inverting glycosidase to have transferase activity, it must, by the principle of microscopic reversibility, act on both anomers of their substrate.

If no transfer products are formed, determining the reaction stereochemistry for a glycosidase requires, first, that the chemical flux from enzyme catalysis is adequate to maintain an anomeric composition significantly different from the position of mutarotational equilibrium, and, second, that the analytical method is fast compared with mutarotation. Stereochemistry determination is therefore easiest with glucosidases and glucanases, where the new reducing ends mutarotate slowly – so slowly, indeed, that anomers can be separated by HPLC on a commercial Dextropak column, so that regiochemistry and stereochemistry for glucanases can be determined from a single HPLC trace.[56]

NMR is the most powerful general method for determining glycosidase stereochemistry. In the case of aldopyranosidases, it is most convenient to follow the anomeric ^{1}H resonances as a function of time – an example is given in Figure 5.14. It is necessary to dissolve the substrate in buffered D_2O and to concentrate highly and/or exchange the enzyme before the experiment, as axial anomeric protons can be masked by large HOD peaks. Since mutarotation is catalysed by buffers, particularly ambident types such as phosphate, it is useful to keep the buffer concentrations as low as possible and use sterically hindered systems such as MOPS,[xi] which are poor catalysts of mutarotation. The use of

[xi] 3-(*N*-Morpholino)propane-1-sulfonic acid.

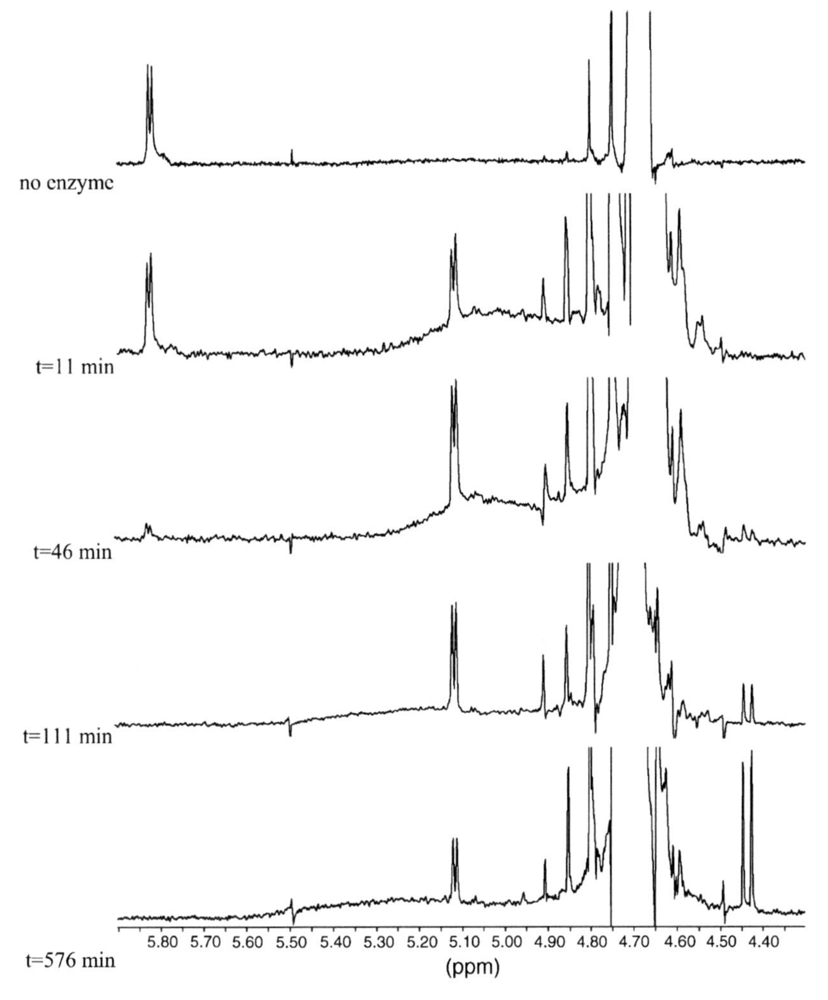

Figure 5.14 Demonstration of retention of configuration by a GH 27 α-galactosidase from *Phanerochaete chrysosporium* by ^{1}H NMR.[59] 1-Naphthyl α-D-galactopyranoside (2.7 mM) was hydrolysed by 0.022 mg mL^{-1} of enzyme in 50 mM sodium formate buffer in D_2O, pD 3.75, at 20 °C. The ^{1}H spectrum shows the anomeric region, peaks centred on δ 5.82 being H1 of the substrate, on δ 5.12 being H1 of α-D-galactopyranose and on δ 4.44 being H1 of β-D-galactopyranose. The large peak centred on δ 4.70 is residual HOD.

D_2O has the further advantage that the solvent isotope effect on mutarotation is usually greater than that on glycosidase catalysis. The temperature coefficient of most enzymes, moreover, is less than that of mutarotation, so it is often advantageous to cool the sample. Finally, the more powerful the magnet, the fewer the FIDs and hence the shorter the time that is required to obtain a good spectrum. *Tours de force* of NMR-based glycosidase stereochemistry determination, were those of an polygalacturonidase, which had to cope with the rapid mutarotation of galacturonic acid residues (presumably because of the acid group acting as an intramolecular general acid catalyst),[57] and an arabinanase, which had to cope with the fast mutarotation of furanose residues.[58]

Polarimetry is the traditional method of determining reaction stereochemistry, but unless the hydrolysis reaction is also monitored simultaneously by another method, it can be misleading.[60] The traditional method of determining whether α- or β-glucopyranose was formed was to add sodium hydroxide, an efficient catalyst of mutarotation; solutions containing excess α-glucopyranose showed a step decrease in rotation and those containing excess β-glucopyranose a step increase.

If the appropriate glycosyl fluorides can be made in the first place, they are excellent substrates with which to determine enzyme stereochemistry, since they have a very good leaving group whose small size makes them excellent substrates for inverting, in addition to retaining, enzymes. Additionally, for reactions followed polarimetrically, since –F and –OH are isoelectronic, the specific rotations of fluorides are similar to those of the corresponding reducing sugar, so rotation time courses are immediately informative.

A particular problem is the demonstration of inversion with sialidases, since they act on α-*N*-acetylneuraminides, but the mutarotational equilibrium is 10% α:90% β.[61] The chemical shifts of the two H3 protons have different chemical shifts in the α- and β-forms of *N*-acetylneuraminic acid and its mutarotation is not inconveniently fast, so demonstrating retention is straightforward. However, to demonstrate inversion, it is necessary to match the rate of liberation of aglycone with that of a retaining sialidase and plot time courses for both enzymes, in order to distinguish inversion from uninformative conversion of substrate to the equilibrium mixture;[62] *i.e.* it has to be demonstrated that the chemical flux through the system is easily adequate to maintain non-equilibrium proportions of α- and β-*N*-acetylneuraminic acid.

5.4.4 Kinetic Isotope Effects

Whereas alteration of enzyme or substrate structure causes large perturbations to the potential energy functions describing enzyme–substrate interactions, a consequence of the Born–Oppenheimer approximation[xii] is that isotopic substitution causes no perturbation of potential energy functions at all. Isotope effects are therefore amongst the most powerful methods of determining enzyme mechanism. If they are measured by any sort of competition method (for example, isotope enrichment in the unreacted starting material or product) then, because one isotopomer acts as a competitive inhibitor of the

[xii] A fundamental assumption of chemical physics, that the motions of electrons and nuclei can be treated independently, since the latter are at least 1000-fold heavier than the former. If it breaks down, description of ions and molecules becomes an insoluble many-body problem.

transformation of the other, the observed isotope effect is always that on k_{cat}/K_m. A problem with small secondary hydrogen effects on k_{cat}/K_m, though, is that there can be significant isotope effects on the binding of the labelled substrate to the protein, as has been demonstrated for human 2′-deoxythymidine arsenolysis catalysed by human thymidine phosphorylase,[63] and is probable for a 5′-tritium effect of 1.028 for arsenolysis of inosine by purine nucleoside phosphorylase.[64] It should always be borne in mind that non-specific hydrogen isotope effects on apolar binding interactions can be comparable to small "chemical" secondary effects of interest – baseline separation of benzophenone and its decadeutero derivative in GLC has been observed![65]

Isotope effects on k_{cat}/K_m are almost always determined by competition, with the isotope ratio in the substrate and product being determined by isotope ratio mass spectrometry or the simultaneous counting of two radioisotopes (one of them being in a hopefully "remote" position). In the case of glucosidases, Berti and co-workers have solved the major problem in the use of the Singleton method (Section 3.10) for enzyme reactions, removal of a large excess of product, by oxidising the glucose product of a glucosidase with glucose oxidase and removing the gluconate ion with an anion exchanger. To measure isotope effects on k_{cat}, it is necessary to compare directly zero-order rates at $[S] \gg K_m$ for highly labelled and unlabelled substrate. Likewise, measurement of isotope effects on individual constants requires direct comparison of pre-steady-state data with labelled and unlabelled substrate.

The Northrop–Cleland nomenclature system for isotope effects greatly simplifies their discussion:[66] the non-abundant isotope and, in the case of secondary effects, site of substitution are written as superscripts to V or V/K in parentheses, so that ${}^{\alpha D}(V/K)$ refers to an α-deuterium kinetic isotope effect on k_{cat}/K_m and ${}^{18}(V)$ refers to an ^{18}O effect on k_{cat}. The effects on individual rate and equilibrium constants are written as superscripts: ${}^{\beta T}K$ is the β-tritium effect on an equilibrium constant and ${}^{14}k_{+2}$ is the ^{14}C effect on the k_{+2} step. Although in principle potentially ambiguous (*e.g.* ^{18} could in principle refer to ^{18}O or the relatively short-lived ^{18}F), in practice any ambiguity is resolved from the context.

The major problem in interpreting enzyme kinetic isotope effects is the intervention of non-chemical steps, which can mask the chemical ones. The problem can be dealt with explicitly by using the concept of "commitment to catalysis", c, $0 < c < 1.0$, which measures the tendency of any ES complex to go on to products. An external commitment of 1.0, for example, would, in the case of a single-step hydrolase following Scheme 5:1, arise from $k_{+2} \gg k_{-1}$, so that every encounter of enzyme and substrate was "sticky" and isotope effects on k_{+2} would not appear in isotope effects on (V/K). More generally, ${}^{i}(V/K)$ is related to the effect on the isotope-sensitive step ${}^{i}k$ and the ${}^{i}K$ by eqn. (5.42), where c_f, the forward commitment, measures the probability of the ES complex before the isotope-sensitive step going on to product and c_r measures the probability of the ES complex after the isotope-sensitive step reverting back to its predecessor:

$$ {}^{i}(V/K) = \frac{{}^{i}k + c_i + c_r\,{}^{i}K}{{}^{i}k + c_i + c_r} \tag{5.42} $$

If c_r is unity, the expression simplifies to ${}^{i}K$, the isotope effect on the equilibrium constant.

All commitments greater than zero tend to reduce the observed isotope effect, and as most effects of mechanistic interest in glycosyl transfer are small, they compound the experimental problem. Rather than trying to deconvolute a system with non-zero commitments, it is often easier to use a poor substrate and/or a non-optimal pH. Obviously, external commitments arising from diffusion are independent of pH, and in practice protein conformational changes or product release steps giving rise to internal commitments are less sensitive than bond-breaking steps to pH.

A useful method of determining isotope effects in reactions which are appreciably reversible and can be followed instrumentally (usually by absorbance), such as phosphorolysis, is equilibrium perturbation[67] (Figure 5.15). A solution containing equilibrium concentrations of unlabelled reactants and products is made up and concentrations fine-tuned so that on addition of enzyme there is no change in absorbance. Identical solutions are then made up, with one of the participants replaced by its labelled counterpart. Enzyme is then added. If there is a kinetic isotope effect, then the forward reaction will proceed faster than the reverse reaction (or *vice versa*) by virtue of the KIE until the isotope itself achieves its equilibrium distribution. This results in a perturbation in the signal, which is described by the difference between two exponentials with very similar exponent, *i.e.* the signal shape is the same as for KIE measurements of non-enzymic reactions by the isotope quasi-racemate method. As with the isotope quasi-racemate method, fitting the whole time course to an expression with all variables completely unknown results in very ill-conditioned fits, and the effect is better estimated from the size of the perturbation. For small kinetic effects not resulting in an equilibrium effect, this is given by eqn. (5.43), where $\delta k = k_{\text{light}} - k_{\text{heavy}}$ and [A] and [B] are the equilibrium concentrations of the

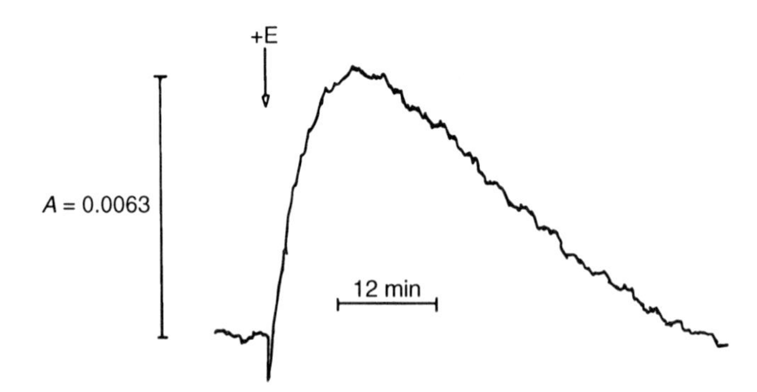

Figure 5.15 Equilibrium perturbation. Absorbance changes at 250 nm consequent upon the addition of *E. coli* purine nucleoside phosphorylase (150 μg in 5 μL) to a solution (3 mL) containing [1′-^{2}H]inosine (1.99 mM), hypoxanthine (0.100 mM), ribose-1-phosphate (0.50 mM) and inorganic phosphate (11.5 mM) in 0.2 M glycine–HCl buffer, pH 9.4, in a 1 cm pathlength cuvette.[68]

isotope-containing species on each side of the equilibrium and P is the size of the perturbation, in concentration units:

$$\frac{1}{[A]}+\frac{1}{[B]}=\frac{\delta k}{2.72kP} \tag{5.43}$$

Obtaining mechanistic information from solvent isotope effects on enzyme reactions is complicated by isotope effects on critical ionisations. Most acids that ionise in a pH range measurable by the glass electrode are about three times weaker in D_2O than in H_2O[xiii] – the exception being thiols, which have very low fractionation factors. [Ordinary glass electrode-based pH meters can be used in D_2O, where they read low by a constant 0.4 units (*i.e.* pD = pH meter reading + 0.4).] This cannot be assumed to apply to enzymes, for two reasons: enzyme pK_as are very often blends of microscopic pK_as, and enzyme structures can have short, tight hydrogen bonds in which the shared hydron has a very low fractionation factor (sometimes as low as 0.3).[69] It is generally advisable, therefore, to measure solvent isotope effects (and proton inventories) at a pH/pD on a horizontal part of the rate profiles, remote from any ionisations. If this is not possible, and the enzyme ionisable catalytic groups are known with reasonable confidence, then proton inventories can be measured at a constant ratio of a chemically similar buffer. Thus, for a retaining *O*-glycosidase with a pH optimum around 5.0, one would use acetic acid buffers.

5.4.5 Structure–Reactivity Correlations

5.4.5.1 Variation of Substrate Structure. Whereas linear free energy relationships play a central role in the determination of non-enzymic mechanisms, they are much less important for enzymes, for two reasons. First, enzymes have evolved to bind their natural substrates, and substituents introduced in an attempt merely to alter electron demand at the transition state may have many other interactions with the enzyme protein. The result is very noisy Hammett and Brønsted plots. Whereas conclusions can be drawn from non-enzymic rates varying over a factor of ~3, with enzyme reactions, to see any trend above the noise it is usually necessary to have rates ranging over several orders of magnitude.

The second problem arises from the multi-step nature of enzyme reactions: by the time electronic demand has been altered sufficiently to see a trend, the rate-determining step may have been changed.

Nonetheless, β_{lg} values for aryl glycosides hydrolysed by glycosidases continue to be measured: the first such, for aryl β-glucosides hydrolysed by almond emulsin (which measured the effect of leaving group acidity from tabulated Hammett σ constants, rather than directly from the leaving group pK_a) dates

---

[xiii] Fractionation factors for the hydrons on the right-hand side of the reaction $AL + L_2O = A^- + L_3O^+$ are roughly unity, on the left-hand side they are 0.69, so change from pure water to D_2O disfavours ionisation by $(0.69)^3$.

from 1954.[70] They can be measured either by plotting $\log(k_{cat}/K_m)$ or $\log k_{cat}$ against leaving group pK_a (Figure 5.16); by analogy with the Northrop–Cleland nomenclature for isotope effects (see Section 5.4.4), these will be termed $^{\beta lg}(V/K)$ and $^{\beta lg}(V)$, respectively.

k_{cat}/K_m contains no terms for steps after the departure of a leaving group, so for a two-step (retaining) glycohydrolase it can be a more informative parameter. However, plots can level off to a constant value at low leaving group pK_a; this probably represents diffusion together of substrate and product becoming rate-determining (*i.e.* the substrate becoming sticky); if this is the case, the absolute value of k_{cat}/K_m should approach $\sim 10^7\,M^{-1}\,s^{-1}$ and the plateau rate should be reduced by inert substances which increase the microscopic viscosity.[71] A rarely recognised snag with $^{\beta lg}(V/K)$ values is that they include any polar effect on binding [for example, charge-transfer interactions between electron-deficient nitrophenol leaving groups and electron-rich protein aromatic groups such as the phenol of tyrosine or (especially) the indole of tryptophan].

Plots of $\log k_{cat}$ against leaving group pK_a do not incorporate polar effects on binding in the derived β_{lg} values, but they can be complicated by a change in rate-determining step, which leads to substrates with better leaving groups all having the same k_{cat}, the value of k_{+3}, the rate of hydrolysis of the glycosyl enzyme.

The behaviour of a GH 1 β-glucosidase from *Agrobacterium* sp. with aryl glucosides illustrates these points. The plot of $\log(k_{cat}/K_m)$ against phenol pK_a exhibits a slope of -0.7 at $pK_a > 8$, but below this $\log(k_{cat}/K_m)$ reaches a plateau value of 7, when enzyme–substrate diffusion becomes rate determining. The analogous plot of $\log k_{cat}$ is horizontal below a phenol pK_a of 8.5, indicating rate-determining hydrolysis of the glycosyl-enzyme.[72] The less efficient GH 39 β-xylosidase of *Thermoanerobacterium saccharolyticum* acting on aryl xylosides shows a similar behaviour of k_{cat}, with dexylosylation rate determining for xylosides of aglycones of $pK_a < 9$, but with xylosylation governing k_{cat}/K_m throughout the accessible range. As with a number of enzymes showing +1 site specificity, acid catalysis does not appear to be applied to the departure of unnatural phenol leaving groups, since $^{\beta lg}(V/K) = -0.97$.[73]

The β_{lg} value itself, once it can be shown to be derived from a single chemical step, is a function both the degree of C–O cleavage and the degree of proton donation; if some independent estimate of the degree of bond cleavage can be made, for example from secondary isotope effects, then the degree of proton transfer can be qualitatively estimated. The GH 10 *exo*-cellobiohydrolase of *Cellulomonas fimi* acting on aryl glucosides exhibits $^{\beta lg}(V/K) = {}^{\beta lg}(V) = -1$, indicating little or no proton donation to the leaving group and nearly complete glycone–aglycone bond fission. With cellobiosides as substrates, k_{cat} was essentially invariant, but $^{\beta lg}(V/K) = -0.3$. Pre-steady-state measurements (Section 5.9.2) confirmed that β_{lg} on the directly measured k_{+2} step was indeed -0.3 and that k_{cat} was determined by k_{+3} for most substrates. In a nice demonstration of an induced fit mechanism, the contrast between the behaviour of the two sets of substrates demonstrated that occupancy of the -2 subsite by a glucose unit was required before the acid catalytic machinery could operate.[74]

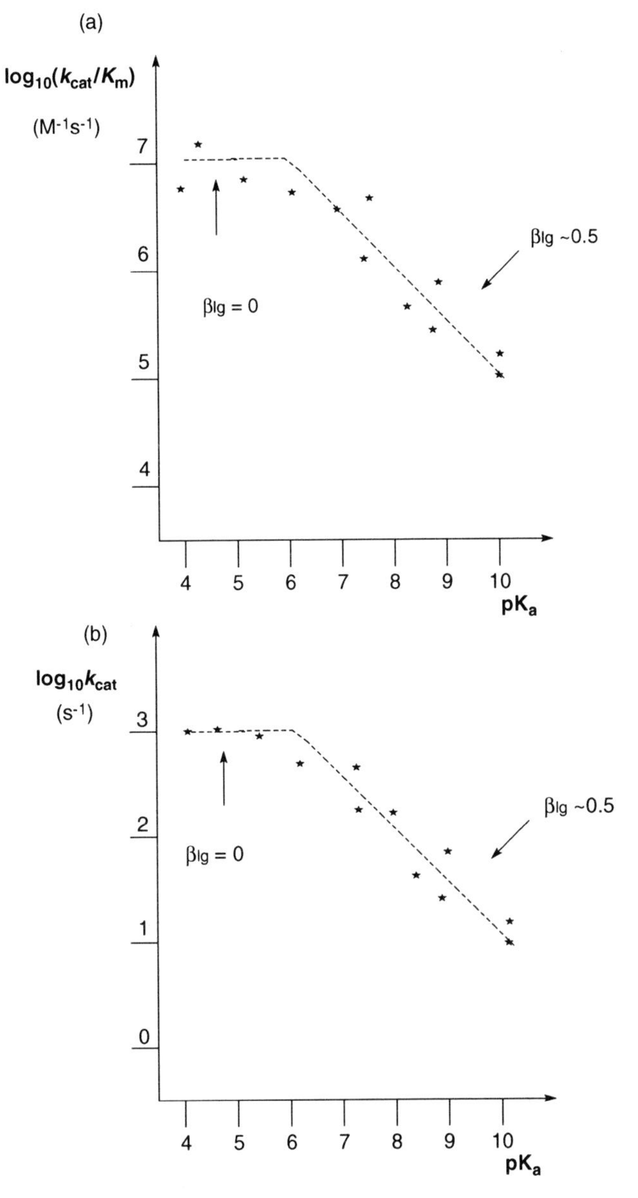

Figure 5.16 Schematic plots of (a) $\log(k_{cat}/K_m)$ against leaving group pK_a, where good substrates become "sticky" (external commitment = 1) and (b) log k_{cat} against leaving group pK_a for a double-displacement glycosidase, where the hydrolysis of the glycosyl-enzyme becomes rate determining. The scatter of the points is an impressionistic rendering of the author's experience with several enzymes. k_{cat} for very good substrates for a retaining glycosidase for which $k_{+2} \gg k_{+3}$, should be exactly k_{+3} – the scatter from the line $\beta_{lg} = 0$ should represent only the error of measurement, not "chemical noise".

The closer to zero the "true" β_{lg} value is, the more Brønsted plots become difficult to differentiate from random scatter caused by non-chemical rate-limiting events.

Some *O*-glycosidases can act on *N*-glycosides and *vice versa* and the contrast between the two types of substrate can be mechanistically useful. Many *O*-glycosidases will act on the corresponding glycosylpyridinium salts, whose structure prohibits the operation of electrophilic or acidic catalysis.[75] $^{\beta lg}(V/K)$ and $^{\beta lg}(V)$ values are in the region −0.7 to −1.0, closer to zero than for the spontaneous hydrolyses (see Section 3.5.2), so that rate enhancements decrease with increasing leaving group ability, even though they are generally in the range 10^8–10^{10} for both inverting and retaining enzymes. A useful feature of a pyridinium leaving group is that it is an equatorial anchor, and may introduce qualitatively different, and very informative, behaviour in *O*-and *N*-glycosides. Thus, yeast α-glucosidase (GH 13) acting on aryl glucosides shows a random scatter in both $^{\beta lg}(V/K)$ and $^{\beta lg}(V)$ plots, which the absence of a leaving group ^{18}O effect shows to be due to a rate-limiting conformation change. However, α-glucosylpyridinium salts give good $^{\beta lg}(V/K)$ and $^{\beta lg}(V)$ plots, confirmed to involve bond breaking by appropriate kinetic isotope effects.[76] The reason appears to be that the reactive conformation is on the skew/boat pseudo-rotational itinerary (*i.e.* on the equator of Figure 2.6), on which the pyridinium salts were, but that a conformation change had to take place with substrates in the $^4C_1$ conformation. By contrast, an initial conformation change in the action of influenza neuraminidase on aryl *N*-acetylneuraminides[77] was not rate limiting where the $^2C_5$ conformation was disfavoured by the anomeric effect, but was completely rate limiting for the much slower hydrolyses of *N*-acetylneuraminylpyridinium salts,[78] where it would have removed the pyridinium group from the equatorial orientation.

Because of the strong and specific interactions made with the glycone by glycosyl-transferring enzymes, successful attempts to manipulate electronic demand by substitution in the glycone have been few. A 1980 study on the *N*-acetylhexosaminidase of *Aspergillus niger* (probably GH 20) recorded $^{\beta lg}(V/K) \approx -0.2$ and $^{\rho^*}(V/K)$ of −1.4 for the effect of successive fluorination in the acetyl group of the substrate[xiv]; the authors correctly concluded that such a large polar effect showed immediately that the substrate amide group acted as the nucleophile in this retaining enzyme.[79] A series of analogues of NAD^+ (Figure 3.8) substituted at the 2′-position of the fissile ribose, gave a $^{\rho I}(V)$ value of −9.4, somewhat greater than that for spontaneous hydrolysis (∼−7; see Section 3.5.2); if the σ_I value for alcoholate was interpolated on the plot, the V_{max} of NAD^+ was correctly predicted, although whether this was just a lucky coincidence, or whether the enzyme does indeed completely ionise the 2-substituent without using it as a nucleophile, remains to be seen.[80]

The existence of a linear free energy relationship between two enzymes acting on the same library of substrates can give an indication of common mechanism. Thus, a plot of $\log(k_{cat}/K_m)$ for hydrolysis of a series of *p*-nitrophenyl glycosides and aryl glucosides by the GH 1 enzyme from the mesophile *Agrobacterium*

[xiv] The Taft ρ^* is the equivalent of the Hammett ρ, for aliphatic systems.

faecalis against its counterpart from the thermophile *Pyrococcus furiosus* gave a correlation coefficient of +0.97.[81]

5.4.5.2 Variation of Enzyme Structure – Site-directed Mutagenesis.[82] It is now possible specifically to change one amino acid to another in an enzyme, provided that the gene for the enzyme is available and can be placed on a plasmid.[xv] The detailed molecular biology of the many current variants of the basic technology is beyond the scope of this text, particularly as many such variants are commercially available in kit form. However, kits do not always work as the instructions say they will and a basic knowledge of the technique will alert the enquirer after mechanism to possible malfunction. The following steps are involved:

(i) Chemical synthesis of an oligonucleotide probe with base replacement(s) that will result in the desired sequence change – *e.g.* GCU, coding for Ala, in place of GAU, coding for Asp. If the sequence is long enough (~40 bases), it will still form a double helix with single-stranded DNA (although if it is too long it may be involved in other interactions).
(ii) Separation of the complementary strands of the plasmid.
(iii) Binding ("annealing") of the primers to the single-stranded DNA: sometimes two primers, binding to each complementary strand of the DNA, are used.
(iv) Extension of the primers, using deoxynucleoside triphosphates and polymerase.
(v) Amplification of the DNA by the polymerase chain reaction (PCR), a technique involving thermal cycling, which leads to a geometric increase in the number of copies of the mutant DNA (but not wild-type DNA).
(vi) Expression in a suitable host.

There are various ways of ensuring the digestion of the wild-type (wt) strand, based on recognition of supercoiling or methylation, but other techniques rely on the outnumbering of the wt gene by the mutant gene possible after many cycles of PCR. PCR, however, is not 100% accurate and can introduce mutations other than the desired one. It is always advisable to have the structure of the mutant gene confirmed by experimental re-sequencing.

In the literature, mutant proteins are often described using the one-letter amino acid code, with the residue number separating wt and mutant residue. Thus D216A represents a mutant in which the wt aspartate has been replaced with alanine.

[xv] A relatively small closed circular piece of DNA which is transferred between bacteria, even between bacteria of different species, without affecting the main chromosome. Transfer of plasmids containing antibiotic-detoxifying enzymes, such as β-lactamases, is a main pathway for the spread of antibiotic resistance.

The mechanistic conclusions that one can draw from simple kinetic measurements on mutant proteins are limited by a number of factors:

(i) A mutant may be inactive because the mutated residue is structural and the whole protein has misfolded. Ideally, X-ray crystal structures for the mutant proteins should be obtained, but failing that, spectroscopic probes of secondary structure, such as circular dichroism, should be the same for mutant and wt.[xvi]
(ii) A catalytically inactive mutant may be apparently active because of the use of a buffer which can provide "catalytic rescue".[83] Thus, acetate and formate buffers can rescue mutants involving active site carboxylates, particularly if the side-chain is truncated (the most extreme case being E → G), as the buffer component occupies the hole in the structure left by the mutation.
(iii) Assays are linear, free energy contributions are logarithmic. Typical k_{cat} values for glycosidase-catalysed hydrolyses acting on their natural substrates are around 10^{16}–10^{17}-fold faster than the spontaneous hydrolysis of the same compounds,[84] corresponding to a lowering of the free energy barrier by between 91 and 97 kJ mol^{-1}. If a mutation lowers k_{cat} 100-fold, 1% activity may not be detectable in an assay, but the catalytic power of the enzyme has been reduced only 12%.
(iv) Mutation of a residue remote from the active site may have fairly large kinetic consequences because of subtle changes in protein conformation. Thus the "site-misdirected mutant" (L9H) of the large subunit of GH 2 *ebg*^{0} β-galactosidase of *Escherichia coli*, L9H, arising from a PCR-introduced T → A mutation in the gene, had an 8-fold lower k_{cat} for aryl galactosides than wt, despite the mutation being remote from active site.[85]
(v) Wild-type contamination (*e.g.* from the use of the same chromatographic columns as used to isolate wt) may result in spuriously high catalytic activities for inactive mutants. The only certain refutation of wt contamination is an active site titration (see Section 5.9.2).

Asp → Asn and Glu → Gln mutations are inadvisable, as chemical deamidation back to the wt residue is a possibility, particularly with thermostable enzymes.

5.4.5.3 Large Kinetic Consequences of Remote Changes in Enzyme or Substrate Structure: Intrinsic Binding Energy and the Circe Effect. During the first quarter-century during which physical-organic ideas were applied to enzyme mechanism (around 1950–75), it was widely thought that the chemical functionality at the enzyme active site was the major contributor to enzyme catalysis and that the rest of the protein served only to place the catalytic group in the

[xvi] Wild-type, *i.e.* proteins isolated from organisms found "in the wild" or those of the same sequence, as distinct from those produced by organisms carrying a mutant gene for this particular protein.

right environment. In a review which revolutionised the field, Jencks[13] pointed out that K_m values were usually very much higher than the dissociation constants of small molecules binding to non-functional proteins, and introduced the concept of intrinsic binding energy. This was the maximum free energy of binding of a small molecule to a protein and could be expressed in non-catalytic proteins. In a catalytic protein, by contrast, the intrinsic binding was only expressed at the transition state and manifested itself as a lowering of the free energy barrier to reaction (Figure 5.6). This intrinsic binding energy often involves interactions at sites remote for the reaction centre and explains the general trend "the smaller the substrate, the bigger the enzyme". With much intrinsic binding energy available, say by the binding of many units of a polysaccharide, the molecular machinery does not have to be very elaborate in order to express enough of it at the transition state for efficient catalysis, whereas more subtleties are required with only the intrinsic binding energy of one monosaccharide unit available. Only in the limit, with very small substrates such as hydrogen peroxide, the substrate of peroxidase, does catalysis rely exclusively on active site chemistry.

Removal of remote transition state binding interactions, whether by alteration of the structure of the enzyme or of the substrate, will have kinetic consequences from which the free-energy contribution of the interaction to catalysis can be estimated. One approach is to deoxygenate all the positions of the sugar ring in turn. Of course, this will also alter the intrinsic reactivity of the glycoside by electrostatic and inductive effects, but this can be estimated by fluorination, which has the same inductive effect as deoxygenation, but in the opposite sense. Thus, a comparison of $\log(k_{cat}/K_m)$ for hydrolysis of 2-, 3-, 4- and 6-deoxygenated and fluorinated 2′,4′-dinitrophenyl β-D-galactopyranosides by GH 2 *E. coli* (*lacZ*)-β-galactosidase led to the conclusion that the transition state was stabilised by $>17\,kJ\,mol^{-1}$ by hydrogen-bonding interactions at positions 3, 4 and 6, and by $>34\,kJ\,mol^{-1}$ by such interactions at position 2.[86] The figure for the 6-OH interaction was strikingly confirmed when His540, which appeared in the X-ray structure of the enzyme to hydrogen bond to O6 of the substrate, was mutated to Phe: (k_{cat}/K_m) for *p*-nitrophenyl galactoside fell by 10^3 [$RT\ln(1000) = 17\,kJ\,mol^{-1}$], but (k_{cat}/K_m) for *p*-nitrophenyl L-arabinopyranoside and D-fucopyranoside remained largely unchanged.[87] With the GH 1 β-glucosidase of *Agrobacterium faecalis*, a similar approach led to the conclusion that hydrogen bonds at positions 3 and 6 stabilised the transition state by $\sim 9\,kJ\,mol^{-1}$ and at position 2 by $18–22\,kJ\,mol^{-1}$.[88]

5.4.6 The Use and Misuse of X-Ray Crystallographic Data in the Determination of Enzyme Mechanism

It is very rare for the X-ray crystal structure of a small molecule to be in any sense incorrect – the handful of cases often involve an incorrect space group and a missed hydrogen in the presence of a very heavy metal. Most protein structures determined by X-ray crystallography, however, lack such absolute, detailed authority. The X-ray diffraction experiment involves measuring the intensities of 10^5–10^6 diffraction spots of protein crystals in a mother liquor mixed with glycerol or similar antifreeze so it can be cooled to 100 K. The best modern practice involves data collection at a synchroton X-ray source and measurements of all spots in an array detector. The minimisation of exposure to X-rays and subsequent radiation damage of the crystals mean that structures can now be determined on milligrams of protein, whereas the structure of lysozyme required 1 kg of crystalline enzyme, as the crystals were exposed to the emissions of X-ray generators for long periods.

Since X-rays are scattered by the electrons in a molecule, the outcome of an X-ray crystal structure determination of a protein is a map of electron density in the unit cell, averaged over the $> 10^{13}$ such cells in an individual crystal. The structure arises from a fit of a chemical model to the electron density – the chemical model is determined by the known sequence of the protein (usually) and common bond lengths and bond angles. The goodness of fit of the model to the data is described by an *R* factor, given by eqn. (5.44), where the structure factors (*F* values) are related to the intensity of the diffraction spot; the sum is the sum over all measured spots:

$$R = \frac{\sum_{hkl} (F_{obs} - F_{calc})}{\sum_{hkl} F_{obs}} \tag{5.44}$$

R factors are commonly 20–30% for proteins, whereas for small molecules they are 2–5%. A rule of thumb is that the *R* factor (in %) should be around 10 times the resolution (in Å). One of the contributors to the *R* factor is thermal motion (even at 100 K); this motion is reported as a *B* factor (sphere in which the atom can be found); some structures may be accurate enough to determine separately anisotropic *B* factors, describing thermal motion in three orthogonal directions. Since protein crystals are up to 70% water, assumptions have to be made about the distribution of these waters. Certain regions of the protein may be disordered, with some crystals having one protein conformation and others another. The occupancy of various sites is then an important consideration.

It is rare for hydrogen atoms to be explicitly located in a protein structure and therefore it is usually impossible to distinguish isoelectronic F, OH and NH_2 groups (therefore Glu and Gln, Asp and Asn usually cannot be distinguished). The most mechanistically informative structures are those in which crystals have been soaked in solutions of various ligands and the protein structure includes structures of ligands bound at the active site. An important parameter here is site occupancy, particularly if the enzyme is a polysaccharidase. With low occupancy, a structure supposedly detailing the binding of a hexasaccharide to a catalytically inactive mutant may well be a blend of some enzymes in the crystal binding a trisaccharide in the −3, −2, −1 subsites and others in the +1, +2, +3 subsites (see Section 5.5).

Even mutant enzymes are rarely completely catalytically inactive over the period of days or weeks with which they are in contact with substrate. Over a period of days, glycosyl transfer enzymes can often catalyse elimination reactions. Thus, sialidases will catalyse the elimination of NANA to DANA (Figure 3.24). The flattening of the sugar ring that results in these eliminations can result in the elimination product being mistaken for a stable, enzyme-bound oxocarbenium ion: resolution is inadequate to distinguish the $^{6}H_5$ conformation of DANA and the $^{4}H_5$ conformation of the oxocarbenium ion.[89]

5.5 ENZYMES WITH MULTIPLE SUBSITES SUCH AS POLYSACCHARIDASES

Topologically, the active sites of simple monosaccharidases without aglycone specificity tend to be pits, like those of other enzymes acting on simple molecules. The active sites of enzymes acting on polysaccharides are of three types. Those of enzymes acting at the ends of chains usually have pits like simple monosaccharidases. Those of enzymes acting in the middle of the chains of soluble polysaccharides or in the middle of chains in the amorphous regions of insoluble ones have clefts, like the typical such enzyme lysozyme. Enzymes acting on the crystalline regions of insoluble polysaccharides, often processively, intriguingly have tunnels as active sites, through which the sugar chain is removed from the surface.

The active sites of enzymes acting on polysaccharides are considered to be made up of a series of subsites, each of which binds a single saccharide residue. Although letters were in the past used to denote these subsites, the use of numbers is now generally accepted.[90] Numbers increase from non-reducing to reducing end and a polysaccharide-chain is, by definition, cleaved between the −1 and +1 subsites, so that the glycosyl residue at which the substitution is taking place is −1 and the leaving group is +1. The idea is illustrated in Figure 5.17.

The terms *exo*- and *endo*-action are often encountered in descriptions of the actions of enzymes acting on polysaccharides. The term "*exo*" originally meant action on a single sugar residue at the non-reducing end of the chain, *i.e.* the enzyme had only one negative subsite. *Endo*-acting glycosidases cleaved the polysaccharide-chain in the middle and could have any number of positive and negative subsites. The *exo*-terminology was extended to enzymes, such as the GH 6 cellobiohydrolases involved in cellulose degradation, which had −1 and −2 subsites and released a disaccharide from the non-reducing end. The classification was further confused by the discovery of GH 7 (retaining) cellobiohydrolases, which had two positive subsites, but up to 10 negative ones, and released cellobiose from the reducing end of a glucan chain. *Exo*-acting enzymes with a significant number of positive subsites tend to be inverting, for obvious reasons. Were they to be retaining, after the first chemical step the polysaccharide bound in the +1, +2, +3, ... sites would have to diffuse away before the glycosyl-enzyme could be hydrolysed.

The commonest simple way to distinguish *exo*- from *endo*-acting enzymes is to measure the generation of reducing sugar equivalents and the reduction of substrate viscosity as the enzyme acts on a soluble polymer. The intrinsic viscosity of a linear polysaccharide is proportional to the molecular weight

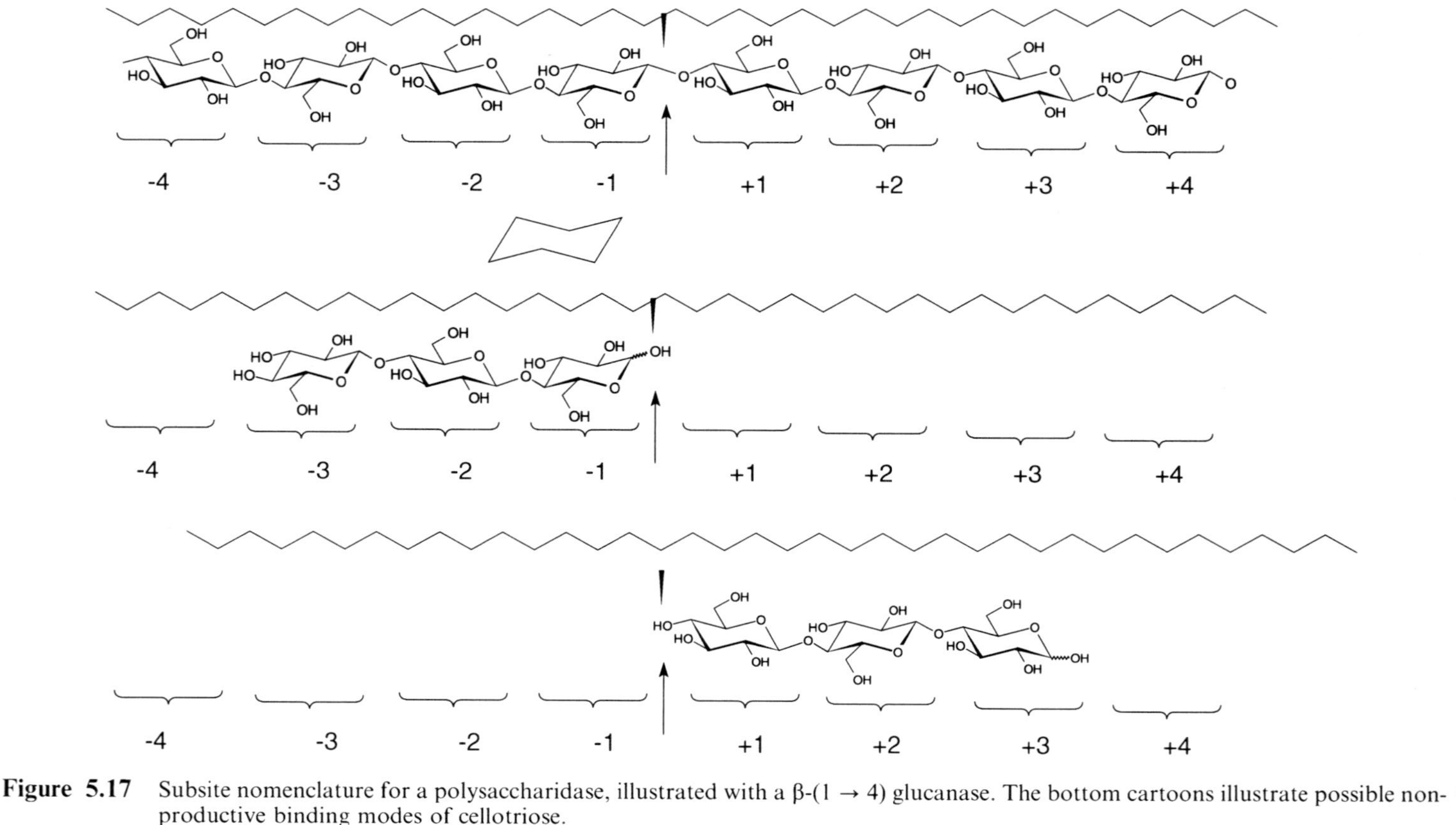

Figure 5.17 Subsite nomenclature for a polysaccharidase, illustrated with a β-(1 → 4) glucanase. The bottom cartoons illustrate possible non-productive binding modes of cellotriose.

raised to a fractional power, usually around 0.7 (Section 4.5.2). So, an *endo*-acting enzyme will bring about a large decrease in molecular weight and hence the viscosity, with each catalytic event. If each link cleaves with a rate constant k, then DP at time t, DP_t is given by the af Eckenstam equation,[472] equation 5.45. This reduces to equation 5.46 if DP_t and DP_0 (the DP at time zero) are both large.

$$\ln(1 - [1/DP_t]) - \ln(1 - [1/DP_0]) = -kt \tag{5.45}$$

$$1/DP_t - 1/DP_0 = kt \tag{5.46}$$

On the other hand, an *exo*-acting enzyme will bring about only a very marginal reduction in the DP of the polymer for each catalytic event. In tests of cellulases, it is usual to monitor both reducing sugar and viscosity as the enzyme in question hydrolyses carboxymethylated cellulose.

A common feature of the action of catenases such as polysaccharidases on short substrates is that the latter bind non-productively. Non-productive binding has no effect on k_{cat}/K_m, as can be seen intuitively, since under conditions of very dilute substrate, where the rate is determined by k_{cat}/K_m, there will still be a large excess of the unliganded enzyme over both productively and non-productively liganded forms. However, the observed $k_{cat(obs)}$ and $K_{m(obs)}$ are obtained by dividing both k_{cat} and K_m for productive binding by $1 + \Sigma(K_m/K_{np})$, where K_{np} is the dissociation constant of the substrate in each particular non-productive binding mode.

Figure 5.18 illustrates a further problem with the use of oligomeric substrates, against retaining polysaccharidases: transglycosylation may increase the degree of polymerisation, as shown, so that good substrates may be built up from poor ones.

The idea that polysaccharide–polysaccharidase interaction could be treated simply as the sum of the interactions of various subsites is intuitively attractive. Hiromi first tackled the algebra,[91] but made the assumption, now known to be implausible, that the microscopic rate constant for bond cleavage in the ES complex was independent of subsite occupancy. This is very unlikely for enzymes in general terms, since it is now accepted that they exploit interactions with the substrate remote from the site of bond cleavage to lower the free energy for the catalysed reaction (the Circe effect,[13] Section 5.4.5.3). The Hiromi assumption was shown to be incorrect by direct experiments with an endoxylanase.[92] A better fit to experimental data was obtained if the subsite affinities were calculated for the first irreversible transition state (*i.e.* on k_{cat}/K_m and cleavage patterns of oligosaccharides), but careful analysis in some systems indicated that even this approximation failed.[93]

Processivity may be encountered with enzymes acting on polysaccharides, that is, the substrate remains bound to the enzyme for more than one catalytic event. In the case of enzymes acting on insoluble substrates, processivity is revealed by a high proportion of soluble to insoluble reducing sugar, as was the case with the GH 48 cellobiohydrolase of *Clostridium thermocellum* acting on

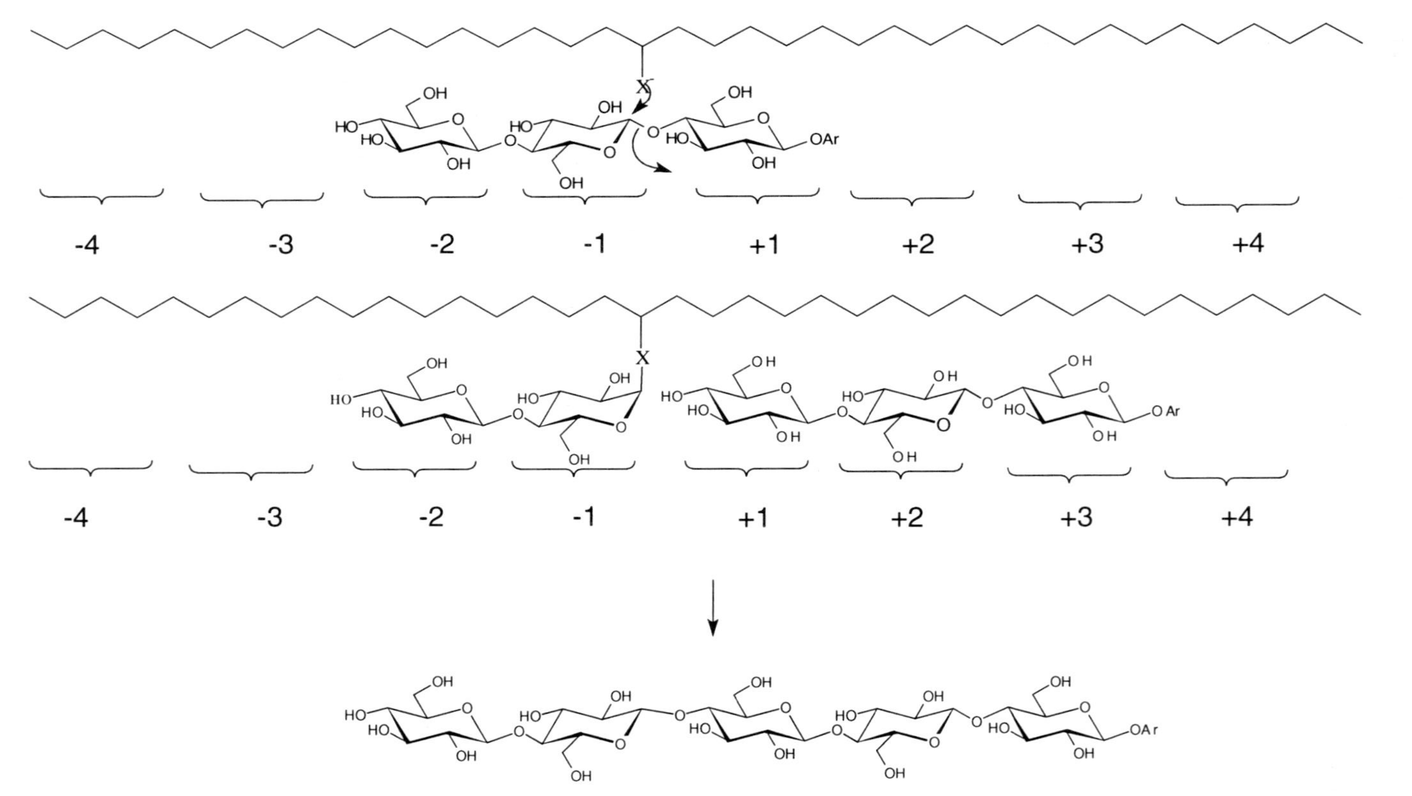

Figure 5.18 Illustration of how a retaining catenase may build up a good substrate from a poor one by transglycosylation. The example shows a retaining cellulose acting on a cellotrisaccharide glycoside.

amorphous cellulose.[94] Processivity can also be demonstrated with soluble substrates: the very low levels of cellohexaose, cellopentaose and cellotetraose that appeared during the hydrolysis of cellooctaose by the GH 7 cellobiohydrolase of *T. reesei* indicated that once the octameric substrate was bound it could be subject to multiple attack; at the same time, the presence of significant quantities of glucose in the products indicated the enzyme was not a classical *exo*-acting cellobiohydrolase.[95] We now know that this enzyme is *exo*-acting from the reducing end: use of cellooligosaccharides ^{18}O-labelled at the reducing end, in conjunction with HPLC/MS, indicated that the initial cellobiose was produced from the reducing end[96] and that the glucose arises from cellooctaose binding with its seventh and eighth monosaccharide units in the +1 and −1 sites.

If the enzyme in question has an *exo*-action from the reducing end, processivity can be measured by chemically derivatising all reducing ends and measuring the release of the reducing ends. Thus, reaction of various types of cellulose with anthranilic acid and reduction of the glycosylamine (as its open-chain Schiff base) gave cellulose labelled with a fluorescent tag. On hydrolysis of this derivatised cellulose with the cellobiohydrolase I (Cel7A) of *T. reesei*, a "burst" of labelled cellobiose was produced in the first catalytic acts of the enzyme, followed by a steady-state rate, as the enzyme completed its catalytic cycle, came off the cellulose and then attacked a fresh labelled reducing end. Comparison of the size of the "burst" with the steady-state rate (the algebra parallels that for a "burst" and steady-state rate when the "burst" arises from formation of the glycosyl-enzyme intermediate [eqns (5.48) and (5.49)] revealed that this enzyme liberated 88, 42 and 34 molecules of cellobiose for every encounter with bacterial cellulose, bacterial microcrystalline cellulose and endoglucanase-treated bacterial cellulose, respectively.[97]

5.6 GENERAL FEATURES OF *O*-GLYCOHYDROLASES

Whether retaining or inverting, *O*-glycosidases have a number of common features:

(i) The catalytic proton donor is *syn* or *anti*, never directly above the plane of the sugar ring.[37,38]

(ii) Pyranoside substrates with equatorial leaving groups in the preferred chair conformation, where they can be observed in the −1 subsite, are often distorted from a perfect chair and the same is true for three of the four glycosyl-enzyme intermediates for which a structure is known (as at summer 2006), which would have an equatorial leaving group in a perfect chair. This distortion is a requirement of ALPH (Figure 3.10) but is also a requirement for bimolecular nucleophilic attack on an equatorial substituent in a chair-conformer six-membered ring, which is prohibited by the axial hydrogens (analogous to H5 and H3 of $^{4}C_{1}$ glucopyranose). 2-Adamantyl *p*-toluenesulfonate, in which the cyclohexyl ring is held rigidly in a chair conformation, is so insensitive to

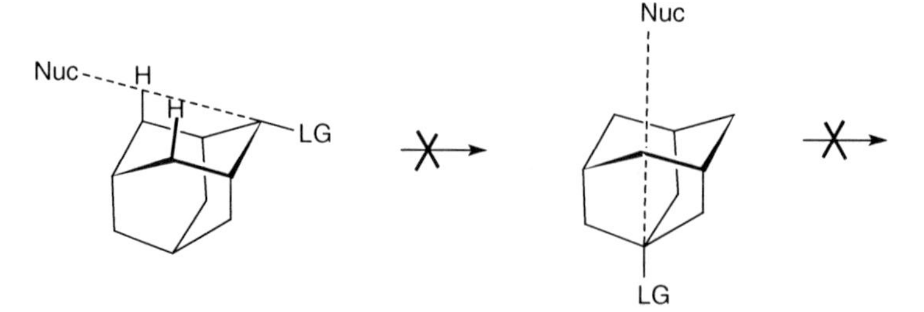

Figure 5.19 2-Adamantyl derivatives are as inert to nucleophilic displacements as 1-adamantyl derivatives.

nucleophiles[99] that it makes as good a standard for zero nucleophilic susceptibility as 1-adamantyl *p*-toluenesulfonate (Figure 5.19) and a better one than *tert*-butyl derivatives.[100]

(iii) The requirement for coplanarity of C2, C1, O5 and C5 for maximum conjugation of oxygen lone pair and electron-deficient reaction centre means that the non-chair conformations that have been observed by protein crystallography in various complexes are adjacent to the possible $^{4}H_{3}$, $^{3}H_{4}$, $^{2,5}B$ and $B_{2,5}$ conformations of the cation (Figure 2.6). Thus, enzyme-bound β-glucopyranosyl rings can adopt the $^{1}S_{3}$ conformation (pseudoaxial leaving group, isoclinal CH_2OH and 2-OH, pseudoequatorial 3-OH and 4-OH), at the same longitude as $^{4}H_{3}$ on the Cremer–Pople sphere. Likewise, enzyme-bound β-mannopyranosyl rings can adopt the $^{1}S_{5}$ conformation (pseudoaxial leaving group, pseudoequatorial CH_2OH, 2-OH and 4-OH, isoclinal 3-OH), on the equator and 30° longitude different from $B_{2,5}$ on the Cremer–Pople sphere.

(iv) All glycosidases, whether inverting or retaining, and whether *syn* or *anti* protonating, have a "hydrophobic patch" of residues which makes contact with the hydrophobic area around C5–C6 in the case of pyranosides. These interactions assist the attainment of half-chair or boat conformations.[101]

(v) Glycosidases have electron-rich groups with lone pairs pointing towards the upper and/or lower lobe of the p-type lone pair on the ring oxygen.[98] Such interactions raise the energy of the orbital in the bound state, but electrostatically stabilise the transition state, in a manner similar to the carboxylate in the model system shown in Figure 3.30.

Of the 42 GH families which had a crystal structure with the −1 subsite occupied (in December 2004) there were 15 *syn*-protonators. In 14 of these *syn*-protonators, the acid catalyst was placed to interact with the lobe of the p-type lone pair on the ring oxygen on the same side as the leaving group (Figure 5.20). As it donated its proton, it would become more negative and thus a mutual synergism would develop between acid and electrostatic catalysis. [The exception was GH 47, which acted upon the crowded α-(1 → 2) mannosyl link, where

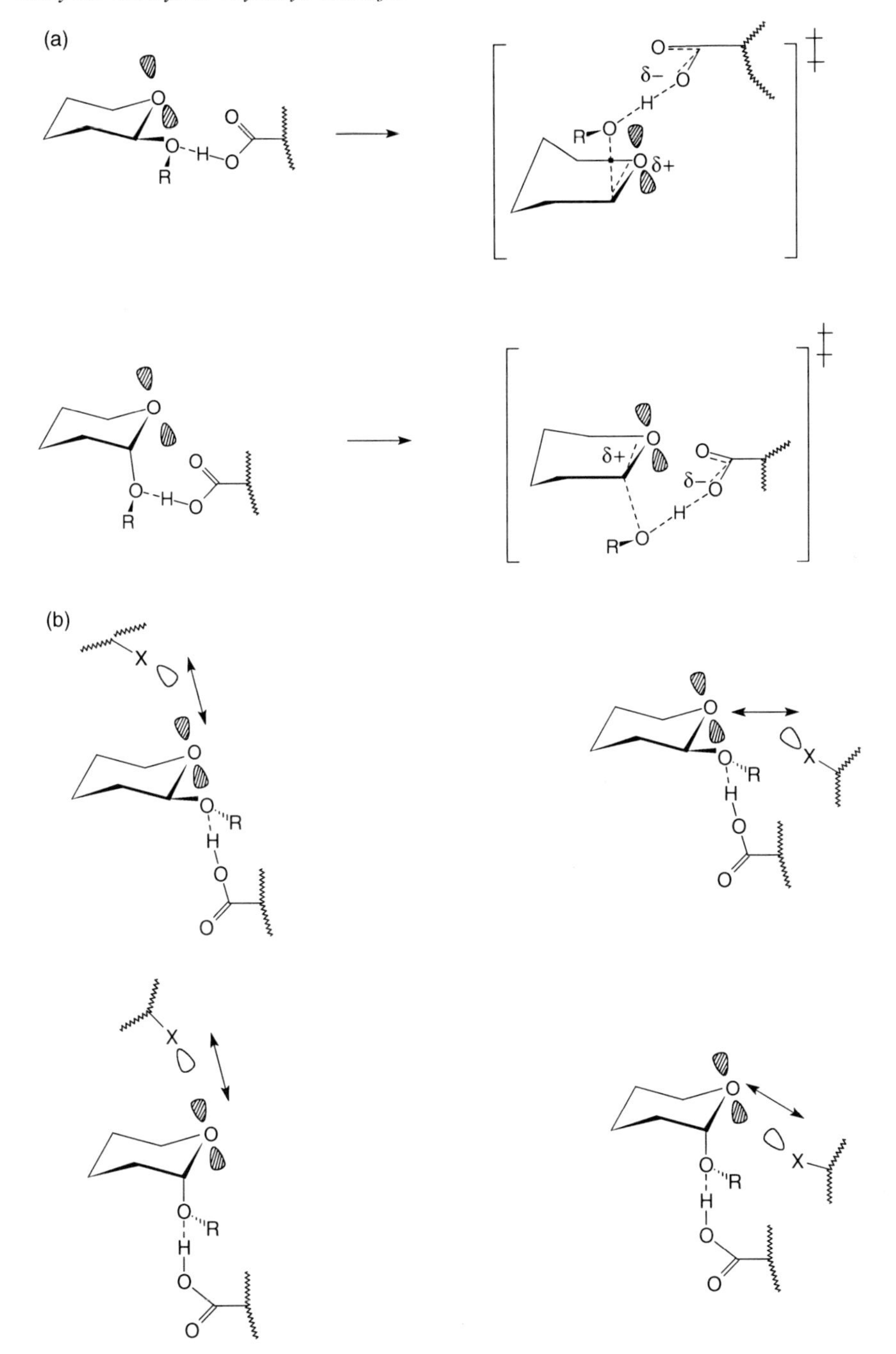

Figure 5.20 Cartoon of interactions with the p-type lone pair on oxygen with electron-rich groups in the active sites of glycosidases: (a) in *syn* protonators, the acid catalyst can act; (b) in *anti* protonators, an auxiliary group, *cis* or *trans* to the leaving group, is involved.

electrostatic assistance was provided by O6 of the substrate.] *Anti*-protonators, by contrast, almost invariably exploited an additional "helper" residue, which could interact with the p-type lone pair lobe either *cis* or *trans* to the leaving group, depending on the family. In several families this "helper" group was strictly conserved. Only in the unusual GH 90 Family are both sides of the C5–O5–C1 (or equivalent) plane accessible to solvent.

5.6.1 Reactions with Enol Ethers

Glycohydrolases will hydrate glycals. The reactions of retaining glycosidases with glycals are well understood:[102,103] a 2-deoxyglycosyl enzyme intermediate is formed, by the microscopic reverse of an *E*1-like *syn E*2 reaction. If the hydration is carried out in D_2O, therefore, in the case of retaining α-glycosidases the proton is added to the β face of C2 and in the case retaining β-glycosidases to the α face. Most of the work with glycals was done before the facts of in-plane protonation by glycosidases were established and failure to observe protonation of glycals by the enzyme acid–base catalyst caused much puzzlement. If the acid catalyst is in the plane of the sugar and well placed to donate a proton to O1, however, donation of a proton to C2 by any trajectory is almost impossible, whereas the hydrophobic patch made by the C1–C2 double bond will probably lower the pK_a of the nucleophilic group. If the enzymes have transferase activity, the 2-deoxyglycosyl residue can be transferred to acceptors rather than to water. The only exception to the "protonation by the reverse-protonated nucleophile" mechanism is the hydration of 2-acetamidoglucal by GH 20 β-glucosaminidases (where the nucleophile is the substrate acetamido group): in this case, the acid catalyst protonates the β-face.[104]

The situation with inverting enzymes is less clear-cut, probably because the catalytic base is further removed and the nucleophilic water molecule stands between it and C2. Nonetheless, various β-amylases (probably GH 14) transferred a hydron to the β face of maltal:[105,106] proton inventory studies showed that the solvent isotope effects [up to ${}^{D_2O}(V)$ of 8] were due to the transfer of a single proton. However, various inverting α-glucosidases (probably GH 15) transferred a proton to the α face of glucal.[107]

Glycosidases will hydrate exocyclic enol ethers, often efficiently (k_{cat} for glycopyranosylidene methylene hydrolysed by GH 2 *lacZ* β-galactosidase is 50 s^{-1} [108]). The double bond is protonated by the acid catalyst; the resulting tertiary cation is captured by the nucleophile of retaining glycosidases or the nucleophilic water of inverting ones; if the inverting enzyme has a + 1 subsite which can accommodate an acceptor, the transfer product is of opposite stereochemistry to the normal substrate.

Enol ethers are reactive molecules, however, and the rate accelerations brought about on them by glycosidases are modest: effective molarities of the catalytic acid in the ES complex are similar to those in model systems (Figure 5.21).[109]

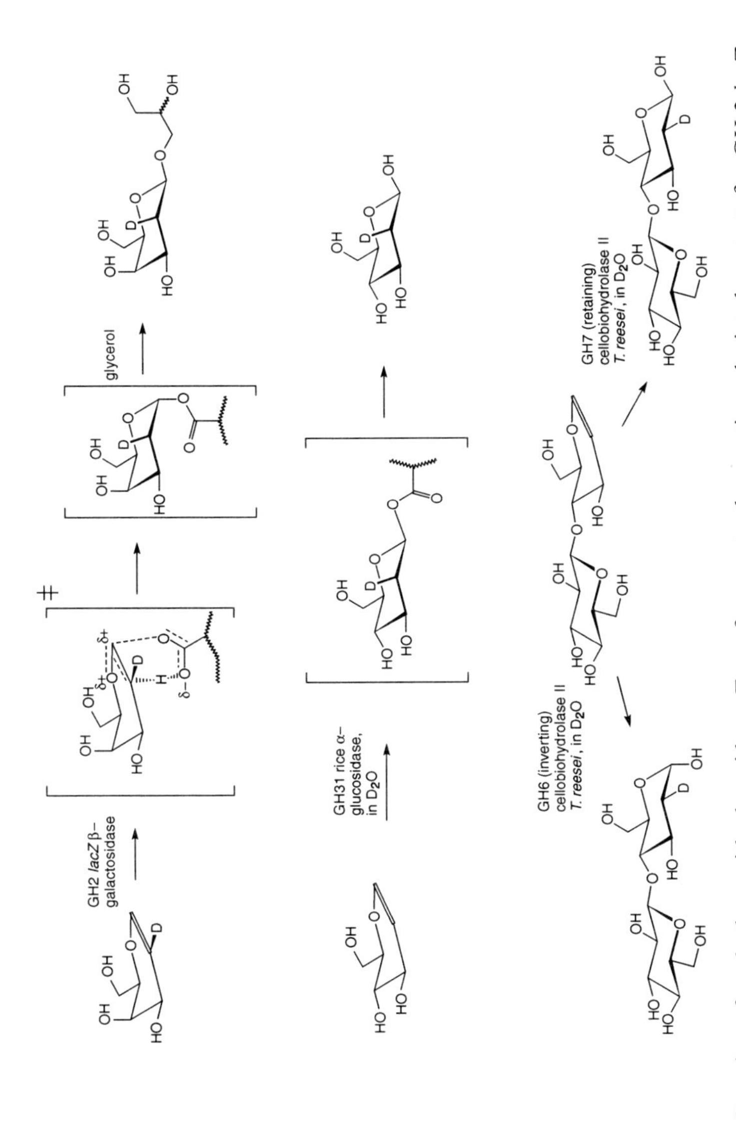

Figure 5.21 Reaction of enol ethers with glycosidases For references to the reactions depicted see text, for GH 2 *lacZ* enzyme Refs. 110 and 111, for hydration of exocyclic gluco compound Refs. 112 and 113 and reaction of cellobial Ref. 114.

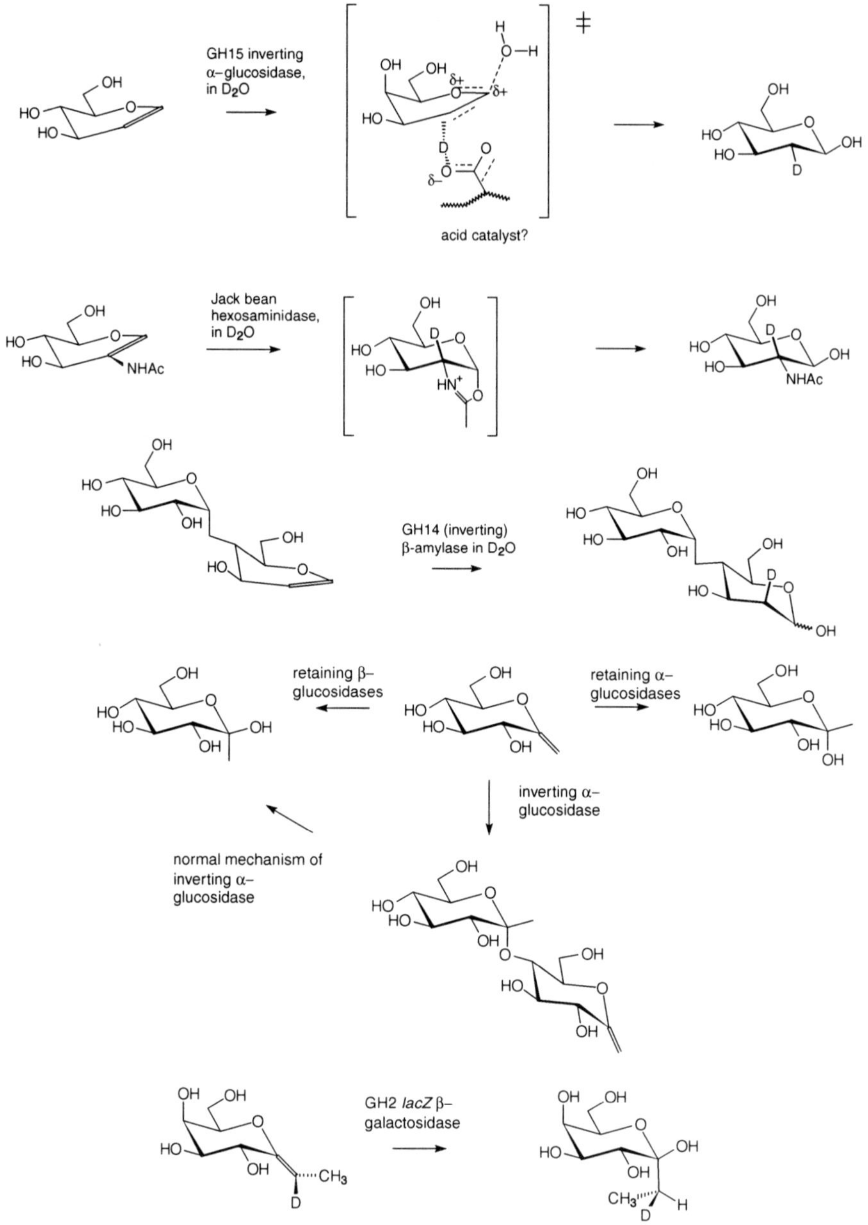

GH15 inverting α-glucosidase, in D_2O

acid catalyst?

Jack bean hexosaminidase, in D_2O

GH14 (inverting) β-amylase in D_2O

retaining β-glucosidases

retaining α-glucosidases

inverting α-glucosidase

normal mechanism of inverting α-glucosidase

GH2 *lacZ* β-galactosidase

Figure 5.21 (Continued)

5.7 INVERTING *O*-GLYCOSIDASES

The reactions catalysed by these enzymes invariably involve a single chemical transition state, of the "exploded" nature described in Chapter 3, with varying mechanisms for activating nucleophile and leaving group, and varying degrees of positive charge build-up on the sugar. They are characterised by the presence in the active site of two protein carboxylate groups – Asp or Glu – about 6–12 Å apart.[4,5,115] One of these carboxylates acts as a general acid for the departure of the leaving group, in the mechanism of Figure 5.1b. In the canonical case, the second carboxylate acts a general base catalyst for the attack of a nucleophilic water molecule. There is some indication that the method and degree and mechanism of activation of the nucleophilic water may vary, with Grotthus (water chain)-like mechanisms and the operation of more than one carboxylate. In accord with this, the "consensual" distance apart of the two catalytic groups widened from 9–10 Å[4] in the first *in silico* survey to 6–12 Å more recently.[5] Families of inverting *O*-glycoside hydrolases where at least one catalytic group is known are given in Table 5.1, together with the trajectory of protonation (*syn* or *anti*), where known.

As might be expected from the fact that the extra methylene group in the glutamate side-chain makes the carboxylic acid group less acidic than that of aspartate, in only one family, where the stereochemistry is not experimentally determined (90), is there any suggestion of a system where glutamate is the base and aspartate is the acid.

There are comparatively few methods of identifying unambiguously catalytic groups involved only in proton transfer. Groups may be labelled by *exo* affinity labels such as 3-epoxybutyl or 2-epoxypropyl glycosides,[122] but that merely locates them in the vicinity of the active site. Generally, heavy reliance has to be placed on X-ray studies and site-directed mutagenesis. Identification of the catalysing acid is generally the easier task, because differential effects are observed, depending on the leaving group. Whereas elimination of the catalytic base should have a similar effect on all substrates, elimination of the catalytic acid should have a greater effect, the more the aglycone needs protonating in order to leave. With enzymes which will tolerate phenol leaving groups in the +1 site, such as the GH 43 β-xylosidases, this can be manifested by dramatically different β_{lg} values: $^{\beta lg}(V/K) = {}^{\beta lg}(V) = -0.36$ for the wt β-xylosidase from *Geobacillus stearothermophilus*, but both values change to ~ -1.8 for the E187G mutant, with rates for wt and mutant enzymes against 3,4-dinitrophenyl xyloside being very similar.[123] That the β_{lg} value is more negative than that for spontaneous hydrolysis of aryl tetrahydropyranyl ethers (−1.18; Section 3.3) is attributed to a hydrophobic +1 site. Even with enzymes which will not tolerate a phenol in the +1 site, they will usually hydrolyse glycosyl fluorides, so that mutation of the acid catalysis will have a far larger effect on the hydrolysis of oligosaccharides than the corresponding glycosyl fluorides.

5.7.1 Evidence from Action on the "Wrong" Fluorides

A powerful piece of evidence for the general mechanism of Figure 5.22 comes from the work of Hehre, who showed with a number of inverting enzymes that glycosyl fluorides of the "wrong" anomeric configuration were in general substrates. The original demonstration was with what is now known to be GH 14 β-amylase, which accepted β-maltosyl fluoride.[124] Likewise, a GH 15

Table 5.1 Inverting *O*–glycosidases, their catalytic groups and stereochemistry of protonation. The substrate stereochemistry is given as ring size (*p* or *f*), with *p* further amplified by whether the leaving group is axial (a) or equatorial (e).

GH Family	*Substrate(s) stereochemistry*	*−1 site conformation*	*Acid*	*Trajectory*	*Base(s)*
6	*p*, e	$^{2}S_{0}$[130,132]	Asp	*syn*	Asp?
8	*p*, e	$^{2,5}B$[116]	Glu	*anti*	Asp
9	*p*, e		Glu	*syn*	Asp
14	*p*, a	$^{4}C_{1}$[117]	Glu	*syn*	Glu
15	*p*, a		Glu	*syn*	Glu
23	*p*, e		Glu	*syn*	?
24	*p*, e		Glu	*syn*	?
28	*p*, a		Asp	*anti*	2 × Asp[118]
43	*p*,a, also *f*		Glu	*anti*	Asp
45	*p*, e		Asp	?	Asp
46	*p*, e		Glu	?	Asp
47	*p*, a (p, e)	$^{1}C_{4}$ (DMJ),[119] $^{3}S_{1}$ (1-thio-substrate)[120]	Arg/Glu/H_2O	*syn*	Glu
48	*p*, e		Glu	?	?
49	*p*, a		Asp	?	Asp
65	*p*, a		Glu	?	(Phosphate nucleophile)
67	*p*, a	$^{2}H_{3}$[121]	Glu	*syn*	Asp + Glu[121]
74	*p*, e		Asp	?	Asp
82	*p*, e		Glu	?	Asp
90	*p*, a		Asp	*anti*	Asp + Glu
94	*p*, e		Asp	*syn*	(Phosphate nucleophile)

Taken from CAZy,[98] except where indicated.

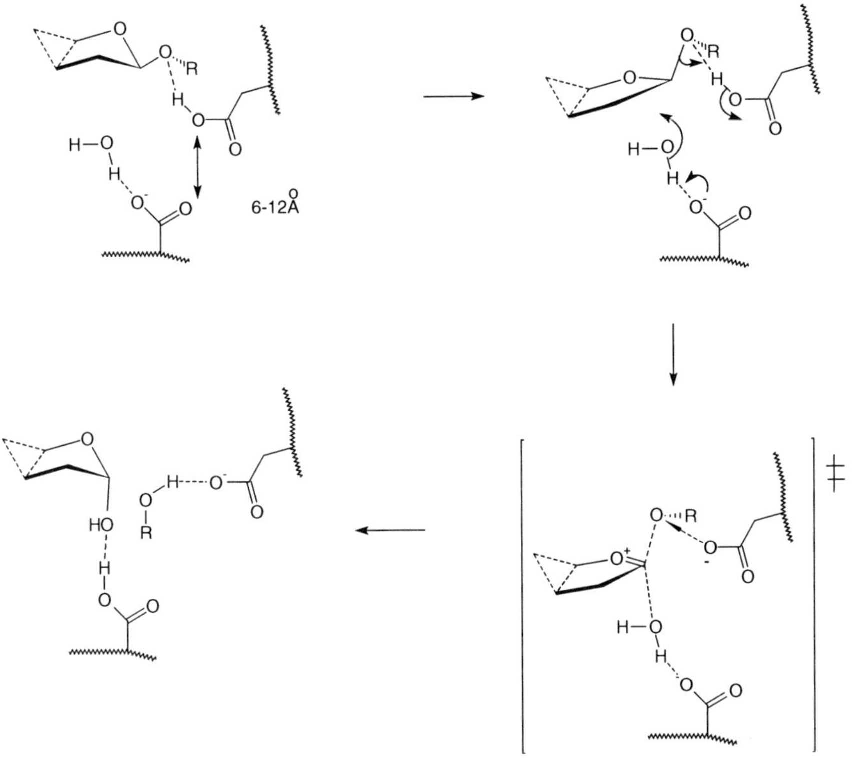

Figure 5.22 Canonical mechanism for an inverting *O*-glycosidase. The reaction proceeds via proton removal from a nucleophilic water molecule and proton donation to the leaving group via a single transition state in which the sugar ring has substantial oxocarbenium ion character. The dotted lines in the sugar ring indicate that the reaction mechanism applies to both pyranosidases and furanosidases: some families contain both, such as (inverting) GH 43, which contains α-L-arabinofuranosidases and β-D-xylopyranosidases. For illustration, *anti* protonation is shown.

glucoamylase [with −1, +1, +2, +3, ... sites for an α(1 → 4) glucan] and a GH 15 glucodextranase [with −1, +1, +2, +3, ... sites for an α(1 → 6) glucan] accepted β-glucosyl fluoride,[125] as did a GH 37 trehalase.[126] Cleavage of α-xylopyranosyl fluoride by a GH 43 β-xylosidase was also observed.[127] The kinetics, however, indicated that two molecules of "wrong" fluoride were bound to the enzyme, although in some cases one of them could be replaced by the unfluorinated glycone (*e.g.* glucose in the case of trehalase). All available evidence points to the mechanism of Figure 5.23, illustrated for GH 6 cellobiohydrolases.[128] The "wrong" fluoride binds in the negative subsites and a second molecule binds in the positive subsites. The mechanism is possible since fluorine and OH are isoelectronic and approximately isosteric, so that the fluorine of the "wrong" fluoride mimics the incoming nucleophilic water in the normal mechanism. A reverse-protonated form of the enzyme, in which

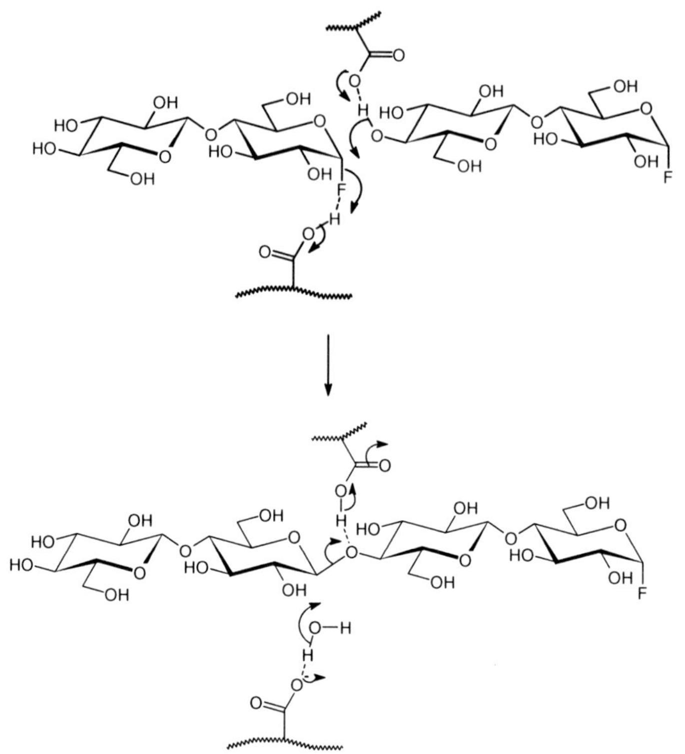

Figure 5.23 Cleavage of α-cellobiosyl fluoride by GH 6 cellobiohydrolases.

the usual base is protonated and the usual acid is deprotonated, is used. One molecule of the "wrong" fluoride binds to the −1 site (or the −1, −2 sites in the case of disaccharide) and another molecule in the +1 (or +1, +2) sites. A new glycosidic linkage is formed. In favourable cases (*e.g.* the GH 43 β-xylosidase), the synthesised saccharide will diffuse away from the active site and can be detected in free solution; in others (including the GH 6 cellobiohydrolases of Figure 5.23) it remains bound and is cleaved by the normal catalytic mechanism of the enzyme.

5.7.2 Mutation of Catalytic Groups

Experiments with various GH 6 cellulases illustrated how site-directed mutagenesis can be used to identify catalytic groups. The D252A mutant of the *CenA* endoglucanase of *Cellulomonas fimi* had wt activity on 2,4-dinitrophenyl β-cellobioside but 5×10^{-6} of wt activity on carboxymethylcellulose, establishing D252 as the acid catalyst; the pK_a of 2,4-dinitrophenol is 4, so that acid catalysis of its departure is not required in the way it is with a sugar. The D392A mutant had no activity on either substrate and so D392 was identified as the general base. As with many glycosidases, there was an additional active site carboxylate whose mutation had large, but not dramatic, effects, D216.[129] D221 in *T. reesei* cellobiohydrolase II was homologous with D252 in *CenA* and, as expected for an acid catalyst, the D221A mutant showed near-wt activity towards β-cellobiosyl fluoride, but none on cellooligosaccharides, since departing fluoride (pK_a of HF = 3.2) needs only some hydrogen bond donation, not protonation.[130] The reverse protonation mechanism of Figure 5.23 predicts that an acid-catalyst mutant should be incapable of hydrolysing the "wrong" fluoride and indeed the *T. reesei* GH 6 cellobiohydrolase D221A mutant showed no activity at all against α-cellobiosyl fluoride. However, the mutation of the "auxiliary" residue to D175A produced similar decelerations in the hydrolyses of both fluorides.

5.7.3 Some Inverting Glycosidase Families

5.7.3.1 GH 6. The structure of a complex of *T. reesei* GH 6 cellobiohydrolase with β-D-Glc*p*-(1 → 4)-β-D-Glc*p*-(1 → 4)-4-thio-β-Glc*p*-(1 → 4)-β-D-Glc*p*-OMe occupying the −2, −1, +1 and +2 sites, *i.e.* with the non-hydrolysable thioglycoside linkage spanning the cleavage site, revealed the glucopyranose unit in the −1 subsite to adopt the $^2S_0$ conformation.[131] The same ligand was also bound in the identical conformation by the analogous *Humicola insolens* GH 6 cellobiohydrolase.[132] This unusual conformation suggests that conjugative electron release from the ring oxygen atom in any oxocarbenium-like transition state would be maximised by the adoption of the adjacent $^{2,5}B$ conformation (Figure 2.6), despite the clashes of the flagstaff hydroxymethyl group with the flagstaff hydrogen on C2. β-D-Glucopyranosylisofagomine, the cellobiose analogue of isofagomine (Figure 5.8), does indeed bind in this conformation.[133] Studies on the endoglucanase (Cel6A) of *Thermobifida fusca* indicated that the adoption of the $^2S_0$ conformation by the −1 sugar was entirely due to clashes with the active site tyrosine, since its replacement with a smaller group resulted in binding of substrate-analogues in the relaxed $^4C_1$ conformation.[134]

GH 6 contains two types of cellulolytic enzymes. The cellobiohydrolases liberate α-cellobiose from the non-reducing ends of the chains of crystalline cellulose. They have an extended active site which is a tunnel and act processively. The tunnel structure may be connected with the work that has to be done removing a cellulose chain from the attractive forces of its neighbours: the only source of energy to do this work is the free energy of hydrolysis of the glucan link. The *T. reesei* cellobiohydrolases can reverse the process and lay down radiolabelled cellulose on a template of bacterial microcrystalline cellulose in the presence of radiolabelled cellobiose.[135] In accord with this idea, the active site tunnel of the *Humicola insolens* enzyme shows structural changes, suggestive of the "tense" and "relaxed" forms of a work-producing enzyme.[136] The two loops of protein forming the roof of the tunnel in GH 6 cellobiohydrolases appear to be able to open, since the substrate in Figure 5.24 is hydrolysed (both ends of which are blocked by bulky groups, so that it cannot thread its way into the tunnel).[137]

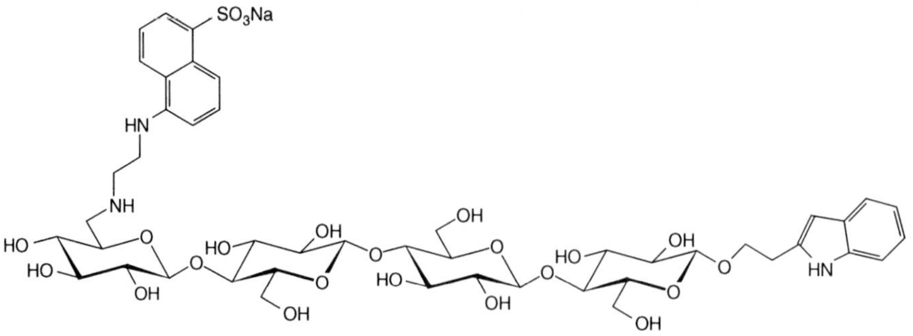

Figure 5.24 Instrumental substrate for endocellulases. The instrumental signal arises from resonance energy transfer: light at 290 nm is used to excite the indole nucleus, which is sufficiently close in space to the amino-naphthalenesulfonyl group that the excitation energy is transferred to it, causing a fluorescence at 490 nm. When the substrate is cleaved, the energy transfer ceases and the fluorescence decreases. Both of the aromatic groups are too large to thread their way into the tunnel of CBHII: the loops of protein forming the roof must open.

The endoglucanases, however, act on amorphous regions of cellulose and have clefts, rather than tunnels, as active sites. It is possible in Family 6 to convert a cellobiohydrolase to an endoglucanase by deleting the loop of protein which forms the roof of the active site tunnel, by site-directed mutagenesis.[138]

5.7.3.2 GH 8. An atomic resolution structure of the complex of an acid-catalyst mutant (Glu95Gln) of *Clostridium thermocellum* endoglucanase with cellopentaose revealed that the ring in the −1 subsite was in a $^{2,5}B$ conformation.[116] This highly strained conformation, although favouring glycosidic bond cleavage because C2, C1, O5 and C5 are already coplanar, seems to require the intrinsic binding energy of a number of sugar residues to offset its inherent strain energy.

5.7.3.3 GH 9. Further clues that the machinery for activation of the nucleophilic water in inverting enzymes may not be that critical come from an endocellulase from *Clostridium stercorarum*, which lost activity against crystalline cellulose, but not soluble oligosaccharides, on mutation of the presumed catalytic base.[139]

5.7.3.4 GH 14. The main members of this family are β-amylases, which yield β-maltose from the non-reducing ends of starch. Four maltose molecules are found in the complex of maltose with the holoenzyme from *Bacillus cereus*, two in tandem in the active site, one in a C-terminal CBM and a fourth in what appears to be an internal CBM.[117]

5.7.3.5 GH 15 Glucoamylase. The GH 15 glucoamylase of *Aspergillus niger* (and other *Aspergillus* species) has been extensively studied because of its commercial importance. It hydrolyses terminal glucose residues off long or short α-(1→4) glucan chains, but also has activity towards α-(1→6) glucan linkages. The former activity explains its use in industrial starch processing, which is assisted by a non-catalytic starch binding domain (see Section 5.10). However, industrial bioreactors work at fairly low water activity and as the equilibrium constant for hydrolysis of the α-(1→6) glucan linkage is lower than that for hydrolysis of the more crowded α-(1→4) glucan linkage, resynthesis of α-(1→6) glucan linkages can be troublesome. Industrial interest led to a very large programme studying the effect of change in the structure of the enzyme (by site-directed mutagenesis) or of the substrate (by deoxygenation).

The catalytic base is Glu400, which is hydrogen-bonded to Glu401 and Tyr48;[140] Tyr48 is largely conserved in GH 15 and the oxygen lone pairs provide the electron-rich region raising the energy of the substrate ring oxygen p-type lone pair, as discussed earlier. The acid catalyst is another glutamate (179).[141] An apparent auxiliary aspartate (175), site directed mutagenesis of which gave kinetic effects ~100-fold,[142] turned out to be too far removed from the −1 site to be directly implicated in bond breaking. Remarkably, mutation of Glu400 to Cys and oxidation of the thiol to a sulfinic acid (R–SO–OH) resulted in an engineered enzyme with a slightly greater activity than wt on both α-(1→4) and α-(1→6) glucooligosaccharides.[143] This *sulfino* enzyme also transformed β-glucosyl fluoride by the Hehre mechanism more efficiently than wt, where under the same conditions reaction could not be detected.

Examination of the pre-steady-state kinetics of oligosaccharide hydrolysis, exploiting the intrinsic fluorescence of the enzyme as a signal, demonstrated that three distinct rate processes were involved. Their identities were unclear, one group[144] favouring fast enzyme–substrate binding, followed by the chemical step and then rate-determining product release and the other[145] fast enzyme–substrate binding, isomerisation of the ES complex and a slow chemical step, with all subsequent steps fast. Both models predict (V/K) isotope effects for α-glucosyl fluoride hydrolysis, since loss of F^- is likely to be irreversible. The effects[146] of anomeric ^{14}C, 1.03, and αT, 1.19, support the usual oxocarbenium ion-type transition state. The kinetic model with two significant ES complexes was supported by a comparison of the pH dependence of the elementary steps and of the Michaelis–Menten parameters.[147]

It is clear that interactions in the +1 site, as with many inverting enzymes, are important: α-glucopyranosyl fluoride is a mere 100-fold better substrate than maltose, despite being intrinsically much more reactive. Interactions with the 4-OH of the +1 sugar of isomaltose appear to account to its susceptibility to the enzyme, as well as maltose.[148] Interactions with the 4′ and 6′ hydroxyls of both isomaltose and maltose in the −1 subsite was crucial for catalysis, as shown both by deoxygenation and site-directed mutagenesis.[149]

The early literature on this enzyme reports many attempts to analyse its interactions with oligosaccharides in terms of the Hiromi subsite analysis. These are now known to be flawed because of the implausible assumptions made.[145,150,151]

5.7.3.6 GH 28. Galacturonases and related pectinases are in this family, which has a distinctive fold consisting almost entirely of β-sheet and a range of basic groups in the active site, to neutralise the negative charges of the substrate on binding. However, the chemistry appears to involve three conserved aspartates, one of which acts as a general acid and the other two activate the nucleophilic water[152] (although one of the two general base catalysts is not strictly conserved, in at least one enzyme being replaced by a glutamate).

5.7.3.7 GH 47. This family, whose paradigmal members are mammalian Golgi α-(1→2) mannosidases (mannosidase I) I acting on N-linked glycoproteins, exhibits a number of unusual features. They are the only Ca^{2+}-dependent glycosidases in which the substrate makes direct contacts with the Ca^{2+}, which is coordinated by O2 and O3 of the −1 sugar. Deoxymannonojirimycin (Figure 5.25a) and the natural product kifunensine Figure 5.25b are bound at the −1 subsite in the $^{1}C_4$ conformation,[119] and the −1-bound ring of the non-cleavable thio analogue α-Man*p*-(1→2)-2-thio-α-Man*p*-OMe, which binds −1, +1, in the $^{3}S_1$ conformation.[120] It therefore appears that the reaction involves motion along a line of longitude on the Cremer–Pople sphere, in which the oxocarbenium ion-like transition state is close to the normally disfavoured $^{3}H_4$ conformation and the pyranose ring describes a least-motion trajectory from the $^{3}S_1$ conformation of the bound substrate, through a $^{3}H_4$ transition state, to a $^{1}C_4$ conformation product.

Although site-directed mutagenesis experiments establish the catalytic base unambiguously as Glu (599 in the human enzyme expressed in the yeast *Pichia pastoris*), mutation of the supposed acid (Glu330) had only modest kinetic effects.[120] The X-ray structure of enzyme–thioglycoside complex revealed a water molecule interposed between Glu330 and the sulfur. With hindsight, it is clear that the approach of a catalytic carboxylic acid to the glycosidic oxygen atom would be sterically restricted, so it is unsurprising that the acid acts through a Grotthus-type mechanism, possibly modulated by an adjacent arginine.

5.7.3.8 GH 48. This family appears to form a clan with GH 8, with the protein fold $(\alpha/\alpha)_6$. The cellobiohydrolase from *C. thermocellum* has the expected tunnel, with eight monosaccharide binding sites and is processive.

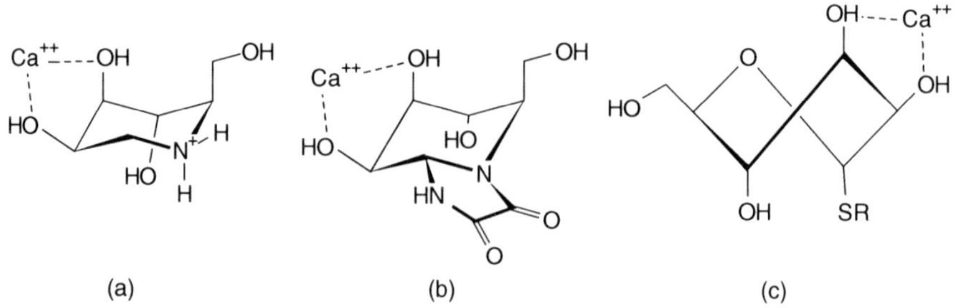

Figure 5.25 Ligand conformations bound to GH 47 enzymes.

The tunnel takes a sharp bend between subsites −1 and +1 and a glucopyranose can be modelled into the −1 subsite only on the assumption that it adopts the $^{2,5}B$ conformation seen experimentally with the other clan member. GH 48 is another inverting enzyme where the base catalyst is not well defined and a Grotthus-type mechanism seems likely.[153] The Cel48F processive endocellulase from another *Clostridium* species has an active site tunnel covering subsites −7 to −1 and a cleft in the four positive subsites; the pyranose ring in the −1 subsite is not in the $^4C_1$ conformation seen with other occupied subsites, but the existence of two sets of binding modes, shifted by about half the length of a sugar residue, in the tunnel made the exact conformation difficult to determine.[154]

5.7.3.9 GH 67. This family includes the enzymes which hydrolyse the 4-*O*-methyl-α-glucuronyl residues off xylan fragments produced by lignocellulose degrading organisms. Crystal structures of the enzyme[121] from *Geobacillus stereothermophilus* reveal a loop of protein carrying the acid catalyst (Glu285) swings into the correct position to donate a proton after substrate binding and that the nucleophilic water was activated by two carboxylates (*cf.* GH 28). Studies with the E285N mutant indicated the sugar ring in the −1 subsite was in the $^2H_3$ conformation; this is not on any likely reaction trajectory and it is possible that the true reactive conformation is the adjacent $^2S_0$.

5.8 REACTION OF *N*-GLYCOSIDES WITH INVERSION

The excision of nucleic acid bases by hydrolases can be both a repair mechanism for the host and an offensive weapon by competitors. There is a whole array of DNA repair glycosylases, some of which are merely hydrolases, others of which have an associated lyase activity.[155] Catalysing a similar reaction are the proteins, formed in the seeds of many plants, which hydrolyse a ribosyl–adenine bond in the sequence GAGA loop region of the 28*S* RNA, thus shutting down protein synthesis.[156] No direct determination of reaction stereochemistry has been carried out for either class of enzyme, nor is an X-ray structure of an enzyme–substrate–analogue complex available, but the crystal structures so far available suggest inversion.

Nucleotide C–N bond hydrolysis and phosphorolysis at the monomer level form part of purine salvage pathways and their mechanisms have been intensively investigated for pharmacological reasons. Humans can biosynthesise purine nucleosides *de novo*, whereas protozoal parasites such as those causing bilharzia and Chagas' disease rely on hydrolysis of preformed nucleosides from the host. Finally, a series of ribosyl transfers from NAD^+ are important in the modification of proteins by pathogens and have been studied extensively.

The most studied RNA *N*-glycosidase is ricin, from the castor oil bean, *Ricinus communis*. This toxin has two component chains, the A chain, containing the *N*-glycohydrolase, and a B chain which is a lectin (a protein which binds carbohydrate structures, in this case galactopyranose), which facilitates

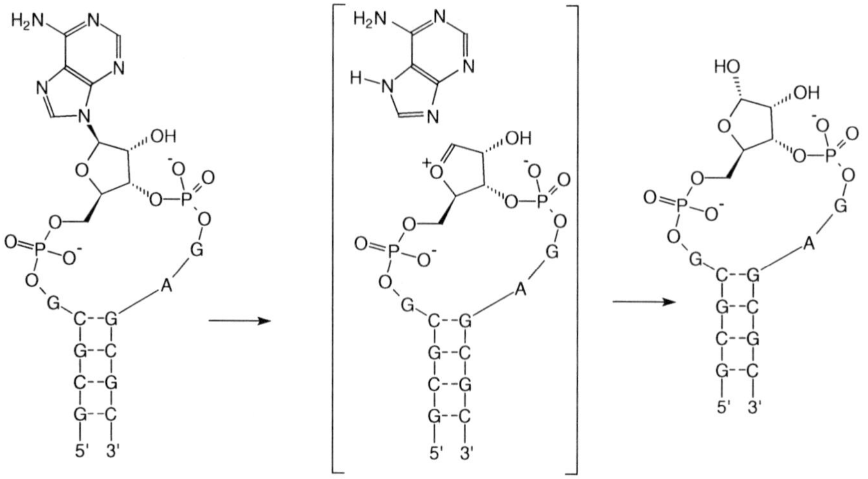

Figure 5.26 Reaction catalysed by ricin A subunit and similar plant toxins.

entry of the complex into the cell. Ricin A chain hydrolyses both DNA and RNA of the correct secondary structure (but DNA fragments several hundred times slower than equivalent RNA fragments[157]) (Figure 5.26) and multiple kinetic isotope effect studies with smallest DNA[158] and RNA[159] substrates possessing the appropriate secondary structure give surprisingly different results. There are no commitments in either case, so the (V/K) effects discussed below (DNA first) refer to bond-breaking steps. Leaving group ^{15}N effects are comparable (1.023 and 1.016), as are α-tritium effects (1.187 and 1.163), both suggesting the expected S_N1-like S_N2 reaction. An inverse effect at N7 [$^{15}(V/K) = 0.981$] for RNA again suggests, very reasonably, that the distal nitrogen of the leaving group is partly protonated. Anomeric ^{14}C effects near unity (0.993 and 1.015), however, led to the claim that in both cases a ribofuranosyl cation was formed reversibly and then collapsed to product in an isotope-insensitive step. However, the mechanism was heavily dependent on the reaction of a ribosyl cation with water being insensitive to isotopic substitution in the sugar, which is only plausible if there is vanishingly small bond formation at the transition state. Moreover, theoretical modelling of the fine details of this and similar reactions were subsequently shown to be dependent on the theoretical methods used.[160] In addition, there were large, unexplained discrepancies between the β-tritium effects in the two systems (*proR* 1.117, *proS* 1.146 for DNA and 1.012 for RNA); since a furanosyl cation has only the ^3E and E_3 conformations available to it, one of the two β effects for DNA should have corresponded to the effect for RNA.

Uracil DNA glycosidase is a DNA repair enzyme which removes the non-DNA base uracil from double-stranded DNA. A pre-steady-state analysis using a fluorescent 2-aminopyridine–uracil mismatch[161] and the *E. coli* enzyme suggested that the enzyme first bound substrate non-specifically with a

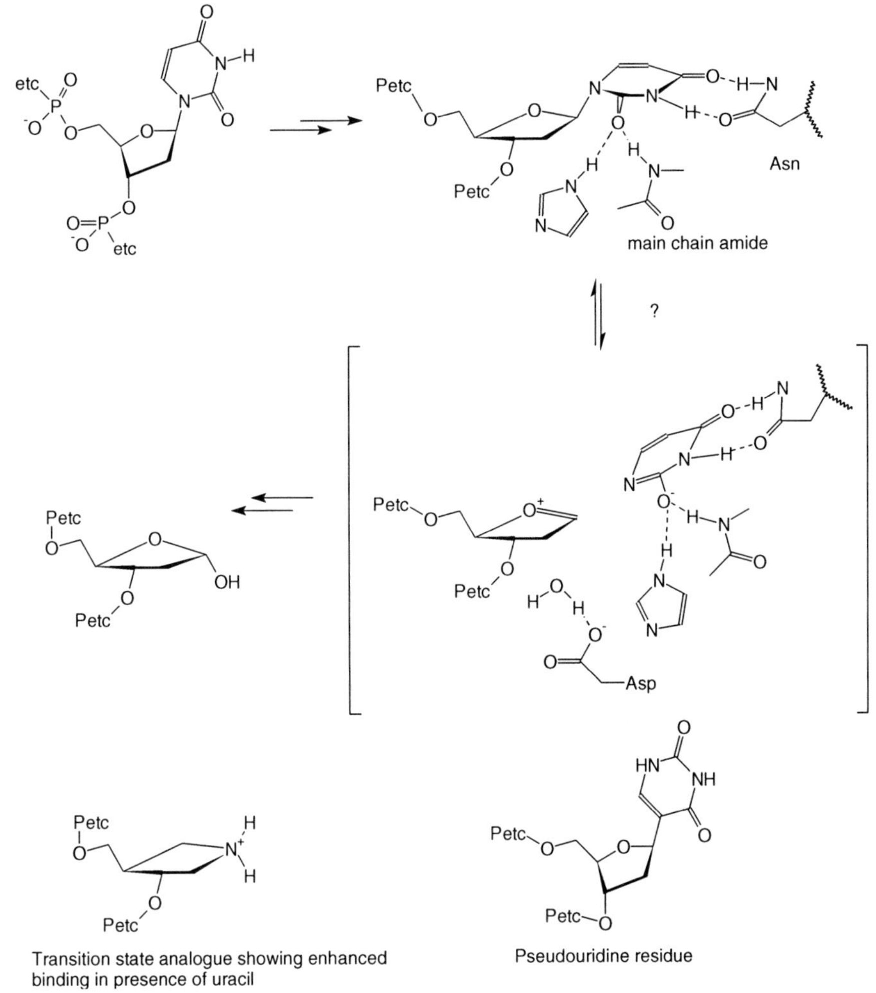

Transition state analogue showing enhanced binding in presence of uracil

Pseudouridine residue

Figure 5.27 Proposed mechanism of uracil DNA glycosylase.

bimolecular rate in excess of $1.8 \times 10^{10}\,M^{-1}\,s^{-1}$,[xvii] then takes the uracil out of the double helix, inserts a leucine side-chain into the resulting hole, cleaves the bond, retracts the leucine side-chain; the rate-determining step is product loss (Figure 5.27). The X-ray structure of the recombinant human enzyme and its complexes with a non-fissile substrate–analogue containing a pseudouridine residue and products suggest a strain mechanism in which the uracil ring is rotated about the C1′–N1 bond and then the angle C1′–N1–C4 bent from 180°

[xvii] This is faster than the diffusion-controlled reaction of azide ion with small cations (see Section 3.2.1), probably because, like many proteins which bind to DNA, the enzyme binds non-specifically to DNA and then undergoes one-dimensional diffusion along the DNA polymer.

to $\sim 130°$, putting the C1′–N1 bond electrons in a better position to be delocalised into the uracil aromatic system. Hydrogen bonding from a histidine and a backbone amide to O2 and an asparagine to O4 and N2 is sufficient to stabilise the anion of uracil in the enzyme active site. Direct NMR evidence indicated that when bound to the enzyme, the pK_a of N1 of uracil was lowered from its value of 9.8 in free solution to 6.4 in the active site.[162] A nucleophilic water molecule poised to attack the α-face of the deoxyribose ring strongly indicated inversion.[163] The idea that uracil left as an ion was firmly established by removal of the key histidine (which abolished activity)[164] and by inhibition studies. A DNA strand constructed with a phosphorylated *trans*-3-hydroxy-4-hydroxymethylpyrrolidine residue in place of uridine and uracil anion bound cooperatively to the enzyme (the K_i for the cationic DNA fell from 2.1 μM to 0.6 nM on addition of uridine).[165] Furthermore, a short, tight hydrogen bond between a neutral histidine residue and bound uracil anion was directly observed by NMR.[166] Kinetic isotope effects [(V/K), under conditions and with a substrate for which they were intrinsic] of $^{\text{anomeric }13}k = 1.010$, $^{\alpha D}k = 1.20$, $^{proR\beta D}k = 1.102$ and $^{proS\beta D}k = 1.106$ were observed;[167] the high secondary deuterium effects and low anomeric ^{13}C effects led to the proposal that the deoxyribofuranosyl cation–uracil anion ion pair was a discrete species with a real lifetime. Such a conclusion, as with ricin catalysis, is very dependent on the fine details of the model used to calculate kinetic isotope effects, and fails to account for the presence of an essential aspartate residue, positioned to act as a general base for the attack of the nucleophilic water and shown to be essential by site-directed mutagenesis.[168] There seems to be no compelling reason to abandon the model of a single transition state at the extreme S_N1 end of the "exploded S_N2" spectrum in favour of a reversibly formed ion pair. Nonetheless, a careful analysis of enzyme-substrate interactions showed that uracil DNA glycosylase is an excellent example of a "Circe effect" enzyme, with remote interactions accounting for a lowering of the transition state free energy by 54 kJ mol^{-1}.[169]

Some DNA glycosylases, in addition to the ability to cleave mismatched or "foreign" bases (uracil, for example, is produced by deamination of cytosine), have the ability to cleave the DNA chain. The initial product of a DNA glycosylase, a β-phosphatoxyaldehyde is labile to $E1_{CB}$ loss of phosphate anyway, but some DNA glycosylases appear to employ an active site lysine residue (or in at least one case an N-terminal proline[170]) to accelerate the loss of phosphate from C3′. The mechanism involves formation of a Schiff base, then Amadori rearrangement to an enamine, followed by expulsion of phosphate (Figure 5.28). The intermediate Schiff base can be reduced with tritiated sodium borohydride, which reduces the Schiff base intermediate, introduces a radiolabel and attaches the DNA to the enzyme.[171] At one point it was thought that the lysine could possibly act as a nucleophile, displacing the base, but this now seems unlikely on the basis of site-directed mutagenesis experiments with various DNA glycosylases. Thus, *MutY* DNA glycosylase has a active site lysine, but does not catalyse strand cleavage,[172] and acts as a glycosylase even if the lysine is mutated to alanine;[173] conversely, conversion of an active site serine to lysine conveys lyase activity.[174] Perhaps the most telling argument

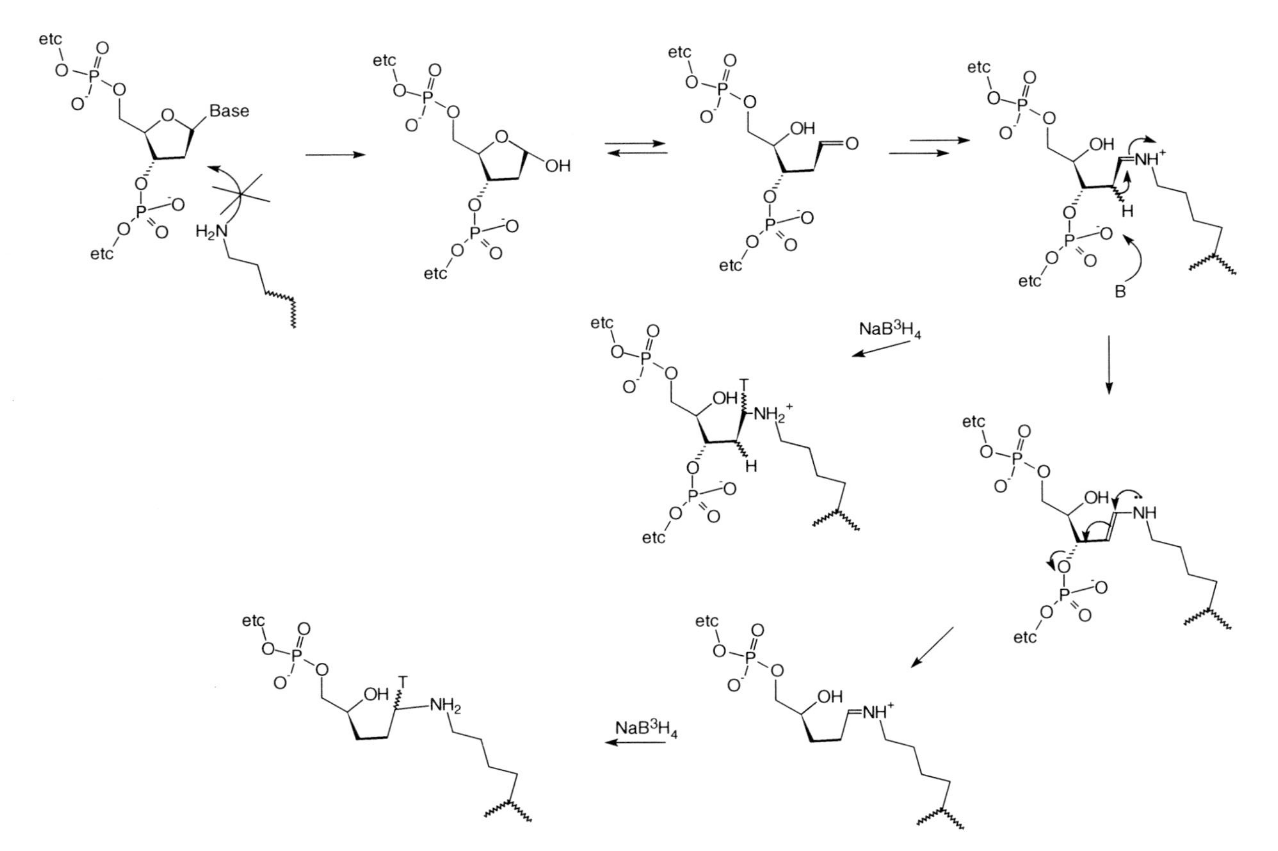

Figure 5.28 Apurinic site lyase mechanism. The probable products of sodium borotritide labelling are also shown.

against lysine as the initial nucleophile, however, was that murine 8-oxoguanine DNA glycosylase hydrolysed the *N*-glycosyl link about 10 times faster than it formed the enamine.[175] The DNA glycosylase that removes the thymidine dimers formed by UV irradiation was said[176] to work by initial nucleophilic attack of the active site lysine, but the key datum (reaction of a methylglycoside apurinic site) could equally be compatible with initial hydrolysis followed by Schiff base formation.

Transition state mapping with multiple kinetic isotope effects has been carried out for a number of nucleoside hydrolases and phosphorylases. Intrinsic effects are given in Table 5.2; for structures of substrates, see Figures 3.14 and 5.31.

It appears that nucleoside hydrolases and phosphorylases can be classified according to their ability to act on *p*-nitrophenyl β-D-ribofuranoside; inosine–uridine nucleosidase (IU-NH) hydrolyses it eight times faster than inosine, whereas inosine–adenine nucleosidase (IAG-NH) hydrolyses it 200 times slower than inosine.[190] Therefore, for (and also guanosine-inosine nucleoside hyarolase, GI-NH) IAG-NH, specific interactions between the protein and the leaving group must operate; partial protonation at N7 is supported by the significant inverse ^{15}N isotope effects at this position, in accord with an increase in bonding. The differences have been exploited to produce powerful isoenzyme-specific inhibitors for these enzymes, with DNJ-type functionality to impart some transition state analogue character and a non-fissile C glycosidic link, but an aglycone which can make many of the interactions of the natural leaving group (Figure 5.29).[191]

Isotope effects for AMP nucleosidase demonstrate a very important feature of transition states for enzymic glycosyl transfer, namely that they are plastic, in the sense that their structure can be altered by comparatively minor changes in the protein. ATP is an allosteric activator and the fact that the relative magnitudes of the various effects alter according to allosteric activation mean that transition state structures must be changing (a change in commitments would have affected all effects equally).

Although some isotope effects were measured on reactions catalysed by *E. coli* purine nucleoside phosphorylase (*e.g.* Figure 5.15), the mammalian (beef) enzyme has been most studied, despite being kinetically inconvenient (it is a trimer and in the absence of phosphate hydrolyses one molecule of inosine to give a species with one molecule of tightly bound inosine per enzyme trimer[192]). Phosphorolysis is reversible (see caption to Figure 5.15), so the isotope data of Table 5.2 were obtained for arsenolysis, since arsenate esters rapidly hydrolyse by As–O fission; the use of arsenate as a nucleophile also made commitments zero. A surprising feature is that N9 isotope effects are low, implying less C–N bond cleavage than usual.

X-ray crystal structures of the beef purine nucleoside phosphorylase enzyme with a substrate–analogue complex (inosine and the non-nucleophilic sulfate), a transition state–analogue complex (immucillin H and phosphate) and a product complex (inosine and ribose-1-phosphate)[193] led to the conclusion that most atomic motion was in the anomeric centre, which swung from attachment to

N9 of hypoxanthine to attachment to phosphate. There were, though tightening and shortening of hydrogen bonds from the protein to the leaving group in the transition state.

The other *N*-glycoside phosphorolysis for which there is a multiple kinetic isotope effect determination of transition state structure is the phosphonoacetolysis of orotidine 5′-phosphate. The reaction of phosphoribosyl pyrophosphate with orotic acid is a key step in the biosynthesis of pyrimidine nucleosides, but the reaction was studied in the non-physiological direction, using phosphonoacetate, rather than pyrophosphate, as a nucleophile to avoid commitments (Figure 5.30). It was a good choice, since the X-ray crystal structures of the *E. coli*[194] and *S. typhimurium*[195] enzymes revealed a complex protein architecture, with the active site constructed from the interface of the two monomers. Pre-steady-state and steady-state kinetic studies with the natural substrates revealed a complex mechanism, in which the most important kinetic event was the movement of a loop of protein to close off the active site from solvent; the chemistry was fast and at equilibrium compared with this.[196,197] The kinetic mechanism was ordered bi–bi with one step effectively irreversible, which gave a pattern of double reciprocal plots similar to ping-pong.

The leaving group ability of a purine nucleoside base (or the nucleophilicity of a purine), can be modulated by proton transfer to and from N7, although it appears that purine nucleoside phosphorylase does not use this mechanism. Both mammalian and bacterial enzymes will accept the adenosine and inosine isomers 3-(β-D-ribofuranosyl)adenine and 3-(β-D-ribofuranosyl)hypoxanthine as substrates, with 3–30% of the activity of the "right" isomer;[198] the X-ray structure of the immucillin H complex shows interaction of a glutamate with N1 and O6, but no interaction with N7. It thus appears that purine nucleoside phosphorylase may adopt a pyrimidine nucleoside-type activation mechanism. At least in DNA uracil glycosylase, this involves a bending of the base (C1′–N1–C4 $\sim$130°) so that the fissile C1′–N bond can interact with the aromatic π system, rather than be orthogonal to it. In the reverse reaction, in which the pyrimidine is a nucleophile, the "bending" mechanism implies that it is a π, rather than a σ, nucleophile, *i.e.* it attacks with the aromatic π system rather than an sp^2 lone pair on N1. The bending mechanism may account for the very low leaving group ^{15}N effects of both orotate phosphoribosyltransferase and purine nucleoside phosphorylase.

There are many transferases catalysing the transfer of a 5-phosphoribosyl group from 5′-ribosyl pyrophosphate to nitrogen, but detailed mechanistic studies are not available. All such enzymes, however, are activated by divalent cations such as Mg^{2+}, which coordinates to the pyrophosphate oxygens and increases leaving group ability.

Many microbial pathogens produce toxins which ADP-ribosylate target proteins; the reaction is a single displacement of nicotinamide from C1′ by a nucleophile on the target protein surface. In the absence of acceptor proteins, however, the catalysing proteins slowly hydrolyse NAD^+. In the case of pertussis (whooping cough) toxin, ^{14}C effects show clearly that the protein reaction has the more S_N2 character, not surprising in view of the thiolate

Table 5.2 (V/K) kinetic isotope effects on enzymic nucleoside bond cleavage (intrinsic + binding effects); isotope as well as position is given.

Enzyme, reaction	*Source*	*C1′*	*N9 (^{15}N)*	*H1′*	*H2′*	*H4′*	*H5′*	*Other*
Inosine–uridine nucleosidase[177]	*Crithidia fasciculata*	1.044/^{14}C	1.026	1.150/T	1.163/T	0.992/T	1.051 av/T	
Guanosine–inosine nucleosidase[178]		1.029/^{14}C	1.030	1.122/T	1.104/T		1.022 av/T	N7 0.997
Inosine–adenine–guanine nucleosidase[178a]	*Trypanosoma brucei*	1.034/^{14}C	1.017	1.125/T	1.101/T		1.026 av/T	N7 0.994
		1.032/^{14}C	1.021	1.104/T	1.107/T		1.026 av/T	N7 0.985
wt AMP nucleosidase[179]–	*Azotobacter vinelandii*	1.035/^{14}C	1.030	1.069/T 1.045/D	1.061/D			
ATP) +ATP		1.032/^{14}C	1.025	1.047/T 1.030/D	1.043D			
Mutant AMP nucleosidase[180]	*Azotobacter vinelandii*							
–ATP		1.040/^{14}C	1.034	1.086/T	1.061/D			
+ATP		1.040/^{14}C	1.021	1.094/T	1.043/D			
NAD^+ hydrolysis[181]	Cholera toxin	1.030/^{14}C	1.029	1.186/T	1.108/T	0.986/T	1.020/T (av)	
NAD^+ hydrolysis[182]	Diphtheria toxin	1.034/^{14}C	1.030	1.200/T	1.142/T	0.9990/T	1.032/T (av)	4′–^{18}O 0.986
NAD^+ hydrolysis[183]	Pertussis toxin	1.021/^{14}C	1.021	1.207/T	1.144/T	0.989/T	1.004/T (av)	
ADP-ribosylation of Cys in $G_{i\alpha}$ (inv)[184]	Pertussis toxin	1.049/^{14}C	1.023	1.199T	1.105/T	0.991/T	1.020/T (av)	
ADP-ribosylation of Cys in $_{\alpha i3}$C20 (inv)[185]	Pertussis toxin	1.050/^{14}C	1.021	1.208T	1.104/T	0.989/T	1.014/T (av)	

Beef purine nucleoside phosphorylase[186]	Inosine arsenolysis	1.022/^{14}C	1.009	1.118/T	1.128/T	1.007/T	1.028/T (av)
Human purine nucleoside phosphorylase[187]	Inosine arsenolysis	1.002/^{14}C	1.029	1.184/T	1.031/T	1.024/T	1.062/T (av)
Malaria purine nucleoside phosphorylase[187]	*Plasmodium falciparum* inosine arsenolysis	0.996/^{14}C	1.019	1.116/T	1.036/T	1.009/T	1.062/T (av)
Human thymidine phosphorylase[188]	2′-Deoxythymidine arsenolysis	1.139/^{14}C	1.022	0.989/T	*proR* 0.974 *proS* 1.036/T	1.020/T	1.061/T (av)
Orotate phosphoribosyl transferase[189]	*Salmonella typhimurium*	1.040/^{14}C	1.006	1.200/T	1.140/T	0.988/T	1.028/T (av)

[a] Because of the uncertainties in the intrinsic effects for studies with inosine, because of high commitments, the experiments were repeated with 2′-deoxyinosine, a slow substrate.

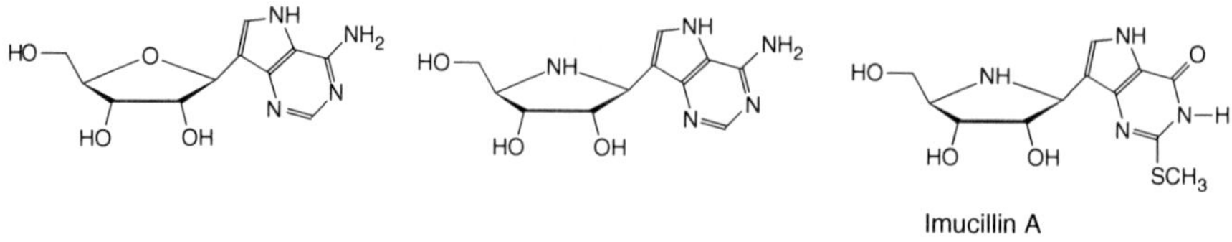

Imucillin A

IU-NH:	1.5 mM	7 nM	0.23 μM
IAG-NH	0.41μM	0.9nM	> 0.05 mM

Figure 5.29 Isoenzyme-specific inhibitors of purine nucleoside hydrolases.

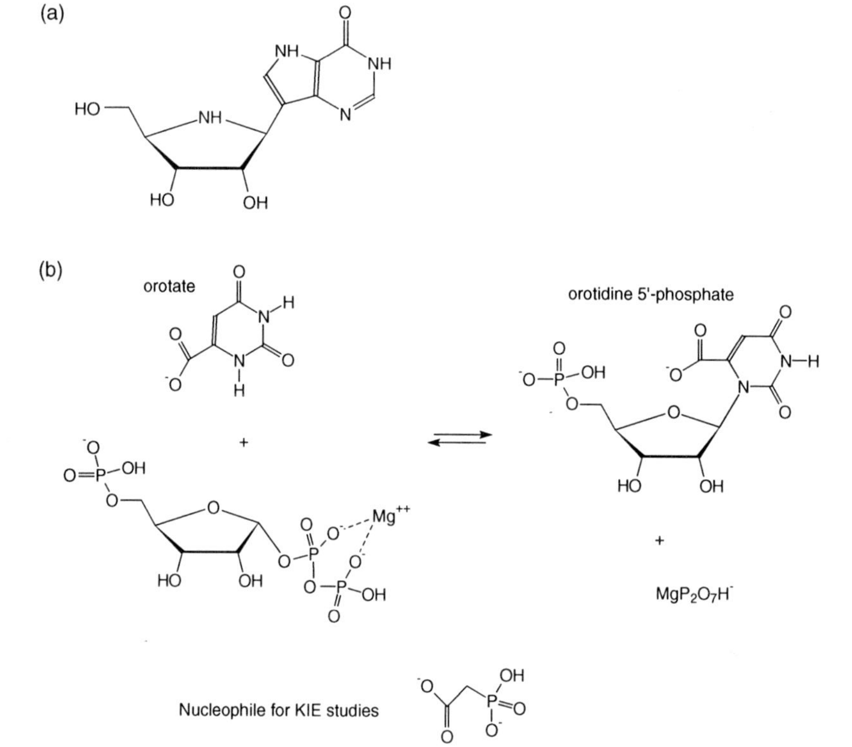

(a)

(b) orotate orotidine 5'-phosphate

$MgP_2O_7H^-$

Nucleophile for KIE studies

Figure 5.30 (a) Immucillin H, a transition state analogue for purine nucleoside phosphorylase. (b) Reaction catalysed by orotate phosphoribosyl transferase.

nucleophile for the protein reaction compared with unactivated water for the hydrolysis. Likewise, the lower β-tritium effects for the protein reaction reflect less charge build-up at C1′; the α-tritium effects are the same for hydrolysis and reaction with thiolate, the increase in effect expected for the softer thiolate nucleophile (see Section 3.3.1) being offset by more S_N2 character.

Proteins can also be ADP ribosylated on arginine residues (Figure 5.31). The crystal structures of two such toxins support the idea of a similar mechanism,

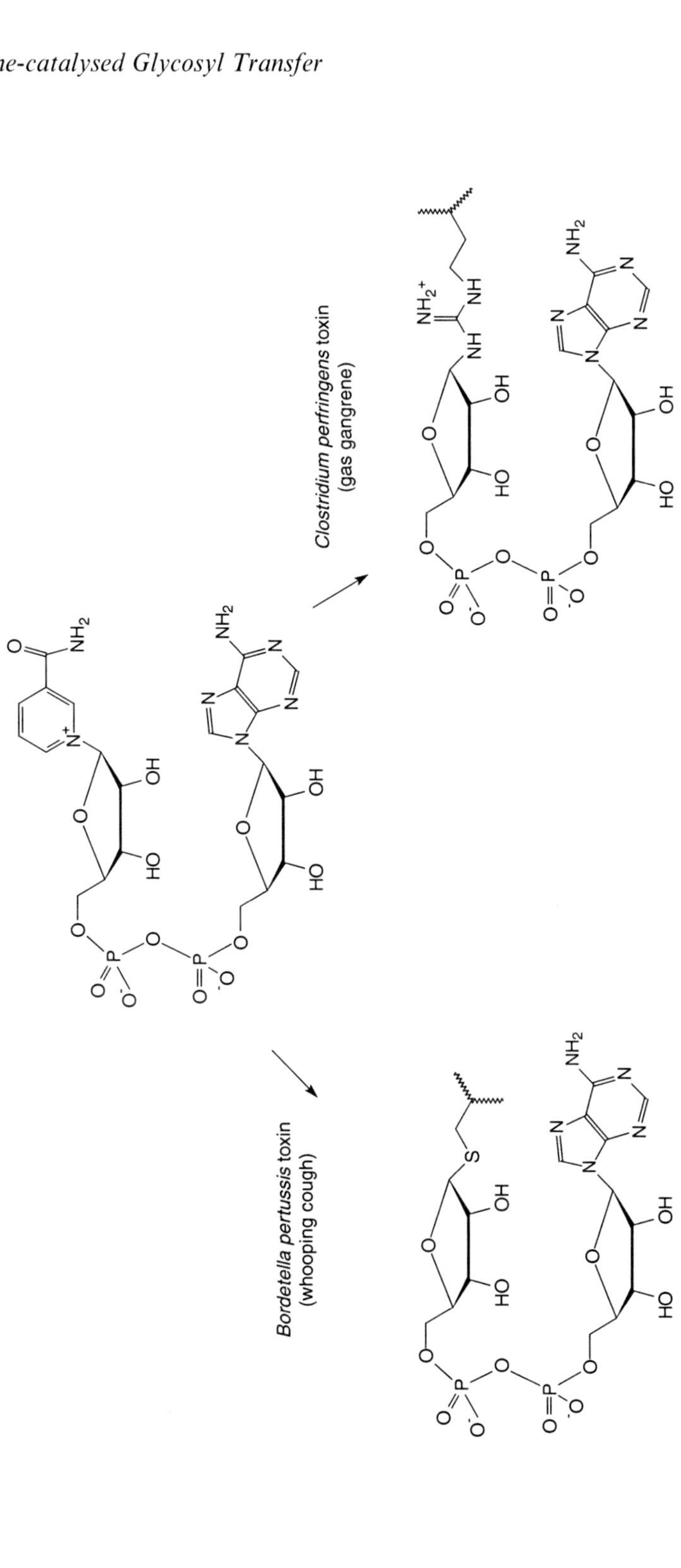

Figure 5.31 ADP ribosylation of proteins.

with the normally cationic guanidino residue being activated by hydrogen bonding to basic groups;[199,200] mutation of one such blocked transfer, but not hydrolysis.[201]

5.9 RETAINING *O*-GLYCOSIDASES AND TRANSGLYCOSYLASES

The canonical mechanism for retaining glycosidases and transglycosylases is given in Figure 5.32. Two transition states, which invariably involve an electron-deficient sugar in the −1 site, are involved, and a covalent intermediate. The covalent nature of the intermediate remained controversial until recently, because of the prestige (rightly) accorded to the first X-ray structure of an enzyme, hen egg-white lysozyme[202] (now known to be GH 22), and by association (wrongly) with a mechanism based on model building into the −1 subsite, which appeared to indicate that the glycosyl-enzyme intermediate could only be a stabilised ion pair on a glycosyl cation and the nucleophilic aspartate, not a covalent ester.[203] The glycosyl-enzyme intermediate in lysozyme itself has only become accessible recently,[204] but starting in 1973 all measurements of α-deuterium kinetic isotope effects (see Section 3.3.1) on the hydrolysis of glycosyl-enzyme intermediates were direct ($k_H/k_D > 1.0$), in the wrong sense for an ion-paired intermediate (Table 5.3).

Although one can write a canonical mechanism for retaining *O*-glycosidases, there are departures from this mechanism in acid-catalytic machinery and the nature of the nucleophile (for some families, not the carboxylate shown). In Table 5.4 are set out the nature of the catalytic groups, the protonation trajectory and the conformations of substrate and glycosyl enzyme, where at least two of these are known.

5.9.1 Inactivation of Glycosidases – *Exo* and Paracatalytic Activation

Active site groups can often be determined chemically, without the need to crystallise the protein. So-called *exo* affinity labels contain intrinsically reactive groups such as halocarbonyl, *e.g.* $-COCH_2Br$,[235] and preferentially alkylate groups in the active site, by virtue of their binding functionality. Thus, ω-epoxyalkyl cellobiosides label active site carboxylates of cellulases:[236] in the GH 5 Family, the most effective reagent is the 4,5-epoxypentyl compound, which labels the nucleophile.[237] The sugar residues bind −2, −3 and the epoxyalkyl group fits across the −1 subsite to reach the nucleophile.

An extended series of investigations by Legler and co-workers on hydroxylated cyclohexene epoxides, starting in the 1960s, was based on the presumption that conduritol B epoxide could mimic either an α- or a β-glucoside, with the epoxide being activated by protonation by the catalytic acid.[238] Elegant experiments with inactivated α- and β-glucosidases, conducted with radiolabelled conduritol B epoxide, appeared to confirm these assumptions. Treatment of inactivated β-glucosidase A3 from *Aspergillus wentii* with hydroxylamine yielded *chiro*-inositol,[239] whereas similar treatment of inactivated mammalian

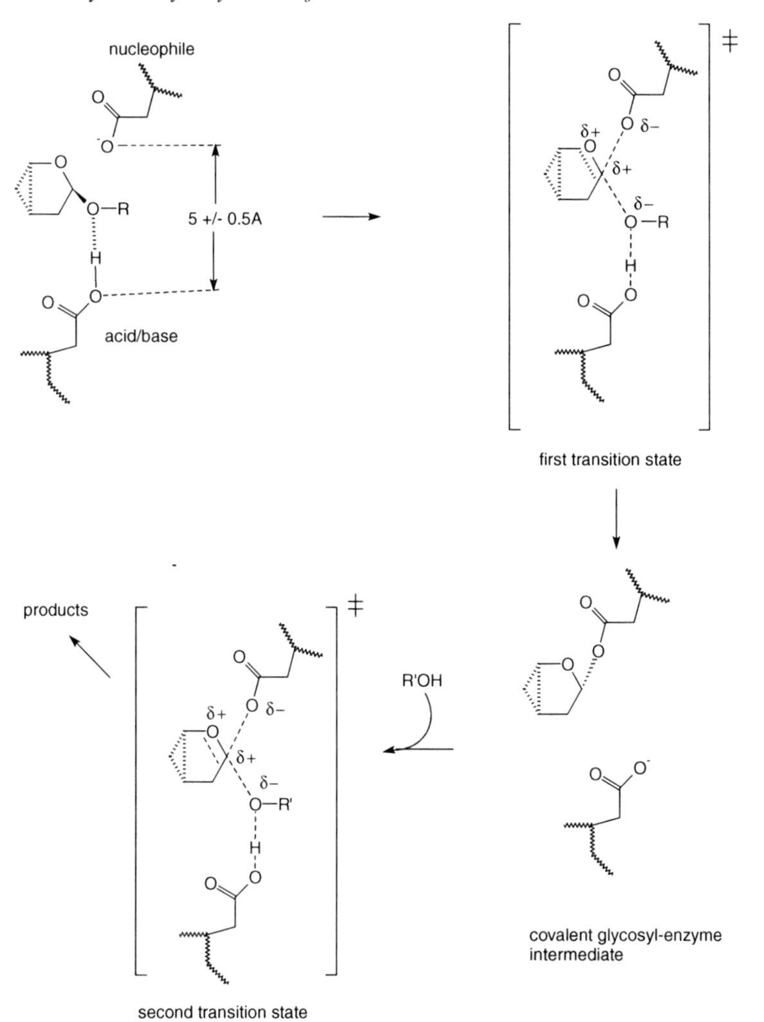

Figure 5.32 Canonical mechanism for a retaining glycosidase or transglycosylase Whether the substrates are pyranosidases or furanosidases, whether the leaving group is axial or equatorial and whether the protonation trajectory is *syn* or *anti* depend on the enzyme. If R = R′, then the principle of microscopic reversibility requires transition states 1 and 2 to be identical. More generally, the group which acts as a general acid in the first step must act as a general base in the second.

sucrase–isomaltase (GH 31) yielded *scyllo*-inositol (Figure 5.33); the L-isomer of conduritol B epoxide was inactive.[240] The peptide sequence around active site aspartate labelled by conduritol B epoxide in the β-glucosidase A3 from *Aspergillus wentii*[241] was subsequently shown to be the same as that around the nucleophilic aspartate of a GH 3 enzyme from another *Aspergillus* species (*niger*)[242] labelled by 2-fluoro-2-deoxy-β-D-glucosyl fluoride; likewise, the

Table 5.3 α-Deuterium kinetic isotope effects on the hydrolysis of the glycosyl-enzyme intermediates of retaining glcyosidases.

Family, enzyme	*Source*	$^{D}k_{+3}$ *(nucleophile water unless indicated)*	*Ref.*
GH 2 β-galactosidase	*E. coli* (*lacZ*)	1.25, 1.35 (MeOH)	205
GH 2 β-galactosidase	*E. coli* (ebg)	1.08, 1.10	206, 207
(Unknown GH) β-glucosidase	*Stachybotris atra*	1.11	208
GH 1 β-glucosidase	*Agrobacterium faecalis*	1.11	72
GH 10 cellulase/xylanase Cex	*Cellulomonas fimi*	1.10	74
GH 27 α-galactosidase	*Phanerochaete chrysosporium*	1.11	59
Unknown GH β-xylosidase	*Trichoderma koningii*	1.02	209
GH 3 β-glucosidase	*Flavobacterium meningosepticum*	1.18	210
GH 2 β-mannosidase, acid–base mutant	*Cellulomonas fimi* E429A	1.12, 1.17 (N_3^-)	211

nucleophile in a GH 31 α-glucosidase from *A. niger* was labelled both by conduritol B epoxide and α-5-fluoroglucosyl fluoride.[243] Therefore, in these cases, conduritol epoxide had labelled the nucleophile; however, when conduritol C epoxide was used against the *lacZ* β-galactosidase of *Escherichia coli*, (GH 2),[244] it did not label the nucleophile but rather Glu461, either the acid catalyst or a coordinator of the electrophilic Mg^{2+}.[245] In the case of human β-glucoceramidase (GH 30), conduritol B epoxide labelled a carboxylate residue comparatively unconnected with catalysis.[246] The mislabelling could arise from the imperfect resemblance of conduritol epoxides and sugars – the naturally occurring cyclophellitols, discovered subsequently, show a better fit.[247] In the first indication of substrate-assisted catalysis with glucosaminidases, however, it was found that acetamido analogue of conduritol B epoxide resembled substrate sufficiently to be slowly converted to an oxazolidine by GH 20 glucosaminidases from jackbean and beef kidney.[248]

Unsurprisingly, the nitrogen analogue of conduritol epoxide inactivated GH 13 yeast α-glucosidase and a GH 1 β-glucosidase from *Agrobacterium faecalis*; the inactivation appeared faster than with conduritol B epoxide itself.[249]

Particularly with overtly reactive labelling reagents such as halocarbonyl compounds, it is desirable to have confirmation that the labelling is active site directed. Only one molecule of the reagent should be incorporated per active site inactivated (these days this readily established by mass spectrometry with "soft" ionisation techniques such as electrospray) and the kinetics of inactivation in the presence of excess inactivator should be first order (if A is activity, then $A_t = A_0\exp(-k_{obs}t)$. In particular, quantitative protection by a reversible

Table 5.4 Substrate stereochemistry, acid/base groups, nucleophiles, protonation trajectory and substrate and glycosyl-enzyme intermediate conformations, where known.

Family	*Substrate stereochemistry*	*Acid/base*	*Trajectory*	*Nucleophile*	*ES conformation*	*Glycosyl-enzyme, conformation*
1	*p,e*	Glu	*anti*	Glu		$^{4}C_{1}$(2-d-2F)[212]
2	*p,e*	Glu	*anti*	Glu	$^{4}C_{1}$	$^{4}C_{1}$(2-d-2F)[213]
3	*p,e*	Glu	*anti*	Asp		
5	*p,e*	Glu	*anti*	Glu	$^{1}S_{3}$[214]	$^{4}C_{1}$[215]
7	*p,e*	Glu	*syn*	Glu	$^{1}S_{3}$[216]	
10	*p,e*	Glu	*anti*	Glu		$^{4}C_{1}$[217]
11	*p,e*	Glu	*syn*	Glu		$^{2,5}B$[218]
12	*p,e*	Glu	*syn*	Glu	$^{1}S_{3}$[219]	$^{4}C_{1}$[220]
13	*p,a*	Glu	*anti*	Asp	$^{4}H_{3}$[221]	$^{4}C_{1}$[222]
16	*p,e*	Glu	*syn*	Glu		
17	*p,e*	Glu		Glu		
18	*p,e*	Glu	*anti*	Subst. $NHCOCH_3$	$^{1,4}B$	$^{4}C_{1}$[223]
20	*p,e*	Glu	*anti*	Subst. $NHCOCH_3$	$^{1,4}B$	$^{4}C_{1}$[224]
22	*p,e*	Glu	*syn*	Asp	"Sofa" intermediate $^{1}S_{3}$ and $^{4}H_{3}$[225]	$^{4}C_{1}$[204]
26	*p,e*	Glu	*anti*	Glu	$^{1}S_{5}$	$^{O}S_{2}$[226]
27	*p,a*	Asp	*anti*	Asp		
29	*p,a*	Glu	*syn*	Asp	$^{1}C_{4}$	$^{3}S_{1}$[227]
30	*p,e*	Glu		Glu		

(*Continued*)

Table 5.4 (*Continued*).

Family	*Substrate stereochemistry*	*Acid/base*	*Trajectory*	*Nucleophile*	*ES conformation*	*Glycosyl-enzyme, conformation*
31	*p,a*	Asp	*anti*	Asp		${}^{1}S_{3}$[228]
32	*f*	Glu		Asp		
33	*p,e* (*gem*, a COOH)	Asp	*anti*	Tyr (σ or π)	${}^{4,0}B$[229]; $B_{2,5}$, ${}^{6}S_{2}$[230]	${}^{2}C_{5}$[230]
34	*p,e* (*gem*, a COOH)	Asp	*anti*	Tyr (σ or π)	$B_{2,5}$[231]	
35	*p,e*	Glu		Glu		
38	*p,a*	Asp	*anti*	Asp		${}^{1}S_{5}$[232]
39	*p,e*	Glu	*anti*	Glu		${}^{4}C_{1}$[233]
42	*p,e*	Glu	*anti*	Glu		
51	*p,e* or *f*	Glu	*anti*	Glu	${}^{4}E$(*Araf*)	${}^{2}T_{1}$(*Araf*)[234]
52	*p,e*	Asp		Glu		
53	*p,e*	Glu		Glu		
54	*p,a*	Asp	*anti*	Glu		
56	*p,e*	Glu	*anti*	Subst. $NHCOCH_3$		
57	*p,a*	Asp	*anti*	Glu		
68	*f*	Glu	*anti*	Asp		
70	*p,a*	Glu		Asp		
72	*p,e*	Glu		Glu		
77	*p,a*	Asp	*anti*	Asp		
79	*p,e*	Glu		Glu		
83	*p,e*	?Arg	*?anti*	Tyr		
86	*p,a*	Glu		Glu		

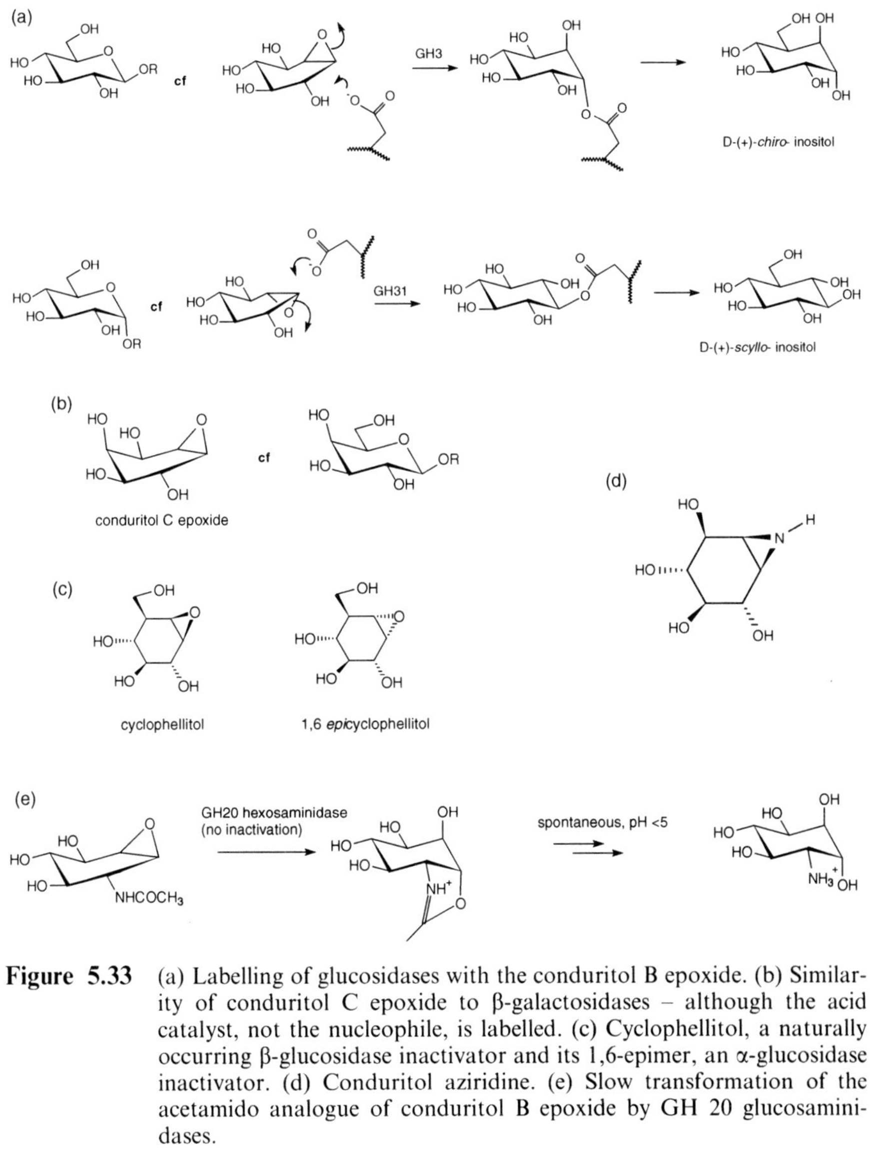

Figure 5.33 (a) Labelling of glucosidases with the conduritol B epoxide. (b) Similarity of conduritol C epoxide to β-galactosidases – although the acid catalyst, not the nucleophile, is labelled. (c) Cyclophellitol, a naturally occurring β-glucosidase inactivator and its 1,6-epimer, an α-glucosidase inactivator. (d) Conduritol aziridine. (e) Slow transformation of the acetamido analogue of conduritol B epoxide by GH 20 glucosaminidases.

inhibitor (or poor substrate) should be observed, according to eqn. (5.47):

$$k_{obs} = \frac{k_{max}[\text{Inact}]}{[\text{Inact}] + K\left(1 + \frac{[\text{I}]}{K_i}\right)} \tag{5.47}$$

Some inactivators – sometimes called "suicide substrates" – are not overtly reactive but rather rely on the catalytic action of the enzyme to generate an electrophilic species. If the electrophilic species has no recognition functionality, it is often free to diffuse away from the active site and alkylate randomly. Such species are generated from the α,α-difluoro-*p*-cresyl[250] or 2-chloromethyl-4-nitrophenyl[251] glycosides (with generate electrophilic quinone methides) and 1,1-difluoroalkyl glucosides,[252] (which generate an acyl fluoride) (Figure 5.34). Salicortin (a constituent of aspen/poplar twigs, apparently produced as an antifeedant against browsing mammals) also generate a quinone methide when hydrolysed by β-glucosidase. The quinone methide attacks several positions in the active site of the GH 1 β-glucosidase from

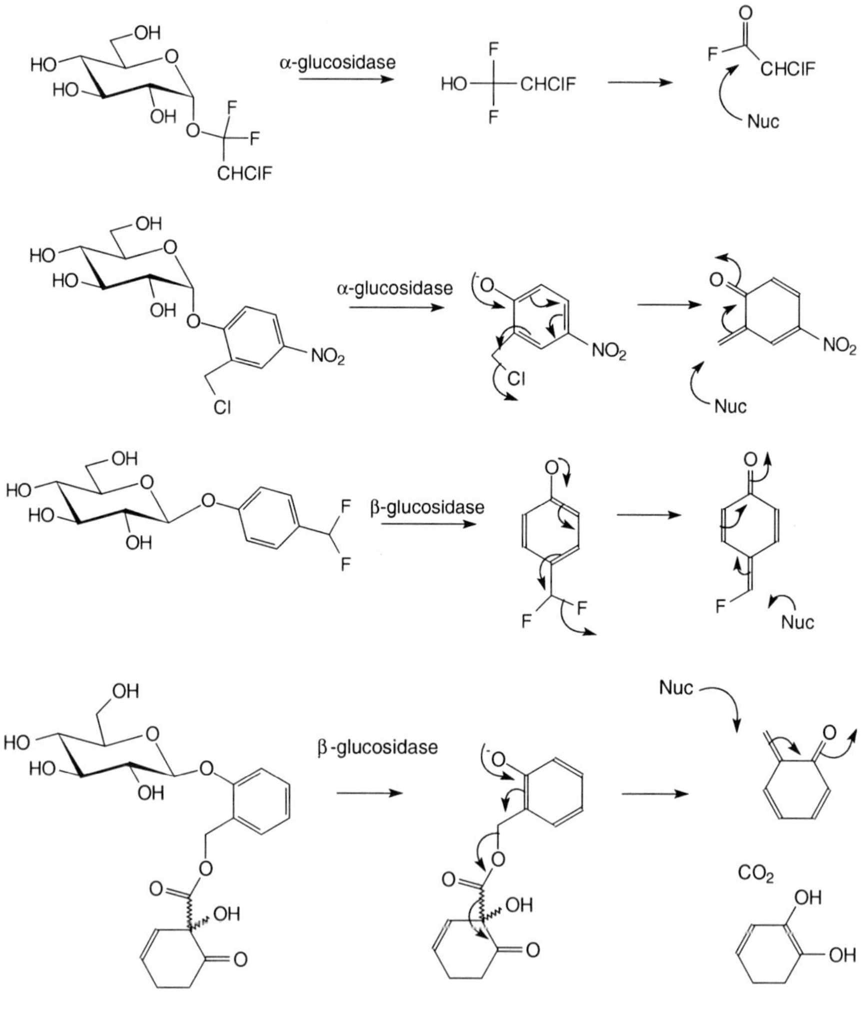

Figure 5.34 Some suicide inactivators generating alkylating species without recognition functionality.

Agrobacterium faecalis,[253] but apparently does not live long enough to diffuse away and attack other proteins, since in a solution of both active and inactive nucleophile mutant of the enzyme, only the active enzyme is hydroxybenzylated.

An important parameter for suicide inactivators is the partition ratio – the number of catalytic events which are necessary for one active site to be inactivated. If the inactivator is stable, it can be estimated by incubating excess enzyme with a limited amount of inactivator; if it is not, and particularly if the partition ratio is low but > 1, numerical solutions to the differential equations involved are needed.[254]

The foregoing suicide inactivators generate avid electrophiles with no recognition functionality, and are therefore prone to modify non-active site groups. One group of suicide reagents generating powerful electrophiles with a sugar recognition functionality still attached are the glycosylmethylaryltriazenes. These generate glycosylmethanediazonium ions in the active sites of some glycosidases, in enzyme-catalysed processes. The glycosylmethyldiazonium ions then either diffuse away and decompose in free solution or alkylate an active site residue (primary alkane diazonium ions have lifetimes of tens or hundreds of milliseconds in aqueous solution – simultaneous fragmentation of the parent triazenes to aniline, nitrogen and a carbenium ion is not an issue). In the case of the GH 2 *E. coli lacZ* β-galactosidase, the residue alkylated is Met502,[255] but in the case of the closely related *ebg* enzyme a glutamate, homologous to the group labelled by conduritol C epoxide in the *lacZ* enzyme and now known to be the acid catalyst or Mg^{2+} ligand, was labelled.[256]

A survey of a limited number of glycosidases failed to find an inverting enzyme which was inactivated by the corresponding glycosylmethyl-*p*-nitrophenyltriazene,[257] although it was later reported that α-mannopyranosylmethyl-*p*-nitrophenyltriazene inactivated Golgi mannosidase I, a GH 47 inverting enzyme, in addition to the (retaining) jackbean enzyme of GH 38.[258] The intuitive expectation that the triggering reaction would involve the acid catalytic group carrying out a reaction analogous to the general-acid-catalysed decomposition of alkylaryltriazenes in free solution appears not be fulfilled. The non-enzymic reaction exhibits β_{lg} values (calculated on the pK_a of $ArNH_3^+$) of ~0.7, depending on the catalysing acid,[259] whereas the β_{lg} values on k_{inact} for inactivation *E. coli lacZ* β-galactosidase[260] are negative, like those on the spontaneous, pH-independent reaction of alkylaryltriazenes (−0.7 calculated on the pK_a of $ArNH_3^+$ or −0.24 calculated on the pK_a of $ArNH_2$, the latter giving a much better correlation). Partitioning ratios of about 4 are observed for *E. coli lacZ* β-galactosidase, but around 80 for inactivation of an α-galactosidase from *Pycnoporus cinnabarinus*[261] (a retaining GH 27 α-galactosidase from *Phanerochaete chrysosporium* is inert to α-galactospyranosylmethyl-*p*-nitrophenyltriazene, however[59]). When these triazenes work, however, they have the advantage over similar reagents of selectively inactivating the target enzyme in whole cultured cells[262] or even perfused organs.[263] Glycosylmethylaryltriazenes as suicide inactivators of retaining glycosidases are shown in Figure 5.35.

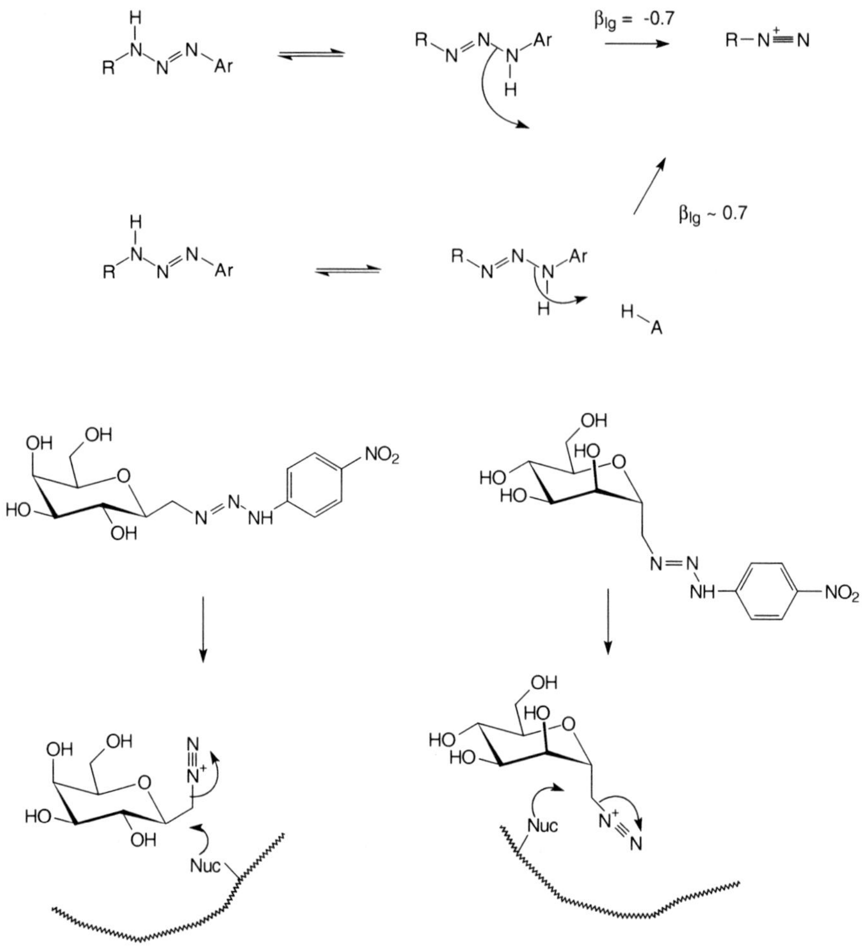

Figure 5.35 Glycosylmethylaryltriazenes as suicide inactivators of retaining glycosidases. Top, pathways for triazene decomposition. Bottom, two triazenes known to work *in vivo*.

5.9.2 Direct Observation of Glycosyl-enzyme Intermediates

The major breakthrough in the paracatalytic inactivation of retaining glycosidases came with the realisation by Withers and coworkers that, according to the canonical mechanism for such enzymes of Figure 5.31, destabilisation of the second transition state by, for example electron-withdrawing groups in the glycone, would make the glycosyl-enzyme turn over slowly, if at all (Figure 5.36). The solution to the problem of also destabilising the first transition state by the same glycone substitutions was to incorporate a good leaving group. In the first attempt, 2,4-dinitrophenol (pK_a 4.2) was used as the leaving group and the 2-OH was replaced by F in an inactivator of GH 1 β-glucosidase of *Agrobacterium faecalis*.[264] The F-for-OH substitution results in a fairly modest change in

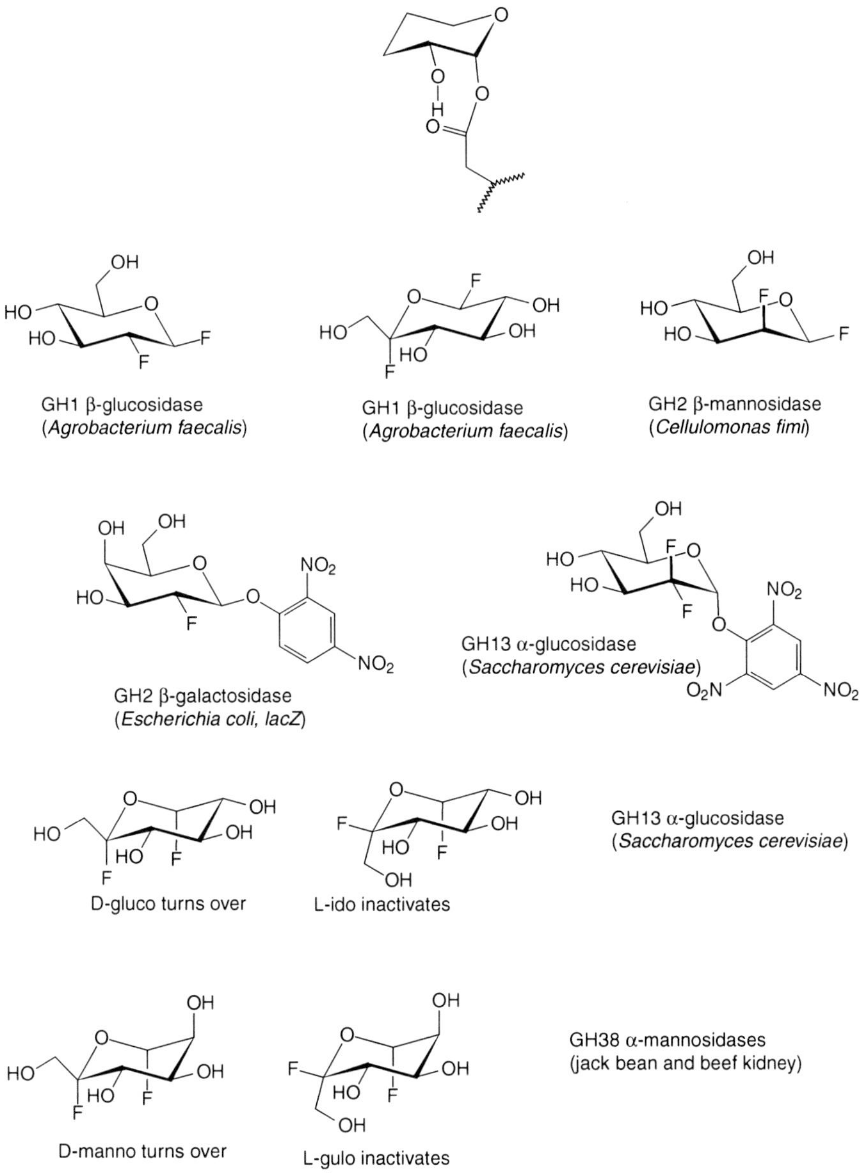

GH1 β-glucosidase (*Agrobacterium faecalis*)

GH1 β-glucosidase (*Agrobacterium faecalis*)

GH2 β-mannosidase (*Cellulomonas fimi*)

GH2 β-galactosidase (*Escherichia coli, lacZ*)

GH13 α-glucosidase (*Saccharomyces cerevisiae*)

D-gluco turns over

L-ido inactivates

GH13 α-glucosidase (*Saccharomyces cerevisiae*)

D-manno turns over

L-gulo inactivates

GH38 α-mannosidases (jack bean and beef kidney)

Figure 5.36 Some inactivators of retaining glycosidases/transglycosylases on the Withers principle. Top: possible 2-OH–nucleophile interaction in retaining β-glycosidases which largely accounts for the success of a single F-for-OH substitution.[273]

electron demand ($\Delta\sigma_I = 0.25$) and it is often found that only partial inactivation is observed with the 2,4-dinitrophenylglycosides, the 2-fluoro-2-deoxyglycosyl-enzyme intermediate turning over slowly. The turnover could often be accelerated by the addition of acceptors which bound to the positive subsites and there reacted faster than water – a typical example is the GH 5 endoglucanase from *Clostridium thermocellum*.[265] This demonstration that the glycosyl-enzyme intermediates were catalytically competent went some way towards stilling initial objections that the reagents were reacting via a mechanism unrelated to the normal mechanisms of retaining glycosidases. Almost all the firm identifications of nucleophiles have been made using reagents based on these principles.

Nucleophiles in GH 1,[266] GH 2[245] and GH 5,[267] GH 7,[268] GH 12,[269] GH 35,[270] and GH 39[271] could be identified using 2,4-dinitrophenyl-2-deoxy-2-fluoroglycosides. Changing the leaving group to the better fluoride (pK_a 3.2) increased the proportion of glycosyl-enzyme during the steady state, but did not, of course, affect the rate of glycosyl-enzyme hydrolysis. It was found[272] that 2-deoxy-2-fluoroglycosyl β-glycosyl fluorides irreversibly inactivated a range of β-gluco- and -galactosidases, gave partial inactivation (with a slow steady-state turnover) with others, but with one exception (an *Aspergillus* α-galactosidase, which gave slow turnover), 2-deoxy-2-fluoroglycosyl α-glycosyl fluorides did not inactivate α-glycosidases. Slow turnover was later also observed with a GH 29 L-fucosidase.[227]

The kinetic equations describing the appearance of leaving groups with the Withers reagents during paracatalytic inactivation are the same as those describing "burst" kinetics with ordinary substrates for which $k_{+2} > k_{+3}$. In both cases the build-up of the glycosyl-enzyme intermediate is monitored directly, by observing the liberation of leaving group on a time-scale short compared with $1/k_{+2}$ and $1/k_{+3}$, before the steady state is established. For "ordinary" substrates, rapid reaction techniques such as stopped-flow are necessary: k_{cat} values for nitrophenylglycosides are commonly 50–1000 s^{-1}, so the faster transients may even be beyond the limits of observation of conventional stopped flow, which has mixing times of the order of milliseconds. A typical stopped flow device is shown in Figure 5.37. If $k_{+2} > k_{+3}$ and the enzyme is saturated with substrate, there is a period over which the enzyme becomes glycosylated. If the appearance of the

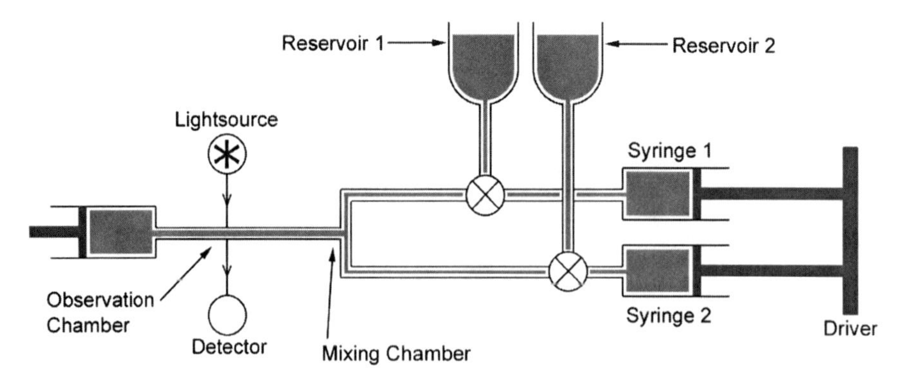

Figure 5.37 Cartoon of a stopped flow kinetic device (Courtesy of TgK Scientific Ltd). Solutions of enzyme and substrate are "pushed" from two syringes into a mixing chamber, then through an observation chamber and then into a third syringe.

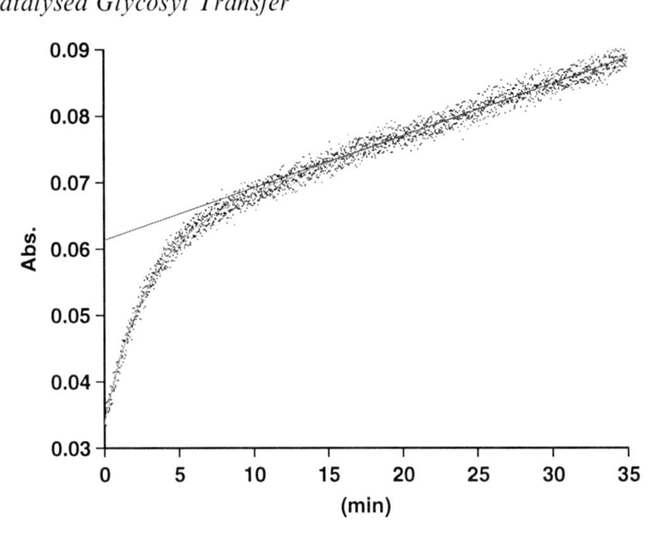

Figure 5.38 Burst kinetics showing liberation of aglycone during the establishment of a steady state during hydrolysis of a substrate by a retaining glycosidase, when the hydrolysis of the glycosyl-enzyme is rate determining. The glycone was fluorinated, so the experiment was performed in a conventional spectrometer (note time-scale). Note the "burst" of 0.03 absorbance units.

aglycone is monitored, by, for example, its UV absorbance or fluorescence, then a burst of aglycone can be seen, as in Figure 5.38.

Three parameters are obtained from such a trace: the steady-state rate, the amplitude of the burst Π and its first-order rate constant $k_{\text{transient}}$, which are given in terms of the microscopic rate constants of Scheme 5.2 by eqns. (5.48) and (5.49):

$$\Pi = \left(\frac{k_{+2}}{k_{+2} + k_{+3}}\right)^2 \left(\frac{[\mathrm{S}]}{K_\mathrm{m} + [\mathrm{S}]}\right)^2 [\mathrm{E}]_0 \tag{5.48}$$

$$k_{\text{transient}} = k_{+3} + \frac{k_{+2}k_{+3}[\mathrm{S}]}{(k_{+2} + k_{+3})K_\mathrm{m} + k_{+3}[\mathrm{S}]} \tag{5.49}$$

The direct measurement of the size of a burst becomes one of the few ways of knowing the concentration of active sites in an enzyme solution, independently of assumptions about the activity or otherwise of the protein. Wild-type enzymes that are 100% active are the exceptions rather than the rule, whereas on the other hand many mutant proteins are reported with low levels of activity which reflect contamination with wild-type.

The squared terms in eqn. (5.48) mean that even with $[\mathrm{S}] = 10K_\mathrm{m}$ and $k_{+2} = 10k_{+3}$, the burst size corresponds to only $0.68[\mathrm{E}]_0$. Further, comparatively large concentrations of enzyme are required (with an extinction coefficient change of 10 000 and a 1 cm pathlength cell, 2 μM enzyme or 0.12 mg mL^{-1} for an "average" glycosidase molecular weight of 60 000, is required to see a burst of 0.02 A, the smallest that can be reliably measured. Use of the more sensitive fluorescent leaving groups can reduce this, but the fluorescence has to be calibrated for the circumstances of the measurement. If Withers inactivators with a fluoride leaving group are used, the sensitivity problem worsens, since the ion-selective electrodes used to monitor fluoride release are at the limit of their sensitivity in the μM range. (The response of fluoride-selective electrodes is so slow that fluoride leaving groups cannot be used to generate observable bursts from natural glycones).

A feature of eqn. (5.49) is that the rate constant for the pre-steady-state phase contains K_m: effectively increasing this parameter by inclusion of a competitive inhibitor {which makes its effective value $K_m(1+[I]/K_i)$} thus slows the rate constant of the burst.

2-Deoxy-2-fluoro-β-glycosyl fluorides identified the nucleophiles in several enzymes, such as sweet almond β-glucosidase (GH 1)[274] and human β-glucuronidase (GH 2)[275] and in GH 30.[246] When they were tested *in vivo*, the activity recovered much faster than when the same enzyme was inactivated by the appropriate triazene. Recovery of enzyme activity in the triazene case was associated with new protein synthesis, but 2-fluoro-2-deoxyglycosyl enzymes slowly reverted to the original active enzyme.[276]

The 2-fluoro-2-deoxy-β-glycosyl fluorides and 2,4-dinitrophenolates owed their success as much to a hydrogen bond between the 2-OH of the glycone and the acylal carbonyl in the α-glycosyl-enzymes, which accelerated their hydrolysis and was not present in the 2-fluoro analogues, as to the electronic effect of the F-for-OH substitution (see Section 5.4.5.3). The absence of such interaction in the β-glycosyl-enzymes of α-glycosidases was a major reason for the failure of the 2-deoxy-2-fluoro compounds as inactivators (although not the only one, since 2-deoxy-2-fluoro-β-D-mannopyranosyl fluoride labelled a GH 2 β-mannosidase).[277] Increasing the electron-withdrawing effect by putting two fluorine atoms at C2 ($\Delta\sigma_I = 0.75$) and converting the leaving group to picrate (2,4,6-trinitrophenolate, $pK_a \approx 1$) permitted the inactivation of GH 13 yeast α-glucosidase and human pancreatic α-amylase[278] and GH 27 α-galactosidase.[59] The inductive effect of two fluorines at position 2, however, made nucleophilic displacement reactions at C1 very difficult (the picrates were made by reaction of the deprotected 1-OH with picryl fluoride, which gave the axial anomers in large excess). Reasoning that an F-for-H substitution at position 5 ($\Delta\sigma_I = 0.5$) would have an effect intermediate between the F-for-OH substitution at position 2 of the first generation of reagents ($\Delta\sigma_I = 0.25$) and the 2,2-fluoro compounds,[xviii] McCarter and Withers showed that α- and β-5-fluoroglucosyl fluorides inactivated GH 13 yeast α-glucosidase and GH 1 *Agrobacterium* β-glucosidase, respectively.[279] The 5-epimeric fluoroglycosyl fluorides were also produced by the synthetic route chosen and it was discovered that β-L-5-fluoroidosyl fluoride

[xviii] This assumes that the ρ_I values for the first step of the enzyme-catalysed reactions is approximately the same for substitution at C2 and C5: at least for the spontaneous hydrolysis of glycosyl 2,4-dinitrophenolates, this is approximately so (Table 3.1).

was the superior reagent for the identification of the nucleophile in GH 13 yeast α-glucosidase.[280] The conformation of this reagent was not discussed, but quoted NMR coupling constants were consistent with the conformational mobility expected from a compound which would have an axial hydroxymethyl group in the $^4C_1$ conformer preferred by its C5 epimer. The better reactivity against the L-idosyl compound then accords with the conformation change suggested by isotope effect and structure–reactivity studies on glucosides.[76] Analogously, 5-fluoro-β-gulosyl fluoride labels the nucleophile in two GH 38 α-mannosidases (from jackbean[281] and beef kidney lysosomes),[282] but here a careful NMR study of the reagent established it was not in a $^4C_1$ chair. With the jackbean enzyme, the "correct" 5-fluoro-β-mannosyl-enzyme intermediate in fact turned over; similarly, the 5-fluoro-β-galactosyl enzyme intermediate generated from GH 27 coffee-bean α-galactosidase and α-5-fluorogalactosyl fluoride turned over, but in this case rapid tryptic digestion of the steady-state concentration of the glycosyl enzyme in the presence of excess reagent enabled the nucleophile to be identified.[283] A similar situation was observed with the GH 38 *Drosophila melanogaster* Golgi α-mannosidase II, which was irreversibly inactivated by 5-fluoro-β-L-gulosyl fluoride, but only partially by the 5-fluoro-α-L-mannopyranosyl fluoride.[232]

In the case of some transglycosylases, it is possible to generate a glycosyl enzyme which turns over slowly by using as substrate a molecule with the acceptor position blocked. This was first done by deoxygenating the non-reducing end of an oligosaccharide acceptor. The GH 13 cyclodextrin glycosyltransferase from *Bacillus* sp.,[xix] which converts starch to cyclodextrins, will accept maltooligosaccharyl fluorides as substrates: 4″-deoxy-α-maltotriosyl fluoride [*i.e.* 4-deoxy-α-D-Glc*p*-(1 → 4)-α-D-Glc*p*-(1 → 4)-α-D-Glc*p*F] cannot of course accept a glycosyl residue, since it has no free 4-position, but does itself form a glycosyl enzyme which hydrolysed sufficiently slowly for the catalytic nucleophile to be identified.[284] To obtain a glycosyl-enzyme stable enough for crystallographic investigation, however, it was also necessary to mutate the acid–base catalyst.[222] Later work extended the concept using a 4,6-cyclic acetal as the "blocking" function of a 4α-glucanotransferase of GH 57.[285]

5.9.3 Effect of Mutation of the Nucleophilic Carboxylate

In general, mutation of the nucleophilic carboxylate to an inert residue (Ala, Gly, *etc.*) leaves an enzyme without catalytic activity. Apparent residual activities are likely arise from wild-type contamination or catalytic rescue by added salts.

Mutation of the nucleophilic glutamate to cysteine in a GH 1 β-glucosidase from *Agrobacterium* sp. (Abg) slowed the formation of the glycosyl-enzyme

[xix] Because of the commercial importance of cyclodextrins, which are used for the encapsulation of many pharmaceuticals and other agents (Section 4.6.2.1), the strains of bacterium producing them in high yield tend to be commercial secrets.

from the 2,4-dinitrophenyl glucoside 10^6-fold, but the thioglycoside intermediate did not turn over detectably.[286] A prescient (but ignored) 1997 paper reported that mutation of the nucleophilic aspartate of lysozyme to glutamate resulted in an isolable covalent intermediate.[287] More systematic studies of shortening and lengthening the chain of the nucleophilic carboxylate were carried out with the small GH 11 xylanase from *Bacillus circulans*. Replacement of the wt glutamate with an aspartate slowed the deglycosylation step around 1600-fold, but had no effect on deglycosylation. Lengthening the chain (by constructing the Glu78Cys mutant and then chemically carboxymethylating) slowed both steps down by 1–2 orders of magnitude.[288]

The E358G mutant of Abg was subject to "catalytic rescue"; in the presence of azide or formate, which could bind in the "hole" left by truncation of the nucleophilic glutamate, it gave α-glucopyranosyl azide and formate from activated substrates, and it also hydrolysed α-glucosyl fluoride.[289] The smaller "hole" left by E/A mutation of the catalytic nucleophile in a GH 10 glycanase from *Cellulomonas fimi* also proved large enough to accommodate an azide ion – the mutant was subjected to catalytic rescue and gave α-glycosyl azides as products.[290] However, catalytic rescue can also be observed with acid–base mutants and also, as with this enzyme, with "helper" residues: the E123A mutant, Glu123 being a not strictly essential "helper" residue, is nonetheless revived in the presence of azide, although the product is reducing sugar, not azide.

The observation of catalytic rescue led Withers and colleagues to the concept of glycosynthases[291]: they reasoned that if formate, *etc.*, could be recruited from free solution and mimic the nucleophile of a nucleophile mutant, a glycosyl derivative of the wrong anomeric configuration with a small, good nucleofuge could also mimic the glycosyl-enzyme intermediate. If the enzyme had positive subsites and if these were occupied by the acceptor, the remaining catalytic apparatus of the enzyme (general base catalysis by the acid–base catalyst and "Circe effect" interactions) would stabilise the transition state of a glycosylation reaction.[292] The absence of an enzyme nucleophile meant that in these reactions the product was stable to subsequent hydrolysis, unlike transglycosylation products made with unmutated retaining glycosidases. These expectations were confirmed with glycosyl fluorides of the wrong anomeric configuration as glycosyl donors. It subsequently proved advantageous to replace the nucleophilic residue with a one which could hydrogen-bond to the departing fluoride, such as serine[293] rather than one which was inert.

In the presence of excess donor, the reaction can sometimes continue indefinitely; thus, in the presence of a *p*-nitrophenyl glucoside primer, the nucleophile mutants (E231G, E231S and E231A) of a GH 17 barley β-laminarase synthesise crystalline laminarin from α-laminarobiosyl fluoride.[294] However, with the same donor and the E134A mutant of a GH 16 β(1→3, 1→4) glucanase from *Bacillus licheniformis*, only one laminarobiosyl residue was transferred to glucose derivatives[295] [to generate a new β-(1→4) link], since the binding of a laminarobiosyl unit across the +1, +2 sites is disfavoured compared with the binding of a cellobiosyl unit.

5.9.4 The Acid–Base Catalytic Machinery

The canonical mechanism shown in Figure 5.32 requires the same carboxylate group which acts as a general acid in the first step to act as a general base in the second. Its pK_a must, therefore, decrease between the free enzyme and the glycosyl-enzyme intermediate. Direct NMR observation of active site carboxylates in the GH 11 β-xylanase of *B. circulans* already discussed in Section 5.4.2 indicated that the acid–base catalytic Glu 172 had a (microscopic) pK_a of 6.7 in the free enzyme and 4.2 in the 2-deoxy-2-fluoro-α-xylobiosyl enzyme generated from the Withers inactivator 2,4-dinitrophenyl 2′-deoxy-2′-fluoro-β-xylobioside.[50]

The acid–base catalytic machinery shows slight variations within families. Thus, the active site glutamate in the GH 3 β-glucosidase labelled by *N*-bromoacetylglucosylamine is not strictly conserved in GH 3,[296] and there individual members of GH 1 and GH 2 which have different acid or base catalysts than the majority.

5.9.5 Effects of Mutation of the Catalytic Acid–Base

Mutation of the acid–base catalyst to a residue with no functionality, such as alanine, has two effects: it removes the base catalyst for the hydrolysis of the glycosyl-enzyme, thus slowing it down, and it removes the acid catalyst for the first step. The overall effect thus depends on leaving group ability. In the case of good leaving groups such as 2,4-dinitrophenolate or fluoride, removal of the acid catalyst will have little effect on k_{+2}, but the removal of the base catalyst for *deglycosylation* will slow it down, probably to the point where it becomes rate limiting. Hence the overall effect will be to lower k_{cat}, and, because of eqn. (5.6), K_m, but to leave k_{cat}/K_m relatively unaffected (equation 5.7). For poor leaving groups, the effect on k_{+2} and k_{+3} will be similar and k_{cat} will be lowered but K_m will relatively unaffected. If a whole range of substrates are studied, the effect of taking out the acid–base catalyst will be to make $^{\beta lg}(V/K)$ close to -1 and to make $^{\beta lg}(V)$ zero (because most likely k_{cat} represents k_{+3}). This behaviour was first seen with the acid/base mutants of a GH 10 exoglucanase/xylanase from *Cellulomonas fimi* (Cex),[297] and subsequently with an E31G mutation in a GH 1 (Abg) β-glucosidase,[286] with a GH 3 β-glucosidase,[298] a G13 α-amylase[299] and an GH 51 α-L-arabinofuranosidase/β-D-xylopyranosidase.[300]

In the case of substitution of the catalytic acid–base by a small, neutral group, "catalytic rescue" of the action on a good substrate by added nucleophiles can often be seen, as the good nucleophiles, which do not require deprotonation, attack the glycosyl-enzyme directly. The commonest such nucleophile is the small, highly nucleophilic azide ion,[286] which gives stable glycosyl azides of the same anomeric configuration as substrate,[296] although in a very favourable case (a GH 1 β-glucosidase), even fluoride ion could effect nucleophilic rescue of an E/A acid catalytic mutant.[301] In an intriguing echo of the work on the spontaneous hydrolysis of salicyl β-glucoside 30 years earlier (Figure 3.17), it was found that k_{cat}/K_m for hydrolysis of this compound by the

E31G mutant of Abg was close to wt, although the deglucosylation rate was of course much reduced.[302]

Careful analysis of the kinetics of acid–base catalytic mutants can reveal "essential" ionisations not readily apparent in the wt. Glycosylation, rather than deglycosylation, was rate limiting for the acid–base E160A mutant of the GH 39 β-xylosidase of *Thermoanaerobacterium saccharolyticum* acting on aryl xylosides.[303] The pH–rate profile was a similar bell-shape to wt, but shifted to more alkaline pH. It was suggested that an "auxiliary" carboxylate was acting as an acid–base catalyst in the mutant and that the pH–rate profile reflected a reverse protonation mechanism.

5.9.6 Retaining Glycosidase Families

5.9.6.1 GH 1. The plant family Cruciferae produce both glucosinolinates, anionic thioglycosides of various structures whose aglycones, once liberated, decompose spontaneously to a range of strong-tasting compounds such as allyl isothiocyanate, and myrosinase, the enzyme to hydrolyse them. Normally myrosinase and glucosinolates are kept in separate compartments of the plant tissue, but when the plant is crushed, for example by a grazing animal, the enzyme and substrates are mixed and the strong-tasting aglycone decomposition products produced. Much is known about the myrosinase from white mustard (*Sinapis alba*): it is a GH 1 enzyme, but one in which the normal acid–base Glu is replaced by Gln. The nucleophilic Glu could be labelled by a Withers inactivator designed to mimic the natural substrate tropeolin (Figure 5.39).[304] The absence of acid catalysis in the first step creates no problems for the loss of an acidic thiolate. The hydrolysis of the glycosyl-enzyme can take place with the hydrogen bonding available from Gln, but *in vivo*, endogenous ascorbic acid (vitamin C) is recruited, which binds in the +1 site and uses its anionic O3 to remove a proton from the nucleophilic water.[305] Because the glucosyl-enzyme can turn over in the absence of ascorbic acid and the ascorbic acid can also bind in the −1 site, the kinetics of ascorbate activation are not of the expected ping-pong type, but display maximum activation at the physiological concentration of ascorbic acid in the plant. Oxocarbenium ion-like transition states are supported by analogy, and the tight binding of glucono-δ-lactone and similar supposed transition state analogues.

5.9.6.2 GH 2. The tetrameric *lacZ* β-galactosidase of *Escherichia coli* has been intensively investigated for over 40 years, initially because of its relationship to Monod's classic work on enzyme induction. A very large set of data accumulated by Wallenfels and Weil[306] unfortunately had to be discarded because it was obtained in inhibitory "Tris" buffers.[307] The first glycosyl-glycosidase intermediate to become kinetically accessible was with this enzyme;[308] the 2-fluoro-2-deoxy-α-D-galactopyranosyl enzyme, the parent enzyme and ES complexes have since been crystallised and its structure determined.[213]

The enzyme is Mg^{2+} dependent, activation of the hydrolysis of *O*-glycosides in both loose and tight complexes being seen.[309] The Mg^{2+} (presumably

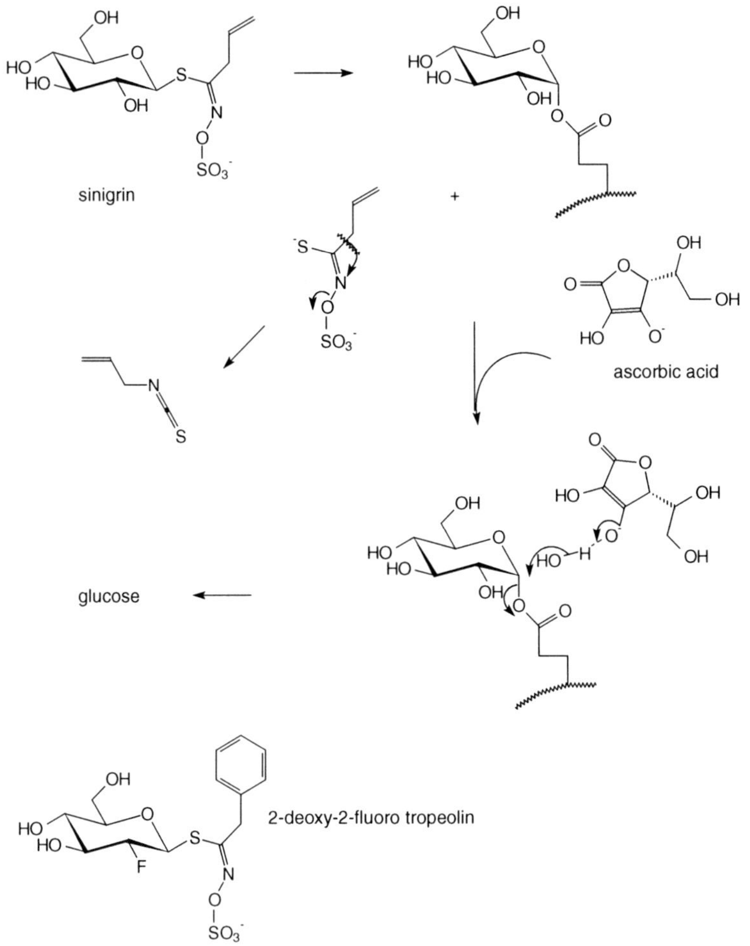

Figure 5.39 Action of myrosinase on sinigrin and the tropeolin-based Dolphin–Withers inactivator.

the tightly bound one) is in the active site, coordinated by Glu461 (homologous to the acid–base catalyst in other GH 2 enzymes), Glu416 and His418.[310] Removal of Mg^{2+} has little effect on the hydrolysis of galactosylpyridinium salts, but lowers k_{cat} for aryl galactosides by 1–2 orders of magnitude, and also makes the Brønsted plot a detectable correlation, rather than a random scatter.[311] The simplest model is that the enzyme is using electrophilic catalysis by Mg^{2+}, rather than general acid catalysis, to increase leaving group ability. On the assumption that loss of the phenol from its enzyme-bound magnesium phenolate can limit the rate of hydrolysis of aryl galactosides by Mg^{2+}-enzyme, a whole range of kinetic measurements, including the otherwise mystifying observation of a significant leaving group ^{18}O KIE on the hydrolysis of *p*-nitrophenyl galactoside when there is no α-deuterium KIE, can be rationalised.[312]

An electrophilic role for Mg^{2+}, however, was considered to be not readily compatible with the X-ray crystal structures, as the Mg^{2+} ion in the observed complexes was considered to be too far away – however, to accommodate both quaternary pyridinium ion leaving groups and *O*-glycosides, the whole electrophilic/acid catalytic apparatus has to be somewhat flexible.

Hydrolysis of a series of substituted-ethyl β-galactopyranosides by Mg^{2+}-enzyme gave $^{\beta lg}(V) = -0.49 \pm 0.13$, $^{\beta lg}(V/K) = -0.75 \pm 0.14$;[313] removal of Mg^{2+} decreased k_{cat} trifluoroethyl galactoside by 10^2-fold but for ethyl galactoside by 10^3-fold. If the same substituted ethanols used as leaving groups were also used to intercept the galactosyl-enzyme, a value of β_{nuc} of -0.19 ± 0.10 was obtained.[314] Combination of this value with the β_{lg} value gave a β_{eq} value of -0.56 ± 0.05 for the equilibrium of eqn. (5.50):

$$\beta\text{Gal}p\text{OEt} + \text{HOCH}_2\text{CH}_{3-n}\text{X}_n \rightleftarrows \beta\text{Gal}p\text{OCH}_2\text{CH}_{3-n}\text{X}_n + \text{EtOH} \qquad (5.50)$$

That the thermodynamic, in addition to the kinetic, stability of glycosides should be so dependent on leaving group ability was surprising.

The E461G mutant had the same reactivity as wt towards galactosylpyridinium ion, a 1300-fold lower reactivity than wt towards *p*-nitrophenyl galactoside (leaving group $pK_a = 7$), but a 500 000-fold lower reactivity to trifluoroethyl galactoside (leaving group $pK_a = 12$). Removal of Mg^{2+} from the E461G mutant had no further effect on hydrolysis of galactosyl azide, but a small further effect on the hydrolysis of *p*-nitrophenyl galactoside.[315] It was argued that, because the effect of mutation of the catalytic acid was much larger than removal of Mg^{2+}, the electrophilic model was implausible. This seems to the author to neglect the possibility that removal of Mg^{2+} by dialysis still leaves general acid catalysis by wt as a possibility: we are comparing electrophilic pull by $Mg^{2+}\cdots{}^{-}O$CO– in the Mg^{2+} wt with that by HO–CO– in the Mg^{2+}-free wt. Complete removal of Glu461 would much more seriously compromise both phenomena. Moreover, if Glu461 is acting as a general acid–base catalyst in the normal way, then the role of Mg^{2+} remains obscure and a geometrically appealing mechanism for the conversion of lactose [β-Gal*p*-(1→4)-Glc] to allolactose [β-Gal*p*-(1→6)-Glc] without release of glucose into solution[316] – simultaneous coordination of O4 and O6 of glucose by the electrophilic Mg^{2+} – has to be abandoned.

Although the E461G mutant undergoes the normal sort of acid–base mutant "catalytic rescue" with nucleophiles, giving β-galactosyl azide, acetate and butyrate in the presence of azide, acetate and butyrate anions, the smaller formate ion, when bound in the "hole" left by truncation of residue 461 from CH_2–CH_2–COO– to H, acts as a general base, rather than a nucleophile and accelerates production of β-galactopyranose.[317]

The single chromosome of *E. coli* has arisen as a result of a double gene duplication, and the presence of a second β-galactosidase, *ebg*, was revealed when a *lacZ*-deleted mutant, which still contained the genes for transport of lactosides into the cell, started grow when plated out on a medium which contained lactose as sole carbon source.[318] Certain colonies began to grow and their ability to hydrolyse lactose was due to the production of mutants of *ebg* (the wt enzyme, *ebg*^{0}, was too catalytically feeble to permit growth on lactose). A limited range of mutations, all involving changes of amino acids in the active site which did not participate in covalency changes, were responsible for changed kinetics, although sometimes these were accompanied by "kinetically silent" mutations elsewhere. The system proved very informative,[319] both about the mechanism of acquisitive evolution[xx] and about the evolution of enzyme energetics and transition state structure.

It can be argued that the ratio of k_{cat}/K_m for the hydrolysis of a glycosyl fluoride to that for the hydrolysis of its 1-fluoro derivative is a measure of charge development at the first transition state, on the grounds that the effective volume occupied by a C–H bond and a C–F bond will not be greatly dissimilar, particularly when the hydrogen (and non-departing fluorine) move considerably with respect to the rest of the molecule during reaction anyway (see Figure 3.13). On this basis, it was found that charge development in the first transition state for wt *ebg* (*ebg*^{0}) was much greater than for evolvants, with second-generation evolvants showing marginally less charge development than first-generation evolvants.[320] In a way parallel to the change in the structure of the transition state in AMP nucleosidase action brought about by the presence of an allosteric activator (Table 5.2), this glycosidase transition state also appears "plastic" and subject to change as a consequence of small perturbations.

The structural possibilities available to the *ebg* gene to make a better β-galactosidase are limited, so that, with hindsight, it is not surprising that the only difference on the free-energy profile of the catalysed reaction to show any sort of uniform behaviour during evolution was that defining k_{cat}/K_m for lactose hydrolysis, which became steadily smaller.[321] Changes in the rest of the profile were random, refuting the thermodynamics-based theory of enzyme evolution of Albery and Knowles.[322] This predicted that the easiest change to bring about would be a uniform increase in binding of all internal states (enzyme-bound intermediates and transition states alike); the second easiest would be a change in the relative energies of intermediates; and the most difficult change to bring about would be the increase in rate of an individual step.

5.9.6.3 GH 7. The active site tunnel of the CBH 1 of *T. reesei* is 50 Å long,[323] and can accommodate ten monosaccharide residues, in accord with the processivity of the enzyme (it releases around 44 cellobiose units per encounter with

[xx] The *ebg* system was one of the first for which "Cairnsian evolution" was shown to apply. In classical Darwinian evolution, the rate of spontaneous mutation is independent of selection pressure. In Cairnsian evolution, the probability of a mutation in a gene increases with the selection pressure it is under: there are chemically reasonable explanations for this (such as the increased likelihood of spontaneous deaminations, *etc.*, in a stretch of DNA which is being continually unwound and wound back, compared with one which is resting). The appearance of growing colonies in *lacZ*-deleted strains plated out on lactose required mutations in both the EBG structural gene and in the EBG repressor, which occurred far more frequently than would be predicted by the "resting" frequency of mutation in *E. coli*. (B. G. Hall, *Curr. Opin. Genet. Dev.* 1992, **2**, 943)

crystalline cellulose).[97] The glucan chain turns itself through almost 180° as it threads its way through this tunnel, with the key "kinks" in the chain being between subsites −4 and −3, and −3 and −2. The dihedral angles φ and ψ (*i.e.* defined in the IUPAC system) are −70° and +123°, and ~−75° and ~+133°, respectively, across the two linkages, compared to $\varphi \approx -90°$ and $\psi \approx +90°$ for the unbound polymer. The action of CBH1 on the highly crystalline *Valonia ventricosa* cellulose results in fibrillation and splitting of the microcrystals, in addition to some sharpening. By use of the Sugiyama "tilt" technique of identifying the crystallographic direction in an individual cellulose crystal by monitoring the changes in its diffraction pattern (Figure 4.10), it was possible to show that the sharpening occurred at the reducing end of the crystals and therefore that Cel7A was processive in from reducing to non-reducing end.[324]

Many studies have been carried out on the synergy in the hydrolysis of cellulose between the components of various cellulose complexes, but most of them are compromised by enzyme impurity – radically different results are obtained if a crucial component is present as a trace impurity, for example an endoglucanase in a cellobiohydrolase. One study where this is not likely to be a problem is that of the action of two recombinant GH 7 cellulases from *T. reesei*, the endoglucanase I and the cellobiohydrolase I.[325] Weight loss from cotton of the two enzymes acting together was twice what would be expected from the loss observed in separate experiments (at a loading of 1 nmol of each enzyme per mg of cellulose). Molecular weight distributions, determined by gel permeation chromatography of the carbanilate, indicated that when the two enzymes acted synergistically there was little change in the molecular weight distribution of insoluble polysaccharide, but when the endoglucanase acted alone there was accumulation of insoluble oligosaccharides of *DP* 10–15.

The four loops of protein forming the roof of the active site tunnel have been shown, by their excision or reinforcement and comparison with the homologous Cel7D of *Phanerochaete chrysosporium* (with three such loops), to be essential for action on crystalline cellulose, with the balance of endoglucanase/exoglucanase action increasing with fewer loops.[326]

5.9.6.4 Non-chair Pyranosyl-Enzyme Intermediates – GH 11, 26, 29, 31 and 38. ALPH suggests that, since an oxygen lone pair is exactly antiperiplanar to an α leaving group in the normally-preferred $^4C_1$ (D) conformation, the glycosyl-enzyme intermediates derived from all retaining β-glycosidases should be in this conformation. They are not. The −1 ring in 2-deoxy-2-fluoro-α-xylobiosyl intermediates from GH 11 β-xylanases is in the $^{2,5}B$ conformation;[219] neither of the oxygen lone pairs approaches antiperiplanarity to the leaving group in this conformation, although one is exactly synperiplanar. No conformational change of the ring, however, is required to attain maximum delocalisation of positive charge on to oxygen. A less marked example of an "anti-ALPH" glycosyl-enzyme intermediate is that from the GH 26 β-mannanase of *Cellvibrio japonicus*. The −1 saccharide unit of the 2-deoxy-2-fluoro-α-mannotriosyl-enzyme intermediate adopts the $^0S_2$ conformation (leaving group and 4-OH isoclinal, 2-OH and CH_2OH pseudoequatorial, 3-OH pseudoaxial). A complex of the unreacted 2,4-dinitrophenyl 2-deoxy-2-fluoro-β-mannotrioside showed the −1 deoxyfluoro sugar in the $^1S_5$ conformation (leaving group pseudoaxial, 2-F, 4-OH and CH_2OH pseudoequatorial, 3-OH isoclinal). The reaction trajectory thus corresponds to passage of the mannopyranose ring along the equator of the Cremer–Pople sphere through 60° of longitude ($^1S_5 \rightarrow {}^{2,5}B \rightarrow {}^0S_2$).[226] However, another GH 26 enzyme, a lichenase from *Clostridium thermocellum*, may adopt a slightly different path. This enzyme, which cleaves β(1 → 3) linkages in polysaccharides containing the structure [β-Glc*p*-(1→4)-βGlc*p*-(1→4)-βGlc*p*-(1→3)]$_n$, binds 3-*O*-β-D-glucopyranosyl isofagomine in a $^4C_1$ conformation.[327]

Conversely, both ALPH and the simple steric difficulty of displacing equatorial leaving groups suggest that glycosyl-enzyme intermediates in retaining α-glycosidases should be in non-chair forms (Figure 5.40). However, at least one GH 13 enzyme, the cyclodextrin-forming transglycosylase, forms a complex in the $^4C_1$ conformation;[222] this may be related to the relative stability of the glycosyl-enzyme and could alter when the +1 subsite is occupied by a sugar which is a good acceptor. The three other retaining β-glycosyl-enzyme intermediates whose structure is known at present appear to have sugar rings in skew conformations, although the data are not especially firm. Inactivation of the GH 29 α-L-fucosidase from *Thermotoga maritima* with 2-deoxy-2-fluoro-α-L-fucosyl fluoride led to the identification of the nucleophile; however, the 2-deoxy-2-fluoro-β-L-fucosyl-enzyme very slowly turned over and the electron density in the −1 subsite of the crystal was interpreted as arising from a mixture of sites containing a covalently-linked 2-deoxy-2-fluoro-β-L-fucosyl group in the $^3S_1$ conformation and 2-deoxy-2-fluoro-L-fucose in the $^1C_4$ conformation (these two conformations being equivalent to $^1S_3$ and $^4C_1$ in the D series).[227] Data for the 5-fluoro-L-gulosyl-enzyme derived from the *Drosophila melanogaster* Golgi mannosidase II (GH 38) show the −1 sugar ring unambiguously in the $^1S_5$ conformation.[232] The objection that the $^4C_1$ conformation of this sugar is especially disfavoured, both by the axial hydroxymethyl group and by the anomeric effect of the equatorial C5–F bond, was met by generating the 2-deoxy-2-fluoromannosyl derivative from the acid–base catalytic mutant. Again, the covalently-attached −1 sugar was in the $^1S_5$ conformation, although partial hydrolysis was also observed. The $^1S_5$ conformation is exactly that seen for the substrate −1 sugar with the GH 26 mannanase, suggesting that it is the optimum for attack on a β-mannosyl derivative. Finally, the YicI α-glucosidase of *E. coli* (GH 31)[228] formed a stable 5-(*R*)-fluoroxylosyl-enzyme intermediate in which X-ray crystallography showed the −1 sugar to adopt the $^1S_3$ conformation. As with the gulosyl-mannosidase, there remains uncertainty about how close the observed conformation is to the one with the natural substrate, since the unnatural structural features necessary to observe the intermediate (in this case an anomeric equatorial C–F bond) of themselves favour non-chair conformers. Whereas the 5-(*R*)-fluoro-β-xylopyranosyl fluoride was an inactivator, its epimer, *meso*-5-fluoro-α-xylopyranosyl fluoride was a competitive inhibitor, for reasons that are not yet understood.

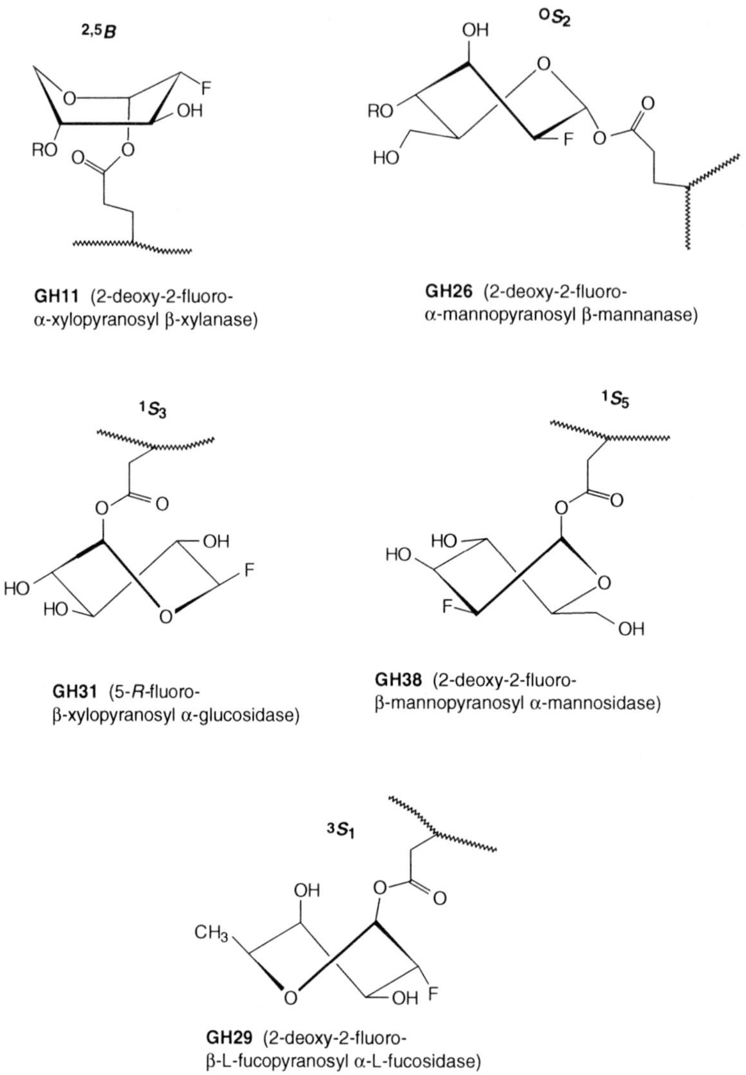

$^{2,5}B$
RO
O
OH
F
GH11 (2-deoxy-2-fluoro-α-xylopyranosyl β-xylanase)
$^{O}S_2$
OH
RO
HO
F
O
GH26 (2-deoxy-2-fluoro-α-mannopyranosyl β-mannanase)
$^{1}S_3$
HO
OH
F
O
GH31 (5-R-fluoro-β-xylopyranosyl α-glucosidase)
$^{1}S_5$
HO
F
O
OH
GH38 (2-deoxy-2-fluoro-β-mannopyranosyl α-mannosidase)
$^{3}S_1$
OH
CH3
O
F
GH29 (2-deoxy-2-fluoro-β-L-fucopyranosyl α-L-fucosidase)

Figure 5.40 Non-chair glycosyl-enzyme intermediates observed by X-ray crystallography.

5.9.6.5 GH 13. This is a very large family, which contains glycosidases and transglycosylases which act on α-glucopyranosyl residues. The protein fold is based on the TIM barrel of GH clan A. They appear to be "anti-ALPH" enzymes: the $^{4}C_1$ ground state of the substrate presents a perfectly disposed antiperiplanar lone pair to the leaving group, yet there is kinetic[76,328] evidence for a change from this conformation in the ES complex of yeast α-glucosidase and structural evidence in the case of cyclodextrin glycosyltransferase.[222] At the same time, the −1β sugar residue of the glycosyl-enzyme of the latter enzyme is in the $^4C_1$ conformation with no antiperiplanar lone pair to the β carboxylate leaving group.[222]

These cyclodextrin glycosyltransferases are particularly intriguing, since they convert (linear) amylose to 6-, 7- and 8-membered (α, β and γ) cyclodextrins. Important interactions occur in the −6 subsites, since the distribution of cyclodextrins can be altered by mutations here, which do not affect the rate of hydrolysis.[329] Remarkably, the cyclodextrins first produced from amylose have between 9 and 60 glucose units in a macrocyclic ring, with the smaller cyclodextrins being produced later.[330]

Interactions with carboxylates in the +1 site seem to modulate transglycosylation activity in the human pancreatic α-amylase: as expected, nucleophile mutants were inactive, acid–base catalyst mutants showed reduced deglycosylation rates, but much reduced glycosylation rates only with oligosaccharide substrates, rather than maltotriosyl fluoride. However, mutation of the "auxiliary" Asp300 impaired transglycosylation, rather than catalysis.

5.9.6.6 The Many Activities of GH 16. Agarases, carrageenanases, lichenases, laminarases and xyloglucan endotranglycosylases (XETs) have so far been found in this family, part of clan B. The XETs are as intriguing as they are biologically important. Xyloglucan binds very tightly to the surface of cellulose and the primary cell wall of plants consists of cellulose microfibrils bound together by xyloglucan. When the cell expands, XET forms a hydrolytically stable glycosyl-enzyme intermediate,[331] cleaving the xyloglucan chain, but remaining attached to the new reducing end. Only when the xyloglucan loose end with a XET molecule attached encounters a non-reducing xyloglucan loose end does the second half of the reaction take place. XETs appear to have no hydrolytic activity whatsoever, even to the point that XLLLG-6-chloro-4-nitrophenol gives no stoichiometric burst as the glycosyl-enzyme is generated.[332] (For xyloglucan nomenclature, see Section 4.6.1.2). Recently solved X-ray crystal structures of κ-carrageenanase,[333] a number of β-(1→3),(1→4) glucanases[334], two β-agarases[335] and poplar XET (as previous) suggest how evolutionary changes in the proteins may have brought about changes in specificity. Figure 5.41 displays the evolutionary tree of the Family 16 enzymes, with the structures of the substrates superimposed.

Between the nucleophilic and acid–base glutamates in the linear sequence lies a catalytically important "assisting" aspartate. GH 16 can be divided into two sub-families. The first, including laminarases and agarases, has an active site linear sequence Glu(nucleophile)–Hph–Asp(assist)–Hph–Met–Glu(acid/base), where Hph is a hydrophobic amino acid such as isoleucine. The second sub-family does not possess the methionine. The X-ray structures of the enzymes indicate that this extra Met residue distorts a β-strand to make a "β-bulge".

5.9.6.7 Nucleophilic Assistance by Vicinal trans-Acetamido Group of Substrate – Mechanisms of GH 18, GH 20, GH 56, GH 84, GH 102, GH 103, GH 104 and Sometimes GH 23. Neighbouring group participation by the *trans*-acetamido group of *N*-acetyl β-glucosaminides and -galactosaminides is intrinsically chemically reasonable (see Section 3.12.1) and was proposed many years ago as a

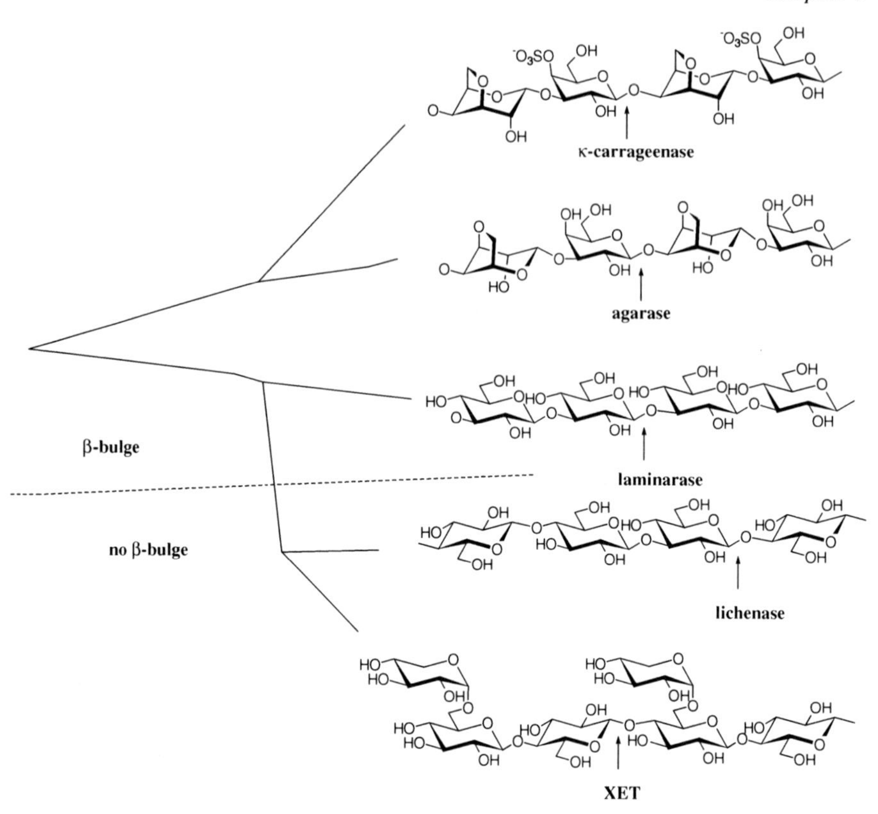

Figure 5.41 Evolution of enzymes of GH 16 and substrate structures.

possible mechanism for hen egg-white lysozyme,[336] before being abandoned on the grounds that β-D-GlcNAc(1→4)-β-D-Glc-*p*-nitrophenol was nearly as good a substrate as β-D-GlcNAc(1→4)-β-D-GlcNAc-*p*-nitrophenol.[337] The mechanism was revived for what is probably a GH 20 hexosaminidase on the basis of the large ρ^* value (−1.4) obtained by progressive fluorine substitution in the acetamido methyl group (see Section 5.4.5.1), but direct X-ray evidence was first obtained for a chitobiase from *Serratia marcescens* which apparently bound its substrate in the +1, −1 subsites, with the −1 residue in a "sofa" conformation (C3 out of the plane of the others).[338] Later work, however, established that in this class of enzymes the ES complex was in a $^{1,4}B$ conformation.[224] A stable analogue of the putative oxazoline intermediate is the corresponding thiazoline ("NAG-thiazoline"), which proved to be a relatively powerful inhibitor of jackbean β-hexosaminidase;[339] the thiazoline could be generated from the action of the enzyme on 4′-methylumbelliferyl 2-thioacetamido-2-deoxy β-glucopyranoside.

Whereas GH 20 contains hexosaminidases and chitobiosidases (*i.e.* enzymes with −1 and possibly +1 subsites only), GH 18 contains chitinases, enzymes which act at the polymer level. X-ray crystallographic evidence for amide participation in GH 18 originally came from the complex of the chitinase/lysozyme hevamine with the naturally occurring inhibitor allosamine, in which the fused ring structure, with obvious similarity to the oxazolidine, bound at the −1 subsite;[340] GH 18 and GH 20 were later shown to have similar intermediate structures and presumed mechanisms.[341] The intermediacy of the oxazolidine in GH 18 as well as GH 20 action was conclusively demonstrated by the conversion of a chitobiose-derived oxazolidine to chitin by a *Bacillus* sp. chitinase.[342]

The protonated oxazolidines are comparatively strong acids (2-methyl-Δ^2-oxazoline has a pK_a of 5.5,[343] which the electron-withdrawing substituents in the sugar will lower), and in GH 18[344,345] and GH 20[224] active site aspartates are observed which hydrogen bond (or perhaps completely deprotonate) the oxazoline nitrogen. The accepted mechanism of GH 18 and GH 20 enzymes is given in Figure 5.42, in all details (including the −1 ring conformation changes) identical with a possible mechanism for lysozyme proposed by Lowe *et al.* in 1967.[336] The mechanism is also followed by GH 56 and GH 84. The usual oxocarbenium ion-like transition states are suggested by the tight binding of the corresponding aldonolactones or nojirimycin analogues.

GH Family 18 contains enzymes which act on crystalline chitin: the presence of hydrophobic, aromatic amino acids at the entrance to a tunnel plays a role, as with the cellobiohydrolases.[346] The GH 18 chitinases A and B of *Serratia marcescens* act on crystalline β-chitin (parallel chains) in opposite senses.[347] In a series of very elegant experiments, Sugiyama and colleagues showed that the digestion of chitin crystals by either enzyme resulted in "sharpening" of one end of the narrow chitin crystals, whereas treatment with both chitinases resulted in sharpening of both ends. To sort out which enzyme was going in which direction, the molecular directionality of the chitin crystals was determined by the tilt microdiffraction technique (Figure 4.10). It was also confirmed by chemical labelling of the reducing end of the sugars. Chitinase A degraded chitin from reducing to non-reducing end, whereas chitinase B was processive in the opposite direction. GH 18 are retentive enzymes in which the nucleophile is on the substrate, rather than the enzyme, so that classical *exo*-acting enzymes can act with retention, without encountering the problem of the rest of the chain in extended positive subsites having to be removed before the second step can take place, at least if the initial product of these GH 18 enzymes is the oxazoline, rather than the sugar.

One type of hyaluronidase catalyses the hydrolysis of the β-GlcNAc-(1→4) GlcA linkage in hyaluronic acid. The crystal structure of the bee venom GH 56 enzyme complex with the hyaluronic acid tetrasaccharide GlcAGlcNAcGlcAGlcNAc bound −4 to −1 showed no obvious enzyme nucleophile and tight control of the acetamido group.[348] A comparison with site-directed mutagenesis results on a human enzyme[349] suggested that in GH 56 also amide participation took place, with Asp111 the polarising group and Glu113 the acid–base catalyst.

The enzymes which hydrolyse the GlcNAc residue attached to Ser or Thr of glycoproteins have no sequence similarity to GH 20, but it appears that GH 84,

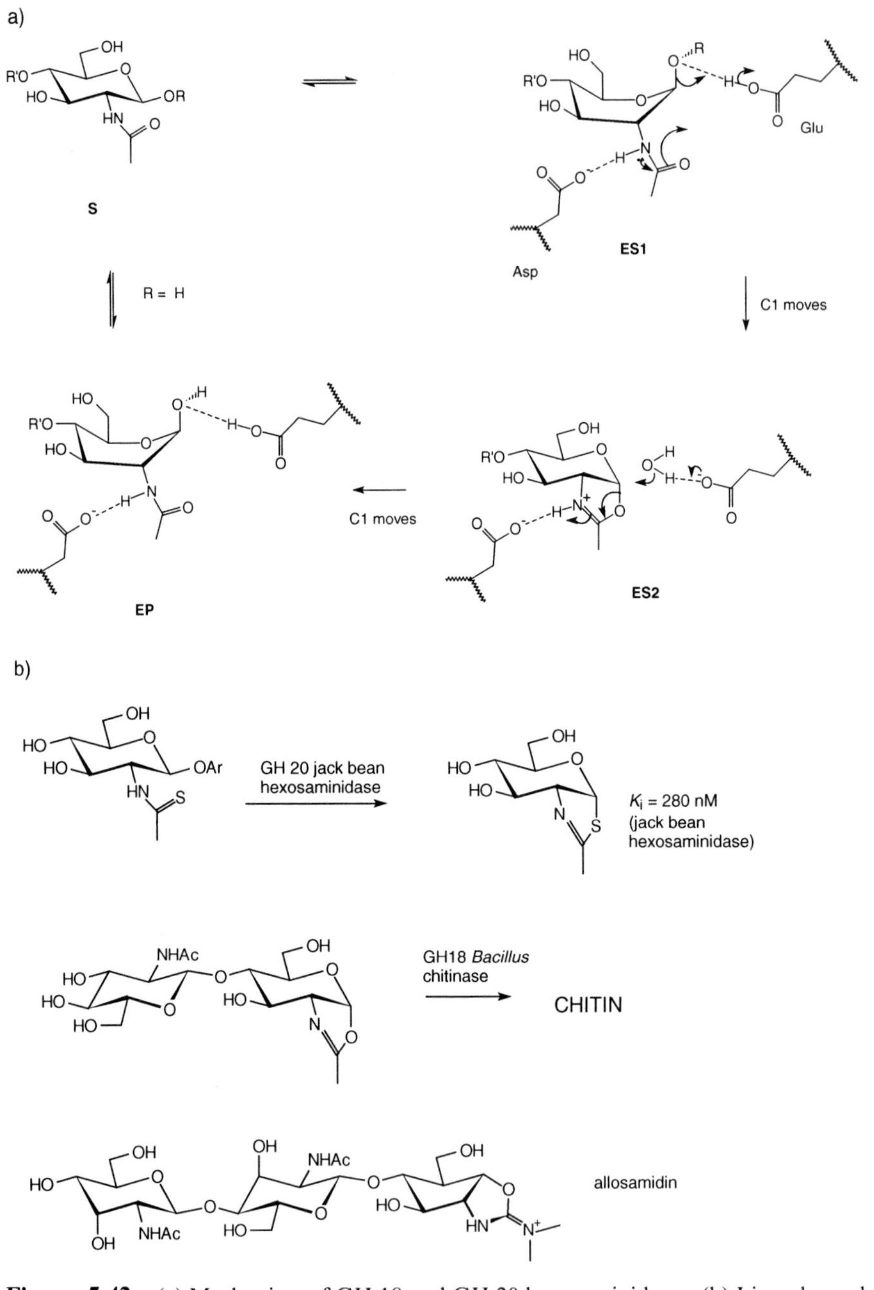

a)

b)

Figure 5.42 (a) Mechanism of GH 18 and GH 20 hexosaminidases. (b) Ligands used in deducing the mechanism. GH 56 (hyaluronidase) and GH 84 follow similar mechanisms, although the acid–base catalyst in GH 84 is Asp.

to which they belong, operates by an amide participation mechanism. The conclusion arose in part from tight binding of the thiazoline and its analogues, and partly from the large effect of successive fluorination at the acetyl methyl group. The value of $\rho^*(V/K)$ obtained (−0.42) is, however, much lower than for the GH 20 case cited earlier or even GH 20 human lysosomal hexosoaminidase (−1.0).[350] It was claimed that in the GH 20 enzymes the apparent value of $\rho^*(V/K)$ was increased by a steric effect, arising from fluorine being larger than hydrogen. Certainly, if the effect of fluorine substitution in the methyl group on the basicity of the acetamido carbonyl is the same as that on the basicity of acetates, then a $\rho^*(V/K)$ of −1.0 corresponds to an unprecendentedly high β_{nuc} of 1. Site-directed mutagenesis studies[351] identified Asp175 as the acid–base catalyst and Asp174 as the group which hydrogen bonded to the nucleophilic amide; in the case of this enzyme class the case was made that in fact classical general base catalysis was involved, with the kinetic pK_a attributed to Asp174 (5.0) being close to the estimated pK_a of the oxazoline. As expected, the D175A mutant had little effect on k_{cat}/K_m for substrates with good leaving groups and a large effect on substrates with poor leaving groups. An interesting feature of this mutant was a biphasic Brønsted plot in both k_{cat} and k_{cat}/K_m, with leaving groups with p$K_a > 7.2$ with the expected β_{lg} of −1 and leaving groups with p$K_a < 7.2$ of around 0.2 (and absolute values of k_{cat}/K_m below diffusion). Below leaving group p$K_a \approx 7.2$ it is suggested that loss of non-covalently bound aglycone is rate determining and is preceded by a rapid and reversible formation of an enzyme-bound oxazoline–leaving group complex.

GH Family 23 is the only family to contain enzymes which act with opposite stereochemical outcomes. Its usual members are goose-type lysozymes, which have a fold similar to hen lysozyme, but lack the nucleophilic aspartate and act with inversion of anomeric configuration. The lytic transglycosylases, like lysozyme, act on murein and have a similar protein fold, but the product is a 1,6-anhydromurein unit (Figure 5.43).[352] They have been classified into their own four families by Blackburn and Clarke (B&C),[353] although recently they have been incorporated into CAZy. Soluble bacterial lytic transglycosidases (B&C1) remain in GH 23, B&C2 (bacterial membrane-bound lytic transglycosylases type A) becoming GH 102, B&C3 (bacterial membrane-bound lytic transglycosylases type B) becoming GH 103 and B&C 4 (bacteriophage lytic transglycosylases) becoming GH 104. Despite differences in sequence, they all appear to act by the same mechanism: thus, the *E. coli* GH 103 enzyme is powerfully inhibited by NAG-thiazolidine in the usual way.[354] The structure of the bacteriophage λ enzyme (GH 104) in complex with hexa-*N*-acetylchitohexaose (in fact cleaved to a tetrasaccharide in −4 to −1 and a disaccharide in +1, +2)[355] suggested a double-displacement mechanism, with an intermediate oxazolinium ion, as in GH 18 and GH 20, as did a similar complex of a peptidoglycan with the GH 23 (soluble) *E. coli* enzyme.[356] The pyranose ring in the 1,6-anhydromurein unit is constrained to the $^1C_4$ and $B_{3,0}$ conformations; the pyranose ring in the ES complex in GH 18 and GH 20 enzymes is in a $^1S_3$ conformation and that in GH 6 inverting cellulase in the $^2S_0$ conformation. These skew/boat conformations are spread over a

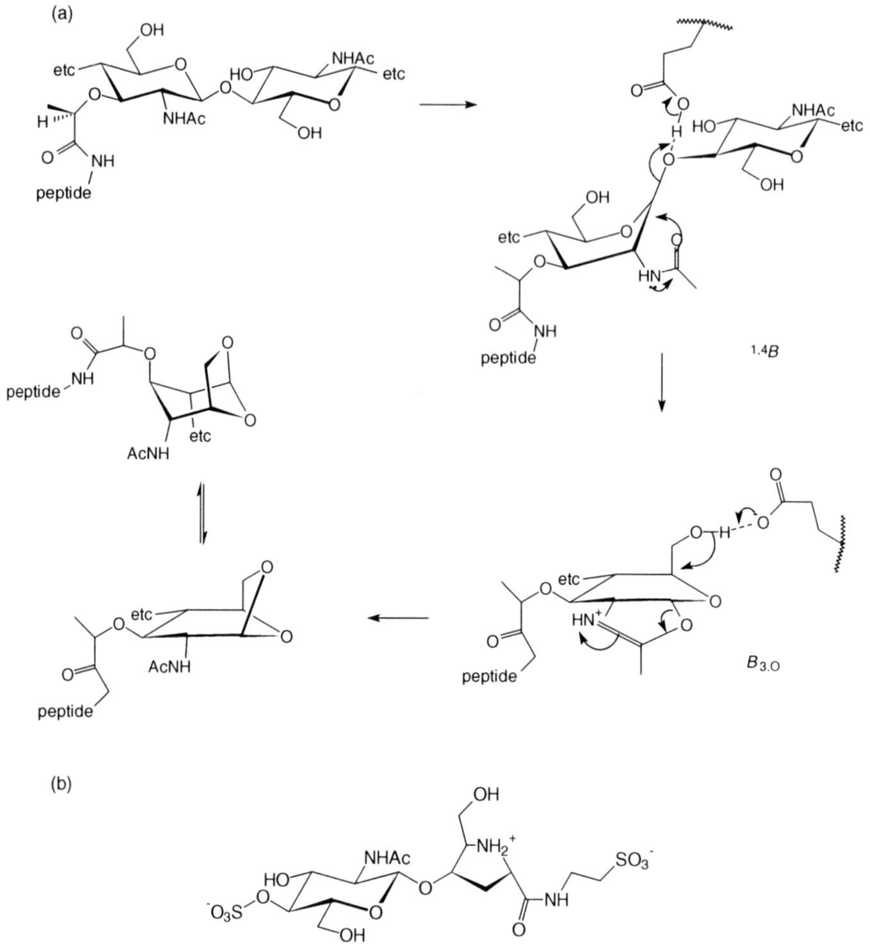

Figure 5.43 (a) Possible mechanism of action of lytic transglycosylases. (b) Bulgecin.

pseudorotational angle of only 60° (Figure 2.6) and it is reasonable that slight shifts of active site structure should tip the balance between a single displacement with a nucleophilic water and participation by amide. In GH 23 the acid catalyst adopts a *syn* trajectory, making it easy for the deprotonated acid catalyst to act as a general base for the attack of the 6-OH on the oxazoline.

The mechanism and likely oxocarbenium-ion-like transition states therein are supported by the powerful inhibition of the lytic transglycosylases by the natural inhibitor bulgecin (Figure 5.43b) and the suggestive disposition of its amide in the complex with a GH 23 enzyme.[357]

5.9.6.8 GH 22 – Lysozyme. The natural substrate of lysozyme is the murein of the cell walls of Gram-positive bacteria (see Figures 5.43 and 4.95(a)). Hen egg-white lysozyme was the first enzyme for which an X-ray structure was available, and the imputation of an acid–base function to Glu35 and substrate distortion as a contributor to catalysis proved well founded. The original mechanism identified clashes of the hydroxymethyl group of the sugar bound in the −1 subsite with the protein as the key determinant of substrate distortion, and the idea was confirmed by the finding that oligosaccharide glycosides where this had been removed from the residue in the −1 subsite were not substrates. Whereas compounds of the type [β-GlcNAc-(1→4)]$_n$-β-GlcNAc–OAr, where $n=2$, 3 or 4 and ArOH was a good leaving group such as 3,4-dinitrophenol, readily liberated ArOH, the corresponding compounds [β-GlcNAc-(1→4)]$_n$-βXylNAc–OAr did not.[358] When an X-ray structure with the −1 subsite occupied finally became available [3-*O*-lactyl-GlcNAc-β-(1→4)-GlcNAc-β(1→4)-3-*O*-lactyl-GlcNAcOH binds −3, −2, −1], the main distortion of the sugar ring was indeed a pushing of the hydroxymethyl group into a pseudoaxial conformation in a ring conformation described as "sofa" and probably, on the Cremer–Pople sphere, at the same longitude, but at an intermediate latitude, between $^1S_3$ and $^4H_3$.[225]

Formation of a glycosyl-enzyme intermediate, with a covalent bond between Asp52 and C1 of the sugar in the −1 subsite, was detected by mass spectrometry with wt enzyme and βGlcNAc(1→4)β-2-deoxy-2-fluoro-GlcF; however to obtain a glycosyl-enzyme stable enough for crystallisation a Glu35 mutant had to be used.[204]

5.9.6.9 GH 31. GH 31 is a standard retaining glycosidase family, acting on axially linked pyranosides; however, an intriguing subgroup within it catalyses an elimination reaction. The enzymes, α-(1→4) glucanlyases, cleave glucose units from the non-reducing end of the polymer, yielding 1,5-anhydrofructose rather than glucose (Figure 5.44).[359] The anhydrofructose arises from spontaneous tautomerisation of the first-formed enol. Surprisingly, this enol arises not from a single-step *E*1-like elimination, in which the catalytic machinery stabilises an oxocarbenium ion-like transition state and the usually nucleophilic group removes a proton from C2, in a *trans* elimination. Rather, there is a normal first step in which a glycosyl-enzyme is formed[360] and the elimination takes place in a *syn* sense in the second step, which can be regarded as the near-microscopic reverse of the first step of the hydration of a glycal. Brønsted plots for aryl glycosides give $\beta_{lg}=-0.32$ or -0.33 on $\log(k_{cat}/K_m)$ or $\log k_{cat}$. The α- and β-deuterium kinetic isotope effects for these substrates are relatively high [$^{\alpha D}(V)=1.19$, $^{\beta D}(V)=1.06$] and indicate a high degree of charge development at the first transition state; taken together with the modest β_{lg} values, a relatively high degree of proton donation to the leaving group is indicated. The only fluoroglycosyl fluoride that acted as a paracatalytic inactivator was the 5-fluoro-L-idosyl β-fluoride; however, decomposition of the 5-fluoroglucosyl enzyme was rate determining during the hydrolysis of 5-fluoroglucosyl α-fluoride, and permitted effects of $^{\alpha D}(V)$ at C1 of 1.23 and a primary deuterium isotope effect at C2 of 1.92 to be measured. These are characteristic of a oxocarbenium ion-like elimination mechanism. The catalytic base is not identified with certainty, but it is probably the other oxygen of the catalytic nucleophile.[361]

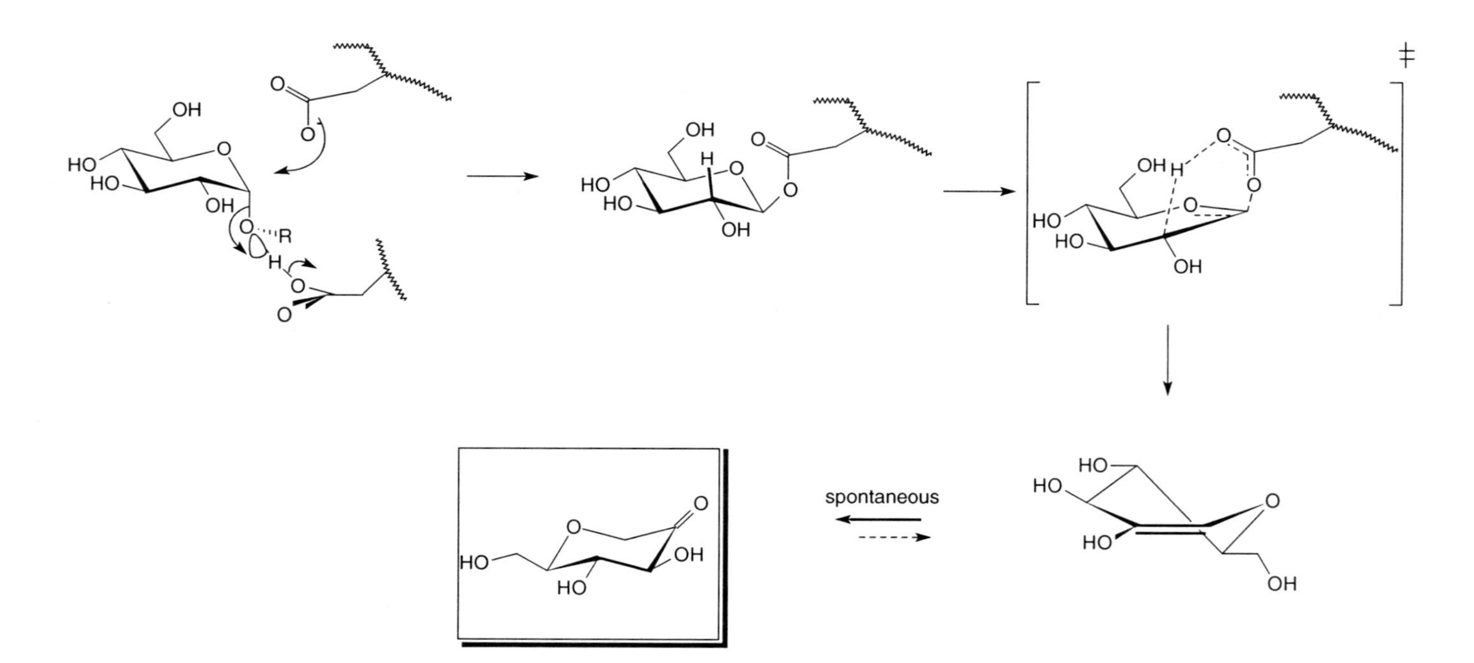

Figure 5.44 Mechanism of GH 31 α-glucan lyase.

5.9.6.10 Sialidases (Neuraminidases). The three sialidase families (GH 33, 34 and 83) all have the same catalytic machinery, an aspartate, which appears to act as a proton donor, and a probable nucleophilic tyrosine, rather than a carboxylate, activated in all likelihood by a glutamate. GH 33 contains all transialidases as well as simple hydrolytic enzymes, whereas GH 34 and GH 83 contain only enzymes from viruses which are mammalian or avian pathogens. GH 33 and GH 34 act with retention of the anomeric configuration and it is currently assumed that GH 83 is similar. Obtaining crystal structures with mechanistically informative ligands bound is complicated by the facility with which sialidases dehydrate *N*-acetylneuraminic acid to its 2,3-dehydro derivative, DANA; the process is most facile with GH 33 enzymes. The influenza sialidase in GH 34 was more amenable and not only bound the minor anomer of NANA, but bound it in the $B_{2,5}$ conformation.[231] Structures of GH 83 sialidases are available only with uninformative ligands such as β-NANA bound.[362]

The use of a phenolic OH as a nucleophile by these enzymes may be related to the difficulty of approach of a carboxylate to a tertiary centre also bearing a carboxylate, even if that carboxylate is hydrogen bonded to no less than three arginine residues in all sialidases so far examined. In the case of the GH 33 *trans*-sialidase of *Trypanosoma cruzi* (the agent of Chagas' disease, a sleeping sickness affecting populations in Central and South America), the nucleophilic Tyr has been labelled by a Withers inactivator,[363] and the pyranose ring in the 3-fluoroglycosyl-enzyme complex subsequently shown to adopt the normal $^{2}C_{5}$ conformation (Figure 5.45)[xxi].[230] Crucially, the intermediate is catalytically competent and will transfer 3-fluorosialic acid residues to lactose, thereby stilling fears that the fluorine may have perturbed the reaction mechanism. The active site glutamate which deprotonates the attacking phenolic OH in the glycosylation step will presumably act as a general acid in the hydrolysis or transglycosylation of the sialosyl-enzyme intermediates.

Mutation of the presumed acid-catalytic Asp to Gly in the GH 33 *Micromonospora virifaciens* enzyme had little effect on the hydrolysis of activated substrates such as 4-methylumbelliferyl or nitrophenyl sialosides, but lowered k_{cat}/K_m for the 3′- and 6′-sialosyllactoses, natural substrates, by around 10^3-fold[364]. This suggested that acid catalysis did not operate during the hydrolysis of aryl sialosides and therefore provided a solution to the puzzle of why the rate-determining step in the hydrolysis of *p*-nitrophenyl-*N*-acetylneuraminides by GH 33 sialidases of *Vibrio cholerae*,[365] *Salmonella typhimurium*[77] and the internal transglycosylase from the leech *Macrobdella decorata*[366] was formation, rather than hydrolysis, of the glycosyl-enzyme intermediate, despite *p*-nitrophenol being a better leaving group than the phenol which formed the

---

[xxi] Since the numbering of carbon atoms in *N*-acetylneuraminic acid starts at the carboxylate group, to obtain the equivalent conformation of a hexopyranoside (Figure 2.6), 1 should be subtracted from the carbon numbers – but not the ring oxygen. Thus, sialic acid $B_{2,5}$ is the counterpart of hexopyranoside $B_{1,4}$, $^{6}S_{2}$ the counterpart of $^{5}S_{1}$ and $^{4,0}B$ the counterpart of $^{3,0}B$.

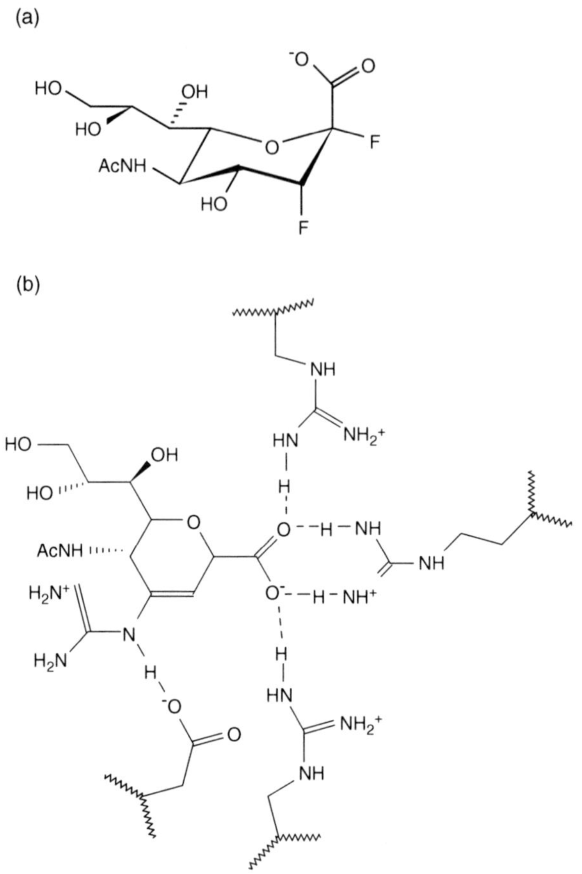

(a)

(b)

Figure 5.45 (a) Dolphin–Withers inactivator for sialidases. (b) Cartoon of Relenza, the anti-influenza drug, showing pattern of hydrogen bonding of 1-carboxylate to arginines and the drug guanidine group to the acid catalyst.

glycosyl-enzyme. Only in the case of hydrolysis of the glycosyl-enzyme was the departure of phenol assisted by partial proton donation.

The exact sugar conformation in the ES complex of GH 33 enzymes seems to vary from substrate to substrate. Thus, complexes of 4-methylumbelliferyl *N*-acetyl-α-neuraminide in complex with the (inactive) acid–base mutant of the *T. cruzi trans*-sialidase reveal a $^{6}S_{2}$ conformation, whereas the complex with sialyllactose has a $B_{2,5}$ ring;[230] these are next to each other on the pseudorotational itinerary (Figure 2.6b). A possible conformational trajectory for the pyranose ring in these enzymes is $B_{2,5}$ (ES complex) → $^{4}H_{5}$ (probably most stable conformation of the sialosyl cation, with NHAc and OH pseudoequatorial) → $^{2}C_{5}$ (glycosyl-enzyme intermediate). This would result in most of the motion of the substrate relative to enzyme in the course of catalysis being in the carbon which is being substituted, in the normal way.

A similar situation to that with the lytic transglycosylases occurs with a leech sialidase (sialidase L), which gives 2,7-anhydro-*N*-acetylneuraminic acid, despite being in GH 33 like all other non-viral sialidases and transialidases. This enzyme binds a *C*-glycoside substrate–analogue (αNeuNAc–CH_2–CH=CH_2) and product in the $^{4,0}B$ conformation.[229] The conformation of glycosyl-enzyme is unknown, but if it is $^2C_5$ like the *T. cruz* transialidase, then a very similar conformation journey, but this time starting from the next boat conformation to $B_{2,5}$, can be envisaged (Figure 2.6b).

Probable mechanisms of a retaining sialidase and of the internal transialidase are shown in Figure 5.46.

Sialidase substrates with the diastereotopic hydrons at C3 separately labelled are available (Figure 3.24) and so conformational information about the transition state is available; this is given in Table 5.5, along with β_{lg} values, for a number of enzymes. From the relative magnitudes of the β-deuterium kinetic isotope effects, it is clear that the precise location of the transition state on the conformational itinerary varies from enzyme to enzyme. In the $^6S_2$ conformation the *proS* hydron and the leaving group are almost *trans*-diaxial, whereas in the $^4H_5$ conformation of the fully formed cation, the *proS* C2-L, makes an angle approaching 90° with the empty p-orbital, whereas the *proR* C2-L bond is much more closely aligned; therefore, if the transition state is close to $^4H_5$, the *proR* effect should be bigger than the *proS* effect, whereas the reverse would be the case if the transition state were closer to $^6S_2$. *V. cholerae* enzyme alone appears to have a transition state closer to the half-chair; that of the influence, leech and *S. typhimurium* enzymes is closer to the skew form. The absolute magnitudes of the effects provide evidence of the usual oxocarbenium ion-like transition states.

The similarity of the β_{lg} values obtained for the acid-catalytic mutant of the *M. viridifaciens* enzyme and wild-type *V. cholerae*, *S. typhimurium*, influenza and *M. decora* enzymes confirm indicate that, at least with unnatural, activated substrates the application of acid catalysis is minimal. Further evidence is the high intrinsic leaving group ^{18}O effects, suggesting that the loss of bonding between C2 and O2 is not offset by partial protonation.

There is a discrepancy between $^{\beta lg}(V)$ and $^{\beta lg}(V/K)$ for most of the enzymes in Table 5.5. In the influenza case, this is clearly due to a non-chemical step (probably the conjoint change of enzyme and substrate back to the $^2C_5$ conformation), as shown by *V* for NeuNAc-αOPNP being isotopically silent. For the *S. typhimurium* enzyme a similar explanation also applies, as isotope effects on *V* are uniformly attenuated as compared with (V/K). However, isotope effects on *V* and (V/K) are the same for the *V. cholerae* enzyme and only some sort of polar effect on binding can explain the result.

Very remarkably, site-directed mutagenesis of the supposedly nucleophilic tyrosine of the GH 33 sialidase from *Micromonospora viridifaciens* yields enzymes which still work, and some of them act with inversion, rather than retention.[367] A detailed examination of the inverting Y 370G mutant revealed that the "hole" left by removal of the tyrosine side-chain was now filled with water molecules; surprisingly, the effect of removing the nucleophilic tyrosine

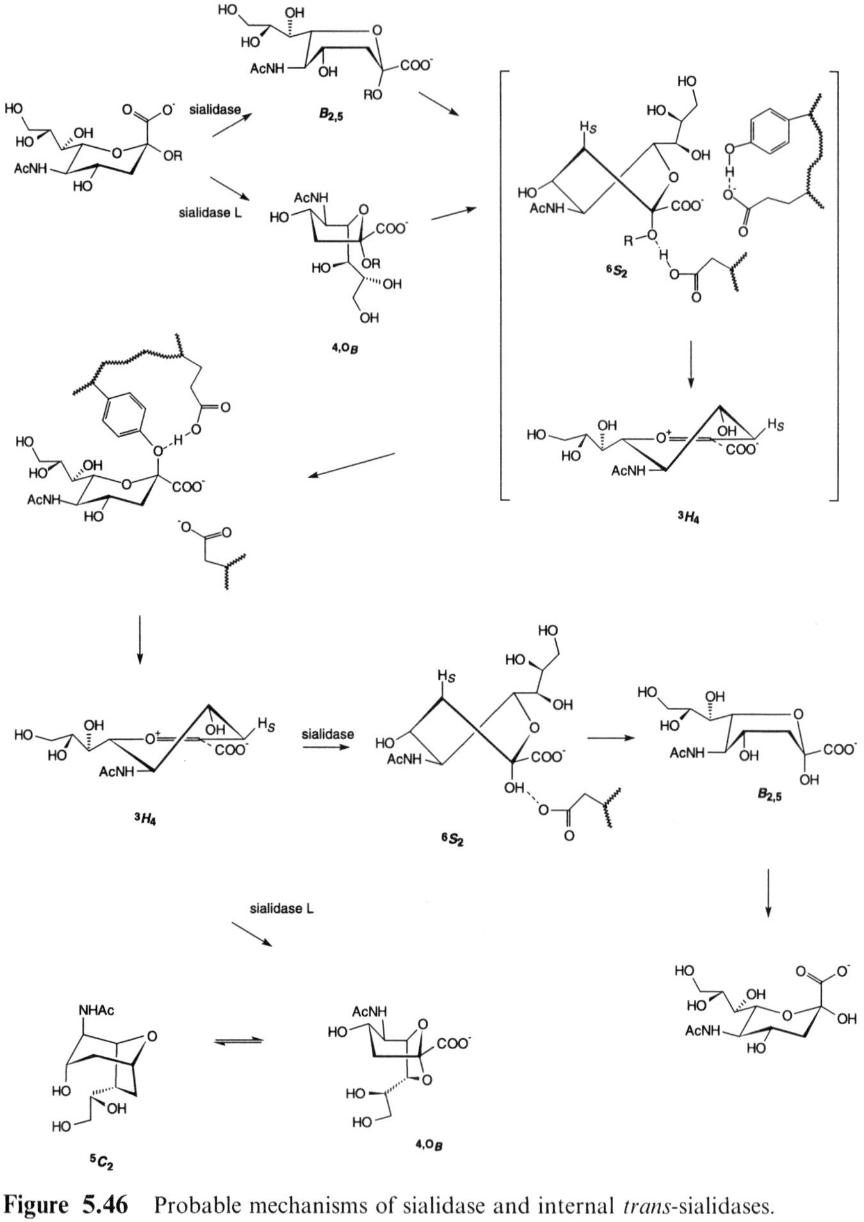

Figure 5.46 Probable mechanisms of sialidase and internal *trans*-sialidases.

was much more pronounced for a pyridinium ion than for *O*-glycosides.[367] To the author, a possible reason could be that the general base catalytic glutamate now directs the nucleophilic attack of water towards the pyridinium ring, in a reversible reaction which would give an inert, tightly bound species.

5.9.6.11 GH 32 and 68: Fructofuranosidases and Transfructofuranosylases. GH 32 and GH 38 includes hydrolases and transglycosylases acting on

Table 5.5 Structure–reactivity parameters and isotope effects for various sialidases (all GH 33 except influenza, GH 34).

Enzyme source	*Intrinsic proR β-DKIE (NeuNAc-αPNP)*	*Intrinsic proS β-DKIE (NeuNAc-αOPNP)*	*Intrinsic 2-^{18}O KIE (NeuNAc-αOPNP)*	$\beta^{lg}(V)$	$\beta^{lg}(V/K)$	*Ref.*
V. cholerae	1.03_7	1.01_8	1.04_6	-0.2_5	-0.7_3	365
M. decora (internal transglycosylase)	1.02_0	1.05_4		-0.5	-0.6	366
S. typhimurium	1.05	1.05	1.05	-0.5_3	-0.8	77
M. viridifaciens D92G (acid-catalyst mutant				-0.3_7	-0.7_2	364
M. viridifaciens Y370G (nucleophile mutant: inverter)				-0.6_3	-0.8_0	367
Influenza	1.05_7	1.07_8	1.06	-0.11	-0.45	77

β-fructofuranosyl residues, including invertase, the enzyme for which the Michaelis–Menten equation [eqn. (5.1)] was first proposed.[368] The name invertase refers not to the stereochemistry of reaction, which is retention, but to the inversion of the sign of the optical rotation of solutions of sucrose, as the dextrorotatory sucrose was converted to a 1:1 mixture of glucose and fructose, which when fully mutarotated is laevorotatory.

Asp23 of the yeast enzyme was labelled by L-conduritol B epoxide (Figure 5.33), which somewhat resembled a β-fructofuranosyl residue (Figure 4.73); the residue was confirmed to be important by site-directed mutagenesis,[369] and the acid catalyst considered to be Glu204 on the basis of site-directed mutagenesis.[370] The crystal structure of the enzyme from *Thermatoga maritime*[371] reveals a five-bladed β-propeller shape with the active site in a deep pocket. The structure is shared, unremarkably, with GH 68,[372] which also contains fructofuranoside hydrolases and transfructofuranosylases, and remarkably[373] with GH 43, which contains both arabinanases and xylanases and is inverting. The three catalytic carboxylates (homologues to Asp23 and Glu204 of the yeast enzyme and an additional "auxiliary" aspartate, presumably acting electrostatically) of the three families can be superimposed,[371] again suggesting the flexibility of the acid–base catalytic machinery in inverting enzymes. The structure of the inulinase from *Aspergillus awamori*[374] is essentially similar, although the complex with β-fructofuranose suggests the "auxiliary" Asp (189) hydrogen bonds to O3 and O4.

Levansucrase is a transfructosylase which converts sucrose to "levan", a largely β(2 → 6)-linked fructofuranose polymer, and has been investigated because of its synthesis by oral bacteria. The levans and dextrans produced from dietary sucrose are sparingly water soluble and contribute to the build-up of dental plaque on the teeth. The enzyme from *Bacillus subtilis* was shown to have ping-pong kinetics,[375] and the intermediacy of a covalent, fructosylated aspartate residue was found to be stable at low pH and was isolated and partly sequenced.[376] Finally, to complete a *tour de force* of mechanistic enzymology, decades ahead of its time (mid-1970s), the complete free energy profile of the enzyme was determined. This indicated that the high energy of the sucrose substrate was largely retained in the glycosyl-enzyme intermediate, ensuring the transfer reaction was downhill.[377] Sequence and structure[372] subsequently confirmed the picture.

5.10 CARBOHYDRATE BINDING MODULES AND THE ATTACK OF GLYCOSIDASES ON INSOLUBLE SUBSTRATES

5.10.1 Occurrence of CBMs

Many glycosidases which attack insoluble or partly soluble polysaccharides have a modular construction, in which catalytic domains and carbohydrate binding modules (CBMs) are produced in the same polypeptide. Although there may be occasional exceptions, the main function of these domains appears to be to increase the local concentration of enzyme by loosely absorbing the enzyme to the substrate. CBMs have been classified into families on the basis of amino acid sequence, like the catalytic domains and are on CAZy,[7] the continuously updated website; in summer 2007 there were 49 families. Early work had concentrated on the cellulose binding domains (CBDs):[378] previous classifications had referred to CBM (or CBD) "Types" and used Roman numerals and, to concur with this usage, CBM Families 1–13 are the same as CBM Types I–XIII.

Although the specificity of the CBMs binding polysaccharides is the same as the catalytic domain more often than not, there are many examples to the contrary from enzymes hydrolysing the plant cell wall. Thus, an acetylxylan esterase *Cellvibrio japonicus* (formerly *P. fluorescens* subsp. *cellulosa*) contains a cellulose binding domain.[379] Moreover the modular structure permits fairly complicated proteins to be built up: one chitinase, for example, from a hyperthermophilic archeon, had two GH 18 catalytic domains and two Family 1 and a Family 5 CBMs.[380]

The molecular complexity increases further in the case of cellulosomes, which had evolved in bacteria to address the problems posed for enzymes which acted synergistically on plant biomass in an environment with a high fluid flow (Figure 5.47). Cellulosomes are extracellular aggregates, M_r around 10^6, with many hydrolases.[381] They are centred around a non-catalytic, large protein called a scaffoldin; this has a number of domains capable of binding other proteins; these domains are termed cohesins. In the first cellulosome to be thoroughly investigated, that from the anaerobe *Clostridium thermocellum*[382] the scaffoldin also carried a CBM which attached the whole aggregate to the cellulose surface. The catalytic proteins of the cellulosome carry dockerin domains which bind to the complementary cohesin domains on the scaffoldin. Sometimes the catalytic proteins also carry CBMs, in addition to the scaffoldin; moreover, the whole cellosome may be attached to the producing bacteria by an attachment protein.

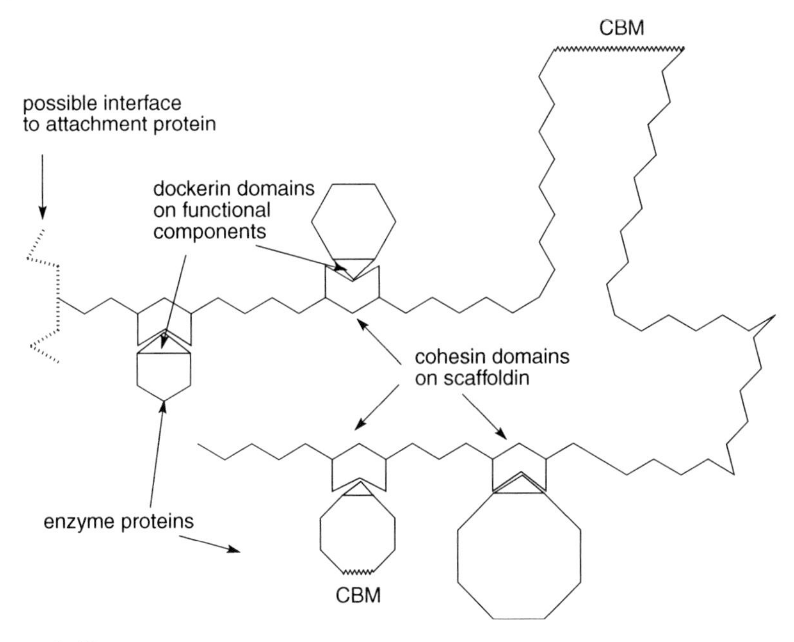

Figure 5.47 Cartoon of a cellulosome.

5.10.2 Methods of Study of CBMs

Discrete CBMs are usually produced by protein engineering methods: however, with some enzymes, such as the cellulase components of *Trichoderma reesei*, the CBMs are connected to the catalytic by long, heavily glycosylated linker regions, which are susceptible to attack by proteases, so that protease hydrolysis of commercial cellulases can be practical method of obtaining CBMs.[383]

The size of CBMs (CBM I being the smallest, with about 35 amino acids; some of the bigger families can range up to 200) makes them the right size for tertiary structure determination by modern NMR methods: the principles of the NMR experiments are much the same as with a polysaccharide, a key role being played by nuclear Overhauser enhancements. What is produced by such studies is a series of conformational constraints, within which the known primary structure has to be fitted, in much the same way as to an electron density map in X-ray crystallography. NMR structures, however, give more information about the degree of molecular motion. X-ray structural studies are also used, particularly of CBMs in complex with oligosaccharides.

Whereas the binding of CBMs to insoluble carbohydrates can be monitored by the UV absorption of the protein, relatively large amounts are required, and in some of the early studies on CBDs they were made radioactive with ^{3}H to increase sensitivity. The general technical problem of monitoring the binding of soluble, non-chromophoric ligands to soluble CBMs, has, however, been solved only in the last decade, with the advent of isothermal titration microcalorimetry

(Figure 5.48).[xxii] Two identical samples of protein, at a concentration greater than the anticipated K_D, are placed in identical compartments which are side by side in the centre of at least two independent systems of thermal insulation. To one of the samples successive constant aliquots of concentrated ligand is added automatically. All reactions take in or give out heat: in the case of exothermic reactions, heat is supplied by an electrical heater to the reference cell until the two temperatures are equal, whereas in the case of endothermic reactions it is applied to the sample compartment. The current and time to re-establish equilibrium are recorded. Initially, at low ligand concentration, every added ligand molecule is bound; however, as the free concentration of ligand approaches K_D, less and less heat is given out with each successive aliquot, until when the protein is saturated the only heat change is that due to the dilution of the ligand.

The whole process is automated, with the aliquots being added at equal time intervals: the output is a set of "spikes" of heat change, which on integration and appropriate correction for dilution gives $\Delta H^{\ominus}$; from the variation of the size of the "spike" with the amount of ligand added, it is possible to calculate (and even estimate roughly by inspection) K_D and hence $\Delta G^{\ominus}$. Therefore, $\Delta H^{\ominus}$, $\Delta G^{\ominus}$ and $\Delta S^{\ominus}$ can all be determined in a single experiment.

5.10.3 Types of CBM

The generalisation that all members of the same CAZy sequence family adopt the same protein fold holds for CBMs in addition to hydrolases. However, the absolute correlation between CAZy sequence family and mechanism that holds for the glycohydrolases breaks down for CBMs.

Functionally, CBMs can be divided into three types:[385] those which bind to the crystalline regions of solid polysaccharides and show weak, if any affinity for soluble carbohydrates (Type A); those which bind polysaccharide-chains, whether as soluble oligosaccharides or as amorphous regions of insoluble polysaccharides (Type B); and those which bind small sugar molecules (Type C). Unsurprisingly, there is considerable overlap between Type C CBMs and lectins, molecules which bind, but do not transform, sugars.

The only CAZy family with a unique protein fold is CBM1 (CBD I), which is confined almost exclusively to fungal enzymes. CBM 1 modules, which are small (33–40 residues) all have a Type A function (although the converse is not true – other CAZy families and protein folds also have Type A function). The CBM1 attached to the GH 7 catalytic domain of the cellobiohydrolase I of

---

[xxii] Previously, the fortunate investigator might have observed a perturbation in the fluorescence of the protein on ligand binding or interpretable NMR signals, although the latter required a very great deal of protein. The unfortunate investigator had to use the only general method, equilibrium dialysis. This required a radioactive ligand and large amounts of protein, which had to be stable for days as equilibrium was reached between the ligand in both dialysis compartments and the protein, which hopefully could not pass through the dialysis membrane, in one.

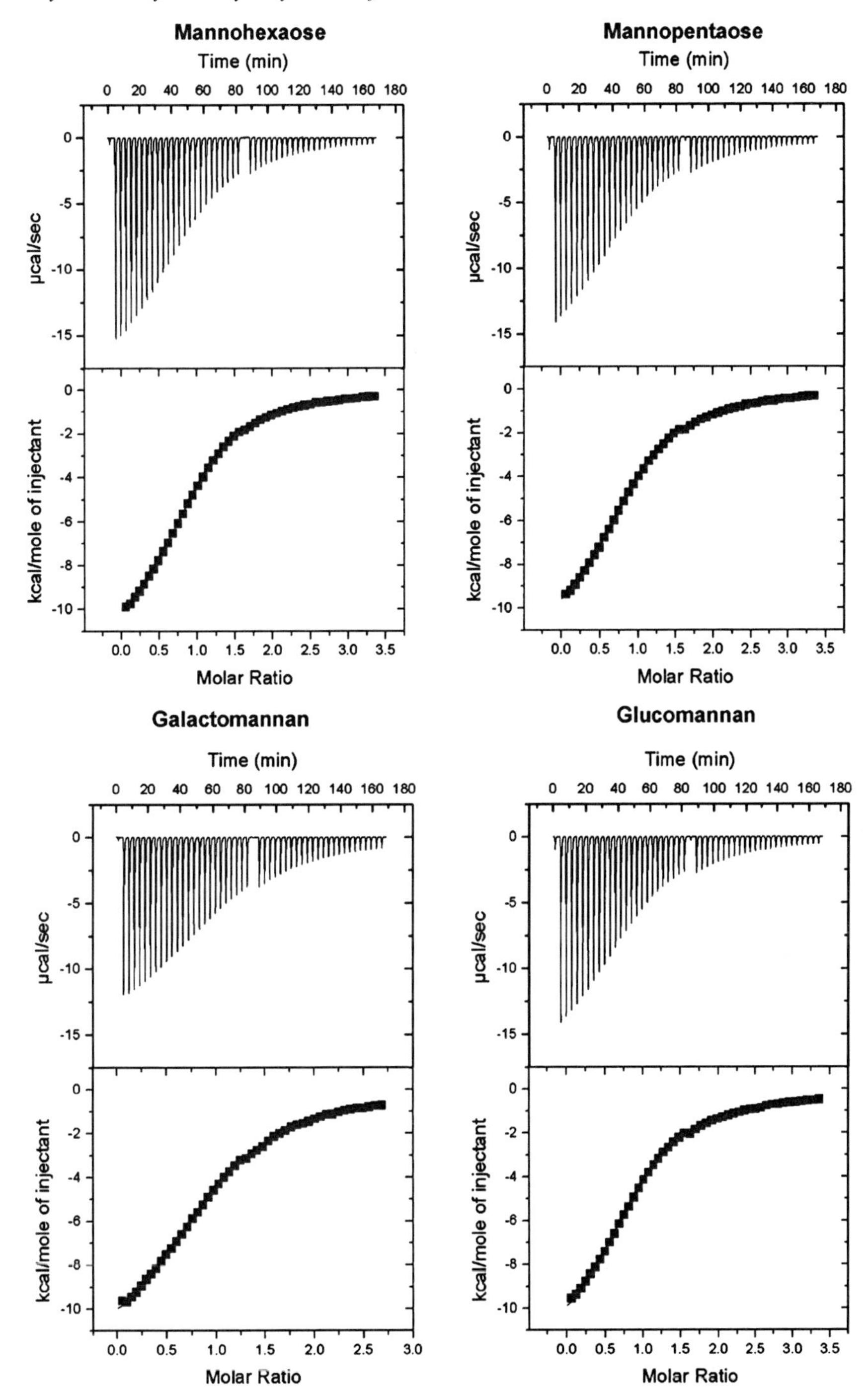

Figure 5.48 Typical output from differential titration microcalorimetry experiments (binding of various polysaccharides to CBM 35[384] reproduced by kind permission of Prof. H. J. Gilbert). The top traces are the rates of heat change, the bottom their integrals.

T. reesei has a wedge shape with one flat face of the wedge being formed from three tyrosine residues.[386] Analogous aromatic residues are a feature of all CBM 1 structures, but Tyr may be replaced with Trp or (less frequently) Phe. Disruption of any one of these residues abolishes binding to cellulose.[387] The protein "wedge" is heavily reinforced by disulfide bridges (three in the case of the *T. reesei* CBHII CBM) – the fold is sometimes termed a "cysteine knot".

All Type B function is associated with a so-called "β-sandwich" fold: this comprises two β-sheets, each consisting of three to six antiparallel β-strands. With one exception (a CBM2a from *C. fimi*) all of the CBMs with a β-sandwich fold have a structural Ca^{2+} ion. The first CBD that bound preferentially to amorphous, rather than crystalline, cellulose (*i.e.* that had Type B functionality) to be discovered was a CBM 4 attached to the N-terminus of a GH 9 endoglucanase from *Cellulomonas fimi*; it bound cellooligosaccharides as well as insoluble cellulose.[388] The isolated module was a 152-residue peptide, whose structure, β-sandwich, was solved by NMR spectroscopy: the active site was a cleft with many potential hydrogen-bonding groups.[389]

All Type B function is associated with a β-sandwich fold, but not all β-sandwich folds are associated with Type B function. The β-sandwich fold is adopted by CAZy sequence Families 2, 3, 4, 6, 9, 11, 15, 17, 22, 27, 28, 29, 32, 34 and 36, but some CAZy Family 2 members (CBM 2a) and all CBM Family 3 members have Type A function. In CBM 2a and CBM 3, three aromatic residues are again disposed to make a planar hydrophobic face sheet, as with CBM 1, even though the overall fold is a β-sandwich. The multiplicity of binding functions that can be supported by the β-sandwich is emphasised from the structure of a CBM6 attached to a *Clostridium thermocellum* xylanase, which had two binding clefts, only one of which appeared to be used in xylan binding; the apparently unused cleft in this structure was, however, the main binding site for a Family 22 and a Family 4 module.[390]

The same disposition of three planar hydrophobic aromatic residues is also adopted by CBM 5 and CBM 10, as other CBMs which have Type A function and unique protein folds. An NMR structure of the 60-residue CBM 5 of an *Erwinia chrysanthemi* endoglucanase revealed a "ski-boot" structure, with a flat hydrophobic face corresponding to the sole and heel of the ski-boot. This hydrophobic face contained the three planar hydrophobic residues (two tyrosines and a tryptophan).[391] Also using a CBM 5, but as a chitin binding domain, is the chitinase B of *Serratia marcescens* (the catalytic domain being GH 18).[223]

The same three-aromatic group arrangement is also produced by the "OB" (oligonucleotide/oligosaccharide binding fold) of CBM 10: it is largely composed of β-sheet and uses this scaffold to provide a flat, planar solvent-exposed face from two tryptophan residues and a tyrosine.[392] Mutation of any one of the three in CBDs greatly weakens the binding to cellulose.[393]

In the chitinase A1 of *S. marcescens*, the two aromatic residues of the CBM 12 form a continuous hydrophobic surface with several aromatic groups of the catalytic site; mutation of any one of them abolishes activity on crystalline chitin.[394]

5.10.4 Type CBM A Function

Although one can have a broad picture of how Type A function comes about, the detailed kinetics and thermodynamics still present unsolved problems. The planar, hydrophobic, aromatic face presented by all the protein folds that have Type A function suggests that the driving force is hydrophobic and that therefore Type A CBDs should bind preferentially to hydrophobic faces of the polysaccharide crystal. Indeed, with gold-labelled CBM 1 and CBM 3 binding to cellulose I_{α} (triclinic) crystals of *Valonia ventricosa*, electron microscopy reveals just that – preferential binding to the (110) face of the crystal.[395] In perfect crystals these faces are fairly small – the major exposed faces along the fibre crystal are 100 and 010, which are hydrophilic – and are not large enough to account for the amount of CBM adsorbed. However, cellulose crystals will be damaged and this damage will be associated with removal of glucan chains from the corners of the crystal, increasing the effective hydrophobic area.

The hydrophobic effect is an entropy-driven one, arising from the ordering of water molecules around a non-polar group or solute as they form tighter, more ordered hydrogen bonds to their near neighbours, to compensate for the absence of hydrogen bonding capacity to the non-polar group, forming the "iceberg". When two "icebergs" come together, ordered water molecules are freed into bulk solution, increasing disorder. Measurements of binding of Type A CBMs to cellulose indicate that the binding is indeed entropically driven.[396]

In the binding direction, it is possible to obtain a Langmuir-type binding isotherms with a Type A CBM. In fact, with bacterial microcrystalline cellulose, saturation is not readily attainable since the material binds CBM1 and CBM 2a at two sets of sites, a tight one accounting for the majority of the binding, and a loose one not saturable at accessible CBM concentrations. The Langmuir isotherm has to be modified according to eqn. (5.51):

$$[\mathrm{CBM}]_{\mathrm{bound}} = \frac{[\mathrm{CBM}]_{\mathrm{free}} N_1}{K_{\mathrm{D}}^1 + [\mathrm{CBM}]_{\mathrm{free}}} + \frac{[\mathrm{CBM}]_{\mathrm{free}} N_2}{K_{\mathrm{D}}^2 + [\mathrm{CBM}]_{\mathrm{free}}} \tag{5.51}$$

where N_1 and N_2 are the effective concentrations of the two types of site in the particular cellulose suspension and K_{D} refers to the dissociation constant from the particular types of sites. In the case of bacterial microcrystalline cellulose binding a Type A CBM2 (CBM2a), about 79% of the available cellulose sites are tight-binding ($K_{\mathrm{D}}^1 \approx 16\,\mathrm{nM}$) and 21% of the sites loose-binding ($K_{\mathrm{D}}^2 \approx 0.9\,\mu\mathrm{M}$).[396]

However, the detailed kinetics and thermodynamics of Type A function present an as yet unsolved problem. With many CBMs, it is possible to obtain apparently conventional binding isotherms against various types of cellulose in the binding direction: the cellulose does not bind more CBM if more is added and the amount of cellulose bound is constant with respect to time. However, if the solution is now diluted with buffer, no loss of CBM from the cellulose is seen, and if bound CBM is centrifuged off and placed in fresh buffer, no CBM is liberated: the CBM binding is apparently irreversible.

The phenomenon was investigated in detail with the two CBMIs from the cellobiohydrolases of *T. reesei*.[397] That from CBHI, attached to a GH 7 catalytic domain, had the planar aromatic face formed by three tyrosines and its binding was reversible;[398] that from CBHII, attached to a GH 6 catalytic domain, had a tryptophan in place of one of the tyrosines and its binding was apparently irreversible. However, conversion of the binding face Trp to Tyr in the CBHII CBD produced a CBD whose binding was reversible, like that of CBHI. Removal of one of the three disulfide bridges had a similar effect. Change of temperature, however, could desorb wt CBHII CBD. Apparent irreversibility arising from binding and desorption being kinetically bimolecular (or higher order) can be ruled out, since no exchange of bound CBM with CBM is solution was seen. The mixed reversible–irreversible binding of two CBM 2a domains to bacterial microcrystalline cellulose has been reasonably successfully treated in terms of an initial, reversible Langmuir-type absorption followed by an irreversible process of trapping of the CBMs in interstices.[399]

The binding does not appear irreversible in two dimensions, however: two-dimensional diffusion on the surface of a *Valonia ventricosa* crystal to two CBM 2as from *C. fimi* was demonstrated.[400] Obviously, having a CBM which bound a catalytic domain not only irreversibly but also immovably would vitiate its biological function.

Making a biomolecular reaction unimolecular confers a maximum entropic advantage of $\sim 10^8$ M.[401] In an attempt to exploit this, a genetically engineered bivalent Type A CBM was constructed from the CBHII and CBHI CBDs from *T. reesei* just discussed. For a thermodynamically well-behaved system, the lower limit on the dissociation constant (M level) of the dimer from cellulose is therefore $10^{-8}K_1K_2$, where K_1 and K_2 are the dissociation constants of the individual monomeric CBMs. The CBHII CBD of this system is not thermodynamically well behaved, but dissociation constants of the monomers are in the μM region and of the dimer only 10–15 times lower.[402]

The role of CBMs may not be entirely passive, although thermodynamics ensure that any interaction has to be stoichiometric rather than catalytic. An isolated CBM 2a, originally from a *Cellulomonas fimi* endoglucanase, formed the usual β-sandwich, but, in solution, when unconstrained by an attached catalytic domain, dimerised.[403] This CBM liberated small particles from cotton linters but not bacterial microcrystalline cellulose; the same behaviour was shown by the holoenzyme which had been inactivated with the appropriate Withers inactivator.[404] However, another CBM 2a, from *Cellvibrio japonicus*, was rigorously shown to act only by increasing substrate proximity.[405]

5.10.5 Type B CBM Function

As with Type A CBMs, the interactions of planar aromatic residues play a role in determining function, but in conjunction with other interactions, specifically hydrogen bonding. The aromatic interactions are often of the "stacking" type in which C–H bonds in electron-deficient systems interact with the cloud of π electrons of an (electron-rich) aromatic system: the well-known chloroform–benzene interaction (their mixing is exothermic) is the simplest example. The interplay between stacking interactions and hydrogen bonding interactions was investigated by site-directed mutagenesis and titration microcalorimetry in the case of a xylan-binding CBM 2b.[406] Mutation of hydrogen-bonding residues in the CBM had no dramatic effect on overall affinity, but did bring about large changes in the balance between entropic and enthalpic contribution to binding. Enthalpic changes were less favourable, but entropic changes, as the polysaccharide became more mobile, became more favourable.

The prevalence of aromatic residues in the active site clefts means that it is often difficult to predict Type A or Type B function from structure. Indeed, the CBM Family 2 supports both Type A (2a) and Type B (2b) function, and switching from one to the other can involve only single amino acid changes. The specificity of CBM 2b from *C. fimi* for xylan depends on two perpendicular tryptophan residues, spaced so as to interact favourably with xylan in its three-fold helical conformation. A single arginine to glycine mutation allowed both tryptophans to lie flat and in the same plane: the function of the mutant was changed from Type B xylan binding to Type A binding of crystalline cellulose.[407]

One of the first studies by titration microcalorimetry revealed that binding of a CBM 4 to cellulose was enthalpy driven, rather than entropy driven like CBM 1.[408] More detailed study revealed that the parent enzyme had two such CBMs, presumably to exploit the entropic advantages of multivalency;[409] the second CBM 4 had a similar structure and enthalpy-driven cellooligosaccharide-binding activity.[410] Enthalpy-driven binding, arising from the formation of direct hydrogen bonds from the CBM to the glycan chain, appears general for Type B CBMs; indeed, disruption of these bonds lowers or even abolishes binding function.[411]

CBM 15 binds largely xylan. The X-ray structure of one such CBM in complex with xylopentaose revealed the expected β-sandwich, in which a concave active site bound xylopentaose in the approximate 3-fold helix of xylan;[412] however, a substantial number of the 2- and 3-OH groups are exposed to solvent. In natural xylan these would be substituted with various "decorations" such as α-L-arabinofuranosyl groups: the xylan could be bound to the CBM even if decorated. Site-directed mutagenesis suggests stacking interactions with aromatic groups and direct hydrogen bonds to the sugar are responsible for xylan-binding activity.[413] The X-ray structure of the holoenzyme (the catalytic domain was GH 10) did not show defined electron density for the linker, suggesting that it was very flexible.[414]

Many isolated catalytic domains are less stable than the holoenzyme from which they are derived, and indeed CBM 22 was first identified as a "thermostabilizing" domain, rather than by its true function of enthalpy-driven binding of xylan and xylan oligosaccharides.[415]

The function of CBM Family 20 (sometimes called SBD, starch binding domains) is particularly well understood in the case where it is attached to a GH 15 catalytic module, as in the *Aspergillus* spp. glucoamylase. The SBD contains two saccharide binding sites, each of which relies on a Trp residue to interact with the hydrophobic portion of the amylose unit. Mutation of either Trp residue gives an enzyme which continues to bind to soluble starch, but no longer acts on insoluble starch.[416] AFM imaging at first revealed a single amylose molecule wound round the SBD, but more recent images showed the SBD inserting itself between the two strands of the amylose helix.[417] Given that glucoamylase is an *exo*-acting enzyme, the proposal that the SBD inserted itself between the two participants at the non-reducing ends of short stretches of straight-chain double helices seemed entirely reasonable.

Although the β-sandwich fold is normally associated with binding of an extended chain, in at least one example, a CBM Family 6 with specificity for laminarin, the active site cleft is closed off so that the module is specific for the non-reducing end.[418]

5.10.6 Type C CBMs

The structures and activities of CBMs which bind only a small number of sugars, or even monosaccharides, overlap with a very large class of sugar-binding proteins, known collectively as lectins, which modulate a whole range of biological activities and are outside the scope of this text. Whereas binding of carbohydrate chains to Type B CBMs occurs both by direct hydrogen bonding and by stacking interactions of aromatic protein residues, the binding of Type C CBMs and lectins appears more dependent on hydrogen bonding.[411]

Notable Type C CBMs are the CBM Family 9 (β-sandwich folded),[419] one member of which has been shown rigorously to bind the reducing end of oligo- and polysaccharides in an enthalpy-driven process.[420]

CBM Family 13 adopt the β-trefoil fold first encountered with the ricin B (carbohydrate-binding) chain.[421] It contains 12 strands of β-sheet arranged in such a way as to provide three independent sugar binding sites with different specificities,[422] and again binding is enthalpy driven.

The fairly small Family 18 CBMs (~40 residues) possess the hevein fold associated with chitin binding in higher plants: despite their small size, they can accommodate chitotetraoside essentially on the "outside" of the domain.[423] Related is another chitin-binding CBM Family, 14.

5.11 RETAINING *N*-GLYCOSYLASES AND TRANSGLYCOSYLASES

These enzymes have broadly similar mechanisms to retaining *O*-transglycosylases, except that the acid catalytic machinery is different. Sensible mechanistic evidence is available for NAD^+-glycohydrolases and a related series of mammalian proteins in which N1 of the adenine acts as a nucleophile towards the glycosyl-enzyme, generating a cyclic ADP-ribose second messenger for calcium signalling in the cell. Other understood enzymes are those which exchange an original G residue in tRNA for unusual modified bases and deoxyribosyl-transglycosylases.

5.11.1 Retaining NAD^+-Glycohydrolases and Cyclases

Various eukaryotic organisms produce proteins which hydrolyse the carbon–pyridinium linkage of NAD^+ in a way reminiscent of the hydrolysis of glycosylpyridinium salts by *O*-glycosidases. Retention was established by the ability of the enzymes from calf spleen[424] and the venom of the banded krait[425] to exchange nicotinamide with other pyridines and for the snake venom enzyme to transfer ADP ribose to methanol.[426] Although the chemistry was masked for hydrolysis of the natural substrate, studies of the selectivity of various pyridines as ADP-ribose acceptors enabled a β_{nuc} value of 0.43 to be obtained for the krait venom enzyme.[426] On the estimate of a β_{eq} of -1.47 for hydrolyses of glycosylpyridinium salts, this indicated a probable β_{lg} in the usual region of -1; indeed, a β_{lg} of -0.9 was obtained for the calf-spleen enzyme for substrates lacking a carbonyl substituent at C3 of the pyridine.[427] The calf-spleen enzyme was also the vehicle for the insightful study of substituents at C2′ of the fissile ribose, discussed in Section 5.4.5.1, which suggested that the 2-OH was effectively deprotonated at the transition state.

Thanks to the sequences becoming available through the human genome project, various human membrane-bound proteins which hydrolase NAD^+ and cyclise it to cyclic ADP-ribose have provided firm structural data for this type of enzyme. The nucleophile in CD 38 was identified as Glu226 by use of the Withers inactivator shown in Figure 5.49; the glycosyl-enzyme intermediate was shown to turn over in the presence of added acceptor. $^{\alpha T}(V/K)$ and $^{1'-14}(V/K)$ values of 1.23 and 1.02 established the usual type of exploded transition state for the first step of the hydrolysis, but not of the cyclisation, for which the chemistry is masked.[428] A crystal structure of CD157 again shows a likely nucleophilic glutamate, as well as another glutamate which could partially deprotonate N2 and thus render N1 more nucleophilic in the second step.[429]

Sir2 protein deacetylases (sirtuins) use an acetyl lysine residue as an oxygen nucleophile to displace nicotinamide from NAD^+; in a second step the imidate intermediate is hydrolysed to 1-O-acetyl ADP ribose and a lysine residue. A study of the effect of acyl substitution in the acyl lysine yielded a ρ^* value of -1.9 for the first step, equivalent to a $^{\beta nuc}(k_{+2})$ of >1,[555] an extraordinarily high value for which there is as yet no explanation.

5.11.2 tRNA Transglycosylases

These enzymes exchange the first-synthesised guanine residues in tRNA[xxiii] for unusual bases, at the polymer level.[430] The nucleophile appears to be Asp in both the *E. coli*[431] enzyme (which exchanges guanine with prequeuine) and the archaeon *Pyrococcus hirikoshi* (which exchanges it with archeosine).[432] The acid–base catalytic machinery remains uncertain: the acceptor ability of various prequeuine derivatives increased slightly with electron-withdrawing

[xxiii] A single strand of RNA which along most of its length forms a double helix with itself, leaving a loop at one end and two loose ends at the other. tRNAs are the carriers of activated amino acids during protein biosynthesis, with each tRNA having a ribose ring *O*-acylated with the amino acid appropriate to the unpaired bases in the loop. These pair up with the mRNA when this specifies a particular amino acid.

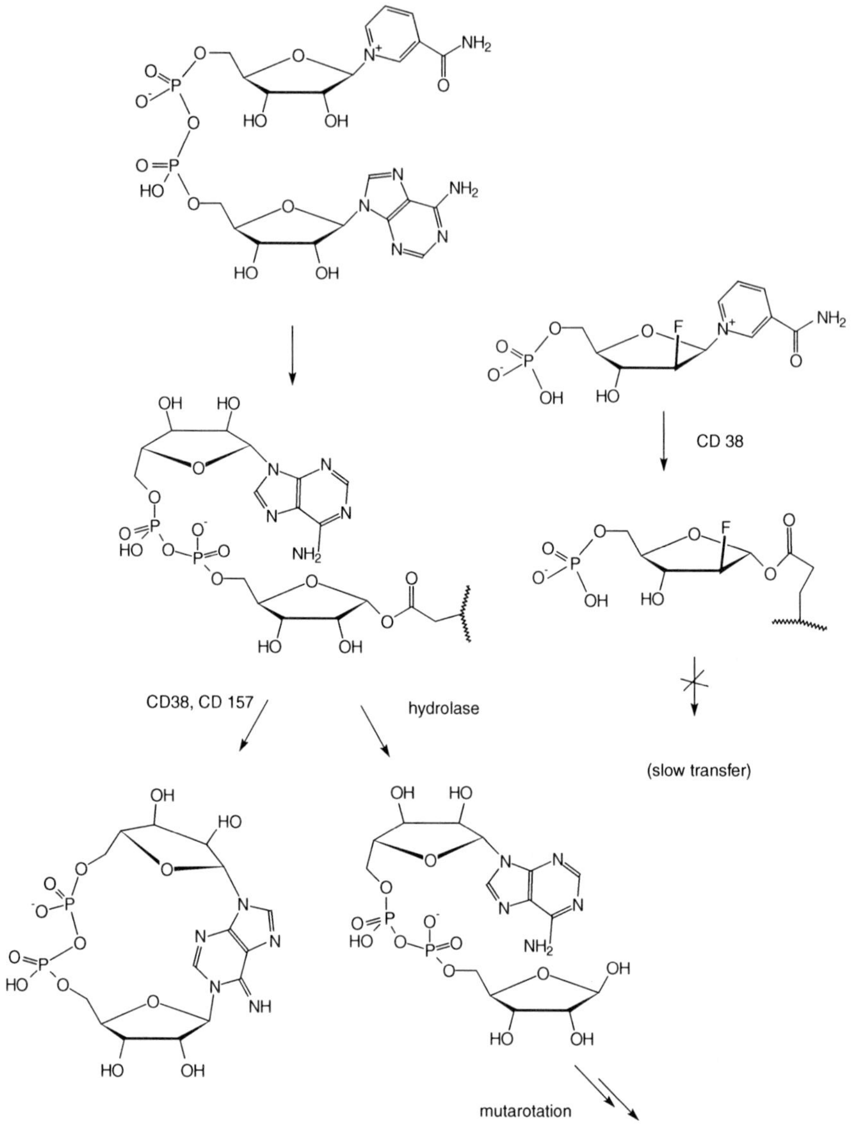

Figure 5.49 NAD$^+$ hydrolysis and cyclisation with retention. The appropriate Withers inactivator is shown.

substituents (although the chemistry was largely masked),[433] suggesting some sort of deprotonation, and a suicide inactivator was obtained when $R = CH_2F$ (Figure 5.50),[434] suggesting the generation of an electrophilic non-aromatic species such as the one shown. This would imply that the acid–base catalytic machinery acts directly at the C1′–N bond, as an acid in the removal of guanine and by activating the acceptor by deprotonating N1. Any sort of synchronicity between C1′–N cleavage or formation and N-deprotonation is, however,

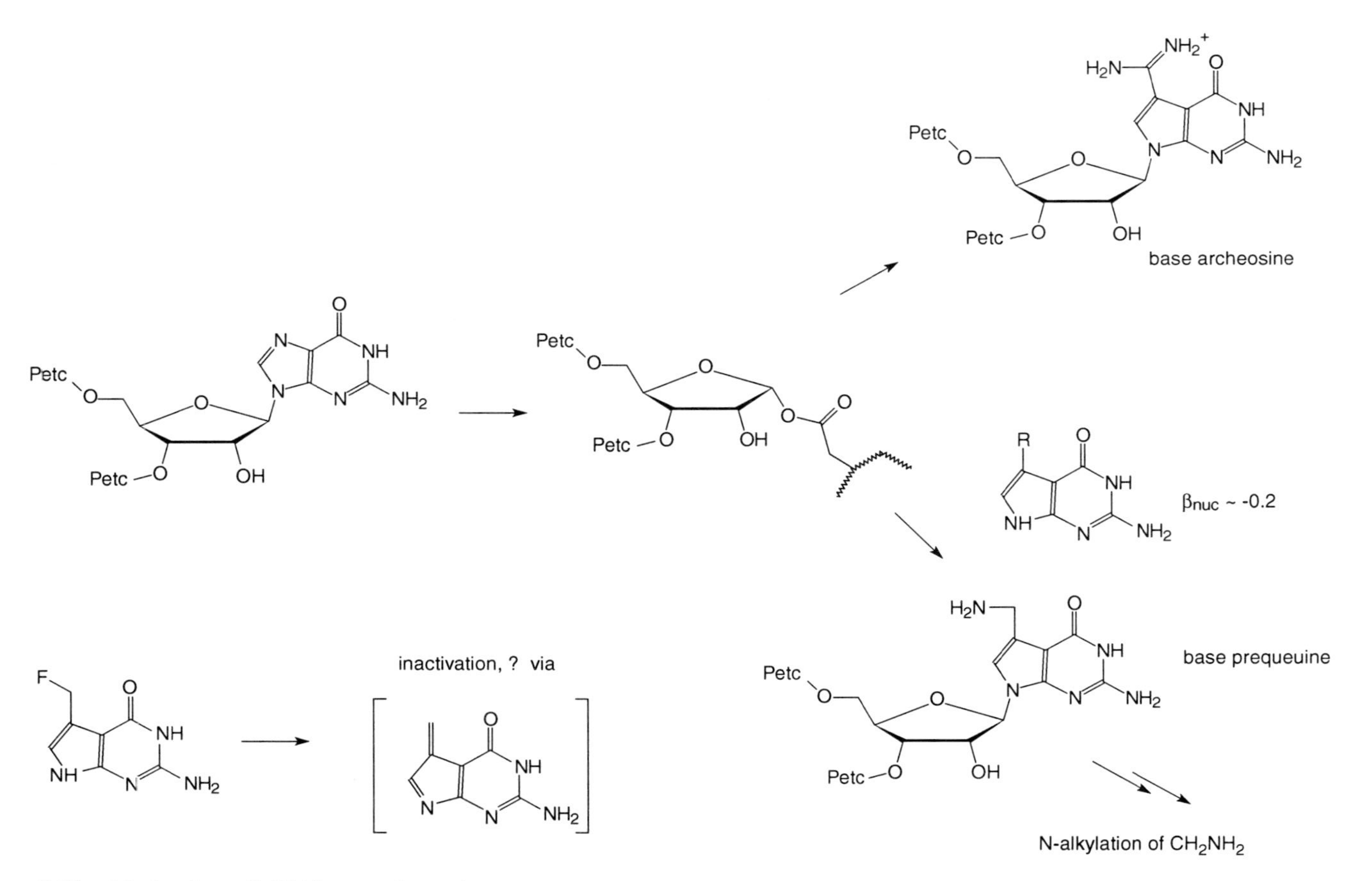

Figure 5.50 Mechanism of tRNA transglycosylases.

stereoelectronically impossible; at best the acid–base group can only act as an acid in C1′–guanine cleavage in the preassociation sense of simply "catching" the incipient anion once it is formed; in the reverse sense, the anion has to be fully formed before C1′–N bond formation.[435]

The isomerisation of specific uridine residues in tRNA to the isomeric *C*-glycosidic pseudouridine appears to be carried out by very similar enzymes, all of which have a conserved, essential and presumed nucleophilic aspartate residue (Figure 5.51).[436] Given the ability of another nucleic acid glycosylase, DNA uracil-glycosylase, to stabilise a uracil anion so that it acts as a leaving group, the proposal that such an ion rotates relative to the glycosyl-enzyme and then attacks with a C5 seems entirely reasonable, and is consistent with X-ray structures of the enzyme in complex with substrates[437] (the uracil anion will certainly act as a π-nucleophile when attacking with carbon). An alternative mechanism involving initial attack of the aspartate on the pyrimidine[438] lacks chemical rationality.

5.11.3 2′-Deoxyribosyl Transferases

Bacterial *Leishmania* spp. produce enzymes which exchange the bases of 2′-deoxyribosylpurines and -pyrimidines, as part of the nucleoside base salvage pathway. There are two classes, Class I,[439] which accepts only purines, and Class II,[440] which accepts both purines and pyrimidines, representative crystal structures of which have been solved. The expected nucleophile is a glutamate, proved directly in the case of a Class II enzyme by glycosylation with the Withers inactivator *N*-9-(2′-deoxy-2′-fluoro-β-D-arabinofuranosyl)adenine,[441] and identified from homology and structural studies in other enzymes. Site-directed mutagenesis studies suggested, but did not prove, that the acid catalyst(s) in the Class II enzyme were aspartate(s). The acid-catalytic machinery (if any) is clearly related to base specificity and the presence of a protein loop in Class II enzymes, which closes off the active site from solvent, but which is missing in Class I enzymes.

As might be expected, in the absence of acceptors the 2′-deoxyribosyl-enzyme slowly hydrolyses; intriguingly, however, the hydrolysis is associated with the production of substantial quantities of ribal,[442] which is then slowly hydrolysed. Formation of ribal is clearly the near microscopic reverse of the first step in the hydration of glycals by retaining glycosidases (Figure 5.21), for some reason favoured with a 2-deoxy substrate.

5.12 GLYCOSYL TRANSFERASES

The glycosyl transferases have, like the glycoside hydrolases and transglycosylases, been classified into families on the basis of sequence similarities and the families are likewise available on CAZy. Most glycosyl transferases are membrane bound and this has limited the number of crystal structures (mostly of engineered enzymes without their membrane anchors) that are available. It has

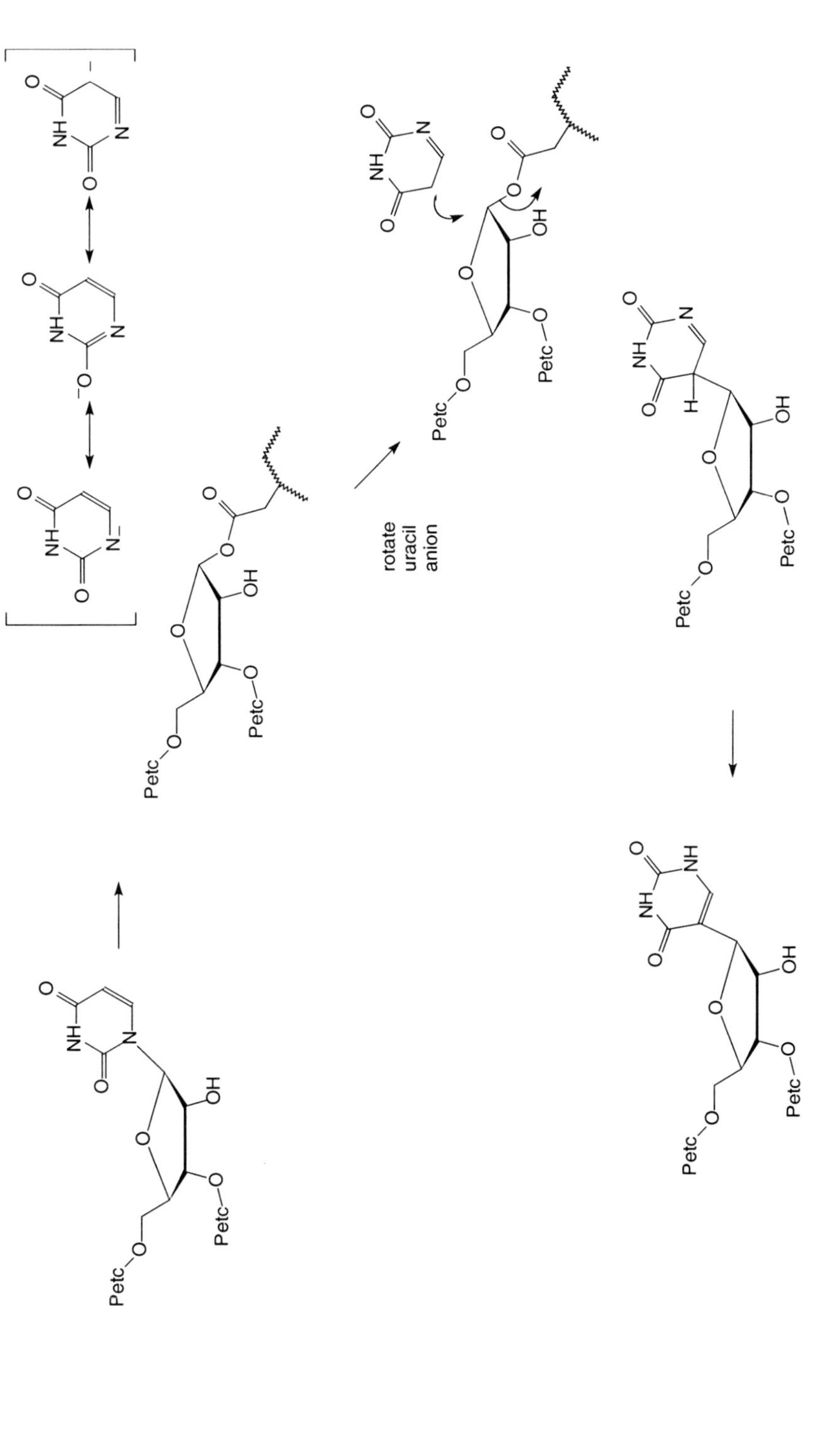

Figure 5.51 Pseudouridine synthetase mechanism.

also limited the scope of kinetic studies. Like glycoside hydrolases, glycosyl transferases can act with either retention or inversion of anomeric configuration.

So far only two protein folds have been characterised, designated A and B, and these are distributed amongst both inverting and retaining transferases (Figure 5.52). The A fold[443] involves two domains: an N-terminal nucleotide binding fold (four parallel β-strands flanked on either side with two α-helices "Rossman domain"), and a mixture of β-sheet and α-helix with several disordered regions, suggesting a conformationally mobile system in which loops of protein can fold over glycosyl donor and acceptor. The B fold has two Rossman domains, with the active site being in a deep cleft between them.[444]

These two protein scaffolds appear to catalyse glycosyl transfer from activated donors with retention and inversion indiscriminately. As discussed in Section 5.12.2, the assumption, made by analogy with the glycoside hydrolases, that retaining glycosyltransferases act by the same sort of double displacement mechanism as retaining glycosidases, founders on the failure to identify the putative enzyme nucleophile, except in GT 6. An S_Ni mechanism similar to that in Figure 3.37 therefore seems elsewhere. The transition states for both retaining and inverting transferases would then lie in the extreme SE corner of Figure 3.2, with almost completely developed oxocarbenium ion character on the sugar ring. Such transition states could then be stabilised indiscriminately by a given protein fold. This is more likely to occur with glycosyltransferases, which have to bring about a rate enhancement between 8 and 10 orders of magnitude less than the glycosidases.

Important evidence that glycosyl transfer from activated donors with retention or inversion can indeed be catalysed by very similar protein folds comes from an α-mannopyranosyl glycerate synthetase with a GTA fold,[445] which according to amino acid sequence belongs to (inverting) GT 2, but catalyses

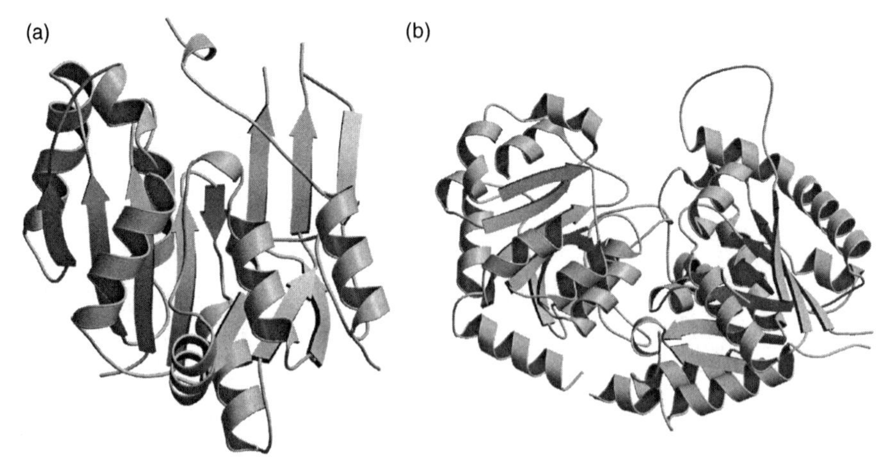

Figure 5.52 The two protein folds associated with glycosyl transferases: (a) GTA[445] and (b) GTB.[446] These greyscale ribbon diagrams were kindly provided by Prof. G. T. Davies.

transfer of an α-pyranosyl residue from a nucleotide diphospho sugar to the 2-OH of D-glycerate, D-lactate or glycollate with retention of configuration. It has been reclassified as GT 78. In addition to being mildly promiscuous with respect to acceptor, the enzyme is grossly indiscriminate with respect to donor, accepting GDPMan, UDPMan, GDPGlc, UDPGlc and β-GDP-L-fucose. It would appear that the main function of the GT 2/78 protein structure is to stabilise an oxocarbenium ion generated from a nucleotide diphospho sugar and that specificity interactions are secondary.

5.12.1 Inverting Glycosyl Transferases

A mechanism along the lines of Figure 5.53, in which the attack of the acceptor hydroxyl is accelerated by general base, accounts for the basic stereochemical facts. With transferases with the A fold, the leaving pyrophosphate is often accelerated by coordination with a divalent metal of similar ionic radius to Mg^{2+} (*e.g.* Mn^{2+}) and this also illustrated; metal dependence of transferases with GTB fold is rare.

5.12.1.1 GT 63. The first glycosyl transferase for which a structure was obtained was an unusual enzyme from bacteriophage T4, in a family by itself (GT 63), which transfers a β-glucosyl residue to a modified cytosine residue (hydroxymethylcytosine) in DNA.[444] It adopts the glycosyl transferase B (GTB) fold.

The wild-type enzyme does not permit a structure with donor UDP-glucose to be obtained (the donor is hydrolysed), but the Asp100A mutant does, so Asp100 is thereby identified as the general base catalyst.[447] Intriguingly, although a metal ion is observed in the complex with UDP,[448] no electron density is seen in the complex with UDP-glucose.[447] The very reasonable suggestion is that the metal assists product release, rather than bond cleavage. The importance of Mn^{2+} (and by analogy Mg^{2+}) in the departure of pyrophosphate has been established for a GDP-mannose mannohydrolase.[449] The X-ray structure of the enzyme in complex with UDP and a 13-mer DNA fragment was obtained in two forms, one with a metal ion at the active site and one with a Tris molecule, making its usual adventitious "sugar cation" contacts.[450]

5.12.1.2 GT 1. This family, which adopts a GTB fold, contains many transferases which glycosylate complex organic structures such as anthocyanines and triterpenes; the structure of the enzyme that glucosylates the anthocyanins in grape skins (and hence red wine) is available (it is the exemplar in Figure 5.52). The other X-ray structures that are available are all connected with the synthesis of the antibiotic vancomycin by *Amycolatopsis orientalis*. The core of the molecule is an oxidised peptide, and the members of the vancomycin family are produced by variable glycosylation. Three donors are used: UDP glucose, dTDP-vancosamine and dTDP-epivancosamine. Structures of the unliganded enzyme which initially glucosylates the 2-OH of the gallic acid

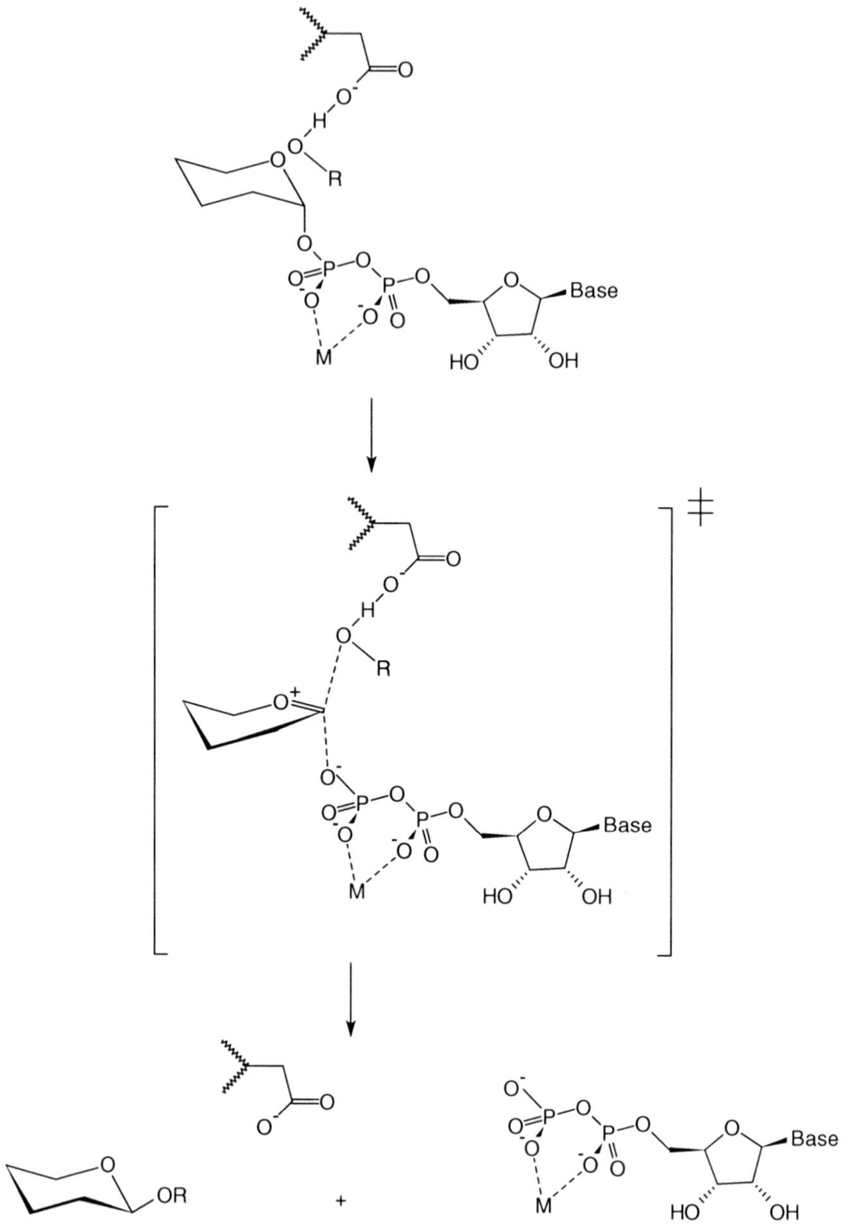

Figure 5.53 Canonical mechanism for inverting glycosyl transferases. A divalent metal ion is shown acting as an electrophilic catalysts: however, data for GT63 (B fold) suggest that any metal-ion dependence of B-fold enzymes arises from metal-promoted product release, not catalysis and metals are rare in transferases with the GTB fold.

residue (GtfB),[451] the liganded enzyme which transfers epivancosamine to a benzylic OH (GtfA)[452] and the liganded enzyme which transfers vancosamine to the 2-OH of the glucose residue added by GtfB (GtfD) are available.[453] The dTDP donors are presumably unstable to spontaneous hydrolysis because of the deoxygenation of the 2 position and destabilising 1,3-*trans* methyl substitution (attempts to obtain donor complexes used dTDP glucose). This 1,3-*trans* dimethyl substitution of vancosamine and epivancosamine, which in either $^{1}C_4$ or $^{4}C_1$ conformation of the pyranose ring results in an axial methyl group, probably leaves the conformation to be determined by the anomeric effect; for this reason, the amino sugar rings of the donors are drawn in the $^{4}C_1$ conformation, even though in the products they are in the $^{1}C_4$ conformation. The acceptor complex of GtfD suggested that Asp13 functioned as the general base and this was confirmed by site-directed mutagenesis (Figure 5.54).

5.12.1.3 GT 2. This, the largest of the GT families, contains the enzymes which biosynthesise the β-(1→4)-linked polymers chitin and cellulose and adopts an A fold. Although the cellulose synthetase enzymes have not been crystallised, there is nothing in their sequences to indicate a second nucleotide binding domain. The synchronous transfer of two glucose units in two adjacent sites (see Section 4.6.1.1.1), very implausible anyway on physical-organic grounds because of the reduced nucleophilicity of the 4-OH in a glucose residue carrying a partial positive charge, can therefore be discarded as a mechanism.

Crystal structures, unfortunately, are confined to an enzyme implicated in bacterial peptidoglycan biosynthesis, whose donor and acceptor sugars are currently unknown. It crystallises with UDPGlc with one divalent metal bridging the two phosphates and a second in the approximate position that would be adopted by a transferred sugar. The cryoprotectant glycerol is found bound and hydrogen bonded to an aspartate.[454] Superposition of this structure[455] on GTA structures from GT 7, GT 13 and GT 43 identified the general-base catalytic residue as an aspartate, a conclusion confirmed by site-directed-mutagenesis on ExoM *Sinorhizobium meliloti*[456] (this GT 2 enzyme transfers a glucosyl residue from UDPGc to a polyprenyl intermediate during the biosynthesis of succinoglycan).

The catalytic mechanism of GT 2 thus appears to be exactly as depicted in Figure 5.53.

5.12.1.4 GT 7. The enzyme which transfers a β-galactosyl residue from UDPGal to the terminal GlcNAc of polylactosamines in cattle belongs to fold GTA.[457] Despite the complex with UDPGal being crystallised in the presence of Mn^{2+}, no metal was coordinated between the two phosphates and density for the galactose residue could not be seen, because of either hydrolysis or disorder. A later structure of the enzyme in the presence of UDPGal[458] indicated that the galactosyl residue was deeply buried and that substantial conformation change after UDPGal binding made the reaction centre inaccessible to water. In the presence of the mammary gland-specific milk protein α-lactalbumin, lactose synthetase is formed; the crystal structure of

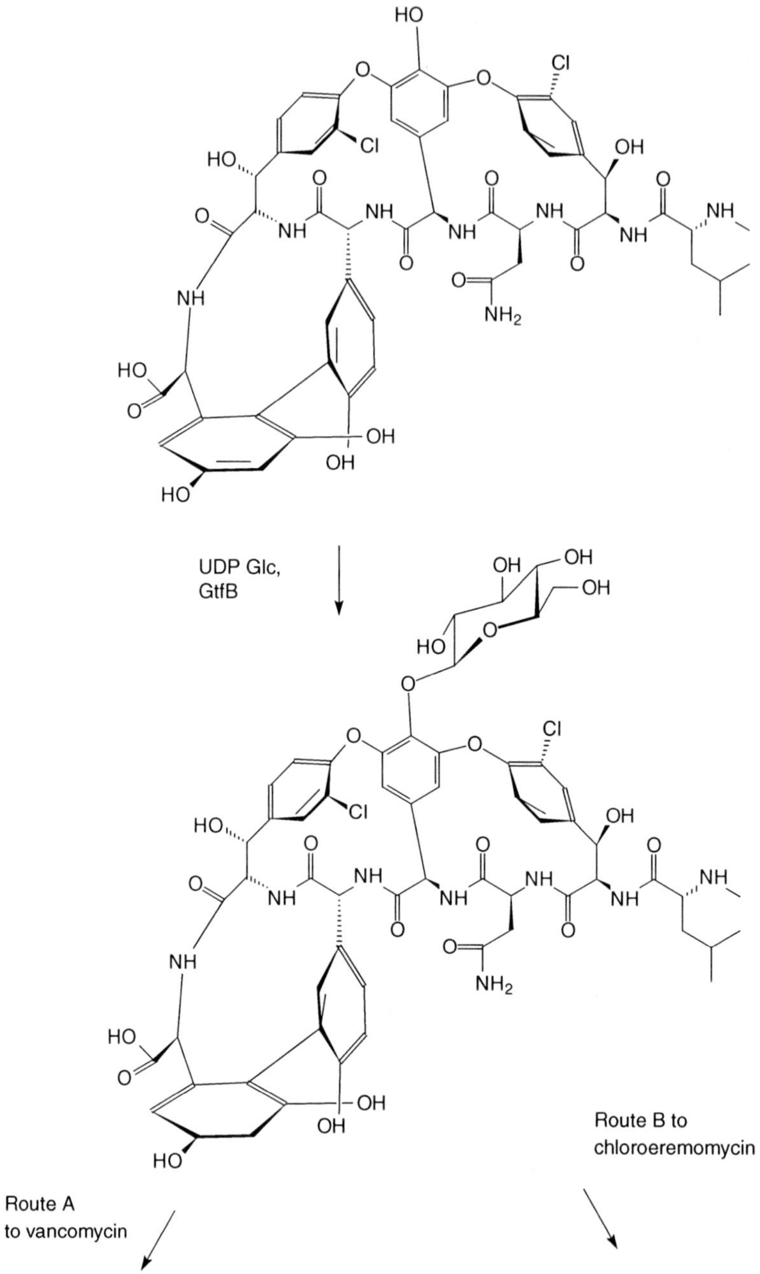
HO
Cl
O
O
Cl
OH
HO
O
O
O
O
NH
NH
NH
NH
NH
NH
O
O
O
NH2
NH
HO
O
OH
OH
HO
UDP Glc,
GtfB
OH
OH
OH
HO
O
O
Route B to
chloroeremomycin
Route A
to vancomycin

Figure 5.54 Reactions carried out by the glycosyltransferases of vancosamine antibiotic biosynthesis.

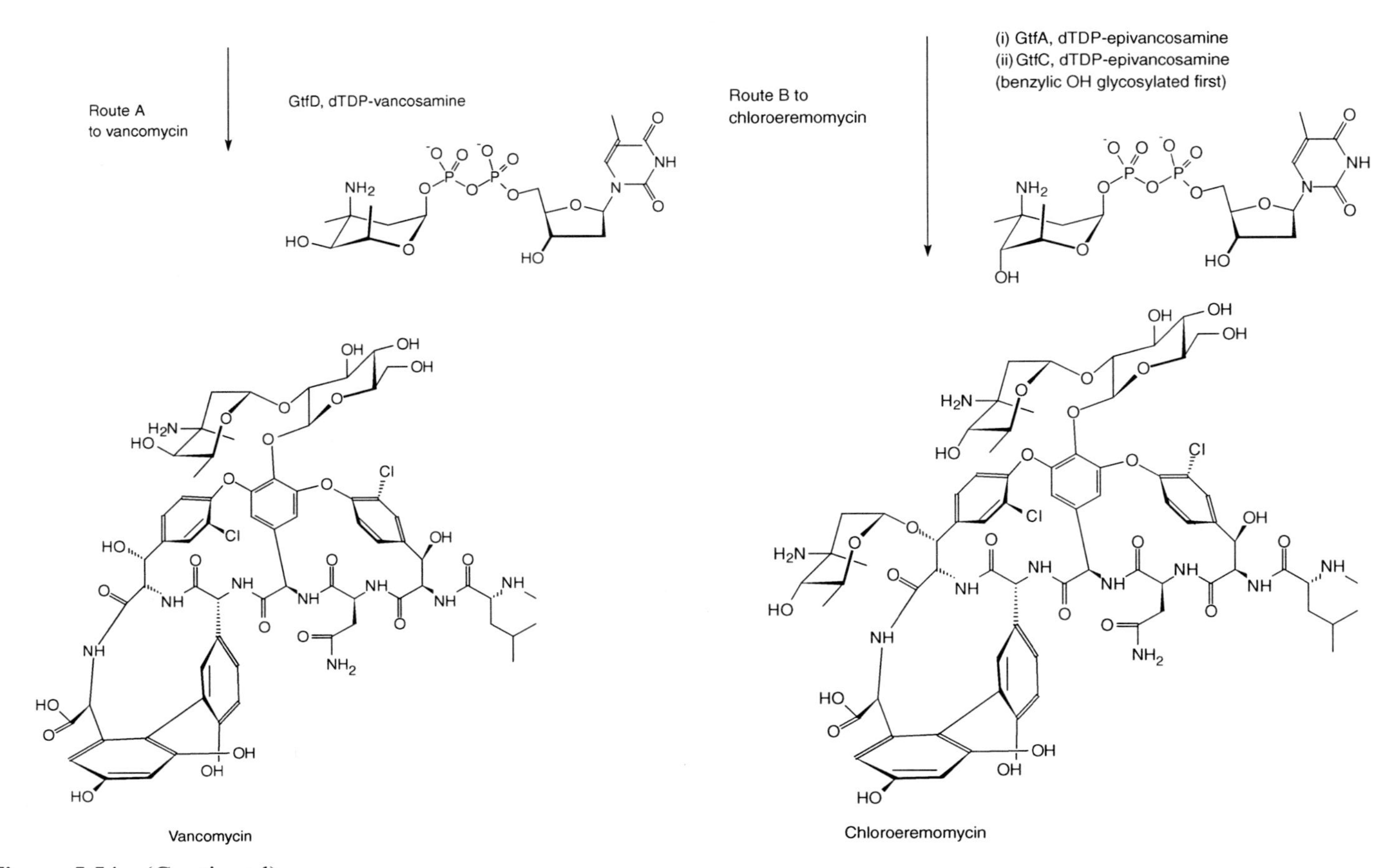

Figure 5.54 (Continued)

UDPGal–lactalbumin complex also reveals large structural changes which "shut off" the active site from the solvent.[459] The presence of α-lactalbumin (a milk protein) alters both donor and acceptor specificity.[460] Both UDPGlc-NAc and UDPGalNAc can be donors and both glucose and GalNAc acceptors. In the tertiary complexes of galactosyl transferase, α-lactalbumin, UDPGlc and *N*-butanoylglucosamine,[460] galactosyl transferase, α-lactalbumin and UDPGalNAc, an Mn^{2+} ion was seen coordinated in the usual way.[461] Structural analyses of the various complexes led to the identification of various minor interactions with the protein which, when eliminated or introduced, alter specificities in a rational way.[461] Kinetic differences between various acceptors and donors were 1.5 orders of magnitude or less, however.

5.12.1.5 GT 9 and GT 13. A structure of the heptosyltransferase II involved in the biosynthesis of lipopolysaccharide in *Escherichia coli* is available on CAZy and establishes the inverting GT9 as the GTB fold, but no discussion of the findings by the crystallographers is as yet available. The β-*N*-acetylglucosaminyl transferase[462] from rabbit, which transfers a β-GlcNAc residue from UDPGlcNAc to the 2-position of an α(1 → 3)-mannopyranosyl residue in the core oligosaccharide of *N*-glycoprotein biosynthesis in GT13, exhibits fold GTA. The acceptor site is only formed after the binding of UDPGlcNAc, by structuring of a 13-residue loop. This explains the kinetics observed with the enzyme, namely a ternary complex with one substrate binding first [eqn. (5.19)].[463] UDP-sugar binds first, followed by oligosaccharide acceptor, then the oligosaccharide product and lastly the nucleotide diphosphate, as expected if the nucleotide diphosphate moiety had to be present to order the 13-residue loop to create the acceptor binding site.

5.12.1.6 GT 28. This family contains one of the enzymes responsible for biosynthesis of the peptidoglycan cell wall in bacteria. It transfers a GlcNAc residue to the 4-position of a peptidylmuramic acid residue attached to a membrane-bound polyprene pyrophosphate (Figure 5.55). After the polyprenyl NAG-NAM structure is biosynthesised on the cytoplasmic side of the cell membranes, it is translocated to the other (cell well) side and the peptidoglycan further elaborated. The structure of the unliganded *Escherichia coli* MurG enzyme established the fold as GTB;[464] the structure of the enzyme complexed with UDPGlcNAc failed to reveal a divalent metal ion.[465] The enzyme is unusually strict with respect to donor specificity, failing to recognise UDPGalNAc or dTDPGlcNAc, but the acceptor specificity and the identity of any base is unknown.

5.12.1.7 GT 42 – Sialyl Transferases. Mechanistic data are available for the soluble form of an enzyme from the human intestinal pathogen *Campylobacter jejuni*, which catalyses the transfer of a NeuNAc residue from CMPNeuNAc (Figure 3.24) to the 3-position of a β-galactosyl residue or the 6-position of another sialyl residue (Figure 5.56).[466] The soluble form of the enzyme was produced by deleting the hydrophobic portion which anchored the enzyme in the membrane, but the soluble enzyme still self-associated as a tetramer. The structure of the soluble enzyme, unliganded and in complex with CMP and with

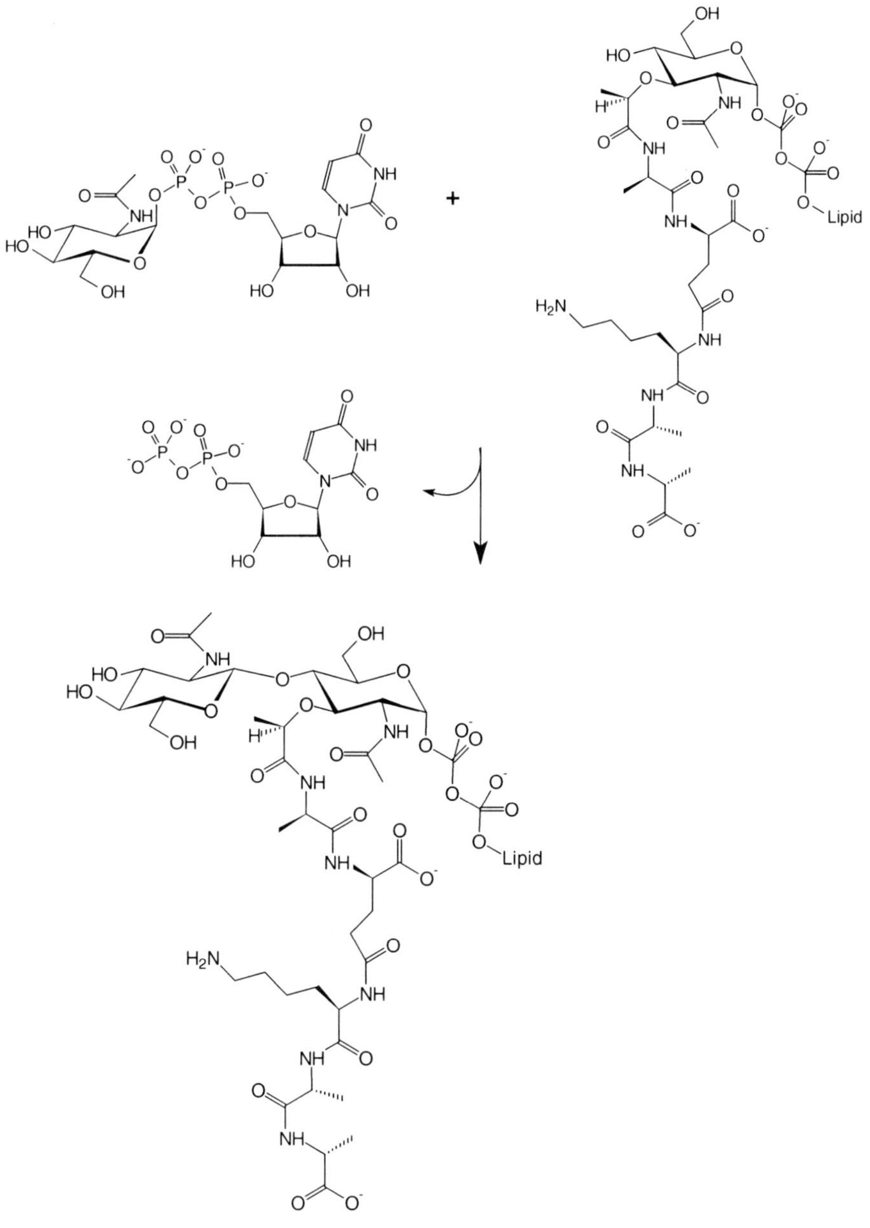

Figure 5.55 MurG enzyme.

CMP-3-(*R*)-fluoro-NeuNAc, has been solved. The overall structure of the monomer showed only one Rossman fold (nucleotide binding domain), with the rest of the molecule showing signs of disorder: it resembled GTA more than GTB, but with marked differences from the "standard" GTA fold. This is not surprising as the glycosyl donor is a monophosphate, not a diphosphate.

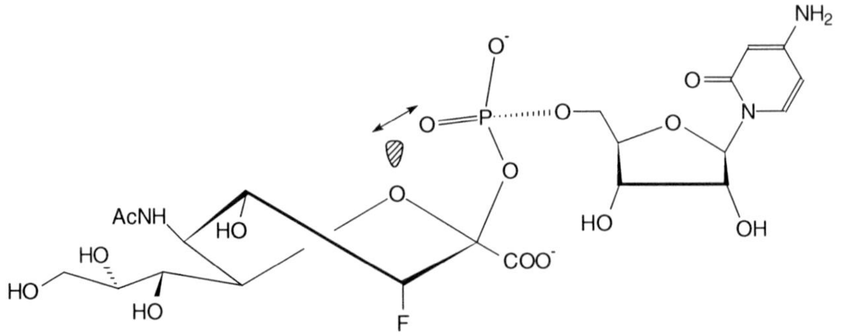

Figure 5.56 Conformation of the glycosyl-donor analogue as bound to sialyltransferase.[466]

The enzyme slowly hydrolyses CMP-NeuNAc in the absence of acceptor, so it is not surprising that attempts to obtain a crystalline CMP-NeuNAc complex failed, with only density for CMP being seen in the electron-density map. Fluorine substitution in the 3-*R* position of the sialic acid ring ($\Delta\sigma_I = 0.5$), however, produced a stable complex in which, remarkably, the pyranose ring of the transferred sialyl residue is not in the $^2C_5$ conformation which would have presented a ring oxygen lone pair antiperiplanar to the leaving group, but in a conformation on the equator of the Cremer–Pople sphere (Figure 2.6b). This conformation is described as $^0S_5$, but in the text C1, C2, C3 and O6 are said to be coplanar (which is not possible in a near-perfect skew or boat conformation), and is also said to be the same as that of NANA bound to influenza sialidase, which is $B_{2,5}$.[231] From the structural formulae in the paper, the bound structure appears to lie on the equator of the Cremer–Pople sphere between $^{3,0}B$ and $^0S_3$. In this conformation, the *proR* oxygen of the phosphate group is only 3.0 Å from the ring oxygen, apparently acting in much the same way as the electron-rich groups seen with glycosidases (Section 5.6) or in the model system of Figure 3.30.

The protein machinery that stabilises the charge on the C1 carboxylate appears very different from that in sialidases – there is not the trio of arginines to balance the C1 carboxylate negative charge, rather it is opposite the positive end of the dipole from a stretch of α helix which is directed towards this carboxylate. The only likely candidate for general base in the attack of acceptor is His188 and mutation of this residue eliminated transferase activity.

5.12.1.8 GT 43 – Glucuronyltransferases. Human β-glucuronyl 1 → 3 transferase I transfers a β-glucuronyl residue from UDPGlcA to the 3-position of the terminal galactose in the structure β-Gal*p*-(1→3)-β-Gal*p*-(1→4)-β-Xyl*p*-1→ serine. This is the reducing-end structure whereby heparan and chondroitin sulfates are attached to the "initiator" protein; subsequently βGlcA and αGlcNAc are added sequentially,[467] and the initial polymer modified as described in Section 4.6.10.1. Structures of the recombinantly expressed human enzyme, with the membrane-anchoring structures deleted, in complex with β-Gal*p*-(1→3)-β-Gal*p*-(1→4)-β-Xyl*p*, UDP and Mn^{2+},[468] established the fold as GTA, with the usual binding of Mn^{2+} chelated by α- and β-phosphate oxygens. A hydrogen bond was discovered between Glu281 and the 3-OH of the acceptor. Subsequently it proved possible to obtain crystals with intact UDPGlcA, which were flash-frozen before structure determination.[469] Superposition of the two structures indicated all the components of the mechanism of Figure 5.53, including base catalysis by Glu281 and α,β-phosphate coordination by Mn^{2+}, were operative.

The structures of human glucuronyltransferase P in complex with UDP and acceptor *N*-acetyllactosamine [β-Gal*p*-(1→4)-β-GlcNAc] have been solved.[470] The enzyme is involved in the biosynthesis of the HNK-1 epitope β-GlcA*p*3S-(1→3)-β-Gal*p*-(1→4)-β-GlcNAc, which is attached to glycoproteins and glycolipids. The glucuronic acid is transferred to an *N*-acetyllactosaminyl residue and then sulfated. Again, a base-catalytic glutamate (in this case Glu284) hydrogen bonding to the acceptor oxygen atom was clearly visible and the D184A mutant was inactive.[471] Mn^{2+} was also clearly seen.

5.12.1.9 GT 66 – Oligosaccharyl Transferase. An important modification of proteins in eukaryotes is the *N*-glycosylation of the side-chain amide of asparagines. The glycosyl donor is the dolichol pyrophosphate derivative (Figure 5.2) of a 4′-glycosylated chitobiose and the acceptor is the asparagine residue in the sequence Asp–X–Ser/Thr, where X can be any amino acid except proline. The chemical reaction itself appears to be a relatively straightforward S_N2 reaction, albeit with very large participants, in which the major point of interest is the enhancement of the nucleophilicity of the amide nitrogen.[31]

However, the spatial and temporal arrangement of the reaction is highly complex, as it forms part of the secretory pathway of proteins.[473] *N*-Glycosylation is a cotranslational process, in which the oligosaccharide is transferred to the target asparagines before the protein is properly folded. The various stages of the reaction occur close to the interfaces of a lipid bilayer and an aqueous environment, in a subcellular structure known as the rough endoplasmic reticulum. This has two sides, the "lumen", a space into which glycoproteins are secreted, and the cytosol, containing the "works" of the cell. The protein translational machinery occurs in the cytosol. Secretory proteins are encoded with a hydrophobic "signal sequence" at the N-terminus, so as they come off the ribosome, the hydrophobic signal sequence "dissolves" in the lipid bilayer and passes through into the lumen. On the luminal side (*i.e.* "outside") of the endoplasmic reticulum, once the N-terminal sequence has cleared the lipid bilayer, it is hydrolysed off by the signal peptidase complex. When enough of the protein proper, minus its signal peptide, has cleared the lipid bilayer on the luminal side of the endoplasmic reticulum, the asparagine in the recognition sequence Asn–X–Ser/Thr is *N*-glycosylated with a tetradecasaccharide. At this point the recognition sequence is 30–40 Å clear of the endoplasmic reticulum, in essentially aqueous medium in the lumen.[474] The oligosaccharyl transferase complex appears to consist of no less than nine proteins, of which STT3 falls in GT 66 and is the probable candidate for the catalytic subunit.[475] The newly synthesised glycosylated and partly folded protein is folded into the correct conformation by a endoplasmic reticulum-bound chaperonin and then exported to the Golgi apparatus, where, according to species and location, some of the existing 14 sugar residues are hydrolysed off the glycoprotein and others, notably sialic acids, are added by nucleotide diphospho sugars or CMP-NeuNAc.

The assembly of the tetradecasaccharide glycosyl donor, shown in Figure 5.57, is remarkable; much of our knowledge of its pathway has come from studies of mutations in the responsible genes in yeast and analysis of the mutant glycosylation patterns resulting, *i.e.* a combination of genetics (including classical genetics) and carbohydrate analysis.[473]

The first seven sugars are sequentially transferred to dolichol pyrophosphate from UDPGlcNAc or GDPMan on the cytosolic side of the membrane. These are labelled "C" in Figure 5.57; it is not known which of the two branches at residue C3 (β-Man) are biosynthesised first. When the heptasaccharyl unit (saccharide units labelled "C" in Figure 5.57) has been assembled on the cytosylic side of the membrane by Leloir glycosytransferases, it is "flipped" to the outside of the membrane (luminal side) by a special "flippase" protein.[476] The remaining seven sugars are added by transfer of Glc or Man from dolichol phosphate (not pyrophosphate) derivatives α-Glc-P-Dol and α-Glc-P-Man, in the order shown in Figure 5.57,[477] in which "L" refers to "lumen" and the number, the order of addition. The organism cannot use Leloir glycosylation with freely diffusible nucleotide donors on the luminal side of the membrane, since these would be lost to the cell, and has to rely on transfer from donors anchored in the membrane by dolichyl groups. Presumably a heavily hydrated heptasaccharide is the largest such species that can be dragged through the lipid bilayer with acceptable efficiency by a "flippase" protein.

It appears that activation of the asparagine nitrogen to electrophilic attack occurs via hydrogen-bonding interactions within the peptide to be glycosylated. Examination of the conformations of asparagine-containing peptides revealed that those constrained to a β-turn conformation were not glycosylated, whereas those which adopted the ASX conformation were excellent acceptors (Figure 5.58).[478] The key interaction was between the carbonyl of the asparagine side-chain and the hydroxyl group of the Ser or Thr residue. Glutamine residues are never glycosylated.[479] There are two mechanistic proposals for the nature of the nucleophile: Imperiali *et al.* prefer a neutral imidol tautomer, albeit one drawn hydrogen bonded to a deprotonated threonine or serine,[478] whereas Bause *et al.* suggest that the amide is fully deprotonated.[480] Both proposals involve a basic group on the enzyme which first deprotonates the amide NH_2 (Imperiali) or Ser/Thr OH (Bause).

N-Nucleophilicity of amides towards glycosyl centres cannot be estimated simply, unlike *O*-nucleophilicity. Protonation of amides occurs on oxygen, rather than nitrogen, and is governed by pK_a values in the range −0.5 to −1.0.[481] The low β_{nuc} values for reactions at acetal centres (0.1–0.2, Section 3.3) mean, therefore, that an amide oxygen is an only a 3–10-fold worse nucleophile towards glycosyl centres than a carboxylate of pK_a 4 and, indeed, as set out in Section 5.9.6.7, acetamido groups do act as nucleophiles in some glycosidase

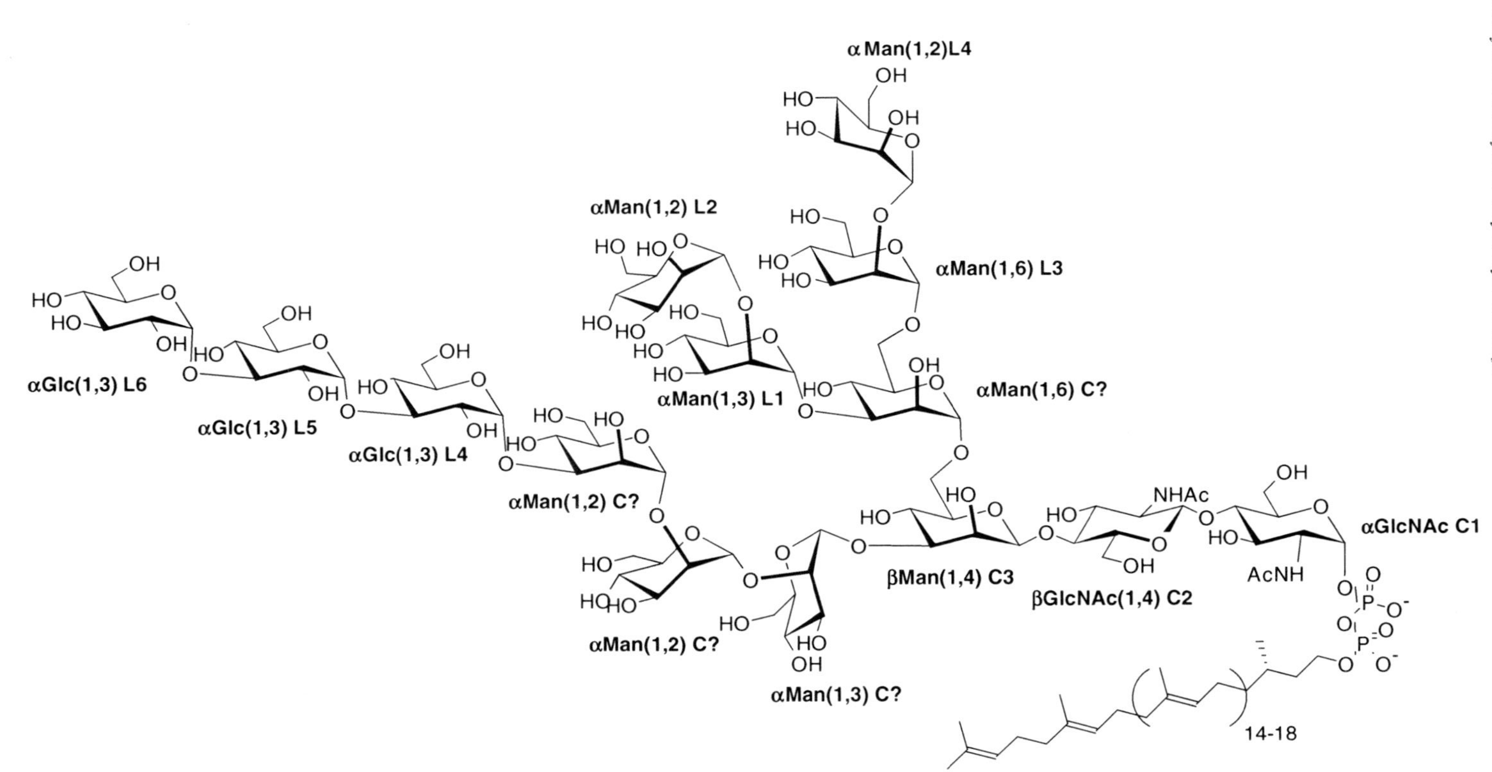

Figure 5.57 Structure of the tetradecasaccharide donor substrate of oligosaccharyl transferase.

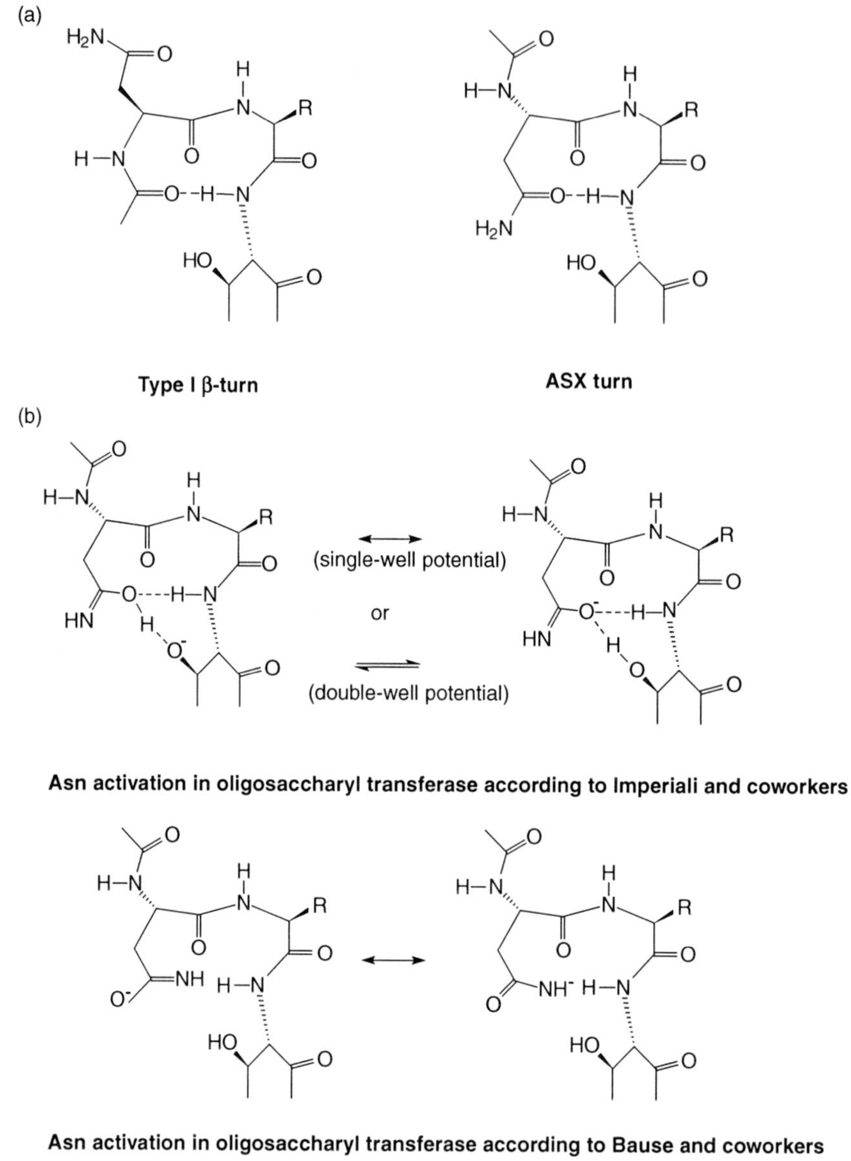

(a)

Type I β-turn

ASX turn

(b)

(single-well potential)

or

(double-well potential)

Asn activation in oligosaccharyl transferase according to Imperiali and coworkers

Asn activation in oligosaccharyl transferase according to Bause and coworkers

Figure 5.58 Conformation and activation of asparagines. (a) β-Turn and ASX turn; (b) activated forms of the amide.

classes. The *N*-nucleophilicity of the unfavourable neutral imidol tautomer of an amide is likely to be much greater than the *O*-nucleophilicity of the favourable tautomer – the pK_a of the imidol nitrogen is likely to be similar to that of an oxazoline (~ 5.5[343]) and nitrogen nucleophiles are usually more effective than oxygen nucleophiles of comparable pK_a. It seems unlikely, therefore, that the active nucleophile has a fully deprotonated amide. Once

Natural Selection had produced a system which stabilised the imidol tautomer of the asparagine $CONH_2$, selection pressure for a more efficient system would likely be diverted elsewhere. Indeed, oligosaccharyl transferase will accept a range of *N*-nucleophilic side-chains in an appropriate peptide, apart from $-CH_2CONH_2$ ($-CH_2CONHNH_2$, $-CH_2ONH_2$, $-CH_2CH_2NH_2$), but pronounced product inhibition is observed, suggesting that product release, rather than any chemical step, can be limiting and that selection pressure would be applied to this step, rather than the chemistry.[482]

The invocation of general base catalysts on the enzyme to transfer protons from amide or serine/threonine hydroxyl seems unnecessary: kinetic barriers to proton transfer from electronegative atoms are small, the difficulty in creating a reasonable *N*-nucleophile from a primary amide lying largely in the unfavourable tautomeric equilibrium.

5.12.2 Retaining Glycosyltransferases

Whereas all enzymic glycosyl transfers with inversion, whether hydrolysis or synthesis, involve a single displacement, such mechanistic uniformity is not apparent for glycosyl transfer with retention. Covalent glycosyl-enzyme intermediates are observed with retaining glycosidases and transglycosidases, but an examination of structures of glycosyl transferases with the GTA fold and of which crystal structures are available with substrate-analogue bound, has revealed that none of them have a potential nucleophile on the β face of the bound sugar.[445] Kinetics, where they have been examined, are ternary complex, like inverting glycosyltransferases, rather than ping-pong, as with retaining glycosylhydrolases and the occasional phosphorylase, such as sucrose phosphorylase,[483] whose sequence places it in a GH, rather than GT, family (in this case, GH 13).

Structures of glycosyl transferases are only now becoming available and kinetic data are sparse. Glycogen phosphorylase (Figure 5.59), which acts in a classical *exo*-fashion on glycogen and starch, producing glucose-1-phosphate, is placed by its sequence in GT 35 (rather than a GH family) and by its structure as GTB; this is fortunate, as much early mechanistic work has been done on this enzyme, now known to be an honorary glycosyl transferase. It is a richly allosteric enzyme, with heterotrophic and homotrophic interactions and covalent modification (phosphorylation of a serine) all involved in conformational changes. When the presumed catalytically active R conformation (R for "relaxed" as distinct from T for "tense"), was eventually observed, in complex with phosphate and the nojirimycin-based tetrazole of Figure 5.11, the only potential nucleophile was a peptide amide carbonyl.[484] Amide carbonyls, as discussed in Section 5.12.1.9, would be only slightly less nucleophilic than carboxylates in glycosyl transfers with the usual "exploded" transition states: by the same token, in the second step, characterised by β_{lg} values close to -1, a protonated imidate would be a very much better leaving group than a carboxylate. If amides are nucleophiles, then Withers inactivators would not be able to trap them because of deglycosylation rates intrinsically 4–5 orders of

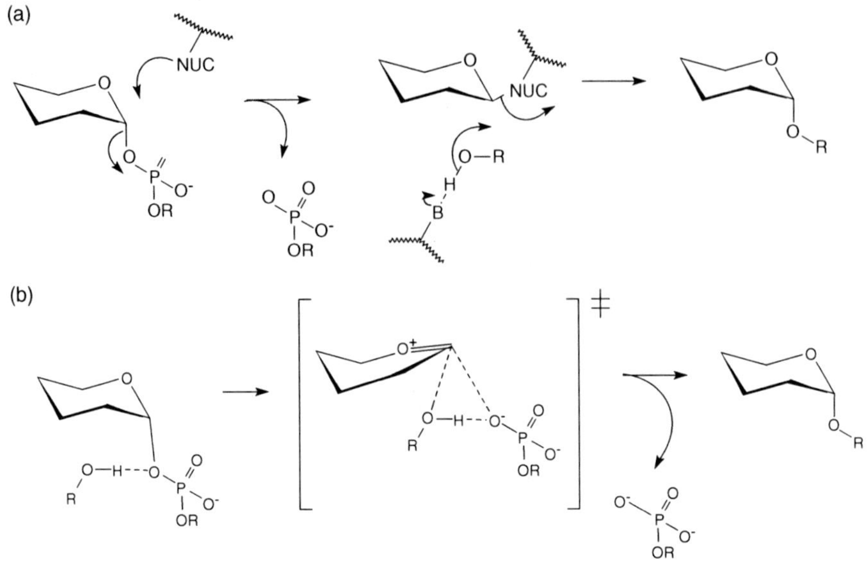

Figure 5.59 Alternative mechanisms for retaining glycosyl transferases and similar enzymes such as glycogen phosphorylase. (a) Classical double displacement. (b) S_Ni. The incoming nucleophile is shown hydrogen bonding to the leaving oxygen, to make an exact parallel with Figure 3.37, but hydrogen bonding to a non-bridge oxygen, and concomitant larger motion of the leaving group/nucleophile hydrogen-bonded pair, remain a possibility.

magnitude faster than deglycosylation of the glycosylated carboxylate nucleophiles of glycosidases. "Internal return" mechanisms for retaining glycosyl transferases are currently favoured, but are by no means proved, and each GT class where there is mechanistic evidence will be examined in turn.

An important tool for testing for any reversible cleavage of an X–OPO(OH)R bond is positional isotope exchange (Figure 5.60). If the substrate is synthesised with the bridge oxygen labelled, as in X–^{18}OPO(OH)R, then once the X–O bond has been cleaved, the $^{-}$OPO(OH) group becomes torsionally symmetrical by virtue of rapid proton transfer and rapid rotation about the R–P bond. Recombination of the X and phosphonyl fragments will thus give the label one-third in the original bridge position and two-thirds in the non-bridge phosphonyl oxygens (neglecting isotope effects, expected to be very small). The distribution of label can be determined very simply by ^{31}P or proton-decoupled ^{13}C NMR spectroscopy. There exists an isotope effect on chemical shifts, with the heavier isotope always giving rise to slightly higher chemical shifts. The heavier isotope has a lower zero point energy, and hence lower amplitudes of the affected vibrations, than the lighter isotope and thus the electrons in the bond are spread over a slightly smaller volume, thereby slightly increasing electron density at the nuclei (*cf.* Figure 3.22). This slight increase in electron density results in slightly more shielding of the nuclei

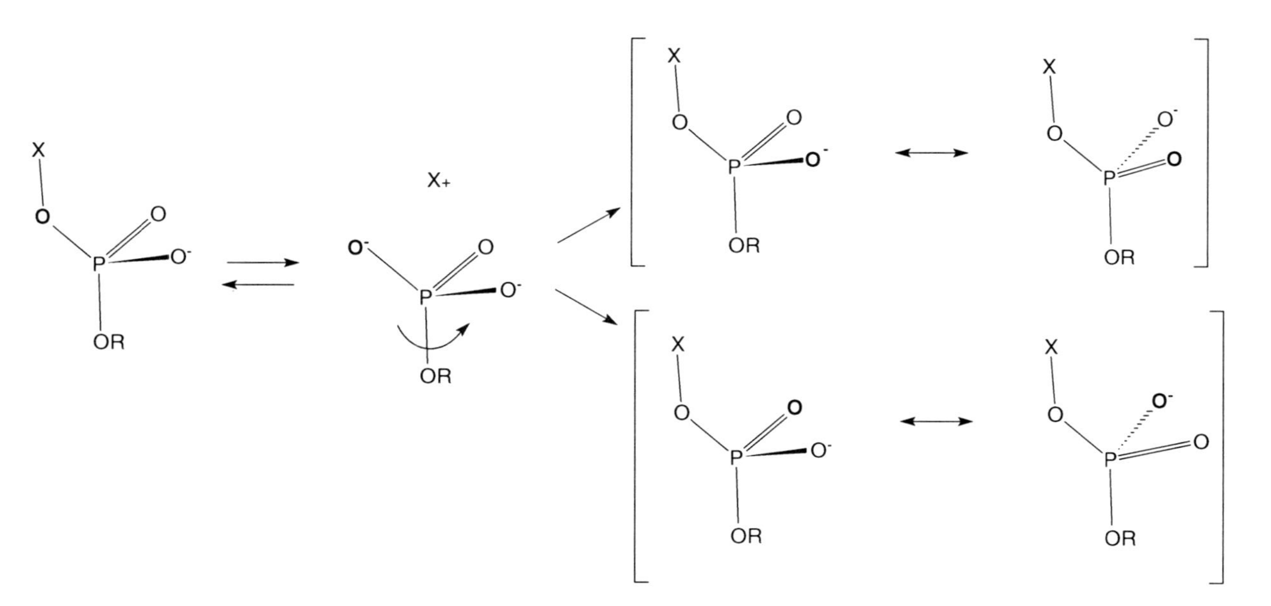

Figure 5.60 Positional isotope exchange. The non-bridge oxygens of a phosphate are prochiral (enantiotopic if X and R are achiral, diastereotopic if they are not).The occurrence of positional exchange can be readily followed by ^{31}P NMR, with the ^{31}P signals increasing in shielding in the order $^{16}O–^{31}P < ^{18}O–^{31}P$ and the isotope shift increasing with bond order.

involved in the vibration: the heavy isotopomer is always the more shielded. Moreover, the amount of this shielding depends, in the first instance, on the vibration frequencies of the bond, so that multiple bonds, with their higher frequencies, give larger isotope shifts. In the present case, ^{18}O in the phosphoryl oxygens (bond order 3/2) gives a larger shift than in the bridge oxygens.

5.12.2.1 GT 5. Mammalian and yeast starch synthetases use UDPGlc as glycosyl donor and occur in GT3, whereas the bacterial and plant types use ADPGlc and occur in GT5. The structure of the *Agrobacterium tumifaciens* starch synthetase in complex with ADP has been solved:[485] it has the GTB fold and, like many transferases with this fold, does not require metals. An essential glutamate was located by site-directed mutagenesis,[486] but its molecular role is uncertain. There is a close similarity between the GT35 (glycogen phosphorylase) and GT5 starch synthetase structures, with the adenosine of the starch synthetase occupying a similar position to the pyridoxal phosphate of glycogen phosphorylase.[485]

5.12.2.2 GT 6. These mammalian enzymes are involved in transfer of α-Gal and α-GalNAc from their UDP derivatives to various antigens. Crystal structures of the human α-Gal and α-GalNAc transferases reveal them to have a GTA fold and the specificity difference between α-Gal transfer and α-GalNAc transfer to arise from the single amino acid making contact with the 2 and 3 substituents of the transferred sugar, Leu266 in the case of the α-GalNAc transferase and Met266 in the case of the α-Gal transferase (there are only four amino acid differences between the enzymes anyway).[487] This difference nonetheless determines whether blood expresses the A antigen or the B antigen. Both enzymes transfer a sugar to the fucose 3OH in the structure β-L-Fuc*p*-(1→2)-β-D-Gal*p*-OR, but their mode of inhibition by structures in which this 3-OH had been replaced by hydrogen or an amino group was different and the inhibitors bound differently.[488] This different behaviour fits the emerging picture of the acceptor site in GTA enzymes being "floppy" and formed after the donor binds. The complexes with UDP and saccharide acceptor show a tightly-bound Mn^{2+}, coordinated in the usual way between the two phosphates. They also suggested that Glu317 might be a catalytic nucleophile, but subsequent site-directed mutagenesis experiments[489] in which the residue in the galactosyltransferase was mutated to Gln revealed that the effect was too small (2400-fold fall in k_{cat} for transfer to lactose, 120-fold fall in k_{cat} for hydrolysis) for a supposedly crucial residue. Importantly, the effects on hydrolysis and transfer were different, indicating that the low levels of activity did not arise from wt contamination. The kinetics of the soluble truncated enzyme were the expected ordered sequential (ternary complex), with UDPGal binding first and UDP being the last to be released.

An enzyme which transfers the α-Gal residue to *N*-acetyllactosamine structures in the glycoproteins of mammals other than humans has attracted attention because the α-Gal*p*-(1→3)-Gal epitope is a major obstacle to the use of organs from other mammals in xenotransplantation: because of the α-Gal*p*-(1→3)-Gal structure on protein and lipid surfaces, the organs are recognised as "foreign" and excite an immune reaction.

The structure of the bovine α-Gal transferase has been solved,[490] and relatively high-resolution structures of complexes[491] with UDP, GalOH and the acceptors β-Gal*p*-(1→4)-Gal*p*-OH and β-Gal*p*-(1→4)-GalNAc*p*-OH obtained (the structures with UDP and GalOH bound resulting from attempts to co-crystallise UDPGal and enzyme). The structures revealed how the acceptor binding site was constructed by conformational changes after the binding of UDPGal, and were augmented by isothermal titration calorimetry measurements. Addition of lactose or *N*-acetyllactosamine to enzyme caused no heat change, but the addition of UDPGal was exothermic (and entropy disfavoured). Exothermic binding of the acceptors could only be detected once the enzyme had been saturated with UDP. These measurements nicely confirm the structural suggestions that the acceptor binding site is only constructed after nucleotide diphosphate binding to an essentially "floppy" protein.

The suggested nucleophile (likewise Glu 317) in the bovine enzyme has been mutated to Ala.

The mutant enzyme has no transferase activity (although it has hydrolase activity) and can be catalytically rescued by azide ion; the product from UDPGal is β-galactopyranosyl azide. This incisive experiment is powerful evidence for the double displacement, rather than S_Ni, mechanism for this family.[497]

5.12.2.3 GT 8. LgtC of the pathogen *Neisseria meningitidis* is an αGal transferase which transfers an αGal residue from UDPGal to the 4-OH of a terminal lactosyl residue on the lipooligosaccharide of the bacterial capsule. The resulting structure thus mimics the sugar moieties of human glycolipids and is thought to be part of the "camouflage" whereby *N. meiningiditis* and related organisms such as *N. gonorrhoea* partly evade the human immune system.[492] The enzyme has been subjected to intensive mechanistic investigation, largely in unsuccessful attempts to trap and identify the supposed enzyme nucleophile (Figure 5.61),[493] and its structure in complex with donor and acceptor analogues has been reported.[494] It has a GTA fold.

K_m values for acceptors do not decrease as sugar residues are added beyond lactose, so presumably the acceptor site only extends to a disaccharide. The ability to use lactose as an acceptor made possible a detailed study of the kinetics,[493] and this was further assisted by the use of UDP-2-deoxy-2-fluoro-Gal. Originally synthesised as a potential Withers inactivator, it proved to be a reversible inhibitor strictly competitive with respect to UDPGal. The plots of $1/v$ *versus* 1/[UDPGal] at various concentrations of lactose were parallel, suggesting a ping-pong mechanism, but also compatible with a ternary complex mechanism if the double reciprocal plots met at values of $-1/v$ and −1/[UDPGal] much more negative than the positive range of the experimental points. The latter was shown to be the true situation by adding UDP-2-deoxy-2-fluoro-Gal, effectively increasing the K_m of UDPGal, when intersecting double reciprocal plots were obtained. A ternary complex mechanism was therefore followed. That this was

UDP-2-deoxy-2-fluoro-Gal
inhibitor, competitive
with UDP-Gal

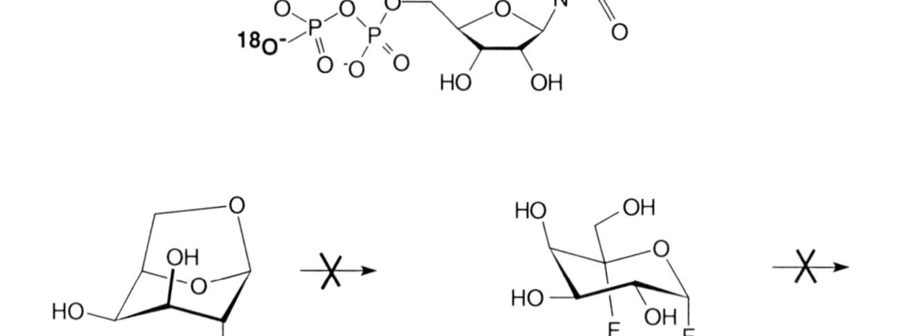

Figure 5.61 Attempts to identify a nucleophile in LgtC.

not rapid equilibrium random was shown by the anticompetitive inhibition exhibited by the product, α-Gal*p*-(1→4)-β-Gal*p*-(1→4)-GlcOH, against UDPGal.

Although the enzyme accepted α-galactosyl fluoride as a glycosyl donor, it did this only in the presence of UDP: UDPGal was first synthesised. Galactosyl fluoride could act both as an acceptor and an precursor of UDPGal, so that if the reaction were followed by liberation of F⁻, it was apparently greater than first order in [GalF], as α-Gal*p*-(1 → 4)-α-GalF was synthesised.[495]

UDP-2-deoxy-2-fluoro-Gal, an incompetent donor, was crystallised with the enzyme in the presence and absence of 4-deoxylactose, an incompetent acceptor, to enable a detailed picture of the reactive conformation to be obtained. The only potentially nucleophilic groups within 5 Å of the reaction centre are the oxygens of Gln189 and the Gal6OH of the lactose acceptor.[494] A double displacement, first by the galactose 6-OH of the lactose acceptor, to generate β-Gal*p*-(1→6)-β-Gal*p*-(1→4)-GlcOH, and then by the galactose 4-OH, to generate the product α-Gal*p*-(1→4)-β-Gal*p*-(1→4)-GlcOH, was considered, despite the 6 → 4 migration involved being a 6-*endo-tet* cyclisation and hence in gross violation of Baldwin's rules.[496] The 6-OH as potential nucleophile was rejected, however, on the experimental grounds that 6′-deoxylactose was not an inhibitor. Any role of the glycosyl donor 6OH as nucleophile had to be abandoned as 1,6-anhydrogalactose was not a substrate.

In the presence of the incompetent acceptor 4′-deoxylactose, there is no positional isotope exchange with UDPGal, nor is there any sign of enzyme inactivation by 5-fluoro-α-galactosyl fluoride, even at high concentrations and in the presence of UDP.

In a final attempt to generate a covalent glycosyl-enzyme intermediate, Gln189 was mutated to Glu, in the hope that the much worse leaving group of any glycosyl-enzyme with this mutation would permit its isolation.[498] In fact, a galactosyl-enzyme intermediate was formed and turned over, but the sugar residue was on Asp190, not Glu189. The crystal structure of the Q189E mutant revealed an active site configuration essentially identical with that of wt enzyme. However, the effects of mutations on k_{cat} for transfer to saturating concentrations of lactose and for hydrolysis were relatively modest – a factor of 8000 for transfer but only 2 for hydrolysis for D190N, a factor of 60 for transfer for Q189A and a factor of 35 for transfer and 7 for hydrolysis for Q189E. These data are consistent with a two-component nucleophile (Figure 5.62), with Asp120 acting as a general base and the oxygen of Gln189 acting as the

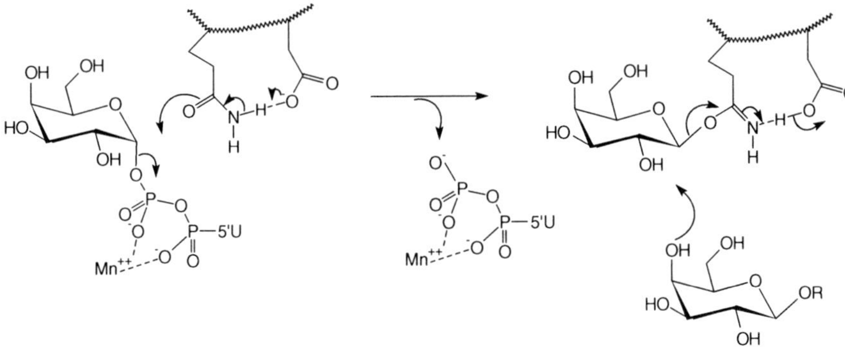

Figure 5.62 Possible nucleophile in LgtC.

nucleophile, in a fashion exactly analogous to the use of intramolecular amide nucleophiles in GH 18, 20 and similar families (Section 5.9.6.7).

Glycogenin, the autocatalytic initiator of starch biosynthesis, catalyses the initial glycosylation of one of its own tyrosine residues, followed by the first few α-glucosyl residues of the starch molecule; the stereochemistry of the transfer to tyrosine is unknown, but that to the 4-OH of glucose residues is retentive. Again, the X-ray crystal structure revealed no likely nucleophile except Asp102, which would require a conformation change to be correctly placed.[499]

5.12.2.4 GT 15. This family contains fungal and yeast enzymes, but no mammalian enzymes. The structures of the α-(1→2)-mannosyltransferase from yeast (*S. cerevisiae*), unliganded and in binary and ternary complexes with GDP and methyl α-mannopyranoside acceptor.[500] The enzyme transfers the mannosyl residues labelled "α-Man(1,2) C?" in Figure 5.57 during the biosynthesis of *N*-linked oligosaccharides and is also involved in *O*-glycoprotein biosynthesis. The fold is GTA, with the Mn^{2+} bridging the α- and β-phosphates in the usual way.

Incisive site-directed mutagenesis experiments demonstrated that the role of two active site glutamates (279 and 247) lay in the provision of the correct active site electrostatics, rather than covalent intervention: their mutation to glutamine abolished activity, whereas their mutation to aspartate reduced k_{cat} by factors of only 3 and 1.3, respectively. The effects of mutation of Tyr220 were more problematic: the Y220F mutant, lacking just the phenolic OH, had a 3000-fold reduced k_{cat}. This comparatively large effect was consistent with Tyr220 being the elusive nucleophile, which the crystal structure permitted. However, transferred mannose could not be observed in complexes obtained with GDPMan, presumably because of the usual slow hydrolytic activity. Nonetheless, in the ternary complex of GDP and acceptor methyl α-mannoside, the 2-OH of the glycosyl donor formed a hydrogen bond with both the β-phosphate and the OH of Tyr220. To the author, it seems that Tyr220 forms part of the vital proton-transfer network for an S_Ni mechanism, rather than acts a nucleophile. There is no chemical logic in an organism going to the metabolic expense of constructing an elaborate, activated glycosyl donor (effective leaving group pK_a 4–7, depending on metal coordination), only to generate a standard aryl glycoside (leaving group pK_a 10) as an enzyme intermediate.

5.12.2.5 GT 20. One representative of this class of transferase, which adopts a GTB structure, has been investigated, trehalose 6-phosphate synthetase, which transfers an αGlc residue to the 1-OH of glucose-6-phosphate. Crystal structures of the native enzyme, a complex of UDP and α-glucose-6-phosphate,[501] and a complex with the inert donor analogue, UDP-2-fluoro-2-deoxyglucopyranose,[502] have been solved. Trehalose [α-Glc*p*-(1↔1)-α-Glc*p*] is an important sugar because it is the reserve carbohydrate of insects, and also because of its unique hydrogen-bonding and solvating properties, which are exploited by many organisms which have to cope with intense dryness. Current thinking suggests that at high concentrations the involatile trehalose can mimic the hydrogen-bonding patterns of water and thus preserve proteins and lipid membranes in their native, functional state when as much as 99% of their water has been removed: trehalose can then be 20% of the weight of the organism.[503] Complexes with the various ligands illustrate the capability of the GTB fold to adopt several conformations by rotation of the otherwise fairly rigid Rossmann domains with respect to each other. Notably, glucose-6-phosphate binding, exclusively as the α-anomer, causes a significant relative shift of the two domains and in the ternary complex with UDP binds with its 1-OH binding to the β-phosphate of UDP, in precisely the manner to suggest an S_Ni reaction.

5.12.2.6 GT 27. The one enzyme in this family with a solved X-ray structure[504] is a mouse UDPGalNAc-dependent *N*-acetylgalactosamine transferase which initiates mucin biosynthesis by transferring α-GalNAc to the hydroxyl of serine or threonine of what becomes the peptide backbone of an *O*-linked glycoprotein. The enzyme has a C-terminal lectin (carbohydrate-binding) domain and a GTA glycosyltransferase domain: bound UDP coordinates Mn^{2+} between the α- and β-phosphates in the usual way. The catalytic domain resembles that of LgtC (GT 8), including an amide–carboxylate pair (Glu319/Asn320) which might act as an enzyme nucleophile in the manner of Figure 5.62. Mutagenesis of Glu319 (E319Q, E319A) results in complete (>99.96%) loss of activity and acid/amide changes at other carboxylates (D156, D209) which presumably contribute to electrostatic transition-state stabilisation, have a similarly drastic effect.[505]

5.12.2.7 GT 35 – Glycogen Phosphorylase. GT 35 contains a number of enzymes which reversibly phosphorolyse α-(1→4) glucan links and which play a key role in energy storage and mobilisation in mammals, fungi and bacteria. The glucose-1-phosphate product is converted to glucose-6-phosphate and then fed into the glycolysis pathway. Investigation of mammalian muscle glycogen phosphorylase has a very long history, the rabbit enzyme having been discovered (and crystallised)[506] in 1942 by Carl and Gerty Cori (Nobel Prize, 1947), who were looking for the enzyme that interconverted the "Cori ester" (α-glucose-1-phosphate) and inorganic phosphate and glycogen.[507]

The mammalian (largely rabbit) muscle enzyme has been the most investigated;[508] X-ray structural work favours a GTB fold. It has control, covalent and allosteric, at several levels and these are usually discussed in terms of a modification of the Monod–Wyman–Changeux model,[509] in which the individual polypeptide monomers can adopt either a non-catalytic T ("tense") state or a catalytic R ("relaxed") state, but mixed oligomers (*e.g.* TR in a dimer) do not occur. In the extensive structural studies, oligomers with mixed conformations have never been observed.

One way of converting an inactive enzyme with T subunits to an active one with R subunits is covalent phosphorylation of Ser14: under physiological conditions this is performed by glycogen phosphorylase kinase and ATP. The negative charge of the phosphate group neutralises the positive charges of the highly basic N-terminal sequence, which in the unphosphorylated enzyme is disordered, so that it curls up into a helix and this drives the T → R transition in the rest of the molecule. The covalently phosphorylated enzyme is known as phosphorylase a and is catalytically active in the presence of substrate alone.

Phosphorylase b, without Ser14 covalently phosphorylated, can be made catalytically active by the binding of AMP, an allosteric activator. Its binding pocket arises from contact with several different regions of the monomer and from across the monomer–monomer interface. However, even in the presence of AMP, catalytic activities are lower than phosphorylase a.

The monomer has a M_r of 97 kDa and it is generally thought that the active form of both phosphorylase a and b is the dimer. In the case of phosphorylase b, in the absence of allosteric effectors the equilibrium lies towards the dimer at accessible protein concentrations. Phosphorylation promotes association to the tetramer by generating surface which becomes the interface of the dimer of dimers.[510] The tetramer of phosphorylase a (and presumably phosphorylase b) is inactive and ligands have only modest effects on the association–dissociation equilibria.[511]

The enzyme is also subject to allosteric inhibition. When glucose binds at the active site, it stabiles the T-state conformation of the enzyme. The T state is also stabilised by bi- or tricyclic aromatic compounds such as caffeine or flavins, which bind at the entrance to the active site tunnel,[512] and by acylated β-glucopyranosylamine derivatives, which bind similarly to glucose, but more tightly.[513] A third allosteric site, formed at the interface of two subunits and normally an internal "pool of water molecules", has recently been discovered in rabbit muscle phosphorylase b[514] and human liver phosphorylase a.[515] Occupancy of this site freezes the enzyme in the T state and inhibitors with $\sim 10^{-8}$ M dissociation constants from the site are being investigated in the treatment of diabetes.

The equilibrium constant, defined as [αGlcP][Glc_n]/[Glc_{n+1}][P_i], where P_i is inorganic phosphate, is 0.29 ($\Delta G^\circ = +3.1\ \mathrm{kJ\ mol^{-1}}$), but under physiological conditions the chemical flux is in the thermodynamically disfavoured direction, with the α-glucopyranosyl phosphate being mopped up by phosphoglucomutase and the enzymes of glycolysis. The Coris originally thought that glycogen phosphorylase was also responsible for the physiological flux to glycogen, but glycogen synthesis is now known to be carried out by Leloir enzymes.

In addition to a covalent activation site, a non-covalent activation site and two allosteric inhibition sites, glycogen phosphorylase also possesses a polysaccharide storage site, distinct from the catalytic site,[516] and occupancy of this site is thought to promote the heterotropic interactions with substrates. It is not, therefore, a carbohydrate binding module. In the direction of glycogen synthesis, the kinetics of rabbit muscle phosphorylase a are hyperbolic with respect to glucose-1-phosphate,[516] but those of rabbit muscle phosphorylase b under presumably saturating concentrations of glycogen (0.1%) are sigmoidal at all concentrations of glucose-1-phosphate and AMP activator, with values of n_H around 1.5 (1.4–1.7) for both substrate and activator.[517] In the direction of glycogen phosphorylolysis, inorganic phosphate binds both in the catalytic site and in the AMP activator site, so sigmoidal kinetics are observed.[518]

The kinetics of both the forward and the reverse reactions are rapid equilibrium random. k_{cat}/K_m values for a series of deoxy and fluorodeoxy analogues of glucose-1-phosphate were transformed by rabbit muscle phosphorylase b[519] 10^2–10^5 times more slowly than the natural substrate: bond cleavage was therefore almost certainly rate determining. LogV_{max} correlated with the logarithmic rates for spontaneous hydrolysis with a correlation coefficient of 0.9 and a gradient of 0.8, indicating an oxocarbenium ion-like transition state with 80% of the charge development of a spontaneous hydrolysis of an α-glucosyl derivative with an anionic leaving group.[520] However, careful attempts to detect an α-tritium kinetic isotope effect on the reaction with the natural substrates, in either the phosphorolysis or synthesis direction, failed to do so.[521] It is possible that some sort of transfer between the polysaccharide storage site and the catalytic site becomes rate determining with natural substrates.

Glycogen phosphorylase contains an active site pyridoxal linked to the protein via a Schiff base to Lys680 (in the usual way of pyridoxal phosphate-dependent enzymes). If the enzyme has evolved from a transferase which used a nucleotide diphospho sugar, it seems possible that pyridoxal phosphate was a molecule which was recruited to fill a "hole" in the protein when a nucleotide was no longer required, and also to provide the electrostatic field of the second phosphate of a nucleotide diphospho sugar: certainly the similar disposition of the pyridoxal in glycogen phosphorylase and a nucleotide diphosphate in starch synthetase (GT 5, Section 5.12.2.1) suggest this, as does the ability of the enzyme reconstituted with "pyridoxaldiphosphoglucose" (Figure 5.63) to transfer an α-glucose residue to the non-reducing end of a glycogen chain.[522] Nonetheless, its catalytic role has been vigorously debated, with one school of thought suggesting it acted as an acid catalyst for phosphate departure.[508a] This seemed intrinsically unlikely, as the leaving group is already a good one. It was finally disproved by two pH-dependence studies. In the first, apo-phosphorylase b reconstituted with pyridoxal and phosphite (HPO_2OH^-, pK_a 6.6) and fluorophosphate (FPO_2OH^-, pK_a 4.8) showed essentially the same dependence of the synthesis reaction on pH as the holoenzyme.[523] In the second, apoenzyme was reconstituted with isosteric analogues of pyridoxal phosphate shown in Figure 5.63 and again the same pH dependences were seen, despite the major changes in electron demand at the phosphate.[524]

Very important information has come from the study of the reaction of glycogen phosphorylase with enol ethers. The enzyme from various sources carries out a *cis* addition of phosphate or arsenate to glucal, with the proton being added to the α-face and the 2-deoxyglucopyranosyl phosphate then slowly reacting to give 2-deoxyglucans (2-deoxyglucopyranosyl arsenate hydrolyses spontaneously by As–O fission).[525] "Heptenitol" (Figure 5.64), by contrast, gives a stable phosphate ester: presumably the methyl group sterically hinders subsequent glycosyl transfers. In an ingenious series of experiments, the reaction has been monitored in crystals of phosphorylase b. Early experiments

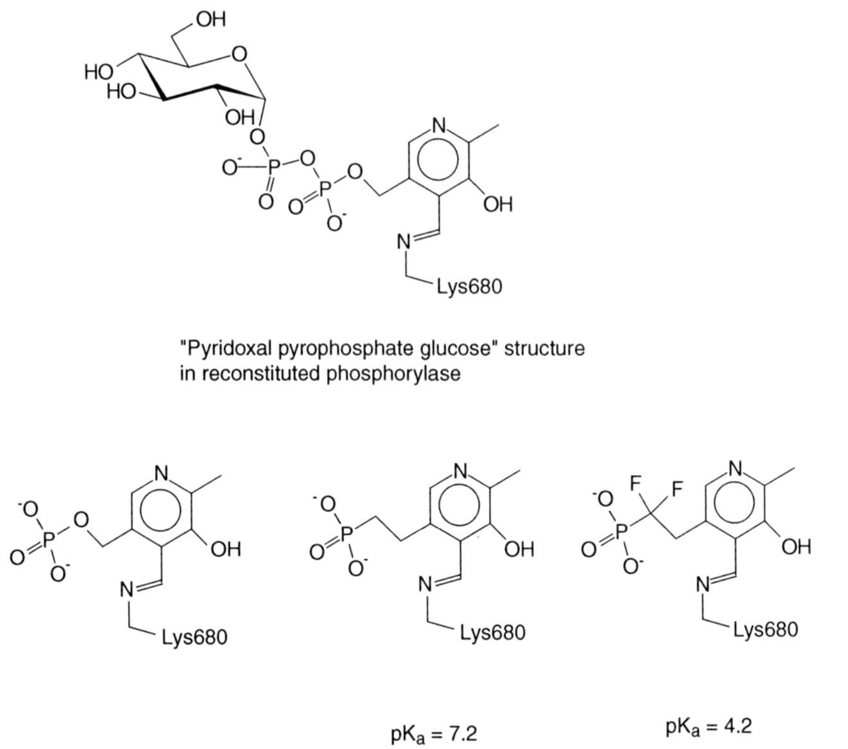

"Pyridoxal pyrophosphate glucose" structure in reconstituted phosphorylase

$pK_a = 7.2$ $pK_a = 4.2$

Figure 5.63 Structures of ligands used to probe the role of pyridoxal phosphate in rabbit muscle glycogen phosphorylase.

used monochromatic X-rays from a synchroton,[526] but these were refined by the use of Laue techniques (see Section 4.2.3),[527] which enabled the first X-ray structures to be obtained within 3 min after the initiation of the reaction (by flash photolysis of crystals soaked in 3,5-dinitrophenyl phosphate, which nearly instantaneously generated high concentrations inorganic phosphate). The 3 min structure showed "heptenitol" with phosphate poised to attack the reaction centre, with little evidence of reaction. In strong evidence for the S_Ni mechanism, electron density from the two reactants slowly changed to electron density from the product, with no suggestion of a covalent intermediate.

In its bound state, the 1-deoxyketoheptosyl phosphate adopts a conformation with $\varphi =$ "224°" (i.e. $-136°$), though, as distinct from $\varphi = 88°$ in the crystal of the substance itself (Figure 5.65).[528] This permits a hydrogen bond to the 2-OH and explains why the 1,2-cyclic phosphate of glucose, in which the relative positions of phosphate and sugar are similar, is a powerful competitive inhibitor.[529]

The conformations in Figure 5.65 again suggest an S_Ni mechanism. Positional isotope exchange in apparently unreacted glucosyl phosphate would not be expected for this mechanism, unless for some reason the glycosylation of the acceptor was reversible in the ternary complex. The evidence from potato starch phosphorylase[530] favours a double displacement mechanism: there is no

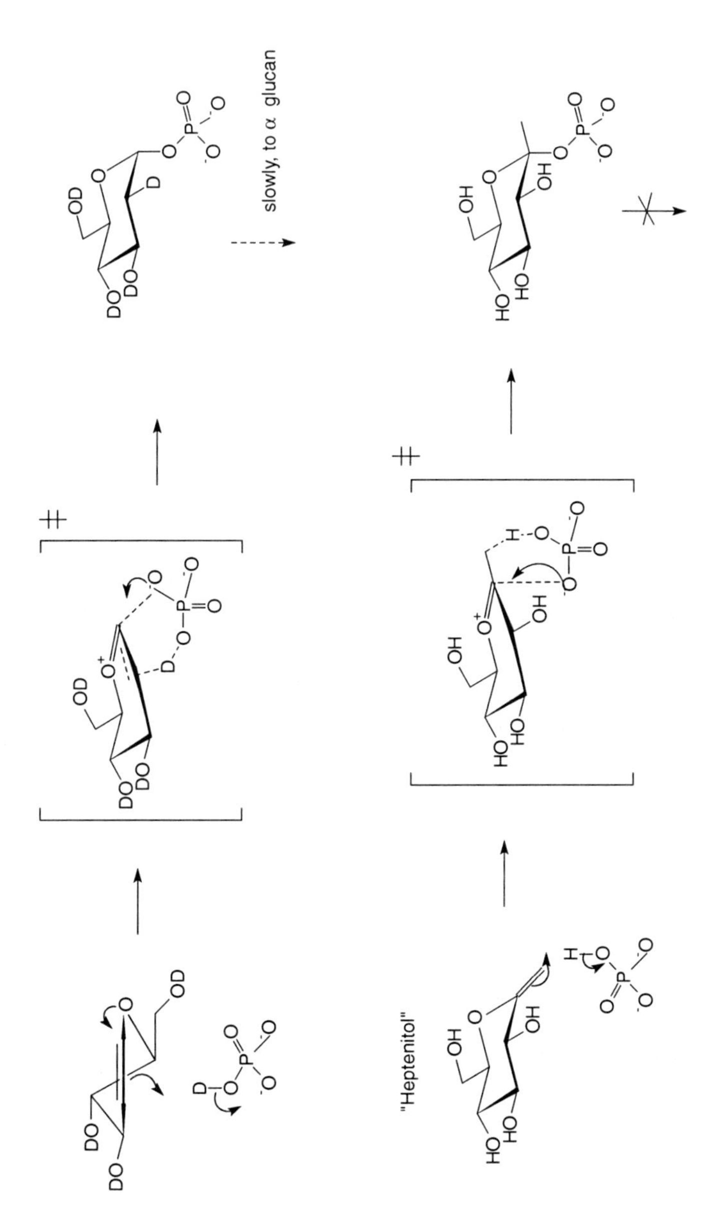

Figure 5.64 Reaction of glucose-derived enol ethers with glycogen phosphorylase.

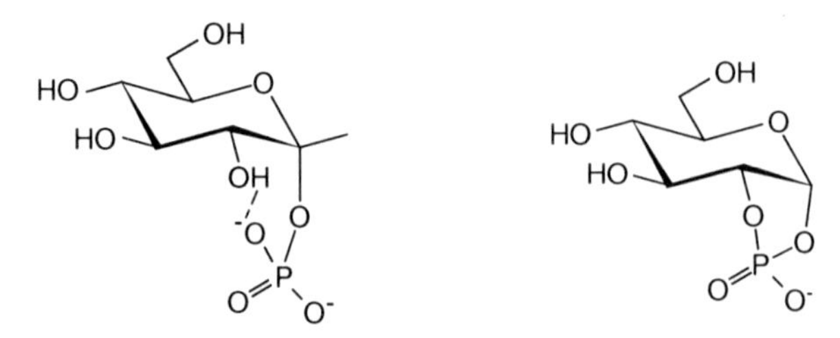

Figure 5.65 Conformations of phosphorylase-bound 1-deoxyketoheptosyl phosphate and 1,2-phosphoglucose.

positional isotope exchange when starch is used as a primer, but extensive scrambling when maltotriose or cyclodextrins are used. This is what would be expected for a glycosyl-enzyme intermediate, with reversibility of glycosyl-pyrophosphate cleavage in the presence of poor acceptors and irreversibility in the presence of good ones. As discussed above, the glucose tetrazole in Figure 5.11 and inorganic phosphate freeze the tertiary complex in the R state and in this state a main-chain peptide carbonyl is placed to be a nucleophile.[484]

Other GT 35 enzymes with less kinetic complexity and structural plasticity have been more recently studied. The starch phosphorylase from *Corynebacterium callunae* exhibits hyperbolic kinetics with respect to glucose-1-phosphate in the synthesis direction and inorganic phosphate in the phosphorolysis direction.[531] Analysis of interactions in crystal structures of the very similar *E. coli* maltodextrin phosphorylase suggested that the His residue whose peptide carbonyl might provide the nucleophile, made interactions though its imidazole group with the 6-OH of the transferred glucose. Mutation of this group indeed dramatically lowered the specificity between glucosyl phosphate and xylosyl phosphate. The other important finding arising from this work was the observation of a linear free energy relationship for wild-type and mutants between gluconolactone binding and k_{cat}/K_m for glucose-1-phosphate, establishing the lactone was a transition state analogue. It was a slow, tight binder, with a "fast" K_i of 0.26 mM tightening to a "slow" K_i of 15 μM. The gradient of the $\log K_i$ *versus* $\log(k_{cat}/K_m)$ plot was ~0.8, suggesting, remarkably, that the lactone had 80% of the character of the transition state.

The bacterial enzyme that performs a physiological function analogous to mammalian glycogen phosphorylase is maltodextrin phosphorylase and a fair amount is known about the enzyme from *E. coli* (MalP gene-product). The dimeric enzyme is 46% identical in amino acid sequence with rabbit glycogen phosphorylase and likewise contains an essential pyridoxal phosphate moiety. Regulation, however, occurs at the level of expression, so the enzyme is non-allosteric and is not activated by covalency changes. X-ray structures of the native enzyme and in complex with maltose and sulfate anion,[532] malt-ooligosaccharides and thiooligosaccharides[533] and acarbose (Figure 5.9)[534] are available, in addition to sequential structures showing catalysis in the crystal.[535] Acarbose binds +1 to +4, *i.e.* not across the site of glycosidic cleavage, as it would do in binding to a glycosidase, and binds weakly (K_i > 5 mM), all in all suggesting a very different mechanism to a glycosidase. Maltopentaose binds −1 to +4, as does its analogue with a non-fissile thioglycoside link, 1-thio-α-Glc*p*-(1→4)-[α-Glc*p*-(1→4)]$_2$-GlcOH; the non-fissile link permits the observation also of a ternary complex with phosphate. The maltooligosaccharide sugars are in the $^4C_1$ conformation, with dihedral angles between subsites +2 and +3 and +3 and +4 little changed from their values in maltooligosaccharides in solution ($\varphi = \Phi = 80$–$90°$; $\Psi = -130$–$-150°$ ($\underline{\psi} \approx 110°$)); between −1 and +1, φ narrows by ~10–20° but Ψ becomes significantly more negative (−173 to −203°, $\underline{\psi} \approx 37$–$67°$).

There is little motion of the protein, the main movement being closure of a loop over the active site after binding phosphate and substrate–analogue. There are indications that when an acceptor maltooligosaccharide and glucose-1-phosphate are both bound, the phosphate is forced down and in an unfavourable torsional angle; however; there is no indication of any direct covalent interaction between the nucleophilic phosphate and the phosphate attached to the pyridoxal moiety, which had been proposed on the basis on NMR studies on glycogen phosphorylase.[536] There is also no indication of any covalent intermediate. Structurally, the S_Ni mechanism is plausible whereas the double-displacement mechanism would create difficulties.

5.12.2.8 GT 44. GT44 contains bacterial retaining transferases which use UDPGlc or UDPGlcNAc as donors. The bacteria concerned are unpleasant human pathogens (*Chlamydia* spp., *Clostridium dificile*, *E. coli* O157:H7) and the transferases are involved in the generation of toxins. The structure of the toxin B from *C. dificile* and its complex with Mn^{2+} and UDPGlc has been solved:[537] it glucosylates a specific Thr of a signalling protein (a Rho GTPase). The enzyme is unusual: in addition to a GTA fold, with the usual UDP and Mn^{2+} interactions, it has a 309-residue domain concerned with target recognition. Its slow hydrolytic action meant that the supposed UDPGlc complex was in fact a complex of UDP, but the presence of both halves of the donor nonetheless indicated that there was no potential nucleophile on the β-face of the sugar. Rather, an array of negatively charged groups seemed poised to stabilise an S_Ni transition state.

The 2-OH of the transferred glucose residue makes contact with Ile383 and Gln385; in another *Clostridium* species (*novyi*), whose toxin B transferred GlcNAc rather than Glc, the equivalent residues were serine and alanine, but the enzymes were otherwise homologous. The Ile383Ser/Gln385Ala mutant of the *C. dificile* enzyme indeed transferred GlcNAc from UDPGlcNAc, rather than Glc from UDPGlc.[538]

5.12.2.9 GT 64. A GlcNAc transferase involved in the early stages of heparan biosynthesis (EXTL2) occurs in this family and the crystal structures of the mouse enzyme in complexes with UDPGalNAc, UDP, Mn^{2+} and acceptor analogue have been solved.[539] In the early stages of heparan biosynthesis, the structure β-Gal*p*-(1→3)-β-Gal*p*-(1→4)-β-Xyl*p*-1→ serine is first biosynthesised on the initiator protein. A GlcA residue is then added and the enzyme responsible is discussed in Section 5.12.1.8. EXTL2 then adds an αGlcNAc residue to the 4-position of this GlcA residue of the reducing-end initiator sequence, which can be mimicked by β-GlcA*p*-(1→3)-β-Gal*p*-(1→O)-CH_2-naphthalene.[xxiv] [The basic polymerisation reaction in the biosynthesis of heparin, and heparan sulfates, is alternate addition of α-GlcNAc-(1→4) and β-GlcA-(1→3) units to the non-reducing end of the growing chain, carried out by bifunctional EXT enzymes.[540]]. EXTL2 is a GTA-fold enzyme, which is Mn^{2+} dependent. In the UDPGalNAc structure there are no plausible nucleophiles on the β-face of the GalNAc, but in the UDP–acceptor analogue structure there is a hydrogen bond between the 4-OH of the GlcA residue of the acceptor–analogue and a β-phosphate oxygen. Mutation of an apparently "electrostatic" carboxylate results in the inert mutant D246. Taken together, the data on this enzyme support the S_Ni mechanism.

5.12.2.10 GT 72. The structure of one enzyme in this class, an α-glucosyltransferase, has been solved.[541] It is a counterpart of the T4 bacteriophage β-glucosyltransferase describe in Section 5.12.1.1, but transfers an α-glucosyl, rather than a β-glucosyl, residue from UDPGlc to the oxygen of hydroxymethylcytosine residues in duplex DNA. It adopts the GTB fold, closely similar to the enzymes of GT 35. Again, there is no evidence of a nucleophile. Of three carboxylate groups in the vicinity of the transferred sugar, only one (Asp115) is close enough to act as a nucleophile: however, k_{cat} for the D115A mutant is 10% of wt. Again, an S_Ni mechanism appears to be indicated.

5.12.3 UDPGlcNAc Epimerase

Bacterial UDPGlcNAc epimerase converts UDPGlcNAc into UDPManNAc, which is used as a mannosamine source. The structure of the enzyme from *E. coli* has been solved and has the GTB fold of the transferases.[542] Previous work with this enzyme[543] had demonstrated incorporation of deuterium into both substrate and product if the reaction was carried out in D_2O, and a primary kinetic isotope effect [in an experiment carried out at [S] = 3.4 × K_m, so for the most part on (V)] of 1.8 in 2-[^{2}H]UDPGlcNAc isomerisation. Oxygen isotope positional exchange was also observed in substrate and product with UDPGlcNAc labelled in the C–O–P bridge as substrate. Taken together, these data establish the mechanism in Figure 5.66, in which 2-acetamidoglucal and UDP are intermediates. Indeed, 2-acetamidoglucal and UDP occasionally come off the enzyme, and at extended reaction times or high enzyme loadings can become the main products, as they are thermodynamically favoured. The GTB fold appears to be stabilising oxocarbenium ion-like transition states both for substitution and elimination, in much the same way as the GH 31 fold.

[xxiv] The paper does not specify the position of substitution of the naphthalene, the synthesis of the analogue is not in the open literature and in the X-ray structures with acceptor the naphthalene is disordered.

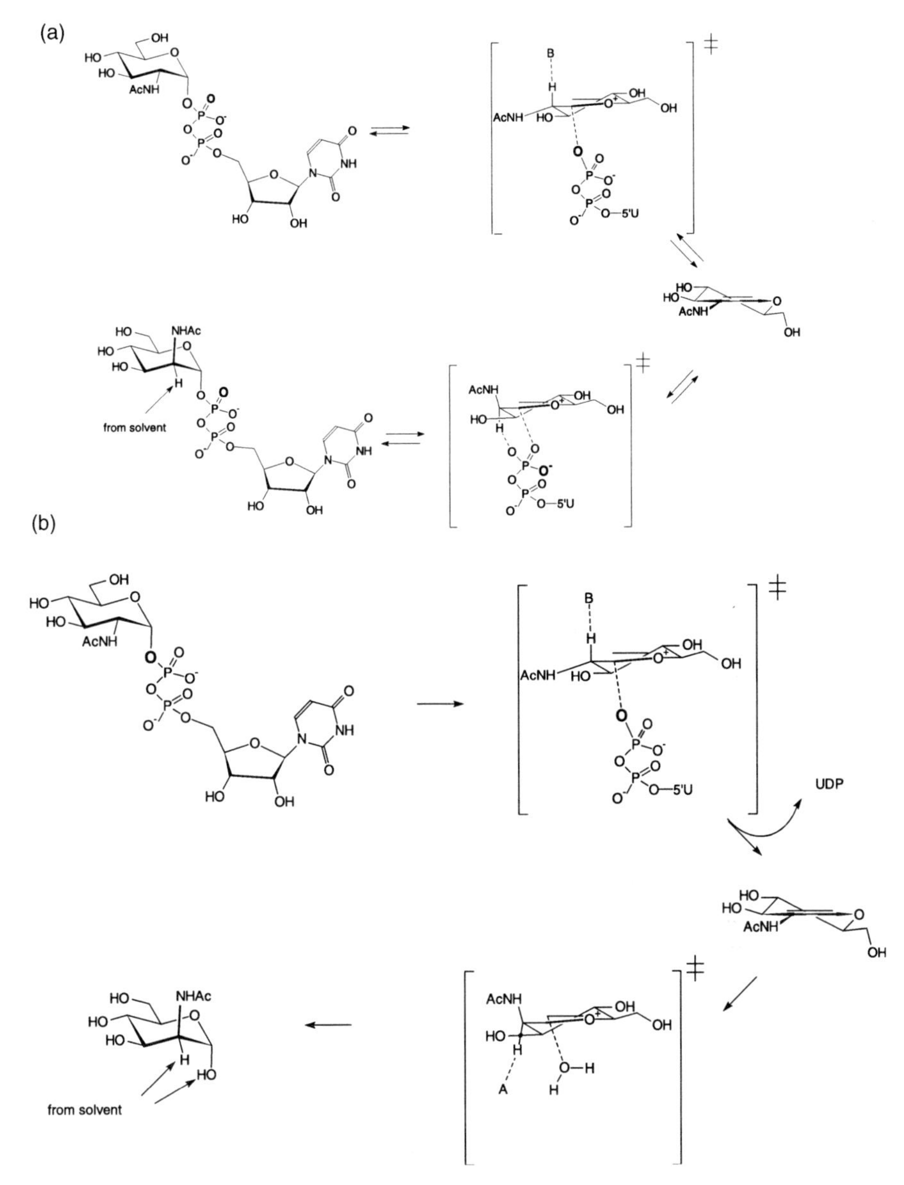

Figure 5.66 Mechanism of action of (a) UDPGlcNAc epimerase and (b) hydrolysing UDPGlcNAc epimerase.

The structure revealed the basis for the pronounced allosteric nature of the enzyme. Initial rates of UDPGlcNAc isomerisation give a Hill coefficient [eqn. (5.20), Figure 5.5] of 2.29 and epimerisation of UDPManNAc does not take place except in the presence of its epimer. The asymmetric unit of the crystals of UDPGlcNAc 2-epimerase contains four copies of the enzyme arranged as two similar copies of the biological homodimer, with each dimer composed of one open and one closed monomer that differ by an approximately 10° interdomain rotation. There is one UDPGlcNAc binding site per monomer. The domains appear to rotate as rigid bodies about the hinge region (residues 169–172 and 356–361) at the domain interface on the backside of the enzyme. It is suggested that only the closed form of the monomer unit is the catalytically active form, and binding at the open form induces a conformation change.

A closely related enzyme is the hydrolysing UDPGlcNAc epimerase, part of a two-domain enzyme which produces *N*-acetylmannosamine 6-phosphate for NeuNAc biosynthesis (the other domain is a kinase, which phosphorylates the ManNAc 6-OH with ATP). The domains can be studied independently,[544] and experiments with the heterologously expressed rat enzyme suggested a mechanism similar to that for UDPGlcNAc epimerase, but with a second step a *syn* hydration of 2-acetamidoglucal, rather than an addition of UDP (Figure 5.66).[545] The initial product was α-ManNAc*p*-OH, in which O1 and H2 arose from water, as shown by reaction in $H_2{}^{18}O$ or D_2O. Labelling of the bridge C–^{18}O–P confirmed that the C–^{18}O bond was broken, with ^{18}O in the UDP product, but also permitted a positional isotope exchange experiment to be done. This was negative – there was no scrambling in "unreacted" UDPGlcNAc, unlike the simple epimerase. Taken together with the absence of a primary H2 deuterium kinetic isotope effect on (V) or (V/K), these data suggested that the *syn* hydration of the 2-acetamidoglucal (which was demonstrated to be a catalytically competent intermediate) was rate determining. Essentially similar work with the enzyme from *N. meningitidis* painted a similar picture, with no positional exchange of the β-phosphate.[546] However, small isotope effects arising from ^{2}H substitution at H2 of the GlcNAc were noticed [$^D(V) = 1.15$ and $^D(V/K) = 1.26$], suggesting that the initial elimination was at least partly rate limiting. Surprisingly, there was little discrimination between hydrogen and deuterium in the hydration of 2-acetamidoglucal.

Three carboxylates were identified as important in both the *E. coli* non-hydrolysing and the *N. meningitis* hydrolysing epimerases (Asp95/100, Glu117/122, Glu131) and mutation to the corresponding amides in all six cases gave proteins with k_{cat} values reduced by $> 10^3$-fold; their role is as yet unclear.

5.12.4 UDP Galactopyranose Mutase

D-Galactofuranose residues occur in cell-wall and extracellular polysaccharides of certain bacteria, but not in mammals. The D-galactofuranose donor is UDP-galactofuranose, generated from UDP-galactopyranose by the remarkable enzyme UDP-galactopyranose mutase. Since the mutase does not occur in mammals, inhibitors are attractive potential antibacterials. The most likely mechanism involves nucleophilic displacements at the anomeric centre and so the enzyme is considered here (Figure 5.67).

The position of equilibrium, as would be expected, lies strongly towards the pyranose form ($[f]/[p] = 0.057$),[547] the furanose form being disfavoured by the *cis* interactions of hydroxyalkyl chain on C4, OH C2 and pyrophosphate on C1. A positional isotope exchange experiment, monitored by the isotope effect

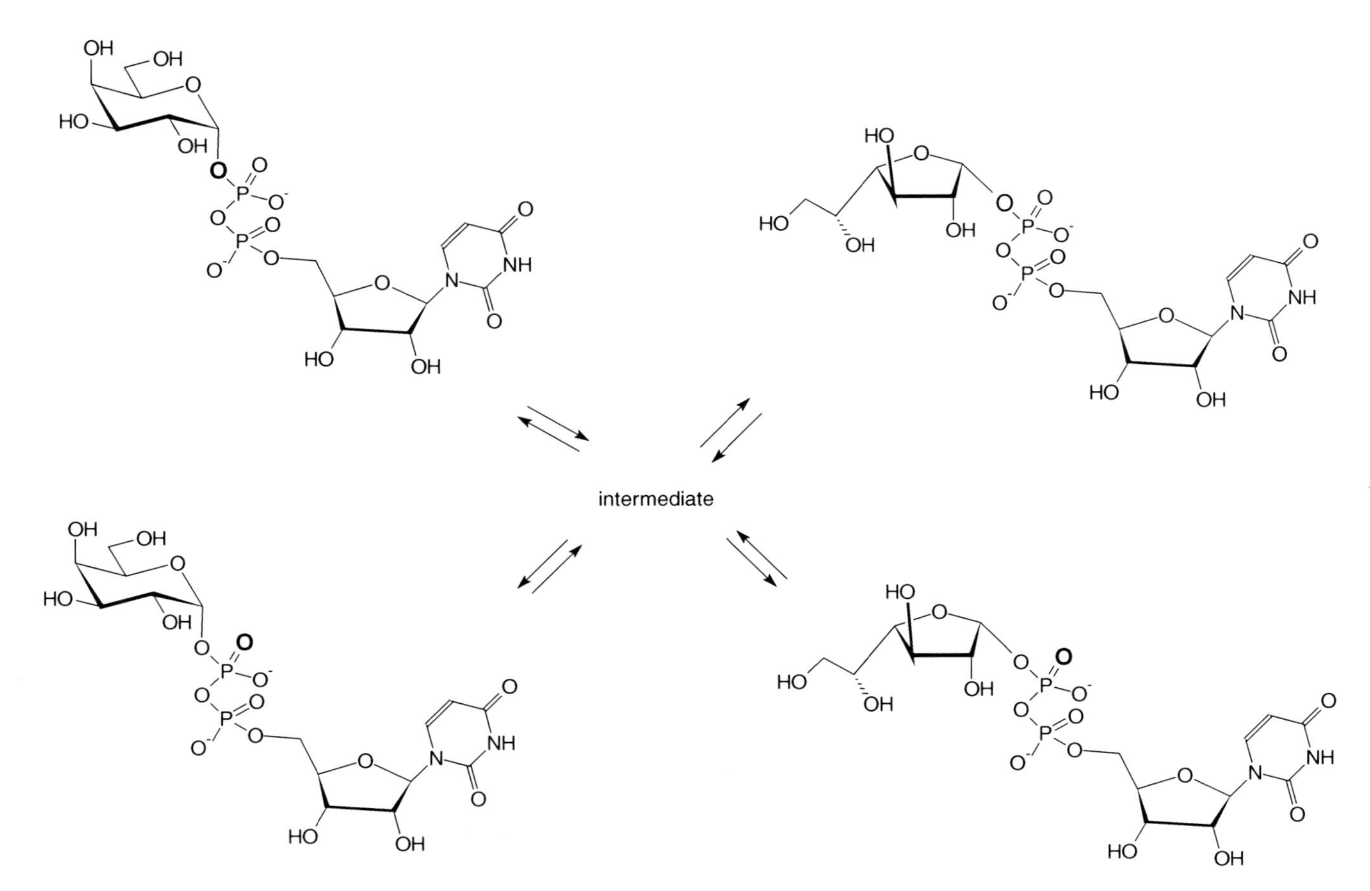

Figure 5.67 Reaction catalysed by UDPgalactopyranose mutase, illustrating positional isotope exchange.

on the anomeric ^{13}C chemical shift rather than the more usual β-^{31}P shift, demonstrated bridge–non-bridge scrambling of the glycosidic oxygen of apparently unreacted UDP-galactopyranose by the *Klebsiella pneumoniae* enzyme.[548] The C1–O1 bond is therefore broken in one ring form of the sugar and re-formed in the other. The suggestion of a bicyclic 1,4-anhydrogalactopyranose intermediate remains a possibility, but cannot explain the role of flavin.

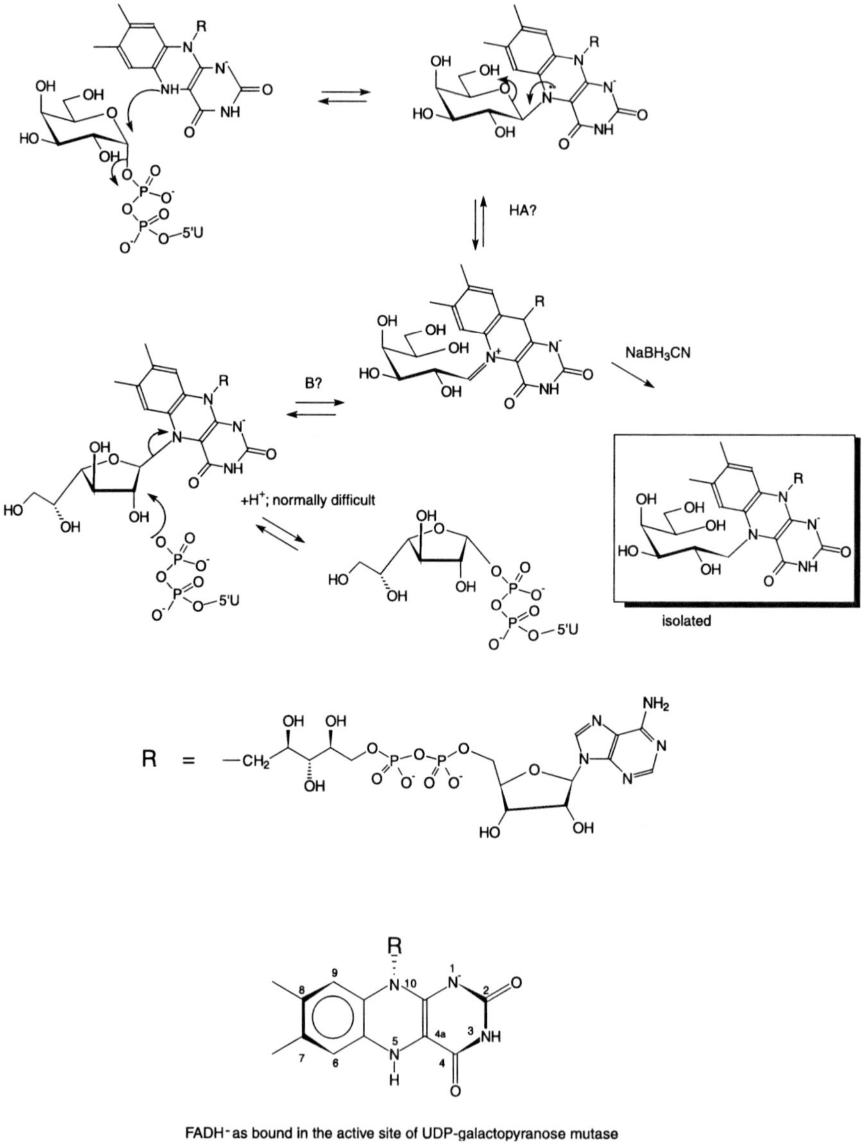

Figure 5.68 Flavin/imine mechanism of UDP galactopyranose mutase. The direction of curvature of bound $FADH^-$ as shown in the X-ray crystal structure is sketched.

The enzymes all contain a tightly, but non-covalently bound molecule of FAD. The enzyme is most active when the flavin is reduced,[547] but the bound form of the enzyme was shown by UV spectroscopy to be $FADH^-$ rather than $FADH_2$.[549]. The neutral semiquinone $FADH^•$ is stabilised by the protein environment, making $FADH^-$ a powerful one-electron reductant. It was therefore proposed that cryptic radical chemistry was involved, with an initial step involving electron transfer and loss of UDP anion, generating an anomeric galactosyl radical. The process seems unlikely because of the negative charges on the electron acceptor.

Rather, it was plausibly proposed that the $FADH^-$ anion acted as a nucleophile at N5 and gave a flavin-derived glycosylamine. By standard and well-precedented glycosylamine chemistry (see Section 1.2.3) the quaternary imine of *aldehydo* galactose, capable of closing to furanose and pyranose forms, was formed and the appropriate ring isomer of the UDP galactose liberated by reverse of the initial nucleophilic displacement (Figure 5.68). The mechanism was further supported by $NaBH_3CN$ reduction of the quaternary imine and isolation of the reduced species,[550] which had the correct molecular ion and UV spectrum for the proposed adduct.

Structures are available for the dimeric enzymes from *E. coli*,[551] *K. pneumoniae* and *Mycobacterium tuberculosis*[552] with FAD in the oxidised state and of the *K. pneumoniae* enzyme in the reduced state.[552] One FAD binds to the side of an active site cleft, but there are as yet no structures with bound nucleotide diphospho sugar. The reduced flavin is bowed (to avoid the central ring being antiaromatic), but apparently bowed to the *si* face, with the *re* face buttressed with fairly rigid-looking protein structures. It was suggested that his bowing is in the wrong sense for N5 to act as a nucleophile, since approach to the electrophile would be hindered, particularly by clashes with the benzene and pyrimidine rings of the reduced flavin. However, the weakest point of the flavin imine mechanism is the last step, in which pyrophosphate has to displace a very poor nitrogen nucleofuge. The difficulty would obviously be tempered if the nitrogen nucleofuge were sterically hindered, so that the glycosylamine were strained.

If the nucleophilicity of N5 were compromised by steric crowding, then the initial step would demand an oxocarbenium ion-like transition state. The very large effects of F-for-OH substitution in UDPGal (k_{cat} for transformation by the *K. pneumoniae* enzyme reduced 70 000-fold by the substitution at C2 and 900-fold at C3),[553] slightly greater than the effects in glycosidases, suggest this. Sterically strained amines have been exploited as good leaving groups by synthetic chemists.[554]

REFERENCES

1. E. Fischer, *Chem. Ber.*, 1894, **27**, 2985.
2. D. E. Koshland, *Proc. Natl. Acad. Sci. USA*, 1958, **44**, 98.
3. D. E. Koshland, *Biol. Rev. (Cambridge)*, 1953, **28**, 416.
4. J. D. McCarter and S. G. Withers, *Curr. Opin. Struct. Biol.*, 1994, **4**, 885.

5. D. Zechel and S. G. Withers, *Acc. Chem. Res.*, 2000, **33**, 11.
6. (a) B. Henrissat, *Biochem. J.*, 1991, **280**, 309; (b) B. Henrissat and A. Bairoch, *Biochem. J.*, 1993, **293**, 781; (c) B. Henrissat and A. Bairoch, *Biochem. J.*, 1996, **316**, 695.
7. http://www.cazy.org/
8. K. Kamata, M. Mitsuya, T. Nishimura, J. Eiki and Y. Nagata, *Structure*, 2004, **12**, 429.
9. The effect of Tris on glycosidases is rediscovered at roughly decade intervals: one of the first such reports was G. Semenza and A.-K. von Balthazar, *Eur. J. Biochem.* 1974, **41**, 149.
10. G. Legler, K.-R. Roeser and H.-K. Illig, *Eur. J. Biochem.*, 1979, **101**, 85.
11. M. J. Snider and R. Wolfenden, *Biochemistry*, 2001, **40**, 11364.
12. B. A. Lyons, J. Pfeifer, T. H. Peterson and B. K. Carpenter, *J. Am. Chem. Soc.*, 1993, **115**, 2427.
13. W. P. Jencks, *Adv. Enzymol.*, 1975, **43**, 219.
14. D. H. Leaback, *Biochem. Biophys. Res. Commun.*, 1968, **32**, 1025.
15. (a) S. Ezaki, *J. Biochem. (Tokyo)*, 1940, **32**, 91; (b) K. Horikoshi, *J. Biochem. (Tokyo)*, 1942, **35**, 39; (c) J. Conchie and G. A. Levvy, *Biochem. J.*, 1957, **65**, 389.
16. J. Baty and M. L. Sinnott, *Can. J. Chem.*, 2005, **83**, 1516.
17. (a) S. Inouye, T. Tsuruoka and T. Niida, *J. Antibiot. (Tokyo)*, 1966, **19**, 288; (b) S. Inouye, T. Tsuruoka, T. Ito and T. Niida, *Tetrahedron*, 1968, **24**, 2125.
18. T. Niwa, T. Tsuruoka, H. Goi, Y. Kodama, J. Itoh, S. Inouye, Y. Yamada, T. Niida, M. Nobe and Y. Ogawa, *J. Antibiot. (Tokyo)*, 1984, **37**, 1579.
19. Y. Miyake and M. Ebata, *Agric. Biol. Chem.*, 1988, **52**, 1649.
20. G. Legler and S. Pohl, *Carbohydr. Res.*, 1986, **155**, 119.
21. *e.g.* G. W. J. Fleet, S. J. Nicholas, P. W. Smith, S. V. Evans, L. E. Fellows and R. J. Nash, *Tetrahedron Lett.*, 1985, **26**, 3127.
22. (a) S. M. Colegate, P. R. Dorling and C. R. Huxtable, *Aust. J. Chem.*, 1979, **32**, 2257; (b) R. J. Molyneux and L. F. James, *Science*, 1982, **216**, 190.
23. W. Dong, T. Jespersen, M. Bols, T. Skrydstrup and M. R. Sierks, *Biochemistry*, 1996, **35**, 2788.
24. F. Morís-Varas, X.-H. Qian and C.-H. Wong, *J. Am. Chem. Soc.*, 1996, **118**, 7647.
25. M. Iwama, T. Takahashi, N. Inokuchi, T. Koyama and M. Irie, *J. Biochem. (Tokyo)*, 1985, **98**, 341.
26. P. A. Fowler, A. H. Haines, R. J. K. Taylor, E. J. T. Chrystal and M. B. Gravestock, *J. Chem. Soc., Perkin Trans. 1*, 1994, 2229.
27. R. A. Field, A. H. Haines, E. J. T. Chrystal and M. C. Luszniak, *Biochem. J.*, 1991, **274**, 885.
28. G. Legler, *Pure Appl. Chem.*, 1987, **59**, 1457.
29. E. Truscheit, W. Frommer, B. Junge, L. Müller, D. D. Schmidt and W. Wingender, *Angew. Chem. Int. Ed. Engl.*, 1981, **20**, 744.

30. L. G. Dickson, E. Leroy and J.-L. Reymond, *Org. Biomol. Chem.*, 2004, **2**, 1217.
31. B. Imperiali, *Acc. Chem. Res.*, 1997, **30**, 452.
32. Comprehensive review: A. Berecibar, C. Grandjean and A. Siriwardena, *Chem. Rev.*, 1999, **99**, 779.
33. See a whole series of papers with the general title "Zur Kenntniss des Süssmandelemulsins" by B. Helferich and collaborators in *Hoppe-Seylers Z. Physiol. Chem.* in the 1960s, eg. B. Helferich and T. Kleinschmidt, *Hoppe-Seylers Z. Physiol. Chem.*, 1968, **349**, 25.
34. See, *e.g.*, P. Suominen and T. Reinikainen, eds., *Trichoderma reesei Cellulases and Other Hydrolases*, Akateeminen Kirjakauppa, Helsinki, 1993.
35. M. T. H. Axamawaty, G. W. J. Fleet, K. A. Hannah, S. K. Namgoong and M. L. Sinnott, *Biochem. J.*, 1990, **266**, 245.
36. D. L. Zechel, A. B. Boraston, T. Gloster, C. M. Boraston, J. M. Macdonald, D. M. G. Tilbrook, R. V. Stick and G. J. Davies, *J. Am. Chem. Soc.*, 2003, **125**, 14313.
37. T. D. Heightman and A. Vasella, *Angew. Chem. Int. Ed.*, 1999, **38**, 750.
38. M. Bols, R. G. Hazell and I. B. Thomsen, *Chem. Eur. J.*, 1997, **3**, 940.
39. T. M. Gloster, J. M. Macdonald, C. A. Tarling, R. V. Stick, S. G. Withers and G. J. Davies, *J. Biol. Chem.*, 2004, **279**, 49236.
40. C. R. Berland, B. W. Sigurskjold, B. Stoffer, T. Frandsen and B. Svensson, *Biochemistry*, 1995, **34**, 10153.
41. R. Mosi, H. Sham, J. C. M. Uitdehaag, R. Ruiterkamp, B. W. Dijkstra and S. G. Withers, *Biochemistry*, 1998, **37**, 17192.
42. P. Ermert, A. Vasella, M. Weber, K. Rupitz, S. G. Withers and A. Vasella, *Carbohydr. Res.*, 1993, **250**, 113.
43. (a) R. Wolfenden and M. J. Snider, *Acc. Chem. Res.*, 2001, **34**, 938; (b) P. A. Bartlett and M. M. Mader, *Chem Rev.*, 1997, **97**, 1281.
44. P. Douzou and C. Balny, *Proc. Natl. Acad. Sci. USA*, 1977, **74**, 2297.
45. T. Barman, F. Travers, C. Balny, G. H. B. Hoa and P. Douzou, *Biochimie*, 1986, **68**, 1041.
46. A. L. Fink and K. Magnusdottir, *J. Protein Chem.*, 1984, **3**, 229.
47. N. More, R. M. Daniel and H. M. Petach, *Biochem. J.*, 1995, **305**, 17.
48. A. Bülow, I. W. Plesner and M. Bols, *J. Am. Chem. Soc.*, 2000, **122**, 8567.
49. R. A. Alberty and V. Bloomfield, *J. Biol. Chem.*, 1963, **238**, 2804.
50. L. P. McIntosh, G. Hand, P. E. Johnson, M. D. Joshi, M. Körner, L. A. Plesniak, L. Ziser, W. W. Wakarchuk and S. G. Withers, *Biochemistry*, 1996, **35**, 9958.
51. M. D. Joshi, G. Sidhu, I. Pot, G. D. Brayer, S. G. Withers and L. P. McIntosh, *J. Mol. Biol.*, 2000, **299**, 255.
52. For a recent glycosidase example, see D. J. Vocadlo, J. Wicki, K. Rupitz and S. G. Withers, *Biochemistry*, 2002, **41**, 9736.
53. R. P. Bell, *The Proton in Chemistry*, 2nd edn., Chapman and Hall, London, 1973.
54. G. Rudnick and R. H. Abeles, *Biochemistry*, 1975, **14**, 4515.

55. A. H. Fielding, M. A. Kelly, M. L. Sinnott and D. Widdows, *J. Chem. Soc. Perkin Trans. 1*, 1981, 1013.
56. C. Braun, A. Meinke, L. Ziser and S. G. Withers, *Anal. Biochem.*, 1993, **212**, 259.
57. P. Biely, J. Benen, K. Henrichová, H. C. M. Kester and J. Visser, *FEBS Lett.*, 1996, **382**, 249.
58. S. M. Pitson, A. G. J. Voragen and G. E. Beldman, *FEBS Lett.*, 1996, **398**, 7.
59. H. Brumer, P. F. G. Sims and M. L. Sinnott, *Biochem. J.*, 1999, **339**, 43.
60. W. Liu, N. B. Madsen, C. Braun and S. G. Withers, *Biochemistry*, 1991, **30**, 1419.
61. J. N. Watson, S. Newstead, A. A. Narine, G. Taylor and A. J. Bennet, *ChemBioChem*, 2005, **6**, 1999.
62. J. N. Watson, V. Dookhun, T. J. Borgford and A. J. Bennet, *Biochemistry*, 2003, **42**, 12682.
63. M. R. Birck and V. L. Schramm, *J. Am. Chem. Soc.*, 2004, **126**, 6882.
64. P. C. Kline and V. L. Schramm, *Biochemistry*, 1993, **32**, 13212.
65. T. Holm, *J. Am. Chem. Soc.*, 1994, **116**, 8803.
66. W. W. Cleland, *CRC Crit. Rev. Biochem.*, 1982, **13**, 385.
67. M. I. Schimerlik, J. E. Rife and W. W. Cleland, *Biochemistry*, 1975, **14**, 5347.
68. P. K. Lehikoinen, M. L. Sinnott and T. A. Krenitsky, *Biochem. J.*, 1989, **257**, 355.
69. W. W. Cleland, *Biochemistry*, 1992, **31**, 317.
70. R. L. Nath and H. N. Rydon, *Biochem. J.*, 1954, **57**, 1.
71. M. P. Dale, W. P. Kofler, I. Chait and L. D. Byers, *Biochemistry*, 1986, **25**, 2522.
72. J. B. Kempton and S. G. Withers, *Biochemistry*, 1992, **31**, 9961.
73. D. J. Vocadlo, J. Wicki, K. Rupitz and S. G. Withers, *Biochemistry*, 2002, **41**, 9727.
74. D. Tull and S. G. Withers, *Biochemistry*, 1994, **33**, 6363.
75. First example, M. L. Sinnott and S. G. Withers, *Biochem. J.*, 1974, **143** 751; examples of hydrolysis by inverting hydrolases: B. Padmeperuma and M. L. Sinnott, *Carbohydr. Res.*, 1993, **250**, 79.
76. L. Hosie and M. L. Sinnott, *Biochem. J.*, 1985, **226**, 437.
77. X. Guo, W. G. Laver, E. Vimr and M. L. Sinnott, *J. Am. Chem. Soc.*, 1994, **116**, 5572.
78. D. T. H. Chou, J. N. Watson, A. A. Scholte, T. G. Borgford and A. J. Bennet, *J. Am. Chem. Soc.*, 2000, **122**, 8357.
79. C. S. Jones and D. J. Kosman, *J. Biol. Chem.*, 1980, **255**, 11861.
80. A. L. Handlon, C. Xu, H. M. Muller-Steffner, F. Schuber and N. Oppenheimer, *J. Am. Chem. Soc.*, 1994, **116**, 12087.
81. M. W. Bauer and R. M. Kelly, *Biochemistry*, 1998, **37**, 17170.
82. Review: H. D. Ly and S. G. Withers, *Annu. Rev. Biochem.* 1999, **68**, 487.
83. *e.g.* D. L. Zechel, S. P. Reid, D. Stoll, O. Nashiru, R. A. J. Warren and S. G. Withers, *Biochemistry*, 2003, **42**, 7195.
84. R. Wolfenden, X. Lu and G. Young, *J. Am. Chem. Soc.*, 1998, **120**, 6814.

85. S. V. Calugaru, B. G. Hall and M. L. Sinnott, *Biochem. J.*, 1995, **312**, 281.
86. J. D. McCarter, M. J. Adam and S. G. Withers, *Biochem. J.*, 1992, **286**, 721.
87. N. H. Roth and R. E. Huber, *J. Biol. Chem.*, 1996, **271**, 14296.
88. M. N. Namchuk and S. G. Withers, *Biochemistry*, 1995, **34**, 16194.
89. W. P. Burmeister, R. W. H. Ruigrok and S. Cusack, *EMBO J.*, 1992 **11**, 49.
90. G. J. Davies, K. S. Wilson and B. Henrissat, *Biochem. J.*, 1997, **321**, 557.
91. K. Hiromi, *Biochem. Biophys. Res. Commun.*, 1970, **40**, 1.
92. M. M. Meagher, B. Y. Tao, J. M. Chow and P. J. Reilly, *Carbohydr. Res.*, 1988, **173**, 273.
93. (a) T. Christensen, B. B. Stoffer, B. Svensson and U. Christensen, *Eur. J. Biochem.*, 1997, **250**, 638; (b) I. Matsui and B. Svensson, *J. Biol. Chem.*, 1997, **272**, 22456.
94. C. Reverbel-Leroy, S. Pages, A. Belaïch, J. P. Belaïch and C. Tardif, *J. Bacteriol.*, 1997, **179**, 46.
95. B. Nidetzky, W. Zachariae, G. Gercken, M. Heyn and W. Steiner, *Enzym. Microb. Technol.*, 1994, **16**, 43.
96. B. K. Barr, Y.-L. Hsieh, B. Ganem and D. B. Wilson, *Biochemistry*, 1996, **35**, 586.
97. K. Kipper, P. Väljamäe and G. Johansson, *Biochem. J.*, 2005, **385**, 527.
98. W. Nerinckx, T. Desmet, K. Piens and M. Claeyssens, *FEBS Lett.*, 2005, **579**, 302.
99. J. L. Fry, J. M. Harris, R. C. Bingham and P. v. R. Schleyer, *J. Am. Chem. Soc.*, 1970, **92**, 2540.
100. F. L. Schadt, T. W. Bentley and P. v. R. Schleyer, *J. Am. Chem. Soc.*, 1976, **98**, 7667.
101. W. Nerinckx, T. Desmet and M. Claeyssens, *FEBS Lett.*, 2003, **538**, 1.
102. G. Legler, *Adv. Carbohydr. Chem. Biochem.*, 1990, **48**, 319.
103. (a) E. J. Hehre, *ACS Symp. Ser.*, 1995, **618**, 66; (b) E. J. Hehre, *Adv. Carbohydr. Chem. Biochem.*, 2000, **55**, 265.
104. E. C. K. Lai and S. G. Withers, *Biochemistry*, 1994, **33**, 14743.
105. E. J. Hehre, S. Kitahata and C. F. Brewer, *J. Biol. Chem.*, 1986 **261**, 2147.
106. S. Kitahata, S. Chiba, C. F. Brewer and E. J. Hehre, *Biochemistry*, 1991, **30**, 6769.
107. S. Chiba, C. F. Brewer, G. Okada, H. Matsui and E. J. Hehre, *Biochemistry*, 1988, **27**, 1564.
108. M. Brockhaus and J. Lehmann, *Carbohydr. Res.*, 1977, **53**, 21.
109. M. L. Sinnott, *Chem. Rev.*, 1990, **90**, 1171.
110. J. Lehmann and B. Ziegler, *Carbohydr. Res.*, 1977, **58**, 65.
111. J. Lehmann and P. Schlesselmann, *Carbohydr. Res.*, 1983, **113**, 93.
112. E. J. Hehre, C. F. Brewer, T. Uchiyama, P. Schlesselmann and J. Lehmann, *Biochemistry*, 1980, **19**, 3557.
113. P. Schlesselmann, H. Fritz, J. Lehmann, T. Uchiyama, C. F. Brewer and E. J. Hehre, *Biochemistry*, 1982, **21**, 6606.

114. M. Claeyssens, P. Tomme, C. F. Brewer and E. J. Hehre, *FEBS Lett.*, 1990, **263**, 89.
115. G. Davies and B. Henrissat, *Structure*, 1995, **3**, 853.
116. D. M. A. Guérin, M. B. Lascombe, M. Costabel, H. Souchon, V. Lamzin, P. Béguin and P. Alzari, *J. Mol. Biol.*, 2002, **316**, 1061.
117. B. Mikami, M. Adachi, T. Kage, E. Sarikaya, T. Nanmori, R. Shinke and S. Utsumi, *Biochemistry*, 1999, **38**, 7050.
118. Y. van Santen, J. A. E. Benen, K.-H. Schröter, K. H. Kalk, S. Armand, J. Visser and B. W. Dijkstra, *J. Biol. Chem.*, 1999, **274**, 30474.
119. F. Vallée, K. Karaveg, A. Herscovics, K. W. Moremen and P. L. Howell, *J. Biol. Chem.*, 2000, **275**, 41287.
120. K. Karaveg, A. Siriwardena, W. Tempel, Z.-J. Liu, J. Glushka, B.-C. Wang and K. E. Moremen, *J. Biol. Chem.*, 2005, **280**, 16197.
121. G. Golan, D. Shallom, A. Teplitsky, G. Zaide, S. Shulami, T. Baasov, V. Stojanoff, A. Thompson, Y. Shoham and G. Shoham, *J. Biol. Chem.*, 2004, **279**, 3014.
122. Y. Isoda and Y. Nitta, *J. Biochem. (Tokyo)*, 1986, **99**, 1631.
123. D. Shallom, M. Leon, T. Bravman, A. Ben-David, G. Zaide, V. Belakhov, G. Shoham, D. Schomburg, T. Baasov and Y. Shoham, *Biochemistry*, 2005, **44**, 387.
124. E. J. Hehre, C. F. Brewer and D. S. Genghof, *J. Biol. Chem.*, 1979, **254**, 5942.
125. S. Kitahata, C. F. Brewer, D. S. Genghof, T. Sawai and E. J. Hehre, *J. Biol. Chem.*, 1981, **256**, 6017.
126. E. J. Hehre, T. Sawai, C. F. Brewer, M. Nakano and T. Kanda, *Biochemistry*, 1982, **21**, 3090.
127. T. Kasumi, Y. Tsumuraya, C. F. Brewer, H. Kersters-Hilderson, M. Claeyssens and E. J. Hehre, *Biochemistry*, 1987, **26**, 3010.
128. D. Becker, K. S. H. Johnson, A. Koivula, M. Schülein and M. L. Sinnott, *Biochem. J.*, 2000, **345**, 315.
129. H. G. Damude, S. G. Withers, D. G. Kilburn, R. C. Miller and R. A. J. Warren, *Biochemistry*, 1995, **34**, 2220.
130. J.-Y. Zou, G. J. Kleywegt, J. Ståhlberg, H. Driguez, W. Nerinckx, M. Claeyssens, A. Koivula, T. T. Teeri and T. A. Jones, *Structure*, 1999, **7**, 1035.
131. A. Koivula, L. Ruohonen, G. Wohlfahrt, T. Reinikainen, T. T. Teeri, K. Piens, M. Claeyssens, M. Weber, A. Vasella, D. Becker, M. L. Sinnott, J.-Y. Zou, G. Kleywegt, M. Szardenings, J. Ståhlberg and T. A. Jones, *J. Am. Chem. Soc.*, 2002, **124**, 10015.
132. A. Varrot, T. Frandsen, H. Driguez and G. J. Davies, *Acta Crystallogr. Sect., D*, 2002, **58**, 2201.
133. A. Varrot, J. McDonald, R. V. Stick, G. Pell, H. J. Gilbert and G. J. Davies, *Chem. Commun.*, 2003, 946.
134. A. M. Larsson, T. Bergfors, E. Dultz, D. C. Irwin, A. Roos, H. Driguez, D. M. Wilson and T. A. Jones, *Biochemistry*, 2005, **44**, 12915.

135. N. S. Sweilem and M. L. Sinnott, in *Carbohydrases from Trichoderma reesi and Other Microorganisms*, ed. M. Claeyssens, W. Nerinckx and K. Piens, Royal Society of Chemistry, Cambridge, 1998, p. 13.
136. (a) W. P. Jencks, *Methods Enzymol.*, 1984, **171**, 145; (b) D. W. Urry, *Angew. Chem. Int. Ed.*, 1993, **32**, 819.
137. S. Armand, S. Drouillard, M. Schülein, B. Henrissat and H. Driguez, *J. Biol. Chem.*, 1997, **272**, 2709.
138. (a) A. Meinke, H. G. Damude, P. Tomme, E. Kwan, D. G. Kilburn, R. C. Miller, R. A. J Warren and N. R. Gilkes, *J. Biol. Chem.*, 1995, **270**, 4383; (b) G. J. Davies, A. M. Brzozowski, M. Dauter, A. Varrot and M. Schülein, *Biochem. J.*, 2000, **348**, 201.
139. K. Riedel and K. Bronnenmaier, *Eur. J. Biochem*, 1999, **262**, 218.
140. T. P. Frandsen, C. Dupont, J. Lehmbeck, B. Stoffer, M. R. Sierks, R. B. Honzatko and B. Svensson, *Biochemistry*, 1994, **33**, 13808.
141. A. E. Aleshin, L. M. Firsov and R. B. Honzatko, *J. Biol. Chem.*, 1994, **269**, 15631.
142. M. R. Sierks, C. Ford, P. Reilly and B. Svensson, *Protein Eng.*, 1990, **3**, 193.
143. H.-P. Fierobe, A. J. Clarke, D. Tull and B. Svensson, *Biochemistry*, 1998, **37**, 3753.
144. M. R. Sierks and B. Svensson, *Biochemistry*, 1996, **35**, 1865.
145. U. Christensen, K. Olsen, B. Stoffer and B. Svensson, *Biochemistry*, 1996, **35**, 15009.
146. Y. Tanaka, W. Tao, J. S. Blanchard and E. J. Hehre, *J. Biol. Chem.*, 1994, **269**, 32306.
147. U. Christensen, *Biochem. J.*, 2000, **349**, 623.
148. T. P. Frandsen, B. B. Stoffer, M. M. Palcic and B. Svensson, *J. Mol. Biol.*, 1996, **263**, 79.
149. M. R. Sierks and B. Svensson, *Biochemistry*, 2000, **39**, 8585.
150. S. Natarajan and M. R. Sierks, *Biochemistry*, 1996, **35**, 15269.
151. S. Natarajan and M. R. Sierks, *Biochemistry*, 1997, **36**, 14946.
152. S. Armand, M. J. Wagemaker, P. Sánchez-Torres, H. C. M. Kester, Y. van Santen, B. W. Dijkstra, J. Visser and J. A. E. Benen, *J. Biol. Chem.*, 2000, **275**, 691.
153. B. G. Guimarães, H. Souchon, B. L. Lytle, J. H. D. Wu and P. M. Alzari, *J. Mol. Biol.*, 2002, **320**, 587.
154. G. Parsiegla, C. Reverbel-Leroy, C. Tardif, J.-P. Belaïch. H. Driguez and R. Haser, *Biochemistry*, 2000, **39**, 11238.
155. Review: H. E. Krokan, R. Standal and G. Slupphaug, *Biochem. J.* 1997, **325**, 1.
156. J. D. Robertus and A. F. Monzingo, *Mini-Rev. Med. Chem.*, 2004, **4**, 477.
157. T. K. Amukele and V. L. Schramm, *Biochemistry*, 2004, **43**, 4913.
158. X.-Y. Chen, P. J. Berti and V. L. Schramm, *J. Am. Chem. Soc.*, 2000, **122**, 6527.
159. X.-Y. Chen, P. J. Berti and V. L. Schramm, *J. Am. Chem. Soc.*, 2000, **122**, 1609.

160. V. L. Schramm, *Acc. Chem. Res.*, 2003, **36**, 588.
161. I. Wong, A. J. Lundquist, A. S. Bernards and D. W. Mosbaugh, *J. Biol. Chem.*, 2002, **277**, 19424.
162. A. C. Drohat and J. T. Stivers, *J. Am. Chem. Soc.*, 2000, **122**, 1840.
163. S. S. Parikh, G. Walcher, G. D. Jones, G. Slupphaug, H. E. Krokan, G. M. Blackburn and J. A. Tainer, *Proc. Natl. Acad. Sci. USA*, 2000, **97**, 5083.
164. M. J. N. Shroyer, S. E. Bennett, C. D. Puttnam, J. A. Tainer and D. W. Mosbaugh, *Biochemistry*, 1999, **38**, 4834.
165. Y. L. Jiang, A. C. Drohat, Y. Ichikawa and J. T. Stivers, *J. Biol. Chem.*, 2002, **277**, 15385.
166. A. C. Drohat and J. T. Stivers, *Biochemistry*, 2000, **39**, 11865.
167. R. M. Werner and J. T. Stivers, *Biochemistry*, 2000, **39**, 14054.
168. A. C. Drohat, J. Jagadeesh, E. Ferguson and J. T. Stivers, *Biochemistry*, 1999, **38**, 11866.
169. T. L. Jiang and J. T. Stivers, *Biochemistry*, 2001, **40**, 7710.
170. D. O. Zharkov, R. A. Rieger, C. R. Iden and A. P. Grollman, *J. Biol. Chem.*, 1997, **272**, 5335.
171. B. Sun, K. A. Latham, M. L. Dodson and R. S. Lloyd, *J. Biol. Chem.*, 1995, **270**, 19501.
172. D. O. Zharkov and A. P. Grollman, *Biochemistry*, 1998, **37**, 12384.
173. S. D. Williams and S. S. David, *Biochemistry*, 1999, **38**, 15417.
174. S. D. Williams and S. S. David, *Biochemistry*, 2000, **39**, 10098.
175. D. O. Zharkov, T. A. Rosenquist, S. E. Gerchman and A. P. Grollman, *J. Biol. Chem.*, 2000, **275**, 28607.
176. A. K. McCullough, A. Sanchez, M. L. Dodson, P. Marapaka, J.-S. Taylor and R. S. Lloyd, *Biochemistry*, 2001, **40**, 561.
177. B. A. Horenstein, D. W. Parkin, B. Estupiñán and V. L. Schramm, *Biochemistry*, 1991, **30**, 10788.
178. D. W. Parkin, A. A. Sauve, P. J. Berti and V. l. Schramm, unpublished results, 2003.
179. D. W. Parkin and V. L. Schramm, *Biochemistry*, 1987, **26**, 913.
180. D. W. Parkin, F. Mentch, G. A. Banks, B. A. Horenstein and V. L. Schramm, *Biochemistry*, 1991, **30**, 4586.
181. K. A. Rising and V. L. Schramm, *J. Am. Chem. Soc.*, 1997, **119**, 27.
182. P. J. Berti and V. L. Schramm, *J. Am. Chem. Soc.*, 1997, **119**, 12079.
183. J. Scheuring and V. L. Schramm, *Biochemistry*, 1997, **36**, 4526.
184. J. Scheuring, P. J. Berti and V. L. Schramm, *Biochemistry*, 1998, **37**, 2748.
185. J. Scheuring and V. L. Schramm, *Biochemistry*, 1997, **36**, 8215.
186. P. C. Kline and V. L. Schramm, *Biochemistry*, 1993, **32**, 13212.
187. A. Lewandowicz and V. L. Schramm, *Biochemistry*, 2004, **43**, 1458.
188. M. R. Birck and V. L. Schramm, *J. Am. Chem. Soc.*, 2004, **126**, 2447.
189. W. Tao, C. Grubmeyer and J. S. Blanchard, *Biochemistry*, 1996, **35**, 14.
190. L. J. Mazzella, D. W. Parkin, P. C. Tyler, R. H. Furneaux and V. L. Schramm, *J. Am. Chem. Soc.*, 1996, **118**, 2111.

191. D. W. Parkin, G. Limberg, P. C. Tyler, R. H. Furneaux, X.-Y. Chen and V. L. Schramm, *Biochemistry*, 1997, **36**, 3528.
192. P. C. Kline and V. L. Schramm, *Biochemistry*, 1992, **31**, 5964.
193. A. Fedeorov, W. Shi, G. Kicska, E. Fedorov, P. C. Tyler, R. H. Furneaux, J. C. Hanson, G. J. Gainsford, J. Z. Larese, V. L. Schramm and S. C. Almo, *Biochemistry*, 2001, **40**, 853.
194. A. Henriksen, N. Aghajari, K. F. Jensen and M. Gajhede, *Biochemistry*, 1996, **35**, 3803.
195. G. Scapin, D. H. Ozturk, C. Grubmeyer and J. C. Sacchetini, *Biochemistry*, 1995, **34**, 10744.
196. G. P. Wang, C. Lundegaard, K. F. Jensen and C. Grubmeyer, *Biochemistry*, 1999, **38**, 275.
197. G. K. Grabner and K. L. Switzer, *J. Biol. Chem.*, 2003, **278**, 6921.
198. A. Bzowska, E. Kulikowska, N. E. Poopeiko and D. Shugar, *Eur. J. Biochem.*, 1996, **239**, 229.
199. H. Tsuge, M. Nagahara, H. Nishimura, J. Hisatsune, Y. Sakaguchi, Y. Itogawa, N. Katunuma and J. Sakurai, *J. Mol. Biol.*, 2003, **325**, 471.
200. H. Ritter, F. Koch-Nolte, V. E. Marquez and G. E. Schultz, *Biochemistry*, 2003, **42**, 10155.
201. H. Barth, J. C. Preiss, F. Hofmann and K. Aktories, *J. Biol. Chem.*, 1998, **273**, 29506.
202. C. C. F. Blake, L. N. Johnson, G. A. Mair, A. C. T. North, D. C. Phillips and V. R. Sarma, *Proc. R. Soc. London Ser. B*, 1967, **167**, 378.
203. C. A. Vernon, *Proc. R. Soc. London, Ser. B*, 1967, **167**, 389.
204. D. J. Vocadlo, G. J. Davies, R. Laine and S. G. Withers, *Nature*, 2001, **412**, 835.
205. M. L. Sinnott and I. J. L. Souchard, *Biochem. J.*, 1973, **133**, 89.
206. B. F. L. Li, D. Holdup, C. A. J. Morton and M. L. Sinnott, *Biochem. J.*, 1989, **260**, 109.
207. A. C. Elliott, S. Krishnan, M. L. Sinnott, J. Bommuswamy, Z. Guo and Y. Zhang, *Biochem. J.*, 1992, **282**, 155.
208. E. Van Doorslaer, O. van Opstal, H. Kersters-Hilderson and C. K. De Bruyne, *Bioorg. Chem.*, 1984, **12**, 158.
209. Y.-K. Li, H.-J. Yao and I.-H. Pan, *J. Biochem. (Tokyo)*, 2000, **127**, 315.
210. Y.-K. Li, J. Chir and F.-Y. Chen, *Biochem. J.*, 2001, **355**, 835.
211. D. L. Zechel, S. P. Reid, D. Stoll, O. Nashiru, R. A. G. Warren and S. G. Withers, *Biochemistry*, 2003, **42**, 7195.
212. W. P. Burmeister, S. Cottaz, H. Driguez, S. Palmieri and B. Henrissat, *Structure*, 1997, **5**, 663.
213. D. H. Juers, T. D. Heightman, A. Vaella, J. D. McCarter, L. Mackenzie, S. G. Withers and B. W. Matthews, *Biochemistry*, 2001, **40**, 14781.
214. A. Varrot, M. Schülein, M. Pipelier, A. Vasella and G. J. Davies, *J. Am. Chem. Soc.*, 1999, **121**, 2621.
215. G. J. Davies, L. Mackenzie, A. Varrot, M. Dauter, A. M. Brzozowski, M. Schülein and S. G. Withers, *Biochemistry*, 1998, **37**, 11707.

216. G. Sulzenbacher, H. Driguez, B. Henrissat, M. Schülein and G. J. Davies, *Biochemistry*, 1996, **35**, 15280.
217. V. A. Notenboom, C. Birsan, R. A. J. Warren, S. G. Withers and D. R. Rose, *Biochemistry*, 1998, **37**, 4751.
218. (a) G. Sidhu, S. G. Withers, N. T. Nguyen, L. P. McIntosh, L. Ziser and G. D. Brayer, *Biochemistry*, 1999, **38**, 5346; (b) E. Sabini, G. Sulzenbacher, M. Dauter, Z. Dauter, P. L. Å. Jørgensen, M. Schülein, C. Dupont, G. J. Davies and K. S. Wilson, *Chem. Biol.*, 1999, **6**, 483.
219. M. Sandgren, G. I. Berglund, A. Shaw, J. Ståhlberg, L. Kenne, T. Desmet and C. Mitchison, *J. Mol. Biol.*, 2004, **342**, 1505.
220. G. Sulzenbacher, L. F. Mackenzie, K. S. Wilson, S. G. Withers, C. Dupont and G. J. Davies, *Biochemistry*, 1999, **38**, 4826.
221. Y. Yoshioka, K. Hasegawa, Y. Matsuura, Y. Katsube and M. Kubota, *J. Mol. Biol.*, 1997, **271**, 619.
222. J. C. M. Uitdehaag, R. Mosi, K. H. Kalk, B. A. van der Veen, L. Dijkhuizen, S. G. Withers and B.W. Dijkstra, *Nat. Struct. Biol.*, 1999, **6**, 432.
223. D. M. F. van Aalten, D. Komander, B. Synstad, S. Gåseidnes, M. G. Peter and V. G. H. Eijsink, *Proc. Natl. Acad. USA*, 2001, **98**, 8979.
224. B. L. Mark, D. J. Vocadlo, S. Knapp, B. L. Triggs-Raine, S. G. Withers and M. N. G. James, *J. Biol. Chem.*, 2001, **276**, 10330.
225. N. C. J. Strynadka and M. N. G. James, *J. Mol. Biol.*, 1991, **220**, 401.
226. V. M.-A. Ducros, D. L. Zechel, G. N. Murshudov, H. J. Gilbert, L. Szabo, D. Stoll, S. G. Withers and G. J. Davies, *Angew. Chem. Int. Ed.*, 2002, **41**, 2824.
227. G. Sulzenbacher, C. Bignon, T. Nishimura, C. A. Tarling, S. G. Withers, B. Henrissat and Y. Bourne, *J. Biol. Chem.*, 2004, **279**, 13119.
228. A. L. Lovering, S. S. Lee, Y.-W. Kinm, S. G. Withers and N. C. J. Strynadka, *J. Biol. Chem.*, 2005, **280**, 2105.
229. Y. Luo, S.-C. Li, Y.-T. Li and M. Luo, *J. Mol. Biol.*, 1999, **285**, 323.
230. M. F. Amaya, A. G. Watts, I. Damager, A. Wehenkel, T. Nguyen, A. Buschiazzo, G. Paris, A. C. Frasch, S. G. Withers and P. M. Alzari, *Structure*, 2004, **11**, 775.
231. J. N. Varghese, J. L. McKimm-Breschkin, J. B. Caldwell, A. A. Kortt and P. M. Colman, *Proteins: Struct. Funct. Genet.*, 1992, **14**, 327.
232. S. Numao, D. A. Kuntz, S. G. Withers and D. R. Rose, *J. Biol. Chem.*, 2003, **48**, 48074.
233. J. K. Yang, H.-J. Yoon, H. J. Ahn, B. I. Lee, J.-D. Pedelacq, E. C. Liong, J. Berendzen, M. Laivenieks, C. Vieille, G. J. Zeikus, D. J. Vocadlo, S. G. Withers and S. W. Suh, *J. Mol. Biol.*, 2004, **335**, 155.
234. K. Hövel, D. Shallom, K. Niefind, V. Belakhov, G. Shoham, T. Baasov, Y. Shoham and D. Schomburg, *EMBO J.*, 2003, **22**, 4922.
235. T. S. Black, L. Kiss, D. Tull and S. G. Withers, *Carbohydr. Res.*, 1993, **250**, 195.
236. G. Legler, *Adv. Carbohydr. Chem. Biochem.*, 1990, **48**, 319.
237. R. Macarron, J. van Beeumen, B. Henrissat, I. de la Mata and M. Claeyssens, *FEBS Lett.*, 1993, **316**, 137.

238. P. Lalégerie, G. Legler and J. M. Yon, *Biochimie*, 1982, **64**, 977.
239. G. Legler, *Hoppe-Seylers Z. Physiol. Chem.*, 1970, **351**, 25.
240. H. Braun, G. Legler, J. Deshusses and G. Semenza, *Biochim. Biophys. Acta*, 1977, **483**, 135.
241. E. Bause and G. Legler, *Hoppe Seylers Z. Physiol. Chem.*, 1974 **355**, 438.
242. S. Dan, I. Marton, M. Dekel, B. A. Bravdo, S. He, S. G. Withers and O. Shoseyov, *J. Biol. Chem.*, 2000, **275**, 4973.
243. S. S. Lee, S. He and S. G. Withers, *Biochem. J.*, 2001, **359**, 381.
244. G. Legler and M. Herrchen, *FEBS Lett.*, 1981, **135**, 139.
245. J. C. Gebler, R. Aebershold and S. G. Withers, *J. Biol. Chem.*, 1992, **267**, 11126.
246. S. Miao, J. D. McCarter, M. E. Grace, G. A. Grabowski, R. Aebershold and S. G. Withers, *J. Biol. Chem.*, 1994, **269**, 10975.
247. K. Tatsuta, Y. Niwata, K. Umezawa, K. Toshima and M. Nakata, *Carbohydr. Res.*, 1991, **222**, 189.
248. G. Legler and R. Bollhagen, *Carbohydr. Res.*, 1992, **233**, 113.
249. G. Caron and S. G. Withers, *Biochem. Biophy. Res. Commun.*, 1989, **163**, 495.
250. S. Halazy, V. Berges, A. Ehrhard and C. Danzin, *Bioorg. Chem.*, 1990, **18**, 330.
251. J. C. Briggs, A. H. Haines and R. J. K. Taylor, *J. Chem. Soc., Chem. Commun.*, 1992, 1039.
252. S. Halazy, C. Danzin, A. Ehrhard and F. Gerhart, *J. Am. Chem. Soc.*, 1989, **111**, 3484.
253. J. Zhu, S. G. Withers, P.B. Reichardt, E. Treadwell and T. P. Clausen, *Biochem. J.*, 1998, **332**, 367.
254. (a) S. G. Waley, *Biochem. J.*, 1985, **227**, 843; (b) T. Funaki, Y. Takanohashi, H. Fukuzawa and I. Kuruma, *Biochim. Biophys. Acta*, 1991, **1078**, 43.
255. M. L. Sinnott and P. J. Smith, *Biochem. J.*, 1978, **175**, 525.
256. A. V. Fowler and P. J. Smith, *J. Biol. Chem.*, 1983, **258**, 10204.
257. P. J. Marshal, M. L. Sinnott, P. J. Smith and D. Widdows, *J. Chem. Soc. Perkin Trans. 1*, 1981, 366.
258. W. McDowell, A. Tlusty, R. Rott, J. N. BeMiller, J. A. Bohn, R. W. Meyers and R. T. Schwartz, *Biochem. J.*, 1988, **255**, 991.
259. C. C. Jones, M. A. Kelly, M. L. Sinnott, P. J. Smith and G. T. Tzotzos, *J. Chem. Soc., Perkin Trans. 2*, 1982, 1655.
260. M. L. Sinnott, G. T. Tzotzos and S. E. Marshall, *J. Chem. Soc., Perkin Trans. 2*, 1982, 1665.
261. R. P. Sewell-Alger, PhD thesis, University of Bristol, 1986.
262. O. P. van Diggelen, A. W. Schram, M. L. Sinnott, Paul J. Smith, D. Robinson and H. Galjaard, *Biochem. J.*, 1981, **200**, 143.
263. P. A. Docherty, M. J. Kuranda, N. N. Aronson, J. N. BeMiller, R. W. Myers and J. A. Bohn, *J. Biol. Chem.*, 1986, **261**, 3457.
264. S. G. Withers, I. P. Street, P. Bird and D. H. Dolphin, *J. Am. Chem. Soc.*, 1987, **109**, 7530.

265. Q. Wang, D. Tull, A. Meinke, N. R. Gilkes, R. A. J. Warren, R. Aebershold and S. G. Withers, *J. Biol. Chem.*, 1993, **268**, 14096.
266. S. G. Withers, K. Rupitz, D. Trimbur and R. A. J. Warren, *Biochemistry*, 1992, **31**, 9979.
267. L. F. Mackenzie, G. S. Brooks, J. F. Cutfield, P. A. Sullivan and S. G. Withers, *J. Biol. Chem.*, 1997, **272**, 3161.
268. L. F. Mackenzie, G. Sulzenbacher, C. Divne, T. A. Jones, H. F. Wöldike, M. Schülein, S. G. Withers and G. J. Davies, *Biochem. J.*, 1998, **335**, 409.
269. D. L. Zechel, S. He, C. Dupont and S. G. Withers, *Biochem. J.*, 1998, **336**, 139.
270. J. F. Blanchard, L. Gal, S. He, J. Foisy, R. A. J. Warren and S. G. Withers, *Carbohydr. Res.*, 2001, **333**, 7.
271. D. J. Vocadlo, L. F. Mackenzie, S. He, G. J. Zeikus and S. G. Withers, *Biochem. J.*, 1998, **335**, 449.
272. S. G. Withers, K. Rupitz and I. P. Street, *J. Biol. Chem.*, 1988, **263**, 7929.
273. M. N. Namchuk and S. G. Withers, *Biochemistry*, 1995, **34**, 16194.
274. S. He and S. G. Withers, *J. Biol. Chem.*, 1997, **272**, 24864.
275. A. W. Wong, S. He, J. H. Grubb, W. S. Sly and S. G. Withers, *J. Biol. Chem.*, 1998, **273**, 34057.
276. J. D. McCarter, M. J. Adam, N. G. Hartman and S. G. Withers, *Biochem. J.*, 1994, **301**, 343.
277. D. Stoll, S. He, S. G. Withers and R. A. J. Warren, *Biochem. J.*, 2000, **351**, 833.
278. C. Braun, G. D. Brayer and S. G. Withers, *J. Biol. Chem.*, 1995, **270**, 26778.
279. J. D. McCarter and S. G. Withers, *J. Am. Chem. Soc.*, 1996, **118**, 241.
280. J. D. McCarter and S. G. Withers, *J. Biol. Chem.*, 1996, **271**, 6889.
281. S. Howard, S. He and S. G. Withers, *J. Biol. Chem.*, 1998, **273**, 2067.
282. S. Numao, S. He, G. Evjen, S. Howard, O. K. Tollersud and S. G. Withers, *FEBS Lett.*, 2000, **484**, 175.
283. H. D. Ly, S. Howard, K. Shum, S. He, A. Zhu and S. G. Withers, *Carbohydr. Res.*, 2000, **329**, 539.
284. R. Mosi, S. He, J. Uitdehaag, B. W. Dijkstra and S. G. Withers, *Biochemistry*, 1997, **36**, 9927.
285. H. Imamura, S. Fushinobu, B.-S. Jeon, T. Wakagi and H. Matsuzawa, *Biochemistry*, 2001, **50**, 12400.
286. S. L. Lawson, R. A. J. Warren and S. G. Withers, *Biochem. J.*, 1998, **330**, 203.
287. R. Kuroki, Y. Ito, Y. Kato and T. Imoto, *J. Biol. Chem.*, 1997, **272**, 19976.
288. S. L. Lawson, W. W. Wakarchuk and S. G. Withers, *Biochemistry*, 1996, **35**, 10110.
289. Q. Wang, R. W. Graham, D. Trimbur, R. A. J. Warren and S. G. Withers, *J. Am. Chem. Soc.*, 1994, **116**, 11594.
290. A. M. Macleod, D. Tull, K. Rupitz, R. A. J. Warren and S. G. Withers, *Biochemistry*, 1996, **35**, 13165.

291. (a) Reviews: (a) S. J. Williams and S. G. Withers, *Aust. J. Chem.* 2002, **55**, 3; (b) D. L. Jakeman and S. G. Withers, *Trends Glycosci. Glycotechnol.*, 2002, **14**, 13.
292. L. F. Mackenzie, Q. P. Wang, R. A. J. Warren and S. G. Withers, *J. Am. Chem. Soc.*, 1998, **120**, 5583.
293. V. M.-A. Ducros, C. A. Tarling, D. L. Zechel, A. M. Brzozowski, T. P. Frandsen, I. von Ossowski, M. Schülein, S. G. Withers and G. J. Davies, *Chem. Biol.*, 2003, **10**, 619.
294. M. Hrmova, T. Imai, S. J. Rutten, J. K. Fairweather, L. Pelosi, V. Bulone, H. Driguez and G. B. Fincher, *J. Biol. Chem.*, 2002, **277**, 30102.
295. C. Malet and A. Planas, *FEBS Lett.*, 1998, **440**, 208.
296. J. Chir, S. G. Withers, C.-F. Wan and Y.-K. Li, *Biochem. J.*, 2002, **365**, 857.
297. A. M. McLeod, T. Lindhorst, S. G. Withers and R. A. J. Warren, *Biochemistry*, 1994, **33**, 6371.
298. Y.-K. Li, J. Chir, S. Tanaka and F. Y. Chen, *Biochemistry*, 2002, **41**, 2751.
299. E. H. Rydberg, C. Li, R. Maurus, C. M. Overall, G. D. Brayer and S. G. Withers, *Biochemistry*, 2002, **41**, 4492.
300. D. Shallom, V. Belakhov, D. Solomon, S. Gilead-Gropper, T. Baasov, G. Shoham and Y. Shoham, *FEBS Lett.*, 2002, **514**, 163.
301. D. L. Zechel, S. P. Reid, O. Nashiru, C. Mayer, D. Stoll, D. L. Jakeman, R. A. J. Warren and S. G. Withers, *J. Am. Chem. Soc.*, 2001, **123**, 4350.
302. Q. Wang and S. G. Withers, *J. Am. Chem. Soc.*, 1995, **117**, 10137.
303. D. Vocadlo, J. Wicki, K. Rupitz and S. G. Withers, *Biochemistry*, 2002, **41**, 9736.
304. S. Cottaz, B. Henrissat and H. Driguez, *Biochemistry*, 1996, **35**, 15256.
305. W. P. Burmeister, S. Cottaz, P. Rollin, A. Vasella and B. Henrissat, *J. Biol. Chem.*, 2000, **275**, 39385.
306. K. Wallenfels and R. Weil, in *The Enzymes*, ed. P. D. Boyer, 3rd edn Wiley, New York, 1972, Vol.7, p. 617.
307. J.-P. Tenu, O. M. Viratelle, J. Garnier and J. M. Yon, *Eur. J. Biochem.*, 1971, **20**, 363.
308. J.-P. Tenu, O. M. Viratelle, J. Garnier and J. M. Yon, *Biochem. Biophys. Res. Commun.*, 1969, **37**, 1036.
309. J.-P. Tenu, O. M. Viratelle and J. M. Yon, *Eur. J. Biochem.*, 1972, **26**, 112.
310. R. H. Jacobsen, X.-J. Zhang, R. F. DuBose and B. W. Matthews, *Nature*, 1994, **369**, 761.
311. M. L. Sinnott, S. G. Withers and O. M. Viratelle, *Biochem. J.*, 1978, **175**, 539.
312. T. Selwood and M. L. Sinnott, *Biochem. J.*, 1990, **268**, 317.
313. J. P. Richard, J. G. Westerfeld and S. Lin, *Biochemistry*, 1995, **34**, 11703.
314. J. P. Richard, J. G. Westerfeld, S. Lin and J. Beard, *Biochemistry*, 1995, **34**, 11713.
315. J. P. Richard, R. E. Huber, S. Lin, C. Heo and T. L. Amyes, *Biochemistry*, 1996, **35**, 12377.
316. A. Jobe and S. Bourgeois, *J. Mol. Biol.*, 1972, **69**, 397.

317. J. P. Richard, R. E. Huber, C. Heo, T. L. Amyes and S. Lin, *Biochemistry*, 1996, **35**, 12387.
318. D. L. Hartl and B. G. Hall, *Nature*, 1974, **248**, 152.
319. B. G. Hall, *Genetica*, 2003, **118**, 143.
320. K. Srinivasan, A. K. Konstantinidis, M. L. Sinnott and B. G. Hall, *Biochem. J.*, 1993, **291**, 15.
321. K. Srinivasan, B. G. Hall and M. L. Sinnott, *Biochem. J.* 1995, **312**, 971, and references therein.
322. (a) W. J. Albery and J. R. Knowles, *Biochemistry*, 1976, **15**, 5627; (b) J. R. Knowles and W. J. Albery, *Acc. Chem. Res.*, 1977, **10**, 105.
323. C. Divne, J. Ståhlberg, T. T. Teeri and T. A. Jones, *J. Mol. Biol.*, 1998, **275**, 309.
324. T. Imai, C. Boisset, M. Samejima, K. Igarishi and J. Sugiyama, *FEBS Lett.*, 1998, **432**, 113.
325. M. Srisodsuk, K. Kleman-Leyer, S. Keränen and T. T. Teeri, *Eur. J. Biochem.*, 1998, **251**, 885.
326. I. von Ossowski, J. Ståhlberg, A. Koivula, K. Piens, D. Becker, H. Boer, R. Harle, M. Harris, C. Divne, S. Mahdi, Y. Zhao, H. Driguez, M. Claeyssens, M. L. Sinnott and T. T. Teeri, *J. Mol. Biol.*, 2003, **333**, 817.
327. E. J. Taylor, A. Goyal, C. I. P. D. Guerreiro, J. A. M. Prates, V. A. Money, N. Ferry, C. Morland, A. Planas, J. A. Macdonald, R. V. Stick, H. J. Gilbert, C. M. G. A. Fontes and G. J. Davies, *J. Biol. Chem.*, 2005, **280**, 32761.
328. X. Huang, K. S. E. Tanaka and A. J. Bennet, *J. Am. Chem. Soc.*, 1997, **119**, 11147.
329. H. Leemhuis, J. C. M. Uitdehaag, H. J. Rozeboom, B. W. Dijkstra and L. Dijkhuizen, *J. Biol. Chem.*, 2002, **277**, 1113.
330. Y. Terada, M. Yanase, H. Takata, T. Takaha and S. Okada, *J. Biol. Chem.*, 1997, **272**, 15729.
331. Z. Sulová, M. Takáčová, N. M. Steele, S. C. Fry and V. Farkaš, *Biochem. J.*, 1998, **330**, 1475.
332. P. Johansson, H. Brumer, M. J. Baumann, Å. M. Kallas, H. Henriksson, S. E. Denman, T. T. Teeri and T. A. Jones, *Plant Cell*, 2004, **16**, 874.
333. G. Michel, L. Chantalat, E. Duee, T. Barbeyron, B. Henrissat, B. Kloareg and O. Dideberg, *Structure*, 2001, **9**, 513.
334. A. Planas, *Biochim. Biophys. Acta*, 2000, **1543**, 361.
335. J. Allouch, M. Jam, W. Helbert, T. Barbeyron, B. Kloareg, B. Henrissat and M. Czjzek, *J. Biol. Chem.*, 2003, **278**, 47171.
336. G. Lowe, G. Sheppard, M. L. Sinnott and A. Williams, *Biochem. J.*, 1967, **104**, 893.
337. G. Lowe and G. Sheppard, *J. Chem. Soc., Chem. Commun.*, 1968, 529.
338. I. Tews, A. Perrakis, A. Oppenheim, Z. Dauter, K. S. Wilson and C. E. Vorgias, *Nat. Struct. Biol.*, 1996, **3**, 638.
339. S. Knapp, D. Vocadlo, Z. Gao, B. Kirk, J. Lou and S. G. Withers, *J. Am. Chem. Soc.*, 1996, **118**, 6804.

340. A. C. Terwisscha van Scheltinga, K. H. Kalk, J. J. Beintema and B. W. Dijkstra, *Structure*, 1994, **2**, 1181.
341. I. Tews, A. C. Terwisscha van Scheltinga, A. Perrakis, K. S. Wilson and B. W. Dijkstra, *J. Am. Chem. Soc.*, 1997, **119**, 7954.
342. S. Kobayashi, T. Kiyosada and S. Shoda, *J. Am. Chem. Soc.*, 1996, **118**, 13113.
343. G. R. Porter, H. N. Rydon and J. A. Schofield, *Nature*, 1958, **182**, 927.
344. E. Bokma, H. J. Rozeboom, M. Sibbald, B. W. Dijkstra and J. Beintema, *Eur. J. Biochem.*, 2002, **269**, 893.
345. K. Bortone, A. F. Monzingo, S. Ernst and J. D. Robertus, *J. Mol. Biol.*, 2002, **320**, 293.
346. T. Watanabe, A. Ishibashi, Y. Ariga, M. Hashimoto, N. Nikaidou, J. Sugiyama, T. Matsumoto and T. Nonaka, *FEBS Lett.*, 2001, **494**, 74.
347. E.-L. Hult, F. Katouno, T. Uchiyama, T. Watanabe and J. Sugiyama, *Biochem. J.*, 2005, **388**, 851.
348. Z. Markovič-Housley, G. Miglierini, L. Soldatova, P. J. Rizkallah, U. Müller and T. Schirmer, *Structure*, 2000, **8**, 1025.
349. S. Arming, B. Strobl, C. Wechselberger and G. Kreil, *Eur. J. Biochem.*, 1997, **247**, 810.
350. M. S. Macauley, G. E. Whitworth, A. W. Debowski, D. Chin and D. J. Vocadlo, *J. Biol. Chem.*, 2005, **280**, 25313.
351. N. Çetinbaş, M. S. Macauley, K. A. Stubbs, R. Drapala and D. J. Vocadlo, *Biochemistry*, 2006, **45**, 3835.
352. N. T. Blackburn and A. J. Clarke, *Biochemistry*, 2002, **41**, 1001.
353. N. T. Blackburn and A. J. Clarke, *J. Mol. Evol.*, 2001, **52**, 78.
354. C. W. Reid, N. T. Blackburn, B. A. Legaree, F.-I Auzanneau and A. J. Clarke, *FEBS Lett.*, 2004, **574**, 73.
355. A. K.-W. Leung, H. S. Duewel, J. F. Honek and A. M. Berghuis, *Biochemistry*, 2001, **40**, 5665.
356. E. J. van Asselt, A.-M. W. H. Thunissen and B. W. Dijkstra, *J. Mol. Biol.*, 1999, **291**, 877.
357. E. J. van Asselt, K. H. Kalk and B. W. Dijkstra, *Biochemistry*, 2000, **39**, 1924.
358. F. W. Ballardie, B. Capon, M. W. Cuthbert and W. M. Dearie, *Bioorg. Chem.*, 1977, **6**, 483.
359. S. Yu, K. Bojsen, B. Svensson and J. Marcussen, *Biochim. Biophys. Acta*, 1999, **1433**, 1.
360. S. S. Lee, S. Yu and S. G. Withers, *J. Am. Chem. Soc.*, 2002, **124**, 4948.
361. S. S. Lee, S. Yu and S. G. Withers, *Biochemistry*, 2003, **42**, 13081.
362. (a) S. Crennell, T. Takimoto, A. Portner and G. Taylor, *Nat. Struct. Biol.*, 2000, **7**, 1068; (b) M. C. Lawrence, N. A. Borg, V. A. Streltsov, P. A. Pilling, V. C. Epa, J. N. Varghese, J. L. McKinn-Breschkin and P. M. Colman, *J. Mol. Biol.*, 2004, **335**, 1343.
363. A. G. Watts, I. Damager, M. L. Amaya, A. Buschiazzo, P. Alzari, A. C. Frasch and S. G. Withers, *J. Am. Chem. Soc.*, 2003, **125**, 7532.

364. J. N. Watson, S. Newstead, V. Dookhun, G. Taylor and A. J. Bennet, *FEBS Lett.*, 2004, **577**, 265.
365. X. Guo and M. L. Sinnott, *Biochem. J.*, 1993, **296**, 291.
366. X. Guo, M. L. Sinnott, S.-C. Li and Y.-T. Li, *J. Am. Chem. Soc.*, 1993, **115**, 3334.
367. S. Newstead, J. N. Watson, T. L. Knoll, A. J. Bennet and G. Taylor, *Biochemistry*, 2005, **44**, 9117.
368. L. Michaelis and M. L. Menten, *Biochem. Z.*, 1913, **49**, 333.
369. V. A. Reddy and F. Maley, *J. Biol. Chem.*, 1990, **265**, 10817.
370. A. Reddy and F. Maley, *J. Biol. Chem.*, 1996, **271**, 13953.
371. F. Alberto, C. Bignon, G. Sulzenbacher, B. Henrissat and M. Czjzek, *J. Biol. Chem.*, 2004, **279**, 18903.
372. G. Meng and K. Fütterer, *Nat. Struct. Biol.*, 2003, **10**, 935.
373. D. Nurizzo, J. P. Turkenburg, S. J. Charnock, S. M. Roberts, E. J. Dodson, V. A. McKie, E. J. Taylor, H. J. Gilbert and G. J. Davies, *Nat. Struct. Biol.*, 2002, **9**, 665.
374. R. A. P. Nagem, A. L. Rojas, A. M. Golubev, O. S. Korneeva, E. V. Eneyskaya, A. A. Kulminskaya, K. N. Neustroev and I. Polikarpov, *J. Mol. Biol.*, 2004, **344**, 471.
375. R. Chambert, G. Gonzy-Treboul and R. Dedonder, *Eur. J. Biochem.*, 1974, **41**, 285.
376. R. Chambert and G. Gonzy-Treboul, *Eur. J. Biochem.*, 1976, **71**, 493.
377. R. Chambert and G. Gonzy-Treboul, *Eur. J. Biochem.*, 1976, **62**, 55.
378. Review: M. Linder and T. T. Teeri, *J. Biotechnol.* 1997 **57**, 15.
379. L. M. A. Ferreira, T. M. Wood, G. Williamson, C. Faulds, G. P. Hazlewood, G. W. Black and H. J. Gilbert, *Biochem. J.*, 1993, **294**, 349.
380. T. Tanaka, S. Fujiwara, S. Nishikori, T. Fukui, M. Takagi and T. Imanaka, *Appl. Environ. Microbiol.*, 1999, **65**, 5338.
381. Y. Shoham, R. Lamed and E. A. Bayer, *Trends Microbiol.*, 1999, **7**, 275.
382. R. Lamed and E. A. Bayer, *Adv. Appl. Microbiol.*, 1988, **33**, 1.
383. M. A. Lemos, M. Teixeria, M. Mota and F. M. Gama, *Biotechnol. Lett.*, 2000, **22**, 703.
384. D. N. Bolam, X. Hefang, G. Pell, D. Hogg, G. Galbraith, B. Henrissat and H. J. Gilbert, *J. Biol. Chem.*, 2004, **279**, 22953.
385. A. Boraston, D. N. Bolam, H. J. Gilbert and G. J. Davies, *Biochem. J.*, 2004, **382**, 769.
386. P. J Kraulis, G. M. Clore, M. Nilges, T. A. Jones, G. Pettersson, J. Knowles and A. M. Gronenborn, *Biochemistry*, 1989, **28**, 7241.
387. M. Linder, M. L. Mattinen, M. Kontteli, G. Lindberg, J. Ståhlberg, T. Drakenberg, T. Reinikainen, G. Pettersson and A. Annila, *Protein Sci.*, 1995, **4**, 1056.
388. P. Tomme, A. L. Creagh, D. G. Kilburn and C. A. Haynes, *Biochemistry*, 1996, **35**, 13885.
389. P. E. Johnson, M. D. Joshi, P. Tomme, D. G. Kilburn and L. P. MacIntosh, *Biochemistry*, 1996, **35**, 14381.

390. M. Czjzek, D. N. Bolam, A. Mosbah, J. Allouch, C. M. G. A. Fontes, L. M. A. Ferreira, O. Bornet, V. Zamboni, H. Darbon, N. L. Smith, G. W. Black, B. Henrissat and H. J. Gilbert, *J. Biol. Chem.*, 2001, **276**, 48580.
391. E. Brun, F. Moriaud, P. Gans, M. J. Blackledge, F. Barras and D. Marion, *Biochemistry*, 1997, **36**, 16074.
392. S. Raghothama, P. J. Simpson, L. Szabó, T. Nagy, H. J. Gilbert and M. P. Williamson, *Biochemistry*, 2000, **39**, 978.
393. T. Ponyi, L. Szabó, T. Nagy, P. J. Simpson, H. J. Gilbert and M. P. Williamson, *Biochemistry*, 2000, **39**, 985.
394. T. Watanabe, Y. Ariga, U. Sato, T. Toratani, M. Hashimoto, N. Nikaidou, Y. Kezuka, T. Nonaka and J. Sugiyama, *Biochem. J.*, 2003, **376**, 237.
395. J. Lehtiö, J. Sugiyama, M. Gustavsson, L. Fransson, M. Linder and T. T. Teeri, *Proc. Natl. Acad. Sci. USA*, 2003, **100**, 484.
396. A. L. Creagh, E. Ong, E. Jervis, D. G. Kilburn and C. A. Haynes, *Proc. Natl. Acad. Sci. USA*, 1996, **93**, 12229.
397. G. Carrad and M. Linder, *Eur. J. Biochem.*, 1999, **262**, 637.
398. M. Linder and T. T. Teeri, *Proc. Natl. Acad. Sci. USA*, 1996, **93**, 12251.
399. H. Jung, D. B. Wilson and L. P. Walker, *Biotech. Bioeng.*, 2002, **80**, 380.
400. E. J. Jervis, C. A. Haynes and D. G. Kilburn, *J. Biol. Chem.*, 1997, **272**, 24016.
401. M. I. Page and W. P. Jencks, *Proc. Natl. Acad. Sci. USA*, 1971, **68**, 1678.
402. M. Linder, I. Salovuori, L. Ruohonen and T. T. Teeri, *J. Biol. Chem.*, 1996, **271**, 21268.
403. G.-Y. Xu, E. Ong, N. R. Gilkes, D. G. Kilburn, D. R. Muhandiram, M. R. Harris-Brandts, J. P. Carver, L. E. Kay and T. S. Harvey, *Biochemistry*, 1995, **34**, 6993.
404. N. Din, H. G. Damude, N. R. Gilkes, R. C. Miller, R. A. J. Warren and D. G. Kilburn, *Proc. Natl. Acad. Sci. USA*, 1994, **91**, 11383.
405. D. N. Bolam, A. Ciruela, S. McQueen-Mason, P. Simpson, M. P. Williamson, J. E. Rixon, A. Boraston, G. P. Hazlewood and H. J. Gilbert, *Biochem. J.*, 1998, **331**, 775.
406. H. Xie, D. N. Bolam, T. Nagy, L. Szabó, A. Cooper, P. J. Simpson, J. H. Lakey, M. P. Williamson and H. J. Gilbert, *Biochemistry*, 2001, **40**, 5700.
407. P. J. Simpson, H. Xie, D. N. Bolam, H. J. Gilbert and M. P. Williamson, *J. Biol. Chem.*, 2000, **275**, 41137.
408. P. Tomme, A. L. Creagh, D. G. Kilburn and C. A. Haynes, *Biochemistry*, 1996, **35**, 13885.
409. P. E. Johnson, P. Tomme, M. D. Joshi and L. P. MacIntosh, *Biochemistry*, 1996, **35**, 13895.
410. E. Brun, P. E. Johnson, A. L. Creagh, P. Tomme, P. Webster, C. A. Haynes and L. P. McIntosh, *Biochemistry*, 2000, **39**, 2445.
411. A. B. Boraston, D. N. Bolam, H. G. Gilbert and G. J. Davies, *Biochem. J.*, 2004, **382**, 769.
412. L. Szabó, S. Jamal, H. Xie, S. J. Charnock, D. N. Bolam, H. J. Gilbert and G. J. Davies, *J. Biol. Chem.*, 2001, **276**, 49061.

413. G. Pell, M. P. Williamson, C. Walters, H. M. Du, H. J. Gilbert and D. N. Bolam, *Biochemistry*, 2003, **42**, 9316.
414. G. Pell, L. Szabo, S. J. Charnock, H. F. Xie, T. M. Gloster, G. J. Davies and H. J. Gilbert, *J. Biol. Chem.*, 2004, **279**, 11777.
415. S. J. Charnock, D. N. Bolam, J. P. Turkenburg, H. J. Gilbert, L. M. A. Ferreira, G. J. Davies and C. M. G. A. Fontes, *Biochemistry*, 2000, **39**, 5013.
416. T. Giardina, A. P. Gunning, N. Juge, C. B. Faulds, C. S. M. Furniss, B. Svensson, V. J. Morris and G. Williamson, *J. Mol. Biol.*, 2001, **313**, 1149.
417. V. J. Morris, A. P. Gunning, C. B. Faulds, G. Williamson and B. Svensson, *Starch/Stärke*, 2005, **57**, 1.
418. A. L. van Bueren, C. Morland, H. J. Gilbert and A. B. Boraston, *J. Biol. Chem.*, 2005, **280**, 530.
419. V. Notenboom, A. B. Boraston, D. G. Kilburn and D. R. Rose, *Biochemistry*, 2001, **40**, 6248.
420. A. B. Boraston, A. L. Creagh, M. M. Alam, J. M. Kormos, P. Tomme, C. A. Haynes, R. A. J. Warren and D. G. Kilburn, *Biochemistry*, 2001, **40**, 6241.
421. V. Notenboom, A. B. Boraston, S. J. Williams, D. G. Kilburn and D. R. Rose, *Biochemistry*, 2002, **41**, 4246.
422. M. Schärpf, G. P. Connelly, G. M. Lee, A. B. Boraston, R. A. J. Warren and L. P. McIntosh, *Biochemistry*, 2002, **41**, 4255.
423. F. A. Saul, P. Rovira, G. Boulot, E. J. M. Van Damme, W. J. Peumans, P. Truffa-Bachi and G. A. Bentley, *Struct. Fold. Des.*, 2000, **8**, 593.
424. F. Schuber, P. Travo and M. Pascal, *Bioorg. Chem.*, 1979, **8**, 83.
425. D. A. Yost and B. M. Anderson, *J. Biol. Chem.*, 1981, **256**, 3647.
426. D. A. Yost and B. M. Anderson, *J. Biol. Chem.*, 1983, **258**, 3075.
427. C. Tarnus and F. Schuber, *Bioorg. Chem.*, 1987, **15**, 31.
428. A. A. Suave, H. T. Deng, R. H. Angeletti and V. L. Schramm, *J. Am. Chem. Soc.*, 2000, **122**, 7855.
429. S. Yamamoto-Katayama, M. Ariyoshi, K. Ishihara, T. Hirano, H. Jingami and K. Morikawa, *J. Mol. Biol.*, 2002, **316**, 711.
430. Review: G. A. Garcia and J. D. Kittendorf, *Bioorg. Chem.*, 2005, **33**, 229.
431. K. Reuter, S. R. Chong, F. Ullrich, H. Kersten and G. A. Garcia, *Biochemistry*, 1994, **33**, 7041.
432. R. Ishitani, O. Nureki, S. Fukai, T. Kijimoto, N. Nameki, M. Watanabe, H. Kondo, M. Sekine, N. Okada, S. Nishimura and S. Yokoyama, *J. Mol. Biol.*, 2002, **318**, 665.
433. G. C. Hoops, L. B. Townsend and G. A. Garcia, *Biochemistry*, 1995, **34**, 15381.
434. G. C. Hoops, L. B. Townsend and G. A. Garcia, *Biochemistry*, 1995, **34**, 15539.
435. This process was characterised in the particularly clear-cut case of carbanion generation by A. Thibblin and W. P. Jencks, *J. Am. Chem. Soc.* 1979, **101**, 4963.
436. L. X. Huang, M. Pookanjanatavip, X. G. Gu and D. V. Santi, *Biochemistry*, 1998, **37**, 344.

437. K. Phannachet and R. H. Huang, *Nucleic Acids Res.*, 2004, **32**, 1422.
438. X. Gu, Y. Liu and D. V. Santi, *Proc. Natl. Acad. Sci. USA*, 1999, **96**, 14270.
439. R. Anand, P. A. Kaminski and S. E. Ealick, *Biochemistry*, 2004, **43**, 2384.
440. S. R. Armstrong, W. J. Cook, S. A. Short and S. E. Ealick, *Structure*, 1996, **4**, 97.
441. S. A. Short, S. R. Armstrong, S. E. Ealick and D. J. T. Porter, *J. Biol. Chem.*, 1996, **271**, 4978.
442. M. Smar, S. A. Short and R. Wolfenden, *Biochemistry*, 1991, **30**, 7908.
443. S. J. Charnock and G. J. Davies, *Biochemistry*, 1999, **38**, 6380.
444. A. Vrielink, W. Rüger, H. P. C. Driessen and P. S. Freemont, *EMBO J.*, 1994, **13**, 3413.
445. J. Flint, E. Taylor, M. Yang, D. N. Bolam, L. E. Tailford, C. Martinez-Fleites, E. J. Dodson, B. J. Davies, H. J. Gilbert and G. J. Davies, *Nat. Struct. Mol. Biol.*, 2005, **12**, 608.
446. W. Offen, C. S. Martinez-Fleites, M. Yang, E. Kiat-Lim, B. G Davis, C. A Tarling, C. M. Ford, D. J. Bowles and G. J. Davies, *EMBO J.*, 2006, **25**, 1396.
447. L. Larivière and S. Moréra, *J. Mol. Biol.*, 2002, **324**, 483.
448. S. Moréra, L. Larivière, J. Kurzeck, U. Aschke-Sonnenborn, P. S. Freemont, J. Janin and W. Rüger, *J. Mol. Biol.*, 2001, **311**, 569.
449. P. M. Legler, H. C. Lee, J. Peisach and A. S. Mildvan, *Biochemistry*, 2002, **41**, 4655.
450. L. Larivière, V. Gueguen-Chaignon and S. Moréra, *J. Mol. Biol.*, 2003, **330**, 1077.
451. A. M. Mulichak, H. C. Losey, C. T. Walsh and R. Garavito, *Structure*, 2001, **9**, 547.
452. A. M. Mulichak, H. C. Losey, W. Lu, Z. Wawrzak, C. T. Walsh and R. M. Garavito, *Proc. Natl. Acad. Sci USA*, 2003, **100**, 9238.
453. A. M. Mulichak, W. Lu, H. C. Losey, C. T. Walsh and R. M. Garavito, *Biochemistry*, 2004, **43**, 5170.
454. S. J. Charnock and G. J. Davies, *Biochemistry*, 1999, **18**, 6380.
455. N. Tarbouriech, S. J. Charnock and G. J. Davies, *J. Mol. Biol.*, 2001, **314**, 655.
456. C. Garinot-Schneider, A. C. Lellouch and R. A. Geremia, *J. Biol. Chem.*, 2000, **275**, 31407.
457. L. N. Gastinel, C. Cambillau and Y. Bourne, *EMBO. J.*, 1999, **18**, 3546.
458. B. Ramakrishnan, P. V. Balaji and P. K. Qasba, *J. Mol. Biol.*, 2002, **318**, 491.
459. B. Ramakrishnan and P. K. Qasba, *J. Mol. Biol.*, 2001, **310**, 205.
460. B. Ramakrishnan, P. S. Shah and P. K. Qasba, *J. Biol. Chem.*, 2001, **276**, 37665.
461. B. Ramakrishnan and P. K. Qasba, *J. Biol. Chem.*, 2002, **277**, 20833.
462. U. M. Ünligil, S. Zhou, S. Yuwaraj, M. Sarkar, H. Schachter and J. M. Rini, *EMBO J.*, 2000, **19**, 5269.
463. Y. Nishikawa, W. Pegg, H. Paulsen and H. Schachter, *J. Biol. Chem.*, 1988, **263**, 8270.

464. S. Ha, D. Walker, V. Shi and S. Walker, *Protein Sci.*, 2000, **9**, 1045.
465. Y. Hu, L. Chen, S. Ha, B. Gross, B. Falcone, D. Walker, M. Mokhtarzadeh and S. Walker, *Proc. Natl. Acad. Sci USA*, 2003, **100**, 845.
466. C. P. C. Chiu, A. G. Watts, L. L. Lairson, M. Gilbert, D. Lim, W. W. Wakarchuk, S. G. Withers and N. C. J. Strynadka, *Nature Stuct. Mol. Biol.*, 2004, **11**, 163.
467. T. Lind, F. Tufaro, C. McCormick, U. Lindahl and K. Lidholt, *J. Biol. Chem.*, 1998, **273**, 26265.
468. L. C. Pedersen, K. Tsuchida, H. Kitagawa, K. Sugahara, T. A. Darden and M. Negishi, *J. Biol. Chem.*, 2000, **275**, 34580.
469. L. C. Pedersen, T. A. Darden and M. Negishi, *J. Biol. Chem.*, 2002, **277**, 21869.
470. S. Kakuda, T. Shiba, M. Ishiguro, H. Tagawa, S. Oka, Y. Kajihara, T. Kawasaki, S. Wakatsuki and R. Kato, *J. Biol. Chem.*, 2004, **279**, 22693.
471. K. Ohtsubo, S. Imajo, M. Ishiguro, T. Nakatani, S. Oka and T. Kawasaki, *J. Biochem. (Tokyo)*, 2000, **128**, 283.
472. A. af Eckenstam, *Ber*, 1936, **69**, 553.
473. E. Dempski and B. Imperiali, *Curr. Opin. Chem. Biol.*, 2002, **6**, 844.
474. I. Nilsson and G. von Heijne, *J. Biol. Chem.*, 1993, **268**, 5798.
475. A. X. Yan and W. J. Lennarz, *J. Biol. Chem.*, 2005, **280**, 3121.
476. J. Helenius, D. T. W. Ng, C. L. Marolda, P. Walter, M. A. Valvano and M. Aebi, *Nature*, 2002, **415**, 447.
477. P. Burda and M. Aebi, *Biochim. Biophys. Acta*, 1999, **1426**, 239.
478. B. Imperiali, K. L. Shannon and K. W. Rickert, *J. Am. Chem. Soc.*, 1992, **114**, 7942.
479. E. Bause, *FEBS Lett.*, 1979, **103**, 296.
480. (a) E. Bause, *Biochem. Soc. Trans.*, 1984, **12**, 514; (b) E. Bause, W. Breuer and S. Peters, *Biochem. J.*, 1995, **312**, 979.
481. B. C. Challis and J. A. Challis, in *Comprehensive Organic Chemistry*, ed. D. H. R. Barton and W. D. Ollis, Pergamon Press, Oxford, 1979, Vol. 2, p. 996.
482. S. Peluso, M. de L. Ufret, M. K. O'Reilly and B. Imperiali, *Chem. Biol.*, 2002, **9**, 1323.
483. (a) J. G. Voet and R. H. Abeles, *J. Biol. Chem.*, 1970, **245**, 1020; (b) J. J. Mieyal, R. H. Abeles and M. Simon, *J. Biol. Chem.*, 1972, **247**, 532.
484. E. P. Mitchell, S. G. Withers, P. Ermert, A. T. Vasella, E. F. Garman, N. G. Oikonomakos and L. N. Johnson, *Biochemistry*, 1996, **35**, 7341.
485. A. Buschiazzo, J. E. Ugalde, M. E. Guerin, W. Shepard, R. A. Ugalde and P. M. Alzari, *EMBO J.*, 2004, **23**, 3196.
486. A. Yep, M. A. Ballicora, M. N. Sivak and J. Preiss, *J. Biol. Chem.*, 2004, **279**, 8359.
487. S. I. Patenaude, N. O. L. Seto, S. N. Borisova, A. Szpacenko, S. L. Marcus, M. M. Palcic and S. V. Evans, *Nat. Struct. Biol.*, 2002, **9**, 685.
488. H. P. Nguyen, N. O. L. Seto, Y. Cai, E. K. Leinala, S. N. Borisova, M. M. Palcic and S. V. Evans, *J. Biol. Chem.*, 2003, **278**, 49191.

489. Y. Zhang, G. J. Swaninathan, A. Deshpande, E. Boix, R. Natesh, Z. Xie, K. R. Acharya and K. Brew, *Biochemistry*, 2003, **42**, 13512.
490. L. N. Gastinel, C. Bignon, A. K. Misra, O. Hindsgaul, J. H. Shaper and D. H. Joziasse, *EMBO J.*, 2001, **20**, 638.
491. E. Boix, Y. Zhang, G. J. Swaminathan, K. Brew and K. R. Acharya, *J. Biol. Chem.*, 2002, **277**, 28310.
492. Y. L. Tzeng and D. S. Stephen, *Microbes Infect.*, 2000, **2**, 687.
493. H. D. Ly, B. Lougheed, W. W. Wakarchuk and S. G. Withers, *Biochemistry*, 2002, **41**, 5075.
494. K. Persson, H. D. Ly, M. Dieckelmann, W. W. Wakarchuk, S. G. Withers and N. C. J. Strynadka, *Nat. Struct. Biol.*, 2001, **8**, 166.
495. B. Lougheed, H. D. Ly, W. W. Wakarchuk and S. G. Withers, *J. Biol. Chem.*, 1999, **274**, 37717.
496. J. E. Baldwin and L. I. Kruse, *J. Chem. Soc. Chem. Commun.*, 1977, 233.
497. A. Monegal and A. Planas, *J. Am. Chem. Soc.*, 2006, **128**, 16030.
498. L. L. Lairson, C. P. C. Chiu, H. D. Ly, S. He, W. W. Wakarchuk, N. C. J. Strynadka and S. G. Withers, *J. Biol. Chem.*, 2004, **279**, 28339.
499. B. J. Gibbons, P. J. Roach and T. D. Hurley, *J. Mol. Biol.*, 2002, **319**, 463.
500. Y. D. Lobsanov, P. A. Romero, B. Sleno, B. Yu, P. Yip, A. Herscovics and P. L. Howell, *J. Biol. Chem.*, 2004, **279**, 17921.
501. R. P. Gibson, J. P. Turkenburg, S. J. Charnock, R. Lloyd and G. J. Davies, *Chem. Biol.*, 2002, **9**, 1337.
502. R. P. Gibson, C. A. Tarling, S. Roberts, S. G. Withers and G. J. Davies, *J. Biol. Chem.*, 2004, **279**, 1950.
503. A. D. Elbein, Y. T. Pan, I. Pastuszak and D. Carroll, *Glycobiology*, 2003, **13**, 17R.
504. T. A. Fritz, J. H. Hurley, L.-B. Trinh, J. Shiloach and L. A. Tabak, *Proc. Natl. Acad. Sci. USA*, 2004, **101**, 15307.
505. F. K. Hagen, B. Hazes, R. Raffo, D. deSa and L. A. Tabak, *J. Biol. Chem.*, 1999, **274**, 6797.
506. A. A. Green, G. T. Cori and C. F. Cori, *J. Biol. Chem.*, 1942, **142**, 447.
507. http://nobelprize.org/nobel_prizes/medicine/laureates/1947/cori-lecture.pdf.
508. Reviews of the first half-century of work on glycogen phosphorylase: (a) E. J. M. Helmreich, *Biofactors*, 1992, **3**, 159; (b) N. B. Madsen, *The Enzymes*, 3rd edn, 1986, Vol. 17, p. 366. (c) C. B. Newgard, P. K. Hwang and R. J. Fletterick, *CRC Crit. Rev. Biochem. Mol. Biol.*, 1989, **24**, 69. (d) D. Palm, H. W. Klein, R. Schinzel, M. Buehner and E. J. M. Helmreich, *Biochemistry*, 1990, **29**, 1099.
509. J. Monod, J. Wyman and J. P. Changeux, *J. Mol. Biol.*, 1965, **12**, 88.
510. D. Barford and L. N. Johnson, *Protein Sci.*, 1992, **1**, 472.
511. Z.-X. Wang, *Eur. J. Biochem.*, 1999, **259**, 609.
512. P. J. Kasvinsky, S. Shechosky and R. J. Fletterick, *J. Biol. Chem.*, 1978, **253**, 9102.
513. N. G. Oikonomakos, M. Kosmopoulou, S. E. Zographos, D. D. Leonidas, E. D. Chrysina, L. Somsák, V. Nagy, J.-P. Praly, T. Docsa, B. Tóth and P. Gergely, *Eur. J. Biochem.*, 2002, **269**, 1684.

514. N. G. Oikonomakos, V. T. Skamnaki, K. E. Tsitsanou, N. G. Gavalas and L. N. Johnson, *Structure*, 2000, **8**, 574.
515. V. L. Rath, M. Ammirati, D. E. Danley, J. L. Ekstrom, E. M. Gibbs, T. R. Hynes, A. M. Mathiowetz, R. K. McPherson, T. V. Olson, J. L. Treadway and D. J Hoover, *Chem. Biol.*, 2000, **7**, 677.
516. P. J. Kasvinsky, N. B. Madsen, R. J. Fletterick and J. Sygusch, *J. Biol. Chem.*, 1978, **253**, 1290.
517. E. A. Sergienko and D. K. Srivastava, *Biochem. J.*, 1997, **328**, 83.
518. R. J. Fletterick and N. B. Madsen, *Annu. Rev. Biochem.*, 1980, **49**, 31.
519. I. P. Street, K. R. Withers and S. G. Withers, *Biochemistry*, 1989, **28**, 1581.
520. S. G. Withers, M. D. Percival and I. P. Street, *Carbohydr. Res.*, 1989, **187**, 43.
521. L. M. Firsov, T. I. Bogacheva and S. E. Bresler, *Eur. J. Biochem.*, 1974, **42**, 605.
522. S. G. Withers, N. B. Madsen, B. D. Sykes, M. Takagi, S. Shimomura and T. Fukui, *J. Biol. Chem.*, 1981, **256**, 10759.
523. S. G. Withers, S. Shechosky and N. B. Madsen, *Biochem. Biophys. Res. Commun.*, 1982, **108**, 322.
524. W. G. Stirtan and S. G. Withers, *Biochemistry*, 1996, **35**, 15057.
525. H. W. Klein, D. Palm and E. J. M. Helmreich, *Biochemistry*, 1982, **21**, 6675.
526. J. Hajdu, K. R. Acharya, D. I. Stuart, P. J. McLaughlin, D. Barford, N. G. Oikonomakos, H. Klein and L. N. Johnson, *EMBO J.*, 1987, **6**, 539.
527. E. M. H. Duke, S. Wakatsuki, A. Hadfield and L. N. Johnson, *Protein Sci.*, 1994, **3**, 1178.
528. L. N. Johnson, K. R. Acharya, M. D. Jordan and P. J. McLaughlin, *J. Mol. Biol.*, 1990, **211**, 645.
529. S. G. Withers, N. B. Madsen and B. D. Sykes, *Biochemistry*, 1981, **20**, 1748.
530. F. C. Kokesh and Y. Kakuda, *Biochemistry*, 1977, **16**, 2467.
531. A. Schwarz, F. M. Pierfederici and B. Nidetzky, *Biochem. J.*, 2005, **387**, 437.
532. M. O'Reilly, K. A. Watson, R. Schinzel, D. Palm and L. N. Johnson, *Nat. Struct. Biol.*, 1997, **4**, 405.
533. K. A. Watson, C. McCleverty, S. Geremia, S. Cottaz, H. Driguez and L. N. Johnson, *EMBO J.*, 1999, **18**, 4619.
534. M. O'Reilly, K. A. Watson and L. N. Johnson, *Biochemistry*, 1999, **38**, 5337.
535. S. Geremia, M. Campagnolo, R. Schinzel and L. N. Johnson, *J. Mol. Biol.*, 2002, **322**, 413.
536. S. G. Withers, N. B. Madsen, B. D. Sykes, M. Tagaki, S. Shimomura and T. Fukui, *J. Biol. Chem.*, 1981, **256**, 10759.
537. D. J. Reinert, T. Jank, K. Aktories and G. E. Schulz, *J. Mol. Biol.*, 2005, **351**, 973.
538. T. Jank, D. J. Reinert, T. Giesemann, G. E. Schulz and K. Aktories, *J. Biol. Chem.*, 2005, **280**, 37833.
539. L. C. Pedersen, J. Dong, F. Taniguchi, H. Kitagawa, J. M. Krahn, L. G. Pedersen, K. Sugahara and M. Negishi, *J. Biol. Chem.*, 2003, **278**, 14420.

540. U. Lindahl, M. Kusche-Gullberg and L. Kjellén, *J. Biol. Chem.*, 1998, **274**, 24979.
541. L. Larivière, N. Sommer and S. Moréra, *J. Mol. Biol.*, 2005, **352**, 139.
542. R. E. Campbell, S. C. Mosimann, M.E. Tanner and N. C. J. Strynadka, *Biochemistry*, 2000, **39**, 14993.
543. P. M. Morgan, R. F. Sala and M. E. Tanner, *J. Am. Chem. Soc.*, 1997, **119**, 10269.
544. K. Effertz, S. Hinderlich and W. Reutter, *J. Biol. Chem.*, 1999, **274**, 28771.
545. W. K. Chou, S. Hinderlich, W. Reutter and M. E. Tanner, *J. Am. Chem. Soc.*, 2003, **125**, 2455.
546. A. S. Murkin, W. K. Chou, W. W. Wakarchuk and M. E. Tanner, *Biochemistry*, 2004, **43**, 14290.
547. Q. Zhang and H. Liu, *J. Am. Chem. Soc.*, 2000, **122**, 9065.
548. J. N. Barlow, M. E. Girvin and J. S. Blanchard, *J. Am. Chem. Soc.*, 1999, **121**, 6968.
549. S. W. Fullerton, S. Daff, D. A. R. Sanders, W. J. Ingledew, C. Whitfield, S. K. Chapman and J. H. Naismith, *Biochemistry*, 2003, **42**, 2104.
550. M. Soltero-Higgin, E. E. Carlson, T. D. Gruber and L. L. Kiessling, *Nat. Struct. Biol.*, 2004, **11**, 539.
551. D. A. R. Sanders, A. G. Staines, S. A. McMahon, M. R. McNeil, C. Whitfield and J. H. Naismith, *Nat. Struct. Biol.*, 2001, **8**, 858.
552. K. Beis, V. Srikannathasan, H. Liu, S. W.B. Fullerton, V. A. Bamford, D. A. R. Sanders, C. Whitfield, M. R. McNeil and J..H. Naismith, *J. Mol. Biol.*, 2005, **348**, 971.
553. J. N. Barlow and J. S. Blanchard, *Carbohydr. Res.*, 2000, **328**, 473.
554. A. R. Katritzky, A. M. El-Mowafy, G. Musumarra, K. Sakizadeh, C. Sana-Allah, S. M. M. El Shafie and S. S. Thind, *J. Org. Chem.*, 1981, **46**, 3823.
555. B. C. Smith and J. M. Denu, *J. Am. Chem. Soc.*, 2007, **129**, 5802.

CHAPTER 6

Heterolytic Chemistry Other than Nucleophilic Attack at the Anomeric or Carbonyl Centre

6.1 REARRANGEMENTS OF REDUCING SUGARS

6.1.1 Types of Rearrangements

Reducing sugars, like other carbonyl compounds, are readily enolised, and the enolisation of, for example, glucose will destroy the chirality at C2. The resulting enediol can be reprotonated on either diastereotopic face at C2, giving glucose and mannose, or on C1, giving fructose. The isomerisation of C2 epimeric aldoses and the corresponding ketose in mild alkali was discovered in 1895,[1] and is known by the names of its high-born discoverers, Cornelis Adriaan Lobry von Trostenburg de Bruyn and Willem Alberda van Ekenstein. Even after truncation of their names, the Lobry de Bruyn–Alberda van Ekenstein transformation is still the longest-named reaction in chemistry.[2] An enolisation mechanism is also employed by key enzymes which catalyse the interconversions of aldose and ketose phosphates, most notably triose phosphate isomerase (interconverting glyceraldehyde-3-phosphate and dihydroxyacetone phosphate)[3,4] and glucose 6-phosphate isomerase (interconverting glucose 6-phosphate and fructose 6-phosphate).[5] The latter enzyme will very occasionally misdirect a proton in the ES complex, resulting in the production of mannose 6-phosphate.[6] A fully-fledged "Lobry de Bruyn–Alberda van Ekenstein-ase", which will efficiently interconvert the 6-phosphates of glucose, mannose and fructose, has recently been discovered in the archeons *Aeropyrum pernix* and *Thermoplasma acidophylum*.[7]

The isomerisation of aldoses and ketoses by hydride transfer, although suggested for enzymic isomerisations of aldoses and ketoses (rather than their phosphates),[8] and having precedent in the metal ion-catalysed isomerisations of α-hydroxy ketones,[9] was considered to be the less common route until the paradigmal aldose–ketose isomerisation, that of glyceraldehyde to dihydroxyacetone, was examined by NMR spectroscopy.[10] Identification and analysis of all products of the reaction in D_2O permitted concentrations to be chosen which minimised side-reactions. A useful feature of the system was that it was possible to integrate the areas of proton resonances in the CHD and CH_2

groups of exchanged and unexchanged dihydroxyacetone separately, since the former was a triplet, split by the deuterium ($I = 1$).

The first-order rate constant for isomerisation of glyceraldehydes (Figure 6.1) increased with [OD^-] to reach a limiting maximum value of $\sim 1.2 \times 10^{-3}$ s^{-1} (at 25.0 °C) above ~80 mM [OD^-]. This was caused by deprotonation of the *gem*-diol of glyceraldehyde hydrate to give an inert species. In 10 mM KOD, the ratio of the CH_2 peak area to the CHD peak area in the proton spectrum of the product dihydroxyacetone was initially around 4, but decreased slightly with increasing degree of reaction because of base-catalysed exchange: it would be 2 for complete reaction by proton transfer and infinity for reaction by hydride shift. In pD 8.4 pyrophosphate buffer, a buffer greatly favouring general base-catalysed processes by statistical factors ($p = 1$, $q = 5$; see Section 1.2.1.3), the peak area ratio was constant at ~2, as expected. The enolate of glyceraldehyde 3-phosphate was known to be protonated on C1 at least 20 times faster than on C2,[11] so that the observation of isomerisation without exchange immediately established the existence of a hydride transfer mechanism. Zn^{II} promotes both the hydride shift and the enolisation mechanism. In acetate buffer of pD 5.7 with 115 mM Zn^{II} the peak area ratio started out in excess of 6.

In the absence of Zn^{II}, the similar dependence of both the hydride-shift pathway and the enolisation pathway on [OD^-] established that the hydrate diol anion was inert to both types of rearrangement, and that the hydride-shift mechanism probably went through the glyceraldehyde anion formed by deprotonation of O2.

In sterically hindered buffers such as hexafluoroacetone hydrate or pivalate in the presence of Zn^{II}, the hydride shift mechanism is completely dominant, and the second-order rate constant is proportional to [OD^-], but in high concentrations of acetate the enolisation mechanism, with $k_{obs} = k_3[Zn^{II}][Base]$, becomes predominant.

In the course of biological evolution, primitive enzymes which bind the appropriate transition states initially fairly poorly are slowly refined as a consequence of Natural Selection. Evolution by Natural Selection thus predicts that, wherever enzymes are found to catalyse equivalent reactions by very different mechanisms, as with aldose–ketose isomerisations by hydride transfer or proton transfer, the spontaneous, uncatalysed mechanisms will be found to proceed at comparable rates under ambient conditions.[i]

The stereochemistry of the enediols and enediolates involved in non-enzymic sugar isomerisations is unknown, but protein X-ray crystallography, particularly of isomerases in complexes with stable analogues of the enediolates, has established the hydroxyl stereochemistry to be *cis*. The argument from Natural

---

[i] The tenets of Darwinian evolution as applied to enzyme catalysis are thus falsifiable in the Popper sense. The existence of radically different enzyme reaction pathways only in those cases where the uncatalysed pathways are comparably facile is also a telling argument against any sort of intelligent design.

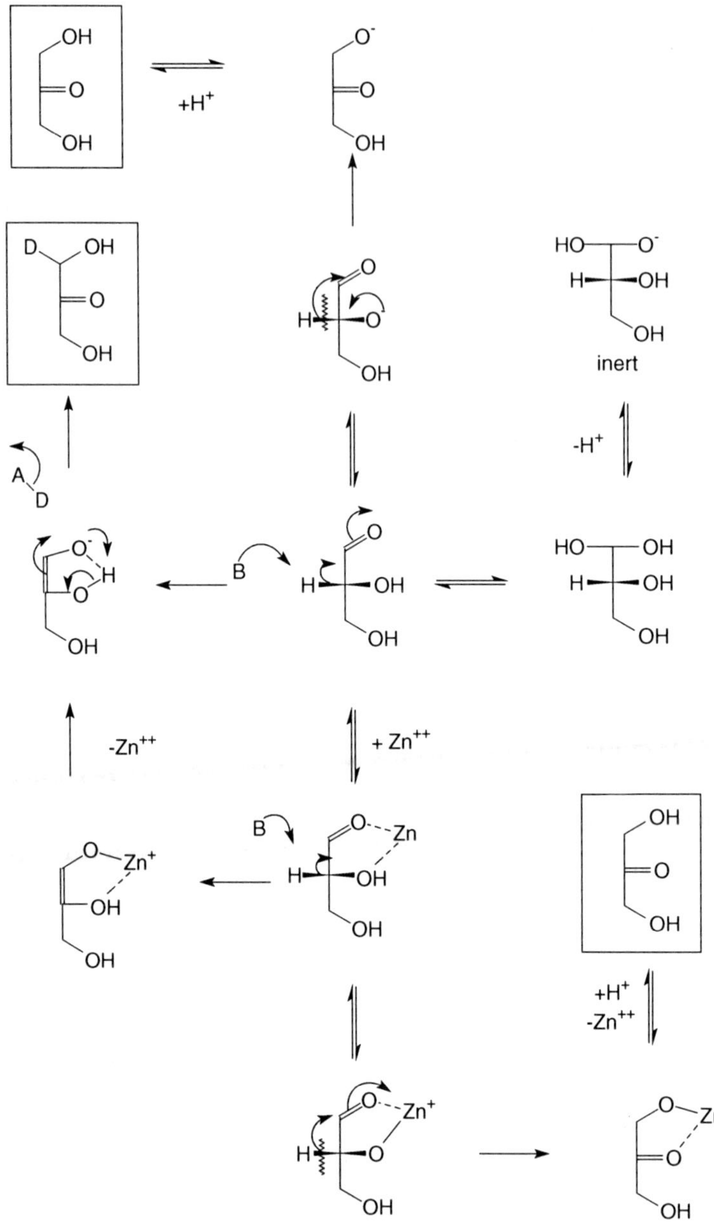

Figure 6.1 Pathways for the isomerisation of glyceraldehydes.

Selection then indicates that the lowest energy pathway for the isomerisation of aldoses and ketoses by the enolisation route likewise involves a *cis*-enediol(ate).

In addition to aldose–ketose interconversions by hydride and by proton transfer, a number of metal-catalysed reactions involving interconversion of epimeric aldoses or of 2-ketoses and 2-branched aldoses are important. If the metal is Mo^{VI} the reaction is the Bílik reaction,[12] but other metals are used, notably Ni^{II},[13] and there is even a possibility that the metal-catalysed route may be important in Lobry de Bryn–Alberda van Ekenstein rearrangements carried out with the classical reagent of calcium hydroxide.[2,14] The key mechanistic discovery was made in the Mo^{VI} system, when it was found that during epimerisation carbons exchanged places, so that [1-^{13}C]glucose gave [2-^{13}C]mannose (but not [2-^{13}C]glucose or any labelled fructose);[15] the Ni^{II}–diamine system was shown to be similar.[13] If the hydrogens are labelled, they migrate with the carbons to which they are attached, so that [2-^{3}H]glucose yields [1-^{3}H]mannose – there is no hydrogen washout. The mechanisms drawn by for both Ni^{II}[16] and Mo^{VI}[17] processes by their discoverers, however, focus very much on the stereochemistry of metal coordination and consider the key step to be a 1,2-shift of an alkyl group with its bonding pair of electrons, similar to those in the pinacol and benzylic acid rearrangements. To the author, the mechanism lacks plausibility since it cannot account for the following features of the reaction, without supplementary hypotheses:

(i) The necessity for a 3-OH–3-deoxymannose and 3-deoxyglucose with molybdate give 3-deoxyfructose,[18] rather than the Bílik product. 3-Deoxyfructose is expected to dominate the Lobry deBruyn–Alberda van Ekenstein equilibrium because of the absence of an electron-withdrawing group adjacent to the carbonyl. The acceleration of the enolisation mechanism by a Zn^{II} has been commented upon above and Mo^{VI} is probably acting similarly.
(ii) Side-products resulting from epimerisation at C3.
(iii) The necessity for a 4-OH.

Rather, the reactions will be interpreted as reverse aldol and then aldol reactions within the coordination shell of a metal, involving the transient existence of a metal complex of a *cis*-enediol and a hydroxyaldehyde (Figure 6.2). The isomerisation occurs when the carbonyl moves over the face of the enediol and recombines with the other carbon. The carbonyl group is a poor ligand and can thus break free of the complex: simple rotation about C3–C4 while still attached to the metal will yield side-products epimerised at C3. If the hydroxyaldehyde it is not held to the metal by other hydroxyls, particularly O4, it can escape completely.

6.1.2 Isomerisation by Enolisation – the Classic Lobry de Bruyn–Alberda van Ekenstein Reaction

6.1.2.1 Non-enzymic Enolisation. The carbanion chemistry of aldoses and ketoses themselves is masked by the ring opening step and mechanistic work is limited. However, a detailed examination has been made of the enolisation of L-glyceraldehyde 3-phosphate,[11] which cannot form monomeric rings. The use of the unnatural (L) enantiomer enabled any racemisation to the natural

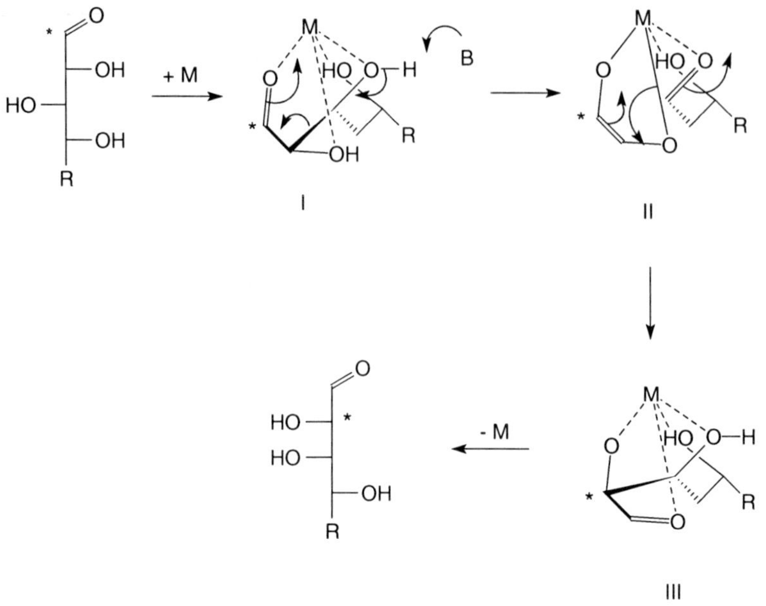

Figure 6.2 Reverse aldol–aldol mechanism for the Bílik and related reactions. In complex II, if the carbonyl group comes loose and rotates about C3–C4, the products will be epimerised at C3.

D-isomer to be monitored by enzyme-linked oxidation of NADH. The same technique was used to monitor the formation of dihydroxyacetone phosphate. The major reaction on formation of the enediolate, however, is loss of phosphate to give the enol of methylglyoxal.[ii] The formation of the enediolate is independent of pH between pH 6 and 10 because of intramolecular proton abstraction by the phosphate dianion. Above pH 10 and below pH 6, the pH–rate profile has a gradient of +1.0, in the former case a reflection of direct attack by OH^- on the phosphate dianion, in the latter a reflection of the proportion of reactive dianionic form of the substrate present.

The enolisation is catalysed by (relatively unhindered) substituted quinuclidine bases with a Brønsted β value of 0.45 and 0.48 for glyceraldehyde 3-phosphate and dihydroxyacetone phosphate. The elimination reaction of the enediolate (Figure 6.3) gives a "clock" whereby the relative efficiency of reprotonation can be measured from slopes of the plot of the ratio of reprotonation to elimination product against buffer concentration. This leads to a Brønsted α value of 0.47 for reprotonation to dihydroxyacetone phosphate (reprotonation to glyceraldehyde phosphate could not be observed in the quinuclidinium-catalysed reaction, but in the water-catalysed reaction,

[ii] The enediolate is stabilised to phosphate loss by the protein of triose phosphate isomerase, but occasionally comes off the enzyme and yields methylglyoxal, which is detoxified by another hydroxy ketone isomerase, glyoxylase.

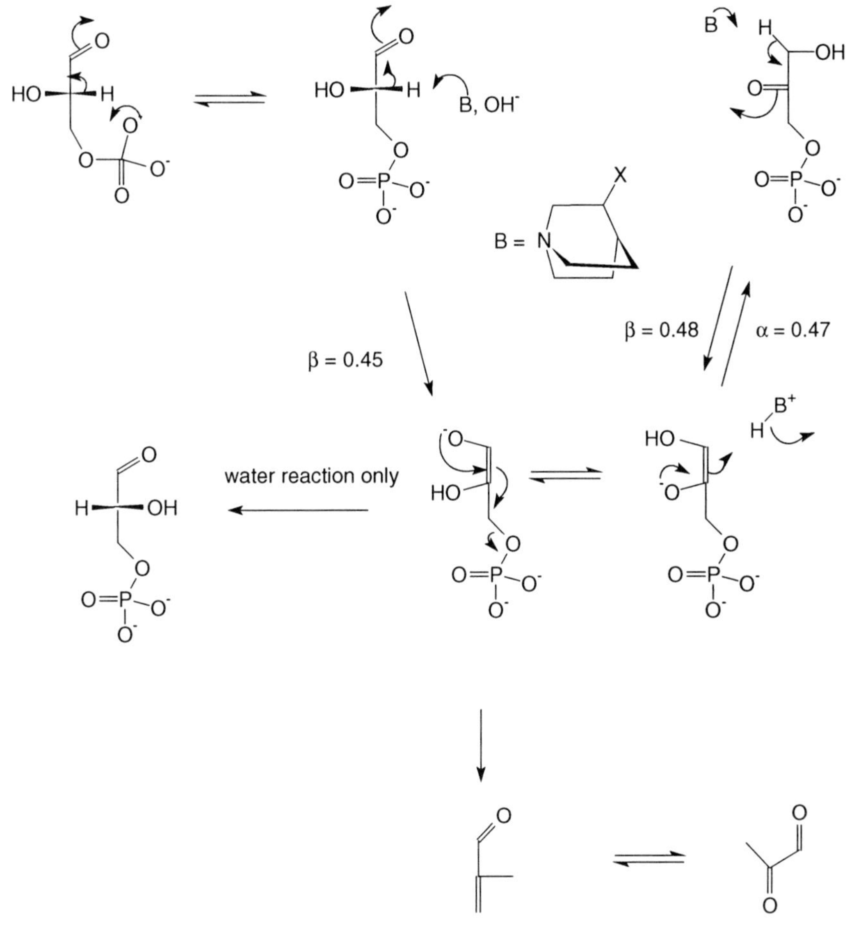

Figure 6.3 Generation and fate of the (probably *cis*) enediolate from dihydroxyacetone and glyceraldehyde phosphates.

racemisation was five times slower than isomerisation to dihydroxyacetone phosphate).

The isomerisation reactions of free sugars are associated with a UV absorbance at 310 nm,[19] attributed to the enediolate. This is only observed in the absence of oxygen, with which enediolates react. Direct monitoring of the 310 nm absorbance led to the suggestion that a key intermediate in the enolisation of reducing sugars was a species in which the sugar chain was in a conformation similar to that which it was constrained to adopt in the ring, and that some interaction between the 5-anion and new carbonyl group persisted (Figure 6.4). Such a model explained the relative rates of enolisation (particularly the slow deprotonation of mannose, with an equatorial H2 pointing away from the O′5) and is reminiscent of the intermediate complex proposed in the mutarotation of 5-thioglucose (see Section 1.2.1.5).

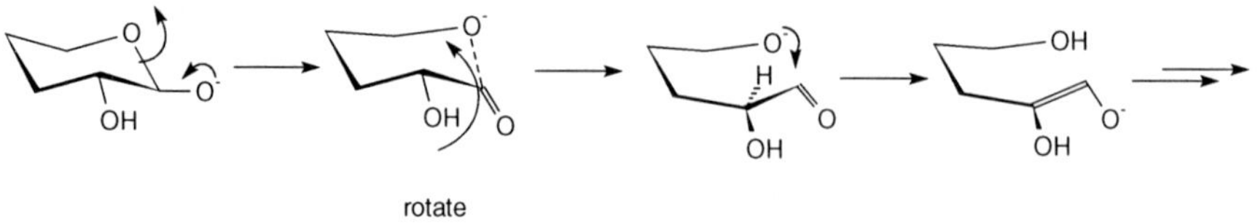

Figure 6.4 Intermediate in the enolisation of sugars?

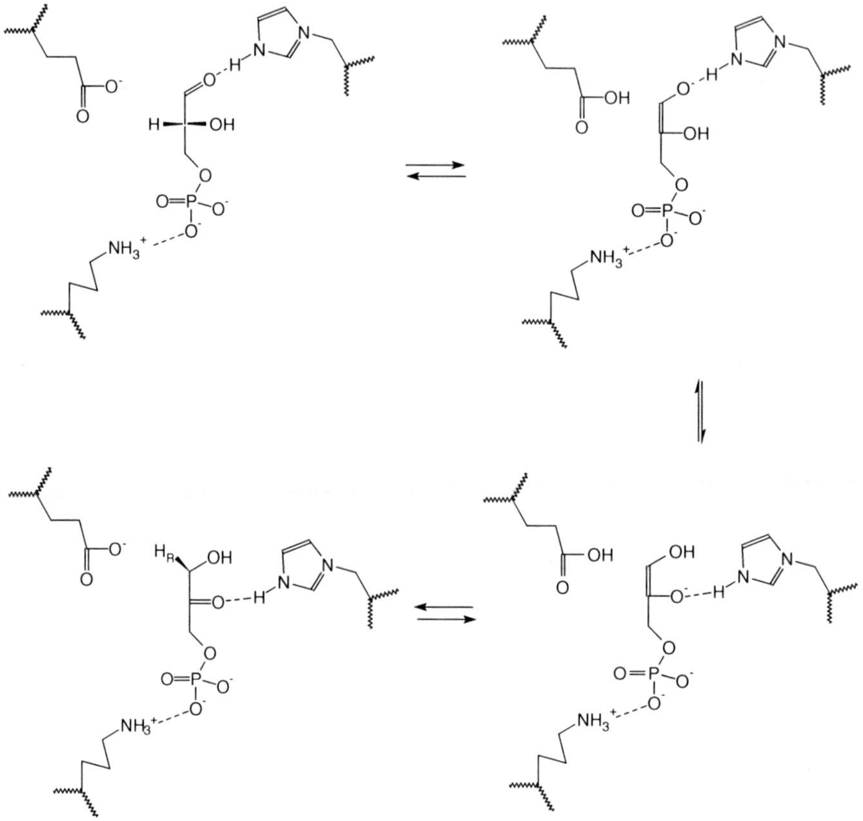

Figure 6.5 Triose phosphate isomerase mechanism.

6.1.2.2 Enzymic Enolisation – the Classic Aldose–Ketose Phosphate Isomerase Mechanism. The paradigmal aldose–ketose isomerase is triose phosphate isomerase (TIM), established by isotope labelling experiments to involve only one catalytic base which conveyed the *proR* H1 of dihydroxyacetone phosphate to C2 of glyceraldehyde phosphate (and vice versa), with partial exchange with solvent[3] (if there had been two enzymic bases, no retention of label would have been observed, and if the reaction had been hydride transfer, no exchange at all with solvent would have been found) (Figure 6.5). Affinity labelling showed

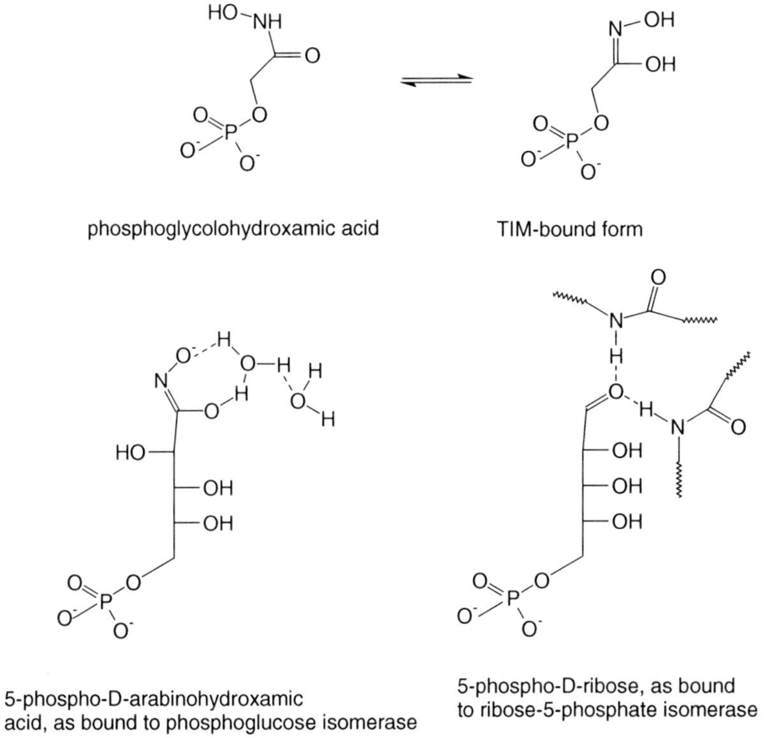

phosphoglycolohydroxamic acid

TIM-bound form

5-phospho-D-arabinohydroxamic acid, as bound to phosphoglucose isomerase

5-phospho-D-ribose, as bound to ribose-5-phosphate isomerase

Figure 6.6 Enediolate analogues and mechanisms of enediolate stabilisation in glucose-6-phosphate (bottom left) and ribose-5-phosphate isomerase (bottom right).

that the base was a Glu165 (in both chicken and yeast enzymes).[20] An analogue of the enediolate intermediate (Figure 6.6) is the imide tautomer of phosphoglycolohydroxamate, and X-ray structures of the yeast[21] and chicken[22] enzymes in complex with this transition state (or high-energy intermediate) analogue (see Section 5.3.2) revealed that after the substrate had bound, a loop of protein had moved to cover the intermediate analogue, and was held in place by a key hydrogen bond from Tyr.[23] The loop keeps the enolate from escaping. Another feature of TIM paralleled in other aldose–ketose isomerases is the binding of the phosphate moiety by a cationic residue [in the case of TIM, Lys12 (yeast)/13 (chicken)]:[24] the use of such phosphate "handles" is common in the enzymes of primary metabolism such as TIM. TIM will in fact catalyse the enolisation of glyceraldehyde itself, but at a bimolecular rate little better that the catalytic constant for enolisation by a general base – some 84% of the free energy barrier lowering comes from "Circe effect" interactions between the phosphate and the enzyme.[25]

A single base mechanism, with glutamate as the base, is common to most aldose–ketose phosphate isomerases, but the electrophilic machinery to stabilise the enolate appears to differ from enzyme to enzyme. TIM uses His95 in its *neutral* form. ^{13}C and ^{15}N NMR spectra establish that the ionisation state of His95 does not alter between pH 9 and 4.5, indicating that the first pK_a (of the imidazolium cation) is decreased by at least 2–2.5 units and the second (of the imidazole, giving the imidazolate anion) presumably likewise.[26] The effect is attributed to the electric field arising from a short stretch of α-helix.[27] This unusually acidic neutral histidine residue can be shown to polarise bound substrate. With efficient editing of Fourier transform infrared spectra taken in D_2O, it is now possible to monitor carbonyl group stretching vibrations of enzyme-bound ligands, and the $\nu C{=}O$ band of dihydroxyacetone phosphate moves from 1732 cm^{-1} in solution or bound to the H95N or H95Q mutants to 1713 cm^{-1} bound to active enzyme.[28]

Ribose-5-phosphate isomerase, which catalyses the interconversion of ribose-5-phosphate and ribulose-5-phosphate, activates the carbonyl differently. The aldose substrate can form a furanose (but not a pyranose) ring, but the furanose forms are so disfavoured (see Table 1.1) that there are likely to be appreciable amounts of open-chain sugar present at equilibrium. Moreover, this equilibrium, involving only furanose forms whose opening is possibly assisted by general acid catalysis from the 5-phosphate, is likely to be achieved rapidly. Although the enzyme from other sources crystallise with ring forms of the substrate bound, that from *Thermus thermophilus* crystallises with both ribose-5-phosphate and arabinose-5-phosphate in the straight-chain forms.[29] The carbonyl group and potential oxyanion are bound in a classic oxyanion hole made up of main-chain amides, whose N–H bonds therefore play the same role as the N–H bond of neutral histidine in TIM. The basic catalytic Glu (108) and phosphate-binding Lys (199), however, parallel TIM.

Polarisation of the substrate carbonyl group appears to be achieved in yet a third way by mammalian glucose 6-phosphate isomerase, which interconverts glucose and fructose 6-phosphates, The crystal structure of the rabbit enzyme in complex with the reactive intermediate analogue D-arabinohydroxamic acid ($K_i = 0.2\ \mu M$) (in its hydroximic form) reveals a cluster of four water molecules hydrogen bonded to each other, to the counterparts of O1 and O2 of the enediolate intermediate and to an active site arginine.[30] Grotthus mechanisms of proton transfer between the two tautomers of the enediolate probably occur.

Spontaneous ring opening of the pyranose substrate is unlikely to be fast enough to generate adequate concentrations of open-chain glucose 6-phosphate. The suggestion[5] that, in the human enzyme, Lys518, close to the 6-phosphate, donates to O5 in the ring opening reaction seems implausible, however, since the substrate already possesses a better intramolecular general acid in the phosphate group.

Triose phosphate isomerase was the subject of a series of elegant and comprehensive isotope exchange experiments, which led to the establishment of the complete free energy profile for the catalysed reaction.[31] In the following discussion, glyceraldehyde 3-phosphate will be abbreviated to G3P, dihydroxyacetone phosphate to DHAP and the enediolate to INT. The enzyme complexes and transition states will be denoted as in Figure 6.7. Reference to one of

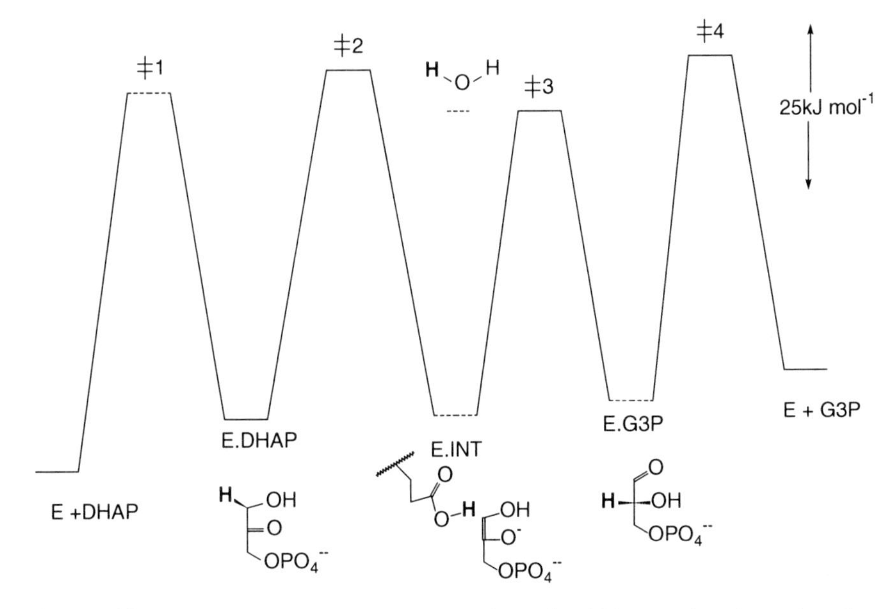

Figure 6.7 Free energy profile for the action of rabbit muscle triose phosphate isomerase. The profile is drawn approximately to scale: the exchanged hydron is shown in bold and the barrier to hydron exchange as a dotted line. Other dotted lines are upper or lower limits.

the participants in the equilibrium as "substrate" means the experiments were carried out with excess over enzyme and initially none of the other isomer.

The key finding was that the proton on Glu165 in the E–INT complex exchanged almost, but not quite, completely with solvent: the tritium in the 1-*R* position of DHAP substrate was incorporated into glyceraldehyde 3-phosphate product only to the level of 3–6%, although the fact that this level increased with the degree of reaction indicated a kinetic isotope effect on the protonation of INT.[32] In the other direction, with tritiated G3P as substrate, virtually no tritium was incorporated into product, indicating complete washout and hence that (in modern terms) that the forward commitment of the enzyme–enediolate complex was low. The idea that DHAP as substrate was enolised only the once before carrying on to product, whereas with G3P as substrate the system went back and forwards several times to INT, was confirmed by the measurement of $^{D}(V) = 2.9$ for (*R*)-[^{2}H]DHAP, but $^{D}(V) = 1.0$ for [2-^{2}H]G3P, as the deuterium was washed out of the substrate before it went over the rate-limiting barrier.[33]

Further information came from use of the exchangeable hydron on Glu165 in the E–INT complex to label substrate and product in tritiated water. With DHAP as substrate, tritium is introduced into DHAP about one-third as frequently as it is introduced into G3P,[34] and with G3P as substrate tritium was introduced into G3P about one-third as frequently as it was introduced into DHAP.[35] (Of course, if ‡2 and ‡3 had been much higher than other transition states, the partitioning of solvent isotope between DHAP and G3P should have been independent of the direction from which the E–INT complex was approached – that it is not gives information about preceding steps). Deconvolution of the system – assisted by a rigorous derivation of the equations governing isotope incorporation in this and similar systems[36] – was simplified by the discovery that solvent deuterium was incorporated into G3P and DHAP exactly as solvent tritium – *i.e.* there was no kinetic isotope effect on the hydron exchange of Glu165–COOH in E–INT.[37]

Construction of the free energy profile (Figure 6.7) for the whole reaction required a choice of thermodynamic standard state, since bimolecular and unimolecular rate constants were being compared. This was taken as 40 μM, the physiological concentration of the triose phosphates.

A key conclusion from the free energy profile in Figure 6.7 was that TIM had reached "evolutionary perfection" – at physiological concentration, all the free energy barriers to reaction, including those of normally diffusion-limited enzyme–substrate combination, were of comparable size. To the author, the idea seems somewhat circular, since evolutionary changes in catalytic activity, particularly in key enzymes of primary metabolism such as TIM, alter physiological concentrations of substrates.[38]

On the basis of the X-ray structure of L-fucose isomerase of *Escherichia coli*,[39] it has been suggested that this Mn^{II}-dependent enzyme acts via a metal ion-stabilised enediolate pathway, similar to Zn^{II} catalysis of dihydroxyacetone enolisation, but the evidence so far is exclusively structural.

6.1.3 Isomerisation of Reducing Sugars by Hydride Shift

Non-enzymic aldose–ketose isomerisations that are acid catalysed appear to involve a 1,2-hydride shift. During acid-catalysed rearrangement of glucose to fructose, the label of [2-^{3}H]glucose substrate is retained in the [1-^{3}H]fructose product, distributed equally between the *proR* and *proS* positions.[40] In the reverse sense retention of the label of tritiated fructose in the glucose and mannose products was not complete. Similar observations were made for the xylose–xylulose interconversion.[41] With an appropriate sugar configuration (ribose),[42] even the base-catalysed reaction proceeds partly with retention of label, presumably by the same mechanism as with trioses.

The isomerisation of glucose to fructose is commercially important in the production of high-fructose corn syrup, and this has led to examination of glucose isomerases from many sources. The enzymes are invariably homotetramers of bacterial origin, with the size of the monomer being around 50 kDa for most bacteria (Group 2), but slightly smaller (43 kDa) for Actinomycetes (Group 1).[43] They require two divalent metal ions of ionic radii less than 0.8 Å to function and to stabilise the protein: Co^{2+}, Mn^{2+} and Mg^{2+} serve.[43] Fe^{2+}, intermediate between Mn^{2+} and Co^{2+} in the first transition series, unsurprisingly can also act,[44] but because of the accessibility of the Fe^{III} state under aerobic conditions is not usually employed.

The enzymes were originally xylose rather than glucose isomerases, the specificity being reflected more in K_m (1–3 mM for xylose, 150–400 mM for glucose)[iii,43] than k_{cat} (1–3 s^{-1} for glucose, 7–8 s^{-1} for xylose).[43] They will tolerate changes at C3 of the substrate (so ribose is a substrate). 3-Deoxy-3-fluoroallose and -glucose are inhibitors, but do not lose F^-. This is telling evidence against the 1,2-enediolate as an intermediate (see Figure 6.3),[45] as is the occurrence of exchange of the transferred hydrogen with solvent less than once per million turnovers.[8]

The enzymes bind the α-pyranose forms of aldose substrates, but available structures with bound ligands all have the ligands in an open chain form, so the ring opening mechanism remains uncertain. Particularly informative is the structure of the *Streptomyces olivochromogenes* enzyme in complex with threonohydroximinate.[46] The ligand is a slow, tight-binding inhibitor ($K_i < 0.1$ μM), of a type which also mimics the enediolate intermediate in proton-transfer isomerases. It would appear that the *cis* stereochemistry of O1 and O2 and coplanarity of O1, O2, C1, C2 and C3 are common to the enolisation transition state and the hydride-transfer transition state. The mechanism of hydride transfer would then appear to be a suprafacial migration above this plane Figure 6.8. The non-linear hydride transfer would explain the low primary kinetic isotope effects [$^D(V) \approx {}^D(V/K) = 2.6$–4.6], which are probably intrinsic. The effects with Mn^{2+}-activated enzymes are greater than those with Mg^{2+},[43] suggesting slight movement of the metal as part of the reaction coordinate, in accord with suggestions from the X-ray structure.[46]

Rhamnose isomerase of *Escherichia coli* appears from the structure of its ligand complexes to have a similar mechanism to glucose isomerase.[47]

6.1.4 The Bílik and Related Reactions

The Bílik reaction[17] occurs readily with molybdate (Mo^{VI}) solutions at mildly acidic pH (0.1–6.0, optimum ~3). As discussed above, C2 epimeric aldoses are brought into equilibrium, with concomitant exchange of H2 and C2 and no crossover of label. A mechanism in which a reverse aldol reaction yields an aldehyde and an enediolate coordinated to the same metal and which can recombine from the same faces of the enediolate or aldehyde carbonyl, is sketched in Figure 6.9. This also explains the observed epimerisation at C3, particularly of aldopentoses. The first step yields a complexed glyceraldehyde which is attached to the template metal by only two OH groups. Since the carbonyl group oxygen is a weak ligand, the carbonyl group can come free and can rotate about C1–C2, allowing side-products from addition of the enediolate to the opposite face of the aldehyde carbonyl. Indeed, all four D-aldopentoses can be produced from D-xylose under the conditions of the Bílik reaction.[48] The alkyl shift mechanism, by contrast, requires the postulation of a process involving two concurrent hydride shifts in competition with the main reaction.[49]

[iii] A 400 mM solution of glucose is 7.2% (w/v) in carbohydrate; it is very doubtful if the separation of k_{cat} and K_m under these conditions is meaningful, as the departure of the V versus [S] plot from proportionality could be an effect of the change of medium.

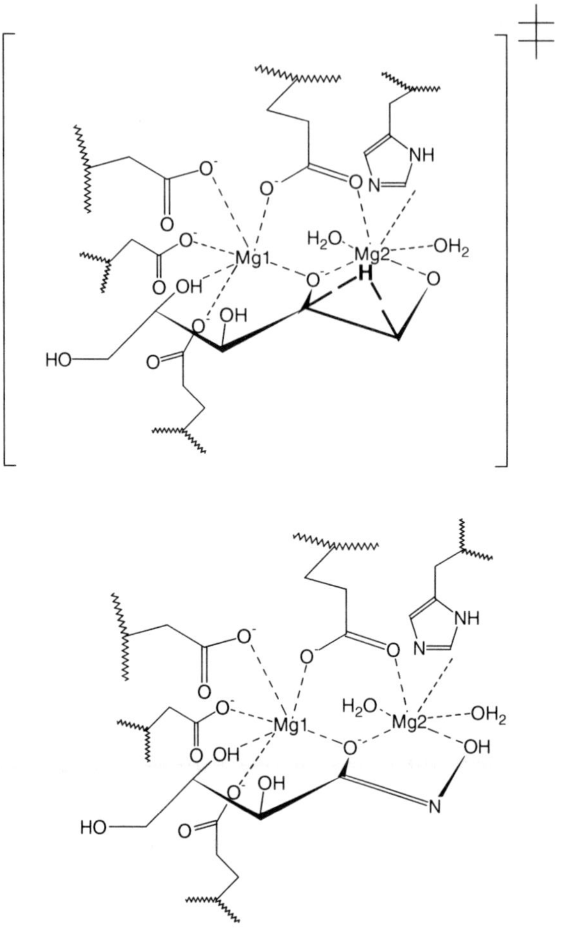

Figure 6.8 Binding of a transition state analogue to (bottom), and proposed reaction mechanism of (top), xylose–glucose isomerases. O2 of the substrate (O1 of the transition state analogue) must be deprotonated in order to bind both Mg^{2+} ions.

The Bílik reaction applied to 2-ketoses yield 2-hydroxymethyl aldoses in which the tertiary carbon originates from C2 of the ketose and the C2 hydroxyl is on the opposite side to the C3 hydroxyl of the ketose (in the Fischer projection). Thus, D-fructose yield D-hamamelose. The position of equilibrium, however, lies towards the straight-chain sugar, although it can be pulled over somewhat towards the branched-chain aldose by the addition of borate. The mechanism in Figure 6.9 again explains the main reaction, but not the formation of sorbose as a by-product, which probably arises from a metal ion-promoted hydride shift, as there is no isotope exchange with solvent.[17] The Bílik reaction can be applied to the production of 1-deoxy-D-xylulose from 2-*C*-methyl-D-erythrose; the reaction is particularly clean and only the two

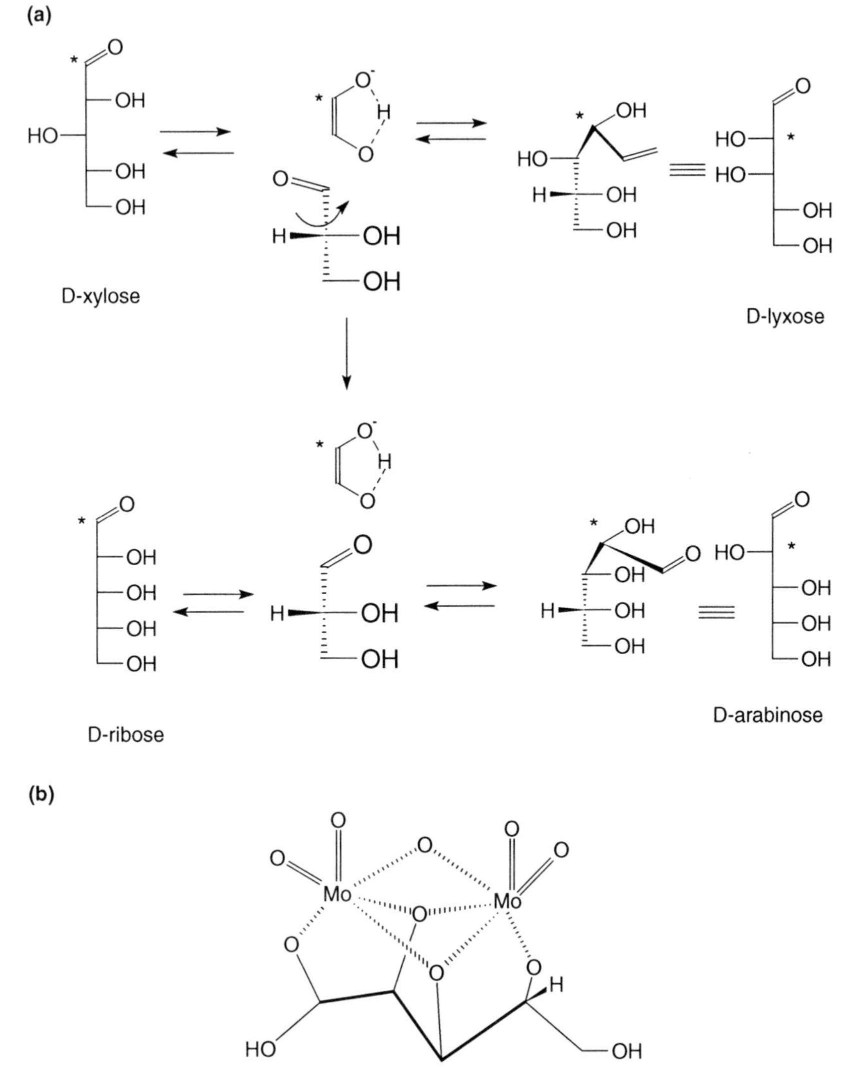

Figure 6.9 (a) Possible reverse aldol–aldol pathways for the Bílik reaction of xylose. Coordination to the metal is not shown. The enediolate in the intermediate is intended to be below the plane of the paper. (b) Dinuclear complex of Mo^{VI} and D-arabinose. Single bonds to molybdenum are drawn as broken. On the alkyl shift mechanism the C3 shifts with its bonding pair of electrons from C2 to C1.

participants in the equilibrium (95% straight-chain 1-deoxyxylulose:5% 2-*C*-methyl D-erythrose) are seen by NMR spectroscopy.[50]

The structures of the complexes formed from sugars and Mo^{VI} under the conditions of the Bílik reaction have been studied by NMR spectroscopy[51] and X-ray crystallography[52] They are based on two oxygen-coordinated Mo^{VI}

octahedra sharing a face, with the sugar providing two of the three shared oxygens and being in a ring form, or a straight chain form with one oxygen of the aldehyde hydrate coordinating the metal. That species can be isolated, of course, does nor mean they are on the reaction pathway. Indeed, the very fact that they *can* be isolated suggests they are not, a suggestion reinforced by the isolation of complexes of ring-form sugars, in which the reactive carbonyl is masked.

Cleaner reactions of the same type are achieved by the use of a nickel–ethylenediamine system in methanol, which uses stoichiometric concentrations of metal.[16] The sugar, $NiCl_2.6H_2O$ and *N*-substituted ethylenediamine in the molar ratio 1:1:2 are refluxed together. There is no epimerisation at C3 of aldopentoses,[13] in accord with the more stable Ni^{II} complexes formed. Nickel coordination by an amine with more than two donor nitrogens results in complete loss of catalytic activity, as the very effective coordination with the nitrogen ligand means that stable complexes with the sugar can now no longer be formed. The proportion of branched-chain 2-aldoses such as hamamelose from 2-ketoses is usefully higher than the Bílik reaction, probably because of a different position of equilibrium when complexed to the metal. The nature of the product metal complexes is unknown, but Ni^{II} can adopt tetrahedral, square planar and octagonal coordination and an octagonal complex with a single diamine ligand *cis* would provide enough sugar coordination sites for the mechanism in Figure 6.9. Osanai's suggestion[16] that an amino alcohol formed from the sugar and diamine, of the type $HO–CHR–NRCH_2CH_2NR'_2$, could be a ligand cannot, however, account for the success of tetramethylethylenediamine as Ni^{II} ligand; species of the type $^{-}O–CH–N^{+}R_3$ are very unstable, sometimes "too unstable to exist".

The enzyme 2-*C*-methyl-D-erythritol-4-phosphate synthetase appears to catalyse a Bílik reaction (Figure 6.10): the substrate 1-deoxyxylulose-5-phosphate is converted to the title compound via an intermediate aldehyde, whose carbonyl derives from C3 of the substrate. The first step is thus a Bílik reaction and the aldehyde is subsequently reduced by the enzyme using NADPH as reductant, The X-ray crystal structure of the *Escherichia coli* enzyme in complex with the promising antimalarial Fosmidomycin (a hydroxamic acid) reveals a bound Mn^{2+} coordinated to oxygens equivalent to the substrate carbonyl and O3.[53] The stereochemistry and regiochemistry follow the normal Bílik course, although the crystallographers favour an alkyl shift rather than a reverse aldol–aldol mechanism. The intermediate aldehyde has been shown to be a catalytically competent intermediate.

6.1.5 Beyond the Lobry de Bruyn–Alberda van Eckenstein Rearrangement – Deep-seated Reactions of Sugars in Base

Under more vigorous conditions, of higher alkalinity or temperature, reducing sugars undergo more deep-seated rearrangements than enediol rearrangements involving only C1 and C2. In principle the enediolate rearrangements can continue down the carbon chain, but in practice they are intercepted by $E1_{CB}$ and reverse aldol reactions. Particularly important are expulsions of leaving groups from enediolates at the end of polysaccharide chains (*cf.* Figure 6.3). If

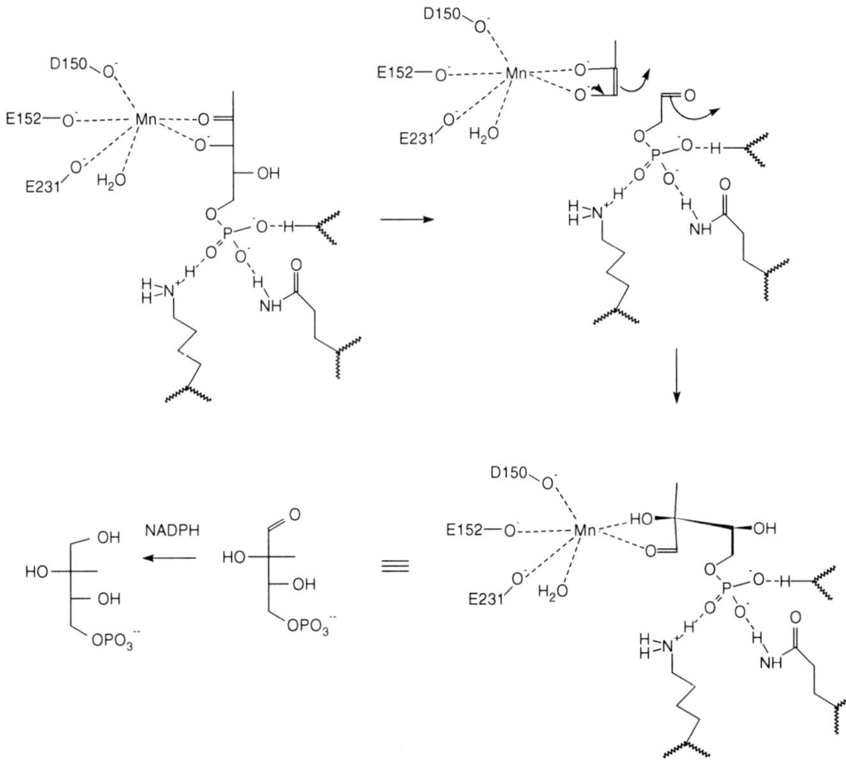

Figure 6.10 Reaction catalysed by 2-*C*-methyl-D-erythritol-4-phosphate synthetase. The reaction is drawn with a metal-templated reverse aldol–aldol mechanism, rather than the alkyl shift mechanism. The prop-2-en-2,3-diolate is below the plane of the glycolaldehyde phosphate.

the leaving group is the anomeric oxygen of another sugar, then its expulsion will generate a new reducing sugar which can degrade similarly. This stepwise, alkaline degradation of polysaccharides from the reducing end via base-catalysed rearrangement of reducing-end residue *n* leading to the expulsion of O1 of residue *n* − 1 is of the highest economic and technical importance. Most of the pulp for making high-quality paper that contains little or no lignin[iv] comes from a century-old chemical treatment of wood chips known as the kraft[v] process, in which the chips are cooked[vi] with a several molar solution of NaOH and NaSH

[iv] Known as "wood-free" pulp. Low-quality pulp for newsprint is made by mechanically defibrillating wood, often with some thermal and/or chemical treatment.

[v] Sometimes known as the sulfate process, since the sulfur is added as sodium sulfate. It is then reduced to S^{-II} by burning with the evaporated "black liquor" (lignin and carbohydrate fragments from a previous cook) in an insufficient supply of air. Water extraction yields "green liquor", a solution of (largely) sodium sulfide and sodium carbonate and which is treated with slaked lime [$Ca(OH)_2$]. Insoluble calcium carbonate precipitates and the resulting solution of sodium hydroxide and sulfide ("white liquor") is used for a new cook.

[vi] The correct technical term.

at 160–185 °C for several hours.[54] During this process the peeling reaction destroys the softwood hemicelluloses, largely galactoglucomannan, but the production of carboxylic acids from carbohydrate fragments causes the alkalinity of the cooking liquor to fall and xylans, which had initially dissolved, partly reprecipitate. The peeling reaction does not continue indefinitely, but rearrangements which convert the reducing end to an alkali-stable species (stopping reactions) eventually take place.

The outlines of the basic mechanisms of alkaline sugar degradation were proposed by Nef,[55] put into modern electronic form by Isbell[56] and confirmed by a series of studies of the products of singly-*O*-methylated glucoses and fructose by Richards and co-workers.[57–59] These studies were performed with paper chromatography as the analytical tool: application of modern methods of analysis to the products of the action of calcium hydroxide at 100 °C on glucose[60a] and fructose[60b] identified over 50 products. The main reactions can be rationalised as follows.[61]

Initial enolisation of glucose yields the 1,2-enediolate. Expulsion of the 3-OH or 3-OR yields a 1-aldehydo-3-deoxy-2,3-enol. This tautomerises to the α-ketoaldehyde and addition of OH^- at C1, followed by shift of H1 as hydride,[vii] yields a pair of straight-chain, C2-epimeric, 3-deoxygluconic acids known as *metasaccharinic acids*. These are the predominant six-carbon products from the peeling reaction of 1→3-linked polysaccharides.

Alternatively, the 1,2-enediolate could accept a proton at C1, yielding fructose, which could then undergo deprotonation at C3. The C2–C3 enediolate could then expel the substituent at C4, yielding a 4-deoxy-3,4-enol, which could tautomerise to a 2,3-diketone. Addition of OH^- at C2, followed by 1,2-migration of CH_2OH,[vii] or addition of OH^- at C3, followed by 1,2-migration of $-CH_2-CHOH-CH_2OH$,[vii] would in both cases yield a pair of C2-epimeric 2-*C*-hydroxymethyl-3-deoxyaldopentonic acids, the *isosaccharinic acids*. The isosaccharinic acids are the predominant six-carbon products of the peeling reaction of 1→4-linked polysaccharides.

Alternatively, the 2,3-enediolate could expel the 1-OH, yielding a 1-deoxy-2-enol. Tautomerisation would now give a 1-deoxy-2,3-diketone. Nucleophilic addition of OH^- at C2, followed by 2,3 migration of the methyl group, or at C3, followed by 3,2-migration of the glycerol group, would in both cases yield a pair of C-2 epimeric 2-*C*-methylaldopentonic acids, the *saccharinic acids*.

Tautomerisation further down the chain, to give a 3,4-enediolate, which then expels the C5 substituent, gives, after tautomerisation, a 5-deoxy-3,4-diketone. Base-catalysed cleavage of these gives *parasaccharinic acids*.[60] These minor products can be detected only with modern analytical techniques, but illustrate the generality of the basic mechanism, which is illustrated in Figure 6.11.

[vii] This is sometimes referred to as a "benzylic acid rearrangement". The paradigmal benzylic acid rearrangement is the conversion of benzil, Ph–CO–CO–Ph, to benzylic acid, $Ph_2COH-COOH$, in concentrated alkali. It seems to the author to stretch the term misleadingly to describe the migration both of alkyl groups (less mobile than phenyl) and hydride (more mobile), as "benzylic acid rearrangements".

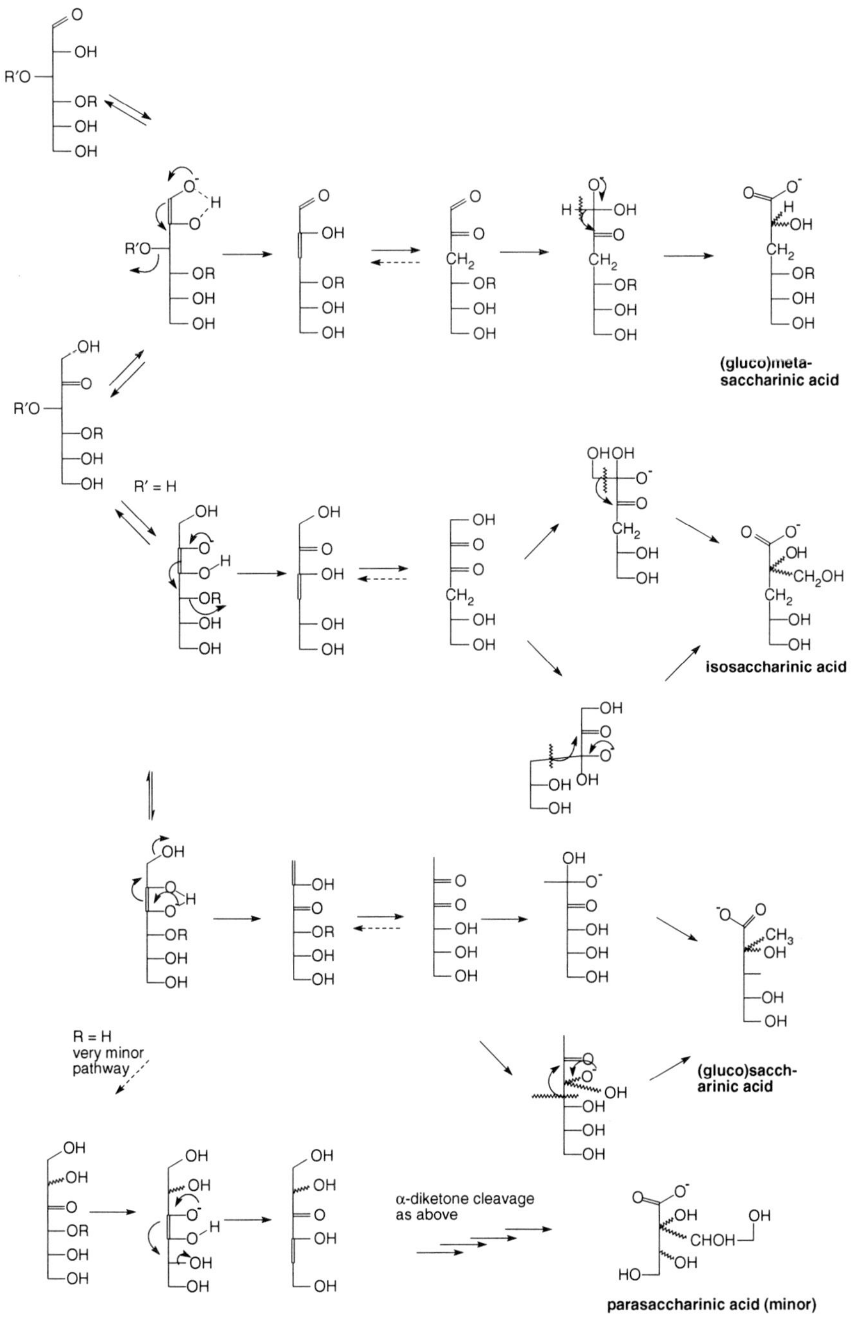

Figure 6.11 Nef–Isbell–Richards mechanisms for degradation of glucose and fructose and the peeling reaction of 1→3- and 1→4-linked polysaccharides. R or R′ can be another sugar residue, when the mechanism of the peeling reaction is illustrated.

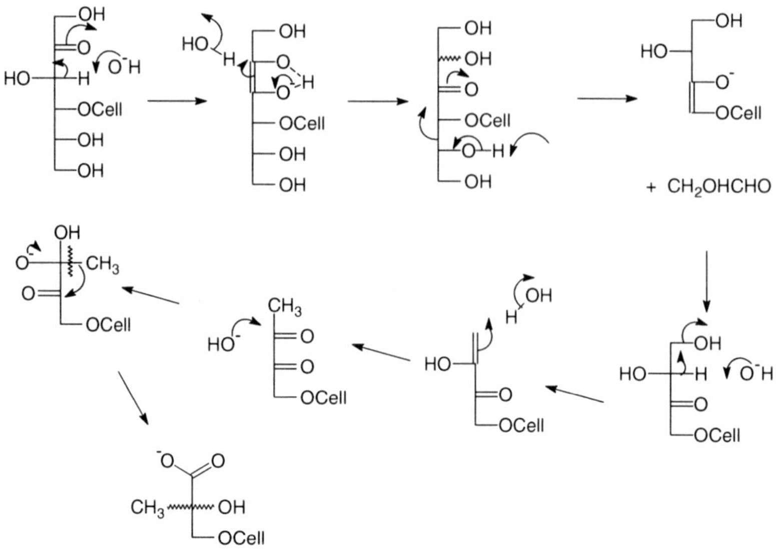

Figure 6.12 Stopping the peeling reaction by formation of 3-*O*-cellulose glycosides of 2-*C*-methylglyceric acids.

The commonest "stopping" reaction in the case of cellulose pulping arises from the pathway which gives glucometasaccharinic acid from glucose (Figure 6.11); the reducing end is converted to an alkali-inert carboxylic acid before OR is expelled. Another important stopping reaction is the formation of derivatives of 2-*C*-methylglyceric acid, probably by a mechanism similar to that in Figure 6.12.[62]

A metasurvey of reports of products from the action of alkali on glucose and cellulose and cellooligosaccharides[61] indicated that salts of formic (HCOOH), acetic (CH_3COOH), glycolic ($CH_2OHCOOH$) and lactic ($CH_3CHOHCOOH$) acids were the major low molecular weight products and glucometasaccharinic acids the major C_6 product. Analyses of kraft black liquors give similar results,[63] but the proportion of glucoisosaccharinic acid for softwoods is around double that for hardwoods. The major hardwood hemicellulose is xylan, whereas the major hemicellulose for softwoods, which is almost completely destroyed during pulping, is galactoglucomannan; it seems likely that glucoisosaccharinic acid is a major product of the degradation of the hexose-based polysaccharides.

2-Hydroxybutanoic acid is a major organic component of kraft black liquors from hardwood pulping, but is formed only in trace quantities from the degradation of glucose polymers. It seems likely that it arises from the xylan by a process along the lines of that illustrated in Figure 6.13.

Interception of the α-dicarbonyl compounds produced in the alkaline degradation of reducing sugars by *p*-hydroxybenzoic acid hydrazide (PAHBAH) (or possibly their production via Maillard-type processes) forms the basis of the relatively recently introduced PAHBAH method for determining reducing sugars.[64] Unlike traditional reducing sugar assays, which depend on

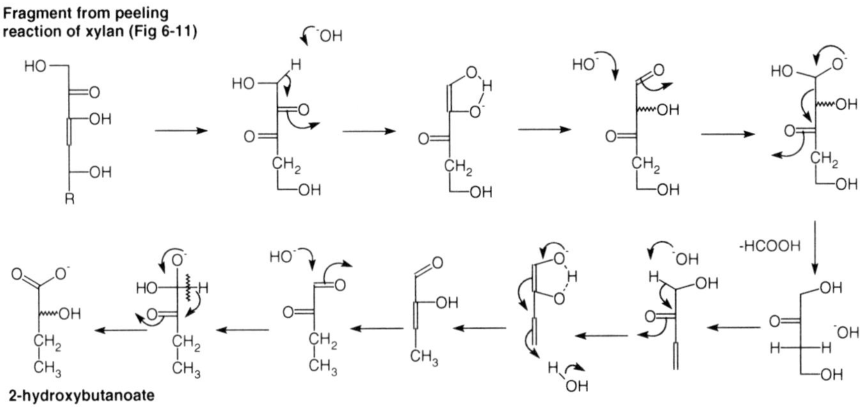

Fragment from peeling reaction of xylan (Fig 6-11)

2-hydroxybutanoate

Figure 6.13 Possible mechanism for the formation of 2-hydroxybutanoate from the peeling reaction of xylan.

reduction of transition metals by enediolates and are thus subject to interference by one-electron reductants, the intense yellow colour, attributed to the ionised form of the bis(hydrazone) (see Figure 7.8), is produced only from sugars.

An enzymic counterpart of these complex base-catalysed rearrangements of sugars may be the reaction catalysed by 4-phospho-3,4-dihydroxy-2-butanone synthetase. The enzyme catalyses the formation of the eponymous intermediate in secondary metabolism from ribulose 5-phosphate. Labelling studies[65] indicated that C1–C3 of the substrate became C1–C3 of the product, that H3 of the substrate derived from solvent and that C4 was lost as formate. X-ray crystal structures of the native enzyme[66] and a partly active mutant in complex with the substrate are available.[67] The active site of the enzyme from *Methanococcus jannaschii* contains two metals, which can be any divalent cations of the approximate radius of Mg^{2+} or Mn^{2+}, the two usually observed. Their disposition is very reminiscent of those in the hydride transfer aldose–ketose isomerases, but also to ribulose-5-phosphate carboxylase, which works by an enolisation mechanism, so the enolisation route suggested by Steinbacher et al.[67] is repeated in Figure 6.14, as is the Bílik-type alkyl group shift, for which an equivalent reverse aldol–aldol mechanism cannot be written.

6.2 FURTHER REACTIONS OF GLYCOSYLAMINES

6.2.1 The Amadori and Heyns Rearrangements[68]

The open-chain forms of glycosylamines (see Section 1.2.3) are Schiff bases and are labilised towards deprotonation of an adjacent carbon in much the same way as the open-chain forms of the sugars themselves. Reprotonation of the first-formed enamine on the carbon bearing the nitrogen atom will give an α-aminocarbonyl compound. The formation of a 1-deoxy-1-amino-2-ketoses

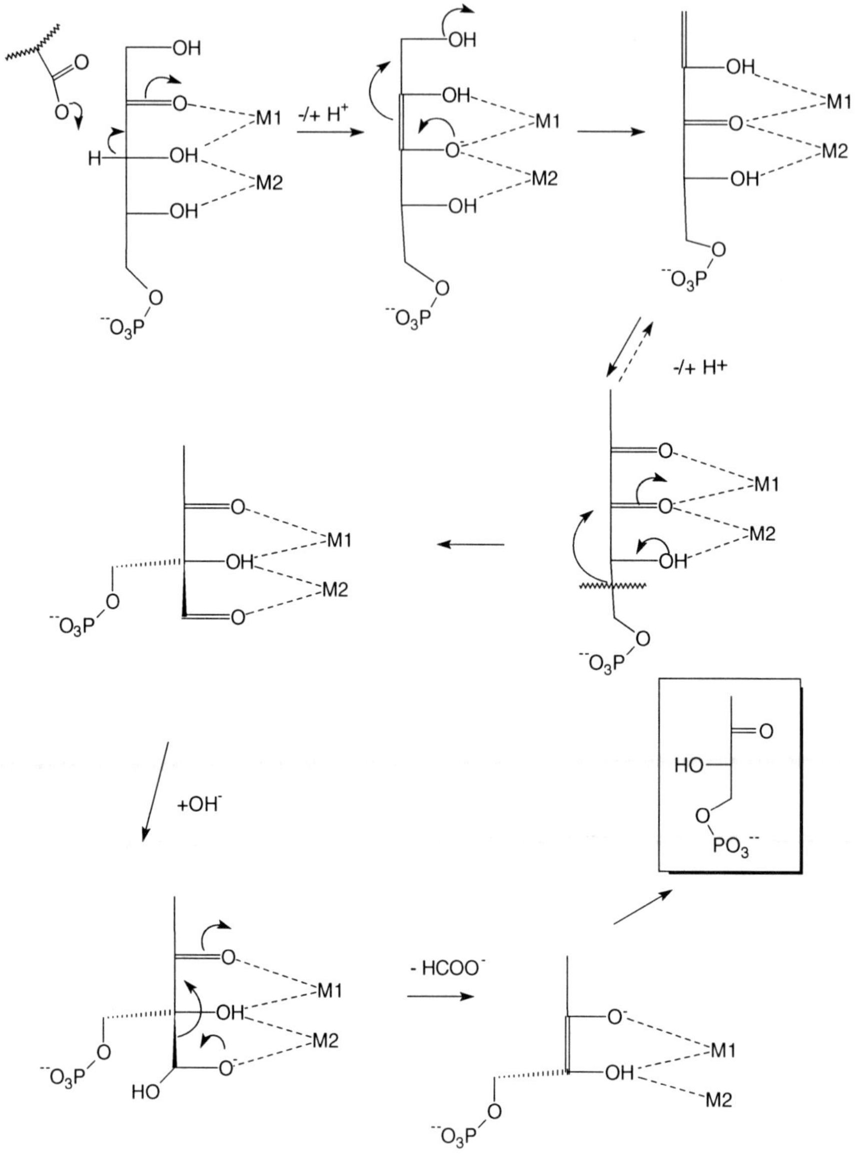

Figure 6.14 Mechanism of action of L-3,4-dihydroxy-2-butanone 4-phosphate according to Steinbacher *et al.*[67]

from aldose-derived glycosylamines is termed the Amadori rearrangement, whereas the formation of 2-deoxy-2-aminoaldoses from glycosylamines derived from 2-ketoses is termed the Heyns rearrangement (Figure 6.15).

In contrast to the reactions of the straight-chain forms of reducing sugars, however, the carbanion-like reactions of glycosylamines are most rapid near neutrality. This is because the conjugate acids of the Schiff bases are much better

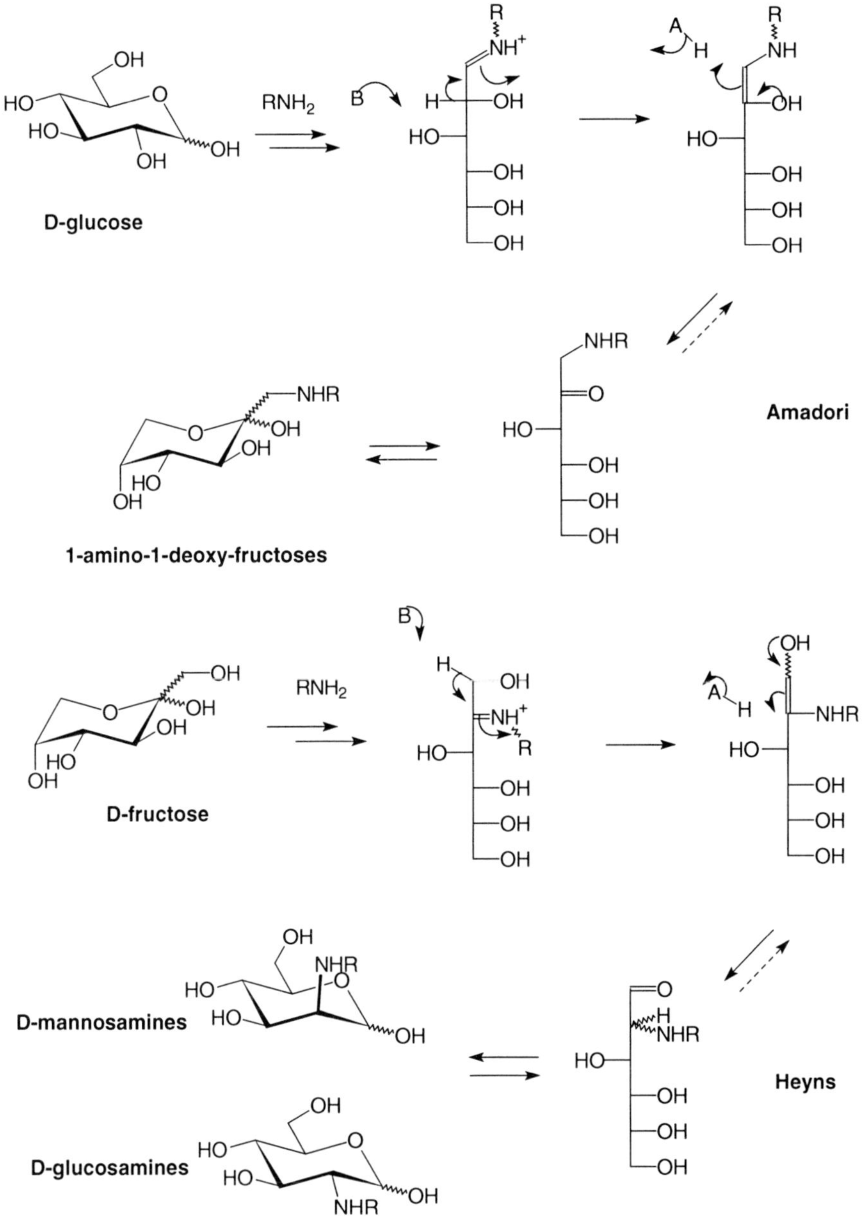

Figure 6.15 Amadori and Heyns rearrangements.

electron acceptors than the Schiff bases themselves, yet bases strong enough to deprotonate a carbon-bound hydrogen are protonated in acid solution.

Reactions of aldoses with amines can be stopped conveniently at the 1-amino-1-deoxy-2-ketose stage if the amine is an arylamine such as aniline (pK_a 4.6) and the reaction is carried out in buffered weakly acid medium (such as glacial acetic acid buffered with sodium acetate or aqueous pyridinium hydrochloride.[69] (Some of Amadori's original "composti" were obtained with neat reagents.)[70] The window of amine pK_a values in which the 1-amino-1-deoxy-2-ketose is readily isolated is fairly narrow – anilines with electron-donating substituents give such products, anilines with electron-withdrawing substituents do not.[71]

For the Heyns rearrangement, it would appear that a more basic amine is required – arylfructosylamines do not rearrange under the conditions where arylglucosylamines give Amadori products.[72] The Heyns reaction was originally discovered with ammonia itself, from the production of glucosamine and mannosamine when fructose was treated with ammonium chloride in the presence phosphate buffer.[73] The early mechanistic work on both the Heyns and Amadori rearrangements, however, focused on yields of various products. It is now difficult to interpret, since yield is determined by several variables, not only the rate of the reaction under study, but also the rate of side-reactions and in this case subsequent reaction of the Amadori or Heyns product down the Maillard cascade (see Section 6.2.3).

6.2.2 Osazone Formation

Reaction of C2-epimeric aldoses and the corresponding 2-ketose with three molecules of phenylhydrazine, to give a common osazone (*i.e.* a bisphenylhydrazone of a 2-ulose), aniline and ammonia, played a crucial role in Fischer's determination of structure of the aldoses (Figure 6.16). The mechanism of the reactions were examined between the mid-1950s and mid-1960s, but not subsequently, as the osazones have lost their importance.

The reaction is agreed to start by the formation of the phenylhydrazone of the reducing sugar, which then enolises. Two mechanisms are then proposed (Figure 6.17). In mechanism A, the lone pairs on the enol oxygen (perhaps with deprotonation) provide enough electronic "push" to cleave the N–N bond of the hydrazine. In mechanism B, the enol tautomerises to the ketone, which then forms a hydrazone. After re-tautomerisation to an enediamine, the N–N bond is cleaved by a push from a nitrogen lone pair.

Compelling evidence that pathway A was operating came from work in which ^{15}N was introduced into the 2-position of *p*-nitrophenylhydrazine[viii],[74] This was then used to make the *p*-nitrophenylhydrazones of glucose, benzoin and 2-hydroxycyclohexanone. These were reacted with excess unlabelled *p*-nitrophenylhydrazone and the ammonia liberated was shown to possess, in the early stages of the reaction, the same isotopic composition as N2 of the parent *p*-nitrophenylhydrazine. In the 2-hydroxycyclohexanone case, the imine intermediate was isolated (as its *N*-acetylenamine) and shown to proceed to osazone after hydrolysis.

As expected from both mechanisms A and B, [1-^{3}H]mannose and -glucose form osazones in which the label is almost completely retained,[75] whereas

[viii] By forming the diazonium compound from *p*-nitroaniline and $Na^{15}NO_2$ and then reducing.

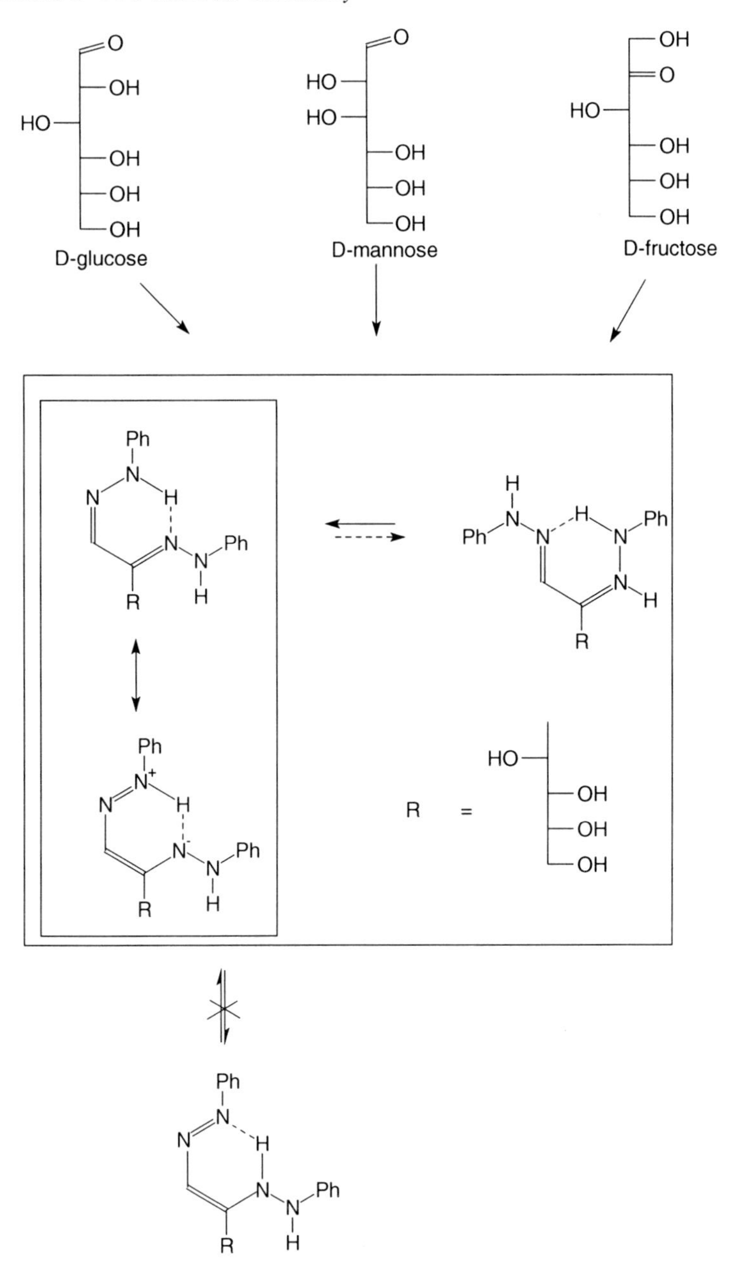

Figure 6.16 Osazone formation. The hydrogen-bonded ring is that established directly by NMR spectroscopy: the chelated proton appears to move in a single potential well.

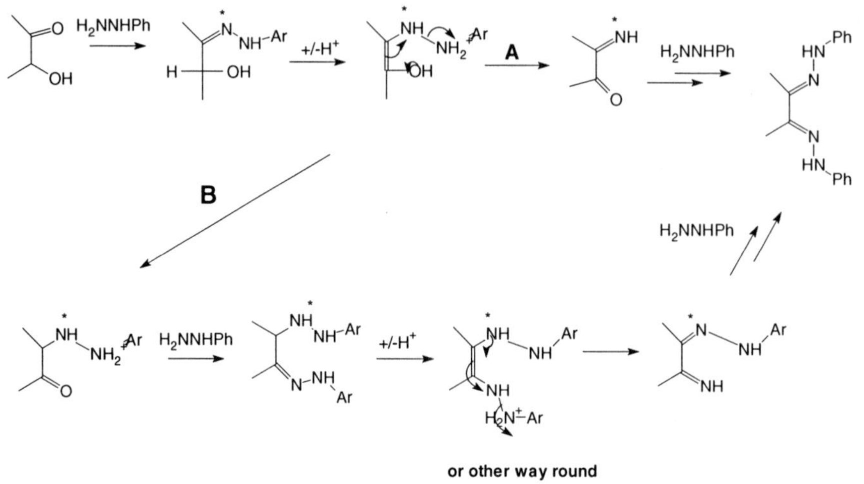

Figure 6.17 Alternative mechanisms of osazone formation. In pathway A all the labelled nitrogen appears as ammonia, and in pathway B less (in the case of symmetrical systems such as benzoin, half).

[2-^{3}H]glucose completely loses the label.[76] The evidence for some reaction via pathway B comes from reactions of [1-^{3}H]-1-deoxy-*p*-toluidinofructose, in which the label is lost on formation of the osazone.[77] The argument, however, rests on the assertion, from work with 4,6-benzylideneglucose, that no Amadori product with an intact N–N bond is an intermediate: the Amadori product with a 1-deoxy-1-arylamino group was isolated.[78]

With either mechanism, however, the question arises as to why the reaction stops at the osazone stage, rather than continues with a series of Amadori rearrangements and N–N cleavages until it has oxidised every hydroxyl group in the chain. The suggestion that the formation of either of a pair of six-membered hydrogen-bonded rings between vicinal hydrazone groups stopped the reaction was confirmed when it was found that 1-methylphenylhydrazine did indeed oxidise every OH in the sugar: xylose, for example, consumed nine equivalents of 1-methyl-1-phenylhydrazine to give an "alkazone" with five phenylhydrazone groups (Figure 6.18).[79] A ^{15}N NMR study of various osozones indicated that in the sugar series the cyclic hydrogen-bonded conformer with the C1=N bond *Z* was the predominant form and that the cyclic structure was stabilised by an aromatic-style ring current.[80] The experimentally established "quasi-aromatic" nature of the hydrogen-bonded six-membered ring explains immediately why the system falls into a free energy "pit" with just one OH oxidised.

6.2.3 The Maillard Reaction

The gamut of complex and deep-seated reactions that occur between reducing sugars and amino acids, proteins and peptides is named after its *Belle Époque*

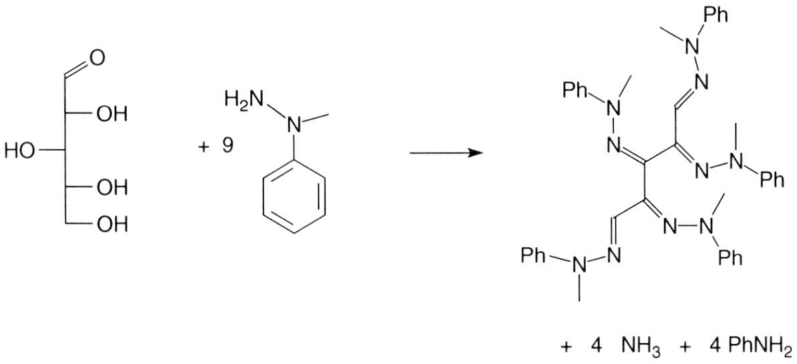

Figure 6.18 Oxidation of every sugar carbon by 1-methyl-1-phenyl hydrazine.

discoverer in the food context, Louis-Camille Maillard.[81] The Maillard reaction gives rise to a whole range of small molecules, many of them volatile, which contribute to the taste and aroma of cooked dishes in which proteins and carbohydrates are combined, and also to the polymeric materials (melanoidins) responsible for their golden-brown colour. Included in the low molecular weight materials is acrylamide ($CH_2{=}CH{-}CONH_2$), said to be a powerful carcinogen[ix].[82] More recently, it has been discovered that *non-enzymic* reactions of reducing sugars with proteins *in vivo* follow broadly similar chemistry and are responsible for many modifications of proteins associated with diabetes and/or ageing.[83] The covalent modification of proteins by *spontaneous* reaction with reducing sugars is termed *glycation*; the link between sugar and protein is never glycosidic, whereas *enzyme-catalysed* attachment of sugars to proteins almost invariably involves a glycosidic linkage and is termed glycosylation.

It is conventional to regard the reaction as taking place in three stages, with initial Maillard products being the Amadori rearrangement products from the first-formed Schiff base, intermediate Maillard products often being transient and highly reactive (*e.g.* methylglyoxal) and advanced Maillard products which are stable.[84] In the medical context these are called "advanced glycation endproducts" (AGEs). Studies of the mechanism of the Maillard reaction usually refer to one of two regimes, those of food chemists referring to temperatures of 100–150 °C and often conditions of low water activity, and those of medical biochemists to dilute solutions near neutrality, at temperatures 25–60 °C.

The available amino groups in proteins are the ε-amino group of lysine and the N-terminus. Many model studies of the Maillard reaction, however, use free amino acids, which can decarboxylate, particularly when the α-nitrogen has combined with a sugar moiety to provide an electron sink. This then produces aldehydes from the alkyl residue of the amino acid (Strecker reaction).[85]

[ix] For 2–3 decades, research biochemists cast their own polyacrylamide gels and handled the material on a regular basis.

The first stage of the Maillard reaction is the formation of glycosylamines by combination of a free amino group and a reducing sugar.[x] The overall browning reaction of various sugars heated with glycine correlates well with the free carbonyl content of their aqueous solutions.[86] The formation of the initial glycosylamine in the reaction of lactose with the major milk protein lactalbumin and lactose has been measured at 50–60 °C[xi].[87] A pH-independent reaction was observed below (room temperature) pH 7.0 and there was evidence for a base-catalysed process above this, which was accounted for by the deprotonation of the ε-amino groups of the lactalbumin (the pK_a of this group, as of all cationic acids, is very temperature dependent: 10.43 at 10.0 °C and 9.84 at 25.0 °C). From analogy with more well-defined glycosylamine formation (see Section 1.2.3), the rate-limiting step in the process formally involving the neutral amino group is probably the initial formation of the carbinolamine, and the pH-independent step its acid-catalysed decomposition. The pH independence could arise from the proportion of reactive deprotonated amine being inversely proportional to $[H^+]$ at pH values well below the pK_a of lysine and the rate of reaction of neutral amine with carbonyl being proportional to $[H^+]$.

After glycosylamine formation, Amadori rearrangement gives the key *N*-substituted 1-aminodeoxy-2-ketose. The reaction of glucose with glycine has been quantitatively modelled according to the reactions in Figure 6.19.[88] The important carbohydrate intermediates are 3-deoxyglucosone (3-deoxy-D-*erythro*-2-aldohexulose) and 1-deoxyglucosone (1-deoxy-D-*erythro*-2,3-hexodiulose); the melanoidins arise from the 3-deoxyglucosone and the methylglyoxal. 3-Deoxyglucosone predominates in acid solution and 1-deoxyglucosone in neutral or basic medium.[89] The outline mechanism had been established.[90] Degradation of glucose without combination with an amino acid has also to be included in the model.

The pathways in Figure 6.19a, complex, however, they are (each rate constant has its own pH and temperature dependence), are an exercise in fitting kinetic data to a model, which can elide several steps and therefore be an oversimplification with respect to molecular mechanism. Instances are:

(i) Spontaneous enolisation of glucose should produce mannose in addition to fructose (steps 2 and 3) and mannose behaves differently than glucose with tryptophan at elevated temperatures (110 and 140 °C).[91] Whereas glucose in excess over tryptophan converts it to *N*-(1-deoxy-D-fructosyl)tryptophan, mannose converts it to *N*,*N′*-bis(1-deoxy-D-fructosyl)tryptophan from a double Amadori rearrangement, which gives melanoidins much more readily.

(ii) 3-Deoxyglucosone would more plausibly arise, not from the Amadori rearrangement product, but its precursor enamine; since Amadori

---

[x] For making light-coloured meringues, therefore, it is essential to use the non-reducing sucrose.

[xi] This Maillard reaction is important in the storage of milk products, since cows' milk contains ~0.3 M lactose and the dietary value of various proteins is often limited by their lysine content.

rearrangement product formation is reversible, this would not affect kinetic modelling. Given that melanoidins are formed from 3-deoxyglucosone, such a route would explain why the formation of melanoidins from glycine and various reducing sugars parallels their equilibrium proportion of carbonyl compounds, but that the most readily formed Amadori rearrangement products gave least browning when heated alone.[86]

(iii) Although some methylglyoxal could be produced from the Amadori rearrangement product, the retro-aldol reactions on 1-deoxyglucosone and 4-deoxyglucosone indicated by ^{13}C labelling studies would be as plausible and more prolific: they would also explain the concurrent production of glyceraldehyde.[89] 1-^{13}C-labelled glucose gives methylglyoxal labelled at both C1 and C3, in accord with its formation from 1-deoxyglucosone and 3-deoxyglucosone, respectively.

(iv) Formic and acetic acids are both formed by retro-Claisen reactions made possible by further enolisations/ketonisations down the sugar chain (Figure 6.19e). While this is easy to recognise in the case of the formation of acetic acid,[92] the formation of formic acid from the Amadori product, as proposed by Hollnagel and Kroh,[93] avoids 3-deoxyglucosone as a discrete intermediate and is more plausible.

(v) The model in Figure 6.19a assumes only specific acid and base catalysis, but glycosylamine formation is subject to general acid–base catalysis and the expectation is that buffer catalysis would also take place in the reactions involving C–H bond formation and cleavage. The recent report that polyatomic anions will accelerate the Maillard browning reaction[94] seems to arise from the simple operation of the statistical factors in general acid–base-catalysed reactions [see eqns (1.4) and (1.5)].

The 1-deoxy-D-hexo-2,4-uloses formed by enolisation of 1-deoxyglucosone can also react by ketone cleavage between C4 and C3, giving glyceric acid and hydroxyacetone (Figure 6.20).[89]

The pathway shown in Figure 6.19e for the elimination of the 4-OH is that followed, *mutatis mutandis*, for the Maillard reaction with 4-linked oligosaccharides, such as the maltooligosaccharides, under aqueous conditions: the products are the 3-deoxypentulose and formic acid shown. The reaction has obvious parallels with the peeling reaction (Section 6.4).[95] Under non-aqueous conditions, however, the product is a 1,4-dideoxyglucosone (1,4-dideoxy-D-hex-2,3-ulose), probably formed by reduction of its 1-amino derivative.

As expected, pentoses react similarly to hexoses in the Maillard reaction, with the formation of 1- and 3-deoxyosones. However, the initial two-carbon fragment formed from the reaction of xylose with alanine is not the glycolaldehyde expected from retro-aldolisation of the Amadori rearrangement product, but glyoxal, for reasons which are not clear.[96]

Enolisation down the chain is a key to understanding the protein cross-linking capacity of Amadori rearrangement products attached to the ε-amino

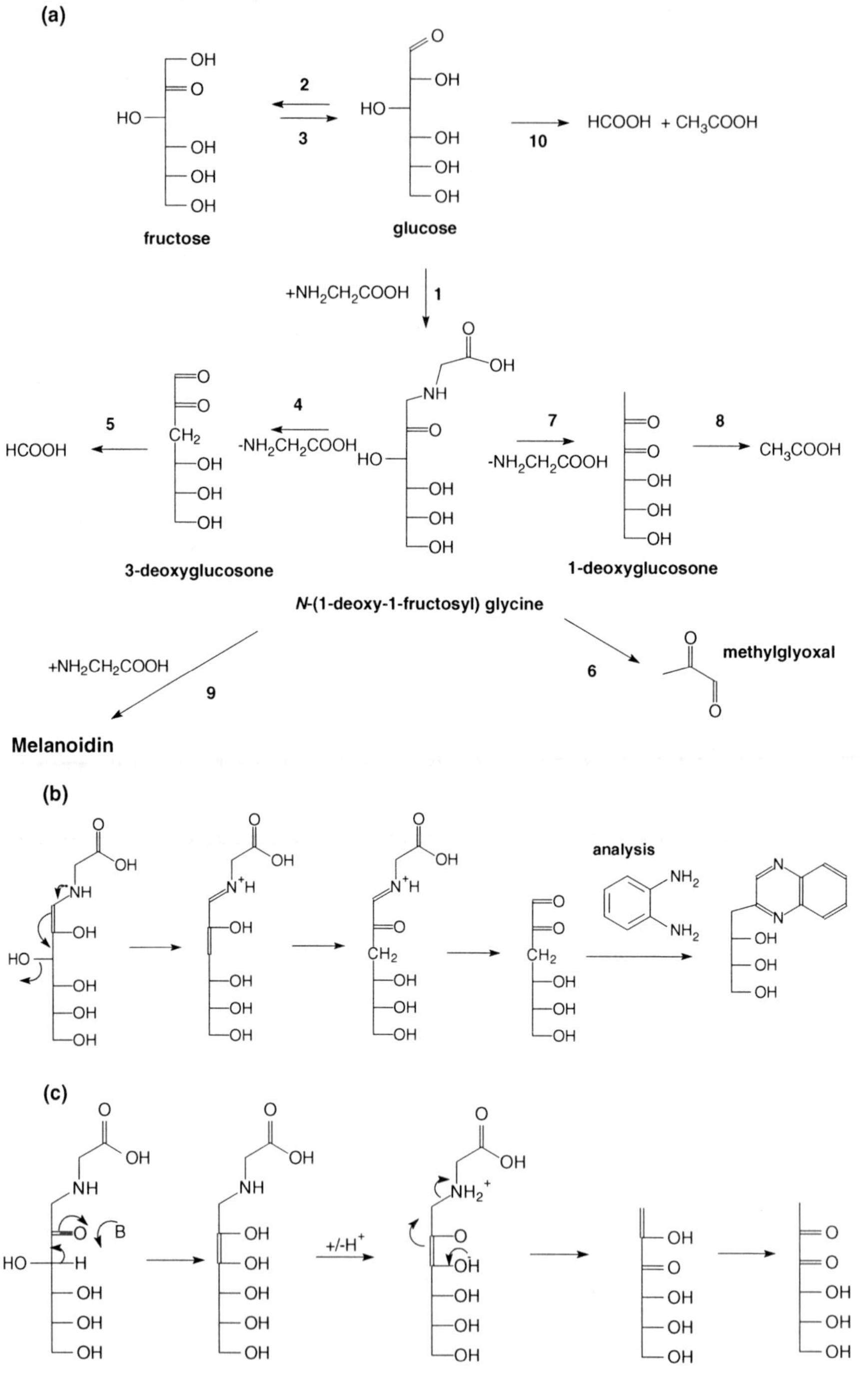

Figure 6.19 (a) Reaction of glucose and glycine: the kinetically significant steps and products. (b) Mechanism of reaction 4 (formation of 3-glucosone), illustrating formation of quinoxalines as stable, analysable derivatives of α-dicarbonyls. (c) Mechanism of reaction 7 (formation of 1-deoxyglucosone).

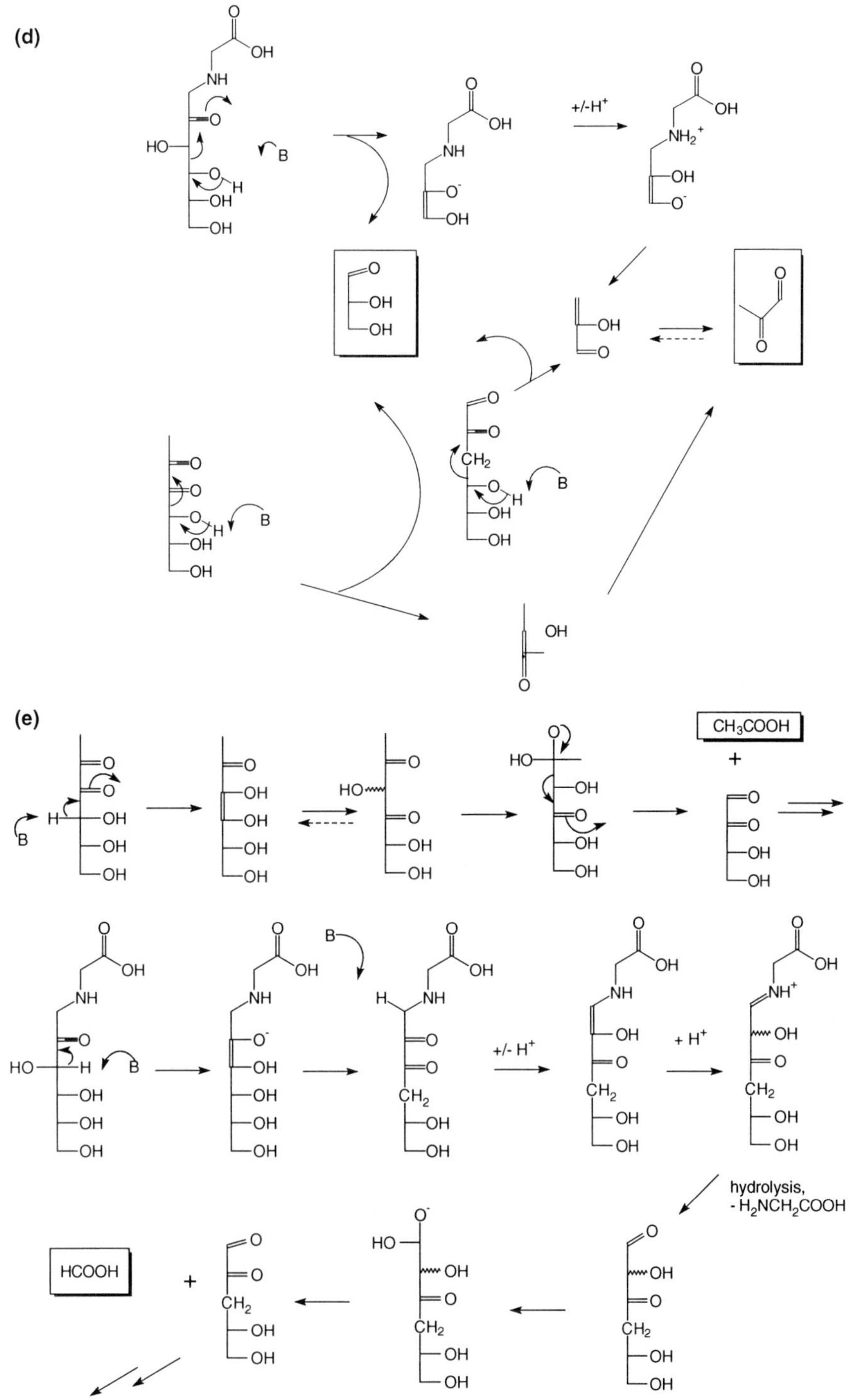

(d)

(e)

Figure 6.19 (d) Pathways for formation of methylglyoxal and glyceraldehyde and possible pathway for reaction 6. (e) Pathways for reactions 5 and 8 (formation of formic and acetic acid).

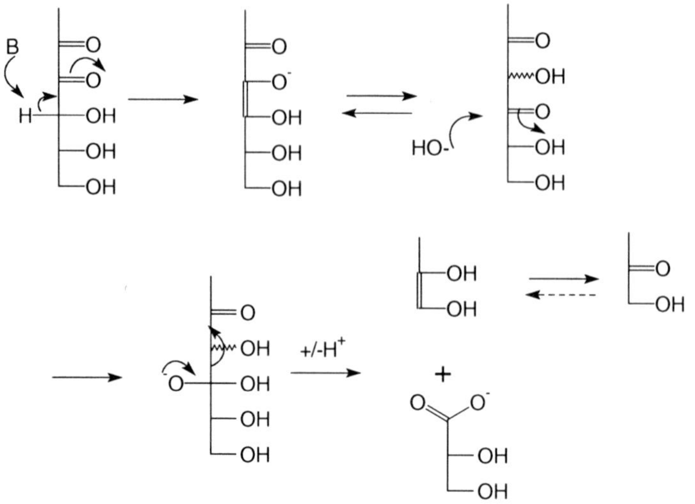

Figure 6.20 Formation of glyceric acid and hydroxyacetone.

group of lysine,[97] which is of such importance in the deterioration of long-lived mammalian connective tissue and in diabetes. The key intermediate is a 4-deoxy-5,6-dicarbonyl compound, for which there are two plausible mechanisms of formation, depending on the enediolate which expels the 4-OH. This can cyclise both on to the lysine and on to the side-chain of an arginine residue to give the key cross-linking structure glucosepane, which is a major protein cross-link in senescent human extracellular matrix.[98] The two proposed[97] pathways for its formation are given in Figure 6.21, with A being more likely, as it does not involve vinylogous enolisations.

Pentoses react similarly, but the structure analogous to glucosepane, pentosinane, has a six-membered ring and is readily oxidised to fluorescent pentosidine.[99] A similar process, however, can take place with glucosepane, to give a likewise aromatic aza-azulene structure.[100] Cross-links similar to glucosepane can also be formed from two- and three-carbon sugar fragments attached to the ε-amino group of lysine. Some of these cross-linking structures are slowly hydrolysed at the guanidino carbon of the arginine to give the non-ribosomal amino acid ornithine [$H_2N(CH_2)_3CH(NH_2)COOH$] in the protein.[101]

Structural studies on melanoidins are in their infancy, but it does appear that they can contain intact carbohydrate fragments, particularly if oligosaccharides are involved.[102] Pyrolysis yields a range of compounds containing aromatic heterocycles, which contribute to the odour of cooked food, with many having furan rings.[103]

The formation of acrylamide in the Maillard reaction appears to involve free asparagine. A detailed study[104] indicated that hydroxycarbonyl compounds, such as hydroxyacetone, were more effective than α-dicarbonyl compounds in catalysing its production. On this basis, it was proposed that acrylamide arose from a 1,3-dipole (Figure 6.22b) rather than the classic Strecker decarboxylation pathway (Figure 6.22a).

6.3 AROMATISATION

The aromatic stabilisation of compounds containing the furan ring means that such compounds are in free energy pits, into which a system undergoing a range of reversible reactions under extreme conditions will fall. Carbohydrates therefore tend to give furans with drastic treatment under neutral and acidic conditions. The acid-catalysed aromatisation of the arabinoglucuronoxylan in maize husks to furfural (Figure 6.23) was a classic entry into chemical feedstocks containing the furan ring;[106] more recently, xylose has been shown to degrade to furfural in neutral superheated water, and also weakly acid solution.[107] In water at around 200 °C, glucose gave 1,6-anhydroglucose at lower temperatures, but under more extreme conditions 5-hydroxymethylfurfural and smaller quantities of furfural were produced,[108] and 5-hydroxymethylfufural was the only product that was not a methylglycoside from supercritical methanolysis of cellulose.[109] The Amadori product from reaction of glucose with the ε-amino group of lysine is degraded to "furosine" when proteins are hydrolysed with concentrated hydrochloric acid, prior to classical amino acid analysis, and the same reaction apparently occurs more slowly under less drastic conditions in the body.[110] However, the N^{α} adduct of glucose and tryptophan gives 5-hydroxymethylfurfural.[111]

6.4 ACIDIC AND BASIC GROUPS IN CARBOHYDRATES

In general, the electron-withdrawing effect of the multiple hydroxyl groups of carbohydrates makes hydroxyl, carboxyl and ammonium groups more acidic than the same groups attached to simple alkyl residues. Other factors, such as hindrance to solvation and the balance between inductive and field effects, also influence pK_a values in detail.

Experimental data on the ionisation of hydroxyl groups are sparse. The most acidic hydroxyl in a reducing sugar is the anomeric OH, which is subject to the inductive/field effect of a geminal OR and has a pK_a value just over 12 (Table 6.1). The equatorial anomer is the more acidic, as the negative charge is more readily solvated. Because the expected pK_a of other hydroxyl groups is appreciably higher, the macroscopic first ionisation of a reducing sugar, measured electrochemically, can usually be equated with the microscopic dissociation constant of the anomeric hydroxyl. The large numbers of potential ionisation equilibria involving non-anomeric hydroxyl groups, however, mean that their deconvolution with current technology is impossible. The titration curve of a methyl hexoside would show four overlapping macroscopic ionisations, but the system has 32 microscopic ionisations, since the pK_a of each OH group would depend on the ionisation states of the other three, for which there are eight possibilities. (For a treatment of microscopic and microscopic pK_a values, see Figure 5.13).

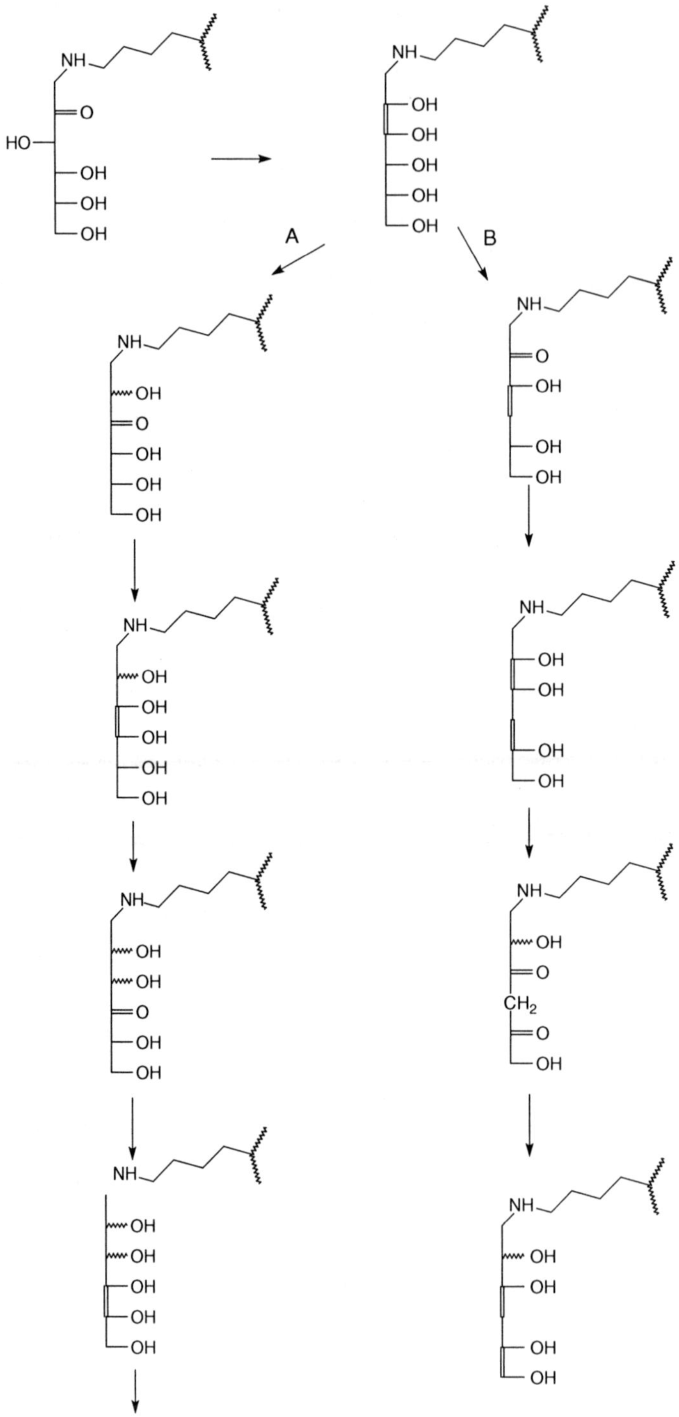

Figure 6.21 Pathways for the formation of glucosepane and related protein cross-links.

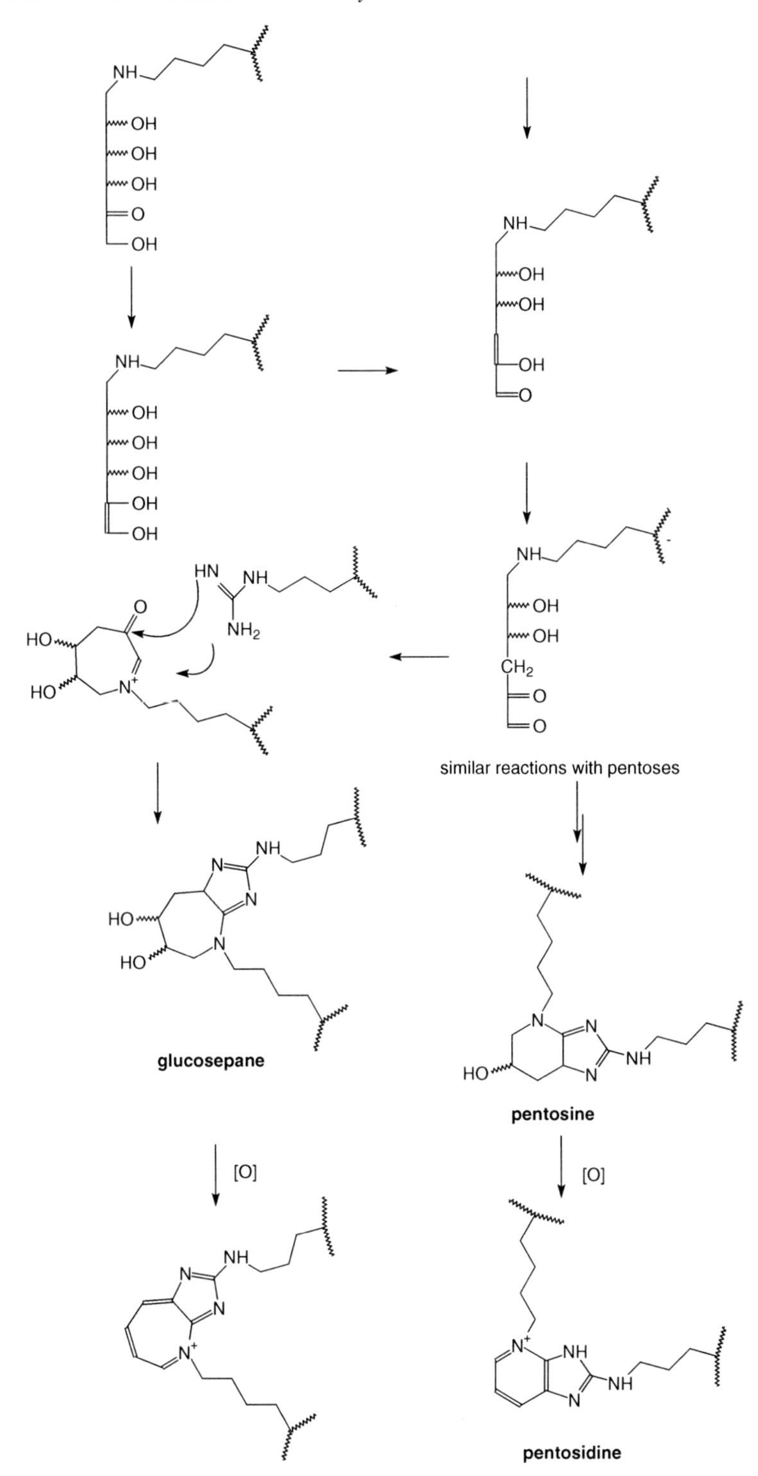

Figure 6.21 (*Continued*)

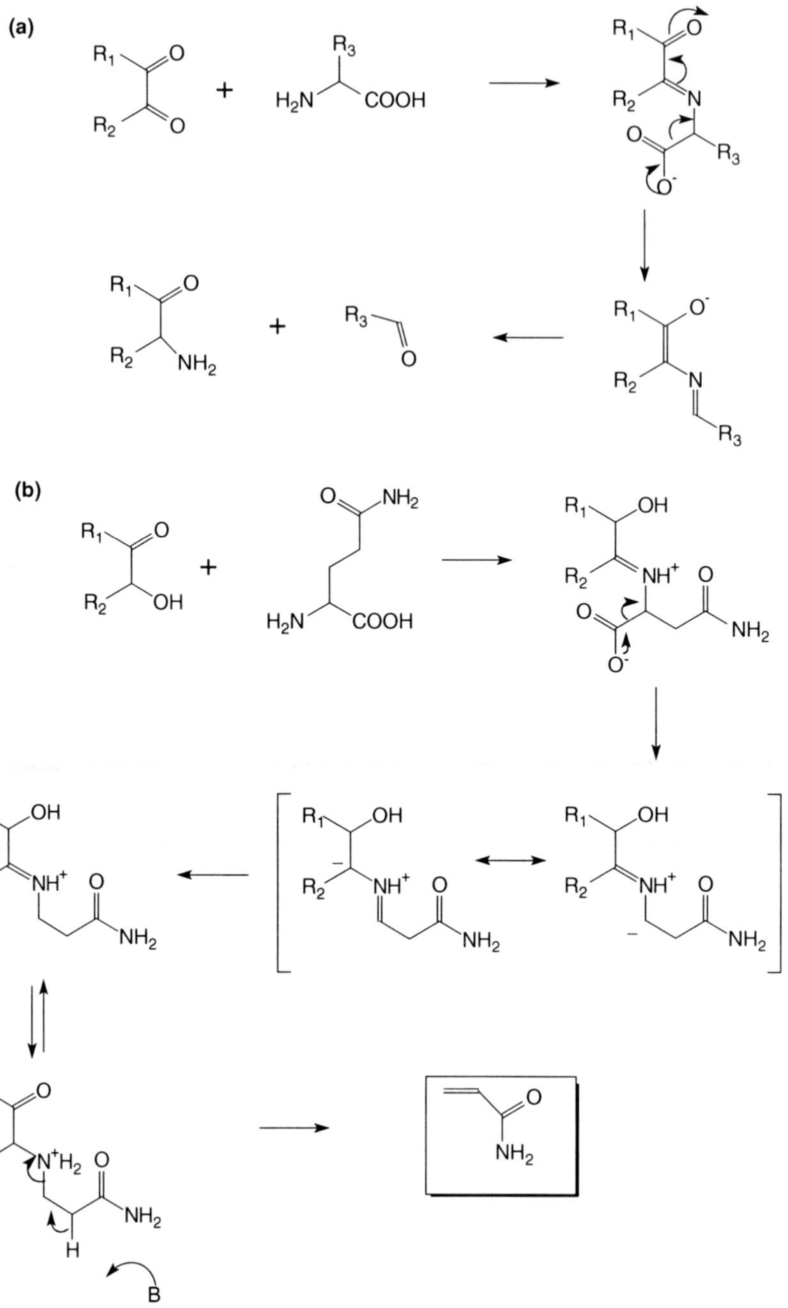

(a)

R1 O R2 O + R3 H2N COOH → R1 O R2 N O R3 O-

R1 O- R2 N R3 → R1 O R2 NH2 + R3 O

(b)

R1 O R2 OH + O NH2 H2N COOH → R1 OH R2 NH+ O O NH2 O-

R1 OH R2 NH+ O NH2 ↔ R1 OH R2 NH+ O NH2

R1 OH R2 NH+ O NH2

R1 O R2 N+H2 O NH2 H B → O NH2

Figure 6.22 (a) Strecker pathway for decarboxylation of α-amino acids (a similar pathway is used by Nature in pyruvate-dependent transaminases and decarboxylases[105]). (b) Production of acrylamide from asparagine by α-hydroxycarbonyl compounds.

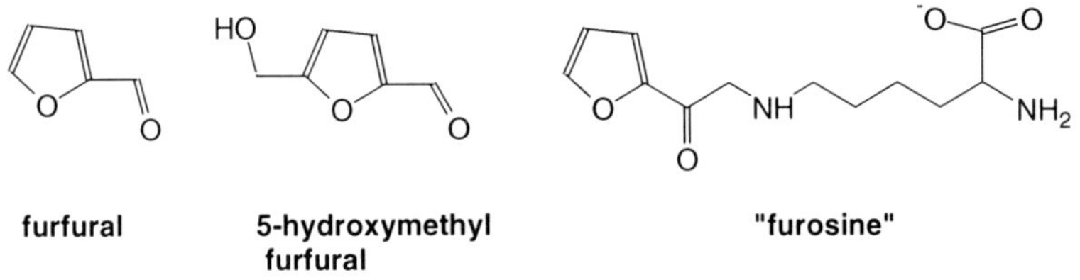

furfural 5-hydroxymethyl furfural "furosine"

Figure 6.23 Aromatisation products of sugar moieties.

Table 6.1 Comparison of carbohydrate and aliphatic aqueous pK_a data (25 °C).

Compound	*pK$_a$*
CH_3OH	15.54[114]
CH_3CH_2OH	16.0[114]
$HOCH_2CH_2OH$	14.77[114] (equivalent to single OH pK_a of 15.07)
$CH_3OCH_2CH_2OH$	14.82[114]
CCl_3CH_2OH	14.24[114]
CF_3CH_2OH	12.37[115]
D-Glucose	α: 12.28;[116] ΔpK_a = 0.57 (β more acidic)[117]
D-Galactose	12.35[116]
D-Mannose	12.08[116]
D-Xylose	12.15[116]
Cyclohexaamylose	12.33 (equivalent to single OH pK_a of 13.59)[112]
Cycloheptaamylose	12.20 (equivalent to single OH pK_a of 13.52)[112]
Cyclooctaamylose	12.08 (equivalent to single OH pK_a of 13.46)[112]
$HOCH_2CH_2NH_2$	9.50[116]
D-Gluco- or -galactosylamine	5.6;[118] ΔpK_a = 0.3 (β more basic)[119]
N-Benzyl-D-gluco- or -galactosylamine	5.3[118]
N-Tolyl-D-gluco- or -galactosylamine	1.5[118]
D-Glucosamine	8.12 (α), 7.87 (β)[117]
D-Galactosamine	8.49 (α), 8.02(β)[117]
D-Mannosamine	7.78 (α), 8.50 (β)[117]
CH_3COOH	4.76[116]
$HOCH_2COOH$	3.83[116]
p-Nitrophenyl-*N*-acetyl-D-neuraminide	2.86 (α), 1.58 (β)[120]
D-Glucuronic acid	2.93 (α), 2.83(β)[121]
β-D-GlcA residues in heparin	2.79[121]
2-Sulfato-α-L-IdoA residues in heparin	3.13[121]

The first pK_a of cycloamyloses is in the accessible range (Table 6.1), because of statistical effects – if all the OH groups had the same microscopic K_a, $K_{a(m)}$, then the macroscopic K_a would be $18K_{a(m)}$ for cyclohexaamylose, $21K_{a(m)}$ for cycloheptaamylose and so forth. When the titration was monitored by ^{13}C

NMR spectroscopy, the chemical shifts of C5, C4 and C1, as expected, did not change, but there was some suggestion of a smaller shift of C6 than C3 or C4, suggesting the microscopic pK_a of the 6-OH group was higher than that of 2-OH or 3-OH.[112]

Calculation can give some idea of the relative acidities of sugar hydroxyls. The deprotonation enthalpies (kJ ol^{-1}) *in vacuo* of the various OH groups of α-D-glucopyranose, relative to that for the anomeric hydroxyl, are 2-OH, 35.5; 3-OH, 33.4; 4-OH, 22.9; and 6-OH, 43.9.[113] Absolutely, these differences are huge (if they were translated completely into pK_a differences, the anomeric OH would be 6–8 pK_a units stronger than the other hydroxyls!), but experiment does suggest the 6-OH is less acidic than the others.

Glycosylamines (NH_2 on C1) are configurationally unstable, but it is possible to measure the pK_a of the mixture and the difference between α- and β-anomers by the technique described by eqn (2.8). The marginally stronger basicity of equatorial glycosylamines is attributed to the more ready solvation of their conjugate acids, but the anomeric effect makes the difference smaller than in carbocyclic systems.

The effect of structure on the basicity of carbohydrate analogues with nitrogen in the ring has been convincingly rationalised with the discovery that the electronic effect of hydroxyl and similar groups is predominantly a field (through-space) effect, which depends on the direction of the C–O dipole. Axial OH groups have a smaller effect than equatorial ones. In Table 6.2 are displayed values of Bols' σ_S parameters,[122] which are simply the decrease in the pK_a of piperidine brought about by the introduction of the group in question (increasing acidity is positive, as with the classical Hammett σ). The values of σ_S for the cyano group do not differ appreciably between axial and equatorial. If CN is treated as a point dipole, then its location is further from the ring than OH and consequently the angle α between the direction of the dipole and the vector joining it to the piperidine nitrogen is similar in equatorial and axial cases. In fact the substituents can be treated by the simple Kirkwood–Westheimer theory, in which the effect of a point dipole substituent is

Table 6.2 Bols σs values for substituted piperidines. The pK_a of piperidines is $10.7 - \Sigma\sigma_S$ and that of pyridazines $7.3 - \Sigma\sigma_S$.

Substituent R	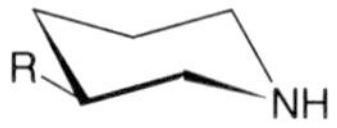 3-equatorial R (piperidine NH)	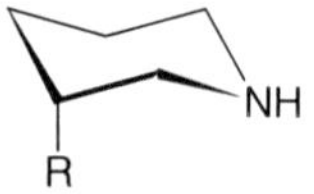 3-axial R (piperidine NH)	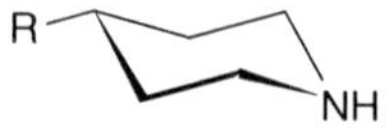 4-equatorial R (piperidine NH)	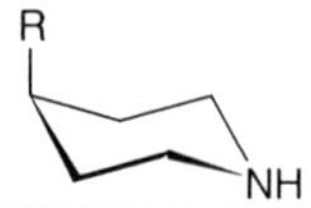 4-axial R (piperidine NH)
OH	1.3	0.5	0.6	0.2
F	2.3	1.5	1.0	
COOMe	1.2	0.2		
$CONH_2$	1.5	1.3		
COO^-	0.5	−0.2	0.2	
CN	2.8	3.0		
CH_2OH	0.4	0.5		

proportional to $\cos\alpha/r^2$, where r is the distance between the point dipole and the reaction centre. For 3-cyanopiperidines this term is identical (0.048 $Å^{-2}$) for axial and equatorial conformers.

In the early days of physical organic chemistry, there was discussion about whether the effects of substituents in saturated systems was through bonds (inductive effects) or through space (field effects),[123] but the consensus view was that it was not useful to try to distinguish them.[xii] These recent data are a decided upset to the consensus and must throw into question any Hammett-type reactivity parameter derived from a conformationally mobile system in any area of physical organic chemistry.

Comparison of the pK_a values of the hexosamines (NH_2 on C2) with that of ethanolamine reveals the same ~1.5 units lowering in the sugar as was seen in the comparison of OH acidities of ethylene glycol and the cycloamyloses. The origin of the variation between anomers of the hexosamines is said to lie in the enhanced solvation of an equatorial NH_3^+ promoted by a vicinal axial OH, so that the α-anomers of glucosamine and galactosamine are the least acidic.[124] When the hexosamine is mannosamine, the acidities of the two anomers are reversed, in accord with this proposal. The higher basicity of galactosamine compared with glucosamine is probably a reflection of the lower σ_S value for an axial hydroxyl.

The equatorial COOH in a uronic acid is acidified by the sugar moiety to below the pK_a of glycolic acid. Attachment directly to the anomeric centre, with an additional electron-withdrawing oxygen, as in the β-neuraminides, lowers the pK_a by a further unit: however, steric hindrance to solvation of the axial carboxylate of neuraminides of the natural (α) configuration raises the pK_a again. One would expect a similar effect in IdoA residues in the $^1C_4$ conformation, but the IdoA residues in heparin have only modestly increased pK_a values, possibly because of the adoption of the less hindered $^2S_0$ conformation.

Attachment of a phosphate group to a sugar lowers the pK_a values of the first two ionisations from those of phosphoric acid itself (pK_{a1} 1.96, pK_{a2} 6.93, pK_{a3} 11.72) by 0.3–0.6 units, as demonstrated by glucose 2-phosphate (pK_{a1} 1.3, pK_{a2} 6.6). Glucose 1-phosphate has a marginally lower value of pK_{a2} (6.5), which may reflect the electron-withdrawing effect of the additional oxygen attached to the carbon bearing the phosphate.[125]

6.5 NUCLEOPHILIC REACTIONS OF OH GROUPS

The basic principles governing nucleophilicity, such as the steric effects making primary OH groups more nucleophilic than secondary OH groups, or, in a

[xii] E.g. " A number of workers refer to electrostatic action transmitted through chains of atoms as *inductive effects* and electrostatic action transmitted either through empty space or through solvent molecules as *field effects*, but the experimental separation of these two 'effects' has thus far proven difficult", E. S. Gould, *Mechanism and Structure in Organic Chemistry*, Holt, Rinehart and Winston, New York, 1959, p. 203.

six-membered ring, equatorial OH groups more nucleophilic than axial OH groups, apply to carbohydrate chemistry. The effect of different pK_a values, however, makes the relative nucleophilicities of OH groups dependent on the acidity or basicity of the reaction medium. If the nucleophile is the alcohol itself, then primary OH will be most nucleophilic for both electronic and steric reasons. However, if the reaction is with the anion at pH values below the OH pK_a, in the absence of countervailing steric effects, the most acidic OH is also the most nucleophilic. (This arises because β_{nuc} values are, maximally, only 0.5 – see Section 3.13.2.1). This means that under conditions of base catalysis, the predominant product from nucleophilic attack by the anomeric OH of an equilibrating mixture of a reducing sugar will be the equatorial (usually β) compound, from the combined effect of the greater steric accessibility and lower pK_a of the equatorial anomer.

6.5.1 Alkylation

Carbohydrate hydroxyl groups are alkylated in three contexts: analysis of polysaccharide sequences, dealt with in Section 4.2.1; protection of OH groups during synthetic transformations, which requires the alkyl residues to be removable; and the modification of cheap polysaccharides, such as starch and cellulose, for more desirable properties.

Widely used alkyl protecting groups are benzyl, *p*-methoxybenzyl, allyl, trityl (triphenylmethyl) and *p*-methoxytrityl. Benzyl, allyl and *p*-methoxybenzyl groups are introduced by reaction of the alkoxide, preformed by reaction of the alcohol with a slight excess of sodium hydride, with the appropriate alkyl halide in an S_N2 reaction (Figure 6.24). Trityl groups are attached by the capture of the carbenium ion, derived from the chloride, by the OH in the presence of an acid acceptor, such as pyridine. The steric requirements of the trityl and substituted trityl groups mean that there is considerable (although not absolute)[126] selectivity for primary OH groups.

Benzyl (Bn) groups are removed by catalytic hydrogenolysis (Pd/C), either with hydrogen gas or by transfer from a molecule which gives an aromatic compound on dehydrogenation, such as cyclohexadiene. *p*-Methoxybenzyl groups can be removed by one-electron oxidation by such species as Ce^{IV} or high-potential quinones,[127] but also by Lewis acids and a carbenium ion trap such as a thiol or dialkyl sulfide.[128] Trityl and methoxytrityl [(CH_3O-*p*-C_6H_4)Ph_2C–] are removed by mild acidic hydrolysis, which generates relatively stable triarylcarbenium ions. The extra lability of the *p*-methoxytrityl group makes it compatible with the phosphate coupling chemistry during oligonucleotide synthesis, and hence *p*-methoxytrityl protection of 5′-OH is standard.[129] Allyl groups are removed by initial treatment with transition metals, which isomerise them to vinyl ethers. The vinyl ethers are then hydrolysed, and by appropriate choice of transition metal complexes the entire deprotection sequence can occur in the same pot.[130]

Production of alkylated celluloses requires material free of other polysaccharides, so "dissolving pulps", produced under acid pulping conditions which remove

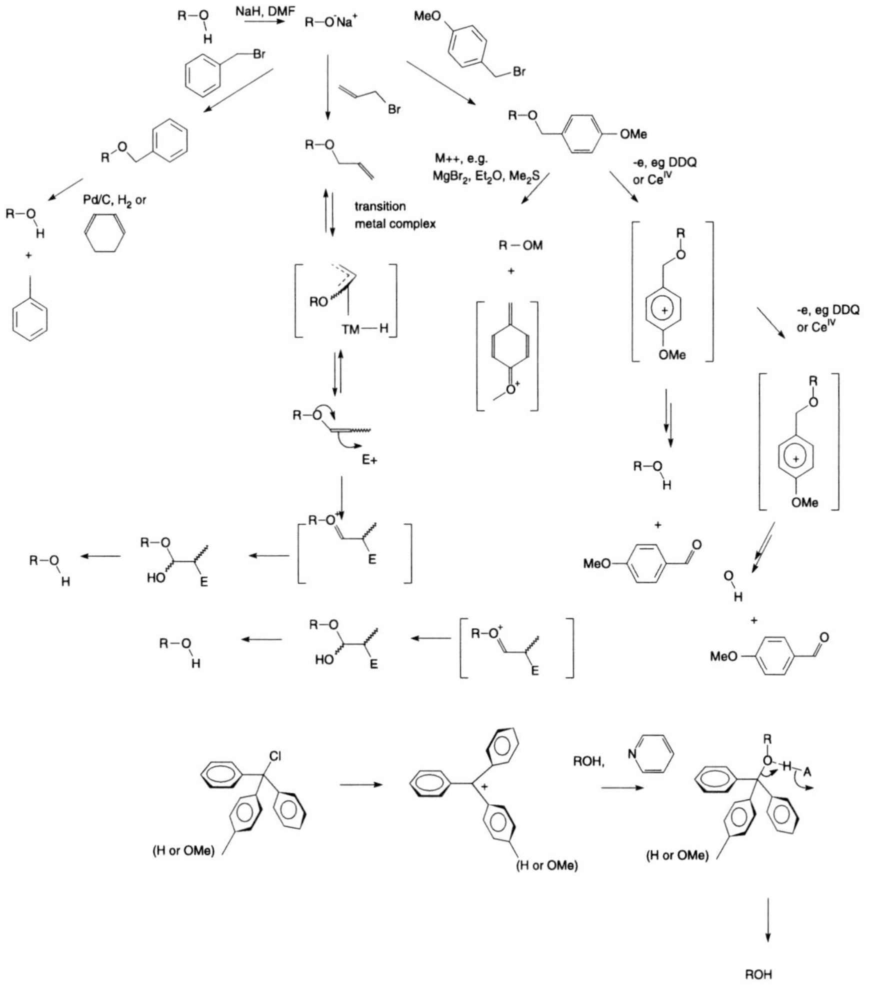

Figure 6.24 Introduction and removal of *O*-alkyl protecting groups.

hemicelluloses, are used. For alkylation, the cellulose is swollen with NaOH and then treated with the appropriate electrophile as shown in Table 6.3.[131] Although accessibility of the solid polysaccharide plays a role in the distribution of substituents, the relative reactivity of hydroxyl groups (6-OH > 2-OH ≫ 3-OH) is what would be expected for reactions of relatively unhindered electrophiles with partly deprotonated nucleophiles. Properties of the commoner alkylated celluloses are set out in Table 6.3, where *DS* is degree of substitution, the number of substituents per glucose unit. Derivatisation with ethylene and propylene oxides may cause this to exceed 3, as short chains of polyethylene and polypropylene glycols grow from the sites of initial substitution, a phenomenon which may have as much to do with the accessibility of the newly created OH groups, away from the packed cellulose microfibrils, as intrinsic chemical reactivity. Also

Table 6.3 Common alkyl derivatives of cellulose.

Electrophile	*Substituent*	*Commercial acronym*	*DS*	*Solubility*	*Uses*
$(CH_3)_2SO_4$ or CH_3Cl	Cell-O-CH_3	MC	1.5–2.4	Water	
CH_3CH_2Cl	Cell-O-CH_2CH_3	EC	2.3–2.6	Organic solvents	Films, lacquers and binders
O	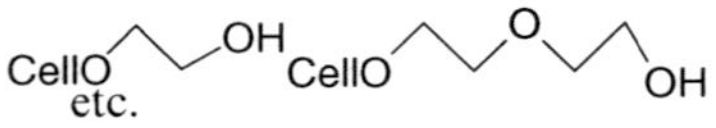 CellO OH CellO O OH etc.	HEC	1.3–3.0 (molar substitution)	Water	Coatings, thickening and emulsifying agents
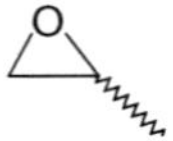 O	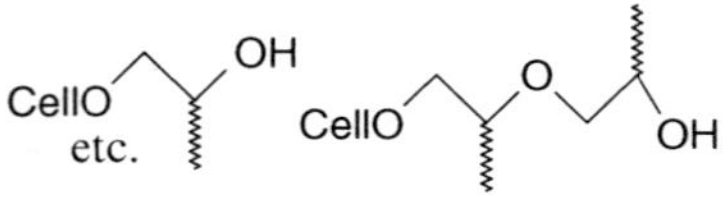 CellO OH CellO O OH etc.	HPC (hydroxypropylcellulose)	2–3 (molar substitution)	Organic solvents, water	Thickenings, coatings, emulsifying agents
$ClCH_2COONa$	Cell-O-CH_2COOH	CMC	0.5–1.2	Water	Anionic polymer, thickener, "poor man's alginate"
$CH_2=CHCN$	Cell-O-CH_2CH_2CN	CEC	0.3–0.5 ~2	Water organic solvents	Fabric softener, electrical insulator

counterintuitive to the organic chemist is that methylation of cellulose, by interfering with the packing of the cellulose chains, makes the polymer water soluble.

Alkylation of starch with ethylene oxide and sodium hydroxide yields hydroxyethyl-starch, which replaces native starch in some industrial applications as it gels more readily. The most frequent alkylations of starch, however, are for the purpose of making it cationic. Cellulose fibres, whether in textiles or paper, carry negative charges because of occasional carboxylate groups, whether from degradation of the cellulose or adherence of uronic acid-containing hemicelluloses. Cationic polymers are used to strengthen the paper sheet by cross-linking negative fibres and to size both paper and textiles. The alkylating agents are the chloride $ClCH_2CH_2NEt_2$ and the epoxide $CH_2(O)CH–CH_2N^+Et_3Cl^-$, which react predominantly at O2:[132] this may be a result of the relative inaccessibility of the 6-OH in the V-amylose helix (see Figure 4.52).

6.5.2 Silylation

Conversion of all the OH groups of a mono- or disaccharide to trimethylsilyl (TMS) ethers [$(CH_3)_3Si–OR$, TMSOR] renders the derivative volatile enough for analysis by gas chromatography. The derivatisation reaction takes place with trimethylsilyl chloride [$(CH_3)_3SiCl$, TMSCl] and hexamethyldisilazane {$[(CH_3)_3Si]_2NH$}, in pyridine, or with neat *N*,*O*-bis(trimethylsilyl)acetamide (bizarrely[xiii] abbreviated as BSA). The reactions appear to be S_N2 reactions on silicon, with the base providing proton abstraction or perhaps some activation by proton abstraction to the attack on TMSCl, and possibly some proton donation from the OH's to the N of BSA. The development of sensitive liquid analytical techniques for which the sugar does not have to be derivatised, such as high-performance anion-exchange chromatography with pulsed amperometric detection (HPAEC/PAD), now confines GLC of carbohydrate TMS ethers to routine analyses where the analysands are known. As with any technique involving derivatisation before chromatography, incomplete derivatisation of sample and/or internal standards is an ever-present concern.

TMS ethers are too labile to acid-catalysed hydrolysis to be preparatively useful, but *tert*-butyldimethylsilyl (TBDMS) ethers are around 10^3 times less susceptible to acid hydrolysis.[133] The steric restriction about the central silicon atom presumably is the cause of the reduced reactivity, which also makes their introduction with *tert*-butyldimethylsilyl chloride (TBDMSCl) in pyridine selective for the primary position, but very slow. With the more basic imidazole (pK_a 7 rather than 5.2 for pyridine) in DMF as base catalyst, the reaction is readier but loses its absolute selectivity for primary positions (Figure 6.25). Reaction of methyl α-D-glucopyranoside with two equivalents of TBDMS

[xiii] In biological chemistry, BSA = bovine serum albumin.

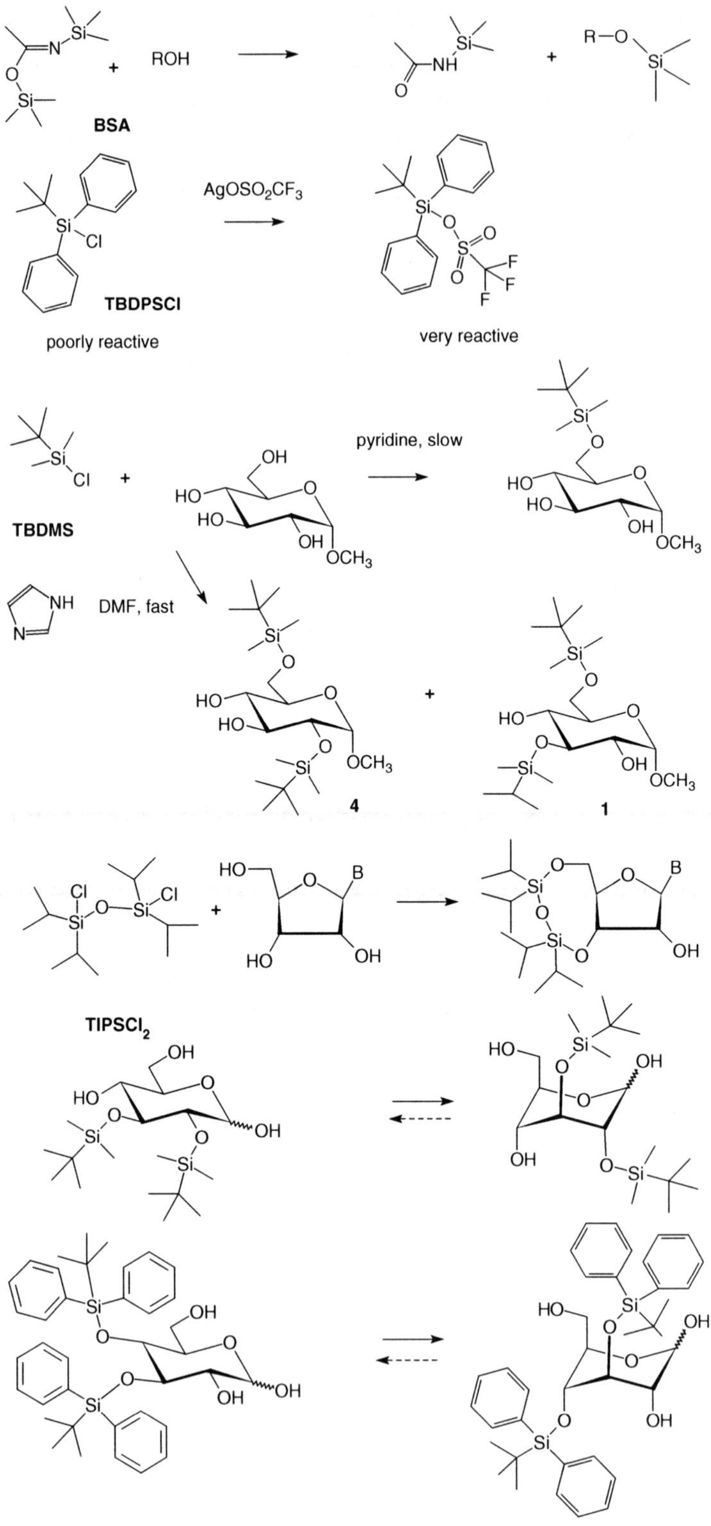

BSA + ROH →

TBDPSCl (poorly reactive) —$AgOSO_2CF_3$→ (very reactive)

TBDMS + (imidazole) — pyridine, slow → ; DMF, fast → **4** + **1**

TIPSCl$_2$

Figure 6.25 Silicon-based protecting groups and their introduction and consequences.

under these conditions, for example, gives the 2,6- and 3,6-bis(TBDMS) derivatives in a 4:1 ratio, whereas the 2,6- and 3,6-bis(TBDMS) derivatives of methyl β-D-glucopyranoside are formed in a 1:1 ratio.[134] Under forcing conditions [TBDMS trifluoromethanesulfonate (triflate)], the TBDMS group can be introduced into both the 2- and 3-positions of a protected D-glucopyranose: the bulk of the substituent forces both anomers of 2,3-bis(TBDMS) glucopyranose compound into the $^{1}C_4$ conformation.[135]

tert-Butyl diphenylsilyl (TBDPS) ethers are a further 100-fold less hydrolytically labile than their TBDMS analogues. They can be introduced by substitution of the chloride,[136] but prior conversion to an agent with a better leaving group, *e.g.* by reaction with its silver salt, is more convenient.[137] The chloride (in DMF with imidazole as a base) can, however, be made very selective, for example silylating the 6-position of completely unprotected β(1→4)-linked mannobiose and mannotriose exclusively and in high yield.[138] Formation of the 3,4-bis(TBDPS) derivative of a protected glucopyranose under forcing conditions and subsequent removal of the other protecting groups, revealed that 3,4-bis(TBDPS) glucopyranose,[139] and its 2,3 isomer,[135] adopted the $^{1}C_4$ conformation.

The distance between the electrophilic centres of the silylation agent $\{[(CH_3)_2CH]_2SiCl\}_2O$ is too long for derivatisation of vicinal OH groups, but useful for converting HO*n*CH_2 and HO*n* – 2 into six-membered cyclic tetraisopropyldisiloxy (TIPS) ethers, particularly in the case of nucleosides ($n = 5$), where 3′, 5′ derivatives are formed.[140]

Because silicon can expand its valence octet, all these silyl derivatives are vulnerable to a greater or lesser extent to migration to other hydroxyl groups under alkaline conditions, via the formation of trigonal bipyramidal intermediates.

Fluoride ion has a particularly high nucleophilicity towards silicon and other electrophilic centres involving second row elements: this is attributed to efficient overlap of two orthogonal filled lone pair orbitals on fluoride with two empty silicon d orbitals. Although the more reactive silyl derivatives can be removed by hydrolysis, therefore, many protocols for mild silyl group removal involve fluoride ion.

Silicon, as silicate, has a large biochemistry in the plant kingdom, which is only beginning to be explored. Silicate appears to be transferred as the ester of a serine OH in the active site of a silicatein, an enzyme with sequence homology to cysteine proteinases, but with the crucial cysteine replaced by serine.[141] Plant cell walls contain considerable amounts of silica[xiv] – dry rice straw can contain 8 wt% silica. Although deposition of silica in diatoms is mediated in the first instance by cationic proteins, a secondary role for a (1→3)-linked mannan has been suggested Figure 6.25.[142]

[xiv] Silica interferes with the alkali recovery cycle in various forms of alkaline pulping and is in consequence the major reason for the neglect of many readily available forms of lignocellulose, particularly cereal straw, for papermaking.

6.5.3 Acylation and Deacylation

6.5.3.1 Non-enzymic Acylation, Deacylation and Migration. Traditionally, acetylation of a reducing sugar with acetic anhydride and fused sodium acetate gives predominantly the fully acetylated equatorial (usually β) pyranose, because of the greater acidity and nucleophilicity of the equatorial OH, whereas acid-catalysed acetylation gives the axial acetate, thermodynamically favoured by the anomeric effect.

Ester protecting groups of varying stability have been developed for use in synthesis. Conjugation of the carbonyl group with the aromatic ring makes benzoates less susceptible to nucleophilic attack than acetates, and the benzoyl (Bz) group can be tweaked further by adding nitro groups (more reactive)[143] or *o*-methyl groups (less reactive: mesitoyl groups have to be removed with AlH_3).[144] Likewise, chlorine or alkoxy substitution into acetyl groups makes them more susceptible to attack at the carbonyl,[145] whereas methyl substitution, particularly as in pivaloyl [$(CH_3)_3CCO$–][146] reduces reactivity.

For synthetic purposes, these acyl groups are introduced by reaction with the chloride or anhydride and a tertiary amine, which means the actual acylating agent is an *N*-acyl derivative. 4-(Dimethylamino)pyridine (DMAP) is often added as a promoter, to effect the acylation of sterically restricted sites, but is much more effective with the acid anhydride than an acid halide. The DMAP-promoted acetylation of cyclohexanol,[147] with triethylamine as acid acceptor, is first order in [DMAP], [cyclohexanol] and [Ac_2O] and zero order in triethylamine, suggesting a rate-determining encounter of the acylpyridinium ion–acetate ion pair with the alcohol. The acetate counterion is thought to deprotonate the alcohol, a role which is supported by the relative ineffectiveness of DMAP promotion in the reactions of acid chlorides, which would give an ion pair with the weakly basic chloride ion. Theoretical calculations suggest that there is concerted transfer of acyl group and proton – any tetrahedral intermediate is "too unstable to exist". In accord with a rate-determining transition state which sees charge annihilation, the reaction is much faster the less polar is the solvent.

The most widely used deprotection is transesterification with methanol, Zemplén deacylation (Figure 6.26). If the methanol is rigorously anhydrous, so that no carboxylic acid is formed, only small, catalytic amounts of sodium methoxide are needed. *N*-Acyl groups, such as the *N*-acetyl groups of hexosamine derivatives, are untouched. Magnesium methoxide is even milder and can be made selective between different *O*-acyl groups.[148] Potassium cyanide in anhydrous methanol generates acyl cyanides, which are readily converted back to cyanide and the methyl ester: both benzoyl and acetyl groups are removed.[149] For substrates which are base-labile but acid-stable, such as 2, 4-dinitrophenyl glycosides, HCl in methanol can be used, although this can never be anhydrous because of the slow conversion of methanolic HCl to methyl chloride and water.[150] Various primary amines, which yield the amide of the protecting acyl group, are also used: ammonia in methanol is a standard debenzoylation reagent.[151] An ingenious method of specifically removing chloroacetyl groups in the presence of other acyl functionality is treatment

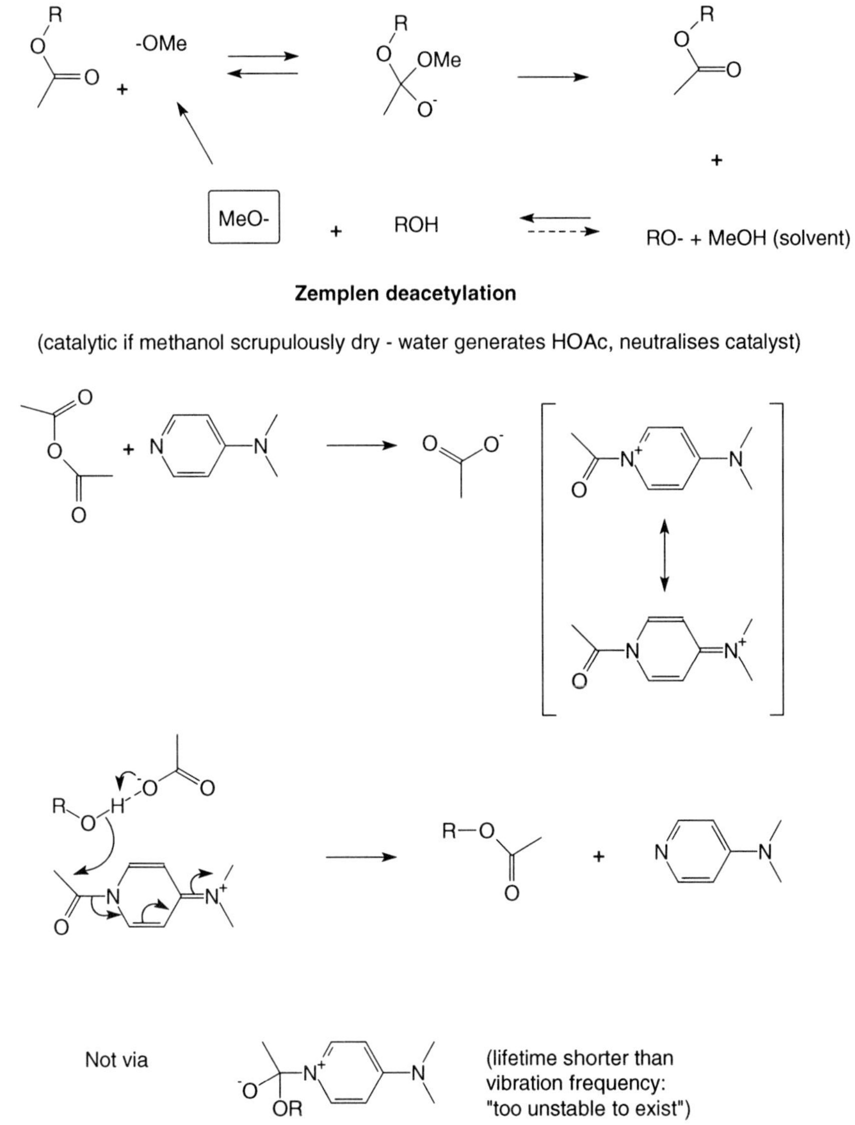

MeO- + ROH ⇌ RO- + MeOH (solvent)

Zemplen deacetylation

(catalytic if methanol scrupulously dry - water generates HOAc, neutralises catalyst)

Not via

(lifetime shorter than vibration frequency: "too unstable to exist")

Figure 6.26 Aspects of the removal and introduction of acyl protecting groups.

with the sterically unhindered tertiary amine 1,4-diazobicyclo[2.2.2]octane (DABCO) in ethanol at room temperature. Presumably the DABCO first displaces the chloride to yield a quaternary ammonium ion whose carbonyl group is, by virtue of the adjacent full positive charge, particularly reactive.[152]

Transesterification takes place under both acidic and basic conditions, so that acyl substituents in partially esterified carbohydrates are prone to migration (Figure 6.27). Presumably for steric reasons, acyl groups tend to migrate to primary positions – the well-known migration of acyl groups from the 4- to the

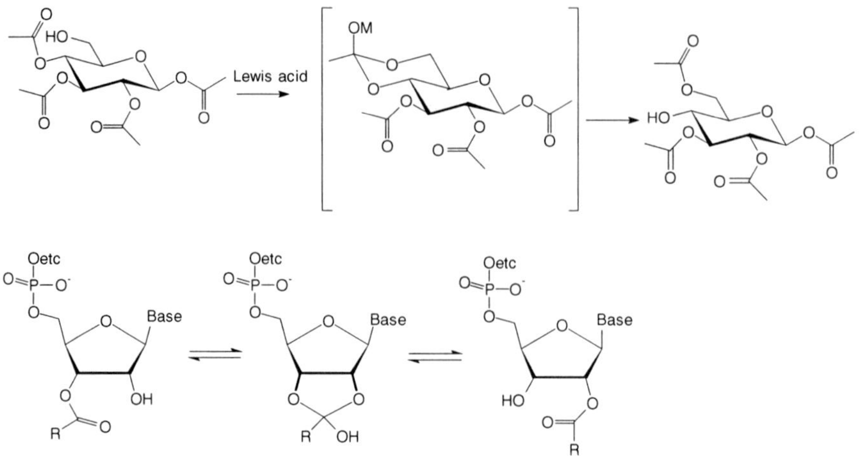

Figure 6.27 Intramolecular transesterification–acyl group migration: (top) 4,6-migration of an acetyl group in glucopyranose; (bottom) equilibration of 2′- and 3′-aminoacyl tRNAs. R = amino acid side-chain.

6-position of glucose is an example. When there is a choice of equatorial secondary positions, as in mono- and diacetylated nitrophenyl β-D-xylopyranoside,[153] at equilibrium the 4-*O*-acetyl compound predominates amongst the monoacetates and the 3,4-diacetyl compound among the diacetates. The 2-*O*-acetyl compounds are the most kinetically labile. Such ready acetyl migration presents problems for the location of *O*-acetyl groups in hemicelluloses, which migrate even under the pH-neutral extraction conditions of superheated steam.[154] Migration between the *cis*-related 2′-OH and the 3′-OH of aminoacyl tRNAs is so rapid that the charged tRNA species involved in ribosomal protein biosynthesis can be isolated only as equilibrating positional isomers. In the complex of valine tRNA with a regulatory protein, the charged tRNA does not exist as either ester, but as the tetrahedral intermediate (hemiorthoester) for their interconversion.[155]

Transesterification of sucrose with triglycerides under alkaline anhydrous conditions gives the non-nutritional fat Olestra™. The sucrose molecule is esterified with 6–8 triglyceride-derived fatty acids to give a material with the texture and taste of cooking oil. However, the cluster of alkyl chains makes the ester linkages inaccessible to the lipases of the human gastrointestinal tract, which work at the interfaces of fat globules and pull one fatty acid chain at a time out of the lipid phase. The material therefore is not digested.[156]

Cellulose acetate is a major commodity chemical.[157] It is made by acid-catalysed acetylation of cellulose with acetic anhydride, often after the cellulose has been acid-swollen and sometimes partly hydrolysed to reduce the *DP* [the acid catalyst is often the (cheap) sulfuric acid]. Material with *DS* 2.2–2.3 is used for lacquers and plastics; the acetone-soluble material of slightly higher *DS* (2.3–2.4) is spun into "acetate rayon" and cellulose acetate of *DS* 2.5–2.6 is used for photographic film. In this application it has replaced nitrocellulose by virtue of its thermal and photochemical stability. The fine control of *DS* required by the various applications is achieved by near-complete acetylation and partial deprotection; steric effects in the deprotection may result in an over-representation of primary OH groups amongst those unsubstituted. Solid cellulose can also be acetylated with preservation of fibre structure – paper tissues made from such material absorb oil, *etc.* An important niche for cellulose acetate films is as dialysis membranes and tubing in practical biochemistry[xv] (and kidney dialysis machines): the pore sizes in appropriately manufactured films are large enough to let small molecules through, but retain macromolecules above a certain MW cutoff.

Cellulose propionate and butyrate, and mixed esters, are made similarly and have uses dictated by their greater solubility in cheap organic solvents such as toluene.

6.5.3.2 Enzymic Acylation, Deacylation and Transesterification. The esterases and transacylases that act on carbohydrates have been classified into families on the basis of their amino acid sequences, in the same way as glycosidases and transglycosylases and glycosyl transferases, although the classification scheme is less successful. At the time of writing (spring 2007) there are 14 carbohydrate esterase (CE) families on CAZy,[158] although CE10 contains few, if any, specifically carbohydrate esterases and has not been updated since 2002. Carbohydrate esterases can be further classified, either by whether the acyl or alcohol moiety is carbohydrate or by catalytic mechanism. The distinction between amidases and esterases is not useful, since several families contain both esterases and amidases such as *N*-deacetylases. Adaptations of three of the four proteinase mechanisms (serine, aspartic and Zn^{2+}-dependent) are found, in addition to an adaptation of the urease mechanism. Serine esterases are found in Families 1, 5, 6, 7, 10 and 12; CE 4 is Zn^{2+}-dependent, CE 8 aspartate–protease-like and CE 9 urease-like. Little is known of the mechanisms of CE 2, 3, 11 and 13.

Enzymes with the same nominal specificity, such as acetyl xylanesterases, have different specificities in the deacetylation of cellulose acetate, according to their CE Family.[159] This parallels, but in a more rational fashion, the use of lipases by synthetic chemists to deprotect sugar acetates regioselectively, or, in non-aqueous suspension, to acylate them selectively;[160] however, in these cases, the choice of lipase is purely empirical.

6.5.3.2.1 Serine carbohydrate esterases and transacylases. The commonest reaction mechanism is the standard serine esterase[161]/protease mechanism, demonstrated paradigmally for chymotrypsin,[162] involving an acyl-enzyme intermediate. The enzyme nucleophile is a serine hydroxyl, which is hydrogen bonded the imidazole of a histidine residue, whose other nitrogen is hydrogen bonded to a buried, but ionised, aspartate residue (Figure 6.28),

---

[xv] They are therefore weakened, if not destroyed, by most cellulases, a fact repeatedly rediscovered by neophyte cellulase researchers.

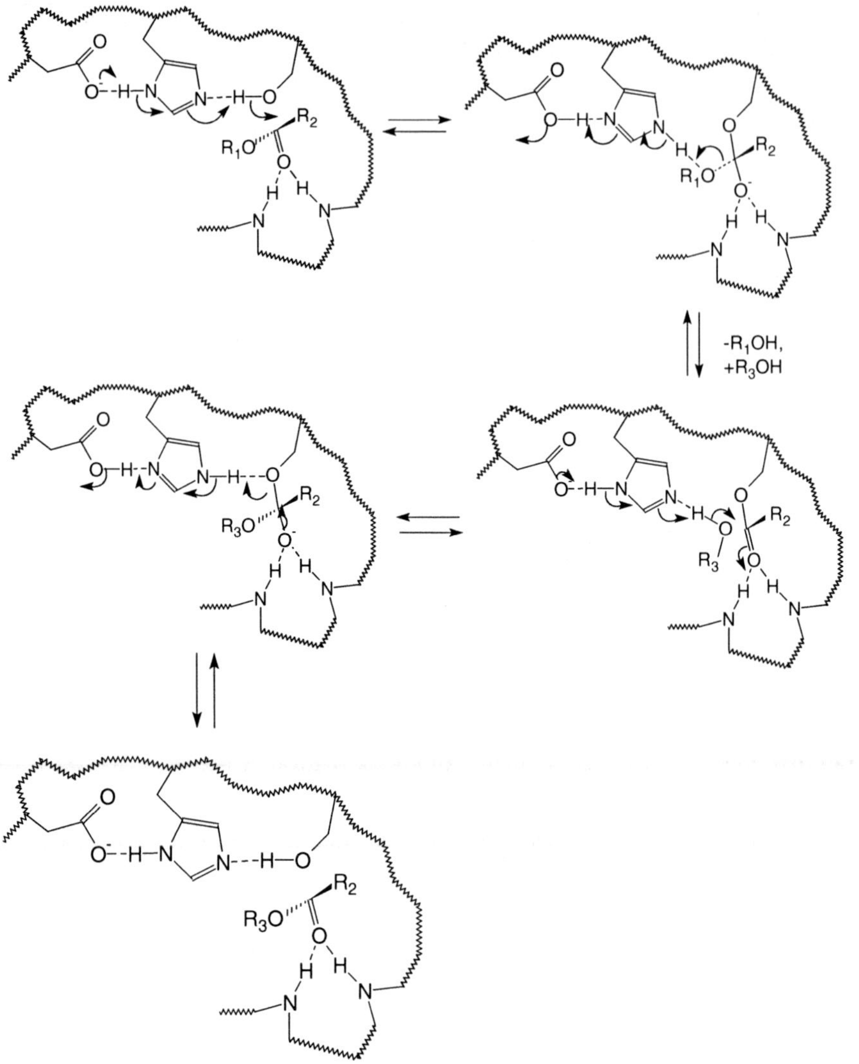

Figure 6.28 Mechanism of serine esterases and transacylases: for a hydrolase $R_3 = H$. The proton transfer steps are speculative, some probes of mechanism indicating that the protonation state of the aspartate does not change during catalysis and that the Asp–His pair has a short, tight hydrogen bond. The tetrahedral intermediates do not accumulate (although acyl-enzymes are well characterised) and there is some suggestion that they have lifetimes too short to have a meaningful existence. For a discussion of the evidence on these points, see Ref. 162.

the Ser–His–Asp sequence being known as a "catalytic triad". The negative charge of the oxygen of the tetrahedral intermediates is stabilised by hydrogen bonding to two or three amide N–H groups, two of which are from the peptide main chain.

Important evidence for the mechanism in Figure 6.28 comes from mechanism-based inhibitors. Serine proteases are irreversibly inhibited by electrically neutral phosphorylating agents such as diisopropyl fluorophosphate (DFP), $[(CH_3)_2CHO]_2POF$ or the nerve gas sarin, $CH_3POF[OCH(CH_3)_2]$. Fluoride is displaced and the active-site serine is irreversibly phosphorylated. The use of these reagents as war gases depends on their ability to phosphorylate irreversibly the active site serine of acetylcholinesterase (which in fact occurs in CAZy CE 10) and interfere with nervous conduction. Organophosphorus insecticides, which do not have fluoride leaving groups, are based on the same principle. It appears that the reagents owe their effectiveness to the recruitment of most of the catalytic power of the enzyme to the phosphorylation, but not the dephosphorylation step.[163] Inactivation by DFP and related agents is a useful indication that an esterase is a serine esterase; phenylmethanesulfonyl fluoride (PMSF), $PhCH_2SO_2F$, is routinely added to crude enzyme solutions to stop proteolysis by endogenous serine proteinases.

Physically safer as a mechanistic criterion is reversible inhibition by boronic acids, $RB(OH)_2$, which add the active site serine to form tetrahedral species $[RB(OH)_2OSer]^-$, which mimic the tetrahedral intermediate/transition state. They also mimic the tetrahedral intermediates in aspartic protease action, however, and are therefore not as definitive.

CE Family 1 is very large and contains members which do not act on carbohydrate-derived substrates. The crystal structure of a CE 1 domain of Xyn10B modular enzyme from *Clostridium thermocellum* has been solved.[164] The CE 1 domain is a feruloyl esterase which hydrolyses the feruloyl groups attached to some arabinofuranosyl O5 groups in native xylan. (The Xyn 10B protein as a whole consists of two CBM 22 domains, a dockerin domain, and a GH 20 xylanase domain, and forms part of a cellulosome – see Section 5.10.) The enzyme has the common α/β hydrolase fold. Studies of ferulic acid complexes of the inactive alanine mutant of the active site serine revealed the classic catalytic triad, and two main-chain peptide NH bonds are in place to form an "oxyanion hole". A remarkable feature is that the enzyme as repeatedly isolated was esterified on the active site serine by phosphate or sulfate.

As families containing the double displacement glycohydrolases can also contain transglycosylases, so CE 1, which contains double-displacement carboxyl esterases, also contains transacylases. Trehalose [α-D-Glc*p*-(1↔1)-α-D-Glc*p*] esterified on O6 with mycolic acid plays a role in maintaining the hydrophobic exteriors of the tubercle bacterium *Mycobacterium tuberculosis*, and the crystal structure of three related mycoloyl transesterases are known.[165,166] These transfer mycoloyl residues between the 6-positions of α,α-trehalose and the cell wall arabinogalactan: one has the ability to disproportionate two molecules of trehalose 6-mycolate into one molecule of trehalose and one of trehalose 6,6′-dimycolate. Mycolic acids have the general formula $R_1CHOHCHR_2COOH$, where R_2 is a long alkyl chain and R_1 is also a long chain containing other functionality. The mycolic acids of *M. tuberculosis* have the general formula $CH_3(CH_2)_{19}$–Cyp–$(CH_2)_y$–Cyp–$(CH_2)_z$–CHOH–CH(n-$C_{24}H_{49}$)–COOH, where Cyp = 1,2-substituted cyclopropane, $y = 10$, 14 or 16 and $z = 11$, 13, 17 or19.[167] They are therefore very hydrophobic, and the acyl transfer appears to take place at a phase boundary, with the transmycolase acting in the scooting mode of lipases, which permits the long alkyl chains to remain in the hydrophobic phase. Nonetheless, the transmycolases have the standard α/β hydrolase fold.

Significant mechanistic and structural work has been done on fungal enzymes in CE 5, the acetylxylan esterase AXEII from *Penicillium purpurogenum*,[168] and the cutinase from *Fusarium solani*, which have the classic α/β fold.[169] The former has the interesting property of having two observable conformations of the active site Ser and His in the crystal. The latter acts on cutin, the water-repellent surface coating of plants, which is a polymer of hydroxy acids through ester linkages. In addition to a well-defined catalytic triad, cutinase has oxyanion hole formed from two main-chain peptide NH bonds. The structure of the unliganded enzyme and that of the enzyme inactivated by *p*-nitrophenyl diethylphosphate and thus with an active–site serine carrying a diethylphosphate [–PO(OEt)$_2$] group, are very similar. This indicates a distinction between cutinases and lipases proper, in which the oxyanion hole is only formed when the enzyme is activated by absorption at the lipid–water interface.

CE 6 apparently contains only acetyl xylan esterases, but catalytic functions of many of the proteins in it have not been experimentally confirmed. One such protein from *Arabidopsis thaliana* proved to have a catalytic triad,[170] and as isolated was found to have the catalytic serine covalently modified by PMSF. This enabled an oxyanion hole formed by two main-chain NH and a side-chain carboxamide to be identified.

CE 7 is remarkable in containing enzymes with very broad specificities: an enzyme identified as a cephalosporin deacetylase proved to have greater acetylxylose esterase activity. Its X-ray structure[171] confirmed it to be a doughnut-shaped hexamer. The active site residues of the six active sites point towards a large central chamber in the centre of the doughnut, whose walls are 60 Å apart at the widest point, although the entrances to the chamber, separated by 36 Å, are smaller. The catalytic triads and oxyanion holes, formed by two main-chain peptide NH bonds, of the six active sites are conventional. It is thought that the broad specificity of the enzyme, specific with respect to the acetyl function but promiscuous with respect to leaving group, arises from the architecture. Because the enzyme had no activity on polymeric substrates, although it completely deacetylated fully acetylated xylose and xylobiose, it was thought to operate intracellularly.

CE 12 contains serine enzymes which are likewise deacetylases. The rhamnogalacturonan I acetyl esterase of *Aspergillus aculeatus* removes acetyl groups from the 2- and 3-positions of the main-chain galacturonic acid residues of rhamnogalacturonan I; it is a serine esterase, with a Ser–His–Asp catalytic triad and an oxyanion hole made from two main-chain peptide NHs and an Asn side-chain,[172] but does not have the α/β fold.

6.5.3.2.2 Zn^{2+}-dependent carbohydrate esterases. CE 4 contains enzymes which are Zn^{2+} dependent, and all enzymes contain a zinc-binding motif of two histidines and an aspartate.[173] Many are involved in the *N*-deacetylation of hexosamine residues at the polymer level. The activity was first discovered in the nodulating factors produced by plant symbiotic bacteria, but a NodB-like domain was later discovered in chitin deacetylases.[174] The peptidoglycan *N*-deacetylase (*Sp*PgdA) is a countermeasure taken by the human pathogen *Streptococcus pneumonia* against the host antibacterial lysozyme, and the only CE 4 enzyme on which both structural and mechanistic work has been done.[173] Deacetylation of GlcNAc in the peptidoglycan of the bacterial cell wall renders it a poor substrate for lysozyme. (A similar enzyme produced by *Bacillus subtilis* acts on the *N*-acetylmuramic acid residues: in this case de-*N*-deacetylation is followed by intramolecular lactamisation, possibly catalysed by additional enzymes).[175]

The crystal structure of native *Sp*PgdA revealed a (possibly product) acetate ion coordinated to the zinc, in addition to the residues of the histidine-binding motif (His330, His326 and Asp275). Potentially catalytic functionality in the region of the active site was probed by site-directed mutagenesis and the D275N, D319N and H417S mutants had no catalytic activity. The X-ray structure of the D275N mutant indicated that a sulfate had replaced the acetate coordinated to zinc in the wild-type (wt) enzyme, and the newly introduced asparagine was in turn hydrogen bonded to this sulfate. The mechanism advanced for these enzymes is essentially the same as that advanced two decades earlier by Christianson and co-workers for the Zn^{2+}-dependent proteinases, such as thermolysin and carboxypeptidase A.[176] Asp275 acts as a general base to a Zn^{2+}-coordinated water molecule and in a single chemical step the tetrahedral intermediate is generated and becomes coordinated to the Zn^{2+}. Collapse comes from general acid catalysis by the protonated form of an active site histidine, with Asp319 acting to maintain the correct protonation state of the enzyme.

The mechanism has a number of difficulties, chief among them the requirement that Zn^{2+} makes the nucleophilic water more nucleophilic and the electrophilic carbon more electrophilic in the same chemical step. Mock has shown convincingly, however, that thermolysin and carboxypeptidase A work by a reverse-protonation mechanism, with the catalytically active form of the enzyme being one in which Zn^{2+} coordinated by water (not hydroxide) and the equivalent to His417 acts a general base for the attack of water (Figure 6.29).[177] Slight modification of this mechanism for *Sp*PgdA would have His417 as the general base (with the role of Asp391 now clarified as increasing the proportion of the deprotonated form), the Zn^{2+} acting purely as an electrophile and Asp275 acting as a weak Zn^{2+} ligand. The Zn^{2+} in the free enzyme would thereby be maintained in an "entactic state", a normally disfavoured coordination environment in which it is particularly electrophilic.

6.5.3.2.3 Aspartic carbohydrate esterases. CE 8 contains only pectin methyl esterases, which demethylate methyl esters of α-(1→4)-polygalacturonan. Two X-ray crystal structures – from the bacterial plant pathogen *Erwinia chrysanthemi*[178] and from carrot (*Daucus carota*), have been solved.[179] The enzyme

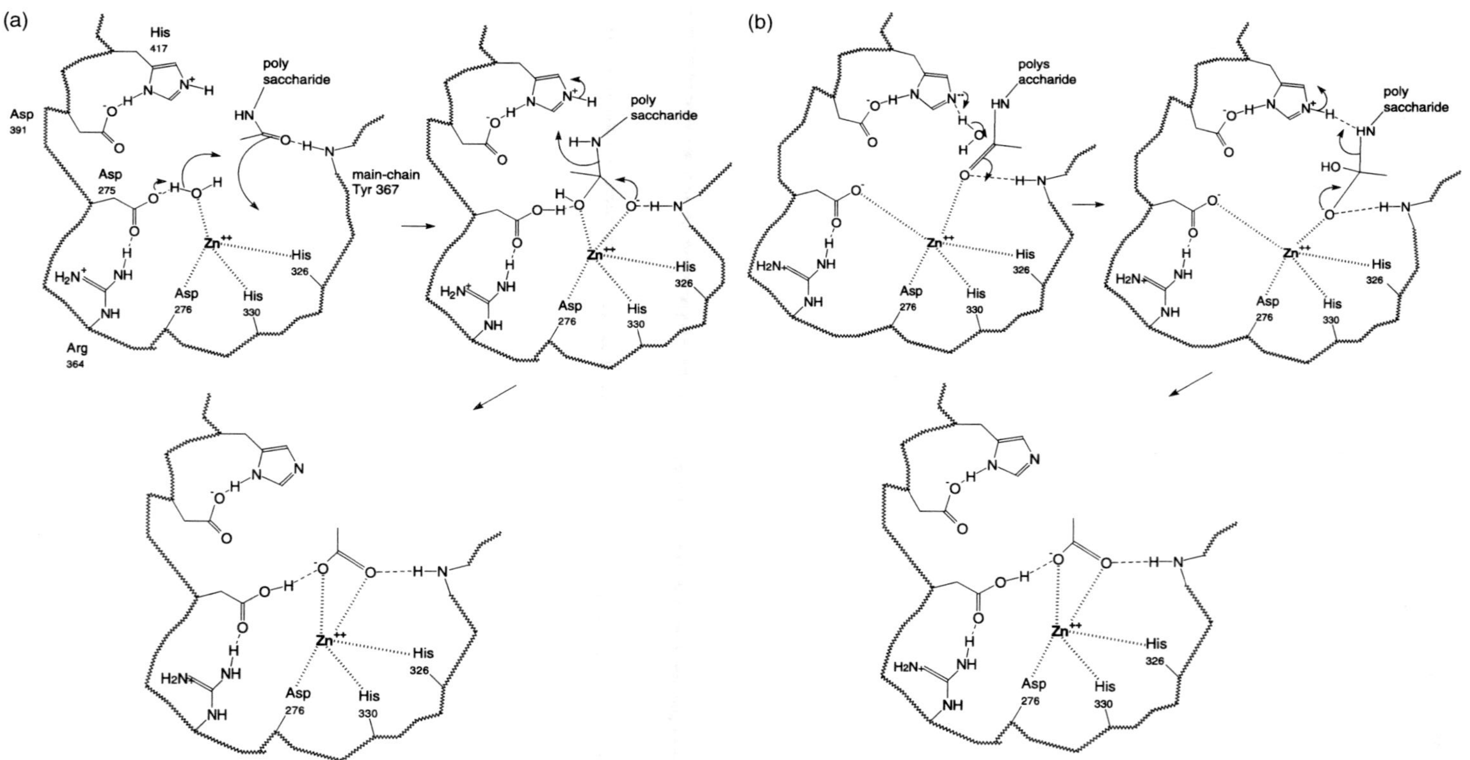

Figure 6.29 Mechanism of the Zn^{2+}-dependent carbohydrate *N*-deacetylases: (a) according to van Aaalten and co-workers;[173] (b) modification of the Mock mechanism for carboxypeptidase.[177]

from *E. chrysanthemi* has a fold consisting of a right-handed helix of β-sheet, totally unlike the α/β hydrolase fold, but resembling both the lyases and glycoside hydrolases which act on pectin. There was no sign of the catalytic triad of a serine protease or the Zn^{2+} binding motif, but two aspartates and an arginine were conserved. It was very reasonably suggested that a close analogue of the mechanism of the aspartate proteases[180] was followed, with one aspartate acting as a general base and the other as a general acid in the generation of a tetrahedral intermediate, which then collapsed to product with no covalent acyl-enzyme intermediate (Figure 6.30). The arginine, not seen in the aspartic proteases, is presumably to modulate the ionisation states of the carboxylates – although the polyelectrolyte nature of the substrate makes pH–rate measurements difficult to interpret, it does appear that the pH optimum (broad, acid limb with $pK_a \approx 5$) of pectin methyl esterase is more neutral than that of pepsin (pH 1–5).

The structure of the carrot CE 8 pectin methyl esterase was very similar to the bacterial enzyme, although for reasons which are not clear a mechanism involving one of the aspartates as a nucleophile, rather than a general base, was preferred to the aspartate protease mechanism. Such mechanisms have been previously proposed for the aspartic proteinases, but were thoroughly disproved when one of this class of enzymes, the HIV proteinase, was found to catalyse ^{18}O exchange from solvent into recovered substrate,[181] a process not possible on the nucleophilic mechanism.

6.5.3.2.4 Twin group VIII metal esterases (urease type). CE 9 contains only *N*-acetylglucosamine 6-phosphate deacetylases. The dimeric enzyme from *Bacillus subtilis* proved to have a modified TIM barrel fold,[182] and to be metal dependent, with two iron atoms in the active site connected by an oxygen bridge. Although the catalytic site was all contained on one subunit, two basic groups on the other subunit were recruited to bind the 6-phosphate in the product complex with glucosamine-6-phosphate. The overall fold resembled that of the urease superfamily, which, however, normally recruits two oxygen-bridged nickel ions rather than iron ions.[183] The mechanism in Figure 6.31 was plausibly advanced, with one iron ion acting as an electrophile towards the carbonyl oxygen and the nucleophile being a hydroxide ion which bridged the two metals. The general acid could not be the histidine as suggested for urease, as the deacetylase possessed no similarly placed residue: Asp281 was, however, well placed so to act.

The NagA enzyme from *E. coli*, unambiguously a member of CE 9, however, on isolation was found to contain one Zn^{2+} per active site and no Fe and was powerfully inhibited ($K_i = 34\,nM$)[184] by the methyl phosphoramidate analogue of the substrate, which accurately mimics the tetrahedral intermediate in both carboxypeptidase- and urease-like mechanisms. The metal dependence of CE 9 appears variable and both urease-like mechanisms and mechanisms analogous to that of CE 4 may operate. The crystal structure of the *E. coli* enzyme reveals only bound Zn^{2+}, in a site equivalent to one of the two Fe sites of the *B. subtilis* enzyme and no potential histidine ligands for the other Fe; the *T. maritime* enzyme appears to be intermediate, with one Fe bound.[185]

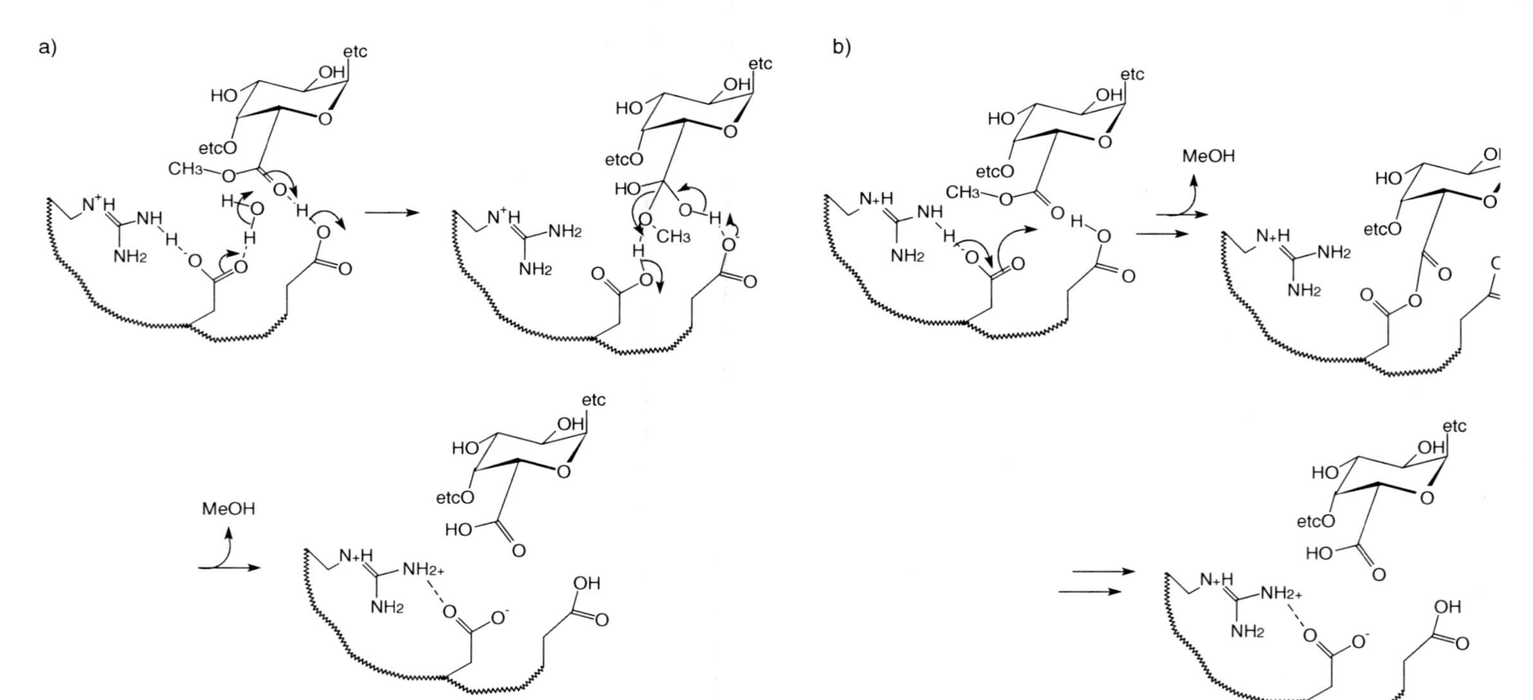

Figure 6.30 Mechanisms for CE 8 aspartate esterases (the anhydride intermediate mechanism (b) has been ruled out for the related aspartic proteinases – see text).

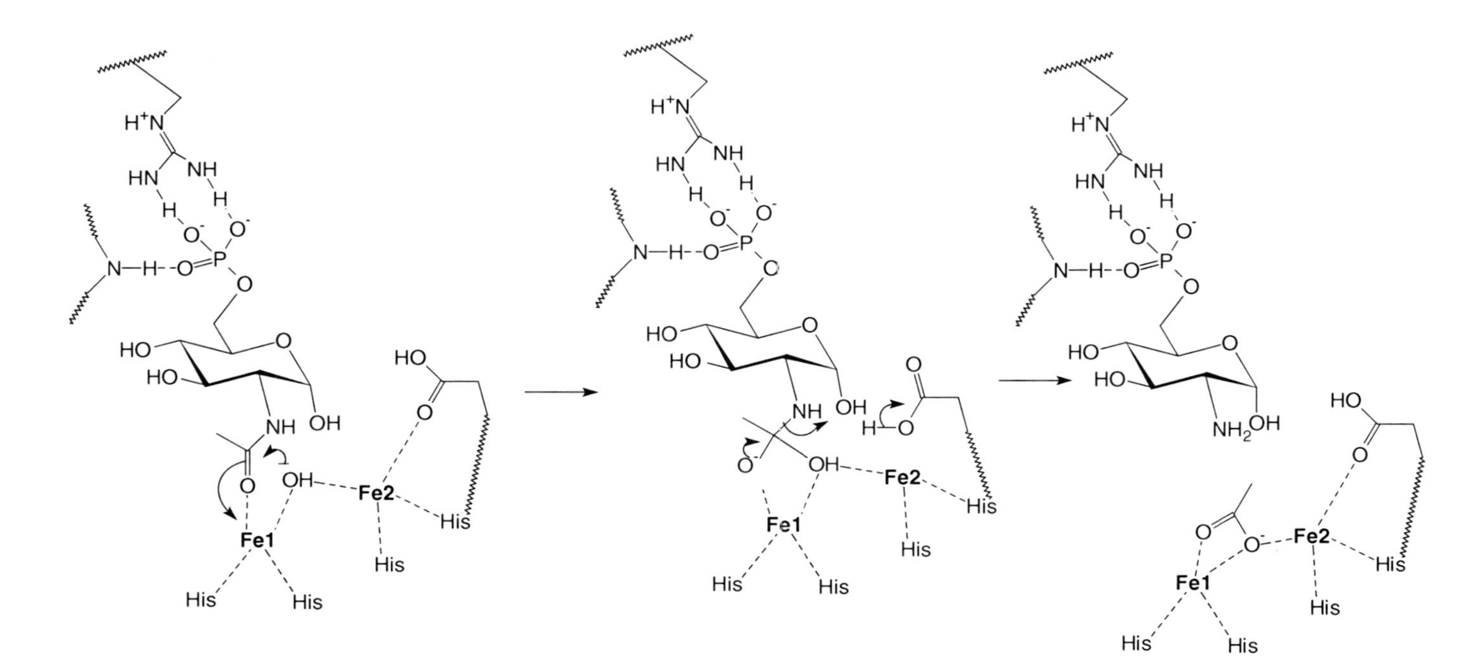

Figure 6.31 Urease-like mechanism for CE 9 carbohydrate esterases.

6.5.4 Carbohydrate Esters of Carbonic Anhydride and Their Nitrogen and Sulfur Analogues

Vicinal diols readily give five-membered cyclic carbonates. The carbonylating reagents can be diaryl carbonates, alkyl carbonates such as triphosgene [bis(trichloromethyl) carbonate], carbonyldiimidazole, chloroformate esters or the World War I war gas phosgene, in the presence of base (stoichiometric for chlorides, catalytic for transesterification from alkyl carbonates).[186] Cyclic carbonates are cleaved readily by base,[187] but are more stable to acid than ordinary esters, possibly because delocalisation of two lone pairs of electrons into the carbonyl group of the protonated carbonate makes it less electrophilic than protonated esters. The alkyl chloroformate route to cyclic carbonates has the advantage that solitary OH groups are also protected, as acyclic mixed carbonates. In the case of vicinal triols, the cyclic carbonate goes on any *cis* pair of hydroxyls. Carbonate is usually regarded as a "non-participating" protecting group in reactions at C1, but participation occasionally occurs, as shown in Figure 6.32, which also illustrates the use of TBDPS protection of primary OH groups in unprotected sugars and the use of "tweaked" acyl groups for selective anomeric and other protection.

Reaction of alkoxides with carbon disulfide yields xanthates, R–O–C(S)S^-;[xvi] reaction of alkali-swollen cellulose with CS_2 yields a solution of cellulose xanthate, "viscose", used in the production of rayon and cellulose film. The viscose process, invented by Cross and Bevan in 1892, is the oldest process for producing synthetic fibre. Dissolving pulp is first mercerised and then allowed to "age" aerobically to reduce the cellulose *DP* to 250–300 (the depolymerisation reactions being probably oxidation of cellulose 2-OH, 3-OH or 6-OH to carbonyls, followed by $E1_{CB}$ eliminations). Reaction with carbon disulfide yields viscose, which is allowed to "ripen". The initial site of substitution is, as expected for a base-catalysed reaction with a small electrophile, predominantly on O2 and during the ripening process the CS_2 substituents redistribute themselves and remaining crystalline regions dissolve as they become loosened from the crystallites and become xanthylated. The ripened viscose solution is then squirted into dilute sulfuric acid through spinnerets to yield rayon fibre and through slits to yield cellulose film, both comprised of cellulose II. Subsequent manipulation of the spinning conditions makes the glucan chains crystallise in the direction of the fibre.

Phenylurethane (carbamate) derivatives of insoluble polysaccharides, produced by reaction with PhNCO in neutral organic solvents, are used in determination of molecular weight distributions (see Section 4.5.1); carbamates are acid stable and base labile and have better alternatives as protecting groups in synthesis.

[xvi] In preparative chemistry, xanthate salts are often methylated with methyl iodide to give methyl xanthates, $R_1R_2CH–R_3R_4C–O–CS–S–Me$, which in simple cases can be pyrolysed smoothly to $R_1R_2C{=}CR_3R_4$, COS and MeSH (the Chugaev reaction). However, the reaction usually fails with carbohydrates, particularly if R_1 or $R_2 = OR_5$, so that a vinyl ether would result. Methyl xanthates are, however, used in radical deoxygenation of sugars – see Chapter 7.

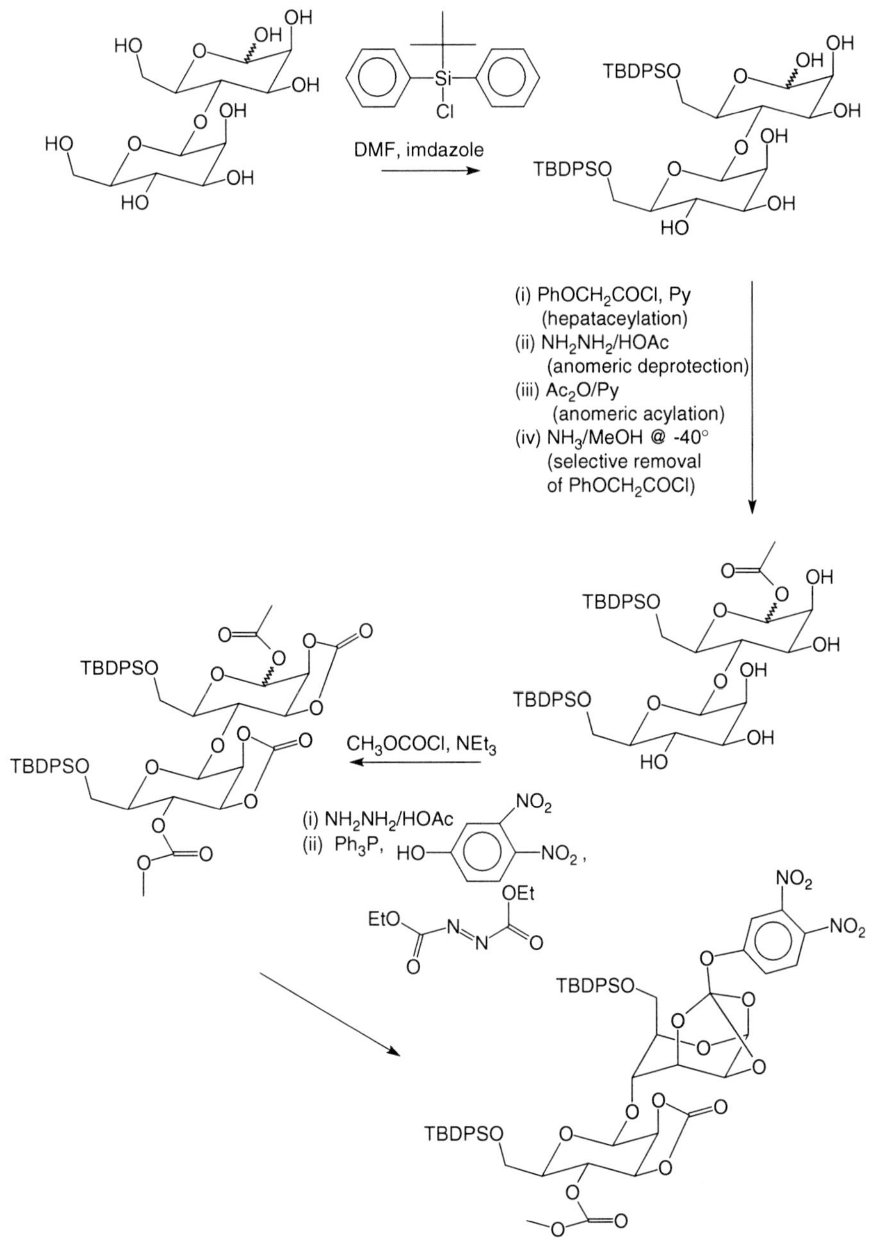

Figure 6.32 Successes and failure of carbonate protection in an attempt to make aryl β-mannobiosides.[188] The formation of the cyclic orthocarbonate must be entirely due to the complex phosphine oxide leaving group in the Mitsunobu reaction, since 4,6-di-*O*-acetyl-α-D-mannopyranosyl bromide 2,3-carbonate gave the expected aryl β-mannosides on reaction with phenoxides.[189]

6.5.5 Acetals of Carbohydrates

Sugar hydroxyl groups can be protected as tetrahydropyranyl or methoxymethyl ethers by reaction with dihydropyran and a catalytic trace of acid[190] or methoxymethyl chloride and acid acceptor, in aprotic solvents. Because of the introduction of a new asymmetric centre at position 2 of the tetrahydropyranyl ring, however, tetrahydropyranyl protection leads to diastereomeric mixtures, with accordingly complex NMR spectra. However, they can be removed under acid conditions which leave other acid-labile functionality such as methoxymethyl or TBDMS untouched ($BF_3 \cdot Et_2O$ in CH_2Cl_2 containing 5% EtSH).

The most important acetalisation and ketalisation reactions of sugar hydroxyls, however, are those where two hydroxyl groups are protected simultaneously (Figure 6.33). By and large, under conditions of thermodynamic control, acetalisation gives rise to six-membered rings (1,3-dioxanes) and ketalisation to five-membered rings (1,3-dioxolanes), but there are frequent exceptions. Acetalisation to 1,3-dioxanes is generally stereoselective, as the new asymmetric centre is generated with the aldehyde R group equatorial. Analogous 1,3-dioxane rings generated by ketalisation necessarily have one of the alkyl groups from the ketone axial, which may partly explain the preference of ketalisation for giving dioxolanes. Dioxolane formation occurs preferentially with *cis*-hydroxyl groups, for obvious reasons in the case of five-membered rings. In pyranose rings, the dihedral angle between *trans*-diequatorial and *cis*-OH groups is the same at approximately 60°. On incorporation into a five-membered ring, the angle has to decrease somewhat. As can be readily illustrated with Dreiding- or Fieser-type models, narrowing the dihedral angle is more difficult for *trans*-diequatorial than *cis* substituents. In terms of the Cremer–Pople sphere (Figure 2.3), narrowing the dihedral angle of *trans*-diequatorial substituents has to alter ring pucker (Q), whereas for *cis* substituents the ring conformation is displaced from the poles.

The reaction of sugars with ketones such as acetone or cyclohexanone or aldehydes such as acetaldehyde or benzaldehyde and a trace of Lewis or Brønsted acid as catalyst has been known for over a century; the classical ketalisation reagent was acetone and the classical acetalisation reagent benzaldehyde. Systematic investigations of sugar acetalisation are, however, few compared with the vast body of preparative anecdotage, which is impossible to rationalise in detail.[191,192] A major problem is that kinetic and thermodynamic products frequently differ and certain catalysts (particularly anhydrous $ZnCl_2$, which becomes hydrated and less electrophilic as acetalisation proceeds) deliver an uncertain mix of the two. Thus, the reaction of benzaldehyde with glycerol catalysed by HCl[193] gives first the *cis*- and *trans*-1,3 dioxolanes and then approaches equilibrium, which in DMF at room temperature is 1.8:1.8:1.2:1.0 *trans*-dioxane:*cis*-dioxane:*cis*-dioxolane:*trans*-dioxolane (Figure 6.34). The equilibrium proportion of *cis*-dioxane increases in carbon tetrachloride, which is poor H-bond acceptor, as the 2-OH of the glycerol forms hydrogen bonds with the ring oxygens. The equilibrium proportion of dioxolanes increases with temperature possibly for the same reason as the proportion of furanoses in a

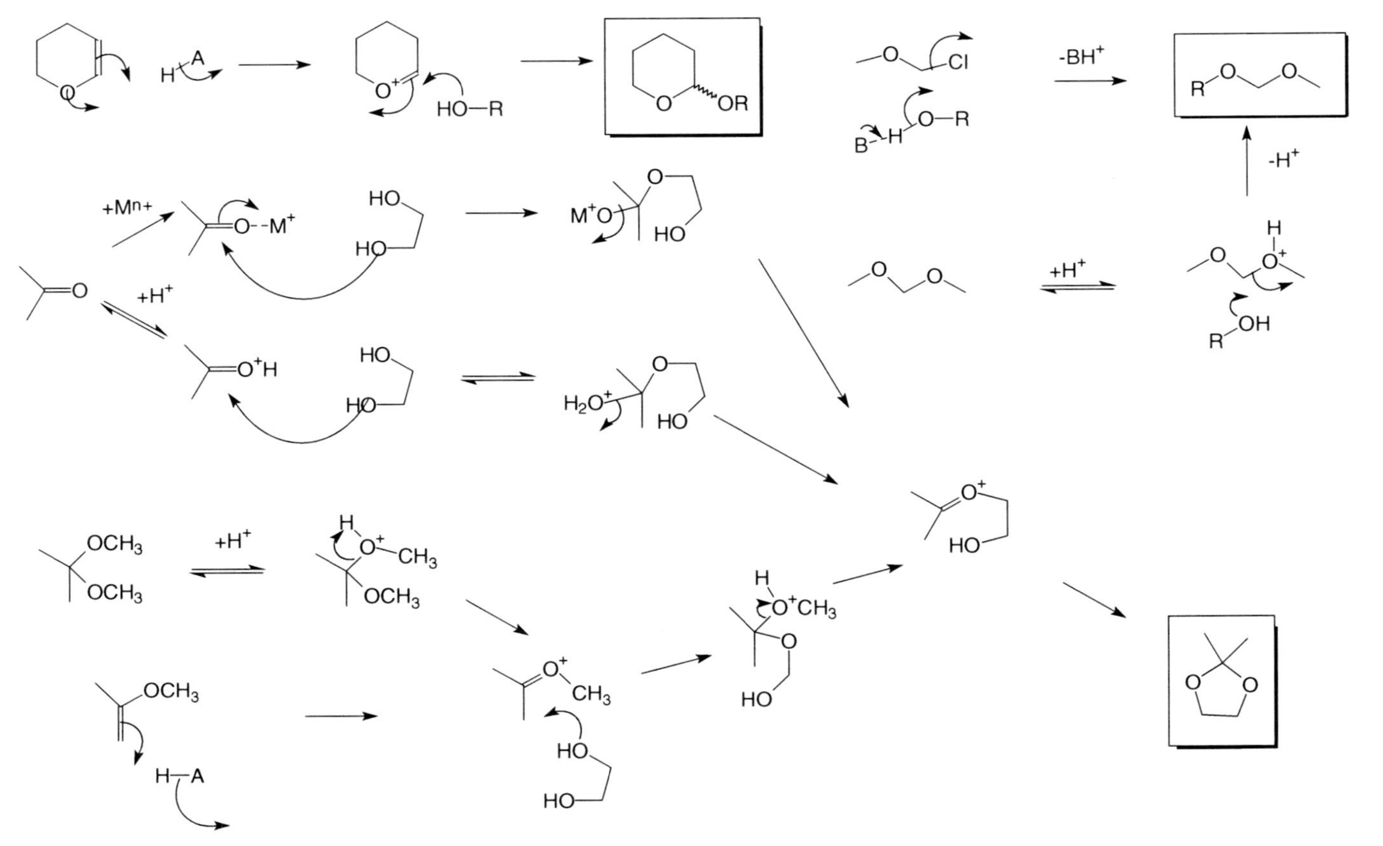

Figure 6.33 Basic reactions of acetal and ketal formation. The generation from the carbonyl compound or other acetal can be catalysed by Lewis or Brønsted acids, that from the enol ether by Brønsted acid only. Reactions are shown as involving oxocarbenium ions or not according to the stabilities in water.

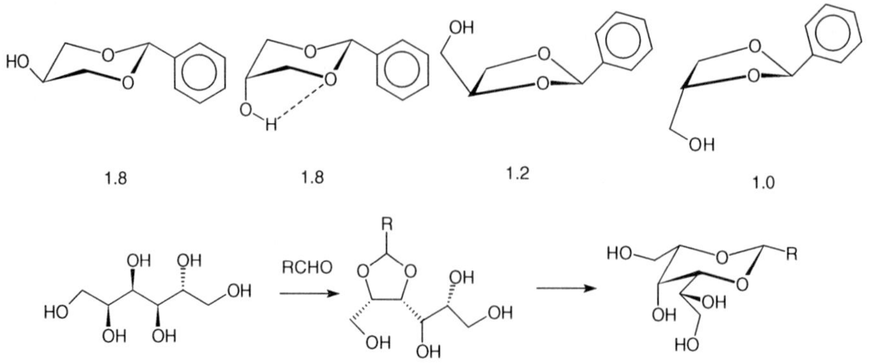

Figure 6.34 (Top) Equilibrium proportions of acetal products from benzalation of glycerol. (Bottom) Major kinetic and thermodynamic product from acetalation of glucitol with aliphatic aldehydes.

reducing sugar: the five-membered rings have an additional degree of free rotation. Acetal groups appear to avoid primary positions, with aliphatic aldehydes reacting with D-glucitol to yield first the 2,3-derivative which rearranges to the 2,4-compound. Similar differences in kinetic and thermodynamic products are observed in ketal formation. Methyl α-D-altropyranoside first gives a mixture of 4,6- and 3,4-isopropylidene derivatives, whereas at equilibrium only the 3,4-derivative (*cis*-dioxolane) is present,[194] and fructose forms first the 1,2;4,5-diisopropylidene β-fructopyranose, which is converted to the 2,3;4,5-isomer. The latter adopts a "$^{6}S_{4}$" ($^{3}S_{0}$) conformation.[192]

When reducing sugars are directly acetalated, the size of the sugar ring can change to accommodate the preference of the acetalation reagent. Because of the preference for dioxolanes to be *cis*-fused to pyranose and *a fortiori* furanose rings, "diacetone glucose", discovered by Emil Fischer, is 1,2:5,6-diisopropylidene α-D-glucofuranose.[195] Galactose, however, in the pyranose form has two pairs of vicinal *cis*-diols, so that "diacetone galactose" is 1,2:3,4-diisopropylidene α-D-galactopyranose. The preference of ketones for forming six-membered rings is not absolute, as can be seen from Figure 6.35, and xylose gives 1,2;3,5-diisopropylidene α-D-xylofuranose with acetone. Likewise, although mannose on reaction with acetone gives a mixture of products, on reaction with cyclohexanone it yields a nicely crystalline 2,3;4,6-dicyclohexylidene mannose. This latter compound establishes cyclic acetals as "non-participating", since Mitsunobu reaction (*cf.* Figure 6.32) with acidic phenols yields aryl 2,3;4,6-dicyclohexylidene β-mannopyranosides,[196] thereby establishing that cyclic ketals, unlike cyclic carbonates, are genuinely non-participating.

From the data in Figure 6.34, it can be seen that at equilibrium, benzaldehyde-derived dioxanes are favoured over dioxolanes by a factor of only 3.3 at room temperature,[xvii] so that dioxolanes can be formed, as glucose forms

[xvii] In the reaction of benzaldehyde with glycerol, the dioxolane products, being racemates, are favoured by a statistical factor of 2 compared with the achiral dioxanes.

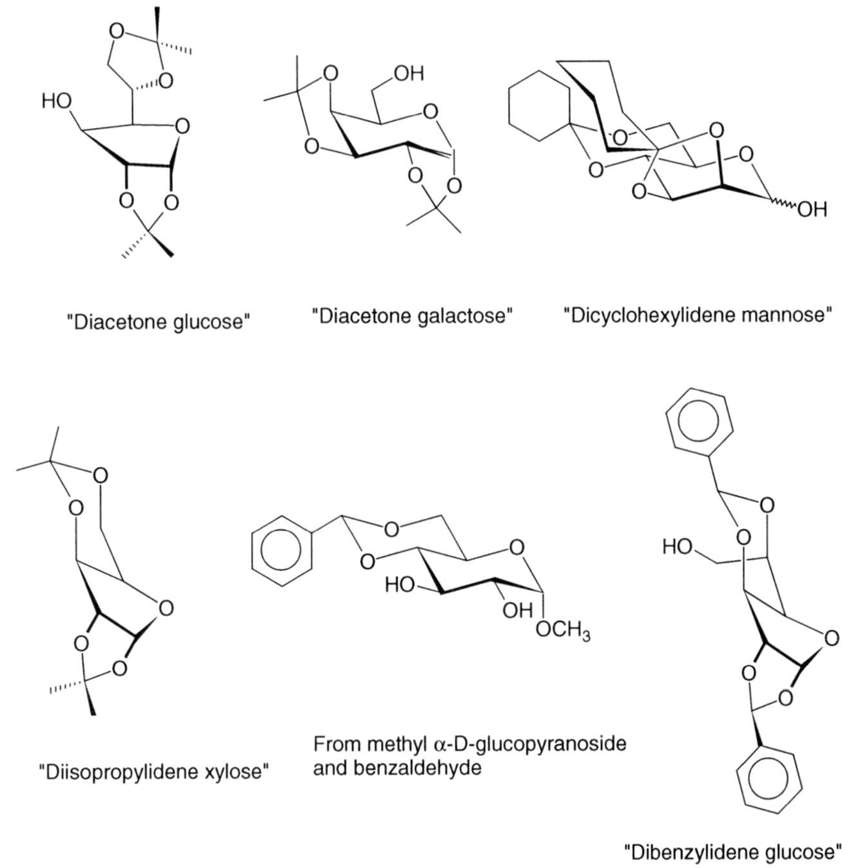

"Diacetone glucose" "Diacetone galactose" "Dicyclohexylidene mannose"

"Diisopropylidene xylose" From methyl α-D-glucopyranoside and benzaldehyde

"Dibenzylidene glucose"

Figure 6.35 Thermodynamic products of acetal and ketal formation from various sugars.

1,2;3,5-dibenzylidene glucofuranose, if the possibilities of forming the favoured dioxane are restricted.

trans-Diequatorial diols can be protected as acid-labile ketals if two adjacent ketal centres are involved (Figure 6.36). The first examples used 2,2′-bis(dihydropyran),[197] but later developments used cyclohexa-1,2-dione in methanol, its tetramethyl ketal or the tetramethyl ketal of butane-2,3-dione as ketalisation agents under acid catalysis.[198] The acetal products are under thermodynamic control. The stereochemistry of the products, in particular the configurations of the two new asymmetric centres, are dictated by the anomeric effect and the strong steric preference for the pyranose ring and the new 1,4-dioxane ring to be *trans* fused. The cyclohexadione derivatives have advantages in terms of crystallinity and availability of ketalisation reagent, but the disadvantage, over the butane-2,3-dione reagents, of sometimes derivatising compounds with the *manno* configuration 2,3, with one of the ketone equivalents acting as cyclohexanone itself.

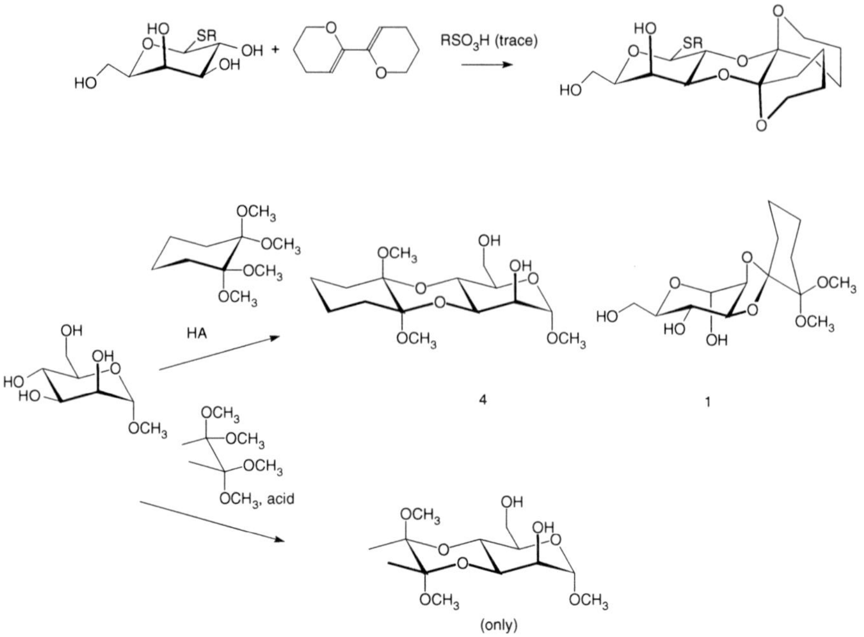

Figure 6.36 1,2-Diketal protection of vicinal *trans*-diequatorial diols.

An intriguing method of introducing cyclic acetals from fully deprotonated diols is to react them with (*E*)- or (*Z*)-1,2-bis(phenylsulfonyl)ethene.[199] The mechanism in Figure 6.37 results in 1,3-dioxolanes or dioxanes. Because of the electron-withdrawing effect of the $PhSO_2$ group on the stability of the oxocarbenium ion formed in their hydrolysis, however, the cyclic acetals are remarkably stable to acid and indeed can be removed only under drastic reducing conditions such as with $LiAlH_4$.

Formation of cyclic acetals by sugar hydroxyls generally retards nucleophilic displacements at the anomeric centre of the same sugar residue. In the case of 4,6-benzylidene derivatives, the mechanism of deactivation appears to be that the dipole of the C6–O6 bond is constrained with its positive end directed towards the anomeric centre.[122] In the case of the 1,2-diketals, the deactivation arises from the increased difficulty of forming half-chair or boat conformations in a six-membered ring, which is part of a *trans*-fused decalin structure.

Hydrolytic removal of acetals can be selective, as in the well-known production of 1,2-isopropylidene α-D-glucofuranose from mild acid-catalysed hydrolysis of 1,2;5,6-diisopropylidene α-D-glucofuranose.[200] The difference in reactivity of the two isopropylidene derivatives lies in the differing stabilities of the oxocarbenium ion intermediates. Under acid-catalysed conditions, a cyclic ketal with different electron demand at the alcohol sites will first open to give the most stable oxocarbenium ion; leaving group ability is not an issue

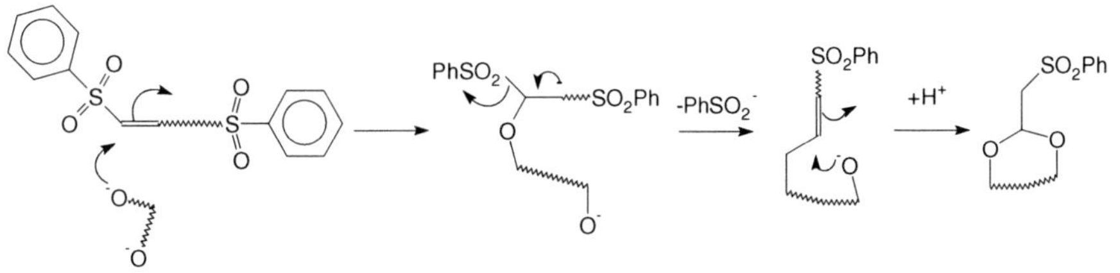

Figure 6.37 Formation of cyclic acetals from 1,2-bis(phenylsulfonyl)ethene.

because of the β_{lg} values of ~ 0 observed with specific-acid-catalysed acetal hydrolysis (see Section 3.7.2). The oxocarbenium ion derived from opening of the 1,2-ketal is destabilised inductively by its proximity to doubly oxygenated C1, compared with analogous ions from the opening of the 5,6-acetal. The reactivity of acetals, as of esters, can be tweaked by manipulation of electron demand in the derivatising agent: thus, *p*-methoxybenzaldehyde yields 4,6-derivatives of hexopyranosides, which are usefully more labile than unsubstituted benzylidene derivatives.

Cyclic acetals and ketals hydrolyse about 1–3 orders of magnitude more slowly than their acyclic counterparts, but this is not due to any reversibility of the initial endocyclic cleavage step.[200,203] A detailed comparison of the rates of acid-catalysed hydrolysis of substituted-benzaldehyde acetals of ethanol and ethylene glycol revealed that electronic effects in the two systems were identical. The 30–35-fold slower hydrolysis of the dioxolanes arose entirely from a less favourable entropy of activation.[201] Because of the dominance of the translational partition function, entropies of activation for reactions giving two fragments are generally greater than those giving only one. The ready hydrolysis of dioxolanes is incompatible with ALPH (see Section 3.5.2), at least in its simple form, as the five-membered ring prohibits the presentation of an antiperiplanar lone pair to the leaving oxygen, although a synperiplanar lone pair is available.

It is unlikely that overall rates of hydrolysis of cyclic acetals and ketals are materially altered by recombination of oxocarbenium ion and leaving group, as was at one time suggested.[200] Re-addition of the newly departed leaving group, compared with addition of water, has been examined in two systems. In the hydrolysis of the substituted-benzylidene acetals of *exo,exo*-2,3-dihydroxynorbornane, recombination becomes more favourable the more stable the oxocarbenium ion[202] (Figure 6.38). Generation of the oxocarbenium ion intermediate in the hydrolysis of cyclic ketals of acetophenone from the vinyl ether enabled its partitioning between cyclisation and addition of water to be monitored directly: water addition was at least an order of magnitude faster than cyclisation.[203] As the oxocarbenium ion intermediates in the hydrolyses of cyclic carbohydrate acetals and ketals will be less stable than the alkylated acetophenones in this study, recombination is not a factor in their hydrolytic reactivity.

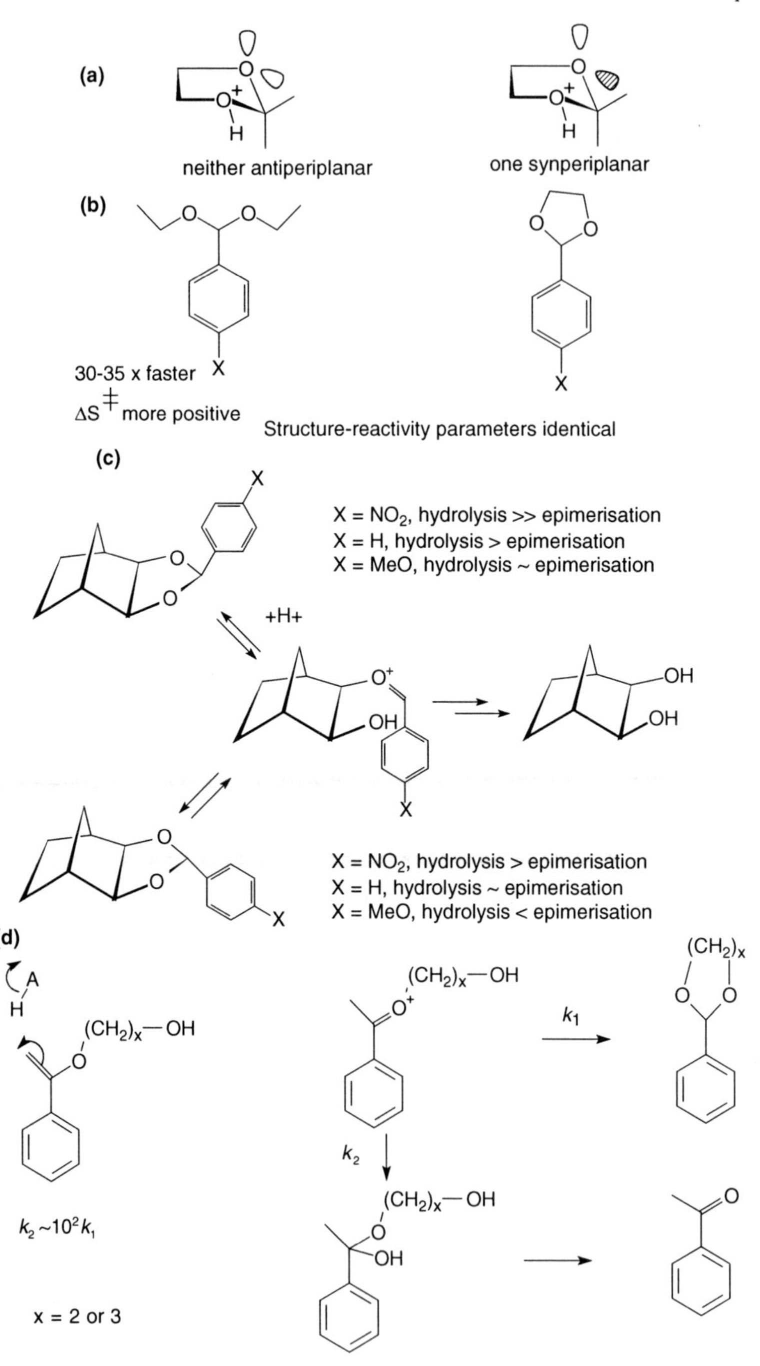

Figure 6.38 Examination of the slower rates of hydrolysis of cyclic acetals compared with their acyclic counterparts: (a) anti-ALPH nature of dioxolane hydrolysis;p[204] (b) two systems with identical structure–reactivity parameters;[201] (c) recombination in a dioxolane hydrolysis; (d) direct measurement of the relative rates of hydrolysis and recyclisation of the acyclic oxocarbenium ion intermediate in cyclic acetal hydrolysis.[202]

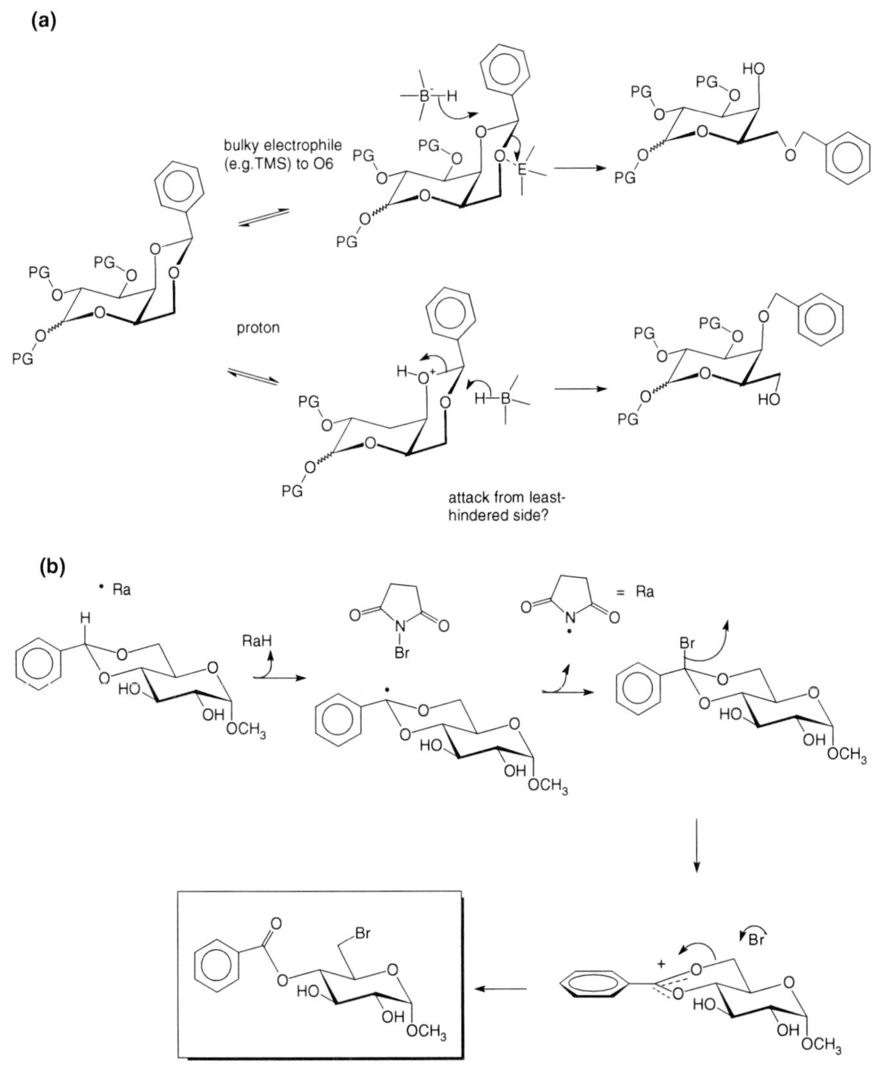

Figure 6.39 (a) Selective cleavages of 4,6-benzylidene acetals. Both *gluco*- and *galacto*-configured acetals are shown, as the reactions are general. (b) Hanessian reaction. (Oxidative opening of 4,6 benzylidene sugars with free radical brominating agents).

In addition to hydrolysis, benzylidene acetals can be catalytically hydrogenolysed off completely or partly reduced off, leaving one hydroxyl free and a benzyl group on the other (Figure 6.39). This last reaction is a nucleophilic displacement at the acetal centre by a complexed hydride, of a hemiacetal anion, which is either protonated or complexed with an electrophile. The bulk of the electrophile governs regioselectivity, with bulky electrophiles coordinating the less hindered O6 and therefore giving the 4-*O*-benzyl compound.[205] Brønsted acids give the 6-*O*-benzyl-substituted product;[206] since protonation of O4 and O6 should be comparable, this selectivity presumably arises because of hindrance of the approach of the bulky hydride nucleophile by substituents on O4.

A related reaction of 4,6-benzylidene acetals is their conversion to 6-bromo-4-*O*-benzoyl compounds by free radical brominating agents such as *N*-bromosuccinimide. The initial product, an *o*-acyl halide, readily ionises (in the non-polar solvents employed, such as carbon tetrachloride, presumably to an ion pair). The bromide ion then acts as a nucleophile at C6.[207]

The only cyclic ketals of sugars which are prominent in Nature are the pyruvate ketals covered in Section 4.6.10.3.1 see particularly Figure 4.92. The conversion of UDPGlcNAc to its 3-*O*-enylpyruvyl ether using phosphoenolpyruvate as vinylating agent, although formally a vinylation, is closely related to acetalations to cyclic pyruvate ketals and so is treated here (see Figure 6.40(a)). In the mechanism in Figure 4.92, rather than the second oxocarbenium ion intermediate (formed by loss of the phosphate) adding a second sugar OH, it loses a proton. The reaction is part of the pathway for the biosynthesis of the cell-wall peptidoglycan of bacteria and the formation of activated *N*-acetylmuramic acid is completed by the reduction of 3-*O*-enylpyruvyl UDPGlcNAc to the 3-lactyl ether UDP-MurNAc, catalysed by the flavoenzyme MurB.

The X-ray structure of the transferase MurA from *Enterobacter cloacae*,[208] highly homologous to the *Escherichia coli* enzyme, is of a two-domain enzyme in an "open" conformation: it was reasonably surmised that when the positive charges of an arginine and a lysine, whose repulsion were keeping the protein in the open conformation, were neutralised by the binding of UDPGlcNAc, the active site would close and eliminate water.

The stereochemistry of the addition/elimination was determined as follows (Figure 6.40(b)). MurA accepts both (*E*)- and (*Z*)-phosphoenolbutyrates and if the reaction is carried out in D_2O, (*E*)-phosphoenolbutyrate is converted to (*E*)-enol ether containing no deuterium and the (Z)-enol ether containing deuterium [at least in the initial stages of the reaction, before the slow conversion of (*E*)- to (*Z*)-phosphoenolbutyrate complicated matters].[209] Since the sugar OH adds to C2 in the first step of the reaction and the phosphate leaves from C2 in the second, if the addition and elimination were of the same stereochemical type, there would have to be a rotation about the C2–C3 bond, resulting in abstraction in the elimination of a different hydrogen from that which had been added. Consequently, the two reactions must be of different stereochemical types. The stereochemistry of the protonation step was determined using (*E*)- and (*Z*)-3-fluorophosphoenolpyruvates, whose kinetics were also very informative.[210] Formation of the tetrahedral intermediate was 10^4 times slower than with phosphoenolpyruvate itself, suggesting a very oxocarbenium ion-like transition state for the addition, but its decomposition was $>10^6$ times slower, which meant that it accumulated (although it bound tightly). By a complex series of stereochemical correlations, it was shown that the proton was added to the C2-*re* face of the

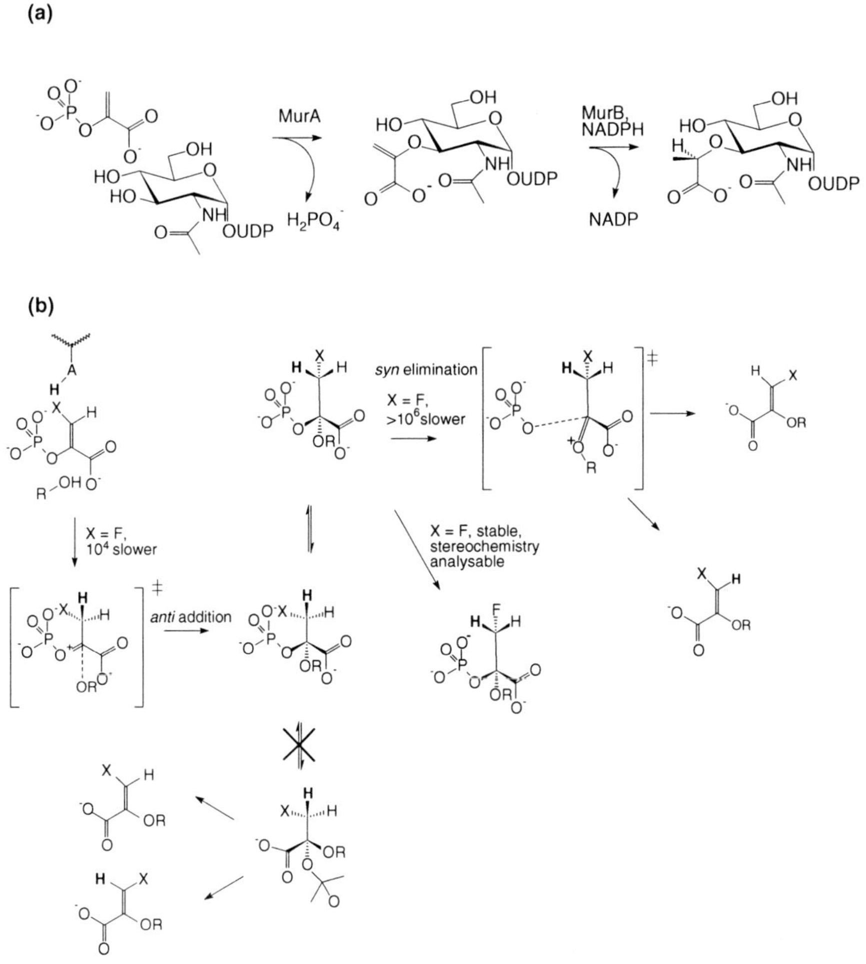

Figure 6.40 Formation of UDP-muramic acid. (a) General reaction; (b) stereochemical course and an indication of how a stereochemical marker and isotopic labelling can detremine the stereochemistry of reaction.

fluorophosphoenolpyruvates (and hence, presumably, to the C2-*si* face of the natural substrate).[xviii] Since the natural tetrahedral intermediate has the *S* configuration,[211] the addition must proceed *anti* and hence the elimination must proceed *syn*.

The fluorophosphoenolpyruvates also act as suicide substrates, alkylating Cys115, also the target of the antibiotic phosphomycin [a simple *exo*-affinity label, (1*R*,2*S*)-epoxypropane-3-phosphonic acid]. It was suggested that an

---

[xviii] COOH has higher priority than CH_2, but lower priority than CHF, in the Cahn–Ingold–Prelog system.

actual oxocarbenium ion was produced on the enzyme, which could partition between capture by substrate OH and by Cys115. It seems to the author, however, that with addition–elimination reactions in the SW corner of the More O'Ferrall–Jencks diagram (see Figure 6.66), slight changes in environment, brought about by interaction with the fluorine atom, could alter the nature of the preassociated nucleophile.

The tetrahedral intermediate formed on the enzyme – but not covalently linked – can be removed by a rapid alkaline quench and its constitution and spontaneous decomposition studied.[211] It decomposes by loss of phosphate to give an ion pair. Rate-determining and product-determining steps are different, with a crucial product-determining ionisation not apparent in the kinetics. The intermediate ion can lose a proton, add O4 of the sugar or add water and revert to pyruvate, in processes governed by the second ionisation of the phosphate, although this is not seen in the pH–rate profile.

Little is known about the catalytic groups, other than Cys115, whose replacement with Asp yielded a fosfomycin-inert but functional enzyme: it was suggested that Cys115 was involved in acid–base catalysis.[212]

The enoylpyruvyl UDPGlcNAc is then reduced to the UDP-muramic acid by MurB, an unusual reductase which uses NADPH as a coenzyme but contains a tightly bound molecule of FAD. The *proS* hydride of NADPH is transferred first to the flavin, and then the flavin N5-H is transferred to C3 of the pyruvate enol ether; a consequence is that the enzyme follows ping-pong kinetics, as the flavin is alternately reduced by NADPH and oxidised by 2-pyruvyl UDPGlc-NAc. The X-ray structure of the native enzyme[213] and the complex with substrate of both wt and Ser229Ala mutants[214] has been solved. The enolate generated by hydride addition appears to be stabilised by interactions with Arg159 and Glu325 (*E. coli* numbering). It is subsequently protonated, probably by Ser229, on the *si* face to give the D-lactyl ether.

6.5.6 Borates and Boronates

The reversible formation of borate esters by carbohydrates in aqueous solution has been known for over a century, and played a role in early conformational studies.[215] Currently it is an ancillary in the separation of carbohydrates, and with its aid very fine separations can be achieved – the secondary tritum isotope effects at H2, H3, H5 and H6 (but not H1) of glucose on borate complexation permit complete isotopic separation to be achieved by ion-exchange chromatography.[216]

The formation and hydrolysis of these borate esters are rapid on the time-scale of bench manipulations (so that rapid equilibration can be assumed for the purposes of chromatography and ion exchange), but slow on the ^{11}B and ^{13}C NMR time-scale, so that separate resonances of various species can be observed.[217] Cyclic borates with both five- and six-membered rings are formed, as are complexes of two carbohydrate ligands with one borate. Measurements made at 25 °C of the association of borate with a range of sugars led to the following generalisations, which are discussed in terms of the negative logarithms of the pH-independent dissociation constants (K_{D1}, M) of the complex between the neutral sugar (L) and the borate anion, $B(OH)_4^-$. The association of neutral sugars with boric acid, $B(OH)_3$, is negligible, so the values of the association constants at any pH can be calculated from the pK_a of boric acid (9.40, expressed conventionally rather than as an association constant for OH^-). The reactivity patterns arise because of the short boron–oxygen bond (1.36 Å), which is most compatible with tetrahedral bond angles, and C–O and C–C bond lengths, if the two C–O bonds of a cyclic borate ester are eclipsed.

(i) The strongest associations (pK_{D1} = 3.40–5.65) are made between vicinal *cis*-hydroxyls of a furanose.
(ii) Exocylic complexes between vicinal hydroxyls in pyranoses, such as that involving 1-OH and 2-OH of fructopyranose show pK_{D1} = 2.40–3.18.
(iii) Exocylic complexes between vicinal hydroxyls in furanoses, such as that involving 1-OH and 2-OH of fructofuranose, have pK_{D1} = 1.70–2.00.
(iv) Vicinal *cis*-diols in pyranoses (axial–equatorial) give rise to comparatively weak complexes (pK_{D1} = 1.00–1.30)
(v) Six-membered cyclic borates are formed between O4 and O6 of pyranoses, whether the O4 and the hydroxymethyl group are *cis* or *trans*, with pK_{D1} = 0.48–0.78.
(vi) No complexes are formed from vicinal *trans*-diols.

These are microscopic pK_{D1} values for a particular mode of borate ester formation: where a sugar forms more than one type of 1:1 complex, the macroscopic association constant ($1/K_{D1}'$) is the sum of all the microscopic association constants [$\Sigma(1/K_{D1})$].

Dimeric complexes of the type L_2B^- are also formed. In favourable cases, such as ribofuranose, where there are no steric clashes between the two ligands to boron, 2pK_{D1} = pK_{D2} ($K_{D2} = [L]^2[B^-]/[L_2B^-]$),[218] and the second carbohydrate binds as tightly to tetracoordinated anionic boron as the first. Such 2:1 complexes, involving the vicinal *cis*-OH groups in a furanose, are employed by Nature to cross-link the side-chains of rhamnogalacturonan II in the plant cell wall (see Section 4.6.3.2.3, Figure 4.67).

Tridentate complexes are observed with β-arabinofuranose, β-fructofuranose and α-galactofuranose, involving a *cis*-1,2 diol and a hydroxymethyl or hydroxyethyl group all on the same side of the ring.[217] They were also discovered in the case of cyclitols, with (all) *cis*-inositol forming an adamantane-type structure (Figure 6.41).[219]

Carbohydrate esters of alkane- or areneboronic acids are made by exchange from their methyl or ethyl esters, *e.g.* $ArB(OEt)_2$ or $RB(OMe)_2$, or by heating the carbohydrate and boronic acid and distilling out the water.[220] The often crystalline phenylboronates have some use as protecting groups in synthesis, surviving conditions for acylation, mild glycosylation and oxidations with

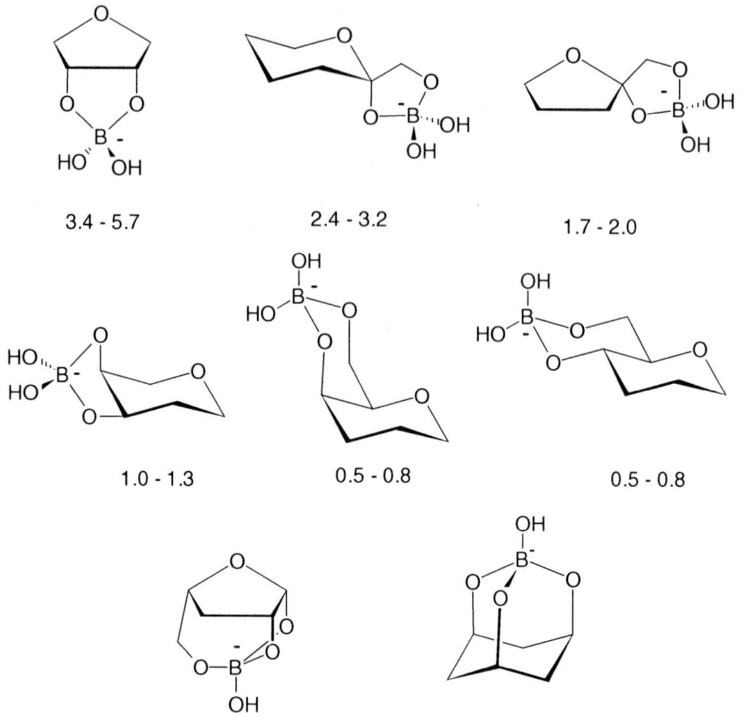

Figure 6.41 Types of borate complexes. The figures are pK_{D1} values at 25 °C.

activated DMSO, but the main current interest in areneboronic acids is as recognition elements for sensors and for affinity chromatography materials for carbohydrates (Figure 6.42). To the former end, fluorescent areneboronic acids such as 6-dimethylaminonaphthalene-2-boronic acid[221] have been constructed, whose fluorescence changes with carbohydrate complexation. Similarly, the original concept of monitoring the electrode potential of the ferrocene/ferricinium ion couple of ferrocene boronic acid as carbohydrate complexes are formed[222] has been elaborated into a receptor having two boronic acids attached through linkers to different rings of the ferrocene, so that discrimination between sugars is improved.[223]

Determination of the structures of the complexes by NMR spectroscopy is more experimentally difficult for boronates than for borates, since each cyclic boronate ester is a pair of stereoisomers, diastereomeric at the newly created asymmetric boron. The association of various sugars with 3-propionidobenzeneboronic acid has been measured as a function of pH.[224] The variation for simple sugars is as expected for their complexation by $ArB(OH)_3^-$ [pK_a $ArB(OH)_2 = 8.6$] and pK_{D1}' values at 25 °C for complexation of glucose (1.93), mannose (2.16) and galactose (2.40) can be calculated from the data given. The dissociation constant of the complex with *N*-acetylneuraminic acid, calculated on total complexed and uncomplexed boronate, is, however,

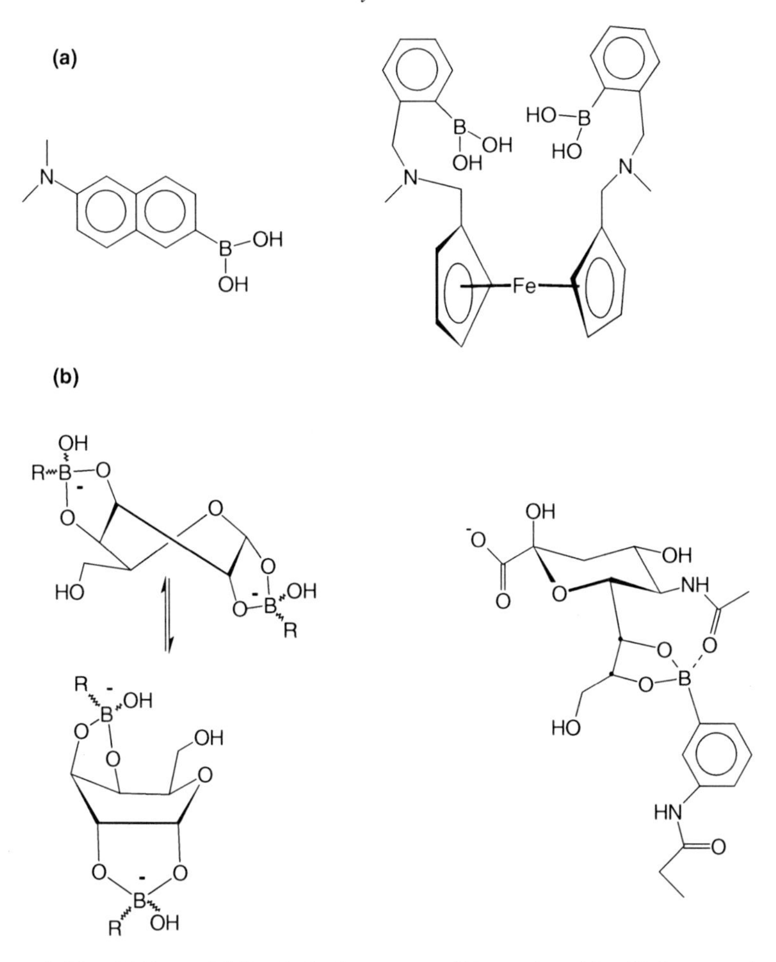

Figure 6.42 (a) Potential "carbohydrate sensor" boronic acids. (b) Suggested structures of areneboronic acid–sugar complexes: left, α-D-galactopyranose; right, β-*N*-acetylneuraminic acid.

independent of pH between pH 4 and 8 and then falls slightly. Such a dependence is compatible with the binding of *N*-acetylneuraminic anion to neutral $ArB(OH)_2$ or its equivalent, binding of neutral NANA (effective $pK_a \approx 1.6$, see Table 6.1) to $ArB(OH)_3^-$. However, extensive NMR data on the complex were more compatible with complexation via O7 and O8 and the acetamido group.

It was suggested, from ^1H NMR studies on the complexation of various sugars by *m*-nitrobenzeneboronic acid and rather low-level calculations (MOPAC and AM1), that complexation of pyranose vicinal *cis*-diols may occur via skew and boat conformations.[225] This was most likely in LB_2^- complexes, where there were two such interactions, such as the 1,2;3,4-diadduct of α-D-galactose, which was calculated to adopt the $^{3,0}B$ and $^0S_2$ conformations.

6.5.7 Nitrites and Nitrates

Simple alkyl nitrites are well known, being formed from alcohols in the presence of nitrosating agents, including, in acidic solution, nitrous acid itself. Alkyl nitrites are much more stable to aqueous alkali than the isoelectronic carboxylic esters. Their alkaline hydrolysis involves a direct displacement of alkoxide at nitrogen by hydroxide, since ^{18}O exchange experiments indicate that any trigonal pyramidal intermediate of the type R–O–N(OH)–O^- does not live long enough for the proton to exchange between the two unsubstituted oxygen atoms.[226] Their acid-catalysed hydrolysis and exchange are, however, much more ready than those of alkyl carboxylates, proceeding through free NO^+ ions:[227] the departure of NO^+ and the transfer of a proton from the (general) acid catalyst is concerted, so that the reaction is an S_E2 reaction on oxygen.[228] The lability to acid means that unprotected carbohydrate nitrites have not been characterised, although cyclodextrin nitrites are formed transiently during cyclodextrin catalysis of hydrolysis of,[229] and *N*-nitrosation by,[230] simple alkyl nitrites under basic conditions. By contrast, under acidic conditions the formation of inclusion complexes with cyclodextrins sequesters simple alkyl nitrites and protects them from hydrolysis. Protected carbohydrate nitrites are sometimes made for the Barton remote functionalisation, in which alkoxy radicals generated by photolysis of nitrites abstract hydrogen atoms from neighbouring, but unactivated, C–H bonds.

In contrast to the obscurity of carbohydrate nitrites, carbohydrate nitrates have been known and exploited for nearly two centuries, long before their chemical constitution was established. During the second quarter of the nineteenth century, the explosive and/or deflagrating substances produced by treatment of various natural materials, now known to be carbohydrate biopolymers, by treatment with nitric acid, with or without sulfuric acid, were investigated;[231] the names coined at the time still persist. Braconnot first produced cellulose nitrate from the action of nitric acid on paper, linen and sawdust in 1819.[232] Unfortunately, he thought the "xyloidin" so produced was the same substance as that he produced in 1833 from the action of nitric acid on starch.[233] The confusion was cleared up a decade later[234] (on the basis of the different solubilities of the exhaustively nitrated materials, nitrostarch being soluble and nitrocellulose crystalline and insoluble), but the name "xyloidin" stuck to nitrostarch. The increase in weight without dissolution of cellulose on nitration was correctly attributed to the formation of nitrate esters by elimination of water by Pelouze in 1838,[235] and a reliable process for "gun-cotton" devised by 1846.[236] Nitrocellulose with lower *DS* (1.5–2.1) was distinguished from gun-cotton by its solubility in mixtures of ethanol and ether and termed variously "collodion" or "pyroxylin". Nitrostarch as an explosive, by then properly distinguished from nitrocellulose, dates from 1905.

Nitration of low molecular weight polyols, rather than polysaccharides, was initiated by Sobrero, who produced glyceryl trinitrate (nitroglycerin) in 1846 and mannitol hexanitrate in 1847, although the explosive properties of hexanitromannitol were not examined until 1878, by Sokolov.[231] During the various arms races between the Crimean War and the end of World War II, a host of polyol nitrates were investigated; some of the compounds that have remained

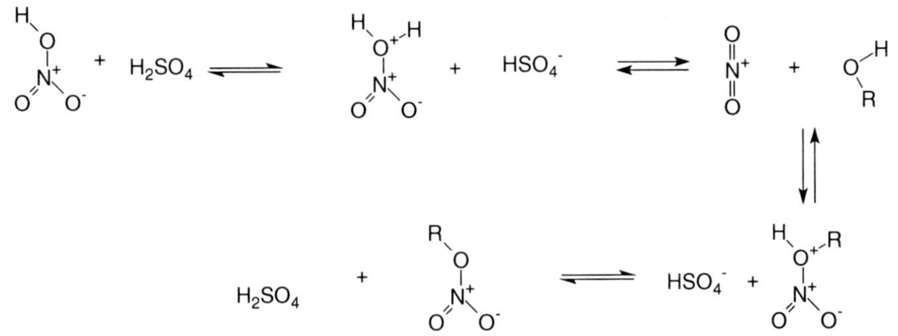

Figure 6.43 Mechanism of formation of nitrate esters. Either formation of the nitronium ion or its reaction with alcohols, can be rate determining. Nitrate ester formation under industrial conditions of nitration is reversible, so products from homogeneous solutions of polyols reflect thermodynamics.

in use are ethane-1,2-diol dinitrate, ethylene di- and triglycol dinitrate, 1,2-propanediol dinitrate, 1,2,4-butanetriol trinitrate and pentaerythritol tetranitrate.[237] Pure nitroglycerin melts at 13 °C, making it unsuitable for blasting in cold weather; later commercial blasting products incorporated other polyol nitrates such as ethane-1,2-diol dinitrate to reduce the melting point.

Popular history regards Alfred Nobel's discovery in 1867 that nitroglycerin could be stabilised by absorption on kieselguhr (diatomaceous earth), to produce dynamite, as a step-change which made possible the safe use of nitroglycerin, but it was, rather, the incremental improvement that proved commercially successful. Stabilisation of nitroglycerin by admixture with an inorganic solid, to damp any initial small shockwaves and neutralise any acid formed by decomposition, was being explored by explosive experts across Europe; for example, Petroshevskii made a magnesium oxide-stabilised product for use in Siberian mines in 1853. Folklore seems also to have exaggerated the instability of liquid nitroglycerin – although a consignment intended for blasting the route of the transcontinental railroad did indeed explode spectacularly at the Wells Fargo shipping office in San Francisco in 1866, it had already survived shipment round Cape Horn.

The mechanism of conversion of alcohols to their nitrates in acidic media involves the simple attachment of the nitronium ion NO_2^+ to the alcohol oxygen, followed by loss of a proton (Figure 6.43). In nitromethane, the reaction of unhindered alcohols with excess nitric acid is zero order (all alcohols react at the same rate) as generation of NO_2^+ becomes rate determining; with hindered alcohols, the reaction is first order, so that nitration of glycerol involves an initial rapid nitration of the primary OHs and a very much slower nitration of 2-OH.[238] Modern studies[xix] of the nitration of cellulose with the nitric acid–sulfuric acid

[xix] There is a large technical literature on nitrocellulose production which ignores Ingold's findings – for details see Short and Munro's 1993 paper.[239]

mixture used to manufacture various grades of nitrocellulose established a similar mechanism.[239] The symmetrical N–O stretching vibration of the linear NO_2^+ ion at $1400\,cm^{-1}$ was detected by Raman spectroscopy in technical acid mixtures, which nitrated cellulose within the time of mixing. Surface nitration, measured by X-ray photoelectron spectroscopy, was shown to be reversible and the regioselectivity ($6 \gg 2 > 3$) shown to be an equilibrium effect, with sulfuric acid catalysing denitration. Oddly, surface nitration was slightly lower than bulk nitration and it was suggested that this was a kinetic effect arising from the fact that the nitronium ion could penetrate between the sheets of glucan chains but acid molecules, to reverse the nitration, could not. The distribution of substituents, however, in bulk and surface nitration was similar. Because of the reversibility of nitrate ester formation and the ability of cellulose trinitrate regions to crystallise, nitrocellulose obtained from sulfuric/nitric acid nitration of cellulose possesses less than the theoretical 14.14%N: complete nitration is achieved with nitric acid and a dehydrating agent.[240]

The structure of crystalline trinitrocellulose is a right–handed single 5_2 helix: the bulky nitrate groups have unwound the helix slightly from the 2_1 form found in cellulose itself. Antiparallel packing of the helices is favoured, but not unequivocally established.[241]

At high degrees of substitution, nitrates of polysaccharides and polyols are "high-energy materials". They decompose to exclusively gaseous products, and, unlike gunpowder (a potassium nitrate–charcoal–sulfur mixture), which produces a thick smoke of inorganics, they are in principle smokeless. These two attributes, of course, account for the 170-year industrial, commercial and military interest in the materials. Nitroglycerin can in principle decompose to completely gaseous products and still have excess oxygen according to

$$4C_3H_5N_3O_9 \rightarrow 12CO_2 + 10H_2O + O_2 \quad (6.1)$$

whereas cellulose nitrate even of *DS* 3 can decompose to gaseous products only by under-oxidising the carbon:

$$2C_6H_7N_3O_{11} \rightarrow 3CO_2 + 9CO + 7H_2O + 6N_2 \quad (6.2)$$

The possibilities offered by these stoichiometries are exploited in two ways, propellants and explosives. The two physicochemical processes involved are deflagration and detonation.[242] Deflagration is initiated by high temperatures and the combustion zone spreads by heat transfer, aided by the gases released. Detonation, by contrast, is spread by a shock wave, involving a nearly instantaneous and massive increase in pressure, which travels through the material at between 2 and $9\,km\,s^{-1}$. The propellant action is required for propulsion charges and detonation for blasting. Although it is possible to convert a deflagration to a detonation by confinement, by and large explosives and propellants have subtly different compositions. Thus, the standard British propellant in World Wars I and II was "cordite", a mixture of 30% nitroglycerin, 65% gun-cotton and 5% petroleum jelly, whereas gelignite, used for blasting, is a solution of "collodion" (nitrocellulose, *DS* 1.9–2.1) in nitroglycerin.

By the use of very fast photodetectors, it has recently become possible to analyse intermediates in the deflagrations and detonations of polyol nitrates. Although data are still limited, it seems as if detonations are initiated by S_N1 departure of nitrate, followed by carbenium ion reactions, whereas deflagrations are initiated by thermal homolysis of the O–N bond.

A classical physical-organic study of the thermolysis of organic nitrates[243] established that under ambient pressure the reaction of all but tertiary nitrate esters was initiated by homolysis of the N–O bond. Thermolysis of 1-butyl nitrate in 1,2,3,4-tetrahydronaphthalene (tetralin) as solvent, which traps radicals efficiently by hydrogen transfer, gave 1-butanol and nitrogen dioxide. It was governed by a positive entropy and volume of activation ($\Delta S^{\ddagger} = 50.8\ \mathrm{J\,mol^{-1}\,K^{-1}}$, $\Delta V^{\ddagger} = 17\ \mathrm{mL\,mol^{-1}}$), which established a dissociative reaction, without the charge separation which could have caused ordering and/or contraction of solvent molecules. The reversibility of the initial homolysis, suggested by the inhibitory effect of added NO_2, was unequivocally established by two further experiments:

(i) Increasing the solvent viscosity reduced the reaction rate, by making it more difficult for the NO_2 and alkoxy radical to escape the solvent cage.
(ii) In the thermolysis of neopentyl nitrate, which gave 2-methyl-2-nitropropane by loss of formaldehyde from the initial neopentyloxy radical, complete crossover was observed when neopentyl-d_9 [^{15}N]nitrate–was thermolysed with unlabelled material: doubly labelled, unlabelled and both d_9 and ^{15}N singly labelled 2-methyl-2-nitropropane were produced.

Loss of formaldehyde from radicals $RCH_2O^{\bullet}$ occurs when $R^{\bullet}$ is relatively stable: it has been implicated as a consequence of O–N fission of the 6-nitrates of nitrocellulose, where the product radical is essentially anomeric.[244] Thermolysis of 2,2-dimethyl-1,3-propanediol dinitrate yields 2,2-dimethyloxirane by N–O homolysis, loss of CH_2O and displacement of NO_2 from oxygen by the tertiary carbon radical (Figure 6.44). A similar mechanism also appears to take place during PETN thermolysis, whose products are NO_2, formaldehyde and a condensation polymer of isolactic acid.

Thermolysis of tertiary nitrates, however, is ionic even in diethyl ether at 70–100 °C. The products are the expected Saytseff elimination products, *e.g.* 2-methyl-2-butene from *tert*-amyl nitrate.

The volume of activation of S_N1 (and $E1$)-like reactions is around −10 to −20 mL mol^{-1}, as the separation of charges polarises solvent molecules and pulls them in to solvate the new ions, a process known as electrostriction. Therefore, increasing pressure disfavours homolytic and favours heterolytic pathways. The plots of rate of decomposition of cyclohexanol nitrate or propane-1,2-diol dinitrate versus applied pressure at 170 °C are U-shaped, with a minimum between 0.4 and 0.8 GPa,[xx] in accord with a change in mechanism

[xx] 1 GPa = 10^9 Pa = 9870 atm.

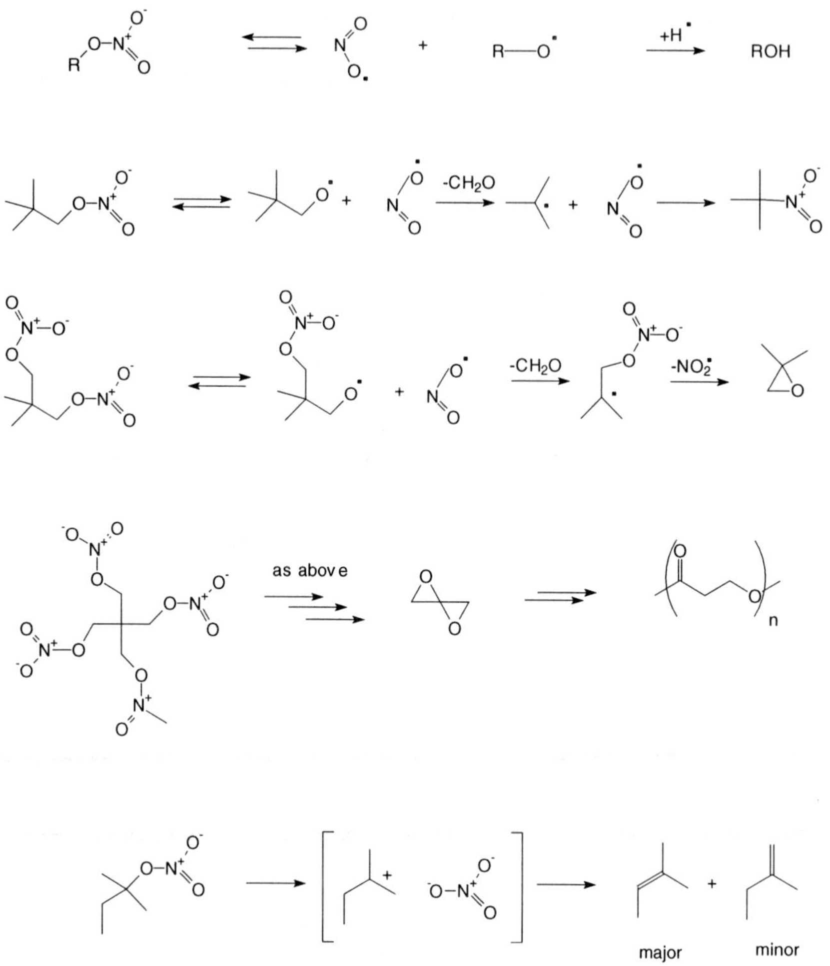

Figure 6.44 Pathways for the thermolysis of nitrate esters.

from homolytic to heterolytic (Figure 6.45). If deuterated tetralin (1,2,3,4-tetrahydronaphthalene) is used as a solvent, there is no kinetic isotope effect at ambient pressure, but one of 2.14 at 0.8 GPa and 2.88 at 1.2 GPa: the tetralin C–H bonds are acting as preassociated concerted nucleophiles, hydride being abstracted by a carbenium ion "too unstable to exist" after nitrate has departed.[245] However, the rate of nitroglycerin decomposition decreases with pressure to the experimental limit of 1.2 GPa, suggesting that destabilisation of a secondary carbenium ion by two vicinal nitrates makes the homolytic mechanism predominant at least to this point. Exposure of solutions of 1-pentyl nitrate in benzene to shock waves producing pressures of 6–16 GPa resulted in ionisation to the 1-pentyl cation, as demonstrated by the production of an array of Wagner–Meerwein-rearranged pentylbenzenes, but neopentyl nitrate still decomposed homolytically.[246] Possibly solvent benzene acted as a

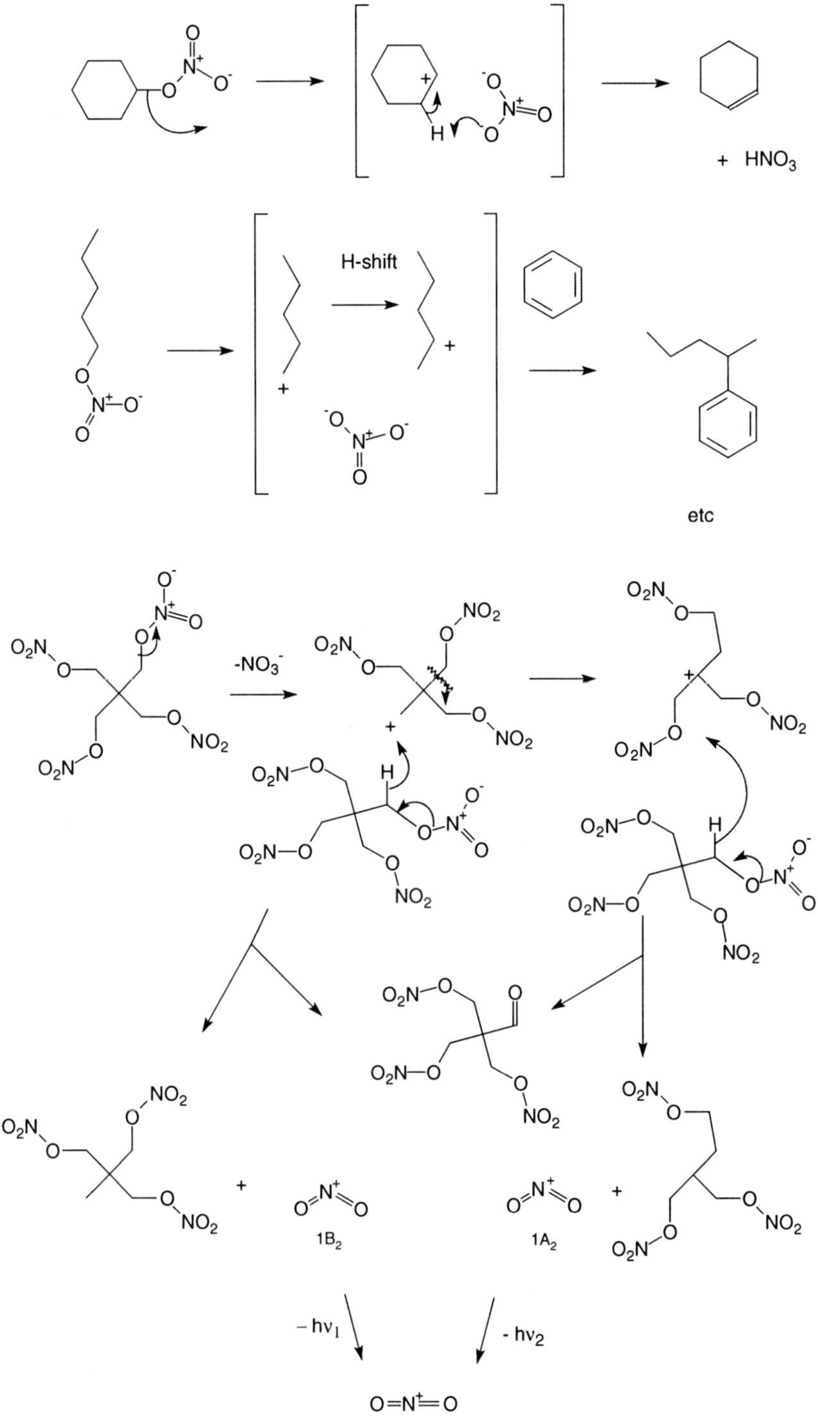

Figure 6.45 Heterolytic fission of nitrates at >1 GPa.

preassociated nucleophile in the reactions of pentyl nitrate, a role sterically prohibited in neopentyl nitrate. Certainly, the evidence that shock waves which produced 4–8 GPa pressures in single crystals of PETN also promoted heterolytic cleavage is strong.[247] The key finding was luminescence from two different electronic states of the nitronium ion, which appeared 200 ns after the shockwave entered the crystal. The (bent) excited states were thought to be produced by abstraction of an α-hydride from a nitrate ester group by a primary and a tertiary carbenium ion, with the abstraction by a primary carbenium ion yielding the higher-energy $^1B_2$ state of the nitronium ion and by the tertiary carbenium ion the lower energy (but still excited) $^1A_2$ state.

Nitrocellulose of *DS* 1.9–2.0 is used as a plastic and material of *DS* 2.0–2.3 for films, including photographic films and cements. Material of *DS* 1.9–2.3 is used for lacquers.[248] Bulk uses (such as plastics and photographic film) have been largely abandoned because of the extreme flammability and instability of nitrocellulose: cellulose nitrate can spontaneously ignite. A recent study of spontaneous ignition of cellulose nitrate of *DS* 2.17 established that the spontaneous ignition arose from heat generated by an autoxidation with oxygen of the air,[249] rather than from a spontaneous decomposition, although the autoxidation is initiated by O–N homolysis. Radical stabilisers act differently, diphenylamine by intercepting NO_2 and 2,6-di-*tert*-butylphenols by intercepting peroxy radicals.[250] The term "safety film" referring to photographic films indicates that the film was made from cellulose acetate, not cellulose nitrate. The "collodion" film of early movies is now disintegrating, partly directly from one-electron chemistry and partly from introduction of carbonyl groups into the cellulose chain and subsequent β-elimination. However, anaerobic treatment of cellulose nitrate with dilute sulfuric acid results in acid-catalysed denitration followed by glycoside hydrolysis at much the same rate as with small molecule models.[251]

Nitrocellulose membranes play important roles in the blotting techniques of biochemistry and molecular biology. Southern discovered that by placing an agarose gel in which DNA fragments had been separated by electrophoresis in contact with a nitrocellulose membrane and then drawing liquid through the agarose gel and nitrocellulose membrane by capillary action (driven by pads of dry paper on the other side of the nitrocellulose membrane), the DNA fragments were adsorbed on the nitrocellulose membrane in the same spatial pattern as the original agarose gel; they could then be identified by various hybridisation techniques.[252] The procedure was named after its discoverer and a very similar one, involving RNA fragments, was waggishly termed "Northern blotting". Extension of the technique to proteins, which could be separated polyacrylamide gel electrophoresis and detected and partly identified and on the nitrocellulose membrane by antibodies, is termed "Western blotting". There is no "Eastern blotting".

Primary and secondary alkyl nitrites under basic conditions eliminate nitrite with concomitant production of a carbonyl group. The nitrite can be estimated by its formation of a diazonium ion with *p*-aminobenzenesulfonamide, followed by formation of an azo dye. The sequence is termed the Griess test

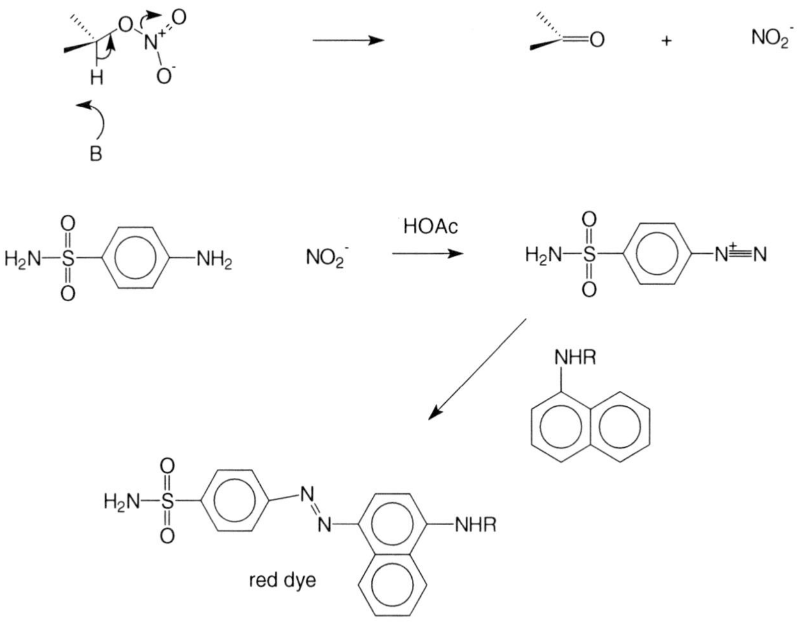

Figure 6.46 Griess test.

(Figure 6.46); in the 1970s, official forensic scientists presented a positive Griess test as proof that the accused had handled materials such as nitroglycerin, but of course a positive test would also be given from handling nitrocellulose harmlessly and legitimately.

Nitroglycerin and similar compounds have been used as a principle therapeutic agent for angina and congestive heart disease for 130 years. It acts as a vasodilator by generating nitric oxide and/or the nitroso compounds, which activate a soluble GTP cyclase, resulting in cyclic GMP. This second messenger brings about the primary physiological effect of vasodilation. Whole organism studies indicated widespread nitrosylation, with *S*-nitrosation, *N*-nitrosation and nitrosation of the iron centre of haem.[253] In principle, a range of endogenous enzymes could play a part in this: the claims of mitochondrial aldehyde dehydrogenase,[254,255] which reduces nitroglycerin to glycerol-1, 2-dinitrate and inorganic nitrite, as a uniquely important enzyme in the process are strengthened by particular features of the enzyme. Normally aldehyde dehydrogenase forms a thiohemiacetal with an aldehyde, from which hydride is removed by NAD^+; the resulting thiol ester is then hydrolysed. There are two further active site thiols in addition to the nucleophile and the suggestion is that after formation of an *S*-nitrate by the nucleophilic thiol, a disulfide is formed with expulsion of nitrite. That such a disulfide is capable of being formed had been established with certain inhibitors. It has to be reduced with each turnover if the enzyme is to act as an NO-generating system, which explains the requirement for exogenous thiols during *in vitro* nitric oxide generation.

However, other enzymes, such as glutathione-*S*-transferase, cytochrome P_{450} enzymes and xanthine oxidase, have also been implicated. Even "old yellow enzyme",[256] a mixture of homodimeric and heterodimeric ~45 kDa gene products each containing an FMN cofactor, first isolated from bottom yeast in 1933 and still of unknown biological function, regioselectively reduces polyol nitrates to nitrite, probably by the mechanism in Figure 6.47. The intermediacy of inorganic nitrite in the generation of NO has, however, been questioned on the grounds that endogenous NO_2^- concentrations are far higher than the

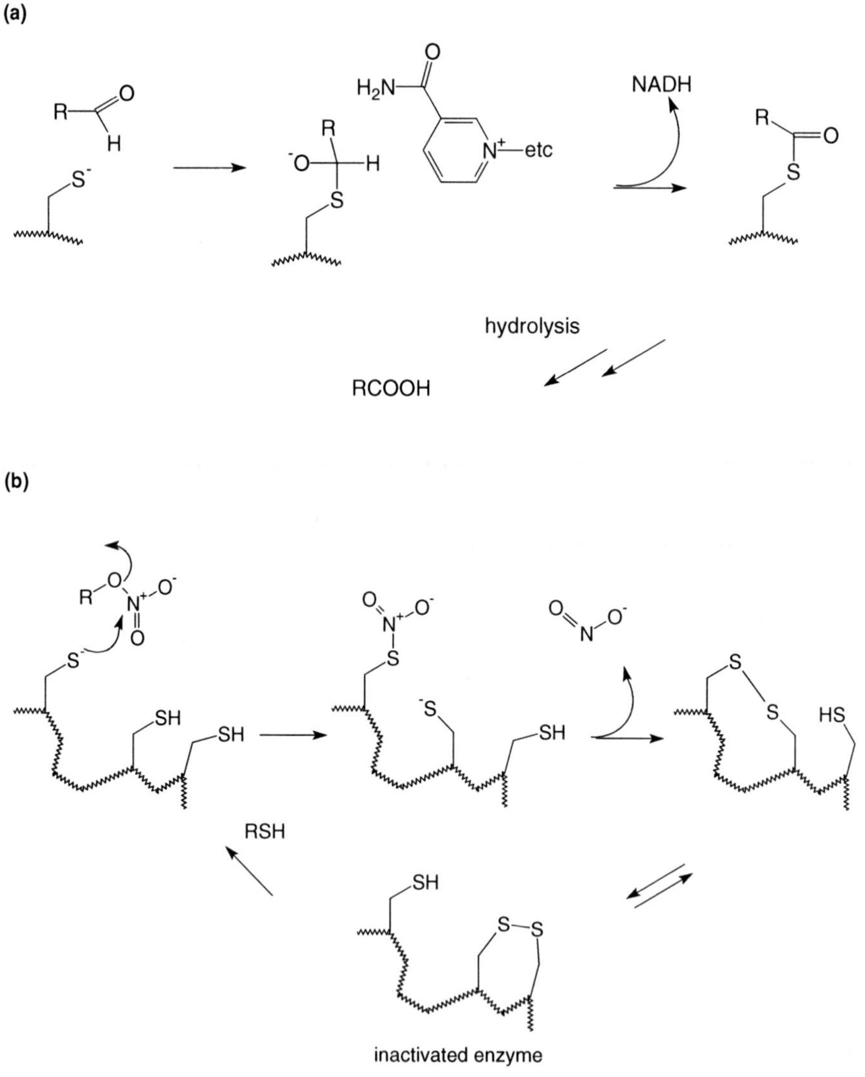

Figure 6.47 Biotransformations of nitroglycerin. (a) Normal mechanism of aldehyde reductase; (b) production of nitrite by mitochondrial aldehyde dehydrogenase.

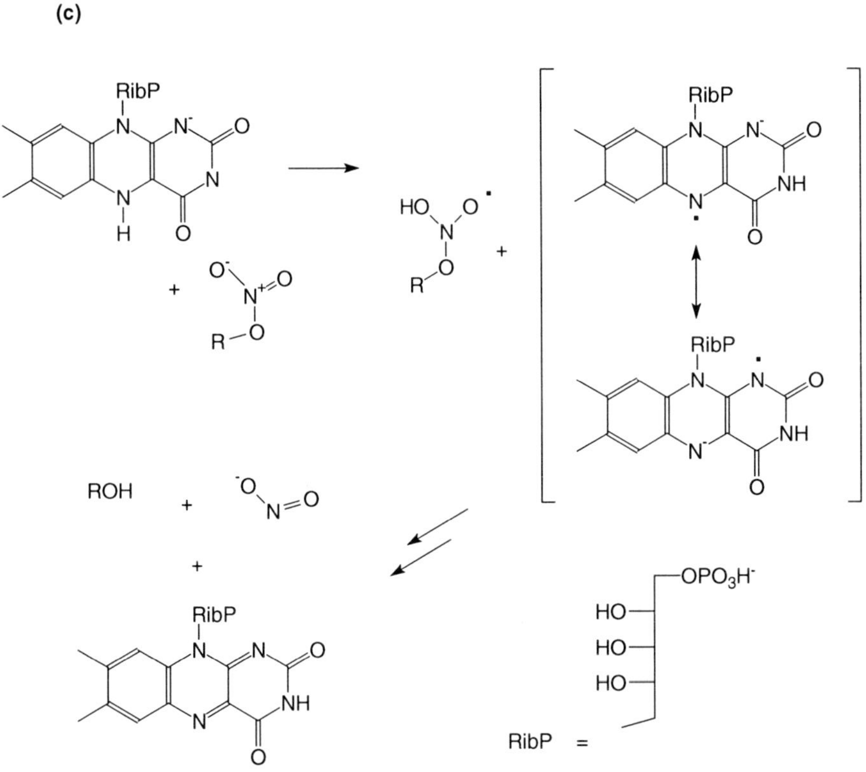

(c)

Figure 6.47 (c) Production of nitrite by old yellow enzyme.

physiological concentrations of nitroglycerin which stimulate guanyl cyclase. When nitroglycerin is reduced by xanthine oxidase under anaerobic conditions, no NO is produced until a thiol is added; with other kinetic measurements, this established that at least this enzyme reduced organic nitrate to organic nitrite, which was the true precursor of nitric oxide and nitroso thiols.[257] The case has also been made that the physiological effects of organic nitrates arise from an array of biotransformations and that emphasis on one particular pathway, such as aldehyde dehydrogenase reduction, is misleading.[258]

6.5.8 Phosphorus Derivatives

6.5.8.1 General Considerations. Phosphorus has low-lying d orbitals and can exhibit covalencies up to 6, as in the anion PF_6^-, which has octagonal coordination. The sizes of the empty 3d orbitals are such that they overlap efficiently with the full 2p orbitals on oxygen and fluorine, with the consequence that P–O and P–F bonds are particularly stable. In synthetic chemistry the affinity of phosphorus for oxygen is exploited in the use of phosphines to turn hydroxy groups into good leaving groups.

Phosphorus can be pentacoordinated (sp^3d hybridised), with trigonal bipyramidal geometry, as in PF_5 and various phosphoranes. A feature of pentacoordination is that there is no way of arranging the ligands around the central atom such that they are equivalent. The closest is the trigonal bipyramid, with two apical and three equatorial bonds. The apical and equatorial bonds can change places rapidly in a process known as pseudorotation because, in the case of PF_5 itself, the changes give the appearance of the molecule having rotated. The simplest pseudorotation is Berry pseudorotation, illustrated in Figure 6.48, in which the two apical bonds move back and behind the paper and two of the three equatorial bonds lengthen and move forward.

The existence of stable phosphoranes suggests that the trigonal bipyramidal species in bimolecular nucleophilic substitution at phosphorus, unlike those in nucleophilic substitution at carbon, could be intermediates, not transition states. The classical investigations of Westheimer's group demonstrated that this was so.[259] The paradigmal experiment showed that the 10^6-fold faster hydrolysis of methyl ethylene phosphate than its acyclic analogues was due to ring strain, yet the ring was preserved in a significant proportion of the products. Rate-determining and product-determining steps in the hydrolysis were therefore different, the classical demonstration of the existence of an intermediate. The conformations of the phosphorane intermediates were governed by the following rules:

(i) Incoming nucleophiles and outgoing leaving groups were always axial.
(ii) Electronegative groups tended to be apical and electropositive ones (alkyl groups, $-O^-$ groups) to be equatorial.
(iii) Five-membered rings were equatorial-apical (angle X–P–Y = 90°) rather than diequatorial (angle X–P–Y = 90°).

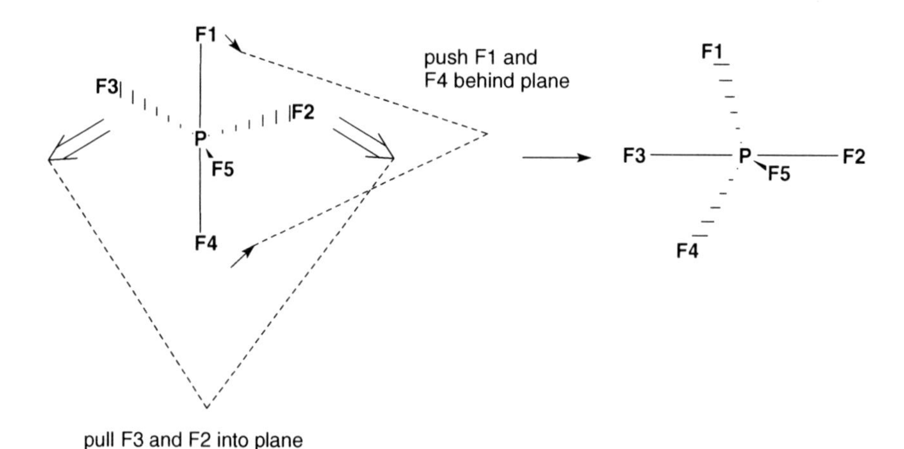

Figure 6.48 Trigonal bipyramidal PF_5, showing Berry pseudorotation. The apical bonds are solid lines, the equatorial bonds wedges. Fluorine atoms are numbered, but if they were not, the pseudorotation would be equivalent to rotation through 90° about the C–F5 bond ("pivot" of the pseudorotation).

Soon after formulation of Westheimer's rules, an example of phosphate ester hydrolysis was found where pseudorotation was rate determining.[260] Pseudorotation is a necessity for some of the examples of neighbouring group participation in RNA hydrolyses and transphosphorylations.[261]

6.5.8.2 Phosphonium Intermediates in the Activation of OH to Nucleophilic Substitution. Triphenylphosphine and dialkyl azidodicarboxylates (the Mitsunobu reaction) are used extensively for activating sugar hydroxyls to nucleophilic substitution.[262] A common protocol uses carboxylic acids to give an ester of the inverted configuration, and involves addition of the dialkyl azidodicarboxylate, to a mixture of alcohol, acid and triphenylphosphine (Figure 6.49). Mechanistic studies of such a reaction[263] indicated that there was a rapid formation of a phosphine–azidocarboxylate adduct, which could be attacked at phosphorus by the carboxylate anion unless a second equivalent of acid was added to form the H-bonded dimer. Then formation of an *O*-alkylphosphonium salt took place. The key displacement took place with a β_{nuc} value of 0.1, less than that of competing elimination reactions. ^{31}P NMR examination of the reaction with an unreactive alcohol revealed the presence of dialkoxyphosphoranes in equilibrium with an alkoxyphosphonium salt–carboxylate ion pair.[264] Although collapse of this ion pair to the mixed phosphorane was rapid on the NMR time-scale, the >100 ppm changes in the ^{31}P chemical shift that could be induced by changing solvent were taken as evidence for two rapidly equilibrating species. A theoretical investigation suggested that, although the precursor of the inverted product is an alkoxyphosphonium salt, an array of phosphoranes and phosphonium salts could be formed, some of which could form retained product under the appropriate conditions.[265] A possible recent example of the latter is an intramolecular displacement of OH by carboxylate which can occur with retention or inversion, as the conditions are tweaked.[266] As with all methods of activating secondary OH groups to nucleophilic displacements, particularly in sugars, feeble intramolecular nucleophiles may compete successfully with relatively powerful external ones – a recent example being competition of internal benzyloxy with external *p*-nitrobenzoate.[267]

A probably similar reaction, particularly useful for the production of primary deoxyhalo sugars under mild conditions, is the reaction of triphenylphosphine with an alcohol and carbon tetrachloride or bromide. The mechanism is not known. Nucleophilic attack by triphenylphosphine on the halogen, to form Ph_3P^+–Cl//$^-CCl_3$, then displacement by alcohol on phosphorus to give the usual Ph_3P^+OR species, followed by S_N reaction of chloride (or bromide), may appear reasonable, but is thrown into question by the isolation of ylids $Ph_3P{=}CCl_2$[268] and $Ph_3P{=}CBr_2$[269] from the reactions of triphenylphosphine with the respective carbon tetrahalides, and the success of the reaction with phenols.

6.5.8.3 Phosphite Esters. Sugar phosphites do not occur naturally, but play a key role in DNA synthesis, particularly automated DNA synthesis. Their reactions are thus at the heart of much modern biotechnology. Many systems

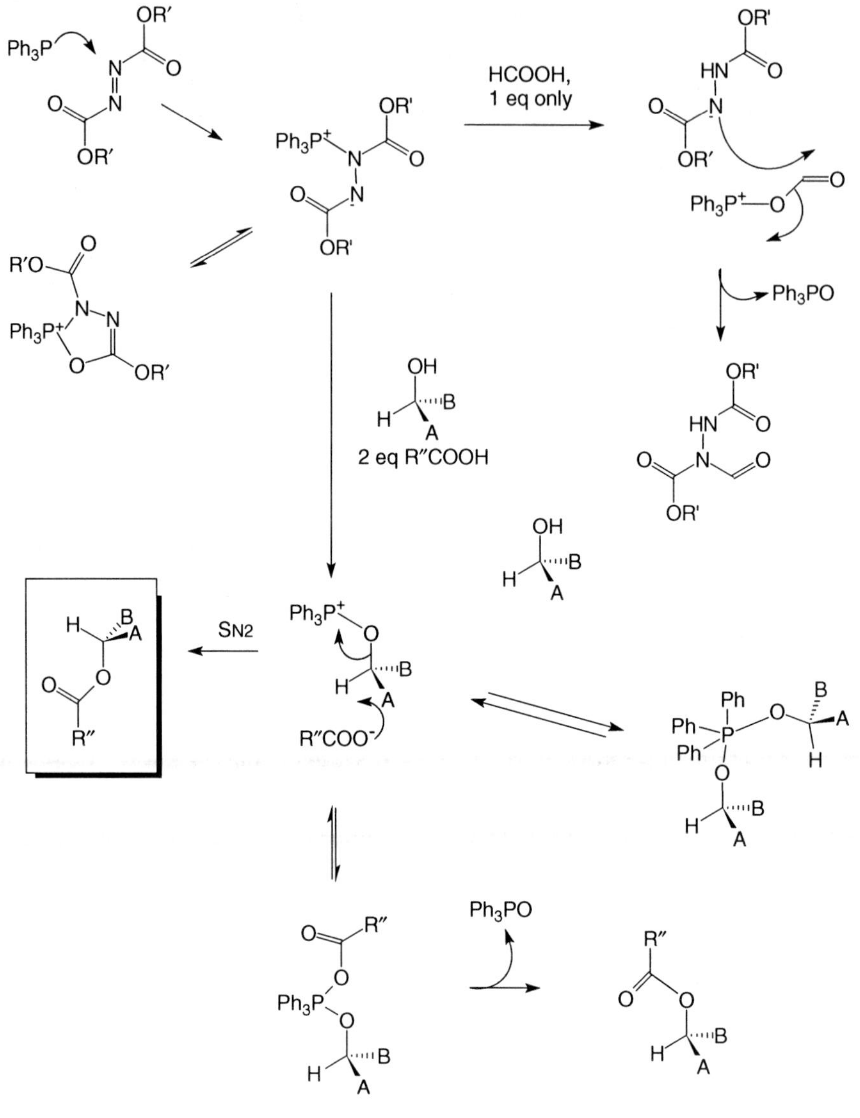

Figure 6.49 Species in the Mitsunobu reaction of an alcohol with carboxylic acids. The formation of retained product via collapse of a phosphorane is usually negligible.

are based on the P^{III} chemistry introduced by McBride and Caruthers[270] in which a deoxynucleotide with the 5′-OH free is attached to a solid support by its 3′-substituent and protected deoxynucleoside 3′-phosphoramidites are added one at a time. The 5′-OH of each monomer is protected as an acid-labile group – methoxytrityl (CH_3O-*p*-$C_6H_4CPh_2$) or dimethoxytrityl [(CH_3O-*p*-$C_6H_4)_2CPh$] are favourites – and can be unmasked with each cycle. Acid catalysts capable of activating the P–N bond to nucleophilic substitution, but not removing the O5′ protection, appear to be largely NH acids such as tetrazole or 4,5-dicyanoimidazole. After the coupling step, there is a 5′ deprotection step, after which the whole sequence can be repeated. When the neutral polyphosphite has reached an appropriate length, the phosphite centres are oxidised to phosphate with oxidants such as H_2O_2. This permits the removal of the third phosphate ligand, frequently a 2-cyanoethyl group removed as acrylonitrile. Alternatively, a combined deprotection/oxidation step after the addition of each nucleoside can be used if the 5′ protection is changed to aryloxycarbonyl, which is removed by the pH 9.5 buffered hydrogen peroxide also used to oxidise the phosphate.[271]

The barrier to inversion about phosphorus in phosphites is large enough for tricovalent phosphites chiral at phosphorus to be configurationally stable, but the conditions for the coupling of phosphoramidites epimerised them at P about 10 times faster than it coupled them,[272] although the oxidation of phosphites to phosphates and thiophosphates proceeded with complete retention of phosphorus stereochemistry.[273]

The problem of the diastereoselective synthesis of phosphite-linked oligonucleosides was eventually solved by incorporating both using 5-isopropylamino-5-deoxy-1,2-cyclopentanylidene D-xylofuranose as a bidentate *O*,*N*-phosphorus ligand in the phosphoramidite (Figure 6.50).[274] The stereospecific reaction can be used to generate DNA stretches with sulfur instead of one of the two diastereotopic oxygen atoms: the sulfuration reaction can be molecular sulfur (S_8) itself or an organic thiosulfonate.[275] These stereospecific introductions of sulfur are one of the few tools available to study the catalytic mechanism of ribozymes.

Trialkyl phosphites undergo the Michaelis–Arbusov reaction with alkyl halides to give dialkyl phosphonates: the initial phosphonium ion is attacked at one the alkyl groups by the newly liberated halide. The reaction is useful in producing non-fissile but isosteric analogues of activated sugars for mechanistic studies (Figure 6.51).[276]

6.5.8.4 Phosphates

6.5.8.4.1 Mechanistic features of phosphoryl transfer and methods of investigation. The chemical phosphorylation of carbohydrates is of importance only in the synthesis of phosphorylated carbohydrates of biological importance. Common phosphorylating agents for monophosphates are dibenzyl or diaryl phosphorochloridates, $ClPO(OAr)_2$ and $ClPO(OBn)_2$; benzyl groups can be removed by catalytic hydrogenation.

The non-enzymic hydrolysis of phosphates has been extensively studied, as a model for the enzymic processes. Many of the tools developed are applicable equally to enzyme-catalysed and spontaneous processes.[277]

Because of rapid protonation and deprotonation, the monophosphate group is torsionally symmetrical. The symmetry can be broken by stereospecific substitution with sulfur, or by stereospecifically labelling with the three stable isotopes of oxygen, ^{16}O, ^{17}O and ^{18}O. An ingenious NMR-based method of analysis, which relies on the isotope shift of the ^{31}P resonance (the heavier the

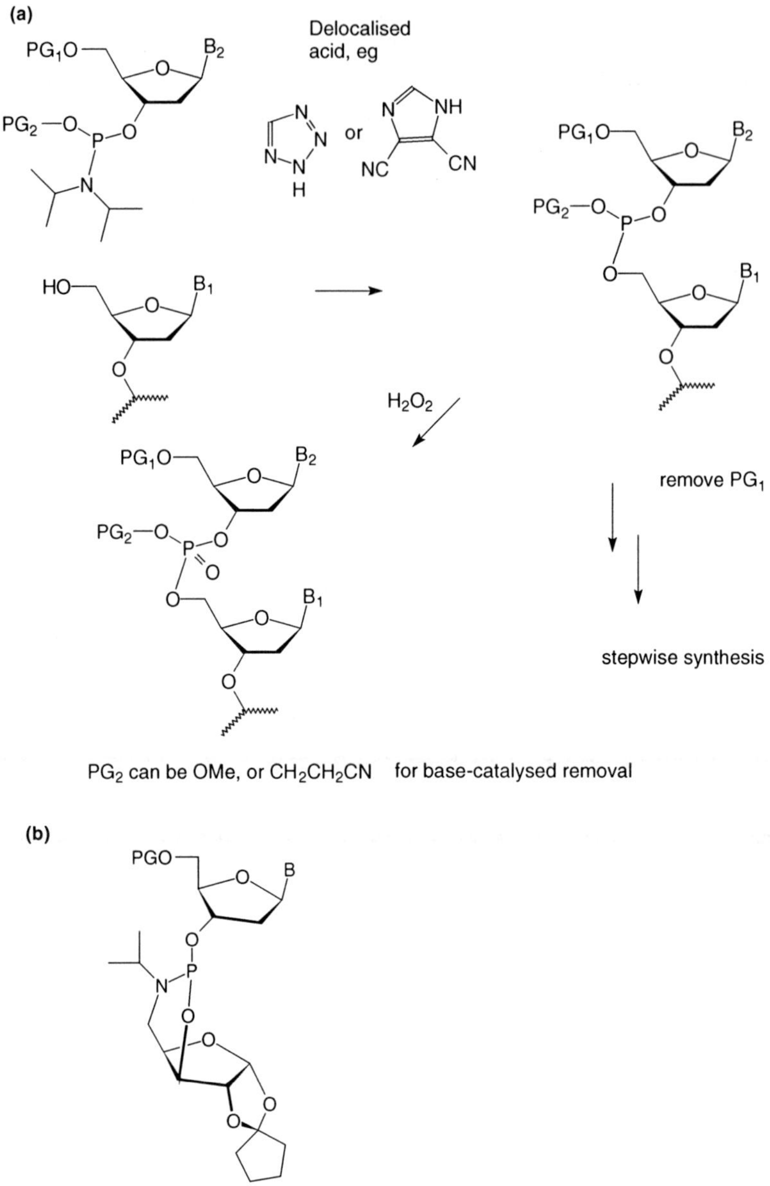
(a)
PG1O
B2
PG2—O
P—O
N
HO
B1
O
Delocalised acid, eg
or
NC
CN
NH
H2O2
remove PG1
stepwise synthesis
PG2 can be OMe, or CH2CH2CN for base-catalysed removal
(b)
PGO
B
O
P
N

Figure 6.50 (a) Outline of the phosphoramidite method of DNA synthesis. (b) A chiral phosphoramidite for P-stereospecific synthesis. PG = protecting group; B = nucleoside base.

isotope, the bigger the shift) made it possible to determine the stereochemistry of a wide range of phosphoryl transfer reactions.

For many years, the metaphosphate ion, isoelectronic with sulfur trioxide, was written as a discrete intermediate. Reactions of monophosphate

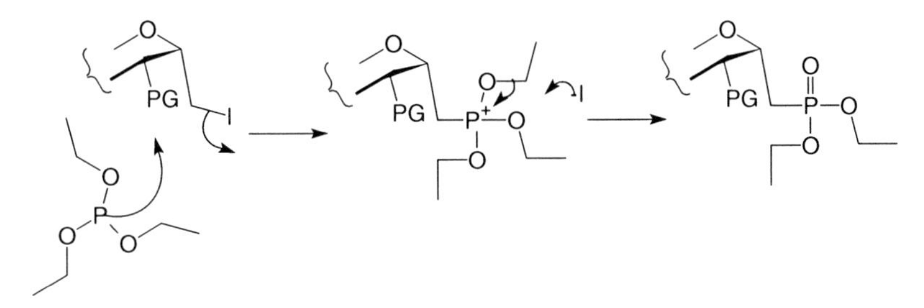

Figure 6.51 Production of a non-fissile glycosyl phosphate analogue via the Michaelis–Arbusov reaction.

monoanions, for example, were considered to react through their tautomer which gave metaphosphate, and dianions of phosphates of acidic phenols to react similarly by S_N1(P) reactions (Figure 6.52). However, of the reactions putatively involving metaphosphate as an intermediate, only those in *tert*-butanol went with appreciable amounts of racemisation.[278] Metaphosphate, like glycosyl cations, is on the borderline of a real existence, with reactions generating it occurring within the same solvent shell or with exploded transition states. However, ethanolysis of ^{18}O-labelled *p*-nitrophenyl phosphorothioate,[279] or its hydrolysis,[280] proceeded with significant amounts of racemisation, indicating that thiomctaphosphatc is a solvent-equilibrated intermediate.

The relationship between mechanistic type and reaction stereochemistry is not *a priori* as clear cut as with reactions at carbon, since the trigonal bipyramidal species involved is an intermediate, which can in principle pseudorotate, rather than a transition state. In practice this appears to happen only in the case of 1,2-migrations, where according to Westheimer's rules, the intermediate phosphorane must pseudorotate.[281]

It has been pointed out that, in the absence of pseudorotation, the molecular motions involved in transfer of a phosphoryl group through a "metaphosphate-like" transition state from an oxygen leaving group (P–O distance ~1.6 Å) to an oxygen nucleophile at the van der Waals contact distance (P–O distance ~3.3 Å) are minimal, with most of the motion being in the phosphorus atom.[282] If the distance between the two oxygens remains the same at ~5.0 Å, a symmetrical transition state would have two P–O distances of ~2.5 Å. Simple bond order–bond length correlations of the Pauling type suggest that this corresponds to bond orders to the two apical ligands of 0.08. If there is some compression of ligands in the ES complex, then bond orders of 0.15 are achievable; in this case there is no motion of non-bridge oxygen atoms at all. Classical S_N2 reactions, with bond orders of ~0.5, require compression of the O_{nuc}–P–O_{lg} distance by ~1.4 Å, completely out of the question (Figure 6.52(d)).

The degree of bonding at the transition state of phosphoryl transfer reactions is accessible by multiple heavy atom kinetic isotope effects. Simple resonance theory indicates that the P–O bond order to the three non-bridging oxygens is 5/3, whereas in the starting ester it is 4/3 (Figure 6.52(b)). The expectation that metaphosphate-like transition states should be associated with inverse ^{18}O

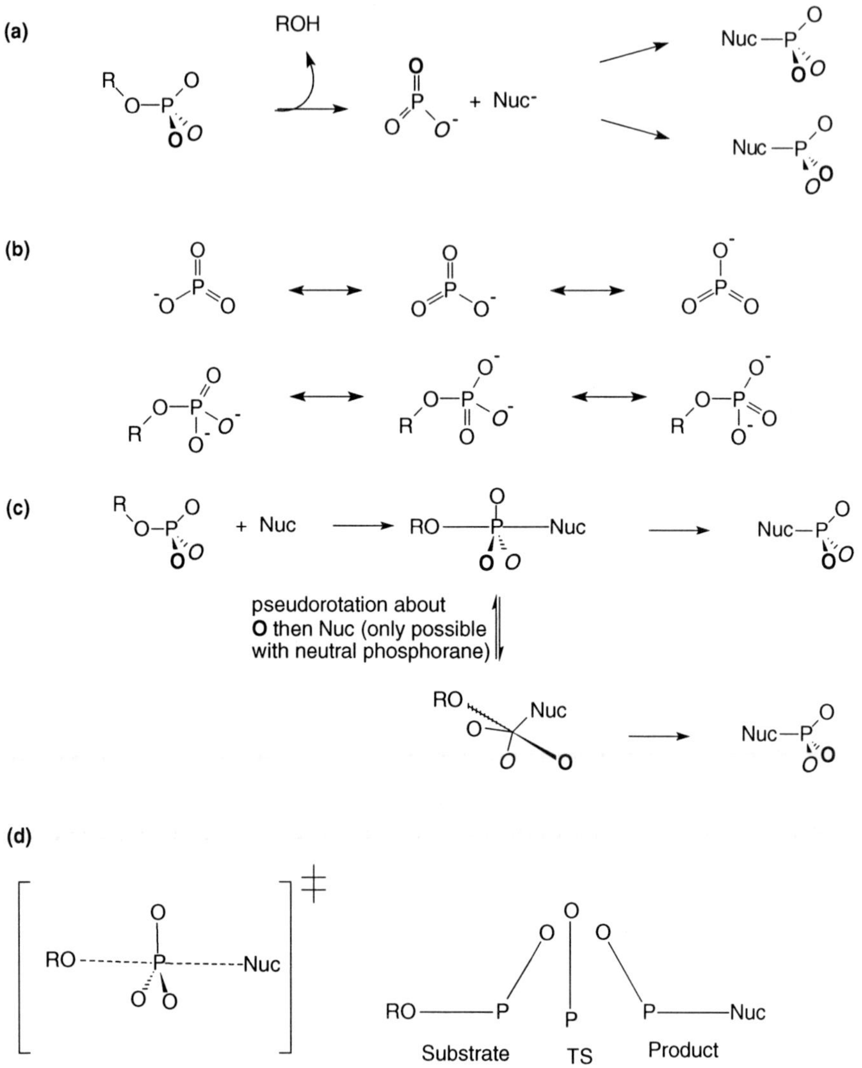

Figure 6.52 Mechanistic possibilities for transfer of a monophosphate. (a) $S_N1(P)$ involving a solvent-equilibrated metaphosphate intermediate. Protonation states are not shown, but different fonts for the phosphorus oxygens, representing different oxygen isotopes, indicate stereochemistry. (b) Canonical structures for metaphosphate and alkyl phosphate dianion. (c) $S_N2(P)$ involving an intermediate which can pseudorotate. (d) "Exploded" $S_N1(P)$-like $S_N2(P)$ transition states, and cartoon of calculated movements of phosphorus and non-bridge oxygens during phosphoryl transfer through such a transition state, where the leaving to nucleophile atom distance is constant.

effects in the non-bridging oxygen atom is fulfilled in the case of *p*-nitrophenolate dianion hydrolysis, for which $k_{16}/k_{18} = 0.9994 \pm 00005$.[283] Leaving group ^{18}O effects resemble those in glycoside hydrolysis, in that the direct effect caused by rupture of the P–O bond (which is expected to be greater than those caused by rupture of the C–O bonds, because the proportional effect of replacing ^{16}O by ^{18}O on the reduced mass is greater) will be offset by any increased bonding to hydrogen as a result of acid catalysis. The two phenomena of P–O cleavage and O–H bond formation affecting the one effect could be deconvoluted by measurement of a β_{lg} value, but also, with a single *p*-nitrophenyl substrate, by measurement of ^{15}N effects, which are significant (and direct) if charge is delocalised into the nitro group.

Mechanistic studies of the hydrolysis of phosphate diesters indicate similar possibilities, with alkyl metaphosphate ($ROPO_2$)-like transition states and $S_N2(P)$ pathways.[284] The hydrolysis of *m*-nitrobenzyl uridine 3′-diphosphate, whose two potential leaving groups have equal pK_as, is associated with isomerisation to the 2′-isomer in weakly acidic solution, which requires pseudorotation of a (probably neutral) phosphorane.[281]

6.5.8.4.2 Mechanisms of biological phosphate transfer to and from carbohydrates. Biological monophosphorylation of carbohydrates is of enormous importance. The monophosphate group constitutes a "handle" which provides ample intrinsic binding energy to proteins, for example by the formation of salt bridges to Lys and Arg. In the case of enzymes such as triose phosphate isomerase, it may be transduced into a lowering of the free energy activation of the reaction.

Although the possibility of pseudorotation during enzymic phosphoryl transfer has been raised many times, there is as yet no unambiguous example of it. If groups enter and leave from apical positions, the trigonal bipyramidal intermediate may have a lifetime less than the period of a vibration and thus be "too unstable to exist." In practice, therefore, the rule "retention = double displacement, inversion = single displacement" appears to hold for phosphoryl transfer at least as well as for glycosyl transfer.

Physiologically, monophosphate groups are added to sugars from ATP and other nucleoside triphosphates by kinases, whereas they are usually hydrolysed off by phosphatases, although sometimes, in examples of metabolic frugality, kinases act physiologically in reverse.[xxi] There is also a third group of monophosphate transfer enzymes, mutases which convert one sugar phosphate to its regioisomer, such as phosphoglucomutase, which converts glucose-1-phosphate to glucose-6-phosphate. Enzymes catalysing phosphoryl transfer exploit the electrophilic properties of metal cations, with both "hard" metals such as

[xxi] Enzymes inevitably catalyse reactions in both directions, but sometimes the physiological flux can be thermodynamically uphill, as subsequent enzymes in the metabolic pathway remove the disfavoured partner in an equilibrium.

Mg^{2+} and "soft"[xxii] metals such as Zn^{2+} being used. The ability to replace a phosphate oxygen by sulfur, in combination with the use of "soft" metals by phosphoryl-transferring enzymes, provides an important probe of stereochemistry and mechanism. If sulfur is incorporated at a site coordinating to Mg^{2+} (or a hydrogen bond donor such as lysine), catalysis is gravely impaired, whereas if the coordination is to Zn^{2+} there is little effect.

In general,[277] kinases act with inversion and mutases act with retention, whereas phosphatases act with either with inversion or retention.

Sugar kinases acting on primary OH groups, of which the best known is hexokinase, which converts glucose to glucose-6-phosphate.[277] It is one of a superfamily[xxiii] of enzymes ("hexokinase–hsp70–actin superfamily") which carry out similar reactions and which have a common protein fold (Figure 6.53).[285] They consist of two large domains with a deep cleft between them, with each domain possessing a five-stranded region of β-sheet. On binding of substrates the domains close, permitting efficient transfer of phosphate without interception of water (*cf.* the glycosyl transferases). In addition to the two domains, each member of the superfamily may possess different substructures which permit oligomerisation and allosteric interactions. The catalytic base is an aspartate and the enzymes are Mg^{2+} dependent. The Mg^{2+} binds between the γ-phosphate and the non-bridging atoms of the β-phosphate, and is also coordinated to a water molecule which hydrogen bonds to the catalytic base. The mechanism is a straightforward in-line phosphoryl transfer, as shown in Figure 6.52d, with the metaphosphate-like transition state being confirmed by inverse non-bridge ^{18}O kinetic isotope effects [$^{18}(V/K)$=0.9965 at the non-optimal pH of 5.3].[286]

The coordination of Mg^{2+} to one of the two diastereotopic oxygens on the γ-phosphate gives rise to two stereoisomers of the complex. Mg^{2+} itself is very labile to substitution, but preparation of the diastereoisomeric complexes of ATP with the substitution-inert transition metal ions Cr^{2+} (d^3) and Co^{2+} (d^6, low spin, high ligand field) enabled the stereochemistry to be determined. Hexokinase has the Λ stereochemistry and phosphofructokinase (which converts fructose 6-phosphate to fructose 1,6-diphosphate) the Δ stereochemistry. The configurational preferences of the γ-phosphate in hexokinase could also be determined by the "thiophilic rescue" technique. ATP in which one of the two diastereotopic non-bridge oxygens of the β-phosphate had been replaced by sulfur exhibited pronounced diastereoselectivity in its reactions with hexokinase, which depended on the divalent metal. The Mg^{2+} complex of the R_P isomer was a much better substrate than the S_P isomer, whereas the preferences were reversed when Mg^{2+} was replaced by Cd^{2+}. If one assumes that Mg^{2+} coordinates preferentially with O and Cd^{2+} with S, then a Λ stereochemistry of the reactive complex is indicated (Figure 6.53(b)).[287]

[xxii] "Hard" metals have no low-lying filled d orbitals and thus preferentially coordinate with first row ligands with tightly -held, non-polarisable electron clouds such as F^-; "soft" metals have filled d orbitals which can overlap with empty d orbitals in ligands possessing them, notably sulfur, which has low-lying unoccupied 3d orbitals.

[xxiii] A term meaning the same as "clan" in the CAZy classifications.

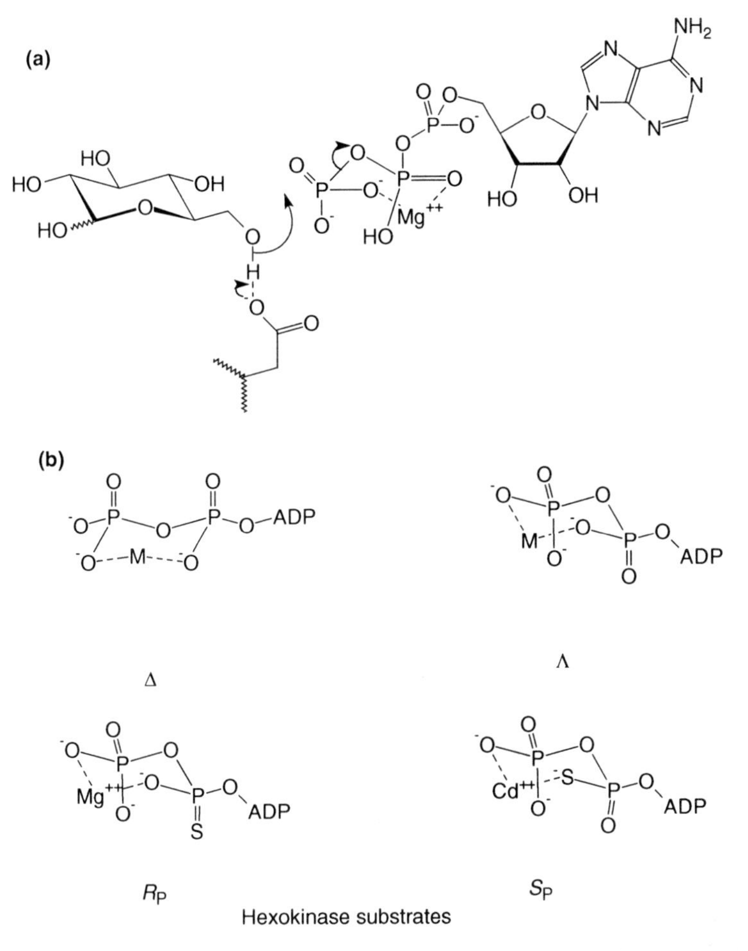

Figure 6.53 (a) Mechanism of enzymes of the hexokinase superfamily. (b) Stereochemistry about the β-phosphate and its determination by sulfur substitution.

The degradation of sugar phosphates is carried out by a range of phosphatases. A paradigmal non-specific phosphatase is the alkaline phosphatase of *E. coli*, which binds two metal ions (usually Zn^{2+}) tightly, about 4 Å apart, and a third metal, usually Mg^{2+}, more loosely. The catalysed reaction is a double displacement, via a phosphorylated serine, with overall retention. An active site arginine interacts with the non-bridge oxygen atoms. Many of the basic kinetic data obtained over a period of 40 years were compromised by the tight binding of the phosphate product ($K_i \approx \mu M$) (micromolar concentrations of phosphate are produced when *p*-nitrophenol phosphate generates an absorbance of 0.01 in a 1 cm cuvette, even in the unlikely event of the substrate being completely pure). A novel radioassay permitted the $^{\beta lg}(V/K)$ value for a series of alkyl phosphates to be redetermined as –0.85; the pH–(V/K) profile exhibited an ionisation at pH 5.5. Extended arguments based on the stability constants of zinc–phenolate

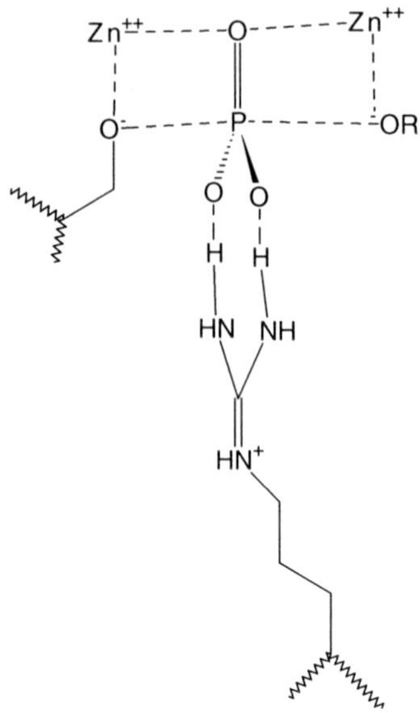

Figure 6.54 Transition states for alkaline phosphatase. In the first, R is derived from the substrate; in the second, R = H.

and zinc–alcoholate complexes led to the conclusion that the transition state (Figure 6.54) was an exploded one in which the leaving group was the zinc phenolate and the nucleophile the serine alkoxide, whose protonation resulted in the pH 5.5 ionisation.[288] The metaphosphate-like transition state was in accord with inverse non-bridge ^{18}O kinetic isotope effects measured earlier; being measured by a competition method, they are uncompromised by phosphate inhibition.[289]

The nucleophile in retaining phosphatases can also be histidine. Glucose-6-phosphatase was early demonstrated to go through a phosphohistidine intermediate;[290] the mammalian enzyme has several isoenzymes and the nucleophile appears general to them.[291] It can also be cysteine. The purple acid phosphatases also use two metal catalysis, but one of the Zn^{2+} ions is replaced by Fe^{2+} (the purple colour arising from the charge transfer band of a tyrosine ligand), the more electrophilic metal possibly accounting for the lower pH optimum.

Fructose 1,6-diphosphatase hydrolyses off the 1-phosphate in the step which commits the substrate to gluconeogenesis. The mammalian enzyme, a homodimer, is richly allosteric, as befits an enzyme at a metabolic branchpoint. It catalyses a single displacement, mediated by no less than four metal ions: three Mg^{2+} sites and one K^+ site have been observed in the crystal structure.[292] The nucleophilic water may be activated by proton transfer from Glu98. The very large claim has been made that by alteration of the conditions of crystallisation, the on-enzyme equilibrium can be switched from fructose-6-phosphate and inorganic phosphate to fructose-6-phosphate and metaphosphate.[293] Given the reactivity of metaphosphate, however, it seems likely that some other phenomenon is giving rise to the observed electron density patterns.

Inositol monophosphatase in the brain is the presumed target of Li^+ therapy for manic depression, and has been intensively investigated in consequence. It catalyses the hydrolysis of all *myo*-inositol monophosphates except the 2-isomer (axial) and has similarities with fructose 1,6-bisphosphatase. The stereochemistry was established as inversion by the use of ^{17}O-labelled phosphorothioate in $H_2{}^{18}O$.[294] This was in accord with the mechanism proposed on the basis of crystal structures of the human enzyme complexed with one Gd^{2+} or three Mn^{2+} ions. Modelling suggested a single transition state like the two in alkaline phosphatase action, with one metal coordinating the inositol leaving group (possibly through O6 as well) and a second the nucleophilic water, which was partly deprotonated by Glu70.[295] The anticompetitive inhibition by Li^+ was thought to arise from the stabilisation of the complex of enzyme and inorganic phosphate.

Phosphoglucomutase interconverts glucose 1-phosphate and glucose 6-phosphate (Figure 6.55). The overall retention of configuration implies a double displacement, and in principle either the enzyme could have two nucleophiles, with one displacement being between enzyme and substrate and the second between the two sites of phosphorylation of the enzyme, or it could have the one nucleophile and the substrate could rotate with respect to the nucleophile. The latter mechanism is the correct one, and of its two possible variants (the free enzyme being phosphorylated and 1,6–diphosphate rotating when bound to the dephosphorylated enzyme or the free enzyme being dephosphorylated and glucose rotating when bound to the phosphoryl enzyme), the former is correct.

The group phosphorylated in free enzyme is Ser; Ser is therefore the nucleophile which attacks the bound glucose 1,6-diphosphate. The enzymes exist in "open" states which can bind either substrate and "closed" states with bound glucose 1,6-diphosphate; with advanced mammalian enzymes the "closure" of the active site, which allows the glucose 1,6-diphosphate to rotate, is very efficient, but for more primitive bacterial enzymes the diphosphate occasionally escapes and glucose 1,6-diphosphate has to be added as a cofactor to rephosphorylate the enzyme. The enzyme is Mg^{2+} dependent, although the Mg^{2+}can be replaced with Cd^{2+}. There appears to be no single catalytic acid–base, with mutants at all likely positions still possessing 4–12% activity. It was suggested that an array of positive groups (Arg, Lys, His) create an electrostatic field sufficient to deprotonate the nucleophile fully, in accord with the slightly different presentation of the serine OH to the transferred phosphate in the two orientations of the glucose 1,6-diphosphate–enzyme complex.[296] Remarkably, mutants of the active site serine were also partly active, raising the possibility of a breakdown of site-directed mutagenesis technology. However, pre-steady-state kinetic analysis revealed a second intermediate, glucose 4,6-diphosphate, in addition to glucose 1,6-diphosphate, which was not directly converted to either glucose 1-phosphate or glucose 6-phosphate.[297] This is in accord with previous ^{19}F NMR studies of fluorinated substrates bound to the enzyme, in which two signals, corresponding to the two orientations of fluorinated glucose 1,6-diphosphate, were observed for all ligands except 4-fluorinated ones, which only exhibited one signal.[298]

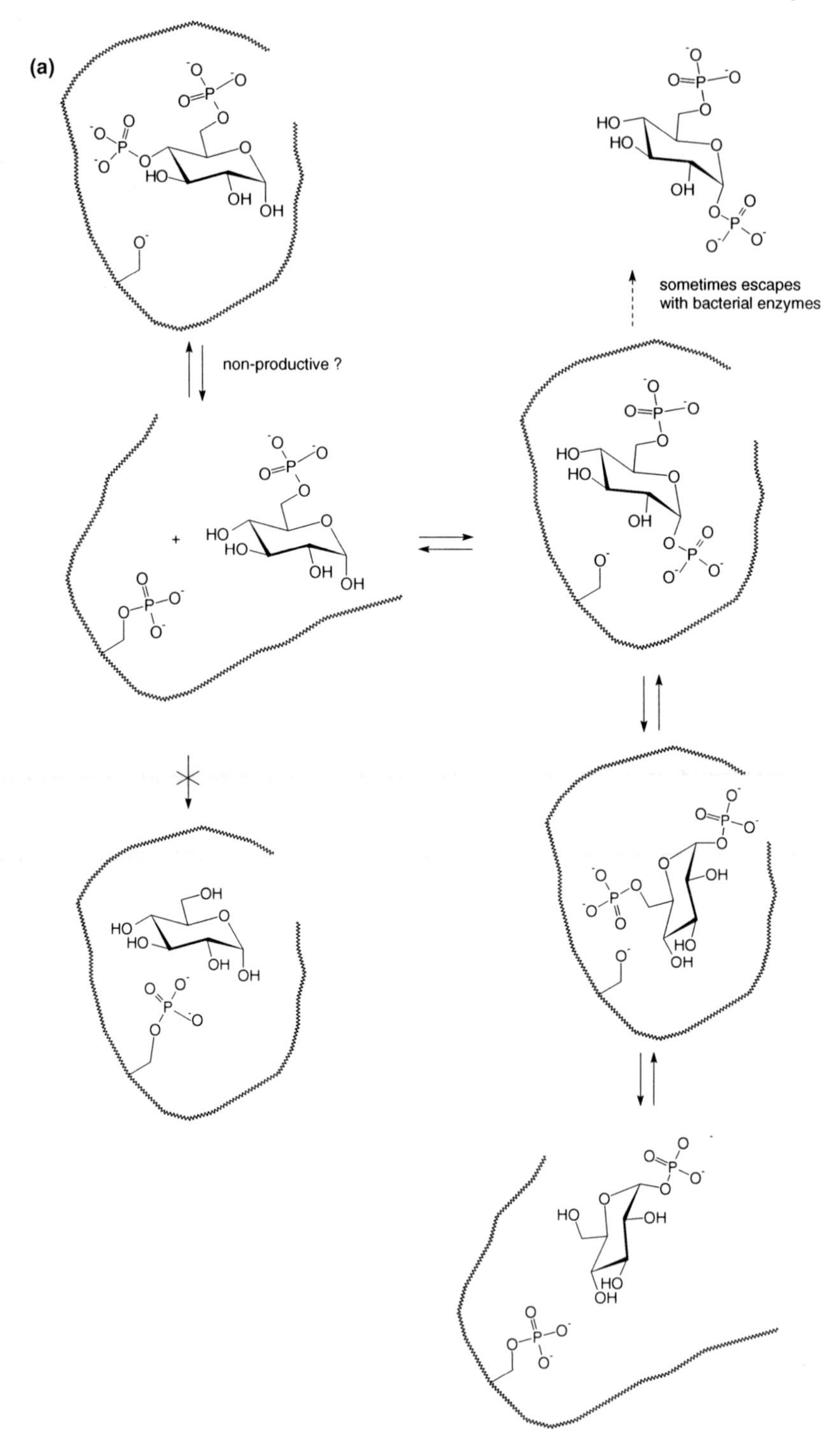

Figure 6.55 (a) α-Phosphoglucomutase mechanism.

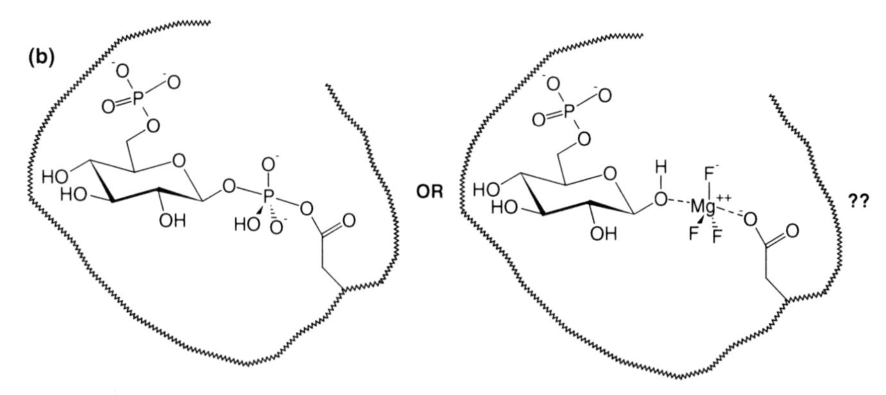

Figure 6.55 (b) Suggested phosphorane intermediate (or artefact) in β-phosphoglucomutase.

A functionally related enzyme in bacteria interconverts β-glucosyl phosphate to 6-phosphoglucose. The gross mechanism is similar, with the phosphorylation of a single residue and rotation of substrate in a closed form of the enzyme.[299] However, the protein fold is that of the haloalkane dehalogenases and the nucleophile is an aspartate, as it is in these enzymes. The crystal structure of the *Lactococcus lactis* enzyme in the presence of either substrate proved to have a phosphate on O6 and what appeared to be a phosphorane, formed from the nucleophilic aspartate residue and a phosphate on O1.[300] However, in an instructive example of some of the weaknesses of X-ray crystallography as a mechanistic tool, it was pointed out that the "phosphorane" could in fact be a complex of MgF_3^-, the enzyme nucleophile and sugar O1 arranged in a trigonal bipyramid with the three Mg–F bonds equatorial (the crystallisation solutions contained 10 mM Mg^{2+} and 100 mM F^-)[xxiv].[301]

The paradigmal sugar diphosphatase is ribonuclease A, which depolymerises RNA (Figure 6.56a).[302] It is small and robust and accordingly well studied. The accepted mechanism involves general acid catalysis of leaving group departure by His119 and general base catalysis of the attack of the 2′-OH by His12, to yield a cyclic phosphate intermediate, which is hydrolysed, but much more slowly, in a subsequent step. Lys41, which hydrogen bonds to the non-bridge oxygen, is also crucial: mutation of any of the three key residues reduces the rate by 10^4-fold. Further evidence for the mechanism includes crystal structures of enzymes from several sources complexed with several ligands, and the inertness of 2′-deoxy nucleosides. The use of the phosphodiester of 3′-uridine and *p*-nitrophenol as substrate showed that His119 was the general acid and His12 the general base, rather than the other way round, since the activity against the His119Ala mutant was nearly the same as wt.

[xxiv] The original authors later replied (L. W. Tremblay, G. Zhang, J. Dai, D. Dunaway-Mariano and K. N. Allen, J. Am. Chem. Soc. 2005, 127, 5298) by analysing the protein/phosphate ratios of the crystals, but using imperfect and potentially non-specific colour tests.

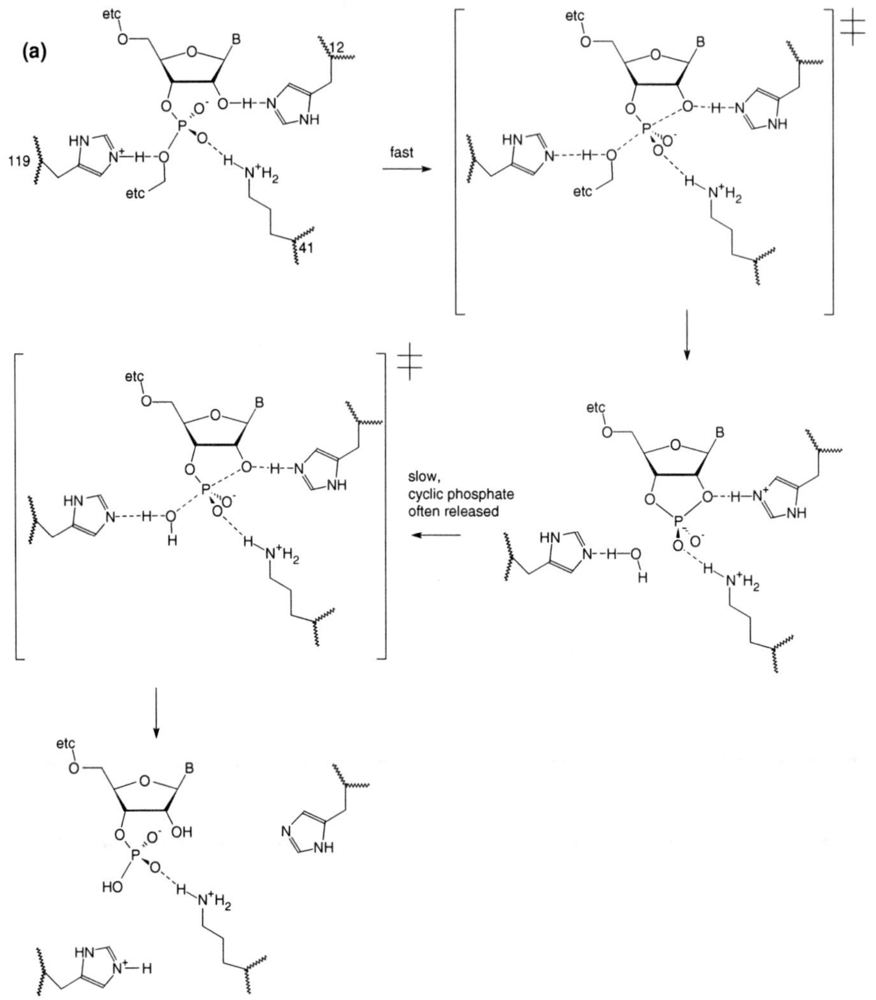

Figure 6.56 (a) Ribonuclease A mechanism.

A mechanistically closely similar enzyme is phosphatidylinositol-specific phospholipase C, which hydrolyses the diphosphate linkages between inositol and a diacylglycerol, the inositol 1,2-cyclic phosphate being released into aqueous solution and the lipid into the lipid bilayer (Figure 5.56b). The cyclic phosphate is hydrolysed by the enzyme some 10^4 times slower than the phosphatidylinositol substrate. The enzyme from *Bacillus thuringensis* acting on monophosphorylated inositol has been most extensively studied,[303] with the crystal structure showing that the lysine of ribonuclease was replaced by an arginine.[304] The role of the arginine in hydrogen bonding to a specific non-bridge oxygen was ingeniously solved by the use of phosphorothioate substrates in which the sulfur had been introduced stereospecifically. R_P phosphorothioate substrates were transformed at about the rate of phosphate substrates, but their

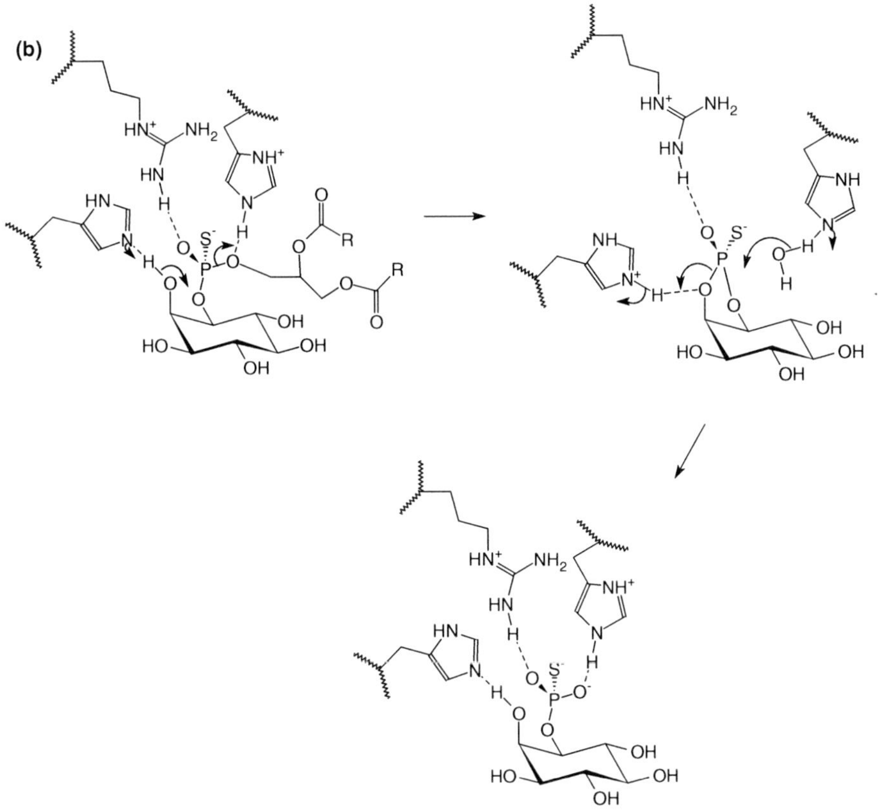

Figure 6.56 (b) Phosphatidylinositol-specific phospholipase C mechanism, shown with the R_P phosphorothioate substrates (note that the final product is 1-D-*myo*-inositol phosphate, the enantiomer of the product formed by cyclisation of 6-phosphoglucose).

S_P epimers were 10^5 slower; if the active site arginine was mutated to a lysine, the stereoselectivity largely disappeared, there being only a 10-fold difference between phosphorothioates epimeric at phosphorus.[305] His32 and Asp274 formed a catalytic dyad which putatively deprotonated O2, with 10^5-fold and 10^4-fold loss of activity in alanine mutations, but 2% residual activity in Asp274Asn mutations; the suggestion was that hydrogen bonding oriented the basic His correctly. Mutations in the leaving-group catalytic dyad (His82–Asp33) were selective, with little effect on the departure of leaving groups which did not require protonation, such as thiolates or *p*-nitrophenolate, but large effects on the activity against substrates with basic oxygen leaving groups. The Asp33Asn mutation had a detectable electronic effect, with β_{nuc} for ester formation from the cyclic phosphate falling from –0.17 in wt to –0.32. The mammalian counterparts of this enzyme act on variously phosphorylated phosphatidylinositols and are Ca^{2+} dependent. Crystal structures are available which are compatible with the mechanism of the bacterial enzyme, as suggested by sequence homologies within the catalytic region, but despite the pronouncements of the crystallographers themselves.[306]

All the very many enzymes involved in DNA and RNA synthesis and many of those involved in degradation and repair involve reactions of sugar diphosphates, but such enzymes are outside the scope of this text, both because of the vast amount of material that would have to be covered and also because "Circe effect" remote interactions with the nucleotide bases or local structural elements play an important role in catalysis, and are the focus of biological interest. Likewise, catalytic RNA,[307] whether self-splicing or promiscuous, is of the highest importance to our ideas of the origin of life. As molecules which can carry information and perform catalysis, they may be the "missing link" between prebiotic self-replication chemistry and the modern division of labour in which DNA carries information and proteins perform functions such as catalysis. Their basic chemistry is that of sugar phosphates, but the catalytic site is assembled in complex ways which depend on interactions between remote bases,[308] so, like the protein DNA and RNA polymerases, detailed treatment is outside the scope of this text. Catalytic RNA molecules assemble the catalytically active site from several metal ions in addition to RNA, and the catalytically active form is often in a comparatively shallow free energy minimum, so that dominant conformations observed by, say, NMR spectroscopy or crystallography may not be the catalytically relevant ones. However, thiophilic metal rescue of chemically synthesised thio analogues of the natural substrate (*e.g.* 2′- and 3′-SH, *pro*S_P of the non-bridging phosphate oxygen)[309], is a valuable technique for delineating which metal binds where and how much it contributes to catalysis.

6.5.9 Sulfites, Sulfates and Sulfonates

Carbohydrate oxyacid esters of sulfur, in which the sulfur covalency is less than 6, are not encountered except as transient intermediates in Pfitzner–Moffatt and related oxidations (Section 6.6.1.1) or in nucleophilic fluorination by DAST (diethylamino sulfotrifluoride, Et_2NSF_3) and related compounds. The activation of the OH group and its nucleophilic displacement by fluoride are carried out in the same pot, usually in aprotic solvents such as dichloromethane. The active intermediate is thought to be $ROSF_2NEt_2$, with SOF_2 being displaced by fluoride or internal nucleophiles; in the simple case the OH group is replaced by fluoride with inversion of configuration.[310] Whatever its precise nature, the S^{IV} leaving group seems to lead to complications from neighbouring group participation and alkyl and hydride shift in reactions of carbohydrates, similar to those encountered with trifluoromethanesulfonates.[311]

Conversion of alcohols to sulfonate esters is a way of labilising them to nucleophilic substitution and elimination. The common leaving groups are arenesulfonates, particularly *p*-toluenesulfonate (tosylate), methanesulfonates (mesylate) and trifluromethanesulfonate (triflate); they are introduced by reaction of the acid chlorides (or, in the case of the trifluoromethanesulfonates, acid anhydrides) in a basic solvent such as pyridine. Traditionally, the reactions are carried out in pyridine as solvent, but both this solvent and the liberated chloride ion are good nucleophiles, potentially converting primary tosylates to chlorides or *N*-alkylpyridinium salts. A relatively recent tosylation protocol uses 1.5–2.5 equivalents of triethylamine and 0.1–1.5 equivalents of trimethylamine hydrochloride: trimethylamine, but not triethylamine or pyridine, reacts to form an *N*-tosyltrimethylammonium ion, which is a true and efficient tosylating agent in a range of aprotic solvents (Figure 6.57(a)).[312]

Alkanesulfonyl chlorides can also react by an initial elimination: in water, mesyl chloride hydrolyses below pH 6.7 by an S_N2 reaction of water at sulfur, between pH 6.7 an 11.8 by capture of the sulfene by water and above pH 11.8 by capture of the sulfene with hydroxide ion.[313] An extended study of mesylations of phenols catalysed by various pyridines revealed a competition between the sulfene pathway and a general base-catalysed attack of the phenol, which was dependent on the pK_a of both phenol and pyridine, [314] with no mention of an *N*-mesylpyridinium ion. However, formation of an *N*-triflylpyridinium ion occurs immediately pyridine and triflic anhydride are mixed and this is probably the triflating species.[315] Reaction via an *N*-sulfonyl quaternary ammonium ion therefore becomes important with more basic amines and more electrophilic sulfonylating reagents.

In the 1960s and 1970s, much synthetic effort went into making and reacting sugar *p*-toluenesulfonates and methanesulfonates, and attempting to displace them with nucleophiles, despite evidence in the physical organic literature that such attempts would give only patchy success at best. S_N2 displacement of equatorial substituents in a cyclohexane ring is impossible: if the alternative chair conformation is prohibited, then reaction occurs through non-chair conformers and if those are also prohibited, as in the 2-adamantyl system (Figure 5.19), the reaction is rigorously unimolecular, goes through a non-classical carbenium ion and occurs with *retention* of configuration.[316] Indeed, attempts to displace sulfonate leaving groups even from simple acyclic secondary centres with acetic acid as a nucleophile results in products derived from delocalised, non-classical carbenium ions with C–C and C–H bridging.[317] The many available lone pair-bearing functionalities of a carbohydrate (including the ring oxygen) are therefore likely to intervene in any attempt to carry out a "difficult" displacement of a sulfonate, although the problems are minimised by the use of the very reactive trifluoromethanesulfonates. The pK_a of trifluoromethanesulfonic acid is −13 to −14, compared with around −2.5 for methanesulfonic acid and −3 for benzenesulfonic acid;[318] with a maximum β_{lg} value for S_N2 reactions of ~ -0.5 (for methyl transfer), the triflates will be at least 10^5 times more reactive than tosylates, *etc.*

Driguez's syntheses of thio cellooligosaccharides with non-fissile thioglycosidic linkages are examples of how many factors have to be favourable before an S_N2 displacement of a sulfonate from a pyranose ring gives acceptable yields. A 75% yield of substitution product (still with 20% elimination product) was obtained only when the nucleophile was a powerful, but non-basic, thiolate ion, generated *in situ* by *S*-deacetylation with diethylamine of pentaacetyl 1-thio-β-D-glucopyranose, the electrophile methyl 2,3-6-tri-*O*-benzyl 4-*O*-trifluoromethanesulfonyl-β-D-galactopyranoside and the solvent the most powerful common dipolar aprotic solvent, hexamethylphosphoramide (Figure 6.57(b)).[319]

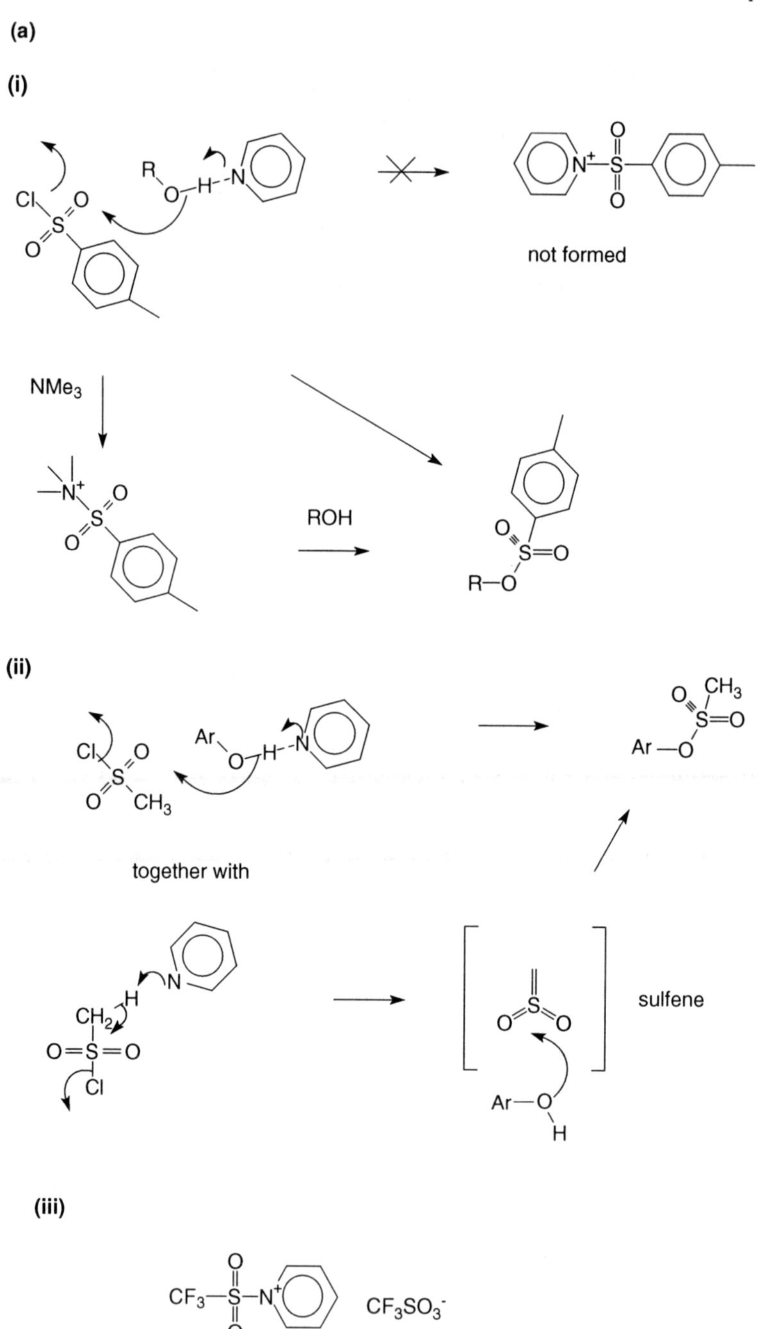

Figure 6.57 (a) Mechanisms of sulfonylation: (i) tosylation, (ii) mesylation and (iii) triflylation.

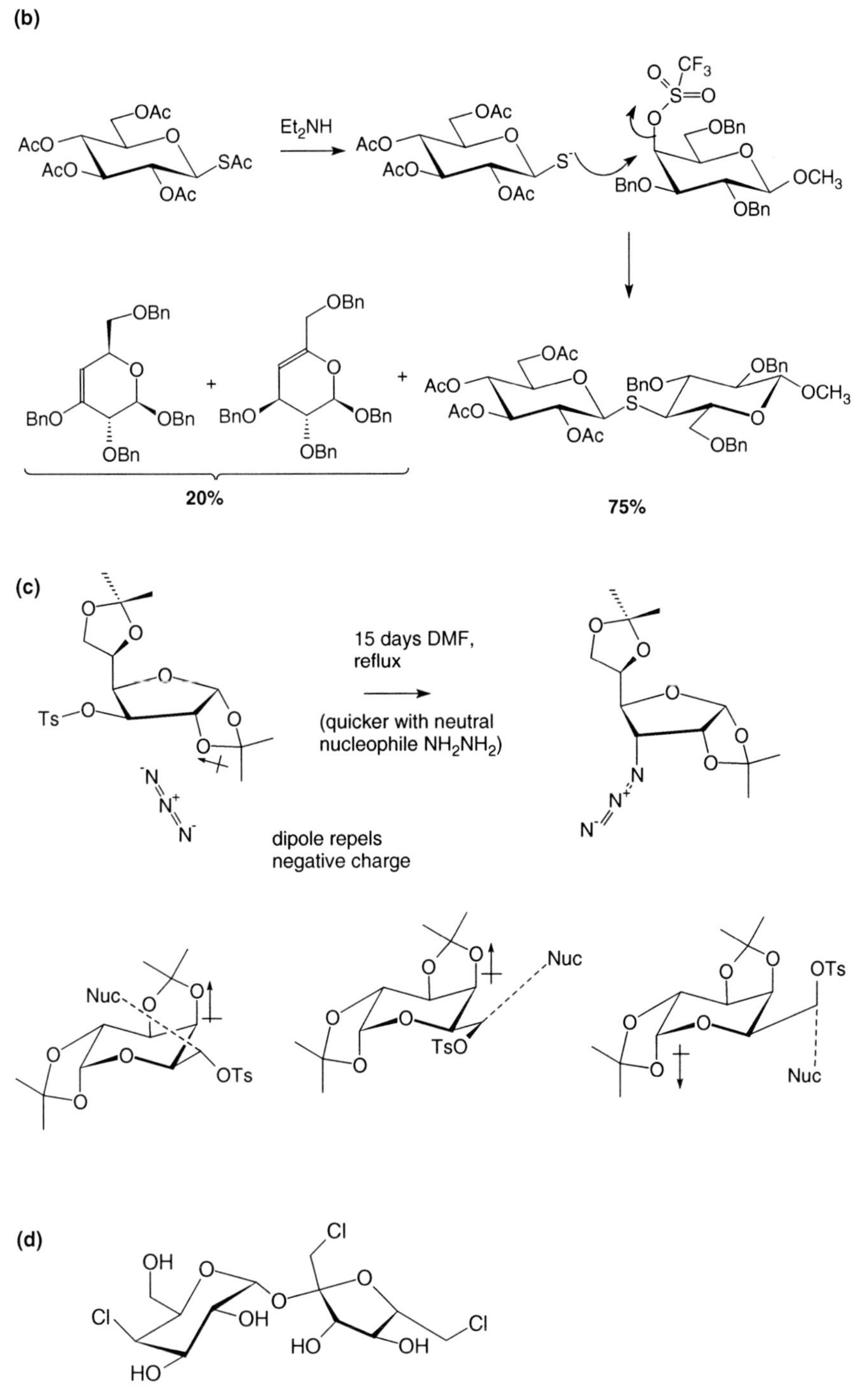

(b)

Et$_2$NH

20%

75%

(c)

15 days DMF, reflux

(quicker with neutral nucleophile NH_2NH_2)

dipole repels negative charge

(d)

Figure 6.57 (b) A successful sulfonate displacement. (c) Dipole hindrance of tosylate displacements. (d) Sucralose.

S_N2 displacements from a five-membered ring or a primary position can also be disfavoured, by repulsion of anionic nucleophiles by the negative ends of substrate dipoles. Notorious examples are 3-*O*-tosyldiisopropylideneglucose and 6-*O*-tosyldiisopropylidenegalactose, whose tosylate groups are displaced only in refluxing dipolar aprotic solvents (Figure 6.57(c)).[320]

Sucrose is the most readily and cheaply available pure organic chemical and was the object of valorisation efforts in the 1970s and 1980s. Reaction with sulfuryl chloride in pyridine gave a range of chlorodeoxysucroses and their stereoisomers, via nucleophilic displacements of chlorosulfate esters R–O–SO_2–Cl. First the 6-*gluco*-position and then the 6-*fructo*-position were chlorinated.[321] The reactions are displacements of chlorosulfate by chloride ion. Formation of chlorodeoxy sites on the ring also occurred, however, via the intermediacy of epoxides.[322] Some of the products proved to have dramatically increased sweetness, with 4-chloro-4-deoxy-α-D-galactopyranosyl-1′,6′-dichlorodideoxy-β-D-fructofuranose being 650 times sweeter than sucrose (Figure 6.57(a)). It is not transformed by intestinal sucrase, and so is non-caloric and is marketed as Sucralose.[323]

Nature is generally reluctant to carry out classical, "central" S_N2 reactions, with the exception of methylations – glycosyl- and prenyl-transfer reactions are very S_N1-like. A much preferred route for displacements of sugar substituents is oxidation by a tightly bound nicotinamide cofactor of a β-hydroxyl, elimination of the leaving group, introduction of the substituent in a Michael addition and reduction, as dealt with in Section 6.8.

The only biological S_N2 reactions of sugars away from the anomeric centre seem to be the apparently enzyme-catalysed ring closures of 6-*O*-sulfated galactans to give 3,6-anhydro residues (see Section 4.6.10.1.3) and the displacement of methionine from the 5′-position of *S*-adenosylmethionine by reduced arsenic species in some marine algae, which gives rise to 5′-arsenolated ribosides such as those in Figure 6.58.[324]

Ease of alkyl–oxygen fission in the reactions of oxyacid esters of second-row elements increases in the order $(RO)_4Si < (RO)_3PO < (RO)_2SO_2 < ROClO_3$ and ease of central atom–oxygen fission increases in the order $ROClO_3 < (RO)_2$-$SO_2 < (RO)_3PO < (RO)_4Si$, so reactions of sulfate esters at sulfur are rare, alkyl esters generally reacting by substitution and elimination at carbon. Where they can be induced to react at sulfur, as with *p*-nitrophenyl sulfate, the mechanisms appear to parallel those of the corresponding phosphate, with SO_3-generating reactions associated with tightening of S–$O_{non\text{-}bridge}$ bonds and, in this case, clearly inverse $^{18}O_{non\text{-}bridge}$ isotope effects.[325]

6.5.10 Stannylene Derivatives

The formal tin counterpart of the dioxolanes formed from vicinal diols by reaction with ketones are stannylenes, formed by reaction with dialkyltin oxides. The commonest reagent is dibutyltin oxide, Bu_2SnO, which converts diols to stannylenes when refluxed in an aprotic solvent such as benzene, with continuous distillation out of the water generated. The partly ionic nature of

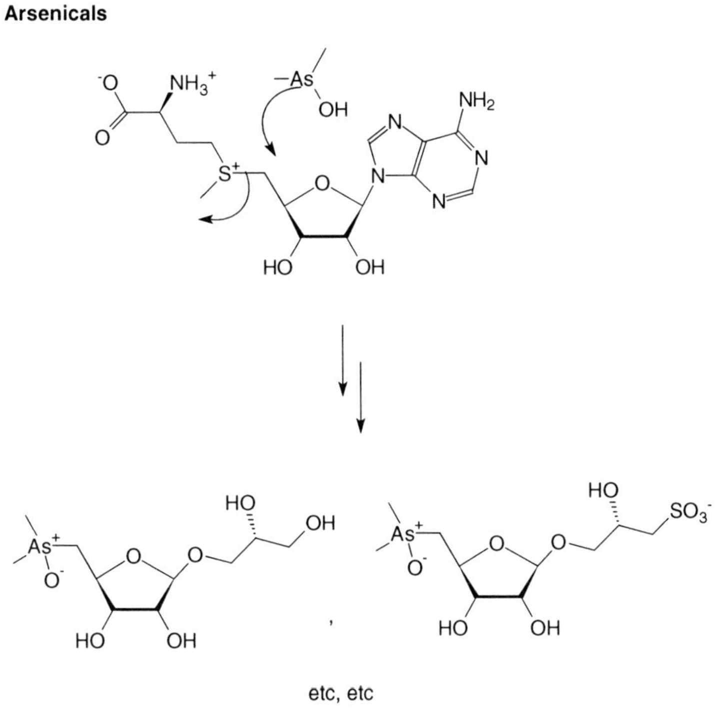

Arsenicals

etc, etc

Figure 6.58 Representative arsenic-containing ribosides and the proposed biosynthetic C–As bond-making step.

the Sn–O bond of stannylenes makes the oxygen atom a good nucleophile and this has even been exploited in glycosylations.[326] However, neither reagent nor derivatives are monomers. Dibutyltin oxide is polymeric, whereas most stannylenes are dimeric, with the tin becoming pentacoordinate in a four-membered ring of alternating tin and oxygen atoms. To form such rings, one of the two oxygens of the diol must act as a ligand to two tin atoms and thus become non-nucleophilic: in the case of the dibutylstannylene derivative of methyl 4,6-benzylidene-α-D-glucopyranoside, this is O3.[327] Stannylene formation thus inherently confers selectivity on the nucleophilic reactions of diols (and also their oxidation by bromine to hydroxy ketones),[328] but as yet there is no reliable way of predicting which way such selectivity will operate.

6.6 OXIDATIONS

6.6.1 Oxidations of Individual OH Groups

6.6.1.1 By Valence Change of an Oxyacid Ester or Related Species. Many of the two-electron oxidations of alcohols of use to preparative chemists involve formation of an ester-type intermediate of general formula >CH–O-X″<, followed by formation of the carbonyl group in what is essentially an *E*2 reaction, yielding $X^{n-2}<$ (Figure 6.59). The notorious Griess test for alkyl nitrates, based on such an elimination with $X = N^{V}$, is dealt with in Section 6.6.7 (Figure 6.46). Preparatively useful examples include S^0, Cr^{VI}, Sn^{II}, Br^0 (but only at the anomeric centre), Ru^{VIII} and Ru^{VII}. Oxidations mediated by TEMPO and related species, whilst involving one-electron chemistry to generate an oxoammonium ion (and which are therefore treated in Chapter 7, especially Figure 7.18), are similar.

The direct oxidation of aldoses by bromine water was one of the first carbohydrate reactions to be studied mechanistically.[329] The products are largely the δ-lactones, although these may open above pH 2. The reaction above pH 1 is first order in bromine and sugar and inversely proportional to $[H^+]$; other Br^{I} species, such as tribromide ion or hypobromous acid, are inert, except in so far as they alter $[Br_2]$. The reactions of α-aldopyranoses correlate with their rate of mutarotation, both when different sugars are compared in the same buffer and the same sugar in different buffers of different effectiveness as mutarotation catalysts. The main reaction therefore involves a bimolecular reaction of the anion of the equatorial aldopyranose with bromine. However, there is a primary hydrogen isotope effect ($k_H/k_T \approx 5$)[330] at H1 of the sugar and the value of the rate constant calculated for attack of the anion of β-glucose on bromine (10^6–$10^7\ M^{-1}\ s^{-1}$) is well below that measured directly for reaction of Br_2 with ^-OH, which is diffusion -controlled ($k \approx 10^{10}\ M^{-1}\ s^{-1}$). The formation of the β-glucose anion–bromine complex giving rise to the isotope effect then has to be at least partly reversible. The complex therefore cannot be the glucosyl hypobromite, but rather its complex with bromide ion.

In strongly acidic solution there is evidence for a slow, direct reaction of a neutral species, possibly the acyclic *gem*-diol.

Reactions of carbohydrates with chlorine have been studied largely in the context of the bleaching of papermaking pulp. The product of any pulping operation is brown, even when, as with kraft pulp, it is nearly pure cellulose, because of chromophores from residual lignin and carbohydrate degradation products. For most purposes pulp is bleached, but this causes a loss of paper strength by lowering the cellulose *DP*.[xxv] The mechanisms of such *DP* lowering are threefold. Hydrogen atom abstraction, followed by recombination with Cl and loss of HCl, will give carbonyl groups, which can then promote elimination reactions similar to the peeling reaction. In neutral to alkaline solution, however, the predominant reaction appears to be formation of hypochlorite esters on O2, O3 and O6 of the cellulose, which on elimination of HCl gives carbonyl compounds, in the same way as β-pyranoses react with bromine to form aldonolactones; certainly, $NaBH_4$ oxidation of chlorine-oxidised cellulose eliminated the subsequent production of metasaccharinic-type-species.[331]

Oxidations by dimethyl sulfoxide and an electrophile[332] involve the formation of a sulfoxonium ion, which then collapses to a carbonyl group. The

---

[xxv] Brown wrapping paper and the brown envelopes used by official bodies are made from unbleached kraft pulp, which is particularly strong.

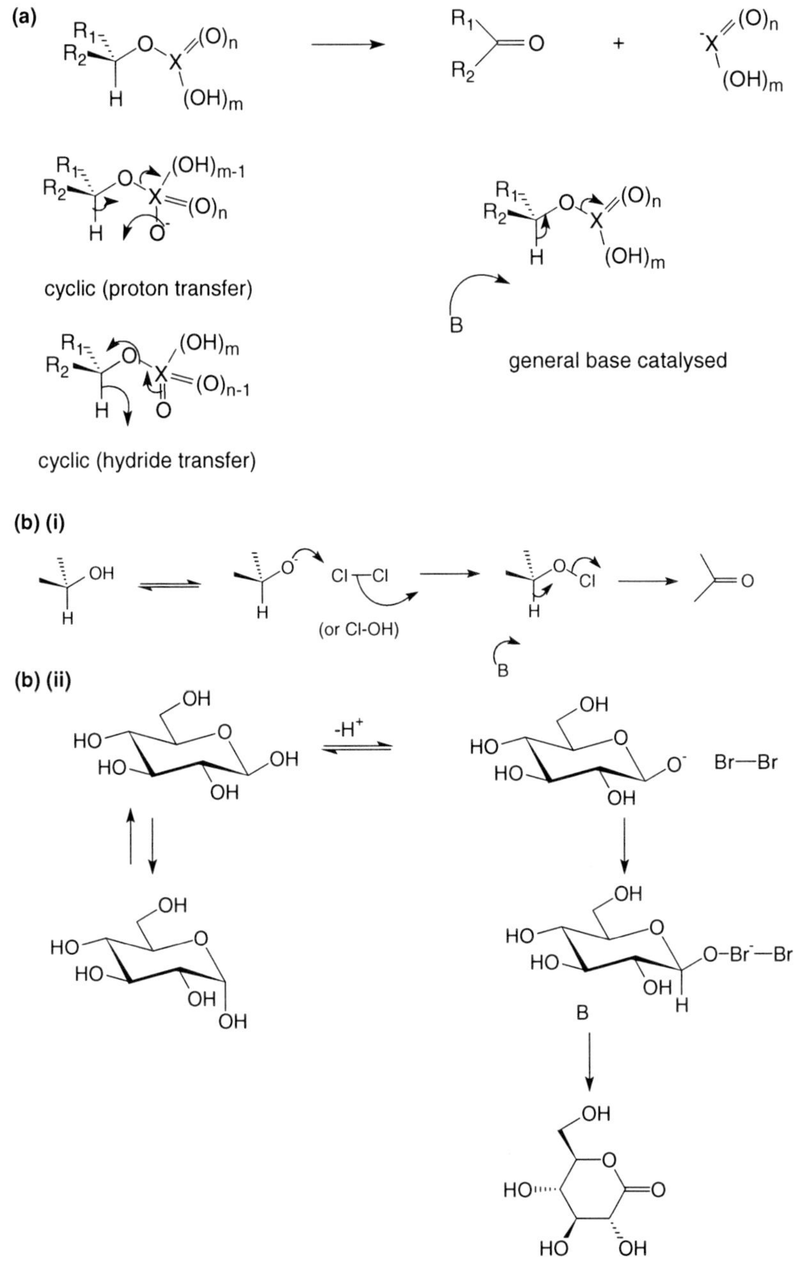

Figure 6.59 (a) General mechanism of two-electron oxidation of alcohols via formation of oxyacid esters; for $X = N^+ <$, $m = 1$, $n = 0$, see Figure 7.18, for $X = N^+$, $m = n = 1$, see Figure 6.48. (b) $m = n = O$, $X = Hal$: (i) chlorine oxidations; (ii) bromine reactions of aldoses. The evidence for the dibromo complex is weak and the intermediate could be a simple hypobromite.

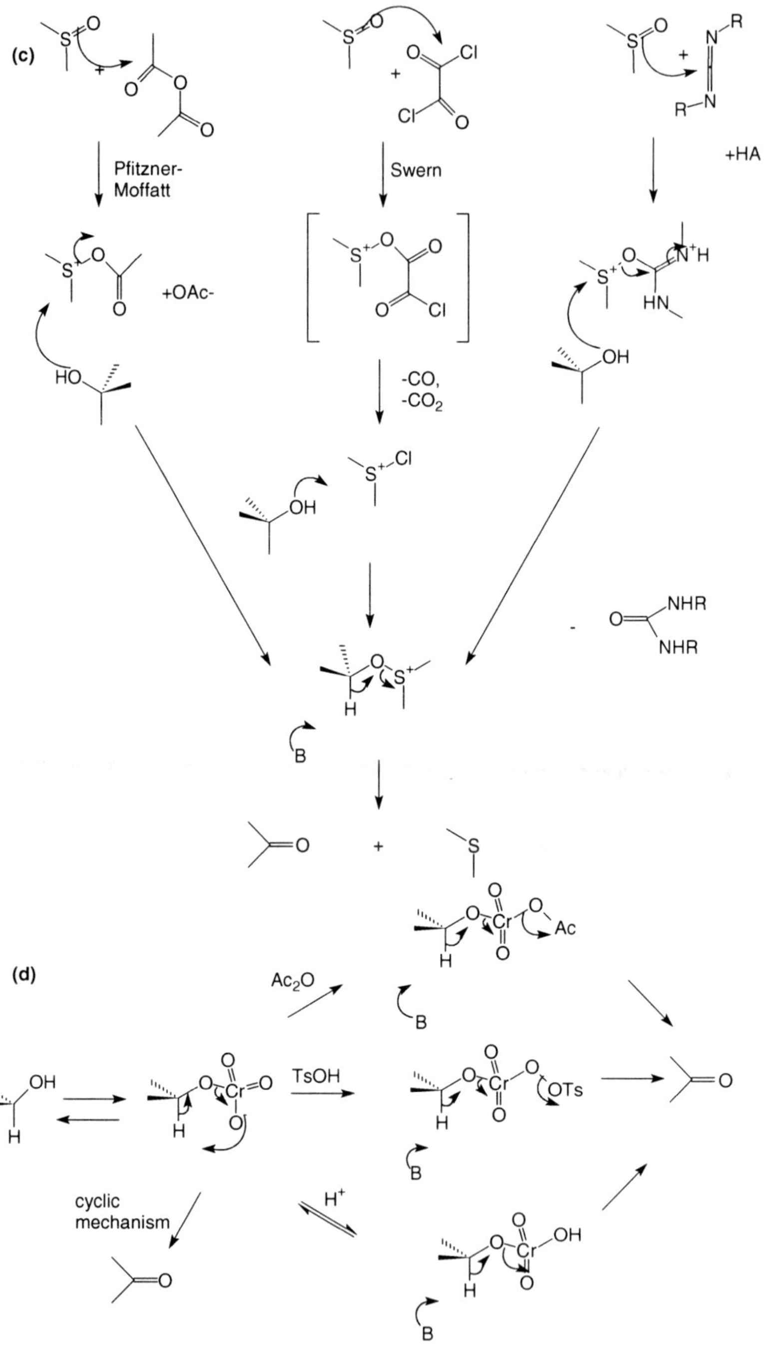

Figure 6.59 (c) Activated DMSO reactions, $m = 1$, $n = O$, $X = S^+ <$. (d) Chromium oxidations, $m = 1$, $n = 2$, $X = Cr$.

reaction is carried out under anhydrous conditions, so further oxidation of an aldehyde hydrate to a carboxylic acid cannot take place. The activators are various: the classical Pfitzner–Moffatt protocol uses acetic anhydride, the Swern protocol oxalyl chloride, and a third common variant is the use of dicyclohexylcarbodiimide and an acid. A feature of the use of acetic anhydride is that it generates acetate ion in DMSO, which, because the latter is a dipolar aprotic solvent, makes the former a powerful base. Pfitzner–Moffatt oxidations with potential leaving groups (sulfonate, pK_a 1, of course, but also various carboxylates, pK_a 4–5) β to the new carbonyl therefore tend to give products from $E1_{CB}$ eliminations.[333] This is less of a problem with the Swern protocol, where the active species is thought to contain chlorine and only chloride ions are present as potential bases.

The well-known oxidations of primary and secondary alcohols with Cr^{VI} species proceed through chromate esters. The definitive mechanistic experiments[334] demonstrating that previously observed chromate esters were indeed on the reaction pathway showed that either formation or decomposition of the ester could be rate-determining. The rates of oxidation of cyclohexanol and the secondary hydroxyl group of a very sterically hindered steroid in aqueous acetic acid were measured as a function of the solvent composition. The former increased radically as the acetic acid concentration increased whereas the latter remained invariant. The former exhibited a primary deuterium kinetic isotope effect of 5,[335] whereas there was no KIE on the oxidation of the crowded steroid. Therefore, the rate-determining step in the oxidation of the cyclohexanol was decomposition of the chromate ester and in the oxidation of the steroid it was its formation, with the ester an obligate intermediate in both reactions.

In the carbohydrate area, the use of neutral or weakly basic variants, such as chromium trioxide–pyridine, $CrO_3.2C_5H_5N$, pyridinium dichromate, quinolinium or quinoxalinium[xxvi] dichromate or pyridinium chlorochromate, $C_5H_5NH^+.ClCrO_3^-$, is advisable to avoid cleavage of acid-labile groups; addition of molecular sieve to remove water is also recommended.[336] The complexes with chromic trioxide can be activated by acetic anhydride, probably making the ester $R–OCrO_2OAc$ rather than $R–OCrO_2OH$, and thereby assisting the elimination.[337] The reaction of quinoxalinium dichromate with alcohols in the presence of *p*-toluenesulfonic acid was likewise thought to proceed through $R–OCrO_2OTs$.[338]

The details of hydrogen transfer from carbon are unclear. Analogy with other reactions of this type, particularly those of TEMPO, would suggest a cyclic transition state, characterised by primary hydrogen effects of ~2, under basic conditions and a general base-catalysed process under acidic conditions, characterised by isotope effects around 5. However, the cyclic mechanism was also proposed for the standard oxidation conditions of aqueous acetic acid, originally[334] on the basis of a rate decrease in aqueous acetic acid as sodium acetate was added, despite such an addition possibly deprotonating the

---

[xxvi] Quinoxaline is 1,4-diazanaphthalene.

chromate half-ester and thus reducing its reactivity. Oxidation of Ph_2CDOH with pyridinium chlorochromate in basic conditions indeed gives a KIE of 2.93[336] (which is suppressed by the addition of molecular sieve desiccant), but the reaction was interpreted as a hydride transfer, like the oxidation of ring-substituted benzyl alcohols with quinoxalinium dichromate, which gave a Hammett ρ value of -1.0.[338] However, negative ρ values could have just as easily arisen from a proton transfer mechanism in which O–Cr cleavage was somewhat ahead of C–H cleavage.

In either mechanism, a low value for the KIE is expected for non-linear hydrogen transfer,[xxvii] with non-linear transfer associated with similar enthalpies and different energies of activation for the two isotopomers. The oxidation of benzyl alcohol by quinoxalinium dichromate is acid catalysed, being first order in dichromate, substrate and added *p*-toluenesulfonic acid, and the KIE for $PhCD_2OH$ oxidation is 6.78. The latter is the product of a primary effect and a secondary α-deuterium effect: if we assume the latter is around 1.2 (see Section 3.3.1), the reaction is a typical acid chromate oxidation with a linear

---

[xxvii] As an approximation, consider the ABH system, in which hydrogen is transferred from A to B, as a triatomic molecule. This has $3n-5$ ($=4$) normal modes of vibration when it is linear and $3n-6$ ($=3$) modes when it is non-linear. These are sketched in the diagram, with modes 1–4 referring to the linear and modes I–III to the non-linear arrangement. Modes 1 and I correspond to the reaction coordinate and have no zero point energy. Modes 2, 3 and II are bending vibrations, which do not alter much during reaction. In mode 4 there is little motion, indeed none at all in the case where the transition state is symmetrical and a maximal KIE is observed. In mode III, however, there is still a large amount of motion of the transferred hydrogen at the transition state and so KIEs are lowered.

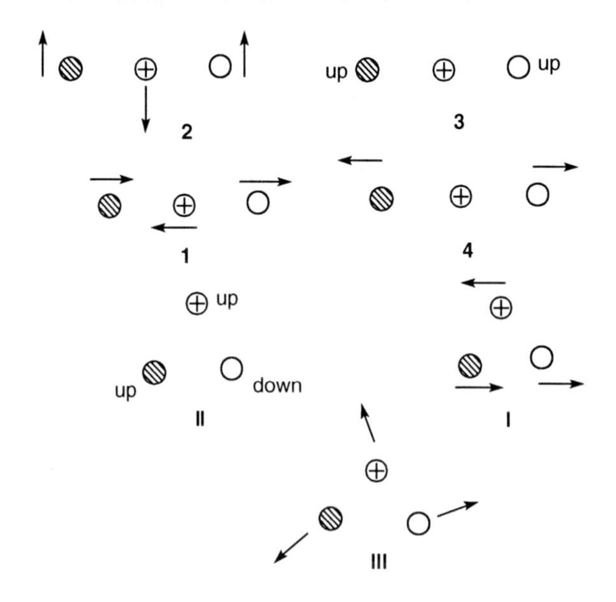

transition state for proton transfer. The chromic acid oxidation of the highly sterically crowded bis(*tert*-butyl)carbinol had $\log_{10}A$ and activation energies of 9.09 and 59 kJ mol^{-1}, respectively, for the protiated and 10.26 and 71 kJ mol^{-1}, respectively, for the deuterated compound.[335] The figures were taken to support the non-linear transfer model, despite values of k_H/k_D larger than normal (around 10). To the author, however, they strongly suggest hydrogen transfer in a linear transition state by quantum mechanical tunnelling in a base-catalysed process, which is well known to be favoured by hindered substrates, and results in lower A values for the protiated compound and high absolute effects.

It seems to the author that the balance of evidence points to a general base-catalysed process in acid and a cyclic one in neutral or alkaline solution, as with other reactions of this type.

Chromate oxidations are further complicated by the ability of other valencies of chromium to oxidise alcohols. Cr^{IV}, in acidic aqueous solution, probably as CrO^{2+}, oxidises alcohols probably by hydride transfer, the chromous (Cr^{II}) ion product reacting with Cr^{IV} to give two equivalents of Cr^{III} with a near-diffusional rate constant of 2×10^9 M^{-1} s^{-1}.[339] Cr^V will also oxidise alcohols; however, the process is slower than the oxidations with Cr^{VI}.[340] There is also an extended literature from the 1970s on oxidations in which Cr^{VI} supposedly removed three electrons to organic substrates in a single chemical step, yielding Cr^{III} directly, which is in error because of mistaken choices of model systems.[341]

6.6.1.2 By Hydride Transfer. An important pathway of oxidation is the removal of hydride attached to an oxygen atom: if the oxygenated centre is OH, the reaction is accelerated by partial or complete removal of the hydroxyl proton.

6.6.1.2.1 Non-enzymic hydride abstraction from carbohydrates. Avid electrophiles, such as ozone[342] or chlorine,[343] will remove an axial anomeric hydride from a glycoside, and ozone will remove an axial or pseudoaxial hydride from five- and six-membered cyclic acetals.[344] The observation that methyl β-glucopyranoside reacted with ozone to give glucono-δ-lactone, whereas methyl α-glucopyranoside appeared inert, was one of the seminal observations that led to the formulation of ALPH.[345] The abstraction of the axial hydride was supposedly promoted by an antiperiplanar lone pair on O5 in addition to one on O1, in a way not possible with the α-glucosides, where only a lone pair on O1 could become antiperiplanar. As with the hydrolysis of conformationally restricted orthoesters, the preference for axial departure can be better rationalised by recognising that it required less motion in the rest of the molecule to attain the C2–C1–O1–O5–C5 coplanarity of the dioxocarbenium ion intermediate, than by invoking ALPH as a fundamental principle (see Figure 3.13b). As expected for a least motion effect, quantitatively it is not large, with methyl α-glucopyranoside reacting with ozone at the acetal oxygen about half as fast as its anomer reacts by hydride abstraction.[342]

Abstraction of hydride by ozone results in an intimate ion pair which was originally thought to collapse exclusively to the covalent hydrotrioxide,[346] which has been well characterised in the case of dioxolane and dioxane substrates.[344] It is now suggested that the ion pair can collapse directly to hemiorthoester with emission of singlet oxygen. The hemiorthoester can also be formed by a water-mediated decomposition of the covalent hydrotrioxide, also with the production of singlet oxygen. This reaction has been shown, in the case of oxidised dioxanes and dioxolanes, to involve an intramolecular proton transfer. Singlet oxygen and the hydroxy ester can also be formed by simple hydrolysis and loss of HOOOH. The hemiorthoesters derived from dioxanes and dioxolanes collapse to hydroxy ester in a process involving multiple proton transfers, as shown by $k_{H_2O}/k_{D_2O} = 4.6$. The hemiorthoester derived from methyl β-glucopyranoside collapses only to glucono-δ-lactone, with little or no acyclic methyl gluconate, as predicted both by ALPH and least motion arguments.

Although, under neutral and alkaline conditions, cellulose, methyl β-glucopyranoside and methyl β-cellobioside react with chlorine predominantly via formation of hypochlorites (see Section 6.6.1.1), in acid (pH < 2), when radical reactions are suppressed by addition of chlorine dioxide, the products are exclusively glucono-δ-lactone (and some cellobionolactone from the cellobioside), in accord with hydride abstraction followed by hydrolysis (Figure 6.60).[343]

Because of the multiplicity of potential coordination sites for the metal, the classical metal ion-promoted hydride transfer between >C=O and >CHOH[xxviii] is little applied in carbohydrate chemistry, but an intramolecular version mediated by anhydrous Sm^{III} in organic solvents has been reported. This converts 2,3,4,6-tetrabenzylaldohexoses to the 1,3,4,6-tetrabenzylketohexoses by an apparently direct 5→1 hydride shift (so that tetrabenzyl-D-glucose gives tetrabenzyl-L-sorbose and tetrabenzyl-D-galactose gives tetrabenzyl-L-tagatose; see Figure 1.6 and 1.7).[347]

6.6.1.2.2 NAD(P)-linked enzymic oxidations[348]. Several biological oxidations of the OH groups of carbohydrates involve hydride transfer to position 4 of the quaternary pyridinium (nicotinamide) moiety of coenzymes NAD^+ and NADP (Figure 6.61). The reactions are formally removal of hydrogen and the enzymes concerned are termed dehydrogenases. The reduced nicotinamide coenzymes are oxidised *in vivo* in a variety of different ways, but *in vitro* the absorption of the reduced nicotinamide ring at 340 nm (6220 M^{-1} cm^{-1}), being well clear of protein absorption at 280 nm, has become a cornerstone of biochemical analysis. Assay for reducing sugars by the appropriate aldose dehydrogenases, for example, is an attractive alternative to chemical methods.

The two positions of position 4 of the reduced coenzyme NAD(P)H are diastereotopic, and $NAD(P)^+$-linked dehydrogenases are absolutely stereospecific for one or the other. The stereospecificity was denoted A or B, before there

---

[xxviii] Meerwein–Ponndorf–(sometimes Verley) reduction, Oppenauer oxidation.

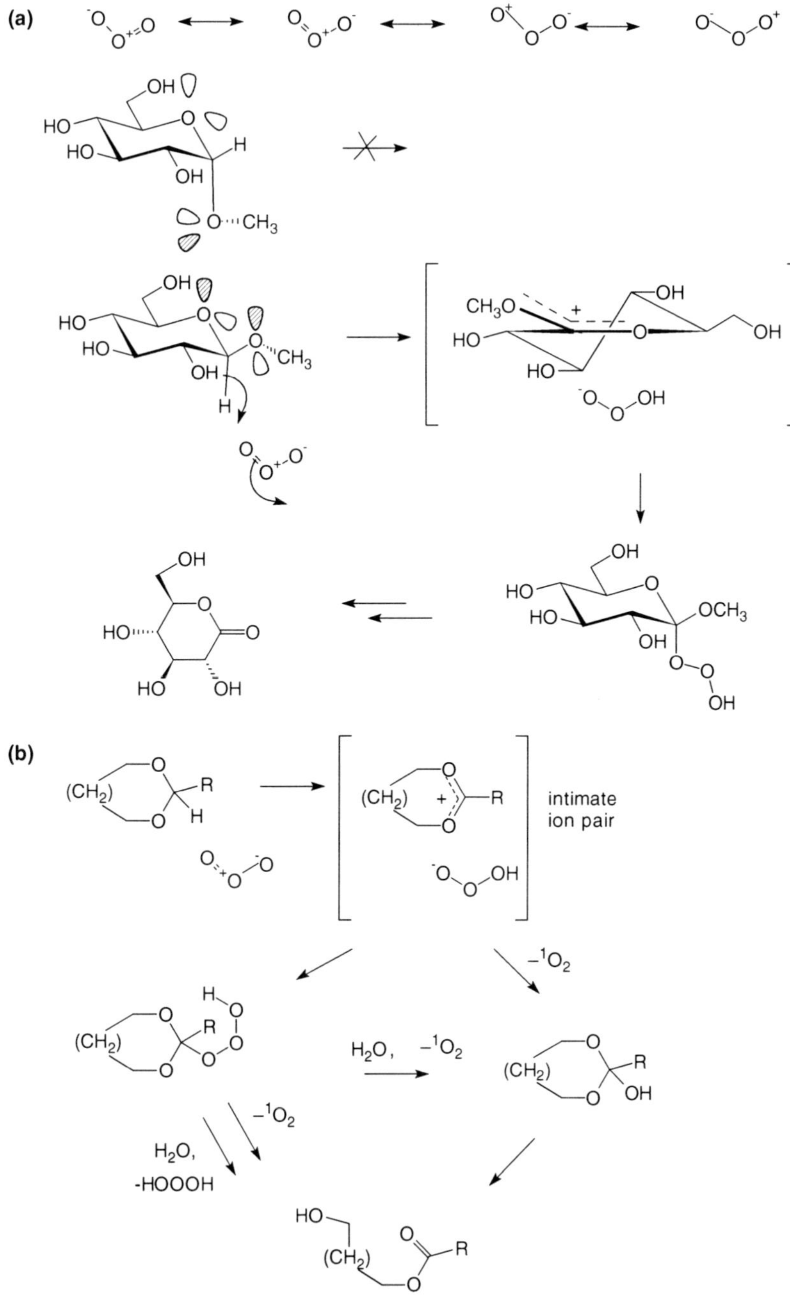

Figure 6.60 (a) Reaction of methyl glucosides with ozone. Important canonical structures for ozone are given. The ALPH-based rationale for the relative reactivities of anomers is given: for a least motion rationale, see Figure 3.13b. (b) Pathways for ozone oxidation of dioxanes and dioxolanes, with detail of the pathways from dioxocarbenium ionhydrotrioxide ion pair.

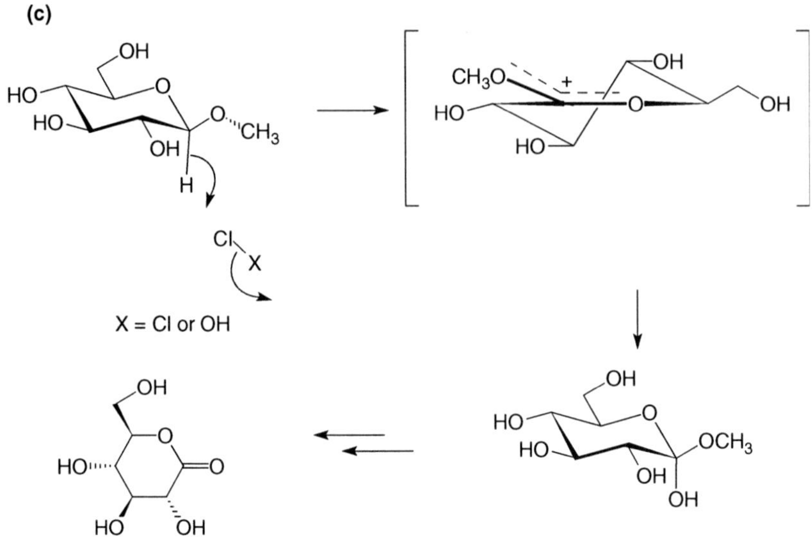

(c)

X = Cl or OH

Figure 6.60 (c) Reaction of cellulose models with chlorine in acidic aqueous solution.

were techniques for telling which was which. There are thus three ways of describing the specificity:

- transfer of hydride to the *re* face of C4 of $NAD(P)^+$ = transfer of *proR* hydrogen of NAD(P)H = A side specificity;
- transfer of hydride to *si* face of C4 of NAD(P) = transfer of *proS* hydrogen of NAD(P)H = B side specificity.

Reaction specificity appears to correlate with the conformation of the bound NAD(P) about the C–N bond. There appear to be two minima in free solution, both apparently dictated by the need to have interactions between large substituents staggered, and hence the pyridine ring eclipsing the anomeric hydrogen, *syn* and *anti*. (Figure 6.61). X-ray studies of a range of enzymes in complex with coenzyme showed that B-side-specific enzymes bind the coenzyme in the *syn* conformation and A-side-specific enzymes bind in the *anti* conformation.

About 25 years ago, it was suggested that reaction stereospecificity was determined by reaction thermodynamics.[349] Examination of the stereochemical outcomes of about 130 NAD(P)-linked dehydrogenases led to the generalisation that if the equilibrium constant, defined by eqn. (6.3), is greater than 0.1 nM, B-side specificity is shown, whereas if it is less than 1 pM, A-side specificity results, and if it is in the 10 pM region, either specificity can be observed.

$$K_{eq} = \frac{[\mathrm{NAD(P)H}][\mathrm{H}^+][>\mathrm{C{=}O}]}{[\mathrm{NAD(P)}][>\mathrm{CHOH}]} \tag{6.3}$$

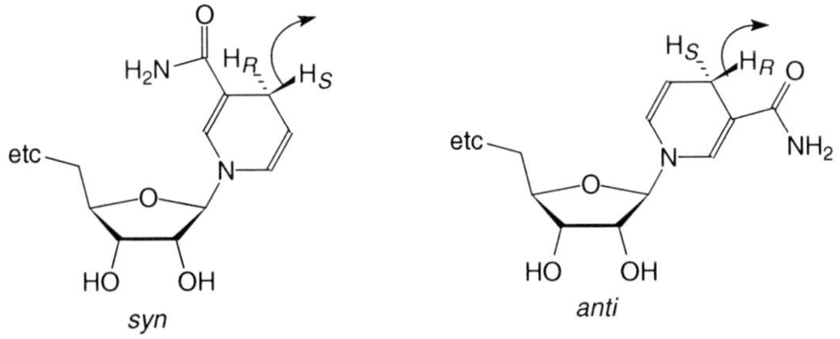

Figure 6.61 Discrimination between diastereotopic hydrogens of nicotinamide cofactors. A and B side specificity are shown with the C–N glycosidic bond in the conformation in which such specificity is usually observed.

In NAD^+ solutions there is a 10-fold greater proportion of the *syn* conformation than in NADH. Therefore, *syn*-NAD^+ is a weaker oxidising agent than *anti*-NAD^+. Hence, if the thermodynamic correlation holds, easier oxidations (higher K_{eq}) will have B-side stereochemistry.

The correlation appeared to be a major advance in the debate about the relative importance, in the evolution of an enzyme mechanism, of the limited structural possibilities available to an ancestral protein and thermodynamic constraints of the catalysed reaction. Objections that the correlation was in fact an artefact of the choice of dataset[350] were supported when the coenzyme stereochemistry of a series of aldose dehydrogenases,[351] which oxidise the β-anomer of aldoses, was examined. Because the product of such reactions is a δ-lactone, not an aldehyde or ketone, the reaction is much more thermodynamically favoured than oxidations to ketones, with K_{eq} values in the μM region or even higher. The reactions should be uniformly B; however, the xylose dehydrogenase of pig liver and the glucose dehydrogenase of *Thermoplasma acidophilum* had A-side stereospecificity, despite K_{eq} being five orders of magnitude in the B-side region in the proposed correlation. The "historical" picture of the evolution of enzymes appeared confirmed, as with the *ebg* system (see Section 5.9.6.2).

In addition to the stereospecificity, NAD(P)H-linked dehydrogenases that convert >CHOH to >C=O differ in the apparatus that has evolved to catalyse proton transfer from substrate oxygen. That apparatus can centre round a zinc ion or an acid–base histidine, tyrosine or lysine.

The short-chain dehydrogenases, originally so called because the linear sequences then available rarely exceeded 250 residues, use tyrosine as the acid–base. As longer sequences with the same fold have become available, the name is something of a misnomer and the M_r range is now 250–350 residues for the core structure.[352] They have B-side (*proS*) stereospecificity for the coenzyme, which they bind in the *syn* conformation. Structurally, they adopt the same protein fold, much like a GH family, even though the degree of similarity of the linear sequences is low. A typical example is the sorbitol dehydrogenase of *Rhodobacter spheroides*,[353] which catalyses the interconversion of sorbitol (D-glucitol) and D-fructose, of L-iditol and L-sorbose and of galactitol and D-tagatose (Figures 1.6 and 1.7), with a preference for the substrate pair sorbitol–D-fructose. The structure is in accord with the proposed general mechanism for short-chain dehydrogenases,[352] which involves a hydrogen-bonded chain through the 2′-OH of the coenzyme to a completely conserved active site lysine, whose importance has also been established by site-directed mutagenesis experiments.

The aldose reductases are part of the aldo–keto reductase superfamily. The classifications of aldose–ketose reductases are vastly confused,[354] but aldose reductases differ from the short-chain alcohol dehydrogenases in the cofactor stereochemistry, transferring the *proR* (A-side) hydrogen of NAD(P)H and binding the cofactor in the *anti* configuration. Hydride is transferred and to the *re* side of the carbonyl of the aldose substrate (*i.e.* in the oxidation of the alditol the *proR* hydrogen is removed) and the active site acid–base is a tyrosine, but there appears to be no pyranose ring-opening catalytic machinery: the enzyme binds the open-chain form of the sugar.

Much attention has been paid to the mammalian enzyme, since its operation in hyperglycaemic patients contributes to complications of diabetes. Its structure in complex with cofactor does not have the usual nucleotide binding fold (Rossman fold).[355] The crystal structure of the apoenzyme supports kinetic measurements which suggest that after binding of coenzyme the binary complex undergoes a large conformational change before binding of reductant: comparison of the NADP- liganded enzyme with apoenzyme reveal that a 12-residue loop of protein closes over the cofactor after binding, by moving up to 17 Å as a body.[356] The loss of NADPH is the rate-determining step in the physiological direction of aldose reduction, which is identified as the opening movement of the loop. The crystal structure of the monomeric human enzyme[357] in complex with an aldose reductase inhibitor supports the identification of an acid–base catalytic residue as Tyr (rather than His).[358]

A considerable medicinal chemistry effort has gone into devising inhibitors of aldose reductase: their inhibition shows a dominant or even exclusive anticompetitive component when measured as inhibitors of aldose reduction, since they form tight complexes of the form E.NAD.I;[359] the inhibition is competitive when alditol oxidation is measured.

Another enzyme in the same family is xylose reductase of *Candida tenuis*, for which a structure is available,[360] and which can use either NAD^+ or $NADP^+$.[361] Site-directed mutagenesis of this enzyme suggests an important, but not critical, role for an active site histidine.[362] The His113Ala mutant had 10^3-fold lower activity than wt, but like wt had a sigmoid pH–rate profile, with the decrease in k_{cat}/K_m[xxix] to higher pH governed by a pK_a of 8.8 (wt) and

[xxix] K_m values refer to the concentration of carbohydrate substrate to give half-maximal rate at saturating concentrations of the coenzyme.

7.6 (mutant). Use of NADD surprisingly gives a larger primary KIE [$^{D}(V/K)$] with wt (2.5) than mutant (1.6). There is no solvent isotope effect on either wt or mutant, immediately indicating that the elaborate sequence of proton transfers proposed on structural grounds does not take place in the chemical step, if it takes place at all. The enzyme has broad specificity and the $^{\rho}(V/K)$ obtained with *meta*-substituted benzaldehydes for the mutant (0.8) is lower than that for wt (1.7). Taken together, these data indicate that the mutation has altered transition-state structure, probably making it earlier. It was suggested that the role of the active site His was to hydrogen bond to the other lone pair of the oxygen of the carbonyl substrate, not to the acid–base tyrosine.

In the very diverse long-chain alcohol dehydrogenase family, the base appears to be the conserved lysine, since the enzymes are Zn^{2+}-independent but the tyrosine is not fully conserved.[363] The enzymes have chain lengths of 340–550 amino acids. The family includes glyceraldehyde-3-phosphate dehydrogenases and a number of enzymes which carry out four-electron oxidations, such as UDP-glucose dehydrogenase, which converts UDPGlc to UDPGlcA. The structures of binary and tertiary complexes of the *Pseudomonas fluorescens* mannitol dehydrogenase with mannitol and NAD^{+} have been solved;[364] the nicotinamide conformation is *syn* and the hydride transfer to the *si* face (B-side specificity). Mannitol is the C2 epimer of the glucitol shown in Figure 6.62 and it is easy to see how the configuration change could alter the stereochemistry. An ingenious series of pH and isotope studies delineated the mechanism.[365,366] The pH–k_{cat}/K_m profiles for mannitol oxidation and fructose reduction are

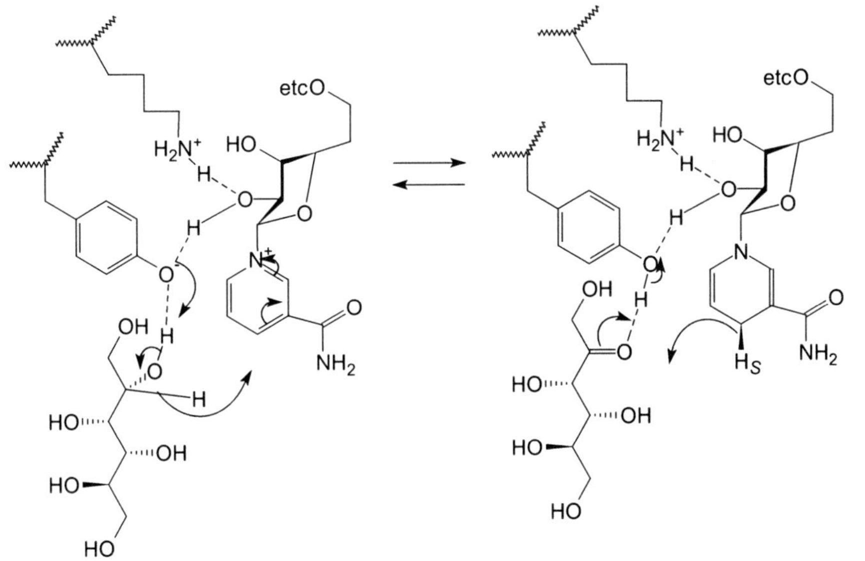

Figure 6.62 Mechanism of short-chain alcohol dehydrogenases, illustrated with bacterial sorbitol dehydrogenase (the proposed mechanism[352] also involves structured proton transfers beyond the lysine, for which there is no evidence beyond structural suggestion).

almost mirror-images of each other, sigmoidal with a pK_a of 9.3 ± 0.2, with mannitol oxidation increasing and fructose reduction decreasing as the pH increases. Oxidation of [2-^{2}H]mannitol exhibited a value of $^D(V/K) = 1.8$ between pH 6 and pH 9.5, which fell to 1.0 above that pH, whereas that for fructose reduction, measured with NADD, was about 2 between pH 6 and 9, falling to 0.83, the equilibrium isotope effect for carbonyl reduction at pH > 10. There were significant solvent isotope effects on both fructose reduction and mannitol oxidation. The ionisation in the pD–rate profile of mannitol oxidation was shifted by about 0.4 units more alkaline compared with the pH–rate profiles, as expected for the ionisation of an enzyme group with a fractionation factor around 1.[xxx] There was no solvent isotope effect on substrate isotope effects. Linear proton inventories for mannitol oxidation at pL 10 (where there was no substrate isotope effect) showed that the significant solvent isotope effect on both k_{cat} and k_{cat}/K_m arose from a single proton, $\varphi = 0.43$. These data indicate that hydride transfer and proton transfer are distinct, with the hydride transfer being partly reversible.

Use of histidine as an acid–base is exemplified by glucose-6-phosphate dehydrogenase (Figure 6.63); in mammals this is the first enzyme of the pentose phosphate pathway. It oxidises the β-anomer of the substrate, transferring the anomeric hydride to the B-side of the coenzyme. The existing sequences are substantially homologous, so the dimeric *Leuconostoc mesenteroides* has been the vehicle for mechanistic study. The bacterial enzyme is relatively indifferent to phosphorylation of the coenzyme, rates with NAD^+ being about an order of magnitude lower than for $NADP^+$. As expected, the coenzyme is bound in the *syn* conformation.[367] Mutation to Asn of the catalytic His240 reduces the rate by 10^5-fold and of the hydrogen-bonded Asp177 by 10^2-fold.[368] The pK_a of the wt enzyme was measured by NMR as 6.4,[369] suggesting that the Asp–His dyad seen in the structure does not affect pK values.

The so-called medium-chain alcohol dehydrogenases are Zn^{2+}-dependent. Examples include the medium-chain sorbitol dehydrogenases (the short-chain ones use Tyr as acid–base; see above) (Figure 6.63). A crystal structure of the enzyme from *Bemisia argentifolii* (a silverleaf whitefly that uses sorbitol as protectant against heat stress) indicates a tetrahedral Zn^{2+}, with three protein ligands (Cys, His and Glu) and a water which is displaced on binding straight-chain sorbitol or fructose. The coenzyme was not observed in the structure, but was modelled into the active site in the *anti* configuration.[370] A structure of the tetrameric human enzyme complexed with NAD^+ was solved, however, and the coenzyme indeed adopts an *anti* configuration about the C–nicotinamide bond.[371] On the basis of the X-ray structure of the complex with a commercial sorbitol dehydrogenase inhibitor, a mechanism was proposed in which the Zn^{2+} became transiently pentacoordinate, with O1 and O2 of the substrate acting as ligands and the carboxylate group released, but there is no other evidence for this. Both enzymes had well-developed Rossman folds for

---

[xxx] The shift arises from the low fractionation factor for H_3O^+ ($= 0.7$), $3\log_{10}(0.7) = -0.46$.

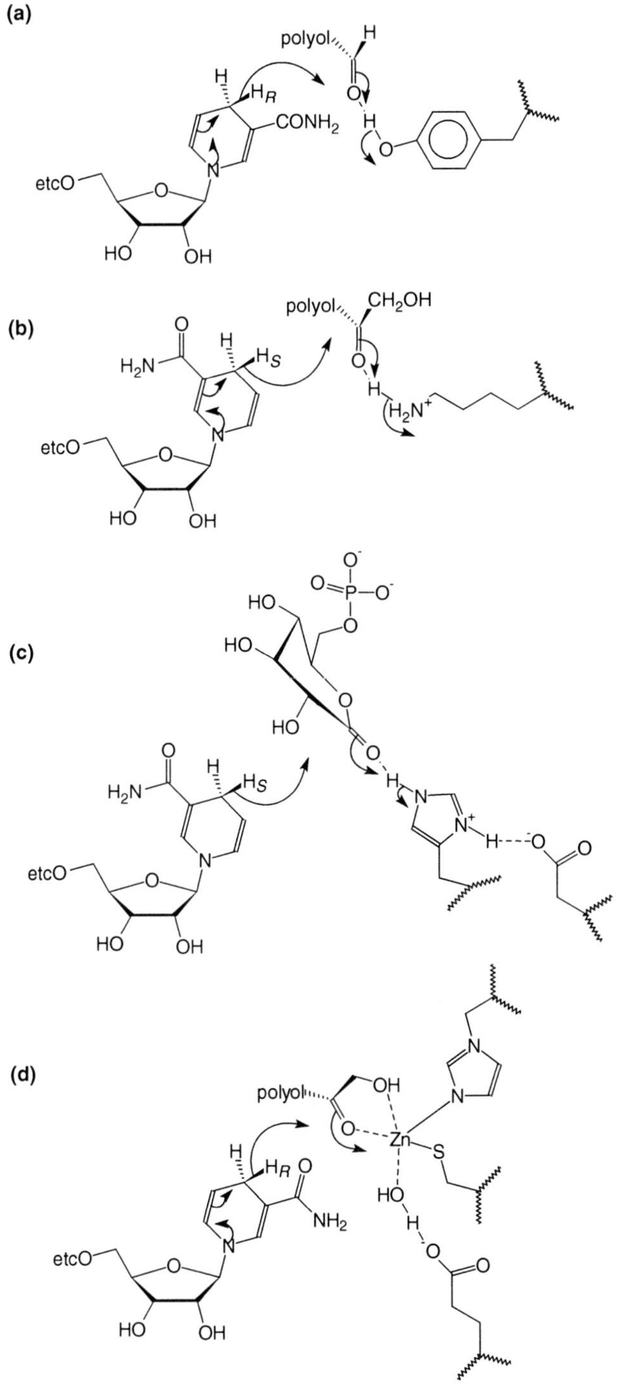

(a)

(b)

(c)

(d)

Figure 6.63 Substrate activation mechanisms and stereochemistry (shown in the sense of ketone reduction) for (a) aldose reductases, (b) long-chain dehydrogenases, (c) glucose-6-phosphate dehydrogenase and (d) sorbitol dehydrogenase. The glutamate ligand to Zn^{2+} is shown removed in the complex with fructose, but this is speculative.

coenzyme binding. There was some indication from molecular modelling studies that Zn^{2+} might assist ring opening and closing.

Xylitol dehydrogenase converts xylitol to the 2-ketopentose xylulose and the tetrameric enzyme from *Galactocandida mastotermitis* has been shown to possess one essential Zn^{2+} per monomer. As expected, binding is ordered with the cofactor binding first; however, binding of carbohydrate is so weak that a "Theorell–Chance" kinetic mechanism obtains (*i.e.* one in which there is a bimolecular reaction between E.NAD$^+$ and xylitol, without detectable E.NAD$^+$.xylitol or E.NADH.xylulose complexes).[372]

A specialised member of the short-chain reductase family is UDP-galactose 4-epimerase.[373] This interconverts UDP-glucose and UDP-galactose and contains one mole of tightly bound NAD$^+$ per active site. The *E. coli* enzyme is dimeric and the NAD$^+$ cannot be removed without denaturing the enzyme, whereas it can be removed from the human enzyme. Another arrangement is observed in yeast, where the galactose 1-epimerase forms part of the same polypeptide chain as the UDP-galactose 4-epimerase, with the dimeric enzyme forming a V-shape with the two active sites are 50 Å apart.[374]

All mechanistic evidence points to a remarkable oxidation–reduction–oxidation mechanism, in which a C4 of the hexose is transiently oxidised to a ketone and then reduced again from the other side. If [4-^{3}H]UDP-glucose is incubated with the enzyme, the tritium is distributed between enzyme and substrate, but tritium is lost from [3-^{3}H]UDP-glucose neither to enzyme nor solvent. These data exclude a potential C3 oxidation, C4 enolisation mechanism. The NAD$^+$ and the uridine diphosphate portion of the UDP sugar are tightly held with the NAD$^+$ in the *syn* conformation. The importance of the binding of uridine diphosphate in assembling the active site can be gauged from the fact that in the binary complex E.NADH the NADH is in the *anti* conformation and only adopts the same *syn* conformation expected for a short-chain aldehyde dehydrogenase in the ternary E.UDP.NADH complex. The cofactor can be reduced in the E.NAD$^+$.UDP complex by hydride transfer to the *si* face of C4 (B-side specificity, as expected for short-chain alcohol dehydrogenases) by reducing agents such as $NaBH_3T$; likewise, the E.NADH complex can be oxidised by 4-ketohexopyranoses. These data compel a mechanism in which the hexose can rotate about the pyrophosphate linking it to the nucleotide and thereby present either face of the transiently formed 4-keto-UDP sugar to the NADH cofactor. The X-ray structure of a mutant of the *E. coli* enzyme which bound both UDPGlc and UDPGal in abortive complexes with E.NADH showed the two sugars accurately placed for hydride transfer from position 4, using the standard Tyr–2′-OH–Lys proton relay of a short-chain dehydrogenase (Figure 6.63) (Lys153 and Tyr149 in the case of the *E. coli* isomerase). The sugars were held in different positions by Asn179 and Tyr299, which H-bonded to 2OH in the *galacto* complex and 6OH in the gluco complex.

Other epimerases using this mechanism appear to be UDPGlcNAc 4-epimerase (indeed, the UDPGal 4-epimerase from some species has a small amount of this activity), UDPXyl 4-epimerase and UDPGlcA 4-epimerase. The mechanism is not confined to transiently oxidising position 4 – an enzyme

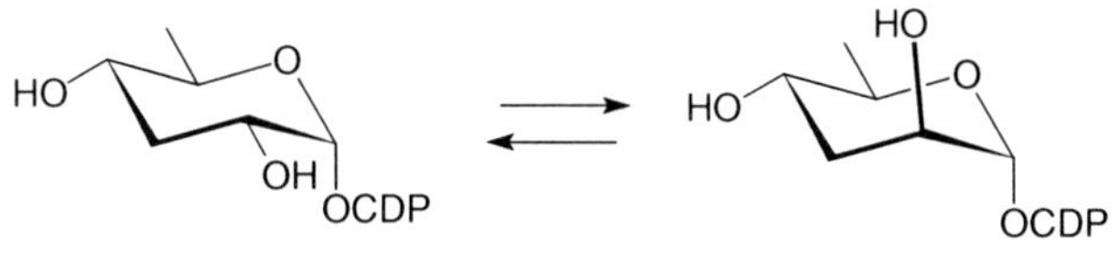

CDP-paratose CDP-tyvelose

Figure 6.64 Reaction catalysed by CDP 2-epimerase.

involved in lipopolysaccharide biosyntheses, CDP tyvelose 2-epimerase, which interconverts the eponymous substrate and CDP paratose, proceeds by the same mechanism (Figure 6.64).

6.6.2 Oxidations of Diols

The oxidation of diols $R_1R_2C(OH)–C(OH)R_3R_4$ to two carbonyl fragments R_1R_2CO and R_3R_4CO by aqueous periodate has long been employed productively in carbohydrate chemistry (Figure 6.65). The reaction is carried out in weakly acidic medium ($1 < pH < 8$, generally pH 3–5) at room temperature or below, and is usually monitored by measuring the decrease in periodate absorbance at 223 nm. Until a couple of decades ago, it played a major role in the elucidation of the structures of novel sugars, since in addition to periodate consumption, small molecules such as formaldehyde (from $HOCH_2–CHOH–R$ groups) and formic acid (from the central carbon of structures of the type R–CHOH–CHOH–CHOH–R′) could be assayed by sensitive colour reactions. It is still extensively used in the modification of polysaccharides, since the polymeric structure is preserved, but reactive aldehyde groups are introduced.

The observed first-order rate constant for the disappearance of the periodate absorbance in the presence of excess glycol show a hyperbolic dependence on glycol concentration, suggesting the formation of an intermediate present in appreciable concentration, *i.e.* eqn. (6.4) holds[xxxi],[375] where K is the dissociation constant of the complex.

$$k_{obs} = k_{max}[\text{glycol}]/(K + [\text{glycol}]) \quad (6.4)$$

The parameter K can also, like K_m values in enzyme kinetics, be an agglomeration of rate constants that refers to a steady state, rather than an equilibrium situation. Rapid reaction kinetics, in the case of propane-1,2-diol, established that the intermediate was kinetically competent.[376] The reaction with pinacol, where K is very large and the reaction is cleanly second order over a wide range of pinacol concentration,[377] exhibited general acid and general

---

[xxxi] The form of this equation has been altered to bring out the parallel with enzyme kinetics: the papers by Bunton and co-workers used association constants, rather than the dissociation constants, which conceptually resemble K_m values.

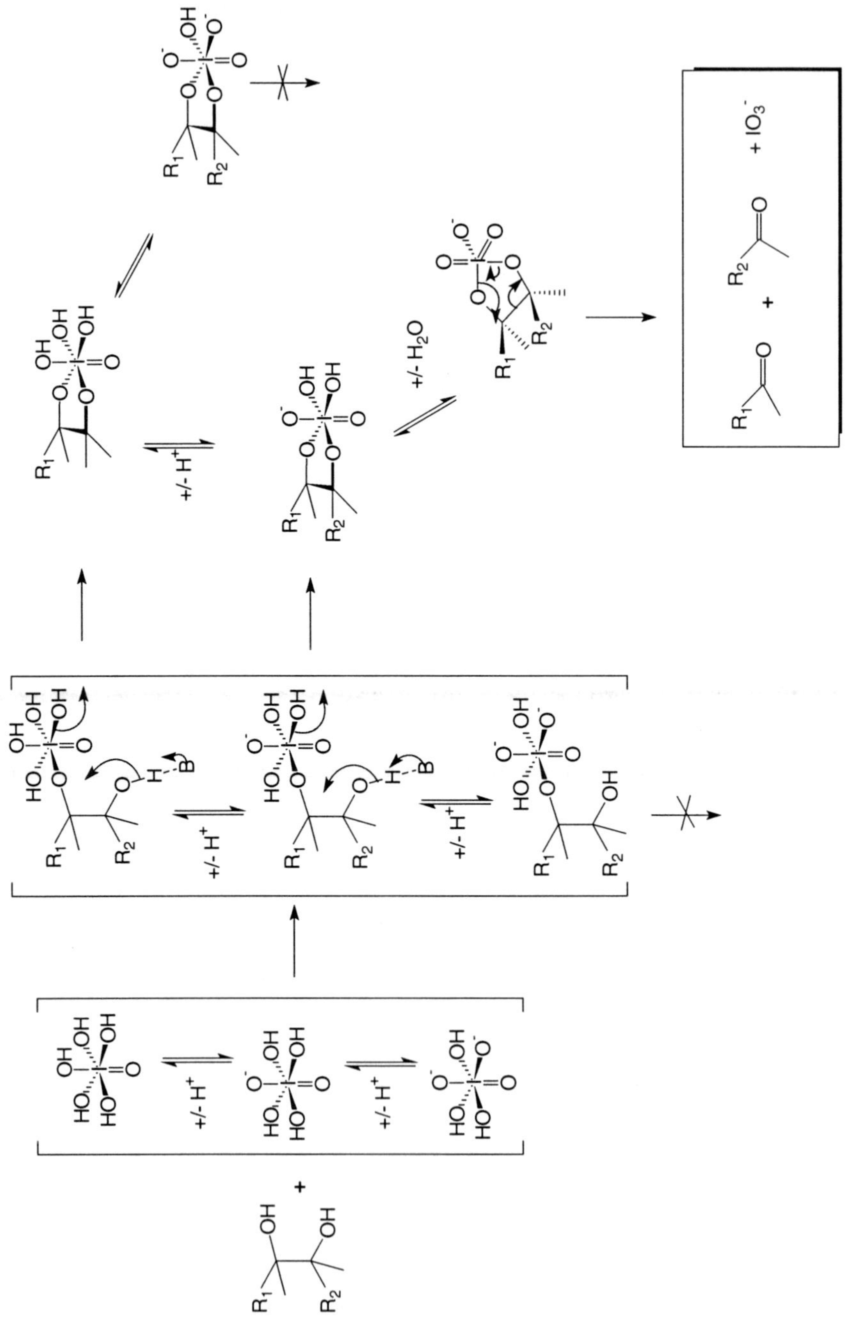

Figure 6.65 Mechanism of periodate oxidation of diols.

base catalysis.[378] The dependence of the second-order rate constant, k_{max}/K, on pH is bimodal, with maxima around pH 2 and 8: reasonably, this requires two intermediates.[379]

The reaction also proceeds with amines of the form >C(OH)–C(NHR)< and >$C(NHR_1)$–$C(NHR_2)$<, where R can be hydrogen, although the reaction takes place through the neutral amine rather than its cation, and therefore requires a higher pH (9–11).[380] The initially formed imines >C=NR hydrolyse spontaneously to the carbonyl compounds. The kinetics of the reaction parallel those of diols, with intermediate complexes.

The mechanism that was advanced to accommodate these observations involved acyclic monoester and cyclic diester intermediates of periodic acid. Under mildly acidic conditions, the conversion of the monoester to the diester was largely rate determining. In molecular terms, the observed general base catalysis was true general base catalysis of the cyclisation of the monoester anion, and the general acid catalysis was the kinetically equivalent general base-catalysed cyclisation of the neutral monoester. At high concentrations of general bases, either formation or decomposition of the cyclic ester could become rate determining. The decomposition of the cyclic ester monoanion is the rate-determining step at pH > 9 and the steep fall-off of rate with pH in this region indicates that the cyclic ester monoanion, rather than the dianion, is the reactive species.

I^{VII} in aqueous solution, free or coordinated to alkoxy ligands, can adopt trigonal bipyramidal or octahedral geometry. The cyclic ester dianion, which is necessarily hexacoordinate, is stable enough for NMR studies; moreover, cyclic triesters, such as that from 1,2-isopropylideneglucofuranose,[381] which cannot readily adopt a trigonal bipyramidal geometry, are also stable. These data confirm that the reactive species in periodate oxidations is the pentacoordinate cyclic ester monoanion.

The point has recently been made that the kinetic data which indicate that the cyclic intermediates are involved are also compatible with mechanisms in which the true reaction is a bimolecular one and the complexes of reactants are inert.[382] In support of this reinterpretation is the observation of an induction period in the reaction with propane-1,2-diol. Although the point is valid, there are two reasons why the Bunton mechanism of Figure 6.65 should be maintained. The first is the rapid-reaction work,[376] which clearly showed the kinetic competence of an absorbing complex. The second is Occam's razor, which would reject the "parasitic equilibria plus unspecified bimolecular process" in favour of the simpler Bunton proposals.

6.7 ELIMINATIONS AND ADDITIONS

6.7.1 General Considerations

More O'Ferrall–Jencks diagrams were devised to deal with the varying effects of the same substituent on differing types of elimination reaction,[383] only later being applied to acid–base catalysis.[384] Figure 6.66 represents such a diagram,

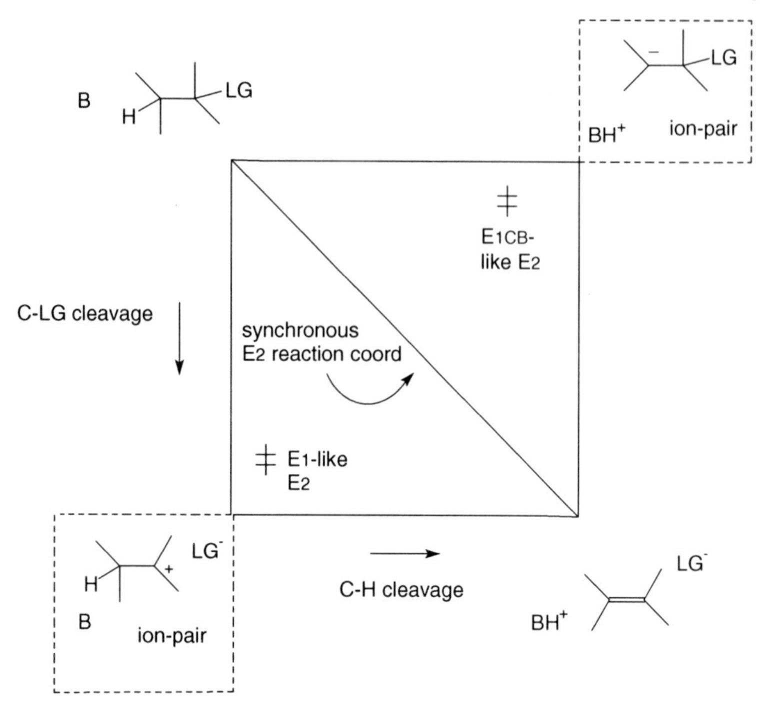

Figure 6.66 More O'Ferrall–Jencks diagram for elimination reactions (like the general acid catalysis of acetals, this is a class e reaction and is drawn in the same way as Figure 3.16).

in which energy can be represented as contours, although in qualitative discussions contours serve no purpose and are usually omitted. "Wings" (the outer dotted lines) are added to represent the barrier to intermediates leaving the same solvent shell (or enzyme active site).

Ingold's two classical elimination reactions are little represented in carbohydrate chemistry. The *E*2 reaction, with C–H and C–LG cleavage synchronous, involves an exactly NW–SE reaction coordinate. The *E*1 reaction, where C–LG is completely broken and the carbenium ion is solvent equilibrated, involves the system travelling the West edge and then exiting the solvent cage. Eliminations from glycosyl derivatives and electrophilic additions to glycals, however, although not involving solvent-equilibrated 2-deoxyglycosyl cations (even in water: see Section 3.6) involve a reaction coordinate along the Western edge but remaining within the solvent cage.

Reversibility of the transits of the system along the Western edge, but within the same solvent shell or active site, is "internal return", as discussed in Chapter 3. Leaving group atom scrambling with a multidentate leaving group such as carboxylate is a sufficient, but not a necessary, condition for internal return: isolation of R–^{18}O-CO–Ar from partial reaction of R–O–C^{18}O–Ar requires the ion pair to live long enough for the carboxylate group partly to rotate, and also for return to occur. Reversibility of the transits along the Northern edge of the diagram, within the same solvent shell/enzyme active site, can likewise be demonstrated by isotope exchange, with hydrogen incorporated from solvent if the base is polyprotic, such as a primary amine, or if the complex of carbanion and conjugate acid lives long enough that the conjugate acid can exchange its single proton with solvent. Where the deprotonation is reversible, the term $E1_{CB}$(r) is used. Both the irreversible $E1_{CB}$(irr) reaction and $E1_{CB}$-like $E2$ reaction exhibit primary hydrogen isotope effects [unlike $E1_{CB}$(r) reactions], but the $E1_{CB}$(irr) reaction is characterised by the absence of any leaving group effects, such as kinetic isotope effects or large electronic effects (although there will be a small electronic effect of the leaving group on the pK_a of the CH bond).

The More O'Ferrall–Jencks diagram does not take into account reaction stereochemistry. For proton removal, orbital overlap is maximised when the carbon sp^3 and base lone pair have a common axis going through the centre of the s orbital of the transferred proton and for leaving group departure the leaving group orbital and the carbon sp^3 orbital are likewise collinear. If these two processes are to be coupled to produce an olefin, *i.e.* if there is to be a partial formation of a π bond, then these two axes should be parallel. Qualitative MO considerations thus indicate that *syn* and *anti* eliminations and additions are favoured and give some indication that the *anti* pathway is the more likely. If we approximate the breaking C–H bond by an sp^3 orbital containing a lone pair and the breaking C–X bond by a σ^* orbital derived from carbon and leaving group sp^3 orbitals, then HOMO–LUMO overlap is maximised in an *anti* transition state, where the large lobe of the HOMO can overlap with the large lobe of the LUMO (Figure 6.67).

Anti elimination is also favoured because it is the least motion pathway and because the *syn* pathway can also be disfavoured by steric repulsion between base and leaving group. Despite these *a priori* reasons why *anti* elimination is favoured, experimentally, there seems to be a only slight preference for *anti* elimination, even for classical $E2$ reactions with central transition states [such as Hofmann eliminations of (acyclic) alkyltrimethylammonium hydroxides].[385] It was at one time thought that $E1$-like reactions were *anti* and $E1_{CB}$-like reactions were more likely to be *syn*,[386] but electrophilic additions to glycals are rarely *trans* and they go through transition states in the SW corner of the More O'Ferrall–Jencks diagram.

The direction of approach of nucleophile or electrophile to a double bond – or the microscopic reverse in an elimination reaction – has a preference for taking place with the "behind and above" approach – at the Bürgi–Dunitz angle in the case of Michael additions. It predicts a *trans*-diaxial addition in the case of six-membered rings. Again, it appears to be a least-motion effect, which is easily overridden. There is some evidence for its reality in the case of ammonia and azide Michael addition to 1-nitroglycals, where protected nitroglucals yield mannosamine derivatives exclusively from variously protected nitroglucals, whereas 3-acetyl-4,6-benzylidene nitrogalactal gave a *galacto* derivative with azide but a *talo* derivative with ammonia.[387] The attacks are, however, *syn*, in the sense that the proton is added equatorially, but this

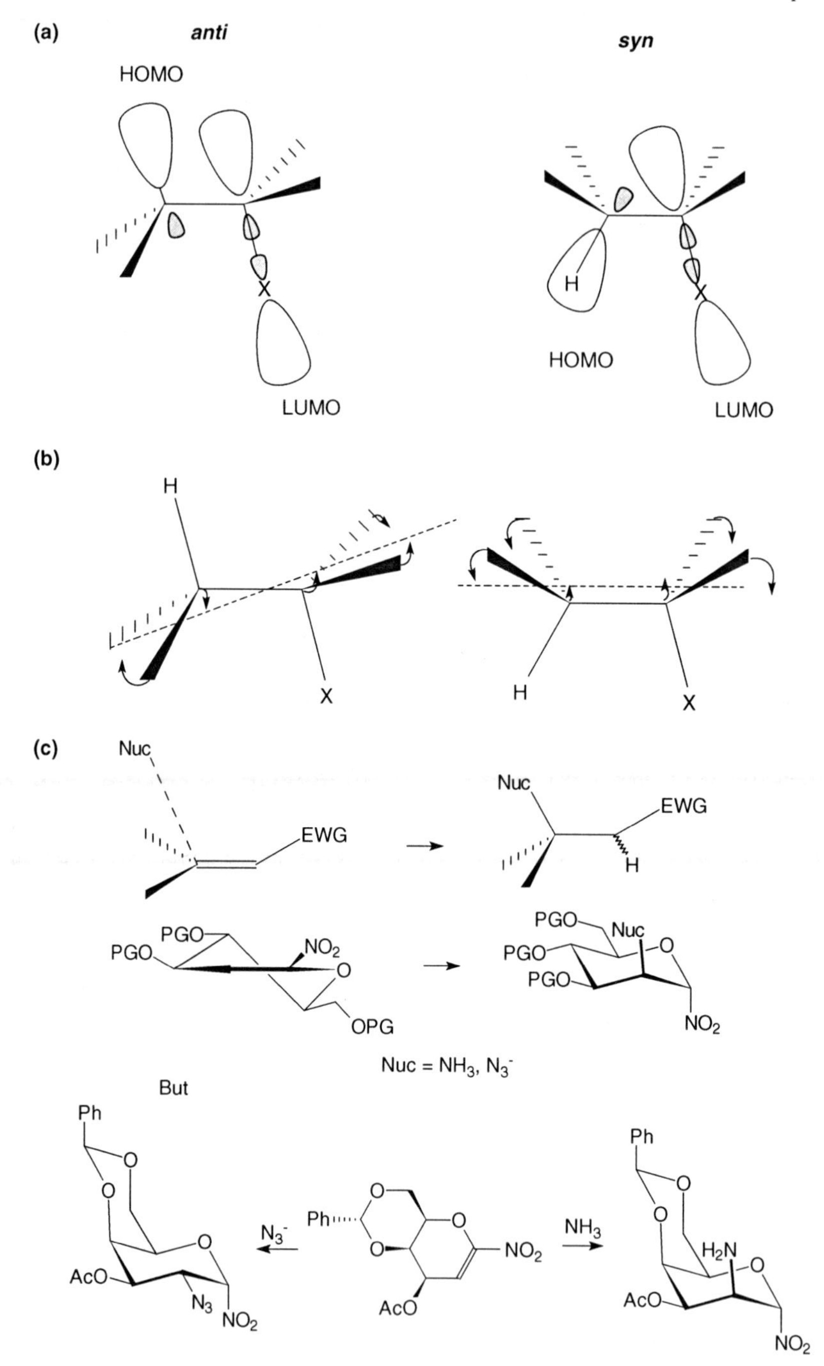

Figure 6.67 Stereoelectronics of eliminations and additions. (a) Orbital overlap favouring *anti* elimination/addition. (b) Cartoon of least-motion arguments favouring *anti* elimination. The dotted line represents the reference plane, which is perpendicular to the plane of the paper. (c) Bürgi–Dunitz-like approach in Michael additions and examples.

could be a reflection of thermodynamic control and the strong anomeric effect of the nitro group.[388]

6.7.2 Electrophilic Additions to Glycals

Addition of an electrophile E–X to a glycal occurs with attachment of E to C2, with generation of an oxocarbenium ion–X^- encounter complex. There are therefore four stereoisomers of the adduct possible. The norm is for mixtures of stereoisomers to be formed, with any predominance of the *trans*-diaxial product, as dictated by elementary texts, a rarity.

Under strongly acidic conditions, glycals add alcohols. The acid of choice is triphenylphosphine hydrobromide. The stereochemistry at C1 appears to be dictated by the position of anomeric equilibrium, with axial products predominating, except in the case of tribenzylallal, where they would be disfavoured by 1,3-diaxial interactions. Running the reaction in deuterated solvents gave a mixture of axial and equatorial deuterium at C2, with between 5:1 and 2:–1 predominance of equatorial protonation, except for dibenzyl-3-deoxyglucal.[389]

Careful analysis of the products of the addition of halogens and pseudohalogens to glycals has led to a definitive picture[390] of the factors yielding various stereochemistries and products (Figure 6.68), as follows.

Iodine azide (from ICl and excess NaN_3) forms an initial iodonium ion in acetonitrile, so that only *trans* products are formed, with between a 4:1 and 5:1 preference for the *trans*-diaxial product (*e.g.* 87% tribenzyl-2-iodo-2-deoxy-α-mannopyranosyl iodide to 13% 2-iodo-2-deoxy-β-glucopyranosylazide from tribenzylglucal).

Bromine in dichloroethane yields an unbridged bromodeoxyglycosyl bromide ion pair, since bromine is less capable of bridging than the less electronegative iodine. The ion pair collapses exclusively to α-products, but the initial electrophilic attack can occur from either face in a way which varies with substitution. Initial attack from the β-face is strongly disfavoured in tribenzylgalactal (>99:1) for obvious steric reasons, but to a lesser extent in triacetyl- and tribenzylglucals (80% and 60% initial α attack), compared with 2-phenyl- and 2-benzyloxydihydropyran (95% and 86% attack *cis* to the substituent). Field effects of the 3- and 4-substituents are obviously involved, but it is not known exactly how. Added tribromide ion has the effect of suppressing initial electrophilic α attack at C2, while leaving the exclusive specificity of nucleophilic α attack at C1 unchanged.

Addition of bromine in methanol gives exclusive *trans* addition (to 2-deoxy-2-bromomethyl glycosides) in approximately equal amounts, except for tribenzylgalactal, which, as expected from the steric restrictions imposed by O4, gives tribenzyl-2-bromo-2-deoxymethyl β-galactoside and tribenzyl-2-bromo-2-deoxymethyl α-L-talopyranoside in a 2.3:1 ratio. Only in the case of the bromomethoxylation of triacetylglucal was it possible to intercept the oxocarbenium ion with azide ion. The Jencks clock, using a value of $7 \times 10^9\,M^{-1}\,s^{-1}$ for the diffusional anion–cation recombination in methanol, then gave a lifetime of 6 ns for the triacetyl-2-bromo-2-deoxy-glycosyl cation in methanol.

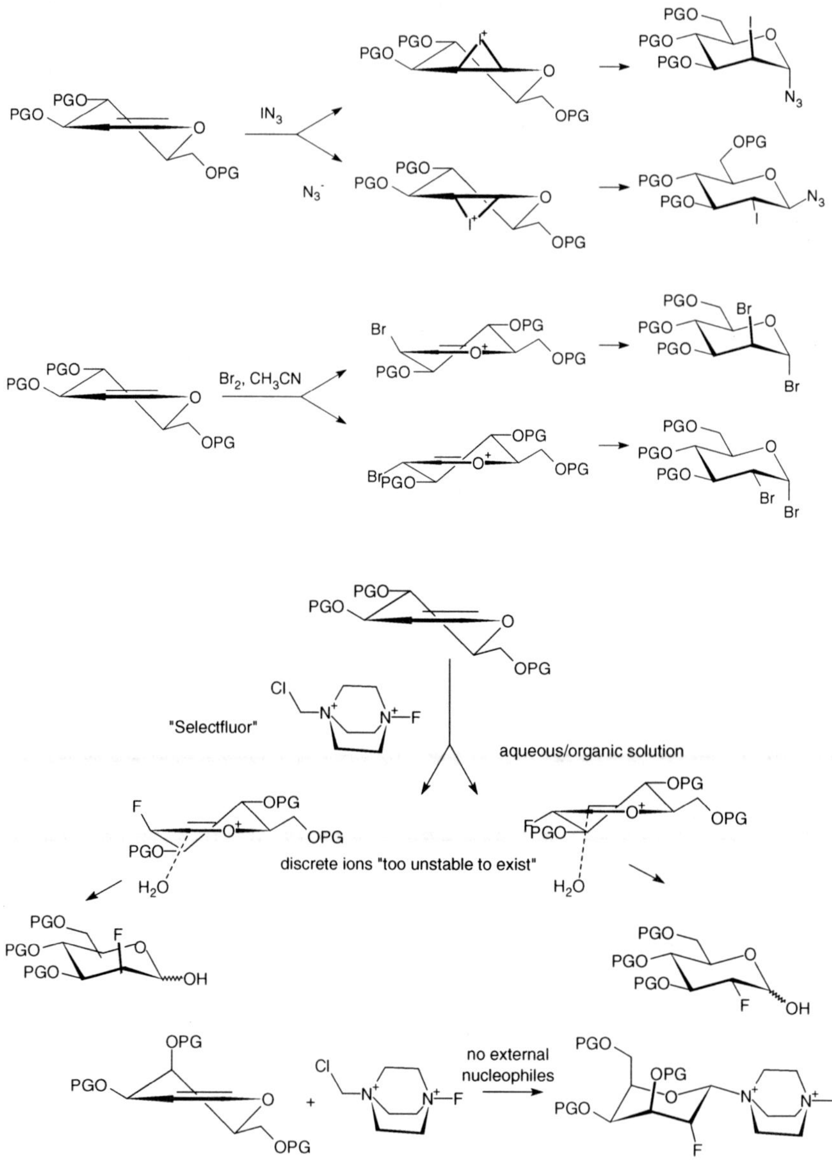

Figure 6.68 Some electrophilic additions of glycals. For details of relative proportions of products, see text.

Why the lifetime of the acetyl-substituted species should be longer than that of the benzyl-substituted species, so that it alone can be timed by the Jencks clock, is not clear.

Electrophilic fluorination of glycals is a key step in the syntheses of Withers glycosidase inactivators and related potential pharmaceuticals.

N-Fluoro-quaternary nitrogen species such as Selectfluor™ are now[xxxii] used. Selectfluor can be used in partly aqueous solutions, and the products are then sugars, a 3:2 mixture of 3,4,6-protected 2-fluoro-2-deoxymannose and 3,4,6-protected 2-fluoro-2-deoxymannose from protected glucal, and exclusively 3,4,6-protected 2-fluoro-2-deoxygalactose from 3,4,6-protected galactal, *i.e.* the axial O4 in the *galacto* series directs equatorial fluorination.[391] In the *galacto* series, a quaternary ammonium intermediate formed by clean *syn* addition to the α-face has been characterised: it appears that the tertiary amine is displaced in a subsequent step is displaced by ambient nucleophiles[xxxiii].[392] If the ambient nucleophiles are not present, the quaternary amine can anomerise (presumably via the fluorides: use of BF_4^-, rather than triflate, as the Selectfluor counterion can result in the formation of glycosyl fluorides in good yields).[393]

Although Lemieux *et al.*'s use of nitrosyl chloride additions to glycals to introduce nitrogen at C2 dates from 1968,[394] the basic idea of simultaneously introducing functionality at C2 and stereodirecting glycoside formation at C1 was not elaborated until many years later, when a plethora of such protocols appeared.[395] In addition to the additions of iodine azide, the addition of ArSCl gives intermediates which can be intercepted to give 2-thioglycosides, which can then be desulfurised to 2-deoxyglycosides.[396]

6.7.3 S_N' Reactions at the Anomeric Centre – the Ferrier Rearrangement[397]

The nucleophilic substitution with allylic rearrangement of 3-substituents of glycals is known to authors other than Ferrier himself[397] as the Ferrier rearrangement (Figure 6.69), since the first modern example was the production of a *p*-nitrophenyl glycoside from tri-*O*-acetyl D-glucal.[398] However, the production of D-*threo*-hex-2-enose from hydrolysis of triacetylglucal (itself readily prepared from "acetobromoglucose" and zinc dust[399]) in hot water was reported in 1925.[400] Although similar transformations can be achieved by other routes, in its classical formulation the reaction probably involves a delocalised oxocarbenium ion formed by departure of the 3-substituent and its capture by a nucleophile or "exploded" S_N2' transition states similar to the "exploded" S_N2 transition states in displacements at the anomeric centre. The question of whether oxocarbenium ions delocalised over four atoms are "too unstable to exist" has not been addressed, but compared with a glycosyl cation, the presumed intermediates in the Ferrier rearrangement are stabilised by the absence of an electron-withdrawing 2-substituent and additional conjugation into the vinyl group. Such conjugation may additionally increase barriers to reaction with nucleophiles disproportionately to its effect on ion stability (see Section 3.4.2). Proton loss from the intermediate cation in the furanose series,

---

[xxxii] The previous reagent, acetyl hypofluorite, CH_3COOF, had to be freshly made as a dilute gas by a procedure involving elemental fluorine.

[xxxiii] Although glycosyl quaternary ammonium salts – particularly with a 2-fluoro substituent – are stable, these derivatives have a leaving group pK_a of around 3 and undergo a reaction which is accelerated in polar solvents.

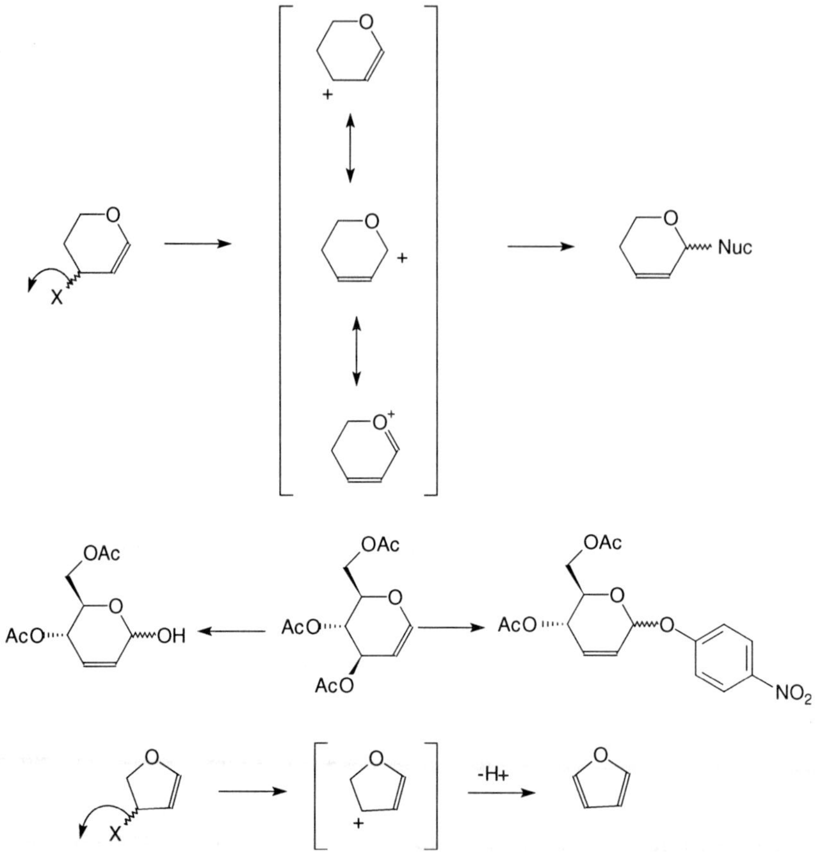

Figure 6.69 The Ferrier rearrangement. Displayed are the basic reaction, the two first examples and the diversion of the reaction to furans in the furanose series.

favoured by the aromaticity of the furan product, explains why the classical Ferrier rearrangement is confined in practice to the pyranose series.

For maximum delocalisation of the positive charge, the delocalised oxocarbenium ion must have all the ring atoms coplanar. The starting material can adopt the $^4H_5$, $^5H_4$, $^{3,0}B$ and $B_{3,0}$ conformations and the product the $^0H_5$, $^5H_0$, $^{1,4}B$ and $B_{1,4}$ conformations. Product conformations will be dominated by the anomeric effect in the usual way.

The observation that L-arabinal adopted the $^5H_4$ conformation exclusively and that the $^4H_5$ and $^5H_4$ conformations of D-glucal were comparably populated led Curran and Suh[401] to propose a "vinylogous anomeric effect", similar to the anomeric effect. Additional stabilisation of an electronegative substituent in an axial orientation was available from a p-type lone pair on oxygen, separated by a double bond, via overlap of the C–X σ* orbital and ψ2 of the enol ether (Figure 6.70). No experimental estimates of its magnitude are available, but it is likely to be smaller than the anomeric effect itself, not least because any electrostatic component is much smaller.

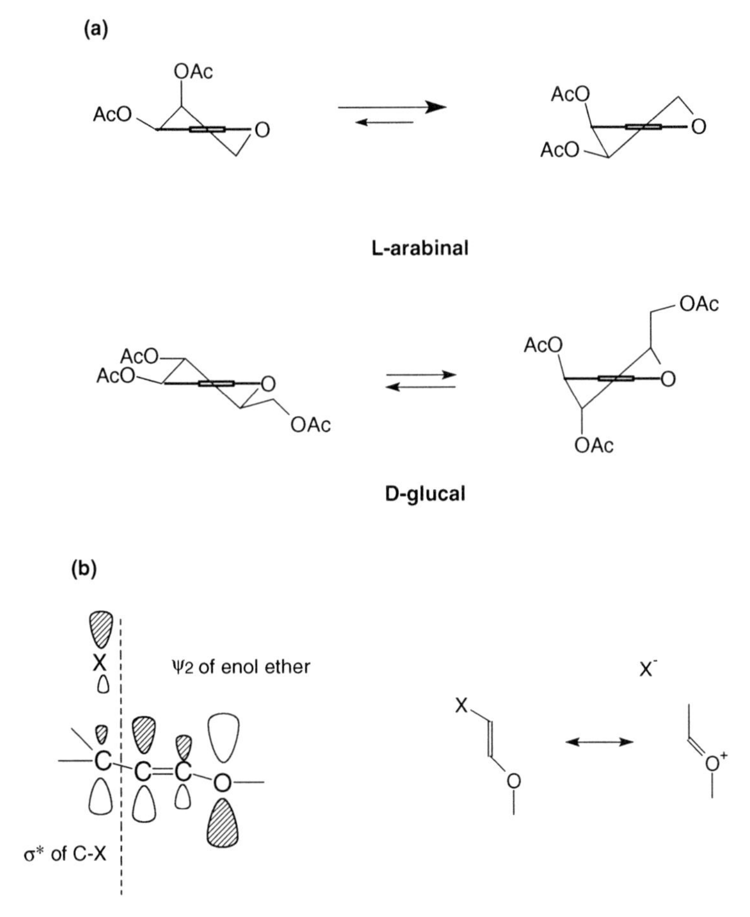

Figure 6.70 The vinylogous anomeric effect. (a) Experimental evidence. (b) Cartoon of overlap of ψ2 for an enol ether with the C3-X σ*. Different sizes of orbitals give an indication of coefficients. A valence bond representation of the effect is also shown.

Carboxylate leaving groups at C3 ($pK_a = 4$–5) or *a fortiori* sulfonate leaving groups ($pK_a \approx 0$) are sufficiently good that they will react with heating, but it is conventional to use Lewis acid catalysts such as boron trifluoride or stannic or titanium chlorides, and work at lower temperatures to avoid side-reactions. Protic acids are not generally used, because of the competing reaction of proton addition to C2, a normal reaction of enol ethers. With a Lewis acid in the mixture, there is some evidence that allal derivatives (C3 epimers of glucal derivatives) are the more reactive, possibly because the leaving group is pseudoaxial in the $^4H_5$ conformation which puts the bulky protected hydroxymethyl group equatorial.[402]

There are a series of reports that Ferrier reactions in the *galacto* series are more difficult than in the *gluco* series, although in the absence of kinetic data it is not clear whether the *galacto*-configured substrates are intrinsically less reactive or merely more prone to side-reactions.[403]

The reaction produces an anomeric mixture in which there is a preponderance of the pseudoaxial glycoside (α in the D-series). The product glycosides are much more labile than ordinary aldopyranosides, and so can equilibrate under the conditions of the reaction: the α predominance of *O*-glycoside products is a reflection of thermodynamic control. *C*-, *N*- and *S*-glycoside products show similar α-selectivity, however. Allyltrimethylsilane reacts with triacetylglucal and triacetylgalactal with titanium tetrachloride as the Lewis acid to give the allyl *C*-glycosides in >10:1 α:β ratio.[404] With a suitably vigorous Lewis acid, even deprotected glycals react: glucal, galactal, allal and rhamnal reacted with trimethylsilyl trifluoromethanesulfonate and phenols to give aryl *C*-glycosides and with allyltrimethylsilane to give allyl *C*-glycosides.[405] The allyl products in particular are most unlikely to be thermodynamic products and the α selectivity can be explained on least-motion arguments. On conversion from the planar cation to the $^{0}H_5$ conformation of the product, α attack requires the anomeric hydrogen to bend only ~30° out of the cation plane and β attack around ~90°. Alternatively, ALPH could be applied in a microscopic reverse sense, favouring the transition state for addition yielding a lone pair antiperiplanar to the incoming group.

Whereas the rearranged products from *O*-nucleophiles are formed under conditions of both thermodynamic and kinetic control, there is evidence that with *S*- and *N*-nucleophiles, the unrearranged, 3-substituted glycals are the thermodynamic products (Figure 6.71). Under normal conditions thiols give the expected 2,3-unsaturated α-glycosides,[406] but under forcing conditions the initial products rearrange to the C3-pseudoaxial ($^{0}H_5$) 3-alkyl- or arylthioglycals.[407] The eventual stereochemistry at C3 (triacetylglucal gives 3-alkylthioallals and triacetylgalactal, 3-alkylthiogulals), is clearly a manifestation of the vinylogous anomeric effect. The original workers[407] ascribed the final regiochemistry to the soft nature of the thiol nucleophile; however, why this should not also be manifest kinetically is not clear. Rather, it seems to the author that the absence of efficient geminal stabilisation of the 2,3-unsaturated glycoside (Section 3.4.1) in the sulfur case tips the equilibrium towards the glycal.

The reaction with azide[408] or 2,6-dichloropurine[409] shows similar features, with the initial 2,3-deoxyglycosyl derivative undergoing isomerisation under more drastic conditions to the 3-substituted glycal.

The reactions of organopalladium compounds with glycals are formally like Ferrier rearrangements, but the intermediates appear to be σ- and π-bonded palladium organometallics rather than stabilised oxocarbenium ions, so the reactions can be carried out on furanoid glycals without yielding furans.[397]

6.7.4 Epimerisations α and Eliminations α,β to the Carboxylates of Uronic Acids

6.7.4.1 Non-enzymic Epimerisation and Elimination. The non-enzymic removal of a carbon-bound proton α to an ionised carboxylate of a uronic acid

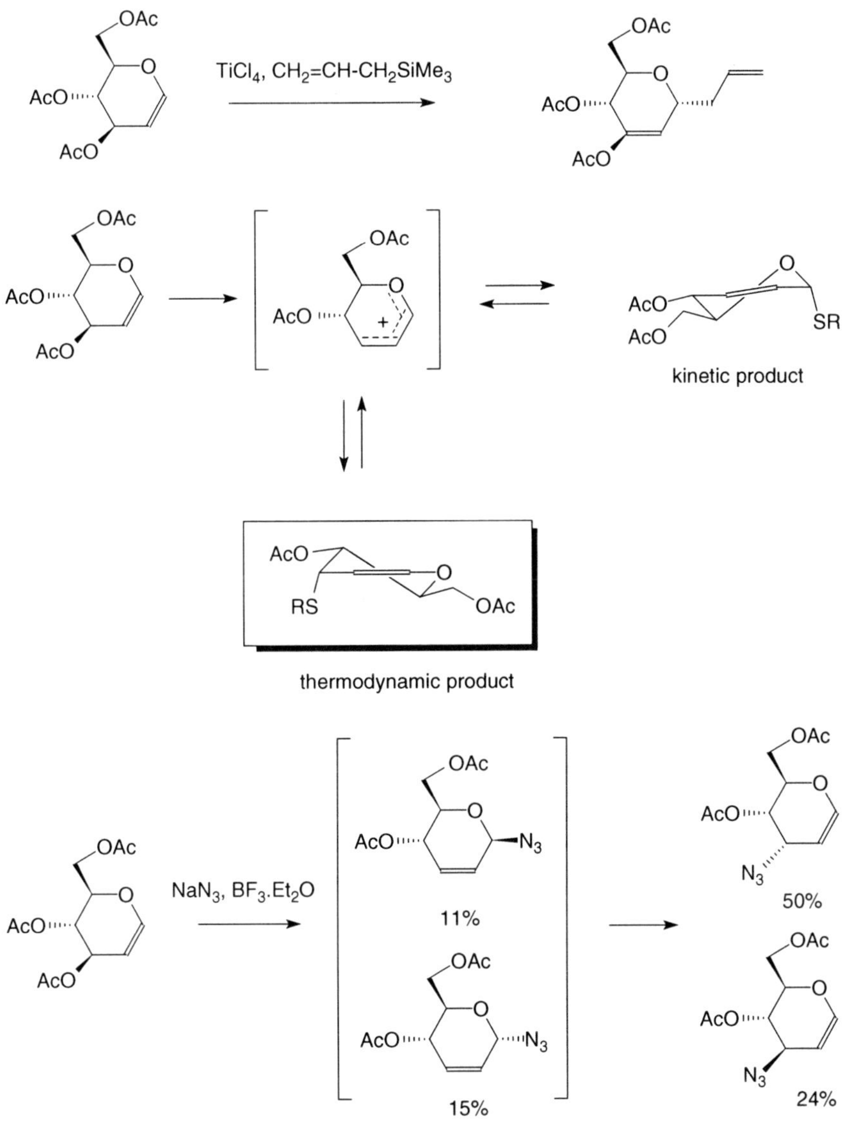

Figure 6.71 Examples of thermodynamic and kinetic control in the Ferrier rearrangement.

was first observed in the reaction of 2-*O*-(4-*O*-methyl-α-D-glucuronopyranosyl)-D-xylitol with 1.0 M NaOH at 150 °C. The substrate was a model for the removal of 4-*O*-methyl-α-D-glucuronyl residues from xylan during kraft cooking.[410] Epimerisation to 2-*O*-(4-*O*-methyl-β-L-iduronopyranosyl)-D-xylitol and elimination to 2-*O*-(4-deoxy-β-L-*threo*-hex-4-eneuronopyranosyl)-D-xylitol were observed (Figure 6.72). The epimerisation was somewhat faster than the elimination, and from the data a second-order rate constant of $9 \times 10^{-4}\ s^{-1}$ can

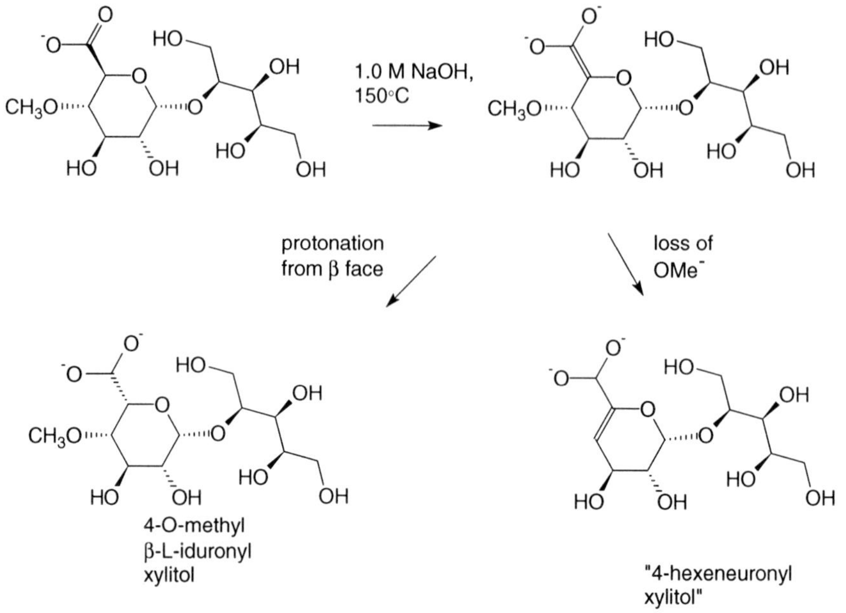

1.0 M NaOH, 150°C

protonation from β face

loss of OMe⁻

4-O-methyl β-L-iduronyl xylitol

"4-hexeneuronyl xylitol"

Figure 6.72 Non-enzymic generation of a glucuronate dianion.

be calculated for the ionisation, with the elimination occurring at a rate of $2.5 \times 10^{-4}\,s^{-1}$. The p$K_a$ of the C5–H bond in a uronate residue can be estimated as around 32.[411]

Analysis of these rate and pK_a data show that the C5–H5 bond of a uronate is a kinetically normal acid site, like OH and NH sites. There is essentially no barrier to proton transfer between such sites, thermodynamically favoured deprotonation occurring at the diffusion rate ($10^{10}\,M^{-1}\,s^{-1}$ at 25 °C), thermodynamically disfavoured deprotonation occurring at a rate of $10^{10-\Delta pK_a}$.[412] Therefore, the rate of deprotonation of a "normal" acid of pK_a 32 by OH^- (pK_a of the conjugate acid = 15.5) is predicted to be $10^{-6.5}\,M^{-1}\,s^{-1}$ at 25 °C. Arrhenius extrapolation of the experimental data for deprotonation of 2-*O*-(4-*O*-methyl-α-D-glucuronopyranosyl)-D-xylitol, using a range of activation energies encompassing all reasonable values of E_a (40–80 kJ mol^{-1}) gives rates $10^{-5.5}$–$10^{-7.5}\,M^{-1}\,s^{-1}$ at 25 °C. Therefore, C5–H5 ionisation of a uronate anion is that of a kinetically normal acid, which is surprising. Carbon acids which require extensive delocalisation of negative charge into electronegative substituents are usually deprotonated far more slowly than normal acids of the same pK_a – the effect is particularly pronounced with nitroalkanes. The effect, particularly the "nitroalkane anomaly", is generally attributed to imbalance in the transition state, with delocalisation and solvation of charge lagging behind proton removal; carbon acids whose anions do not have such delocalised charge (such as acetylenes or HCN) are kinetically normal.[413] By treating proton transfer and rearrangement/solvation in the rest of the molecule as separate processes, in multidimensional More O'Ferrall–Jencks diagrams, rates of deprotonation of carbon acids can now be predicted.[414] That the C5–H5 bond of a uronate is a kinetically normal acid, therefore, implies that the additional solvation and geometric movement in the carboxylate is minimal (possibly because it is already solvated in the required sense).

Analysis of reprecipitated xylan on fibres from kraft pulping of hardwoods by modern methods indicated that the model system of 20 years earlier had been very perceptive: both hexeneuronic acid residues [415] and β-L-iduronyl residues were found[416] (The finding caused a stir in the pulp industry, since a $KMnO_4$ titre ("kappa number") had been used traditionally as measure of the residual lignin in the pulp, but in hardwood pulps these titres were too high, by factors of ≤ 2, because of the reaction of permanganate with the hexeneuronic acids[417]).

6.7.4.2 Polysaccharide Lyases. The results on the modification of 4-*O*-methylglucuronic acid indicate that C5 epimerisation and elimination of O4 are comparably facile processes *in vitro*. Natural selection therefore predicts that where enzymes have evolved in the same organism, working on the same substrate, to perform both epimerisation and elimination, they should be structurally and evolutionarily related. A case in point is the alginate lyases and epimerases,[418] although the sequences of modern lyases and epimerases are not similar, possibly because of the different stereochemical requirements of the two reactions.[419]

Since the barrier to C5 deprotonation is entirely thermodynamic, both lyases and epimerases can work by stabilising the carbanion thermodynamically. This can be achieved by stabilising the negative charges on carboxylate oxygens by coordination to metal ions such as Ca^{2+}, or by hydrogen bonding. The idea that hydrogen bonds of a "short, tight" type are particularly effective in stabilising anionic intermediates in active sites was advanced just over a decade ago.[420] Their characteristics were a length of less than 2.5 Å, high bond energies (for hydrogen bonds $> 20\,\mathrm{kJ\,mol^{-1}}$ in aqueous solution), low deuterium fractionation factors (<0.4) and large ^{1}H NMR chemical shifts ($\delta \geq 14$). Their strength was at a maximum when the basicities of X and Y in the system X···H···Y were matched. In some cases the energy barrier between two positions of the hydrogen was smaller than the zero point energy of the hydron, so that the system X···H···Y was governed by a single potential well and X···D···Y by a double potential well. The ideas were perhaps too enthusiastically taken up, and the claim was then made that the phenomenon did not exist at all and that short, tight hydrogen bonds were largely an artefact of model systems, in which atoms bearing two lone pairs were in closer than van der Waals contact before protonation, which relieved the consequent strain.[421] In the enzyme context such objections seem to the author immaterial, the "inbuilt strain" argument an example of the well-known mechanism in enzymic catalysis of destabilising the ES complex with respect to a bound transition state.

Although most polysaccharide lyases and epimerases act on the polyuronic acids, a number of those acting on pectin do so on the methyl esters. Whereas the second pK_a of acetic acid is around 35–36,[411] the pK_a of ethyl acetate is 25,[422] so clearly pectin lyases have a much easier catalytic task than pectate lyases.

Polysaccharide lyases have been grouped into a number of CAZy families and mechanistic and structural data are available for a number of them. As with the glycoside hydrolases, enzymes of the same CAZy family have the same protein fold and similar, if not identical, mechanisms. In June 2007 there were 18 polysaccharide lyase families.

6.7.4.2.1 Polysaccharide lyase Family 8 (PL 8). This family encompasses lyases acting upon mammalian connective tissue glycosaminoglycans, and for this reason is the best investigated, with a number of crystal structures available. It has a two-domain fold with an $(\alpha/\alpha)_5$ toroid connected to a domain almost exclusively of β-sheet. The reactions catalysed are *syn* eliminations, with available evidence indicating an irreversible $E1_{CB}$ mechanism, in which the intermediate enolate is stabilised by hydrogen bonding, rather than metal ions, and both deprotonation of C5 and protonation of O4 are carried out by the same tyrosine residue.[423]

The chondroitin AC lyase of *Arthrobacter aurescens* in fact has a relaxed specificity, cleaving hyaluronan with slightly greater efficiency than chondroitin-4-sulfate, thereby demonstrating the unimportance of the hexosamine C4 configuration or substitution. It has a two-domain structure,[424] with the N-terminal α-helical domain containing 13 α-helices, 10 of which form an incomplete double-layered $(\alpha/\alpha)_5$ toroid. The C-terminal domain is composed very largely of antiparallel β-strands arranged into four β-sheets. There is a long, deep groove on one side of the $(\alpha/\alpha)_5$ toroid where electron density from the chondroitin 4-substrate tetrasaccharide β-($\Delta^{4,5}$-4-deoxy)-Glc*p*A-(1→3)-β-GalNAc*p*4S-(1→4)-β-Glc*p*A-(1→3)-β-GalNAc*p*4S, obtained by lyase action, is observed. This is cleaved slowly in the crystal and snapshots of the reaction were obtained. At short reaction times a subsite occupancy of 0.6 is obtained for the unreacted substrate and on this basis the glucuronate residue at subsite +1 (the fissile carbohydrate residue in a lyase, to preserve the convention that cleavage takes place between +1 and −1 subsites) is assigned a distorted conformation $B_{3,0}$.[xxxiv] This ensures that the +1 substrate carboxylate forms close hydrogen bonds to Asn183 (possibly in the imido form) and His183 H5, a probable anion-stabilising arrangement. H5 is also in close proximity to Tyr242, the active site tyrosine. The structure resembles a hyaluronate lyase from *Streptococcus agalactiae*,[425] which has, however, an additional small domain of β-sheet on the opposite side of the toroid from the C-terminal domain. The enzyme from *Streptococcus pneumoniae* has a processive action on hyaluronate but not on chondroitin or chondroitin sulfates;[426] it has the usual $(\alpha/\alpha)_5$ toroid plus β-domain structure. Mutation of the base catalyst, Tyr408, produced an inactive enzyme, but mutation of the supposedly carbanion-stabilising Asn349 and His399 left 6% and 12% activity, respectively, indicating that the residues were not crucial.[427] The high activity of the Asn349 mutant

---

[xxxiv] Described as $^{0,3}B$ in the cited paper, but the designation is clearly incorrect, because of the reference to a pseudoaxial carboxylate in the text; the mechanism sketched has the sugar ring in what appears to be the $^{2,5}B$ conformation.

argues strongly against any sort of crucial enolate stabilisation by "short, tight" hydrogen bonds, which would make interactions with this residue absolutely crucial and the high activity of the His399 mutant eliminates it as a base. Mutation of a series of hydrophobic amino acids (Trp291, Trp292 and Phe343), supposedly a "hydrophobic platform" associated with processivity, did not alter the product distribution and had moderate overall kinetic effects.[428] The xanthan lyase of *Bacillus* sp., which removes the terminal pyruvylated mannose from the xanthan side-chain (Figure 4.92), leaving a dehydrated uronyl structure, again has the same overall fold,[429] but groups, particularly Arg612 in the −1 subsite, determine the specificity for terminal pyruvylated mannose.

Measurement of the loss of ArO^- from a series of 4-*O*-benzyl phenyl-β-D-glucuronides, catalysed by the chondroitin AC lyase of *Flavobacterium heparinum*, resulted in a β_{lg} value of zero.[430] The 4-deoxy-4-fluoro substituent was readily introduced, with Walden inversion, by the action of modified DAST on a *galacto* configured substrate, so isotope effects were measured for the transformation of phenyl 4-deoxy-4-fluoro-β-glucuronide (cf. Figure 6.57(b)). That bond cleavage was rate determining for these poor substrates was confirmed by a primary kinetic isotope effect [$^{5D}(V/K)$] of 1.67. The β_{lg} value of zero, together with a value of $^{4D}(V/K)$ of 1.01, confirmed that even though C5 was being deprotonated, there were no changes at C4, and that therefore the reaction was $E1_{CB}$. No solvent deuterium incorporation into unreacted phenyl 4-deoxy-4-fluoro-β-D-glucuronide was observed during its transformation, but the experiment was considered to be uninformative, as the expulsion of F^- from the carbanion was probably faster than reprotonation. To provide evidence that the enzyme could protonate the C5 anion, the protonation of the nitronate anion from phenyl 5-*S*-nitro β-xylopyranoside, isoelectronic and isosteric with the dianion of phenyl β-D-glucuronide, was examined: it was an excellent substrate ($k_{cat}/K_m = 3 \times 10^6\, M^{-1}\, s^{-1}$).[431] The Tyr234Phe mutant (Tyr234 being the basic catalyst) has no activity against the natural substrate, the nitronate or phenyl 4-*O*-(2,4-dinitrophenyl) β-D-glucuronide, establishing its crucial role in the initial stages of the reaction. The mechanism is set out in Figure 6.73.

6.7.4.2.2 Pectin and pectate lyases. Pectate and pectin lyases (Figure 6.74) catalyse the *trans* elimination of O4 and H5 from galacturonic acid residues in pectins, with pectate lyases acting on the stretches of polygalacturonate in pectin and pectin acting on the methyl esters. Structures of enzymes in PL 1, PL 3, PL 9 and PL 10 are available. Several structures are available for PL 1 pectin or pectate lyases (listed in Ref. 432); they are remarkable in being composed almost exclusively of β-strands, wound into a helix and then into a large shallow coil.

PL 1 pectate lyases are Ca^{2+} dependent, but pectin lyases are not, although some stimulation of activity at high pH is observed; the pH optimum of pectate lyases is about 8.5 and that of pectin lyases about 5.5. Comparison of the structures of the PL 1 pectin lyase A[433] and B[434] from *Aspergillus niger* with PL 1 pectate lyases reveals that a Ca^{2+}-binding carboxylate is replaced by an arginine. The probable ~10 units difference in the pK_a values of C5–H5 in the

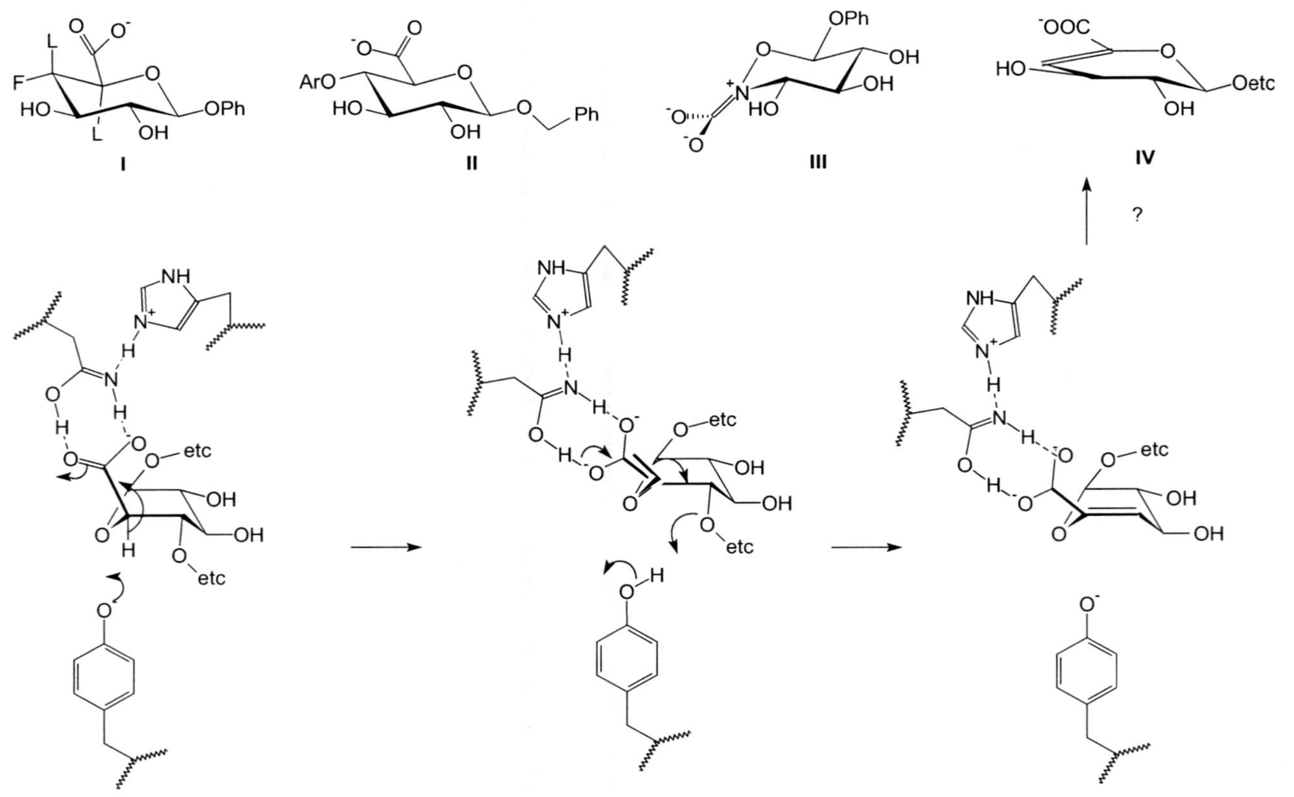

Figure 6.73 Chondroitin AC lyase mechanism, showing reactive conformation (bottom). Top: (I) substrates for KIE (L = H or D) and (II) β_{lg} measurements; (III) nitronate anion whose protonation is catalysed; (IV) *E*1 conformation of ΔUA residues.

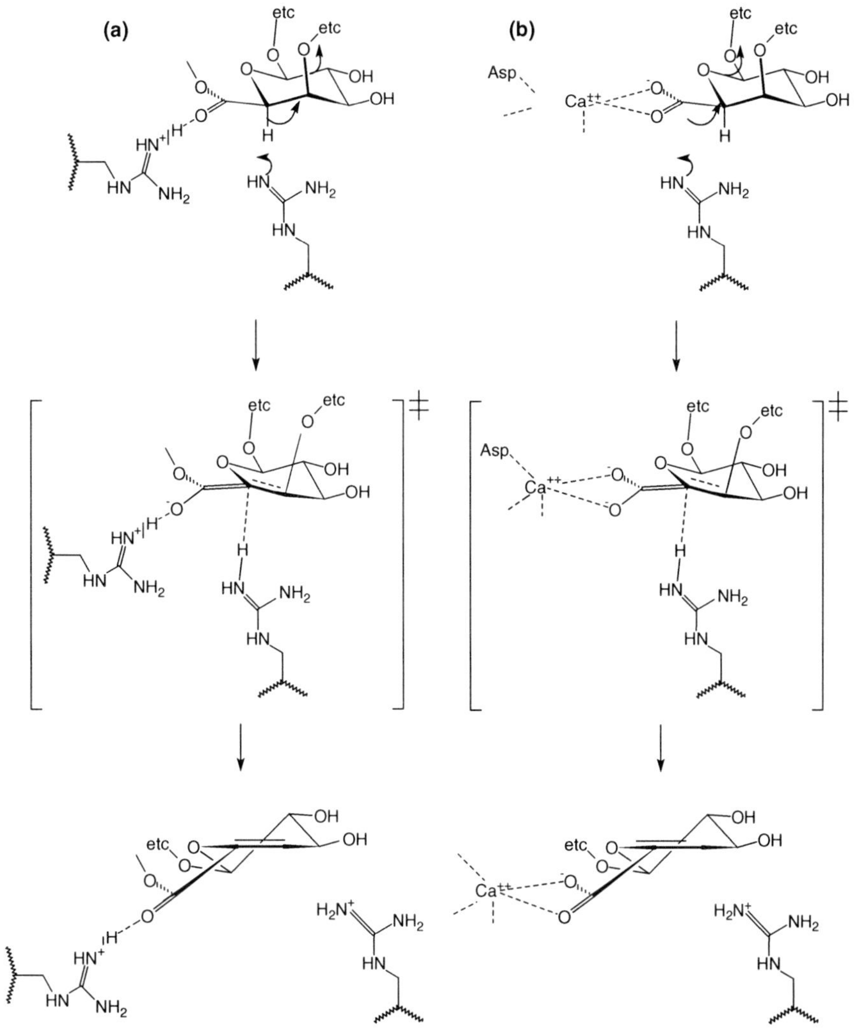

Figure 6.74 Proposed mechanism of pectin and pectate lyases. (a) Pectin lyases. (b) Pectate lyases: in PL the base is a lysine. Note that detailed mechanistic investigations, similar to PL 8, are lacking.

esterified and unesterified residues makes the generation of a carbanion much readier. Four amino acid residues in *A. niger* pectin lyase A (Asp154, Arg176, Arg236 and Lys239) were identified as potential catalytic residues by sequence comparison. Asp154 mutants were reasonably active and Arg236 mutants completely inactive whereas Arg176 and Lys239 mutants had activities reduced 150–300-fold. Model building suggested that Arg236 was the base; in the structure it was sandwiched between Arg176 and Lys239 and it is suggested that these positive groups stabilise the enolate.

The Family 3 structure is similar in structure to Family 1 pectate lyases, the pectate lyase of a *Bacillus* species being a parallel eight-turn β-helix.[435] Site-directed mutagenesis experiments supported a role for an arginine as the active site base that deprotonated C5, with a role for the second Ca^{2+} in stabilising the enolate (the first Ca^{2+} being structural). It is possible that the very existence of a Family 3 PL, separate from Family 1, is an artefact of the many basic residues incorporated into this enzyme that ensure its alkaline pH optimum.

The Family 9 pectate lyase from the plant pathogen *Erwinia chrysanthemi* has again a structure composed largely of β-strands, being a 10-turn helix of parallel β-strands. Analogy with the closely similar PL 1 enzymes suggests that the enzyme binds a catalytic Ca^{2+} between enzyme and substrate, in addition to a "structural", tightly bound Ca^{2+} observed without addition of Ca^{2+} to solutions. A similar mechanism to PL 1 and PL 9 is proposed, with the exception that, there being no arginines in the putative active site, the base is a lysine.[436]

PL 10 has an open two-domain structure, with the two domains composed largely of α-helix and the active site at the intersection of the two domains. There are two members of this family for which structures are available, *Azospirillium irakense*[437] and *Cellovibrio japonicus*.[438] For the latter enzyme, it was possible to obtain a structure with a α-Gal*p*A-(1→4)-α-Gal*p*A-(1→4)-α-GalAOH trisaccharide bound in the −1, +1 and +2 subsites of the inactive Asp389Ala mutant. All three sugars were in the $^4C_1$ conformation. In this structure the −1 and +1 carboxylate coordinated a single calcium, but it was suggested that in active enzyme a second Ca^{2+} would have been coordinated to Asp389 and the carboxylate of the +1 residue. Arg524 is well placed deprotonate C5 in a mechanism similar to that of PL 1 pectate lyases. An $E1_{CB}$-like $E2$ reaction is the likely mechanism of PL 1, PL 3 and PL 10.

No dedicated acid catalyst which donates a proton to O4 has as yet been found in any pectin or pectate lyase: transfer from solvent or from more remote protein acids and bases via a Grotthus mechanism appears to be all that is required.

PL 4 contains rhamnogalacturonan lyases which cleave the [α-D-GalA*p*-(1→2)-α-L-Rha*p*-(1→4)]$_n$ backbone of rhamnogalacturonan I. The one crystal structure, of the enzyme from *Aspergillus aculeatus*, reveals a three-domain structure, with the catalytically active domain likely residing in the largest domain, composed of β-strands: there is some suggestion that lysine, rather than arginine, is the catalytic base, since there is no wholly conserved Arg in PL 4.[439]

6.7.4.2.3 Alginate lyase. It is to be expected there will be at least two distinct types of alginate lyases, those acting on guluronate (G) residues in the +1 subsite, carrying out a *trans* elimination, and those acting on mannuronate (M) residues, carrying out a *syn* elimination. Subdivisions might occur according to whether the −1 subsite preference is for mannuronate or guluronate and according to the degree of acetylation. Specificities are denoted by the occupants of the −1 and +1 sites, MM, GG, GM and MG.

Sphingomonas A1 produces three alginate lyases, ALI (63 kDa), ALII (23 kDa) and ALIII (40 kDa), the last two being formed by autocatalytic processing of ALI. ALII has specificity for GG and ALIII specificity for MM, and ALI presumably both specificities. The structure of ALIII, which belongs to PL 5, has been solved in complex with the trisaccharide.[440] The overall structure is a toroid of α-helices in which the active site forms a deep cleft. The trisaccharide $\Delta^{4,5}$-4-deoxy-β-Man*p*A-(1→4)-β-Man*p*A-(1→4)-Man*p*AOH bound with the unsaturated unit deepest in the tunnel (−1), with the −3 residue largely sticking out. The −1/−2 and −2/−3 glycosidic linkages adopted approximately the same conformation as in solution, $\varphi = \Phi$ and Ψ being, respectively, −63.3 and −177.2° and −68.3 and −109.6°. From the residues in the vicinity of the +1 site, it appeared that a mechanism very similar to that of PL 8 applied, with Tyr246 acting as base towards H5 and an acid to O4 and the carboxylate dianion being stabilised by Asn, His and Arg residues. Mutation of the His residue gave an activity $10^{-4.5}$ of that of wt.[441] Given the similarity of the protein fold and identity of reaction-type, this is very reasonable.

ALII, which cleaves between guluronate residues, is a member of PL 8 and is one[442] of the three structures in this family which have been solved.[443] They are small enzymes (M_r = 20–30 kDa), Pacman-like molecules with the two "jaws", which close over the substrate, being constructed from β-strands. There are no clues as to its mechanism, the structure-based guesswork being (probably erroneously) based on the assumption that the reaction is similar to the *syn* eliminations catalysed by M–M specific alginate lyases.

6.7.4.2.4 PL 9 – Both chondroitin and alginate lyase. PL 9 contains both M–M alginate lyases and chondroitin lyases, which both catalyse *syn* eliminations. The one structure available (that of a chondroitinase from *Flavobacterium heparinum*)[444] is of a helix of β-strands typical of enzymes acting on pectate, although three short α-helices are also in the structure. Despite the existence of a complex with −2 and −1 sites occupied, assignment of catalytic groups is speculative.

A structure of a PL 18 alginate lyase is in CAZy – it is a helix of β-strands – but no paper has yet appeared.

6.7.4.2.5 PL 18 – Hyaluronan lyase. An unusual hyaluronate lyase occurs as part of the protein "tail" of one of the bacteriophages responsible for converting non-pathogenic *Streptococcus pyogenes* into a "flesh-eating bacterium". The crystal structure of this protein, HylP1,[445] revealed three intertwined polypeptides, giving rise to a structure with threefold symmetry, with stretches of α-helix at the "nose" and "tail", but the central portion consisting of three intertwined β-strands, one from each polypeptide. They form a single extended right-handed helix in the form of an irregular triangular tube, where each of the three faces of the tube is composed of alternating β-strands from each of the three polypeptides, with the β-strands orthogonal to the long axis of the enzyme. There is a 200 Å long substrate-binding groove. The initial products of enzyme action on hyaluronan are largely octa- and decasaccharides, with significant quantities of dodeca- and tetradecasaccharides and small amounts of hexasaccharide. Extended incubation with large amounts of enzyme yields largely hexa- and octasaccharides, with minor amounts of tetra- and decasaccharide. The *endo* action of this enzyme is therefore very marked, with kinetically significant subsites extending from −6 to +6, in accord with the extended polysaccharide binding site. It was suggested that the biological function of the hyaluronidase was reduction in viscosity to permit bacteriophage invasion; this is most effectively achieved with few, widely spaced cleavages.

There are three catalytic sites per molecule. Site-directed mutagenesis of Tyr142 and Asp137 produced inactive protein in which structural integrity had been preserved. Therefore, it was reasonably suggested that a mechanism similar to PL 8, with Tyr142 as the base/acid, operated.

6.7.5 C5 Uronyl Residue Epimerases[373]

Both the glucuronyl epimerases involved in the biosynthesis of heparin[xxxv,446] and the mannuronyl epimerases involved in the biosynthesis of alginate exchange H5 with solvent.[447] The enzymes have no cofactors, so mechanisms involving simple deprotonation of substrate and protonation of carbanions from both directions are likely.

Most is known about bacterial enzymes which modify the newly synthesised β-mannuronan in alginate-producing bacteria. The enzymes in the periplasmic space, products of the *AlgG* gene, appear to be independent of Ca^{2+} (the structure of AlgG enzyme from *Pseudomonas aeruginosa*[448] is a right-handed β-helix of the usual polyuronate-binding type, but catalytic residues are unidentified). The products of the *AlgE* genes (seven, AlgE1–AlgE7, in the case of *Azotobacter vinelandii*), which are found in the culture fluid, require Ca^{2+} for full activity.

AlgE enzymes are modular, with one or two catalytic (A) domains and between one and seven R domains, with the structural complexity of the linear protein sequence varying from AR (AlgE4) to $A_1R_1R_2R_3A_2R_4R_5R_6R_7$ (AlgE3).[449] The distribution of products from the enzymes differs, with the simple AlgE4 producing alternating GM sequences almost exclusively and Alg E6 ($A_1R_1R_2R_3$) long G blocks. AlgE2 and AlgE5, both with the linear modular sequence $A_1R_1R_2R_3R_4$, produce G blocks, but shorter stretches than AlgE6. The enzymes with two A domains, AlgE3 and AlgE1 ($A_1R_1R_2R_3A_2R_4$), produce both G blocks and alternating GM sequences, presumably one sort in each active site, although this has not been demonstrated. AlgE7, despite its simple structure ($A_1R_1R_2R_3$), is unique in having both epimerase and lyase activity.[450]

The R modules are not carbohydrate-binding modules, but appear to be the opposite. Direct atomic force microscopy (AFM) measurements found that polymannuronan bound more tightly to the AlgE4 A subunit than to the AR holoenzyme,[449] and that the isolated R module did not bind polymannuronate at all. From the various force–extension curves it is possible to calculate the

[xxxv] The kinetics of approach to chemical and isotopic equilibrium were inadequately analysed in Ref. 446, so the supposed tritium KIE of 20 on protonation of the enolate should be discounted.

probability of observing a bond-cleavage event in the absence of applied force: the molecule is desorbed from the enzyme by thermal motion. This is the "off" rate of enzyme kinetics and proved to be ~40 times lower than k_{cat} values for AlgE4, indicating processivity. The function of the R modules appears to make the enzyme less processive. An analysis[451] of the evolution of the products during epimerisation indicated that with the holo enzyme about 10 residues were epimerised per encounter between enzyme and substrate, with a minimum substrate size of the hexamer; this and the heptamer and octamer appeared to bind −2 to +4, +5 and +6 (by analogy with the lyases, the carbanion site is designated +1).

The structure of the R module from AlgE 4 from *A. vinelandii* has been solved by NMR spectroscopy; it is an elongated molecule consisting of right-handed parallel β-roll, a similar to several other polyuronate-binding proteins.[452]

6.8 BIOLOGICAL OXIDATION–ELIMINATION–ADDITION AND RELATED SEQUENCES

A number of carbohydrate-processing enzymes use the transient generation of a ketone group in the carbohydrate as a way of accessing carbanion chemistry to replace, reorganise or epimerise substituents. The first such enzyme to be recognised was *S*-adenosylhomocysteine hydrolase (although transient oxidation is involved in UDPGal epimerase action, carbanion chemistry is not).

6.8.1 *S*-Adenosylhomocysteine Hydrolase

One of the few biological S_N2 reactions is methylation by *S*-adenosylmethionine, but this leaves *S*-adenosylhomocysteine to be recycled into the cell metabolism by an apparent hydrolysis to adenosine and homocysteine. Even though the C5′ position is primary, this is not an S_N2 reaction, and in a series of classical experiments Palmer and Abeles[453] established the central features of the mechanism (Figure 6.75). The tetrameric enzyme isolated from mammalian sources had a molecule of tightly bound NAD^+ per active site and catalysed the exchange of H4′ (ribose) of *S*-adenosylhomocysteine, adenosine and 5′-deoxyadenosine with solvent: if the reaction with 5′-deoxyadenosine was stopped by denaturation and $NaBH_3T$, radioactivity was found predominantly in the 5′-deoxy-β-D-xylofuranosyladenosine produced, as expected for hydride transfer to the least hindered, α-face of the 5′-deoxy-3′-ketoadenosine. The enzyme also converted 4,5-dehydroadenosine to adenosine (or to *S*-adenosylhomocysteine, in the presence of homocysteine). Finally, the transformations were accompanied by changes in the absorbance spectrum of the enzyme, with a new maximum at 327 nm, close to that of unbound NADH at 340 nm.

X-ray crystallography of the recombinant rat liver[454] and human[455] enzyme have fleshed out some details of Palmer–Abeles mechanism. The wt enzyme crystallises in an open conformation, which on binding of

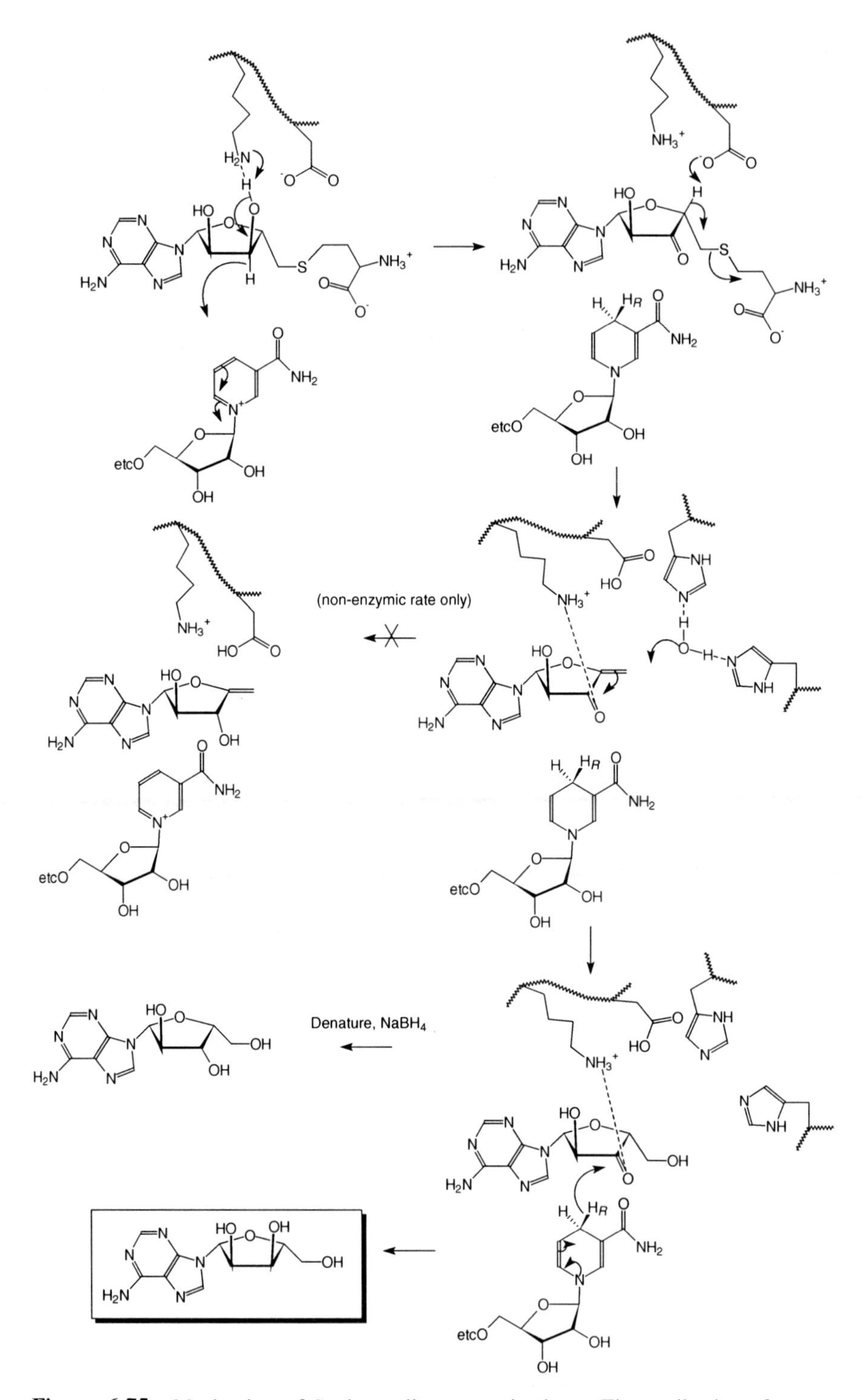

Figure 6.75 Mechanism of *S*-adenosylhomocysteine lyase. The attribution of proton transfer to various groups is as in Ref. 455, on the assumption that residual activities of mutants are largely due to wt contamination, catalytic rescue or chemical deamidation.

adenosylhomocysteine closes: the same closed structure is also seen in the Asp224Glu mutant of the rat liver enzyme substrate. The NAD^+ binds *anti*, as expected for the experimentally determined[456] A-side specificity. The enzyme from the malaria parasite *Plasmodium falciparium*,[457] a potential target for anti-malarial drugs, is similar, with the addition of a loop of protein away from the active site.

The stereochemistries of the reactions in the adenosyl moiety were determined by ^1H NMR spectroscopy of the H5′ resonances of C5 chirally deuterated substrates.[458] First, the C5′ stereochemistries of the *S*-adenosylhomocysteine produced from chirally deuterated adenosine and homocysteine were compared with samples of known stereochemistry; the overall reaction proceeded with retention of configuration. Then the stereochemistry of the addition reaction was examined by examining the H5′ resonances of the homocysteine produced from (*E*)- and (*Z*)-5′-deutero-4,5-dehydro-3-ketoadenosine: the stereochemistry of addition was *syn*. Therefore, because of the overall retention of configuration, the stereochemistry of both elimination and addition is *syn*.

Pre-steady-state kinetic studies established that the appearance of the NADH chromophore on addition of substrate was a two-step process, and these steps can now be identified as closure of the active site and hydride transfer.[459] This study indicated that the on-enzyme equilibrium for addition of water or homocysteine to the enone was close to unity (and the value in free solution), whereas the equilibrium for oxidation of NAD^+ by bound adenosine was 10^5 times more favourable than in free solution. The focusing of the catalytic power of the enzyme on the oxidation step avoids the formation of abortive complexes by hydride transfer between enone and NADH, yielding 4,5-dehydroadenosine and NAD^+. This happens about 10^{-4} times faster than productive hydride transfer at the beginning and end of the catalytic cycle,[460] with the slow rate (close to that of model reactions) apparently arising from a conformationally modulated increase in the distance the hydride has to be transferred.

Site-directed mutagenesis[461] and the crystal structure of the closed form of the enzymes suggested Asp130 as the base which deprotonates C4′, Lys185 (H-bonded to Asp189) as the base which deprotonates O3′ and His54 and His300 as joint base catalysts for the attack of water (an immobilised water is seen so hydrogen bonded). However, the relative residual activities in mutants (0.7% Asp130Asp; 0.5% Lys185Asn; 0.1% Asp189Asn) are at variance with the proposed functions: mutation of the *C*-deprotonating residue should eliminate activity, whereas mutation of a pK_a-modulating group should have relatively modest effects.

6.8.2 Biosynthesis of Nucleotide Diphospho 6-Deoxy Sugars

The donors of the 6-deoxy sugars are biosynthesised from the nucleotide diphospho derivatives of common sugars, GDP-L-Fuc from GDP-D-Man and dTDP-L-Rha from dTDP-D-Glc (Figure 6.76). Hexoses deoxygenated at

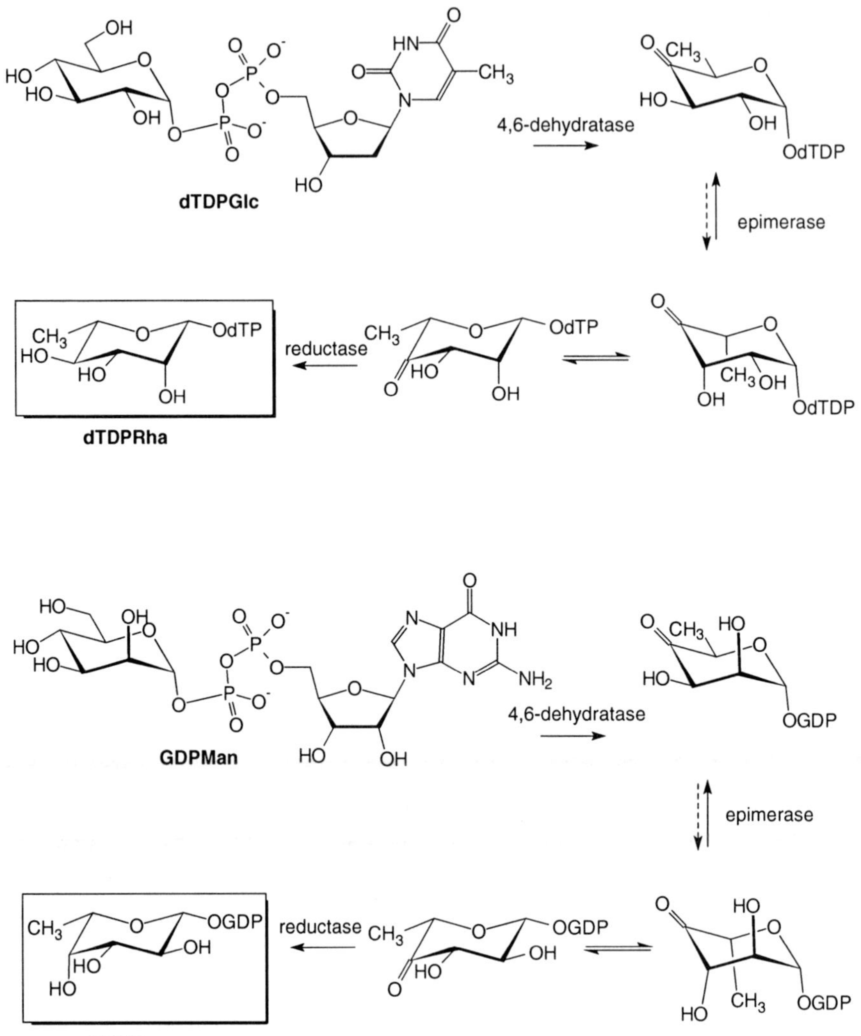

Figure 6.76 Pathways for the biosynthesis of nucleotide diphospho sugars of common, 6-deoxy sugars.

the 2-, 3- and 4-positions are known – the mechanism of 2-deoxygenation is similar to that of 6-deoxygenation, that of 3- and 4-deoxygenation involves pyridoxal chemistry and that of 3-deoxygenation involves iron–sulfur cluster chemistry[462] – but in view of their secondary biochemical role and the limited claim of highly deoxygenated molecules such as desosamine (3,4,6-trideoxy-3-dimethylamino-D-glucose) to be considered carbohydrates, they will not be treated further.

6-Deoxygenation involves three enzyme activities: the 4,6-dehydratase, the 3,5-epimerase and the reductase. The 4,6-dehydratases work by a broadly similar mechanism to the *S*-adenosylhomocysteine hydrolase. They contain one tightly bound $NAD(P)^+$ molecule per active site. This oxidises the 4-position to a ketone and an $E1_{CB}$ reaction drives off the 6-OH. The glucosenone is then reduced by the NAD(P)H to the 4-keto-6-deoxyhexose. The intermediacy of the glucosenone has been verified by rapid quench/soft ionisation mass spectrometry.[463]

The stereospecificity (*proS*, B) of the NAD^+ cofactor chemistry of the glucose 4,6-dehydratases is opposite to that of *S*-adenosylhomocysteine hydrolase, however.[464] (The nucleotide diphosphoglucose 4,6-dehydratase most amenable to study is the CDP-glucose 4,6-dehydratase of *Yersinia pestis*, which in fact commits the glucose to the 3,6-deoxy sugar pathway.) The stereochemistry of the replacement of the 6-OH was determined to be inversion by making the product methyl group chiral by virtue of isotopic substitution with ^2H and ^3H: in the case of the *E. coli* dTDP-glucose 4,6-dehydratase the C6 of the sugar was stereospecifically labelled with ^2H and ^3H,[465] whereas the *Y. pestis* dehydratase substrate had ^3H stereospecifically introduced at C6 and ^2H at C4.[466] On the assumption that, as the reaction was intramolecular, the overall migration of the hydride from C4 to C6 was suprafacial, the overall inversion at C6 means that the loss of H4 and 6-OH was a *syn* elimination, whereas the Michael addition of hydride was *anti*.

The structures of the dTDP-glucose dehydratases from *Streptococcus suis* and *Salmonella enterica* serovar *typhimurium* in complex with dTDP, dTDPGlc and dTDPXyl permit the proton transfers of the basic mechanism to be identified (Figure 6.77).[467] The fold of is that of a short-chain dehydrogenase and the base that deprotonates glucose O4 is clearly identified as Tyr, as expected from this analogy. The initial oxidative step could be studied using dTDPXyl as substrate: mutation of the acid–base Tyr made hydride transfer partly rate determining in its action.[468] The pK_a of the Tyr OH was brought down to 6.7 by key interactions with a conserved Lys, as expected in the short-chain dehydrogenases. O4 is also hydrogen bonded to a Ser, which may play a role in stabilising the subsequent enolate. The *syn* elimination of water requires an acid and a base. The acid was elegantly identified as an Asp residue (135 in the *E. coli* enzyme) by the demonstration that although Asp135 mutants were catalytically crippled against the natural substrate, they transformed 6dTDP-6-deoxy-6-fluoroglucose, a substrate whose leaving group did not require full protonation, at a similar rate to wt.[469] The base that deprotonated C5 was identified as Glu both from the X-ray and site-directed mutagenesis and also by its ability to exchange H5 of the product with solvent.

By monitoring exchange of the C5 proton and the C6 oxygen with solvent by MALDI-TOF mass spectrometry in both reactant and product, it was possible to show that the reaction was concerted.[470] With the acid catalytic Asp mutated to Asn, however, deprotonation at C5 could occur without loss of O6.

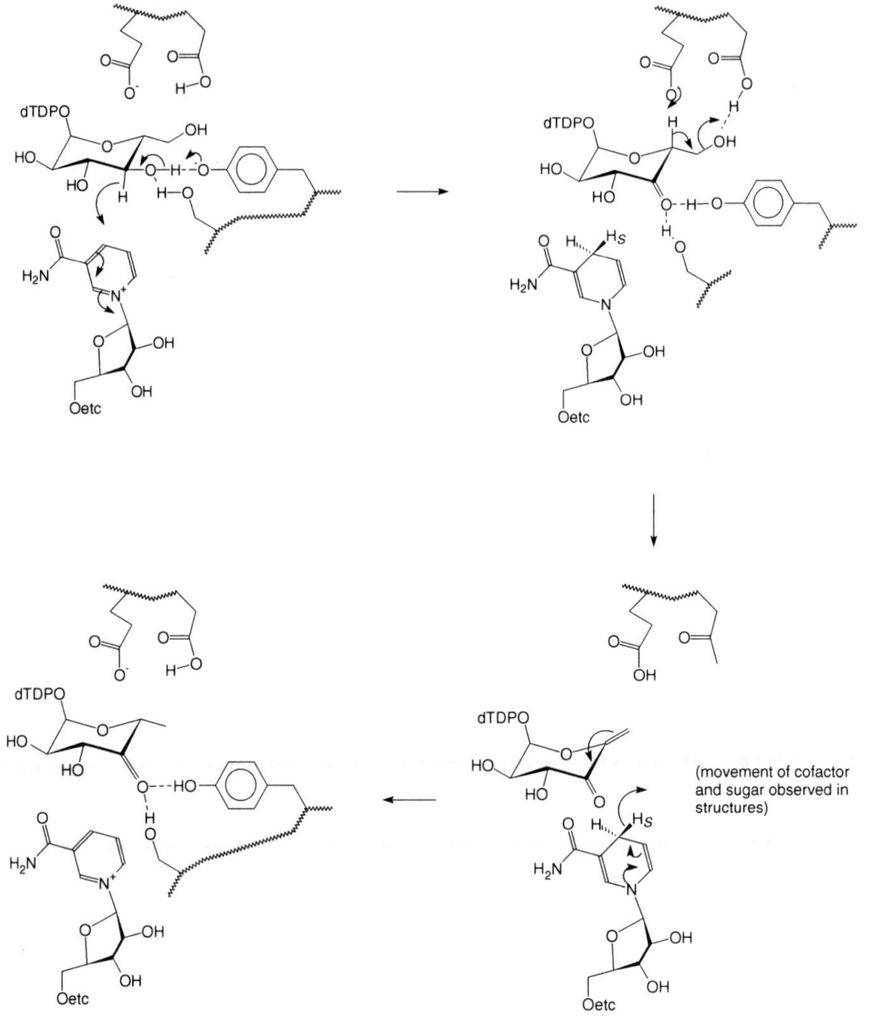

(movement of cofactor and sugar observed in structures)

Figure 6.77 Mechanism of nucleotide diphosphoglucose 4,6-dehydratase.

The mechanism of the epimerase involves enolisation at C3 and C5, since proton exchange with solvent is observed.[471] The structure of the epimerase has also been solved in complex with dTDP-glucose and dTDP-xylose;[472] slight movement is observed between the sugars, but sufficiently strong interactions are observed to rule out any mechanism involving the sugar coming partly loose and rotating, in the manner of UDPGal 4-epimerase. Therefore bases are required to deprotonate C3 and C5 in the sugar and protonate them in the C3-enolate and the C5-enolate, on both sides of the ring (*i.e.* both *cis* and *trans* to the OdTDP group) The structure seems reasonably clear with His76 (*S. suis* numbering), hydrogen bonded to Asp180 in a "catalytic dyad", operating *cis* to the OdTDP group. The situation on the "top" of the ring as normally drawn is less clear, but an acid–base tyrosine is plausible, together with a protonated lysine to stabilise the enolate oxygen. There is no indication as to which enolisation occurs first (or, indeed, whether there is a preference at all).

The structures of GDPMan 4,6-dehydratases from *E. coli*,[473] *Arabidopsis thaliana*[474] and *Pseudomonas aeruginosa*[475] are very similar, despite the first enzyme being tetrameric and the second and third dimeric. They are short-chain alcohol dehydrogenases, but use a molecule of tightly *syn*-bound NADP, rather than NAD^+, per active site. The acid–base groups involved in catalysis (Tyr H-bonded to Lys for the oxidation, Glu for C5 deprotonation) appear to be the same as for dTDPGlc 4,6-dehydratase and a similar mechanism is proposed, supported by basic site-directed mutagenesis experiments supporting the role of the assigned catalytic Glu, Tyr and Lys in the mechanism of Figure 6.62 and 6.77.[473]

The GDP-6-deoxy-4-keto-D-mannose then can either be reduced, giving GDP-D-rhamnose (the unusual stereoisomer of rhamnose), or be isomerised at positions 3 and 5 and reduced to L-fucose at the active site of an unusual bifunctional enzyme, which is both reductase and isomerase. The reduction occurs with *proS* (B-side) specificity of the NADPH coenzyme and the enzyme will catalyse epimerisation and exchange of H3 and H5 in the absence of NADPH.[476] The available structures of the *E. coli* enzyme[477,478,] show NADPH bound in the expected *syn* conformation and counterparts to the His and Lys suggested to play a role in the action of the dTDP-6-deoxy-4-ketoglucose isomerase in the correct position.

6.8.3 GH Family 4

The oxidation–elimination–Michael addition–reduction route for what is formally a nucleophilic displacement appears to be taken by one family of glycoside hydrolases, GH 4. The enzymes are dependent on divalent metal ions, particularly Mn^{2+}, and contain a molecule of tightly bound NAD^+. Uniquely for a glycohydrolase family, GH 4 contains both α- and β-glycosidases.

The GH 4 6-phospho-β-glucosidase of *Thermotoga maritima* yielded trideuteromethyl 6-phospho-2-[^{2}H]-β-glucoside when reacted with *p*-nitrophenyl 6-phospho-β-glucoside in a mixture of D_2O and CD_3OD, establishing that the reaction proceeded with retention of configuration, by a process which labilised H2 to exchange with solvent.[479] 6-Phospho-1,5-anhydroglucitol likewise exchanged H2 when incubated with the enzyme in D_2O, thereby establishing that anomeric chemistry was not the cause of the exchange. Using *p*-nitrophenyl 6-phospho-β-glucoside as substrate, primary deuterium kinetic isotope effects $^{3D}(V/K) = 1.91$, $^{3D}(V) = 1.63$, $^{2D}(V/K) = 2.03$, $^{2D}(V) = 1.63$ were observed, clearly indicating that both C2–H and C3–H cleavage contributed to a virtual rate-determining transition state.

A crystal structure of this enzyme[480] showed that the NAD^+ was ideally placed to remove a hydride from C3 to the *proR* (A) face of the NAD^+. The structure of a 6-phospho α-glucosidase from *Bacillus subtilis* revealed a very

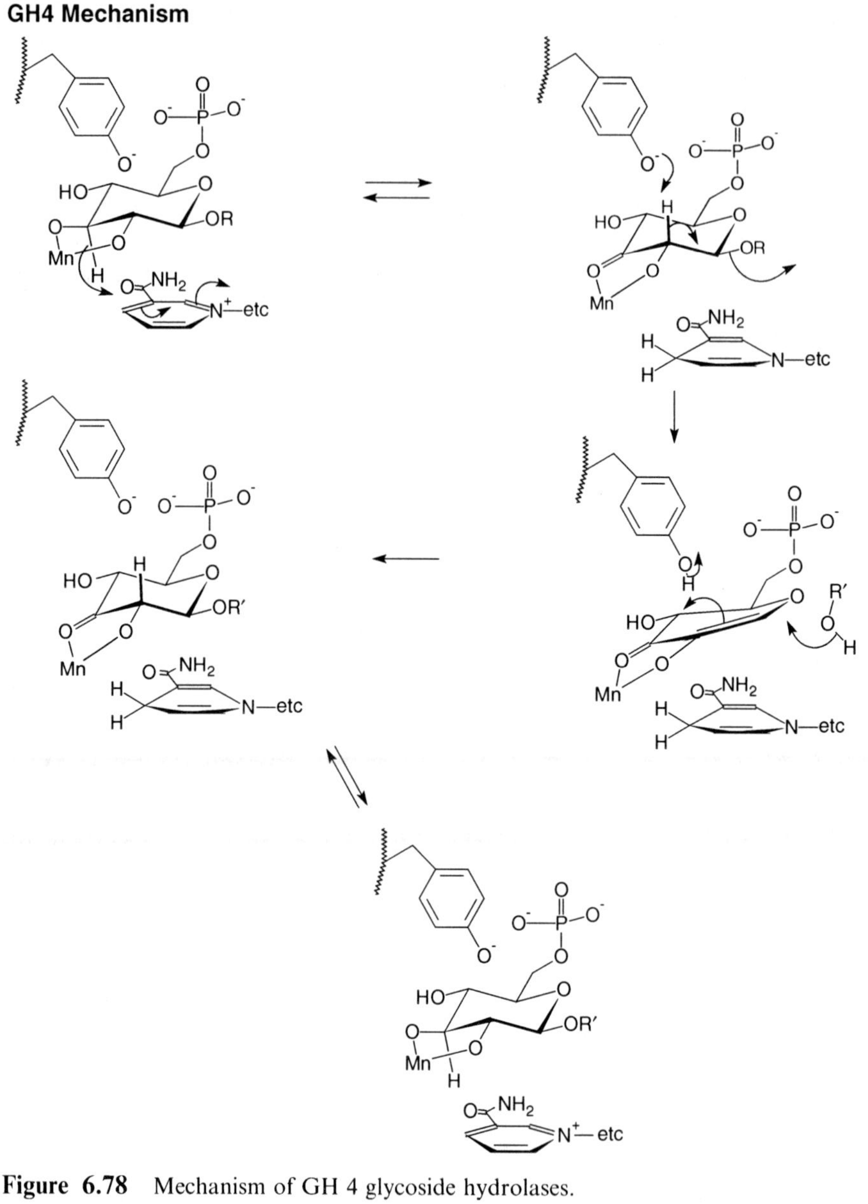

GH4 Mechanism

Figure 6.78 Mechanism of GH 4 glycoside hydrolases.

similar active site architecture.[481] In both cases Mn^{2+} is chelated by O2 and O3 of the sugar. These data are therefore adequate to establish the broad outlines of the mechanism (Figure 6.78):

(i) Hydride is removed from C3 by NAD^+, leaving a keto sugar.
(ii) In an $E1_{CB}$-like reaction, the aglycone is lost to generate a 2-hydroxyenone.

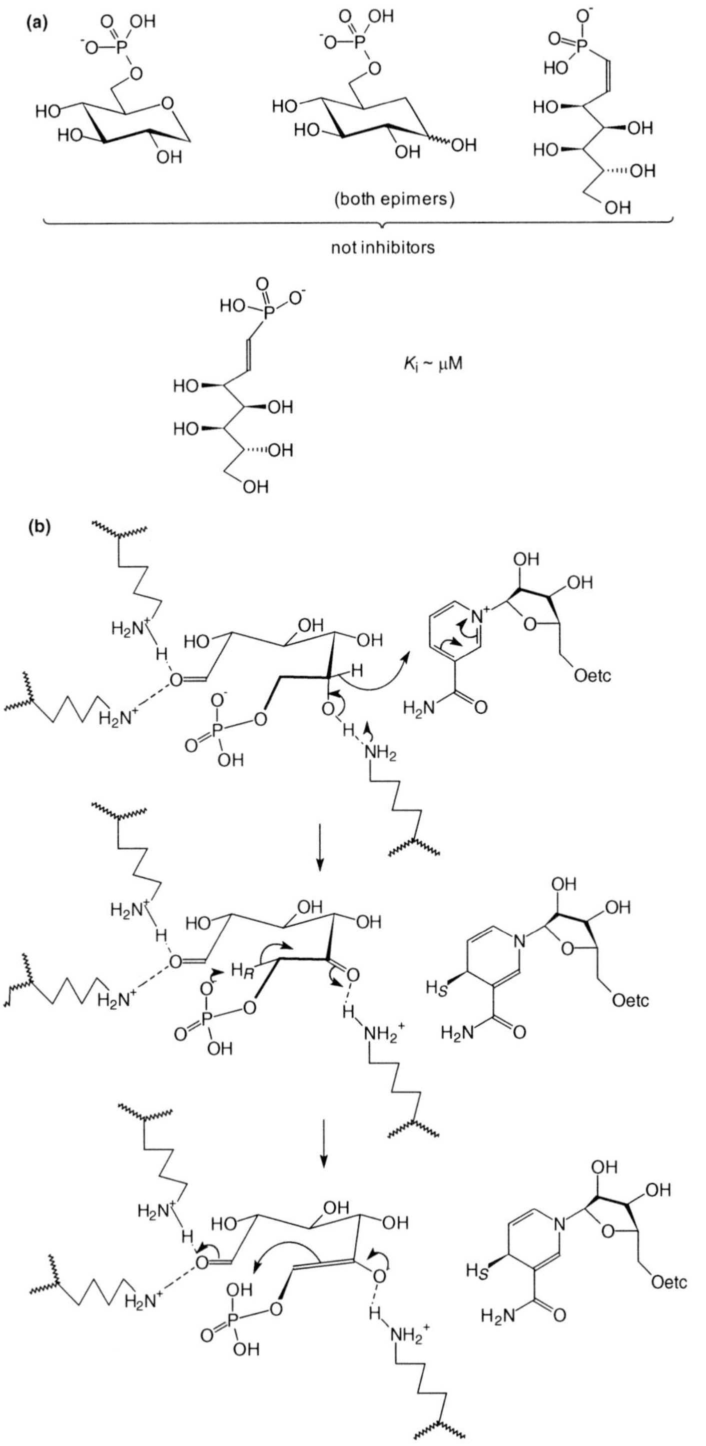

Figure 6.79 Mechanism of 1-L-*myo*-inositol 1-phosphate. (a) Structures of potential inhibitors; (b) mechanism.

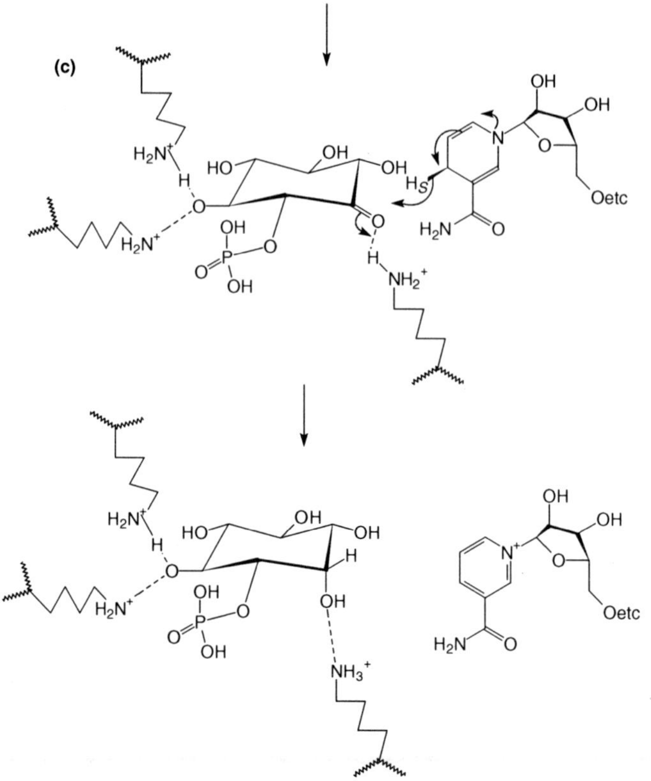

Figure 6.79 (*Continued*)

(iii) Water or alcohols add in a Michael fashion to the unsaturated ketone.
(iv) The keto group at position 3 is reduced and the product released.

The groups involved in proton transfer are as yet unclear: in both 6-phospho α- and β-glucosidases a tyrosine is clearly positioned to act as catalytic base in the elimination reaction and then general acid in the Michael addition, but the tyrosine does not appear to be conserved in GH4 α-glucosidases whose substrates are not phosphorylated. General acid catalysis of aglycone departure in the elimination reaction appears to be not strictly necessary, as *p*-nitrophenyl 6-phospho-1-thio-α-glucoside is as good a substrate for the phospho-α-glucosidase as its oxygen analogue.

6.8.4 L-Myoinositol 1-Phosphate Synthetase[482]

L-Myoinositol 1-phosphate synthetase converts glucose 1-phosphate to myoinositol phosphate. Its structure is largely conserved in eukaryotes from yeast to humans, although its oligomeric structure differs from species to species. It contains one tightly bound NAD^+ per active site and during the cyclisation the *proR* hydrogen is lost from C6[483] of glucose 1-phosphate. $^T(V/K)$ isotope effects of 5.7–2.0, depending on (crude) enzyme source, were observed with 5-tritiated substrate, although the tritium does not exchange with water; the C5–H5 bond is broken in a rate-determining step, but H5 remains in the product.[484] The tetrameric yeast enzyme is the most studied: it binds NAD^+ in a *syn* conformation with the amide hydrogen bonded to the sugar diphosphate;[485] in the *Mycobacterium tuberculosis* structure the amide and pyrophosphate are connected by a Zn^{2+}.[486]

The outlines of the mechanism thus involve transient oxidation of C5 to a ketone, removal of the *proR* hydrogen to yield an enolate and an intermolecular aldol condensation (Figure 6.79). The intermediate L-myosose-1-phosphate was independently synthesised and shown to bind to the NAD^+ form of the yeast enzyme with $K_i = 3.6\,\mu M$.[487] Further details were provided by studies of reversible inhibitors. Analogues of pyranose forms of glucose-6-phosphate which were incapable of ring opening, either because 1-OH had been removed or because O5 had been replaced by a CH_2 group, were not inhibitors.[488] Replacement of the 6-CH_2O–PO_3H_2 part of the substrate with a –CH = CH–PO_3H_2 grouping was particularly informative: the *Z*-isomer failed to inhibit, but the *E*-isomer gave μM inhibition. It was therefore concluded that the enzyme bound the small proportion (0.4%) of glucose 1-phosphate in the acyclic form directly and that the phosphate, not an enzyme group, deprotonated C6.

The structure of the yeast enzyme in complex with NAD^+ and "2-deoxy-D-glucitol 6-(*E*)-vinylhomophosphonate"[489] suggested that the carbonyl activation was achieved by hydrogen bonding to two lysines, 373 and 412 in the case of C1 and 369 and 489 in the case of the transient ketone at C5; certainly mutation of Lys369, 412 and 489 abolished activity.

REFERENCES

1. C. A. Lobry de Bruyn and W. Alberda van Ekenstein, *Recl. Trav. Chim. Pays-Bas* 1895, **14**, 156, and succeeding papers in the same journal.
2. S. J. Angyal, *Top. Curr. Chem.*, 2001, **215**, 1.
3. S. V. Rieder and I. A. Rose, *J. Biol. Chem.*, 1959, **234**, 1007.
4. J. R. Knowles and W. J. Albery, *Acc. Chem. Res.*, 1977, **10**, 105.
5. J. Read, J. Pearce, X. Li, H. Muirhead, J. Chirgwin and C. Davies, *J. Mol. Biol.*, 2001, **309**, 447.
6. S. H. Seeholzer, *Proc. Natl. Acad. Sci. USA*, 1993, **90**, 1237.
7. T. Hansen, D. Wendorff and P. Schönheit, *J. Biol. Chem.*, 2004, **279**, 2262.
8. (a) I. A. Rose, E. L. O'Connell and R. P. Mortlock, *Biochim. Biophys. Acta*, 1969, **178**, 376; (b) K. J. Schray and I. A. Rose, *Biochemistry*, 1971, **10**, 1058.
9. C. I. F. Watt, S. N. Whittleton and S. M. Whitworth, *Tetrahedron*, 1986, **42**, 1047.
10. R. W. Nagorski and J. P. Richard, *J. Am. Chem. Soc.* (a) 1996, **118**, 7432; (b) 2001, **123**, 794.
11. J. P. Richard, *J. Am. Chem. Soc.*, 1984, **106**, 4926.

12. V. Bílik and Š. Kučár, *Carbohydr. Res.*, 1970, **13**, 311.
13. T. Tanase, F. Shimizu, S. Yano and S. Yoshikawa, *J. Chem. Soc., Chem. Commun.*, 1986, 1001.
14. T. Yamauchi, K. Fukushima, R. Yanagihara, S. Osanai and S. Yoshikawa, *Carbohydr. Res.*, 1990, **204**, 233.
15. M. L. Hayes, N. J. Pennings, A. S. Serianni and R. Barker, *J. Am. Chem. Soc.*, 1982, **104**, 6764.
16. S. Osanai, *Top. Curr. Chem.*, 2001, **215**, 43.
17. L. Petruš, M. Petrušová and Z. Hricovíniová, *Top. Curr. Chem.*, 2001, **215**, 15.
18. V. Bílik, L. Petruš and J. Zemek, *Chem. Zvesti*, 1978, **32**, 242.
19. G. de Wit, A. P. G. Kieboom and H. van Bekkum, *Carbohydr. Res.*, 1979, **74**, 157.
20. S. G. Waley, J. C. Miller, I. A. Rose and E. L. O'Connell, *Nature*, 1970, **227**, 181.
21. R. C. Davenport, P. A. Bash, B. A. Seaton, M. Karplus, G. A. Petsko and D. Ringe, *Biochemistry*, 1991, **30**, 5821.
22. Z. D. Zhang, S. Sugio, E. A. Komives, K. D. Liu, J. R. Knowles, G. A. Petsko and D. Ringe, *Biochemistry*, 1994, **33**, 2830.
23. N. S. Sampson and J. R. Knowles, *Biochemistry*, 1992, **31**, 8488.
24. P. J. Lodi, L. C. Chang, J. R. Knowles and E. A. Komives, *Biochemistry*, 1994, **33**, 2809.
25. T. L. Amyes, A. C. O'Donoghue and J. P. Richard, *J. Am. Chem. Soc.*, 2001, **123**, 11325.
26. P. J. Lodi and J. R. Knowles, *Biochemistry*, 1991, **30**, 6948.
27. P. J. Lodi and J. R. Knowles, *Biochemistry*, 1993, **32**, 4338.
28. E. A. Komives, L. C. Chang, E. Lolis, R. F. Tilton, G. A. Petsko and J. R. Knowles, *Biochemistry*, 1991, **30**, 3011.
29. K. Hamada, H. Ago, M. Sugahara, Y. Nodake, S. Kuramitsu and M. Miyano, *J. Biol. Chem.*, 2003, **278**, 49183.
30. D. Arsenieva, R. Hardré, L. Salmon and C. J. Jeffrey, *Proc. Natl. Acad. Sci. USA*, 2002, **99**, 5872.
31. W. J. Albery and J. R. Knowles, *Biochemistry*, 1976, **15**, 5627.
32. J. M. Herlihy, S. G. Maister, W. J. Albery and J. R. Knowles, *Biochemistry*, 1976, **15**, 5601.
33. P. F. Leadlay, W. J. Albery and J. R. Knowles, *Biochemistry*, 1976, **15**, 5617.
34. S. G. Maister, C. P. Pett, W. J. Albery and J. R. Knowles, *Biochemistry*, 1976, **15**, 5607.
35. S. J. Fletcher, J. M. Herlihy, W. J. Albery and J. R. Knowles, *Biochemistry*, 1976, **15**, 5612.
36. W. J. Albery and J. R. Knowles, *Biochemistry*, 1976, **15**, 5588.
37. L. M. Fisher, W. J. Albery and J. R. Knowles, *Biochemistry*, 1976, **15**, 5621.
38. H. Kacser and R. Beeby, *J. Mol. Evol.*, 1984, **20**, 38.
39. J. E. Seeman and G. E. Schulz, *J. Mol. Biol.*, 1997, **273**, 256.

40. D. W. Harris and M. S. Feather, *J. Am. Chem. Soc.*, 1975, **97**, 178.
41. S. Ramchander and M. S. Feather, *Arch. Biochem. Biophys.*, 1977, **178**, 576.
42. W. B. Gleason and R. Barker, *Can. J. Chem.*, 1971, **49**, 1433.
43. C. I. F. Watt, in *Comprehensive Biological Catalysis*, ed. M. L. Sinnott, Academic Press, London, 1997, Vol. 1, p. 253.
44. M. Callens, H. Kersters-Hilderson, O. Van Opstal and C. K. De Bruyne, *Enzyme Micobiol. Technol.*, 1986, **8**, 696.
45. K. N. Allen, A. Lavie, G. A. Petsko and D. Ringe, *Biochemistry*, 1995, **34**, 3742.
46. K. N. Allen, A. Lavie, G. K. Farber, A. Glasfeld, G. A. Petsko and D. Ringe, *Biochemistry*, 1994, **33**, 1481.
47. I. P. Korndörfer, W.-D. Fessner and B. W. Matthews, *J. Mol. Biol.*, 2000, **300**, 917.
48. V. Bílik, L. Petruš and V. Farkaš, *Collect. Czech. Chem. Commun.*, 1978, **43**, 1163.
49. V. Bílik, L. Petruš and V. Farkaš, *Chem. Zvesti*, 1975, **29**, 690.
50. S. Zhao, L. Petruš and A. S. Serianni, *Org. Lett.*, 2001, **3**, 3819.
51. M. Matulová and V. V. Bílik, *Carbohydr. Res.*, 1993, **250**, 203.
52. G. E. Taylor and J. M. Waters, *Tetrahedron. Lett.*, 1981, **22**, 1277.
53. S. Steinbacher, J. Kaiser, W. Eisenreich, R. Huber, A. Bacher and F. Rohdich, *J. Biol. Chem.*, 2003, **278**, 18401.
54. E. Sjöström, *Wood Chemistry*, Second Edition, Academic Press, London, 1993, p. 140.
55. J. U. Nef, *Justus Liebigs Ann. Chem.*, 1910, **376**, 1.
56. H. S. Isbell, *J. Res. Natl. Bur. Stand. (U. S.)*, 1944, **32**, 45.
57. J. Kenner and G. N. Richards, *J. Chem. Soc.*, 1955, 1810.
58. J. Kenner and G. N. Richards, *J. Chem. Soc.*, 1957, 3019.
59. G. Machel and G. N. Richards, *J. Chem. Soc.* 1960, (a) 1924; (b) 1932; (c) 1938.
60. B. Y. Yang and R. Montgomery, *Carbohydr. Res.*, 1996, **280**, (a) 27; (b) 47.
61. C. J. Knill and J. F. Kennedy, *Carbohydr. Polym.*, 2003, **51**, 281.
62. E. Sjöström, *TAPPI J.*, 1977, **60**, 151.
63. R. Alén, in *Forest Products Chemistry*, ed. P. Stenius, Vol. 3 of *Papermaking Science and Technology*, Series Eds J. Gullichsen and H. Paulapuro, Fapet Oy, Helsinki, 2000, p. 59.
64. For a discussion of mechanism and intereference in connection with a modern application, see P. Hartmann, S. J. Haswell and M. Grasserbauer, *Anal. Chim. Acta*, 1994, **285**, 1.
65. R. Volk and A. Bacher, *J. Biol. Chem.* (a) 1990, **265**, 19479; (b) 1991, **266**, 20610.
66. D. I. Liao, Y. J. Zheng, P. V. Viitanen and D. B. Jordan, *Biochemistry*, 2002, **41**, 1795.
67. S. Steinbacher, S. Schiffman, G. Richter, R. Huber, A. Bacher and M. Fischer, *J. Biol. Chem.*, 2003, **278**, 42256.

68. T. M. Wrodnigg and B. Elder, *Top. Curr. Chem.*, 2001, **215**, 115.
69. L. Rosen, J. W. Woods and W. W. Pigman, *Chem. Ber.*, 1957, **90**, 1038.
70. M. Amadori, *Atti Accad. Naz. Lincei*, 1925, **2**, 337.
71. F. Micheel and B. Schleppinghoff, *Chem. Ber.*, 1956, **89**, 1702.
72. C. P. Barry and J. Honeymoon, *J. Chem. Soc.*, 1952, 4147.
73. K. Heyns and W. Koch, *Z. Naturforsch., Teil B*, 1952, **7**, 486.
74. M. M. Shemyakin, V. I. Maimind, K. M. Ermolaev and E. M. Bamdas, *Tetrahedron*, 1965, **21**, 2771.
75. H. Siman, K. -D. Keil and F. Weygand, *Chem. Ber.*, 1962, **95**, 17.
76. D. Palm and H. Simon, *Z. Naturforsch., Teil B*, 1963, **18**, 419.
77. F. Weygand, H. Simon and J. F. Klebe, *Chem. Ber.*, 1958, **91**, 1567.
78. F. Micheel and I. Dijong, *Justus Liebigs Ann. Chem.*, 1963, **669**, 136.
79. O. L. Chapman, W. J. Welstead, T. J. Murphy and R. W. King, *J. Am. Chem. Soc.*, 1967, **89**, 7005.
80. L. Mester, G. Vass, A. Stephen and J. Panello, *Tetrahedron Lett.*, 1968, **9**, 4053.
81. L.-C. Maillard, *C. R. Acad. Sci.*, 1912, **154**, 66.
82. R. H. Stadler, I. Blank, N. Varga, F. Robert, J. Hau, P. A. Guy, M.-C. Robert and S. Riediker, *Nature*, 2002, **419**, 449.
83. J. McPherson, B. H. Shilton and D. J. Walton, *Biochemistry*, 1988, **27**, 1901.
84. M. A. J. S. Van Boekel, *Nahrung/Food*, 2001, **45**, 150.
85. D. R. Cremer and K. Eichner, *Food Chem.*, 2000, **71**, 37.
86. H. Hashiba, *Agric. Biol. Chem.*, 1982, **46**, 547.
87. M. N. Lund, K. Olsen, J. Sørensen and L. H. Skibsted, *J. Agric. Food. Chem.*, 2005, **53**, 2095.
88. S. I. F. S. Martins and M. A. J. S. Van Boekel, *Food Chem.*, 2005, **92**, 437.
89. H. Weenen, *Food Chem.*, 1998, **62**, 393.
90. S. I. F. S. Martins, A. T. M. Marcelis and M. A. J. S. Van Boekel, *Carbohydr. Res.*, 2003, **338**, 1651.
91. V. A. Yaylayan and N. G. Forage, *Food Chem.*, 1992, **44**, 201.
92. T. Davidek, S. Devaud, F. Robert and I. Blank, *Ann. N. Y. Acad. Sci.*, 2005, **1043**, 73.
93. A. Hollnagel and L. W. Kroh, *J. Agr. Food. Chem.*, 2002, **50**, 1659.
94. G. P. Rizzi, *J. Agric. Food. Chem.*, 2004, **52**, 953.
95. L. W. Kroh and A. Schulz, *Nahrung/Food*, 2001, **45**, 160.
96. T. Hofmann, *Eur. Food Res. Technol.*, 1999, **209**, 113.
97. O. Reihl, T. M. Rothenbacher, M. O. Lederer and W. Schwack, *Carbohydr. Res.*, 2004, **339**, 1609.
98. D. R. Sell, K. M. Biemel, O. Reih, M. O. Lederer, C. M. Strauch and V. M. Monnier, *J. Biol. Chem.*, 2005, **280**, 12310.
99. D. R. Sell and V. M. Monnier, *J. Biol. Chem.*, 1989, **264**, 21597.
100. O. Reihl, K. M. Biemel, W. Eipper, M. O. Lederer and W. Schwack, *J. Agric. Food. Chem.*, 2003, **51**, 4810.

101. D. R. Sell and V. M. Monnier, *J. Biol. Chem.*, 2004, **279**, 54173.
102. B. Cämmerer, W. Jalyschko and L. W. Kroh, *J. Agric. Food Chem.*, 2002, **50**, 2083.
103. A. Adams, R. C. Borrelli, V. Fogliano and N. De Kimpe, *J. Agric. Food Chem.*, 2005, **53**, 4136.
104. R. H. Stadler, F. Robert, S. Riediker, N. Varga, T. Davidek, S. Devaud, T. Goldmann, J. Hau and I. Blank, *J. Agric. Food Chem.*, 2004, **52**, 5550.
105. M. L. Hackert and A. E. Pegg, in *Comprehensive Biological Catalysis*, ed. M. L. Sinnott, Academic Press, London, 1998, Vol. 2, p. 201.
106. M. S. Feather and J. F. Harris, *Adv. Carbohydr. Chem. Biochem.*, 1973, **28**, 161.
107. P. J. Oefner, A. H. Lanziner, G. Bonn and O. Bobleter, *Monatsh. Chem.*, 1992, **123**, 547.
108. M. Watanabe, Y. Aizawa, T. M. Iida, T. Aida, C. Levy, K. Sue and H. Inomata, *Carbohydr. Res.*, 2005, **340**, 1925.
109. Y. Ishikawa and S. Saka, *Cellulose*, 2001, **8**, 189.
110. R. Krause, K. Knoll and T. Henle, *Eur. Food Res. Technol.*, 2003, **216**, 277.
111. V. Yaylayan and N. G. Forage, *J. Agric. Food Chem.*, 1991, **39**, 364.
112. R. I. Gelb, L. M. Schwartz and D. A. Laufer, *Bioorg. Chem.*, 1982, **11**, 274.
113. M. E. Brewster, M. Huang, E. Pop, J. Pitha, M. J. S. Dewar, J. J. Kaminski and N. Bodor, *Carbohydr. Res.*, 1993, **242**, 53.
114. P. Ballinger and F. A. Long, *J. Am. Chem. Soc.*, 1960, **82**, 795.
115. P. Ballinger and F. A. Long, *J. Am. Chem. Soc.*, 1959, **81**, 1050.
116. J. A. Dean, *Handbook of Organic Chemistry*, McGraw-Hill, New York, 1987.
117. A. Blaskó, C. A. Bunton, S. Bunel, C. Ibarra and E. Moraga, *Carbohydr. Res.*, 1997, **298**, 163.
118. G. Legler, *Pure Appl. Chem.*, 1987, **59**, 1457.
119. C. L. Perrin and R. Armstrong, *J. Am. Chem. Soc.*, 1993, **115**, 6825.
120. V. Dookhun and A. J. Bennet, *J. Am. Chem. Soc.*, 2005, **127**, 7458.
121. H.-M. Wang, D. Loganathan and R. J. Linhardt, *Biochem. J.*, 1991, **278**, 689.
122. H. H. Jensen and M. Bols, *Acc. Chem. Res.*, 2006, **39**, 259.
123. Eg, R. D. Topsom, *Prog. Phys. Org. Chem.*, 1976, **12**, 1.
124. A. Neuberger and A. P. Fletcher, *J. Chem. Soc. B*, 1969, 178.
125. G. M. Ullmann and E.-W. Knapp, *Eur. Biophys. J.*, 1999, **28**, 533.
126. T. V. Tyrtysh, N. E. Byramova and N. V. Bovin, *Bioorg. Khim.*, 2000, **26**, 460.
127. A. Cappa, E. Marcantoni, E. Torregiaim, G. Bartoli, M. C. Bellucci, M. Bosco and L. Sambri, *J. Org. Chem.*, 1999, **64**, 5696.
128. (a) T. Onoda, R. Shirai and S. Iwasaki, *Tetrahedron Lett.*, 1997, **38**, 1443; (b) A. Bouzide and G. Sauve, *Synlett*, 1997, 1153.
129. G. L. Greene and R. L. Letsinger, *Nucleic Acids Res.*, 1975, **2**, 1123.

130. Recent protocols: (a) T. Taniguchi and K. Ogasawara, *Angew. Chem. Int. Ed.* 1998, **37**, 1136; (b) S. Tanaka, H. Saburi, Y. Ishibashi and M. Kitamura, *Org. Lett.*, 2004, **6**, 1873.
131. E. Sjöström, *Wood Chemistry – Fundamentals and Applications*, Academic Press, San Diego, 2nd edn 1993, p. 214.
132. R. Manelius, A. Buléon, K. Nurmi and E. Bertoft, *Carbohydr. Res.*, 2000, **329**, 621.
133. E. J. Corey and A. Venkataswarlu, *J. Am. Chem. Soc.*, 1972, **94**, 6190.
134. H. Brandstetter and E. Zbiral, *Helv. Chim. Acta*, 1978, **61**, 1832.
135. H. Yamada, K. Tanigakiuchi, K. Nagao, K. Okajima and T. Mukae, *Tetrahedron Lett.*, 2004, **45**, 9207.
136. S. Hanessian and P. Lavallee, *Can. J. Chem.*, 1975, **53**, 2975.
137. S. A. Hardinger and N. Wijaya, *Tetrahedron Lett.*, 1993, **34**, 3821.
138. M. Underwood and M. L. Sinnott, unpublished work, 1999.
139. H. Yamada, K. Tanigakiuchi, K. Nagao, K. Okajima and T. Mukae, *Tetrahedron Lett.*, 2004, **45**, 5615.
140. W. J. Markiewicz, B. Nowakowska and K. Adrych, *Tetrahedron Lett.*, 1988, **29**, 1561.
141. J. N. Cha, K. Shimuzu, Y. Zhou, S. C. Christiansen, B. F. Chmelka, G. D. Stucky and D. E. Morse, *Proc. Natl. Acad. Sci. USA*, 1999, **96**, 361.
142. A. Chiovitti, R. E. Harper, A. Willis, A. Bacic, P. Mulvanry and R. Wetherbee, *J. Phycol.*, 2005, **41**, 1154.
143. E.g. K. Walczak, J. Lau and F. B. Pedersen, *Synthesis* 1993, 790.
144. S. Cai, S. Hakomori and T. Toyokuni, *J. Org. Chem.*, 1992, **57**, 3431.
145. E.g. R. F. Helm, J. Ralph and L. Anderson, *J. Org. Chem.*, 1991, **56**, 7015.
146. M. Osswald, U. Lang, S. Friedrich-Bochnitschek, W. Pfrengle and H. Kunz, *Z. Naturforsch., Teil B*, 2003, **58**, 764.
147. S. Xu, I. Held, B. Kempf, H. Mayr, W. Steglich and H. Zipse, *Chem. Eur. J.*, 2005, **11**, 4751.
148. Y. C. Xu, A. Bizuneh and C. Walker, *J. Org. Chem.*, 1996, **61**, 9086.
149. J. Herzig, A. Nudelman, H. E. Gottlieb and B. Fischer, *J. Org. Chem.*, 1986, **51**, 727.
150. F. Ballardie, B. Capon, J. D. Sutherland, D. Cocker and M. L. Sinnott, *J. Chem. Soc. Perkin Trans. 1*, 1973, 2418.
151. L. Q. Chen and F. Z Kong, *Carbohydr. Res.*, 2002, **337**, 2335.
152. D. J. Lefeber, J. F. Kamerling and J. F. G. Vliegenthart, *Org. Lett.*, 2000, **2**, 701.
153. M. Mastihoubová and P. Biely, *Carbohydr. Res.*, 2004, **339**, 1353.
154. M. A. Kabel, P. de Waard, H. A. Schols and A. G. J. Voragen, *Carbohydr. Res.*, 2003, **338**, 69.
155. C. Förster, S. Zimmer, W. Zeidler and M. Sprinzl, *Proc. Natl Acad. Sci USA*, 1994, **91**, 4254.
156. J. M. Jones, *Chem. Ind. (London)*, 1996, **13**, 494.
157. E. Sjöström, *Wood Chemistry – Fundamentals and Applications*, Academic Press, San Diego, 2nd edn, 1993, p. 211.

158. http://www.cazy.org.
159. C. Altaner, B. Saake, M. Tenkanen, J. Eyzaguirre, C. B. Faulds, P. Biely, L. Viikari, M. Siika-aho and J. Puls, *J. Biotechnol.*, 2003, **105**, 95.
160. M. Therisod and A. M. Klibanov, *J. Am. Chem. Soc.*, 1986, **108**, 5638.
161. D. M. Quinn and S. M. Feaster, in *Comprehensive Biological Catalysis*, ed. M. L. Sinnott, Academic Press, London, 1998, Vol. 1, p. 455.
162. C. W. Wharton, in *Comprehensive Biological Catalysis*, ed. M. L. Sinnott, Academic Press, London, 1998, Vol. 1, p. 345.
163. I. M. Kovach, M. Larson and R. L. Schowen, *J. Am. Chem. Soc.*, 1986, **108**, 5490.
164. J. A. M. Prates, N. Tarbouriech, S. J. Charnock, C. M. G. A. Fontes, L. M. A. Ferreira and G. J. Davies, *Structure*, 2001, **9**, 1183.
165. D. H. Anderson, G. Harth, M. A. Horwitz and D. Eisenberg, *J. Mol. Biol.*, 2001, **307**, 671.
166. D. R. Ronning, V. Vissa, G. S. Besra, J. T. Belisle and J. C. Sacchettini, *J. Biol. Chem.*, 2004, **279**, 36771.
167. J. Asselineau and G. Lanéelle, *Frontiers Biosci.*, 1998, **3**, 164.
168. D. Ghosh, M. Sawicki, P. Lala, M. Erman, W. Pangborn, J. Eyzaguirre, R. Gutiérrez, H. Jörnvall and D. J. Thiel, *J. Biol. Chem.*, 2001, **276**, 11159.
169. C. Martinez, A. Nicolas, H. van Tilbeurgh, M. P. Egloff, C. Cudrey, R. Verger and C. Cambillau, *Biochemistry*, 1994, **33**, 83.
170. E. Bitto, C. A. Bingman, J. G. McCoy, S. T. M. Allard, G. E. Wesenberg and G. N. Phillips, *Acta Crystallogr. Sect. D*, 2005, **61**, 1655.
171. F. Vincent, S. J. Charnock, K. H. G. Verschueren, J. P. Turkenburg, D. J. Scott, W. A. Offen, S. Roberts, G. Pell, H. J. Gilbert, G. J. Davies and J. A. Brannigan, *J. Mol. Biol.*, 2003, **330**, 593.
172. A. Mølgaard, S. Kauppinen and S. Larsen, *Structure*, 2000, **8**, 373.
173. D. E. Blair, A. W. Schüttelkopf, J. I. MacRae and D. M. F. van Aalten, *Proc. Natl. Acad. Sci. USA*, 2005, **102**, 15429.
174. D. Kafetzopoulos, G. Thireos, J. N. Vournakis and V. Bouriotis, *Proc. Natl. Acad. Sci. USA*, 1993, **90**, 8005.
175. T. Fukushima, T. Kitajima and J. Sekiguchi, *J. Bacteriol.*, 2005, **187**, 1287.
176. D. A. Whittington, K. M. Rusche, H. Shin, C. A. Fierke and D. W. Christianson, *Proc. Natl. Acad. Sci. USA*, 2003, **100**, 8146.
177. W. L. Mock, in *Comprehensive Biological Catalysis*, ed. M. L. Sinnott, Academic Press, London, 1998, Vol. 1, p. 425.
178. J. Jenkins, O. Mayans, D. Smith, K. Worboys and R. W. Pickersgill, *J. Mol. Biol.*, 2001, **305**, 951.
179. K. Johannsson, M. El-Ahmad, R. Friemann, H. Jörnvall, O. Markovič and H. Eklund, *FEBS Lett.*, 2002, **514**, 243.
180. T. D. Meek, in *Comprehensive Biological Catalysis*, ed. M. L. Sinnott, Academic Press, London, 1998, Vol. 1, p. 327.
181. L. J. Hyland, T. A. Tomaszek, G. D. Roberts, S. A. Carr, V. W. Magaard, H. L. Bryan, S. A. Fakhoury, M. L. Moore, M. D. Minnich, J. S. Culp, R. L. DesJarlais and T. D. Meek, *Biochemistry*, 1991, **30**, 8441.

182. F. Vincent, D. Yates, E. Garman, G. J. Davies and J. A. Brannigan, *J. Biol. Chem.*, 2004, **279**, 2809.
183. M. A. Halcrow in *Comprehensive Biological Catalysis*, ed. M. L. Sinnott, Academic Press, London, 1998, Vol. 1, p. 506.
184. C. Xu, R. Hall, J. Cummings and F. M. Raushel, *J. Am. Chem. Soc.*, 2006, **128**, 4244.
185. F. M. Raushel *et al.*, unpublished work, 2006; Dr Raushel is thanked for information prior to publication and his helpful e-mail about CE 9 mechanisms.
186. T. Desai, J. Gigg and R. Gigg, *Carbohydr. Res.*, 1995, **277**, C5.
187. E.g. J. G. Randolph, K. F. McClure and S. J. Danishefsky, *J. Am. Chem. Soc.*, 1995, **117**, 5712.
188. M. L. Sinnott and M. Underwood, unpublished data, 1999.
189. H. P. Kleine and R. S. Sidhu, *Carbohydr. Res.*, 1988, **182**, 307.
190. S. S. Bhattacharjee, J. A. Schwarcz and A. S. Perlin, *Carbohydr. Res.*, 1975, **42**, 259.
191. Comprehensive reviews of cyclic acetals and ketals of aldoses and aldosides of the "classical" period: A. N. de Belder, *Adv. Carbohydr. Chem.* 1965, **20**, 219; updated, A. N. de Belder, *Adv. Carbohydr. Chem. Biochem.* 1977, **34**, 179.
192. Review of cyclic acetals and ketals of ketoses: R. F. Brady, *Adv. Carbohydr. Chem. Biochem.*, 1971, **26**, 197.
193. N. Baggett, J. M. Duxbury, A. M. Foster and J. M. Webber, *Carbohydr. Res.*, 1966, **2**, 216.
194. J. G. Buchanan and R. M. Saunders, *J. Chem. Soc.*, 1964, 1796.
195. E. Percival, *Structural Carbohydrate Chemistry*, E. Garnett Miller, London, 1962, p. 68.
196. P. J. Garegg and T. Iversen, *Carbohydr. Res.*, 1979, **70**, C13.
197. S. V. Ley, R. Leslie, P. D. Tiffin and M. Woods, *Tetrahedron Lett.*, 1992, **33**, 4767.
198. S. V. Ley, D. K. Baeschlin, D. J. Dixon, A. C. Foster, S. J. Ince, H. W. M. Priepke and D. J. Reynolds, *Chem. Rev.*, 2001, **101**, 53.
199. F. Chéry, P. Rollin, O. De Lucchi and S. Cossu, *Tetrahedron Lett.*, 2000, **41**, 2357.
200. B. Capon, *Chem. Rev.*, 1969, **69**, 407.
201. T. H. Fife and L. K. Jao, *J. Org. Chem.*, 1965, **30**, 1492.
202. B. Capon and M. I. Page, *J. Chem. Soc. D, Chem. Commun.*, 1970, 1443.
203. R. A. McClelland, B. Watada and C. S. Q. Lew, *J. Chem. Soc., Perkin Trans. 2*, 1993, 1723.
204. M. L. Sinnott, *Adv. Phys. Org. Chem.*, 1988, **24**, 113.
205. R. Johansson and B. Samuelsson, *J. Chem. Soc., Perkin Trans. 1*, 1984, 2371.
206. P. J. Garegg, H. Hultberg and S. Wallin, *Carbohydr. Res.*, 1982, **108**, 97.
207. Initial discovery: S. Hanessian, *Carbohyd. Res.* 1966, **2**, 86; scope and limitations: S. Hanessian and N. R. Plessas, *J. Org. Chem.*, 1969, **34**, 1035, 1045 and 1053. The very many subsequent variants can be followed in the

preparative review literature, *e.g.* K. Jarowicki and P. Kocienski, *J. Chem. Soc., Perkin Trans. 1*, 2001, 2109.
208. E. Schönbrunn, S. Sack, S. Eschenburg, A. Perrakis, F. Krekel, N. Amrhein and E. Mandelkow, *Structure*, 1996, **4**, 1065.
209. W. J. Lees and C. T. Walsh, *J. Am. Chem. Soc.*, 1995, **117**, 7329.
210. D. H. Kim, W. J. Lees and C. T. Walsh, *J. Am. Chem. Soc.*, 1995, **117**, 6380.
211. B. Byczynski, S. Mizyed and P. J. Berti, *J. Am. Chem. Soc.*, 2003, **125**, 12541.
212. H. Kim, W. J. Lees, K. E. Kempsell, W. S. Lane, K. Duncan and C. T. Walsh, *Biochemistry*, 1996, **35**, 4923.
213. T. E. Benson, C. T. Walsh and J. M Hogle, *Structure*, 1996, **4**, 47.
214. T. E. Benson, C. T. Walsh and J. M. Hogle, *Biochemistry*, 1997, **36**, 806.
215. S. J. Angyal, *Angew. Chem. Int. Ed.*, 1969, **8**, 157.
216. C. R. Shoaf, W. D. Heizer and M. D. Caplow, *Carbohydr. Res.*, 1982, **103**, 195.
217. R. van den Berg, J. A. Peters and H. van Bekkum, *Carbohydr. Res.*, 1994, **253**, 1.
218. S. Chapelle and J.-F. Verchere, *Tetrahedron*, 1988, **44**, 4469.
219. S. G. Angyal, D. Greeves and V. A. Pickles, *Carbohydr. Res.*, 1974, **35**, 165.
220. P. J. Wood and R. Siddiqui, *Carbohydr. Res.*, 1974, **36**, 247.
221. X. Gao, Y. Zhang and B. Wang, *Tetrahedron*, 2005, **61**, 9111.
222. A. N. J. Moore and D. D. M. Wayner, *Can. J. Chem.*, 1999, **77**, 681.
223. J. C. Norrild and I. Søtofte, *J. Chem. Soc., Perkin Trans. 2*, 2002, 303.
224. H. Otsuka, E. Uchimura, H. Koshino, T. Okano and K. Kataoka, *J. Am. Chem. Soc.*, 2003, **125**, 3493.
225. M. P. Nicholls and P. K. C. Paul, *Org. Biomol. Chem.*, 2004, **2**, 1434.
226. S. Oae, N. Asai and K. Fujimori, *J. Chem. Soc., Perkin Trans. 2*, 1978, 571.
227. M. J. Crookes and D. L. H. Williams, *J. Chem. Soc., Perkin Trans. 2*, 1989, 1319.
228. E. Eglesias, I. García Río, J. R. Leis, M. E. Peña and D. L. H. Williams, *J. Chem. Soc., Perkin Trans. 2*, 1992, 1673.
229. E. Eglesias, *J. Am. Chem. Soc.*, 1998, **120**, 13057.
230. E. Eglesias, *New J. Chem.*, 2000, **24**, 1025.
231. T. F. Sikora, revised R. L. Dilworth and M. E. Heldberg, *Military Explosives*, Technical Manual, TM9-1300-214, United States Army Headquarters, Washington, DC., 1990, p. 2-5.
232. H. Braconnot, *Ann. Chim Phys.*, 1819, **12**, 185.
233. H. Braconnot, *Ann. Chim Phys.*, 1833, **52**, 290.
234. C. H. D. Buijs Ballat, *J. Prakt. Chem.*, 1844, **31**, 209.
235. T.-J. Pelouze, *J. Prakt. Chem.*, 1838, **16**, 168.
236. C. F. Schönbein, *Philos. Mag.*, 1846, **31**, 7.
237. T. F. Sikora, revised R. L. Dilworth and M. E. Heldberg, *Military Explosives*, Technical Manual, TM9-1300-214, United States Army Headquarters, Washington, DC., 1990, pp. 8-1*ff.*

238. E. L. Blackall, E. D. Hughes, C. Ingold and R. B. Preston, *J. Chem. Soc.*, 1958, 4366.
239. R. D. Short and H. S. Munro, *Polymer*, 1993, **34**, 2714.
240. T. F. Sikora, revised R. L. Dilworth and K. E. Heldberg, *Military Explosives*, Technical Manual, TM9-1300-214, United States Army Headquarters, Washington, DC., 1990, p. 8-2.
241. D. Meader, E. D. T. Atkins and F. Happey, *Polymer*, 1978, **19**, 1371.
242. T. F. Sikora, revised R. L. Dilworth and M. E. Heldberg, *Military Explosives*, Technical Manual, TM9-1300-214, United States Army Headquarters, Washington, DC., 1990, pp. 4-1*ff.*
243. M. A. Hiskey, K. R. Brower and J. C. Oxley, *J. Phys. Chem.*, 1991, **95**, 3955.
244. J. K. Chen and T. B. Brill, *Combust. Flame*, 1991, **85**, 479.
245. D. L. Naud and K. R. Brower, *J. Org. Chem.*, 1992, **57**, 3303.
246. L. L. Davis and K. R. Brower, *J. Phys. Chem.*, 1996, **100**, 18775.
247. Z. A. Dreger, Y. A. Gruzdkov, Y. M. Gupta and J. J. Dick, *J. Phys. Chem. B*, 2002, **106**, 247.
248. E. Sjöström, *Wood Chemistry*, Academic Press, San Diego, 2nd edn, 1993, p. 210.
249. K. Katoh, L. Le, M. Kumasaki, Y. Wada, M. Arai and M. Tamura, *Thermochim. Acta*, 2005, **431**, 161.
250. K. Katoh, L. Le, M. Kumasaki, Y. Wada and M. Arai, *Thermochim. Acta*, 2005, **431**, 168.
251. B. A. Lurye, Z. T. Valishina and B. S. Svetlov, *Vysokomol. Soyedin., Ser. A*, 1991, **33**, 100.
252. E. M. Southern, *J. Mol. Biol.*, 1975, **98**, 503.
253. D. R. Janero, N. S. Bryan, F. Saijo, V. Dhawan, D. J. Schwalb, M. C. Warren and M. Feelisch, *Proc. Natl. Acad. Sci. USA*, 2004, **101**, 16958.
254. Z. Chen, J. Zhang and J. S. Stamler, *Proc. Natl. Acad. Sci. USA*, 2002, **99**, 8306.
255. Z. Chen, M. W. Foster, J. Zhang, L. Mao, H. A. Rockman, T. Kawamoto, K. Kitagawa, K. I. Nakayama, D. T. Hess and J. S. Stamler, *Proc. Natl. Acad. Sci. USA*, 2005, **102**, 12159.
256. Y. Meah, B. J. Brown, S. Chakraborty and V. Massey, *Proc. Natl. Acad. Sci. USA*, 2001, **98**, 8560.
257. H. Li, H. Cui, X. Liu and J. L. Zweier, *J. Biol. Chem.*, 2005, **280**, 16594.
258. G. R. J. Thatcher, in E. van Faassen and A. Vanin, Eds., *Radicals for Life: the Various Forms of Nitric Oxide*, Elsevier, Amsterdam, 2007, p. 347.
259. F. H. Westheimer, *Acc. Chem. Res.*, 1968, **1**, 70.
260. R. Kluger, F. Covitz, E. Dennis, L. D. Williams and F. H. Westheimer, *J. Am. Chem. Soc.*, 1969, **91**, 6066.
261. Recent theoretical work on the pseudorotation in this context: C. Silva López, O. Nieto Faza, A. R. de Lera and D. M. York, *Chem. Eur. J.*, 2005, **11**, 2081.

262. O. Mitsunobu, *Synthesis*, 1981, **1**, 1.
263. D. L. Hughes, R. A. Reamer, J. J. Bergan and E. J. J. Grabowski, *J. Am. Chem. Soc.*, 1988, **110**, 6487.
264. D. Camp and I. D. Jenkins, *J. Org. Chem.* 1989 **54**, (a) 3045; (b) 3049.
265. S. Schenk, J. Weston and E. Anders, *J. Am. Chem. Soc.*, 2005, **127**, 12566.
266. A. B. Hughes and M. M. Sleebs, *J. Org. Chem.*, 2005, **70**, 3079.
267. R. Persky and A. Albeck, *J. Org. Chem.*, 2000, **65**, 3775.
268. R. Rabinowitz and R. Marcus, *J. Am. Chem. Soc.*, 1962, **84**, 1312.
269. N. B. Desai and N. McKelvie, *J. Am. Chem. Soc.*, 1962, **84**, 1745.
270. L. J. McBride and M. H. Caruthers, *Tetrahedron Lett.*, 1983, **24**, 245.
271. A. B. Sierzchala, D. J. Dellinger, J. R. Betley, T. K. Wyrzykiewicz, C. M. Yamada and M. H. Caruthers, *J. Am. Chem. Soc.*, 2003, **125**, 13427.
272. W. J. Stech and G. Zon, *Tetrahedron Lett.*, 1984, **25**, 5279.
273. P. M. Cullis, *J. Chem. Soc., Chem. Commun.*, 1984, 1510.
274. E. Marsault and G. Just, *Tetrahedron*, 1997, **53**, 16945.
275. R. P. Iyer, W. Egan, J. B. Eagan and S. L Beaucage, *J. Am. Chem. Soc.*, 1990, **112**, 1253.
276. F. Casero, L. Cipolla, L. Lay, F. Nicotra, L. Panza and G. Russo, *J. Org. Chem.*, 1996, **61**, 3428.
277. A. Hengge, in *Comprehensive Biological Catalysis*, ed. M. L. Sinnott, Academic Press, London, 1998, Vol. 1, p. 517.
278. (a) S. L. Buchwald, D. H. Pliura and J. R. Knowles, *J. Am. Chem. Soc.*, 1984, **106**, 4911; (b) J. M. Friedman, S. Freeman and J. R. Knowles, *J. Am. Chem. Soc.*, 1988, **110**, 1268.
279. P. M. Cullis and A. Iagrosso, *J. Am. Chem. Soc.*, 1986, **108**, 7870.
280. J. Burgess, N. Blundell, P. M. Cullis, C. D. Hubbard and R. Misra, *J. Am. Chem. Soc.*, 1988, **110**, 7900.
281. B. Gerratana, G. A. Sowa and W. W. Cleland, *J. Am. Chem. Soc.*, 2000, **122**, 12615.
282. W. W. Cleland and A. A. Hengge, *FASEB J.*, 1995, **9**, 1585.
283. A. C. Hengge, W. A. Edens and H. Elsing, *J. Am. Chem. Soc.*, 1994, **116**, 5045.
284. A. C. Hengge, A. F. Tobin and W. W. Hengge, *J. Am. Chem. Soc.*, 1995, **117**, 5919.
285. For a recent structural example, see D. Grueninger and G. E. Schulz, *J. Mol. Biol.*, 2006, **359**, 787.
286. J. P. Jones, P. M. Weiss and W. W. Cleland, *Biochemistry*, 1991, **30**, 3634.
287. E. K. Jaffe and M. Cohn, *J. Biol. Chem.*, 1978, **253**, 4823.
288. P. J. O'Brien and D. Herschlag, *Biochemistry*, 2002, **41**, 3207.
289. P. M. Weiss and W. W. Cleland, *J. Am. Chem. Soc.*, 1989, **111**, 1928.
290. R. Parvin and R. A. Smith, *Biochemistry*, 1969, **8**, 1748.
291. A. Ghosh, J.-J. Shieh, C.-J. Pan and J. Y. Chou, *J. Biol. Chem.*, 2004, **279**, 12479.
292. J.-Y. Choe, S. W. Nelson, H. J. Fromm and R. B. Honzatko, *J. Biol. Chem.*, 2003, **278**, 16008.

293. J.-Y. Choe, C. V. Iancu, H. J. Fromm and R. B. Honzatko, *J. Biol. Chem.*, 2003, **278**, 16015.
294. C. M. J. Fauroux, M. Lee, P. M. Cullis, K. T. Douglas, M. G. Gore and S. Freeman, *J. Med. Chem.*, 2002, **45**, 1363.
295. S. J. Pollack, J. R. Atack, M. R. Knowles, G. McAllister, C. I. Ragan, R. Baker, S. R. Fletcher, L. L. Iversen and H. B. Broughton, *Proc. Natl. Acad. Sci. USA*, 1994, **91**, 5766.
296. L. E. Naught, C. Regni, L. J. Beamer and P. A. Tipton, *Biochemistry*, 2003, **42**, 9946.
297. L. E. Naught and P. A. Tipton, *Biochemistry*, 2005, **44**, 6831.
298. M. D. Percival and S. G. Withers, *Biochemistry*, 1992, **31**, 505.
299. G. Zhang, J. Dai, L. Wang, D. Dunaway-Mariano, L. W. Tremblay and K. N. Allen, *Biochemistry*, 2005, **44**, 9404.
300. S. D. Lahiri, G. Zhang, D. Dunaway-Mariano and K. N. Allen, *Science*, 2003, **299**, 2067.
301. G. M. Blackburn, N. H. Williams, S. J. Gamblin and S. J. Smerdon, *Science*, 2003, **301**, 1184.
302. N. H. Williams, in *Comprehensive Biological Catalysis*, ed. M. L. Sinnott, Academic Press, London, 1998, Vol. 1, p. 543.
303. R. J. Hondal, Z. Zhao, A. V. Kravchuk, H. Liao, S. R. Riddle, X. Yue, K. S. Bruzik and M.-D. Tsai, *Biochemistry*, 1998, **37**, 4568.
304. D. W. Heinz, M. Ryan, T. L. Bullock and O. H. Griffith, *EMBO J.*, 1995, **14**, 3855.
305. R. J. Hondal, S. R. Riddle, A. V. Kravchuk, Z. Zhao, H. Liao, K. S. Bruzik and M.-D. Tsai, *Biochemistry*, 1997, **36**, 6633.
306. L.-O. Essen, O. Perisic, M. Katan, Y. Wu, M. F. Roberts and R. L. Williams, *Biochemistry*, 1997, **36**, 1704.
307. J. A. Grasby, in *Comprehensive Biological Catalysis*, ed. M. L. Sinnott, Academic Press, London, 1998, Vol. 1, p. 563.
308. M. A. Engelhardt, E. A. Doherty, D. S. Knitt, J. A. Doudna and D. Herschlag, *Biochemistry*, 2000, **39**, 2639.
309. S. Shan, A. V. Kravchuk, J. A. Piccirilli and D. Herschlag, *Biochemistry*, 2001, **40**, 5161.
310. D. Haigh, L. J. Jefcott, K. Magee and H. McNab, *J. Chem. Soc., Perkin Trans. 1*, 1996, 2895.
311. P. Borrachero, F. Cabrera-Escribano, A. T. Carmona and M. Gómez-Guillén, *Tetrahedron Asymmetry*, 2000, **11**, 2927.
312. Y. Yoshida, Y. Sakakura, N. Aso, S. Okada and Y. Tanabe, *Tetrahedron*, 1999, **55**, 2183.
313. J. F. King, J. Y. L. Lam and S. Skonieczny, *J. Am. Chem. Soc.*, 1992, **114**, 1743.
314. S. Lyashchuk, Y. G. Skrypnik and V. Besrodnyi, *J. Chem. Soc., Perkin Trans.*, 1993, **2**, 1153.
315. T. Umemoto, G. Tomizawa, H. Hachisuka and M. Kitano, *J. Fluorine Chem.*, 1996, **77**, 161.

316. M. C. Whiting and H. J. Storesund, *J. Chem. Soc., Perkin Trans. 2*, 1975, 1452.
317. C. N. Cooper, P. J. Jenner, N. B. Perry, J. Russell-King, H. J. Storesund and M. C. Whiting, *J. Chem. Soc., Perkin Trans. 2*, 1982, 605.
318. www.chem.wisc.edu/areas/reich/pkatable/index.htm. This site displays the p*K* data measured by F. W. Bordwell.
319. V. Moreau, J. C. Norrild and H. Driguez, *Carbohydr. Res.*, 1997, **300**, 271.
320. U. G. Nayak and R. L. Whistler, *J. Org. Chem.*, 1969, **34**, 3819.
321. J. M. Ballard, L. Hough, A. C. Richardson and P. H. Fairclough, *J. Chem. Soc., Perkin Trans. 1*, 1973, 1524.
322. C. K. Lee, *Carbohydr. Res.*, 1987, **162**, 53.
323. L. Hough, T. Suami and T. Machinami, *Pure Appl. Chem.*, 1997, **69**, 693.
324. (a) K. A. Francesconi, J. S. Edmonds, R. V. Stick, B. W. Skelton and A. J. White, *J. Chem. Soc., Perkin Trans. 1*, 1991, 2707; (b) J. S. Edmonds, K. A. Francesconi and R. V. Stick, *Nat. Prod. Rep.*, 1993, **10**, 421.
325. R. H. Hoff, P. Larsen and A. C. Hengge, *J. Am. Chem. Soc.*, 2001, **123**, 9338.
326. E.g. E. Kaji and N. Harita, *Tetrahedron Lett.*, 2000, **41**, 53.
327. T. S. Cameron, P. K. Bakshi, R. Thangarasa and T. B. Grindley, *Can. J. Chem.*, 1992, **70**, 1623.
328. S. David and Thieffry, *J. Chem. Soc., Perkin Trans. 1*, 1979, 1568.
329. Ref. 200. No mechanistic work of significance has been done on the bromine oxidation of aldoses since this review.
330. H. S. Isbell and L. T. Sniegoski, *J. Res. Natl. Bur. Stand. (U. S.)*, 1964, **A68**, 145.
331. B. Alfredsson and O. Samuelson, *Sven. Papperstidn.*, 1974, **77**, 449.
332. (a) A. J. Mancuso and D. Swern, *Synthesis*, 1981, 165; (b) T. T. Tidwell, *Synthesis*, 1990, 857.
333. N. L. Holder, *Chem. Rev.*, 1982, **82**, 287.
334. J. Roček, F.H. Westheimer, A. Eschenmoser, L. Moldoványi and J. Schreiber, *Helv. Chim. Acta*, 1962, **45**, 2554.
335. H. Kwart and J. C. Nickle, *J. Am. Chem. Soc.*, 1973, **95**, 3394.
336. J. Hersovici, M.-J. Egron and K. Antonakis, *J. Chem. Soc., Perkin Trans. 1*, 1982, 1967.
337. P. J. Garegg and B. Samuelsson, *Carbohydr. Res.*, 1978, **67**, 267.
338. B. Özgün and N. Değirmenbaşi, *Monatsh. Chem.*, 2004, **135**, 483.
339. S. I. Scott, A. Bakac and J. H. Espenson, *J. Am. Chem. Soc.*, 1992, **114**, 4205.
340. M. Krumpolc and J. Roček, *Inorg. Chem.*, 1985, **24**, 617.
341. S. Ramesh, S. N. Malpatro, J. H. Liu and J. Roček, *J. Am. Chem. Soc.*, 1981, **103**, 5172.
342. G. Y. Pan, C. L. Chen, J. S. Gratzl and H. M. Chang, *Res. Chem. Intermed.*, 1995, **21**, 205.

343. P. S. Fredricks, B. O. Lindgren and O. Theander, *Sven. Papperstidn.*, 1971, **74**, 597.
344. T. Tuttle, J. Cerkovnik, B. Plesničar and D. Cremer, *J. Am. Chem. Soc.*, 2004, **126**, 16093.
345. P. Deslongchamps, C. Moreau, D. Fréhel and P. Atlani, *Can. J. Chem.*, 1972, **50**, 3402.
346. F. Kovac and B. Plesničar, *J. Am. Chem. Soc.*, 1979, **101**, 2677.
347. M. Adinolfi, G. Baroni and A. Iadonisi, *Tetrahedron Lett.*, 1998, **39**, 7405.
348. A. R. Clarke and T. R. Dafforn, in *Comprehensive Biological Catalysis*, ed. M. L. Sinnott, Vol. 3, p. 1.
349. S. A. Benner, *Experientia*, 1982, **38**, 633.
350. N. J. Oppenheimer, *J. Am. Chem. Soc.*, 1984, **106**, 3032.
351. S. B. Mostad, H. L. Helming, C. Groom and A. Glasfeld, *Biochem. Biophys. Res. Commun.*, 1997, **233**, 681.
352. C. Filling, K. D. Berndt, J. Benach, S. Knapp, T. Prozorovski, E. Nordling, R. Ladenstein, H. Jörnvall and U. Oppermann, *J. Biol. Chem.*, 2002, **277**, 25677.
353. A. Philippsen, T. Schirmer, M. A. Stein, F. Giffhorn and J. Stetefeld, *Acta Crystallogr., Sect. D*, 2005, **61**, 374.
354. J. M. Petrash, *Cell. Mol. Life Sci.*, 2004, **61**, 737; this clears up much confusion and quotes a CAZy-like website (www.med.upenn.edu/akr/), last updated in November 2006.
355. (a) J. M. Rondeau, F. Tête-Favier, A. Podjarny, J. M. Reymann, P. Barth, J. F. Biellmann and D. Moras, *Nature*, 1992, **355**, 469; (b) D. K. Wilson, K. M. Bohren, K. H. Gabbay and F. A. Quiocho, *Science*, 1992, **257**, 81.
356. K. M. Bohren, J. M. Brownlee, A. C. Milne, K. H. Gabbay and D. H.T. Harrison, *Biochim. Biophys Acta*, 2005, **1748**, 201.
357. F. Ruiz, I. Hazemann, A. Mitschler, A. Joachimiak, T. Schneider, M. Karplus and A. Podjarny, *Acta Crystallogr., Sect. D*, 2004, **60**, 1347.
358. P. Várnai and A. Warshel, *J. Am. Chem. Soc.*, 2000, **122**, 3849.
359. K. M. Bohren and C. E. Grimshaw, *Biochemistry*, 2000, **39**, 9967.
360. K. L. Kavanagh, M. Klimacek, B. Nidetzky and D. K. Wilson, *Biochemistry*, 2002, **41**, 8785.
361. D. K. Wilson, K. L. Kavanagh, M. Klimacek and B. Nidetzky, *Chem. Biol. Interact.*, 2003, **143**, 515.
362. R. Kratzer, K. L. Kavanagh, D. K. Wilson and B. Nidetzky, *Biochemistry*, 2004, **43**, 4944.
363. M. Klimacek, K. L. Kavanagh, D. K. Wilson and B. Nidetzky, *Chem. Biol. Interact.*, 2003, **143**, 559.
364. K. L. Kavanagh, M. Klimacek, B. Nidetzky and D. K. Wilson, *J. Biol. Chem.*, 2002, **277**, 43433.
365. M. Slatner, B. Nidetzky and K. D. Kulbe, *Biochemistry*, 1999, **38**, 10489.
366. M. Klimacek and B. Nidetzky, *Biochemistry*, 2002, **41**, 10158.
367. M. S. Cosgrove, S. Gover, C. E. Naylor, L. Vandeputte-Rutten, M. J. Adams and H. R. Levy, *Biochemistry*, 2000, **39**, 15002.

368. M. S. Cosgrove, C. E. Naylor, S. Paludan, M. J. Adams and H. R. Levy, *Biochemistry*, 1998, **37**, 2759.
369. M. S. Cosgrove, S. N. Loh, J.-H. Ha and H. R. Levy, *Biochemistry*, 2002, **41**, 6939.
370. M. J. Banfield, M. E. Salvucci, E. N. Baker and C. A. Smith, *J. Mol. Biol.*, 2001, **306**, 239.
371. T. A. Pauly, J. L. Ekstrom, D. A. Beebe, B. Chrunyk, D. Cunningham, M. Griffor, A. Kamath, S. E. Lee, R. Madura, D. Mcguire, T. Subashi, D. Wasilko, P. Watts, B. L. Mylari, P. J. Oates, P. D. Adams and V. L. Rath, *Structure*, 2003, **11**, 1071.
372. R. Lunzer, Y. Mamnun, D. Haltrich, K.D. Kulbe and B. Nidetzky, *Biochem. J.*, 1998, **336**, 91.
373. Reviewed to mid-2001 by J. Samuel and M. E. Tanner, *Nat. Prod. Rep.* 2002, **19**, 261.
374. J. B. Thoden and H. M. Holden, *J. Biol. Chem.*, 2005, **280**, 21900.
375. G. J. Buist and C. A. Bunton, *J. Chem. Soc.*, 1954, 1406.
376. G. J. Buist and C. A. Bunton, *J. Chem. Soc. B*, 1971, 2117.
377. G. J. Buist, C. A. Bunton and J. Lomas, *J. Chem. Soc. B*, 1966, 1094.
378. G. J. Buist, C. A. Bunton and J. Lomas, *J. Chem. Soc. B*, 1966, 1099.
379. G. J. Buist, C. A. Bunton and W. C. P. Hipperson, *J. Chem. Soc. B*, 1971, 2128.
380. L. Maros, I. Molnár-Perl, E. Schissel and V. Szerdahelyi, *J. Chem. Soc., Perkin Trans. 2*, 1980, 39.
381. A. S. Perlin and E. von Rudloff, *Can. J. Chem.*, 1965, **43**, 2071.
382. (a) J. E. Taylor, *J. Phys. Chem.* 1995, **99**, 59; (b) J. E. Taylor, *J. Phys. Chem. A*, 1998, **102**, 2172; (c) J. E. Taylor and H. Masui, *J. Phys. Chem. A*, 2001, **105**, 3532.
383. R. A. More O'Ferrall, *J. Chem. Soc. B*, 1970, 274.
384. W. P. Jencks, *Chem. Rev.*, 1972, **72**, 705.
385. J. Sicher, *Angew. Chem. Int. Ed. Engl.*, 1972, **11**, 200.
386. T. H. Lowry and K. S. Richardson, *Mechanism and Theory in Organic Chemistry*, Harper Collins, New York, 1987, p. 610.
387. F. Baumberger, D. Beer, M. Christen, R. Prewo and A. Vasella, *Helv. Chim. Acta*, 1986, **69**, 1191.
388. L Somsák, *Chem. Rev.*, 2001, **101**, 81.
389. N. Kaila, M. Blumenstein, H. Bielawska and R. W. Franck, *J. Org Chem.*, 1992, **57**, 4576.
390. A. Boschi, C. Chiappe, A. De Rubertis and M. F. Ruasse, *J. Org Chem.*, 2000, **65**, 8470.
391. M. D. Burkart, Z. Zhang, S.-C. Hung and C.-H. Wong, *J. Am. Chem. Soc.*, 1997, **119**, 11743.
392. M. Albert, K. Dax and J. Ortner, *Tetrahedron*, 1998, **54**, 4839.
393. S. P. Vincent, M. D. Burkart, C.-Y. Tsai, Z. Zhang and C. -H. Wong, *J. Org. Chem.*, 1999, **64**, 5264.
394. R. U. Lemieux, T. L. Nagabhushan and I. K. O'Neill, *Can. J. Chem.*, 1968, **46**, 413.

395. S. J. Danishefsky and M. T. Bilodeau, *Angew. Chem. Int. Ed. Engl.*, 1996, **35**, 1381.
396. *E.g.* (a) T. Ercegovic and G. Magnusson, *J. Chem. Soc., Chem. Commun.* 1994, 831; (b) I. P. Smolyakova, W. A. Smit, E. A. Zal'chenko, O. S. Chizhov, A. S. Shashkov, R. Caple, S. Sharpe and C. Kuehl, *Tetrahedron Lett.* 1993, **34**, 3047.
397. R. J. Ferrier, *Top. Curr. Chem.*, 2001, **215**, 153.
398. R. J. Ferrier, W. G. Overend and A. E. Ryan, *J. Chem. Soc.*, 1962, 3667.
399. E. Fischer, *Chem. Ber.*, 1914, **47**, 196.
400. M. Bergmann, *Justus Liebigs Ann. Chem.*, 1925, **443**, 223.
401. D. P. Curran and Y.-G. Suh, *Carbohydr. Res.*, 1987, **171**, 161.
402. K. Takeda, H. Nakamura, A. Ayabe, A. Akiyama, Y. Harigaya and Y. Mizuno, *Tetrahedron Lett.*, 1994, **35**, 125.
403. G. Grynkiewicz, W. Priebe and A. Zamojski, *Carbohydr. Res.*, 1979, **68**, 33.
404. S. J. Danishefsky and J. F. Kerwin, *J. Org. Chem.*, 1982, **47**, 3803.
405. K. Toshima, G. Matsuo, T. Ishizuka, Y. Ushiki, M. Nakata and S. Matsumura, *J. Org. Chem.*, 1998, **63**, 2307.
406. A. de Raadt and R. J. Ferrier, *Carbohydr. Res.*, 1991, **216**, 93.
407. W. Priebe and A. Zamojski, *Tetrahedron*, 1980, **36**, 287.
408. R. D. Guthrie and R. W. Irvine, *Carbohydr. Res.*, 1980, **82**, 207.
409. R. J. Ferrier and M. M. Ponpipom, *J. Chem. Soc. C*, 1971, 553.
410. M. H. Johansson and O. Samuelson, *Carbohydr. Res.*, 1977, **54**, 295.
411. W. W. Cleland, in *Comprehensive Biological Catalysis*, ed. M. L. Sinnott, Academic Press, London, 1998, Vol. 2, p. 1.
412. M. Eigen, *Angew. Chem. Int. Ed. Engl.*, 1964, **3**, 1.
413. A. C. Lin, Y. Chiang, D. B. Dahlberg and A. J. Kresge, *J. Am. Chem. Soc.*, 1983, **105**, 5380.
414. J. P. Guthrie, *J. Phys. Org. Chem.*, 1998, **11**, 632.
415. A. Teleman, T. Hausalo, M. Tenkanen and T. Vuorinen, *Carbohydr. Res.*, 1996, **280**, 197.
416. A. Teleman, M. Siika-aho, H. Sorsa, J. Buchert, M. Perttula, T. Hausalo and M. Tenkanen, *Carbohydr. Res.*, 1996, **293**, 1.
417. J. Li and G. Gellerstedt, *Carbohydr. Res.*, 1997, **302**, 213.
418. P. Gacesa, *FEBS Lett.*, 1987, **212**, 199.
419. D. S. Feingold and R. Bentley, *FEBS Lett.*, 1987, **223**, 207.
420. W. W. Cleland and M. M. Kreevoy, *Science*, 1994, **264**, 1887.
421. J. P. Guthrie, *Chem. Biol.*, 1996, **3**, 163.
422. E. P. Serjeant and B. Dempsey, Eds, *Ionization Constants of Organic Acids in Solution*, IUPAC Chemical Data Series No. 23, Pergamon Press, Oxford, 1979.
423. Y. Maruyama, W. Hashimoto, B. Mikami and K. Murata, *J. Mol. Biol.*, 2005, **350**, 974.
424. V. V. Lunin, Y. Li, R. J. Linhardt, H. Miyazono, M. Kyogashima, T. Kaneko, A. W. Bell and M. Cygler, *J. Mol. Biol.*, 2004, **337**, 367.
425. S. Li and M. J. Jedrzejas, *J. Biol. Chem.*, 2001, **276**, 41407.

426. D. J. Rigden and M. J. Jedrzejas, *J. Biol. Chem.*, 2003, **278**, 50596.
427. S. Li, S. J. Kelly, E. Lamani, M. Ferraroni and M. J. Jedrzejas, *EMBO J.*, 2000, **19**, 1228.
428. M. Nukui, K. B. Taylor, D. T. McPherson, M. K. Shigenaga and M. J. Jedrzejas, *J. Biol. Chem.*, 2003, **278**, 3079.
429. W. Hashimoto, H. Nankai, B. Mikami and K. Murata, *J. Biol. Chem.*, 2003, **278**, 7663.
430. C. S. Rye and S. G. Withers, *J. Am. Chem. Soc.*, 2002, **124**, 9756.
431. C. S. Rye, A. Matte, M. Cygler and S. G. Withers, *ChemBioChem*, 2006, **7**, 631.
432. P. Sánchez-Torres, J. Visser and J. A. E. Benen, *Biochem. J.*, 2003, **370**, 331.
433. O. Mayans, M. Scott, I. Connerton, T. Gravesen, J. Benen, J. Visser, R. Pickersgill and J. Jenkins, *Structure*, 1997, **5**, 677.
434. J. Vitali, B. Schick, H. C. M. Kester, J. Visser and F. Jurnak, *Plant Physiol.*, 1998, **116**, 69.
435. M. Akita, A. Suzuki, T. Kobayashi, S. Ito and T. Yamane, *Acta Crystallogr., Sect. D*, 2001, **57**, 1786.
436. J. Jenkins, V. E. Shevchik, N. Hugouvieux-Cotte-Pattat and R. W. Pickersgill, *J. Biol. Chem.*, 2004, **279**, 9139.
437. H. Novoa de Armas, C. Verboven, C. De Ranter, J. Desair, A. Vande Broek, J. Vanderleyden and A. Rabijns, *Acta Crystallogr., Sect. D*, 2004, **60**, 999.
438. S. J. Charnock, I. E. Brown, J. P. Turkenburg, G. W. Black and G. J. Davies, *Proc. Natl. Acad. Sci. USA*, 2002, **99**, 12067.
439. M. A. McDonough, R. Kadirvelraj, P. Harris, J.-C. N. Poulsen and S. Larsen, *FEBS Lett.*, 2004, **565**, 188.
440. H.-J. Yoon, W. Hashimoto, O. Miyake, K. Murata and B. Mikami, *J. Mol. Biol.*, 2001, **307**, 9.
441. H. J. Yoon, Y. J. Choi, O. Miyake, W. Hashimoto, K. Murata and B. Mikami, *J. Microbiol. Biotechnol.*, 2001, **11**, 118.
442. M. Yamasaki, K. Ogura, W. Hashimoto, B. Mikami and K. Murata, *J. Mol. Biol.*, 2005, **352**, 11.
443. T. Osawa, Y. Matsubara, T. Muramatsu, M. Kimura and Y. Kakuta, *J. Mol. Biol.*, 2005, **345**, 1111.
444. W. Huang, A. Matte, Y. Li, Y. S. Kim, R. J. Linhardt, H. Su and M. Cygler, *J. Mol. Biol.*, 1999, **294**, 1257.
445. N. L. Smith, E. J. Taylor, A.-M. Lindsay, S. J. Charnock, J. P. Turkenburg, E. J. Dodson, G. J. Davies and G. W. Black, *Proc. Natl. Acad. Sci. USA*, 2005, **102**, 17652.
446. A. Hagner-McWhirter, U. Lindahl and J.-P. Li, *Biochem. J.*, 2000, **347**, 69.
447. B. Larsen and H. Grasdalen, *Carbohydr. Res.*, 1981, **92**, 163.
448. S. A. Douthit, M. Dlakic, D. E. Ohman and M. J. Franklin, *J. Bacteriol.*, 2005, **187**, 4573.
449. M. Sletmoen, G. Skjåk-Bræk and B. T. Stokke, *Carbohydr. Res.*, 2005, **340**, 2782.

450. B. I. G. Svanem, G. Skjåk-Bræk, H. Ertesvåg and S. Valla, *J. Bacteriol.*, 1999, **181**, 68.
451. C. Campa, S. Holtan, N. Nilsen, T. M. Bjerkan, B. T. Stokke and G. Skjåk-Bræk, *Biochem. J.*, 2004, **381**, 155.
452. F. L. Aachmann, B. I. G. Svanem, P. Güntert, S. B. Petersen, S. Valla and R. Wimmer, *J. Biol. Chem.*, 2006, **281**, 7350.
453. J. L. Palmer and R. A. Abeles, *J. Biol. Chem.* (a) 1976, **251**, 5817; (b) 1979, **254**, 1217.
454. Y. Hu, J. Komoto, Y. Huang, T. Gomi, H. Ogawa, Y. Takata, M. Fujioka and F. Takusagawa, *Biochemistry*, 1999, **38**, 8323.
455. M. A. Turner, C. S. Yuan, R. T. Borchardt, M. S. Hershfield, G. D. Smith and P. L. Howell, *Nat. Struct. Biol.*, 1998, **5**, 369.
456. D. J. T. Porter, *J. Biol. Chem.*, 1993, **268**, 66.
457. N. Tanaka, M. Nakanishi, Y. Kusakabe, K. Shiraiwa, S. Yabe, Y. Ito, Y. Kitade and K. T. Nakamura, *J. Mol. Biol.*, 2004, **343**, 1007.
458. R. J. Parry and L. J. Askonas, *J. Am. Chem. Soc*, 1985, **107**, 1417.
459. D. J. T. Porter and F. L. Boyd, *J. Biol. Chem.*, 1991, **266**, 21616.
460. X. Yang, Y. Hu, D. H. Yin, M. A. Turner, M. Wang, R. T. Borchardt, P. L. Howell, K. Kuczera and R. L. Schowen, *Biochemistry*, 2003, **42**, 1900.
461. Y. Takata, T. Yamada, Y. Huang, J. Komoto, T. Gomi, H. Ogawa, M. Fujioka and F. Takusagawa, *J. Biol Chem.*, 2002, **277**, 22670.
462. X. He and H.-W. Liu, *Curr. Opin. Chem. Biol.*, 2002, **6**, 590.
463. J. W. Gross, A. D. Hegeman, M. M. Vestling and P. A. Frey, *Biochemistry*, 2000, **39**, 13633.
464. T. M. Hallis and H.-W. Liu, *Tetrahedron*, 1998, **54**, 15975.
465. C. E. Snipes, G. U. Brillinger, L. Sellers, L. Mascaro and H. G. Floss, *J. Biol. Chem.*, 1977, **252**, 8113.
466. Y. Yu, R. N. Russell, J. S. Thorson, L.-D. Liu and H.-W. Liu, *J. Biol. Chem.*, 1992, **267**, 5868.
467. S. T. M. Allard, K. Beis, M.-F. Giraud, A. D. Hegeman, J. W. Gross, R. C. Wilmouth, C. Whitfield, M. Graninger, P. Messner, A. G. Allen, D. J. Maskell and J. H. Naismith, *Structure*, 2002, **10**, 81.
468. B. Gerratana, W. W. Cleland and P. A. Frey, *Biochemistry*, 2001, **40**, 9187.
469. J. W. Gross, A. D. Hegeman, B. Gerratana and P. A. Frey, *Biochemistry*, 2001, **40**, 12497.
470. A. D. Hegeman, J. W. Gross and P. A. Frey, *Biochemistry*, 2002, **41**, 2797.
471. R. J. Stern, T. Y. Lee, T. J. Lee, W. Yan, M. S. Scherman, V. D. Vissa, S. K. Kim, B. L. Wanner and M. R. McNeil, *Microbiology*, 1999, **145**, 663.
472. C. Dong, L. L. Major, A. Allen, W. Blankenfeldt, D. Maskell and J. H. Naismith, *Structure*, 2003, **11**, 715.
473. J. R. Somoza, S. Menon, H. Schmidt, D. Joseph-McCarthy, A. Dessen, M. L. Stahl, W. S. Somers and F. X. Sullivan, *Structure*, 2000, **8**, 123.
474. A. M. Mulichak, C. P. Bonin, W. D. Reiter and R. M. Garavito, *Biochemistry*, 2002, **41**, 15578.

475. N. A. Webb, A. M. Mulichak, J. S. Lam, H. L. Rocchetta and R. M. Garavito, *Protein Sci.*, 2004, **13**, 529.
476. S. Menon, M. Stahl, R. Kumar, G.-Y. Xu and F. Sullivan, *J. Biol. Chem.*, 1999, **274**, 26743.
477. M. Rizzi, M. Tonetti, P. Vigevani, L. Sturla, A. Bisso, A. De Flora, D. Bordo and M. Bolognesi, *Structure*, 1998, **6**, 1453.
478. W. S. Somers, M. L. Stahl and F. X Sullivan, *Structure*, 1998, **6**, 1601.
479. V. L. Y. Yip, A. Varrot, G. J. Davies, S. S. Rajan, X. Yang J. Thompson, W. F. Anderson and S. G. Withers, *J. Am. Chem. Soc.*, 2004, **126**, 8354.
480. A. Varrot, V. L. Y. Yip, Y. Li, S. S. Rajan, X. Yang, W. F. Anderson, J. Thompson, S. G. Withers and G. J. Davies, *J. Mol. Biol.*, 2005, **346**, 423.
481. S. S. Rajan, X. Yang, F. Collart, V. L. Y. Yip, S. G. Withers, A. Varrot, J. Thompson, G. J. Davies and W. F. Anderson, *Structure*, 2004, **12**, 1619.
482. A. L. Majumder, M. D. Johnson and S. A. Henry, *Biochim. Biophys. Acta*, 1997, **1348**, 245.
483. M. W. Loewus, F. A. Loewus, G. U. Brillinger, H. Otsuka and H. G. Floss, *J. Biol. Chem.*, 1980, **255**, 11710.
484. M. W. Loewus, *J. Biol. Chem.*, 1977, **252**, 7221.
485. A. Stein and J. H. Geiger, *J. Biol. Chem.*, 2002, **277**, 9484.
486. R. A Norman, M. S. B. McAlister, J. Murray-Rust, F. Movahedzadeh, N. G. Stoker and N. Q. McDonald, *Structure*, 2002, **10**, 393.
487. M. E. Migaud and J. W. Frost, *J. Am. Chem. Soc.*, 1995, **117**, 5154.
488. F. Tian, M. E. Migaud and J. W. Frost, *J. Am. Chem. Soc.*, 1999, **121**, 5795.
489. X. Jin, K. M. Foley and J. H. Geiger, *J. Biol. Chem.*, 2004, **279**, 13889.

CHAPTER 7

One-electron Chemistry of Carbohydrates

7.1 CLASSES OF RADICAL REACTIONS OF CARBOHYDRATES

Reactions of carbohydrates involving intermediates with unpaired electrons occur in two main contexts, oxidations and reductions.

Most of the classic tests for "reducing sugar" depend on one-electron oxidations by transition metals of anions derived from the open-chain form of the sugar, but modern investigations have focused on oxidations with atmospheric oxygen. The ground state of molecular oxygen is a triplet, with two unpaired electrons, not a singlet, with a closed shell of valence electrons; there is therefore a rich chemistry deriving from the interaction of carbohydrate radicals with atmospheric oxygen.

Elementary molecular orbital cartoons rationalise why the O_2 ground state is a triplet (Figure 7.1) with two unpaired electrons. The term "triplet" arises because the two electronic spins of $S = 1/2$ are not independent, but add to give a spin of 1, which can align itself in three directions with respect to an external axis (+1, 0, −1). Consider first the closed-shell, singlet configuration. If each oxygen atom in O_2 were sp^2 hybridised, an electronic configuration similar to ethylene results, with a σ bond formed from a singly occupied sp^2 orbital on each atom, a π bond from a singly occupied p orbital on each atom and the remaining four electrons on each atom occupying two sp^2 lone-pair orbitals. Such a species, singlet oxygen, is known – it can be made via an electrocylic reaction from 9,10-diphenylanthracene 9,10-peroxide, itself made from triplet oxygen and a triplet excited state of 9,10-diphenylanthracene.[1] It is, however, more energetic by 96 $kJ\,mol^{-1}$ than the triplet ground state, possibly since the axes of its four lone-pair orbitals are in the same plane and repel each other (*cf.* the low bond-energy of F_2): its electron distribution is not cylindrically symmetrical about the O–O bond. Singlet oxygen is a vigorous two-electron oxidant.[2]

Consider now the consequences of having the oxygen atoms sp hybridised. An O–O σ bond is formed from each singly occupied sp orbital, with the other being doubly occupied (lone pair). The remaining six valence electrons of the O_2 molecule are distributed between the two degenerate, orthogonal π orbitals and the two degenerate, orthogonal π* orbitals formed from the p_x and p_y atomic orbitals on each atom. According to Hund's rule of maximum

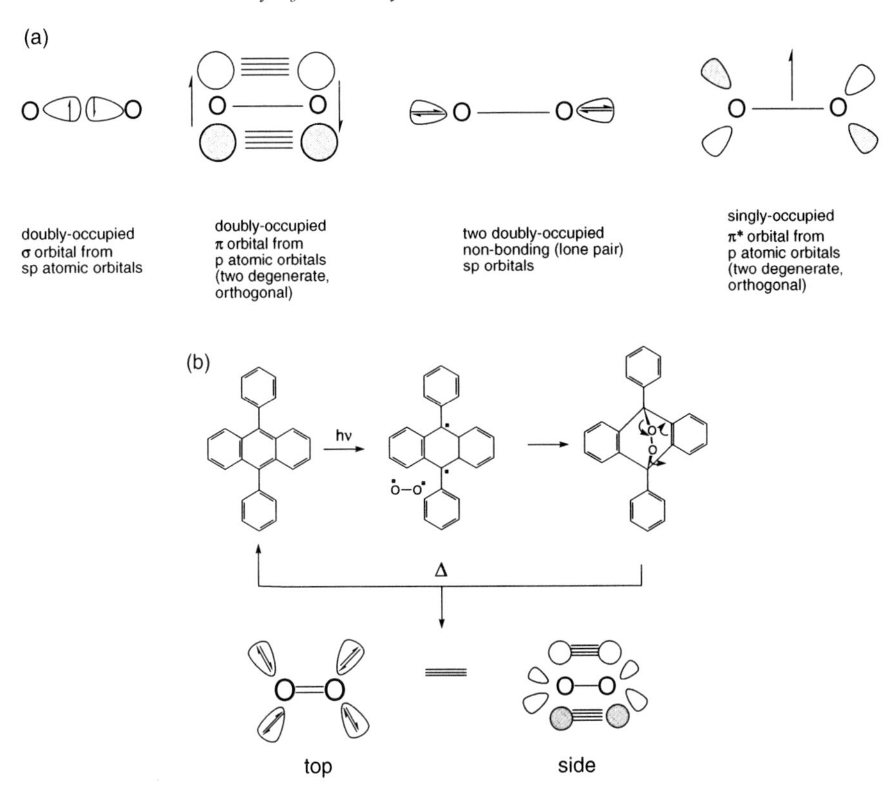

Figure 7.1 Cartoons of singlet and triplet molecular oxygen. (a) Triplet oxygen. Sections through the bonding electrons showing (left to right) bonding σ, bonding π, non-bonding (lone pair) and antibonding π orbitals. (b) Singlet oxygen, showing a method of generation and top and side views of the bonding electrons.

multiplicity, therefore, the π* orbitals are each singly occupied. The distribution is cylindrically symmetrical about the O–O bond.

Addition of a single electron to one of the half-occupied π* orbitals of triplet oxygen yields the superoxide radical anion, $O_2^{\bullet -}$. This is a weak base (pK_a of $HOO^{\bullet}$ = 4.8[3]) and a surprisingly inactive oxidant of closed-shell organic molecules. It is, however, highly cytotoxic, attacking the iron–sulfur clusters of various enzymes. Its usual fate is to disproportionate by electron transfer to oxygen and hydrogen peroxide, either spontaneously or catalysed by the various superoxide dismutases that have evolved to detoxify it *in vivo*. Addition of a second electron to the superoxide radicals yields the peroxide dianion. On protonation this gives hydrogen peroxide (pK_a 11.75), which in the presence of transition metals with two sequential, accessible valence states is reduced to hydroxide ion and hydroxy radical. The latter is a voracious oxidant (see below) and the array of reduced oxygen species ($O_2^{\bullet -}$/$HOO^{\bullet}$, HOO^-/H_2O_2, $HO^{\bullet}$) can be termed, particularly in the biological literature, "ROS" [reduced (or reactive) oxygen species].

Abiotic air (or oxygen) oxidation of sugars is important in the degradation of polysaccharides, particularly cellulose, and in some synthetic transformations, such as the catalysed oxidation of hydroxymethyl groups to uronic acids. Biological oxidation of glucose or galactose with molecular oxygen usually involves enzymes termed oxygenases, since the reduction product is hydrogen peroxide rather than water; the cofactor of microbial glucose oxidase is a flavin[4] and that of galactose oxidase a "Type 2" copper site.[5] Cellobiose oxidase/dehydrogenase appears to play a role in the biological degradation of lignocellulose, possibly by generating hydrogen peroxide for the ligninases and lignin peroxidases: it is unusual in having both flavin and haem cofactors. Additionally, some glucose dehydrogenases have the orthoquinone PQQ as cofactors and use ubiquinone as a coenzyme.[6]

The other type of radical chemistry of importance in the carbohydrate field is one-electron reductions. A handful of these reactions (such as the metallic Zn reduction of acetobromoglucose to triacetylglucal) have been used in synthesis for decades, but, starting with the Barton–McCombie deoxygenation of sugars in the mid-1970s[7] there has been an explosion of interest, as increasingly sophisticated cascades of elementary radical steps have been devised.[8] Such reactions are driven by the homolysis of weak bonds such as Sn–H or N–O under conditions of photolysis or mild thermolysis. Nature uses a similar basic principle in Type II ribonucleotide reductases, where the weak bond in question is the cobalt–carbon σ bond in the corrin cofactor.[9]

7.2 METHODS OF INVESTIGATION OF RADICALS IN CARBOHYDRATE CHEMISTRY AND BIOCHEMISTRY

Many of the basic kinetic techniques of physical organic chemistry lose their usefulness when applied to radical reactions. Because of quantum-mechanical restrictions on spin multiplicity, reaction of a radical with a closed-shell molecule or ion generates another radical, with the radical species themselves being present in low concentration. For this reason, radical reactions are often chain reactions, whose kinetics are dominated by initiation or termination steps, with the rates of reactions generating most product frequently having little influence on the overall rate law.

In deriving the rate law for a particular radical mechanism, the "steady-state assumption" in respect of all species with unpaired electrons, *i.e.* $d[Ra^{\bullet}]/dt = 0$, can usually be made, since the absolute concentrations are so low. The assumption parallels that made in steady-state enzyme kinetics ($d[ES]_n/dt = 0$), but whereas, in the absence of cooperativity, the steady-state approximation in enzyme kinetics always leads to the Michaelis–Menten equation (whatever the significance of k_{cat} or K_m in terms of individual rate constants), the steady-state assumption in radical kinetics often leads to complex expressions. Dimerisations and dissociations in key steps can lead to fractional powers in the rate law: a simple, old example is the thermal decomposition of dibenzoyl peroxide, which proceeds by the following mechanism:

$$(Ph - CO - O)_2 \rightarrow 2PhCOO^{\bullet}$$
$$PhCOO^{\bullet} \rightarrow CO_2 + Ph^{\bullet}$$
$$Ph^{\bullet} + (Ph - CO - O)_2 \rightarrow Ph - O - CO - Ph + Ph^{\bullet}$$

but which nonetheless produces the rate law

$$-\frac{d[(Ph - CO - O)_2]}{dt} = k\left[(Ph - CO - O)_2\right] + k'\left[(Ph - CO - O)_2\right]^{\frac{3}{2}} \quad (7.1)$$

Even conceptually simple reactions – such as the textbook anti-Markovnikov addition of HBr to isobutene, catalysed by dibenzoyl peroxide, can lead to sets of simultaneous equations which may need computer algebra such as *Mathematica* to solve.

Direct or indirect measurements of rate constants for elementary steps therefore play a much greater role in the determination of mechanisms of one-electron than of two-electron reactions. However, there are some simplicities in radical kinetics not available to the investigator of two-electron reactions. Because radicals are electrically neutral, solvent effects are small to negligible, and absolute rate constants determined in one solvent can sometimes be applied to another, in a way impossible in two-electron chemistry. The following generalisations can be made:[10]

(i) There are no solvent effects on the rates of hydrogen atom abstraction from a C–H bond.
(ii) There may be small solvent effects on unimolecular radical scission reactions but these are generally fairly small (see entry for *tert*-butoxy fragmentation, by the standards of radical fission reactions exhibiting a relatively large solvent effect, in Table 7.1).
(iii) There are large solvent effects on the rates of hydrogen atom abstraction from O–H bonds and, to a lesser extent, from N–H bonds.

Another feature of elementary radical reactions which makes it possible to apply absolute rate data across a range of systems is that substituent effects are relatively small: abstraction of $H^{\bullet}$ from tri-*n*-butyltin hydride by $CH_3^{\bullet}$, $CH_3CH_2^{\bullet}$, $(CH_3)_2CH^{\bullet}$ and $(CH_3)_3C^{\bullet}$ ranges over a factor of only 6.9 (at 25 °C),[11] in stark contrast to the effect of alkyl substitution on carbenium ion stability, where a single α-methyl group increases rates of formation by $\sim 10^8$-fold.[12] Consequently, absolute rate constants determined for one substrate can be confidently applied to all substrates of that type, so that the rate constant for $CH_3CH_2^{\bullet} + Bu^n{}_3SnH$ can be applied to all reactions of the type $RCH_2^{\bullet} + Bu^n{}_3SnH$.

The physics of the intermediates provides a further tool with no analogue in two-electron chemistry. Their unpaired spin makes their direct observation by electron spin resonance (ESR) possible. The physical principles governing ESR of organic radicals are exactly the same as those for NMR of nuclei with the same spin (1/2), but the technique is, in concentration terms, more sensitive.

7.2.1 Electron Spin Resonance – Aspects of Importance to Carbohydrates[13]

The basic physics of the ESR experiment is similar to that of NMR. When placed in a magnetic field H_0, the magnetic dipole of the electron associated with its spin of magnitude β (the Bohr magneton) introduces an energy difference ΔE, depending on whether the (quantised) spin is with or against the applied field. The transition between the two states results in absorption or emission of electromagnetic radiation of frequency ν. Equation (7.2) gives the relation between these quantities:

$$h\nu = \Delta E = g\beta H_0 \tag{7.2}$$

The parameter g, the Lande g factor, has a value close to 2 for electrons in free space and in practice for most organic radicals.

The 1835-fold smaller mass of the electron than the proton, offset by a 2.789-fold greater g value, results in ΔE being 658 times larger for the electron at the same H_0; if $\Delta E \ll RT$, therefore, ESR is 658 times more sensitive than ^{1}H NMR. Good spectra for $g = 2.0$ organic radicals are obtained with solutions with radical concentrations in the 0.1–1 μM range.

ESR spectra are split by atomic nuclei with spins, in the same way as NMR spectra. The coupling constant depends on the overlap of the orbital containing the unpaired electron with an occupied orbital with a non-zero electron density at the nucleus in question. The spectrum of a single radical will therefore consist of a single resonance with multiple splittings. The multiplet structure arising from splitting by equivalent $I = 1/2$ nuclei (^{1}H, ^{13}C, ^{19}F and ^{31}P) is exactly the same as in NMR, a series of equally spaced lines of intensities derived from Pascal's triangle. Thus, two equivalent hydrogens give a 1:2:1 triplet, 3Hs a 1:3:3:1 quartet, 4Hs a 1:4:6:4:1 quintet, 5Hs (*e.g.* cyclopentadienyl radical) a 1:5:10:10:5:1 sextet, 6Hs a 1:6:15:20:15:6:1 septet and 7Hs (*e.g.* cycloheptatrienyl radical[14]) a 1:7:21:35:35:21:7:1 octet; obviously, experimentally it becomes increasingly difficult to detect and identify the outlying low-intensity components of the multiplet. With the $I = 1$ nucleus ^{14}N, there are three orientations of the nuclear magnet with respect to the electronic magnet ($m = +1$, 0 and -1), so a single ^{14}N splits the ESR signal into three equally spaced, equal intensity subunits.[i]

All the structural information in the ESR spectrum of an organic radical lies in the splitting pattern, since the similarity of g values means that there is no

[i] The higher energies involved in electron magnetic resonance make the *proportional* uncertainty in the energy of the ^{14}N spin states, imposed by the Uncertainty Principle and their short lifetime, brought about by efficient quadrupolar relaxation, relatively less important. ESR resonance lines showing ^{14}N splittings are therefore not noticeably broadened.

real counterpart to NMR chemical shift information. Because of this and the technology available when ESR was developed, ESR spectra are presented differently to NMR spectra. [Most instruments still use microwave radiation and permanent magnets and the application of (the) "sophisticated pulse and Fourier transfer techniques" (of NMR) is thought to lie in the future by Mile,[13] writing in 2000]. Usually a trace of the first differential of the absorbance (or emission) with respect to field is recorded: this makes multiplets visually clearer at the cost of losing intensity information. In the case of $g = 2.0$ organic radicals, the amplitudes of the various components of the multiplets when present as a differential are approximately proportional to their intensities, but with other systems this is far from the case. Splittings, also called hyperfine coupling constants, are quoted in millitesla (mT), gauss (G) or sometimes cm^{-1} (a spectroscopists' unit of energy), rather than hertz (Hz), although like NMR coupling constants, quoted values are independent of the external field strength, H_0. Tesla and gauss are the SI and cgs units of magnetic field strength, respectively (10 G = 1 mT), and splittings of $g = 2.0$ organic radicals in cm^{-1} are, for practical purposes, 1000 × the splitting in mT. Hyperfine splitting constants are usually represented by upper or lower case A or a.

The trends in magnitude of splittings of an organic radical ESR signal by an α-^{13}C are readily understood. If the electron is in a pure p orbital, there is a node at the ^{13}C nucleus and splitting is minimised: as an alkyl radical becomes increasingly pyramidal, the s character of the orbital it occupies increases and with it the possibility of interaction between nuclear and electronic spin and hence the splitting.[15] For radicals in which the spin density resides exclusively on the carbon in question, eqn. (7.3) applies, where φ[ii] is the angle between the plane defined by two of the bonds to spin-carrying carbon and the third. $a_C(0)$ is the value of the coupling for the methyl radical (3.85 mT), which probably arises from spin polarisation.[16]

$$a_C(\varphi)/\text{mT} = a_C(0)/\text{mT} + 119(2\tan^2\varphi) \qquad (7.3)$$

Splitting by β-protons likewise occurs by a readily grasped mechanism, overlap of the full σ orbital of the β-C–H bond with the half-filled orbital at the spin site; the mechanism closely parallels the weakening of β-C–H bonds next to a carbenium ion centre which gives rise to β-hydrogen kinetic isotope effects (see Section 3.10.3, particularly Figure 3.23). The splitting is geometry dependent in exactly the same way, and an equation identical in form with eqn. (3.12), eqn. (7.4), governs the relationship between observed splitting, maximum splitting and the dihedral angle θ between the axis of the electron-deficient orbital and the C–H bond:[17]

$$a_{\beta H}(\theta)/\text{mT} = a_{\beta H}(0)/\text{mT} + 4.9\cos^2\theta \qquad (7.4)$$

[ii] The literature cited uses the symbol θ, inviting confusion with the dihedral angle between the half-filled orbital and the β-CH bond.

$a_{\beta H}(0)$ usually has the value 0.1–0.5 mT. In the case of the three equivalent hydrogens of a freely rotating methyl group, the effective value of $\cos^2\theta$ for all three hydrogens is 0.5.

The right-hand side of both eqns. (7.3) and (7.4) has, in the case of delocalised radicals, to be pro-rated for spin density on the carbon in question. Thus, the maximum value of $a_{\beta H}$ for splitting by a delocalised radical with a spin density of 0.5 is 2.7 mT.

The origin of the splitting by an α-C–H is not as straightforward (it is usually attributed to spin polarisation), but for delocalised π radicals (such as those formed by adding a single electron to the π* orbital of an aromatic compound such as naphthalene) the spin density, *i.e.* probability of finding the spin on a particular carbon, is approximately $a_{\alpha H}/3$, where $a_{\alpha H}$ is the hyperfine splitting in mT.[18] The spectrum of the methyl radical is a 1:3:3:1 quartet, with $a_{\alpha H} = 2.3$ mT, suggesting that it is essentially a π radical. The spectrum of the ethyl radical combines $a_{\alpha H}$ of 2.24 mT, which splits the signal into a 1:2:1 triplet, and $a_{\beta H}$ of 2.69 mT, which splits each component of the triplet into a 1:3:3:1 quartet.[19] The resulting pattern appears complex because of the similarity of the two hyperfine splittings.

Small splittings by γ-hydrogens are also observed; these are maximised when the axis of the p orbital, the H–Cγ bond, the Cγ–Cβ bond and Cβ–Cα bond form a coplanar W arrangement (*cf.* W couplings in NMR); their maximum value in carbohydrate radicals appears to be about 0.4 mT; however, with unfavourable geometry they fall to zero.

α-, β- and γ-H splittings of the tetra-*O*-methylglucopyranosyl radical are shown in Figure 7.2.

ESR as applied to the active sites of metal-dependent enzymes, such as those catalysing various one-electron reactions of sugars, is much more complicated. Because of slow tumbling of proteins, isotropic spectra are a rarity and even when they can be observed, complications not experienced with organic free radicals intrude.[21] Thus, vanadyl acetoacetate, $VO(CH_3COCHCOCH_3)_2$, has a single unpaired electron, $S = 1/2$, and a single vanadium isotope, ^{51}V, $I = 7/2$, but the ESR spectrum is not the array of eight $(2I + 1)$ equally spaced, equal-amplitude lines one might expect: the lines have equal intensities but, because of the differential way of presenting ESR spectra, not equal amplitudes, some lines being broader than others in a way which depends on field strength. The separation of the lines increases slightly with field strength, because of the failure of the first-order perturbation assumption, that interaction energies, in this case between electron and nuclear spins, are small compared with the energy of spectral transitions. Cu^{II} likewise has $S = 1/2$, but two common isotopes, ^{63}Cu and ^{65}Cu, with $I = 3/2$. However, the a values for the two isotopes are close together, so the expected eight-line spectra, two superimposed sets of four lines in the ratio of isotopic abundances, are not observed: the ESR spectrum of $Cu(CH_3COCHCOCH_3)_2$ in chloroform is four unequal bumps.

If isotropic conditions are abandoned – a necessity with large redox proteins – then the differing values of g in the three orthogonal directions in the crystal field have to be taken into account and spectra display additional complexities.

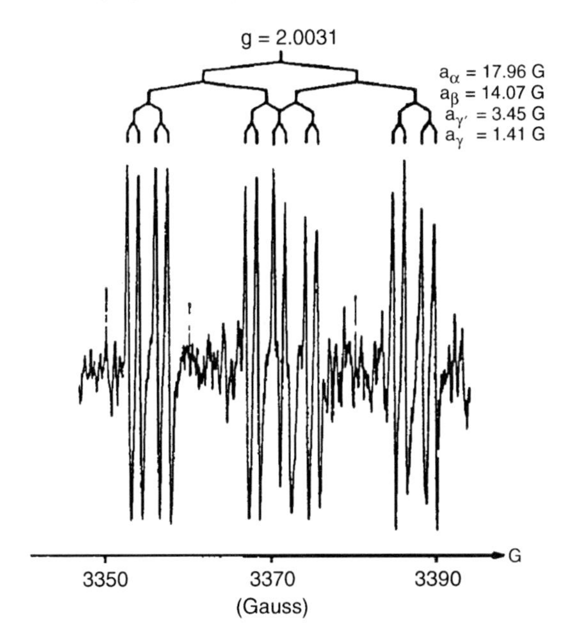

Figure 7.2 ESR spectrum of the tetra-*O*-methylglucopyranosyl radical at 20°C (10 G = 1 mT).[20] Subsequent ^{2}H labelling at H5 identified Hγ' as H5 and Hγ as H3.[17] Thanks are due to Prof. Dr Bernd Giese, University of Basel, for kindly providing this spectrum from his laboratory and allowing its reproduction.

If there is more than one unpaired electron spin on the same atom, the spins combine: two unpaired electrons give a system with $S = 1$. This can adopt three orientations with respect to an external direction, corresponding to M_S (the projection of the spin on this direction) $= +1$, 0 and -1. The resonance condition is always $\Delta M_S = \pm 1$, so that if the three orientations are of equal energy, only one resonance is observed. Such isotropic systems are observed in rapidly tumbling liquids and certain rigorous symmetries (octahedral, tetrahedral and cubic). All isotropic systems with $S > 1/2$ behave in this way. However, in non-isotropic systems a parameter called the zero field splitting (D), essentially a measure of the interaction of the electrons with each other, can be deduced from the multiple lines observed. The same applies to organic diradicals as to transition metal ions, and the zero field splitting is used as a measure of the interaction between two radical sites in an organic molecule. As they get further apart and are more insulated from each other electronically, the D value decreases.[22]

7.2.2 Conformations of Carbohydrate Radicals as Determined by ESR

The methyl radical is effectively planar, although the force constant for distortion is low. As the radical becomes pyramidal, the orbital containing the unpaired electron (SOMO)[iii] changes from p to sp^3, as the carbon orbitals forming the C–H σ bond change from sp^2 to sp^3. With a substituent of the same electronegativity as carbon, this has only minor energetic consequences. If, however, the substituents are electronegative, the change in carbon hybridisation as the radicals are pyramidalised means that the electrons forming the σ bonds move away from carbon, and hence closer to the substituent. Electronegative substituents thus favour pyramidalisation of a radical centre, and this can be monitored from a_C values [eqn. (7.3)], which for increasingly fluorinated methyl radicals are $^{\bullet}CH_3$ 3.85, $^{\bullet}CH_2F$ 5.48, $^{\bullet}CHF_2$ 14.88 and $^{\bullet}CF_3$ 27.16 mT.[16]

Substituent electronegativity promoting pyramidalisation is offset, where the substituent has lone pairs of electrons, by the partial transfer of spin to the heteroatom by overlap of the half-filled an lone-pair orbitals. For this to be optimal, both the SOMO and lone-pair orbitals should be of the p type. Therefore, progressive oxygenation of the methyl radical has a smaller effect on a_C than progressive fluorination, since p–p orbital overlap is more efficient with oxygen lone pairs than fluorine lone pairs: a_C values are $^{\bullet}CH_3$ 3.85, $^{\bullet}CH_2OH$ 4.74, $^{\bullet}CH(OCD_3)_2$ 9.79 and $^{\bullet}C(OCH_3)_3$ 15.27 mT.[16] These values are lowered, both by the lesser pyramidalisation and by removal of spin density by the conjugation in question. In the case of 2-alkoxy-2-tetrahydropyranyl radicals, the fraction of the spin density on C2 was estimated from both $a_{\beta H}$ splittings, assuming $a_{\beta H}(90°) = 0$, as 0.51, with $\varphi = 11°$ (Figure 7.3).[16]

Protected glycopyranosyl radicals could be generated for ESR study by photolysis of hexaalkyldistannanes in the presence of glycosyl–X, where the glycosyl–X bond was weak (X = Br, SePh) or readily photolysed [$X = CO(CH_3)_3$]; $R_3Sn^{\bullet}$ displaced the glycosyl radicals from Br, Se, *etc.*[17] For the tetra-*O*-acetylglucopyranosyl radical, $a_C = 4.73$ mT, compared with 4.14 mT for the cyclohexyl radical, demonstrating the slightly more pyramidal nature of radicals with electronegative substituents, but also that the radical was fundamentally of the π type. Application of a parameterised version of eqn. (7.3) indicates that φ is about 3.9°, *i.e.* the degree of s character is ~1%.

The values of proton hyperfine coupling constants for various pyranosyl radicals varied with sugar, protecting group and temperature. $a_{\beta H}$ for the tetra-*O*-acetyl- (2.76 mT) and tetra-*O*-methylgalactopyranosyl radicals (2.76 and 2.60–2.54 mT over a temperature range from 66 to 90 °C, respectively), together with an absence of coupling to H3 and a 0.23–0.25 mT coupling to H5, indicate a $^4H_3$ conformation, with a substantial spin delocalisation to O5 (Figure 7.4). By contrast, the $^4C_1$ conformation appears largely preserved in the tetra-*O*-acetylmannopyranosyl radical, as the value of $a_{\beta H}$, 0.35 at 20 °C and 0.42 at 90 °C, is too small for a half-chair (for a chair, $\theta \approx 60°$, $\cos^2\theta = 0.25$, but for a half-chair, $\theta \approx 30°$, $\cos^2\theta = 0.75$). Similar large values of $a_{\beta H}$ to those of the protected galactopyranosyl radicals are observed with glucopyranosyl radicals protected either with a 4,6-benzylidene group or both a 4,6-benzylidene and a 2,3-isopropylidene group, in accord with a conformation of high latitude at the

[iii] SOMO = singly occupied molecular orbital.

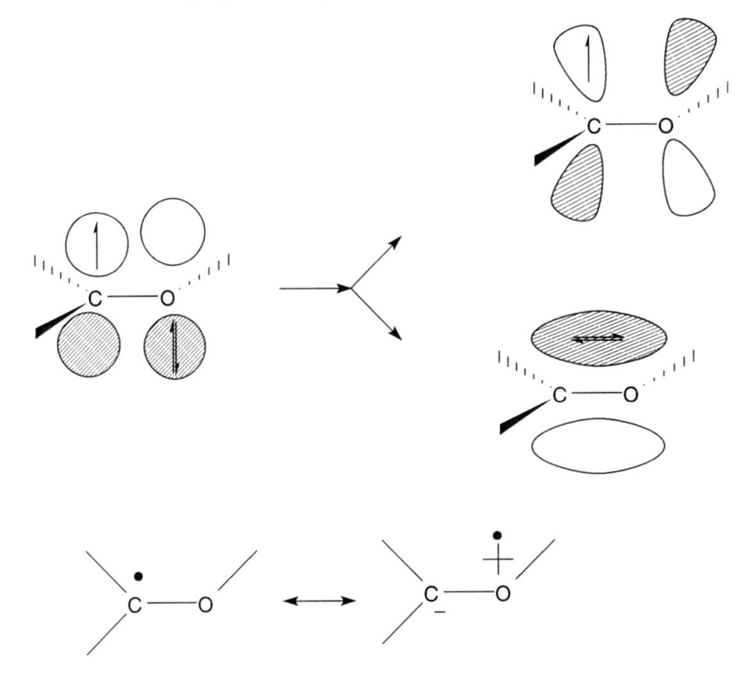

Figure 7.3 Orbital overlap in oxygenated radicals.

longitude of $^{4}H_3$ on the Cremer–Pople sphere (Figure 2.6). However, the spectra of the tetra-*O*-acetyl- and tetra-*O*-methylglucopyranosyl radicals themselves indicate a conformation radically different to $^{4}H_3$ or $^{4}C_1$: $a_{\beta H}$ values are in the 1.1–1.4 mT region, too small for either of these conformations, and are steeply temperature dependent (*e.g.* between −22 and +44 °C its value for the tetramethylglucopyranosyl radical changes from 1.06 to 1.27 mT), suggesting the adoption of more than one conformer. Pronounced splittings from H5 (~0.35 mT) and H3 (~0.15 mT) suggest an layout of the hydrogens, the SOMO and the intervening bonds in a W-arrangement.

The $a_{\beta H}$ values are compatible with $^{1}C_4$, $^{1,4}B$ and $B_{2,5}$ conformations. The $^{1}C_4$ conformation can be ruled out, because the spectrum of the 2,3-diacetyl-4,6-benzylideneglucopyranosyl radical has the same hyperfine splittings as that of the tetra-*O*-acetylglucopyranosyl radical, despite the $^{1}C_4$ conformation prohibited by the benzylidene bridge. The original workers[17,23] discussed the conformation in terms of a single $B_{2,5}$ conformation somewhat distorted to a skew boat, $B_{2,5}$ being preferred to $^{1,4}B$ since it permitted W-arrangements of bonds more compatible with observed $a_{\gamma H}$ values. It seems to the author that the conformational preference of the glucopyranosyl radical is best described as occupancy of a small range of conformations between $B_{2,5}$ and $^{1}S_5$ on the equator of the Cremer–Pople sphere (Figure 2.6). The conformations of the tri-*O*-acetylxylopyranosyl and -lyxopyranosyl radicals are, unsurprisingly, the same as those of their glucopyranosyl and mannopyranosyl homeomorphs.

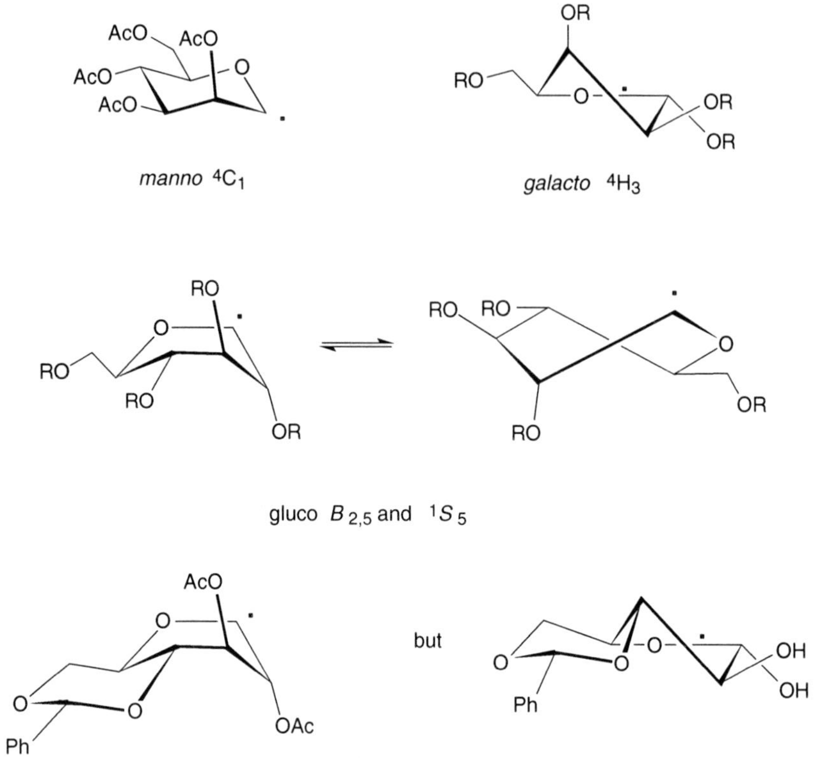

Figure 7.4 Conformations of pyranosyl radicals. R = Ac or Me.

The reason for the adoption of these normally disfavoured conformations is mutual overlap of the oxygen lone-pair orbital, the SOMO and the C2–O2 σ^* orbital (Figure 7.5). The overlap is most effective if the two occupied orbitals are regarded in the first instance as atomic p orbitals, when both dihedral angles are zero. In valence-bond terms, no-bond resonance, similar to that producing the anomeric effect, operates. In the case of mannopyranosyl radicals, because of the ground-state axial configuration of the C2–O2 bond, the effect can operate in the $^4C_1$ conformation and distortion to the $^4H_3$ conformation is not necessary. Galactopyranosyl radicals are disfavoured from adopting the $B_{2,5}$ conformation by the vicinal di-pseudoxial interactions that result and the $^{1,4}B$ conformation by interactions of the now flagstaff O4 and accordingly remain in the $^4H_3$ conformation. If O2 carries an electron-withdrawing substituent, the strength of the effect should be increased if the no-bond resonance model puts negative charge in addition to spin on O2. Such a mechanism may be responsible for the different conformations adopted by the 4,6-benzylideneglycopyranosyl radical and its 2,3-di-*O*-acetyl derivative, the former adopting the $^4H_3$ and the latter the $B_{2,5}$ conformation.

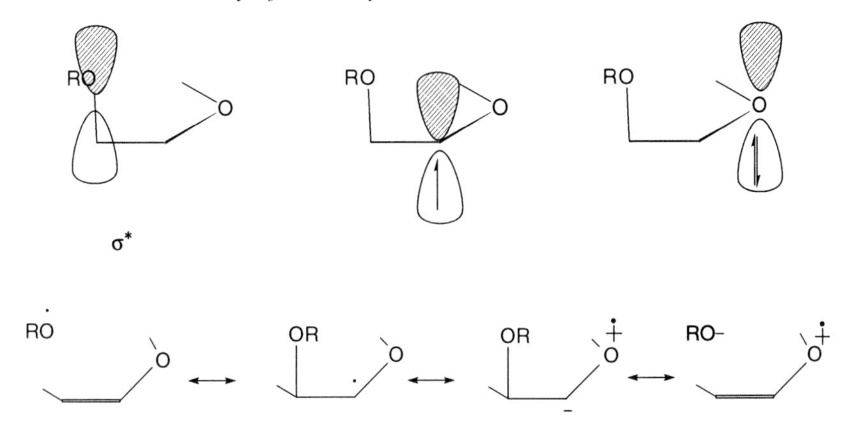

Figure 7.5 Orbital alignment in the glucopyranosyl radical. The top shows the unoccupied, singly occupied and doubly occupied atomic orbitals involved, the bottom a pictorial valence bond representation of the same phenomenon.

There is some evidence that the tetra-*O*-acetylglycopyranosyl radical may be initially generated in a $^4C_1$ conformation, which is a local energy minimum: if it is generated from the action of ^{60}Co γ-rays on acetobromoglucose in a methanol glass at 77 K, a 4.15 mT splitting can be discerned in the resultant, somewhat indistinct, ESR spectrum.[24]

Acetylated 2-, 3- and 4-deoxypyranosyl radicals can be generated similarly, by photolysis of hexamethylditin (hexamethyl distannane, $(CH_3)_3Sn\text{-}Sn(CH_3)_3$) in the presence of the appropriate bromodeoxy or iododeoxy precursor, the reaction being a bromine atom abstraction by $(CH_3)_3Sn^{\bullet}$. Their splitting patterns were entirely as expected for π-type radicals with the trivalent carbon sp^2 hybridised, the unpaired electron in a p orbital and the sugar in a $^4C_1$ conformation.[25]

Remarkably, the tetra-*O*-acetylglucopyranosyl radical rearranged into the 2-deoxy-1,2,4,5-tetra-*O*-acetyl-α-2-glucopyranosyl radical, *i.e.* there was a 1,2-migration of the acetoxy group.[26] The migration proved quite general and was always suprafacial (*i.e.* the acetoxy group remained on the same face of the ring). Thus, acetobromomannose, yielding the 2-deoxy-1,2,4,5-tetra-*O*-acetyl-β-2-glucopyranosyl radical from and 2-deoxy-1,2,4,5-tetra-*O*-acetyl-α-2-galactopyranosyl radical from acetobromogalactose. The 2 → 1 migration also occurred in the furanose series; when used preparatively with tributyltin hydride as a hydrogen donor, the rearrangement gave rise to a simple and useful preparation of 2-deoxy sugars.[27] The reaction also proceeds with phosphatoxy migration in much the same way (Figure 7.6), with the bulky axial phosphatoxy group subsequently dictating approach of the reductant from the β-face of the sugar.[28]

Rearrangement of an alkoxy radical to a secondary alkyl radical is normally contrathermodynamic and the driving force for the rearrangement is still unclear. When the rearrangement was first discovered, the driving force was thought to be the anomeric stabilisation of the acyloxy group, but this explanation now has the

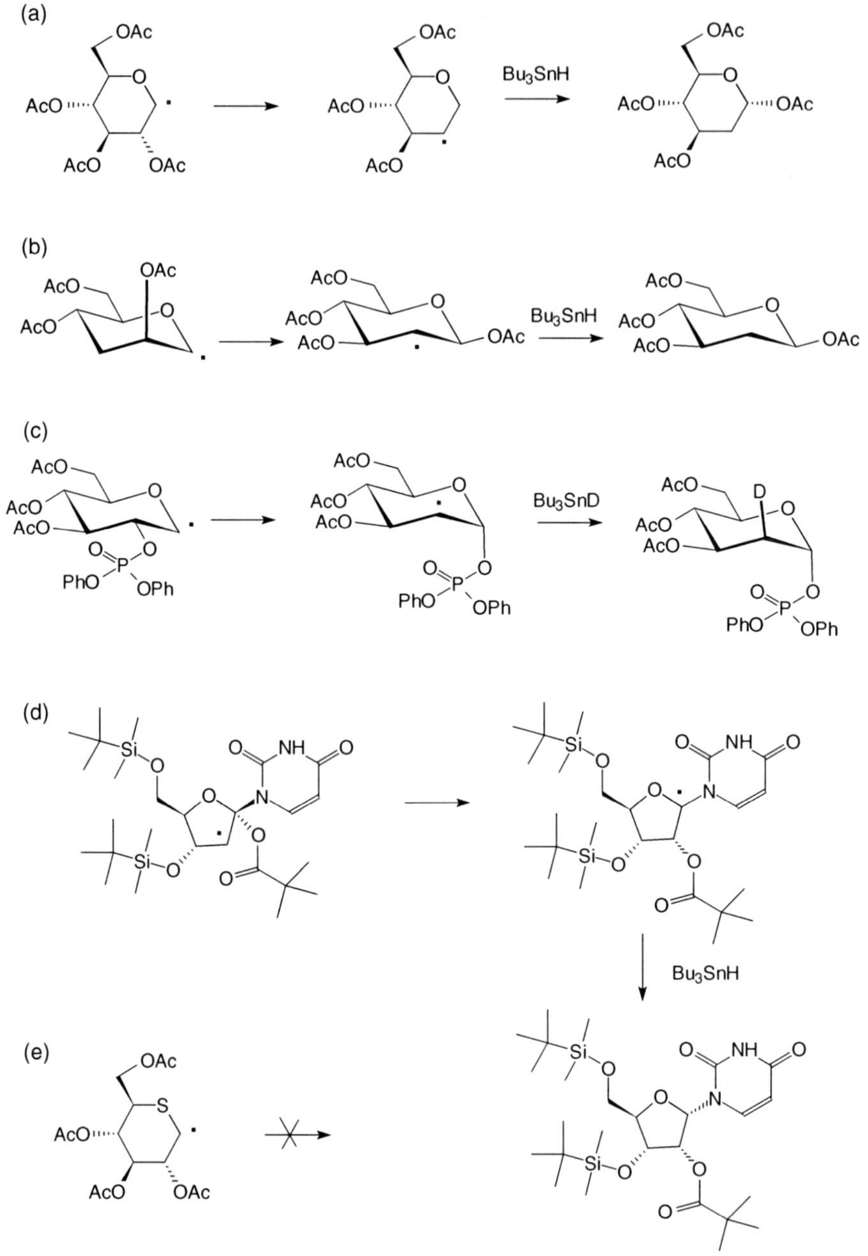

Figure 7.6 Rearrangement of 2-acyloxy and related glycopyranosyl radicals: (a) tetra-*O*-acetylglucopyranosyl, (b) tetra-*O*-acetylmannopyranosyl, (c) a phosphatoxy migration; note how the α-face of the sugar is (d) reverse migration where the anomeric centre is stabilised, (e) failure of migration of a 5-thio radical.

difficulty that the rearrangement also occurs with the tetra-*O*-acetylmannopyranosyl radical, to yield, unless a non-chair conformation is adopted, an equatorial acyloxy group in the product. Certainly the driving force is small, with the tetra-*O*-acetyl-5-thioglucopyranosyl radical, stabilised by a sulfur rather than oxygen, failing to rearrange,[25] and a pivalate group migrating in the opposite $1 \rightarrow 2$ sense when the anomeric centre is stabilised by an additional lone pair, even one involved in aromatic bonding (Figure 7.6d).[29]

The carbohydrate radicals formed by hydrogen abstraction from carbohydrates are still oxygenated at the radical site; the syllable "an" is introduced into the name, so that the glucopyranosan-1-yl radical is the product of abstraction of the anomeric hydrogen of glucopyranose.[17] They have been studied in an ESR flow system after $H^{\bullet}$ abstraction by $HO^{\bullet}$, generated from a modified Fenton system (see Section 7.2) of H_2O_2 and Ti^{III}. The abstraction is statistical, with all C–H bonds in *myo*-inositol[30] and glucose[31] being attacked equally.

The ESR signals from deprotected sugar radicals occasionally show hyperfine splitting from OH groups, as proton exchange is not fast on the ESR timescale (*cf.* NMR in DMSO rather than water). These β-splittings are small ($\sim$0.1 mT)[30] and often not observed, presumably because for optimal overlap between the SOMO and the oxygen lone pair (Figure 7.3) the lone pair orbital will be close to pure p, the oxygen sp^2 hybridised and the O–H bond therefore roughly orthogonal to the SOMO [eqn. (7.4), $\theta \approx 90°$].

It proved possible to deconvolute the superimposed ESR spectra of the six radicals generated from α- or β-glucopyranose, produced under conditions where mutarotation was slow. The spectra of α-glucopyranosan-3- and -4-yl radicals and β-glucopyranosan-2-, -3- and -4-yl radicals, are dominated by two 2.5–3.0 mT β-couplings to the two adjacent axial C–H protons; that of the α-glucopyranosan-2-yl radical has a 3.0 mT coupling to H3 and a 1.3 mT coupling to equatorial H1. The conformation of the pyranose ring in α- and β-glucopyranosan-2-, -3- and -4-yl radicals is therefore $^4C_1$, with the radicals being of the π type. ESR spectra of α- and β-glucopyranosan-5-yl radicals show 3.3 and 3.7 mT splittings, respectively, from H4, but only small splittings (1.0 and 0.7, and 0.7 and 0.5 mT, respectively) to the two diastereotopic hydrogens on C6. The C6–O6 bond is therefore eclipses the SOMO, to obtain the stabilisation as in Figure 7.7. The same phenomenon dictates that the two favoured conformations of the α- and β-glucopyranosan-6-yl radicals about C5–C6 have the SOMO eclipsing the C5–O5 bond and, consequently, a low $a_{\beta H}$ value of 0.75 mT.

The glucopyranosan-1-yl radical has the expected $a_{\beta H}$ value of 2.47 mT, with an additional $a_{\gamma H}$ of 0.26 mT to H5 and a splitting of 0.16 mT from the OH, probably a reflection of the non-planar dioxygenated radical centre and the lesser importance of an individual lone pair–SOMO overlap in a radical stabilised by two such interactions.

7.2.3 Kinetics of Radical Elementary Steps

The first direct measurements of radical reactions were made by kinetic ESR, in continuous flow arrangements of the Hartridge–Roughton type, with the time

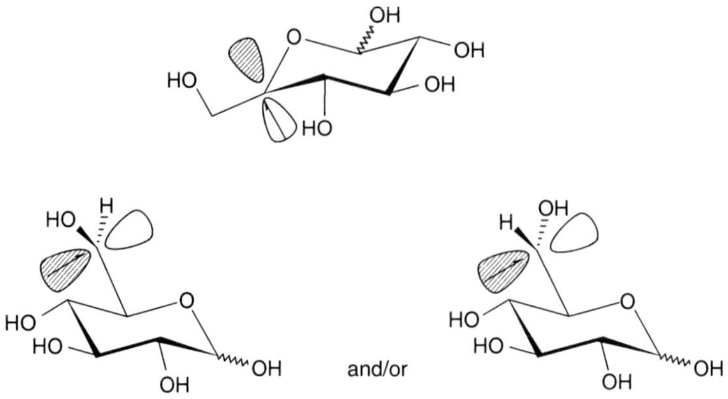

Figure 7.7 Conformations of glucopyranosan-5- and -6-yl radicals about C5–C6.

after mixing depending on the length from the mixing chamber. More recently, kinetic ESR has exploited a combined photolysis–ESR layout.[32] Deoxygenated solutions of a photochemical radical precursor flow slowly through a thermostated cell in the ESR cavity and are irradiated by UV light. In the absence of reactive substrates, the spectrum of the initial radical is observed; in the presence of interceptors, product radicals are also observed. The photolysing light can be interrupted by a rotating sector[iv] so that radicals accumulate during the "on" period and decay during the "off" period. The time courses are accumulated and added together electronically. Modern ESR spectrometers can have a response time as low as 10 μs, so that decay curves with half-times on the order of milliseconds can be successfully accumulated.

Most radicals in solution react with each other at about the diffusion limit, but because of a quantum mechanical restriction only one of the four possible electron spin states of radical pair (↑↑, ↑↓, ↓↑ and ↓↓) is productive. For most solvents, the recombination rate at room temperature is $5 \times 10^9\,M^{-1}\,s^{-1}$, with a range of $1 \times 10^9 - 1 \times 10^{10}\,M^{-1}\,s^{-1}$.[11] Radicals at the 0.1 μM concentration required for ESR quantitation thus have (second-order) half-lives of milliseconds, which confines the technique to very rapid reactions. In the direct analysis of radical kinetics by this method, therefore, the reaction of interest has compete with self-destruction of the initiating radicals. However, in a study of the $(CH_3)_3C–O^\bullet$ (*tert*-butoxyl) radical, which decomposes[33] rapidly to $CH_3^\bullet$ and acetone, it proved possible to monitor the methyl radical and deduce the rate of decomposition of its precursor *tert*-butoxyl radical from the decay curve for the methyl radical (in the limit where the decomposition of *tert*-butoxyl is much faster than other processes, the curve for the decay of the methyl radical is shifted to longer time by the reciprocal of the first-order rate constant for *tert*-butoxyl decomposition).

[iv] Compare the "choppers" in double beam UV/Vis and IR spectrometers.

$$\begin{array}{c} \text{generation} \\ \downarrow \\ \text{R–X} \xleftarrow{k_2[\text{XY}]} \text{R}^{\bullet} \underset{k_{-1}}{\overset{k_1}{\rightleftharpoons}} \text{Я}^{\bullet} \xrightarrow{\kappa_2[\text{XY}]} \text{Я–X} \end{array}$$

Scheme 7.1

Direct measurement of the reaction of interest is sometimes possible using rapid reaction techniques. In laser flash photolysis, an intense, short-lived pulse of light irradiates the sample and the products are monitored by a variety of techniques, from basic UV/Vis spectroscopy to techniques – such as laser-excited fluorescence – which require a second, analytical pulse of radiation.[34] In pulse radiolysis, a short (1–10 ns) pulse of high-energy (1–10 MeV) electrons irradiates the sample and the decay of the fragments can be analysed in the same way as the fragments from flash photolysis.[35] The equipment for pulse radiolysis is even more complex and costly than that for flash photolysis, and tends to be concentrated in national facilities.

In these circumstances, where routine kinetic measurements are uninformative and direct measurements of the product-forming steps difficult, comparative methods, involving competition between a calibrated and a non-calibrated reaction, come into their own. Experimentally, ratios of products from reaction cascades involving a key competition between a first-order and a second-order processes are measured as a function of trapping agent concentration. Relative rates are converted to absolute rates from the rate of the known reaction. The principle is much the same as the Jencks clock for carbenium ion lifetimes (see Section 3.2.1). However, in radical chemistry Newcomb prefers to restrict the term "clock" to a calibrated unimolecular reaction of a radical,[36] but such restriction obscures the parallel with the Jencks clock, where the calibrated reaction is a bimolecular diffusional combination with N_3^- and the unknown reaction a pseudounimolecular reaction of carbenium ion with solvent. Whatever the terminology, the practical usefulness of the method stems from the possibility of applying the same absolute rate data to all reactions of the same chemical type, as discussed in Section 7.1.

Consider a radical $R^{\bullet}$ which is converted to another radical Я$^{\bullet}$; both $R^{\bullet}$ and Я$^{\bullet}$ can react with a compound with a weak bond, X–Y, by transfer of X, generating R–X and Я–X, according to Scheme 7.1.

If the conversion of $R^{\bullet}$ to Я$^{\bullet}$ is irreversible ($k_{-1} = 0$) and [XY] is present in large enough excess to ensure first-order conditions, the ratio of the products is simply related to the ratio k_1 and k_2 by eqn. (7.5):

$$\frac{[\text{RX}]}{[\text{ЯX}]} = \frac{k_2[\text{XY}]}{k_1} \tag{7.5}$$

If the isomerisation step is significantly reversible, then the ratio of products is given by eqn (7.6):

$$\frac{[\text{RX}]}{[\text{ЯX}]} = \frac{k_1 k_2}{k_1 \kappa_2} + \frac{k_2[\text{XY}]}{k_1} \tag{7.6}$$

generation

$$\downarrow$$

$$\text{R–X} \xleftarrow{k_2[\text{XY}]} \text{R}^\bullet \xrightarrow{\kappa_2[\text{ЖY}]} \text{R–Ж}$$

Scheme 7.2

Therefore, k_2/k_1 is the gradient of a plot of the ratio of the two capture products versus the concentration of the capturing agent. With precise enough data, κ_2/k_{-1} can be obtained from the ratio of slope to intercept. If interconversion of $R^\bullet$ and $Я^\bullet$ is sufficiently fast that equilibrium is maintained, the first term dominates eqn. (7.6) and the plot is horizontal.

In practice, the analytical system that distinguishes R–X and Я–X may not be compatible with a large excess of [XY], but the equations governing [R–X]/[Я–X] become very unwieldy if XY is significantly depleted over the course of the reaction. Equation (7.7), where subscript 0 refers to initial and subscript f to final concentration, is such an expression for the simplest case of an irreversible isomerisation of $R^\bullet$ ($k_{-1} = 0$):

$$\frac{[\text{ЯX}]}{[\text{RX}](1+[\text{ЯX}]/[\text{RX}])} = \frac{k_1}{k_2}\{\ln([\text{XY}]_0 + k_1/k_2) - \ln([\text{XY}]_f + k_1/k_2)\} \quad (7.7)$$

[XY] can be kept constant under certain conditions by the slow addition of a precursor. Benzeneselenol, PhSeH, transfers hydrogen atoms to radicals about 10^3 times faster than tributyltin hydride and is therefore useful as a clock reagent for faster reactions inaccessible to Bu_3SnH. PhSeH can be generated *in situ* from equal amounts of PhSeSePh (diphenyl diselenide) and Bu_3SnH, with slow addition of a stoichiometric amount over the course of the reaction. Even though at any moment the concentration of PhSeH is sub-stoichiometric, it is effectively constant and first-order conditions prevail.[37] Using the first-order rate constant for the rearrangement of the 2,3,4,6-tetra-*O*-acetylglucopyranosyl radical (Figure 7.6a) measured by ESR ($46 \times 10^2\,s^{-1}$ at 75 °C)[25] as a clock, the rate constant for quenching of the anomeric radical with PheSeH was shown to be the surprisingly slow $3.6 \times 10^6\,M^{-1}\,s^{-1}$ at 78 °C, possibly because of the boat conformation of the radical.

Occasionally, it is useful to trap the same radical with two different quenchers (Scheme 7.2), rather than relying on an unimolecular isomerisation of the radical.

Equation (7.8) gives the product ratio for the case when both XY and ЖY are in large excess, so that their concentrations are invariant over the course of the reaction: such simple competition experiments are common throughout physical organic chemistry.

$$\frac{[\text{RЖ}]}{[\text{RX}]} = \frac{\kappa_2/[\text{ЖY}]}{k_2[\text{XY}]} \quad (7.8)$$

Some unimolecular clock reactions are given in Table 7.1 and bimolecular clock reactions in Table 7.2, together with their kinetics. Rate constants at any

Table 7.1 Unimolecular clock reactions.

Reaction	*$Log_{10}(A/s^{-1})$*	*$E_a/kJ\,mol^{-1}$*	*k_{25}/s^{-1}*	*Ref.*
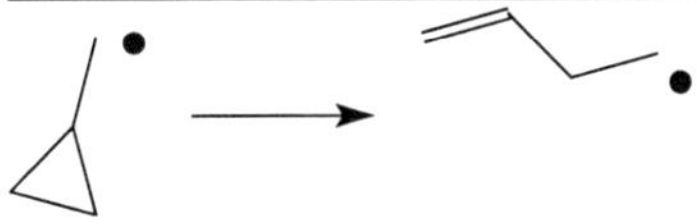	13.04	29.5	6.7×10^7	38
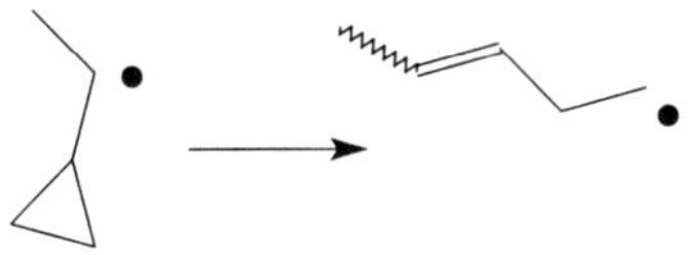	13.15	31.3	4.0×10^7	39
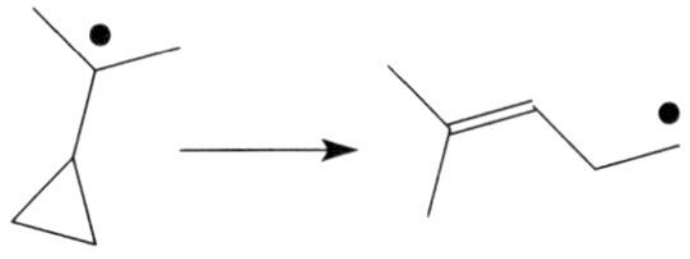	13.15	30.9	5.0×10^7	39
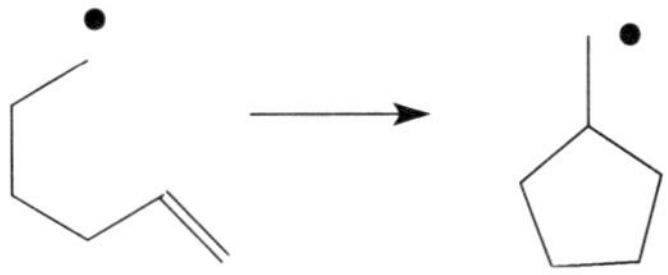	9.3	20.9	2.3×10^5	40
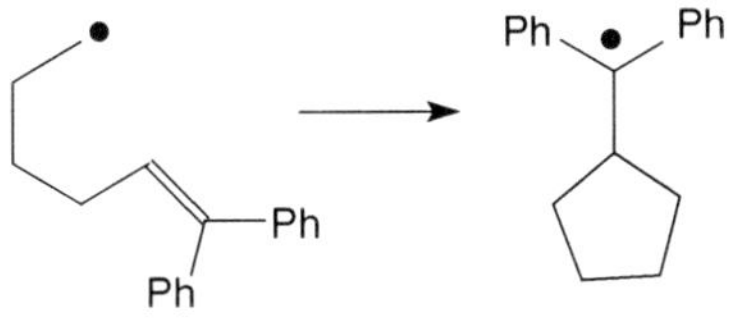 Ph, Ph → Ph, Ph	10.17	14.6	4×10^7	38
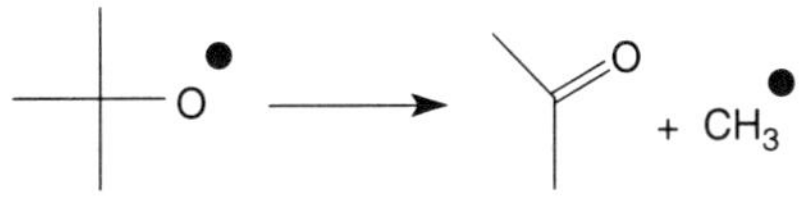 O → O + CH_3	12.8	48.7	2.0×10^4	33
in benzene *tert*-Butoxy fragmentation in CF_2Cl–$CFCl_2$ (Frigen 113)	13.2	52.7	8.0×10^3	33

given temperature are, like those for all reactions, given by the parameterised Arrhenius eqn. (7.9), where T is the absolute temperature and $0.0191 = 2.303R$ in kJ mol^{-1}.

$$\log_{10}k = \log_{10}A - E_a/(0.0191\,T) \qquad (7.9)$$

The lower A values in Table 7.1 for ring-closing reactions, compared with ring-opening reactions, is as expected, since the ring-closing reactions require loss of internal degrees of freedom, as the radical site and the double bond attain the correct disposition in which to react. *tert*-Butoxy fragmentation is modestly solvent dependent, in accord with the radical's slightly more polar nature than the acetone product.

Table 7.2 Bimolecular clock reactions.

Reaction	$Log_{10}(A/M^{-1}s^{-1})$	$(E_a/kJ\,mol^{-1})$	$k_{25}/M^{-1}s^{-1}$	*Ref.*
$Bu_3Sn\text{-}H + CH_3^{\bullet}$	9.4	13.4	1.1×10^7	40
$Bu_3Sn\text{–}H + RCH_2^{\bullet}$	9.1	15.5	2.4×10^6	40
$Bu_3Sn\text{–}H + (CH_3)_2CH^{\bullet}$	8.7	14.6	1.4×10^6	40
$Bu_3Sn\text{–}H + (CH_3)_3C^{\bullet}$	8.4	12.5	1.6×10^6	40
$PhSH + RCH_2^{\bullet}$	9.4	7.1	1.4×10^8	41
$PhSH + (CH_3)_2CH^{\bullet}$	9.3	7.1	1.0×10^8	41
$PhSH + (CH_3)_3C^{\bullet}$	9.2	6.7	1.1×10^8	41
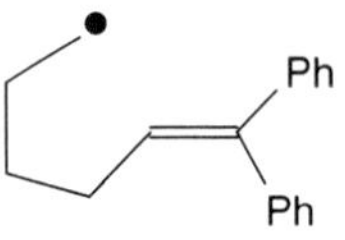 Ph Ph + PhSeH	10.7	9.3	1.2×10^9	38
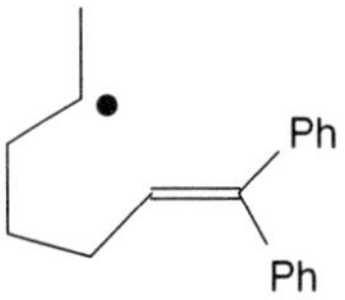 Ph Ph + PhSeH	9.9	4.8	1.2×10^9	38
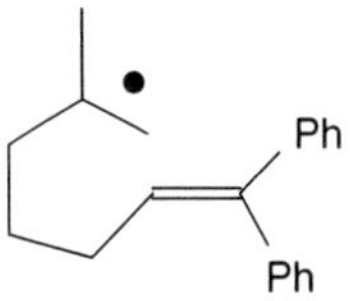 Ph Ph + PhSeH	10.0	4.8	1.2×10^9	38

7.3 GENERATION OF RADICALS

Carbohydrate-derived radicals are generated by direct electron transfer, hydrogen abstraction or fission of weak bonds. Direct electron transfer from the enediolate of reducing sugars is the basis of most reducing sugar assays.

7.3.1 Direct Transfer of Electrons

Studies of the simple removal of electrons from reducing sugars are sparse, but the kinetics of reaction of D-glucose in alkaline solution with anthraquinone-2-sulfonate (a redox mediator in alkaline pulping of wood chips) were in accord with a rate-determining electron transfer between glucose enolate and the quinone. The product was almost exclusively glucosone and isomerisation of glucose (to fructose, mannose and further products) competed with oxidation.[42] Moreover, 2-deoxyglucose reacted 10^2–10^3-fold slower than glucose itself. Whether the reaction involved a two-electron transfer or two single-electron transfers could not be determined, although the semiquinone product of one-electron transfer was detected.

7.3.1.1 Reducing Sugar Assays. Many traditional tests for reducing sugars involve the one-electron reduction of a transition metal complex by the sugar in alkaline solution, and indeed the term "reducing sugar" arose from such tests. Such tests are not standard tests for aldehydes – fructose is a reducing sugar and in its open chain form possesses no aldehydic CH. Traditional tests included Fehling's solution (alkaline Cu^{II}, coordinated by tartrate, whose deep blue colour is discharged and replaced by red cuprous oxide), Tollen's reagent (alkaline Ag^{I}, coordinated by ammonia, which on reduction gives the famous "silver mirror") and the more useful, because easily quantitated, assay of Hagedorn and Jensen, in which $[Fe^{III}(CN)_6]^{3-}$ is reduced to $[Fe^{II}(CN)_6]^{4-}$ and then the two combine with cations in solution to give colloidal Prussian blue. Rates of reaction of an extensive series of reducing sugars with alkaline Cu^{II} tartrate[43] and alkaline ferricyanide[44] were zero order in oxidant and first order in sugar and hydroxide ion. Rate constants for a wide range of sugars correlated well with the rate of C2 deprotonation: once the enediolate was formed, it reacted rapidly by electron transfer to the transition metal ion.

Such reactions, leading to intense colours which can be measured in a spectrometer or even a colorimeter, form the basis of most reducing sugar assays. 2,2′-Bicinchoninate (2,2′-biquinoline 4,4′-bicarboxylic acid) (Figure 7.8) forms stable, intensely blue complexes with Cu^{I} and this is now widely used in place of the traditional colour tests. The Cu^{II}-bicinchoninate reacts with single reducing ends in oligosaccharides[45] on the surface of insoluble polysaccharides and this makes it very useful in investigations of cellulolytic, *etc*., enzymes.

If reducing sugar assays are based on the colour reactions of the reduced oxidant, as all the foregoing are, then there is the ever-present possibility that the reductant may have been a contaminant and not a carbohydrate at all. This possibility is avoided by the PAHBAH assay,[46] which is based on the formation of bis-hydrazones with *p*-hydroxybenzoic acid hydrazide, of small α-dicarbonyl compounds formed from alkaline degradation of sugars, or possibly their generation via Maillard-type reactions (see Section 6.2.3).[47] Fragments from the sugar are incorporated into the chromophore.

Because the products of oxidation reactions are osones, they can tautomerise again to new ene-diolates, which can then in their turn be oxidised. This means that the exact titre depends on the sugar: effective extinction coefficients for monosaccharides assayed by the 2,2′-bicinchoninate assay range over a factor of 3 for common sugars.[48] Similar differences between sugars are found with PAHBAH.

The traditional method of assay of 2-acetamido sugars is the Morgan–Elson reaction. This involves treatment of the sugars in base, which produces furans, and reaction of these furans *in situ* with Ehrlich's reagent (*p*-dimethylaminobenzaldehyde in acid), which generates delocalised chromophores, as shown in Figure 7.8.[49]

7.3.1.2 Ascorbic Acid, the Natural Antioxidant. L-Ascorbic acid,[v] vitamin C, is a biological antioxidant, arguably *the* water-soluble one to complement the

[v] Its structure so thwarted the classical degradative reactions of carbohydrate chemistry that for a time it was called "godnose"!

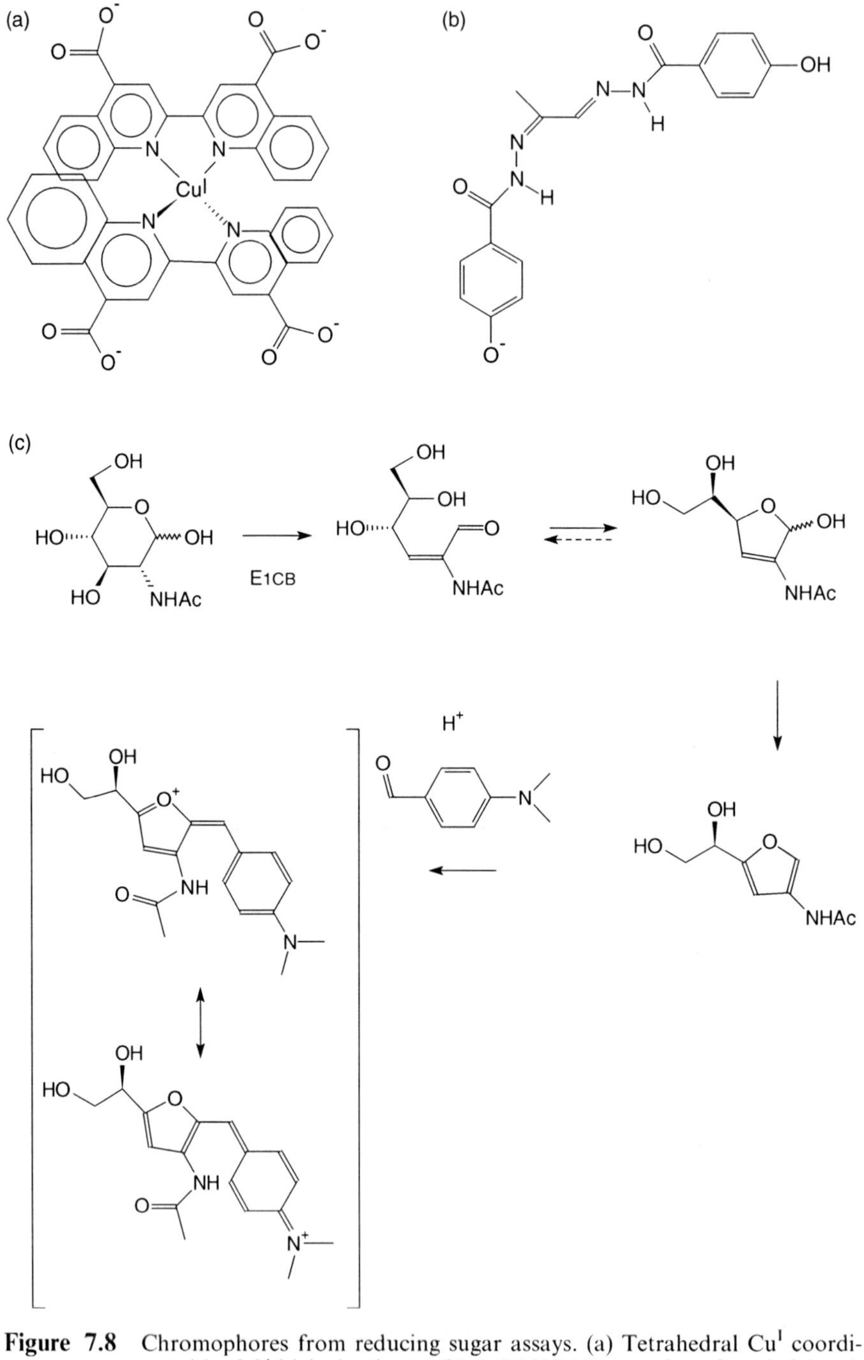

(a) (b) (c)

Figure 7.8 Chromophores from reducing sugar assays. (a) Tetrahedral Cu^{I} coordinated by 2,2′-bicinchoninate. (b) A PAHBAH complex of methylgloxal. (c) The Morgan–Elson test for hexosamines. The Morgan–Elson and PAHBAH reactions are of course heterolytic, but are included here for completeness.

α-tocopherol in membranes.[50] It possesses an enediol function conjugated with a lactone carbonyl, which lowers the first pK_a to 4.17, similar to a carboxylic acid, and the second to 11.57.[51] It has several biosynthetic routes: in mammals it is biosynthesised from glucuronic acid via lactonisation to O3 and reduction of C1 (GlcA numbering), yielding L-gulono-γ-lactone. The key enzyme in its biosynthesis is then the FAD-dependent L-gulono-γ-lactone oxidase, which oxidises the 3-position (Gul numbering, lactone carbonyl being 1) to a ketone; the hydroxy ketone spontaneously isomerises. In animals which can normally get enough vitamin C in their diet (such as guinea pigs and humans), the gene for L-gulono-γ-lactone becomes non-functional and rapidly accumulates mutations.[52] In plants there are several parallel routes, starting from GDP mannose and/or L-galactose.[53]

Ascorbate anion acts as a scavenger of ROS and other radicals by single-electron transfer to yield a stable, persistent ascorbic anion-radical. Ascorbate anion reacts with most radicals at the diffusion limit and even reduces various semiquinone anions, which are usually stable and persistent, by electron transfer at a rate around $2 \times 10^8\,M^{-1}\,s^{-1}$ at room temperature and water-soluble models for the tocopherol radical marginally more slowly (Figure 7.9). The product is a weak oxidant: ascorbate radical anion oxidises semiquinone anions much more slowly ($\sim 10^5\,M^{-1}\,s^{-1}$).[54]

7.3.1.3 Glucose Oxidase[4] and Related Enzymes. The FAD[vi]-dependent enzyme glucose oxidase converts β-D-glucopyranose and oxygen to D-gluconolactone and hydrogen peroxide.[vii] The fungal enzymes (*Aspergillus niger* and *Penicillium amagasakiensis*) have been intensively studied, with crystal structures for the former, a 90 kDa dimer with one molecule of FAD per subunit. The two half-reactions, oxidation of glucose, with concomitant reduction of the cofactor and O_2 re-oxidation of the reduced enzyme, are best considered separately, as the overall kinetics are ping-pong (although a structurally related enzyme, cholesterol oxidase, exhibits sequential kinetics and its X-ray structure reveals a narrow tunnel through which only oxygen can reach the active site).[55]

The oxidative half-reaction is strongly pH dependent, being dependent on a single ionisation of pK_a 8.1, with a low-pH limit of $1.5 \times 10^6\,M^{-1}\,s^{-1}$ and a high pH limit of $6 \times 10^2\,M^{-1}\,s^{-1}$.[56] The ionisation was that of His516, as was shown by the His516Ala mutant, which had the "high pH" activity across the whole range. The reaction was unaffected by change of solvent viscosity and showed no solvent isotope effects, eliminating diffusion-limited processes or combined electron and proton transfer processes as rate determinants. At both the high and low pH extremes, though, a value of $^{18}(V/K)$ of 1.02_8 in reactions of $^{16}O{-}^{18}O$ was observed, indicating a reduction of bonding between the oxygens in the rate-determining step. The oxidation is therefore initiated by an electron transfer from the reduced flavin anion (pK_a 6.7 in free solution,[57] but lowered

---

[vi] Flavin adenine dinucleotide.

[vii] The term "oxidase" normally refers to enzymes which use molecular oxygen as an electron acceptor, generating hydrogen peroxide.

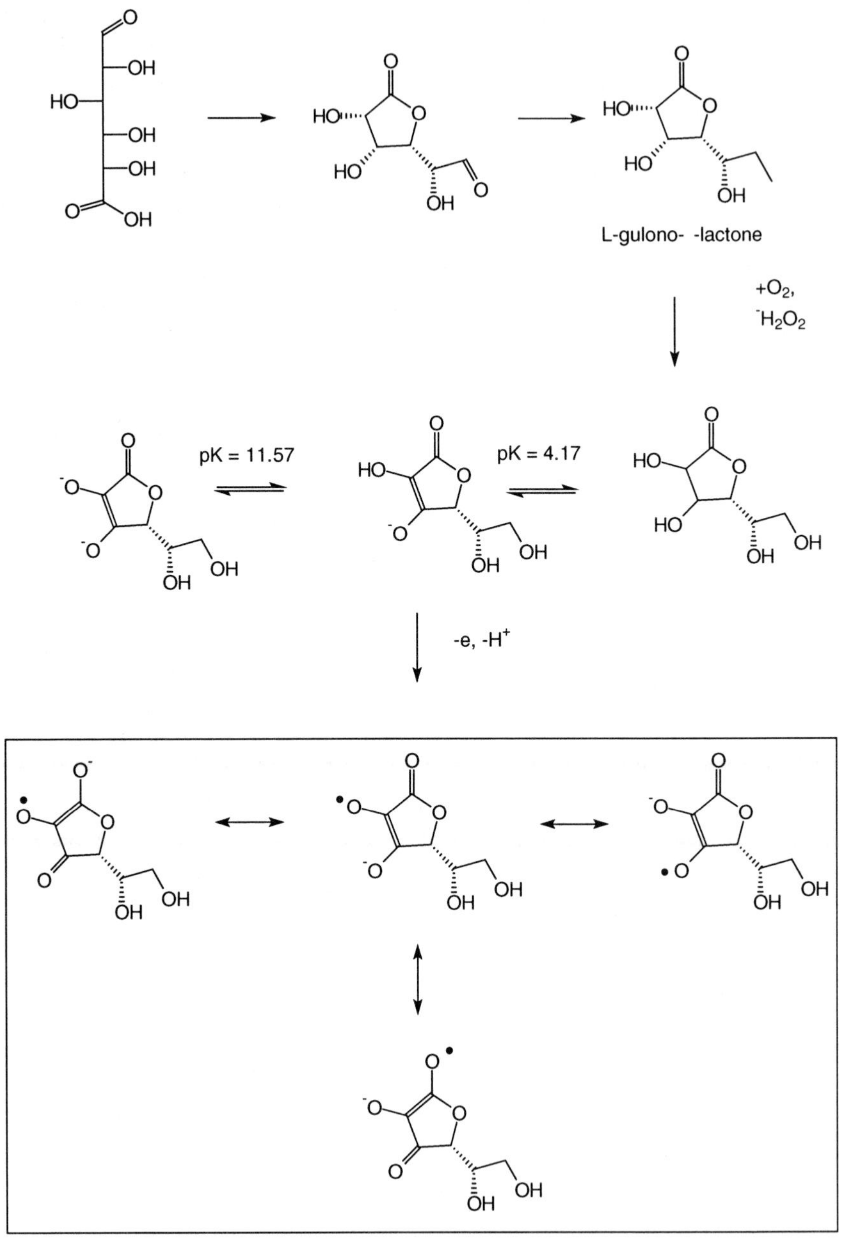

L-gulono- -lactone

$+O_2$, $^-H_2O_2$

pK = 11.57

pK = 4.17

-e, -H^+

Figure 7.9 Ascorbic acid biosynthesis and action as a reductant.

by about 3 units in the enzyme complex) to yield a semiquinone radical. Marcus theory-based calculations support the transfer being facilitated by the protonation of His516. The semiquinone radical has a pK_a of 7.3, but pulse radiolysis studies have shown the neutral molecule, rather than the anion, to react with superoxide at the diffusion rate. The oxidative half-reaction can therefore be written as in Figure 7.10, since the 4a hydroperoxide, generated by pulse radiolysis, is not catalytically competent, decomposing to hydrogen peroxide and flavin slower than the overall reaction.

Although K_m values for sugars can readily be measured under steady-state conditions, the individual reductive half-reaction, measured directly by stopped flow, is first order in sugar at all accessible concentrations. Only the β-anomer of glucose is accepted. A primary kinetic isotope effect is observed with [1-^{2}H]glucose in the reductive half-reaction, whose rate now saturates at high concentrations, establishing that a rapid step precedes an isotope-sensitive step. Studies in which FAD is replaced by a carbon analogue at position 5, whose reduction product is configurationally stable, indicate that the substrate hydrogen is added to the *re* face of the cofactor (*i.e.* from below as drawn in Figure 7.10). Modelling of glucose binding with these constraints suggests that His516 is ideally placed to deprotonate O1 as H1 is transferred to N5 of the cofactor, possibly without intermediates, *i.e.* as a hydride. It was possible to obtain a crystal structure of a complex between a closely related enzyme, cellobiose dehydrogenase of *Phaneroochaete chrysoporium*, and an analogue of the carbohydrate substrate, cellobionolactam (Figure 7.11).[58] C1 of the carbohydrate was on the *re* face of the flavin only 2.9 Å from N5 and the lactam oxygen was hydrogen bonded to His689, the counterpart to His516. Although radical and carbanion mechanisms for flavin-catalysed oxidations could not be disproved from the structure, the geometry is perfect for a hydride transfer. Site-directed mutagenesis of His689 produced enzymes with k_{cat} values 10^{-3} times those of wild-type or less, implicating the residue in catalysis.[59] "Cellobiose dehydrogenase" appears to be identical with the protein previously called cellobiose quinone oxidoreductase, a proteolytic fragment of cellobiose oxidase containing only the flavin and carbohydrate domains.[60] Cellobiose oxidase, which produces hydrogen peroxide, contains a haem-containing domain as well as a flavin domain; electrons are presumably transferred to oxygen from the flavin via the haem.

7.3.1.4 Pyrroloquinoline Quinone (PQQ)-dependent Glucose Dehydrogenase.[6,61] The oxidation of glucose and other sugars to the δ-lactone is sometimes carried out in bacteria by enzymes, which contain the cofactor PQQ[viii] (Figure 7.12a), which, like FAD, mediates both one- and two-electron chemistry. Like flavin, PQQ has one- and two-electron-reduced forms and stands at the interface of one- and two-electron chemistry, with reoxidation of the cofactor a one-electron process and hydride transfer apparently the mechanism of carbohydrate oxidation. The structure of the soluble PQQ-dependent glucose dehydrogenase from *Acinetobacter calcoaceticum* has been solved.[62] The cofactor is in its planar reduced state and the β-D-glucopyranose is in a $^4C_1$ conformation (the enzyme has an absolute specificity for the β-anomer). O1 is

---

[viii] In fact, three other minor structural variants are also known, but mechanistic work has been confined to enzymes using PQQ itself.

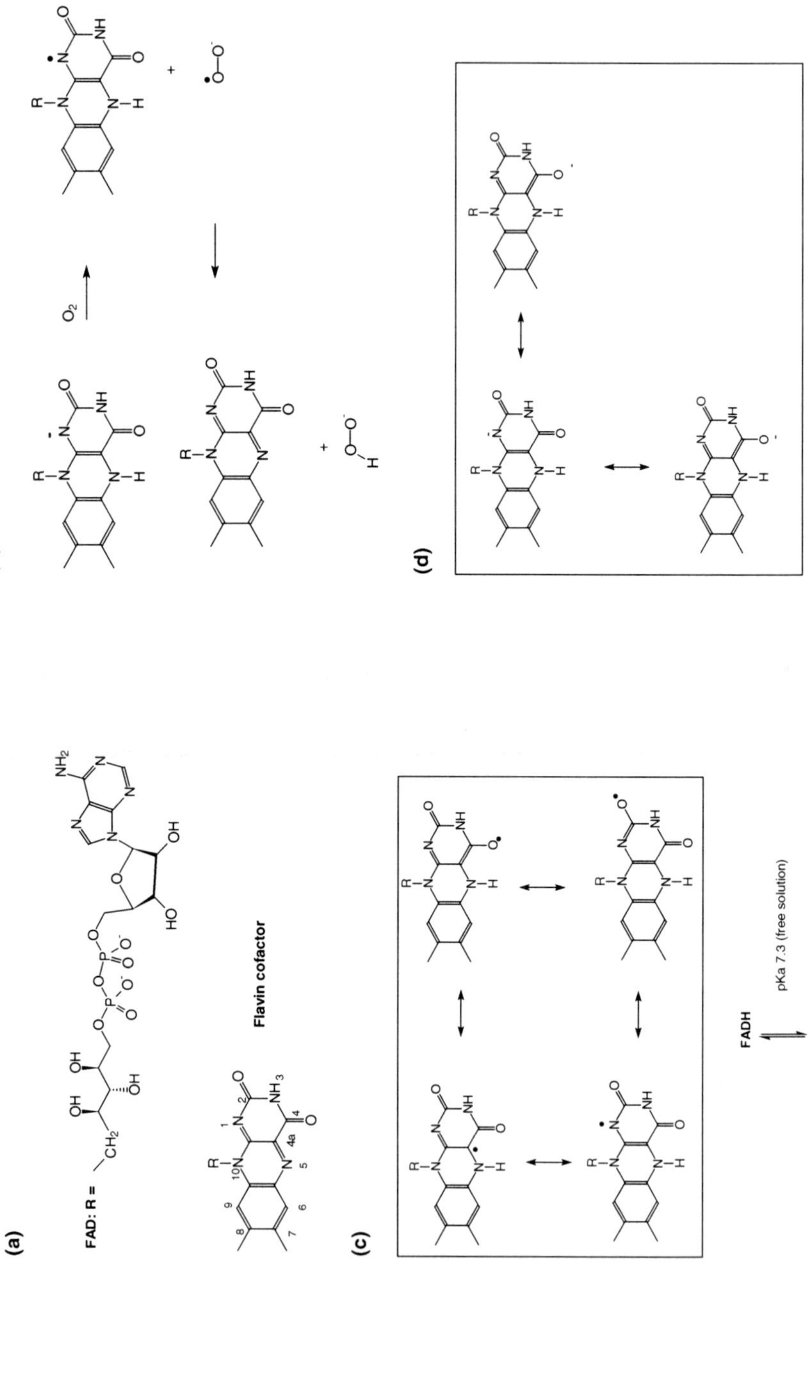

(a)

FAD: R =

Flavin cofactor

(b)

O_2

(c)

(d)

FADH

pKa 7.3 (free solution)

Figure 7.10 *(Continued)*

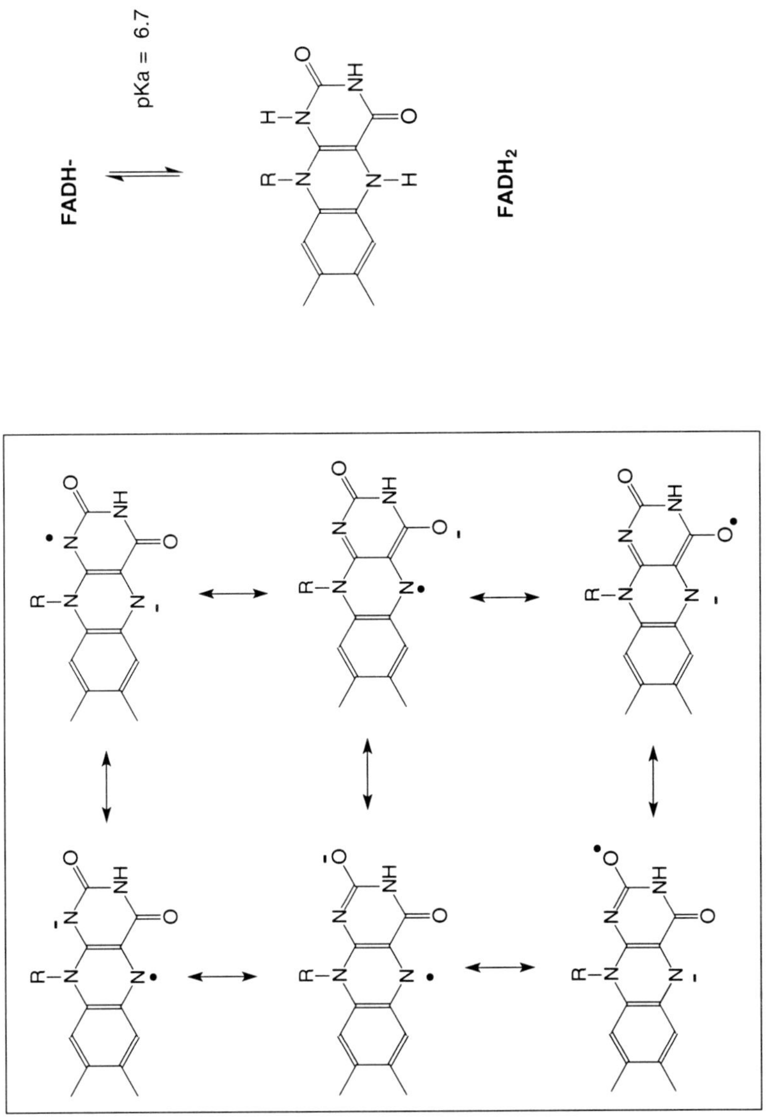

Figure 7.10 (a) Structure of flavin and of FAD. (b) Oxidative half-reaction of flavin enzymes. (c) Half-reduced flavin in two ionisation states: only reasonably important canonical structure, not involving separation of charge, are drawn. (d) Fully reduced flavin in two ionisation states.

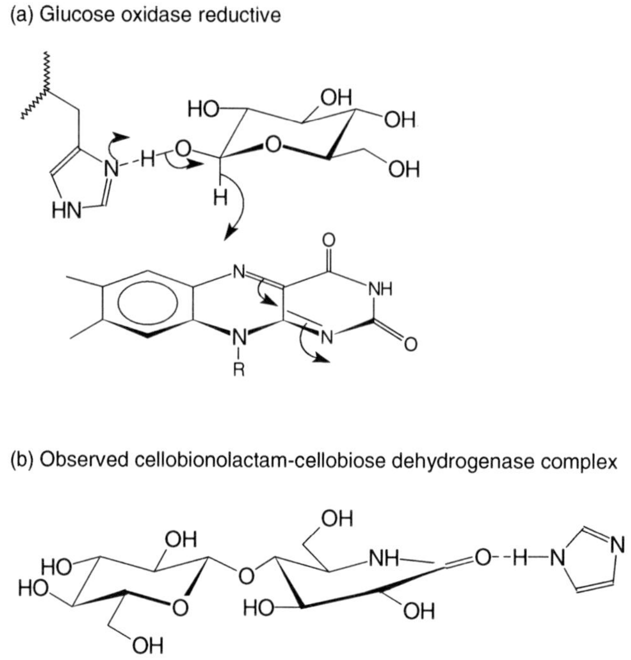

(a) Glucose oxidase reductive

(b) Observed cellobionolactam-cellobiose dehydrogenase complex

Figure 7.11 (a) Probable mechanism of the reductive half-reaction of glucose oxidase. (b) Cellobionolactam, as hydrogen bonded in the complex with *P. chrysosporium* cellobiose dehydrogenase.

hydrogen bonded to His144 and O5, N6 and O7a of the cofactor are coordinated to the essential active-site Ca^{2+} ion. Coupled with pre-steady-state measurements which indicate a fast reaction of the chromophore preceding a slow step, the structure supports a mechanism very similar to the reductive half-reaction of glucose dehydrogenase (Figure 7.12b), with the hydride adding to C5 of the cofactor from the *si* face (*i.e.* from the top as the cofactor is drawn in Figure 7.12a), being pushed off C1 of glucose by partial deprotonation of 1-OH by His144 and being pulled on to C5 of the cofactor by the latter's coordination to calcium. A subsequent slow step aromatises the cofactor.

The reduced cofactor can be reoxidised by two one-electron-transfers, for example the common oxidant ubiquinone.

7.3.2 Hydrogen Abstraction

7.3.2.1 By Hydroxyl and Alkoxyl and Related Species, and Reactive Oxygen Species. The Autoxidation of Carbohydrates. Fenton's reagent is the name given to a mixture of ferrous ions and peroxides: a single-electron transfer from the metal to the coordinating peroxide results in a hydroxy or alkoxy ferric complex and a free hydroxy radical (Scheme 7.3, top), As indicated in Section 7.2.2, the hydroxy radical is an avid hydrogen abstractor, removing H from all

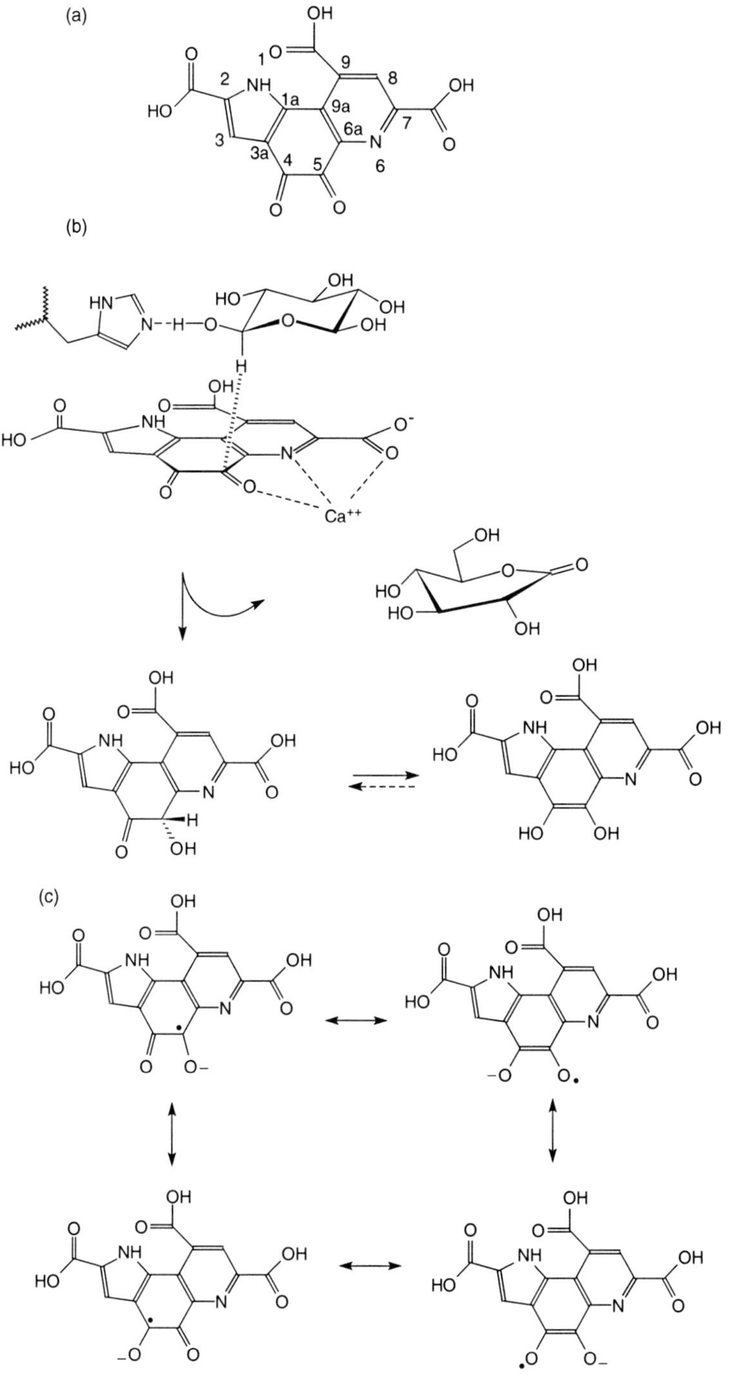

(a)

(b)

(c)

Figure 7.12 (a) PQQ structure. (b) Reductive half reaction of glucose PQQ-dependent dehydrogenase. (c) Important resonance structures for the semiquinone radical anion.

$$Fe^{2+} + H\text{–}O\text{–}O\text{–}H \rightarrow [Fe^{III}OH]^{2+} + HO^{\bullet}$$

$$Ti^{3+} + H\text{–}O\text{–}O\text{–}H \rightarrow [Ti^{IV}OH]^{3+} + HO^{\bullet}$$

Scheme 7.3

carbons in glucose. A very similar system involving Ti^{3+} is used to generate organic radicals for ESR study, as the Ti^{IV} product does not have unpaired electrons. There has been some question as to whether Fenton chemistry is indeed the chemistry of the hydroxyl radical, which cannot be observed directly in solution, rather than ferryl ($Fe^{III}{=}O$) and similar species.[63] However, the balance of opinion[13,64] seems to be that metal complexes are not involved, largely because the putative $HO^{\bullet}$ fulfils the classical criterion for an intermediate, that its reactivity is independent of its method of generation, whether by Ti^{III}, Fe^{II} or in a metal-free system such as radiolysis.

In the gas phase, $HO^{\bullet}$ behaves as a σ-radical with a high spin density on oxygen (*i.e.* the oxygen is sp hybridised, one forming the OH bond and the other the SOMO.[65] In solution, the uncertainty principle broadening of ESR lines, arising from the two degenerate lone-pair p orbitals, makes them unobservable. Alkoxy radicals appear similar.

$HO^{\bullet}$ shows little selectivity in $H^{\bullet}$ abstractions from pyranose rings, but shows enhanced reactivity towards the 4′-positions of aldofuranosides and the 5′-positions of ketofuranosides such as sucrose.[66] This appears concordant with studies on $H^{\bullet}$ abstractions by $(CH_3)_3C\text{–}O^{\bullet}$, where an enhanced reactivity of C–H bonds antiperiplanar to a p-type lone pair on oxygen was noted.[67] As discussed in Section 7.5.1, this could be a least-motion effect as much as a manifestation of ALPH. The labilisation of hydrogens α to an ether linkage – the cause of the well-known peroxidation of ethers – was apparent even in hydrogen abstraction by the normally avid and unselective *tert*-butoxy radical, with reactivities, relative to the C–H bonds in cyclohexane, of 2.7, 0.14 and 0.21 for the α, β and γ hydrogens of tetrahydropyran.[68]

Fenton chemistry is exploited by Nature in its degradation of lignocellulose. Iron is, of course, ubiquitous and at least two brown rot[ix] fungi, *Gloeophyllum trabeum* and *Postia placenta*, produce 2,5-dimethoxy-1,4-hydroquinone (Figure 7.13). The quinol and semiquinone forms serve to reduce oxygen to hydrogen peroxide and Fe^{III} to Fe^{II}, thus ensuring a continuous supply of both components of the reagent.[69] The hydroxyl radicals then indiscriminately attack the carbohydrates.

Under aerobic conditions, radicals react with oxygen to give peroxy radicals, $ROO^{\bullet}$ (Figure 7.14). Peroxy radicals are less reactive than alkoxy and hydroxy radicals, possibly, as was deduced from ESR splittings with specifically

[ix] Brown rot fungi, most notoriously *Merulius lacrymans*, which causes dry rot of timber, normally attack only the carbohydrate components of lignocellulose, leaving the lignin as a brown powder, hence the name. White rot fungi produce lignin-degrading oxidative species and degrade all lignocellulose components.

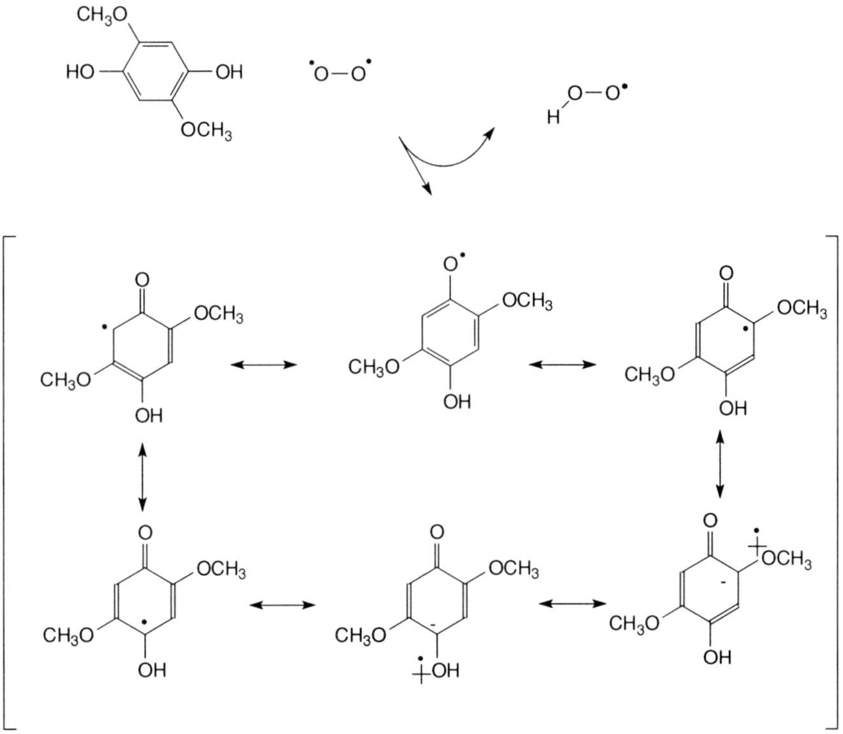

Figure 7.13 2,5-Dimethoxyhydroquinone. The protonated form of the semiquinone and hydroquinone is shown, as rotting fungi work in acidic media.

^{17}O-labelled species, because the spin is 30% on the internal oxygen and only 70% on the external oxygen.[70] The removal of electrons from the central oxygen in $HOO^•$ is presumably the reason why its pK_a is 7 units lower than that of H_2O_2 (see Section 7.1).

The generation of peroxy radicals by combination of carbon radicals with oxygen is very fast – even a stabilised cyclohexadienyl radical reacts with oxygen at close to diffusion ($1.2 \times 10^9\ M^{-1}\ s^{-1}$).[71] Kinetic isotope effects on the addition of oxygen to carbon radicals are surprising: addition of $^{17}O–^{16}O$ is 1.1% faster than addition of $^{16}O–^{16}O$, but addition of $^{18}O–^{16}O$ is 1% slower,[72] possibly because of contributions of classical zero point energy and moment of inertia effects in opposite senses.

Peroxy radicals dimerise at diffusion rates to tetroxides, which in turn, above 150 K, decompose into alkoxy radicals and oxygen. Low-temperature ESR has established the linear tetroxides as real molecules, with mM to μM dissociation constants at 110–150 K.[13] The combination is subject to a very large magnetic isotope effect, caused by the coupling to ^{17}O to the electronic spin, which relaxes some of the spin-symmetry prohibitions on radical recombination; a value of k_{17}/k_{16} of 1.8 is observed for combination of $R–^{16}O–^{17}O^•$ with of $R–^{16}O–^{16}O^•$.[72] If, however, the peroxy radical is α to a hydroxyl

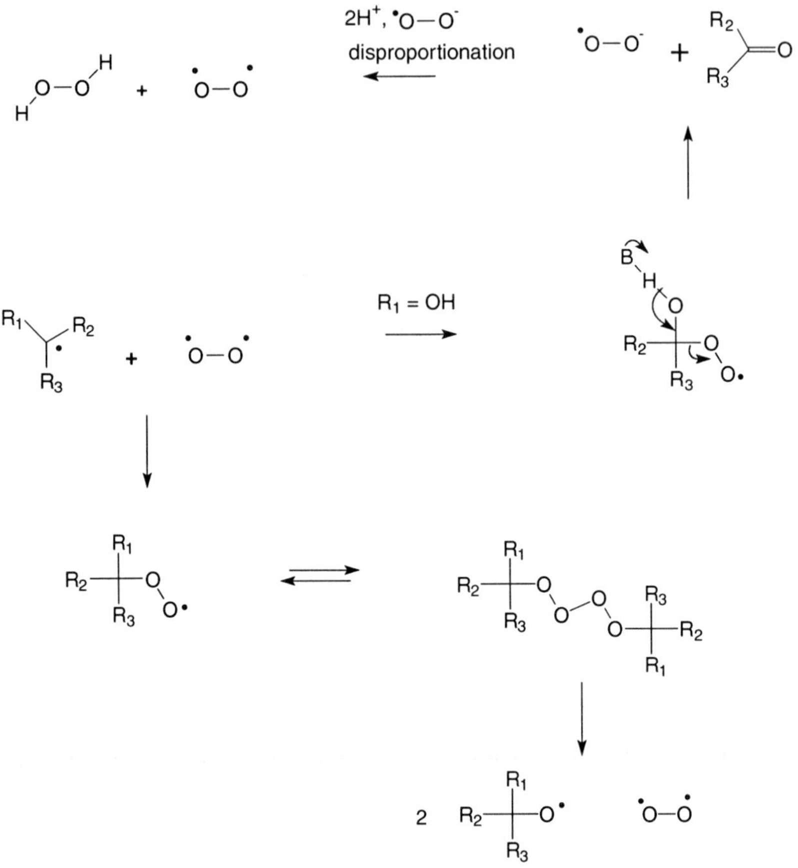

Figure 7.14 Reactions of peroxy radicals under aerobic conditions.

group, then superoxide is expelled in a base-catalysed reaction.[73] This generates a carbonyl group, which, in the case of most polysaccharides, can promote the cleavage of glycosidic linkages by $E1_{CB}$-like eliminations, either directly (*e.g.* oxidation of the 2-OH or 6-OH of starch and cellulose) or after forming the appropriate enediolate (as with the 3-OH of starch and cellulose) (see Section 6.2.4). On protonation, superoxide dismutates to oxygen and hydrogen peroxide, a reaction which is fastest at pH 4.8, the pK_a of the hydroperoxy radical. (Electron transfer between $HOO^\bullet$ and $O_2^{\bullet-}$ is faster than between two $HOO^\bullet$ radicals, and transfer between two $O_2^{\bullet-}$ ions does not take place).[3]

Action of ROS on plant cell wall polysaccharides, in addition to various glycosidases and transglycosylases, has been suggested as a mechanism for plant cell-wall expansion.[74] As discussed earlier, $HO^\bullet$ is very fleeting and anyway difficult to observe by ESR, so the technique of spin trapping is often used to demonstrate its presence in various, particularly biochemical, systems. A spin trap is a stable molecule which can add a reactive radical to yield a

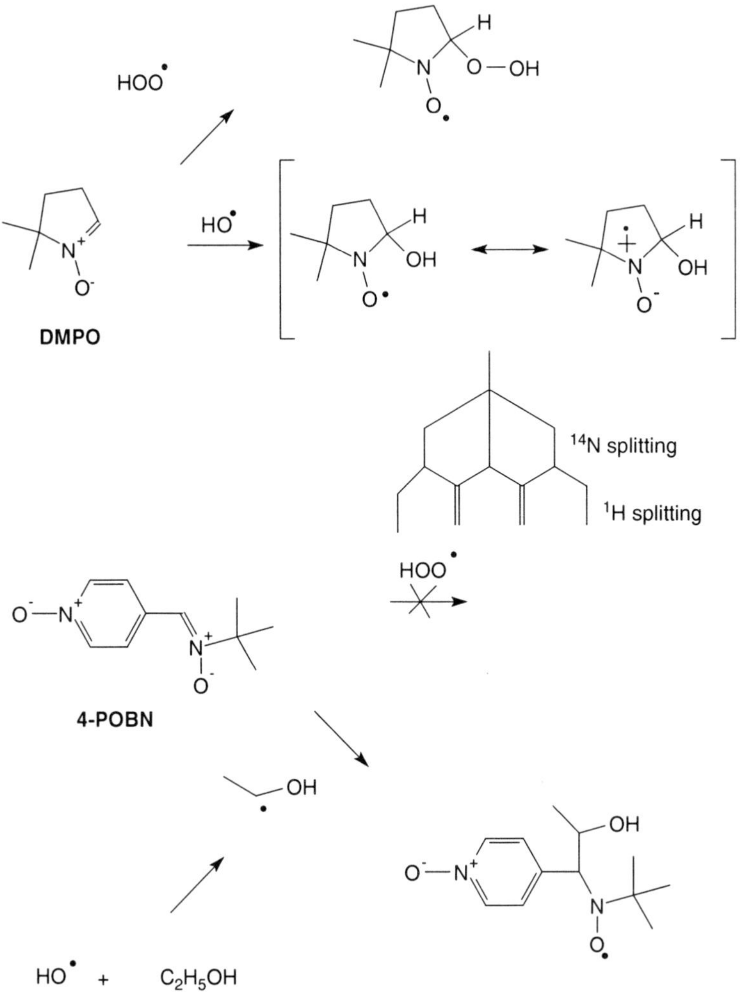

Figure 7.15 Spin trapping by DMPO and 4-POBN.

stable one which can build up and be examined by ESR. Thus, the spin trap 5,5-dimethyl-1-pyrroline *N*-oxide (DMPO) adds $HO^{\bullet}$ to give a nitroxide radical, characterised by an apparent 1:2:2:1 quartet (in fact a triplet of doublets which, because of the fortuitous near-equality of a_N and $a_{\beta H}$, gives this appearance). The adduct with $HOO^{\bullet}$ is more complex, probably because a_N and $a_{\beta H}$ are now no longer equal and $a_{\gamma H}$ no longer negligible. The more stabilised α-(4-pyridyl *N*-oxide)-*N*-*tert*-butyl nitrone (4-POBN) does not add superoxide or hydroperoxide, but does add $CH_3CH^{\bullet}OH$ formed by H-abstraction from ethanol by $HO^{\bullet}$ and this provides what is considered a good test for the presence of $HO^{\bullet}$ (Figure 7.15). In this way it was shown that $HO^{\bullet}$ degradation of polysaccharides did indeed occur in isolated plant cells.[75]

In mammalian cells, the oxidative damage caused by ROS to the components of living cells is thought to be connected with the onset of many ailments and even of ageing itself. The protective effects of many polysaccharides have been documented, as polysaccharides form the basis of many "traditional" medicaments. Any effects probably arise because certain radical sites from which $H^{\bullet}$ is abstracted are stabilised and thus serve as chain breakers in the same sort of way as butylated hydroxytoluene or ascorbate. For example, pectin was shown to protect against damage to the rat small intestine by hydroxy and peroxy radicals,[76] and agarose oligosaccharides to act as antioxidants in respect of $HO^{\bullet}$ and $O_2^{\bullet -}$.[77] Compared with other polysaccharides, agarose (Figure 4.87) has no obvious sites whence hydrogen abstraction would yield a particularly stable residue, except C4 of the 3,6-anhydro-L-galactose, where the radical would be stabilised by the axial C3–O3 bond, which permits the n–SOMO–σ* overlap depicted in Figure 7.5. Such stabilisation would be in accord with ESR studies of the radicals formed from polysaccharides with sulfate radical anion ($SO_4^{\bullet -}$), where the dominant determination of selectivity was the possibility of SOMO–σ* overlap of this type.[78] However, abstraction of H5 from the uronic acids in pectin (and the uronic acid components of glycosaminoglycans) would yield a radical potentially stabilised by the captodative effect (Figure 7.16).[79] The radical site is substituted with a π donor, oxygen, and a π-acceptor, a carbonyl group, and a ~40 kJ mol^{-1} stabilisation has been calculated for such systems.[80] Other than a comment in a paper addressing other matters, that pyranosyl radicals with a carbonyl group at C2 were remarkably unreactive, probably because of their captodative stabilisation,[81] data on the operation of the captodative effect in carbohydrate radicals are lacking.

ROS will generate secondary oxidising species whose interactions with carbohydrates are significant to *in vivo* degradative reactions and chemical technology. Thus, superoxide anion reacts with nitric oxide, the mammalian second messenger, at diffusion-controlled rates to yield peroxynitrite.[82] The parent acid (pK_a 6.8) can be made by acid-catalysed nitrosation of H_2O_2 with nitrous acid, followed by alkali quenching, but is unstable to O–O homolysis to $HO^{\bullet}$ and NO_2 ($k = 0.8\ s^{-1}$ at 20 °C). Peroxynitrite also reacts with CO_2 to yield the carbonate radical ion, $CO_3^{\bullet -}$, at a rate of $3 \times 10^4\ M^{-1}\ s^{-1}$.[83]

$CO_3^{\bullet -}$, generated from peroxynitrite, has been investigated as the basis of a pulp bleaching system.[84] The traditional bleaching reagent has been chlorine, but this produces environmentally persistent chlorinated aromatic compounds from the lignin. Chlorine dioxide, ClO_2, is a radical oxidant which reacts readily with lignin by electron abstraction to yield lignin radical cations, but hardly at all with

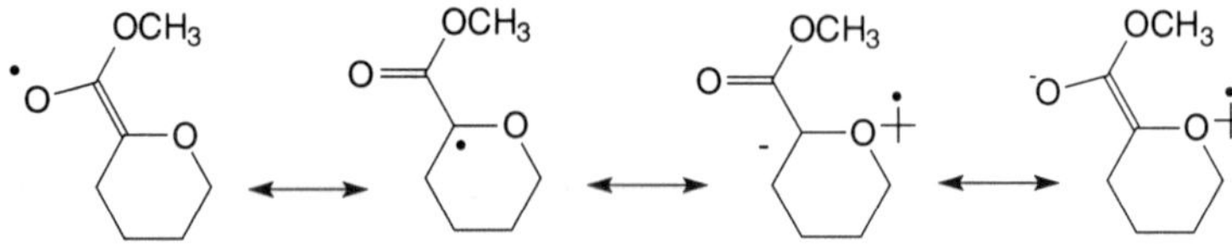

Figure 7.16 Potential captodative stabilisation of radicals at position 5 of uronic derivatives.

cellulose and other carbohydrates,[85] although it attacks the hex-4-eneuronyl residues produced from glucuronoxylan-4-*O*-methylglucuronyl residues in the kraft process.[86] Technically, it is the reagent of choice, but environmentally it still produces small quantities of chlorinated aromatics, so that oxygen, ozone or hydroperoxide bleaching are increasingly important.[x] All three reagents can give rise to the indiscriminate OH radical and loss of cellulose *DP*.

$CO_3^{\bullet-}$ shows the expected selectivity for the anomeric hydrogens of methyl β-D-glucoside and cellobioside, reacting with the latter at a rate of $7 \times 10^4\ M^{-1}\ s^{-1}$ at 20 °C (*i.e.* about five orders of magnitude slower than OH).[84] However, when cellulose (as "cotton linters") was used as a substrate, hydrogen abstraction was indiscriminate, possibly because intrinsic chemical reactivities were counterbalanced by relative accessibility.[87]

7.3.2.2 Hydrogen Abstraction by Halogens. The reaction of cellulose and models such as methyl β-glucopyranoside and -cellobioside with chlorine in aqueous solution has been studied in connection with the loss of cellulose *DP* and hence fibre strength during pulp bleaching. Most of the reactions appear to be heterolytic and are treated in Chapter 6, but in dilute acid (pH 2), a homolytic $H^{\bullet}$ abstraction from all positions, suppressed by chlorine dioxide, is observed. The resulting $>C^{\bullet}OH$ sites abstract chlorine atoms from hypochlorous acid or chlorine, giving transiently >CClOH and then a carbonyl group, leading to loss of *DP* from $E1_{CB}$ reactions.[88]

Classical photochemical bromination [*N*-bromosuccinimide (NBS), light] can be applied to carbohydrates if the anomeric centre is sufficiently activated. Photochemical bromination of tetraacetyl β-D-gluco-, -galacto- and -mannopyranosyl chlorides gives largely the retentive replacement of H1 by Br (Figure 7.17). Increasing the electronegativity of X (*e.g.* to F) or the avidity of the halogenating agent (to *N*-chlorosuccinimide or SO_2Cl_2), results in an increase in H5 abstraction. Substitution of both H1 and H5 occurs with complete retention of configuration, *i.e.* the halogen atom approaches from the α-face.[89] As might be expected, this reaction can be applied only to monosaccharides; attempts to apply it to cellooligosaccharides lead to exclusive replacement of internal anomeric hydrogens.[90]

7.3.2.3 Selective Oxidation of Hydroxymethyl Groups. Platinum on charcoal in the presence of air is the conventional method for selectively oxidising hydroxymethyl groups to carboxylic acids, and thus hexoses to uronic acids; anomeric protection is required, however, and the mechanism of this heterogeneous reaction is unknown. The reaction fails with polymers, probably for mass transport reasons.[91]

A widely used mediator of the oxidations of carbohydrates, particularly polysaccharides, is the stable nitroxide radical TEMPO (2,2,6,6-tetramethylpiperidine-1-oxyl) and its 4-substituted derivatives.[92] Removal of a single

---

[x] In paper industry jargon, chlorine dioxide bleaching is "ECF bleaching" (elemental-chlorine-free) and bleaching with oxygen species "TCF bleaching" (totally-chlorine-free).

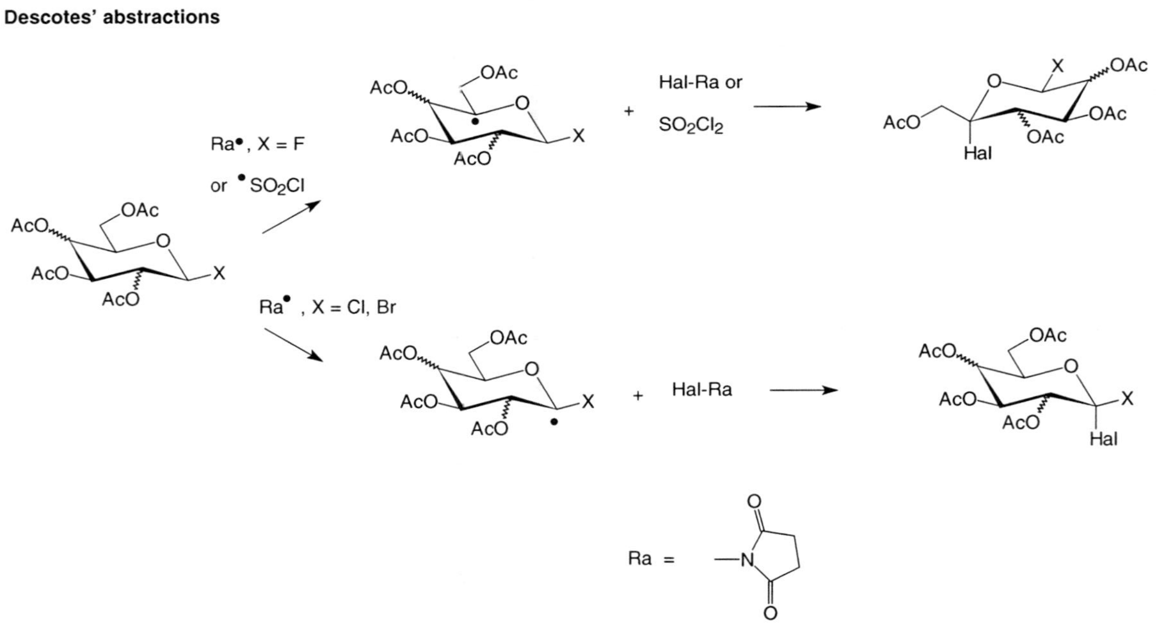

Descotes' abstractions

Figure 7.17 Photochemical halogenation of monosaccharide derivatives (*gluco*, *manno* and *galacto*, but not *talo*, compounds were investigated).

electron generates a cationic species variously called a nitrosonium ion[92] or oxoammonium ion;[93] the author prefers the latter as the nitrosonium ion is NO^+. TEMPO can be used with a wide variety of primary oxidants, including atmospheric oxygen via enzymes such as laccase, and even electrochemically, with direct transfer of electrons from TEMPO to an anode.

Although the generation of the oxoammonium ion from TEMPO involves one-electron chemistry, the actual oxidation of the substrate is heterolytic, as is shown by the oxidation of cyclopropylcarbinol to cyclopropylaldehyde without ring opening.[94] Substituent effects of X on p-$XC_6H_4CH_2OH$ oxidation are modest, suggesting little charge build-up on the oxidised carbon.

Nucleophilic addition of the alcohol to the trigonal oxoammonium ion yields a tetrahedral species, which collapses to the carbonyl compound and hydroxylamine by two mechanisms. Under basic conditions, the N–OH group is deprotonated and can act as an internal base in a cyclic transition state, but if it is protonated, a slower reaction involving an extraneous base has to occur (Figure 7.18). The two transition states are supported by different hydrogen isotope effects for the oxidation of $PhCD_2OH$: 1.7 under basic[94] and 3.1 under acidic conditions[xi],[95] in accord with the non-linear hydrogen transfer giving lower primary kinetic isotope effects (see footnote xxii, Chapter 6).

In accord with its crowded cyclic transition state, the basic pathway has a very high selectivity for oxidation of primary OH groups, whereas the slower acid pathway, with the more open transition state, has lower selectivity, oxidising secondary alcohols only modestly slower than primary ones.

The oxoammonium ion (or perhaps the primary oxidant) will oxidise reducing sugars, but otherwise alkaline (pH 8–10) TEMPO oxidations smoothly oxidise sugars to uronic acids and neutral polysaccharides to polyuronic acids. No definitive reason has been advanced for such oxidations proceeding to the carboxylic acid, when in organic solvents TEMPO oxidations of primary hydroxyls stop cleanly at the aldehyde. It is very reasonably assumed that the substrate for the second oxidation is the aldehyde hydrate, as the formation of which water is required, yet TEMPO- and laccase-mediated air oxidation of benzyl alcohol, in water, yields benzaldehyde. It seems to the author that the presence of water is a necessary, but not a sufficient, condition for the second stage of the oxidation, with a further requirement being that the aldehyde be substantially hydrated in water (which sugar 6-aldehydes, because of the electron-withdrawing effect of the sugar, will be).

If the anomeric centre is not protected, TEMPO-mediated oxidations of monosaccharides with Cl_2 or Br_2 yield aldaric acids, *i.e.* ω-diacids in which the all secondary hydroxyl groups are preserved;[96] the use of the 4-acetamido derivative of TEMPO is to be preferred because of its greater stability and lower volatility.[97] This method of making aldaric (saccharic) acids supersedes the use of nitric acid, used in Emil Fischer's classical work to make both ends of

---

[xi] Interpretation is complicated by the superimposed primary and secondary α-deuterium effects.

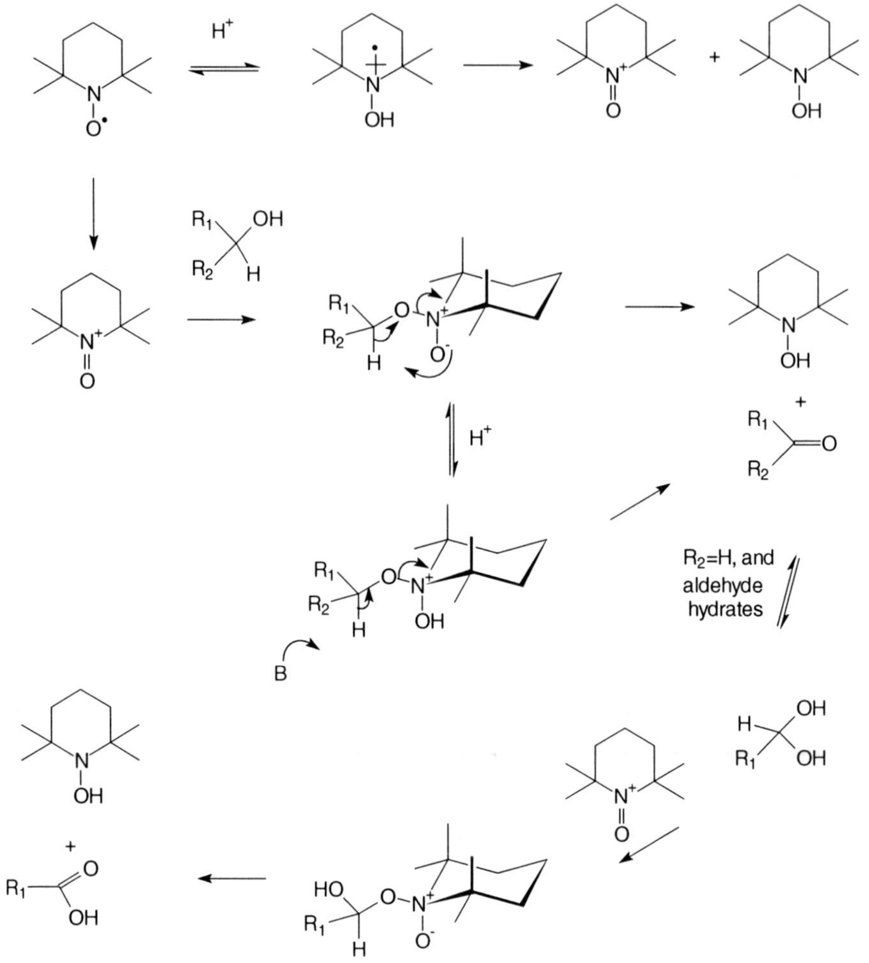

Figure 7.18 Mechanisms of TEMPO oxidations. The top line shows the disproportionation of TEMPO in acid and the rest of the figure the acidic and basic mechanisms for oxidation of alcohols, with the reaction of hydrated aldehydes also indicated.

an aldose structure identical. The apparent specificity of the nitric acid oxidation of glucose to glucaric acid seems to be an artefact of late-nineteenth century isolation procedures: the yield is maximally 44% and a range of smaller oxidised fragments are also produced.[98]

TEMPO-mediated reactions of solid polysaccharides, such as rayon (cellulose II) also result in clean conversion of C6 hydroxymethyl groups to carboxylic acids.[99]

7.3.2.3.1 Galactose oxidase[5]. Galactose oxidase is a fungal enzyme that oxidises C6 of D-galactose to the corresponding aldehyde, reducing molecular oxygen to H_2O_2. Although it does not act on D-glucose or L-galactose, it will oxidise a range of smaller primary alcohols; indeed, the best substrate, four times better than galactose, is dihydroxyacetone. Model building from the structure of the native enzyme suggests a "lock and key"-type active site, with larger molecules such as D-glucose being prohibited from binding by steric clashes with the protein, but smaller molecules being transformed indiscriminately.

The active site contains a "Type II" copper, in which a single copper ion is coordinated by two histidines and two tyrosines, one of which is covalently modified to form a thioether bond with a cysteine residue, *ortho* to the phenolic hydroxyl. The enzyme is biosynthesised with a 17-residue propeptide on the N-terminus.[100] The X-ray structure of the precursor reveals electron density that might indicate the cysteine sulfhydryl is oxidised to a sulfenic acid, CH_2SOH; in any event the peptide bond between Arg-1 and Ala-1 is cleaved and the active site configured in its resting state in the presence of copper ions an oxygen. No ESR signal is obtainable for the resting enzyme and this is now known to be because the spin on the phenoxy radical of the modified tyrosines and the spin on the Cu^{II} are antiferromagnetically coupled.

The currently accepted mechanism[101] involves direct binding of O6 (galactose as substrate), to the copper, with concomitant release of the unmodified tyrosine from the coordination sphere (Figure 7.19). $H6_S$ is abstracted by the oxygen radical of the modified tyrosine and an electron is transferred to Cu^{II} from O6: the relative timing of these processes is unknown, but they generate a complex of aldehyde and a two-electron reduced form of the enzyme, with the metal as Cu^I, and the active tyrosyl radical as a closed-shell species. After loss of the oxidised substrate, oxygen binds and is reduced to hydrogen peroxide by electron transfer from Cu^I and hydrogen transfer from tyrosine.

7.3.3 Fission of Weak Bonds

7.3.3.1 Radical Deoxygenation. The carbon–oxygen bond of a sugar becomes labilised to homolysis if it is attached to thioncarbonyl derivative, which in its turn is attacked by a thiophilic metalloid free radical, such as tri -*n*-butyltin. The original Barton–McCombie reaction[7] used a methyl xanthate in the presence of tri-*n*-butyltin hydride, but the thioncarbonyl group can be thiocarbonylimidazolide or even sometimes thioncarbonate.[102] The slow step in the reaction is, unsurprisingly, C–O homolysis, which results in the initial addition of the tri-*n*-butyltin radical being reversible (Figure 7.20). Possible competing reactions make the reaction work best with secondary hydroxyls. Except at bridgehead positions, Chugaev elimination of tertiary methyl xanthates competes favourably with C–O homolysis. Generation of a primary radical is more difficult than generation of a secondary radical, so that in deoxygenation of primary methyl xanthates, Me–S fission competes with C–O fission, and it is advisable to use other derivatives such as thiocarbonylimidazolides and somewhat higher temperatures than for secondary deoxygenations. The reaction is usually carried out in a refluxing organic solvent such as

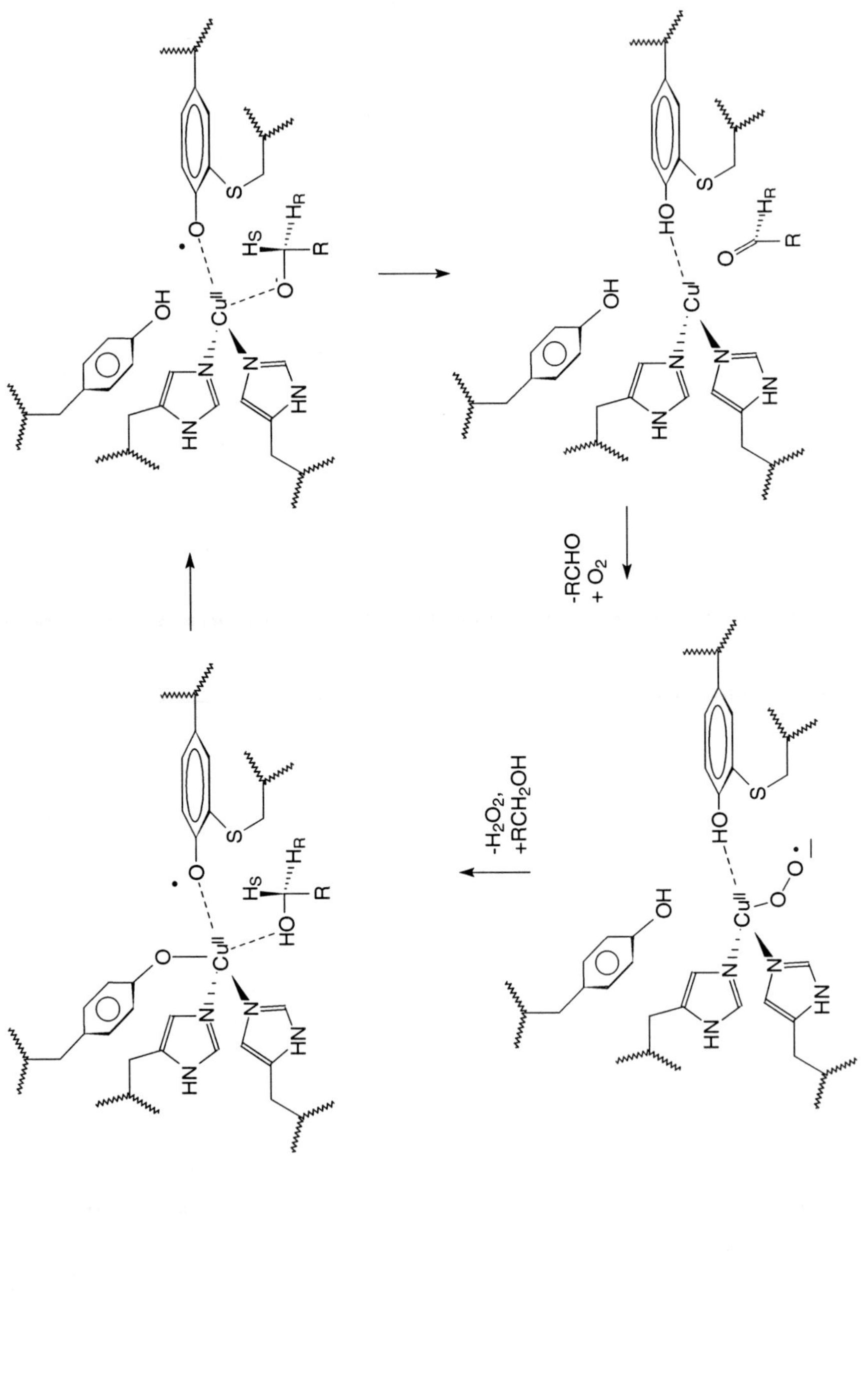

Figure 7.19 Reaction mechanism of galactose oxidase.

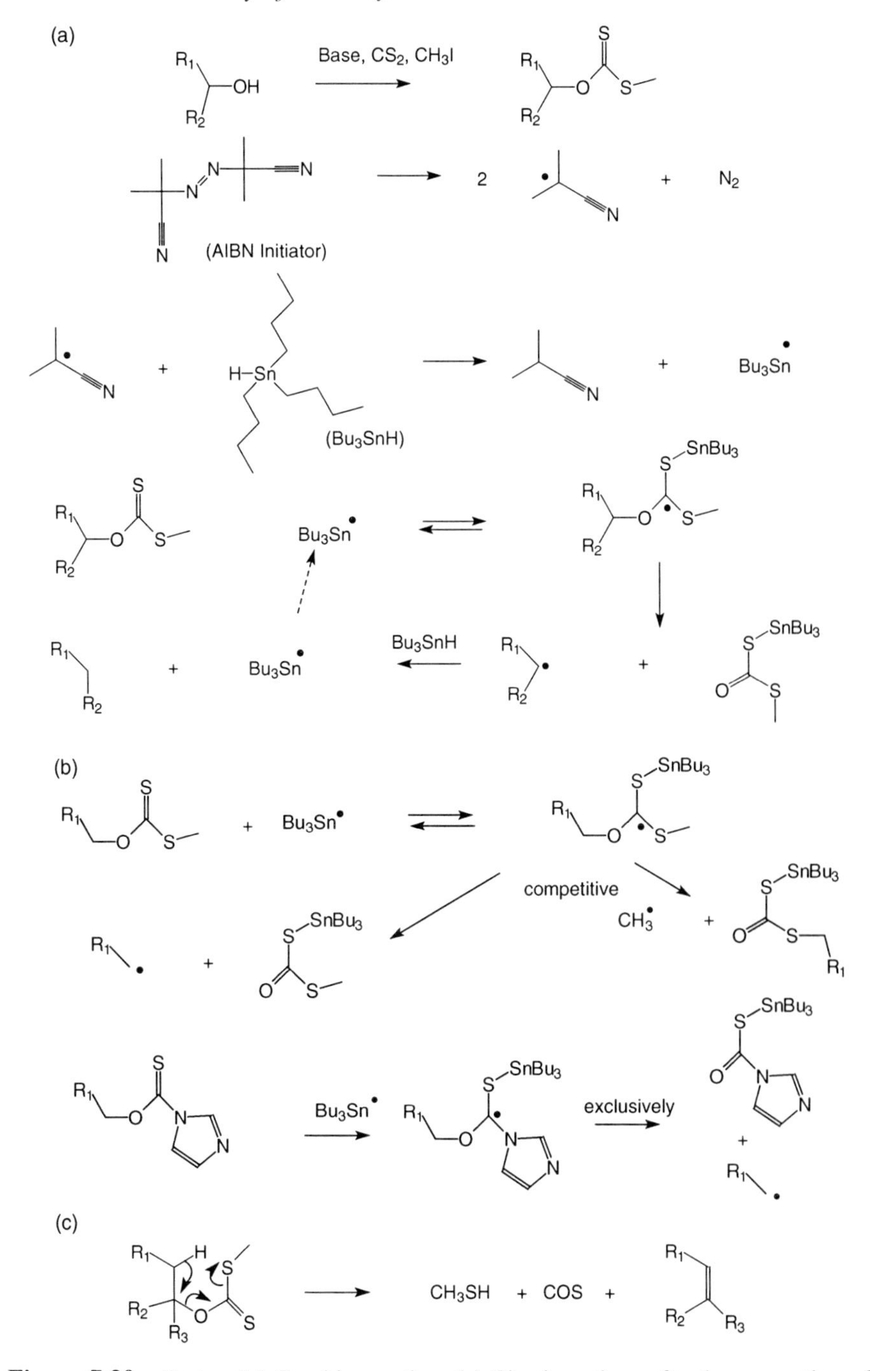

Figure 7.20 Barton–McCombie reaction. (a) Classic pathway for deoxygenation of secondary alcohols; $Bu_3Sn\ SCOSCH_3$ loses COS under the reaction conditions to yield Bu_3SnSCH_3 (b) Competing reactions in the deoxygenation of primary xanthates, and their avoidance (c) Chugaev reaction of tertiary methyl xanthates.

tetrahydrofuran or benzene, with a small amount of azobisisobutyronitrile (AIBN) as an initiator. The rate constants for quenching of alkyl radicals are such (Table 7.2) that with the molar concentrations of Bu_3SnH used, dimerisation of radicals is not a problem.

If two vicinal OH groups are converted to methyl xanthates and then subjected to the Barton–McCombie reaction, the product is the olefin rather than the completely deoxygenated sugar (Figure 7.21).[103] It is thought that the initially formed radical loses MeS(COS)$^{\bullet}$. By contrast, early work with the generally less satisfactory thionbenzoate esters in the Barton–McCombie deoxygenation revealed that a vicinal thionbenzoyloxy was not lost from the radical centre, but rather cyclised to yield a benzylidene thio sugar (Figure 7.22).[104]

An intriguing combination of ketal cleavage and radical deoxygenation has recently been reported for the 2-(2-iodophenyl)-1-cyanoethylethylidene group protecting O4 and O6.[105] Introduced by acetal exchange from triethyl

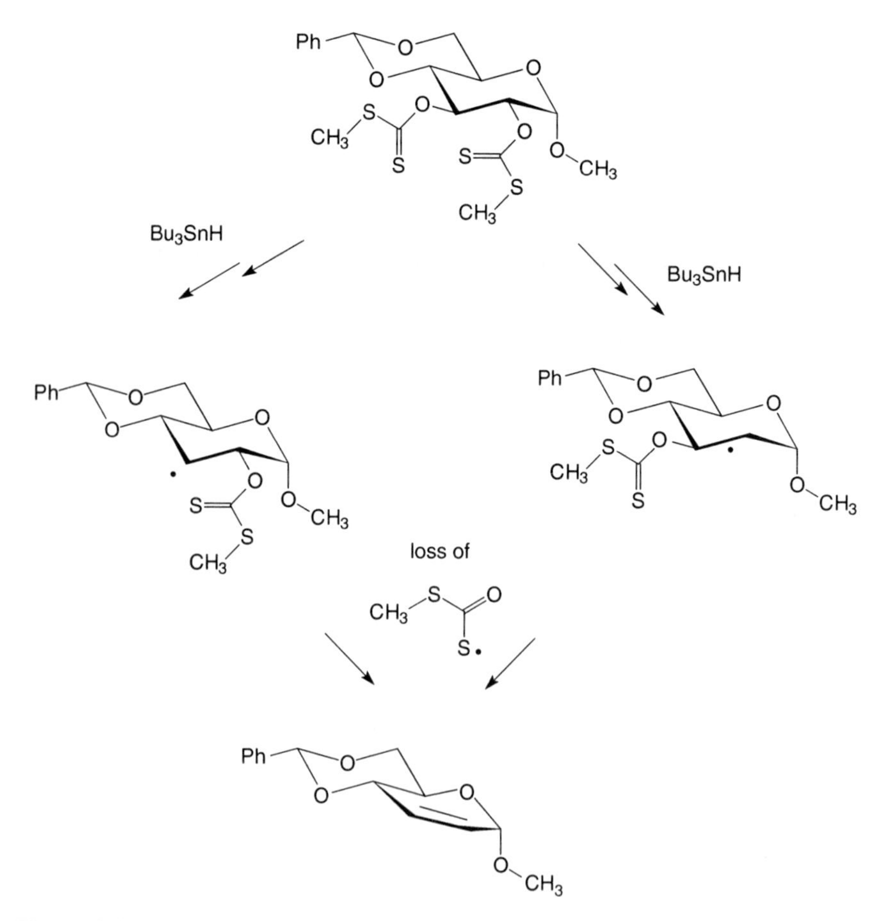

Figure 7.21 Application of the Barton–McCombie reaction to diols.

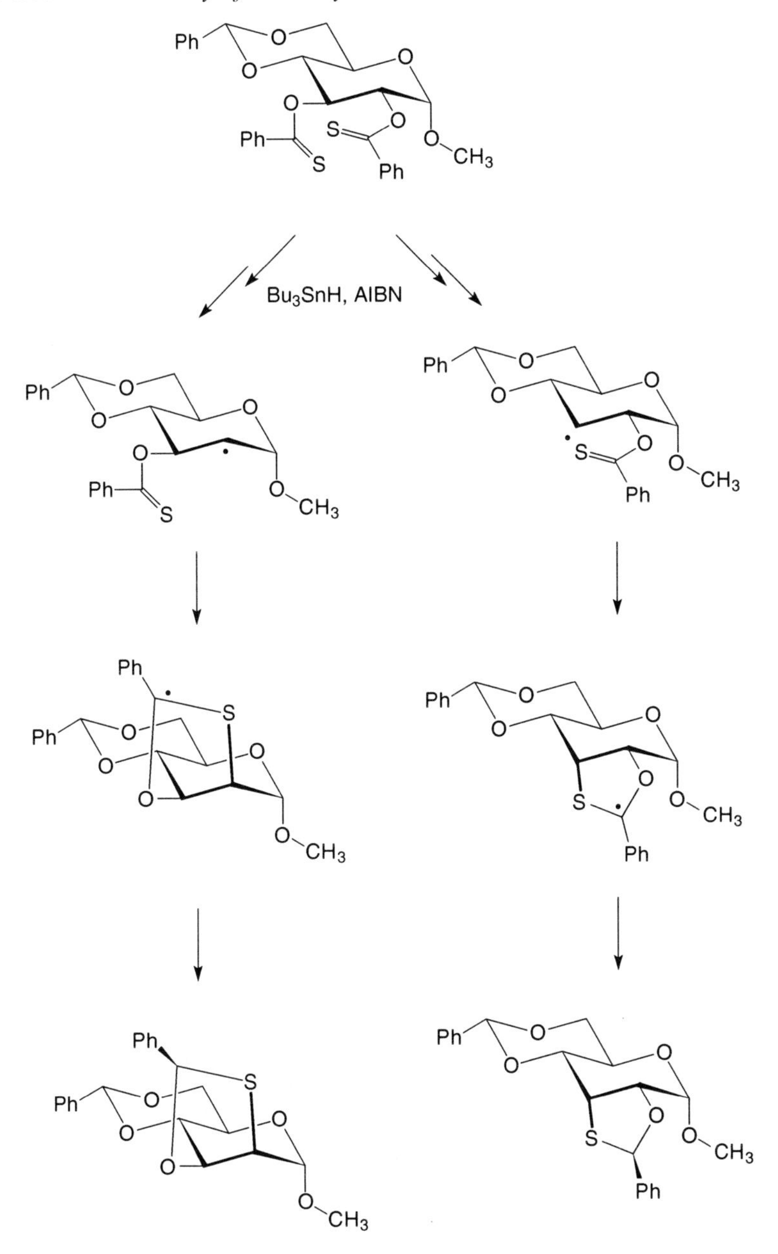

Figure 7.22 Cyclisation of vicinal thionbenzoates. Note the preferred axial approach of the sulfur.

2-iodophenylorthoacetate, followed by displacement of the remaining ethoxide with trimethylsilyl cyanide, the group fragments into a 6-deoxy sugar when heated with AIBN initiator and Bu_3SnH. The selectivity for reduction of the primary position is not preserved, however, when the 4,6-benzylidene derivative

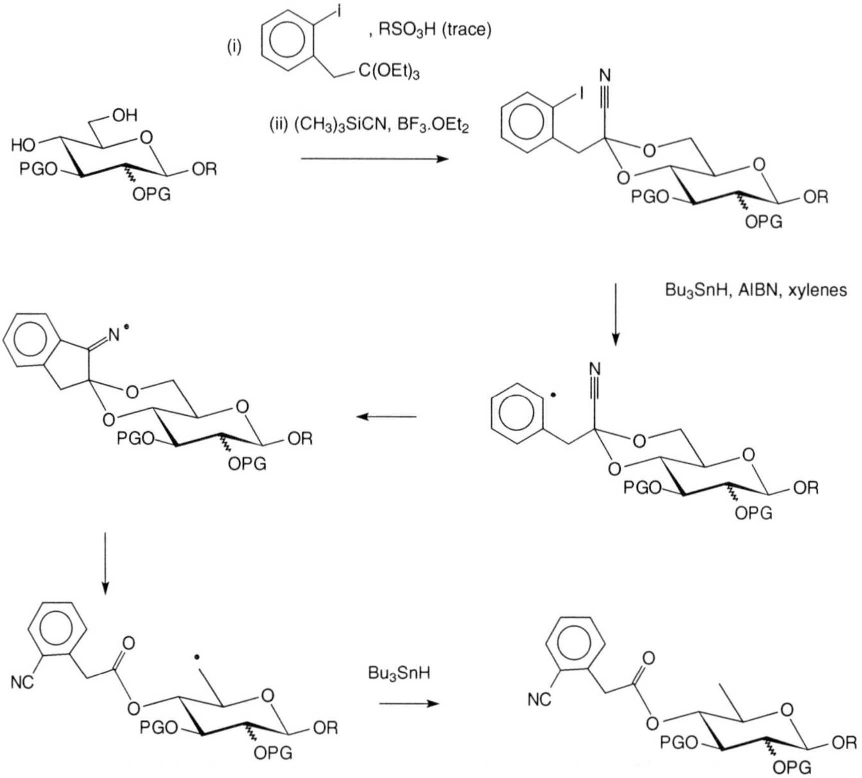

Figure 7.23 Combined deoxygenation and deprotection at position 6.

in the *galacto* series is made, which gave comparable quantities of 4- and 6-reduced product (Figure 7.23).

7.3.3.2 Radicals from Carboxylic Acids. Carboxy radicals appear to be involved in three venerable reactions in preparative carbohydrate chemistry, the Ruff degradation, the Hunsdiecker reaction and the Kolbe electrolysis.

The Ruff degradation, in its classical version, conversion of the calcium salt of an aldonic acid to the aldose of one fewer carbon atoms by treatment with Fe^{III} and hydrogen peroxide, was one of the reactions used by Emil Fischer in his determination of the structure of the aldoses. Its success with Fe^{III} is mysterious, as one would expect Fenton chemistry involving $HO^{\bullet}$ to give molecular rubble, rather than good yields of a single product. However, the reaction is catalysed by transition metals in general (even Ti^{IV})[106], with soluble Cu^{II} giving better yields than Fe^{III}.[107] Retention of deuterium at C2 of the aldonic acid in the product indicates that a α-keto acid is not an intermediate.[108] Scheme 7.4 sets out a mechanism in accord with the known mechanistic facts. The Fe^{III} reaction succeeds possibly because the ability of Fe^{II} to reduce hydrogen peroxide is suppressed by its strong coordination during this often heterogeneous reaction.

$$\text{R–CHOH–COO}^- + \text{Cu}^{II} \rightarrow \text{R–CHOH–COO}^{\cdot} + \text{Cu}^{I}$$

$$\text{R–CHOH–COO}^{\cdot} \rightarrow \text{RCH}^{\cdot}\text{OH} + \text{CO}_2$$

$$\text{RCH}^{\cdot}\text{OH} + \text{Cu}^{II} \rightarrow \text{RCH=O}^{+}\text{H} + \text{Cu}^{I} \rightarrow \text{RCHO} + \text{H}^{+}$$

$$\text{H}_2\text{O}_2 + 2\text{Cu}^{I} + 2\text{H}^{+} \rightarrow 2\text{Cu}^{II} + 2\text{H}_2\text{O}$$

Scheme 7.4

$$\text{R–CHOAc–COOAg} + \text{Br}_2 \rightarrow \text{R–CHOAc–COO–Br} + \text{AgBr}$$

$$\text{R-CHOAc–COO–Br} \rightarrow \text{R–CHOAc–COO}^{\cdot} + \text{Br}^{\cdot}$$

$$\text{R-CHOAc–COO}^{\cdot} \rightarrow \text{RCH}^{\cdot}\text{OAc} + \text{CO}_2$$

$$\text{RCH}^{\cdot}\text{OAc} + \text{Br}^{\cdot} \rightarrow \text{RCH(OAc)Br}$$

Scheme 7.5

The Hunsdiecker reaction, a bromodecarboxylation classically involving treatment of the silver salt of a carboxylic acid with molecular bromine, results in the transient generation of an acyl hypobromite, followed by O–Br homolysis. It works with protected silver aldonates,[109] as shown in Scheme 7.5.

In the Kolbe electrolysis, an electron is removed from a carboxylate ion at an electrochemical anode. Although carboxylic radicals lose carbon dioxide rapidly, they have a real lifetime, just, and the modern consensus is that transfer of the electron to the electrode and decarboxylation are separate events: the decarboxylation of $PhCH_2COO^{\bullet}$ has a rate constant of $5 \times 10^9\,s^{-1}$ at 25 °C in acetonitrile.[110] The same rate constant for decarboxylation is obtained by photolysis of a naphthalene methyl esters in methanol, which also yielded decarboxylation rates ($10^{-9}k/s^{-1}$) for $RCOO^{\bullet}$ of <1.3 ($R = CH_3$), 2.0 ($R = CH_3CH_2$), 6.5 ($R = (CH_3)_2CH$) and 11 [$R = (CH_3)_3C$], all at 25 °C.[111] The rate of decarboxylation thus shows a modest increase with product radical stability when the product radical is of the π type. Radicals of the σ type, such as the phenyl radical, in which the SOMO is a carbon sp^2 atomic orbital, are generated more slowly, with $PhCOO^{\bullet}$ losing CO_2 at a rate of $2.0 \times 10^6\,s^{-1}$ at 25 °C in carbon tetrachloride ($\log_{10}A = 12.6$, $E_a = 35.9\,kJ\,mol^{-1}$),[112] and radicals from oxalic half-esters decarboxylating according to $\log_{10}k = 12.2–40.1/(0.01917)$.[113]

Carboxylic radicals can also be generated from the esters of *N*-hydroxy-2-thiopyridone.[114] Strictly, IUPAC would have these as mixed anhydrides of a carboxylic ester and a thionhydroxamic acid, but they are generally known as PTOC (pyridinethiocarbonyl) or Barton esters (Figure 7.24).[115] Readily made by reaction of the acid chloride with the commercially available sodium salt of 2-mercaptopyridine *N*-oxide, they are pale yellow in colour due to the long-wavelength tail of a strong absorption band at 350 nm. This can be exploited to avoid the use of an initiator, since an adequate concentration of radicals is produced by photolysis with an ordinary tungsten filament light bulb. In the

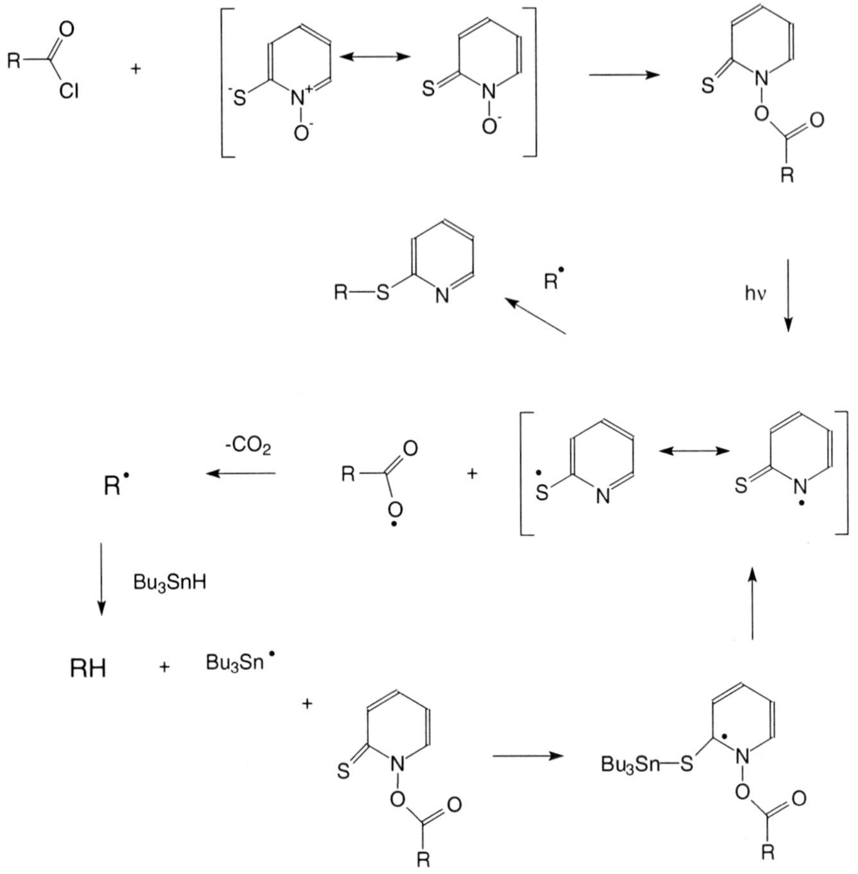

Figure 7.24 PTOC (Barton) esters and their generation of radicals.

absence of an atom donor, the decomposition of PTOC esters yields the 2-alkylthiopyridines, but the alkyl radical can be intercepted by hydrogen donors, typically $HSnBu_3$, to give an overall reduction. It can also abstract chlorine from CCl_4, bromine from $BrCCl_3$ and iodine from CHI_3 to give the decarboxylated halide and can be induced to form C–P and C–S bonds with appropriate reagents.

7.4 REACTIONS OF RADICALS

7.4.1 Stereochemistry of Atom Transfer to Oxygenated Radicals

Ab initio molecular orbital calculations suggest that hydrogen abstraction by α-oxygenated radicals takes place along a trajectory similar to the Bürgi–Dunitz trajectory for addition of nucleophiles to carbonyl groups.[116] This is not surprising, as oxygenated radicals are still largely planar, with $\varphi = 3.9°$ for monooxygenated and 11° for dioxygenated radicals (see above), and such an approach represents the least-motion pathway. In the case of glycopyranosyl or glycopyranosan-1-yl radicals, axial approach of the radical donor requires least nuclear motion in the rest of the system (see Figure 3.13 for the microscopic reverse in the case of a dioxocarbenium ion). Advocates of ALPH have advanced this preference for axial attack as evidence for some sort of kinetic anomeric effect in radicals, but the preference is perfectly well explained by the least-motion arguments of Figure 3.13. Moreover, it is not clear why homolytic ALPH should be more in evidence than its heterolytic cousin, when the simple electron density-transfer model on which ALPH is based would predict the reverse.

Support for "homolytic ALPH" being also a least-motion effect comes from the relative rates of *intramolecular* anomeric hydrogen abstraction: the anomeric ratio of a mixture of tetra-*O*-benzylphenacyl glucopyranosides remained accurately constant ($< \pm 1\%$) during competitive photolyses (Figure 7.25).[117] Unlike the supposedly fundamental ALPH, least-motion effects can be expected to be over-ridden by other effects. In this case, the absence of any preference for axial hydrogen abstraction in the Norrish Type II reaction of the carbonyl triplet was confirmed to be a true reflection of organic reactivity by a second competition experiment with the tetrahydropyranyl compound. This was consumed much faster (the transition state for hydrogen abstraction by the carbonyl triplet being destabilised by the inductive/field effect of the sugar alkoxy substituents).

The preference of glycopyranosan-1-yl radicals for axial attack by an intermolecular hydrogen donor has been exploited in a β-mannoside synthesis (Figure 7.26);[118] the ulosonic acid was obtained by oxidative destruction of a furan ring.

7.4.2 Heterolysis of Carbohydrate Radicals

The chemistry of carbohydrate radicals is dominated by loss of vicinal substituents which can leave as relatively stable anions and molecules. The attack of $OH^{\bullet}$ from the Ti^{III}–H_2O_2 system on *myo*-inositol at pH 4 yields the radicals **1**, **2**, **3** and **4** shown in Figure 7.27 in the expected 2:2:1:1 ratio, as judged from ESR signal intensities. At lower pH values, acid-catalysed loss of OH became apparent, with the reaction of radical **1** occurring with a second-order rate constant of $2.8 \times 10^6\,M^{-1}\,s^{-1}$ at 20 °C, 11 times faster than the loss of equatorial OH from, for example, radical **3**.[30] The preferential loss of the axial OH is probably a least-motion effect of the type held to support ALPH.

In neutral solution, two types of radical dominate the ESR-detectable products from glucopyranosanyl radicals,[119] acyclic semidiones ($R\text{–}CO^{\bullet}\text{=}CH\text{–}O^{-}$) incorporating all the carbons of the sugar and semidiones derived from C4–C6. The former are plausibly considered to arise from base-catalysed ring opening of the pyranosan-2-yl radicals, with α-glucopyranose and α-mannopyranose giving a mixture of the same (*E*)- and (*Z*)-semidiones and β-glucopyranose only the (*Z*)-semidiones (Figure 7.28). This product distribution is readily rationalised by a preference for the vicinal leaving group,

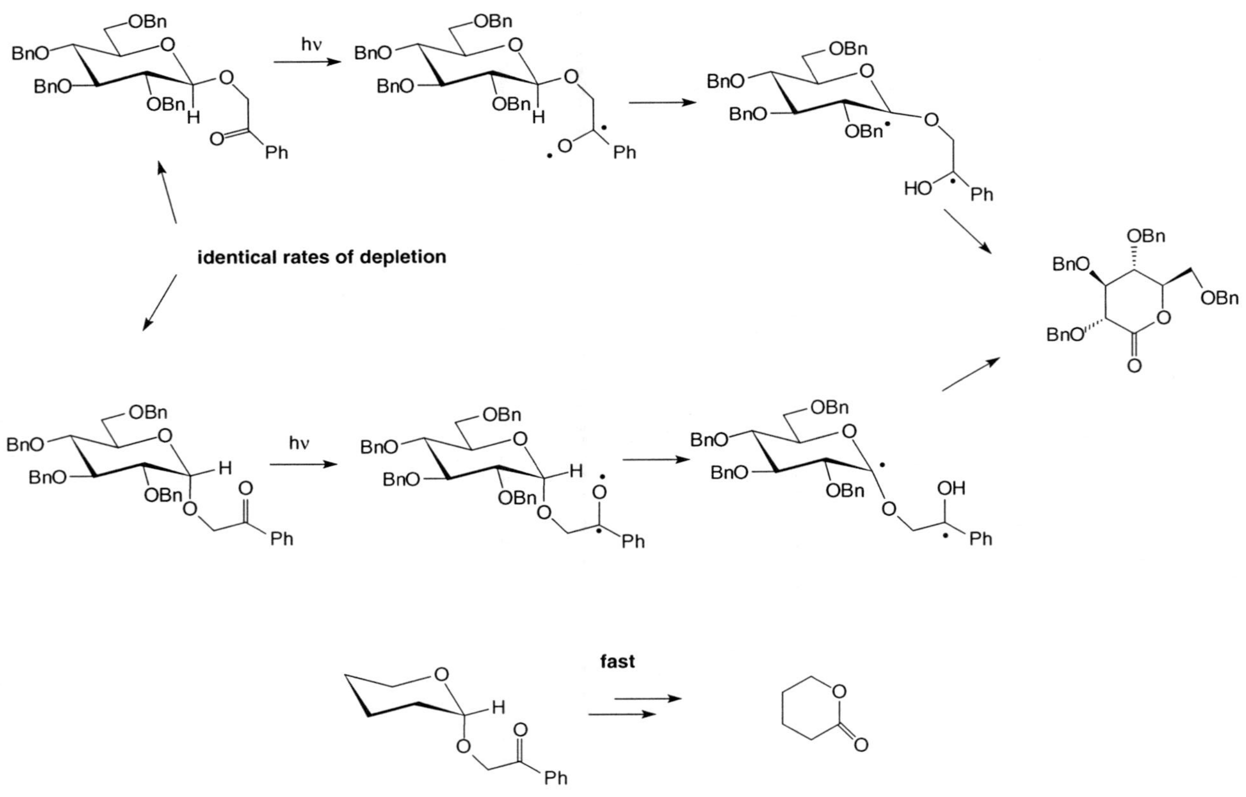

Figure 7.25 Competitive photolysis of phenacyl glycosides.

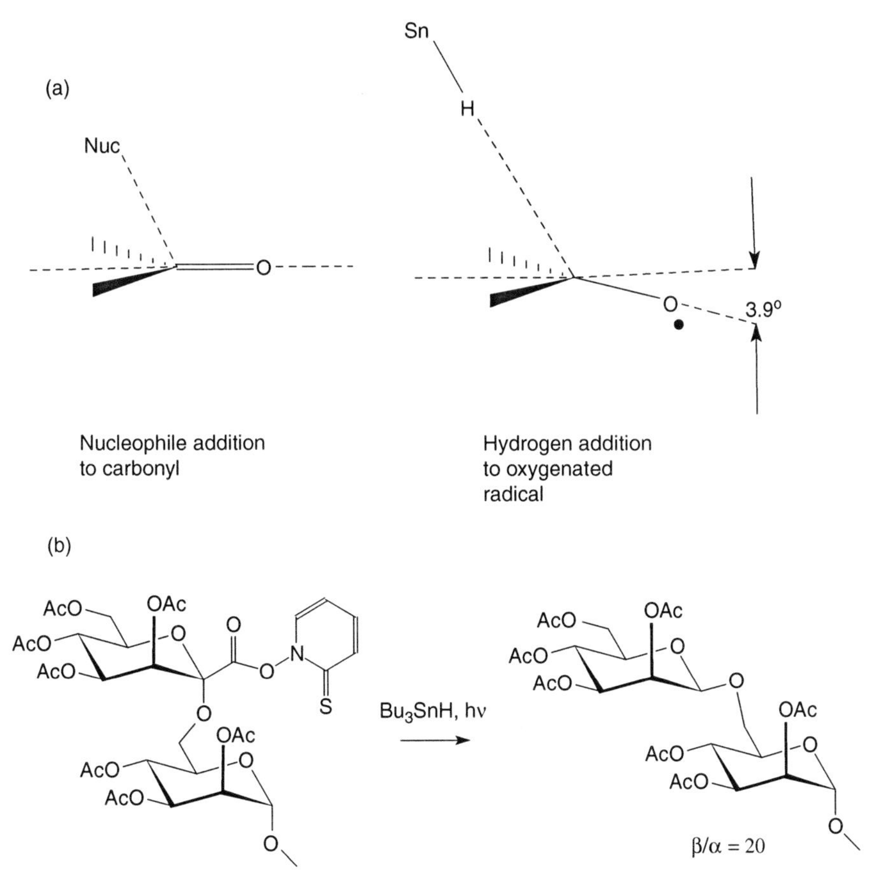

Figure 7.26 (a) Approach of a hydrogen donor. (b) Exploitation of this stereochemistry in the synthesis of a β-mannoside.

in this case O5, to depart approximately orthogonally from the plane defined by the product anion radical. The semidiones may dominate the ESR spectra because they are greatly stabilised and therefore long-lived, rather than because of their importance in the overall reaction. Thus, the radical products of the departure of the 2-OH from the glucopyranosan-1-yl radical or of the 1-OH from the glucopyranosan-2-yl-radical are observed only with difficulty. Such lack of correlation between the intensity of radical signals from a reaction and the chemical flux supported by a particular intermediate is an inherent weakness of mechanism determination by ESR.

These losses of OH^- or OR^- can occur in principle with or without acid assistance to leaving group departure and deprotonation of the hydroxyl group: the presence of a radical centre on the carbon carrying an OH group lowers the latter's pK_a by about 5 units.[9] So regarded, they from part of a continuum of radical fragmentations whose importance (and ubiquity) has only recently been recognised.[120] The key reaction is the departure of a leaving group β to a tricovalent carbon bearing an unpaired electron, to give a π cation

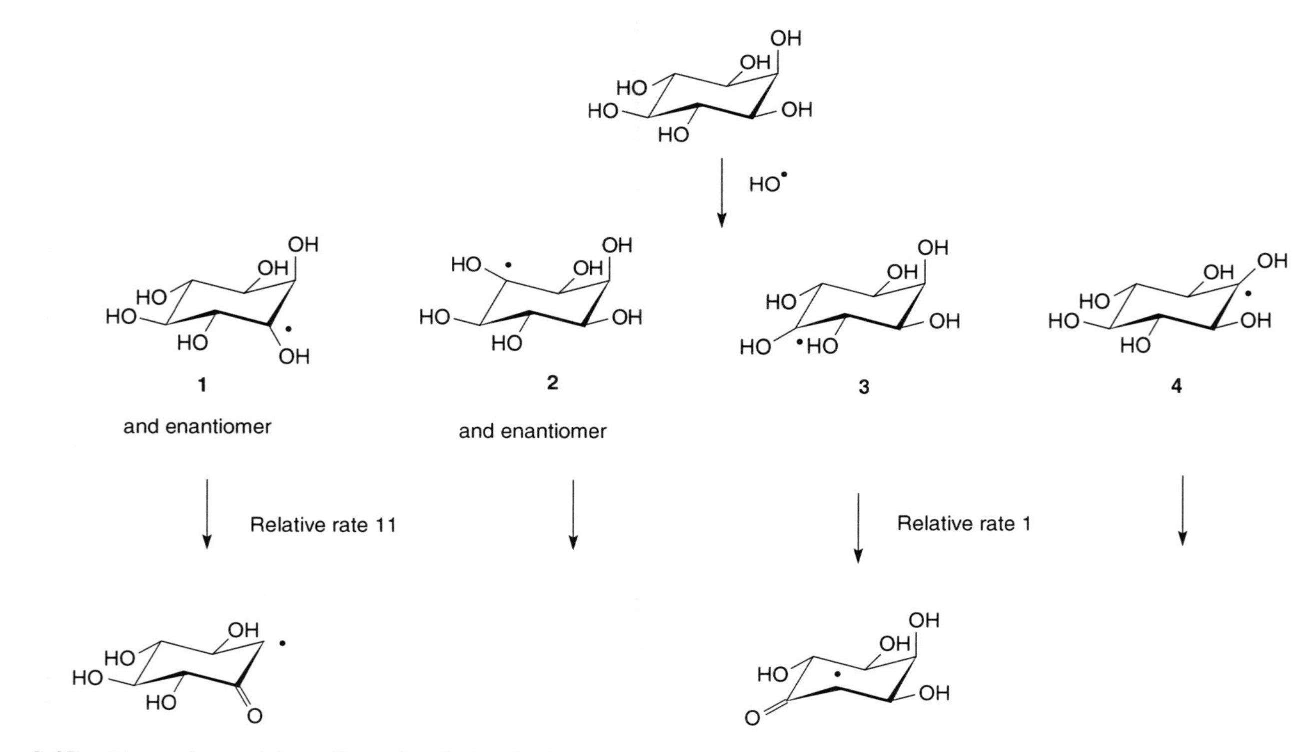

Figure 7.27 Generation and fate of *myo*-inositol radicals.

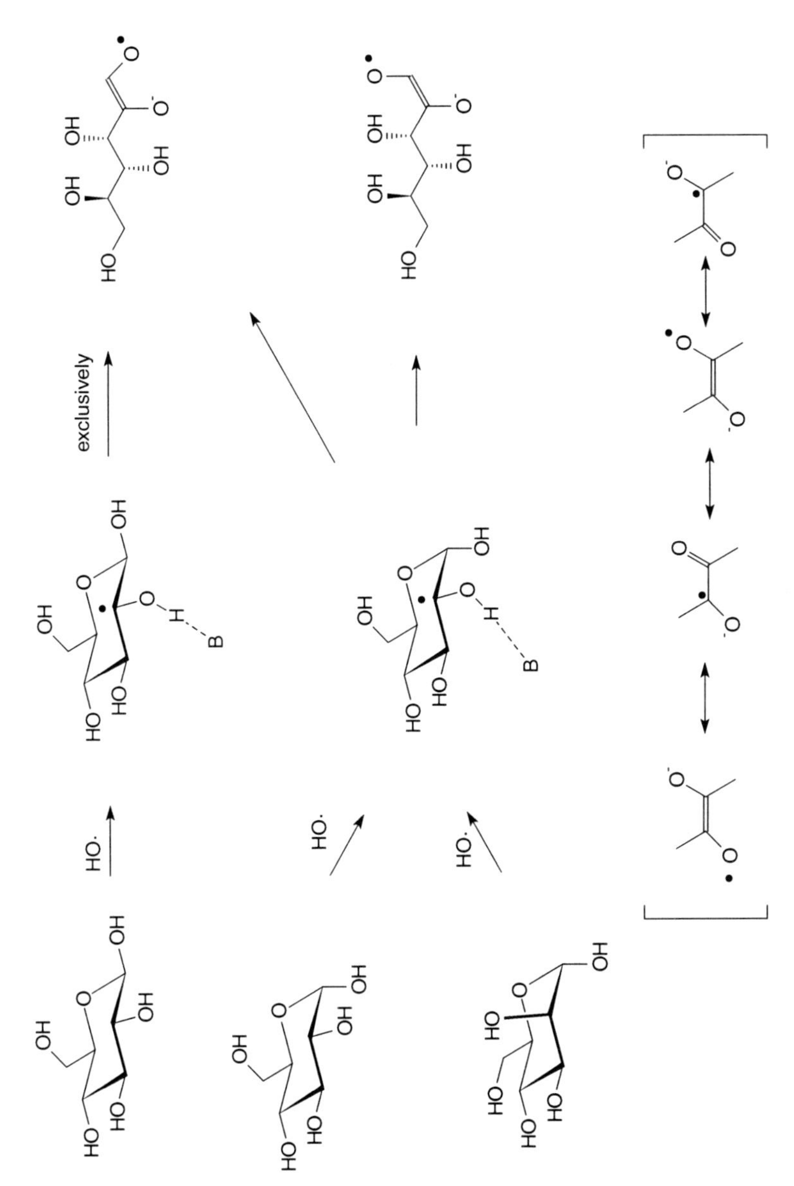

Figure 7.28 Formation of semidiones from oxidation of reducing sugars.

radical. Obviously closely analogous to the S_N1 reaction of closed-shell molecules, it is very many orders of magnitude faster.[3] The greater reactivity of C–H bonds which make a dihedral angle of close to zero with a p-type lone pair on an adjacent oxygen to hydrogen abstraction[67] make the 4′-position of DNA particularly vulnerable to such abstraction, since it is also tertiary. The resulting radical is susceptible to loss of the phosphate; *in vivo* this would lead to DNA strand cleavage. Because of such biological significance, therefore, much of the data for the heterolysis of radicals comes from models for this process.

In groundbreaking work, radicals derived from various 2-methoxyethyl phosphates were generated by photolytically-mediated $H^\bullet$ abstraction (*e.g.* via photolysis of acetone, *tert*-butyl peroxide or hydrogen peroxide), and the kinetics of their decomposition at 3 °C were followed by ESR.[121] It proved possible to monitor the departure of $H_2PO_4^-$, HPO_4^{2-} and PO_4^{3-} from the $CH_3OCH^\bullet CH_2O\mathbf{P}$ radical (where **P** denotes a phosphoryl group of indeterminate ionisation state). They were lost with rate constants of $\sim 3 \times 10^6$, 10^3 and 0.1–$1\ s^{-1}$ at ~ 20 °C, respectively. These data clearly indicate a β_{lg} value of ≈ -1, as would be expected for an S_N1 reaction. The same set of experiments generated CH_3O–CH_2–$CH^\bullet$–O**P**, and it proved possible to measure the second ionisation of the phosphate groups in both radicals. The pK_a value of 6.5 in the closed-shell compound was unchanged by formation of the β-radical, but lowered to 5.3 in the α-radical, clearly demonstrating the electron-withdrawing effect of oxygen lone pair donation to the radical site.

The complete generality of the reaction of Figure 7.29 (b) was demonstrated using pulse radiolysis with rapid-response conductimetric monitoring (which, however, is "blind" to any elimination to give neutral species within an ion-pair). Typical rate constants (s^{-1}) at 19 °C are, for $R_1 = CH_3$, $R_2 = R_3 = R_4 = H$, LG = Br, 7×10^3; for $R_1 = CH_3$, $R_2 = R_3 = R_4 = H$, LG = OSO_2CH_3, 2×10^5; and for $R_1 = R_2 = CH_3$, $R_3 = R_4 = H$, LG = OSO_2CH_3, $\geq 10^6$.[122]

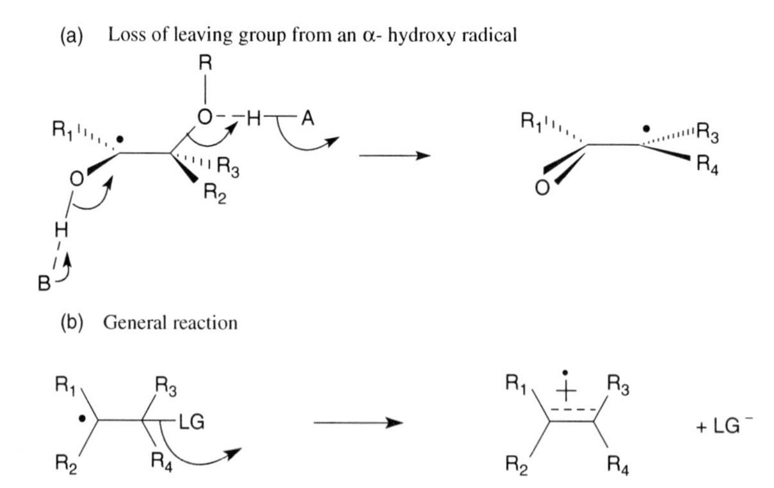

Figure 7.29 Comparison of (a) loss of OH^- from a polyhydroxy radical and (b) the general reaction.

Detailed quantitation of the loss of phosphate, which also unequivocally established the relevance of the reaction for DNA degradation, was carried out with the model compounds shown in Figure 7.30.[123] In the first set of experiments, shown in Figure 7.30a, the 4′-radical derived from *N*-benzoyl-2′-deoxyadenine derivatives, phosphorylated on the 3′-OH with a series of dialkyl phosphates, was generated by addition of PhS• to a pre-formed 4′,5′-double bond. These 4′-radicals could either lose the phosphate diester monoanion, yielding a radical and eventually *N*-benzoyl-*S*-phenyl-2′-deoxy-3′,4′-dehydro-5′-thioadenosine, or be trapped by hydrogen transfer from PhSH. The rate of the trapping reaction was independent of leaving group. The ratio of trapping product to dephosphorylation product showed an excellent correlation with the leaving group pK_a, with a β_{lg} value of -1.3 in both 4:1 aqueous acetone and in toluene. As expected, in the far less polar solvent toluene, loss of dialkylphosphate was slower by around 100-fold, a small factor for such a large solvent change by the standards of two-electron chemistry, possibly because the radical cation is delocalised. In a second set of experiments, the relative leaving group abilities of the 5′- and 3′-diethylphosphate monoanion was compared using the 4′-radical of a thymidine 3′,5′-diphosphate generated from the 4′-phenylseleno derivative, abstracted by Bu_3SnH and light. In this case a methanolic solution was used, to act as a nucleophilic trap, together with Bu_3SnH as a radical trap. The loss of the 5′-diethylphosphate anion from the original radical was so slow compared with the loss of the 3′-diethylphosphate anion that estimates of the relative rate of 5′-diethylphosphate departure could be made only from the methanol adduct of the cation radical from 3′-diethylphosphate departure. In Figure 7.30b, assuming all processes are irreversible, that no other radical processes are involved and that reduction of all neutral C4′ radicals by Bu_3SnH proceeds with a constant bimolecular rate constant k_2, $k_1/(k_2[Bu_3SnH]) = [\mathbf{1}]/([\mathbf{2}]+[\mathbf{3}]+[\mathbf{4}]+[\mathbf{5}])$ and $k_3/(k_2[Bu_3SnH]) = ([\mathbf{4}]+[\mathbf{5}])/[\mathbf{3}]$. Product analysis gave a value of k_1/k_3 of about 60, indicative of a heterolytic process.

The reactions of nucleophiles with oxygen-substituted radical cations seem to be slower than the reactions of closed-shell molecules. In trifluoroethanol–acetonitrile–water mixtures the tetrahydropyranyl 2,3-radical cation[xii] reacts with water at ambient temperature with a rate constant of $1.0 \times 10^4\,M^{-1}\,s^{-1}$.[124] This extrapolates to a lifetime in pure water of 18 ns, whereas that of the tetrahydropyranyl cation,[xiii] probably similar to that of the

xii

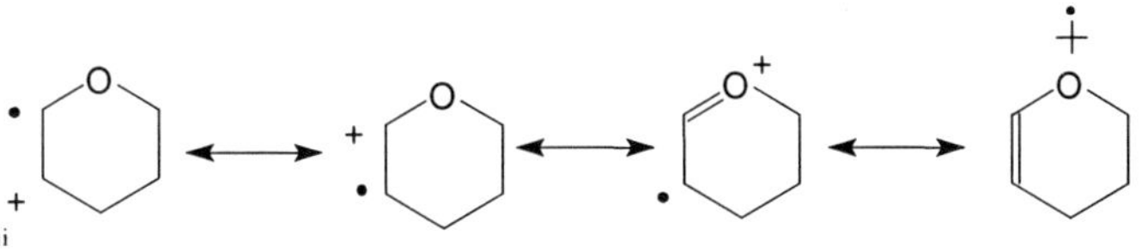

xiii

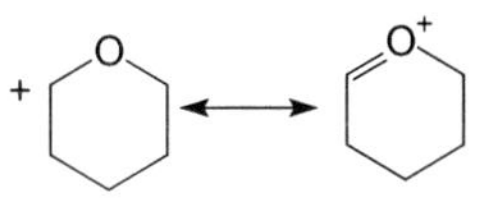

2-deoxyglucosyl cation, is around 10 ps (see Section 3.6) (the radical cations are UV transparent, but can be monitored by a diffusion-controlled electron transfer from triarylamines to give coloured $Ar_3N^{+\bullet}$). The 2-(benzoyloxymethyl)tetrahydrofuranyl radical cation, a closer model for DNA-derived radical cations, has a longer lifetime of 90 ns. Conjugation seems to be more effective than alkylation in stabilising radical cations of this type to oxygen nucleophiles: an upper limit of 1×10^3 M^{-1} s^{-1} was obtained for bimolecular methanol attack on (*Z*)-$[Ph–CH{=}C(OMe)Me]^{+\bullet}$,[126] suggesting a lifetime in water on the order of 10^{-5} s.

A feature of undelocalised oxocarbenium radical cations, such as those formed by 3′-phosphate departure from nucleotide 4′-radicals, is that they are strong enough oxidants to react with the electron-rich purine guanine by electron transfer, creating a positive hole in the π cloud. This hole can then be transmitted along the DNA double helix via stacking interactions.[125]

The classical demonstration of the real existence of a solvent-equilibrated intermediate was finally performed with the radical Ph–CHX–C$^{\bullet}$(OMe)Me, which was generated by flash photolysis of the Barton esters in tetrahydrofuran. The loss of X^- monitored by the change of UV absorbance; at 20 °C loss of Br^- and $^-O_2P(OPh)_2$ occurred with rate constants of 2.8×10^5 and

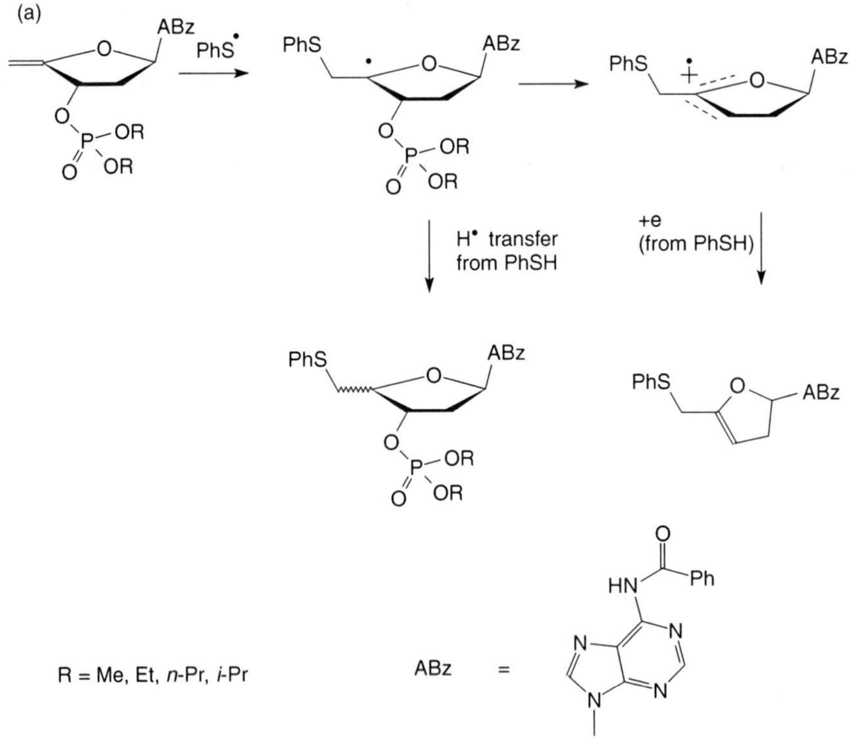

Figure 7.30 (*Continued*)

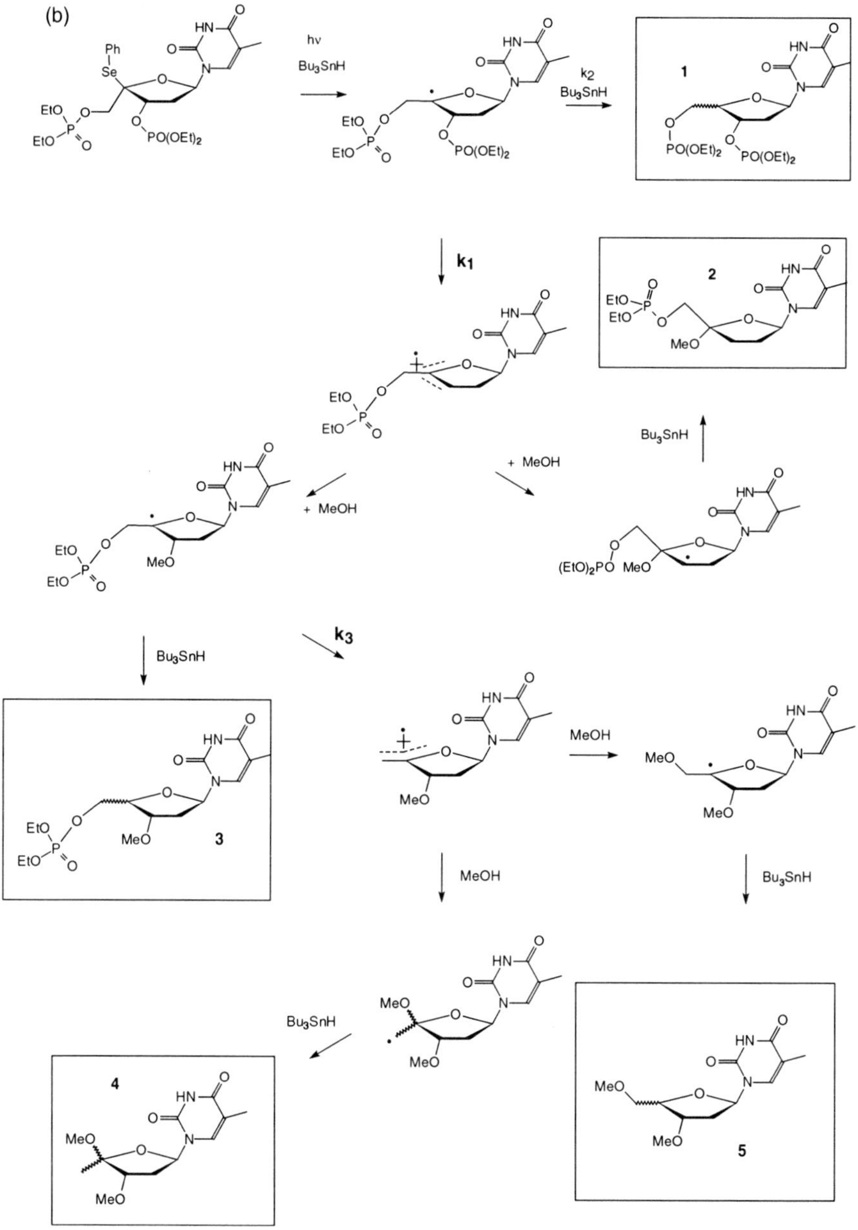

Figure 7.30 (a) System for investigating effect of 3′-phosphate acidity. (b) Comparative leaving abilities of 3′- and 5′-phosphate.

$3.4 \times 10^6\ s^{-1}$, respectively.[126] Loss of diphenyl phosphate in pure acetonitrile, or of both leaving groups in acetonitrile–methanol mixtures, was too rapid to follow even with equipment with 10 ns temporal resolution, but loss of Br^- occurred at about $8 \times 10^7\ s^{-1}$ in pure acetonitrile. The expected large increase in rate with solvent polarity expected for an elementary reaction involving separation of charge is thus observed.

The product is largely the (*Z*)-enol ether shown in Figure 7.31 (the *E*-isomer is not found, and is calculated to be significantly less stable). This arises from hydrogen atom transfer to the oxygen atom, followed by proton loss. A less favoured transfer to carbon, followed by nucleophilic attack of methanol on the oxocarbenium ion, is thought to be the origin of the dimethyl ketal. These are the two products whose ratio is independent of leaving group.

7.4.2.1 Ribonucleotide Reductase.[9,127] Ribonucleotide reductase catalyses the production of 2′-deoxynucleotides from their oxygenated precursors. As there are no other pathways for generating deoxynucleotides, the flux for all DNA repair and synthesis has to pass through this enzyme, and in consequence the enzyme has been intensively investigated for biomedical reasons. Like Sir Derek Barton, Nature uses sulfur radical chemistry for sugar deoxygenation. There are three mechanistic types of ribonucleotide reductase, Type I in mammals and DNA viruses, Type II in aerobic prokaryotes and Type III in anaerobic prokaryotes (Figure 7.32). The preferred substrates for Type I and III enzymes are the nucleotide 5′-diphosphates, and for the Type II enzymes, the nucleotide 5′-triphosphates. All three mechanistic types feature an abstraction of H3′ of the ribose by a cysteinyl ($-CH_2S^\bullet$) radical, but differ markedly in the initiation step whereby this radical is generated. The propagation steps differ somewhat between Type III, which uses formate as reductant and Types I and II which use the thiol exchange reactions of the glutaredoxin and thioredoxin systems and eventually NADPH as reductant (Type I enzymes in fact have a second redox-active cysteine/cystine system in the C-terminus of the large subunit containing the active site).

In the first step of both types of propagation sequence, the active site cysteinyl residue abstracts the 3′-hydrogen. This, of course, is the reverse of the hydrogen transfers in free solution, where a thiol would be expected to quench an α-hydroxy radical: polarisation of the substrate – observed in Type I enzymes, where an active site glutamate hydrogen bonds to 3′-OH – must be involved. Loss of the 2′-OH, assisted by general acid catalysis from a second active site cysteine and by general base catalysis from the aforementioned glutamate in Types I and II enzymes (the base is unclear in Type III enzymes) leads to a keto-radical. This then accepts a hydrogen atom from a second cysteine. In the case of Type I and II enzymes, the product is a disulfide radical anion, formed from a third cysteine and the 2′-deoxy-3-keto sugar. It is thought that a combined electron and proton transfer as depicted in Figure 7.32a results in a 3′-hydroxy-3′-radical, which is quenched by the original thiol, regenerating the key active-site cysteinyl radical. Substrate binding to oxidised Type I and Type II enzymes is prohibited by clashes between 2′-OH and the disulfide bridge. In Type III enzymes, the second cysteinyl radical is quenched by formate, generating $CO_2^{\bullet-}$. A combination of hydrogen and electron transfers is speculatively depicted in Figure 7.32b. In accord with this scheme, the formate hydrogen is lost to solvent, the second 2′-hydrogen derived from it,

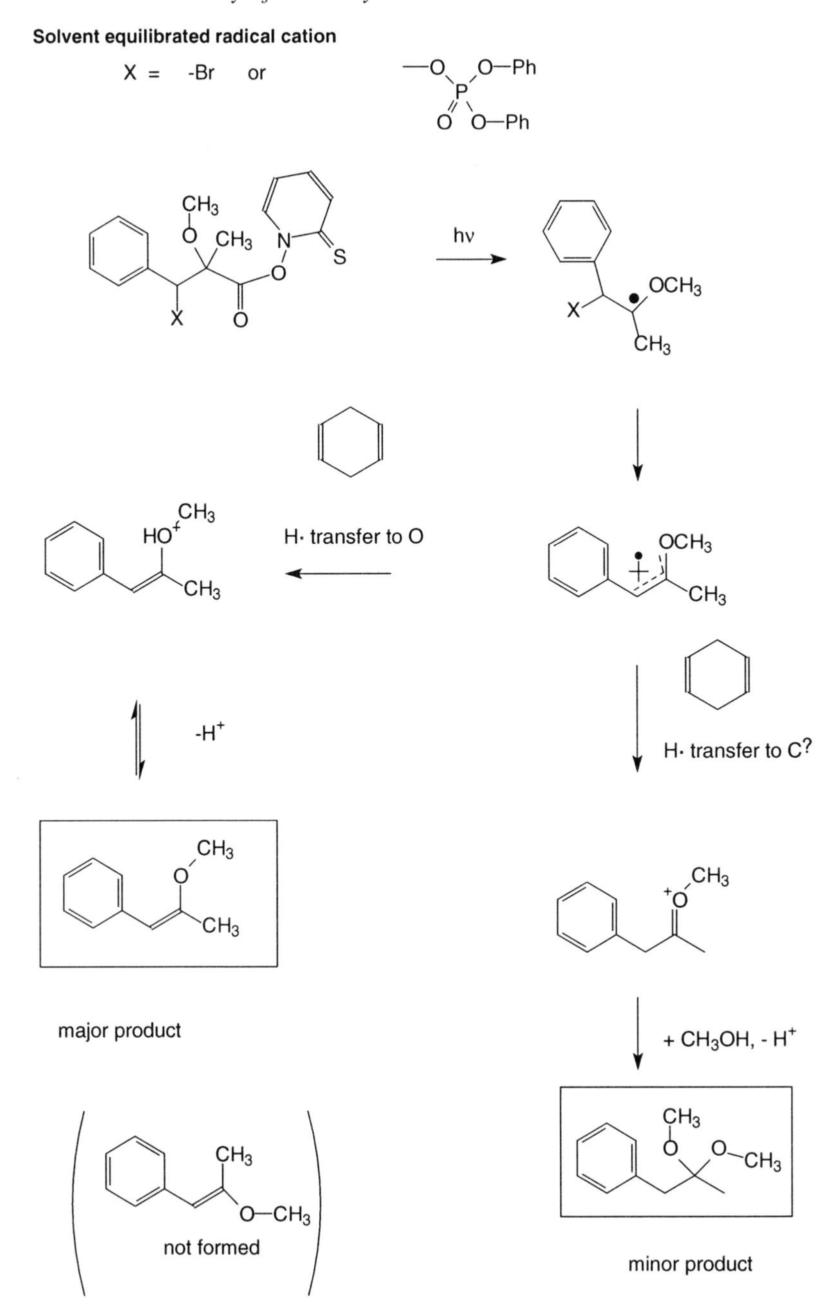

Figure 7.31 Generation of solvent-equilibrated radical cations. The origin of the small amount of saturated dimethyl ketal is not entirely certain – the route shown is most probable, but attack of methanol on the radical cation cannot be completely ruled out.

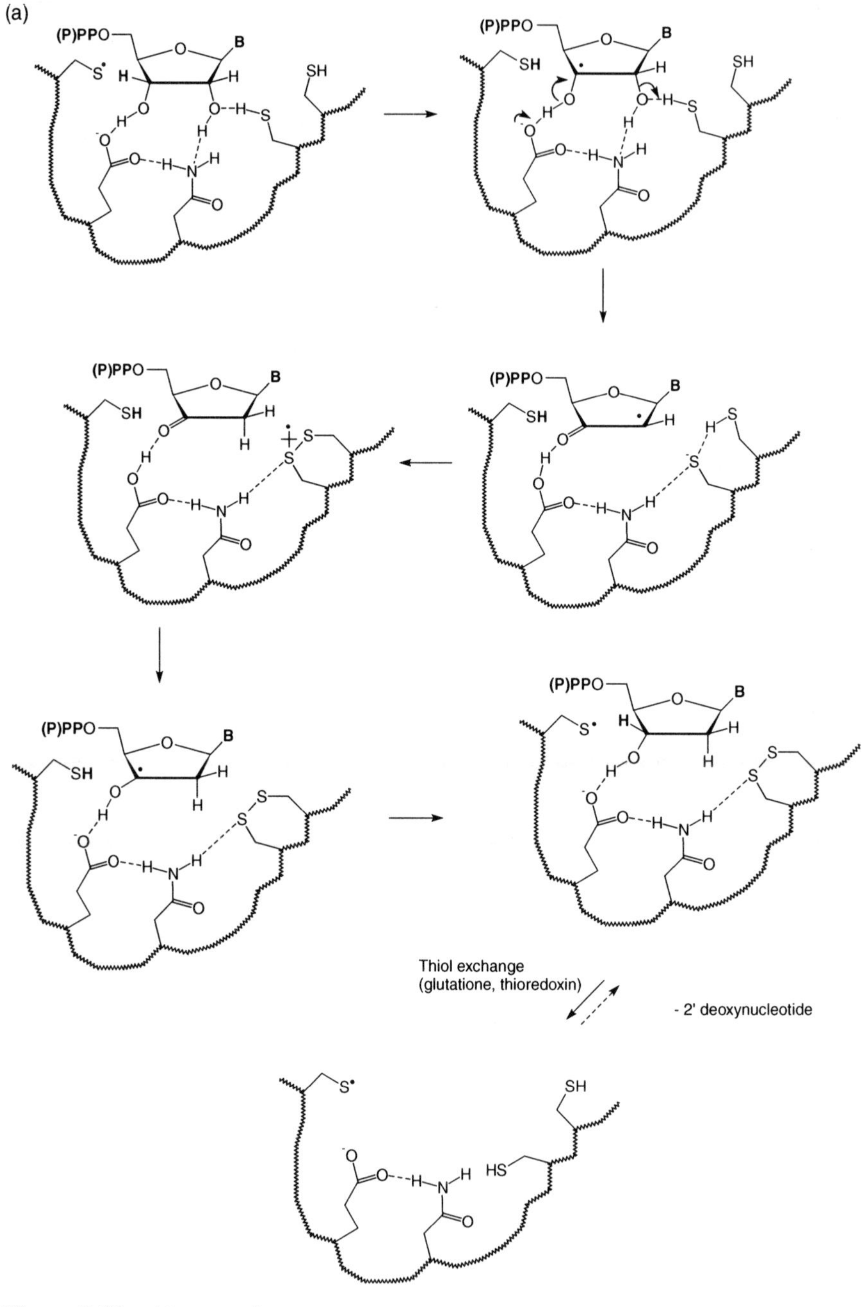

(a)

Thiol exchange
(glutatione, thioredoxin)

- 2' deoxynucleotide

Figure 7.32 (*Continued*)

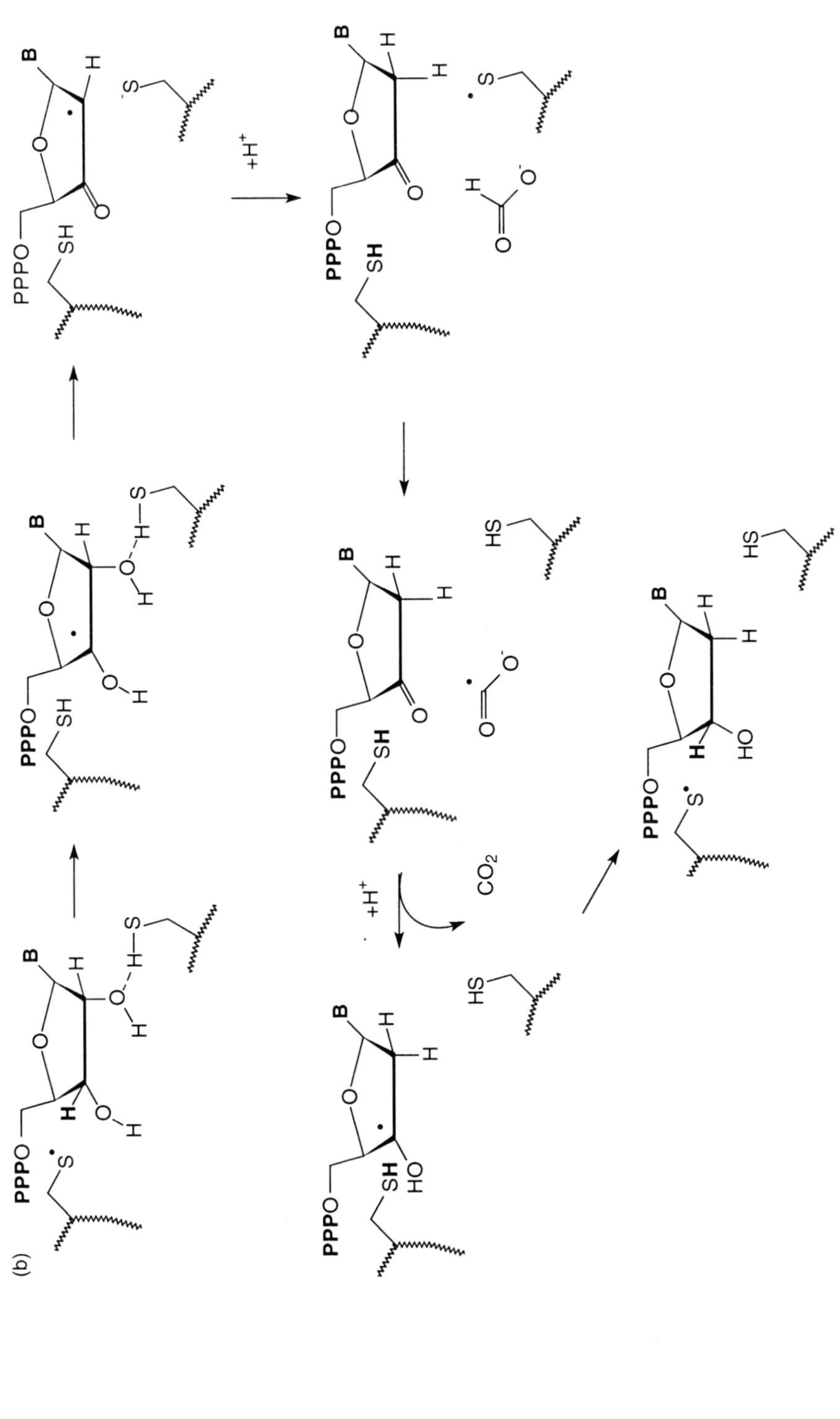

Figure 7.32 *(Continued)*

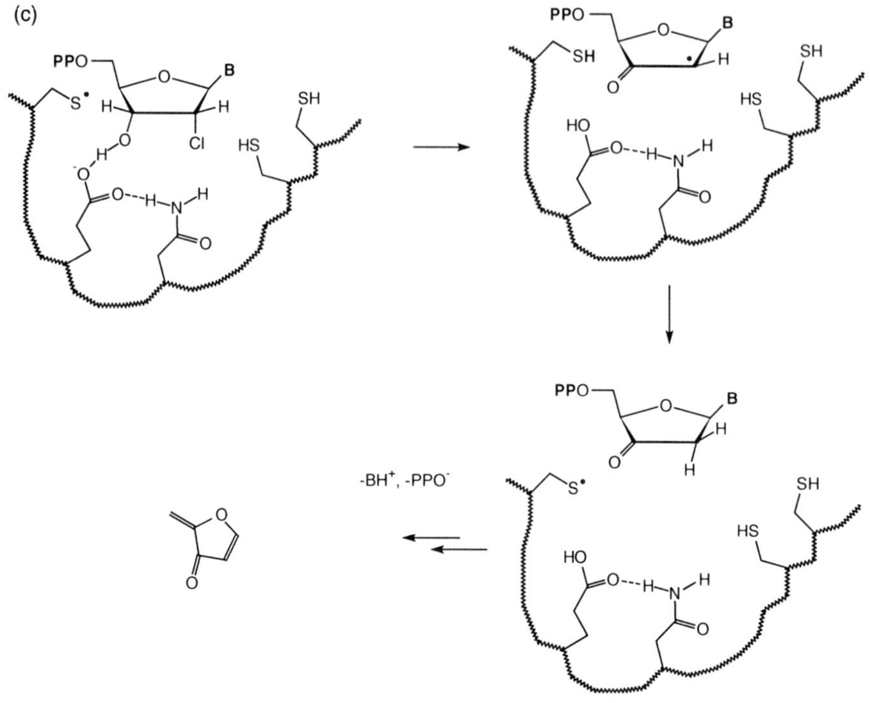

Figure 7.32 Propagation steps of (a) Types I and II ribonucleotide reductases. (b) Type III ribonucleotide reductase.[128] (c) "Suicide" enzyme inactivation by 2′-deoxy-2′-halo- or -pseudohaloribonucleotides.[129]

and the 3′-hydrogen is very largely retained in the product, although there is a couple of percent exchange with solvent.

Substrate analogues in which the 2′-OH is replaced by chlorine or azide are converted by Type I and II enzymes into an avid Michael acceptor, which gains aromatic stabilisation on addition of nucleophiles at C5′.[131] The process apparently involves production of the fully reduced ketone, followed by loss of the 5′-pyrophosphate and aglycone in $E1_{CB}$-like reactions. Possibly because loss of chloride or azide does not require general acid catalysis, the hydrogen originally at C3′ is delivered to C2′ (Figure 7.32c).

The initiation step involves the generation of a radical site away from the active site which transforms substrate. Early ESR studies of Type I holoenzymes revealed a persistent tyrosyl radical in the holoenzyme, and for a long time this was erroneously thought to participate in the catalytic reaction. Type I holoenzymes have an $\alpha_2\beta_2$ quaternary structure, with the active site in the so-called R1 subunit (the large α_2 homodimer). The tyrosyl radical in fact resides on the R2 subunit (the small β_2 homodimer), the C-terminus of which fits in a hydrophobic pocket in the R1 subunit.[130] From the tyrosyl radical, the unpaired electron is transferred through a chain of hydrogen-bonded tyrosines. The tyrosyl residue is generated by a redox centre with two octahedrally-coordinated non-haem iron atoms (Figure 7.33), with the ultimate oxidant being molecular oxygen [in some Type I enzymes, the two α or β subunits are not identical; in the yeast enzyme the second tyrosine radical-generating (β) subunit is in fact devoid of iron binding activity, which resides wholly in the first β subunit].[131]

In Type II enzymes, which are monomers or homodimers, the radical site is generated by homolytic fission of the Co–C bond of adenosylcobalamin.

Type III enzymes operate anaerobically, and rather than $\alpha_2\beta_2$ quaternary structure of the Type I enzymes, have an $\alpha_2+\beta_2$ structure, in which the permanent radical site, presumably stabilised capto-datively, lies on a main chain glycine residue of the dimeric α unit. The site is generated by a separate small iron–sulfur protein (β_2 subunit) which brings about a homo-olytic cleavage of *S*-adenosylmethionine, whose fragments generate the glycyl radical.

7.4.3 Acyloxy and Related Rearrangements

A cousin of the β-heterolysis of radicals is the rearrangement of carbon-centred radicals of the type $R_1R_2C^{\bullet}$–CR_3R_4–O–X=O to radicals of the type O=X–O–CR_1R_2–$C^{\bullet}R_3R_4$.[27] The most investigated case is the acyloxy migration (X=R_5C), examples of which are shown in Figure 7.6, but migrations of phosphate [X = P(OR_5)$_2$], nitrate (X = NO) and sulfonate (X = OSOR) are also known.[132] In cyclic systems, the migrating fragment remains on the same side of the ring. No intermolecular version of this rearrangement has ever been found. Moreover, any mechanism involving separation of a radical and an olefin cannot explain the facilitation of the rearrangement by polar substituents, or, in the case of the acyloxy rearrangement, the absence of decarboxylation products. Together with other evidence that the rearrangement involves charge separation, the mechanistic possibilities are limited to the following intermediates or transition states (Figure 7.34(a)):

(i) cyclic radical;
(ii) cation-radical/leaving group intimate ion pair;
(iii) five-membered cyclic transition state, polarised in the sense of the intimate ion pair;
(iv) three-membered cyclic transition state, polarised in the sense of the intimate ion pair.

The cyclic radical pathway can be dismissed in general, since side-reactions from the fully formed radical do not take place as expected. In a study of the rearrangement of RCOOC(CH_3)$_2$C–$CH_2^{\bullet}$ to (CH_3)$_2C^{\bullet}$–CH_2OOCR,[133] no cyclopropyl ring opening in the rearrangement was observed when R = cyclopropyl, despite independent generation of the 2-dioxolanyl radical resulting in rapid ring opening (Figure 7.34(b)). Likewise, rearrangements of species that could generate a phosphoranyl radical do not yield the products of other modes of fission expected of such an intermediate (Figure 7.34(c)).[134] This is despite, in some cases, the putative intermediate being, in theory, kinetically

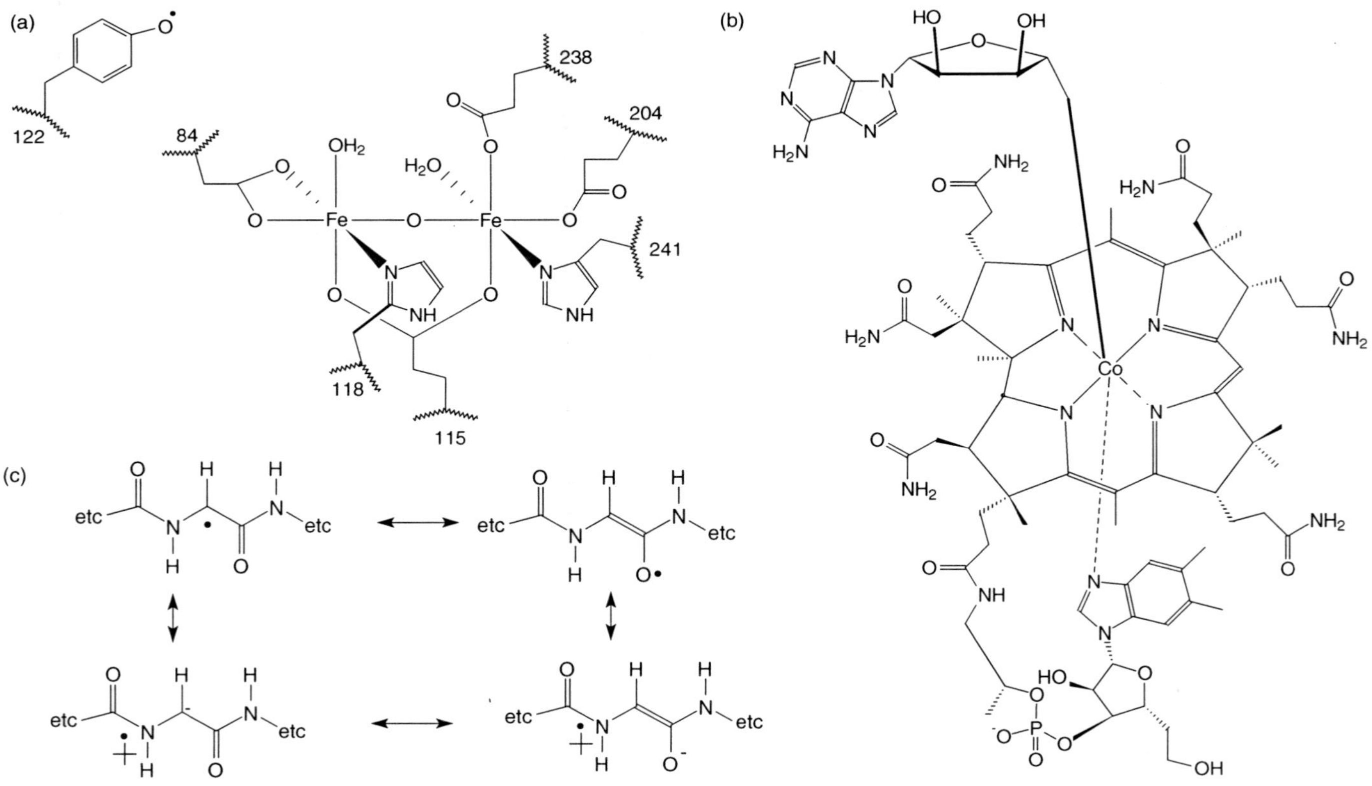

Figure 7.33 (a) Radical-generation cofactor in the R2 (β) subunit of Type I ribonucleotide reductases. Numbering is for the *Escherichia coli* enzyme. (b) Adenosylcobalamin, initiator for Type II ribonucleotide reductases. The exaggerated C–Co bond is homolysed to act as a radical source. (c) Glycyl radical, initiator for Type III ribonucleotide reductase.

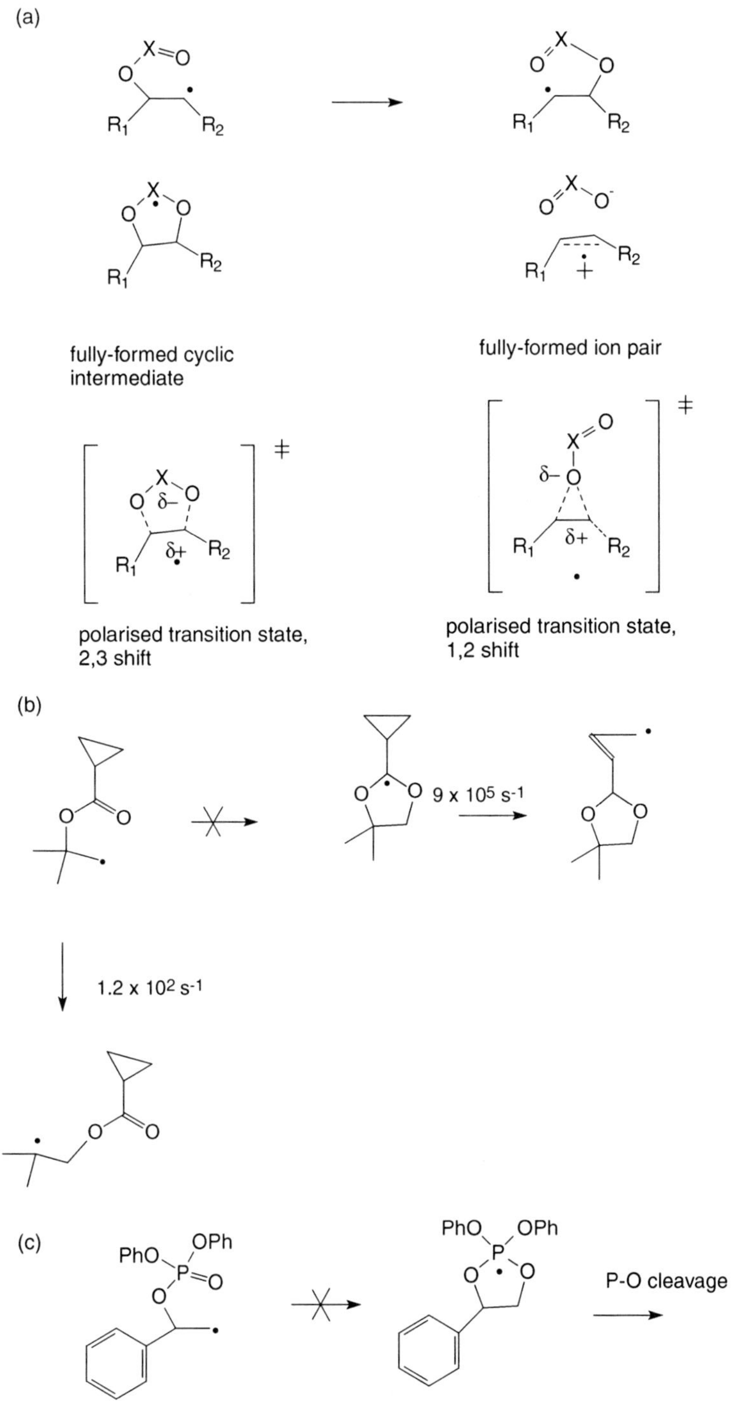

(a)

X=O O R$_1$ R$_2$ → O=X O R$_1$ R$_2$

X O O R$_1$ R$_2$ O=X O$^-$ R$_1$ + R$_2$

fully-formed cyclic intermediate

fully-formed ion pair

[X O δ– O R$_1$ δ+ R$_2$]‡ [X=O δ– O R$_1$ δ+ R$_2$]‡

polarised transition state, 2,3 shift

polarised transition state, 1,2 shift

(b)

O O O O $9 \times 10^5 \text{ s}^{-1}$ O O

$1.2 \times 10^2 \text{ s}^{-1}$

O O

(c)

PhO OPh P=O O PhO OPh P O O

P-O cleavage

Figure 7.34 (a) Potential mechanisms for acyloxy and related shifts. Evidence against cyclic intermediates for (b) acyloxy migration and (c) phosphatoxy migration.

competent. Kinetic ESR studies of the $CH_3COOC(CH_3)_2C–CH_2^•$ to $(CH_3)_2C^•–CH_2OOCCH_3$ rearrangement[133] in hydrocarbon solvents at 75 °C indicated that the opening of the 4,4,2-trimethyl-1,3-dioxolan-2-yl radical (at $7.6 \times 10^3 s^{-1}$) was easily fast enough to support the rearrangement (at $5.1 \times 10^2 s^{-1}$) (although earlier studies of a similar rearrangement in water suggested that here the dioxolan-2-yl radical was kinetically incompetent).[135]

The cyclic radical pathway predicts that complete interchange between carbonyl and alkoxy oxygens occurs in the case of acyloxy shift. Acyloxy 2 → 1 migrations in protected sugars, as in several aliphatic systems, when monitored by the ^{18}O isotope shift on ^{13}C NMR resonances, did indeed show complete interchange of carbonyl and carboxyl oxygens.[136] Similar results had been obtained in several aliphatic systems, but in the 2 → 1 rearrangement of the 2-propionoxytetrahydropyr-1-anyl radical, ^{17}O NMR spectroscopy indicated that 66% of the label remained in the carbonyl group, *i.e.* removing all the electron-withdrawing substituents from a sugar had made the 2,3- and 1,2-migrations proceed at comparable rates. That a 1,2-shift, compared with a 2,3-shift, is favoured by factors stabilising transition states with charge separation was also shown nicely with ^{17}O NMR in the rearrangement of p-$XC_6H_4CMe(OC^{17}O$-n-$C_3H_7)CH_2^•$ to $ArC^•Me–CH_2OOC$-n-C_3H_7.[xiv] The starting radical was derived from the action of AIBN and Bu_3SnH on the corresponding bromide, p-$XC_6H_4CMe(OC^{17}O$-n-$C_3H_7)CH_2Br$ and the shift was timed by competition with hydrogen quenching of the primary radical by Bu_3SnH, which was assumed to have the same rate constant as the directly determined quenching of $(CH_3)_3CH_2^•$. Complete ^{17}O interchange was observed for X = p-CN and H in benzene, but with X = MeO in benzene there was a 39% 1,2-shift and 25% with X = H in methanol.[137]

In the case of the phosphatoxy migrations, once the possible intermediacy of a phosphoranyl radical is discounted, 1,2- and 2,3-shifts can be quantitated from their effect on the stereochemical designation of an asymmetric phosphorus atom. 1,2-Shifts do not alter the priorities of atoms around phosphorus and so apparently result in retention of configuration, whereas 2,3-shifts exchange the priorities of two of the phosphorus ligands and result in apparent inversion.[xv] This stereochemical probe gave the same result as isotopic labelling, that the phosphatoxy migration occurs predominantly (50–80%) by a 1,2-route.[134]

In the cases of nitrate and sulfonate migration, the 2,3-pathway can occur to either of two equivalent oxygen atoms (unless, in the sulfonate case, the alkyl residue within which the migration takes place contains an asymmetric centre, so that sulfonyl oxygens are diastereotopic) and correction for this statistical advantage is problematic. Nonetheless, even after correction, rearrangement occurs predominantly by 1,2-shift (Figure 7.35).[134]

[xiv] A competing neophyl rearrangement to $MeC^•(OCOPr)–CH_2Ar$ had to be corrected for.

[xv] The original papers refer simply to retention and inversion at phosphorus, but no P–O bond cleavages or phosphoranyl pseudorotations are involved.

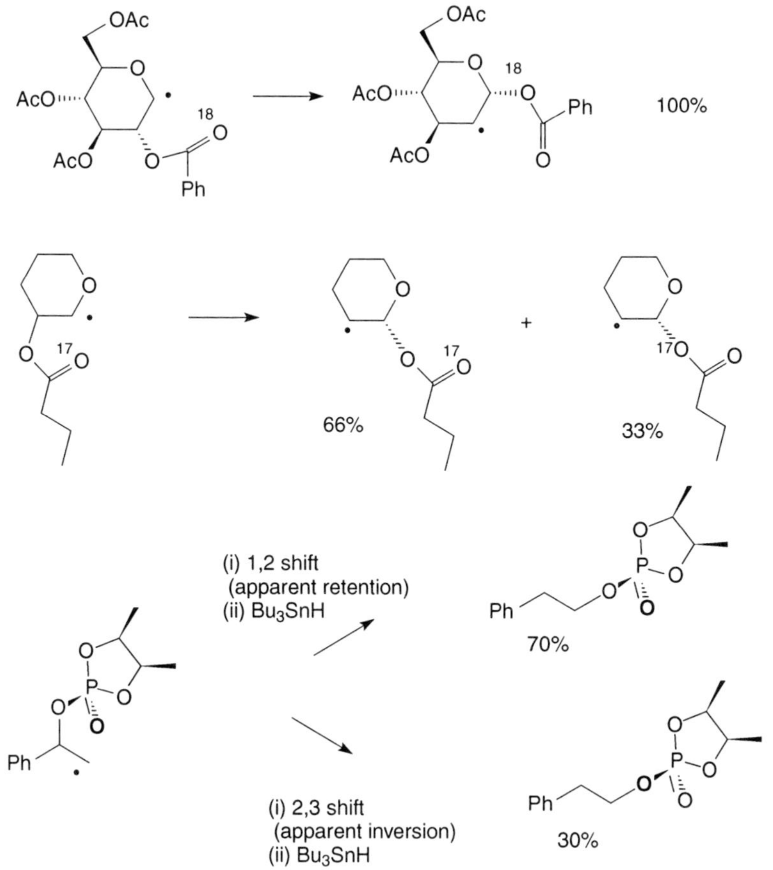

Figure 7.35 Stereochemical and labelling evidence for concurrent 1,2- and 2,3-shifts.

The phosphatoxy rearrangement occurs about $\sim 10^3$ times faster than the acyloxy rearrangement, in accord with its polar nature and the $\sim$4 unit difference in the pK_a values of diaryl phosphates and carboxylic acids. The rates of the rearrangements of PhCH[OPO(O-*p*-$C_6H_4X)_2]CH_2^{\bullet}$ to $PhCH^{\bullet}$–CH_2–O–PO(O-*p*-$C_6H_4X)_2$, measured by competition with PhSeH trapping of the primary radical, correlated with σ_p of X with a ρ value of 2.1, indicating a largely anionic character for the phosphate moiety in the rate-determining transition state (Figure 7.36).[138] The 2 → 1 diphenyl phosphate migration in anomeric radicals[139] shows a similar reactivity patterns to the S_N1 hydrolysis of glycosides with anionic leaving groups, including the enhanced reactivity of 6-deoxy sugars and sensitivity to the electron-withdrawing nature of the protecting group (benzoyl being more electron-withdrawing than acetyl). The only disparity with 2-electron chemistry is the slower reaction of the *galacto* than the *gluco* compound, probably related to the boat conformation of

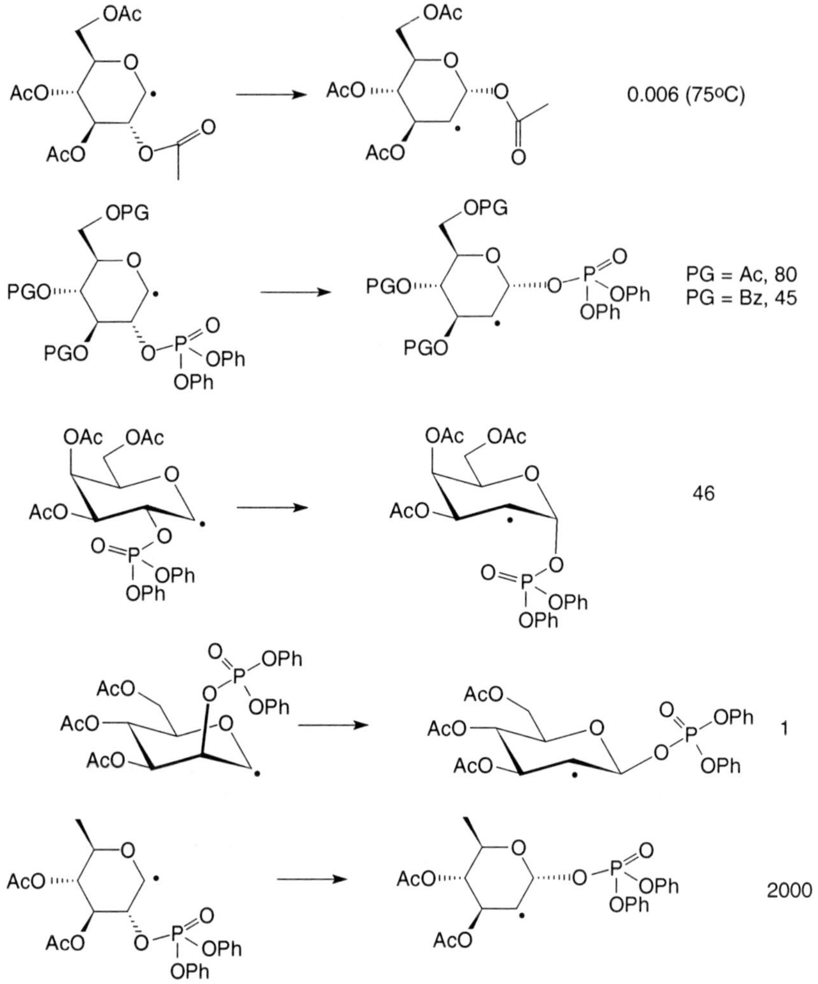

Figure 7.36 Rates of rearrangement ($10^{-5}k$/s^{-1}) of anomeric radicals at 27 °C in benzene.[139]

the *gluco* anomeric radical, and the recently recognised field effects in Cl reactivity.

Further evidence that the alkyl moiety in these migrations carries a positive charge comes from the kinetics of the rearrangement of p-$XC_6H_4CMe(OCO$-n-$C_3H_7)CH_2^{\bullet}$ to $ArC^{\bullet}Me$–CH_2OOC-n-C_3H_7. For X = H in benzene, $\log_{10}(A/s^{-1}) = 11.7$ and $E_a = 46$ kJ mol^{-1}. From the limited data (X = MeO, H and CN), a correlation of rearrangement rate in benzene with σ_p^+ could be discerned, yielding $\rho = -0.7$.

A unifying mechanistic description of these rearrangements has yet to advanced: a More O'Ferrall–Jencks diagram, with the cyclic radical in the NW corner and the radical cation–anion ion pair in the SE corner, is tempting, but such a description does not explain the origin of the 1,2-shifted product, which is often predominant. The existence of 1,2-shifted product, in excess of that expected from the randomisation of the oxygen atoms of the leaving group, demands a specific transition state for both concurrent 1,2- and 2,3-pathways: intimate ion pairs will give, maximally, the statistical amount of apparent 1,2-shift (50% from a carboxylate or dialkyl phosphate, 33% from a sulfonate). Intimate ion pairs can, however, in some circumstances be intermediates, as shown by the existence of products from a phosphatoxy rearrangement of an allylic radical derived from the recombined, allylically shifted phosphatoxy radical.[140] Likewise, if the radical $Ph–CH[OPO(OPh)_2]–C^{\bullet}(CH_2Ph)_2$ is generated (by photolysis of the Barton ester), it both rearranges to the $PhCH^{\bullet}–C(CH_2Ph)_2OPO(OPh)_2$ and eliminates, via the intimate (contact) ion pair to the allyl radical $Ph–CH^{\bullet}–C(CH_2Ph){=}CH–Ph$.[141] Rearrangement of $Ph–CH[OPO(OPh)_2]–C^{\bullet}(CD_2Ph)_2$, monitored by rapid-response UV spectroscopy, proceeded at the same rate as undeuterated material, but the product-determining step in the formation of the allyl radical was subject to the expected primary kinetic isotope effects. Solvent effects were in accord with a process involving charge separation and it was even possible, in aqueous acetonitrile, to detect the radical cation $Ph–CH^{\bullet}–C(CH_2Ph)–C^{+}H–Ph$ as a solvent-equilibrated intermediate.

A unified mechanism for 1,2- and 2,3-acyloxy shifts could perhaps be formulated in terms of More O'Ferrall–Jencks diagrams sharing a common edge (C_{origin}–O cleavage). Alternatively, if a degree of concurrent bonding to both oxygens is permitted, a three-dimensional More O'Ferrall–Jencks diagram as shown in Figure 7.37 could be drawn. The internal energy distribution would then be such that at low energies of the ion pair, pathways to 1,2-shift are favoured.

7.5 CARBOHYDRATE CARBENES

Methylene, CH_2, the simplest carbene, can exist in two electronic states (Figure 7.38). Triplet carbene, sometimes written $CH_2^{:}$, has two unpaired electrons in two degenerate p orbitals and the H–C–H nuclei co-linear. It behaves like a diradical, so that, for example, additions to olefins are non-stereospecific. Singlet carbene, sometimes written $CH_2^{\pm}$, has the central carbon sp^2 hybridised and its additions to olefins are stereospecific.[142] The third sp^2 orbital contains a lone pair, while a p orbital is empty, so that the H–C–H system is non-collinear. In carbene itself the triplet is the ground state, with the singlet some 37 kJ above it,[143] but any substitution, particularly with polar substituents, stabilises the singlet with respect to the triplet. The stabilising effect of substitution of the carbene carbon with inductively withdrawing and conjugatively donating substituents is such that bis(dimethylamino)carbene[144] and bis(diisopropylamino)carbene[145] are stable species; the latter can be sublimed and its X-ray structure revealed an N–C–N bond angle of 121°.

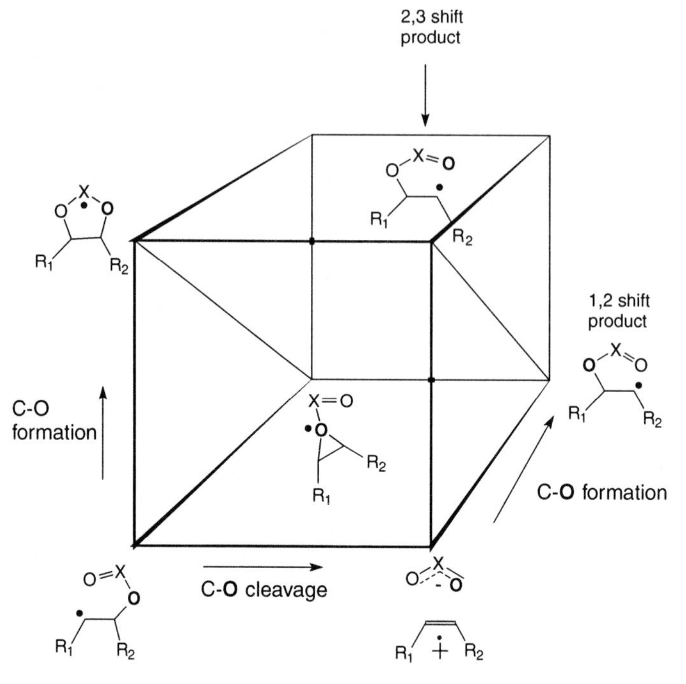

Figure 7.37 Three-dimensional More O'Ferrall–Jencks diagram for radical acyloxy and related reactions.

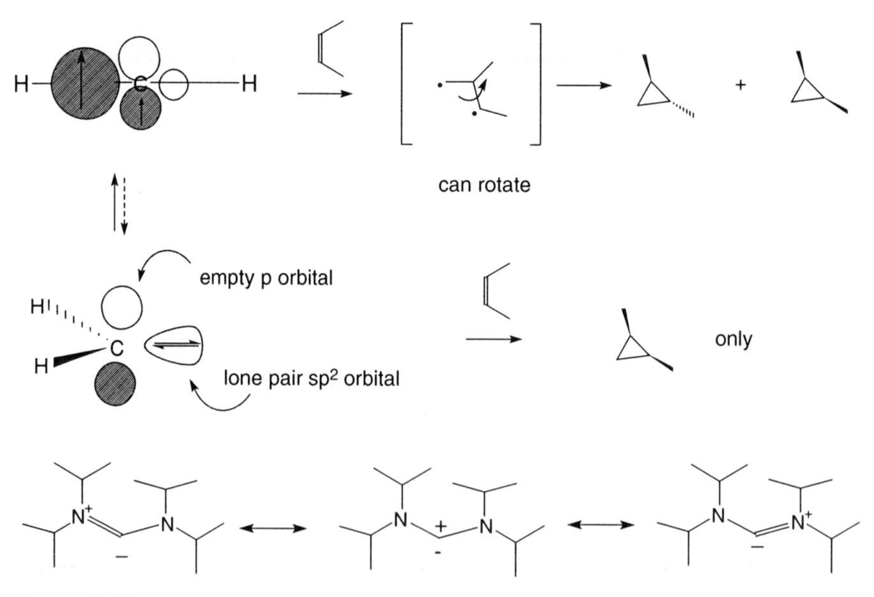

Figure 7.38 Atomic orbital cartoons of singlet and triplet carbenes, illustrating a diagnostic cycloaddition. Canonical structures for the stable bis(diisopropylamino)carbene are also shown.

Carbenes at position 1 of various protected sugars can be generated by thermolysis of anomeric diazirines[146] or photolysis of the sodium salts of tosylhydrazones of aldonolactones[147] or geminal diazides.[148] The lactone tosylhydrazones are prepared by oxidation of *N*-glycosyltosylhydrazines, the diazides by the action of silver azide on glycosyl geminal dihalides[xvi] and the diazirines by a sequence involving oxime formation from the reducing sugar, oxidation to the lactone oxime, *O*-sulfonation, displacement of sulfonate with ammonia and finally oxidation of the diaziridine to the diazirine with, *e.g.*, iodine (Figure 7.39).

The same glycosidenecarbenes seem to be generated both from the thermolysis of the diazirine and photolysis of the sodium tosylhydrazone, since the same distribution of cyclopropanes was produced from various electron-deficient olefins and glycopyranosylidenes generated by either method. This appeared to be the singlet, since tetra-*O*-benzylglucopyranosylidene added stereospecifically to dimethyl maleate and dimethyl fumarate, giving cyclopropanes with COOMe groups *cis* and *trans*, respectively (Figure 7.40) (although there was little discrimination between the two possible stereochemistries at the anomeric carbon).[149] Generated in solution in acetone or cyclohexanone, tetrabenzylglucopyranosylidene gave the two possible spiroepoxides, plus the hydride-transfer product tribenzylbenzyloxyglucal.[150]

Glycosylidenecarbenes were investigated largely for their potential as glycosylation agents, since they apparently inserted into the OH bond of alcohols and phenols to give anomeric mixtures of glycosides.[151] The reaction of tetra-*O*-benzylglucopyranosylidene with alcohols was monitored directly in a series of laser flash photolysis experiments in acetonitrile.[152] The carbene had no useful absorbance above the solvent cutoff, but by monitoring the first-order build-up of the pyridine ylide, which absorbed at 500 nm, after laser flash photolysis of tetra-*O*-benzylglucopyranosylidenediazirine, it was possible to measure the rate of reaction of the carbene both with pyridine and with various alcohols. Equation 7.9 held, Py being pyridine:[xvii]

$$k_{obs} = k_0 + k_{py}[\mathrm{Py}] + k_{\mathrm{ROH}}[\mathrm{ROH}] \qquad (7.10)$$

At ambient temperature, capture by pyridine was fast ($7.9 \times 10^8\,M^{-1}\,s^{-1}$), as was reaction with alcohols. Plots of k_{obs} against alcohol concentration (at constant [Py]) were not linear, as hydrogen-bonded oligomeric aggregates reacted faster than monomeric alcohols. From the four alcohols studied, it is possible to derive a very approximate Brønsted α of ~0.4 for the reaction with monomeric alcohols, with the rate for hexafluoro-2-propanol ($2.9 \times 10^9\,M^{-1}\,s^{-1}$) approaching the diffusion limit.

[xvi] Silver azide is a powerful explosive.

[xvii] The interception of the carbene by alcohols increases the rate at which the pool of carbene is depleted, and so increases the rate of the reaction, although of course it decreases the final absorbance.

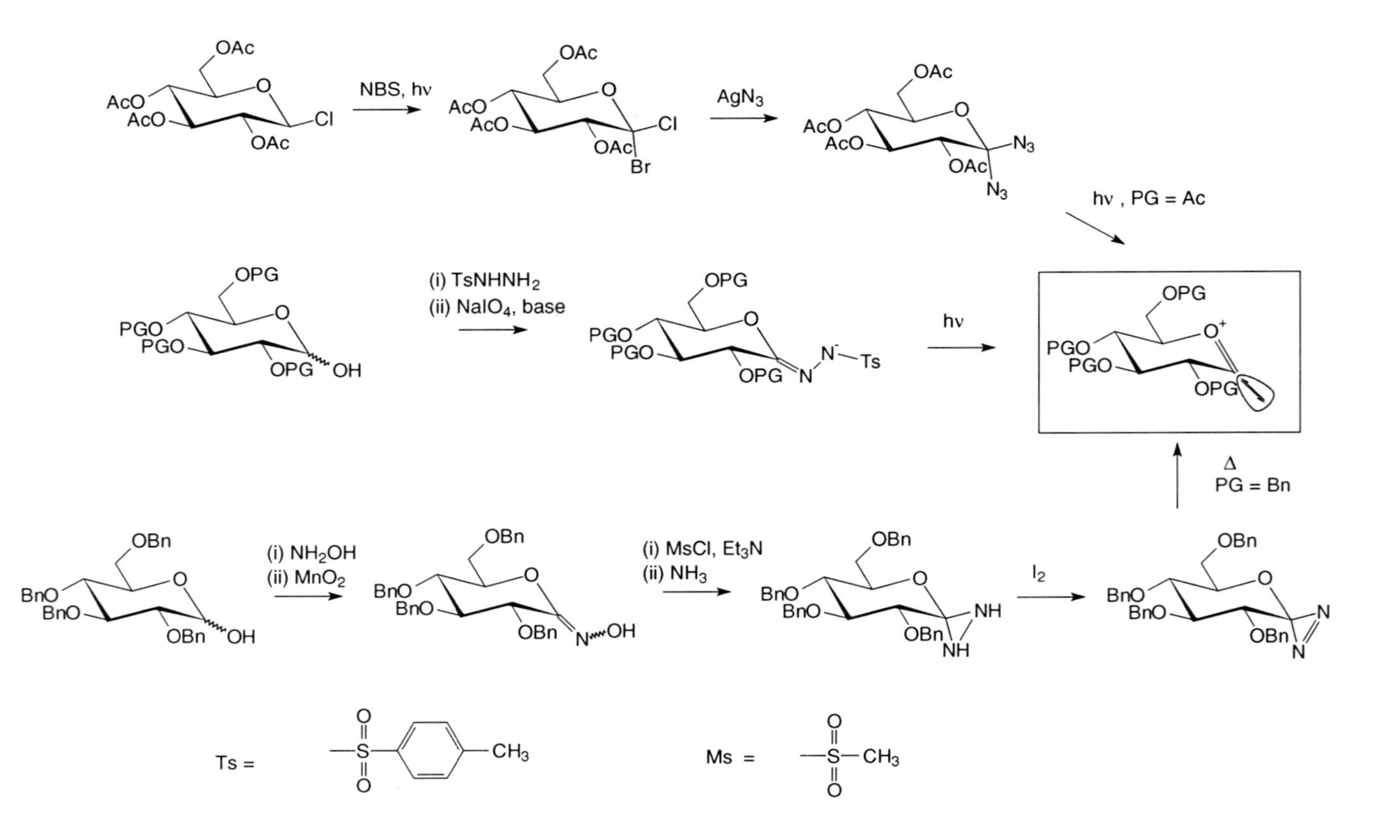

Figure 7.39 Formation of glycopyranosylidenecarbenes. Conjugation to the oxygen lone pair is shown as not being fully developed in the carbene, but there is no evidence on the point.

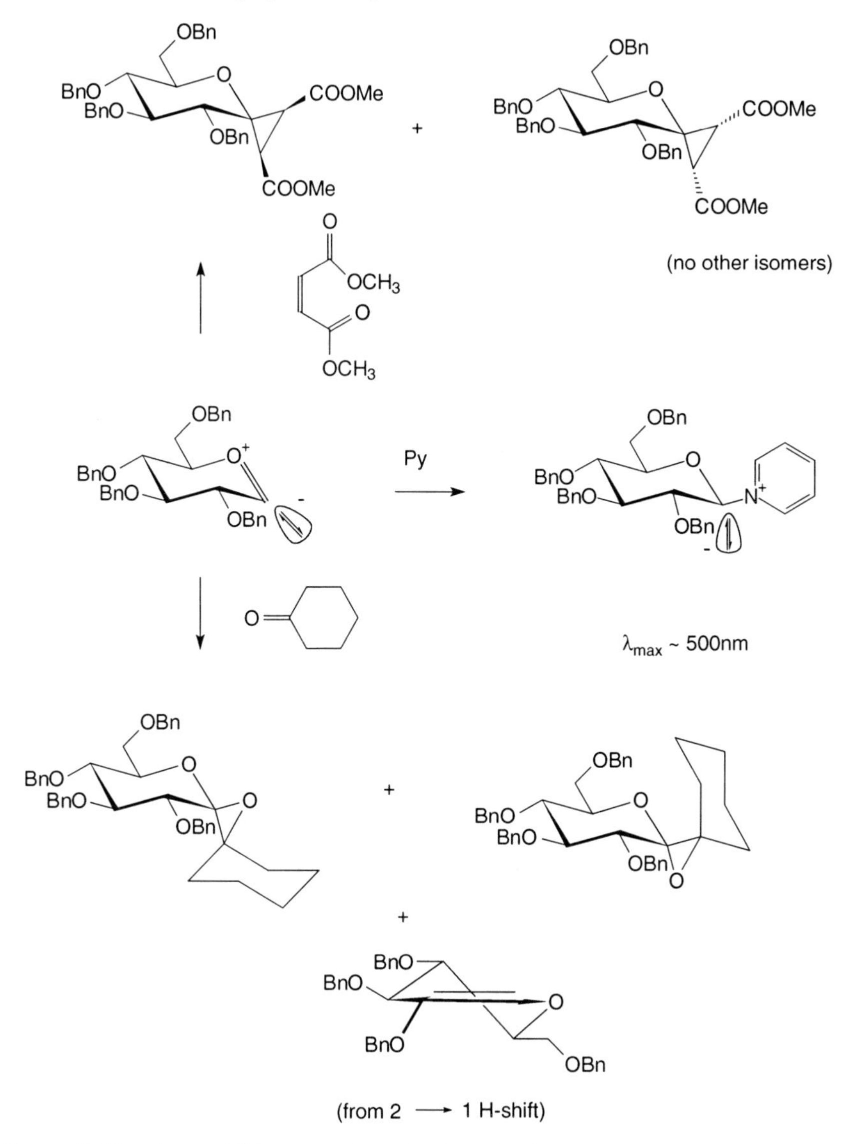

Figure 7.40 Cycloaddition reactions showing the singlet nature of glycopyranosylidene carbenes. The formation of the chromophoric pyridinium ylide is also shown.

The initial step in the reaction of glycosylidenes with alcohols was therefore thought to consist of protonation of the lone pair in the plane of the sugar ring (the empty p orbital interacting with the ring oxygen) to yield an ion pair, followed by nucleophilic combination of the alkoxide with the oxocarbenium ion. This occurs in the π plane (*i.e.* probably at close to the Bürgi–Dunitz angle) (Figure 7.41). In general, therefore, the reaction yields anomeric mixtures. The

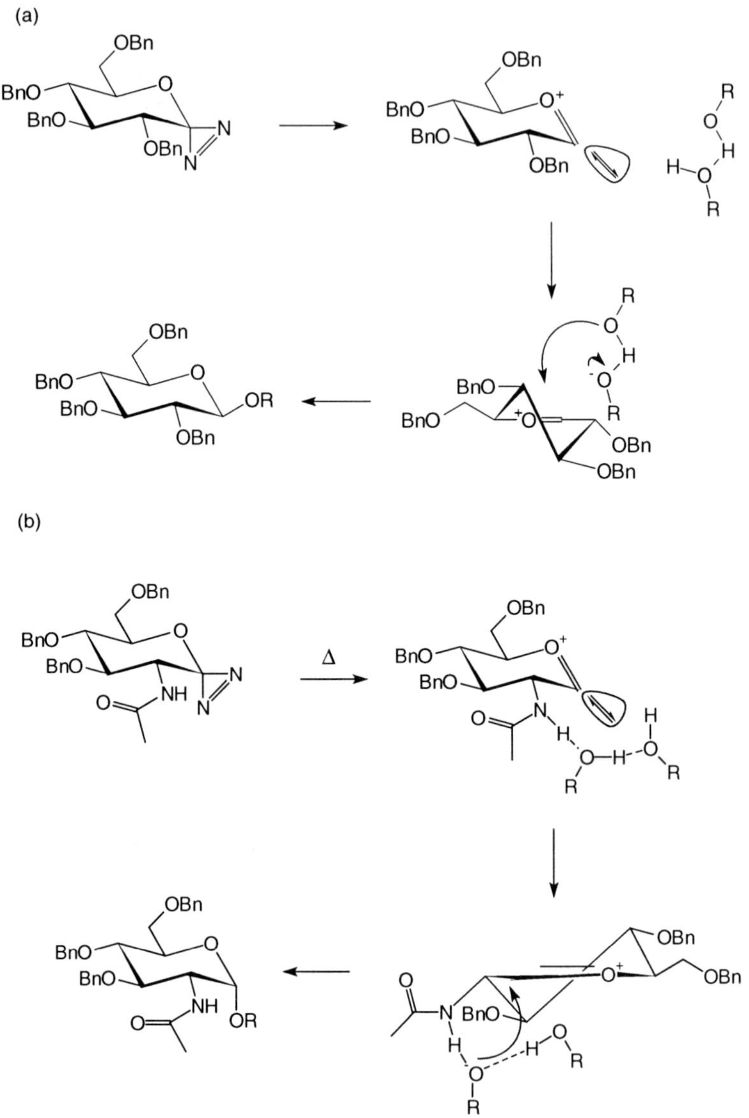

Figure 7.41 Mechanisms of reaction of glycosylidene carbenes with alcohols. (a) Illustrating how hydrogen-bonded alcohol clusters may protonate in plane and add at the Bürgi–Dunitz angle: in the case shown turning the alcohol dimer "upside down" would result in the other anomer: both are observed. (b) Direction of glycosidation by a 2-acetamido group.

possibility that intramolecular hydrogen bonding patterns of polyols might be exploited to achieve regioselectivity and stereopecificity, with protonation by the most acidic OH bond and nucleophilic attack by the most nucleophilic and/or closest oxygen in the hydrogen-bonded cluster, was explored, but the

(a) Successful:

gluco and *galacto*,
red blood cell hexose transporter

Part of tetravalent ligand,
antibody

β galactosidase

hexosaminidase

(b) Unsuccessful:

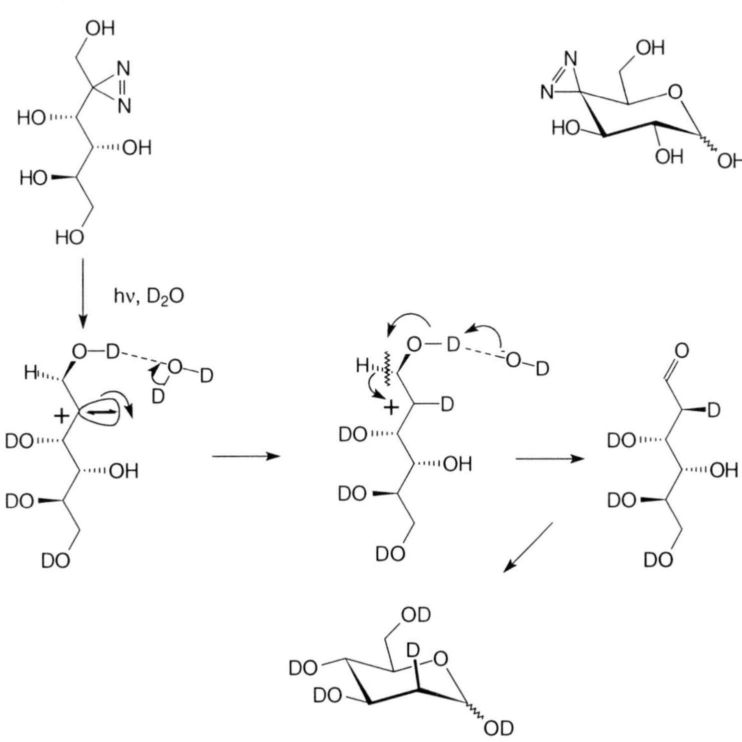

Figure 7.42 Successful and unsuccessful photoaffinity labels from carbohydrate diazirines. The mechanistic details for the observed stereochemistry of reaction of the 2-azi-mannitol in D_2O are speculative.

glycosylation specificities observed were modest.[153] Glycopyranosylidenes with an equatorial 2-acetamido group yield largely or even exclusively α-glycosides, presumably by virtue of hydrogen bond donation from NH to alcohol oligomers: specificity increases with increasing alcohol pK_a, presumably as the key N–H···O hydrogen bond increases in strength.[154]

Sugars with diazirine functions away from the anomeric centre have been used as photoaffinity labels for carbohydrate-active transport and binding proteins and enzymes (Figure 7.42). Irradiation generates carbenes at the active site and these can in principle react with the protein. Against enzymes, successes have been reported where the diazirine is in the +1 site, such as the labelling of an active site Glu in human hexosaminidase B with β-D-GalNAc*p*-S–CH_2–CH_2–C(<N_2)CH_3,[155] or active site peptides of *E. coli* β-galactosidase with β-D-Gal*p*C(<N_2)CH_3.[156] Against transport and binding proteins elaboration at C6 of the reagent is successful.[157,158] The site of the diazirine within the reagent is important: incorporation of the diazirine vicinal to a CH_2OH resulted in production of a –CH_2CHO group in the ligand and no labelling of protein. Presumably the (singlet) carbene produced on photolysis is first protonated, possibly via a water chain from the vicinal OH, and then there is an intramolecular hydride shift which competes successfully with any intermolecular insertions of the carbene into protein. Similar, but less marked, intramolecular reactivity is observed if the diazirine is vicinal to a secondary OH.

REFERENCES

1. H. H. Wasserman, J. R. Scheffer and J. L. Cooper, *J. Am. Chem. Soc.*, 1972, **94**, 4991.
2. Reviews: (a) N. Getoff, *Radiat. Phys. Chem.*, 1995, **45**, 609; (b) R. P. Wayne, *Res. Curr. Intermed.* 1994, **20**, 395; (c) J. K. McCullough and M. Nojima, *Curr. Org. Chem.* 2001, **5**, 601; (d) M. C. De Rosa and R. J. Crutchley, *Coord. Chem. Rev.* 2002, **233**, 351.
3. D. H. J. Bielski, D. E. Cabelli, R. L. Arudi and A. B. Ross, *J. Phys. Chem. Ref. Data*, 1985, **14**, 1041.
4. B. A. Palfrey and V. A. Massey, in *Comprehensive Biological Catalysis*, ed. M. L. Sinnott, Academic Press, London, 1998, Vol. 3, p. 83.
5. A. Messerschmidt, in *Comprehensive Biological Catalysis*, ed. M. L. Sinnott, Academic Press, London, 1998, Vol. 3, p. 401.
6. C. Anthony, in *Comprehensive Biological Catalysis*, ed. M. L. Sinnott, Academic Press, London, 1998, Vol. 3, p. 155.
7. D. H. R. Barton and S. W. McCombie, *J. Chem. Soc., Perkin Trans. 1*, 1975, 1574.
8. P. Renaud and M. P. Sibi, Eds, *Radicals in Organic Synthesis*, Wiley-VCH, Weinheim, 2001.
9. B. C. Golding and W. Buckel, in *Comprehensive Biological Catalysis*, ed. M. L. Sinnott, Academic Press, London, 1998, Vol. 3, p. 239.

10. K. U. Ingold, *Pure Appl. Chem.*, 1997, **69**, 241.
11. M. Newcomb, *Tetrahedron*, 1993, **49**, 1151.
12. J. L. Fry, E. M. Engler and P. v. R. Schleyer, *J. Am. Chem. Soc.*, 1972, **94**, 4628.
13. B. Mile, *Curr. Org. Chem.*, 2000, **4**, 55.
14. P. J. Barker, A. G. Davies and M.-W. Tse, *J. Chem. Soc., Perkin Trans. 2*, 1980, 941.
15. J. K. Kochi, *Adv. Free-Rad. Chem.*, 1975, **5**, 189.
16. R. D. McKelvey, T. Sugawara and H. Iwamura, *Magn. Reson. Chem.*, 1985, **23**, 330.
17. H.-G. Korth, R. Sustmann, J. Dupuis and B. Giese, *J. Chem. Soc., Perkin Trans. 2*, 1986, 1453.
18. D. B. Chestnut and G. J. Sloan, *J. Chem. Phys.*, 1960, **33**, 637.
19. R. W. Fessenden and R. H. Schuler, *J. Chem. Phys.*, 1963, **39**, 2147.
20. J. Dupuis, B. Giese, D. Rüegge, H. Fischer, H.-G. Korth and R. Sustmann, *Angew. Chem. Int. Ed. Engl.*, 1984, **23**, 896.
21. C. D. Garner, *et al.*, in *Lexicon of Terms and Concepts in Mechanistic Enzymology*, ed. M. L. Sinnott, C. D. Garner, E. First and G. Davies, Vol. 4 of *Comprehensive Biological Catalysis*, ed. M. L. Sinnott, Academic Press, London, 1998, p. 14.
22. W. Adam, C. van Barneveld, O. Emmert, H. M. Harrer, F. Kita, A. S. Kumar, W. Maas, W. M. Nau, S. H. K. Reddy and J. Wirz, *Pure Appl. Chem.*, 1997, **69**, 735.
23. R. Sustmann and H.-G. Korth, *J. Chem. Soc., Faraday Trans. 1*, 1987, 95.
24. H. Chandra, M. C. R. Symons, H.-G. Korth and R. Sustmann, *Tetrahedron Lett.*, 1987, **28**, 1455.
25. H.-G. Korth, R. Sustmann, K. S. Gröninger, M. Leisung and B. Giese, *J. Org. Chem.*, 1988, **53**, 4364.
26. H.-G. Korth, R. Sustmann, K. S. Gröninger, T. Witzel and B. Giese, *J. Chem. Soc., Perkin Trans. 2*, 1986, 1461.
27. A. L. J. Beckwith, D. Crich, P. J. Duggan and Q. Yao, *Chem. Rev.*, 1997, **97**, 3273.
28. A. Koch and B. Giese, *Helv. Chim. Acta*, 1993, **76**, 1687.
29. Y. Itoh, Z. Haraguchi, H. Tanaka, K. Matsumoto, K. T. Nakamura and T. Miyasaka, *Tetrahedron Lett.*, 1995, **36**, 3867.
30. B. C. Gilbert, D. M. King and C. B. Thomas, *J. Chem. Soc., Perkin Trans. 2*, 1980, 1821.
31. B. C. Gilbert, D. M. King and C. B. Thomas, *J. Chem. Soc., Perkin Trans. 2*, 1981, 1186.
32. T. Zytowski and H. Fischer, *J. Am. Chem. Soc.*, 1997, **119**, 12869.
33. M. Weber and H. Fischer, *J. Am. Chem. Soc.*, 1999, **121**, 7381.
34. See, for example, the Annual Reports of the UK Central Laser Facility: www.clf.rl.ac.uk/reports.
35. D. J. Holder, D. Allan, E. J. Land and S. Navaratnam, in *Proceedings of the 2002 European Particle Accelerator Conference*, 2002, p. 2804; available at http://epac.web.cern.ch/EPAC/Welcome.html.

36. M. Newcomb, in *Radicals in Organic Synthesis. Volume 1: Basic Principles*, ed. P. Renaud and M. P. Sibi, Wiley-VCH, Weinheim, 2001, p. 317.
37. D. Crich, X.-Y. Jiao, Q. Yao and J. S. Harwood, *J. Org. Chem.*, 1996, **61**, 2368.
38. M. Newcomb, S.-Y. Choi and J. H. Horner, *J. Org. Chem.*, 1999, **64**, 1225.
39. V. W. Bowry, J. Lusztyk and K. U. Ingold, *J. Am. Chem. Soc.*, 1991, **113**, 5687.
40. C. Chatgilialoglu, K. U. Ingold and J. C. Scaiano, *J. Am. Chem. Soc.*, 1981, **103**, 7739.
41. J. A. Franz, B. A. Bushaw and M. S. Alnajjar, *J. Am. Chem. Soc.*, 1989, **111**, 268.
42. T. Vuorinen, *Carbohydr. Res.*, 1983, **116**, 61.
43. S. V. Singh, O. Saxena and M. P. Singh, *J. Am. Chem. Soc.*, 1970, **92**, 537.
44. N. Nath and M. P. Singh, *J. Phys. Chem.*, 1965, **69**, 2038.
45. L. W. Doner and P. L. Irwin, *Anal. Biochem.*, 1992, **202**, 50.
46. M. Lever, *Anal. Biochem.*, 1972, **47**, 273.
47. P. Hartmann, S. J. Haswell and M. Grasserbauer, *Anal. Chim. Acta*, 1994, **285**, 1.
48. S. Waffenschmidt and L. Jaenicke, *Anal. Biochem.*, 1987, **165**, 337.
49. L. Rodén, H. Yu, J. Jin, G. Ekborg, A. Estock, N. R. Krishna and P. Livant, *Anal. Biochem.*, 1997, **254**, 240.
50. S. J. Padayatty, A. Katz, Y. Wang, P. Eck, O. Kwon, J.-H. Lee, S. Chen, C. Corpe, A. Dutta, S. K. Dutta and M. Levine, *J. Am. Coll. Nutr.*, 2003, **22**, 18.
51. Quoted by A. Hara, M. Shinoda, T. Kanazu, T. Nakayama, Y. Deyashiki and H. Sawada, *Biochem. J.* 1991, **275**, 121.
52. M. Nishikimi, R. Fukuyama, S. Minoshima, N. Shimizu and K. Yagi, *J. Biol. Chem.*, 1994, **269**, 13685.
53. N. Smirnoff, P. L. Conklin and F. A. Loewus, *Annu. Rev. Plant Physiol. Plant Mol. Biol.*, 2001, **52**, 437.
54. V. Roginsky, C. Michel and W. Bors, *Arch. Biochem. Biophys.*, 2000, **384**, 74.
55. P. I. Lario, N. Sampson and A. Vrielink, *J. Mol. Biol.*, 2003, **326**, 1635.
56. J. P. Roth and J. P. Klinman, *Proc. Natl. Acad. Sci. USA*, 2003, **100**, 62.
57. G. N. Yalloway, S. G. Mayhew, J. P. Malthouse, M. E. Gallagher and G. P. Curley, *Biochemistry*, 1999, **38**, 3753.
58. B. M. Hallberg, G. Henriksson, G. Pettersson, A. Vasella and C. Divne, *J. Biol. Chem.*, 2003, **278**, 7160.
59. F. A. J. Rotsaert, V. Renganathan and M. H. Gold, *Biochemistry*, 2003, **42**, 4049.
60. N. Habu, M. Samejima, J. F. D. Dean and K.-E. L. Eriksson, *FEBS Lett.*, 1993, **327**, 161.
61. A. Oubrie, *Biochim. Biophys. Acta*, 2003, **1647**, 143.
62. A. Oubrie, H. J. Rozeboom, K. H. Kalk, A. J. J. Olsthoorn, J. A. Duine and B. W. Dijkstra, *EMBO J.*, 1999, **18**, 5187.
63. D. T. Sawyer, A. Sabkowiak and T. Matsushita, *Acc. Chem. Res.*, 1996, **29**, 409.

64. (a) C. Walling, *Acc. Chem. Res.*, 1998, **31**, 155; (b) P. A. MacFaul, D. D. M. Wayner and K. U. Ingold, *Acc. Chem. Res.*, 1998, **31**, 158.
65. J. M. Brown, M. Kaise, C. M. L. Kerr and D. J. Milton, *Mol. Phys.*, 1978, **36**, 553.
66. B. C. Gilbert, D. M. King and C. B. Thomas, *J. Chem. Soc., Perkin Trans. 2*, 1983, 675.
67. V. Malatesta and K. U. Ingold, *J. Am. Chem. Soc.*, 1981, **103**, 609.
68. W. K. Busfield, I. D. Grice and I. D. Jenkins, *J. Chem. Soc., Perkin Trans. 1*, 1994, 1079.
69. R. Cohen, K. A. Jensen, C. J. Houtman and K. E. Hammel, *FEBS Lett.*, 2002, **531**, 483.
70. S. D. Wetmore, R. J. Boyd and L. A. Eriksson, *J. Chem. Phys.*, 1997, **106**, 7738.
71. E. Mvula, M. N. Schuchmann and C. von Sonntag, *J. Chem. Soc., Perkin Trans. 2*, 2001, 264.
72. A. L. Buchachenko, L. L. Yasina and V. A. Belyakova, *J. Phys. Chem.*, 1995, **99**, 4964.
73. C. von Sonntag and H.-P. Schuchmann, in *Peroxyl Radicals*, ed. Z. Alfassi, Wiley, Chichester, 1997, p. 173.
74. S. C. Fry, *Biochem. J.*, 1998, **332**, 507.
75. C. Schweikert, A. Liszkay and P. Schopfer, *Phytochemistry*, (a) 2000, **53**, 565; (b) 2002, **61**, 31.
76. R. Kohen, V. Shadini, A. Kakunda and A. Rubinstein, *Br. J. Nutr.*, 1993, **69**, 789.
77. J. Wang, X. Jiang, H. Mou and H. Guan, *J. Appl. Phycol.*, 2004, **16**, 333.
78. B. C. Gilbert, J. R. L. Smith, P. Taylor, S. Ward and A. C. Whitwood, *J. Chem. Soc., Perkin Trans. 2*, 2000, 2001.
79. H. G. Viehe, R. Merényi and Z. Janousek, *Pure Appl. Chem.*, 1988, **60**, 1635.
80. E. R. Davidson, S. Chakravorty and J. J. Gajewski, *New. J. Chem.*, 1997, **21**, 533.
81. F. W. Lichtenthaler, M. Lergenmüller and S. Schwidetzky, *Eur. J. Org. Chem.*, 2003, 3094.
82. M. Kirsch and H. de Groot, *J. Biol. Chem.*, 2002, **277**, 13379.
83. D. Stenman, M. Carlsson and T. Reiberger, *J. Wood Chem. Technol.*, 2004, **24**, 83.
84. M. Carlsson, J. Lind and G. Merényi, *Holzforschung*, 2006, **60**, 130.
85. S. Lemeune, H. Jameel, H.-M. Chang and J. F. Kadla, *J. Appl. Polym. Sci.*, 2004, **93**, 1219.
86. O. Dahlman, A. Jacobs and J. Sjöberg, *Cellulose*, 2003, **10**, 325.
87. M. Carlsson, D. Stenman, G. Merényi and T. Reiberger, *Holzforschung*, 2005, **59**, 132.
88. P. S. Fredricks, B. O. Lindgren and O. Theander, *Sven. Papperstidn.*, 1971, **74**, 597.
89. J.-P. Praly, L. Brard, G. Descotes and L. Toupet, *Tetrahedron*, 1989, **45**, 4141.
90. A. K. Konstantinidis, PhD Thesis, University of Illinois at Chicago, 1992.

91. L. A. Edye, G. V. Meehan and G. N. Richards, *J. Carbohydr. Chem.*, 1994, **10**, 273.
92. P. L. Bragd, H. van Bekkum and A. C. Besemer, *Top. Catal.*, 2004, **27**, 49.
93. I. W. C. E. Arends, Y.-X. Li, R. Ausan and R. A. Sheldon, *Tetrahedron*, 2006, **62**, 6659.
94. M. F. Semmelhack, C. R. Schmid and D. A. Cortés, *Tetrahedron Lett.*, 1986, **27**, 1119.
95. V. A. Golubev, V. N. Borislavskii and A. L. Aleksandrov, *Izv. Akad. Nauk SSSR, Ser. Khim.* 1977, 2025; *Bull. Acad. Sci. USSR, Div. Chem. Sci.*, 1977, 1874.
96. N. Merbouh, J. M. Bobbitt and C. Brückner, *J. Carbohydr. Chem.*, 2002, **21**, 65.
97. N. Merbouh, J.-F. Thaburet, M. Ibert, F. Marsais and J. M. Bobbitt, *Carbohydr. Res.*, 2001, **336**, 75.
98. C. L. Mehltretter and C. E. Rist, *J. Agric. Food Chem.*, 1953, **1**, 779.
99. B. Sun, C. Gu, J. Ma and B. Liang, *Cellulose*, 2005, **12**, 59.
100. S. J. Firbank, M. S. Rogers, C. M. Wilmot, D. M. Dooley, M. A. Halcrow, P. F. Knowles, M. J. McPherson and S. E. V. Phillips, *Proc. Natl. Acad. Sci. USA*, 2001, **98**, 12932.
101. L. Xie and W. A. van der Donk, *Proc. Natl. Acad. Sci. USA*, 2001, **98**, 12863.
102. S. Z. Zard, in *Radicals in Organic Synthesis. Volume 1: Basic Principles*, ed. P. Renaud and M. P. Sibi, Wiley-VCH, Weinheim, 2001, p. 90.
103. A. G. M. Barrett, D. H. R. Barton and R. Bielski, *J. Chem. Soc., Perkin Trans. 1*, 1979, 2378.
104. D. H. R. Barton and S. W. McCombie, *J. Chem. Soc., Perkin Trans. 1*, 1975, 1574.
105. D. Crich and A. A. Bowers, *J. Org. Chem.*, 2006, **71**, 3452.
106. G. Hourdin, A. Germain, C. Moreau and F. Fajula, *Catal. Lett.*, 2000 **69**, 241.
107. G. Hourdin, A. Germain, C. Moreau and F. Fajula, *J. Catal.*, 2002, **209**, 217.
108. H. S. Isbell and M. A. Salam, *Carbohydr. Res.*, 1981, **90**, 123.
109. F. A. H. Rice and A. R. Johnson, *J. Am. Chem. Soc.*, 1956, **78**, 428.
110. C. P. Andrieux, F. Gonzalez and J.-M. Savéant, *J. Electroanal. Chem.*, 2001, **498**, 171.
111. J. W. Hilborn and J. A. Pincock, *J. Am. Chem. Soc.*, 1991, **113**, 2683.
112. J. Chateauneuf, J. Lusztyk and K. U. Ingold, *J. Am. Chem. Soc.*, 1988, **110**, 2886.
113. P. A. Simakov, F. N. Martinez, J. H. Horner and M. Newcomb, *J. Org. Chem.*, 1998, **63**, 1226.
114. D. H. R. Barton, D. Crich and W. B. Motherwell, *J. Chem. Soc., Chem. Commun.*, 1983, 939.
115. W. B. Motherwell and C. Imboden, in *Radicals in Organic Synthesis. Volume 1: Basic Principles*, ed. P. Renaud and M. P. Sibi, Wiley-VCH, Weinheim, 2001, p. 109.

116. (a) W. Damm, J. Dickhaut, F. Wetterich and B. Giese, *Tetrahedron Lett.*, 1993, **34**, 431; (b) J. E. Eksterowicz and K. N. Houk, *Tetrahedron Lett.*, 1993, **34**, 427.
117. J. Brunckova and D. Crich, *Tetrahedron*, 1995, **51**, 11945.
118. D. Crich, J. T. Hwang and H. Yuan, *J. Org. Chem.*, 1996, **61**, 6189.
119. B. C. Gilbert, D. M. King and C. B. Thomas, *J. Chem. Soc., Perkin Trans. 2*, 1982, 169.
120. A. L. J. Beckwith, D. Crich, P. J. Duggan and Q. Yao, *Chem. Rev.*, 1997, **97**, 3273.
121. G. Behrens, G. Koltzenburg, A. Ritter and D. Schulte-Frohlinde, *Int. J. Radiat. Biol.*, 1978, **33**, 163.
122. G. Koltzenburg, G. Behrens and D. Schulte-Frohlinde, *J. Am. Chem. Soc.*, 1982, **104**, 7311.
123. B. Giese, X. Beyrich-Graf, J. Burger, C. Kesselheim, M. Senn and T. Schäfer, *Angew. Chem. Int. Ed.*, 1993, **32**, 1742.
124. M. Newcomb, N. Miranda, M. Sannigrahi, X. Huang and D. Crich, *J. Am. Chem. Soc.*, 2001, **123**, 6445.
125. (a) E. Meggers, D. Kusch, M. Spichty, U. Wille and B. Giese, *Angew. Chem. Int. Ed.*, 1998, **37**, 460; (b) B. Giese, *Acc. Chem. Res.*, 2000, **33**, 631.
126. B. C. Bales, J. H. Horner, X. Huang, M. Newcomb, D. Crich and M. M. Greenberg, *J. Am. Chem. Soc.*, 2001, **121**, 3623.
127. H. Eklund, U. Uhlin, M. Färnegårdh, D. T. Logan and P. Nordlund, *Prog. Phys. Mol. Biol.*, 2001, **77**, 177.
128. H. Eklund and M. Fontecave, *Struct. Fold. Des.*, 1999, **7**, R257.
129. J.-A. Stubbe, *J. Biol. Chem.*, 1990, **265**, 5329.
130. M. Uppsten, M. Färnegårdh, V. Domkin and U. Uhlin, *J. Mol. Biol.*, 2006, **359**, 365.
131. W. C. Voegtli, J. Ge, D. L. Perlstein, J.-A. Stubbe and A. C. Rosenzweig, *Proc. Natl. Acad. Sci. USA*, 2001, **98**, 10073.
132. D. Crich and G. F. Filzen, *Tetrahedron Lett.*, 1993, **34**, 3225.
133. L. R. C. Barclay, D. Griller and K. U. Ingold, *J. Am. Chem. Soc.*, 1982, **104**, 4399.
134. D. Crich, Q. Yao and G. F. Filzen, *J. Am. Chem. Soc.*, 1995, **117**, 11455.
135. A. L. J. Beckwith and P. L. Tindal, *Aust. J. Chem.*, 1971, **24**, 2099.
136. H.-G. Korth, R. Sustmann, K. S. Gröninger, M. Leisung and B. Giese, *J. Org. Chem.*, 1988, **53**, 4364.
137. A. L. J. Beckwith and P. J. Duggan, *J. Am. Chem. Soc.*, 1996, **118**, 12838.
138. D. Crich, X.-Y. Jiao, Q. Yao and J. S. Harwood, *J. Org. Chem.*, 1996, **61**, 2368.
139. (a) A. Koch and B. Giese, *Helv. Chim. Acta*, 1993, **76**, 1687; (b) A. Koch, C. Lamberth, F. Wetterich and B. Giese, *J. Org. Chem.*, 1993, **58**, 1083.
140. D. Crich, J. Escalante and X.-Y. Jiao, *J. Chem. Soc., Perkin Trans. 2*, 1997, 627.
141. M. Newcomb, J. H. Horner, P. O. Whitted, D. Crich, X. Huang, Q. Yao and H. Zipse, *J. Am. Chem. Soc.*, 1999, **121**, 10685.

142. For a recent theoretical discussion of the cycloadditions of methylene, see P. Pérez, J. Andrés, V. S. Safont, R. O. Contreras and O. Tapia, *J. Phys. Chem. A* 2005, **109**, 4178.
143. Recent theoretical work also reviewing experimental and theoretical values: O. Demel, J. Pittner, P. Čársky and I. Hubač, *J. Phys. Chem. A* 2004, **108**, 3125.
144. R. W. Alder, L. Chaker and F. P. V. Paolini, *Chem. Commun.*, 2004, 2172.
145. R. W. Alder, P. R. Allen, M. Murray and A. G. Orpen, *Angew. Chem. Int. Ed.*, 1996, **35**, 1121.
146. K. Briner and A. Vasella, *Helv. Chim. Acta*, 1989, **72**, 1371.
147. S. E. Mangholz and A. Vasella, *Helv. Chim. Acta*, 1991, **74**, 2100.
148. J.-P. Praly, Z. El Kharraf and G. Descotes, *J. Chem. Soc., Chem. Commun.*, 1990, 431.
149. A. Vasella and C. A. A. Waldraff, *Helv. Chim. Acta*, 1991, **74**, 585.
150. A. Vasella, P. Dhar and C. Witzig, *Helv. Chim. Acta*, 1993, **76**, 1767.
151. Review: A. Vasella, *Pure Appl. Chem.* 1993, **65**, 731.
152. A. Vasella, K. Briner, N. Soundarajan and M. S. Platz, *J. Org. Chem.*, 1991, **56**, 4741.
153. P. Reddy Muddasani, B. Bernet and A. Vasella, *Helv. Chim. Acta*, 1994, **77**, 334.
154. A. Vasella and C. Witzig, *Helv. Chim. Acta*, 1995, **78**, 1971.
155. B. Liessem, G. J. Glombitza, F. Knoll, J. Lehmann, J. Kellermann, F. Lottspeich and K. Sandhoff, *J. Biol. Chem.*, 1995, **270**, 23693.
156. C.-S. Kuhn, J. Lehmann, G. Jung and S. Stevanović, *Carbohydr. Res.*, 1992, **232**, 227.
157. J. Lehmann and S. Petry, *Carbohydr. Res.*, 1993, **239**, 133.
158. J. Lehmann and M. Scheuring, *Carbohydr. Res.*, 1992, **225**, 67.

APPENDIX

Elements of Protein Structure

1 AMINO ACID STRUCTURES

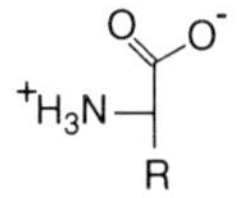

$^{+}H_3N$–CH(R)–CO_2^-

R = H, glycine, Gly, G (achiral)
R = CH_3, alanine, Ala, A
R = $CH(CH_3)_2$, valine, Val, V
R = $CH_2CH(CH_3)_2$, leucine, Leu, L
R = $CH_2CH(CH_3)CH_2CH_3$, isoleucine, Ile, I (*erythro* configuration)

R = CH_2OH, serine, Ser, S
R = $CHOHCH_3$, threonine, Thr, T
(*threo* configuration).
R = CH_2SH, cysteine, Cys, C
R = $CH_2CH_2SCH_3$, methionine, Met, M

R = $CH_2CH_2CH_2CH_2NH_3^+$, lysine, Lys, L
R= $CH_2CH_2CH_2NHC(=NH^+)NH_2$, arginine, Arg, R
R = —CH_2–(imidazolyl: NH, N) histidine, His, H

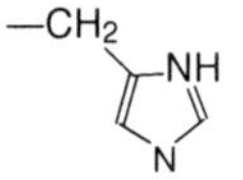

R = CH_2–(phenyl)
phenylalanine, Phe, F

R = CH_2–(phenyl)–OH
tyrosine, Tyr, Y

R = –CH_2–(indolyl: N–H)
tryptophan
Trp, W

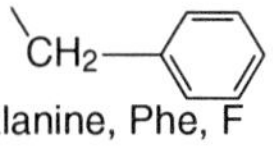

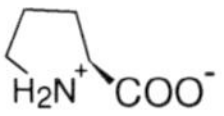

(pyrrolidine ring) H_2N^+ COO^-
proline, Pro, P
(imino-acid: complete structure shown)

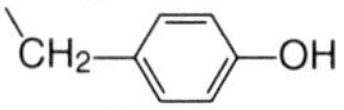

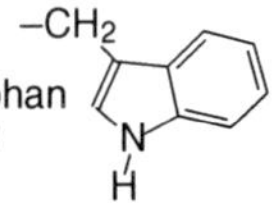

R = CH_2COOH, aspartic acid, Asp, D
R = CH_2CONH_2, asparagine, Asn, N

R = CH_2CH_2COOH, Glu, E
R = $CH_2CH_2CONH_2$, Gln, Q

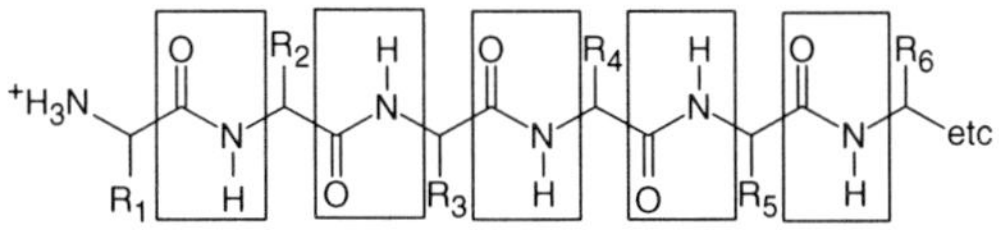

$^{+}H_3N$–CH(R_1)–[C(=O)–N(H)]–CH(R_2)–[C(=O)–N(H)]–CH(R_3)–[C(=O)–N(H)]–CH(R_4)–[C(=O)–N(H)]–CH(R_5)–[C(=O)–N(H)]–CH(R_6)–etc

2 ELEMENTS OF PROTEIN STRUCTURE

Note that the α-carbons are shown in the L-configuration.

Antiparallel β-Strand

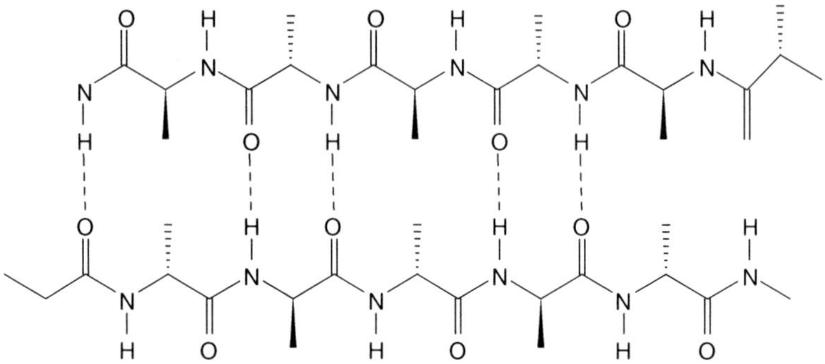

Antiparallel β-Sheet

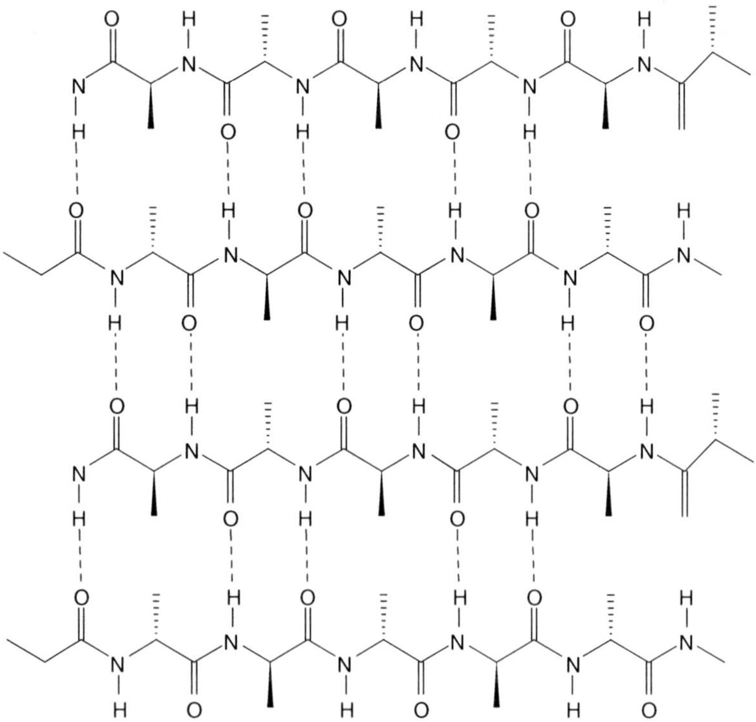

Parallel β-Strand

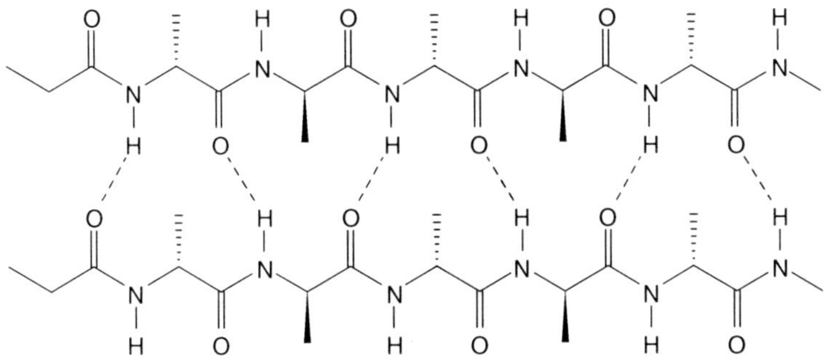

Parallel β sheet

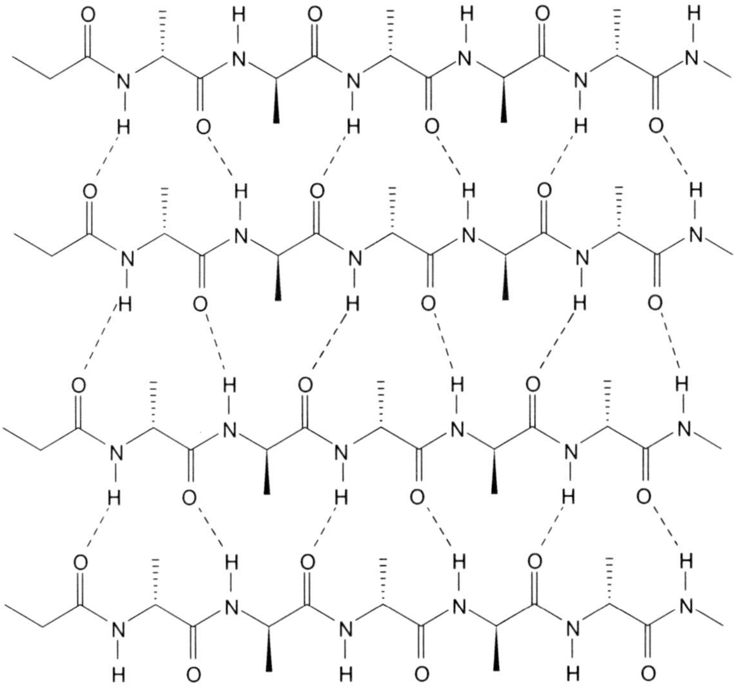

α-Helix

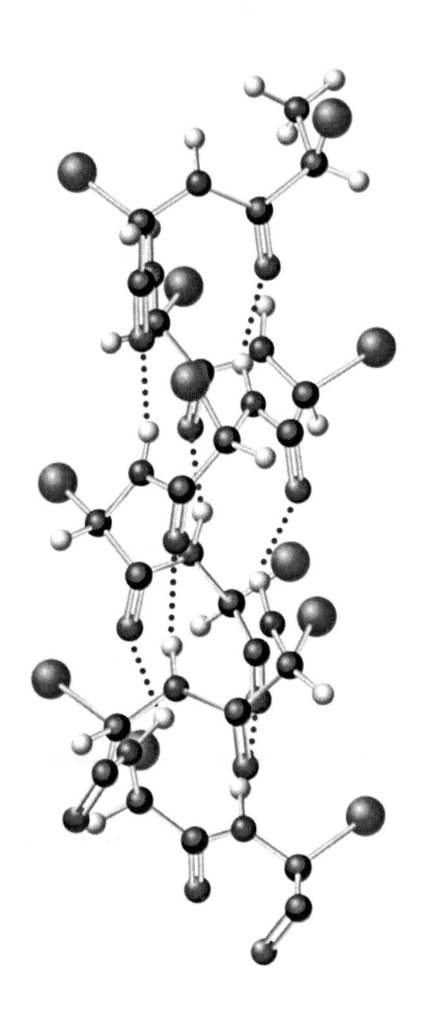

Subject Index